DAIRY CATTLE SCIENCE

(Animal Agriculture Series)

D0164548

by

M. E. Ensminger, B.S., M.S., Ph.D.

Formerly: Assistant Professor in Animal Science
University of Massachusetts

Chairman, Department of Animal Science
Washington State University

Consultant, General Electric Company
Nucleonics Department (Atomic Energy Commission)

Currently: President
Consultants-Agriservices
Clovis, California

President
Agriservices Foundation
Clovis, California

Collaborator
U.S. Department of Agriculture

Adjunct Professor
California State University-Fresno

Adjunct Professor
The University of Arizona-Tucson

Distinguished Professor
University of Wisconsin-River Falls

Third Edition

INTERSTATE PUBLISHERS, INC.
Danville, Illinois

Editions:
First 1971
Second 1980
Third 1993

Order from

Interstate Publishers, Inc.
510 North Vermilion Street, P.O. Box 50
Danville, IL 61834–0050
Phone: (800) 843–4774
FAX: (217) 446–9706

Library of Congress Catalog Card No. 91–74155

ISBN 0–8134–2930–7

3 4 5 6 7 8 9 10   05 04 03 02 01 00

DAIRY CATTLE SCIENCE

(Animal Agriculture Series)

by

M. E. Ensminger, B.S., M.S., Ph.D.

Dr. M. E. Ensminger is President of Agriservices Foundation, a nonprofit foundation serving world agriculture. Also, he is Adjunct Professor, California State University-Fresno; Adjunct Professor, The University of Arizona-Tucson; Distinguished Professor, University of Wisconsin-River Falls; Collaborator, U.S. Department of Agriculture; and Honorary Professor, Huazhong Agriculture College-Wuhan, People's Republic of China. Dr. Ensminger (1) grew up on a Missouri farm; (2) completed B.S. and M.S. degrees at the University of Missouri, and the Ph.D. at the University of Minnesota; (3) served on the staffs of the University of Massachusetts, the University of Minnesota, and Washington State University; and (4) served as Consultant, General Electric Company, Nucleonics Department (Atomic Energy Commission).

Dr. Ensminger is the author of 21 widely used books that are translated into several languages and used throughout the world.

Among Dr. Ensminger's honors and awards are: Distinguished Teacher Award, American Society of Animal Science; Washington State University named and dedicated the *Ensminger Beef Cattle Research Center*, in recognition of his contributions to the University; and an oil portrait of him was placed in the 300-year-old gallery of the famed Saddle and Sirloin Club, which is recognized as the highest honor that can be bestowed on anyone in the livestock industry. In 1991, he was the recipient of the Outstanding Achievement Award of the University of Minnesota.

Cover Pictures: Starting at the lower left with
the Jersey bull and cow, and continuing clockwise—

Jerseys. (Courtesy, The American Jersey Cattle
Club, Reynoldsburg, OH)

Milking Shorthorns. (Courtesy, American Milking
Shorthorn Society, Beloit, WI)

Brown Swiss. (Courtesy, Brown Swiss Cattle
Breeders' Assn., Beloit, WI)

Holsteins. (Courtesy, Ernst Peterson,
Hamilton, MT)

Guernseys. (Courtesy, American Guernsey
Assn., Reynoldsburg, OH)

Ayrshire. (Courtesy, Ayrshire Breeders'
Assn., Brandon, VT)

To Ella Ensminger,
my mother
and
Missouri's Mother-of-the-Year in 1956,
who encouraged me to
keep on keeping on

PREFACE TO THE THIRD EDITION OF *DAIRY CATTLE SCIENCE*

Milk and milk products have played an important and vital role in American history since 1611 when the first cows were brought to the Jamestown Colony, in Virginia. But the Pilgrims did not bring any cows with them on the Mayflower, which arrived at Plymouth, Massachusetts in 1620, with the result that the lack of milk is said to have contributed to the high death rate of the colonists, particularly of the children.

Today, the United States dairy industry includes the farmers who produce the milk, processors and manufacturers who provide the facilities and services needed to produce a variety of nutritious and delicious products, and the retailers who make the dairy products available to consumers.

Dairy products account for about 12% of total U.S. farm income from all farm commodities. In 1989, cash farm receipts from dairy products totaled $19.4 billion, which ranked it second only to cattle and calves with $36.7 billion. That same year, feed crops and poultry/eggs followed dairy products in cash receipts with $16.7 and $15.3 billion, respectively.

Although milk is produced and processed in every state, over half of the total U.S. milk production comes from Wisconsin, California, New York, Minnesota, and Pennsylvania. Wisconsin is, by far, the leading dairy state of the nation, with the most milk cows on farms.

Substantial structural changes are taking place in the dairy industry, in both dairy farms and manufacturing levels. During the 39-year period 1950–1989, the number of dairy farms declined from 3.6 million to 205,000; the average herd size increased from 5.8 cows to 49 cows; the number of milk cows declined from 23.9 million to 10.1 million; and the milk production per cow increased from 5,314 lb to 14,642 lb. The increase in milk production per cow enabled milk production to more than keep pace with commercial needs throughout the 1950 to 1989 period. The structural changes detailed above will continue in the decades to come. In the 1930s, 54% of the milk supply was used for butter, compared to 16.9% in 1990. Today, more milk is used for the manufacture of cheese than for butter (16.9% for butter vs 31.9% for cheese).

Federal and state dairy programs play an important role in the pricing and marketing of milk in the United States. The major government dairy programs are: dairy price supports, provided by the Commodity Credit Corporation (CCC) through the purchase of butter, cheese, and nonfat dry milk; federal milk marketing orders; import restrictions; and state regulations.

Imports of a number of dairy products are restricted by specific import quotas to prevent lower cost and subsidized dairy products from undercutting U.S. dairy price supports. For example, New Zealand has a clear advantage over the United States in milk production cost, due to its pasture-based system. The import quotas on manufactured dairy products limit imports to about 2.5 billion pounds milk equivalent, just under 2% of U.S. milk production in 1989. Exports of dairy products are rather small; historically, only about 2% of U.S. milk production has been disposed of by concessional sales or food aid donations from government supplies.

Per capita consumption of milk and most dairy products appears to have reached a rather stable stage. The most important factors affecting the future demand for dairy products will continue to be changes in population, income, consumer preference, and new products.

It is expected that per capita consumption of various dairy products will follow the trend of recent years; products high in fat will decline in per capita consumption, while those low in fat will increase. The proportion of milk consumed in fluid form will decrease; per capita consumption of butter and evaporated milk will likely decline on a gradual basis; and low fat fluid milk, nonfat dried milk, low fat frozen desserts, cheese, and sour cream dressings are likely to increase in per capita consumption.

Without doubt, new low fat dairy products and nondairy substitutes will replace some of the consumption of similar products higher in fat. The use of synthetic milk will increase.

Increasingly, the U.S. pricing system will (1) differentiate and price milk on a component basis, and (2) pay a premium for higher quality milk.

The operations involved in producing and marketing milk and dairy products will continue the current trend to bigness, fewer numbers, more mechanization, environmental control, and higher quality products.

Market forces, rather than federal or state support programs, will likely determine farm milk prices in the future.

The impact of biotechnology will gradually increase to the year 2000 — and well beyond.

Hopefully, *Dairy Cattle Science* will give a big assist, not only in adapting to change, but in shaping it.

Like all my works, this has been a team approach. I am especially grateful to Mrs. Ensminger, who supervised the processing of *Dairy Cattle Science* from manuscript to camera ready; to Janetta Shumway, assisted by Jo Schepers and Joan Wright, who deciphered my Missouri hieroglyphics and typed the manuscript; to Margo Williams who did the art and pasteup work; and to Randall and Susan Rapp, who typeset the camera-ready copy. Additionally, a host of individuals, associations, and companies provided pictures and subject matter information, or made other notable contributions, which are gratefully acknowledged at appropriate places throughout the book.

Clovis, California
1993

M. E. Ensminger

Other books by M. E. Ensminger
available from Interstate Publishers, Inc.:

The Stockman's Handbook
Stockman's Handbook Digest
Animal Science
Animal Science Digest
Beef Cattle Science
Sheep and Goat Science
Swine Science
Poultry Science
Horses and Horsemanship

The Stockman's Handbook and *Stockman's Handbook Digest* present the "why" as well as the "how." They contain, under one cover, the pertinent things that a livestock producer needs to know in the daily operation of a farm or ranch. They cover the broad field of animal agriculture, concisely and completely, and wherever possible in tabular and outline form.

Animal Science and *Animal Science Digest* present a perspective or panorama of the far-flung livestock industry; whereas each of the other books presents specialized material pertaining to the specific class of farm animals indicated by its respective title.

CONTENTS

Holstein-Friesian cows. (Courtesy, Holstein-Friesian Assn. of America, Brattleboro, VT)

Fig. 1–1. Eight registered Ayrshire cows on pasture. (Courtesy, Ayrshire Breeders Assn., Brandon, VT)

Fig. 1–2. Dairy products—the point and purpose of dairying. (Courtesy, United Dairy Industry Assn., Rosemont, IL)

Contents

CHAPTER

1

THE DAIRY INDUSTRY

Contents   Page

Under natural conditions, wild mammals produce only enough milk for their offspring. However, long before recorded history, people found that milk was good—and good for them—with the result that they domesticated milk-producing animals and began using and selecting them for higher production for their own use. For the most part, this included the cow, the buffalo, and the goat—although the ewe, the mare, the sow, and other mammals have been used for producing milk for human consumption in different parts of the world. The importance of the cow in milk production is attested to by her well-earned designation as, "the foster mother of the human race."

POSITION OF DAIRY CATTLE IN THE ZOOLOGICAL SCHEME

Domesticated cattle belong to the family *Bovidae*, which includes ruminants with hollow horns. Members of this family possess one or more enlargements along the esophagus for food and chew their cuds. In addition to what we commonly call cattle or oxen, the family *Bovidae* (and the subfamily *Bovinae*) includes the true buffalo, the bison, musk-ox, banteng, gaur, gayal, yak, and zebu.

The following outline shows the basic position of the domesticated cow in the zoological scheme:

Kingdom *Animalia*: Animals collectively; the animal kingdom.

Phylum *Chordata*: One of approximately 21 phyla of the animal kingdom in which there is either a backbone (in the vertebrates) or the rudiment of a backbone (in the chorda).

Class *Mammalia*: Mammals or warm-blooded, hairy animals that produce their young alive and suckle them for a variable period on a secretion from the mammary glands.

Order *Artiodactyla*: Even-toed, hoofed mammals.

Family *Bovidae*: Ruminants having polycotyledonary placenta; hollow, non-deciduous, up-branched horns; and nearly universal presence of a gallbladder.

Genus *Bos*: Ruminant quadrupeds, including wild and domestic cattle, distinguished by a stout body and hollow, curved horns standing out laterally from the skull.

Species *Bos taurus* and *Bos indicus*: *Bos taurus* includes the ancestors of the European cattle and of the majority of the cattle found in the United States; *Bos indicus* is represented by the humped cattle (Zebu) of India and Africa and the Brahman breed of America.

HISTORY AND DEVELOPMENT OF THE DAIRY INDUSTRY

Cattle are the most important of all the domesticated animals, and next to the dog the most ancient. In 1989, there were 222,846,000 milk cattle in the world.[1]

The word *cattle* seems to have the same origin as chattel, which means possession. This is a very natural meaning; for when Rome was in her glory, a person's wealth was often computed in terms of cattle possessions. That the ownership of cattle implied wealth is further attested by the fact that the earliest known coins bear an ox head; and the Roman word *pecunia* for money (preserved in our adjective pecuniary) was derived from the Latin word *pecus,* meaning cattle.

To the nomadic Masai of Africa, cattle are a status symbol and like money in the bank. A person's wealth and station in life are measured by the number of cattle owned and a young man cannot get married without paying the asking price by the father of the girl—usually 10 to 50 cows, along with a few goats and 5 to 10 gallons of beer. It follows that cattle numbers are more important than quality.

ORIGIN AND DOMESTICATION OF CATTLE

It seems probable that cattle were first domesticated in Europe and Asia during the New Stone Age. In the opinion of most authorities, today's cattle bear the blood of either or both of two ancient ancestors—*Bos taurus* or *Bos indicus*. Other species or sub-species were frequently listed in early writings, but these are seldom referred to today. Perhaps most, if not all, of these supposedly ancestral species were also descendants of *Bos taurus* or *Bos indicus* or crosses between the two. All U.S. dairy breeds belong to *Bos taurus*.

USE OF CATTLE IN ANCIENT TIMES

Like other animals, cattle were first hunted and used as a source of food and other materials. As civilization advanced and people turned to tillage of the soil, it is probable that the domestication of cattle was first motivated because of their projected value for draft purposes.

As populations became more dense, feed became more abundant, and cattle became more plentiful, people became more interested in larger production of meat and milk. The pastoral people adopted a more settled life and began selecting out those animals that possessed the desired qualities—including rapid growth, fat storage, and milk production. Records exist of cows being milked as early as 9000 B.C. The Bible contains many references to milk, one of the best remembered of which is Exodus 3:8—"milk and honey." Also, Sanskrit writings, thousands of years old, relate that milk was one of the most essential of all foods. Hippocrates recommended milk as a medicine five centuries before Christ.

In contrast with the very great importance of cattle in western Asia and Europe in both ancient and modern times, it is noteworthy that cattle were never very highly valued in China, Japan, or Korea. The people of these countries have never used much beef, milk, butter or cheese. In India, on

---

[1]*FAO Production Yearbook*, Vol. 43, 1989, p. 271, Table 99.

the other hand, cattle play as important a role as in our western civilization; and, additionally, they retain a great religious significance.

CATTLE IN MEDIEVAL FARMING

The best of medieval farms would excite the scorn or contempt of a modern farmer. Very little cow's milk was available, most of it being produced during the grazing season. In fact, more goat's milk than cow's milk was consumed in liquid form. Even in the 13th century, when farming methods had improved, one writer indicated that three cows could be expected to produce only 3½ lb of butter per week. Most cow's milk was used in cheese making.

HISTORY OF THE DAIRY INDUSTRY IN THE UNITED STATES

When Christopher Columbus first arrived in America, there were no cows on the American continent. On his second voyage, in 1493, be brought four calves and two heifers, along with other farm animals, to the West Indies. Coronado later brought about 150 head of cattle to North America during his expedition to Mexico. These cattle were the forerunners of the Longhorns of the Southwest.

The first cows were brought over to the Jamestown colony in 1611. The Pilgrims did not bring any cows with them on the Mayflower, which arrived at Plymouth, Massachusetts (about 40 miles south of Boston), in 1620. As a result, the lack of milk is said to have contributed to the high death rate of the colonists, particularly of the children. The Colony received its first cows in 1624.

Throughout the colonial period, and until the middle of the 19th century, dairying was limited to a few cows cared for by the family. A farm generally had one or two cows to supply fresh milk. The availability of milk was largely seasonal, with the peak coming in the spring and early summer, when pastures were lush. Urban populations in this period received almost no fluid milk because of its highly perishable nature and the lack of rapid transportation and adequate storage facilities. If city dwellers did receive milk, it was usually obtained from two sources: (1) shipped in from neighboring rural areas, or (2) from local cows fed distillers' by-products. In either case, milk was not subjected to health or quality control regulations and, consequently, was often diluted with water and of low quality. Excesses of milk in this period were converted to cheeses to prevent spoilage.

The middle of the 19th century proved to be the turning point for the dairy industry. New techniques in handling, storage, and processing of milk were developed at a rapid pace that was to continue up to the present. The government recognized the value of agricultural research in the mid and latter half of the 19th century and provided the needed funds to establish agricultural colleges (the Land Grant Act) and to foster agricultural research (the Hatch Act).

The early years of the 20th century ushered in an era where quality of food was given considerable attention. Laws regulating the quality and preparation of foods were passed and new techniques were developed that improved the acceptability and nutritive value of milk.

Immediately following World War II, there arose a keen interest in artificial insemination, a technique of farm production that was to revolutionize the dairy industry. Since World War II, numerous other developments have arisen that have increased both the yields of dairy cattle and improved the handling and quality control of milk and milk products; among them, (1) flavor control equipment, (2) adaptation of computers to production record keeping and prediction of genetic merit, (3) new milk containers, (4) electronic testing for milk fat, and (5) embryo transfer and biotechnology.

Table 1–1 lists the dates, in the order in which they happened, of important developments in the U.S. dairy industry.

TABLE 1–1
CHRONOLOGY OF THE DEVELOPMENT OF THE DAIRY INDUSTRY

Year	Event
1611	First cows arrived at the Jamestown Colony.
1624	Cows arrived at the Plymouth Colony.
1841	Milk was first shipped by rail (unrefrigerated) on a regular basis from Orange County, New York to New York City.
1851	First cheese factory was established in the United States, at Oneida, New York.
1855	Michigan established the first agriculture college in the United States.
1856	Pasteur began his experiments that marked the beginning of bacteriology. Gail Borden received the first patent for condensed milk in the United States and Great Britain.
1857	Gail Borden established America's first successful condensery at Burrville, Connecticut.
1861	Mechanical refrigeration was developed.
1862	Land Grant Act was passed, providing funds for the development of land grant colleges.
1868–1880	Breed associations were established for the five major breeds of dairy cattle.
1878	Dr. Gustav De Laval invented the continuous cream separator.
1884	Dr. Harvey Thatcher invented the milk bottle.
1885	*Hoard's Dairyman* was first published.
1887	Hatch Act was passed, providing additional funds for agricultural research.
1890	Dr. S. M. Babcock introduced his fat test for milk and cream. Tuberculin testing of cattle was initiated.
1892	Certified milk was originated by Dr. H. L. Coit in Essex County, New York.
1895	Commercial pasteurizing machines were introduced.
1905	First U.S. cow-test program was established in Newaygo County, Michigan.
1908	First compulsory pasteurization law for milk in the United States was passed in Chicago.
1914	Tank trucks were introduced as alternative transportation to the traditional milk wagon.
1919	Homogenized milk was first marketed.
1932	Vitamin D fortification of milk was introduced on a practical basis.
1936	Artificial insemination was begun in dairy cattle.
1938	First bulk tanks for milk were introduced.

(Continued)

TABLE 1–1 *(Continued)*

Year	Event
1946	Vacuum pasteurization was perfected.
1948	Ultra-high temperature pasteurization was first used.
1950	Plastic-coated paper milk cartons were first used.
1964	Plastic milk containers were introduced.
1967	Nondairy milk substitutes were introduced.
1968	Official acceptance was given for the electronic testing for milk fat.
1975	Metric conversion first began in the United States.
1977	Yogurt annual sales topped 600 million lb, a 496% increase over the previous 10 years.
1981	UHT (ultra high temperature) milks gains national recognition.
1988	Lower fat dairy products gain widespread acceptance. Low fat plus skim milk sales exceed whole milk sales for the first time.

MILK—THE FOUNDATION OF GOOD NUTRITION

Hippocrates, often referred to as the father of medicine, described milk in his writings as "the most nearly perfect food." Indeed, this must be so. The newborn mammal relies almost totally on its mother's milk for food. Its digestive tract is largely undeveloped, and the food (milk) must be nutritionally complete and easily digested and absorbed. The body of the newborn mammal grows rapidly and must cope with a great deal of stress. If milk were anything less than complete, survival of young would be difficult.

NUTRITIVE VALUE OF MILK

Milk plays an important role in the diet of the average American, as is shown in Table 1–2. The major nutrients provided by dairy products in the American diet are: food energy (10%), protein (20%), fat (12%), calcium (77%), phosphorus (36%), vitamin A (16%), and riboflavin (36%).

PROTEIN IN MILK

The protein in milk is of extremely high quality, having a biological value of 85 as compared to 50 to 65 for the cereal proteins. This high value indicates that the protein of milk is highly digestible and contains a well-balanced amino acid profile. Except for the sulfur-containing amino acids, 1 pt of milk supplies the recommended daily allowance of all the essential amino acids.

Additionally, the protein to calorie ratio is very favorable in milk, assuring consumers that they are not ingesting "empty" calories. Rather, they are drinking a highly nutritious source of protein and energy.

CARBOHYDRATES IN MILK

Lactose is, by far, the most abundant sugar in milk, constituting up to 36% of the dry matter. This disaccharide, consisting of glucose and galactose, is only one-fifth as sweet as sucrose (table sugar) and is less soluble in water. In the small intestine, lactose is cleaved to form its constituent monosaccharides. Glucose is readily absorbed, while galactose is slowly absorbed, thereby serving as growth promotant for intestinal bacteria that synthesize vitamin K and the B complex vitamins. Lactose is also converted to lactic acid by intestinal bacteria. When lactic acid is produced in the intestines, growth of pathogenic bacteria is inhibited.

Lactose has been shown to facilitate the absorption of several minerals; among them, calcium, phosphorus, magnesium, and barium.

FATS IN MILK

Milk fat consists primarily of short-chain fatty acids. These fatty acids are more readily digested and absorbed than the long-chain fatty acids found in vegetable protein. Milk fat is the fraction of milk that carries vitamin A. Additionally, it has been shown to facilitate calcium absorption.

TABLE 1–2
FOOD NUTRIENTS: U.S. PERCENTAGE OF TOTAL CONTRIBUTED BY LIVESTOCK AND POULTRY PRODUCTS[1]

	Food Energy	Protein	Fat	Carbo-hydrates	Calcium	Phos-phorus	Iron	Vitamin A Value	Thiamin	Ribo-flavin	Niacin
	(%)	(%)	(%)	(%)	(%)	(%)	(%)	(%)	(%)	(%)	(%)
Dairy products, excluding butter	10.0	20.0	12.0	6.0	77.0	36.0	3.0	16.0	9.0	36.0	2.0
Meat, fish and poultry	20.0	44.0	34.0	[2]	4.0	30.0	27.0	30.0	26.0	25.0	46.0
Eggs	2.0	4.0	3.0	[2]	2.0	4.0	5.0	4.0	2.0	5.0	[2]
Total	32.0	68.0	49.0	6.0	83.0	70.0	35.0	50.0	37.0	66.0	48.0

[1]*Agricultural Statistics*, USDA, p. 471, Table 668. Data for 1985.

[2]Less than 0.5%.

MINERALS IN MILK

Of all the foods commonly eaten in the American diet, milk is the best and most commonly consumed source of calcium. The diet of the average American is based largely on meat and cereal grains—foods that are deficient in calcium. Most meats contain only 5 to 15 mg of calcium per 100 g, whereas milk contains about 120 mg of calcium per 100 g. Thus, milk should be included in the diet. Additionally, the calcium to phosphorus ratio (1.4:1) of milk is very conducive to providing a balanced profile of the two minerals.

Cow's milk is deficient in iron and copper; hence, it is important that these minerals be supplemented in the diet. Since these two minerals are stored by the developing calf fetus during gestation, a deficiency of them is not critical to the newborn calf.

VITAMINS IN MILK

The diet of the average American is generally deficient in vitamin A and riboflavin. Milk provides an excellent source of these two vitamins. Additionally, vitamin D fortified milk is widely marketed.

MILK AS A MEANS OF FEEDING THE HUNGRY

It is not easy to present a precise picture of the extent of undernutrition and malnutrition in the world. Intake of energy and protein over a period of time appears to be the best indicator of dietary sufficiency. This information is presented in Fig. 1–3, which shows the per capita calories and proteins per day of countries.

Through the ages, milk has been a basic food. In the future, it will become increasingly important in meeting the challenge for feeding the hungry people of the world. Nevertheless, the food vs feed argument will wax hot. Can we afford to feed grains to animals while denying them to hungry people? There are many arguments for and against the practice of feeding animals to produce human food. However, dairy cows have proven to be extremely valuable as a source of food for the following reasons:

1. **They convert relatively low-protein feeds into a high-quality food very efficiently.** Table 1–3 (next page) lists the feed to food efficiency rating by species of animal. On a feed efficiency basis (pounds of feed to produce 1 lb of product), the dairy cow ranks number two. On a calorie of feed consumed to calorie of product produced basis, the dairy cow has the highest efficiency rating; and on a protein efficiency basis, the dairy cow ranks number two, being exceeded only by broilers.

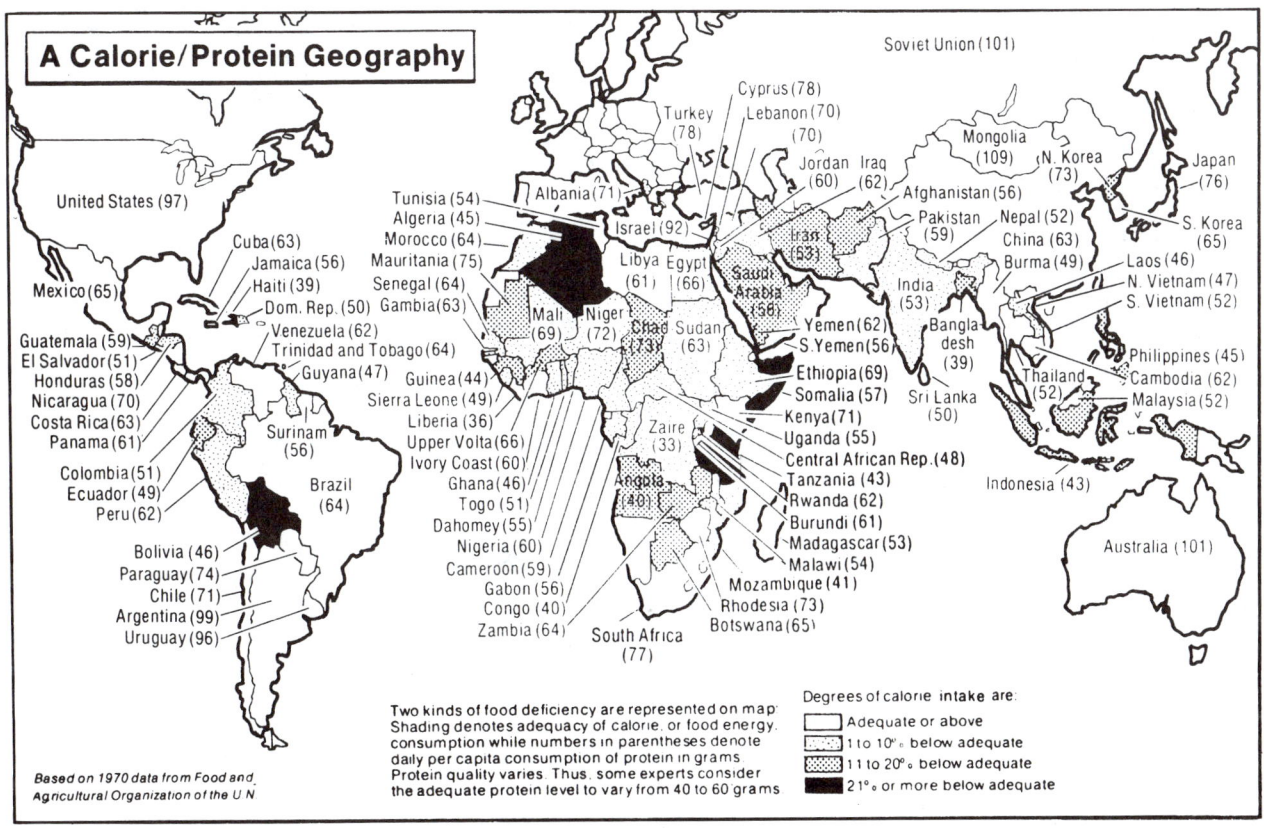

Fig. 1–3. World geography of calories and proteins. (Courtesy, the *New York Times*)

TABLE 1–3
FEED TO FOOD EFFICIENCY RATING BY SPECIES OF ANIMALS, RANKED BY PROTEIN CONVERSION EFFICIENCY
(Based on Energy as TDN or DE and Crude Protein in Feed Eaten by Various Kinds of Animals Converted into Calories and Protein Content of Ready-to-Eat Human Food)

Species	Unit of Production (on foot)	Feed Required to Produce One Production Unit — Pounds	TDN[2]	DE[3]	Protein	Dressing Yield — Percent	Net Left	Ready-to-Eat; Yield of Edible Product (meat and fish deboned and after cooking) — As % of Raw Product (carcass)	Amount Remaining from One Unit of Production	Calorie[4]	Protein[4]	Feed Efficiency[1] (lb feed to produce one lb product) (%)	(ratio)	Calorie Efficiency[5] (%)	(ratio)	Protein Efficiency[6] (%)	(ratio)
		(lb)	(lb)	(kcal)	(lb)	(%)	(lb)	(%)	(lb)	(kcal)	(lb)	(%)	(ratio)	(%)	(ratio)	(%)	(ratio)
Broiler	1 lb chicken	2.47	1.94[8]	3,880	0.218	72[9]	0.72	54[10]	0.39	274	0.11	41.7	2.4:1	7.1	14.2:1	52.4	1.9:1
Dairy cow	1 lb milk	1.11[7]	0.90[8]	1,800	0.108	100	1.00	100	1.00	309	0.037	90.0	1.11:1	17.2	5.8:1	37.0	2.7:1
Turkey	1 lb turkey	5.2[7]	4.21[8]	8,420	0.46[8]	79.7[9]	0.797	57[11]	0.45	446	0.146	19.2	5.2:1	5.3	18.9:1	31.7	3.2:1
Layer	1 lb eggs (8 eggs)	4.6[7]	3.73[8]	7,460	0.41[8]	100	1.00	100[12]	1.00[12]	616	0.106	21.8	4.6:1	8.3	12.1:1	25.9	3.9:1
Rabbit	1 lb fryer	3.0[13]	2.20	4,400	0.48	55[13]	0.55	79[13]	0.43	301	0.08	35.7	2.8:1	6.8	14.6:1	16.7	6.0:1
Fish	1 lb fish	1.6[14]	0.98	1,960	0.57	65[15]	0.65	57[16]	0.37	285	0.093	62.5	1.6:1	14.5	6.9:1	16.3	6.1:1
Hog (birth to 200 lb)	1 lb pork	4.9[17]	3.67	7,340	0.69	70[18]	0.70	44[19]	0.31	341	0.088	20.4	4.9:1	4.6	21.5:1	12.7	7.8:1
Beef steer (yearling finishing period in feedlot)	1 lb beef	9.0[17]	5.85	11,700	0.90	58[18]	0.58	49[19]	0.28	342	0.085	11.1	9.0:1	2.9	34.2:1	9.4	10.6:1
Lamb (finishing period in feedlot)	1 lb lamb	8.0[17]	4.96	9,920	0.86	47[18]	0.47	40[19]	0.19	225	0.052	12.5	8.0:1	2.3	44.1:1	6.0	16.5:1

[1]Feed efficiency as used herein is based on pounds of feed required to produce 1 lb of product. Given in both percent and ratio.

[2]TDN pounds computed by multiplying pounds feed (column to left) times percent TDN in normal rations. Normal ration percent TDN taken from M. E. Ensminger's books and rations, except for following: dairy cow, layer, broiler, and turkey from *Agricultural Statistics 1974*, p. 358, Table 518. Fish based on averages recommended by Michigan and Minnesota Stations and U.S. Fish and Wildlife.

[3]Digestible Energy (DE) in this column given in kcal, which is 1 Calorie (written with a capital C), or 1,000 calories (written with a small c). Kilocalories computed from TDN values in column to immediate left as follows: 1 lb TDN = 2,000 kcal.

[4]*Lessons on Meat*, National Live Stock and Meat Board, 1965.

[5]Kilocalories in ready-to-eat food = kilocalories in feed consumed, converted to percentage. Loss = kcal in feed ÷ kcal in product.

[6]Protein in ready-to-eat food = protein in feed consumed, converted to percentage. Loss = pounds protein in feed ÷ pounds protein in product.

[7]*Agricultural Statistics 1974*, p. 358, Table 518. Pounds feed per unit of production is expressed in equivalent feeding value of corn.

[8]Since pounds feed (column No. 3) per unit of production (column No. 2) is expressed in equivalent feeding value of corn, the values for corn were used in arriving at these computations. No. 2 corn values are TDN, 81%; protein, 8.9%. Hence, for the dairy cow 81% × 1.11 = 0.9 lb TDN; and 8.9% × 1.11 = 0.1 lb protein.

[9]*Marketing Poultry Products*, 5th ed., by E. W. Benjamin, *et al.*, John Wiley & Sons, 1960, p. 147.

[10]*Factors Affecting Poultry Meat Yields*, University of Minnesota Sta. Bull. 476, 1964, p. 29, Table 11 (fricassee).

[11]*Ibid.*, p. 28, Table 10.

[12]Calories and protein computed basis per egg; hence, the values herein are 100% and 1.0 respectively.

[13]Based on information in *Commercial Rabbit Raising*, Ag. Hdbk. No. 309, USDA, 1966, and *A Handbook on Rabbit Raising*, by H. M. Butterfield, Washington State College Ext. Bull. No. 411, 1950.

[14]Data from report by Dr. P. J. Schaible, Michigan State University, *Feedstuffs*, April 15, 1967.

[15]*Industrial Fishery Technology*, edited by M. E. Stansby, Reinhold Pub. Corp., 1963, Ch. 26, Table 26-1.

[16]*Ibid.* Reports that, "Dressed fish averages about 73% flesh, 21% bone, and 6% skin." In limited experiments conducted by A. Ensminger, it was found that there was a 22% cooking loss on filet of sole. Hence, these values—give 57% yield of edible fish after cooking, as a percent of the raw, dressed product.

[17]Estimates by the author.

[18]Ensminger, M. E., *The Stockman's Handbook*, 5th ed., Sec. XIV.

[19]Allowance made for both cutting and cooking losses following dressing. Thus, values are on a cooked, ready-to-eat basis of lean and marbled meat, exclusive of bone, gristle, and fat. Values provided by National Live Stock and Meat Board (personal communication of June 5, 1967, from Dr. W. C. Sherman, Director, Nutrition Research, to the author), and based on data from *The Nutritive Value of Cooked Meat*, by R. M. Leverton and G. V. Odell, Misc. Pub. MP–49, Appendix C, March 1958.

Fig. 1–4. Cows are efficient converters of feed to food. This shows Holstein-Friesian cows on good pasture.

2. **They consume mostly roughages.** As shown in Table 1–4, nearly 60% of the feed consumed by dairy cattle is roughage.

TABLE 1–4
PERCENTAGE OF FEED FOR DIFFERENT CLASSES OF U.S. LIVESTOCK DERIVED FROM (1) CONCENTRATES, AND (2) FORAGES, INCLUDING PASTURE[1]

Class of Animal	Concentrates	Roughages
	(%)	(%)
Beef cattle	15.5	84.5
Dairy cattle	41.3	58.7
Sheep and goats	6.2	93.8
Swine	95.7	4.3
Horses and mules	27.0	73.0
Poultry	100.0	0.0
All livestock	38.3	61.7

[1]USDA, Economics Research Service. Data for the feed year 1983–1984.

3. **They provide outlets for the utilization of by-products from food processing and manufacturing.** Cull vegetables, nuts, hulls, and numerous other products considered unsuitable for human consumption are routinely fed to dairy cattle.

4. **They provide a continuous source of food.** Unlike meat animals that must be fed for a period of time before a usable product can be attained, lactating dairy cows produce milk throughout the year. Additionally, the dairy cow can produce many times her weight in milk throughout her life, while the products derived from meat animals are restricted to their final market weight.

5. **Milk consumption is permitted in vegetarian societies.** In countries such as India and Pakistan, the eating of meat is taboo by religious precept. But the consumption of milk is acceptable and constitutes the vast majority of the animal protein consumed by the people of these countries.

Fig. 1–5. Cattle will continue to furnish most of the milk of the world. But milk shortages in the developing countries will limit the nutritional program, even for children and pregnant-lactating mothers. This shows a woman milking a cow in North Senegal, Africa. In order to induce milk-letdown, cows are first suckled briefly by a calf; then the woman goes ahead with milking. (Courtesy, FAO, Rome, Italy)

HEALTH ASPECTS OF MILK CONSUMPTION

Milk has always been associated with good nutrition. As is shown in Table 1–5, the selected countries with high life expectancies consume considerable quantities of milk on a per capita basis. Numerous advantages and some disadvantages should be considered when incorporating milk into a sound nutrition program.

TABLE 1–5
PER CAPITA MILK CONSUMPTION OF SELECTED COUNTRIES HAVING HIGH LIFE EXPECTANCY AT BIRTH FOR MALES

Longevity Rank	Country	Life Expectancy[1]	Per Capita Fluid Milk Consumption[2]	
		(years)	(lb)	(kg)
1	The Netherlands	74.0	301.2	136.9
2	Spain	74.0	238.7	108.5
3	Sweden	74.0	317.0	144.1
4	Switzerland	74.0	233.4	106.1
5	Australia	73.0	226.6	103.0
6	Canada	73.0	236.3	107.4
7	Italy	73.0	160.2	72.8
8	Norway	72.7	512.2	232.8
9	United States	72.7	228.1	103.7

[1]*The World Almanac and Book of Facts 1991*, World Almanac, Pharos Books, NY.

[2]*World Dairy Situation*, USDA, Foreign Agricultural Service, Circular Series, FD 1–91, May 1992, p. 41. Data for 1990.

BENEFITS OF MILK

In addition to milk's high nutrient content and digestibility, it serves several functions in the promotion of good health:

1. **It has a tranquilizing effect on the body.** Soon after consumption of milk, a general mild tranquilizing effect can be felt. While the exact cause of this effect has not been determined, some researchers feel that it may be due to the high concentration of calcium in milk.

2. **It is commonly used in ulcer therapy.** Because lactose is not absorbed rapidly, it is less irritating to the gastric and intestinal mucosa than many foods.

3. **It may help to reduce the incidence rate of stomach cancer.** Several studies have shown that populations consuming small amounts of milk have higher incidence rates of stomach cancer than populations that consume large amounts of milk. Since this trend is largely based on statistical surveys of populations, and not physiological experiments, it cannot be accepted as conclusive.

HEALTH PROBLEMS ASSOCIATED WITH MILK CONSUMPTION

Several metabolic disorders are found in certain individuals when milk is introduced into the diet. Some people are allergic to it, while others have an inherent inability to digest milk properly as evidenced in individuals with lactose intolerance. Also, in recent years, a vigorous debate has ensued concerning the role of animal fats in atherosclerosis.

ALLERGIES

Occasionally human infants are hypersensitive to cow's milk and must be fed a diet completely free of it. In these cases, the infant is generally allergic to the protein lactalbumin. Severe cases of milk allergy are rare; but in extreme cases of hypersensitivity, a mere drop of cow's milk can cause a rash, asthma, acute gastrointestinal disorders, and shock. However, most milk allergies are not severe, with the symptoms of a rash and diarrhea predominating in most cases. Where allergies exist, quite often goat's milk can be substituted for cow's milk.

LACTOSE INTOLERANCE

Milk has one major limitation that must be considered when supplemental milk programs are initiated in certain countries where milk consumption has been traditionally limited. Certain people have what is referred to as a lactose intolerance. In this condition, susceptible individuals lack the enzyme lactase which splits lactose (milk sugar) into the readily digestible sugars, galactose and glucose. Thus, if such a person ingests milk or certain dairy products, his or her system cannot properly digest the materials. A reaction, generally in the form of cramps and diarrhea, occurs when large amounts of water are drawn into the lumen of the gut due to the osmotic effect exerted by the uncleaved lactose molecule.

This metabolic disorder interests the anthropologist as well as the physiologist because certain populations throughout the world show extremely high rates of lactose intolerance. Scandinavian and Western Europeans exhibit the lowest rate of lactose intolerance, about 2 to 8%, while about nine out of every ten Japanese, Thais, and Filipinos are intolerant to lactose. In the United States, it has been estimated that about 70% of the African-American population is lactose intolerant.[2] The question posed by anthropologists is whether the intolerance to lactose evolved due to the inadequate consumption of milk over a period of generations or whether the low consumption of milk was a result of low gut lactase activity originally. That is, did these people once have the ability to digest lactose and eventually lose it, or have they always had this lactase insufficiency?

Fortunately, many people having lactose intolerance can digest cultured dairy products, such as yogurt, where the lactose is converted to lactic acid by a bacterial innoculum.

ATHEROSCLEROSIS AND ANIMAL FAT

In recent years, the attention of the American public has been focused on heart disease — the leading killer in the United States. Daily, the nation's newspapers and airwaves devote space and time, respectively, to its causes, treatment, and prevention.

Numerous types of heart disease contribute to the yearly death toll of about one million persons in the United States; among them, hypertension, cerebrovascular disease (stroke), congestive heart failure, and atherosclerosis. Diet has been implicated in a number of these diseases, and much attention has been given to the role of animal fats in atherosclerosis. Atherosclerosis is a form of heart disease wherein a buildup of soft, amorphorus lipids and connective tissue develops on the walls of the arteries of the heart. If these deposits become sufficiently large, clots may form and subsequently decrease the diameter of the arterial lumen — in some cases, totally blocking blood flow. When blood flow to the heart is greatly impaired, a heart attack will ensue.

The animal fat-heart disease debate first began in 1953, when Dr. Ancel Keys, at the University of Minnesota,[3] reported a positive correlation between the consumption of fat and the occurrence of atherosclerosis. Subsequent research indicated that individuals with high serum cholesterol levels had a higher rate of atherosclerosis than people with normal levels. Increased serum cholesterol levels can be induced in susceptible individuals when saturated fats are consumed. Thus, the hypothesis that animal fats (highly saturated in nature) cause heart disease became accepted by many as fact. In recent years, research has clearly indicated that this position is entirely too simplistic. Studies have shown that certain African tribes whose diets consist almost entirely of animal products do not have elevated serum cholesterol levels. Additionally, it has been shown that consumption of dietary cholesterol and saturated fats causes a temporary elevation of serum cholesterol in healthy individuals, but, after a short time, the level returns to normal.

[2]Goodhart, R. S., and M. E. Shils, *Modern Nutrition in Health and Disease*, Lea & Febiger, Philadelphia, 1973.

[3]Keys, Ancel, *Journal of Chronic Diseases*, 1956, Vol. 4, p. 364.

That the form of dietary fat is implicated in atherosclerosis is fairly well documented by unbiased, controlled experiments; but it must be emphasized that research indicates that it is not the sole cause. Rather, a number of factors enter into the cause of heart disease, many of which play far greater roles than cholesterol; among these, stress, heredity, hypertension, diabetes mellitus, smoking, lack of exercise, and obesity.

Research has demonstrated that unsaturated fats prevent the elevation of serum cholesterol. This has spurred animal scientists to develop practices which will alter the levels of unsaturated fats in animals. By protecting dietary unsaturated fats, the content of unsaturated fatty acids in goat's and cow's milk has been increased dramatically—from levels of 2 to 35%.

When evaluating the risks of heart disease posed by the consumption of animal products, one must weigh the benefits against the risks. As shown in Table 1–5, the selected countries with the high life expectancies are noted for their milk consumption. The nutrients supplied by milk and meat provide well-balanced nutrition; hence, animal products must not be eliminated from the diet lest nutrient deficiencies arise. Rather, a well-planned diet, along with exercise and a minimum of stress, constitutes the best prevention against heart disease.

WORLD DISTRIBUTION AND PRODUCTION OF DAIRY CATTLE

Cattle furnish most of the milk of the world; only about 8.6% (less than one-twelfth) of the world's milk supply comes from buffalo, goats, and sheep. Table 1–6 shows the ten leading milk cow producing countries of the world, ranked by cow numbers. As noted, the U.S.S.R. has a commanding lead.

TABLE 1–6
TEN LEADING COUNTRIES IN (1) MILK COWS ON FARMS, AND (2) MILK PRODUCTION PER COW

Country	Number Milk Cows on Farms[1]	Country	Average Milk Production Per Cow[2]	
	(1,000 head)		(lb)	(kg)
U.S.S.R.	41,800	Israel	18,768	8,531
India	29,000	Japan	16,650	7,568
Brazil	14,650	United States . .	14,214	6,461
United States . . .	10,127	Denmark	13,686	6,221
Germany	6,960	Sweden	13,658	6,208
Mexico	6,300	The Netherlands	13,182	5,992
France	5,820	Norway	12,206	5,548
Poland	4,981	Canada	12,115	5,507
United Kingdom .	3,142	Finland	11,442	5,201
Italy	2,973	Switzerland . . .	10,602	4,819

[1]From *Agricultural Statistics 1990*, p. 310, Table 475.

[2]Figures for Israel were obtained from *FAO Yearbook, Production*, Vol. 43, 1989, p. 272. The rest of the milk production per cow figures were obtained from *Agricultural Statistics 1990*, p. 310, Table 475.

Table 1–6 also shows the proficiency of cows in the principal countries of the world. It is noteworthy that the United States ranks third in average milk production per cow, being outproduced by both Israel and Japan.

Some countries produce more dairy products than they can use, whereas others are importers. New Zealand, which has low milk production cost because of being pasture-based, accounts for about 20% of the total world dairy exports. The United Kingdom is an especially heavy importer of butter, cheese, and condensed milk.

Among the factors that determine the present development of the dairy industry in different countries are the character and preferences of the people; the adaptation of the country to dairying; the relative size of urban and rural population; and the extent and effectiveness of dairy research and education.

THE DAIRY INDUSTRY IN THE UNITED STATES

As shown in Table 1–6, the United States ranks fourth in number of milk cows and third in average production per cow, among the leading dairy countries of the world. But it should be noted that the full potential of the dairy industry in the United States has yet to be realized.

COMPONENTS OF THE DAIRY INDUSTRY IN THE UNITED STATES

The process of getting milk to the consumer involves the coordinated efforts of many people, from production through marketing. This process can be broken down into three primary divisions, or stages: (1) producers, (2) processors, and (3) market specialists.

PRODUCERS

It is impossible to describe the "average" American dairy farm, for producers can choose to tailor their operations in an infinite number of ways. Herd size, breeds, milking equipment, frequency of milking, and intensity of production all enter into the makeup of the dairy farm.

In general, however, dairy operations can be classified in the following manner:

1. **Grade or mixed herds.** These herds may consist of a mixture of registered and unregistered cows, and they may range in size from about 20 to 5,000 lactating cows. Production is the sole determinant of whether a cow is kept or culled and is the primary basis for providing income for the dairy operator. If a cow becomes a health problem or fails to meet the production standards set by the operator, she is generally sent to slaughter.

2. **Registered herds.** Operators of registered dairy herds are interested in two aspects of dairying: (a) production, and (b) improvement of genetic potential. Through careful record keeping and well-planned breeding programs, registered dairy herd operators provide the industry with top bulls and heifers. Thus, premium prices are paid for young stock from a registered herd, thereby supplementing the income received from milk production.

3. **Dairy herds as sideline.** In many areas which are not conducive to large, intensified dairy production, farmers maintain a small dairy herd for supplemental income. In addition to the dairy cows kept on the farm, the farm operator may grow cash crops or raise other livestock.

4. **Part-time dairying.** Some people keep a few milk cows in addition to holding a full-time job off the farm.

5. **Dairy beef.** When prices are favorable, dairy steers are fed for beef or young dairy bulls are raised for early slaughter as veal calves. With the increasing consumer demand for leaner beef, more and more dairy steers have gone into the feedlot for finishing. In addition, many virgin heifers are bred to beef bulls so that the resulting calves will be smaller and pose fewer calving problems. Thus, a good market for dairy × beef calves is available.

Through the years, dairy farms have become larger and more specialized. The practices of separating cream, making butter, processing market milk, growing and mixing concentrates, and keeping bulls have largely disappeared from the average dairy farm; and this trend will continue. Dairy producers tend to buy more of their feed and replacements, instead of investing in more land. A few produce part or all of their forage requirements. Some have specialized heifer-raising operations.

Most dairy farms in the major dairy areas—especially the Lake States, Corn Belt, and Northeast—continue to be operated as family enterprises. However, as dairy farms become multiple units, operating agreements—such as parent-child agreements—become more important. Through a partnership, or by incorporating, it is possible (1) to assure continuity of the enterprise by transfer of ownership from one generation to the next without excessively heavy tax penalty, and (2) to arrange for the necessary capital for prospective producers who cannot finance their own operations.

A limited number of dairy farms are owned and operated by milk-processing plants, or are under corporations, or in cow pools and cooperatives.

The trends in the U.S. dairy industry are as follows:

1. **Decline in dairy farms; increase in herd size.** The number of farms reporting dairy cows has declined sharply and will, in all likelihood, continue to decline in the future. During the 39-year period 1950 to 1989, the number of dairy farms declined by 90%, but herd size increased by 1,020% (Table 1–7).

TABLE 1–7
TOTAL NUMBER OF FARMS AND FARMS WITH MILK COWS[1,2]

Census of	Farms Reporting Milk Cows	
	Total	Average Herd Size
(year)	(1,000)	(number)
1900	4,514	3.8
1910	5,141	3.3
1920	4,461	4.4
1925	3,729	4.7
1930	4,453	4.6
1935	—	—
1940	4,644	5.2
1945	—	—
1950	3,648	5.8
1955	2,936	6.9
1960	1,792	9.2
1965	1,134	12.9
1970	568	19.7
1975	414	25.6
1989	205	49.0

[1]Data through 1975 from *1978 Dairy Producers Highlights*, National Milk Producers Federation, p. 2.

[2]Data for 1989 from *Dairy Background for 1990 Farm Legislation*, USDA, ERS, Commodity Economics Division, p. 4, Table 1. In 1989, commercial dairy farms (farms with 10 or more milk cows) were estimated at around 160,000 with an average of around 65 cows per farm.

2. **Small dairy farms went out of business; large dairy farms got bigger.** The number of farms with 19 or fewer cows decreased by 96% from 1959 to 1987 (see Table 1–8). The number of farms with 20 to 49 dairy cows also decreased, while the number of farms with over 50 cows dramatically increased in the United States.

TABLE 1–8
FARMS REPORTING MILK COWS BY HERD SIZE[1]

Herd Size	Farms Reporting Milk Cows			
	1959	1969	1978	1987
	(1,000)	(1,000)	(1,000)	(1,000)
1–19	1,706	402	168	66
20–49	243	172	101	68
50–99	30	42	48	48
100 plus	7	11	16	20
Total	1,986	627	333	202
	Percent Distribution			
1–19	85.9	64.1	50.3	32.5
20–49	12.2	27.4	30.4	33.5
50–99	1.5	6.7	14.4	23.9
100 plus	0.4	1.8	4.9	10.1
Total	100.0	100.0	100.0	100.0

[1]*Dairy Background for 1990 Farm Legislation*, USDA, ERS, Commodity Economics Division, p. 5, Table 2.

3. **Cow numbers decreased.** During the 15-year period of 1975 to 1990, the number of milk cows in the United States declined from 11,220,000 to 10,149,000.

4. **Both herd size and production per cow have increased.** Tables 1–8 and 1–15 (pp. 10, 15), illustrate these trends well.

PROCESSORS

Fig. 1–6. Ice cream—a longtime favorite dairy product. (Courtesy, United Dairy Industry Assn., Rosemont, IL)

The dairy processing industry has undergone marked change in recent decades, with substantial gains in efficiency and reductions in real costs. Changes in the industry in recent decades include: fewer but larger plants, increased importance of producer cooperatives, and regional shifts precipitated by population shifts and shifts in milk production in excess of fluid sales. The number of plants producing cottage cheese and butter dropped over 90% from 1950 to 1988. Hard cheese and ice cream plants declined by approximately three-fourths and nonfat dry milk plants by over 80% (Table 1–9). In 1988, average output per plant was over 15 times the 1950 level for butter and cheese, about 7 times for nonfat dry milk and ice cream, and 18 times for cottage

TABLE 1–9
NUMBER OF DAIRY PRODUCT
MANUFACTURING PLANTS, SELECTED YEARS[1]

Product	1950	1970	1980	1983	1988
Hard cheese	2,158	963	737	696	573
Butter	3,060	622	258	222	165
Nonfat dry milk (human food)	459	219	113	101	76
Hard ice cream	3,269	1,628	949	862	765
Cottage cheese curd . . .	1,900	593	269	240	185

[1]Source: *Dairy Products, Annual Summary.* USDA, National Agricultural Statistics Service, various issues.

cheese. Automation and technological advances, such as continuous churns, have increased economies of size in processing. However, changes in assembly and distribution costs were probably of equal importance.

A brief relative to each of the branches of the dairy processing segment follows:

1. **Fluid milk.** This type of processing requires only a minimum of product alteration. Milk is routinely pasteurized and homogenized and perhaps modified in such a way as to enhance its marketability; e.g., low-fat skim milk, cream, or vitamin D fortified milk. This branch of processing is centered primarily around population centers where the highly perishable product can be readily transported to the consumer.

2. **Butter.** The three leading butter producing states by rank are: Wisconsin, California, and Minnesota. This branch of the processing industry serves as an outlet for excess milk. When the volume of milk is great, much of it is diverted to butter—a commodity that is not as perishable as milk in fluid form.

3. **Cheese.** The cheese industry is located in the same general region as the butter industry, with Wisconsin the leading cheese-producing state, manufacturing one-third of the cheese in the United States. Minnesota ranks second in cheese production, California ranks third, and New York ranks fourth. With cheese consumption continuing to increase in the United States, the relative importance of this branch of the processing industry will increase.

Fig. 1–7. Cheese, the fastest growing dairy product. (Courtesy, United Dairy Industry Assn., Rosemont, IL)

4. **Ice cream and frozen dairy products.** Ice cream consumption has continued to expand in recent years. Additionally, ice milk, frozen yogurt, and other frozen deserts have been developed and marketed successfully. With improved manufacturing techniques and the development of new frozen products, this branch of the processing industry will continue to expand.

5. **Nonfat dry milk.** Perishability and bulk of fluid milk have been major drawbacks to transporting it over long distances. However, recent developments of the nonfat dry milk industry have made it possible to export milk abroad. Also, dry milk may be transported over relatively long distances domestically (1) to consumers who desire a cheap source of milk, and (2) to processors of other dairy products in areas where milk for manufacturing may not be readily available. The leading states in the production of nonfat dry milk for human food, by rank, are: California, Minnesota, Iowa, and Wisconsin.

MARKET SPECIALISTS

Once milk has been processed for marketing, the finished product must be packaged, distributed, and promoted in such a way as to achieve maximum consumer demand. Marketing of milk is a diverse field with each aspect requiring the services of a specialist in that particular area. The following market specialists are involved in the marketing of milk:

1. **Market forecasters.** These specialists accumulate a large volume of data to predict what future trends in supply and demand will ensue. If there is too great a volume of grade A milk on the market, thereupon depressing prices, these specialists recommend possible alternatives to decrease the volume, such as diverting fluid milk to manufacturing channels.

2. **Research and development specialists.** These specialists are responsible for improving the quality and acceptability of present dairy products and for developing new dairy products which will have consumer appeal.

3. **Advertisers.** These specialists study the wants and desires of consumers and tailor the marketing of dairy products to meet these needs. Packaging and promotion can turn a relatively unknown product with low demand into one with great appeal.

4. **Public relations and education experts.** These specialists are playing an ever-increasing role in marketing. Their primary function is to promote the dairy industry as a whole rather than individual products. Their main responsibility is to educate the public relative to the nutritional and economic merits of dairy products.

IMPORTANCE OF THE U.S. DAIRY INDUSTRY

Milk is an important constituent of the American diet in terms of per capita consumption (Table 1–10).

In 1988, the total milk equivalent consumption of all dairy products was 585 lb, a 21% decline from 1950. But the growth in human population offset per capita decline, with the result that total consumption increased. From Table 1–10, several trends are apparent. Americans have become weight conscious in recent years leading to a decreased per capita consumption of fluid whole milk and cream and an increased per capita consumption of low-fat milk. Per capita fluid whole milk consumption declined 64% from the period of 1950 to 1988, while per capita low-fat milk consumption increased 746%. Of the manufactured dairy products, per capita consumption of butter declined 59%, and evaporated and condensed milk 62%, in the period 1950 to 1988. Per capita total cheese consumption increased 206% in this same period. Per capita consumption of ice cream has remained relatively constant since 1950.

TABLE 1–10
DAIRY PRODUCTS: U.S. PER CAPITA CIVILIAN CONSUMPTION, SELECTED YEARS[1, 2]

| | Fluid Milk Products | | | | | | | | | | Manufactured Milk Products | | | | | | | | | | | | |
| | Fluid Whole Milk | | Cream | | Low-fat Milk | | Total Product | | Whole Milk Equiv. of Butterfat | | Butter | | Total Cheese | | Cottage Cheese | | Ice Cream | | Evaporated and Con-densed Milk | | Nonfat Dry Milk | | All Milk Equivalent | |
	(lb)	(kg)	(lb)	(kg)	(lb)	(kg)	(lb)	(kg)	(lb)	(kg)	(lb)	(kg)	(lb)	(kg)	(lb)	(kg)	(lb)	(kg)	(lb)	(kg)	(lb)	(kg)	(lb)	(kg)
1950	278	126	11.1	5.0	15.6	7.1	304	138	321	146	10.7	4.9	7.7	3.5	3.1	1.4	17.2	7.8	20.5	9.3	3.7	1.7	740	336
1955	290	132	9.6	4.4	20.0	9.1	320	145	326	148	9.0	4.1	7.9	3.6	3.9	1.8	18.0	8.2	16.2	7.3	5.5	2.5	706	320
1960	276	125	9.1	4.1	23.8	10.8	309	140	309	140	7.5	3.4	8.3	3.8	4.8	2.2	18.3	8.3	13.7	6.2	6.2	2.8	653	296
1965	264	120	7.6	3.4	34.0	15.4	306	139	294	133	6.4	2.9	9.6	4.4	4.7	2.1	18.5	8.4	10.7	4.9	5.6	2.5	620	281
1970	229	104	5.6	2.5	57.5	26.1	292	132	260	118	5.3	2.4	11.5	5.2	5.1	2.3	17.7	8.0	7.1	3.2	5.4	2.4	561	254
1975	195	88	5.9	2.7	84.7	38.4	286	130	244	111	4.8	2.2	14.5	6.6	4.7	2.1	18.6	8.4	5.3	2.4	3.3	1.5	546	248
1980	146	66	5.0	2.3	95.0	43.5	246	112	194	88	4.5	2.0	17.8	8.1	4.5	2.0	17.3	7.9	7.1	3.2	3.0	1.4	543	247
1985	120	55	6.0	2.7	107.0	48.6	239	109	186	84	4.9	2.2	22.6	10.3	4.1	1.9	18.0	8.2	7.9	3.6	2.2	1.0	593	269
1988	100	45	6.0	2.7	132.0	59.9	238	108	184	84	4.4	2.0	23.6	10.7	3.8	1.7	17.8	8.1	7.9	3.6	2.7	1.2	585	266

[1]1950 to 1975 data from *1978 Dairy Producer Highlights*, National Milk Producers Federation, Washington, DC, p. 18. Hawaii and Alaska were included beginning in 1960.

[2]1980, 1985, and 1988 data from *1989 Dairy Producer Highlights*, National Milk Producers Federation, Arlington, VA, p. 17.

Although the per capita consumption of milk has declined in recent years, milk and other dairy products still remain one of the best food buys. In fact, the real price of milk, butter, cheese, and ice cream, as shown by minutes of work required to buy these products has actually declined (see Table 1–11). In 1970, it took the average worker 10.3 minutes to earn enough money to buy ½ gal of milk. In 1990, this figure was only 7.9 minutes.

Table 1–12 shows the quantity of milk going into different channels, and Fig. 1–10 (next page) shows the relative importance of each outlet.

About 37.3% of the milk and cream produced by farmers in 1990 was consumed in fluid form. In the 1930s, 54% of the milk supply was used for butter, compared to 16.9% in 1990.

TABLE 1–11
MINUTES OF WORK REQUIRED TO PURCHASE MILK AND OTHER DAIRY PRODUCTS AT STORES, 1970–1990[1, 2]

Year	Milk	Butter	American Cheese	Ice Cream
	(½ gal)	(1 lb)	(½ lb)	(½ gal)
1970	10.3	15.5	9.0	15.1
1975	9.8	12.7	9.5	15.2
1976	9.5	14.5	9.9	14.6
1977	8.9	14.1	9.4	14.3
1978	8.6	14.4	9.3	14.2
1979	8.9	15.0	9.6	14.6
1980	8.7	15.5	9.7	15.0
1981	8.4	15.0	9.6	15.1
1982	7.9	14.4	9.3	14.8
1983	7.7	14.0	9.0	14.7
1984	7.4	13.8	8.9	14.5
1985	7.1	13.4	8.7	14.5
1986	6.9	13.3	8.0	14.5
1987	7.0	13.2	8.2	14.9
1988	7.0	12.7	8.2	14.5
1989	7.3	12.2	8.4	14.9
1990	7.9	11.0	N/A	14.4

[1]From *1991 Milk Facts*, Milk Industry Foundation, Washington, DC, p. 7.

[2]Based on hourly wage rates of all manufacturing industries. Source: USDA

TABLE 1–12
HOW THE U.S. MILK SUPPLY IS USED[1]

Product	Milk Equivalent	
	(mil lb)	(mil kg)
Fluid milk and cream sales (25.8 billion qt)	55,370	25,138
Cheese.	47,368	21,505
Creamery butter	25,043	11,370
Frozen dairy products[2].	12,307	5,587
Used on farms where produced	2,048	930
Evaporated and condensed milk	1,926	874
Other uses	4,413	2,004

[1]*1991 Milk Facts,* Milk Industry Foundation, p. 29. Data for 1990.

[2]Plus 2,014 mil lb of milk equivalent in other manufactured dairy products used in production of frozen dairy products.

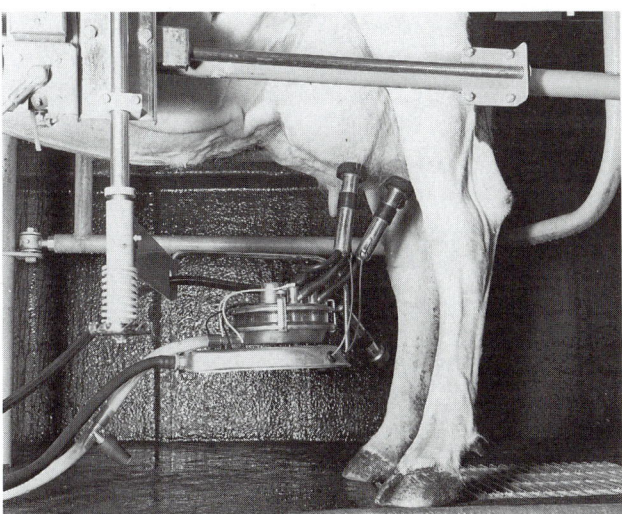

Fig. 1–8.   Milk is a good buy in the United States because the modern dairy industry is automated and very efficient. (Courtesy, Babson Bros., Oak Brook, IL)

Fig. 1–9.   Frozen yogurt, a popular dairy product. (Courtesy, United Dairy Industry Assn., Rosemont, IL)

Today, more milk is used for the manufacture of cheese than for butter (see Fig. 1–10).

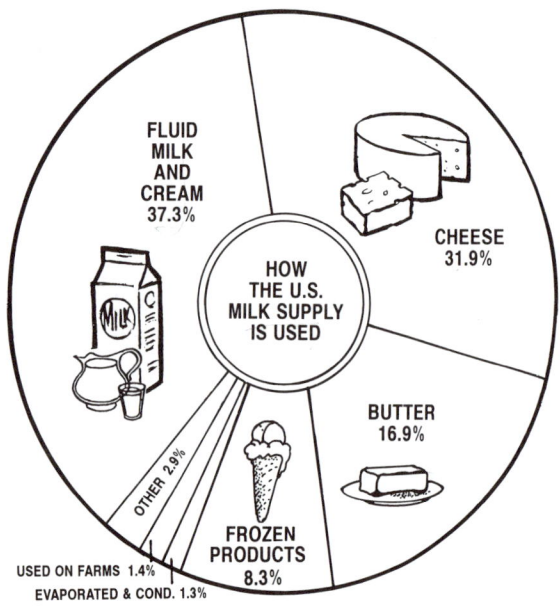

Fig. 1–10. The uses of milk, in percentages. (From *Dairy Situation and Outlook Report*, USDA, ERS, DS-430, July 1991, p. 13. Data for 1990.

DAIRY FARM INCOME

Table 1–13 shows the importance of the U.S. dairy industry. As noted, cash farm income from milk and cream totaled almost $20 billion in 1990.

Table 1–14 shows the importance of dairy products as

TABLE 1–13
U.S. DAIRY FARM DATA[1]

Milk cows on farms (not including heifers not fresh) . . . . (no.)	10,127,000
Farm milk production . . . . (lb)	145,300,000,000
. . . . (kg)	65,966,200,000
Average production per cow . . . . (lb)	14,642
. . . . (kg)	6,647
Cash farm income from milk and cream . . . . . . ($)	19,952,300,000

[1]*1991 Milk Facts*, Milk Industry Foundation, pp. 23, 27. Data for 1990.

compared to the other sources of farm and ranch income. Dairy products rank second only to cattle and calves in percentage of livestock receipts. Note, too, that the income from dairy products exceeds the income from feed crops. The importance of dairy cattle as a source of farm revenue is further magnified when one considers that the income from

the sale of cows, heifers, and calves from dairy herds amounts to about $2 billion annually—a figure that is included in the total income from the sale of cattle and calves.

TABLE 1–14
U.S. FARM INCOME BY COMMODITY[1]

Commodity	Rank by Commodity Group	Income	Percentage of Commodity Group
		(mil $)	(%)
Livestock and poultry:			
Cattle and calves	1	36,676	43.8
Dairy products	2	19,401	23.2
Poultry and eggs	3	15,346	18.3
Hogs	4	9,426	11.3
Sheep and lambs	5	489	0.06
Other	6	2,386	2.8
Total		83,724	100.0
Crops:			
Feed crops	1	16,656	22.1
Oil-bearing crops	2	12,172	16.1
Vegetables and melons . . .	3	11,340	15.0
Fruits and tree nuts	4	9,020	12.0
Food grains	5	6,139	10.7
Cotton (lint and seed)	6	4,740	6.3
Tobacco	7	2,381	3.2
Other	8	11,068	14.7
Total		75,449	100.0

[1]Source: *Agricultural Statistics 1990*, USDA, p. 391, Table 575. Data for 1989.

LEADING DAIRY STATES

Fig. 1–11. Guernsey cows at *Hoard's Dairyman* Farm, Ft. Atkinson, Wisconsin, the leading dairy state of the nation. (Courtesy, The American Guernsey Cattle Assn., Reynoldsburg, OH)

Milk is produced in every state of the Union. However, the greatest concentration of dairy cows is found in those areas with the densest human populations. This is as one would expect from the standpoint of the demand for, and the marketing of, fresh milk—a highly perishable product.

Table 1–15 lists the ten leading states in dairy cattle numbers by rank. Human population centers, which provide

improved breeding, feeding, disease control, and management. With this increased production, efficiency has increased, also.

The cows which were in record-keeping dairy herd improvement associations, and which obviously were the more efficient ones, produced an average of 17,612 lb of milk and 652 lb of butterfat in 1989.

TABLE 1–15
TEN LEADING STATES IN (1) MILK COWS ON FARMS, (2) MILK PRODUCTION PER COW, AND (3) TOTAL MILK PRODUCTION[1]

Rating	State	Number Milk Cows on Farm	State	Milk Production per Cow[2]	State	Total Milk Production[2]
		(thousands)				(mil lb)
1	Wisconsin	1,753	New Mexico	18,815	Wisconsin	24,400
2	California	1,135	Washington	18,557	California	20,953
3	New York	768	California	18,461	New York	11,102
4	Minnesota	710	Arizona	17,447	Minnesota	10,006
5	Pennsylvania	683	Colorado	17,182	Pennsylvania	9,933
6	Texas	386	Nevada	16,600	Texas	5,539
7	Michigan	344	Idaho	16,475	Michigan	5,233
8	Ohio	342	Oregon	16,273	Ohio	4,495
9	Iowa	305	Utah	15,838	Washington	4,398
10	Washington	237	Rhode Island	15,455	Iowa	4,330

[1]From *1991 Milk Facts*, p. 22. Data for 1990.

[2]To convert pounds to kilograms, multiply by 0.454.

a large market for milk, have been a major factor in determining the intensity of dairying. Also land, climate, and feed exert a considerable influence. Wisconsin is, by far, the leading dairy state in the United States, having the most milk cows on farms. Some states have a well-managed dairy industry; but because of their small size, they do not have a large total dairy cattle population. Therefore, proficiency of dairy herd management is best indicated by average production per cow. Additionally, several states, notably those of the Southwest, have highly intensified dairy farms due to the limitations of arable land and cost of irrigation. They must have high-producing cows in order to meet high overhead. New Mexico leads the United States in milk production per cow, with an average of 18,815 lb in 1990.

MILK COW PRODUCTION AS RELATED TO HUMAN POPULATION

The average annual production per cow in the United States has steadily increased, from 3,138 lb in 1920 to 14,642 lb per cow in 1990; simultaneously, cow numbers have declined while the human population has increased (see Table 1–16 and Fig. 1–12, next page). This increase in average production per cow can be attributed to recent progress in

TABLE 1–16
U.S. MILK PRODUCTION PER COW, MILK COW NUMBERS, AND HUMAN POPULATION

Year	Milk Production Per Cow[1]		Number of Milk Cows[2]	Human Population[3]	Ratio of Milk Cows to Humans
	(lb)	(kg)			(Cows:Humans)
1920	3,138	1,423	21,455,000	105,710,620	1:4.9
1930	4,508	2,045	23,032,000	122,775,046	1:5.5
1940	4,625	2,098	24,940,000	131,669,275	1:5.6
1950	5,314	2,410	23,853,000	150,697,361	1:6.6
1960	7,002	3,176	19,527,000	180,684,000	1:9.3
1970	9,385	4,257	12,483,000	202,711,000	1:16.2
1980	11,891	5,399	10,799,000	226,545,805	1:21.0
1990	14,642	6,655	10,127,000	248,082,000	1:24.5

[1]USDA sources.

[2]USDA sources.

[3]*World Almanac and Book of Facts 1991*.

MILK PRODUCTION, NUMBER OF COWS, AND MILK PER COW

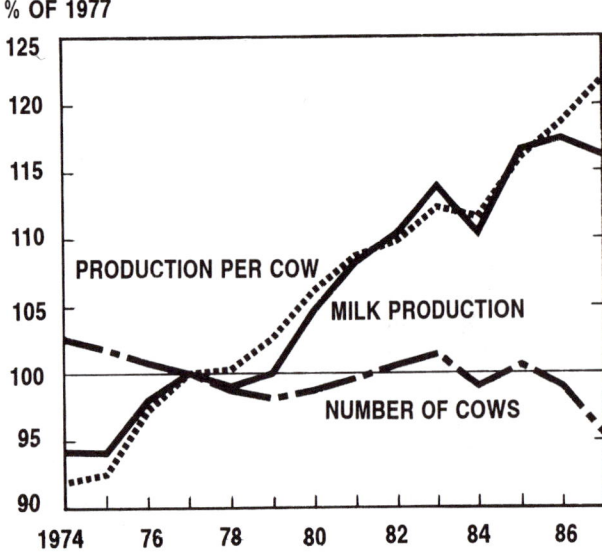

% OF 1977

Fig. 1–12. Although cow numbers have gone down, milk production has increased due to higher production per cow. (Source: *1988 Agricultural Chartbook*, USDA, Agricultural Handbook No. 673, p. 90, Chart 197)

DAIRY FARM BUSINESSES

On a typical dairy farm, the business may consist of one or more of the following enterprises: (1) the milking herd, (2) raising heifers for replacement purposes, (3) dairying and production of forage, and (4) dairying and production of cash crops.

Fig. 1–13. Good Jersey cows on good pasture. Dairying is a stable business. (Courtesy, The American Jersey Cattle Club, Reynoldsburg, OH)

As can be seen, dairy farm businesses are seldom comparable to one another because the proportions of the enterprises vary. Hence, from the standpoint of records and analyses, it is important that each enterprise of the dairy farm be considered separately. Only by doing so, is it possible to determine which enterprises are most profitable, and which are least profitable. Such information makes it feasible for the owner to make important managerial decisions as to the future course—how much and what feed crop to grow, whether to enlarge or reduce the number of cows in the milking herd, whether to raise replacement heifers, and what to do to reduce costs or improve income in each enterprise in the dairy farm business.

FACTORS FAVORABLE TO DAIRY PRODUCTION

Some of the special advantages of dairy production as compared to other livestock on the farm are:

1. **Dairying is a stable business.** Total milk production does not vary from year to year as much as the output of most other agricultural products; the change is often less than 1%, and usually not more than 2%.

2. **The dairy cow is unequaled as an efficient producer of human food.** A cow producing 20,000 lb of milk per year supplies as much food nutrients as are produced by four 1,250-lb steers. Additionally, she is still available for more productive years. See Table 1–3 for the feed to food efficiency rating of the dairy cow compared to other species of animals.

3. **Steady income is assured.** A grain farmer, a fruit producer, or a vegetable grower receives income only when products are sold, usually once per year. Likewise, beef cow-calf producers secure most of their income when the calf crop is marketed, and, here again, this is generally once per year. On the other hand, a dairy producer receives a regular milk check at frequent intervals (bi-weekly or monthly) throughout the year.

4. **Steady employment is provided for labor.** Many types of agricultural work are highly seasonal, with the result that a labor force must be increased and decreased at such intervals as necessary, particularly during harvest. In the dairy enterprise, however, fairly uniform labor needs exist throughout the year. This makes it possible to keep better quality employees on a permanent basis.

5. **Dairy cows use much unsalable roughage.** Each year, considerable amounts of roughage are produced on the farms and ranches of America which would have little value if not used by dairy cattle and other ruminants. Also, much of the rolling land on which such feeds are grown is unsuited to the production of grain or other crops.

6. **Soil fertility is maintained.** By returning the manure to the land, the fertility and physical condition of the soil are preserved.

FACTORS UNFAVORABLE TO DAIRY PRODUCTION

Among the factors which, under certain conditions, may be unfavorable to dairy production are:

1. **Considerable capital is required.** The investment

for land, buildings, equipment, and cows is double, or more, what is normally required in investment per cow in a beef cow-calf enterprise.

2. **Successful dairy management necessitates superior training.** The larger the dairy enterprise, the more important this becomes. A superior manager must be knowledgeable in the basic sciences, business administration, animal physiology, and nutrition; and possess the necessary personality and ability to weld this knowledge into a smooth functioning, efficient production unit.

3. **The bewildering number and kinds of regulatory programs.** The federal, state, and local regulatory programs necessitate that the dairy producer be familiar with, and follow, those that are applicable to the operation. No matter how noble the objectives of these regulatory programs, this takes considerable time, and often it is a frustrating experience.

4. **Dairying is confining.** Unlike many agricultural endeavors, dairying must be done regularly, and without fail, particularly where it is a market milk enterprise. Thus, an owner or a manager who is interested in vacation time, short work weeks, or even short days, should not enter into a one-worker dairy enterprise.

5. **Hourly returns are low.** Hourly returns to dairy farmers have, on the average, been below returns in many other types of farming, and well below the average for all U.S. manufacturing industries.

6. **The threat of imitations.** Milk—long extolled as the *perfect food*—is being threatened by imitations. New *nondairy products*, such as coffee creamers, are proving to be stiff competition to the dairy industry.

FUTURE OF THE DAIRY INDUSTRY IN THE UNITED STATES

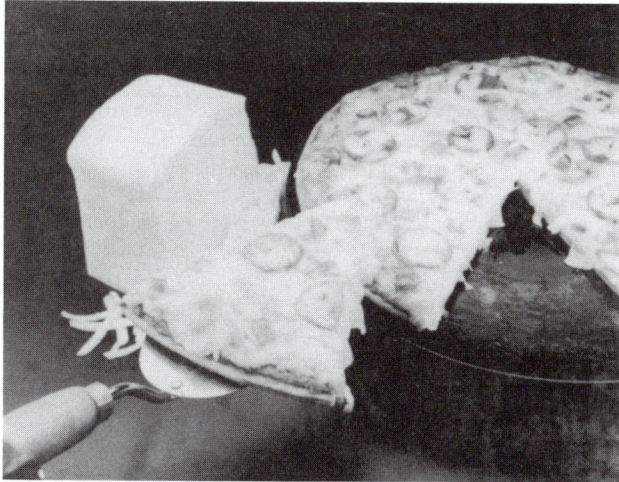

Fig. 1–14. Pizza. Old reliable dairy products, and new and useful products, assure a bright future for the dairy industry. (Courtesy, United Dairy Industry Assn., Rosemont, IL)

One factor, above all others, assures U.S. dairy production a bright future: the efficiency of the dairy cow from the standpoint of transforming plant materials into animal food products for humans.

The following developments will likely characterize the dairy production of the future:

1. **Larger units and increased automation.** Table 1–8 shows that units have become bigger and fewer. This trend will continue, and with it there will be more automation.

2. **Higher production and greater efficiency per cow.** It is evident that, as profit margins narrow and knowledge becomes greater, we shall continue to move, at an accelerated pace, to higher production and greater efficiency per cow. This will be achieved through the application of modern breeding, feeding, milking, management, marketing, and biotechnology.

3. **Fewer dairy producers, fewer cows.** To the year 2,000 and beyond, it is expected that the number of dairy producers and the number of milk cows will decline.

4. **Milk output per worker will increase.** Milk output per worker will increase as a result of increased herd size, higher milk production per cow, and the application of labor-saving machinery and equipment.

5. **Free-stall housing and mechanized milking will increase.** More and more producers will use free-stall housing and mechanized milking systems.

6. **Family farms will continue.** It is expected that the majority of dairy farms will continue to be family owned and operated.

7. **New and useful products.** Intensive research attempts are being made to provide products (a) with the characteristics of fresh milk, but which will last longer in storage, and (b) which can be transported at lower cost.

8. **More dairy beef.** The dairy industry provides an estimated one-fourth of the beef consumed in this country, with these animals being marketed as veal calves, cull dairy cows and bulls, and finished dairy heifers and steers. Improvements in the science and technology of feeding and processing, along with consumer preference for less fat and more lean beef, favor the growing and finishing of more dairy beef.

9. **More training, experience, and business acumen.** The successful dairy manager of the larger and more highly specialized units of the future must have more know-how and experience, and the ability to operate the dairy establishment as a big business.

10. **Yearly production fluctuations will be minimized.** In order to maximize the use of processing equipment, a continuous, even volume of milk must be available. Today, there is an average 12% difference between the high and low months of production. In the future this differential will be decreased.

11. **All milk will be subject to the same quality standards.** In 1990, about 90% of the nation's milk supply was Grade A. In the future, practically all milk, whether used for fluid milk (beverage) or manufactured products, will Grade A.

12. **Changes will be made.** Many factors will determine to what extent the above changes will occur. Feed costs will be a factor. Also, uncertainties about government policies relative to dairy product imports and grain exports will help to determine the degree and direction of change.

QUESTIONS FOR STUDY AND DISCUSSION

1. Why are some mammals other than milk cows—such as water buffalo, goats, and mares—used for milk production in different parts of the world?

2. Outline the basic position of the domesticated cow in the zoological scheme.

3. Discuss (a) the use of cattle in ancient times, (b) cattle in medieval farming, and (c) the history of the dairy industry in the United States.

4. List six important developments of the U.S. dairy industry (Table 1–1) in the approximate order in which they happened.

5. Discuss (a) the nutritive value of milk, and (b) the important role of milk in the diet of the average American.

6. Discuss the role of milk in feeding the hungry people of the world.

7. From a health standpoint, (a) what are the benefits of milk, and (b) what are the health problems associated with milk consumption?

8. Why does the U.S.S.R. hold such a commanding lead in milk cow numbers?

9. Israel and Japan must be doing a lot of things right in order to rank first and second, respectively, in average milk production per cow. Explain.

10. List the components of the dairy industry in the United States. Classify U.S. dairy producers.

11. What are the trends in the U.S. dairy industry?

12. List and discuss the major segments of dairy processing.

13. List and discuss the market specialists involved in the marketing of milk.

14. Discuss (a) trends in dairy product consumption, (b) why milk and other dairy products are a good buy in terms of work required to purchase them, and (c) how the U.S. milk supply is used.

15. Discuss the importance of the U.S. dairy industry in terms of (a) farm income, (b) leading states, and (c) milk production as related to human population.

16. List and discuss the factors *favorable* to dairy production.

17. List and discuss the factors *unfavorable* to dairy production.

18. List and discuss the developments that will likely characterize the dairy production of the future.

19. In the future, how is dairy production likely to fare in comparison with beef production, sheep production, swine production, and poultry production, from the standpoint of increasing or decreasing as cereal grains become scarcer and higher in price?

20. What factors have prompted the trend toward more and more specialization in the dairy enterprise, in contrast to diversification?

SELECTED REFERENCES

Title of Publication	Author(s)	Publisher
Animal Science, Ninth Edition	M. E. Ensminger	Interstate Publishers, Inc., Danville, IL, 1991
Dairy Cattle in American Agriculture	A. R. Porter J. A. Sims C. F. Foreman	Iowa State University Press, Ames, IA, 1965
Dairy Cattle Feeding and Management, Sixth Edition	W. M. Etgen P. M. Reaves	John Wiley & Sons, Inc., New York, NY, 1978
Dairy Cattle Management	J. M. Wing	Reinhold Publishing Corp., New York, NY, 1963
Dairy Cattle: Principles, Practices, Problems, Profits, Third Edition	D. L. Bath, *et al.*	Lea & Febiger, Philadelphia, PA, 1985
Dairy Farm Management	T. Quinn	Delmar Publishing, Inc., Albany, NY, 1980
Large Dairy Herd Management	C. J. Wilcox, *et al.*	University Presses of Florida, Gainesville, FL, 1978
Principles of Dairy Farming, The, Sixth Edition	K. Russell S. Williams	Farming Press Ltd., Fenton House, Ipswich, England, 1972
Science of Providing Milk For Man, The	J. R. Campbell R. T. Marshall	McGraw-Hill Book Company, New York, NY, 1975
Stockman's Handbook, The, Seventh Edition	M. E. Ensminger	Interstate Publishers, Inc., Danville, IL, 1992

Fig. 2–1. The world record milk champion, *Beecher Arlinda Ellen*, a Holstein-Friesian cow. In 1975, this great cow completed a 365-day production record of 55,661 lb of milk in twice-a-day milking. *Ellen* reached a peak of 194.5 lb, or 22.6 gal, per day. Bred and owned by Harold L. Beecher, Rochester, IN. (Courtesy, Mr. Beecher)

Contents

CHAPTER

2

BREEDS OF
DAIRY CATTLE

Fig. 2–2. Holsteins and Jerseys involved in research at the Ohio Agricultural Research and Development Center. (Courtesy, Ohio State University, Columbus)

There are six major breeds of dairy cattle in the United States, all of which have proven satisfactory in different sections of the country.

POPULARITY OF BREEDS

Table 2–1 shows (1) when each of the breed registries was formed, and (2) the 1990 and total registrations to date of the breeds of dairy cattle. Even though data for one year fail to show trends, the recent annual figures reflect the current popularity and numbers of the respective breeds more

than do the total registrations since establishing the breed registry.

CHARACTERISTICS OF BREEDS

Table 2–2 gives in summary form the place of origin and the characteristics of each of the breeds of dairy cattle. Figs. 2–3 to 2–8 (p. 22) show outstanding representatives of each of the breeds.

TABLE 2–1
DAIRY BREEDS: (1) YEAR IN WHICH BREED REGISTRY WAS FORMED,
AND (2) 1990 AND TOTAL REGISTRATIONS

Breed	Breed Association Formed		1990 Registration	Total Registration
	Abroad	In U.S.		
Holstein-Friesian	1873	1871	395,906	16,377,790
Jersey	1833	1868	53,547	4,359,094
Guernsey	1814	1877	14,088	3,620,195
Brown Swiss[1]	1911	1880	11,756	996,319
Ayrshire	1877	1875	8,020	1,123,285
Milking Shorthorn	1822	1882[2]	2,596	246,343

[1]Not including Brown Swiss beef registrations.

[2]The Milking Shorthorn Society of the U.S. was organized in 1910, but both beef and Milking Shorthorns continued to be registered by the American Shorthorn Assn. until 1949.

TABLE 2–2
BREEDS OF DAIRY CATTLE AND THEIR CHARACTERISTICS[1]

Breed	Place of Origin; First U.S. Importation	Color	Distinctive Head Characteristics	Other Distinguishing Characteristics	Disqualifications; Comments
Ayrshire	County of Ayr in southwestern Scotland, prior to 1800. Made by H. W. Hills of Windsor, CT; about 1822.	Light to deep cherry red, mahogany, brown, or a combination of these colors with white, or white alone. Black or brindle is objectionable.	Horns are widespread and tend to curve upward and outward. There are some naturally polled Ayrshires.	The udders are especially symmetrical and well attached to the body. The breed is noted for its style and animation, good feet and legs, and grazing ability.	
Brown Swiss	The Alps of Switzerland. Brought to America in 1869.	Solid brown, varying from very light to dark. White markings are objectionable.	The nose and tongue are black, and there is a characteristic light-colored band around the muzzle. Medium-length horns.	Strong and rugged, with some tendency toward the heavy muscling characteristic of the beef breeds. Calm and unexcitable.	Spotting is undesirable. In 1971, the Brown Swiss Cattle Breeders' Assn. formed the Brown Swiss Beef International, Inc., for registration of beef-type Brown Swiss.
Guernsey	Isle of Guernsey. First U.S. importation was made by an American sailing captain, Captain Prince, of Boston, in 1831.	Fawn with white markings clearly defined; preferably a clear (buff) muzzle.	Good length of head; horns incline forward, are refined and medium in length, and taper toward the tips.	The milk is especially yellow in color; golden yellow skin pigmentation; the unhaired portions of the body are light or pinkish in color (whereas in the Jersey they are near black); calves are relatively small at birth.	
Holstein-Friesian	The Netherlands and northern Germany. The first cattle from Holland were brought to the U.S. by the early Dutch settlers. There is reference to black-and-white cattle in this country as early as 1621.	Black and white or red and white.	Clean-cut broad muzzle, open nostrils, strong jaw, broad and moderately dished forehead, straight bridged nose.	Large angular animal; females should weigh 1,500 lb (mature); males in breeding condition, 2,200 lb.	Black and white animals are disqualified if (1) solid black, (2) solid white, (3) black in switch, (4) solid black belly, (5) one or more legs circled with black that touches the hoof, (6) black on one or more legs from the hoof to above knees or hocks, or (7) grayish color. Suffix "Red" is added to name of red and white animals.
Jersey	Island of Jersey. The first registered Jerseys were imported into the U.S. by John A. Taintor, of Connecticut, in 1850.	Jerseys vary greatly in color, but the characteristic color is some shade of fawn, with or without white markings.	Forehead broad and moderately dished with large, bright eyes. Clean-cut head, proportionate to body.	Jerseys are especially known for their well-shaped udders, strong udder attachments, and ease of calving. They are also very angular and refined.	Total blindness, permanent lameness that interferes with normal function, blind quarter, freemartin heifers and animals showing signs of being operated on or tampered with.
Milking Shorthorn	England; the breed traces to a milking strain of Shorthorns developed by Thomas Bates. First importations to America were made in the 1790s.	Red, white, or any combination of red and white.	Fine horns that are rather short.	Very adaptable.	Tattoo required for registration. No calf is eligible for registration unless its dam was at least 18 months of age at the birth date of the calf, and the sire was at least 9 months of age at the time of service. In 1949, the Milking Shorthorn breed split off from the American Shorthorn Assn. and formed a separate breed registry—the American Milking Shorthorn Society. In 1968, the American Milking Shorthorn Society designated the Milking Shorthorn as a dairy breed, rather than a dual-purpose breed. In 1973, the American Shorthorn Assn. made provision to register Milking Shorthorns in their Herd Book. But there is no reciprocal arrangement for acceptance of Shorthorn (beef) blood in the American Milking Shorthorn Herd Book. All Illawarra cattle imported to this country are eligible for registration in the American Milking Shorthorn Society Herd Book, but they are given an American number.

[1]In addition to the six major dairy breeds listed in Table 2–2, there are limited numbers of two other dairy breeds; namely, (a) Dutch Belted, which originated in Holland, and (b) Red and White (also known as Red and White Holsteins), which evolved from Hostein-Friesian cattle carrying the recessive red color.

Fig. 2–3. Ayrshire cow, *Ardrossan R Com Kate Try It*, the highest scoring cow of the breed in 1990. (Courtesy, Ayrshire Breeders' Assn., Brandon, VT)

Fig. 2–6. Holstein-Friesian cow, *Lawcrest Rotate Mindy*, No. 1 in the United States for protein record. (Courtesy, *Holstein World*, Sandy Creek, NY)

Fig. 2–4. Brown Swiss cow, *Idyl Wild Improver Jinx*, the record selling Brown Swiss cow—at $60,000. (Courtesy, The Brown Swiss Cattle Breeders' Assn. of the U.S.A., Beloit, WI)

Fig. 2–7. Jersey cow, *Rocky Hill Favorite Deb*, breed leader for both milk and fat. (Courtesy, The American Jersey Cattle Club, Reynoldsburg, OH)

Fig. 2–5. Guernsey cow, *Marfred Ron Nadia*, classified as *Excellent*. (Courtesy, American Guernsey Assn., Reynoldsburg, OH)

Fig. 2–8. Milking Shorthorn cow, *Innisfail Wild Rose*, 1990 Class Leader. (Courtesy, American Milking Shorthorn Society, Beloit, WI)

MILK AND BUTTERFAT PRODUCTION OF BREEDS

There are breed differences in milk and butterfat production. Table 2–3 summarizes the averages by breeds. On the basis of milk production, they rank in the following order: Holstein, Brown Swiss, Ayrshire, Milking Shorthorn, Guernsey, and Jersey. However, on the basis of the butterfat test, they rank: Jersey, Guernsey, Brown Swiss, Ayrshire, Holstein, and Milking Shorthorn.

The U.S. milk and butterfat record holders for each of the breeds are given in Table 2–4.

PROGRAMS OF THE REGISTRY ASSOCIATIONS

In addition to registering animals, most of the breed associations promote the following programs:

1. **Production testing in Dairy Herd Improvement Registry (DHIR).** Each breed registry association has a program for testing registered cows under the Unified Rules for Official Testing as adopted by the Purebred Dairy Cattle Association and the American Dairy Science Association. This program is conducted cooperatively by the breed association and the Division of Dairy Herd Improvement Investigations of the U.S. Department of Agriculture (the division which, in cooperation with the states, is responsible for DHIR records). Records recognized as official by both groups are included in one program.

Under DHIR, milk testing is conducted once each month, with the tester obtaining a 24-hour milk weight and butterfat test. Also, the tester secures data on each cow that has freshened, feed consumption and quality, labor, price of milk, etc. All this data is sent to a central laboratory for analysis, with many states cooperating on a regional basis in the Electronic Data Processing Method (EDPM). The machine-processed records are then returned to the producers, giving them current information on milk yield, income over feed costs, milk produced per cow, and other pertinent information to help them in making culling and managerial decisions.

In 1966 and 1967, two breed registry production-testing programs—(a) Advanced Registry (AR), and (b) Herd Improvement Registry (HIR)—were discontinued, and the Dairy Herd Improvement Registry (DHIR) became the official milk-recording program of all the breeds.

TABLE 2–3
BREED DIFFERENCES IN MILK AND
BUTTERFAT TEST AND PRODUCTION[1]

Breed	Milk Production	Butterfat Test	Butterfat Production
	(lb)	(%)	(lb)
Holstein	15,528	3.65	566
Brown Swiss	13,063	3.97	519
Ayrshire	11,730	3.90	457
Milking Shorthorn	10,596	3.64	386
Guernsey	10,573	4.61	487
Jersey	10,150	4.90	497

[1]Data for Table 2–3 was compiled by the Animal Improvement Programs Laboratory, Science and Education Administration, USDA, from records submitted by the 11 dairy records processing centers serving the National Cooperative Dairy Herd Improvement Program.

TABLE 2–4
U.S. PRODUCTION RECORDS FOR EACH BREED[1]

Breed	Name of Cow	Year	Days; Times Milked	Milk Production[2]	Butterfat Test	Butterfat Production
				(lb)	(%)	(lb)
Milk production:						
Holstein	Beecher Arlinda Ellen	1975	(365–X2)	55,661	2.8	1,573
Guernsey	Royal Paulson Champion	1987	(365–X2)	40,240	3.6	1.440
Brown Swiss	Century Acres Liz-C	1980	(365–X2)	37,846	4.4	1,667
Ayrshire	Leete Farms Betty's Ida	1976	(365–X2)	37,170	.4.3	1,592
Jersey	Rocky Hill Favorite Deb	1978	(365–X2)	35,880	5.4	1,923
Milking Shorthorn . .	Washita Ann's Bonnie-EXP	1986	(365–X2)	30,790	2.9	922
Butterfat production:						
Holstein	Breezewood Patsy Bar Pontiac	1976	(365–X2)	47,500	4.7	2,230
Brown Swiss	LPS Telstar Sally May Twin	1990	(365–X2)	33,520	6.2	2,067
Jersey	Rocky Hill Favorite Deb	1978	(365–X2)	35,880	5.4	1,923
Guernsey	Fauver Hill Tel Odette-ET	1989	(365–X2)	35,090	5.1	1,781
Ayrshire	Lette Farms Betty's Ida	1976	(365–X2)	37,170	4.3	1,592
Milking Shorthorn . .	Ashgrove Golden Lily 38th	1982	(365–X2)	25,750	3.7	1,122

[1]Currently, the dairy breed registries are also recording and recognizing protein production, in percentage of protein produced and in total pounds of protein produced in the lactation.

[2]To convert pounds to kilograms, multiply by 0.454.

2. **Type classification.** Since many animals are not exhibited at cattle shows, the associations started a voluntary program of herd classification whereby a qualified classifier, selected by the association at the request of the owner and on a nominal charge basis, comes to the farm and classifies, for type or conformation, each milking animal in the herd. Each animal is rated and placed in the corresponding category, as shown in Table 2–5.

special milk-merchandising programs. These are summarized in Table 2–6.

These awards vary somewhat between breeds, and they are revised from time to time so as to maintain them as worthwhile goals. For breeder recognition, the general factors taken into account are ownership of a certain minimum number of (a) registered females, and (b) animals bred by owner; meeting established minimum production requirements

TABLE 2–5
TYPE CLASSIFICATION BY BREEDS

Nomenclature	Ayrshire	Brown Swiss	Guernsey	Holstein[1]	Jersey	Milking Shorthorn
E–Excellent	90 & over	90& over	90& over	90& over	90& over	90& over
VG–Very Good	85–89	85–89	80–89	85–89	85–89	85–89
GP–Good Plus (D–Desirable in Guernseys, Jerseys, and Milking Shorthorns)	80–84	80–84	70–79	80–84	80–84	80–84
G–Good (A–Acceptable in Guernseys, Jerseys, and Milking Shorthorns)	75–79	75–79	60–69	75–79	75–79	75–79
F–Fair	65–74	65–74	50–59	65–74	Under 74	65–74
P–Poor	50–64	50–64	Under 50	50–64	——	50–64
Year when initiated	1941	1944	1947	1929	1932	

[1]The Holstein Association *classifier* rates each animal on the basis of 17 functional, or Linear Descriptive Traits, following which the classifier uses the following guide to compute a final score: *For cows* —(a) General Appearance, 30%; Dairy Character, 20%; Body Capacity, 20%; and Mammary System, 30%. *For bulls* — General Appearance, 45%; Dairy Character, 30%; and Body Capacity, 25%. Then, the final score represents the degree to which an animal resembles physical perfection. It is expressed in numbers (scores) given in the column headed "Holstein," and in the nomenclature of the first column of this table.

The program has been increasingly utilized by breeders, and it has been highly effective in the general improvement of conformation and in merchandising cattle.

3. **Recognition awards.** Some of the breed associations have established certain programs and recognition awards for breeders, cows, and bulls; and two of them have

under one of the breed testing programs; meeting established minimum type classification requirements; and evidence of a healthy herd, especially with respect to tuberculosis and brucellosis. Sire and dam awards are based on their apparent ability to transmit a high level of production and/or type to a specified percentage of offspring.

TABLE 2–6
BREED RECOGNITION AWARDS

Breed	Breeder Award	Sire Award	Dam Award	Merchandising Program
Ayrshire	Constructive Breeder	Approved Sire	Approved Dam	
Brown Swiss	Honor Roll Protein Award	Superior Sire Qualified Sire	Elite Cows Certified Cows Superior Brood Cows	
Guernsey	Gold Star Breeder Gold Star Herd Double Gold Star Breeder and Herd		Gold Star Cow	Golden Guernsey Milk (started in 1923).
Holstein-Friesian	Progressive Breeders Registry Award Progressive Genetics Herd Award	Gold Medal sire	Gold Medal Dam Dam of Merit	International livestock marketing.
Jersey	Master Breeder	None—the Jersey breed uses USDA Sire Summaries.	Leading living lifetime production for milk, fat, and protein—class leaders.	All-Jersey milk. Jersey marketing service, domestic and international.
Milking Shorthorn	Progressive Breeder Award	None	Medal Certificates Memorial Awards	

QUESTIONS FOR STUDY AND DISCUSSION

1. How do you account for the fact that, currently, more than four times as many Holstein-Friesian cattle are being registered in the United States than all other breeds combined? Is it good or bad that one breed should occupy such a dominant position?

2. Of what importance are the (a) distinguishing characteristics, and (b) disqualifications of the breeds of dairy cattle?

3. Justify any preference or bias that you may have for one particular breed of dairy cattle.

4. Is there a need for more breeds of dairy cattle than we now have in the United States?

5. Dutch Belted dairy cattle are almost an *endangered breed*. (See footnote No. 1 of Table 2–2.) Is it important that a breed of such minor importance be perpetuated?

6. Are breeds of dairy cattle likely to decline in importance as happened in the poultry industry?

7. How do you account for the difference between breeds in milk and butterfat production?

8. Do milk and butterfat records reflect efficiency of feed utilization? Why are the dairy breed registry associations also recording and recognizing protein production?

9. What factors make for a world record milk champion, such as *Beecher Arlinda Ellen*?

10. Why is the U.S. Department of Agriculture involved in milk production records?

11. Of what value are breed type classifications?

12. Of what value are breed recognition awards?

13. Obtain breed registry association literature for one breed of dairy cattle. Evaluate the soundness and value of the material that you receive.

SELECTED REFERENCES

Title of Publication	Author(s)	Publisher
Dairy Cattle Breeds	R. B. Becker	University of Florida Press, Gainsville, FL, 1973
Modern Breeds of Livestock, Fourth Edition	H. M. Briggs D. M. Briggs	The Macmillan Company, New York, NY, 1980
World Dictionary of Livestock Breeds	I. L. Mason	C.A.B. International, Wallingford, Oxon, U.K., 1988

Fig. 2–9. Dutch Belted, which originated in Holland, a less populous breed of dairy cattle. (Courtesy, Dutch Belted Cattle Assn. of America, Inc., Venus, FL)

Fig. 2–10. Ayrshire cow at 19 years of age and her calf.

Fig. 2–13. Holstein-Friesian cows. (Courtesy, Holstein-Friesian Assn. of America, Brattleboro, VT)

Fig. 2–11. Brown Swiss cattle on pasture. (Courtesy, Brown Swiss Cattle Breeders' Assn. of the U.S.A., Beloit, WI)

Fig. 2–14. Jersey cows on pasture. (Courtesy, Jersey Cattle Club, Reynoldsburg, OH)

Fig. 2–12. Guernsey cows. (Courtesy, American Guernsey Assn., Reynoldsburg, OH)

Fig. 2–15. Milking Shorthorns. (Courtesy, American Milking Shorthorn Society, Beloit, WI)

Fig. 3–1. These four Ayrshire cows collectively produced 1,000,000 lb of milk during their lifetime—and they were still living and producing at the time this photo was taken. Selecting and culling influence production and profits! (Courtesy, Ayrshire Breeders' Assn., Brandon, VT)

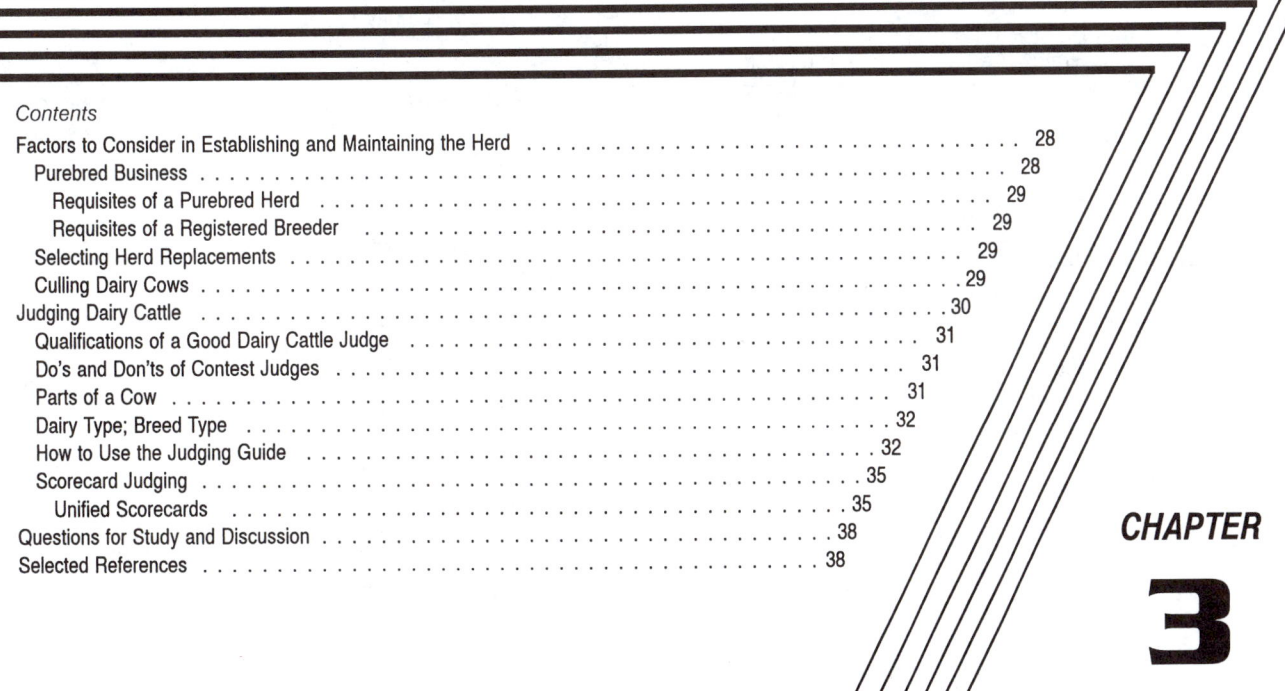

Contents

CHAPTER

3

ESTABLISHING
THE HERD;
SELECTING
AND JUDGING

Whether establishing a new herd or maintaining an old one, dairy producers must constantly appraise or evaluate their animals—they must select, retain, cull, and judge.

There are four bases of selections: (1) selection based on type or individuality, (2) selection based on pedigree, (3) selection based on show-ring winnings, and (4) selection based on production testing. Progressive producers make use of all four bases, with increased emphasis on production testing. They evaluate cows on the basis of Dairy Herd Improvement (DHI) testing, and they select sires on the bases of U.S. Department of Agriculture—Dairy Herd Improvement Association (USDA-DHIA) sire summaries, Predicted Difference (PD) and Repeatability (R), all of which are covered in Chapter 5, Breeding Dairy Cattle.

This chapter is devoted to selection based on type or individuality—commonly called judging.

FACTORS TO CONSIDER IN ESTABLISHING AND MAINTAINING THE HERD

In establishing a dairy herd, the following factors must be considered:

1. **Dairy or dual-purpose type.** Approximately 20% of the cows of the United States are kept for milk production. Of these, about 70% belong to the six major dairy breeds, and the remainder to the dual-purpose breeds, beef cattle, or nondescript breeding. Where dairying is highly specialized, most milk cows are of strictly dairy breeding, rather than of dual-purpose type.

2. **Grades or purebreds.** Technically, a purebred animal is one that can meet ancestry requirements in one of the breed registry associations, whereas a registered animal is a purebred which has been recorded in one of the registry books. Grade animals are those that are not registered or eligible for registry; however, such animals frequently approach purebred status as a result of several generations of breeding up by using sires of one breed. Thus, if a registered sire is used successively for seven generations, the final offspring will, mathematically speaking, consist of 99% registered parentage. From this, it can be reasoned that, from an inherited production potential, the gap between registered and grade animals is often very small.

In general, the dairy producer who is inexperienced in handling dairy animals, or who has a limited amount of capital, should start with grade animals, then improve them by the use of good purebred sires through artificial insemination programs. However, the person who is experienced, and who has adequate capital, may well consider the purebred business.

3. **Choice of a breed.** The choice of a breed is usually made on the basis of personal preference, prior association with the breed and the breeders, and the availability of the breed in the immediate area. Perhaps it is well to add that the choice of a breed is likely of less importance than the choice of good individuals within the breed, simply because there is more difference within breeds than between breeds.

4. **Buying cows, heifers, or calves.** In starting a herd, three methods are available: buying cows, buying heifers, or buying calves. Also, cows and heifers of breeding age may be either open or bred at the time of purchase. The choice between the alternatives should be determined primarily by (a) the time when it is desired to be in production, (b) available capital, and (c) experience.

Where there is no question pertaining to the honesty and integrity of the seller, and good cows with production records back of them can be acquired at what is considered to be a fair and reasonable price, this usually constitutes the best buy. In acquiring cows, it is well to keep in mind that, on the average, they remain in the milking herd for about 4 years only, and for the most part, they are culled from, or leave, the herd before they are 7 years of age.

The purchase of heifers at breeding age, or as "springers," to start a herd is a very popular method. When buying heifers, consideration should be given to the caliber of their dams and the record of their sires.

The purchase of calves requires the least initial capital of any of the methods, but it also takes more time to get into production. In many ways, however, the purchase of calves offers the best opportunity to get high-quality breeding animals.

5. **Production records.** Production records should be used in selecting individual animals, with proper consideration given to the environmental factors under which the records were made.

6. **The disease problem.** In establishing a herd, one should take every possible precaution to avoid bringing in a disease, especially such diseases as tuberculosis, brucellosis, leptospirosis, trichomoniasis, and mastitis. Despite all precautions, however, it is well that newly acquired animals be isolated for a period of 30 to 60 days, and that they be retested at the end of that period before being placed in the herd.

Of course, there are other factors to consider in establishing a herd; among them, longevity. Since it normally takes the profit from the first lactation plus about one-half of the second lactation to cover the initial investment in the animal before she freshens, it becomes crystal clear that longevity is important to the owner. Other factors of importance are uniformity, condition, reproductive ability, temperament, size, and price.

PUREBRED BUSINESS

It is estimated that approximately 10% of the dairy cattle of the United States are purebred. However, these animals exert a powerful influence through supplying seed stock to the dairy industry—to other purebred breeders and to those who have grade dairy cattle.

REQUISITES OF A PUREBRED HERD

The primary requisites of a purebred herd are:

1. **High production.** If purebred animals are to bring about further improvement through the dissemination of seed stock, it is imperative that they be top producers.

2. **Good type.** There is such controversy over the value of type and its relationship, or lack thereof, to production. Nevertheless, no one has proved that, on the average, cattle with good conformation produce less than cattle with poor conformation. Further, it is generally acknowledged that, among top producers, one can select animals with superior type. Also, people instinctively desire and appreciate beauty, no matter what the object; and usually they are willing to pay extra for it. It follows, therefore, that good type is important in the purebred herd because animals must be sold.

3. **Attractive surroundings.** Although it is not necessary that the physical plant of a purebred herd be elaborate, it should be neat and attractive.

Fig. 3–2. One cow in four in the average U.S. dairy herd is replaced each year.

REQUISITES OF A REGISTERED BREEDER

As is true of the commercial producer with grade cows, to be successful the purebred breeder must possess a love for the dairy business and the necessary knowledge and experience in the production, handling, and marketing of milk. Additionally, the registered breeder should (1) have knowledge of breeding and pedigrees, (2) pay attention to details, (3) be able to promote and sell, (4) be able to withstand disappointment, and (5) practice honesty and integrity in all dealings.

SELECTING HERD REPLACEMENTS

The average dairy cow in the United States remains in the milking herd about 4 years. This means that about 25% of the average milking herd must be replaced each year. These replacement animals must be either raised or purchased. In either event, the animals brought in should be of the herd-improving kind.

In this chapter, selection of herd replacements will be limited to females. Bull selection is covered in Chapter 5, Breeding Dairy Cattle.

When the major source of income is from the sale of milk, selection is simplified. The cows can be ranked from high to low on the basis of milk production, and the most profitable ones used as replacements. Where purebred breeding stock is involved, the breeder usually invokes an additional point—type or conformation.

Where heifers are being selected for eventual replacement purposes, pedigree information is very important.

CULLING DAIRY COWS

Successful dairy producers are constantly checking their herd and evaluating individual cows, then culling those that do not make money. With high milk prices and low salvage prices, they often find it difficult to decide which cows to cull from the herd, and how many they should cull. Tables 3–1 and 3–2 (next page) are designed to assist in this regard. Such forms facilitate establishing minimum standards in each area of concern and importance.

Table 3–1, Culling Guide, is simply a form on which the butterfat production of each cow can be listed (list each cow by number in the proper category), with provision for first lactation cows, second lactation cows, and mature cows. By using this form, culling decisions can be made without having to compute age-conversion factors. After listing cows in the proper columns, the producers can decide where to draw the line. For example, if they decide to cull all mature cows that are below the 425-lb level, they will take out all cows in columns, A, B, C, and D. Similarly, they might decide to draw the line on second lactation cows at 376 lb and first lactation cows at 340 lb.

Where it is desired to cull cows on the basis of milk production or protein production, producers can develop a form similar to Table 3–1, with the culling levels based on pounds of milk or pounds of protein, rather than pounds of butterfat.

After culling on the basis of productivity, the producers must also cull for other reasons. The cow evaluation form, Table 3–2, is designed for this purpose. It can be applied, and those that do not measure up can be culled.

TABLE 3–1
CULLING GUIDE (BASED ON 305-DAY BUTTERFAT PRODUCTION)[1]

	A	B	C	D	E	F	G	H	I	J	K	L
Pounds of butterfat	Under 350	350–374	375–399	400–424	425–449	450–474	475–499	500–524	525–549	550–574	575–599	600–Over
Mature cows (3rd lactation or more)												
Pounds of butterfat	Under 310	310–331	332–353	354–375	376–397	398–419	420–441	442–464	465–486	487–508	509–530	531–Over
Second lactation cows												
Pounds of butterfat	Under 280	280–299	300–319	320–339	340–359	360–379	380–399	400–419	420–439	440–459	460–479	480–Over
First lactation cows												
Herd total												

[1]From *A Guide to Culling Dairy Cows*, AXT-11, by Robert D. Appleman, University of California Agricultural Extension Service.

TABLE 3–2
COW EVALUATION[1]

Production	Retain		Cull
	High	**Average**	**Low**
Mature cows	10% above mature herd average.	Within 10% of mature herd average.	10% or more below mature herd average.
Second lactation cows	Above mature herd average.	Within 20% of mature herd average.	20% or more below mature herd average.
First lactation cows	Above or within 10% of mature herd average.	10 to 30% below mature herd average.	30% or more below mature herd average.
Miscellaneous factors: Expected dry period	2 months.	3–5 months.	6 months or more.
Health and injuries	Good.	Temporary.	Chronic.
Milking qualities	Fast.	Medium.	Slow and hard.
Disposition	Quiet.	Not easily excitable.	Nervous or dull.
Type, particularly udder conformation	Highly desirable.	Sound.	Undesirable.

[1]From *A Guide to Culling Dairy Cows*, AXT-11, by Robert D. Appleman, University of California Agricultural Extension Service.

JUDGING DAIRY CATTLE

The main function of the dairy cow is to produce milk. Yet, only about 32% of the cows in lactation are tested to determine their milk-yielding capacity. Even fewer dairy animals are subjected to the scrutiny of an experienced judge in the show-ring. Thus, the only method, other than pedigree, available to evaluate the great bulk of dairy animals is by what is commonly known as *judging*.

Judging—as practiced in shows, in contests, or on the farm—*is an attempt to place or rank animals in the order of their excellence in body type.* Scoring, or type classifying, an animal accomplishes the same thing, in that the individual being scored is classified and compared to an animal that is theoretically perfect, and a rating is assigned on this basis.

Admittedly, there is considerable question as to the degree of correlation between type and production, for appearance is not always indicative of a cow's productive ability. Yet, it is generally recognized that desirable type (1) is indicative of a cow's wearing ability, and (2) does not negate functional value. Moreover, it is generally recognized that attractiveness, and what we think of as desirable type, enhances the market value of purebred animals. Also, well-attached udders are less subject to injury and mastitis infection, and strong legs hold up longer than weak legs and feet.

It is noteworthy that good and successful owners and managers are generally good judges of dairy cattle.

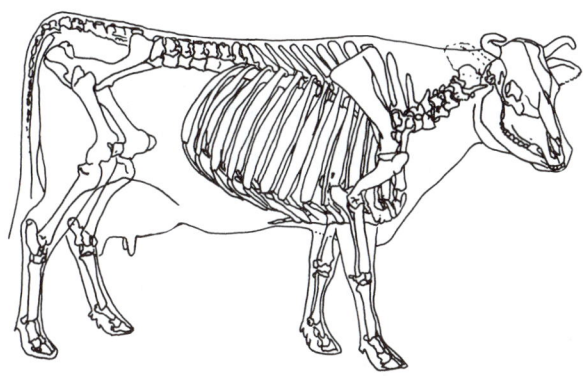

Fig. 3–3. General appearance is determined by the animal's bone structure; and proper bone structure is indicative of a cow's wearing ability and useful life. (From *Dairy Cattle Judging*, Oklahoma State University, Stillwater)

QUALIFICATIONS OF A GOOD DAIRY CATTLE JUDGE

The essential qualifications which a good dairy cattle judge must possess are:

1. **Knowledge of the parts of an animal.** This consists of mastering the language that describes and locates the different parts of a dairy animal.

2. **A clearly defined ideal or standard of perfection.** The successful dairy cattle judge must know what to look for.

3. **Keen observation and sound judgment.** The good judge possesses the ability to observe both good conformation and defects, and to weigh and evaluate the relative importance of the various good and bad features.

4. **Honesty and courage.** The good judge of any class of livestock must possess honesty and courage, whether it be in making a show-ring placing or in conducting a breeding and marketing program. For example, it often requires considerable courage to place a class of animals without regard to (a) placings in previous shows, (b) ownership, and (c) public applause. It may take even greater courage and honesty with oneself to discard from the herd a costly animal whose progeny has failed to measure up.

5. **Logical procedure in evaluating.** There is always great danger of the beginner making too close an inspection. Good judging procedure consists of the following three separate steps: (a) observing at a distance (20 to 30 ft) and securing a panoramic view where several animals are involved, (b) using close inspection (and handling), and (c) moving the animal in order to observe action. Also, it is important that a logical method be used in viewing an animal from all directions, as for example (a) side view, (b) rear view, and (c) front view; thus avoiding overlooking anything and making it easier to retain the observations that are made.

6. **Tact.** In discussing either (a) a show-ring class, or (b) animals on a farm or ranch, it is important that the judge be tactful. Owners are likely to resent any remarks which indicate that their animals are inferior.

DO'S AND DON'TS OF CONTEST JUDGES

Members of 4-H Clubs, FFA students, college judging classes, and other prospective dairy cattle judges should first become thoroughly familiar with the six qualifications of a good judge as outlined in the previous section. Next, they should observe the following do's and don'ts:

1. **Do's:**

a. Make certain how the class is numbered, and keep the numbers straight.

b. Get a clear picture of the class and of each individual animal in mind, so that they will be remembered when giving reasons.

c. Keep in a position of vantage where the class can be seen at all times; usually this means some distance away rather than too close.

d. Make placings on the basis of the big things.

e. Make certain that the card is filled out completely and correctly, and that the correct members are kept in mind.

f. If permissible, make concise notes that will assist in recalling each individual in the class; record such things as distinctive color markings, outstanding faults, etc.

g. When giving reasons, use good poise and look the judge in the eye.

h. Talk reasons clearly, and with conviction and confidence.

i. Give reasons in logical sequence; give the major reason first.

j. Use terms appropriate to the class of animals (for example, mammary system for lactating cows).

k. Use comparative and descriptive terms in giving reasons. Avoid such vague terms as "good," "better," and "best."

l. Concede or grant good points and faults, regardless of the placing of the animal.

2. **Don'ts:**

a. Don't act on hunches; if the first placing is arrived at after due consideration and in a logical manner, stick to it.

b. Don't place animals on the basis of small relatively unimportant characters.

c. Don't destroy self-confidence and self-respect by discussing the class with others before giving reasons.

d. Don't pay attention to what you overhear others say about a class; be an independent judge.

e. Don't give wordy and meaningless reasons.

f. Don't bluff; if you don't know the answer to a question, say so.

PARTS OF A COW

One of the characteristics of good judges is that they possess a thorough knowledge of animals. In speaking of the characteristics of a dairy animal, we usually refer to parts rather than to the individual as a whole. It is important, therefore, to become familiar with the names of the parts. Fig. 3–4 shows an animal in outline form and identifies by name the various parts of the animal. This figure should be studied until each part of the animal can be easily and quickly identified by location and name. Nothing so quickly sets a real dairy producer apart from a novice as a thorough knowledge of the parts and the language commonly used in describing them.

major dairy breed associations has an active committee on dairy type.

There is no conflict between dairy type and breed type; the latter merely adds certain distinctive breed characteristics. The breed registry associations promulgate breed type through shows, type classification programs, models, and paintings.

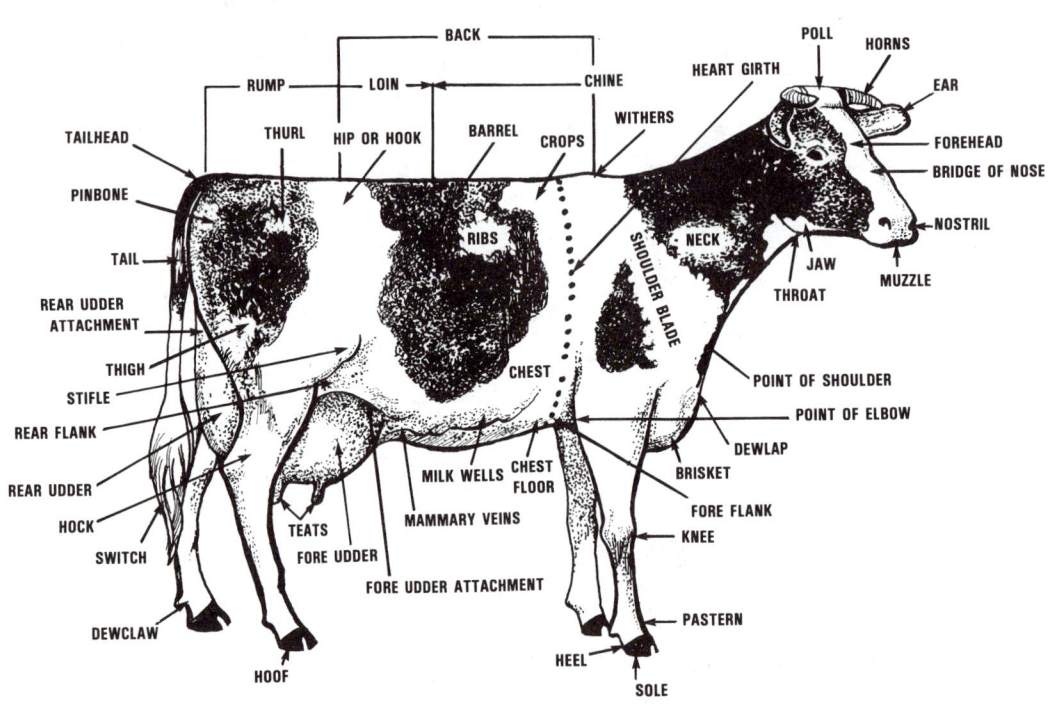

Fig. 3–4. Parts of a dairy cow.

DAIRY TYPE; BREED TYPE

Webster defines *type* as *the combination of characters appropriate to a special kind of use.* Certainly, this definition is adequate for distinguishing beef-type from dairy-type cattle, or perhaps even for distinguishing Jersey type from Holstein type. However, it lacks the necessary specificity for those desiring to differentiate type within a breed. Hence, the following definition is proposed:

Type refers to an ideal or standard of perfection combining the physical characteristics which contribute to an animal's usefulness for a specific purpose.

Additionally, to practical dairy producers, the word *type* has come to express the kind of cow that is adapted to modern herd management, with emphasis placed on profitability.

Type has become more meaningful in dairy cattle since the establishment of the type classification program, first introduced in the Holstein breed in 1929. Today, each of the

A monetary value is placed on type by most producers, particularly by purebred breeders; hence, type may be said to be of market importance.

HOW TO USE THE JUDGING GUIDE

Table 3–3 is a handy guide for judging dairy cattle. It is suggested that it be used as follows:

1. Examine animals in the order indicated; namely, (a) side view, (b) rear view, and (c) front view.

2. Study the points listed under *ideal type,* and know *the common faults.*

3. Rank or place the animals according to their consistent rating on all points, especially the most important ones; or if preferred, the scorecard method which follows in Figs. 3–15 and 3–16 (pp. 36–37) may be used.

TABLE 3–3
JUDGING GUIDE FOR DAIRY CATTLE

Procedure for Examining, and What to Look For	Ideal Type	Common Faults
Side view:  Fig. 3–5	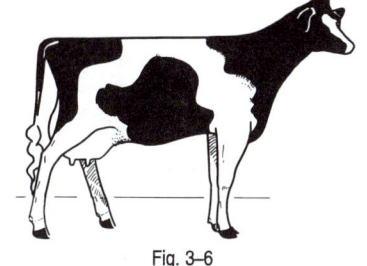 Fig. 3–6	 Fig. 3–7

Procedure for Examining, and What to Look For

Side view:

1. **General appearance.** Style and attractiveness, size, relative length and depth throughout, straightness of back and levelness of rump, attachment and shape of udder, and quality as denoted by fineness of hair, smoothness of joints, fineness of bone, and absence of coarseness throughout.

At the walk, observe the style and carriage, straightness of the legs, and the blending of the parts.

a. **Breed characteristics.** Breed type and size.

b. **Head.** Relative proportion to body; clean-cutness.

c. **Shoulder blades.** Smoothness.
d. **Back.** Straightness and strength.
e. **Rump.** Length, width, and levelness.

f. **Legs and feet.** Set to the legs, strength of pasterns, cleanness of joints, and size of feet.

2. **Dairy character.** Evidence of milking ability; angularity; openness, without weakness; not coarse.

a. **Neck.** Length, leanness, smoothness, cleanness about the throat, dewlap, and brisket.

b. **Withers, ribs, flanks, thighs.** Shape of withers; width between ribs, and shape and length of ribs; shape of thigh.

3. **Body capacity.** Relatively large in proportion to size of animal; ample capacity, strength and vigor.

a. **Barrel.** Support of barrel; length and depth.

b. **Heart girth.** Spring of forerib; fullness of crops; width of chest floor.

4. **Mammary system.** Attachment; balance, capacity, texture.

a. **Udder.** Shape, attachment, texture.

b. **Fore udder.** Length, width, depth, and attachment.

c. **Rear udder.** Height, width, shape, and attachment.

d. **Teats.** Size and placement.

e. **Mammary veins.** Size, length, crookedness, branching.

Ideal Type

1. Great style and beauty.

a. True to the breed as evidenced by the conformation of the head and color markings.
Adequate size for the breed.

b. Head in proportion to body, moderate length, clean-cut, and alert.

c. Shoulder blades blend smoothly into the body.
d. Back straight from withers to tailhead; strong.
e. A long and nearly level rump, from the hip bones to the pin bones; high thurls; tailhead set level with the back bone and free of coarseness.

f. Proper set to the legs; hind legs nearly perpendicular from hock to pastern as viewed from the side; front legs straight as viewed from the side; standing well on the pasterns; clean joints.

2. A triangular body; loose, without being weak; and showing quality.

a. A long lean neck, which blends smoothly into the shoulders; clean-cut throat, dewlap, and brisket.

b. Withers sharp and free of excessive flesh; ribs wide apart, and rib bones wide, flat, and long; thighs rather thin and flat.

3. Great capacity, primarily achieved through length of body rather than extreme depth.

a. Strongly supported; long, with adequate depth; well sprung ribs; greater depth of barrel toward rear.
b. Well sprung forerib; full crops; wide chest floor.

4. Udder strongly attached, well balanced, capacious, and fine textured.

a. Symmetrical; moderately long, wide, and deep; strongly attached; noticeable but not deep division between left and right halves; division between front and rear quarters not marked, and right and left quarters evenly developed; soft, pliable, and collapsed after milking.

b. Moderate length, uniform width from front to rear, and strongly attached.

c. The rear udder should extend high up, and be wide, slightly rounded, of fairly uniform width from top to floor, and strongly attached.

d. Teats of convenient size and squarely placed.

e. The milk veins are large, long, crooked, and branching; the milk wells are large and numerous; and the veins on the udder are large, crooked, and numerous.

Common Faults

1. Plain; lacking in attractiveness, in symmetry, and in balance.

a. Lacking in breed character.
Undersized.

b. Head either too big or too little for the body; plain headed.

c. Loose or out at the shoulders.
d. Weak, sway backed; high tailhead.
e. Short, sloping rump; low thurls; high tailhead.

f. Sickle hocks (hind legs set too far forward); back at the knees (calf-kneed); over at the knees (buck-kneed); down on the pastern; puffy around the joints.

2. A round, thick body; and sluggish temperament.

a. A short, thick neck; open shoulders; heavy, wasty brisket.

b. Excessive flesh over the withers; close ribbed, and rib bones narrow, round, and short; and full fleshy thighs.

3. Lacking capacity; short bodied; and lacking depth.

a. Lacking capacity; short bodied; and lacking depth in the middle.
b. Lacking spring of forerib; slack in the crops; and narrow chest floor.

4. Udder broken away from the body; different size halves and quarters; small; coarse.

a. Quarters not evenly developed; deeply cut between quarters or halves; excessive depth of udder; meaty textured; weakly attached.

b. Fore udder not extended well forward, lacking width, and weakly attached.

c. Not attached high, and narrow, flat, not uniform in width from top to floor, and weakly attached.

d. Teats too large or too small, and poorly placed.
e. Small milk veins and wells.

(Continued)

TABLE 3–3 (Continued)

Procedure for Examining, and What to Look For	Ideal Type	Common Faults
Rear view:  Fig. 3–8	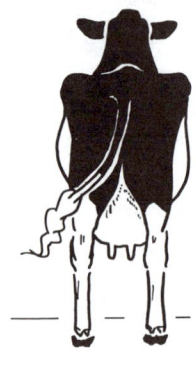 Fig. 3–9	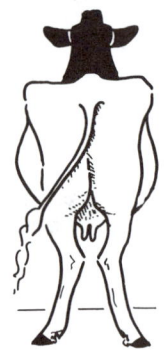 Fig. 3–10

Rear view:

1. Back.

2. Width over loins and rump.

3. Width between the hips and pin bones.
The width between the hips and pins and the levelness of the rump are believed to be associated with the size and shape of the udder.

4. Height and width of thurls.

5. Straightness of hind legs.

6. Size and shape of hind feet.

7. Rear udder attachment.

Ideal Type:

1. Straight and strong, nearly level, vertebra well defined.

2. Wide over the loins and rump.

3. Wide between the hip and pin bones.

4. Thurls high and wide apart.

5. Hind legs straight as viewed from the rear.

6. Feet short, compact, and well rounded with deep heel and level sole.

7. Rear udder that extends high and is wide and strongly attached.

Common Faults:

1. Sagging, roached, weak, vertebra not well defined.

2. Narrow over the loin and rump.

3. Narrow between the hip and pin bones.

4. Low thurls.

5. Cow hocked; puffiness and swelling of hock joints.

6. Small feet; toes not of equal size.

7. Rear udder not extended high up; weakly attached.

Front view:

Fig. 3–11

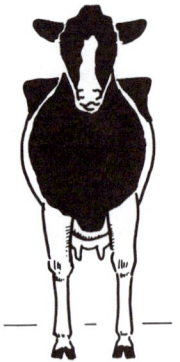

Fig. 3–12

Fig. 3–13

Front view:

1. Shapeliness of head.

2. Sex character.

3. **Neck.** Length, leanness; blending of neck and shoulders; throat, dewlap, and brisket.

4. **Chest.** Width of floor.

5. **Front legs and feet.** Set to front legs; size and shape of front feet.

Ideal Type:

1. A shapely head, with a broad muzzle, large nostrils, and a strong jaw.

2. Cows show femininity about the head and front end; bulls show masculinity and have a well-developed crest.

3. Neck long, lean, and blended smoothly into the shoulders, with clean-cut throat, dewlap, and brisket.

4. Wide chest floor, indicating constitution and chest capacity.

5. Forelegs medium in length, straight, wide apart, and squarely placed.

Common Faults:

1. A plain head.

2. Cows lacking femininity; bulls lacking masculinity.

3. Short, thick neck; neck not blending smoothly into shoulders; leathery and wasty about the throat, dewlap, and brisket.

4. A narrow chest.

5. Crooked front legs; puffiness and swelling at knee joints; curled toes.

SCORECARD JUDGING

A scorecard is a listing of the different parts of an animal, with a numerical value assigned to each part according to its relative importance. It is a standard of excellence. The use of the scorecard involves studying each part, then assigning a score to it.

Show-ring judging or actual selection on the farm are not accompanied by the aid of a scorecard. But a scorecard is a valuable teaching aid in acquainting students and beginners with the various parts of an animal and the relative importance of each. It systematizes judging and avoids any part of the animal being overlooked. However, a scorecard has the following limitations: (1) A near worthless animal may score quite high—for example, an animal that is so structurally unsound that it can hardly walk may have a rather high total score; and (2) it evaluates each part of an animal, rather than the system—the mammary system, the skeletal system, etc.

UNIFIED SCORECARDS

Fortunately, there was a common meeting ground by the purebred dairy cattle associations in the development of the Unified Scorecards herewith reproduced as Figs. 3–15 and 3–16 (pp. 36–37). Breed characteristics may be, and are, considered in the use of the Unified Scorecard—among them are the picturesque style of the Ayrshire; the traditional strength and ruggedness of the Brown Swiss; the tractable disposition and milk color of the Guernsey; the size, scale, and color markings of the Holstein; and the refinement of the Jersey.

AYRSHIRE COW   BROWN-SWISS COW   GUERNSEY COW

HOLSTEIN-FRIESIAN COW   JERSEY COW   MILKING SHORTHORN COW

Fig. 3–14. Breed characteristics portrayed by breed registry ideals.

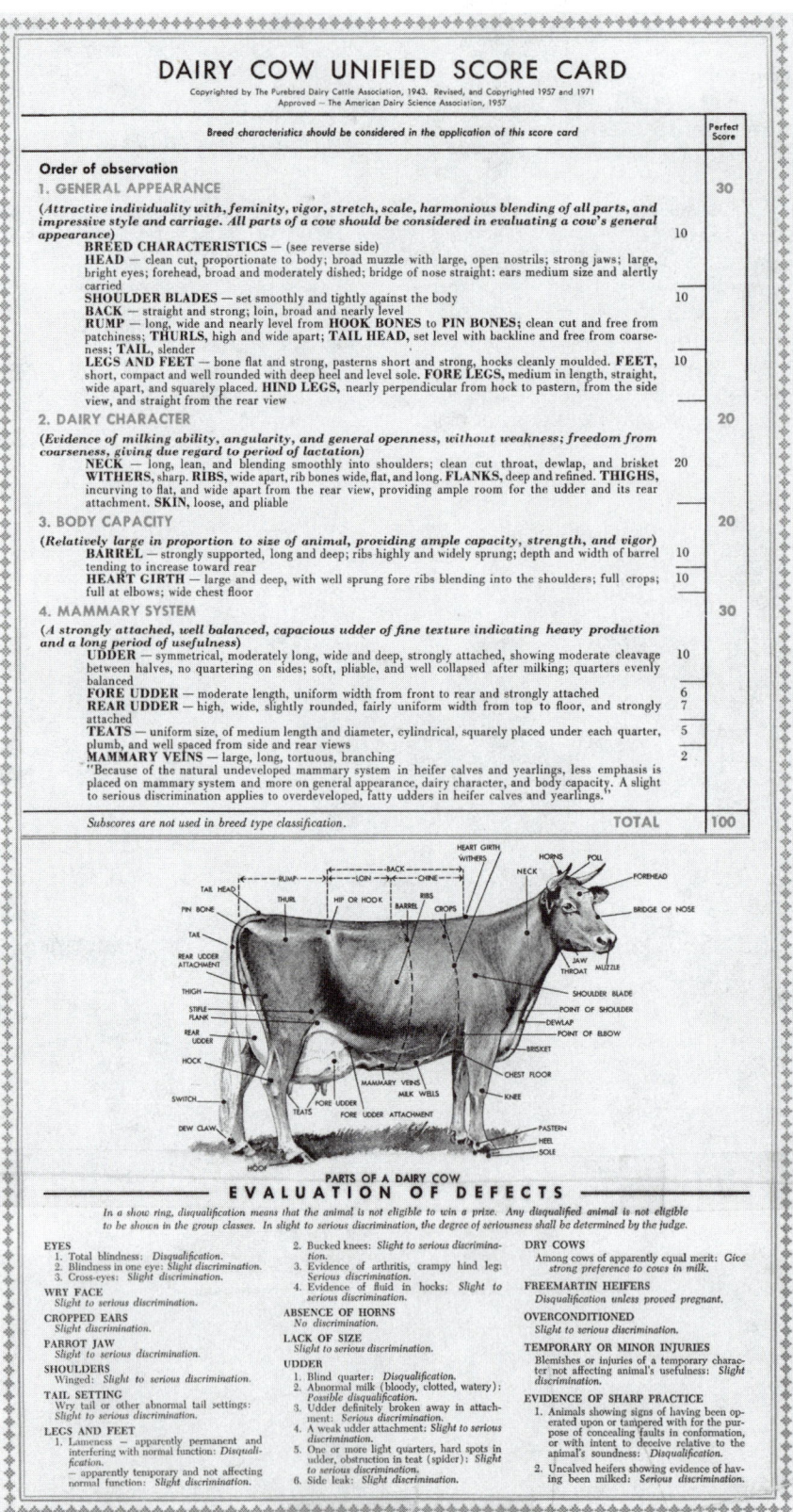

Fig. 3–15. Dairy cow unified scorecard. (Courtesy, The Purebred Dairy Cattle Assn.; approved by the American Dairy Science Assn.)

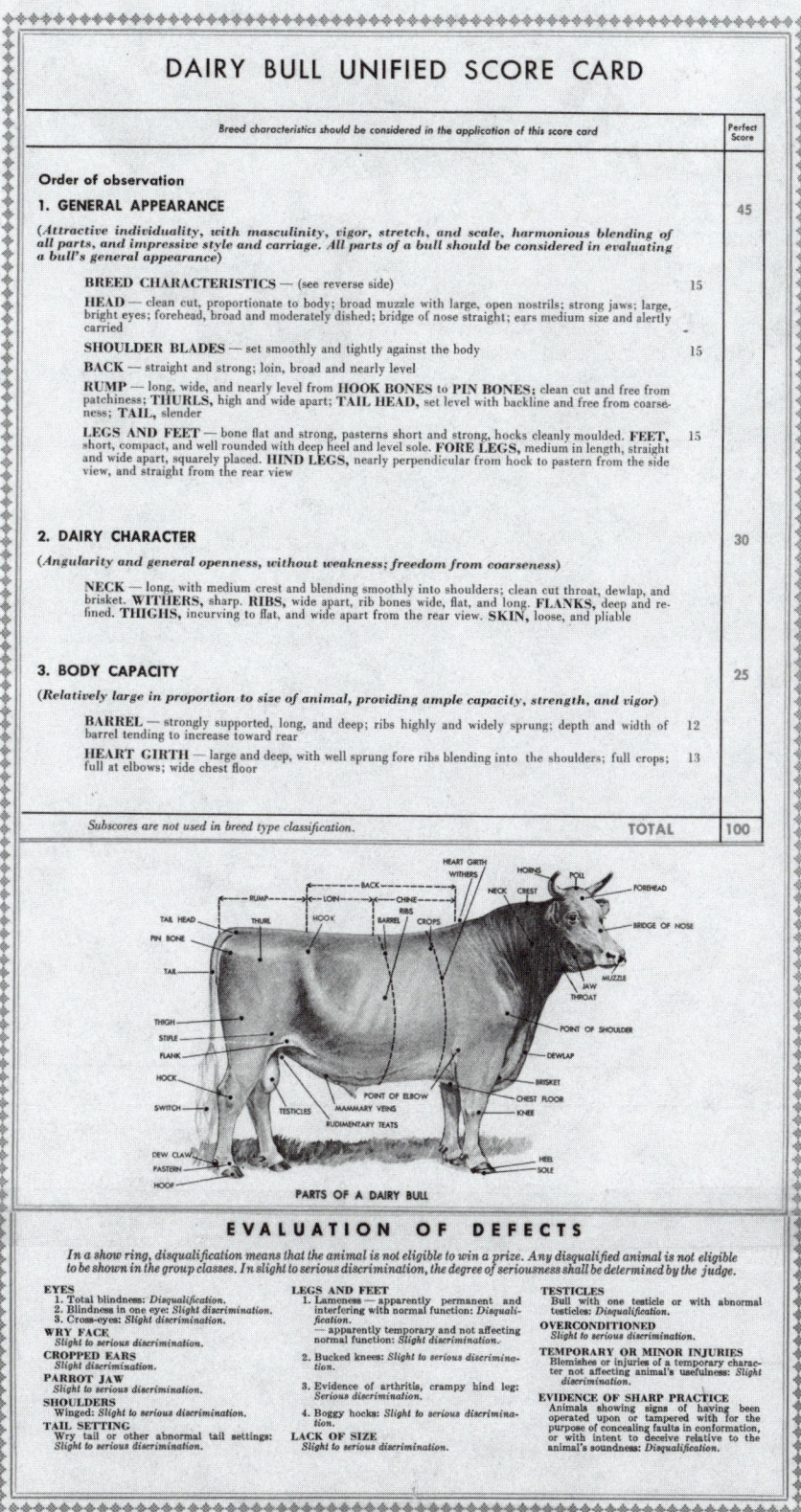

DAIRY BULL UNIFIED SCORE CARD

Breed characteristics should be considered in the application of this score card	Perfect Score
Order of observation	
1. GENERAL APPEARANCE	45
(Attractive individuality, with masculinity, vigor, stretch, and scale, harmonious blending of all parts, and impressive style and carriage. All parts of a bull should be considered in evaluating a bull's general appearance)	
BREED CHARACTERISTICS — (see reverse side)	15
HEAD — clean cut, proportionate to body; broad muzzle with large, open nostrils; strong jaws; large, bright eyes; forehead, broad and moderately dished; bridge of nose straight; ears medium size and alertly carried	
SHOULDER BLADES — set smoothly and tightly against the body	15
BACK — straight and strong; loin, broad and nearly level	
RUMP — long, wide, and nearly level from **HOOK BONES** to **PIN BONES**; clean cut and free from patchiness; **THURLS**, high and wide apart; **TAIL HEAD**, set level with backline and free from coarseness; **TAIL**, slender	
LEGS AND FEET — bone flat and strong, pasterns short and strong, hocks cleanly moulded. **FEET**, short, compact, and well rounded with deep heel and level sole. **FORE LEGS**, medium in length, straight and wide apart, squarely placed. **HIND LEGS**, nearly perpendicular from hock to pastern from the side view, and straight from the rear view	15
2. DAIRY CHARACTER	30
(Angularity and general openness, without weakness; freedom from coarseness)	
NECK — long, with medium crest and blending smoothly into shoulders; clean cut throat, dewlap, and brisket. **WITHERS**, sharp. **RIBS**, wide apart, rib bones wide, flat, and long. **FLANKS**, deep and refined. **THIGHS**, incurving to flat, and wide apart from the rear view. **SKIN**, loose, and pliable	
3. BODY CAPACITY	25
(Relatively large in proportion to size of animal, providing ample capacity, strength, and vigor)	
BARREL — strongly supported, long, and deep; ribs highly and widely sprung; depth and width of barrel tending to increase toward rear	12
HEART GIRTH — large and deep, with well sprung fore ribs blending into the shoulders; full crops; full at elbows; wide chest floor	13
Subscores are not used in breed type classification. **TOTAL**	**100**

PARTS OF A DAIRY BULL

EVALUATION OF DEFECTS

In a show ring, disqualification means that the animal is not eligible to win a prize. Any disqualified animal is not eligible to be shown in the group classes. In slight to serious discrimination, the degree of seriousness shall be determined by the judge.

EYES
1. Total blindness: *Disqualification.*
2. Blindness in one eye: *Slight discrimination.*
3. Cross-eyes: *Slight discrimination.*

WRY FACE
Slight to serious discrimination.

CROPPED EARS
Slight discrimination.

PARROT JAW
Slight to serious discrimination.

SHOULDERS
Winged: *Slight to serious discrimination.*

TAIL SETTING
Wry tail or other abnormal tail settings: *Slight to serious discrimination.*

LEGS AND FEET
1. Lameness — apparently permanent and interfering with normal function: *Disqualification.*
  — apparently temporary and not affecting normal function: *Slight discrimination.*
2. Bucked knees: *Slight to serious discrimination.*
3. Evidence of arthritis, crampy hind leg: *Serious discrimination.*
4. Boggy hocks: *Slight to serious discrimination.*

LACK OF SIZE
Slight to serious discrimination.

TESTICLES
Bull with one testicle or with abnormal testicles: *Disqualification.*

OVERCONDITIONED
Slight to serious discrimination.

TEMPORARY OR MINOR INJURIES
Blemishes or injuries of a temporary character not affecting animal's usefulness: *Slight discrimination.*

EVIDENCE OF SHARP PRACTICE
Animals showing signs of having been operated upon or tampered with for the purpose of concealing faults in conformation, or with intent to deceive relative to the animal's soundness: *Disqualification.*

Fig. 3–16.　Dairy bull unified scorecard. (Courtesy, The Purebred Dairy Cattle Assn.)

QUESTIONS FOR STUDY AND DISCUSSION

1. Were you to enter the dairy business, what breed would you select? Would you start with (a) grades or pure-breds; or (b) cows, heifers, or calves? Justify your choices.

2. List and discuss the primary requisites of (a) a pure-bred herd, and (b) a registered breeder.

3. Discuss the merit of a culling guide based on butter-fat production (Table 3–1). Would a culling guide based on (a) milk production, or (b) protein production be preferable?

4. Discuss the merit of a cow evaluation based on each of the miscellaneous factors listed in Table 3–2.

5. Is it important that a modern dairy producer be a good judge of dairy cattle?

6. List and discuss the essential qualifications of a good dairy cattle judge.

7. Why should a good dairy cattle judge know the parts of an animal?

8. Of what value is (a) dairy type, and (b) breed type?

9. Why should a dairy cattle judge have in mind the ideal type, yet be able to recognize the common faults?

10. Of what value is a scorecard?

11. How can breed characteristics be considered when using the Unified Scorecard?

12. Should production records replace body type evaluation?

13. Under what circumstance would you evaluate an animal by (a) use of the scorecard, (b) show-ring record, and (c) type classification sponsored by the breed registry?

14. How can a person become an expert dairy cattle judge?

SELECTED REFERENCES

Title of Publication	Author(s)	Publisher
Dairy Cattle Judging Techniques	G. W. Trimberger	Prentice-Hall, Inc., Englewood Cliffs, NJ, 1977
Livestock Judging, Selection and Evaluation, Third Edition	R. E. Hunsley W. M. Beeson	Interstate Publishers, Inc., Danville, IL, 1988
Stockman's Handbook, The, Seventh Edition	M. E. Ensminger	Interstate Publishers, Inc., Danville, IL, 1992

Plate 1. A modern dairy, owned and operated by the Gale Haase family. (Courtesy, *Holstein World*, Sandy Creek, NY)

The Dairy Industry

7

Plate 2. Indoor elevated calf pens. (Courtesy, Dr. Tom Schultz, Farm Advisor, Visalia, CA)

Plate 5. Yogurt. (Courtesy, American Dairy Assn., Rosemont, IL)

Plate 3. Fenceline feeding lactating cows. (Courtesy, Dr. Tom Schultz, Farm Advisor, Visalia, CA)

Plate 6. Assorted cheeses. (Courtesy, American Dairy Assn., Rosemont, IL)

Plate 4. Bou-Matic Milking Center. (Courtesy, *Holstein World*, Sand Creek, NY)

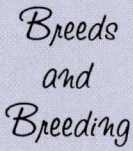

*Breeds
and
Breeding*

Plate 10. Guernsey cow, *Penn Del Vicuvius Glenna*, scored Excellent 98, and was All-American and All-Canadian aged cow in 1989. (Courtesy, American Guernsey Assn., Reynoldsburg, OH)

Plate 7. Artificial insemination cart used on the Souza Dairy, Fresno, California. (Courtesy, Dr. G. E. Higginbotham, Dairy Farm Advisor, Fresno, CA)

Plate 11. Jersey cow, *Billings Top Rosanne*, 1988 and 1989 National Grand Champion, owned by Billings Farm and Rosanne Syndicate, Pottersville, New Jersey. (Courtesy, The American Jersey Cattle Club, Reynoldsburg, OH)

Plate 8. Brown Swiss cow with five of her calves produced by embryo transfer. (Courtesy, Brown Swiss Cattle Breeders' Assn. of the U.S.A., Beloit, WI)

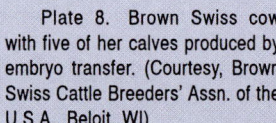

Plate 9. Ayrshire cow, *Maple Dell Hi Kick Sweet Pea*, the breed's 4-year-old *protein champion* in 1990. (Courtesy, Ayrshire Breeders Assn., Brandon, VT)

Plate 12. Milking Shorthorn cow. (Courtesy, American Milking Shorthorn Society, Beloit, WI)

Plate 13. Guernsey cows being judged in the Harrisburg, Pennsylvania, show, in 1990. (Courtesy, American Guernsey Assn., Reynoldsburg, OH)

Plate 16. *Right:* Holstein-Friesian cow, *Crescentmead Rotate Amanda*, the Number 1 cow in the U.S. for protein production. (Courtesy, *Holstein World*, Sandy Creek, NY)

Plate 14. Ayrshire cow, *Sycamore Meade Star Cathy*, a maternal sister to *Sycamore Meade L. L. Olympic*, the top AI Ayrshire sire of the breed in 1991. (Courtesy, Ayrshire Breeders' Assn., Brandon, VT)

Plate 17. *Right:* Milking Shorthorn champion steer, being sold at auction. (Courtesy, American Milking Shorthorn Society, Beloit, WI)

Plate 15. *Below:* Brown Swiss cow, *LPS Telstar Sally May*. (Courtesy, Brown Swiss Cattle Breeders' Assn. of the U.S.A., Beloit, WI)

Plate 18. *Below:* Junior exhibitors showing in the National Junior Jersey Show, Louisville, Kentucky. (Courtesy, The American Jersey Cattle Club, Reynoldsburg, OH)

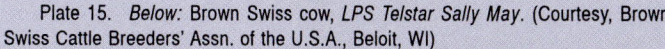

Feeds

Plate 19. Concrete and oxygen-limiting upright (tower) silos for ensiling forages at the USDA Dairy Forage Research Center, Prairie du Sac, Wisconsin. (Courtesy, Dr. D. J. Schingoethe, South Dakota State University, Brookings)

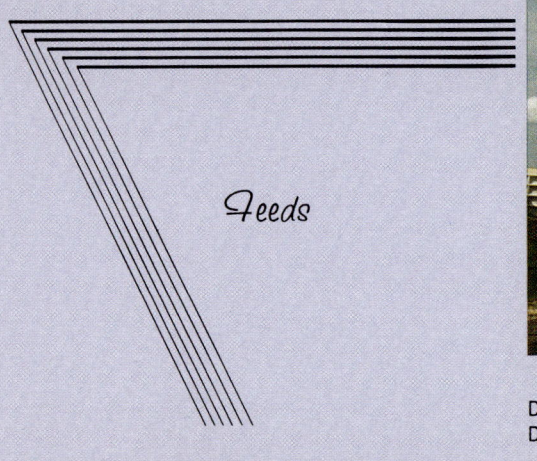

Plate 20. Stanchioned Guernsey cows eating concentrate. (Courtesy, American Guernsey Assn., Reynoldsburg, OH)

Plate 22. Lactating cows on pasture.

Plate 23. Feed ingredients stored at a large dairy for use in ration formulation and mixing. (Courtesy, Dr. D. J. Schingoethe, South Dakota State University, Brookings)

Plate 21. *Below:* Brown Swiss cows eating silage. (Courtesy, Brown Swiss Cattle Breeders' Assn. of the U.S.A., Beloit, WI)

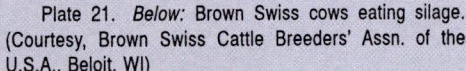

Plate 24. Bulk bins for storage of grains and protein supplements for dairy cattle at the USDA Dairy Forage Research Center, Prairie du Sac, Wisconsin. (Courtesy, Dr. D. J. Schingoethe, South Dakota State University, Brookings)

Plate 25. Proper nutrition and superior sanitation are essential for raising dairy calves successfully. (Courtesy, Dr. D. J. Schingoethe, South Dakota State University, Brookings)

Plate 28. Lactating cows eating alfalfa cubes. (Courtesy, Dr. Tom Schultz, Farm Advisor, Visalia, CA)

Plate 26. Replacement heifers, in lock-up gates, fenceline feeding. (Courtesy, Dr. D. J. Schingoethe, South Dakota State University, Brookings)

Plate 29. Jersey cows on grass-clover pasture. (Courtesy, Dr. D. L. Bath, Extension Dairy Nutritionist, University of California, Davis)

Plate 27. Guernsey cows eating alfalfa hay. (Courtesy, American Guernsey Assn., Reynoldsburg, OH)

Plate 30. *Right:* Cow feeding at magnetic cow feeder. (Courtesy, Dr. Tom Schultz, Farm Advisor, Visalia, CA)

Behavior/
Environment/
Management/
Health &
Disease

Plate 31. Lactating cows in a large corral at the Del Rio Dairy, in Arizona. Note the shade with thermostatically controlled fans for cooling. (Courtesy, Dr. D. J. Schingoethe, South Dakota State University, Brookings)

Plate 32. Solids (manure) separated for bedding. (Courtesy, Dr. Tom Schultz, Farm Advisor, Visalia, CA)

Plate 33. *Left:* Dairy lagoon. Note the ducks. (Courtesy, Dr. Tom Schultz, Farm Advisor, Visalia, CA)

Plate 34. Guernsey cows on pasture, with shade provided by trees. (Courtesy, American Guernsey Assn., Reynoldsburg, OH)

Plate 35. Jersey cows fenceline feeding, under shade, in a corral. (Courtesy, The American Jersey Cattle Club, Reynoldsburg, OH)

Plate 36. Controlled environmental calf facility. (Courtesy, Brown Swiss Cattle Breeders' Assn. of the U.S.A., Beloit, WI)

Plate 37.   Headquarters of Bur-Oak Farm (Dairy), Stockton, Minnesota.

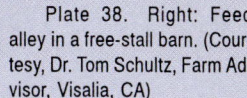

Plate 38.   Right: Feed alley in a free-stall barn. (Courtesy, Dr. Tom Schultz, Farm Advisor, Visalia, CA)

Plate 39.   Cows in a tie stall barn. Note grates over the gutters which keep the cows cleaner. (Courtesy, Dr. D. J. Schingoethe, South Dakota State University, Brookings)

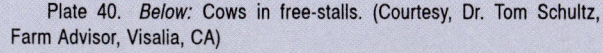

Plate 40.   *Below:* Cows in free-stalls. (Courtesy, Dr. Tom Schultz, Farm Advisor, Visalia, CA)

Plate 41.   Twenty-four stall polygon milking parlor with automatic detachers and clean-in-place washup system, at Stotz' Dairy, Buckeye, Arizona. (Courtesy, Dr. D. J. Schingoethe, South Dakota State University, Brookings)

Plate 42.   *Below:* Portable multi-individual calf hutches, at Alberts Brothers Dairy, Pine Island, Minnesota. (Courtesy, Dr. D. J. Schingoethe, South Dakota State University, Brookings)

Dairy Products/
Business Aspects

Plate 43. Dairy products.
(Courtesy, USDA)

Plate 44. Butter and freshly baked bread. (Courtesy, American Dairy Assn., Rosemont, IL)

Plate 46. Ice cream. (Courtesy, American Dairy Assn., Rosemont, IL)

Plate 47. Cheese. (Courtesy, USDA)

Plate 45. Cottage cheese. (Courtesy, American Dairy Assn., Rosemont, IL)

Plate 48. Computerized dairy. (Courtesy, Dr. Tom Schultz, Farm Advisor, Visalia, CA)

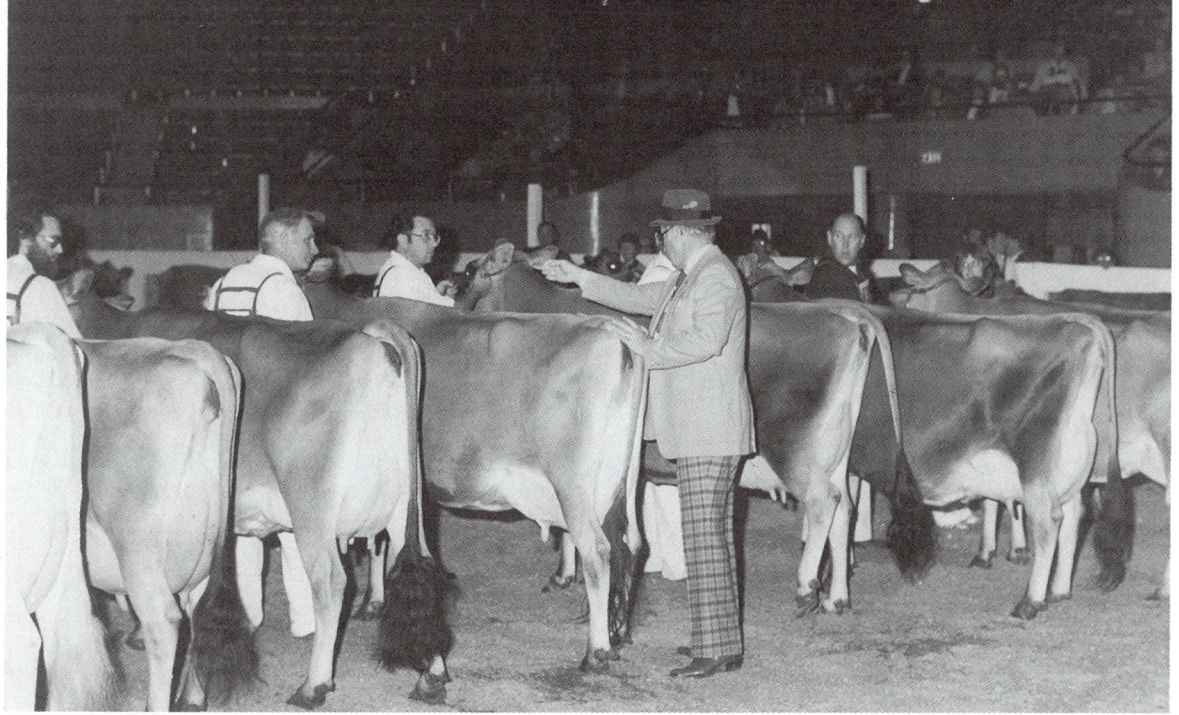

Fig. 4–1. Jersey cows showing at the All American Jersey Show. (Courtesy, The American Jersey Cattle Club, Reynoldsburg, OH)

Contents

CHAPTER

4

FITTING
AND SHOWING
DAIRY CATTLE

There is no higher achievement than that of breeding and fitting a champion; an animal representing an ideal which has been produced through intelligent breeding and then fitted to the height of perfection. Also, it is realized that most of the advertising value of the show-ring accrues to those who exhibit winners, thus behooving exhibitors to select, fit, and show their animals to the best possible advantage.

The selection of the prospective show animals is the first and most important assignment. The second requisite for successful showing is to fit the animals to the peak of perfection.

Assuming that show animals have been carefully selected and properly fed, there yet remains the assignment of parading before the judge. In order to present a pleasing appearance in the show-ring, animals must be well trained, thoroughly groomed, and properly shown. Competition is keen, and often the winner will be selected by a very narrow margin. Close attention to details may, therefore, be a determining factor in the decisions.

Fig. 4–2. Milking Shorthorn cow being posed in the show-ring. (Courtesy, American Milking Shorthorn Society, Beloit, WI)

ADVANTAGES AND DISADVANTAGES OF SHOWING

Dairy shows have both advantages and disadvantages. Human nature being what it is, not all exhibitors share equally in the many advantages that may accrue from showing. In general, however, dairy cattle shows offer the following **advantages**:

1. They afford the best medium yet discovered for molding breed type.

2. They provide an incentive to breed better cattle, for breeders can determine how well they are keeping pace with their competitors only after securing an impartial appraisal of their entries in comparison with others.

3. They offer an opportunity to study the progress being made within other breeds and classes of livestock.

4. They serve as one of the very best advertising or promotional mediums for both the breed and the breeder.

5. They give breeders an opportunity to exchange ideas, thus serving as an educational event.

6. They offer an opportunity to sell a limited number of breeding animals.

7. They may set sale values for the animals back home, for such values are often based on the sale of show animals.

8. They direct the attention of new breeders to those types and strains of cattle that are meeting with the approval of the better breeders and judges.

9. They provide instruction, training, and encouragement for young people to join the dairy industry.

Like many good things in life, dairy shows also have some **disadvantages**, among which are the following:

1. Breed fancy points—involving such things as color and markings, shape of head, and shape of horns (if present)—may overshadow utility value.

2. The desire to win sometimes causes exhibitors to resort to *surgical means* and *filling* in order to correct defects. Admittedly, such artificially made corrections are not hereditary, and their effects are often not too durable—as is belatedly discovered by some innocent purchaser.

3. Valuable animals are sometimes kept out of reproduction in order to enhance their likelihood of winning in the show-ring.

4. Their educational value has not been exploited. Many people who attend a dairy cattle show for the first time are bewildered by the procedure and the many breeds and classes that they see. Others have never gone to a cattle show because they feel that their lack of knowledge of showing and judging cattle would prevent them from enjoying it.

SELECTING SHOW CATTLE

The breeder who is planning to show has the difficult job of selecting animals which, following fitting, will meet the show-ring ideal. Requiring as it does a projection into the future, no judging assignment is quite so difficult. But the success or failure attained in the show-ring depends to a very great extent on the animals selected.

The selection of animals for the show should be made as early in the season as possible, and from among animals that are healthy and vigorous and have good dispositions.

The essential factors to be considered when selecting individual show prospects are (1) type, (2) breeding, and (3) age and show classification.

TYPE

When they make their placings, judges endeavor to select dairy animals that are, or will be, most efficient at producing milk. This calls for dairy type. Such animals are characterized by an angular form and a well-developed mammary system. They are especially adapted to converting feed efficiently into the maximum of milk.

Signs of fertility and reproductive efficiency are extremely important in breeding cattle. Bulls should have two testicles that are of normal size, and that are well defined in the scrotum, rather than surrounded by excess fat. Also, bulls should be masculine.

When selecting females for show, it is well to look for femininity at all ages. Feminine females are trim in the jaw, throat, and dewlap, and smooth shouldered. Also, the mammary system is extremely important. The udder should be strongly attached, well balanced, capacious, and of fine texture — indicating heavy production and long usefulness.

Structural soundness is also important in breeding cattle. Proper structural soundness involves a big foot, a deep heel, toes that are the same size and point straight ahead, clean joints, and correctly set legs. Both bulls and cows must be able to travel well. Also, bulls must be able to mount for service, and cows must be able to calve easily.

(Also see Chapter 3, Establishing the Herd; Selecting and Judging.)

BREEDING

Animals selected for show should always be from good ancestry, for this is added assurance of satisfactory future development. This is especially true of young animals. Also, they should be true to breed type and characteristics, as indicated by the color, conformation of the head, shape of the horns (if present), and size.

AGE AND SHOW CLASSIFICATION

Most successful exhibitors attempt (1) to select animals as old as possible within the respective age classifications, so that they may show to the best possible advantage; and (2) to fill as many show classes as possible, thereby increasing their chances of winning. In order to meet these requisites, the exhibitors must be fully versed relative to the usual show classifications.

The show classifications recommended for 1991 local, district, and state fairs (the rules and classes for each breed's regional and national shows should be recommended by the breed involved), prepared by the Purebred Dairy Cattle Association,[1] P. O. Box 160, Cabool, MO 65689, with the classes arranged in recommended order of judging for a one-day judging program, follow:[2]

Class 1. Bull Calf, born after August 31, 1990 and over four months of age.

Class 2. Yearling Bull, born after August 31, 1989 and before September 1, 1990.

Class 3. Champion Bull, winners of Classes 1 and 2.

Class 4. Intermediate Heifer Calf, born after November 30, 1990 and before March 1, 1991.

Class 5. Heifer Calf, born after August 31, 1990 and before December 1, 1990.

Class 6. Summer Yearling Heifer, born after May 31, 1990 and before September 1, 1990.

Class 7. Junior Yearling Heifer, born after February 28, 1990 and before June 1, 1990.

Class 8. Winter Yearling Heifer, born after November 30, 1989 and before March 1, 1990.

Class 9. Senior Yearling Heifer, born after August 31, 1989 and before December 1, 1989

Class 10. Junior Champion Female, winners of Classes 4, 5, 6, 7, 8 and 9 (Rosette).

Class 11. Reserve Junior Champion Female, except Junior Champion, winners of Classes 4, 5, 6, 7, 8 and 9 and second place winner in Junior Champion's open individual class (Rosette).

Class 12. Junior Best Three Females. This group to consist of three (3) females from heifer classes, none of which have freshened, all bred by and at least one owned by exhibitor.

Class 13. Junior Two Year Old Cow, born after February 28, 1989 and before September 1, 1989.

Class 14. Senior Two Year Old Cow, born after August 31, 1988 and before March 1, 1989.

Class 15. Three Year Old Cow, born after August 31, 1987 and before September 1, 1988.

Class 16. Four Year Old Cow, born after August 31, 1986 and before September 1, 1987.

Class 17. Five Year Old Cow, born after August 31, 1985 and before September 1, 1986.

Class 18. Aged Cow, six years old and over, born before September 1, 1985.

Class 19. Senior Champion Female, winners of Classes 13, 14, 15, 16, 17 and 18 (Rosette).

Class 20. Reserve Senior Champion Female, except Senior Champion, winners of Classes 13, 14, 15, 16, 17 and 18 and second place winner in Senior Champion's open individual class (Rosette).

Class 21. Grand Champion Female, winners of Classes 10 and 19 (Rosette).

Class 22. Reserve Grand Champion Female, except Grand Champion, winners of Classes 10, 11 19 and 20 (Rosette).

Classes 23, 24 and 25 Best Three Females, Breeder's Get of Sire and Produce of Dam, respectively, given below are optional classes or may be replaced with only the Breeder's Herd of Five Females class.

Class 23. Best Three Females.

Class 24. Breeder's Get of Sire or Senior Get of Sire (choose the one desired).

Class 25. Produce of Dam.

OR

Class 23. Breeder's Herd of Five Females. This group, all owned by the exhibitor, consists of two females over two years of age, two females under two years of age and one any age. Three must be bred by the exhibitor. This group, when used as an optional class, replaces all of the former group classes except Junior Best Three Females. It is strongly recommended that the total monies allocated to the group classes in the past be placed on this Breeder's Herd of Five Females class.

---

[1]*Member Associations*: The American Guernsey Assn., The American Jersey Cattle Club, The Ayrshire Breeders' Assn., The Brown Swiss Cattle Breeders' Assn. of the U.S.A., The Holstein-Friesian Assn. of America, and the American Milking Shorthorn Society.

[2]The number of places to receive premium money and the percentage of total monies for each class should be based on an average of the number of head shown in each class during the past 3 years.

(Class number of following classes dependent on option chosen on previous page.)

Class 26. Premier Sire. Recognition to be awarded to the owner, or last recorded owner of the sire, whose progeny accumulates the most points on not less than four and not more than eight progeny in the open individual classes. This will not be a lead out class. The management will calculate the points for every sire having four or more progeny exhibited in the show. The scale of points will be the same as that for determining Premier Breeder.

Class 27. Premier Exhibitor's Award. The exhibitor winning the most points on not to exceed six animals owned and exhibited by self in the open individual classes shall be designated the Premier Exhibitor. The scale of points will be the same as that for determining Premier Breeder.

Class 28. Premier Breeder's Award. The Breeder winning the most points on not to exceed six animals in the open individual classes, exhibited by self and/or other exhibitors shall be designated the Premier Breeder. The point system for determining Premier Breeder is as follows:

	Placings	1	2	3	4	5	6	7	8	9	10
Points	Senior Females	20	18	16	14	12	10	8	6	4	2
Points	Bulls and Junior Females	10	9	8	7	6	5	4	3	2	1

■ **Best Udder Award.** It is recommended that in each of the milking cow classes the judge select the first and second best uddered cows in each class. The selection shall be made on the basis of udder alone and a milk-out will be at the discretion of the judge. It is recommended that the two cows in each class be given special recognition just prior to the final placing of their respective class. It is not recommended that the best uddered cows from each of the classes compete for an overall best udder award of the show.

■ **State Herd Class for Regional and National Breed Shows.** A State Herd shall consist of eight animals from one state as follows:

Two females, 2 years old or over, three females, under 2 years of age; three other females any age or one bull and two females any age.

The eight animals in the state herd must be owned by not less than three exhibitors from one state with no exhibitor furnishing more than three animals. All animals in which a breeder owns a partnership interest will be counted in the three allowed. All animals from a herd or farm, regardless of ownership, shall be counted in the three head allowed one exhibitor.

Animals owned by state institutions will be allowed to show in the state herd class, but all institutions in a state will be considered as one exhibitor and allowed to show not more than three head in a state herd group.

Note well: Most dairy cattle breeders have discontinued showing bulls at their national shows. Young bulls are shown at some local and state shows.

FEEDING FOR SHOW OR SALE

Dairy animals intended for show or sale should be fed so as to achieve a certain amount of finish and bloom, but they should not be too fat.

At the beginning of show preparation, check for both internal and external parasites. If any are present apply the recommended treatment. (See Chapter 15, Dairy Cattle Health, Disease Prevention, and Parasite Control.)

RULES OF FEEDING SHOW CATTLE

Some general rules of feeding may be given, but it must be remembered that most successful cattle fitters have worked out systems of their own through years of practical experience and close observation—they do not follow set rules. Nevertheless, the beginner may well profit by the experience of successful fitters, and it is with this hope that the following general rules of feeding show cattle are presented.

1. **Practice economy, but avoid false economy.** Although the ration should be as economical as possible, it must be remembered that rapid growth and ideal condition are primary objectives, even at somewhat additional expense.

2. **Use care in getting animals on feed.** In starting animals on feed, use extreme caution to see that they do not get digestive disturbances.

3. **Provide a variety of feeds.** A good variety of feeds increases the palatability of the ration and makes it easier to supply the proper balance of nutrients. Furthermore, a fitting ration consisting of only one or two feeds may lose its palatability during a long feeding period.

4. **Feed a balanced ration.** A balanced ration will be more economical and will result in better gains. That is to say, the ration should contain the proper balance of energy, protein, minerals, and vitamins. It must also be remembered that growing animals require a higher proportion of grain than mature ones. Then, too, because of its high cost, it is not economical under most conditions to feed more protein than is required.

5. **Do not overfeed.** Feed plenty, but do not overfeed. Overfeeding is usually caused by the desire of the inexperienced caretaker to push the animal too rapidly.

6. **Keep the feed box clean.** Never leave uneaten feed in the box or trough. It may become sour and cause the animal to go off feed.

7. **Do not underfeed.** It never pays to underfeed an animal. When animals are consuming too little grain, the caretaker may look for several causes, such as the consumption of too much roughage, unpalatability of the ration, or discontentment on the part of the animal.

8. **Supply palatable and light feeds.** In order to consume the maximum amount, the animal must relish the feed. Unpalatable feeds may be fed in limited quantities, provided that they are mixed with more palatable ingredients. Black-strap molasses is relished by animals and is excellent for increasing the palatability of the ration (although blackstrap molasses is preferable, beet molasses is satisfactory). Usually the molasses is added by diluting it with water (warm water in winter) and mixing it with the grain ration just before feeding. One-half to one pint of molasses diluted with an equal volume of water will be entirely satisfactory for this purpose. Most commercial feeds contain some molasses in the mixture. Beet pulp, oats, barley, and wheat bran are also popular feeds in fitting and showing rations. Linseed meal is commonly used to impart bloom to the hair. Soaking certain feeds before feeding also makes for increased palatability. Good-quality forages are always very important.

9. **Provide succulent feeds.** Succulence is provided in such feeds as silage, root crops, and grasses. These feeds have a beneficial effect in the ration. They increase the palatability and produce a laxative effect on the animal's digestive system.

10. **Do not feed damaged feeds.** Moldy, musty, or spoiled feeds may cause digestive disturbances and should not be fed to animals being fitted for show or sale.

11. **Prepare grains.** The grain ration of cattle intended for show purposes is almost always coarsely ground or rolled. Most caretakers prefer steamed flaked grains. The preparation of hay is neither necessary nor advisable.

12. **Feed regularly.** Animals intended for show should be fed with exacting regularity.

13. **Avoid sudden changes.** Sudden changes in either the kind or amount of feed are apt to cause digestive disturbances. Any necessary changes should be gradual.

14. **Provide minerals.** All animals should be given free access to salt at all times. Stabilized iodized salt should be used in iodine-deficient areas; and trace mineralized salt should be used in areas where other trace mineral deficiencies exist.

When a nonlegume roughage is provided, it is especially likely that there will be a deficiency of calcium and phosphorus. Certain trace elements are usually needed, too.

15. **Keep the animal quiet and contented.** Quiet and contentment are important when fitting animals for show. Uncomfortable quarters, isolation from other animals, annoyance by parasites, sudden changes in quarters or feeds, improper handling, and unnecessary noise are the most common causes of discontentment.

16. **Provide exercise.** A certain amount of exercise is necessary in order to promote good circulation and to increase the thrift and vigor of the animal. Exercise also tends to stimulate the appetite and makes for greater feed consumption. Animals can usually be kept in condition by turning them into a paddock at night.

17. **Avoid scouring.** If the droppings are too thin or if there is scouring, (a) decrease the grain allowance, and (b) clean up the quarters. If trouble still persists, decrease the legume roughage and the protein supplement (especially linseed meal).

18. **Avoid sudden water changes.** Frequently, show cattle fail to drink enough water while at the fair. As a result, they become gaunt and show at a disadvantage. Usually this situation is caused by (a) the sudden change in drinking from a trough or tank at home to drinking out of a bucket at the show, and (b) the different taste of the water, primarily due to chlorine or mineral content. This problem can be alleviated by (a) getting the animal accustomed to drinking from the same bucket that will be used at the fair, beginning about a week before leaving home; and (b) adding a tiny bit of molasses to each bucket of water, from the time bucket watering is started until the show is over, thus avoiding any flavor or taste change in the water.

RATIONS

Variations can and should be made in the rations, depending upon the individual animal, the relative prices of feeds, and the supply of homegrown feeds. To secure the correct condition, a suitable ration must be selected and the animal or animals must be fed with care over a sufficiently long period.

Any of the rations listed in Chapter 12, Dairy Feeding Programs, are suitable for use in fitting show animals of similar classification, or a commercial feed may be used.

EQUIPMENT FOR FITTING AND SHOWING

Every exhibitor should have the following equipment:

■ **Halters**—A rope halter, a heavy leather halter for use in the barn, and a narrow leather show halter with one leather and one chain-leather lead strap.

■ **Blanket**—Proper size and weight.

■ **Grooming equipment**—Rubber curry comb, dandy brush, and body brush.

■ **Soap or shampoo**—Either a mild bar of soap or a shampoo concentrate.

■ **Oil and polish**—Oil for the hair and polish for horns and hoofs.

■ **Foot trimming tools**—The necessary foot trimming tools should include a rasp, farrier's knife, and nippers.

■ **Clippers and scissors**—For clipping and shearing.

TRAINING AND GROOMING

Table 4–1 is an illustrated guide for training and grooming dairy cattle.

TABLE 4–1
TRAINING AND GROOMING GUIDE FOR DAIRY CATTLE

The Essential Steps in Training, Grooming, and Showing	Why	How
1. Selecting	Selection of the prospective show animal is the first and most important assignment. Unless the right kind of animal is selected, no amount of fitting and showing can make a champion.	Select show animals and begin fitting at least 2 months ahead of the show. Separate show animals from the rest of the herd and keep them in the barn—out of the sun and away from flies.
2. Feeding and watering	The second requisite is proper fitting so as to enhance the attractiveness of the animal, without excessive fatness.	Animals with excess condition will be placed down in the show because of lack of dairy character. So, if the animal is carrying excess condition (coarse at the withers, patchy over the pinbones, throaty, or fat), place it on a low maintenance ration. If the animal is thin and in poor condition, feed extra grain daily. Some young animals will grow faster than others. So observe their growth pattern carefully. Feed plenty of hay since this will develop capacity and body depth. The grain mixture can be any homegrown grain (corn, oats, barley, etc.) plus minerals, or a commercial feed may be used. Provide plenty of clean, fresh water. A water-taste problem may arise at the show. So, teach the animal to drink from the kind of bucket that will be used at the show barn. To help cover up off-flavors in water, add ½ cup of molasses to each pailful. Condition the animal to the bucket and molasses-flavored water about a week before the show.
3. Bedding	To keep the animal and the blanket clean.	Use long, clean bright straw for bedding. Remove droppings and soiled bedding as often as possible. If the animal becomes stained from manure or dirty bedding, washing the stained spots with water containing some bleach will be helpful.
4. Training	Proper training is necessary if the animal is to be shown to advantage.	Practice makes for perfection. So, begin training the animal well ahead of the show. The exhibitor and the animal must work as a team, each knowing what to expect of the other. Get the animal used to the halter that will be used in the show. Train it to walk slowly one half step at a time in a clockwise direction. The exhibitor should walk backward on the left side of the animal, with the lead strap in the left hand; and the animal should always be betweeen the exhibitor and the judge. When the animal is stopped, it should pose with the topline straight, the head held high enough to appear alert and natural looking, and the feet properly positioned.
5. Washing	To make and keep the animal clean.	Wash the animal about 6 weeks prior to the show. Get it clean, then keep it clean. Wash the stained areas and switch frequently. Check the poll and head area since dirt accumulates there quickly. A second washing may be needed about 2 days before the show, but do not wash an animal too often. Washing removes the natural hair oils and makes the hair look loose and fuzzy, just the opposite of what is desired. Use a mild soap (detergent can cause skin irritation), and scrub with a coarse brush. Avoid getting water in the ears and slowly accustom the animal to water (avoid cold water and high pressure hoses). Remove all soap with a good rinse. After rinsing, rub down the animal with the flat of your hand to remove excess water and smooth the hair coat. Blanket the animal. Remove wax and dirt from inside the ears with a clean cloth dampened with rubbing alcohol.
6. Blanketing	To protect the animal from dust, dirt, and flies; to keep the hair smooth; to loosen the dead, dull hair and bring out the natural oils in the skin.	Begin blanketing the animal 6 weeks before the show. In warm weather, use a light blanket. In cold weather, use a heavy blanket. Be sure the blanket is clean. Smooth the hair before putting on the blanket. The blanket should fit properly, covering the animal from the tailhead to the neck, including the withers. It should be long enough to cover the flank and extend below the belly line.
7. Grooming	To remove hair and dust, loosen the hide, make the hair lie flat, and bring out the natural oils and add gloss.	Curry, brush, and rub down the animal daily. Use a rubber curry comb and work it in a vigorous, circular motion to loosen the hair and hide. (Never use a steel curry comb. It won't loosen the hair as well as rubber. Besides, it will cut and break the ends off new hair and make them stiff and bristlelike.) After the hair has been loosened; and, take a stiff long-bristled brush; and, with quick, flicking action of the wrist, remove the hair and dust from the animal. Follow this procedure with a good rubdown, using the palm of your hands. This method helps to keep the hair in place, and, at the same time, removes excess loose hairs. Also, it brings out the natural oils and adds gloss to the animal's hair coat. Don't comb the switch during the early grooming period. The object is to keep all the long hairs intact in the switch. Combing will pull hair out. Several washings before show day should be enough to clean the switch. If stains are present, bleaching may help remove them. To remove excess water following washing, briskly flip the switch several times. Separate any hair twists with your hands. To fluff the switch, take the long hairs of the switch in your hand and brush upward on them, teasing the hair. Release a few hairs with each stroke. This makes the switch very fluffy. Added curl to the fluff can be attained by braiding the switch a few hours before show time, then brushing out the braids just before going to the show-ring.

(Continued)

TABLE 4–1 (Continued)

The Essential Steps in Training, Grooming, and Showing	Why	How
8. Clipping Fig. 4–3. How to clip legs and tail. Fig. 4–4. How to clip head and neck.	To accentuate quality and dairy character—to attract attention to the animal's strong points and minimize the weak ones. Clip the *tail* to accentuate the switch and make the tail appear slender. Clip the *head and neck* to impart a clean-cut appearance. Clip the *legs* to accentuate quality of bone and correct stance. Clip the *udder and belly* to bring out veining and show quality.	Do not clip the entire animal (1) unless it has an extremely rough hair coat, has not lost the winter hair coat, has stained areas, or shows excess sun bleaching; or (2) if the show is less than 2 months away. Clip the animal on the tail, head and neck, legs, and udder-belly a day or two before each show. Start at the top of the switch or as much as 4 in. above the long hairs (let your eyes determine the starting point; after clipping, the switch should look like a well-blended part of the tail, rather than the business end of a broom), and clip up the tail against the grain of the hair to within 4 to 5 in. of the tailhead. Clip the tailhead with the grain of the hair (referred to as blending). If the tailhead is higher than desired, clip the hair short on top, then blend with the sides. Clipping the rump is confined to "touching up"; to make the rump appear as level as possible and to blend the tailhead and the rump area. Be careful not to call attention to defects in the rump region. If there is a high area, clip closely. If a low point exists, leave the hair. Clip the head and neck closely against the grain of the hair. Clip the area forward of a line formed by the point of the shoulders and the front of the withers. The natural crease formed by the neck and shoulders can be used to blend the long hair. Care should be exercised when clipping about the head because animals are sensitive to the clipper and may throw the head. Hair should be removed from both outside and inside the ears. If the animal will not allow the clipper, use scissors to clip the long hair in the ears. Be careful with the scissors! Clipping the front end of Milking Shorthorns is optional. Clip the sides and back of the rear legs from the hock down closely against the grain of the hair. Take advantage of natural lines and attempt to correct the legs by carefully removing hair. The blood vein in the hock region makes an excellent point for blending. The back of the legs should be clipped in a straight line to the point of the pinbones. On milk cows, clip the entire udder, and clip the underline from the udder to the navel, so as to show the milk veins to advantage. But do not clip the belly of calves or heifers since it makes the animals appear shallow bodied. Extremely long, wooly body hair can be removed by holding the clippers away from the body when clipping. At intervals, lubricate the clipper by dipping the blades in a shallow widemouthed can of light oil or kerosene to remove dirt, dust, and minimize wear and dulling. When a good job of clipping has been done, no lines are visible where blending has occurred.
9. Trimming hooves Fig. 4–5. *Top*: normal foot. *Bottom*: foot that needs attention.	So that the animal will stand squarely and walk properly, and that the feet will be clean and attractive. Long toes increase chances of foot rot, punctures from the sole, and broken or cracked toes. Each of these leads to lameness in the animal. The bottom or sole of the foot should be trimmed so that the animal stands squarely on its feet with the wall (outside shell) supporting the weight.	The normal foot should appear as shown in Fig. 4–5. It should be well rounded (see A), have short toes (see B), and be deep at the heel (see C). The proper leg formation puts the leg directly under the weight it is to support (see D). Be familiar with the bone structure in the foot, and see that the weight of the animal is carried properly by that structure (see E). The bottom drawing of Fig. 4–5 shows a foot that needs attention. The dark area shows the amount of growth that needs to be removed. When the toes are too long, it causes the animal to carry too much weight in an unfavorable position; too much weight is on the heel and not enough on the toes. If the hooves are too long, shorten them with a hoof trimmer, nipper, or a chisel and mallet. A hoof knife (tool with a U-shaped tip) will remove excess growth and is safer than other tools. The foot can be smoothed with a rasp. Wear leather gloves to protect your hands. The following methods can be used to work on the feet: Restrain the animal in a set of stocks or a restraining table, or throw it. Work on one foot at a time. Proceed with care to avoid injuring yourself or your animal.
10. Oiling	To impart "bloom" to the animal. Oil is not needed unless the hair is unusually dry and lifeless.	Several good commercial oils are on the market. However, a homemade oil can be prepared by mixing equal parts of glycerine, sweet oil, and rubbing alcohol. If the animal is oiled, use it only on dark hair because it may yellow white hair. If the hair is dry, apply oil lightly with a rag just before entering the ring.
11. Cleaning	To enhance the attractiveness.	Before the animal enters the ring, clean the hooves and dewclaws with a stiff brush and water, then polish. Use colorless fingernail polish or colorless furniture polish to shine light-colored hooves. Use black shoe polish on black hooves.
12. Filling	To ensure best appearance.	Observe the animal to see how much it should eat or drink to ensure its best appearance at show time. Too much fill may make it look unnatural and stand awkwardly; not enough fill may make it appear gaunt and narrow.

HOW TO GROOM FOR THE SHOW

It requires time and work to have a clean, glossy, eye-catching hair coat and a pliable hide for the show. Figs. 4–6 to 4–10 show how to groom a dairy animal for the show. (Figs. 4–6 to 4–10 courtesy, The Pennsylvania State University, University Park.)

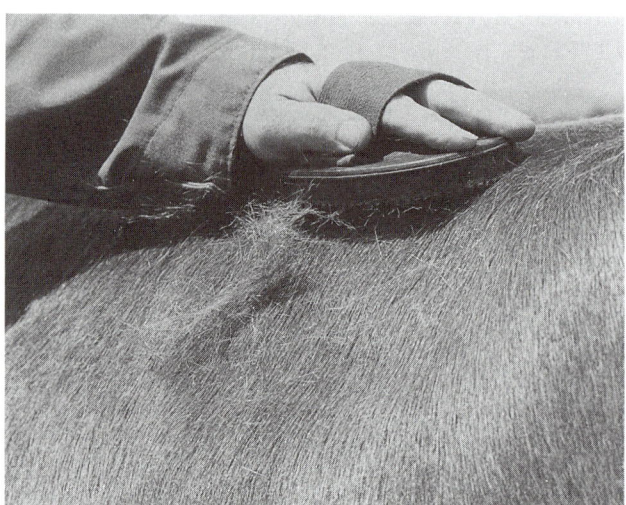

Fig. 4–8. Use a rubber curry comb daily. Work it in a vigorous, circular motion to loosen the hair and hide.

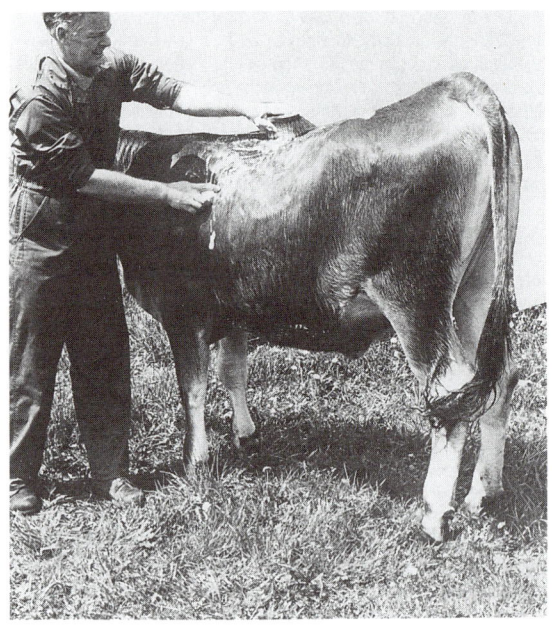

Fig. 4–6. Wash with a mild soap, work up a good lather, and scrub with a stiff brush to remove dead, dull, sun-bleached hair and to stimulate new hair growth.

Fig. 4–7. Use a properly fitted blanket to protect the animal from dirt and flies, smooth the hair coat, and bring out the natural oils of the skin.

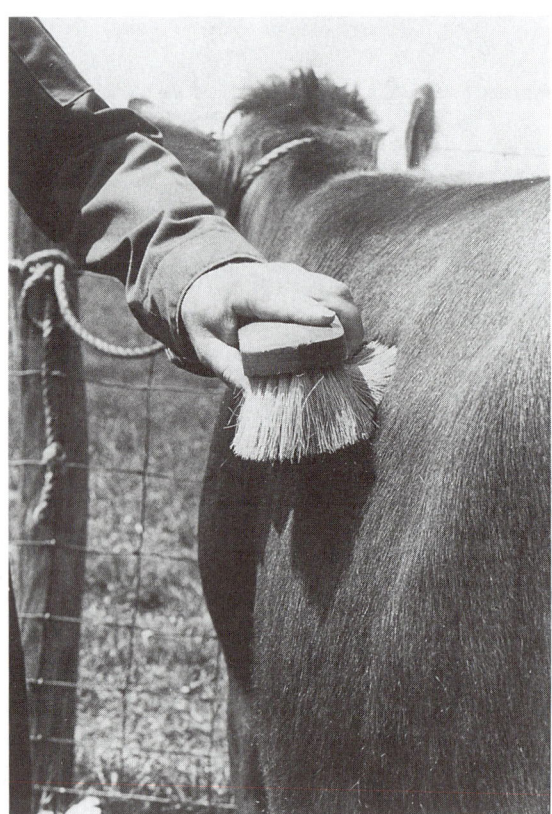

Fig. 4–9. Use a long-bristled brush. With a quick, flicking action of the wrist, remove the hair and dust from the animal.

Fig. 4–10. Wash the switch several times before the show. Use lots of soap and work up a good lather. Then, fluff it by brushing against the lay-of-the-hair and releasing a few hairs with each stroke.

HOW TO CLIP FOR THE SHOW

A properly clipped animal has a definite advantage in the show-ring. It attracts attention to the animal's strong points and minimizes its weak points. Clipping is an art that requires patience, a good eye, practice, and sharp clipper blades. Figs. 4–11 to 4–16 show how to clip a dairy animal for the show. (Figs. 4–11 to 4–16 courtesy, The Pennsylvania State University, University Park.)

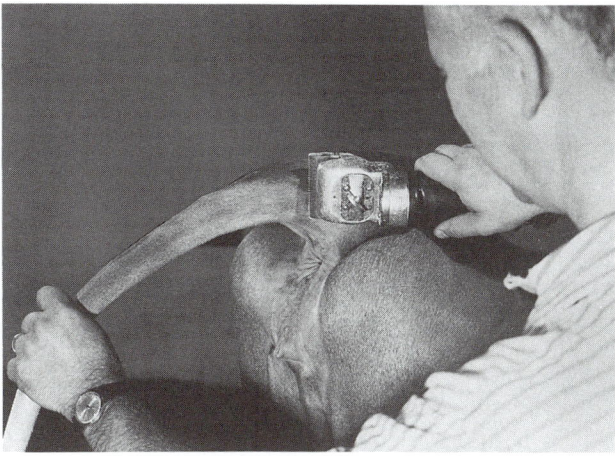

Fig. 4–12. Clip the tailhead with the grain of the hair. Keep a straight top line and blend the tailhead with the sides.

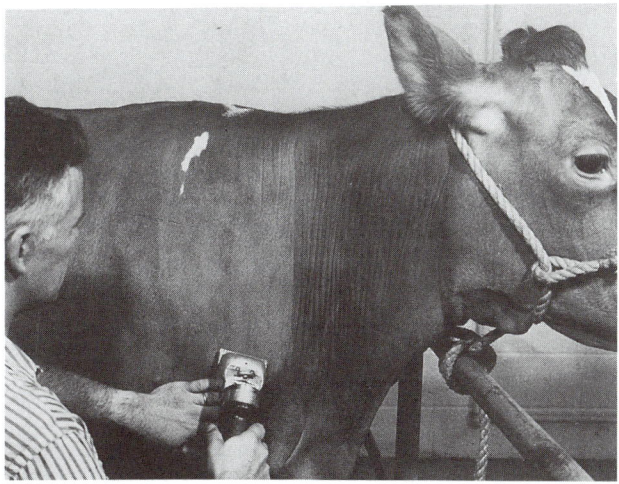

Fig. 4–13. Clip the neck forward from a line formed by the point of the shoulders and the front of the withers.

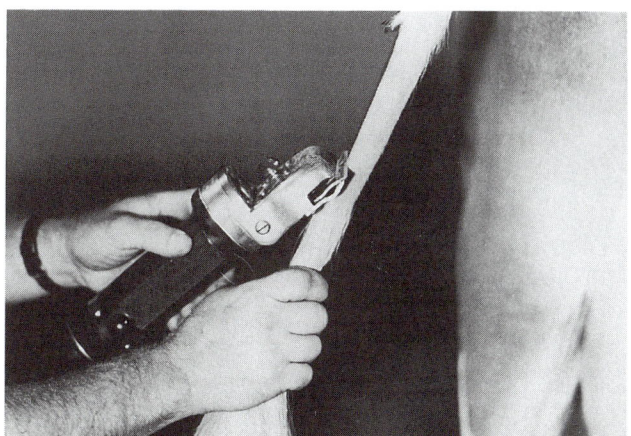

Fig. 4–11. Clip the tail from the top of the switch, or from as much as 4 in. above the long hairs, with the exact starting point determined by your eyes.

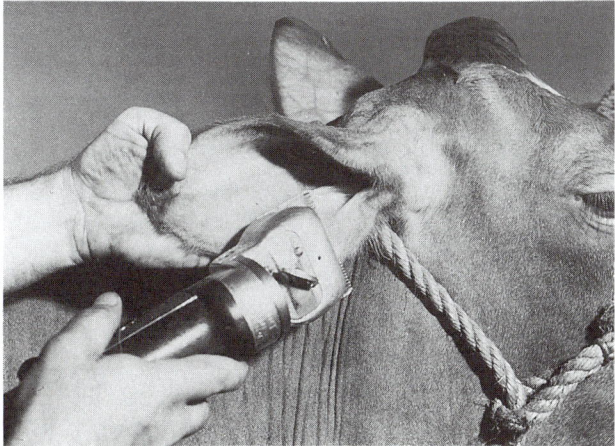

Fig. 4–14. Clip the hair from both inside and outside the ears, using your free hand to brace the ear.

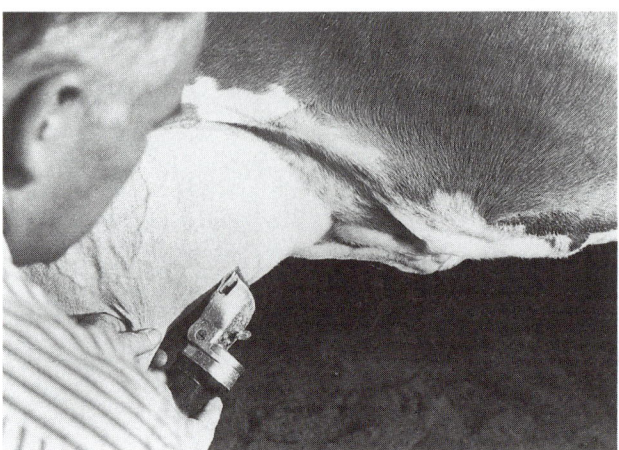

Fig. 4–15.   Clip the udder of lactating cows to accentuate quality.

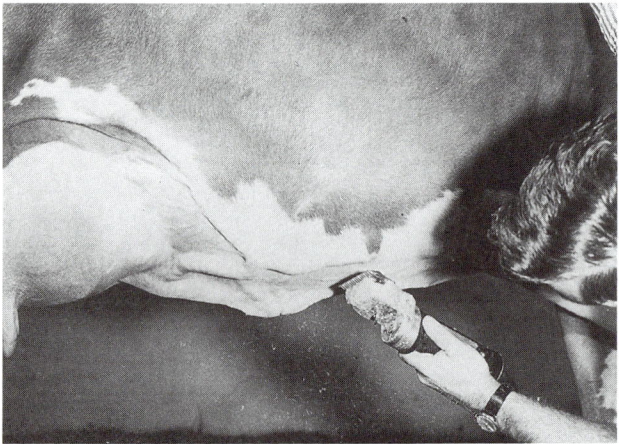

Fig. 4–16.   Clip the underline of lactating cows from the udder to the navel to show veining.

MAKING FAIR ENTRIES

Well in advance of the show, the exhibitor should request that the show manager or secretary provide a premium list and entry blanks. All rules and regulations of the show should be studied carefully and followed to the letter—including requirements relative to entrance, registration certificates, vaccinations, health certificates, stall fees, exhibitor's and helper's tickets, and other matters pertaining to the show.

Generally, entries must be filed with the show about 30 days in advance of the opening date. Most shows specify that entries be made out on printed forms and in accordance with instructions thereon. The class, age, breed, registry number, and usually the name and registry number of the sire and dam must be given. Entries must be made in all individual and group classes in which it is intended to show, but no entries are made in the championship classes, the first place winners being eligible for the latter.

PROVIDING HEALTH CERTIFICATES

Health certificates, signed by an accredited veterinarian of the state of origin, are always required for show animals. For cattle, most shows specify that this certificate indicate that within 30 days prior to entry and within 90 days prior to the date of exhibiting, the veterinarian has examined each animal offered for entry and has found it free of tuberculosis, brucellosis, and other infectious or contagious diseases. This provides reasonable assurance that diseases are not being spread. In addition, some states require that a special permit issued by the state veterinarian must accompany cattle on their trip home from the show.

SHIPPING TO THE FAIR

Show animals are usually shipped via truck; rail shipments are seldom made anymore. It is important that the following details receive consideration:

1. Schedule the transportation so that the cattle will arrive within the limitations imposed by the show and at least 2 to 3 days in advance of the date that they vie for awards.

2. Before using, thoroughly clean and disinfect any public conveyances.

3. Use long, clean, bright straw for bedding in order not to soil the hair or introduce foreign matter into it. It is also a good plan to sand the floor so that cattle will not slip.

4. In transporting by truck, cattle are generally stood crosswise in the truck, with the largest animal near the cab and tied facing to one side. The direction of facing the remaining animals is alternated; the second animal is faced in the opposite direction from the first, and so on. Some prefer to tie the animals so that all of them face the same direction. It takes more space, but they stay cleaner that way. In either case, it is best to tie animals fairly short and near enough together so that they will not lie down.

5. If space is at a premium, place the feed supply, bedding, and show equipment on a deck or platform in the truck, preferably at least $5\frac{1}{2}$ ft above the floor. Allow for air circulation and tying of smaller animals under the deck.

6. When mixed feeds are used, as is usually the case in fitting rations, a supply adequate for the entire trip should be taken along in the truck. This will reduce the hazard of animals going off feed because of feed changes.

7. Limit show cattle to a half feed at the last feeding before loading out and while in transit.

8. In transit, animals should be handled quietly and should not be allowed to become hot nor to be in a draft.

STALL SPACE, FEEDING, AND MANAGEMENT AT THE FAIR

As soon as the show is reached, the animals should be unloaded and placed in clean stalls that are freshly bedded with clean straw. The cattle should be arranged in order of size so as to make the exhibit as attractive as possible.

While at the show, it is preferable that the cattle receive the same ration to which they were accustomed at home. Usually only a half ration is allowed for the first 24 hours after arrival at the show, and a normal ration is provided thereafter. So that the animals will maintain their appetite, however, it is necessary that they receive exercise while at the show. It is usually best to exercise the animals one-half hour or more in the cool of the evening and morning, when the animals are being led to and from their nightly tieouts. This also is a convenient time to clean out the stalls.

It is customary for exhibitors to identify their respective exhibits by means of a neat and attractive sign, the size of which must be within the limitations imposed by the show. This sign usually gives the name of the breed of cattle and name and address of the exhibitor.

SHOWING DAIRY CATTLE

Fig. 4–17. Showing Milking Shorthorn cows. (Courtesy, American Milking Shorthorn Society, Beloit, WI)

Expert showing cannot be achieved through reading any set of instructions. Each show and each ring will be found to present unusual circumstances. However, there are certain guiding principles which are always adhered to by the most successful cattle exhibitors. Some of these are:

1. Train the animal long before entering the ring.

2. Be prepared to give your animal's birthdate, last freshening date, and date due to calve, should the judge ask.

3. Have all health, registration, and entry papers in order and checked in. Know what time the show begins, what breeds will be shown first, and when your class is scheduled.

4. Follow the same feeding, watering, bedding, and grooming routine on the morning of the show as was established at home. Just before going into the ring, give the animal a final drink of water, but watch the sides (good spring of rib, but not rounded). If the animal dislikes the water, add a little molasses to cover up chlorine, mineral, and other tastes. You may do this at home to get your animal used to the molasses-tasting water.

5. Have the animal carefully groomed and ready for the parade before the judge.

6. Dress neatly for the occasion. White clothes are not imperative, but they are preferable.

7. Enter the ring promptly when the class is called.

8. Lead the animal from the left hand, while walking backward and moving the animal slowly so as to show its best form. Move in a clockwise direction, so as to allow you to watch both the judge and your animal. When circling, allow 2 to 4 ft between you and the animal ahead of you.

9. Line up when asked. Go quickly, but not brashly. Line up closely to the animal next to you; by so doing, there is less chance of the judge placing another animal above you.

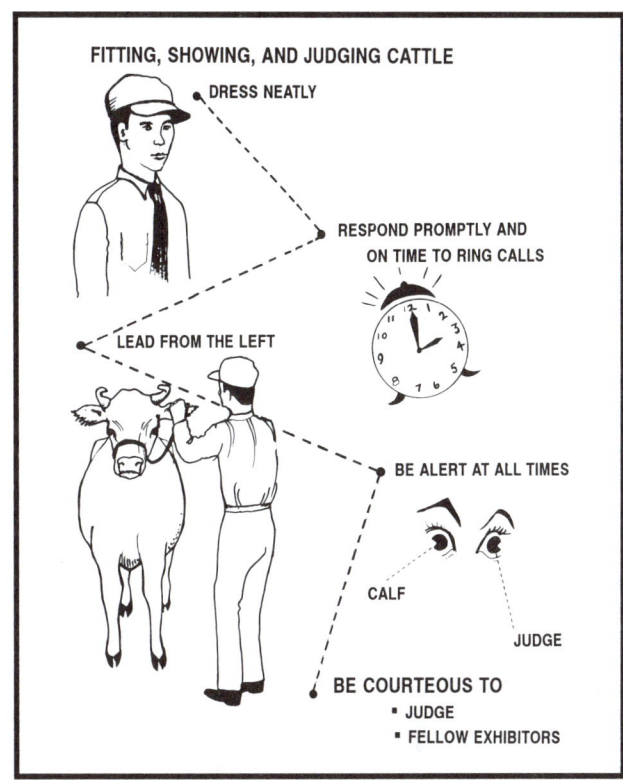

Fig. 4–18. Some of the guiding principles observed by the most successful showmen.

10. Pose the animal correctly when stopped, and so as to minimize faults. Take the strap in the left hand and set the animal up. If possible, select a slightly higher spot for your animal's front feet, but keep it in line with the other animals in the class. Position the animal and its feet by means of halter commands (not with your feet or by body pushing) as follows:

Calves and heifers: The front feet should be parallel or straight across from each other. The hind foot *nearest* the judge should be one half step *BACK*. Usually this is the right rear leg (the heifer appears longer and more stretchy this way).

Cows: Again the front feet are parallel, but the hind foot *nearest* the judge is one half step *FORWARD* (this is normally the right hind leg). This allows the judge to see the rear and fore udder attachments.

Keep the animal's head up and its back straight. This may be accomplished by (a) holding the head up, (b) pushing backward on the shoulder and halter, (c) flattening the tail with the hand, (d) stroking the topline gently with the open hand, and/or (e) pinching the loin slightly about 4 in. in front of the hips to raise low pinbones.

11. Keep one eye on the judge and the other on the animal. Center your attention entirely on showing the animal. The animal may be under the observation of the judge at a time when you least suspect it.

12. Let the animal stand "at ease" if you are in a big class and the judge is working at the other end of the ring.

13. Lead the animal out of the lineup and around the end of the line when it is nervous or restless. But don't make this a habit, and don't lead the animal between the judge and an animal that the judge is viewing.

14. Do not stand so that you block the judge's view; the judge is interested in seeing your animal—not you.

15. Move to a location of vantage if you find that you are hemmed in and that the judge cannot see your animal, unless, of course, the judge has asked you to hold your position.

16. Keep calm and collected. Remember that the nervous exhibitor creates an unfavorable impression.

17. Work in close partnership with the animal.

18. Be courteous and respect the rights of other exhibitors.

19. Do not enter into conversation with the judge. Speak to the judge only when the judge asks you a question; and never question the judge's placings.

20. Be a modest winner and a gracious loser.

21. Listen carefully to the reasons the judge gives for placings. By so doing, the exhibitor can learn and study the defects of an animal and try to select future show animals with fewer faults.

AFTER THE FAIR IS OVER

Most shows have regulations requiring that all exhibits remain on the grounds until a specified time, after which signed releases must be secured from the superintendent of the show. Because most exhibitors are anxious to travel when the show is over and there is considerable confusion at this time, it is usually advisable to load all equipment, leftover feed, and other articles before the release of animals is secured. Then all that remains to be done is to load out the animals.

Upon returning to the farm, it is usually good policy to isolate the show herd for a period of 3 weeks. This procedure reduces the possibility of spreading diseases or parasites to the balance of the herd.

It is important that young stock to be developed for show purposes the following year continue to receive an adequate, though lighter, grain ration.

Where the herd is being exhibited on a circuit, the caretaker must use great expertise in keeping the animals in show condition at all times. The peak condition should be reached at the strongest show. In order to be successful, showing on the circuit requires great skill on the part of the caretaker, especially from the standpoint of feeding and exercising the cattle.

Fig. 4–19. Champion Guernsey cows at the All-American Dairy Show. (Courtesy, American Guernsey Assn., Reynoldsburg, OH)

QUESTIONS FOR STUDY AND DISCUSSION

1. Under what circumstances would you recommend that a purebred breeder, a commercial producer, and a 4-H club member (a) should show, and (b) should not show dairy cattle?

2. Defend either the affirmative or the negative position of each of the following statements:

a. Fitting and showing does not harm dairy cattle.

b. Livestock shows have been a powerful force in dairy cattle improvement.

c. Too much money is spent on dairy shows.

d. Unless all cattle are fitted, groomed, and shown to the same degree of perfection, show-ring winnings are not indicative of the comparative quality of animals.

3. List the advantages and the disadvantages of showing dairy animals.

4. List and discuss the three most important factors to be considered when selecting individual show prospects. Are some of the same factors important when selecting commercial dairy animals? Why is it important that exhibitors be knowledgeable relative to the "age and show classifications"?

5. How does feeding cattle for the show differ from feeding cattle in a commercial dairy operation?

6. What types of halters are used on show animals? Why are two types of lead straps used?

7. How may an exhibitor avoid possible off-flavors in drinking water, which may cause the animal to refuse to drink and become gaunt?

8. Describe the posture of a well-trained show animal.

9. Why isn't a steel curry comb used in grooming dairy animals? Why isn't the switch combed during the early grooming period?

10. Why is the clipping of the front end of Milking Shorthorns optional, whereas clipping the front end is a must for Ayrshires, Brown Swiss, Guernseys, Holsteins, and Jerseys?

11. How may an animal be "filled" before entering the show-ring? Why is this important?

12. Why do dairy cattle exhibitors generally walk backward when leading animals in the show-ring, whereas beef cattle exhibitors usually walk forward?

13. Do exhibitors lead animals clockwise or counterclockwise around the ring?

14. Explain how to maneuver a nervous animal while moving around the show-ring and in the lineup.

15. Why should the exhibitors listen to the judge's reasons for placings?

SELECTED REFERENCES

Title of Publication	Author(s)	Publisher
Dairy Cattle Judging Techniques	G. W. Trimberger	Prentice-Hall, Inc., Englewood Cliffs, NJ, 1977
Livestock Judging, Selection, and Evaluation, Third Edition	R. E. Hunsley W. M. Beeson	Interstate Publishers, Inc., Danville, IL, 1988
Stockman's Handbook, The, Seventh Edition	M. E. Ensminger	Interstate Publishers, Inc., Danville, IL, 1992

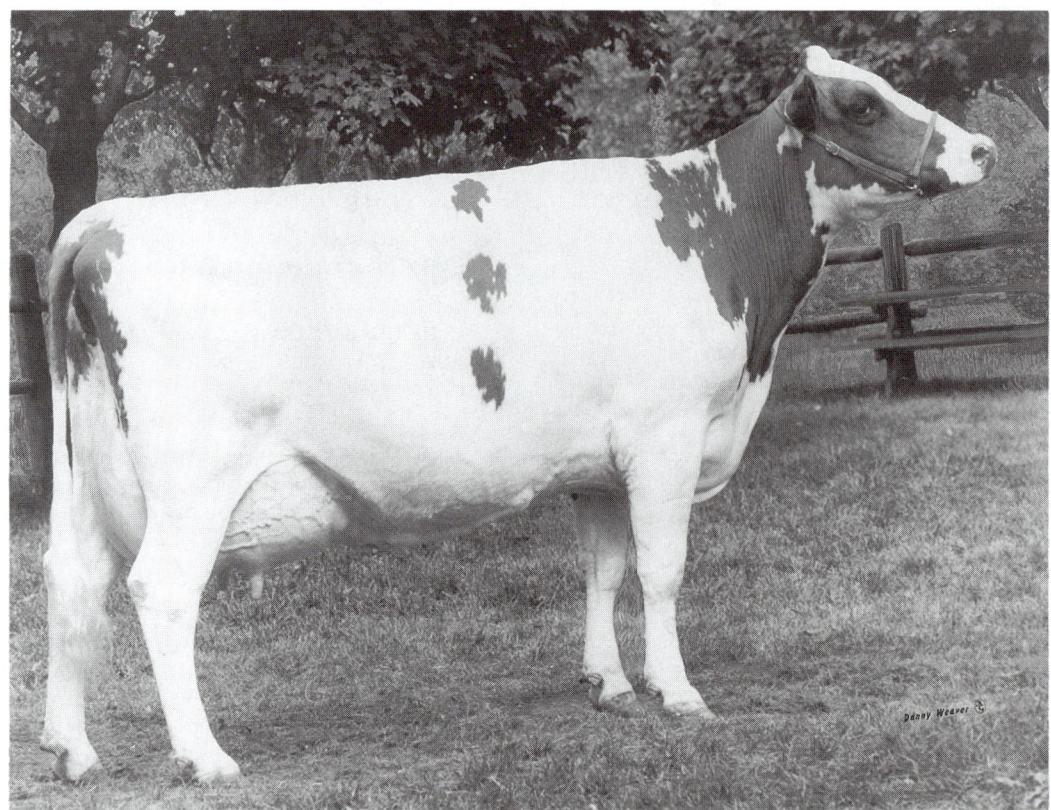

Fig. 4–20.   A beautiful Ayrshire cow beautifully fitted and shown. (Courtesy, Ayrshire Breeders' Assn., Brandon, VT)

Fig. 4–21.   A beautiful Milking Shorthorn cow beautifully fitted and shown. (Courtesy, American Milking Shorthorn Society, Beloit, WI)

Fig. 5–1. Mother cow and her "litter" of calves produced by embryo transfer. Biotechnology will dominate and pace the changes in dairy cattle breeding in the decades ahead. (Courtesy, Holstein-Friesian Assn. of America, Brattleboro, VT)

Contents

CHAPTER

5

BREEDING

DAIRY CATTLE

The objective of dairy cattle breeding is to mate individuals whose offspring will possess the necessary heritability to (1) produce the maximum amount of milk of the desired composition, and (2) develop the desired body type; then to feed and manage these animals so that their maximum genetic potential will be expressed. This recognizes the fact that dairy cattle are products of heredity and environment, as are all other animals.

The economic justification for improved breeding is that good cows make more profit. This fact is clearly illustrated in Fig. 5–2. There are three main explanations for the increase in rate of production per dairy animal in recent years: (1) The productive ability of milk cows has been increased through the selection of better producing animals, (2) cows have been better fed and managed, and (3) with the decline in milk cow numbers, most of the culling has come about among the low producing cows and in the marginal herds, with the result that the remaining cows are usually among the higher producers.

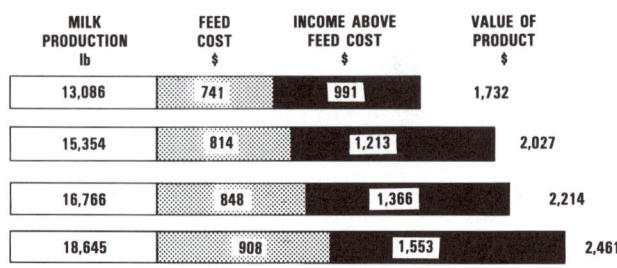

MILK PRODUCTION lb	FEED COST $	INCOME ABOVE FEED COST $	VALUE OF PRODUCT $
13,086	741	991	1,732
15,354	814	1,213	2,027
16,766	848	1,366	2,214
18,645	908	1,553	2,461

Fig. 5–2. It costs more to feed high-producing cows, but the practice pays handsome dividends. The reason: feed and overhead costs for maintenance are practically the same, regardless of the level of production. (Based on data from New York Dairy Records Laboratory; Courtesy, L. R. Brown, Extension Dairyman, The University of Connecticut, Storrs)

As noted in Fig. 5–2, it costs more to feed high producers; it costs $167 more to feed cows producing 18,645 lb of milk than cows producing only 13,088 lb. However, the income above feed costs more than compensates for the higher feed costs; the income above feed costs is $562 more for cows producing 18,645 lb of milk than for cows producing only 13,088 lb. The reason: feed and overhead costs for maintenance are practically the same, regardless of the level of production.

Today, all dairy cattle breeding may be classed as two kinds: (1) breed improvement, and (2) herd improvement, both of which come about through improvement of individual animals. Of course, dairy producers throughout the country as a whole are greatly interested in breed improvement. However, each individual dairy farmer is primarily interested in herd improvement. Naturally, if the owner/operator is able to improve the herd so that it makes for greater net return, it enhances the welfare of the owner/operator. Also, any permanent herd improvements made by an individual producer inevitably contribute to permanent breed improvement. In fact, breed improvement through the years has largely come about through the progress made by master breeders in their own herd improvement programs. It is important, therefore, that modern dairy breeders understand and follow a constructive breeding program.

RECORDS ARE NECESSARY

The foundation for a good breeding program is production records on every cow in the herd, year after year. Additionally, for the purebred breeder, type evaluation records are important. Use of these two tools — production and type evaluation records — through a careful culling and selection program tends to get rid of the undesirable genes and concentrate on those which are superior. Milk and butterfat records are also the key to scientific dairy cattle feeding.

KIND OF RECORDS; CHOOSING A SYSTEM

As is true in bookkeeping or cost accounting, a number of different systems and forms of dairy records are available. The important thing is to choose that system, or those systems, which will give the desired information, then stick to it.

MILK AND BUTTERFAT RECORDS

A number of production record systems have been developed (see Chapter 14, Dairy Cattle Management).

BREEDING, HEALTH, AND LIFETIME RECORDS

Fig. 5–3. Guernsey cow, *Zabels FWP Dixie*, is Excellent 90 at 10 years of age, with a lifetime production of 204,880 lb of milk and 9,103 lb of fat. (Courtesy, American Guernsey Assn., Reynoldsburg, OH)

Breeding, health, and lifetime records are as essential as milk production records. Among other things, they are the only way in which to diagnose and reduce infertility.

Surveys show that about 5% of the dairy cows in the United States become infertile each year. Further, about half of these represent a serious threat to the reproductive health of the rest of the herd, simply because they are "spreaders" of whatever trouble exists.

With proper records, the following are reasonable goals for which to strive:

1. At least 70% of the cows should conceive at first breeding.
2. At any given time, there should be no more than 10% of the cows with reproductive difficulties.

3. At the end of a year, the herds should average no more than 1.3 services per conception.
4. Calving interval should be no longer than 12½ months.

But to achieve these goals, or to be able to diagnose and reduce infertility, good records are essential. The breeding records should serve the following purposes:

1. Indicate when to start breeding.
2. Aid the feeding program.
3. Indicate breeding efficiency.
4. Suggest disease problems or the need for veterinary services.
5. Indicate (a) the fertility of the bull, and (b) the effectiveness of natural and artificial insemination.
6. Indicate when to turn each cow dry.
7. Indicate approximate date of calving.
8. Show parentage and disposal of calves.

An example of a good breeding, health, and lifetime record form is given in Figs. 5–4a and 5–4b (pp. 56–57).

In addition to the monthly record provided by a testing program, for genetic progress it is important to have a lifetime record on each individual cow. This should provide (1) complete identification of the individual animal, (2) individual lifetime lactation summaries, (3) breeding record, (4) calving record, and (5) health and veterinary record. Figs. 5–4a and 5–4b will serve this purpose. With this information available, each cow may be evaluated.

RECORDS AFFECTED BY ENVIRONMENT

The effect of environment on dairy cattle was clearly demonstrated in an experiment in New Zealand. It involved the selection of 20 calves from low-producing herds and 20 calves from high-producing herds. All of them were sired artificially by outstanding bulls. The 40 head were assembled at the Rurakura Experiment Station, raised and milked together for the first lactation. Under these conditions, no significant difference between the production of the two groups was observed. Then, they were sent back to the respective herds from whence they came, whereupon their production was comparable to that of the cows with which they were being milked. Then, for a second time, they were returned to the Rurakura Experiment Station, where again there was no significant difference in their production. The Rurakura Station then went one step further; they confirmed these results by using identical twins, with both twins milked at the Rurakura Station, and then later divided between high- and low-producing herds for subsequent lactation.

The New Zealand experiment points up the importance of management. No matter how good the genetics, a good environment is essential to obtain high production.

HERD NO.

	JAN.	FEB.	MAR.	APRIL	MAY	JUNE	JULY	AUG.	SEPT.	OCT.	NOV.	DEC.

NAME OF COW	REG. No.	DATE OF BIRTH		BIRTH WEIGHT		CONDITION AT BIRTH	

SIRE	REG. No.	TATTOO OR EAR TAG	CALFHOOD VACCINATION	RUMEN MAGNETS

RIGHT   LEFT   NO.   DATE

DAM	REG. No.	DEHORNED	EXTRA TEATS REMOVED	DATE & REASON FOR REMOVAL FROM HERD

MASTITIS & UDDER HEALTH						BREEDING RECORD						CALF RECORD		
DATE	RF	LF	LR	RR	TREATMENT	HEAT DATES	DATE BRED	SERVICE BULL	PREG. CHECK	DATE DUE	DATE FRESH	SEX	NAME OR NO.	CALF DISPOSAL

PRODUCTION RECORDS								COMMENTS, NOTES & MISCELLANEOUS	SKETCH OR PICTURE
	AGE	DAYS IN MILK	MILK — LBS.		TEST	FAT — LBS.			
			ACTUAL	M. E.		ACTUAL	M. E.		
1.									
2.									
3.									
4.									
5.									
6.									
7.									
8.									
9.									
10.									

Fig. 5–4a.   An example of a good breeding, health, and lifetime record form, prepared by Utah State University. This shows the front side. See Fig. 5–4b for reverse side.

DAIRY BREEDING AND HEALTH RECORD

Prepared by Utah State University
Extension Services
Logan, Utah

DIAGNOSIS AND TREATMENT BY VETERINARIAN

REPRODUCTIVE DISEASES			OTHER DISEASES		
DATE	CONDITION	TREATMENT	DATE	CONDITION	TREATMENT

ADDITIONAL INFORMATION

Fig. 5–4b. An example of a good breeding, health, and lifetime record form, prepared by Utah State University. This shows the reverse side of Fig. 5–4a.

Among the environmental and other factors affecting milk production records are the following:

1. **Feeding.** The most important of all factors in determining the productivity of a cow is the quantity and quality of feed provided.

2. **Milking practices.** Good milking practices and a properly functioning milking machine are necessary for high milk production.

3. **Age of animal.** On the average, production increases each year from the time the first calf is born until the cow reaches 5 to 8 years of age, after which it declines.

4. **Size of animal.** Within the same breed and age group, the larger animals, as measured by their capacity to consume more feed, usually produce more milk than the smaller animals.

5. **Season of freshening.** Cows that freshen in the spring and summer months usually produce less than cows freshening in the winter months. Of course, this variation differs among herds and areas.

6. **Calving interval.** Cows that calve within 12 months to 14 months from last calving produce more for that lactation than cows calving at shorter intervals. Lifetime production will usually be reduced if calving intervals are longer than 12 to 14 months.

7. **Length of dry period.** Cows with a dry period of 6 to 8 weeks produce more during the following lactation than those cows with a dry period of less than 4 weeks' duration.

8. **Freedom from disease, parasites, and injury.** Any one of these may depress production, with the degree of depression determined by the severity of the ailment.

9. **Rate of maturity.** Some strains or families of cows mature at a slower rate than others.

10. **Yearly differences.** There are important yearly differences within a given area, primarily due to weather conditions and the general quality of feed available. Nevertheless, when an attempt is being made to evaluate the breeding worth of an individual animal, all of these factors may be playing an important part in the production records.

CORRECTION FACTORS

It is frequently desirable to compare the performance of individuals or groups of animals. To do so, it is necessary to correct all records to a comparable basis. For this purpose, correction factors have been developed for each breed for (1) length of lactation, (2) the number of milkings per day, (3) age and month of calving, and (4) fat content of milk. These four adjustments are important for comparing milk and fat of cows in different environmental conditions. Each of these factors will be discussed. At the outset, however, the following point is pertinent: Although correction factors are usually necessary in order to reduce two or more records to a common basis, it is recognized that records that are comparable without factors are more reliable.

LENGTH OF LACTATION

The most generally accepted standard length of lactation records is 305 days. When a cow is milked longer than 305 days, her yield for the first 305 days is used as the standard lactation yield. Partial lactations (those terminated in less than 305 days because of environmental influences having no relation to the cow's genetic ability to complete normal length lactations) are considered legitimate measures of the cow's performance up to the time they were terminated and are used with a correction factor to 305 days. The factors commonly used for this projection are:

Days Milked	Factor
95	2.82
125	2.16
155	1.77
185	1.51
215	1.32
245	1.18
275	1.08

For comparing a 365-day record, reduce this to a 305-day record equivalent by taking 85% of it.

Total lactation records, or 365-day records, are often quoted verbally and in promotional literature, with or without an adequate definition of the lactation length. Care should be taken to clarify the length of the lactation when comparing or evaluating production records.

NUMBER OF MILKINGS PER DAY

Most cows are milked twice daily (usually referred to as 2X); hence, for most lactations no adjustment is necessary.

To convert 3-times-a-day milking to 2-times-a-day basis, multiply by 83% (0.83). For purposes of illustrating how this works, let's assume that we have a 4-year-old Holstein cow that has a 3-times-a-day, 305-day record of 14,000 lb of milk and 610 lb of fat, and that it is desired to convert it to a 2-times-a-day basis. Simply multiply the cow's record by 0.83. Hence—

14,000 lb milk × 0.83 = 11,620 lb milk on 2X basis
610 lb fat × 0.83 = 506.3 lb fat on 2X basis

AGE AND MONTH OF CALVING

The age of a cow is always based on her age when she calved, which is when her record begins. It is estimated, on a rule-of-thumb basis, that at 2 years of age a cow produces approximately 70–80% of her mature production; at 3 years, 80–90%; at 4 years, 90–95%; 5 years, 96–100%; and at 6 years, her mature record.

Age adjustment factors have been developed to standardize 305-day lactation records to a mature equivalent

basis and to minimize environmental variation due to the month of the year in which the record began. These age and month-of-calving factors are based on a total of 4,452,332 official DHI and DHIR lactations between 1964 and 1968. These factors are more accurate than any others available for milk and fat, because they more accurately remove recent environmental effects from age and month of calving in individual breeds and regions.

Table 5–1 shows the milk and fat age adjustment factors for cows in the United States calving in the month of May, by breed, for selected ages. A complete list of adjustment factors for milk and fat by breeds, by regions (and for the United States), by month of calving and by age, is given in the following report: *USDA-DHIA Factors for Standardizing 305-Day Lactation Records for Age and Month of Calving*, ARS-NE-40, USDA.

The standardized yield is obtained by multiplying yield for the first 305 days of lactation by the factor corresponding to the age at calving for the appropriate breed, region of the country, season of the year, and trait (milk or fat production). For example, let's assume that we have a Guernsey cow that was 20 months old when she calved and began her lactation record in the month of May; that she was milked 2 times daily; and that her 305-day record was 11,510 lb of milk and 508 lb of fat. By referring to Table 5–1, it is observed that the age adjustment factors for a 20-month-old Guernsey cow are 1.29 for milk and 1.28 for fat. Hence—

11,510 lb milk × 1.29 = 14,847 lb milk on ME basis
508 lb fat × 1.28 = 650.2 lb fat on ME basis

FAT CONTENT OF MILK (FCM)

For comparative purposes, the fat content of milk is usually based on calculating the milk and fat production to 4% fat (4% FCM), but it may be calculated to any desired fat basis. The formula for 4% FCM is:

4.0% FCM = (0.4 × milk weight) + (15 × fat weight)

GENETICS OF DAIRY CATTLE BREEDING

In all animal breeding (and dairy cattle are no exception), no new genetic material is created. Rather, it is simply a matter of sorting or rearranging the many factors already present in the male and female gametes. These factors, called genes, are contained in the chromosomes of the sperm of the male and the egg of the female.

Dairy animals have 30 pairs of chromosomes in each cell. The number of genes per chromosome is not definitely known; estimates are that there may be as many as 30,000 genes for dairy cattle. These genes are responsible for how the animal looks and produces.

When the sperm and egg unite, the new cell formed contains 30 pairs of chromosomes, or a total of 60 chromosomes, half of which came from the sperm (male) and half from the egg (female). What determines which genes and which chromosomes are to be passed on to the new cell is still a relatively dark secret.

TABLE 5–1
MILK AND FAT AGE ADJUSTMENT FACTORS FOR COWS IN THE UNITED STATES
CALVING IN THE MONTH OF MAY, BY AGE AND BY BREED[1]

Age	Ayrshire		Brown Swiss & Red Poll		Guernsey		Holstein & Red Dane		Jersey		Milking Shorthorn	
(months)	(milk)	(fat)	(milk)	(fat)	(milk)	(fat)	(milk)	(fat)	(milk)	(fat)	(milk)	(fat)
20	1.33	1.31	1.54	1.51	1.29	1.28	1.40	1.39	1.37	1.36	1.49	1.47
24	1.23	1.21	1.40	1.37	1.21	1.21	1.30	1.29	1.27	1.26	1.31	1.29
30	1.16	1.13	1.29	1.27	1.13	1.12	1.21	1.20	1.17	1.16	1.19	1.18
36	1.13	1.12	1.20	1.18	1.09	1.08	1.15	1.15	1.12	1.11	1.16	1.16
42	1.09	1.08	1.14	1.13	1.06	1.05	1.10	1.10	1.07	1.06	1.14	1.15
48	1.06	1.06	1.10	1.09	1.04	1.04	1.07	1.07	1.04	1.04	1.12	1.12
54	1.04	1.03	1.06	1.06	1.02	1.02	1.04	1.04	1.02	1.02	1.09	1.10
60	1.02	1.02	1.04	1.04	1.01	1.02	1.02	1.03	1.00	1.01	1.06	1.07
66	1.01	1.02	1.03	1.03	1.01	1.02	1.01	1.02	0.99	1.00	1.04	1.05
72	1.01	1.02	1.02	1.03	1.01	1.02	1.01	1.01	0.98	1.00	1.02	1.03
84	1.00	1.02	1.01	1.02	1.01	1.03	1.01	1.02	0.98	1.00	1.00	1.01
95	1.00	1.02	1.00	1.02	1.01	1.04	1.01	1.02	0.98	1.01	0.98	1.00
110	1.02	1.04	1.01	1.03	1.03	1.06	1.03	1.05	1.00	1.03	0.98	1.01
120	1.03	1.05	1.02	1.05	1.04	1.08	1.05	1.07	1.01	1.04	1.00	1.02
135	1.04	1.07	1.04	1.08	1.06	1.11	1.08	1.10	1.03	1.07	1.01	1.04
140	1.06	1.09	1.05	1.09	1.06	1.11	1.10	1.12	1.04	1.08	1.02	1.05
150	1.08	1.11	1.07	1.11	1.08	1.14	1.12	1.15	1.06	1.10	1.03	1.06
160	1.10	1.14	1.09	1.13	1.10	1.16	1.15	1.18	1.08	1.12	1.04	1.07

[1]*USDA-DHIA Factors for Standardizing 305–Day Lactation Records for Age and Month of Calving*, ARS-NE-40, Agricultural Research Service, USDA, pp. 80–91.

HERITABILITY OF CHARACTERS

The expression of a trait, such as milk production, depends upon two factors: (1) inheritance, or the ability to produce, and (2) environment, or the opportunity to express the inherited ability. Heritability is 100% when the expression of the trait varies solely because of inheritance. A trait that varies solely because of environment has a heritability of zero. Variations in most traits are neither wholly environmental nor completely hereditary. The heritability of some common dairy cattle traits is given in Table 5–2.

TABLE 5–2
HERITABILITY ESTIMATES FOR SOME DAIRY CATTLE TRAITS

Trait	Heritability
	(%)
Milk yield	30
Fat yield	25
Fat percentage	50
Protein percentage	50
Solids not fat percentage	50
Milking rate	30
Longevity (length of life)	5
Type (final rating)	30
Breeding efficiency	5

As shown in Table 5–2, milk yield is about 30% heritable, whereas fat percentage is 50% heritable.

The following example will show how heritability can be computed: Let us assume that we have a herd that averages 440 lb of butterfat on a mature level. Further, let us select a young sire that we estimate is capable of transmitting inheritance for 500 lb of butterfat production. He is mated to select cows in the herd with production records averaging 500 lb of butterfat in a normal 305-day lactation period. Because heritability is 30%, we expect only three-tenths of the apparent superiority of the parents expressed in the offspring. The selected parents averaged 60 lb of butterfat higher than the herd. Three-tenths of the 60 equals 18 lb of butterfat. Thus, the offspring would be expected to average 458 lb of butterfat in this herd when given the same opportunity as the parents. The additional butterfat of the parents resulted from influences such as better feeding and management.

NUMBER OF CHARACTERS
SELECTED FOR SIMULTANEOUSLY

Most rapid progress can be made in a breeding program by selecting for one trait only. However, when two characteristics are inherited in close relationship, considerable progress may be achieved in both of them. This is true, for example, in total milk production and total production of fat. On the other hand, body type and high production of milk are not closely associated in inheritance, with the result that selection for both of them will result in relatively slow progress in either one. Under these circumstances, herd owners must decide which of the traits in the herd need improvement most and make their decision on the characteristics to be emphasized in selection accordingly. If good type will bring more monetary return than increased milk production, it should be emphasized. On the other hand, if higher production will increase income to a greater degree than improved body type, then it should be given greater importance in selection.

In many herds, particularly purebred herds, both type and production are important and selection for both should be made. In other words, selection for either type or production should not be made at the expense of serious loss in the other.

As a general rule, milk production and butterfat percentage are the characters usually considered in the breeding of dairy cattle, but other characters should also be considered, as many of them have economic value.

GENETIC EVALUATION OF DAIRY CATTLE IN THE PAST

For many years, genetic improvement of the U.S. dairy cattle was focused on identifying and selecting superior cows, with little attention given to the evaluation of bulls. As improved methods were developed, dairy producers who used them gained some advantage in rate of genetic improvement over their competitors. In order, the following methods of genetic evaluation were used in the past: (1) lactation record, (2) daughter average, (3) daughter-dam comparison, (4) herdmate comparison, and (5) contemporary comparison. Each of these is briefed in the sections that follow.

LACTATION RECORD

Because of environmental influences, knowledge of a cow's lactation record tells very little about her breeding value for milk production. A cow's production may vary as much as 50% between a well-managed herd and a poorly managed herd. On the average, about 80% of the difference between herds is due to environment and 20% is genetic.

DAUGHTER AVERAGE

This index is based on the premise that the average production of a bull's daughter is indicative of his transmitting ability for milk production. Its main weakness is that it does not consider the production of their dams.

DAUGHTER-DAM COMPARISON

This is one of the oldest methods of evaluating bulls; it was widely used from about 1930 to 1960. It consists in comparing the production of a bull's daughters with that of their dams. The disadvantages of a Daughter-Dam Comparison are (1) the dams and daughter may not be milking at the same time, which means that environmental differences are inevitable; (2) comparisons are usually made in one, or only a few herds, which means that only a limited range of

environmental conditions prevail; and (3) many dams do not have records, with the result that there is nothing with which to compare their daughters.

In the early 1960s, the Daughter-Dam Comparison was largely abandoned in favor of the Herdmate Comparison, which was superior in eliminating environmental effects from estimates of breeding value.

HERDMATE COMPARISON

The Herdmate Comparison, which was developed by the U.S. Department of Agriculture and first used in 1954, was the first genetic procedure developed specifically for bulls. It was used extensively from the early 1960s until 1974, when the U.S. Department of Agriculture introduced the Modified Contemporary Comparison.

The term herdmates refers to other cows in the herd that were calved in the same year and season as the daughters of the sire being compared. This method of sire evaluation has two objectives: (1) to compare performance of one animal with another which is not paternally related, and (2) to compare records made at the same time under the same environment. Under this system, all records are converted to twice-a-day milking, 305 days, and mature (5 years) age. All animals in the herd, regardless of age, which freshen within a period from 3 months before to 3 months after each daughter of the particular sire in question freshens, are included in the comparison.

CONTEMPORARY COMPARISON

The Contemporary Comparison is based on the same principles as the Herdmate Comparison except that only first records are used. The name stems from the fact that all animals used are *contemporaries*; that is, they commenced their first records at approximately the same time. Proponents of the contemporary comparison claimed two primary advantages: (1) use of first records minimized many of the possible inaccuracies that might arise from the use of age-correction factors; and (2) the use of first records, which were made only by cows that escaped being culled; hence, were a select group. The primary disadvantage of the Contemporary Comparison is that in many herds there were too few cows that qualified as true contemporaries.

GENETIC EVALUATION OF DAIRY CATTLE AT PRESENT

The U.S. dairy industry and consumers have benefitted greatly from the national research program on genetic improvement of dairy cattle conducted by the Animal Improvement Programs Laboratory (AIPL), Agricultural Research Service, U.S. Department of Agriculture, and from the research at the land grant universities. Genetic evaluation of bulls (Sire Summaries) and cows (Cow Indexes) emanating from this research has been the primary source of information for identifying animals with superior genetic merit for yield.

Since 1974, national genetic evaluations of bulls and cows for milk and fat have been calculated by AIPL-USDA using the Modified Contemporary Comparison. National Sire

Summaries for protein and solids-not-fat have been calculated by the Mixed Model Method since 1976. In 1983, the AIPL-USDA evolved with the first National Buck Summary for dairy goats.

Since there is no measure of a sire's individual performance, his evaluation is based on the performance of his daughters. An estimate of a sire's ability to transmit may be arrived at in two steps: (1) an overall average is calculated from the average difference from herdmates for all of his daughters; and (2) the overall average difference is weighed according to repeatability of that amount of information. Two terms are used in expressing the resulting estimate of genetic transmitting of the sire: Predicted Difference (PD), and Repeatability.

■ **Predicted Difference (PD)** — This is the term applied to the genetic values which rank bulls for production traits. *Predicted Difference is the expected extra production per daughter per year, when compared to a zero PD bull.*

Fig. 5–5. *Favorite Saint*, the highest PD milk bull in the history of the Jersey breed. His PD is +2,248 lb milk and +3.2 for PD type. (Courtesy, The American Jersey Cattle Club, Reynoldsburg, OH)

■ **Repeatability** — It is important to realize that PD values are not absolute. They change, up and down, as more daughters are added in more herds. *Repeatability is an estimate of how sure we are that the Predicted Difference reflects an individual bull's true transmitting ability.*

MODIFIED CONTEMPORARY COMPARISON (MCC)

Since 1974, national genetic evaluations of bulls and cows for milk and fat have been calculated by AIPL-USDA using a procedure called the Modified Contemporary Comparison. To comprehend the value of the MCC over a simple average of daughter-herdmate differences, it is necessary to understand some problems that have arisen in sire evaluation.

Many owners have used high Predicted Difference sires year after year and have developed entire herds sired by these top sires. These herds have far surpassed the average dairy herd in their genetic ability for production. Any bull with daughters competing in herds of this caliber has a difficult time coming up with a high PD.

By contrast, other owners have used mediocre sires,

and in many cases have done little culling. Any bull whose daughters are competing in these herds looks better than he should.

The Modified Contemporary Comparison (MCC) takes into account the genetic level of herdmates competing with the daughters. This is done by the use of PDs of the sires of these herdmates.

Another frequent problem in sire evaluation is the bias due to culling that the older cows have survived. This may be a problem when a bull's summary includes older daughters as well as when the competition the heifer faces includes older cows. The MCC contains an adjustment for bias that occurs due to culling for production after the first lactation.

The MCC takes its name from the use of contemporary groups or herdmates that are of similar age. It makes use of two contemporary groups: *Contemporary Group 1* includes first lactation only; *Contemporary Group 2* includes second and later lactations. The greatest emphasis in the sire evaluation is given to the comparisons of daughters' records with contemporary information — with Group 1 records in the case of daughter's first lactation or with Group 2 records for second and later lactations of daughters. A daughter's non-contemporary herdmates are also included but are counted the equivalent of only one additional contemporary.

Another characteristic of the USDA-MCC sire summary is the use of pedigree information on the sire. The pedigree information used is the PD of the bull's sire and that of his maternal grandsire. Each bull is put into a *genetic group* based on this pedigree evaluation. The average superiority (or inferiority) of the daughter's of the bulls in this genetic group is then used in the sire's summary. This pedigree group average is weighted along with the daughter's information according to the genetic worth of each source. As daughter information (Repeatability) increases, the weight given to the pedigree group average decreases. This relationship is R to 1-R where R is the Repeatability and 1-R is the weight given to the pedigree group average.

The PD computed by USDA is adjusted to a constant genetic base such as PD 82. This means that a breed average bull in 1982 had a PD of zero. As the breed improves genetically, more and more bulls will have higher and higher PDs. The breed average of sire may be +400, +500 or even higher after a period of years. This genetic improvement of the breed may necessitate changing the base to a more recent year base so that an average bull is again zero and a "+" sign will again denote a breed improver.

■ **Pedigree evaluation** — The sire and dam contribute equally to the genetic makeup of the offspring. The estimates of these contributions are the PD of the sire and the Cow Index (CI) or *Estimated Average Transmitting Ability* of the dam. The pedigree estimate of breeding value (EBV) of a heifer or bull would therefore be:

Pedigree EBV = Sire's PD plus Dam's CI.

As an example, suppose a heifer calf is sired by a bull with PD = +1,200 and is from a cow with CI = +1,000.

Pedigree EBV = +1200 plus +1000 = +2200

The best estimate of this heifer's future producing ability, above or below her competition, is +2,200. Since she is expected to transmit only a sample half of her genetic superiority to her offspring, the best estimate of her future transmitting ability would be half her pedigree EBV, or +1,100. The offspring is expected to be intermediate in genetic merit to her parents.

MIXED MODEL METHOD

The term *Mixed Model* refers to the statistical properties of the effects in these models where some of the effects are random and some are fixed.

The Mixed Model Method possesses all the attributes of the Modified Contemporary Comparison—and more. The Mixed Model Method uses fixed seasons; AIPL-USDA uses two 6-month seasons—(1) January through June, and (2) July through December.

The AIPL-USDA continues to use the Modified Contemporary Comparison for national genetic evaluation, rather than the Mixed Model Method, because of the prohibitive cost of the latter method.

In addition to the Mixed Model Method being more costly to conduct than the Modified Contemporary Comparison, it is of questionable accuracy for daughters of non-AI bulls with most or all daughters in a single herd.

SUMMARY OF DAIRY GENETICS

The ultimate goal for the dairy herd breeding program is to produce replacement heifers that will:

1. Produce large quantities of milk with a protein, fat and total solids content that will command a premium price.
2. Reproduce regularly and without problems, calving every 12 or 13 months.
3. Have a minimum of health problems, such as mastitis, milk fever, and ketosis.
4. Be completely mobile and require minimal care of feet.
5. Have a disposition that allows them to fit the facilities and management routine for the herd.
6. Milk out quickly, cleanly, and without special labor requirements at milking time.
7. Wear well and do the above for a long lifetime.

The most rapid avenue for attaining this goal is by selection of the best sires available through artificial insemination. Top pedigreed young sires are also a good buy and can contribute to herd improvement if a number of them are used, each sparingly. If natural services are used, they should be highly selected, with a high pedigree estimate of breeding value, and should be used sparingly and no more than one season.

Milk production is the primary trait for which to select. Fat, protein, or other total solids are also highly important and likely to increase in importance when and if nutritional pricing becomes widespread. Some attention to conformation is necessary to avoid serious problems, especially in the udder traits. The best approach is probably to select the highest production sires and then eliminate the very poorest ones based on total score of their daughters. This should avoid creating conformation problems that cannot be controlled by light culling of females in the herd.

Common sense dictates that the breeder should be aware of how production is being sacrificed each time a lower PD sire is accepted in hopes of obtaining an improvement in some other trait. If the possible improvement in that other trait is worth more than the potential loss in production, the sacrifice should be made. If too many sacrifices in production are made, the genetic improvement for production in that herd will grind to a stop.

■ **Ranking of estimates of transmitting ability of bulls**—Most dairy scientists and producers rank different kinds of estimates of transmitting ability of bulls in descending order of accuracy as follows:

1. Modified Contemporary Comparison or Mixed Model Method.
2. Herdmate Comparison.
3. Daughter-Dam Comparison.
4. Daughter Average.
5. Outstanding records, show-ring winnings, distant relatives, or other selected information.

■ **Ranking of estimates of transmitting ability of cows**—Most dairy scientists and producers rank different kinds of phenotype information for estimating the transmitting ability of cows in descending order of accuracy as follows:

1. Modified Contemporary Comparison or Mixed Model Cow Index.
2. Herdmate Comparison Cow Index.
3. Daughter-Dam Comparison.
4. Lactation Record.
5. Highest Record or other selected information.

■ **Further information**—All dairy farmers can improve the genetic ability of their herds to produce. Modern cow evaluations are available through the national Cooperative Dairy Herd Improvement Program in every state. Accurate and useful sire evaluations are available from the U.S. Department of Agriculture through the extension dairy specialists, artificial insemination (AI) associations, and breed associations. Service to top sires is available through AI Breed associations provide additional programs of benefit in the breeding and management of dairy herds especially through their type classification programs. The Cooperative Extension Service in every state stands ready to help with explanations and suggestions for use of all available information.

The ingredients for a successful breeding program to improve the efficiency and profitability of the herd are available. It is up to dairy producers to take the necessary time to plan, to make decisions, and to use the cow and sire evaluations for the improvement of their herds.

QUALITATIVE TRAITS

In the simplest type of inheritance, known as qualitative traits, only one pair of genes is involved. Examples of qualitative traits in dairy cattle are hair color, horned vs polled, inherited abnormalities, and blood antigens.

Coat color is important from the standpoint of breed requirements. However, in recent years there has been a tendency in most of the breed registry associations to relax color or color-pattern requirements.

A pair of genes is responsible for hair color in cattle. This situation is illustrated in Fig. 5–6. Thus, a Milking Shorthorn having two genes for red (RR) is actually red in color, whereas an animal having two genes for white (rr) is white in color. On the other hand, a Milking Shorthorn which has one gene for red (R) and one for white (r) is neither red nor white but roan (Rr), which is a mixture of red and white. Thus, red × white matings in Milking Shorthorn cattle usually produce roan offspring. Likewise, white × white matings generally produce white offspring; but it must be remembered that white in Milking Shorthorns is seldom pure, for the face bristles, eyelashes, and ears usually carry red hairs. Roans, having one gene for red and one for white on the paired chromosomes will never breed true and, if mated together, will produce calves in the proportion of one red, two roans, and one white. If one wishes to produce roans, the most certain way is to mate red cows with a white bull or vice versa, for then all the calves will be roan. If a roan animal is bred to a red one, one-half the offspring will be red, whereas the other half will be roan. Likewise when a roan animal is bred to a white one, approximately an equal number of roan and white calves will be produced.

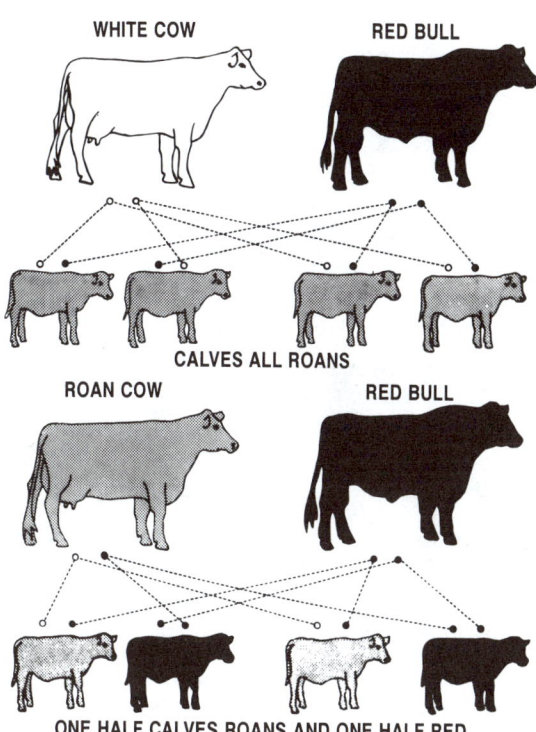

WHITE COW   **RED BULL**

CALVES ALL ROANS

ROAN COW   **RED BULL**

ONE HALF CALVES ROANS AND ONE HALF RED

Fig. 5–6. Diagrammatic illustration of the inheritance of color in Milking Shorthorn cattle. White × red matings in Milking Shorthorn cattle usually produce roan offspring, whereas roan × red matings produce one half red offspring and one half roan offspring.

Environment has little effect on hair color except for extreme circumstances such as molybdenum toxicity, copper deficiency, long exposure to tropical sun, or freeze branding.

In addition to color of hair, other examples of simple gene inheritance in animals (sometimes referred to as qualitative traits) include the presence or absence of horns, type of blood, and lethals.

DOMINANT AND RECESSIVE FACTORS

In the example of Milking Shorthorn colors, each gene of the pair (R and r) produced a visible effect, whether paired as identical genes (two red or two white) or as two different genes (red and white).

This is not true of all genes; some of them have the ability to prevent or mask the expression of others, with the result that the genetic makeup of such animals cannot be recognized with perfect accuracy. This ability to cover up or mask the presence of one member of a set of genes is called dominance. The gene which masks the one is the dominant gene; the one which is masked is the recessive gene.

In cattle, the polled character is dominant to the horned character. Thus, if a *pure polled* bull is used on horned cows (or vice versa), the resulting progeny are not midway between two parents but are of polled character.[1] It must be remembered, however, that not all hornless animals are pure for the polled character; many of them carry a factor for horns in the hidden or recessive condition. In genetic terminology, animals that are pure for a certain character—for example the polled characteristic—are termed *homozygous,* whereas those that have one dominant and one recessive factor are termed *heterozygous.* A simple breeding test can be used in order to determine whether a polled bull is homozygous or heterozygous, but it is impossible to determine such purity or impurity through inspection. The breeding test consists of mating the polled sire with a number of horned females. If the bull is pure or homozygous for the polled character, all of the calves will be polled; whereas if he is impure or heterozygous, only half of the resulting offspring will, on the average, be polled and half will have horns like the horned parents.

It is clear, therefore, that a dominant character will cover up a recessive. Hence, an animal's breeding performance cannot be recognized by its phenotype (how it looks), a fact which is of great significance in practical breeding.

---

[1]It is noteworthy, however, that when a homozygous polled animal is crossed with a homozygous horned animal, some *scurs* or small, loosely attached horns usually appear. There are conflicting reports and opinions concerning the inheritance of scurs, with the following theories prevailing:

a. That the gene for scurs is recessive and independent of the major genes for horns. According to this theory, scurs appear only in individuals homozygous for the scurred gene (sc sc).

b. That scurs are a sex-influenced character. According to this theory, scurs will occur in males either homozygous (Sc Sc) or heterozygous (Sc sc) for the character, but only in females homozygous (Sc Sc) for the character; in other words, it acts as a dominant in polled males and a recessive in polled females.

c. That the major gene (P) for polled condition is only partially dominant, with heterozygous individuals (Pp) tending to be scurred, especially in bulls.

Horns prevent the expression of any genes an animal may have for scurs, and thus complicate studies designed to determine the exact mode of inheritance of scurs.

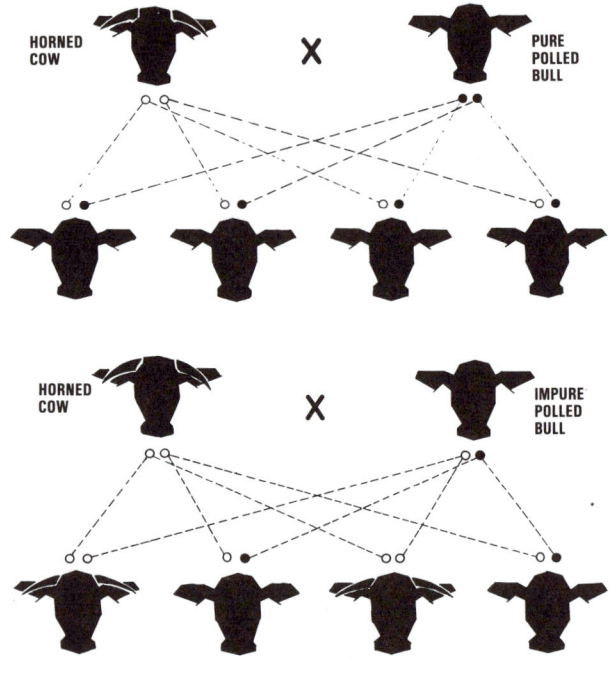

KEY: ○ = *GENE FOR HORNED CHARACTERISTIC*
     ● = *GENE FOR POLLED CHARACTERISTIC*

Fig. 5–7. Diagrammatic illustration of the inheritance of horns in cattle. Although there may be a very occasional exception, if a bull that is considered pure or homozygous for the polled character is mated with a number of horned females, all of the calves will be polled; whereas if a bull that is impure or heterozygous for the polled character is mated with a number of horned females, only half of the calves will, on the average, be polled. (Drawing by R. F. Johnson)

In general, black is dominant to the other colors, although in some cases the dominance is incomplete. For example, black is completely dominant to red in Holsteins. But crossing Holsteins with other breeds usually results in a mixture of black and colored hairs on the nonwhite areas of the offspring.

As can be readily understood, dominance often makes the task of identifying and discarding all animals carrying an undesirable recessive factor a difficult one. Recessive genes can be passed on from generation to generation, appearing only when two animals, both of which carry the recessive factor, happen to mate. Even then, only one out of four offspring produced will, on the average, be homozygous for the recessive factor and show it.

INCOMPLETE OR PARTIAL DOMINANCE

The results of crossing polled with horned cattle are clear-cut because the polled character is completely dominant over its allele (horned). If, however, a cross is made between a red and a white Milking Shorthorn, the result is a roan (mixture of red and white hairs) color pattern. In the latter cross, the action of a gene is such that it does not cover the allele, which is known as incomplete dominance; or, stated differently, the roan color is the result of the action of a pair of genes (joint action) neither of which is dominant. This explains the futility of efforts to develop Milking Shorthorns pure for roan.

The above discussion also indicates that there are varying degrees of dominance—from complete dominance to an entire lack of dominance. In the vast majority of cases, however, dominance is neither complete nor absent, but incomplete or partial. Also, it is now known that dominance is not the result of single-factor pairs but that the degree of dominance depends upon the animal's whole genetic makeup together with the environment to which it is exposed, and the various interactions between the genetic complex (genotype) and the environment.

LETHALS AND OTHER ABNORMALITIES

Lethals are genetic factors which cause death of the calves carrying them prior to or shortly after birth. Documented cases of such abnormalities are numerous. Most lethals are recessive and may, therefore, remain hidden for many generations. The prevention of such genetic abnormalities requires that the germ plasm be purged of the "bad" genes. This means that, where recessive lethals are involved, the dairy producer must be aware of the fact that both parents carry the gene. For the total removal of the lethals, test matings and rigid selection must be practiced.

Fig. 5–8. "Limber leg" calf. This abnormality appears to be inherited as a simple autosomal recessive. (Courtesy, Robert C. Lamb, Utah State University)

In addition to hereditary abnormalities, there are certain abnormalities that may be due to nutritional deficiencies, or to accidents of development—the latter include those which seem to occur sporadically and for which there is no well-defined reason. When only a few defective individuals occur within a particular herd, it is often impossible to determine whether their occurrence is due to (1) defective heredity, (2) defective nutrition, or (3) merely to accidents of development. If the same defect occurs in any appreciable number of animals, however, it is probably either hereditary or nutritional. In any event, the diagnosis of the condition is not always a simple matter.

The following conditions would tend to indicate a hereditary defect:

1. If the defect had been reported previously as hereditary in the same breed.
2. If it occurred more frequently within certain families or when there has been inbreeding.
3. If it occurred in more than one season and when different rations had been fed.

The following conditions might be accepted as indications that the abnormality was due to a nutritional deficiency:

1. If previously it has been reliably reported to be due to a nutritional deficiency.
2. If it appeared to be restricted to a certain area.
3. If it occurred when the ration of the mother was known to be deficient.
4. If it disappeared when an improved ration was fed.

If there is suspicion that the ration is defective, it should be improved, not only from the standpoint of preventing such deformities, but from the standpoint of good and efficient management.

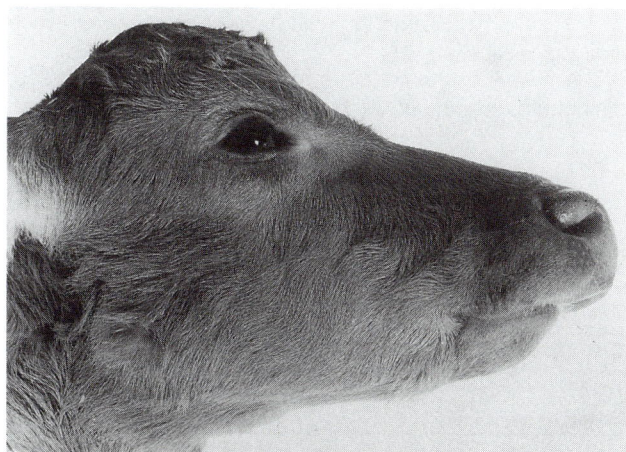

Fig. 5–9. Agnathia (short lower jaw) in a calf. (Courtesy, Professor H. W. Leipold, Kansas State University, Manhattan)

If there is good and sufficient evidence that the abnormal condition is hereditary, the steps to be followed in purging the herd of the undesirable gene are identical to those for ridding the herd of any other undesirable recessive factor. An inbreeding program, of course, is the most effective way in which to expose hereditary lethals in order that purging may follow.

Assuming that a hereditary defect or abnormality has occurred in a herd and that it is recessive in nature, the breeding program to be followed to prevent or minimize the possibility of its future occurrence will depend somewhat on the type of herd involved—especially on whether it is a commercial or purebred herd. In a commercial herd, the breeder can usually guard against further reappearance of the undesirable recessive simply by using an outcross (unrelated) sire within the same breed or by crossbreeding with a sire from another breed. With this system, the breeder is fully aware of the recessive being present, but the action taken is designed to keep it from showing up.

On the other hand, if such an undesirable recessive appears in a purebred herd, the action should be more drastic. Reputable purebred breeders have an obligation not only to themselves but to their customers among both the purebred and the commercial herds. Purebred animals should be purged of undesirable genes and lethals. This can be done by:

1. Eliminating those sires and dams that are known to have transmitted the undesirable recessive character.
2. Eliminating both the abnormal and normal offspring produced by these sires and dams (approximately half of the normal animals will carry the undesirable character in the recessive condition).
3. Breeding a prospective herd sire to a number of females known to carry the factor for the undesirable recessive, thus making sure that the new sire is free from the recessive.

Such action in a purebred herd is expensive, and it calls for considerable courage. Yet it is the only way in which purebred cattle can be freed from such undesirable genes.

BLOOD TYPING CATTLE (BLOOD ANTIGENS)

Cattle blood typing was developed by the University of Wisconsin during the 1940 to 1950 decade. It involves a study of the components of the blood, which are inherited according to strict genetic rules that have been established in the research laboratory. By determining the genetic "markers" in each sample and then applying the rules of inheritance, parentage can be determined. To qualify as the offspring of a given female or male, an animal must not possess any genetic markers not present in its alleged parents. If it does it constitutes grounds for illegitimacy.

Blood typing is used for the following purposes:

■ **To verify parentage**—The test is used in instances where the offspring may bear some unusual color or markings or carry some undesirable recessive characteristic. It may also be used to verify a registration certificate. Since an estimated 5% of all registered animals in the United States are illegitimate, there is need to use blood typing as a bulwark of breed integrity. Through blood typing, parentage can be verified with 90% accuracy.[2] Although this means that 10% of the cases cannot be settled, it is not possible to do any better than that in human blood typing.

■ **To determine which of two sires**—When a female has been served by two or more males during one breeding season, blood typing can exclude the incorrect male and include the correct male in over 91% of the cases.

---

[2]In a personal communication to the author, Dr. Clyde Stormont, Professor of Immunogenetics, Department of Reproduction, School of Veterinary Medicine, University of California, Davis, reported that in the California laboratory they have been able to solve approximately 91% of all cattle parentage cases.

■ **To provide a permanent blood type record for identification purposes** — Two samples of blood are required for each animal to be studied; and the samples must be taken in tubes and in keeping with detailed instructions provided by the laboratory. In parentage cases, this calls for blood samples from the offspring and both parents; in paternity cases, samples must be taken from the offspring, the dam, and all the likely sires.

■ **To substitute for fingerprinting** — Much attention is now being given to the idea of utilizing blood typing as a positive means of identification of stolen animals, through proving their parentage.

■ **To detect fertile heifers born co-twin with bulls** — About 15% of all heifers born twin with a bull are potentially fertile; the other 85% are sterile, or freemartins. Blood typing alleviates the need to wait until such heifers reach breeding age in order to ascertain their breeding potentialities. Instead, a blood sample from each of the twins (the bull and the heifer) can be submitted to a service-typing laboratory and a diagnosis made. If the bull and heifer have *like* blood types (except possible differences in the J system), the heifer is diagnosed as a freemartin and non-breeder. If the bull and the heifer have *unlike* blood types (except possible differences in the J system alone), the heifer is diagnosed as potentially fertile.

The basis for this remarkable method of diagnosing the breeding potentialities of a heifer born twin with a bull goes back to the early events in the embryology of cattle twins. In about 85% of the cattle embryos, some of the chorionic blood vessels become anastomosed, or joined together. This results in a communal blood vascular system. Hence, the twins come to share each other's blood-forming tissues. As a result, they have like blood types.

■ **Blood typing laboratories** — The following laboratories are capable of determining bull parentage:

Cattle Blood Typing Laboratory, Ohio State University, Columbus, OH 43210 (cattle blood typing only)

ImmGen, Inc., P.O. Box 10135, Rt. 4, Box 539C, College Station, TX (cattle blood typing only)

Serology Laboratory, School of Veterinary Medicine, University of California, Davis, CA 95616 (both cattle and horse blood typing)

Stormont Laboratories, Inc. 1237 E. Beamer St., Suite D, Woodland, CA 95695 (provides a variety of blood typing services involving several species: cattle, bison, cat, dog, horse)

In Canada

Bovine Blood Testing Laboratory, Saskatchewan Research Council, 15 Innovation Blvd., Saskatoon, Saskatchewan, S7N 2X8 (cattle blood typing only)

QUANTITATIVE TRAITS (MULTIPLE GENE INHERITANCE)

Relatively few characters of economic importance in farm animals are inherited in as simple a manner as the coat color or polled conditions described. Important characters — such as milk yield and composition, conformation, efficiency, and disease resistance — are due to many genes; thus, they are sometimes called multiple-factor characters or multiple-gene characters. Because such characters show all manner of gradation — from high to low performance, for example — they are known as quantitative traits.

Estimates of the number of pairs of genes affecting each economically important characteristic vary greatly, but the majority of geneticists agree that for most such characters 10 or more pairs of genes are involved. Growth rate in a dairy heifer replacement, therefore, is affected by the following: (1) the animal's appetite; (2) feed consumption; (3) feed utilization — that is, the proportion of the feed eaten that is absorbed into the bloodstream; (4) feed assimilation — the use to which the nutrients are put after absorption; and (5) feed conversion — whether used for muscle, fat, or bone formation. This should indicate clearly enough that such a characteristic as growth rate is controlled by many genes and that it is difficult to determine the mode of inheritance of such characters.

In addition to being influenced by many pairs of genes, quantitative traits differ from qualitative traits in that they are frequently influenced by the environment.

MULTIPLE BIRTHS

Multiple births among cattle have been observed since their domestication.

A review of the literature reveals that, on the average, such multiple births occur at the frequencies shown in Table 5–3.

TABLE 5–3
FREQUENCY OF TWINS IN CATTLE

Breed	Total Number of Births	Percentage of Twin Births
Brown Swiss	14,111	2.70
Holstein	18,736	3.08
Jersey	87,926	1.02

Selection for natural twinning does not appear to hold much promise because the heritability of twinning is low. It is noteworthy, however, that the repeatability of twinning is estimated to be three to four times higher than the average of the population, once a cow has given birth to the first set of twins.

Twins may be produced in any of the following five ways:

1. By two eggs being produced at the same heat period, with both fertilized and carried to term.

2. By two eggs being shed at the same heat period, but the cow being bred to two different bulls with a sperm from each of the bulls uniting with an egg.

3. By a cow coming in heat and being bred, then 3 weeks later coming in heat again and being rebred, with both matings resulting in viable offspring.

4. By a single fertilized ovum splitting during the early stage of development.

5. By the use of hormones to induce superovulation.

Twins may be either fraternal (dizygotic) or identical (monozygotic). Fraternal twins are produced from two separate ova that were fertilized by two different sperm. Identical twins result when a single fertilized egg divides very early in its embryology, into two separate individuals.

In humans, nearly half of the like-sexed twins are identical, whereas in cattle only 5 to 12% of such births are identical. Such twins are always of the same sex, a pair of males or a pair of females, and alike genetically—their chromosomes and genes are alike; they are 100% related. When identical twins are not entirely separate, they are known as Siamese twins.

Genetically, fraternal twins are no more alike than full brothers and sisters born at different times; they are only 50% related. They usually resemble each other more, however, because they were subjected to the same intrauterine environment before birth and generally they are reared under much the same environment. Also, fraternal twins may be of different sexes.

Distinguishing between identical and fraternal twin calves is not easy, but the following characteristics of identical twins will be helpful:

1. Identical twins are usually born in rapid succession, and frequently there is only one placenta.
2. The calves are necessarily of the same sex.
3. The coat colors are identical; i.e., if there is a broken color, there must be a strong degree of resemblance in this respect.
4. There is little variation in birth weights, general conformation, and, more particularly, the shape of the head, position of the horns, and occurrence of skin pigmentation, rudimentary teats, etc.
5. Muzzle prints show a degree of resemblance.
6. The shape, twisting, and position of the horns and the behavior of the twins can be observed at a later stage. Identical twins are inclined to keep together when grazing, walking, lying down, or ruminating.
7. Identical twins have the same blood group.

Most dairy producers prefer single births to twins, for the following reasons:

1. The high incidence of stillbirths in twins.

2. About 85% of all heifers born twin with bulls are apt to be freemartins (sterile heifers).
3. Twin calves average 20 to 30% lighter weights at birth than singles.
4. The tendency of cows that have produced twins to have a lowered conception rate following twinning.

HOW SEX IS DETERMINED

The possibility of sex determination and control has fascinated humans since time immemorial.

On the average, and when considering a large population, approximately equal numbers of males and females are born in all common species of animals. To be sure, many notable exceptions can be found in individual herds.

Sex is determined by the chromosomal makeup of the individual. One particular pair of the chromosomes is called the sex chromosomes. In cattle, the female has a pair of similar chromosomes (usually called X chromosomes), whereas the male has a pair of unlike sex chromosomes (usually called X and Y chromosomes).

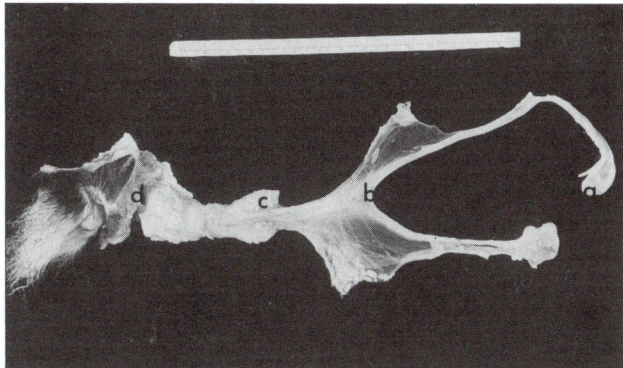

Fig. 5–10. Freemartin calf born co-twin to a male partner. Gonads (a), cordlike Mullerian ducts (b), seminal vesicles (c), and short vagina (d). (Courtesy, Professor H. W. Leipold, Kansas State University, Manhattan)

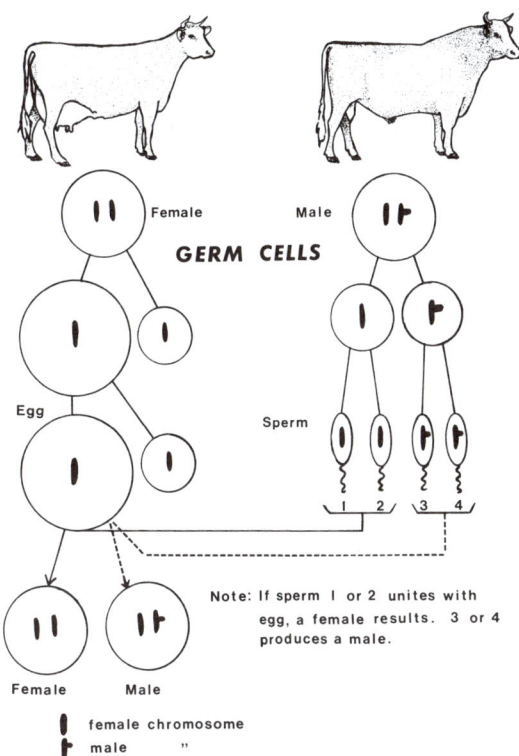

Fig. 5–11. Diagrammatic illustration of the mechanism of sex determination in cattle showing how sex is determined by the chromosomal makeup of the individual. The cow has a pair of like sex chromosomes, whereas the bull has a pair of unlike sex chromosomes. Thus, if an egg and sperm of like sex chromosomal makeup unite, the offspring will be female; whereas if an egg and sperm of unlike sex chromosomal makeup unite, the offspring will be a male.

The pairs of sex chromosomes separate out when the germ cells are formed. Thus, the ovum or egg produced by the cow contains the X chromosome; whereas the sperm of the bull are of two types, one-half containing the X chromosome and the other half the Y chromosome. Since, on the average, the egg and sperm unite at random, it can be understood that half of the progeny will contain the chromosomal makeup XX (females) with the other one-half XY (males).[3]

SEX CONTROL

Sex ratios of dairy calves at birth show that there is a slight deficiency in heifers; out of each 100 calves, on the average, 49 are heifer calves and 51 are bull calves.

Obviously, some method of controlling sex of offspring would have tremendous significance in the dairy field. For example, dairy producers wishing to build up a herd could then secure a high percentage of heifer calves.

It appears that nature is about to yield to sex control. Predetermination of the sex of 6- to 12-day-old embryos is a reality, and progress is being made in the separation of sperm cells containing X chromosomes from those containing Y chromosomes.

SELECTING BREEDING STOCK

For most rapid progress, the dairy producers should have an organized program of selecting and breeding. The first step in such a program consists of (1) establishing goals — goals in milk production, body conformation and wearability, longevity, freedom from hereditary defects, etc.; (2) analyzing where the herd is now — recording the pertinent information on each animal; and (3) determining how to get from "hither to yon" — from where the herd is now to the goals that have been set.

BASES OF SELECTION

The success of any breeding program depends primarily on the ability of the breeder to select properly the animals that are to be parents of the next generation, with primary emphasis on milk production per cow, since, on the average, dairy producers obtain about 90% of their income from the sale of milk.

Three methods of selecting animals are recommended:

1. **Individual merit.** This consists of selecting cows on the basis of their milk production and/or body type. It must be recognized that this basis of selection is materially affected by environment. For this reason, it is most effective if based on more than one lactation period, or even on the basis of a lifetime average, although, admittedly, the latter is too slow for most conditions. Judgment in the use of individual merit records as a basis of selection may be as essential as the records themselves.

2. **Pedigree.** The usefulness of a pedigree depends to a large extent on its completeness, and upon the understanding of the descriptive material available. There is no generally accepted method of reporting information on a pedigree. However, a trend is rapidly developing toward reporting information on twice-a-day milking for a 305-day lactation, and either listing the actual age at time of freshening of each lactation or figuring all records to a mature equivalent basis of six years of age. When the actual cow age for each record is listed, it permits an appraisal of the frequency of calving, calving interval, and to some extent, the breeding efficiency.

Of course, the ancestors close-up in the pedigree are the most important ones in attempting to evaluate the breeding worth of an animal from its pedigree.

It is generally agreed that pedigree selection should be used as an accessory to individual selection. It is particularly useful when selecting young animals for traits that are sex-limited or that are exhibited only after sexual maturity; for example, udder shape and attachment, milk production, etc. Also, for purebred breeders, bloodline information has monetary value if the production is high and the animals are of good conformation.

3. **Progeny testing.** This is the most desirable method of selection. It involves evaluation of the individual's offspring. Progeny testing is particularly valuable for selecting for quantitative traits such as milk production and body conformation (type). When properly used, progeny tests prevent the breeder from being deceived by the effects of environment. It is emphasized, however, that progeny testing should be used in combination with, rather than to replace, the other two bases of selection.

SYSTEMS OF SELECTION

In addition to arriving at a basis of selection (usually a combination of individuality, pedigree, and progeny test), and the traits for which selection is to be made, the dairy cattle breeder must determine what method of selection shall be used. The following three general methods are available:

1. **Cull simultaneously, but independently, for each character.** This means that culling levels are established for each trait, with emphasis on milk yield, below which all individuals are culled, no matter how good they may be in other respects.

2. **Tandem method of selection.** In this method, one characteristic is selected for at a time until it is improved, then selection is made for a second trait, and later a third, and so on. Tandem selection will result in improving one trait faster than can be achieved through any other method; but while that is being done, other traits may deteriorate.

3. **Establish a selection index.** This consists of totaling the animal's score for its merits in each characteristic, then retaining those with the highest total score. The selec-

---

[3]The scientists' symbols for the male and female, respectively, are ♂ (the sacred shield and spear of Mars, the Roman God of War), and ♀ (the looking glass of Venus, the Roman Goddess of Love and Beauty).

tion index is looked upon more favorably than the tandem method because it permits unusually high merit in one characteristic to make up for deficiencies in some other trait.

In practice, a combination of all three methods of selection is usually most desirable and effective.

SELECTING COWS

Fig. 5–12. Jersey cow, *OSB E Settler Shadow Maggie*, E 93%, 29,720 lb of milk, 4.3% fat, and breed leader for protein. In the decades ahead, increased selection pressure will be exerted for protein. (Courtesy, American Jersey Cattle Club, Reynoldsburg, OH)

When milk is the major source of income, selection is simplified. Cows in production are ranked from high to low on the basis of milk production, and the most profitable milk cows are retained and the least profitable ones sold.

The USDA annually compiles and publishes estimates of the genetic transmitting ability of cows identified (through the Dairy Herd Improvement Testing Program) as having demonstrated superior genetic merit for milk production. These cow index values are listed in the USDA-DHIA Cow Performance Index and are based on production records of the cow (modified contemporary deviation), her paternal half-siblings (sire's PD), and her dam's cow index.

A purebred breeder usually finds it desirable and profitable to select animals for type as well as production.

Where grade cows are involved, and replacement heifers are not being raised, cows can be selected or culled primarily on the basis of milk production. Where breeding animals are involved—that is, where replacement heifers or bulls are being selected for retention in the herd or for sale purposes—cows should be selected on the bases of their milk production, pedigree, and progeny, provided all three are available.

SELECTING REPLACEMENT HEIFERS

The number of heifer replacements needed each year to maintain herd size will depend upon the number of cows eliminated from the herd because of disease, injury, low production, or poor type. Normal turnover in DHIA herds is

Fig. 5–13. Replacement heifers. (Courtesy, Holstein-Friesian Assn. of America, Brattleboro, VT)

about 25% each year. To meet this, and to allow some opportunity for culling undesirable first-calf heifers, it is necessary to raise approximately one-third as many heifer calves each year as there are milking animals in the herd.

Producers who raise their replacements are in a better position to evaluate the animals genetically than operators who buy replacements, simply because their dams and close relatives are available in the same herd under similar feed and management conditions.

SELECTING PROVED SIRES

The selection of a sire is extremely important because he becomes the parent of many more offspring than an individual cow. A superior sire may be responsible for 80% or more of the genetic improvement in a herd.

Generally speaking, the producer has three sources of herd sires: (1) artificial insemination service, (2) purchase of a herd sire, or (3) raising a herd sire. The producer must also decide between a proven sire and a young sire. The challenge is to select the herd sire that will maintain a higher level of production than the present herd average. This is essential if there is to be improvement in the herd.

Sire evaluation has become a very sophisticated procedure as a result of experience, the development of larger and faster computers, improved statistical methods, and research.

The most reliable source of superior germ plasm for the breeding program of a herd is bulls that have been accurately evaluated for a large number of traits, including yield, conformation, and calving ease. Basically, this means bulls available through AI.

Twice annually, the USDA publishes a USDA-DHIA Sire Summary. These genetic evaluations are based on information on the bulls' daughters in herds participating in official production testing programs (DHIA and DHIR).

The USDA sire summaries are publicized and interpreted by the Cooperative Extension Service, state associations, dairy breed registries, AI organizations, dairy magazines, and other channels.

The *Hoard's Dairyman* Bull List of top active AI bulls, which was started in 1967, is excellent and widely used. Pertinent details relative to it follow.

Hoard's Dairyman uses three official sources: (1) USDA Sire Summaries (all production information and colored breed type information), (2) the Holstein Association (Holstein type information), and (3) the National Association of Animal Breeders (calving ease information). Bulls are ranked according to Predicted Transmitting Ability Milk-Fat Protein Dollars (PTA MFP$). To further emphasize the importance of the ranking, the bulls are grouped according to their percentile rankings. (Those in the 90th percentile have higher MFP$ than 90% of the other active AI bulls of the same breed.) The bulls are listed under the following headings:[4]

■ **PTA LB P, %P, LB M, LB F, and %F** — PTAs for pounds protein, percent protein, pounds milk, pounds fat, and percent fat indicate how much more (or less) performance to expect from an average daughter of a bull with a PTA of zero for the same trait. The new genetic base, called PTA90, was selected so half the cow population will have positive PTAs. It was established by setting to zero the weighted average PTAs of all cows born in 1985.

■ **PTA MF$** — Predicted Transmitting Ability Milk-Fat Dollars weights the PTA Milk and Fat to reflect the gross income per lactation future mature daughters of bulls will earn in excess of herdmates sired by bulls having a MF$ equal to zero. It is based on prices of $12.50 per hundred of 3.5% fat and 14.8 cents per point fat differential. This was the U.S. average milk price for 1989 minus the average hauling and assessments for promotion.

			Predicted Transmitting Abilities												
	Reg.	NAAB	Protein					Milk and Fat				Type Data		TPI	Calving Ease
Name of Bull	Number	Code	Rel.	MFP$	lb P	%P	CY$	Rel.	lb M	lb F	%F	Rel.	PTAT	PTI	Rel. %DBH

■ **Name of Bull** — Bull's registered name.

■ **Reg. Number** — Bull's registration number.

■ **NAAB Code** — A three-part code. The number before the letter indicates the stud from which the bull's semen can be purchased. The letter indicates the breed. The number following the letter is an individual bull identification number assigned by the bull stud.

■ **Rel.** — Reliability replaces Repeatability as a measure of the accuracy of the genetic evaluations. Four Reliabilities are listed . . . one for milk, fat, and proteins; one for milk and fat; one for type; and one for calving ease.

The closer Reliability is to 100, the more reliable the Predicted Transmitting Abilities (PTAs). Reliabilities on Active AI sires range from 45 to 99%. They do not measure the bull's conception rate.

■ **PTA MFP$** — Weights the Predicted Transmitting Abilities (PTAs) for Milk, Fat, and Protein to reflect the gross income per lactation future daughters of bulls will earn in excess of herdmates sired by bulls having a MFP$ equal to zero when milk receives a protein premium. It is based on prices of $12.50 per hundred for milk with 3.5% fat and 3.2% protein. Differentials are 14.8 cents per point of fat and 14.3 cents per point of protein.

■ **CY$** — Predicted Transmitting Ability Cheese Yield Dollars reflects the income per lactation future mature daughters of the bull will earn if their milk is priced according to its value in Cheddar cheese.

■ **PTAT** — Predicted Transmitting Ability-Type is the expected difference in final score between daughters of the bull and breed average.

■ **TPI (Holsteins)** — Type-Production Index is a value which is determined by placing an emphasis of 2 for PTA Protein, 2 for PTA Fat, 1 for PTA Type, and 1 for udder composite traits.

■ **PTI (all breeds except Holsteins)** — The Production Type Index is a value which is determined by placing an emphasis of 3 for PTA$, 3 for CY$, and 1 for Predicted Transmitting Ability-Type (PTAT).

■ **Calving Ease, %DBH** — This is the estimate of the Percentage of Difficult Births in Heifers when they calve the first time. Producers use this information when choosing bulls to breed heifers.

SELECTING YOUNG BULLS

Generally speaking, producers can avail themselves of the use of proven sires through artificial insemination. However, good proven sires are not always available at a price

---

[4]Source: *Hoard's Dairyman*, February 25, 1990, by W. D. Hoard and Sons Company, Fort Atkinson, WI.

that the individual breeder can afford to pay, particularly when it is desired to use them in natural service. Further, they are even expensive for artificial insemination associations to purchase. Additionally, it is recognized that proven sires are generally 7 to 8 years old, and that they have a remaining life expectancy of only 2 to 3 years. For these reasons, there is increasing interest in young sires.

Generally, young bulls are highly selected on pedigree and should, on the average, have high PDs when proven.

The young, sampled sires are priced lower and can be useful to the dairy farmer if used properly. It is recommended (1) that a few cows be bred to each of several young bulls, and (2) that not more than 25% of all matings be to young sires. One disadvantage with young sires is that their calving ease is an unknown factor. So, young sires should be bred to cows rather than to heifers.

SYSTEMS OF BREEDING

In dairy cattle breeding, one of the following systems of breeding is generally followed: (1) inbreeding, which embraces (a) close breeding, or (b) line breeding; (2) outcrossing; or (3) crossbreeding.

Close breeding (the mating of animals that are closely related) is usually limited to those dairy cattle breeders who are particularly good students of their animals, who recognize their weak and strong points, and who cull ruthlessly when the situation demands such action. When successful, it concentrates the most desirable traits and produces some outstanding individuals who transmit fairly uniformly. Close breeding is best left to the breeder who has complete records of production and type, and whose herd is at a high average level of production, and has been so for a number of years.

Line breeding is practiced much more extensively by cattle breeders than close breeding. It is a conservative type of breeding program which the vast majority of average and small dairy breeders can safely follow to their advantage.

Outcrossing (the mating of unrelated animals) is the most widely used system of breeding by the majority of dairy producers. It offers considerable opportunity to introduce new genes into the herd, simply because a wide choice of animals can be made. It often results in producing animals which are highly desirable within themselves, but they may not transmit uniformly. However, this system does not carry the dangers that often go with inbreeding, such as possible reduction in size and scale, lack of vigor, development of possible recessive factors which may become undesirable, and a greater concentration of any undesirable trait.

Crossbreeding cannot be practiced in registered herds because the offspring cannot be registered. For those milk producers who have no preference as to color and general appearance, crossbreeding may be followed with good results provided good proved sires are used. The aims of this system of breeding are to use the best sires available regardless of breed, and to gain hybrid vigor in the offspring (reduced calfhood mortality) and possibly more economical milk production.

DEVELOP AND FOLLOW A BREEDING PROGRAM

Where replacement animals are raised, in either a purebred or a grade herd, a breeding program must be developed and followed if herd progress and breed progress are to be made. The following steps are pertinent to such a program:

1. Choose a suitable breed of cattle; with consideration given to breed preference, the market for milk, and the sale of surplus animals.

2. Select or purchase the best cows available, based primarily upon their milk production records, but with due consideration given to type and pedigrees of the individuals.

3. Decide on the breeding system—inbreeding, vs outcrossing, vs crossbreeding—that shall be followed.

4. Evaluate the strong points and the weak points of the cows in the herd, known as herd analyses.

5. Obtain the services of the sire(s) which offer the greatest promise of improvement in production and in type, based on predicted difference and repeatability of the sire's proof, but with due consideration given to the price of the sire, along with the age and health of the individual if he is to be used in natural service.

6. Enroll in the particular testing program (DHIA or DHIR) which best meets the breeding and management programs that will be followed and which achieves your goals.

7. Follow that type evaluation which best meets the needs of the breeding and herd-building program you are attempting.

8. Arrive at the method(s) of selection that shall be followed; choosing between individual culling levels, the tandem method, or a selection index.

9. Establish and maintain reasonable standards for freedom from disease, temperament, fertility and sterility, ease and completeness of milking, and such other factors as are considered important.

10. Follow a feeding and management program which will permit the animals in the herd to express the maximum genetic potential which they possess.

PHYSIOLOGY OF REPRODUCTION

Dairy producers encounter many reproductive problems, a reduction of which calls for a full understanding of physiology and the application of scientific practices therein. In fact, it may be said that reproduction is the first and most important requisite of successful dairy cattle breeding, for if animals fail to reproduce, the breeder is soon out of business. Simply stated, milk production is a by-product of the reproductive process.

Table 5–4 (next page) gives the reasons cows leave herds. As shown, low productivity results in 32.5% of cows being culled. Reproductive problems are the second most important reason for eliminating cows from dairy herds, accounting for 26.6% of cows culled.

TABLE 5–4
REASONS FOR COWS LEAVING DAIRY HERDS[1]

Reasons for Culling	Percent
	(%)
Cow production	32.5
Reproductive problems	26.6
Mastitis	10.4
Disease or inabilities	7.7
Teat or udder injury	7.2
Udder conformation	5.0
Accident and injury	4.0
Type	3.1
Disposition and milking ease	2.7

[1]*Dairy Guide*, Cooperative Extension Service, The Ohio State University, Leaflet DG 300, by B. J. Conlin, Univesity of Minnesota.

REPRODUCTIVE ORGANS OF THE BULL

The reproductive organs of the bull are designed to produce semen and to convey it to the female at the time of mating. Semen consists of two parts: (1) the sperm (genetic portion) which are produced by the testes; and (2) the liquid (energy portion), or semen plasma, which is secreted from the seminiferous tubules, the epididymis, the vas deferens, the seminal vesicles, the prostate, and the Cowper's glands. Actually, the sperm make up only a small portion of the ejaculate. On the average, at the time of each service, a bull ejaculates 4 to 7 cubic centimeters of semen, containing about 6 to 10 billion sperm. The sperm concentration is about 1½ billion per cubic centimeter.

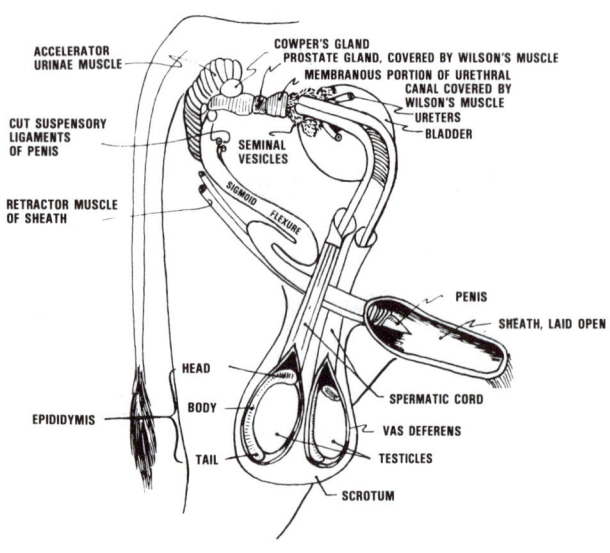

Fig. 5–14.   Diagram of the reproductive organs of the bull.

REPRODUCTIVE ORGANS OF THE COW

The cow's functions in reproduction are (1) to produce the female reproductive cells, the eggs or ova; (2) to develop the new individual, the embryo, in the uterus; (3) to expel the fully developed young at time of birth or parturition; and (4) to produce milk for the nourishment of the young. Actually, the part played by the cow in the generative process is much more complicated than that of the bull. It is imperative, therefore, that the modern dairy producer have a full understanding of the anatomy of the reproductive organs of the cow and the functions of each part.

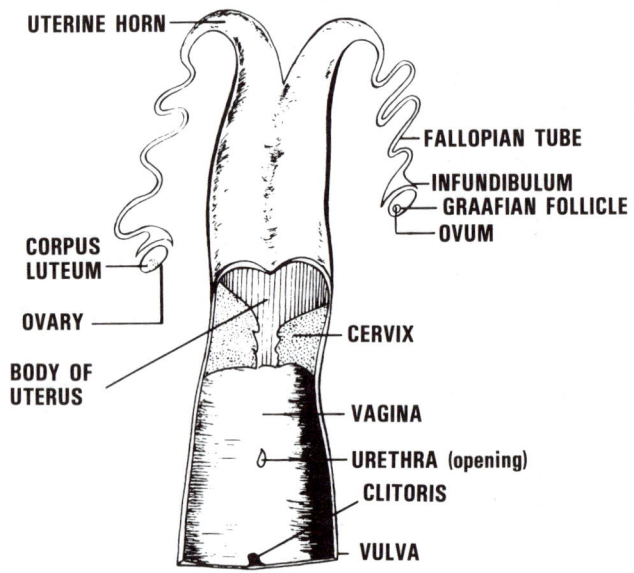

Fig. 5–15.   The reproductive organs of the cow.

NORMAL HEAT AND GESTATION PERIODS

After heifers reach puberty, the normal heat period recurs at approximately 21-day intervals, but it may vary from 18 to 23 days. Best conception is obtained when cows are bred near the end of standing heat.

The normal gestation period of the dairy cow is 283 days. Based on studies made by the Ohio Experiment Station, however, it is evident that there is variation among breeds. The following figures show the gestation period by breeds, as found in this experiment: Ayrshire, 278 days; Brown Swiss, 288 days; Guernsey, 283 days; Holstein, 279 days; and Jersey, 278 days. The Ohio Station also reported (1) that the gestation period of first-calf heifers is about 2 days less than that of older cows of the respective breeds, and (2) that the gestation period where bull calves are born is about 1 day longer than where females are born.

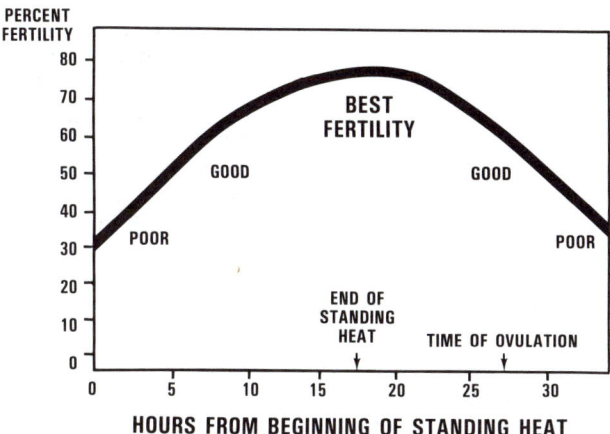

Fig. 5–16. For best conception, breed near the end of standing heat.

COW HEAT DETECTION METHODS AND DEVICES

The problem of heat detection becomes more important as (1) herds get larger, (2) good hired help is more difficult to come by, (3) cows produce more milk, and (4) animal value increases.

Under ordinary farm conditions, caretakers miss an estimated 25 to 50% of the heat periods. On the average, a missed heat period prolongs the calving interval by 30 to 40 days and means a loss of more than $20 in a dairy herd. Some owners pay their employees a bonus for catching a cow in heat. For these reasons, producers are interested in heat detection methods. Among them are the following:

1. **Chin-Ball Marker.** This device was developed in New Zealand. It is similar to a ball-point pen attached to a halter under the chin of a surgically modified teaser bull, often called a "Gomer." (One of the first ranches in North America to use the Chin-Ball Marker gave this name to the bull on which it was used.) During preservice sex play, it is usual for a bull to place his head over the shoulders, back, and rump of the cow. This causes a smearing of the colored ink from the ball-point onto the cow.

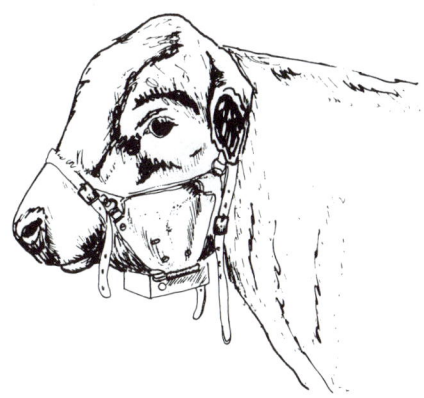

Fig. 5–17. Chin-Ball marking device. (Courtesy, American Breeders Service, De Forest, WI)

One filling of the stainless steel container is sufficient to mark 15 to 25 cows. Experience indicates that one Gomer bull can work approximately 80 cows. In large pastures and in larger-sized herds, it is best to have two bulls.

This method of heat detection is a most dependable management tool.

2. **The KaMaR Heat-Mount Detector.** The heat-mount detector is a 2- by 4½-in. fabric base to which is attached a white plastic capsule. Inside the capsule is a small plastic tube containing red dye. The tube is constructed so the dye is released slowly by moderate pressure. When enough dye is released from the tube (after about 4 to 5 seconds of pressure), it spreads over the inner lining of the capsule, causing it to turn red.

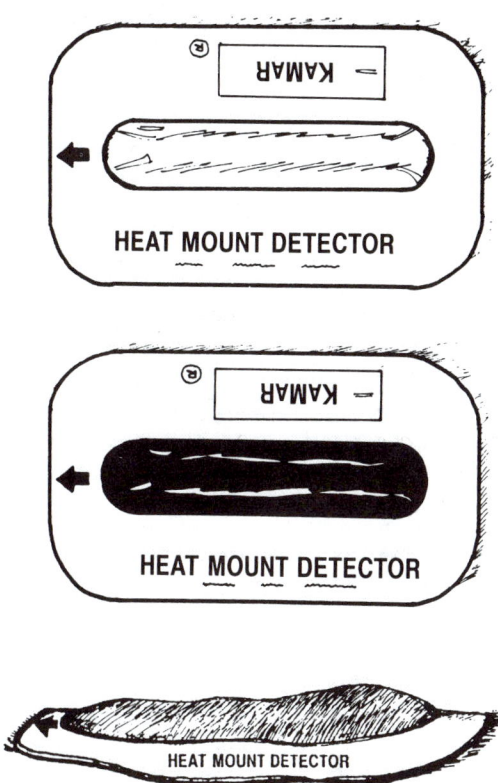

Fig. 5–18. Device for heat detection, as an aid in the artificial insemination of cows. At the top, the KaMaR Heat-Mount Detector is shown before activation. Center shows detector bright red after activation, indicating that the cow is in heat. Lower view shows side or profile view of the device, which is applied to the cow by an adhesive. (Courtesy, KaMaR, Inc., Steamboat Springs, CO)

The detector relies on the natural bovine instinct of bulling or mounting during estrus. The pressure from the brisket of a mounting animal causes the dye to be released and the detector to turn red. If the cow does not stand for the mounting animal, there will not be enough pressure to release the dye and turn the detector red. This device has resulted in catching 95% of the heat periods.

3. **Pen-O-Block.** The Pen-O-Block is a plastic tube placed within the bull's sheath and held in place with a stainless steel pin. The bull can detect cows in heat and

mount them in a normal way, but the device mechanically prevents him from making contact with the cow.

The Pen-O-Block consists of a white plastic tube, the pen or cannula, two washers, and a cotter pin. The device is inserted within the bull's sheath and held in place by the cannula. The procedure is best carried out by a veterinarian, as it requires skill.

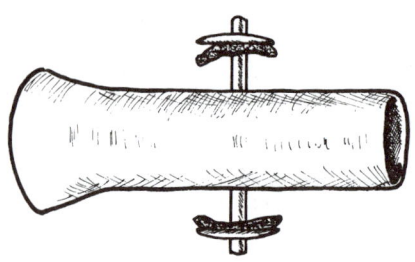

Fig. 5–19. Pen-O-Block marking device. (Courtesy, American Breeding Service, De Forest, WI)

Properly used, these three aids will improve heat detection. They are by no means replacements for visual heat detection; nor will they solve all the problems in breeding a dairy herd artificially. Other factors that need attention are:

1. **Nutrition.** Cows must have adequate nutrition to cycle at a satisfactory rate for successful breeding.

2. **Rest interval.** This is very important, as cows must have calved at least 50 to 60 days prior to breeding for satisfactory performance.

3. **AI facilities.** Facilities should be adequate for handling and breeding the cow herd.

4. **Personnel.** Trained personnel are needed to do heat detection, gather the in-heat cows, and inseminate the herd.

HORMONAL CONTROL OF HEAT

Today, hormonal control of heat and induced parturition are much sought as management tools. Most phases of dairy management have shifted from individual care to herd or flock care, primarily to save labor. Thus, fewer and fewer lactating cows are individually fed. Instead, they are group fed. But no such progress has been made in breeding and parturition. Animals still receive individual attention at these times, despite the fact that labor costs no longer permit such luxury. So, producers need to move toward mass handling of animals at breeding and parturition, primarily through estrous synchronization and induced parturition. Controlled estrus greatly facilitates both artificial insemination and embryo transfer, for which purposes it is now in limited use — albeit unperfected. Controlled parturition would make for the several advantages enumerated under "Induced Calving," which is presented later in this chapter.

Planned parenthood is not new. It has long been practiced among females of all species, women included. Livestock producers have long "tampered with" the breeding and parturition season that was common in the wild state. Prior to domestication, animals brought forth their young in the fields and glens, inhibited only by age and feed, and influenced somewhat by seasons. But people changed all this — even without the use of hormones. Sly controls have been exercised over breeding for a very long time. For example, farm flock owners controlled reproduction in chickens by the simple act of putting eggs under an old setting hen — unless she hid out. Modern poultry producers regulate chick hatchings by controlling when, and how many, eggs go into the incubator. It is more difficult to accomplish the same thing in four-footed animals.

Purebred dairy cattle breeders who show, plan their breeding programs to take maximum advantage of show classifications; commercial milk producers are concerned with weather and feed supply; and all producers want the largest flow of milk at a time when the product is likely to bring the highest price.

Livestock producers have altered nature's way in farm animals (1) by confining the male at certain times, or hand mating; (2) by emulating spring conditions — through providing better feed and shelter (and/or blankets) when breeding at other times of the year; (3) by flushing — through feeding females more liberally 2 to 3 weeks ahead of the breeding season; and (4) by artificially controlling the hours of light per day — through use of ordinary electric lights, which activate hormone production. Each of these methods has been used with varying degrees of success. All have fallen short of achieving the hoped-for goal under most commercial conditions — that of bringing females in heat at will, followed by a high conception rate. Hormonal control may be the answer.

Many different drugs have been administered (either orally, by injection, or by implantation) in attempts to control the estrous cycle of females, with progestagens and prostaglandins, heading the list. These products are availabale under different trade names, such as Syncro-Mate B (SMB).

■ **Progestagens** — These are compounds that mimic the hormone progesterone, which is produced naturally by the corpus luteum on the female's ovary.

Feeding, injecting, or implanting nonpregnant cows with progestagens for a sustained period of 14 to 20 days, followed by withdrawal of the dose, will result in animals exhibiting heat 2 to 8 days after withdrawal. Essentially this method of control places animals in a type of false pregnancy; they are fooled by the high levels of the progesterone-mimicking progestagen. Upon withdrawal of the compound, progestagen levels in the blood drop and new follicles mature, bringing the animals similarly treated into heat synchronistically. Three problems have been encountered with progestagens: (1) Heats and ovulations in animals so treated are not sufficiently synchronized to breed them successfully at a prescribed time; (2) conception at the first synchronized heat is subnormal; and (3) the progestagens must be administered over 14 to 20 days, which involves much labor.

■ **Prostaglandins** — These are hormonelike substances, found in almost every cell and tissue, that are believed to play a key role in regulating cellular metabolism. The name *prostaglandin,* which is a misnomer, was given to these substances because, initially, they were believed to have originated in the male's prostate gland. Later research revealed that they actually come from the seminal vesicles (another accessory sex gland of the male).

Several natural compounds have been identified as prostaglandins. For example, the marine animal—the sea whip— which is found off the Florida coast, is a rich source. Also, all prostaglandins are now being synthesized from commercially available materials.

These highly potent substances have been called local hormones or tissue hormones because they do their work in the immediate area in which they are produced, as distinguished from circulating hormones which aim at distant targets. Contrary to some opinions, prostaglandin is not a heat-producing drug. Rather, it is used in females to regress the *corpus luteum* (the growth on the ovary that prevents ovulation). This allows the natural estrous cycle to begin again. The time interval between injection with a single prostaglandin dosage and the onset of estrus and ovulation (1) is very short (in cows, about 90 hours after injection), and (2) is predictable.

In 1979, the Upjohn Company, Kalamazoo, Michigan, received FDA approval on the use of *LUTALYSE*, a prostaglandin product, for synchronization of estrus in nonlactating dairy heifers and beef cows and heifers.

■ **Syncro-Mate-B (SMB)**— This is a trade name for an estrous synchronization product which was approved by FDA in 1982. It contains *NORGESTOMET*, a patented, potent synthetic progestin, and estradiol valerate, a synthetic estrogen. SMB, which is used as an ear implant, is designed to cause cows or heifers to ovulate in a predictable period of time.

The major criteria for measuring the success of hormone induced estrus are (1) the percentage of females that come in heat, and (2) the percentage of conception.

Researchers in both colleges and industries are in general agreement that hormone controlled estrous synchronization will work, and that it offers promise of good returns when properly used in a well-managed dairy herd. Scientists also realize that we do not know all the answers—that further research work is necessary. Nevertheless, it appears that planned parenthood is here to stay—that its wide use only awaits getting the technique perfected and lowering costs, both of which will come. In the meantime, dairy cattle breeders are admonished to keep abreast of developments and to rely on well-informed advisers.

FERTILIZATION

Fertilization is the union of the male and female germ cells, sperm and ovum. The sperm are deposited in the vagina at the time of service and from there ascend the female reproductive tract. Under favorable conditions, they meet the egg and one of them fertilizes it in the upper part of the oviduct near the ovary.

In cows, fertilization is an all or none phenomenon, since only one ovum is ordinarily involved. Thus, the breeder's problem is to synchronize ovulation and insemination; to ensure that large numbers of vigorous, fresh sperm will be present in the fallopian tubes at the time of ovulation.

A female is fertile only when an egg is present which can be fertilized. Moreover, an egg can live for only a short time after being shed from an ovary unless it is fertilized. The optimal time for insemination is in advance of the time of ovulation, which varies according to species.

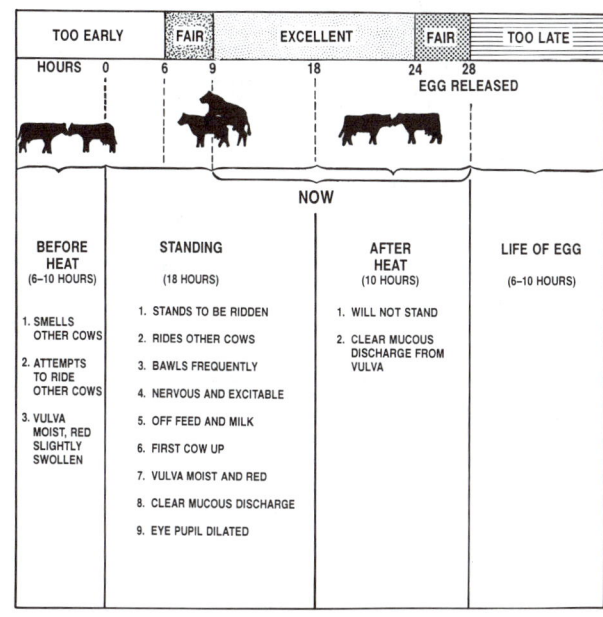

Fig. 5–20. Breeding time guide for the average cow.

A cow does not shed her egg from the ovary until about 10 hours after the close of standing heat, and the egg lives only 6 to 10 hours. Thus, for optimal results from insemination, cows should be bred between the middle of standing heat and 6 hours after the end of standing heat.

From a practical standpoint, few dairy producers are able to check for heat every 2 hours. Normally, cows are in standing heat 18 hours, but it may vary from 4 to 40. They may show other heat signs, such as mucus, swollen vulva and/or nervousness, from 6 to 36 hours before standing heat to 36 hours after standing heat. It varies with individual cows.

Considering all factors, the best guide is (1) to breed cows that are in standing heat in the morning that same afternoon, and (2) to breed cows that are in standing heat in the afternoon the next morning.

SUPEROVULATION

Superovulation consists of injecting the female with drugs which cause the larger follicles (each of which contain one oocyte, or egg) to mature, rupture, and release the egg. Commercial employment of superovulation is limited primarily to cattle at the present time.

The bull is capable of producing from several thousand to millions of sperm daily whereas the cow normally produces one ovum (occasionally two ova) every 17 to 21 days. Now it is possible, through the administration of hormones, to obtain several ova (5 to 50) from a cow at one estrous cycle. It is also feasible to obtain a large number of eggs from very young heifer calves, by injection of hormones.

Eggs which are shed from the ovaries are stored in large follicles. The basic principle of superovulation is to stimulate extensive follicular development through the use of a hormone preparation, given intramuscularly or subcutaneously, with follicle-stimulating hormone (FSH) activity. The most common sources of such a hormone are pregnant mares' serum (PMSG) and FSH extracts from pituitaries of slaughtered animals. Many animals so treated will come into estrus about five days after initiation of treatment and ovulate, through release of their own luteinizing hormone (LH). However, to help assure that multiple ovulations occur, the ovulating LH from pituitaries or from human chorionic gonadotropin (HCG) is injected. The multiple ovulations occur at about the same time the cow would have normally ovulated one egg (21 days after the previous ovulation).

Studies have shown that FSH should be administered twice daily over a period of 4 to 5 days. PMSG has a longer biological life than FSH and a single subcutaneous injection is normally used. Then, 5 to 6 days after the original FSH or PMSG shot, LH or HCG is given intravenously.

The heifers should ovulate by the seventh day after starting hormone treatment.

Since ovulation occurs over a period of time, not all the eggs are fertilized unless the donor is inseminated repeatedly. A yield of 5 to 12 good fertilized eggs per donor cow may be expected.

Of course, the real economic value of superovulation lies in the successful transfer of excess eggs from more valuable donor cows to less valuable recipient cows.

EMBRYO TRANSFER

Fig. 5–21. Brown Swiss cow with her five embryo transfer calves. (Courtesy, Brown Swiss Cattle Breeders' Assn. of the U.S.A., Beloit, WI)

Embryo transfer is the placing of an embryo into the lumen of the oviduct or uterus.

Artificial insemination has given a means for the widespread distribution of desirable genes via the sperm. Similar genetic selection through high-quality females has, however, been limited since, normally, one cow will produce one calf per year and the average number of offspring per female will seldom exceed five in a lifetime. Out of the latter arose the idea that a marked increase in the production of offspring from desirable cows might be effected by superovulation, followed by transfer of the fertilized ova to less desirable cows, with the latter serving as host-mothers or foster-mothers to the developing embryo. Embryo transfer is a seven-step process as follows: (1) synchronize heat cycles of donor and recipient cows; (2) obtain a large number of ova from the donor cow by giving her a drug so that she superovulates; (3) breed donor cow (AI or natural); (4) collect ova from donor cow, 5 days after breeding; (5) examine eggs, make sure that they are normal and fertilized; (6) prepare foster mothers, by synchronizing (usually by hormone control) their ovulation with the donor; and (7) transfer eggs to recipient (see Fig. 5–22).

Steps in Conventional Embryo Transfer

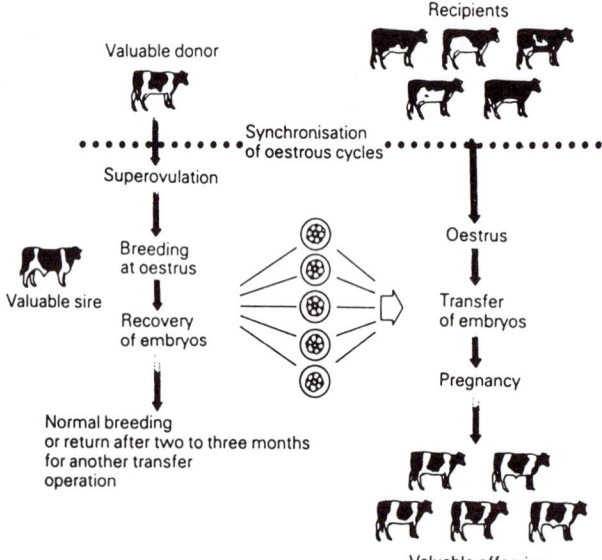

**but each is different -
18, or 20, or 23, or 25, or 30,000 # milk**

Fig. 5–22. Steps in conventional embryo transfer. (Prepared by Dr. Robert Walton, American Breeders Service, a division of W. R. Grace & Co., De Forest, WI)

Pregnancy in the recipients can be diagnosed in about 30 days. Full-term pregnancies result in full sibs (brothers and sisters) with the genetic traits of the donor cow and the bull to which she was bred. Recipients have no genetic influence on the calves they carry—they merely serve as incubators.

The following **advantages** accrue from extensive use of embryo transfer:

1. A dozen calves may be obtained from a valuable cow during a year's time.

2. The rate of progress in genetic improvement is

speeded up, because of the increased number of progeny from valuable cows.

3. Valuable cows that produce normal ova but fail to conceive due to some hormonal or anatomical defects need not be culled because of sterility; such animals may be used as donors for supplying ova for transplantation.

4. Heifers may be effectively progeny tested at an early age if large numbers of fertilized eggs are procured from heifers and transplanted to sexually mature recipients. The generation time of cattle may be reduced by 1 year or more.

PREGNANCY TESTS

Pregnancy tests can make for increased profits. It is expensive business to keep a cow that will not produce a calf, necessitating feed, interest, labor, and other costs.

Barren cows can either be treated or marketed following testing. Most packers will actually pay a premium for cows known to be open. Where valuable purebred animals are involved, bred cows can be sold with a more certain guarantee of being safely in calf.

Absence of heat is not always a sign of pregnancy, but a positive diagnosis can be made by palpating the uterine horn on the side of the corpus luteum. If the cow is pregnant, the blastodermic vesicle may be felt sliding past the fingers. In the hands of an experienced technician, a positive diagnosis can be made as early as 35 days following breeding. This is the most common test of pregnancy. It is popular because it affords early diagnosis, and there is little hazard when performed by experienced operators (see Fig. 5–23).

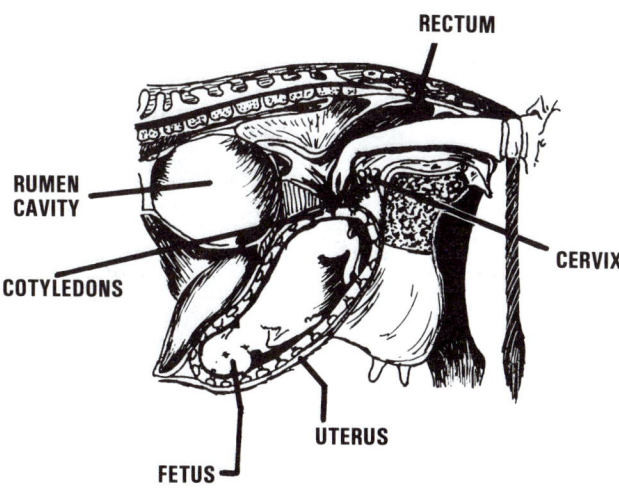

Fig. 5–23. Rectal method for determining pregnancy in the cow.

Other less-used tests for pregnancy in cows are:

1. **Abdominal ballottement** may be used from the fifth to the seventh month of pregnancy. This consists in feeling the fetus by the following technique: (a) place the hand or fist against the abdomen in the lower right flank region; and (b) execute a short, vigorous, inward-upward thrust in this region and retain the hand in place. The hard fetus may be felt. Because of the amniotic fluid, the technique described above will make the fetus recede, but it will fall back in place almost immediately.

2. **The fetal heartbeat** can sometimes be detected after the sixth month of pregnancy, though this method is not as certain in the cow as in other classes of farm animals. Use of a stethoscope is preferred, though good results are sometimes secured by merely placing the ear against the right lower abdominal region and listening. The fetal heartbeat can be distinguished from that of the mother because of its greater frequency and lighter and higher pitch.

3. **Fetal movements** can sometimes be observed through the abdominal wall during the latter half of pregnancy. This method of detecting pregnancy requires much patience. The observer simply must wait until voluntary movement of the fetus on the right side of the cow is observed. The practice of trying to induce movement of the fetus by allowing a very thirsty cow to take on a fill of cold water is cruel and is to be condemned.

INDUCED CALVING

Instead of letting nature take its course, scientists are now synchronizing parturition. Currently, the technique is being used to a limited extent in cattle. The objectives of induced early calving are (1) lowering birth weight of calves, thereby lessening parturition difficulty; (2) predicting calving dates in order to pool labor and concentrate watching; (3) gaining a longer period from calving until rebreeding; and (4) shortening the calving interval and obtaining more offspring in the lifetime of the animal.

Females that have passed the 269th day of pregnancy will calve within 24 to 72 hours following intramuscular injection with an adrenal steroid. Experimental work indicates that such induced calving will result in (1) 5 to 8 days earlier than normal calving, and (2) calves 6 to 8 lb lighter in birth weight than those carried to term. However, a higher incidence of retained placentas and temporarily lowered milk production accompany early calving. Antibiotics are indicated where the fetal membrane remains attached to the uterus longer than normal. Failure to expel the membranes after induced calving appears to have no effect on fertility as cows suffering this problem usually have no trouble breeding back.

Induced calving should be limited to healthy cows, free from disease, because the steroid "knocks out" the animal's immune body system for several days. Also, until the problem of retained placentas is solved, it should not be used on dairy cows in commercial milk production, because of (1) restrictions in the use of antibiotics in dairy cows in lactation; (2) aesthetic reasons, where cows are housed and milked in the usual manner; and (3) the high incidence of metritis following retained placentas, which will likely infect herdmates and make for difficult breeding.

CARE OF THE COW AT CALVING TIME

The careful and observant caretaker will be ever alert and make definite preparations for calving in ample time. It is especially important that first-calf heifers be watched at calving time, for frequently they will need some assistance. Older cows that habitually have trouble in parturition may well be culled from the herd.

SIGNS OF APPROACHING PARTURITION

Perhaps the first sign of approaching parturition is a distended udder, which may be observed some weeks before calving time. Near the end of the gestation period, the content of the udder changes from a watery secretion to a thick, milky colostrum. As parturition approaches, there generally will be a marked shrinkage or falling away of the muscular parts in the region of the tailhead and pinbones, together with a noticeable enlargement and swelling of the vulva.

The immediate indications that the cow is about to calve are extreme nervousness and uneasiness, separation from the rest of the herd, and muscular exertion and distress.

PREPARATION FOR CALVING

At the time the signs of approaching parturition seem to indicate that the calf may be expected within a short time, arrangements for the place of calving should be completed.

During the seasons of the year when the weather is warm, the most natural and ideal place for calving is a clean, open pasture away from other livestock.

Under pasture conditions, there is decidedly less danger of either infection or mechanical injury to the cow and calf. A good procedure consists in having a small pasture adjoining headquarters into which heavy springing cows are placed a few days before calving.

During inclement weather or when no pasture is available, the cow should be placed in a roomy, well-lighted, well-ventilated, comfortable maternity stall, which should first be carefully cleaned, disinfected, and bedded for the occasion.

NORMAL PRESENTATION

Labor pains in a mild form usually start some hours before actual parturition. After a time, the water bag appears on the outside, usually increasing in size until it ruptures from the weight of its own contents. This is closely followed by the appearance of the amniotic bladder (the second water bag), with the fetus. With the rupture of the second water bag, the straining becomes more violent, and presentation soon follows. Most commonly in presentation, the front feet come first followed by the nose which is resting on them, then the shoulders, the middle, the hips, and then the hind legs and feet. Fig. 5–24 shows normal presentations.

With posterior presentation (hind feet first), there is likely to be difficulty in calving. Moreover, there is considerably more danger of having the calf suffocate through rupture of the umbilical cord and strangulation.

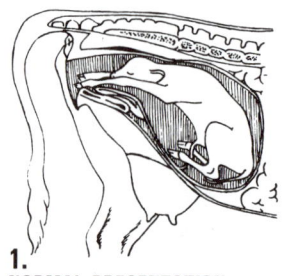
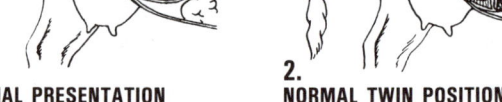
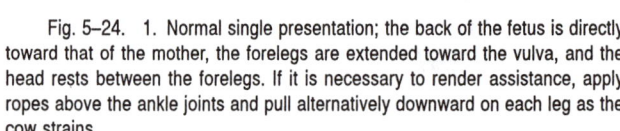

1. NORMAL PRESENTATION   **2. NORMAL TWIN POSITIONS**

Fig. 5–24. 1. Normal single presentation; the back of the fetus is directly toward that of the mother, the forelegs are extended toward the vulva, and the head rests between the forelegs. If it is necessary to render assistance, apply ropes above the ankle joints and pull alternatively downward on each leg as the cow strains.

2. Normal twin positions. If delivery does not proceed normally, this is a case for a veterinarian.

RENDERING ASSISTANCE

Figs. 5–25 and 5–26 show some abnormal presentations.

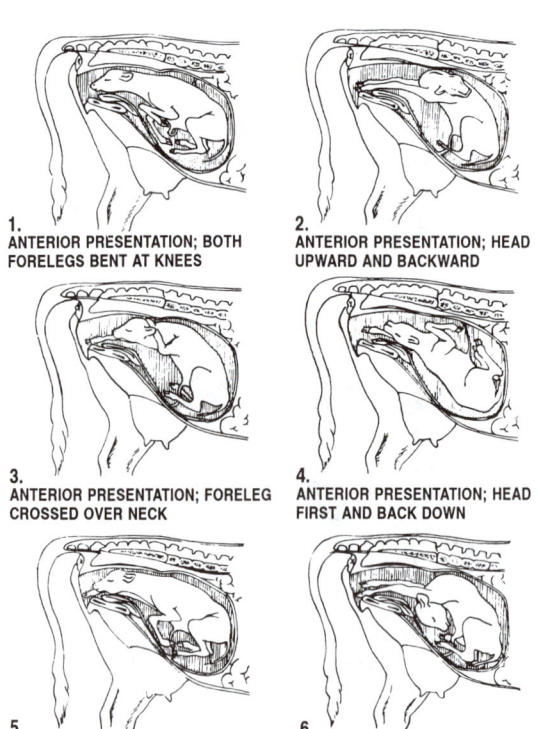

1. ANTERIOR PRESENTATION; BOTH FORELEGS BENT AT KNEES   **2. ANTERIOR PRESENTATION; HEAD UPWARD AND BACKWARD**

3. ANTERIOR PRESENTATION; FORELEG CROSSED OVER NECK   **4. ANTERIOR PRESENTATION; HEAD FIRST AND BACK DOWN**

5. ANTERIOR PRESENTATION; ONE FORELEG BENT AT KNEE   **6. ANTERIOR PRESENTATION; HEAD TURNED BACK**

Fig. 5–25. Some abnormal presentations with suggestions for correction:

1. Extend the legs so that delivery can be accomplished.

2. Push back the fetus, which will often bring the head into its normal position.

3. Grasp the crossed leg a little above the ankle, raise it, draw it to the proper side, and extend it in the genital canal.

4. Rotate the fetus, extend the forelegs, and deliver by traction.

5. Lift the head and draw up and rope the leg so that it does not slip back again.

6. Rope the forelegs and then push them forward; place the head in normal position.

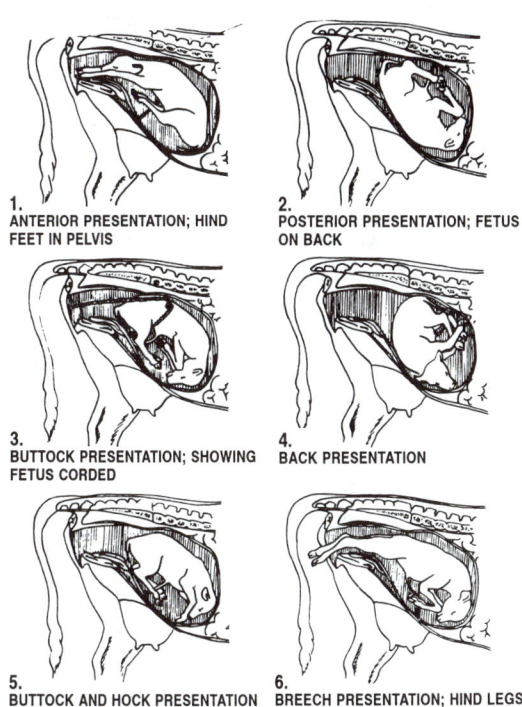

1.
ANTERIOR PRESENTATION; HIND FEET IN PELVIS

2.
POSTERIOR PRESENTATION; FETUS ON BACK

3.
BUTTOCK PRESENTATION; SHOWING FETUS CORDED

4.
BACK PRESENTATION

5.
BUTTOCK AND HOCK PRESENTATION

6.
BREECH PRESENTATION; HIND LEGS FIRST

Fig. 5–26. Some abnormal presentations with suggestions for correction:

1. Force back the hind feet. This is a very serious malpresentation, in which it is generally impossible to save the fetus if delivery is far advanced.

2. Rotate the fetus, extend the rear legs, and deliver by traction.

3. Push the fetus forward, and bring the legs properly into the genital passage.

4. Turn the fetus so that either the head and forelegs or the rear legs can be started through the pelvis.

5. Push the fetus forward and bring the legs properly into the genital passage.

6. Usually delivery is normal but traction may facilitate; beware of prolonged labor because calf may suffocate due to rupture of the naval cord.

If presentation is normal and within an hour or two after the onset of signs of calving, no assistance will be necessary. However, if the cow has labored for some time with little progress or is laboring rather infrequently, it is usually time to give assistance. Such aid will usually consist of fastening small ropes around the pasterns and pulling the young outward and downward as the cow strains. This should be done by an experienced caretaker or a competent veterinarian. It is always well to be reminded that rough, careless, or unsanitary methods at such a time may do more harm than good.

THE NEWBORN CALF

If parturition has been normal, the cow can usually take care of the newborn calf, and it is best not to interfere. However, in unusual cases, it may be necessary to wipe the mucus from the nostrils to permit breathing; or, more rarely yet, artificial respiration methods must be applied to some calves. This may be done by blowing into the mouth, working the ribs, rubbing the body rather vigorously, and permitting the calf to fall gently. The cow should be permitted to lick the calf dry.

With calves born in sanitary quarters or out on clean pastures, there is little likelihood of navel infection. To lessen the danger of such infection, the navel cord of the newborn calf should be treated at once with a 2% solution of tincture of iodine.

A vigorous calf will attempt to rise in about 15 minutes and usually will be nursing in half an hour to an hour. The weaker the calf, the longer the time before it will be able to be up and nursing. Sometimes it may even become necessary to assist the calf by holding it up to the cow's udder.

The colostrum (the milk yielded by the mother for a short period following the birth of the young) is most important for the well-being of the newborn calf. Experiments have shown that it is very difficult to raise a calf that has not received any colostrum. Aside from the difference in chemical composition, compared with later milk, the colostrum seems to have the following functions:

1. It contains antibodies which temporarily protect the calf against certain infections, especially those of the digestive tract.

2. It serves as a natural purgative, removing fecal matter which has accumulated in the digestive tract.

3. It contains a very high content of vitamin A, from 10 to 100 times that of ordinary milk. This provides the young

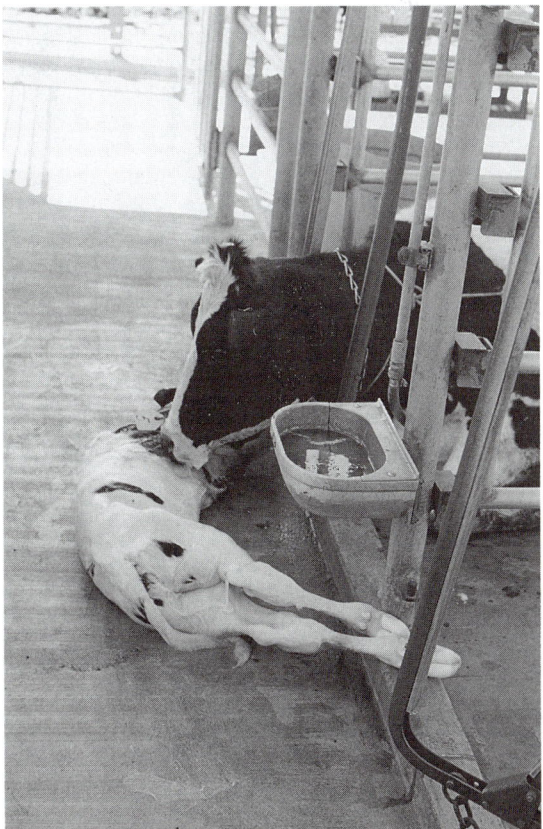

Fig. 5–27. Newborn calf. (Courtesy, James Tappan, Higley, AZ)

calf, which is born with little body storage of this vitamin, with as much vitamin A on the first day as it would secure in some weeks from normal milk.

THE AFTERBIRTH

Under normal conditions, the fetal membranes (placenta or afterbirth) are expelled from 3 to 6 hours after parturition. Should they remain as long as 24 hours after calving, competent assistance should be given by an experienced caretaker or a licensed veterinarian. The operation of removing a retained afterbirth requires skill and experience; and, if improperly done, the cow may be made a nonbreeder. Furthermore, before doing this, the fingernails should be trimmed closely; the hands and arms should be thoroughly washed with soap and warm water, disinfected, and then lubricated with Vaseline or linseed oil. In no case should a weight be tied to the placenta in an attempt to force removal.

As soon as the afterbirth is ejected, it should be removed and burned or buried in lime, thus preventing the development of bacteria and foul odors.

ARTIFICIAL INSEMINATION

Artificial insemination is, by definition, the deposition of spermatozoa in the female genitalia by artificial rather than by natural means. Legend has it that this method had its origin in 1322, at which time an Arab chieftain used artificial methods to impregnate a prized mare with semen stealthily collected by night from the sheath of a beautiful stallion belonging to an enemy tribe. However, the first scientific research relative to the artificial insemination of domestic animals was conducted with dogs by the Italian physiologist, Lazarro Spallanzani, in 1780.

In the United States today, artificial insemination is more extensively practiced with dairy cattle than with any other class of farm animals; more than 65% of the nation's dairy cattle are artificially inseminated.

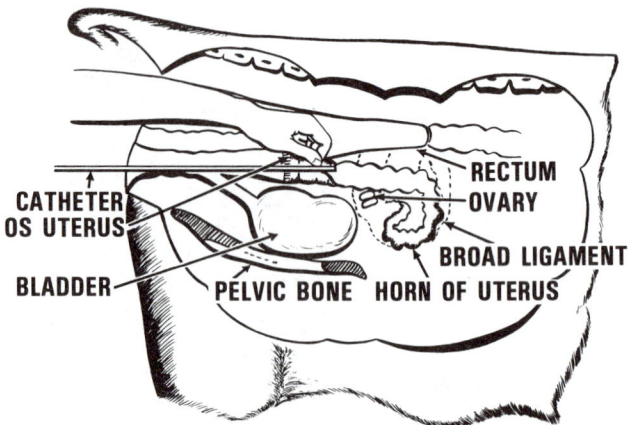

Fig. 5–28. Deep uterine insemination of the cow. The cervix is grasped per rectum and the inseminating tube is carefully worked into and through the cervical canal. (Courtesy, Dr. H. A. Herman, Columbia, MO 65211)

ADVANTAGES OF ARTIFICIAL INSEMINATION

Some of the advantages of artificial insemination are:

1. **It increases the use of outstanding sires.** Through artificial insemination, many breeders can avail themselves of the use of outstanding sires, whereas the services of such males were formerly limited to a relatively few females of one owner, or, at the most, a small group of owners.

2. **It alleviates the danger and bother of keeping a bull.** Some hazard and bother are usually involved in keeping a bull. Usually, the dairy producer may choose from the breeding programs of one or more established artificial insemination organizations and eliminate the necessity of maintaining a sire.

3. **It makes it possible to overcome certain physical handicaps to mating.** Artificial insemination is of value in (a) mating animals of greatly different sizes—for example, in using heavy, mature bulls on young heifers; and (b) using stifled or otherwise crippled bulls that are unable to perform natural service.

4. **It makes it possible to use a sire that is not alive at the time.** Since frozen semen can be stored for many years, it is possible to use a sire far beyond his lifetime.

5. **It lessens sire costs.** In most herds, artificial insemination is usually less expensive than the ownership of a worthwhile bull together with the accompanying housing, feed, and labor costs.

6. **It reduces the likelihood of costly delays through using infertile bulls.** Because the breeding efficiency of sires used artificially is constantly checked, it reduces the likelihood of breeding females to a bull that is of low fertility, or even sterile, for an extended period of time.

7. **It helps to control diseases.** Artificial insemination is a valuable tool in preventing and controlling the spread of certain types of diseases, especially those associated with the organs of reproduction, such as vibriosis and trichomoniasis, *provided it is properly done.* For disease control, it is essential (a) that all males be carefully examined for symptoms of transmissible disease, (b) that bacterial contamination be avoided during the collection and storage of semen, and (c) that clean, sterile equipment be used in insemination. Those artificial insemination organizations that are members of The National Association of Animal Breeders, Inc., are expected to follow the rigid Sire Health Code approved by The American Veterinary Medical Association and adopted by The National Association of Animal Breeders.

8. **It makes it feasible to prove more sires.** Because of the small size of the herds in which they are used, many bulls used in natural service are never proved. Still others are destroyed before their true breeding worth is known. Through artificial insemination, it is possible to determine the genetic worth of a bull at an earlier age and with more certainty than in natural service. The best of the bulls proved at an early age are put into heavy use and have a longer period of usefulness than is possible under natural breeding methods.

9. **It creates large families of animals.** The use of artificial insemination makes possible the development of a large number of animals within a superior family, thus providing uniformity and giving a better basis for a constructive breeding program.

10. **It increases pride of ownership.** The ownership of progeny of outstanding sires inevitably makes for pride of ownership, with accompanying improved feeding and management.

11. **It alleviates distance and time as limiting factors.** The male and the female may be separated by thousands of miles, and, with frozen semen, years may pass between the time of collection of the semen and insemination of the female.

12. **It increases profits.** The offspring of outstanding sires are usually higher and more efficient producers, and thus more profitable. AI provides a means of using such sires more widely.

LIMITATIONS OF ARTIFICIAL INSEMINATION

Like many other wonderful techniques, artificial insemination is not without its limitations. A full understanding of such limitations, however, will merely accentuate and extend its usefulness. Some of the limitations of artificial insemination are:

1. **It must conform to physiological principles.** One would naturally expect that the practice of artificial insemination must conform to certain physiological principles. Unfortunately, much false information concerning the usefulness of artificial insemination has been encountered—for example, the belief that females will conceive if artificially inseminated at any time during the estrous cycle. Others have even accepted exaggerated claims that the quality of semen may be improved through such handling, only to be disappointed.

2. **It requires trained technicians.** To be successful, artificial insemination must be carried out by skilled technicians. This means training, preferably training augmented by experience. While the cow breeding process is not complicated, it has been found that a small percentage of people who attempt to learn it never succeed in doing so.

3. **It may accentuate the damage of a poor sire.** It must be realized that when a male sires the wrong type of offspring his damage is merely accentuated because of the increased number of progeny possible. For this reason, untried or untested males are seldom used extensively in a stud. Fortunately, suitable standards for evaluating sires of dairy animals have evolved through progeny testing. Thus, it is noteworthy that 60% of the dairy sires are proved, and that these sires account for about 80% of the matings made. This precautionary measure virtually eliminates the possibility of using a genetically inferior dairy sire.

4. **It restricts the sire market.** It has been argued that the widespread adoption of artificial insemination has greatly decreased the market for poor or average sires. Such an argument is shortsighted. The principal thrust of artificial insemination is in the direction of maximum improvement through maximum use of superior sires. Obviously, this eliminates the necessity for using poor or average sires; hence, it must be regarded as an attribute, rather than a limitation.

5. **It may increase the spread of disease.** As previously indicated, the careful and intelligent use of artificial insemination will lessen the spread of disease. To date, no outbreaks of disease traceable to the use of artificial insemination have been reported in the United States. However, it must be recognized that carelessness or ignorance may result in the rapid spread of disease. Thus, semen should always be obtained from a source known to observe recommended health and sanitation procedures.

6. **It may be subject to certain abuses.** If semen is transported from farm to farm, the character of the technician must be above reproach. Trained workers can detect differences in the spermatozoa of the bull, ram, boar, stallion, or cock; but even the most skilled scientist is unable to differentiate between the semen of a Holstein and a Jersey, to say nothing of the difference between two bulls of the same breed. However, it appears that such abuse is more suspicioned than real. In a blood type study with cattle, Rendel of Sweden found 4.2% family records in error out of 615 animals by natural service, compared to 4.0% family records in error out of 199 sired by artificial insemination.[5]

7. **It is not yet fully practical to bring cows in "true heat" at will.** Many advantages would accrue from bringing females of all species in heat and ovulation when desired and with certainty. By using hormones, planned parenthood may be imminent; perhaps we shall soon be able to breed a cow on the day desired instead of waiting for the natural occurrence of the estrual cycle. With such a development, (a) breeding artificially would be simplified; and (b) it would be possible to have the young born exactly when desired.

PREPARATION OF AI EQUIPMENT

The success of artificial insemination is directly proportionate to the cleanliness of equipment, operators, and animals. This is because spermatozoa are highly sensitive to, and quickly killed by, dirt, water, urine, excess heat, cold shock, or light, and by just about any other substance foreign to their natural environment.

It is recommended that all equipment that comes in contact with semen during the collection process be washed in a mild detergent after use. It should then be thoroughly rinsed in tap water to remove all detergent residues. The next step is rinsing with distilled water to remove any and all impurities from the tap water.

Rubber equipment should be boiled in distilled water, then rinsed in ethanol. Better yet, disposable sterile plastic should be substituted for rubber wherever possible.

At present, the inseminating tubes, or catheters used in breeding cows are disposable plastic. So are the gloves used by inseminating technicians. This means that all such equipment can be thrown away after one use. The costly, time-consuming sterilization techniques needed with the glass equipment used in the early days of artificial insemination has been eliminated.

---

[5]Rendel, J., "Studies of Cattle Blood Groups. II. Parentage Tests," *Acta. Agric. Scand.*, Vol. 8, No. 131, p. 140.

COLLECTION AND HANDLING OF SEMEN

The method of collecting semen should be adapted to the bull, should be easy for the operator to use, and should permit the collection of a sample of normally ejaculated semen free from contamination with dirt, bacteria, or excess secretions.

There are four recognized methods of collecting semen: (1) using an artificial vagina; (2) using an electro-ejaculator or electrical stimulation (a necessity for crippled bulls that are unable to mount); (3) massaging the accessory genital organs (seminal vesicles and ampulla) per rectum; and (4) collecting the semen from the vagina that has just been bred by natural service.

■ **Artificial vagina** — Many techniques have been developed for the collection of semen, but the most satisfactory one consists of using the artificial vagina, a Russian invention. Males usually respond to this method with little previous training.

The artificial vagina consists of an outer tube or casing — which is usually constructed of heavy rubber, metal, or one of the plastics — and an inner tube, or lining, of thin rubber. The space between the two tubes is usually filled with warm water and a little air. One end of the apparatus is open to allow the entrance of the penis, and other end is attached to a glass tube or breaker to receive the ejaculated semen.

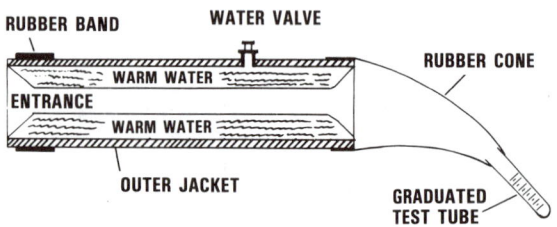

Fig. 5–29. Diagrammatic artificial vagina for the collection of bull semen. (Drawing by R. F. Johnson)

Most males, especially young ones, can be trained to use dummies. Training consists in exposing the male to an in-heat female, without permitting service. After two or more experiences of this kind, the male is introduced to the dummy at the same location; thereby being taught to anticipate service when brought to the breeding quarters. The dummy should be strongly built and firmly anchored.

Many commercial semen-producing businesses have replaced dummies with live *jump stock* — steers as well as cows. It has long since been demonstrated that bulls will mount cows not in heat and stand to be collected with an artificial vagina. More recently, it was discovered that most bulls could be collected in the same manner if a steer replaced the cow as the *jump animal*. The advantage is that steers are bigger and stronger than cows — better able to bear the weight of the bull from which collection is being made.

■ **Electrical stimulation** — This method first found limited use with rams. It has since been adapted to, and widely used

in, collecting semen from bulls that will not or cannot mount jump animals or that lack in libido. It is also useful on farms where there are no facilities for normal semen collection and in cases where time does not permit training bulls to artificial vaginas.

Electro-ejaculator apparatuses are available commercially. The simplest equipment introduces a weak alternating current to the sacral and pelvic nerve via electrodes placed in the rectum on a probe or by hand. For bulls, a single electrode, delivering 5 to 30 volts, is placed in the rectum; then a series of stimulations is applied until erection and ejaculation occur.

■ **Rectal massage of ampullae** — This system has been used successfully in bulls. It consists of the rectal massage of the ampullae and the collection of semen in a funnel leading to a glass tube. This method has the disadvantage of (1) requiring considerable experience on the part of the operator, (2) necessitating two persons for collection, and (3) resulting in abnormal semen. Hence, it should be resorted to only on bulls that are unable or unwilling to mount or ejaculate.

■ **Collecting semen from the vagina** — Semen may be collected from the vagina of a cow with a sponge, pipette, spoon, or similar object. This method is not recommended, however, because the semen is contaminated with mucus and urine, and there is danger of spreading genital diseases.

VOLUME OF SEMEN/CONCENTRATION OF SPERM

The volume of semen ejaculated at one service and the concentration of sperm vary according to species and individuals. Table 5–5 gives average figures by classes of animals.

TABLE 5–5
SEMEN VOLUME AND SPERM CONCENTRATION OF FARM ANIMALS

Class of Animal	Average Volume of Semen per Ejaculate	Average Concentration of Sperm	Number of Females That Can Be Bred per Ejaculate
	(ml)	(millions/ml)	
Bull	5–6	800–1,200	300–500
Ram	1	800–4,000	40–100
Boar	200–300	25–1,000	15–25
Stallion	50–150	30–800	8–12

As a rule, the smaller ejaculates of high sperm concentration, such as found in bull semen, are the most suitable for artificial insemination; for they can be diluted into a large volume with special dilutes that retain the life of the sperm to a higher degree than the natural seminal fluid.

In cattle, sperm numbers are of more importance than semen volume.

SEMEN EXTENDERS

Addition of extenders to freshly collected semen is almost imperative because (1) they provide needed volume; and (2) they exert a beneficial effect on the sperm.

Table 5–6 summarizes the pertinent facts relative to the most widely used semen extenders in the United States. A recent survey showed that about half of the bull studs in America use the yolk-citrate extender and most of the remainder use boiled milk. Other extenders are (1) glucose plus tartrate, sulfate, or phosphate salts (developed by Milovanov, a Russian); (2) modified Ringer solution (developed by Bonnier and Trulsson); (3) skim milk; (4) dry skim milk powder; (5) homogenized whole milk; (6) synthetic pabulum (developed by Phillips and Spitzer); (7) Krebs solution (developed by Lardy and Phillips); (8) IVT (developed by Illinois); (9) Tris buffered extenders; (10) saline solution extenders (for short-time durations); (11) Caprogen; and (12) certain commercially prepared extenders.

TABLE 5–6
COMMON SEMEN EXTENDERS

Type of Extender	Formula of Extender	Classes of Animals for Which It Is Satisfactory	Comments
Egg yolk-citrate (developed by Salisbury, Fuller and Willet)	1. Prepare a 2.9% solution of sodium citrate Na$_3$C$_6$H$_5$O$_7$•2H$_2$O. 2. Prepare egg yolk. 3. To prepare 100 ml of the egg yolk-citrate diluent, mix 20 ml of the egg yolk and 80 ml of the sodium citrate buffer solution. Antibiotic should be added to the diluent.	Bull Ram	Only the yolk portion of the egg is used because egg white contains *lysozyme* which is toxic to spermatozoa. The main advantage of this extender is that it disperses the fat globules of egg yolk, thus facilitating the microscopic examination of extended semen.
Milk extender	The following procedure is recommended for preparing the milk extender: 1. Use either fresh, pasteurized, homogenized milk or fresh pasteurized skim milk. 2. Heat the milk in a double boiler (Pyrex glass preferred) at 198° to 203°F for 10 minutes. 3. Add antibiotics before extending the semen.	Bull Ram Boar Stallion	The main objection to whole milk as an extender is that the fat globules make microscopic examination difficult. The problem is avoided when skim milk is used and can be partially alleviated by the use of homogenized milk, or by the use of goat milk which is "naturally" homogenized.
CWE extender (recommended by Cornell University workers)	**For liquid bull semen:** 1. 80% of a specified liquid salt buffer solution. 2. 20% egg yolk. 3. Plus penicillin and streptomycin. **For frozen bull semen:** 1. Buffer of raffinose and distilled water, 75.3% by volume. 2. Plus 4.7% glycerol (by volume), 20% egg yolk (by volume), and penicillin and streptomycin.	Bull	
Egg yolk-phosphate (developed by Lardy and Phillips)	1. To 100 cc of boiling glass-distilled water add: 0.2 g KH$_2$PO$_4$ (chemically pure) 2.0 g Na$_2$HPO$_4$, 12H$_2$O (chemically pure) Cool to room temperature. 2. Add an equal volume of *fresh egg* yolk which has been carefully separated from the whites. Solution No. 1 can be stored, but a fresh mixture (Nos. 1 and 2) should be made up each time.	Bull Ram Stallion	When using this extender for stallion semen, add 10 g dextrose or glucose (chemically pure) per 100 cc of boiling distilled water. Currently, this extender is little used in cattle breeding in the United States.
Glycine-containing diluents	1. 50% of a 4% glycine solution. 42.5% skim milk, heated to 92°C and then cooled. 7.5% egg yolk. —or— 2. 15% of a 4% glycine solution. 72.25% skim milk, heated to 92°C and then cooled. 12.75% egg yolk.	Bull Stallion	Glycerol depresses the fertility of both boar and stallion semen. However, studies at Texas A&M University indicate that evaporated milk can be used with glycerol as an extender in freezing stallion semen.
Milk-glycerol diluent (developed by Almquist of Penn. State)	1. Dilute the fresh semen about one-half the final desired sperm concentration with fresh, previously heated and cooled homogenized or skim milk containing the recommended antibiotics. 2. Cool slowly to 41°F. 3. Add an equal volume of cooled milk diluent containing 20% glycerol by volume. 4. The final product, which contains 10% glycerol, is ready for packaging and shipping.	Bull	Milk-glycerol diluent is especially well adapted to packaging and shipping liquid semen.
Minnesota GO extender	1. The basic Minnesota GO buffer consists of 10 substances, varying from 1 g to 21 g dissolved in 1,000 ml distilled water. 2. Dispense 50 g of spray dried egg-yolk solids in 1,000 ml precooled (41°F) Minnesota GO buffer.	Bull	Studies show that Minnesota GO extender is equal or superior to egg yolk-citrate or milk-glycerol.

Most fluid semen extenders (listed in Table 5–6) presently used in the United States for cattle breeding contain added glycerol. This has improved fertility, particularly in older semen. When semen is frozen, glycerol must be included in the extender.

■ **Antibiotics and sulfanilamide** — These products have been successfully used to prolong the keeping quality of semen and to increase conception rate 5 to 12%. Penicillin and streptomycin are the commonly used antibiotics, and sulfanilamide is the most commonly used sulfa. Bovine vibriosis may be controlled by treating semen with antibiotics.

STORAGE AND HANDLING OF SEMEN

Fig. 5–30.   Transferring semen from shipping tank to regular storage tank. (Photo by T. O. and J. Jaggers, Leitchfield, KY)

Semen may be used as (1) fresh liquid semen, or (2) frozen semen.

■ **Fresh liquid semen** — If semen is to be used within 1 or 2 hours following collection, it may be kept at room temperature. For longer storage (1 to 4 days) as liquid semen, and delayed use, it should be properly diluted and gradually cooled (avoiding temperature shock) and stored at a temperature of 35 to 40°F.

It is noteworthy that the greatest use of liquid semen stored at ambient temperature is made in New Zealand. Liquid semen is used to inseminate most of the cattle on AI because the main breeding season is short. Of the approximately 1 million head of cows inseminated annually, 90% are bred from mid-September to December, during the lush grazing season of their spring months. New Zealand workers have developed an ambient temperature storage extender by gassing the citrate-buffered egg yolk extender with nitrogen and adding capronic acid. The extender is called *caprogen*.

■ **Frozen semen** — In 1949, British scientists reported that the addition of glycerine to semen diluters permitted them to freeze certain semen at temperatures much below zero (they used dry ice to freeze at a temperature of –79°C or –110°F), and still retain a high degree of fertility following thawing.

Today, semen is being frozen for storage in liquid nitrogen at –196°C, or –320°F. Liquid nitrogen, which is the fourth coldest known substance, is the universally used refrigerant in the United States because uniform temperatures may be maintained for long periods of time and the method is more convenient for shipping and storing frozen semen. Dry ice at –101°F is not cold enough, and mechanical refrigerators are not reliable.

Frozen bull semen has been stored for more than 30 years and conception obtained. There is a small reduction in fertility of semen stored long periods, but it is usual to find semen stored one year or longer in routine use.

Frozen semen, refrigerated with liquid nitrogen, can be shipped to all parts of the world. The thawing of the semen is accomplished by placing the vial in a container of water containing thawing ice (34-38°F) immediately prior to insemination.

Frozen semen is potentially the most valuable breeding technique yet known. Through it, the following may be achieved:

1. The usefulness of outstanding sires can be extended far beyond their lifetime; also, it ensures the proven sire should he die.

2. Outstanding sires can be used, nationwide and worldwide.

3. A multiherd progeny test can be completed at a much earlier age.

4. A stock of semen can be built up while waiting for progeny record assessment.

5. Long-term storage of semen lessens semen wastage, and facilitates long-distance transport.

6. Semen from valuable sires may be fully utilized.

7. A herd owner can usually obtain the sire of choice at any time.

■ **Custom freezing** — On a fee basis, many established bull studs collect, freeze, process, store, and dispense semen from bulls owned by private breeders.

Such semen is generally used to service cattle belonging to the owner of the sire. More than likely, this practice will increase in the future.

PACKAGING SEMEN

The world trend is toward packaging and freezing semen in concentrated form. Among the methods are ampules, pellets, straws, shell-freezing, and lyophilization of semen.

■ **Ampule** — This is a glass container with enough semen for only one service. Usually, each ampule contains 0.5 to 1 ml of diluted semen. In comparison with straws, ampules take much more time to thaw. Also, in going (1) from ampule, (2) to breeding tube, (3) to cow, there is loss of some sperm in semen which clings to the ampule and breeding tube.

■ **Pellet** — The pelleting of semen was developed by the Japanese in 1962. Pellets are thawed in physiological saline (0.9% NaCl). Pellets are a concentrated method of storing semen, but they present problems in identification, automation, and sanitation.

■ **Straws**—Plastic straws range from about 2½ to 5 in. in length and commonly hold about 0.25 to 0.5 ml of semen. In comparison with glass ampules, plastic straws require less storage space, yield a higher recovery rate of motile sperm when thawed, and result in fewer sperm being lost in the insemination process (e.g., fewer sperm adhere to the container). Today, 95% of all units of frozen semen marketed in the United States are in straws.

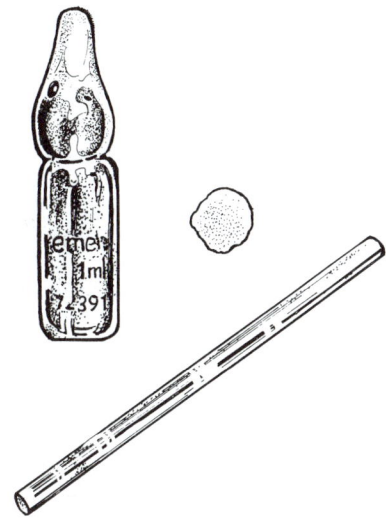

Fig. 5–31. The three most common methods of packaging and storing frozen semen (left to right): ampule, pellet, and straw.

■ **Shell-freezing**—This method was developed by Graham of Minnesota in 1966. It results in a higher percentage of live sperm than other methods of packaging; the dead and abnormal sperm are removed by passing the semen through a glass-fiber filter.

■ **Lyophilization of semen**—The goal in the use of this technique is the preservation of semen in the dry state at low temperature. Progress along these lines is encouraging, but the process is still in the experimental stage.

WHEN TO BREED

A female is fertile only when an egg is present which can be fertilized. Moreover, an egg can live for only a short time after being shed from an ovary unless it is fertilized. The optimal time for insemination is in advance of the time of ovulation.

A cow does not shed her egg from the ovary until about 10 hours after the close of standing heat, and the egg lives only 6 to 10 hours. Thus, for optimal results from insemination, cows should be bred in the latter two-thirds of heat or within a few hours after having gone out of heat; roughly, this means that a cow should be bred within a 24-hour period after she is first noticed in standing heat. To accomplish this, it is recommended that cows first observed in standing heat in the morning be bred during the afternoon or evening of the same day, and that those observed in heat in the evening be bred the next morning.

THAWING FROZEN SEMEN

With both straws and ampules, there is a wide variation in thawing recommendations and techniques. This is due to the interrelationships between freezing and thawing rates associated with the type of semen extender used, the type and size of semen unit package, and the temperature involved. Thus, it is no surprise that some AI organizations recommend relatively high temperatures while others recommend lower temperatures for semen thawing. So, in all cases, it is best to follow the recommendations and instructions of the AI organization that actually processed and packaged the semen. Instructions for thawing are drawn up to fit the particular brand of semen involved.

INSEMINATION OF THE FEMALE

Cleanliness during all insemination manipulations is essential and is a crucial point for success or failure. This applies to the instruments used, to the hands of the operator, and to the animals.

The inseminating tube should be passed just through the cervix, stopping at the location where the cervix ends and the uterus begins. Precisely at this point the semen should be deposited—slowly (see Fig. 5–32)

Cows are usually inseminated with 0.5 to 1.5 ml of extended sperm.

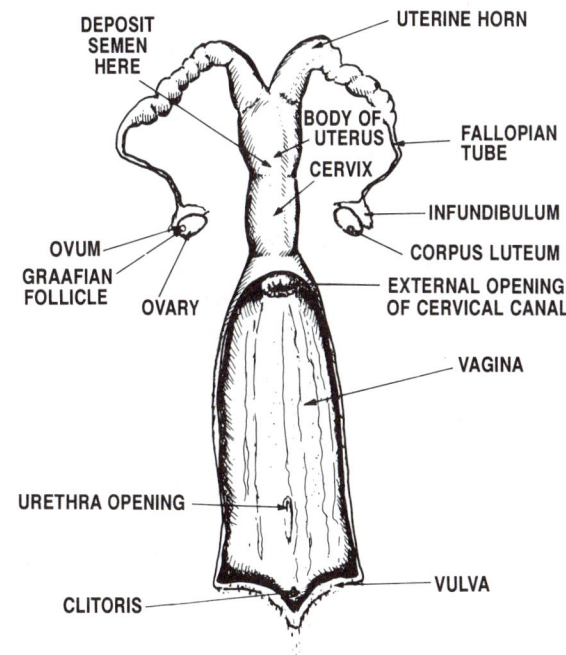

Fig. 5–32. Where to deposit semen.

PRACTICAL APPLICATION

Today, artificial insemination is more extensively practiced with dairy cattle than with any other class of farm animals. In 1938, when the program first began in America, only

7,359 cows and 646 herds were bred by this means in organized groups in the United States.

In 1988, artificial insemination programs involved 65% of milking cattle and more than 90% of the dairy herds in the United States.

Based on present knowledge, gained through research and practical observation, it may be concluded that livestock producers can make artificial insemination more successful through the following:

1. Give the female a reasonable rest following parturition and before rebreeding; in cows this should be 50 to 60 days.

2. Keep records of heat periods and note irregularities.

3. Watch carefully for heat signs, especially at the approximate time.

4. Where an association is involved, notify the insemination technician promptly when an animal comes in heat.

5. Avoid breeding diseased females or females showing cloudy mucus. The latter condition indicates an infection somewhere in the reproductive tract.

6. Have the veterinarian examine females that have been bred three times without conception or that show other reproductive abnormalities.

7. Follow a proper nutrition program at all times.

ARTIFICIAL INSEMINATION OF DAIRY CATTLE

In 1937, the North Central School of Agriculture and Experiment Station, Grand Rapids, Minnesota, conducted the first large-scale demonstration of artificial insemination of dairy cattle in the United States. Subsequent growth was rapid. To dairy producers, AI offered a way in which to enhance the genetics of the herd, control infectious diseases, and eliminate keeping dangerous bulls on the farm. These advantages were further enhanced by the positive economics of AI. Return over investment for the dairy producer averages about 50% on typical semen sales, and may be as high as 100% for herds that are just starting AI.

Today, new and improved techniques, knowledge, and computers impart confidence in predicting the genetic transmitting ability of AI sires.

Genetic progress in AI sires has been phenomenal in recent years. The AI bulls are now about 750 lb of milk genetically superior to their daughters. These bulls are pulling the cows' genetics up by about 200 lb per year. This makes AI attractive to dairy farmers.

BULL COSTS VS AI COSTS

With the increase in artificial insemination in recent years, a frequently asked question is: What's the cost of AI vs natural service?

Many cost figures seen in various publications show bull costs by natural service ranging from $6 to $12 per cow. The Nebraska Station reported that the cost of keeping a bull is much higher than this, even if the bull is depreciated over a 4-year period (see Table 5–7).

TABLE 5–7
ANNUAL COSTS OF OWNING AND MAINTAINING A BULL[1]

	If Depreciated Over 2 Years		If Depreciated Over 4 Years	
	Total Costs	Direct Costs	Total Costs	Direct Costs
Hay, 1.5 tons	$ 75.00	$ 75.00	$ 75.00	$ 75.00
Winter pasture	25.00	—	25.00	—
Summer pasture	108.00	—	108.00	—
Salt and mineral	4.00	4.00	4.00	4.00
Veterinary and medicine	14.00	14.00	14.00	14.00
Death loss	6.00	6.00	6.00	6.00
Depreciation ($1,500 purchase price, $840 selling price)	330.00	330.00	165.00	165.00
Interest on bull (12%)	140.00	140.00	140.00	140.00
Interest on feed and operating expense (12%)	6.00	6.00	6.00	6.00
Labor (10 hr)	55.00	—	55 .00	—
Use of buildings and equipment .	9.00	3.00	9.00	3.00
Miscellaneous expense	6.00	6.00	6.00	6.00
Total	$778.20	$584.00	$613.00	$419.00
30 cows	26.00	19.50	20.50	14.00

[1]Estimates by Guyer & Jose, University of Nebraska, Lincoln.

As shown in Table 5–7, one of the largest bull cost items is depreciation. These figures are based on a purchase price of $1,500 and a selling price of $840. On this basis, bull depreciation alone would amount to more than $10 per cow. Direct costs, excluding pasture, amount to $13.97 per cow (depending on how long the bull is kept), and $14.87 to $21.27 per cow if the charge for summer pasture is included.

Artificial insemination eliminates the expense and problems associated with maintaining bulls. But of course, it involves some different expenses, primarily for the cost of semen and the added labor.

DAIRY ASSOCIATION RULES RELATIVE TO AI

All dairy breed registry associations require compliance with the "Requirements Governing Artificial Insemination of Purebred Dairy Cattle," adopted by The Purebred Dairy Cattle Association and The National Association of Animal Breeders. A copy of the requirements may be obtained from each breed registry association. One requirement common to all the breed dairy associations is that the blood type of a bull and his living unblood-typed parents be on file with the respective breed registry associations.

BIOTECHNOLOGY IN DAIRY CATTLE BREEDING

During the decades ahead, biotechnology will dominate and pace the changes in dairy breeding. Scientists will develop genetically altered animals that require less feed to produce more milk—and much more. DNA, genetic engineering, patented animals, and cloning will become house-hold words. In the next century, artificial insemination may be replaced by *in vitro* (test tube) artificially fertilized embryos.

DNA (DEOXYRIBONUCLEIC ACID)

Many years ago, the gene was usually defined as the *smallest unit of inheritance*. In recent years, research has shown that the gene is a portion of a deoxyribonucleic acid (DNA) molecule; and much information is now available on the chemistry and function of genes. The DNA in animal cells is found in the nucleus and extends the length of the chromosome, near its center. It is a long, helical (spiral) structure, resembling a long, twisted ladder with two sides or strands joined together by rungs, or steps (Fig. 5–33). The fact that chromosomes occur in pairs also causes genes to occur in pairs. Members of the homologous pairs of chromosomes, have a *loci*, or the same position where similar genes (alleles) are carried. So, *alleles are genes that occupy the same location on homologous chromosomes.*

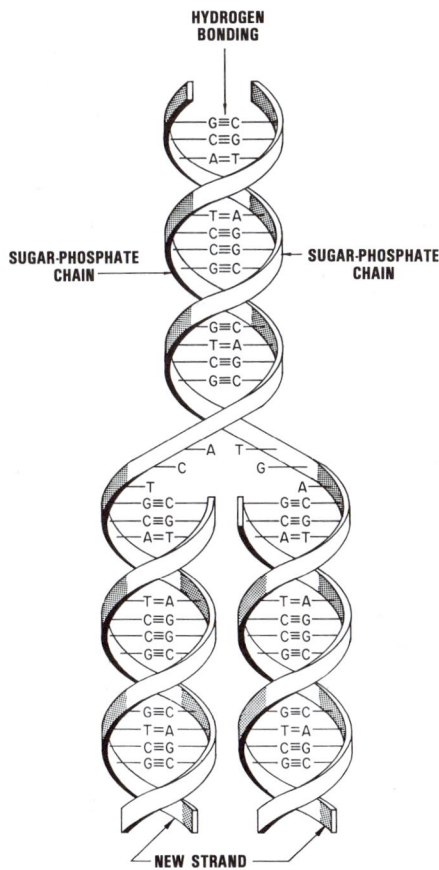

Fig. 5–33. The spiral structure of deoxyribonucleic acid, or DNA—the basic building block of life on earth. It is a double helix (a double spiral structure), with the sugar (deoxyribos)—phosphate (phosphoric acid) *backbone* represented by 2 spiral ribbons. Connecting the *backbone* are 4 nitrogenous bases (a base is the nonacid part of a salt): adenine (A) paired with thymine (T), and guanine (G) paired with cytosine (C); with the parallel spiral ribbons held together by hydrogen bonding between these base pairs. Adenine and guanine are purines, while thymine and cytosine are pyrimidines.

More than one alternative form of gene may exist, with each one affecting the same trait but in a different manner. These forms are called *alleomorphos*, or *alleles*.

Each of the paired autosomes in the male and female of the species normally carries the same loci. However, the sex chromosomes do not always carry the same locus, because in mammals the X chromosome is much larger than the Y chromosome. For this reason, the X chromosome carries loci not carried on the Y chromosome. Since female mammals are XX and males are XY, the male possesses less genetic material than the female.

The DNA of the chromosomes in the nuclei of cells carries the coded master plans for all of the inherited characteristics—size, shape, and orderly development from conception to birth to death. DNA is different for each species, and even for each individual within a species. These differences consist of minor rearrangements of sequences among the nitrogenous bases, which constitute a code containing all the information on the heritable characteristics of cells, tissues, organs, and individuals.

The messages carried by DNA are put into action in the cells by the other nucleic acid, RNA. To do this, DNA serves as a template (as the pattern or guide) for the formation of RNA. The genetic message is coded by the sequence of purine and pyrimidine bases attached to the *backbone* of the DNA structure—a long chain of the sugar deoxyribose and phosphoric acid. Purine bases in DNA include cytosine and thymine. One molecule of DNA may contain 500 million bases. The *backbone* of RNA is also a sugar, the sugar ribose, plus phosphoric acid. However, in RNA the pyrimidine base thymine is replaced by uracil, another pyrimidine. RNA molecules are considerably smaller than DNA, containing from less than a hundred to hundreds of bases, not millions.

Thus it may be said that scientists are coming closer and closer to understanding the very essence of life—DNA.

GENETIC ENGINEERING

Building upon the present knowledge and understanding of genes, and the nucleic acids DNA and RNA, scientists are now making such ideas as genetic engineering and cloning into realities.

The development of gene-splicing (also known as recombinant DNA) ushered in a new era of genetic engineering—with all its promise and possible peril.

On May 23, 1977, scientists at the University of California, San Francisco reported a major break-through as a result of altering genes—turning ordinary bacteria into factories capable of producing insulin, a valuable hormone previously extracted at slaughter from pigs, sheep, and cattle, so essential to the survival of 1.8 million diabetics. The feat opened the door to further genetic engineering or splicing—the transferring of a gene from one individual to another. Already, this genetic wizardry has been used in transplanting into bacteria (and recently into yeast cells) genes responsible for many critical biochemicals in addition to insulin; among them, endorphin, somatotropin, interferon, and vaccines.

Genetic manipulations to create new forms of life make biologists custodians of a great power. Despite different schools of thought, scare headlines, and political hearings, molecular biologists will continue recombinant DNA studies, with reasonable restraints, and work ceaselessly away at making the world a better place in which to live. (See Fig. 5–34.)

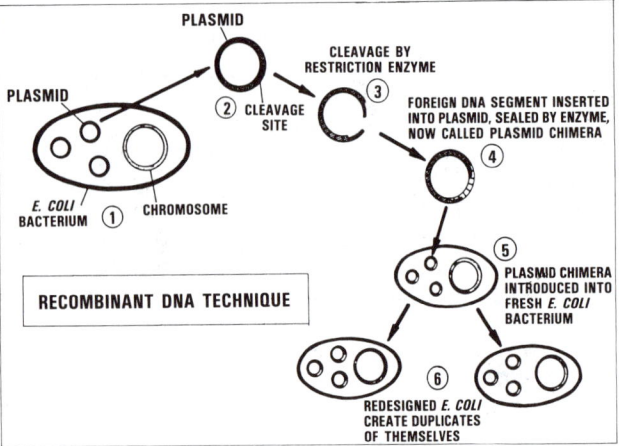

Fig. 5–34. Redesigning *E. coli*, common bacteria of animal and human intestines. The steps:

1. The scientist places the bacterium in a test tube with a detergent. This dissolves the microbe's outer membrane, causing its DNA strands to spill out.

2. The plasmids (the closed loops), which are genetic particles found in bacteria, are separated from the chromosomal DNA in a centrifuge.

3. The plasmids are placed in a solution with a chemical catalyst called a restriction enzyme, which cuts through the plasmids' DNA strips at specific points.

4. The opened plasmid loops are then mixed in a solution with genes—also removed by the use of restriction enzymes—from the DNA of a plant, animal, bacterium, or virus. In the solution is another enzyme called a DNA ligase, which cements the foreign gene into place in the opening of the plasmids. These new loops of DNA are called plasmid chimeras because, like the chimera—the mythical lion-goat-serpent after which they are named—they contain the components of more than one organism.

5. The chimeras are placed in a cold solution of calcium chloride containing normal *E. coli* bacteria. Then the solution is suddenly heated, at which time the membranes of the *E. coli* become permeable, allowing the plasmid chimeras to pass through and become a part of the microbe's new genetic structure.

6. When the redesigned *E. coli* reproduce, they create duplicates of themselves, new plasmids—and DNA sequences—and all.

PATENTED ANIMALS

In April 1987, the U.S. Patent and Trademark Office (PTO) ruled that patents could be issued on genetically engineered animals. Subsequently, the PTO (1) put the new animals on a par with mechanical inventions by decreeing that livestock producers must pay fees to those who patent genetically altered animals, and (2) ruled that livestock producers must pay royalties to the patent holder on each generation of patented animals for the life of the patent—which may be as long as 17 years.

Stud fees and one-time payments for animals have always been a part of livestock farming, but farmers have the rights to, and complete control over, subsequent generations without any additional payments.

In 1988, the U.S. Patent Office granted a patent to Harvard University for a genetically engineered mouse, involving developing a way to add cancer genes to the embryo of mice, thus making them more likely to get cancer thereby facilitating cancer experiments.

Farmers face a dilemma: Generally speaking, they want to reap the benefits of genetically improved livestock, but they do not want to pay royalties on each succeeding generation.

CLONING

Cloning of an animal is the production of an exact genetic copy. In a technical sense, identical twins are clones; they are derived from a single cell, as a result of the embryo splitting early in development to yield what is essentially two carbon copies.

Through cloning, it will be possible for all dairy animals in an entire herd to look alike, be genetically alike, have the same nutritive requirements, and produce the same quantity of milk of the same composition.

The exciting and much sought technological breakthrough in the cloning of mammals involves the manipulation of embryos.

The dream of cloning is based on the following two pieces of scientific evidence:

1. With few exceptions, all cells in the body of an animal appear to contain the same genetic information. This information is contained in the DNA, a molecule that is located in a sac inside cells called the nucleus. Thus, within an animal, the DNA sequence in the nucleus of a liver cell is identical to that in a skin cell. These cells differ in appearance and function because they make use of different parts of the genetic information, not because the total amount of information differs. Further, all of these cells have the genetic information that was present in the one-cell embryo that developed in the animal. Therefore, if the nucleus of any of these cells were used to replace the genetic information in any one-cell embryo, an exact genetic copy of the animal whose cells donated the nucleus would develop. With such an approach, thousands of cloned copies could be made.

2. The second piece of scientific evidence is that nuclear transplantation experiments have been done successfully with several species of animals, especially frogs and fish, which have the big advantage of their eggs being thousands of times larger than mammalian eggs.

Historically, research and development in cloning has passed through the following stages, in order and period of time:

1. Identical twin calves were produced by microsurgically splitting embryos, then transferring half embryos to recipients.

2. A bull calf was born as a result of using the laboratory culturing technique (*in vitro*) of maturing an egg, fertilizing the egg *in vitro*, then transferring the fertilized egg to a surrogate mother.

Nuclei have been taken from 16-cell bovine embryos and placed in one-cell bovine eggs whose nuclei had already been removed. The new one-cell embryos were matured and transferred into recipient cows, which subsequently gave birth to the cloned female calves.

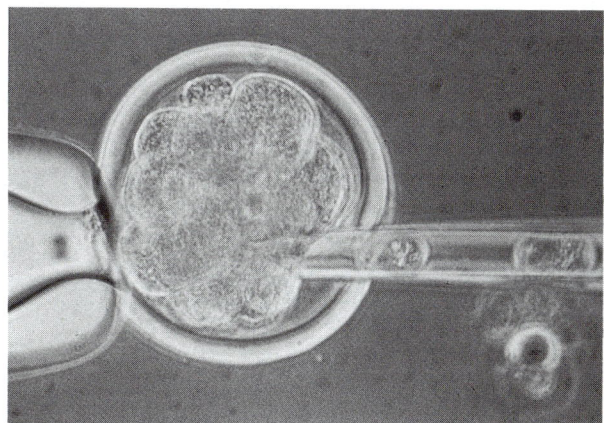

Fig. 5–35. This shows the removal of cells of a valuable embryo to use in cloning, to produce 8 to 20 cloned embryos from this embryo by transfer of each of these cells to an enucleated oocyte. (Courtesy, Dr. Robert Walton, American Breeders Service, a division of W. R. Grace & Co., De Forest, WI)

During the late 1980s, the most advanced cloning procedure consisted of flushing the embryo out of the donor cow at day five of its development (at the 32-cell stage); followed by putting the embryo under a microscope and manually removing one of the cells, then freezing the remaining 31-cell embryo (much like semen is frozen) and putting it away until an order is received. Next, the technician takes an unfertilized egg that has been flushed from a low-grade donor cow, removes the nucleus from it, inserts the borrowed cell into the egg, patches up the hole with a short zap of electricity,

and the cloned embryo divides day after day and develops into a genetically identical duplicate of the heifer that results from the 31-cell embryo.

The goal ahead is (1) to let embryos grow to the 32-cell stage, then split them all into 32 more embryos each, resulting in 1,024 (32 x 32) genetically identical copies of the same pedigree—*the exact same*; (2) to let 31 of the embryos grow up, split apart the 32nd, and keep carrying out the chain for as long as desired; and the first animal born would be identical to the last one—*identical*. The goal ahead is illustrated in Fig. 5–36.

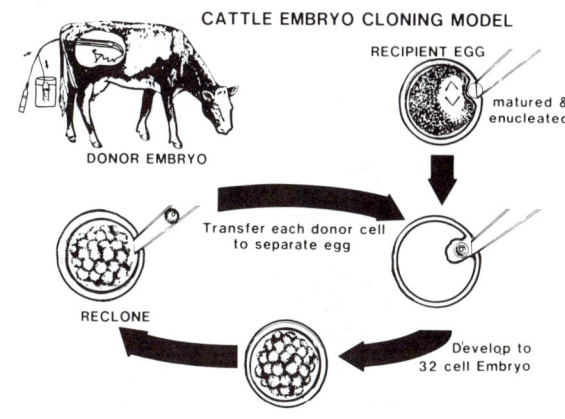

Fig. 5–36. Cattle embryo cloning model. (Prepared by Dr. Robert Walton, American Breeders Service, a division of W. R. Grace & Co., De Forest, WI)

QUESTIONS FOR STUDY AND DISCUSSION

1. What is the economic justification for improved breeding in dairy cattle?

2. What kind of records are most important in a dairy cattle breeding program?

3. The effect of environment on dairy cattle was clearly demonstrated in an experiment conducted in New Zealand. Describe this experiment, including the results.

4. Why are "correction factors" necessary? Describe each of the following correction factors: (a) length of lactation, (b) the number of milkings per day, (c) age and month of calving, and (d) fat content of milk.

5. The expression of a trait, such as milk production, depends on two factors. What are these factors?

6. Describe each of the following genetic evaluations of dairy cattle which were used in the past: (a) lactation record, (b) daughter average, (c) daughter-dam comparison, (d) herdmate comparison, and (e) contemporary comparison.

7. Describe each of the genetic evaluations of dairy cattle which are currently in use: (a) modified contemporary comparison, and (b) mixed model method. Define the following terms which are used in expressing the resulting estimate

of genetic transmitting of the sire: (a) predicted difference, and (b) repeatability.

8. How are chromosome numbers maintained constant from one generation to the next?

9. What is the difference between qualitative and quantitative traits? List some traits of each type.

10. Define dominance, and give an example of it in dairy cattle.

11. Define lethals, and give an example of it in dairy cattle. List the conditions which would indicate that an abnormality is due to a hereditary defect. List the conditions which would indicate that an abnormality may be due to a nutritional deficiency.

12. Of what value is blood typing in dairy cattle?

13. Why don't dairy producers select for multiple births?

14. How is sex determined?

15. Of what practical importance would a successful method of sex control be to a dairy producer?

16. List and discuss the three common bases of selecting dairy animals.

17. List and discuss the three general systems of selection.

18. How would you go about selecting cows or dairy heifer replacements?

19. How would you go about selecting proved dairy sires? What are USDA-DHIA Sire Summaries? Why are they important?

20. Under what circumstances might it be desirable to use a young sire rather than a proved sire?

21. Explain the difference between, and the advantages of (a) inbreeding, (b) outcrossing, and (c) crossbreeding. Under what circumstances would you use each?

22. Outline a sound breeding program for a dairy herd.

23. How much semen and how many sperm does a bull ejaculate on the average?

24. List the cow's functions in reproduction.

25. Why is heat detection so important? List and discuss the common heat detection methods and devices.

26. Why is hormonal control of heat a much sought management tool? List the drugs that head the list in the control of the estrous cycle of cows.

27. For optimal fertilization, at what time should cows be bred in the heat period?

28. Discuss the economic value of (a) superovulation and (b) embryo transfer in dairy cattle.

29. Why and how should a dairy producer pregnancy test cows?

30. What are the objectives of induced early calving?

31. Describe normal presentation of the calf at parturition.

32. What are the functions of colostrum?

33. List (a) the advantages and (b) the limitations of AI.

34. List and describe the four recognized methods of collecting bull semen.

35. How does the volume and concentration of bull semen compare with the ram, boar, and stallion? (See Table 5–5.)

36. Discuss (a) semen extenders, and (b) storage and handling of semen.

37. Discuss (a) packaging semen, (b) when to breed, and (c) thawing frozen bull semen.

38. Compare the cost of keeping a bull vs the cost of AI. (See Table 5–7.)

39. Are the dairy breed registry associations favorable to the use of AI?

40. Describe the application and/or impact of each of the following in the era of biotechnology in dairy cattle breeding: (a) DNA, (b) genetic engineering, (c) patented animals, and (d) cloning.

SELECTED REFERENCES

Title of Publication	Author(s)	Publisher
Artificial Insemination and Embryo Transfer of Dairy and Beef Cattle, The, Seventh Edition	H. A. Herman F. W. Madden	The Interstate Printers & Publishers, Inc., Danville, IL, 1987
Artificial Insemination of Farm Animals, The	E. J. Perry	Rutgers University Press, New Brunswick, NJ, 1968
Cattle Fertility and Sterility	S. A. Asdell	Little, Brown and Company, Boston, MA, 1968
Dairy Cattle Breeding	L. O. Gilmore	J. B. Lippincott Co., Philadelphia, PA, 1951
Dairy Cattle; Principles, Practices, Problems, Profits, Third Edition	D. L. Bath, *et al.*	Lea & Febiger, Philadelphia, PA, 1985
Dairy Cattle Sterility	H. D. Hafs L. J. Boyd	W. D. Hoard and Sons Co., Fort Atkinson, WI
Dairy Guide	Staff	The Ohio State University Cooperative Extension Service, Columbus, OH, 1979
Illinois-Iowa Dairy Handbook	Staff	University of Illinois at Urbana-Champaign, Iowa State University at Ames Cooperative Extension Service, 1983
Improving Cattle by the Millions	H. A. Herman	University of Missouri Press, Columbia and London, 1981
Principles of Dairy Farming, The, Sixth Edition	K. Russell S. Williams	Farming Press Ltd., Ipswich, England, 1972
Principles of Dairy Science	G. H. Schmidt L. D. Van Vleck	W. H. Freeman and Co. Publishers, San Francisco, CA 1974
Stockman's Handbook, The, Seventh Edition	M. E. Ensminger	Interstate Publishers, Inc., Danville, IL, 1992

Fig. 6–1. Newborn calf being greeted by two heifers. Since reproductive ability is fundamental to milk production, it can be readily understood that sterility constitutes a major annual loss in the dairy industry. (Courtesy, USDA)

Contents

CHAPTER

6

STERILITY AND
DELAYED BREEDING
IN DAIRY CATTLE

Sterility (infertility or barrenness) may be defined as temporary or permanent reproductive failure; resulting from anestrus (lack of heat), failure to conceive, or abortion.

Based on number of calves born in comparison with number of cows bred, the U.S. calf crop for all cattle (dairy and beef) is 88%. This means that the other 12% are nonproducers, either temporarily or permanently. Since reproductive ability is fundamental to milk production, it can be readily understood that sterility constitutes a major annual loss in the dairy industry.

It has been estimated that sterility and delayed breeding in dairy cattle account for an average yearly loss of $60 per cow, for a national total loss of $650 million. At least half of these losses could be prevented. Thus, on the average, sterility and infertility may be expected to cause losses of the magnitude indicated in Table 6–1 (center column), at least half of which could be prevented (right column).

TABLE 6–1
HERD SIZE DETERMINES THE SAVINGS POSSIBLE
BY MINIMIZING STERILITY

Herd Size	Expected Yearly Loss	Possible Yearly Saving
	($)	($)
40	2,400	1,200
70	4,200	2,100
100	6,000	3,000
150	9,000	4,500
200	12,000	6,000
500	30,000	15,000

The incidence of sterility varies greatly from herd to herd, and within the same herd from year to year. Despite this fact, dairy producers should establish arbitrary standards by which breeding performance may be gauged. To this end, the following reasonable averages are proposed: not more

than 10% breeding difficulty in the cows at any one time;[1] not more than an average of 1.85 service per conception;[2] and not lower than a 94% calf crop. Of course, the better managed and the more fortunate herds will do better. If these standards are accepted, however, there should be reason for concern if performance falls below these averages—it should then be assumed that something is wrong and that investigation is needed.

Fortunately, comparatively few barren cows and sterile bulls are totally and permanently infertile. Those that are should be sold for slaughter without further delay or expense. Most of the others will regain their breeding abilities with good care and management and appropriate treatment. Mating at the proper stage of estrus and the correct training and use of bulls will do much to maintain a high conception rate. Care in the selection of disease-free breeding stock, isolation of newly purchased animals, and periodic health examinations are effective preventive measures. When breeding irregularities are noted or disease strikes, however, treatment should be prompt. In general, diagnosis and treatment should be left to a veterinarian who possesses training, experience, and skill in handling reproductive failures. Since infertility constitutes one of the major problems with which veterinarians must deal, they will wish to be well informed.

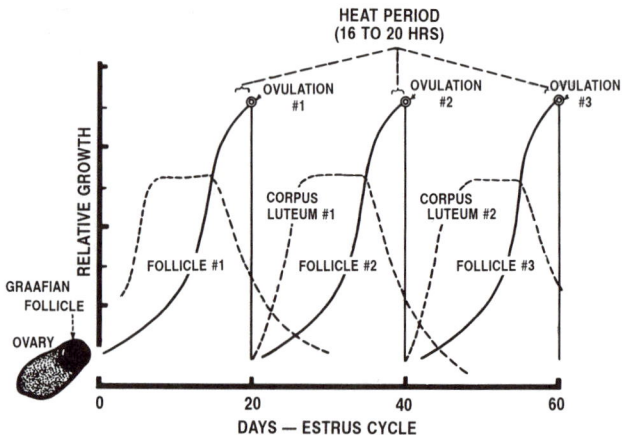

Fig. 6–2. The first requisite to avoiding a high incidence of barrenness and delayed breeding in a dairy herd is an understanding of the cyclic nature of estrus. This figure is a diagrammatic illustration of recurrence of the estrous cycle and the chain of events which take place in the ovary of the nonpregnant cow. Continuous line shows the relative growth of the Graafian follicle. When maximum growth of the follicle is attained, rupture takes place and the ova is released (see double circles on top of diagram). At rupture, the size of the follicle reaches zero and the follicle is replaced by the corpus luteum (dotted line) which increases in size, reaches a maximum, then declines before the subsequent follicle starts to develop. Note that the rupture of the follicles coincides with the end of heat symptoms. The interval between two ovulations is 19 to 20 days on an average.

---

[1]These were the average figures obtained in a survey of dairy cattle in New York, as reported in the Northeastern States Regional Bull. 32, Cornell University Ag. Exp. Sta. Bull. 924, p. 5.

[2]*Ibid.*

Dairy producers should also be well informed relative to reproductive failures, because the enlightened producer will (1) encounter less trouble as a result of the application of preventive measures, (2) more readily recognize serious trouble when it is encountered, and (3) be more competent in carrying out the treatment prescribed by the veterinarian.

Table 6–2 lists the most common causes of infertility in cattle and shows the relative importance of each.

TABLE 6–2
COMMON CAUSES OF INFERTILITY IN CATTLE AND THE RELATIVE IMPORTANCE OF EACH[1]

Cause	Percentage of Infertile Cattle Affected By	Percentage of Affected Animals Having Lowered Fertility	Percentage of All Infertility for Which It Accounts
Vibriosis	25	60	15
Purulent metritis . .	10	100	10[2]
Glandular vaginitis	50	10	5
Leptospirosis	15	20	3
Brucellosis	3	50	1.5
Trichomoniasis . . .	1	100	1
Silent heats	20	36	7.2
Nymphomania . . .	5	100	5
Anestrus	10	30	3
Ovulation failure . .	3	30	0.9
Genetic defects . . .			10
Nutritional deficiencies			10
Improper care at calving			10[2]
Lack of observation			2
Breeding too soon after calving . . .			2
Total			75.6[2]

[1]Estimates by Durward Olds, University of Kentucky. The percentage figures in the first column total more than 100 because of the fact that simultaneous infections occur in many cases. It should be noted that bovine virus diarrhea, infectious bovine rhinotracheitis, and epizootic bovine abortion, all of which cause abortion, are not listed in Table 6–2 because they did not become prevalent in the United States until recent years.

[2]The 10% due to purulent metritis is also listed as improper care at calving.

This table accounts for 75.6% of all infertility and reveals that infectious diseases and poor management are the most common causes of infertility in cattle. Additional causes are listed in the discussion that follows, and, of course, it is recognized that the calf crop may be affected by the season of breeding, and other factors.

STERILITY IN THE COW

Usually the failure of a female to have a heat period before 18 months of age, or to come in heat within 3 months after calving, or to conceive after three matings should constitute sufficient basis for assuming that an abnormal condition exists and that the services of a veterinarian should be obtained for diagnosis and possible treatment. Occasionally

such conditions will correct themselves without treatment; in other cases, they subside for a time only to recur later—they become irregular breeders.

Repeat breeders—cows which exhibit regular or irregular heat periods, but fail to conceive—are most perplexing. The condition may be due to failure of fertilization or to early embryonic death.

When a cow fails to come in heat, she should first be checked for pregnancy (see section entitled "Pregnancy Tests," Chapter 5); approximately one cow in twenty thought to be sterile will be found safely in calf.

For convenience, the common causes of sterility and delayed breeding in dairy cows are herein classified as (1) genital infections and diseases, (2) poor management and feeding, (3) physiological and endocrine disturbances, (4) inherited (genetic) abnormalities, (5) anatomical defects and injuries, and (6) miscellaneous and unknown causes. Of course, at the outset it is recognized that no definite demarcation exists between these classifications; that many of the causes of sterility may be, and are by some authorities, listed under other classifications than those given herein; and that there may be interaction between two or more forces. Also, whatever the cause of sterility, there are no "cure-alls"; rather, each individual case requires careful diagnosis and specific treatment for whatever is wrong.

GENITAL INFECTIONS AND DISEASES

Table 6–2 shows that specific genital diseases account for 35.5% of all sterility in cattle. In addition, nonspecific infections of the cervix and/or uterus are common causes of sterility.

NONSPECIFIC INFECTIONS OF THE GENITAL TRACT

Studies reveal that the invasion of the cervix and uterus by a variety of microorganisms normally follows parturition, but that such infections usually clear up, without treatment, within 40 to 50 days after calving. The return of estrual cycles of normal length and duration is, therefore, a reasonably good indication that the reproductive tract is again normal.

Because bacteria are likely to be present in the reproductive tract immediately following calving, and since breeding at this time may interfere with the normal breeding process and extend the period of infertility, it is recommended that cows not be rebred within 60 days after calving. This will give the cow sufficient time to recover. If the placenta was retained or other calving difficulties were encountered, the cow should be allowed to go through at least one normal heat period before rebreeding.

The veterinarian-prescribed treatment for non-specific genital infections will depend upon their nature and the extent to which they have invaded the reproductive tract; antiseptics, antibiotics, and other drugs may be indicated for local or systemic use, and/or sexual rest may be recommended.

SPECIFIC GENITAL DISEASES

Bovine virus diarrhea (BVD, mucosal-disease complex), brucellosis, epizootic bovine abortion (EBA, foothill abortion), infectious bovine rhinotracheitis (IBR), leptospirosis, metritis, trichomoniasis, vaginitis, and vibriosis are the most troublesome specific genital diseases of cattle.

■ **Bovine virus diarrhea (BVD, mucosal-disease complex)** is an infectious disease of cattle caused by a myxovirus, characterized by diarrhea and dehydration. Also, the BVD virus may cause abortions in pregnant cows. Such abortions usually occur 3 to 6 weeks after infection, and most generally in the first 3 months of pregnancy. The disease can be effectively prevented by immunization with a modified live virus or an inactivated virus vaccine. Immunization is commonly performed when cattle are 6 to 10 months of age. The vaccine manufacturer's recommendations should be followed as to use and inoculation schedule. Pregnant animals should not be vaccinated.

■ **Brucellosis** in cattle is a serious genital disease caused primarily by the *Brucella abortus* bacteria, although the suis and melitensis types are also seen in cattle. It is characterized by (1) abortions at any stage of pregnancy, but most commonly between the fifth and eighth months; (2) above normal incidence of retained placenta; and (3) lowered conception rate. In the United States, screening tests are used to locate infected herds. The Brucella Ring Test (BRT), using milk samples collected on each herd at the milk processing plant or creamery, is used in detecting infected dairy herds; and the Market Cattle Test (MCT), based on blood testing at the time of slaughter, is used in detecting nonlactating infected herds. Supplemental tests, including acidified antigen tests such as the card and buffered plate tests, may be employed to pinpoint infected animals in infected herds detected by a screening test. The disease can be prevented by vaccinating with Strain 19. Dairy heifer calves should be vaccinated at 3 to 6 months of age.

■ **Epizootic bovine abortion (EBA, foothill abortion)** is an infectious disease of cattle, epizootic to California, where it is known as *foothill abortion* because of the high incidence in cows which are pastured on foothill terrain. It is caused by *Chlamydia psittaci*. Infected cows usually abort between the third and sixth months of gestation. In epizootic areas, only first-calf heifers and new cattle introduced from areas free of the disease are affected. The abortion rate varies from 25 to 75%. Animals that have aborted appear to be immune and should be retained in a herd. Keeping pregnant heifers and cows out of the endemic areas during the months of pajaroello tick (the vector) activity, or until after the cows are 6 months pregnant, has proved beneficial. Immunizing agents and antibiotics have not proved useful in preventing or treating EBA.

■ **Infectious bovine rhinotracheitis (IBR)** is an acute contagious viral infection characterized by inflammation of the upper respiratory tract. It was first diagnosed in the United States in 1950. IBR is most prevalent where there are large concentrations of cattle under confinement. The virus may invade the placenta and fetus via the maternal bloodstream, causing abortion or stillbirth from 2 to 3 months subsequent to the respiratory infection. The best method of control of IBR is the routine immunization with modified live or killed vaccines of all potential replacement animals when they are 6 months to 1 year old; after calves have lost their maternally conferred immunity at about 6 months of age, and before breeding age.

The intranasal vaccines are recommended. A single vaccination should be sufficient. However, booster inoculations may be given when the animal is not pregnant.

■ **Leptospirosis** is caused by several species of corkscrew-shaped organisms of the spirochete group. Among other symptoms, it is apt to produce a large number of abortions anywhere from the sixth month of pregnancy to term. In newly infected herds, abortions may approach 30%. The disease can be diagnosed by a blood test. For control, dairy producers rely on annual vaccinations.

■ **Metritis** is an infection of the uterus. Lacerations at the time of calving, wounds inflicted by well meaning but inexperienced operators, and/or retention of afterbirth are the principal predisposing factors. *Escherichia coli* are the most common bacteria isolated from the uterus, although other pathogens are isolated less frequently. Metritis usually develops soon after parturition. It is characterized by a foulsmelling discharge from the vulva that may be brownish or blood-stained and finally becomes thick and yellow. An acute infection may develop into the chronic form, producing sterility. Treatment should be left to the veterinarian. Most cases are treated by the introduction of an antibiotic or sulfa into the uterus.

■ **Trichomoniasis** is a genital-tract infection caused by the protozoan organism, *Trichomonas foetus*. The disease is characterized by (1) irregular sexual cycles; (2) early abortions, usually between 60 to 120 days; (3) a whitish vaginal discharge; and (4) resorption of the fetus, while the uterus becomes filled with a thin grayish fluid. When these symptoms are observed in a herd known to be free of brucellosis,

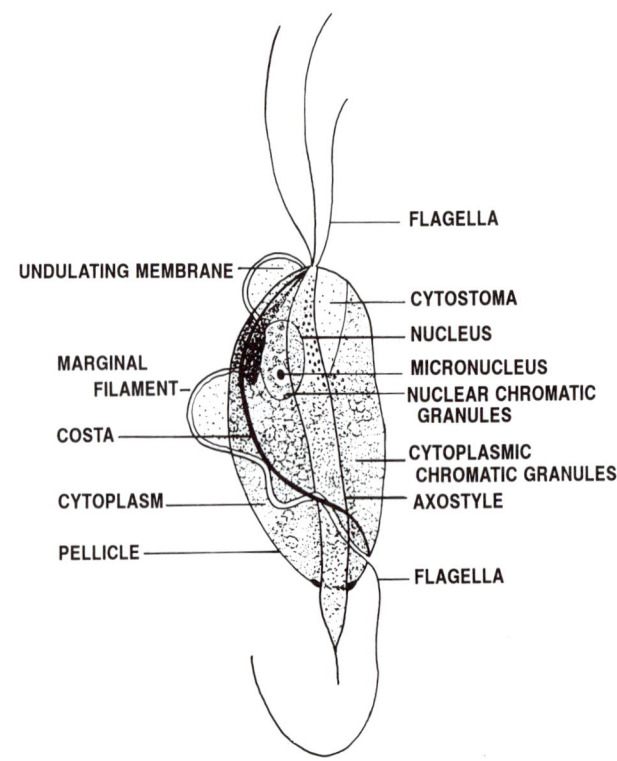

Fig. 6–3. Protozoan *Trichomonas foetus*. (Source: Dr. H. A. Herman, *The Artificial Insemination of Dairy Cattle*, Columbia, MO)

trichomonad infection should be suspected. The diagnosis can be confirmed microscopically by finding the organism. Trichomoniasis may be eliminated from the herd by adoption of a hygienic breeding program. Such a program depends upon using semen from bulls free of trichomoniasis, which can be accomplished by artificial insemination. In cows, the disease appears to be self-limiting; that is, cows appear to acquire an immunity after about 3 months' sexual rest. Slaughter, rather than treatment of bulls, is recommended. If natural service is essential, only young bulls should be used, as they lack susceptibility to trichomonad infection; all bulls over 4 years of age should be eliminated.

■ **Vaginitis** is an infection of the vagina and vulva which causes an inflammation of varying intensity and results in difficult breeders. It may be caused by bruising or laceration of the vagina and vulva at parturition, service from a large and vigorous bull, or prolapse of the vagina. The inflamed vagina is painful, swollen, and often there is an offensive-smelling discharge, indicating infection. The veterinarian may place antibiotics in the uterus and vagina, or he may treat by injecting antibiotics or sulfa drugs. Granular vaginitis is characterized by small spherical nodules on the vulva mucosa of cattle. A similar condition may occur in the lymphatic follicles of the bull's penis. The condition is a response of the lymphatic tissue in the affected area to an irritant or antigen; it is not a disease in the classic sense. Losses in females are in terms of lower percent calf crop and decreased milk production. Bulls may refuse to breed.

Treatment of females is not recommended. It will clear up spontaneously, although it may take several weeks. The condition in the bull is more persistent and should be treated. The prolapsed penis and sheath should be massaged with a suitable antibiotic ointment, repeated sufficiently often to assure elimination of any existing infection.

■ **Vibriosis** is caused by the microorganism *Vibrio fetus*, which is transmitted at the time of breeding. The disease is characterized by (1) several (four or more) services per conception; (2) cows exhibiting irregular heat periods, but finally settling without much difficulty and carrying calves to normal term; and (3) 3 to 5% abortions, usually between the fifth and seventh months of pregnancy. For positive diagnosis, laboratory methods must be used. Infected cows may be treated by injecting drugs into the uterus and/or by allowing sexual rest. Artificial insemination with semen (1) from known non-infected bulls; or (2) which has been antibiotic-treated, is a rapid and practical method of stopping the transmission of infection from cow to cow. The disease tends to be self-limiting in the cow; a cow seems to be free of the disease once she has had a calf following an infection. Vaccination of cows is effective in controlling the disease. Usually vaccination is done about 2 months before breeding and is repeated annually. Artificial insemination is a rapid and practical method of avoiding infection from cow to cow.

(Also see Chapter 15, Dairy Cattle Health, Disease Prevention, and Parasite Control.)

POOR MANAGEMENT AND FEEDING

The term *management* is somewhat elusive and all-inclusive. As used in the discussion that follows, reference is made only to those dairy cattle management practices pertinent to breeding efficiency. It is noteworthy, however, that management is very important in determining breeding efficiency, and that breeders largely determine their own destiny in this regard.

OVULATION AND BREEDING NOT PROPERLY SYNCHRONIZED

If the sperm are introduced into the female reproductive tract much in advance of the egg's release, the chances of fertilization are greatly reduced. It has been shown that the best conception rates are obtained when cows are bred during the final 10 hours of standing heat, or during the first 10 hours after the end of standing heat. To meet these timing relationships, cows that are detected in heat in the morning should be bred during the afternoon of the same day, and those detected in the afternoon should be bred late that evening or early the next morning.

Where artificial insemination or hand mating are practiced, this means that failure to detect accurately when a cow is in heat may account for some breeding failures. Experience and careful observation are the answers to this problem. In addition to recognizing the signs of estrus, where artificial insemination or hand mating are to be followed, the caretaker should keep a record of breeding dates and observed heat periods for each cow.

IMPROPER FEEDING

Improper feeding may imply (1) uncommonly high or low feed intake, or (2) a deficiency of specific nutrients.

Restricted rations often occur during periods of drought, when pastures are overstocked, or when winter rations are skimpy. When such deprivations are extreme, there may be lowered reproductive efficiency on a temporary basis. It is noteworthy, however, that experimental results to date fail to show that the fertility of the germ cells is seriously impaired by uncommonly high or low feed and nutrient intakes.[3] In contrast to these findings, it is equally clear that the onset of heat and ovulation in young heifers is definitely and positively correlated with the level of nutrient intake; thus, if feed or nutrient intake is too low for normal rates of growth and development, the onset of reproductive function is delayed.[4]

Likewise, too liberal feeding and high condition may cause sterility. Sometimes the presence of an excessive amount of fat in the pelvic region leads to a partial protrusion of the vagina and eventual inflammation. Also, where a female remains in an excessively high condition for an extended period of time, degeneration of the ovarian follicles may occur, thereby producing a prolonged state of sterility.

Under natural conditions, a deficiency of vitamin A is the only vitamin likely to be lacking in cattle rations; and they obtain plenty of the precursor (carotene) when they are on green pasture or receive reasonable amount of green hay not over one year old. A severe deficiency of vitamin A may result in a low conception rate, a small calf crop—with many calves

---

[3]*Ibid.*, p. 30.

[4]*Ibid.*

weak or stillborn and with some calves born blind or without eyeballs, but estrus may remain normal.

Under practical conditions, it has been observed that when cows are on a phosphorus-deficient ration, there is a marked inhibition of estrus and a tendency to reproduce every other year. Also, it is believed that heat periods are suppressed when there is a copper deficiency. However, experimental results have failed to show that the fertility of cows is seriously impaired by low trace-mineral content rations.[5]

Although an adequate supply of minerals and vitamins is essential for normal growth and health, adding an excess of these nutrients to a well-balanced dairy cattle ration fed according to recommended practices has no known value in curing breeding troubles.

EXERCISE

Although heat periods are more easily detected when cows are out in the open, exercise is not essential for normal reproduction. It is recognized, however, that exercise is necessary for the normal well-being of the individual.

SEXUAL REST

For maximum reproductive efficiency, cows should not be rebred too soon after calving. Although the reproductive tract usually returns to normal within about 6 weeks, barring infection or other abnormalities, it is not advisable to rebreed a cow earlier than 60 days following parturition. This will still allow time for a second service if needed without exceeding the 12-month calving interval. A New York study revealed that the highest conception rate was obtained when breeding was between 70 and 90 days after calving.[6] In case of retained placenta or other calving difficulties, the cow should be allowed to go through at least one normal heat period before rebreeding.

SEASON AND LIGHT

Seasonal variations in conception rate have been observed in artificial insemination associations; poorest conception is obtained in the winter and maximum in the spring. From this it may be concluded that the amount of daylight (and perhaps the temperature) has an influence on fertility—as the amount of daylight increases, breeding efficiency increases proportionately until temperatures become too hot. In the South, the poorest conception rates occur in the hot summer months.

PHYSIOLOGICAL AND ENDOCRINE DISTURBANCES

The development of the reproductive organs, the production of ova, sexual behavior, the attachment and development of the fetus, parturition, and lactation are primarily regulated by hormones. Many cases of reproductive failure,

particularly of conception and early development, may be due to hormone imbalance.

If neither infection of the genital tract nor any unusual condition is observed, the administration of hormones may be indicated. A wide variety of both natural and synthetic hormones is available, but none should be used except under the direction of a well-informed practitioner. The temptation is always strong to administer a mixture of hormones in the hope that one of them will correct the trouble. A wiser plan is to prescribe the specific hormone which it is believed will produce the desired results; however, this is often difficult or impossible, with the result that "trial and error" methods may be the only alternative.

ANESTRUS (FAILURE TO COME IN HEAT)

Anestrus is the prolonged period of sexual quiescence between the mating seasons of animals. The term is used to describe a cow which is not showing any external signs of heat, which may or may not be associated with ovarian inactivity. This condition is normal in cows immediately following calving. It may also occur in cows in the late winter and early spring when nutritional levels are low. Estrus can be readily induced in such animals by the administration of one of several natural or synthetic estrogens. Although it is unlikely that ovulation will accompany the induced heat, normal cycles are often re-established thereby and conception may occur at the next heat period.

In some cases of anestrus, it may be preferable to administer follicle-stimulating hormones. Frequently, these will induce both estrus and ovulation, but there is a likelihood of undesirable multiple ovulations from treatment.

DISTURBED ESTROUS CYCLES

Numerous variations of the normal estrous cycle occur, all of which are explainable in terms of improper gonadotropin-estrogen-progesterone relationships. Among such conditions are the following:

1. Ovulation without estrus; silent heats; or estrus of such low intensity that recognition is difficult. It has been estimated that 20% of ovulations in cows are not accompanied by external signs of heat.

2. Animals showing estrous cycles, but with a delay in ovulation.

3. Long or short heat periods.

4. Abnormal intervals between heat periods.

5. Estrus without ovulation, known as an anovulatory cycle.

If it is definitely determined that cows with disturbed estrous cycles are not pregnant, and if infections have been ruled out, the judicious and careful use of hormones may be appropriate. Such treatment may restore the endocrine balance, which, in turn, will condition the reproductive system for estrus, conception, and pregnancy.

---

[5]*Ibid.*

[6]*Ibid.*, p. 5.

SEXUAL INFANTILISM

In this condition, the entire reproductive tract remains small. Ovulation may not occur; if it does, affected heifers may exhibit silent heat periods or be irregular in their sexual cycles. Sometimes sexual development is delayed but reproduction is normal after puberty. Heifers with this condition may become excessively fat and resemble spayed heifers or steers.

Sexual infantilism appears to be due to lack of gonadotropic hormone secretion by the anterior pituitary gland, but, unfortunately, treatment with gonadotropins is not often successful. If malnutrition does not appear to account for the condition, the possibility of a heritable factor should be suspected.

RETAINED CORPUS LUTEUM (RETAINED YELLOW BODY)

After the ovarian follicle ruptures and the egg is released, the cells within the follicular cavity change in character and function, forming a corpus luteum or yellow body in the cavity of the ruptured follicle. If the corpus luteum persists, subsequent heat periods do not usually occur. The corpus luteum produces a hormone (progesterone) which suppresses the pituitary output of the follicle-stimulating hormone (FSH). Thus, future follicular development and ripening is inhibited and the estrogens which would induce heat periods are not produced.

A retained corpus luteum may be suspected if a cow is not seen in heat within 60 days after calving. A few years ago these were removed manually, but this procedure usually results in some bleeding. Injections of FSH have replaced the old method.

CYSTIC OVARIES

Cystic ovaries may result when the ovarian follicle fails to rupture. The follicle persists, increases in size and forms a cyst. It is believed that this condition is due to a derangement of gonadotropic-hormone secretion. An excessive secretion of FSH without adequate luteinizing hormone (LH) to produce ovulation causes continued follicular development and estrogen production.

When the condition is allowed to persist in cows, they frequently become chronic *bullers* or nymphomaniacs. Such individuals may show pronounced anatomic and psychologic changes; the pelvic ligaments relax so that there is a sagging of the loin region and elevation of the tailhead. Affected cows may acquire such male characteristics as thickened forequarters and may bellow and behave like bulls.

Recommended treatment for cystic ovaries consists in the administration of a gonadotropin preparation rich in the LH fraction, made either from anterior pituitary glands or from human pregnancy urine. The use of progesterone for 14 consecutive days will clear up this condition. In some cases, a veterinarian may physically rupture the cyst and the cow may return to normal heat periods.

There is evidence that the tendency toward cystic ovaries is inherited in cattle. Also, there is a higher incidence in the dairy breeds than in the beef breeds.

RETAINED PLACENTA (RETAINED AFTERBIRTH)

Normally, the placenta is expelled within 3 to 6 hours after parturition. If it is retained as long as 12 hours after calving, competent assistance should be rendered.

Retained placenta occurs in about 10% of dairy cattle. It is more common following abnormally short or abnormally long pregnancies, among older cows, and following twinning. Experimentally, it has been found that a high incidence of retained afterbirth occurs when premature calving is induced by the administration of glucocorticoid drugs.

While infections such as brucellosis, vibriosis, and others have been associated with abortion and retained afterbirth, these are by no means the only causes. Nutritionally, deficiencies of carotene or vitamin A have been incriminated. Also, it appears that fewer cases of retained placenta occur (1) when calves stay with their dams and nurse for 12 to 24 hours, and (2) when cows are kept on pasture year-round. Among cows which have previously retained the placenta, 20% are likely to do so again.

Calves born when the placenta is retained are likely to be weak. A retained placenta may cause pathological conditions resulting in uterine tissue destruction. This condition may or may not affect milk production, but it very likely will result in 5 to 10% lower fertility than for normal cows.

When a retained placenta is encountered, appropriate treatment should be administered by the veterinarian, who will likely use either antibiotics or sulfonamides, either by direct infusion into the uterus or by other routes (or both).

It is seldom advisable to attempt removal of retained placenta. If the membranes are dragging on the ground, they should be cut off at the hocks. But never, never tie bricks or other objects to it. In most instances, the membranes will fall out by themselves in 1 to 2 weeks.

It is desirable to have all cows which have had retained afterbirth examined at about 30 days after calving. If pus is present, they may be treated with estrogenic hormones to induce heat and then the uterus can be infused with an antibiotic solution or perhaps with a dilute Lugol's (iodine) solution. Such examination and treatment may save considerable time with regard to the onset of normal cycles and may result in a higher conception rate.

INTERSEXES AND HERMAPHRODITES

Intersexes and hermaphrodites may be defined as follows:

■ **Intersex**—*An individual showing both maleness and femaleness, in which the sex differences are not confined to clearly demarcated parts of the body but blend more or less with one another.*

■ **Hermaphrodite**—*An individual whose genital organs have the characters of both male and female in greater or less degree.*

These conditions, which occur only rarely in cattle, are a result of (1) imbalance in the maternal fetal endocrine system, or (2) genetic factors.

INHERITED (GENETIC) ABNORMALITIES

The development of breeds or families within breeds which differ in prolificacy is ample evidence that fertility may have a genetic basis. Thus, it is common knowledge that some once-popular families have become extinct because of the high incidence of infertility. A classical example of a situation of this type occurred in the Duchess family of Shorthorn cattle, founded by the noted pioneer English Shorthorn breeder, Thomas Bates. Bates, and more especially those later breeders who emulated him, followed preferences in blood lines until, ultimately, they were selecting cattle solely on the basis of fashionable pedigrees, without regard to fertility. Ironically, during their heyday, the scarcity of this strain of cattle contributed to their value. Eventually, but all too late, the owners of Duchess Shorthorns came to a realization that indiscriminate inbreeding and lack of selection had increased sterility to the point that the family name was in disrepute; a high incidence of sterility had actually contributed to their scarcity. As a result, Duchess Shorthorns became virtually extinct a few years later.

Of course, reproductive disorders of a heritable nature should not knowingly be perpetuated.

LETHAL GENES

Lethal genes, and other recessive genes causing sterility, belong in the group of highly heritable characters. Among such hereditary lethals are mummification, cystic ovaries, and gonad hyoplasia or gonadless. Generally these abnormalities are easily recognized by their action, and their mode of transmission can often be analyzed and determined. Although lethals are of interest to the geneticist, they are generally no great problem to the dairy producer because their distribution remains under control. Usually recessive, their gene frequency is kept low by the self-destruction of the double recessive; thus, they are self-selective.

WHITE HEIFER DISEASE

This name is a misnomer, for, although the condition is most commonly found in white heifers of the Shorthorn and Milking Shorthorn breeds, it has been reported in roan and red Shorthorns and in colored animals of other breeds. It appears to be due to faulty development of the Mullerian ducts. Some of the more common characteristics are closed hymen or hymen persisting in varying degrees, distention of one or both uterine horns, and uterine body present in rudimentary form, complete absence of cervix, and anterior vagina.

White heifer disease appears to be caused by two pairs of autosomal recessive genes, one pair affecting the left uterine horn and the other the right uterine horn.

ANATOMICAL DEFECTS AND INJURIES

A long list of anatomical defects and injuries to the genital organs has been reported. Some of these are so severe as to cause sterility; others affect the degree of fertility. A Pennsylvania study of repeat breeders—cows which had failed to conceive after four services—revealed that 13% were anatomically abnormal.[7] A brief account of some of the more general anatomical defects follows.

FREEMARTIN HEIFERS

Sterile heifers that are born twin with a bull are known as freemartins.

The origin of the term *freemartin* is obscure. *Free* may be a contraction of the Anglo-Saxon *faer* (meaning empty, or void), or it may come from the Scottish *ferow* (meaning barren, not carrying a calf); while the Gaelic *mart* or *martin* probably signified a spayed bovine female slaughtered because of infertility.

Freemartins occur in about 85 out of 100 twin births when a calf of each sex is involved. The fetal circulations fuse, and the male hormones get into the circulation of the unborn female where they interfere with the normal development of sex and modify the female embryo in the direction of the male. However, it has become increasingly clear that the hormonal theory alone is inadequate to explain the mode of origin of freemartins, and that cellular mechanisms may also be involved. In approximately 15% of twin births of unlike sexes, fusion of the circulation does not occur, and the animal is normal and fertile.

Since only about 15% of such heifers are fertile, it is usually best to assume that they are sterile and market them, unless (1) an experienced person determined at the time of birth that their circulatory systems were not fused; (2) an examination of the vagina reveals that the animal is normal (in freemartin heifers, the vagina is usually about one-third normal length); or (3) skin-grafting[8] or blood-typing[9] techniques show that they are not freemartins and that they may, therefore, be regarded as reproductively normal.

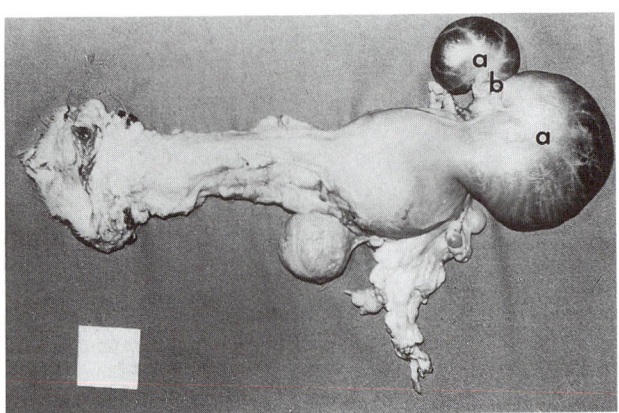

Fig. 6–4. White heifer disease. Note (a) balloonlike distention of uterine horns, and (b) segmental blockage. (Courtesy, Prof. H. W. Leipold, Kansas State University)

[7]*Ibid.*, p. 21.

[8]Billingham, R. E., and G. H. Lampkin, *J. Embryol. Exp. Morph.*, Vol.5, Part 4, Dec. 1957, pp. 351-367.

[9]Stormont, Clyde, *Journal of Animal Science*, Vol. 13, No. 1, Feb. 1954, pp. 94-98.

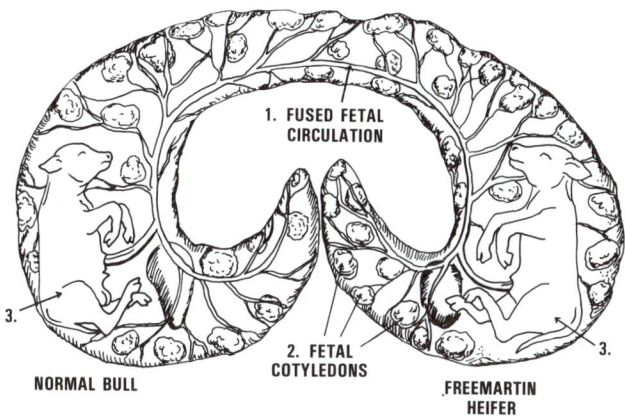

Fig. 6–5. Diagram showing fused fetal circulation of twin calves of opposite sex. Note (1) the fetal circulation of the male fused with that of the female, (2) fetal cotyledon free yolk sac, and (3) normal bull on the left and freemartin heifer on the right.

MECHANICAL INJURIES TO THE GENITAL ORGANS

Mechanical injuries may occur in the female at service and at parturition. A large, vigorous male may inflict injury when breeding; and complicated parturition may result in the loss of the offspring, permanent damage to the reproductive organs, or even death of the cow herself. Perforation of the uterine or vaginal walls, laceration of the cervix, eversion of the vagina, cervix, and uterus or of the rectum may all be sequelae of complicated birth. In some cases, adhesions or secondary infections of the genitalia cause tissue damage which prevents further reproduction.

MISCELLANEOUS AND UNKNOWN CAUSES

Unfortunately, many of the causes of sterility are unrecognized and unknown. A discussion of one of these follows.

EMBRYONIC MORTALITY (FETAL DEATH OR PRENATAL MORTALITY)

It is well known that early embryonic mortality occurs normally in the pig and the rabbit, where there is a surplus production of female gametes. Although it is not so common in the cow, it appears that 20 to 30% of the ova fertilized may meet embryonic death in 2 to 6 weeks. In such cases, the embryo may be absorbed or be expelled unobserved from the female reproductive tract. The cow may assume normal sexual cycles with the conclusion that the fertilization did not take place.

Fetal death, followed by resorption or abortion, may occur at any stage of pregnancy. Such fetuses may range from decomposed masses to bones or dried mummies.

The cause or causes of prenatal death are not known.

STERILITY IN THE BULL

Any bull of breeding age that is purchased should be a guaranteed breeder; in fact, this is usually understood among reputable dairy cattle breeders.

The most reliable and obvious indication of fertility in a bull is a large number of healthy calves from a seasons' service. However, a good evaluation of a bull's fertility may be obtained through a microscopic examination of the semen made by an experienced person. It is recommended that all bulls be semen tested prior to the breeding season; and, where valuable purebred bulls are involved, periodic tests during the breeding season are desirable. Such procedure may alleviate much loss in time, feed, and labor, and avoid delayed and small calf crops.

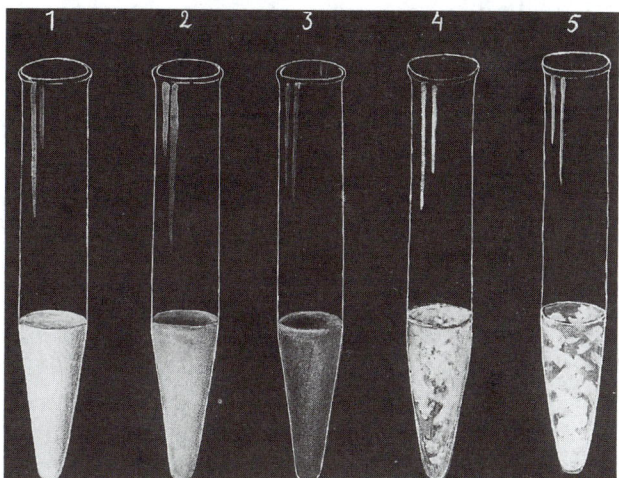

Fig. 6–6. Semen samples of different kinds. 1. Semen with normal appearance, about 1,000,000 spermatozoa per cubic millimeter. 2. & 3. Semen from a bull with hypoplastic testicles. Sample 2 contains about 200,000 spermatozoa per cubic millimeter. Sample 3, which is almost transparent contains about 25,000 spermatozoa per cubic millimeter. 4. & 5. Semen from a bull with inflammation in the seminal vesicles. In the semen, which is almost transparent, there are big, purulent flocci. (Courtesy, Prof. Nils Lagerlof, Department of Obstetrics and Gynecology, Royal Veterinary College, Stockholm, Sweden)

For purposes of convenience, the common causes of sterility and delayed breeding in bulls are herein classified as (1) poor semen, (2) physical defects and injuries, (3) psychological, (4) genital infections and disease, (5) poor management and feeding, (6) physiological and endocrine disturbances, and (7) inherited (genetic) abnormalities.

POOR SEMEN

It is always well to obtain a sample of semen and to make a laboratory examination of the number and condition of the sperm. The four main criteria of semen quality are (1) volume, (2) sperm count, (3) progressive movement, and (4) morphology (shape). Although this technique is not infallible, an experienced person can predict with reasonable accuracy the relative fertility of bulls so examined.

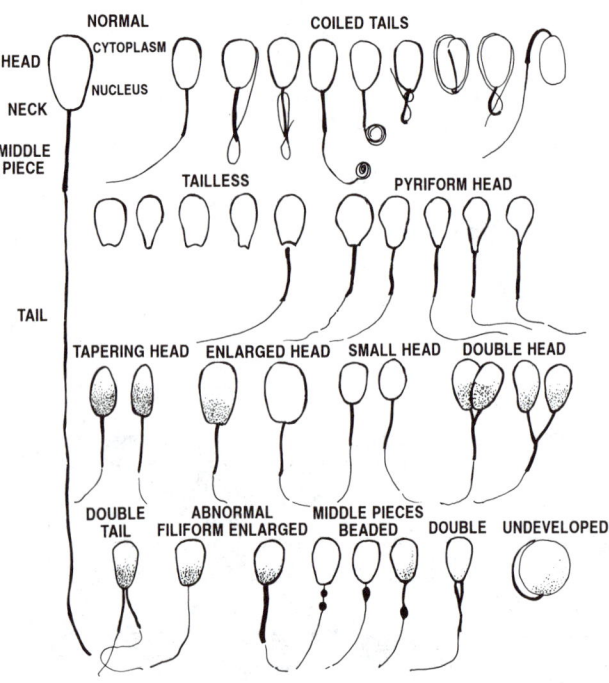

NORMAL AND ABNORMAL TYPES OF BOVINE SPERMATOZOA

Fig. 6–7. Normal and morphologically abnormal spermatozoa of the bull. (Source: Dr. H. A. Herman, *The Artificial Insemination of Dairy Cattle*; courtesy, Dr. Herman)

PHYSICAL DEFECTS AND INJURIES

The most common defects involving the reproductive system are degenerating testes, abscessed testes, fibromas of the penis, broken penis, hematoma, adhesions within the sheath, and paralysis of the retractor muscle. Sometimes certain of these conditions can be corrected by skilled surgery. Infrequent, temporary, or permanent sterility may result from bruising, inflammation, and lacerations of the scrotum and testes. Also, the penis may become bruised or lacerated during service.

Natural service is sometimes interfered with by unsound limbs and feet. This includes such conditions as broken limbs, bad sickle hocks, sore, overgrown or malformed feet, and arthritic or rheumatic joints. The latter condition is more prevalent in old bulls. Defects of the hind limbs are especially troublesome, because they make the bull unstable or cause severe pain when he shifts much of his weight onto them when mounting.

Keratitis (pinkeye) and blindness, especially if in both eyes, will interfere with mating.

An extremely paunchy bull may be unable to mate because the "potbelly" acts as a mechanical obstruction and causes the penis to be directed too low. Usually such inability is temporary and may be corrected by reducing the ration and increasing the exercise.

Improvements can be effected in some of these physical defects and injuries, while others are irreparable.

PSYCHOLOGICAL

Sterile bulls are frequently victims of psychological sterility. Usually, this condition expresses itself in either (1) absence of or lowered sex drive, or (2) faulty reflex behavior during mating or ejaculation.

There is a wide individual difference in the reaction of bulls to their environment; and likewise, these factors are, in part, hereditary. Thus, there are four major types of temperament in bulls: the nervous, the sulky, the placid, and the treacherous. The treacherous types should always be culled at an early age. Then, in selecting from among the other three types, consideration should be given to the herd requirements—whether a commercial or purebred herd is involved, whether pasture or hand mating is to be used, etc. Also, it is generally recognized that bulls lacking in masculinity or secondary sexual characteristics (bulls that are very docile, very fine boned, and lacking development of the crest, etc.) are likely to possess poor psycho-physiological sexual activity.

Young, inexperienced bulls mating for the first time are usually awkward to handle. They approach the cow hesitantly, spend a long time exploring the genitalia, mount hesitantly without erection, descend, and try to mount again. Extreme patience and careful handling should be exercised during this critical first service; otherwise, difficult breeding habits may be established.

CAUSES OF PSYCHOLOGICAL STERILITY

Some common causes of psychological sterility are:

1. **Excitement.** Shouting, noises, distractions during mounting, and the presence of strangers may cause low sex drive. When restrained with a dog, a highly fertile bull may become sexually impotent. Bulls show evidence of such excitement when they urinate more frequently, as they do following the visit of strangers or the introduction of new animals.

Also, it is important to keep sires as quiet as possible in the nonbreeding season. Undue excitement causes sexual impulses and results in the flow of semen into the ampullae.

2. **Transportation.** Psychological sterility is not uncommon in bulls that have been transported long distances by truck or rail.

3. **Animal management.** Young bulls that are isolated from all cows for long periods of time manifest homosexual reflexes and may become impotent.

Also, inadequate sex drive may be due to an attempt to use a young bull on a cow that is too large for him to mount successfully. As a result of his failure to copulate, he develops a mild sense of frustration. Frequently, such a bull can be restored to a high state of breeding efficiency by giving him assistance and trying him on a heifer selected for her small size and willingness to stand quietly; or the same effect may be obtained by standing a large cow in a pit.

Where hand mating is practiced, the bull should not be used immediately following feeding.

4. **Wrong technique during semen collection.** Certain inhibitory reactions may develop from improper use of

the artificial vagina, including wrong timing in applying it, too hot or too cold water, or holding it at a wrong angle.

GENITAL INFECTIONS AND DISEASES

The presence of bacteria in the semen of bulls will not only tend to decrease the viability of the sperm cells but may very easily infect the cows to which they are mated, thereby preventing their conception even when bred to other bulls. If this condition is due to inflammation of the seminal vesicles or prostate, the systematic use of the appropriate antibiotic or sulfa drug may be recommended by the veterinarian. Occasionally, rectal massage of an infected prostate or seminal vesicle will aid the passage of the purulent exudate into the urethra and thus hasten recovery.

Brucellosis, trichomoniasis, vaginitis, and vibriosis are the specific genital diseases of most concern in bulls. Each of these diseases is fully covered in Chapter 15; and, likewise, a pertinent discussion of each disease as it pertains to the cow appears earlier in this chapter. Thus, the ensuing comments will be limited to effects on the bull.

The Brucella organism may localize in the testes, seminal vesicles, or vas deferens, and bulls may spread the disease by copulation. Sex drive is generally reduced if testicular involvement occurs.

In trichomoniasis, the bull is the source of infection. Positive diagnosis of infection in the bull may be made by means of (1) a microscopic examination of smears taken from the prepuce, or (2) mating him with a virgin heifer and checking her cervical and uterine smears for the organism as she approaches the next heat period.

It is believed that granular venereal disease is commonly transmitted by the bull at the time of service, but this is not the only means of transmission since virgin heifers may be infected. Infected bulls should either be treated or sold for slaughter. Treatment consists of massaging the prolapsed penis and sheath with a suitable antibiotic ointment, repeated sufficiently often to assure elimination of infection.

In vibriosis, no clinical lesions are observed in bulls. However, *V. fetus* appears to persist indefinitely in the genital tract of carrier bulls and the bull may transmit the infection. For diagnosis of vibriosis in individual bulls, one should (1) culture the semen, and, where possible, (2) breed to one or more virgin heifers and collect vaginal mucus for culture 10 to 20 days later. Effective control can be obtained by (1) adding antibiotics to semen from infected bulls and breeding artificially, or (2) establishing a new herd of sexually immature animals.

POOR MANAGEMENT AND FEEDING

In all too many cases, little thought is given to the management and feeding of the bull, other than during the breeding season. Instead, the program throughout the entire year should be such as to keep the bull in a vigorous, thrifty condition at all times. Also, lack of fertility in the bull may often be traced back to his early care and feeding. He may have been small and weak at birth, he may have been improperly

fed during the first year of his life, or he may have been prey to infection.

IMPROPER FEEDING

Severe undernutrition (a poor, thin, run-down condition) and vitamin A deficiency are the two most common causes of lowered fertility in the bull, with young bulls more affected than mature bulls. Also, overfat, heavy bulls should be regarded with suspicion, for they may be uncertain breeders.

Vitamin A deficiency inhibits spermatogenesis and adversely affects semen quality and fertility in a manner similar to undernutrition. The common deficiency symptoms are degeneration of the germinal epithelium, reduced diameter of the seminiferous tubules, and testicular atrophy, usually accompanied with cystic changes in the pituitary. Also, there have been responses in sperm production, semen quality, and fertility in cattle to the following trace elements: copper, cobalt, zinc, manganese, and iodine.

Fitting bulls for show and sale results in a variable effect upon their semen-producing ability; in some it has no detectable detrimental effect, while others are severely affected by such practices. Many highly fitted bulls show a complete lack of sperm cells; others produce semen comparable in quality to a bull exhibiting testicular degeneration. Many, perhaps most, such fat bulls eventually reach normal breeding efficiency if they are properly let down—primarily by reducing the grain and by increasing the exercise.

EXERCISE

Most dairy producers feel that regular, daily exercise for the bull is important. Certainly, it is one of the best ways in which to keep the bull in condition. It has also been assumed that forced exercise is important in retaining semen quality. However, studies with dairy bulls have cast considerable doubt on the relationship of exercise to fertility. For example, in one study involving dairy bulls used in artificial insemination, eight bulls which were force exercised were compared with a like number which were kept in box or tie stalls, without forced exercise. The exercised group showed a nonreturn rate of 63.8%, whereas the bulls that were not exercised showed a nonreturn rate of 65%.[10] Hence, the bulls without exercise were actually a little more fertile than the exercised ones.

RETARDED SEXUAL MATURITY

As stated earlier in this chapter, the normal age of puberty in cattle is 12 months. But there is wide variation in the age of sexual maturity of bulls as measured by semen quality. Some yearling bulls are fully equipped to produce a healthy percentage of calves; others are not. At the present time, it is not known whether such retarded sexual development is due to hereditary, nutritional, and/or management factors. There is need, therefore, for additional research on this subject.

---

[10]Snyder, J. W., and N. P. Ralston, "Effect of Forced Exercise on Bull Fertility," *Journal of Dairy Science,* Vol. 38, 1955, pp. 125-130.

THE OVERWORKED BULL

Low fertility of the bull is frequently caused by over-service. In pasture breeding, the bull may copulate four or five times in succession, thereby correspondingly reducing his powers and lessening the size of the herd on which he should be used. However, it is recognized that this situation is compensated for, in part at least, by the fact that bulls on pasture have sexual vigor not possessed by stall-fed bulls.

PHYSIOLOGICAL AND ENDOCRINE DISTURBANCES

The development of the reproductive organs, the production of sperm, and sexual behavior are primarily regulated by hormones; thus, the possibilities of endocrine disturbances are endless. Examples of some of the more common abnormalities follow.

SEXUAL INFANTILISM

In this condition, the entire reproductive tract remains small and the testes are visibly reduced in size. Affected animals lack sex drive. If a low plane of nutrition does not appear to account for the condition, the possibility of a genetic factor should be suspected.

SEX DRIVE (LIBIDO)

Lack of sexual drive (lack of libido) may be due to nutrition, endocrine imbalance, environmental factors, or heredity.

Sex drive, or libido, is important, for if the bull is unwilling to perform and in-heat cows are not bred, there will be no reproduction.

Libido is affected before spermatogenesis. In bulls, both undernutrition and longtime vitamin A deficiency have led to depressed libido. In Europe, the sexual activity of bulls on trace-element deficient soils has been improved by supplementation with iodine, cobalt, copper, and zinc.

Also, overfeeding and obesity in bulls can lead to loss of libido, especially in hot weather.

INHERITED (GENETIC) ABNORMALITIES

Lack of fertility in the bull may often be traced right back to his own sire and dam. He may have been sired by a bull of low vigor and fertility and out of a cow of equally low fertility.

Reproductive disorders of a heritable nature should not knowingly be perpetuated. A discussion of some of the common fertility-affecting hereditary abnormalities in bulls follows.

CRYPTORCHIDISM

When one or both of the testicles of a bull have not descended to the scrotum, the animal is known as a cryptorchid. The undescended testicle(s) is usually sterile because of the high temperature in the abdomen. Since this condition is usually heritable, it is recommended that animals so affected not be retained for breeding purposes.

SCROTAL HERNIA

When a weakness of the inguinal canal allows part of the viscera to pass out into the scrotum, the condition is called scrotal hernia. The abnormality may interfere with the circulation in the testes and result in their atrophy.

TESTICULAR HYPOPLASIA

Hypoplasia of the testicles is an inherited defect in which the potential for development of the spermatogenic epithelium is lacking. It is best known in Swedish Highland cattle, although it occurs in other breeds. It is caused by a recessive autosomal gene with incomplete (about 50%) penetrance.

Testicular hypoplasia is suspected only at puberty, or later, because of reduced fertility or sterility. One or both testicles may be hypoplastic. A hypoplastic testicle is reduced in size, and the semen is watery and contains few or no spermatozoa.

UMBILICAL HERNIA

This condition, which may interfere with breeding efficiency, has been reported as due to (1) a sex-limited dominant gene, or (2) one or more pairs of autosomal recessive factors.

A PROGRAM OF IMPROVED FERTILITY AND BREEDING EFFICIENCY

A program designed to give improved fertility and breeding efficiency in dairy cattle follows:[11]

1. Keep complete breeding records. Maintain complete fertility records on each animal, examine them periodically, and cull low producers. Where hand mating is followed, keep a record of dates bred and observed heat periods of each cow; calculate the expected estrous cycle by adding 21 days to the date of last estrus.

2. At intervals, check the bull for physical defects and quality of semen.

3. Where natural service is used, breed only healthy cows to healthy bulls.

4. Avoid either an overfat or a thin, emaciated condition in all breeding animals; and feed balanced rations.

5. Provide plenty of exercise for bulls and pregnant cows.

6. Do not breed young animals until they are sufficiently mature.

7. Provide an ample rest period between pregnancies; do not rebreed within 60 days after calving. Where the placenta was retained or other calving difficulties were encountered, allow the cow to go through at least one normal heat period before rebreeding.

8. Do not overwork the bull.

9. Observe breeding females carefully during the

---

[11]In this section, special emphasis is placed on increased fertility and breeding efficiency; thus, there is some repetition and there are some additions to the section entitled, "A Program of Dairy Cattle Health, Disease Prevention, and Parasite Control," as given in Chapter 15 of this book.

breeding season; otherwise, heat periods of short duration may be missed. Also, keep a close watch for shy breeders; expose them to the bull often.

10. Diagnose cows for pregnancy.

11. Handle the newborn so that its health shall be assured, and in order that it may have uninterrupted development.

12. Retain as future replacements only those animals which are the progeny of healthy parents, that were carried in utero for a normal gestation period of from 279 to 288 days, and that were born without difficult calving, retained afterbirth, or metritis.

13. Isolate newly acquired animals for a minimum of 3 weeks, during which time they should be tested for brucellosis, leptospirosis, trichomoniasis, and vibriosis. However, first make every reasonable effort to ascertain that they came from herds which are known to be free from these and other diseases.

14. When possible, purchase virgin heifers and bulls. Isolate nonvirgin bulls for a period of 3 weeks, and then turn them with a limited number of virgin heifers; observe these heifers for 30 to 60 days after breeding as an aid in preventing the introduction of breeding diseases.

15. When sterility is encountered, promptly call upon the veterinarian for treatment; do not delay action until the condition is of long standing.

QUESTIONS FOR STUDY AND DISCUSSION

1. Compute the following for your dairy herd, or for a herd with which you are familiar:

a. How much is being lost in annual gross sales of milk and calves when considering the current calf crop percentage vs a 100% calf crop?

b. How much does it cost to maintain all of the barren cows for a year?

2. If a certain dairy producer is experiencing (a) 15% breeding difficulty in the cow herd at a given time, (b) 2.5 services per conception, and (c) a 75% calf crop, outline, step by step, your recommendations for determining the difficulties and improving the situation.

3. What precautions should a dairy producer take to avoid the introduction into the herd of genital infections and diseases?

4. What symptoms characterize each of the following specific genital diseases in cows: bovine virus diarrhea, brucellosis, epizootic bovine abortion, infectious bovine rhinotracheitis, leptospirosis, metritis, trichomoniasis, vaginitis, and vibriosis? What positive diagnosis, if any, can be made of each? What control program should be initiated when the presence of each disease is known?

5. For best results, (a) how many hours should elapse following the known onset of heat before breeding, and (b) how many days should be allowed to elapse following parturition before rebreeding?

6. May sterility be caused by (a) lack of feed as sometimes occurs during droughts, or (b) high condition as when fitted for show?

7. Why should dairy producers call on their veterinarian if hormone injections are given to cattle?

8. On the basis of experimental evidence, should the corpus luteum be removed by an experienced technician by pressure applied through the rectal wall (rectal palpation) if the cow does not come in heat within 60 days after calving?

9. How should a dairy producer handle a case of retained placenta?

10. Define the following: (a) nymphomania, (b) intersex, (c) hermaphrodite, and (d) freemartin.

11. How can you tell whether a heifer is a freemartin?

12. What precautions should be taken when purchasing a bull to avoid sterility and delayed breeding?

13. When bringing in a new bull, what precautions should be taken to avoid psychological sterility?

14. What symptoms characterize each of the following specific genital diseases in the bull: brucellosis, trichomoniasis, vaginitis, and vibriosis; and what positive diagnosis, if any, can be made of each? What control program should be initiated when the presence of each disease is known?

15. Prepare a recommended breeding schedule for bulls of different ages showing the interval between services.

16. Define: (1) cryptorchidism, (b) scrotal hernia, and (c) umbilical hernia.

17. Write out a "program of improved fertility and breeding efficiency" for your herd or for a herd with which you are familiar.

SELECTED REFERENCES

Title of Publication	Author(s)	Publisher
Artificial Insemination and Embryo Transfer of Dairy and Beef Cattle, The, Seventh Edition	H. A. Herman F. W. Madden	Interstate Publishers, Inc., Danville, IL, 1987
Breeding Difficulties in Dairy Cattle, Cornell University Ag. Exp. Sta. Bull. 924	S. A. Asdell	Cornell University, Ithaca, NY, 1957
Cattle Fertility and Sterility	S. A. Asdell	Little, Brown and Co., Boston, MA, 1968
Dairy Cattle Sterility	H. D. Hays L. J. Boyd	*Hoard's Dairyman*, W. E. Hoard and Sons Company, Fort Atkinson, WI, 1964
Factors Affecting Reproductive Efficiency in Dairy Cattle, Kentucky Ag. Exp. Sta. Bull.	D. Olds D. M. Seath	University of Kentucky, Lexington, KY, 1954
Physiology of Reproduction and Artificial Insemination in Cattle, Second Edition	G. W. Salisbury N. L. Van Demark J. R. Lodge	W. H. Freeman & Co., San Francisco, CA, 1978
Prenatal and Post Natal Mortality in Cattle	H. Marsh, *et al.*	National Academy of Sciences, Washington, DC, 1968
Reproduction and Infertility	Centennial Symposium	Agricultural Experiment Station, Michigan State University, Grand Rapids, MI, 1955
Reproduction and Infertility III Symposium	Edited by F. X. Gassner	Pergamon Press, New York, NY, 1958
Reproduction in Dairy Cattle, Ext. Bull. 115	C. H. Boynton	University of New Hampshire, Durham, NH
Reproduction in Domestic Animals	H. H. Cole P. T. Cupps	Academic Press, New York, NY, 1977
Reproduction in Farm Animals, Fourth Edition	Edited by E. S. E. Hafez	Lea & Febiger, Philadelphia, PA, 1980
Reproductive Physiology	A. V. Nalbandov	W. H. Freeman & Co., San Francisco, CA, 1958
Veterinary Endocrinology and Reproduction, Third Edition	Edited by L. E. McDonald	Lea & Febiger, Philadelphia, PA, 1980

Fig. 7–1. Holstein cows on alfalfa pasture. (Courtesy, J. C. Allen and Son, Inc., West Lafayette, IN)

Contents

CHAPTER

7

PASTURE AND GREEN CHOP

Pastures provide a highly nutritious feed that cows relish, especially when fresh growth is maintained. Besides, the cows do the harvesting.

The accent on pastures in the great dairy countries of the world — The Netherlands, Switzerland, and New Zealand, among them — is evidenced by their common practice of measuring dairy production in terms of milk and butterfat produced per acre of pasture, rather than production per cow.

Approximately 50% of the total land area of the United States is devoted to pasture and grazing lands. Much of this area can be utilized only by cattle and sheep.

It is estimated that 58.7% of the nation's milk is produced from forages. Good pasture alone will provide cows with sufficient nutrients for body maintenance and for the production of 20 lb to more than 40 lb of milk daily, depending on the quality of forage.

Fig. 7–2. Good Jersey cows on good pasture. More than half of the U.S. milk supply is produced on forages, including pasture. (Courtesy, The American Jersey Cattle Club, Reynoldsburg, OH)

As larger numbers of lactating cows are concentrated on smaller acreages and as milk production per cow increases, dairy producers depend less on pasture and more on other feeds. A major reason for this is the inability of high-producing cows to consume enough feed to supply their energy requirements when pasture is their main feed source. The physical form and volume of pasture fill the rumen to capacity before nutrient needs of high producers are fulfilled. However, pastures continue to be practical in smaller herds and for heifers and dry cows in both large and small herds.

The general trend among dairy producers, however, is a gradual reduction in use of pasture and increased dependence on stored feeds and green chop as herd size and milk production per cow increase.

To date, in altogether too many cases pastures have been taken for granted. As a result, their improvement has not kept pace with the genetic progress achieved in dairy cows. Thus, it is high time that dairy producers recognize that

the following goals of pasture production are well within the realm of possibility:

1. To produce higher yields of palatable and nutritious forage.

2. To extend the grazing season from as early in the spring to as late in the fall as possible — preferably the year-around.

3. To provide a fairly uniform supply of feed throughout the entire season.

At the outset, it should be recognized that no one plant embodies all the desirable characteristics necessary to meet the above goals. None of them will grow the year-around, or during extremely cold or dry weather. Each of them has a period of peak growth which must be conserved for periods of little growth. Consequently, the progressive producer will find it desirable (1) to grow more than one species, and (2) to plan pastures for each season of the year. In general, a combination of permanent, rotation, and temporary pastures — accompanied by scientific management — will best achieve these ends.

ADVANTAGES OF PASTURES

The following **advantages** may be cited in favor of pastures over confinement production:

1. **They lessen feed costs.** Pasture makes for a saving in feed costs because, when milk cows are on pasture, (a) less grain and protein supplement are required, and (b) grass is cheaper than harvested and stored hay or silage.

Cows will consume from 100 to 200 lb (set basis) of pasture per head daily. Since pasture normally contains 70 to 85% moisture, that is 15 to 60 lb of dry matter per day.

On the average, it is estimated that good pastures result in saving one-half the grain and protein supplement, and most of the hay (even when lactating cows are on good pasture, from 1 to 5 lb of hay per head per day is recommended).

Despite the high regard that most producers have for high-quality pasture, it is doubtful if grass can substitute economically for all of the grain in a milk cow's ration. Without any grain, good cows on excellent pasture can be expected to produce 70 to 80% as much milk as when grain is fed. Medium-quality pasture, when fed alone, will rarely allow cows to exceed 60% as much production as would be obtained with a suitable grain mixture added.

2. **They lessen the hazard of nutritional deficiencies.** Well-managed pastures provide good sources of high-quality protein, certain vitamins and minerals, and unknown factors.

3. **They lessen communicable diseases.** Animals on pasture come in contact with each other less than animals in confinement, with the result that fewer communicable disease problems are encountered.

4. **They lessen costs for buildings and equipment.** Lower cost buildings and equipment can be used in a pasture system than in confinement, with the result that it requires less capital investment on a per cow basis.

5. **They make for greater flexibility.** Pasture operations are more flexible than confinement programs—an important consideration where renters are involved or where other uncertainties exist with a long-range program.

6. **They do not require as high levels of skill and management.** Pasture production does not require as high levels of skill and management as are necessary to make confinement production work. Also, it requires more competent labor to operate a fully automated, highly mechanized, confinement complex than a pasture operation.

7. **They make for approved soil conservation practices on rolling land.** Where there is rolling land and/or need for organic matter, a pasture system may be preferred to confinement. Certainly, pastures conserve the maximum fertility value of the manure and lessen erosion. When animals are on pasture, 80% of the plant nutrients may be returned to the soil.

8. **They favor limited feeding.** Pastures permit limited feeding. This may be important when feed is scarce and high in price.

9. **They improve reproduction.** Pastures provide a desirable way of life for breeding animals, chiefly because of improved nutrition and valuable exercise. Thus, they result in more satisfactory calves being dropped and in a more abundant milk flow. Also, bulls on pasture are more vigorous and more certain breeders.

10. **They maximize noncrop areas.** Pastures permit the maximum utilization of areas not suited to crop production.

DISADVANTAGES OF PASTURES

A number of **disadvantages** of pasturing dairy cattle have caused, and will continue to cause, a shift away from their use to confinement, or dry lot, production. Among the disadvantages sometimes attributed to pastures are:

1. **They mitigate against enlarging dairy production without enlarging the farm.** With high-priced land, this fact must be weighed when it is desired to increase the size of the dairy operation.

2. **They may require more labor.** Running dairy cattle on pasture does not lend itself to automation and labor-saving to the extent of confinement production, primarily in feeding and watering.

3. **They may prevent more remunerative uses of land.** On many dairy farms, operators can make more money from growing corn, soybeans, and other crops than they can from pastures.

4. **They do not facilitate manure handling.** Although less manure has to be handled in a pasture system than in confinement, it is more difficult to automate and handle manure where animals are scattered over a large area. Also, dropping manure on permanent pasture year after year may result in neglect of the other fields.

5. **They waste through trampling.** Pasturing results in considerable waste of the crop through trampling, with lower yield of nutrients per acre than from harvested crops.

6. **They result in cows expending energy in grazing.** The activity of cows on pasture increases the energy required for maintenance in proportion to the amount of energy expended in grazing; up to 40% increase in the maintenance requirement may occur.

7. **They may not provide sufficient nutrients.** Lush pasture may be so high in water content that it supplies inadequate nutrients even though the cows are full.

8. **They do not maintain uniform growth and quality.** Because of variable weather conditions, it is difficult to maintain uniform growth and quality in pastures, with the result that milk production fluctuates. For example, pastures that are not irrigated often become dry and parched during the summer season.

9. **They make for shade and water problems.** In some pastures, availability of shade and water presents a problem.

10. **They make for heat and fly problems.** When cows are on summer pasture, the combination of heat and flies may cause much discomfort to the animals and reduce milk production.

11. **They reflect soil composition.** The nutritive value of pasture is directly related to the composition of the soil. Hence, a soil that is low in certain minerals produces pasture that is low in those minerals. Conversely, toxicities can result in animals grazing on soil with high concentration of certain trace minerals; for example, selenium.

DESIRABLE CHARACTERISTICS OF A GOOD PASTURE

Although it is recognized that no one forage excels all others in all the desired qualities, the following characteristics may serve as criteria in the choice of pasture crops for dairy cattle. Desirable pasture should—

1. Be adapted to local soil and climatic conditions. Although this is a prime requisite, practical producers cannot afford to disregard the grazing qualities.

2. Be palatable and succulent.

3. Be able to withstand tramping and grazing.

4. Be easy to grow, and be grown at a nominal cost.

5. Provide tender and succulent growth for a short period or consistent growth over a long period.

6. Be highly nutritious; rich in proteins, vitamins, and minerals, and low in fiber.

7. Have high carrying capacity.

8. Fit satisfactorily into the crop rotation.

9. Not be contaminated with disease or parasites.

10. Not result in excessive bloat.

Over most of the United States, one or more adapted grass-legume mixtures possess most of these qualities.

CLASSES OF PASTURE

Broadly speaking, all pastures in the United States may be classified as either (1) seeded pastures or (2) native pastures. Although no sharp line of demarcation exists between the two groups in areas of occurrence, seeded pastures are generally established in areas where moisture conditions are favorable. This is usually where rainfall exceeds 20 in. annually or where irrigated. They are the seeded (cultivated) pastures of the Corn Belt, the South, the East, and the irrigated areas, and smaller and scattered moderate to high rainfall areas throughout the West.

Seeded pastures may be further classified as follows:

1. **Permanent pastures.** Those which, with proper care, last for many years. They are most commonly found on land that cannot be profitably used for cultivated crops, mainly because of topography, moisture, or fertility. Also, permanent pastures are generally found in small pastures or where a high concentration of cattle is kept during most of the pasture season. The vast majority of the farms of the United States have one or more permanent pastures, and most range areas come under this classification.

Fig. 7–3. Permanent pasture. Vast areas, such as this rough terrain, are not suited to cultivation. Hence, their only use is for pasture or forest. (Photo by J. C. Allen and Son., Inc., West Lafayette, IN)

2. **Semipermanent or rotation pastures.** Those that are used as a part of the established crop rotation. They are generally used for 2 to 7 years before plowing.

3. **Temporary or supplemental pastures.** Those that are used for a short period, usually annuals such as Sudangrass, millet, rye, wheat, oats, or soybeans. They are seeded for the purpose of providing supplemental grazing during the season when the regular permanent or rotation pastures are relatively unproductive.

Broadly speaking, all U.S. pastures may be further classified as either (1) subhumid (20 to 30 in. rainfall), humid (30 to 40 in. rainfall), and irrigated; or (2) arid (0 to 10 in. rainfall)

and semiarid (10 to 20 in. rainfall) range pastures. Although no sharp line of demarcation exists between the two groups, the former includes those pastures which either receive above approximately 20 in. of rainfall annually or which are irrigated, whereas the latter includes those range pastures which receive less than 20 in. of rainfall annually. The general principles and the objectives sought are the same, but there are considerable differences in the recommended seeding and management practices for the two groups. Most dairy pastures fall in the first group—they are subhumid, humid, or irrigated.

IRRIGATED PASTURE

Well-managed irrigated pastures can enhance the flexibility and add to the stability of dairy forage programs. Throughout the western United States, irrigated pastures have been used successfully to improve carrying capacities, reduce feed costs, lengthen the grazing season, improve milk production, and improve breeding efficiency.

Irrigated pastures provide forage of high quality at a relatively low cost, often on land unsuitable for other crops. Both perennial and annual irrigated pastures are important feed crops.

Successful pasture irrigation involves special decision making relative to (1) irrigation—the method, frequency, and amount of irrigating, and the removal of excess water; and (2) the kind and amount of fertilizer.

■ **Method of water application**—Two basic methods of irrigating pastures are practiced: (1) flood, and (2) sprinkler. The choice of the method for any given pasture should be determined by soil type, topography, water supply, and funds available for irrigation development.

The efficiency of flood irrigation can be improved with the use of borders. The border-flood method is adapted where a large head of water is available and the land is level or requires only minor movement. When properly used, there is no runoff and efficiency of water utilization is high.

Sprinkler irrigation may be preferable to the flood method (1) on land that is not level enough for surface irrigation or where the cost of leveling would be prohibitive; (2) on soils of variable texture where the amount and frequency of application can be adjusted to the water-holding capacity of the soil; or (3) where water cost is high or the supply of water is limited. Sprinklers are on the increase throughout the United States, with laborsaving, center-pivot and wheel move systems making hand-move systems obsolete.

■ **Frequency and amount of irrigation**—It is recommended that in many parts of the United States, especially the west, irrigation water must be applied when it is available, not necessarily when it is desired, and often it may not be available for part of the season. Such restriction may severely limit pasture production. However, when possible and practical, water should be applied: (1) at a rate of frequency to maintain good soil moisture throughout the root profile; (2) immediately following grazing where grazing is rotated; (3) according to the consumptive use rate of the major species; and (4) relative to the content of soluble salt, as soils high in salt may be flushed while irrigation water high in salts should be used sparingly.

Many pasture plants, especially the clovers, are shallow-rooted and require more frequent and lighter irrigations than deep-rooted plants. The Washington Station scientists report that the highest yields per acre from an orchardgrass-ladino clover pasture can be obtained with a summer irrigation frequency of 7 to 11 days, rather than at less frequent intervals, and that more frequent irrigations also give the highest proportion of clover. In important irrigated areas, county extension agents and district conservationists are usually knowledgeable relative to the proper time to irrigate specific crops in the particular area; hence, they should be consulted when developing a schedule.

■ **Excess water** — Excess water in irrigated pastures is caused by either (1) overirrigation and the inability of the excess water to drain from the soil, or (2) subsurface drainage from adjacent and higher land. Allowing excess water to stand on pasture can drown desirable plants, with the resulting area growing up in weeds. Also, standing water is a breeding ground for insects. Surface drains are necessary to remove excess irrigation. Drainage is particularly important wherever there is danger of salt accumulation. Deep drainage ditches spaced at proper intervals help to remove these excess salts and control the level of the water table. In some areas, drainage ditches must be augmented by tile drainage in order to keep the salt content below levels that are harmful to plants.

■ **Plant species** — Selection of species for establishment of irrigated pastures should be dependent on (1) adaptation to the general area, (2) water availability, (3) soils, (4) salinity problems, and (5) forage needs.

■ **Fertilization** — Irrigated pastures require high soil fertility to be productive. The kind and amount of fertilizer should be determined by the level of the productivity desired, and the role of the legumes in the mixture. Production levels of irrigated pastures are increased more by N fertilizer than by other fertilizer elements, with responses also obtained from P and K where soils are deficient in these elements. Nitrogen stimulates grass growth, whereas P increases the legume component. Nitrogen can be supplied by either fertilizer or inoculated legumes. When legumes are a major component of pasture, economic returns from applied N, measured in increased animal production, may not be obtained.

■ **Grazing** — Although continuous grazing has been used effectively in some locations, the potential benefits from irrigated pastures in the West are of such magnitude that some form of rotation grazing should be employed.

FACTORS AFFECTING VALUE OF PASTURE

Many factors affect the value of pasture, including (1) soil and fertilizer, (2) plant species, (3) stage of maturity, (4) rate of growth and season of years, and (5) grazing.

■ **Soil and fertilizer** — Soil and fertilizer affect the growth and composition of pasture crops. Many experiments have been conducted to determine the effect of soil fertility and fertilizer application on pasture. A brief summary of the benefits that generally accrue from pasture fertilization follows:

1. **Increased yields.** The chief benefit to accrue from applying fertilizer to pasture is an increase in yield.

2. **Increased proportion of legumes.** In grass-legume pastures, proper fertilizing can influence the proportion of legumes; in turn, this increases the protein, calcium, phosphorus, and vitamin content of the mixture. Generally legume-grass pastures with about 50% legumes do not require N fertilization. However, on such pastures it is important to maintain adequate levels of lime, P, and K.

3. **Extended grazing season.** Properly fertilized pasture plants grow over a longer period of time than those in infertile soils; they begin growth earlier in the spring, and they continue growth later in the fall.

4. **Increase protein and palatability.** The protein content of young, immature nonlegume pasture is increased appreciably by nitrogenous fertilization, unless of course, there is already plenty of nitrogen in the soil. This increase may be sufficient to add materially to the palatability and the feeding value of grass pasture.

5. **Increased calcium and phosphorus.** Calcium-deficient soils affect pastures in two ways: (a) the percentage of calcium in nonlegume crops is considerably reduced, and (b) the legume crop, if present, will not thrive.

On phosphorus-deficient soils, the phosphorus content of grasses may drop so low that phosphorus deficiency in animals is produced, unless a phosphorus supplement is provided. According to research workers at the Texas Station, pasture forage having less than 0.15% phosphorus on a dry matter basis will not supply enough of the element to meet safely the requirements of beef cattle grazed thereon without supplemental feed. The phosphorus content of legumes is less affected by a deficiency of soil phosphorus than that of nonlegumes. However, most legumes do not thrive or yield well on phosphorus-deficient soils.

■ **Plant species** — Plant species affect the feeding value of pasture. Generally speaking, legumes contain a higher percentage of protein and calcium than nonlegumes. Also, there are marked differences between different kinds of pasture plants as growth advances. For example, bromegrass retains its palatability and nutritive value over a longer period than most grasses,. by contrast, reed canarygrass is readily eaten when young, but becomes woody, high in alkaloids, and unpalatable with maturity. Most pasture legumes retain their palatability and nutritive value as they mature better than most grasses. An exception to the latter rule is *lespedeza sericea*, which becomes bitter and distasteful with maturity due to the accumulation of tannin in the plants. However, plant breeders have developed *sericea* that is low in tannin, thereby overcoming this problem to some degree.

■ **Stage of maturity** — There are great differences in nutritive value between young, immature pasture and the same plants when they are mature or even at the usual hay stage. An example of these wide differences is shown in Fig. 7–4.

MATURITY CHANGES OF ORCHARDGRASS/ LADINO CLOVER PASTURE

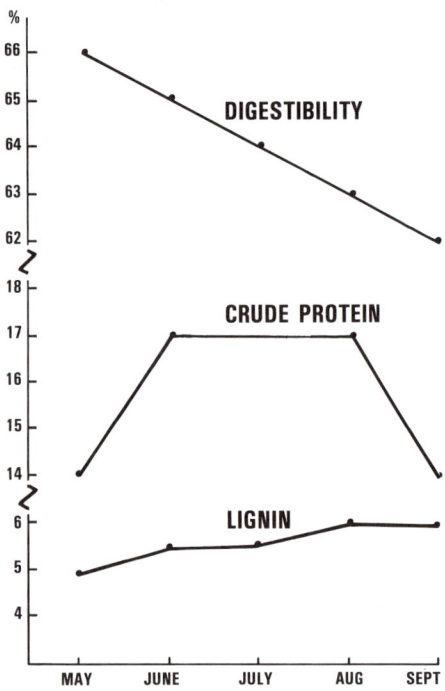

Fig. 7–4. Changes in digestibility, crude protein, and lignin content of orchard-grass (*Dactylis glomerata*)/ladino clover (*Trifolium repens*) pasture in Tennessee, during a 5-month period, from May to September. **Note well:** Digestibility decreased with maturity; it decreased steadily from 66% in May to 62% in September. Protein decreased with maturity; it dropped from 17% in June to 14% in September. Lignin increased with maturity; it increased from 4.8% in May to 5.8% in September. (Source of data: Lane, D. D., K. M. Barth, and J. B. McLaren, *Journal of Animal Science*, Vol. 34, No. 2, 1972, p. 351)

Stage of maturity affects pasture composition as follows:

1. **Protein decreases with maturity.** Young plants are much richer in protein, on a dry basis, than the same plants at maturity. Young actively growing grasses may run 16 to 20% protein content, or better; and young legumes, such as alfalfa and clover, are even higher — 20 to 25% protein or more. The protein content of grasses decreases greatly as they head out. Very mature dormant grasses may contain as little as 2% protein.

The protein content of legumes decreases with maturity, but to a lesser extent than in grasses.

The above underscores the need for producers to take advantage of the high-protein content of young pastures when formulating rations so as to avoid overfeeding costly protein supplements.

2. **Fiber and lignin increase with maturity.** The percentage of fiber and lignin in plants influences both palatability and digestibility. Although there are inherent differences among plant species as well as among plants of the same

species, young plants and regrowth are always lower in fiber and lignin than mature plants. As a result, they are more tender and digestible. On a dry matter basis, new growth in a grass legume pasture will range up to 68% total digestible nutrients in comparison with 51% at the normal hay stage. In humid areas, the nutrient levels of some grasses is leached rapidly following maturity and the digestibility and nutrient value is comparable to poor grade straw.

To a considerable degree, the decrease in digestibility as plants mature is due to an increase in lignification, which lowers digestibility, even by ruminants.

3. **Calcium and phosphorus decrease with maturity.** The calcium content of plants decreases with maturity. However, the percentage change in calcium is much less than occurs in phosphorus.

Young grasses or legumes grown on phosphorus-rich soils usually contain 0.25%, or more, phosphorus. Although phosphorus in pasture decreases with maturity, there is generally plenty left for dairy cattle, unless (a) it is produced on phosphorus-deficient soils, or (b) it is left to cure on the stalk, which makes for weathering and bleaching. Early spring grass (April) may contain five times more phosphorus than leached and weathered grass in the winter (March).

4. **Vitamin value decreases with maturity.** Actively growing green parts of plants are high in carotene. Likewise, such plants are usually rich in most of the B complex vitamins, in vitamin E, in ascorbic acid, and in certain unidentified factors. But the content of vitamins, especially of carotene, decreases as plants mature. When pasture plants are left to cure on the stalk, and to leach and weather, carotene disappears rapidly.

■ **Rate of growth and season of year** — Rapidly growing grass is usually rich in protein and in other nutrients. It is important, therefore, that pasture plants be properly fertilized and, to the extent possible, managed so that they keep growing and that they be prevented from heading out. This can be accomplished by mowing, but it is more practical when done with animals through intensive grazing programs.

Grass is usually higher in protein and other nutrients early in the spring than later in the season. If plant growth is sharply checked in the summer — due to drought, hot weather, and/or lack of available plant food — the protein content and the digestibility will be lower than that of grass at the same stage of maturity earlier in the season.

If pasture resumes growth after the fall rains come, it may be nearly as high in protein and other nutrients as spring growth.

It is also noteworthy that all plants are most digestible and most preferred during periods of rapid growth. Recognizing this, producers are establishing warm-season grass pastures to obtain higher milk production during the normal midsummer growth slum in areas where cool-season grasses predominate.

■ **Grazing** — When pastures are grazed closely throughout the season, the total yield of dry matter is usually 30 to 50% less than when they are allowed to grow to the normal hay stage. This is due to the smaller leaf surface and lowered photosynthesis. This explains why rotational, strip, and green chop grazing yield more than close continuous grazing.

The total effect of frequent grazing will depend on the kind of plants. The yield of tall-growing plants — such as timothy, orchardgrass, alfalfa, and the erect clovers — can be reduced much more than that of low-growing spreading plants, such as bluegrass, Bermudagrass, and white clover.

In contrast to the lowering of the yield of dry matter, frequent grazing usually results in greater total production of protein for the season than when the crop is cut for hay. Also, because immature plants are lower in fiber and more digestible than mature plants, the yield of total digestible nutrients is not reduced as much by frequent grazing as the dry matter yield — dry matter production is lowered by 30 to 50%, whereas digestibility is lowered only by 25 to 40%.

Also, it is noteworthy that when allowed to graze selectively in an extensive grazing program animals pick and choose the leaves and finer parts of stems, which are more tender and more nutritious, and reject the courser, stemmy parts. Thus, the portion consumed under such circumstances will always differ appreciably from the chemical composition of the entire plant.

ADAPTED AND/OR COMMON GRASSES AND LEGUMES OF THE UNITED STATES

The specific grass or grass-legume mixture will vary from area to area, according to differences in soil, temperature, and rainfall. Fig. 7–6 shows the ten generally recognized U.S. pasture areas; and Chart 7–1 (next page) shows the best adapted and/or most common grasses and legumes for each of these areas for dairy cattle.

Fig. 7–5.   Guernsey cows on legume (clover) pasture. (Courtesy, American Guernsey Association, Reynoldsburg, OH)

LEGUMES AND GRASSES ADAPTED TO 10 AREAS OF THE 48 CONTIGUOUS STATES

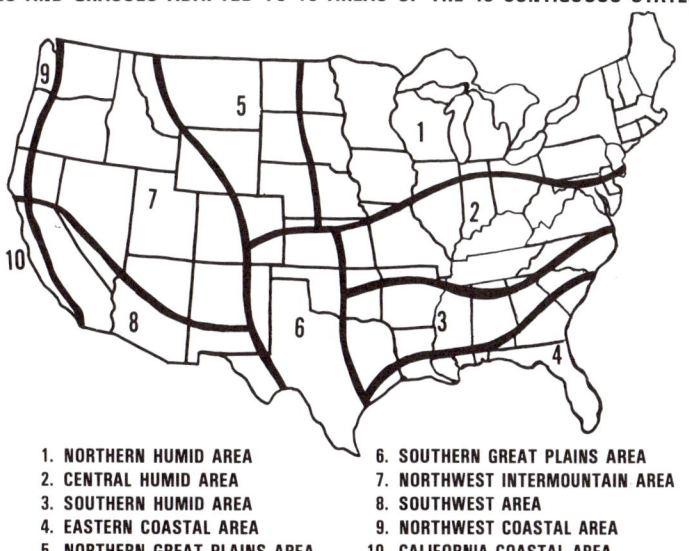

1. NORTHERN HUMID AREA
2. CENTRAL HUMID AREA
3. SOUTHERN HUMID AREA
4. EASTERN COASTAL AREA
5. NORTHERN GREAT PLAINS AREA
6. SOUTHERN GREAT PLAINS AREA
7. NORTHWEST INTERMOUNTAIN AREA
8. SOUTHWEST AREA
9. NORTHWEST COASTAL AREA
10. CALIFORNIA COASTAL AREA

Fig. 7–6.   The ten generally recognized U.S. pasture areas.

CHART 7–1
ADAPTED GRASSES AND LEGUMES (INCLUDING BROWSE AND FORBS) FOR CATTLE PASTURES, BY 10 GEOGRAPHICAL AREAS OF THE UNITED STATES (SEE FIG. 7–6 FOR GEOGRAPHICAL AREAS)

Grasses, shrubs, forbs:	1	2	3	4	5	6	7	8	9	10
Bahiagrass			x	x						
Bermudagrass		x	x	x		x		x		x
Bluegrass, big							x		x	
Bluegrass, Kentucky	x	x	x			x		x		
Bluestem, big	x	x	x	x	x	x				
Bluestem, Caucasian		x	x			x				
Bluestem, little	x	x	x	x	x	x				
Bluestem, sand	x	x			x	x				
Bristlegrass, plains						x		x		
Bromegrass, meadow						x		x	x	x
Bromegrass, smooth	x	x				x		x	x	x
Buckwheat (wild)								x		
Buffalograss					x	x				
Buffelgrass						x				
Canarygrass, reed	x	x					x		x	
Cottontop, Arizona								x		
Curly mesquite						x		x		
Dallisgrass			x	x						
Digitgrass, pangola			x	x						
Dropseed, sand						x		x		
Fescue, tall	x	x	x				x	x	x	x
Foxtail, creeping							x		x	
Galleta						x		x		
Gamagrass, eastern	x	x	x	x	x	x	x			
Grama, black								x		
Grama, blue					x	x	x	x		
Grama, sideoats	x	x	x	x	x	x	x	x	x	
Hardinggrass						x			x	x
Indiangrass	x	x				x	x			
Indianwheat								x		
Johnsongrass			x	x		x				
Kleingrass						x				
Koleagrass, Perla									x	x
Limpograss				x						
Lovegrass, Lehmann							x	x		
Lovegrass, sand	x	x			x	x				
Lovegrass, weeping			x			x				
Maidencane				x						
Millet	x	x	x	x		x				
Muhly, spike							x	x		
Needle-and-thread	x				x					
Needlegrass, green	x				x					
Oatgrass, tall									x	
Oats	x	x	x	x	x	x		x	x	x
Orchardgrass	x	x	x	x	x		x	x	x	x
Paragrass				x						
Pearlmillet		x	x	x		x				
Redtop	x						x		x	
Rescuegrass			x	x					x	
Rhodesgrass			x	x						
Ricegrass, Indian							x	x		
Rye	x	x	x	x	x	x		x	x	x
Ryegrass, annual		x	x	x		x				x
Ryegrass, perennial	x	x	x							x
Sacaton, alkali						x	x	x		

	1	2	3	4	5	6	7	8	9	10
Sage, pitchers	x	x	x		x	x				
Saltbrush, fourwing					x	x	x	x		
Sorghum-Sudan hybrids	x	x	x	x	x	x	x			
Stargrass			x							
Sudangrass	x	x	x	x	x	x	x	x	x	x
Sunflower, Maximilian	x	x	x		x	x				
Switchgrass	x	x	x		x	x				
Three-awn				x	x	x	x	x		
Timothy	x	x					x		x	
Tobosa grass						x				
Wheat	x	x	x	x	x	x	x	x	x	x
Wheatgrass, bluebunch						x		x		x
Wheatgrass, crested						x		x		
Wheatgrass, intermediate	x					x		x	x	x
Wheatgrass, pubescent						x	x	x		x
Wheatgrass, tall	x					x	x	x		x
Wheatgrass, western	x	x				x	x	x		x
Wild-rye, basin						x		x		
Wild-rye, Canada	x	x	x			x	x	x		
Wild-rye, Russian						x		x		
Winterfat (white sage)								x		
Wintergrass, Texas						x				
Legumes:										
Alfalfa (lucerne)	x	x	x	x	x	x	x	x	x	x
Alyceclover			x	x						
Black medic (yellow trefoil)			x			x		x		
Bur-clover		x						x		x
Clover, alsike	x	x			x		x	x	x	
Clover, arrowleaf		x	x							
Clover, crimson		x	x							
Clover, Hubam (white sweet clover)	x	x						x	x	
Clover, Kura	x	x			x		x			x
Clover, Ladino	x	x	x	x			x	x	x	x
Clover, prairie						x		x		
Clover, red	x	x	x	x			x	x	x	
Clover, strawberry					x		x	x		x
Clover, subterranean			x	x				x	x	x
Clover, white	x	x	x	x			x	x	x	x
Cowpeas			x	x						
Crown vetch	x	x								
Flat pea			x	x			x			
Hairy indigo				x						
Lespedeza (annual)		x	x	x						
Lespedeza (perennial, sericea)		x	x	x						
Milk vetch, cicer	x					x		x		
Peas, field									x	
Pea shrub								x		
Prairie clover, purple	x	x	x		x	x	x			
Ratany								x		
Soybeans	x	x	x	x			x			
Sweet clover, white	x	x							x	x
Sweet clover, yellow	x	x				x	x	x	x	
Trefoil, birdsfoot	x	x	x				x		x	x
Velvet bean			x	x						
Vetch		x	x	x	x	x			x	x

In using Fig. 7–6 and Chart 7–1, bear in mind that many species of forages have wide geographic adaptation, but varieties often have rather specific adaptation. Thus, alfalfa, for example, is represented by many varieties which give them species adaptation to nearly all states. Variety then, within species, makes many forages adapted to widely varying climate and geographic areas. Also, some species have very narrow adaptation ranges, but may be important within that limited area. Some species or varieties have adaptations to specific soil conditions, such as salinity, heavy clay, or high water tables. The county agricultural agent or state agricultural college can furnish recommendations for that area that they serve. For more specific and individual farm recommendations, dairy producers are urged to seek the advice of local authorities or to write to their state agricultural college.

The following points are pertinent to the recommendations given in Chart 7–1:

1. **Fertilizer rates.** Because of the high price of fertilizer, along with concern relative to possible pollution of groundwater, fertilizer rates should be based on soil test values.

After legumes have been lost from a grass-legume sward, it is recommended that they be reestablished in the grass sod. If the latter is not feasible, an annual nitrogen application at the rate of 60 to 100 lb of actual nitrogen per acre per year should be applied in split applications or increments.

Where a center-pivot irrigation system is used, nitrogen is usually applied in increments, three to four times during the growing season.

2. **Varieties.** The best guide for the varietal selection of grasses and legumes from the numerous varieties available is the use of certified seed of an adapted variety.

ESTABLISHING NEW PASTURES

The following practices are usually adhered to when planning forage programs and in successfully establishing a new pasture in the subhumid, humid, and irrigated areas:

1. **The species and cultivars are selected.** The selection of species and cultivars to seed should receive first priority. One species or a specific combination best fits the needs of a certain climate, soil, and intended use. A single species may be adequate for one pasture while several species may be best for another. Chart 7–1 gives the general recommendations of competent specialists residing in different areas of the United States. For more specific recommendations for a particular farm or ranch, the dairy producer should consult such local authorities as the county extension agent, district conservationist, or successful neighbors. Often new varieties have been released that may be proving superior to standard varieties in production, palatability, digestibility, or disease resistance.

2. **The soil is tested and limed/fertilized.** Following testing of the soil, it should be limed and/or fertilized according to the needs. In humid regions where periodic application of lime is needed, it is best to work lime into the soil considerably in advance of seeding. Phosphorus levels are especially critical when establishing a new seeding. When the soil

is deficient in nitrogen, the application of a small amount of nitrogen at seeding time may be helpful.

3. **High-quality seed is purchased.** The seed should be of good quality, of high germination and purity as indicated on the tag, and free of noxious weeds. Also, proof of origin is of prime importance when an imported variety is obtained. Certified seed carries more assurance of being high quality than noncertified seed, and gives proof of its origin much as a registration certificate does for a purebred animal.

Pure Live Seed (PLS) has become the standard in seeding recommendations. It is also advisable to purchase seed on a PLS basis. PLS can be determined by % PLS = % germination × % purity × 100. These items should be found on all seed tags.

4. **Scarified legume seed is used.** In the purchase of certain legume seed, it is important that it be scarified, which breaks the seed coat and allows faster moisture penetration—thus assuring quicker and more uniform germination and a better stand in the first year.

5. **Legume seed is inoculated.** Since most legumes can use nitrogen from the air provided they are inoculated with the proper bacteria, it is important that legume seed be inoculated. This is traditionally done at planting time by coating the seed with a water-based slurry consisting of a commercial preparation of bacteria. The expiration date is given on the container. It is important that the seed not be inoculated more than a few hours before seeding because these nitrogen-fixing bacteria are easily killed by drying, heat, sunlight, chemical seed treatment, or by direct contact with acid fertilizers.

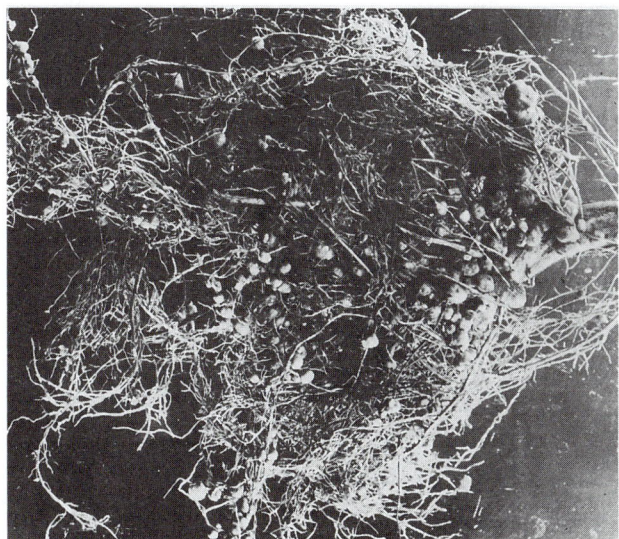

Fig. 7–7. Nodules (small bumps) containing nitrogen-fixing bacteria on a well-inoculated soybean root. (Courtesy, USDA)

On fields or pastures where the legume species has been grown regularly for a number of years, inoculation of seed is not necessary because the soil contains a sufficiently large population of nodule bacteria to fulfill crop requirements.

6. **A good seedbed is prepared.** A good seedbed is free from weeds, fine-textured, firm, and moist.

Weeds are usually destroyed by growing row crops or a small grain the year preceding seeding to pasture and by cultivating frequently following the harvesting of this crop.

There are many different ways in which to prepare a good seedbed. Perhaps as good a method as any consists in (a) plowing as far in advance of seeding as possible, (b) discing, (c) harrowing one or more times to level the field and smooth the surface, and (d) cultipacking or rolling. A properly prepared seedbed should be so firm that one barely leaves a footprint when walking across it; the firmer the better from the standpoint of moisture conservation of small seeds.

7. **The seeding operation is timed and carried out properly.** The seeding time will vary, being determined primarily by the area and by the species or mixture used. Cool-season species should be seeded early enough, in the spring or fall, for seeding development so they can withstand the stress of hot summers and/or avoid winter injury. Conversely, warm-season species may rot before germination if planted before critical soil temperatures are reached in the spring.

The actual seeding operation may be (a) by broadcast, with a whirlwind seeder or by hand; (b) by cultipacker, consisting of two corrugated rollers with seed-metering boxes; or (c) by drilling with one of several types of conventional seeders over a band of fertilizer $\frac{1}{5}$ to $\frac{3}{5}$ in. deep. Drilling is the preferred method, for it ensures more uniform placement of seeds in both depth and amount of seed per acre and results in a more uniform stand. When broadcast methods of seeding are used, rates should be increased 50 to 100% above the amount recommended for drilling.

Seeding rates are dependent on many factors. Normally, only about a third of sown seed produces seedlings, and only half of the seedlings survive the first year. So, high seeding rates appear to be justified. Moreover, high seeding rates reduce weed invasion and increase yields the first year.

Since most grass and legume seeds are very small, they should not be covered deeply. A good rule is that they should not be covered more than 4 to 5 times the diameter of the seed; usually this means not more than $\frac{1}{6}$ to $\frac{1}{2}$ in. deep.

8. **A companion (nurse crop) or preemergence herbicide may be used.** The value of planting a "companion" or nurse crop—usually consisting of annuals or short-lived perennials—with new seed crops varies with the intended crop, soils, and season. The **advantages** are: (a) it furnishes a crop of value while the new seeding is being established, (b) it lessens erosion, and (c) it reduces the weed population. The **disadvantages** are: (a) it may retard the growth of the seedlings for whose protection it is grown and delay establishment, and (b) it may rob the new seedlings of so much moisture and light that it kills them during dry spells unless the companion crop is harvested early as pasture, hay, or silage.

Some perennial grass species are extremely fragile when in the early seedling stage. Establishment is often enhanced by protection and shade afforded either by a nurse crop or the noncompetitive stubble remaining from a previous crop.

Preemergence herbicides may replace the traditional companion crop, especially where a single species is being established. Usually these result in higher forage yields in the seeding year than can be obtained when a companion crop is used.

RENOVATING OLDER PASTURES

Renovation is the rejuvenation of existing pastures. Pastures are renovated in order to regain production potential by (a) reestablishing the desired mix of species; (2) eliminating persistent, undesirable weeds and/or brush; and (3) breaking up soil compaction to improve soil aeration, water infiltration, and root development.

Renovation is accomplished either by partial or complete destruction of the existing sod. The method will vary from one area to another depending on (1) stand and species, (2) soils and topography, and (3) degree or stage of degeneration. Pastures may be renovated by the following methods:

1. **Fertilizing and overseeding.** Often on pastures where a fair, but unproductive, stand of desired plants exists, renovation may simply consist of specially prescribed fertilizer applications, and liming if needed. Occasionally, this may be accompanied with weed and brush control treatments. This condition represents large areas in the south humid and subhumid regions and the occurrence is strangely influenced by the ratio of fertilizer costs and prices received for livestock.

Fig. 7–8. Power-driven seeder for fast, economical pasture renovation. (Courtesy, Deere & Co., Moline, IL)

2. **Complete seedbed preparation with reseeding.** Complete seedbed preparation followed by reseeding has been widely used in accomplishing the objective of renovation. This is especially effective where perennial, weedy grasses have invaded the sod and are otherwise difficult to control. This method facilitates the working of lime and fertilizers into the surface soil and is generally superior to other methods in establishment of legumes such as alfalfa. Occasionally, the use of an intervening crop for one or two years may be desirable in reducing populations of undesirable plants; and a cash crop may help defray renovation costs.

3. **Sod-seeding or overseeding (no-till seeding).** In humid regions, reestablishment of desirable plants, especially legumes, into existing pasture sod is often preferred to

complete seedbed preparation. The basic requirements for successful sod-seeding are: the partial (strip or rows) or complete destruction of existing plants by use of herbicides and close grazing, thereby reducing competition; liming when necessary; fertilizing; proper seed placement; and good pasture management following seeding.

MANAGEMENT OF DAIRY PASTURES

Many good dairy pastures have been carefully established only to be lost in succeeding years through poor management. Efficient and profitable pasture management in the subhumid, humid, and irrigated areas involves the following practices:

1. **Controlled grazing.** Nothing contributes more to good pasture management than controlled (proper) grazing. At its best, it embraces the following:

a. **Protecting first-year seedings.** First-year seedings should be grazed lightly or not at all in order that they may get a good start. Where practical, instead of grazing, it is preferable to mow a new seeding about 3 in. above the ground and to utilize it as hay or silage, provided there is sufficient growth to so justify.

b. **Shifting the location of salt and water.** Where portable salt containers are used, more uniform grazing may be obtained simply by the practice of shifting the location of the salt to less grazed areas of the pasture. Where possible and practical, water locations should be well distributed.

c. **Fencing into smaller units.** Development of more and smaller pastures and employing rotation grazing programs permits greater control of when and where cattle graze. Many more intensive systems are designed with stationary salt and water locations, while the area being grazed may change every 3 to 4 days, or even daily in some cases. Various forms of electric fencing are being used in a number of different pasture arrangements, all designed to facilitate grazing greater numbers of animals on an area for shorter periods of time.

d. **Deferred spring grazing.** When possible allow 6 to 8 in. of growth before turning out to pasture in the spring, thus giving grass a needed start. This is not always practical on year-round pasture programs. However, when two or more pastures are available, early spring grazing can be rotated to the benefit of plants in all pastures.

e. **Avoiding close late-fall grazing.** Pastures that are closely grazed late in the fall start growing late in the spring. Plants should be allowed to replenish root reserves prior to going dormant. With most pastures, 3 to 5 in. of growth should be left for winter cover. An exception to this close grazing rule should be made where winter annual clovers are to be seeded, or are expected to volunteer, especially on Bermudagrass or bahiagrass pastures. Under such circumstances, close grazing or mowing of the Bermudagrass or bahiagrass is recommended.

f. **Avoiding overgrazing.** Never graze more closely than 2 to 3 in. during the pasture season. Continued close grazing reduces the yield, weakens the plants, encourages weeds to invade, and increases run-off and soil erosion. The use of temporary and supplemental pastures, such as Sudan, may "spell off" regular pastures through seasons of drought and other pasture shortages and thus alleviate overgrazing.

g. **Avoid undergrazing.** Undergrazing seeded pastures should be avoided, because (1) rank growth is less palatable and of lower nutritive value; (2) tall-growing grasses may drive out desirable low-growing plants such as white clover due to shading; and (3) weeds, brush, and coarse grasses are more apt to gain a foothold when the pasture is grazed insufficiently. It is a good rule, therefore, to graze the pasture fairly close at least once a year.

2. **Clipping pastures and controlling weeds.** Pastures should be clipped as necessary to control competing weeds (and brush) and to get rid of uneaten clumps and other unpalatable course growth left after incomplete grazing. Good grazing management will reduce the amount of clipping needed. Pastures that are grazed continuously may be clipped at or just preceding the usual haymaking time; rotated pastures may be clipped at the close of the grazing period.

It is noteworthy that mowing can be expensive ($3 to $8 per acre) and should be applied only when results are clearly beneficial. Clipping solely for cosmetic reasons should be critically evaluated.

3. **Topdressing.** Like animals, for best results, grasses and legumes must be fed properly throughout a lifetime. It is not sufficient that they be fertilized (and limed if necessary) only at or prior to seeding time. In addition, in most areas it is desirable and profitable to topdress pastures with fertilizer annually, and, at less frequent intervals, with reinforced manure and lime. Such treatments should be based on soil tests, and are usually applied in the spring or fall.

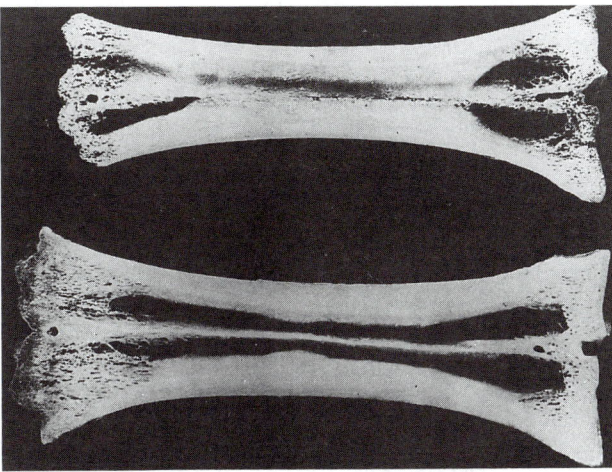

Fig. 7–9. Split bones from two calves of similar breeding and age. Small bone (above) obtained from calf pastured on belly-deep grasses grown on highly weathered soils low in mineral content. Big bone (below) obtained from calf grown on moderately weathered, but highly mineralized, residual limestone soil. (Courtesy, Dr. W. A. Albrecht, University of Missouri)

4. **Scattering droppings.** The droppings should be scattered at the end of each grazing season in order to prevent animals from leaving ungrazed clumps and to distribute the droppings over a larger area. This can best be done by the use of a brush harrow or a chain harrow.

When animals are concentrated under intensive systems such as strip grazing, there is a tendency for droppings to be broken up by hoof action and the need for harrowing is reduced.

5. **Grazing by more than one kind of animal.** Grazing by two or more species of animals makes for more uniform pasture utilization and fewer weeds and parasites, provided the area is not overstocked. Different species of animals have different habits of grazing; they show preference for different plants and they graze to different heights. For example, sheep consume shorter and fine forages and more forbs than cattle.

6. **Irrigating where practical and feasible.** Where irrigation is practical and feasible, it alleviates the necessity of depending on natural precipitation.

7. **Avoiding puddling.** Pastures should be protected during periods of heavy rainfall or when irrigating, by removing cattle until the soil has dried sufficiently to prevent puddling.

GRAZING SYSTEMS FOR SEEDED PASTURES

Several systems of grazing management have been successfully applied to pastures. Generally speaking, the more intensive the system of management on such pasture, the higher the yield of forage and of dairy products.

It is noteworthy that pasture grazing systems have been changed/adapted by both researchers and farmers; and with such changes/adaptations, they have been given different names. Nevertheless, under whatever name, the basic types are continuous grazing, rotation grazing, intensive grazing, strip grazing, and green chop, each of which is covered in a section which follows.

CONTINUOUS GRAZING

The name identifies the practice. *Continuous grazing is the uninterrupted grazing of a specific pasture by animals throughout the year or grazing season.* It can be successful provided moderate stocking is practiced, with some adjustment in animal numbers to reduce the severity of under- or over-grazing.

The **advantages** of continuous grazing as compared to rotational grazing are (1) lower costs for fencing and watering facilities, (2) fewer management decisions when animals are not moved from pasture to pasture, and (3) often slightly better individual animal performance when younger animals, such as heifers, are grazed.

The **limitations** of continuous grazing are (1) animal numbers are seldom flexible; (2) pastures must be stocked lighter than desired when forage growth is maximal to avoid overgrazing during periods of minimal forage growth; (3) animals selectively graze some species in preference to others and return to graze the regrowth of the same plants, thus

selectively reducing plant vigor; and (4) animals often show preference to grazing certain portions of pastures, resulting in uneven fertilization.

ROTATION GRAZING

Rotation grazing is that system in which two or more pastures are grazed and rested in a planned sequence. In this system, pastures are divided into two or more pastures, with the objective to develop a grazing program where major forage species are harvested and then provided a period of rest enabling the plants to regrow and remain thrifty and vigorous.

Rotation grazing involves the concept of time as a management variable for either the grazing period or the regrowth internal of each pasture. Duration of grazing and rest generally are governed by herbage growth rate, which depends primarily on the time of year, moisture, fertility, and species. The number of animals grazing each system may be fixed or variable. Experimental studies comparing continuous and rotation grazing have given inconsistent results. In general, however, rotation grazing has proven beneficial to lactating cows.

The **advantages** of a rotation grazing system are:

1. It permits the dairy farmer to match grazing more adequately to the growth habit of the forage species, condition of the pasture, and animal needs than does continuous grazing.

2. It improves stand persistence and production. Plants are given recovery periods during the growing season for more or less unhampered development of tillers and leaves. This is essential to replenish root reserves. This system of grazing enables the tall-growing legumes and grasses to survive.

3. It increases carrying capacity. Greater amounts of feed nutrients can be removed in the form of herbage with reduced losses due to trampling, fouling, and herbage death and decay.

4. It encourages equalization of grazing. It helps prevent overgrazing and undergrazing, and results in maintaining a better balance of the legumes and grasses. Also, both the palatable and the inferior species are grazed more nearly the same.

5. It often provides more nutritious herbage since the herbage is at the most ideal pasture stage. It is high in protein and low in crude fiber.

6. It helps prevent the grasses from heading out. This is done by concentrating grazing animals or by mowing when animals are shifted to new pastures. This allows new growth to come back uniformly and keeps it more palatable.

7. It helps control animal parasites, especially intestinal worms. Life cycles of worms can be broken by proper planning of grazing and rest periods.

8. It makes it more convenient to harvest surplus forages as hay or silage.

The **limitations** of rotation grazing are:

1. It requires a higher input of capital and management than continuous grazing.

2. There is a continuous day-to-day decline in the quality of the available forage, especially on the more intensive systems. At first turn-on, animals have access to leafy, high-quality forage, but the quality of the forage gets poorer and poorer during the grazing period.

INTENSIVE GRAZING

Several ingenious intensive grazing systems have evolved. All of them are designed to provide and harvest the maximum of high quality forage; to utilize the highest quality pastures for the highest producing animals; and to increase profits.

These systems were first described and used in Europe by Voisin and have since been installed to some extent throughout the United States and much of the world. Many designs are utilized for intensive short duration grazing systems. Generally, dairy producers favor the conventional (rectangular) system, which usually consists of small pastures of about equal size or production, that are fenced in a grid arrangement. Water and salt may be located in each of the pastures or a single source may be used with cattle gaining access by a lane or alley.

■ **First and second grazers.** This short duration grazing system involves two herds: first grazers and second grazers. It calls for using the best quality pastures for lactating cows. Here is how it may work with dairy cows: High-producing lactating cows, which have a high energy requirement, may be first grazers; they are allowed to graze the higher-quality (leafy) portion of pasture No. 1, following which they are moved to pasture No. 2—a fresh pasture. Dry cows or replacement heifers, which have a low energy requirement, may be second grazers; they are turned into pasture No. 1 immediately following the removal of the high producers. This progression is continued through all pastures, then the cycle is repeated.

The chief **advantage** of the system of first and second grazers is the enhanced productivity of the first grazers. The main **limitations** are the necessity of maintaining (1) two groups of animals of different productivity levels, and (2) balanced stocking rates and pasture sizes.

STRIP GRAZING

In this system, cows are allowed access to a strip which may be large enough for several days of grazing or small enough for $\frac{1}{2}$ to 1 day of grazing. Heavy stocking rates of upwards to 50 animal units per acre are used by fencing each strip with movable electric fences both in front and behind the grazing animals. The **advantages** claimed for this method are:

1. Increased utilization of herbage, with wastage reduced to 10 to 20%.
2. Increased milk yields per acre up to 25%.
3. Improved stability of milk yield because the nutritive value of the pasturage consumed is quite constant.
4. Improved utilization of the available forage. Less herbage is soiled by dung, urine, and treading. Under strip grazing, animals are quieter and settle down quickly for

steady grazing rather than roaming about and tramping forage.

5. Increased animal units maintained on a given area, although individual animal productivity may not be increased.

GREEN CHOP

Fig. 7–10. Harvesting alfalfa as green chop. (Courtesy, Ford New Holland, New Holland, PA)

Green chop is fresh herbage that is cut and chopped in the field, then transported and fed to animals in confinement.

Green chop, which is also called *soilage* or *zero grazing*, consists of growing a succession of forage crops, harvesting them with mechanized equipment, and hauling the green feed to the animals rather than allowing the animals to harvest their own forage. Historically, cutting green forage and hauling it to animals developed where land and forage were scarce and labor was plentiful. Under this system of feeding, each animal unit has to be supplied with upwards of 150 lb of green forage daily, depending upon its succulence.

Green chop minimizes the loss of moisture, color, nutrients, and wastage. Alfalfa, ladino clover, orchardgrass, bromegrass, grass-legume mixtures, Sudangrass, corn, sorghum, soybeans, and cereal grains are sometimes used in this manner. With tall-growing crops, more feed value may be realized from a given area than can be obtained by conventional pasturing. However, green chop requires special equipment and harvesting every day. Also, there are harvesting problems in wet weather.

Most green chop is fed to lactating dairy cows, usually in combination with hay or silage because the total intake tends to be greater. Green chop has increased with herd size, with more intensive forms of dairying, with drylotting of cows, and with high grain prices. Also, the use of green chop has been facilitated by the greater mechanization present on larger and more modern dairy farms.

GREEN CHOP CONSIDERATIONS

Experiments and experience indicate that the following points should be thoroughly considered before starting a green chop program:

1. **Daily consumption.** At least 100 lb of green chop per head per day must be harvested for lactating cows.

2. **Emergency reserve.** A supply of hay and/or silage should be available to meet emergencies caused by machinery breakdowns.

3. **Integrate with hay and silage making.** It is important that a green chop program be integrated with hay and silage making, both from the standpoint of amortizing machinery cost and from the standpoint of feed reserves. Surplus forage at peak growth should be made into silage or hay.

4. **Farm application.** Before attempting this system of summer feeding, a careful analysis of the time, machinery, investment, and labor requirements should be made.

5. **Green chop calendar.** By carefully planning a green chop program, it is possible to cut and feed green forage to cows continuously from spring to fall.

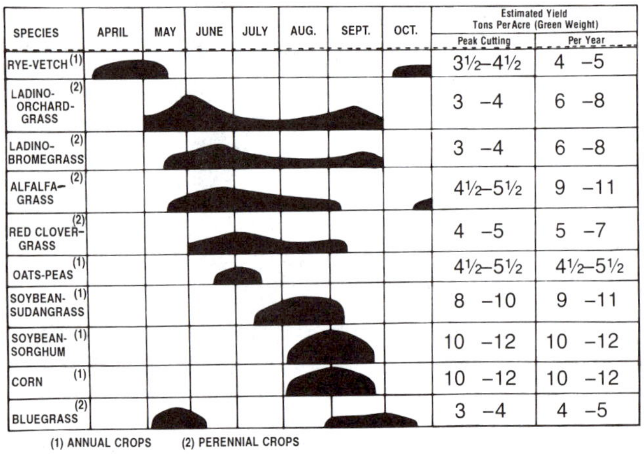

SPECIES	APRIL	MAY	JUNE	JULY	AUG.	SEPT.	OCT.	Estimated Yield Tons Per Acre (Green Weight)	
								Peak Cutting	Per Year
RYE-VETCH [1]								3½–4½	4 –5
LADINO-ORCHARD-GRASS [2]								3 –4	6 –8
LADINO-BROMEGRASS [2]								3 –4	6 –8
ALFALFA-GRASS [2]								4½–5½	9 –11
RED CLOVER-GRASS [2]								4 –5	5 –7
OATS-PEAS [1]								4½–5½	4½–5½
SOYBEAN-SUDANGRASS [1]								8 –10	9 –11
SOYBEAN-SORGHUM [1]								10 –12	10 –12
CORN [1]								10 –12	10 –12
BLUEGRASS [2]								3 –4	4 –5

(1) ANNUAL CROPS   (2) PERENNIAL CROPS

Fig. 7–11. A green chop calendar for northeastern U.S. (Recommended by Rutgers University, the Soil Conservation Service, and the U.S. Department of Agriculture)

FACTORS FAVORABLE TO GREEN SHOP

Among the factors **favorable** to green chop are the following:

1. **It increases carrying capacity.** With tall-growing crops, 30 to 50% greater carrying capacity per acre can be achieved than can be secured from conventional grazing.

2. **It makes it more practical to use distant, small, and scattered pastures.** With large herds, pastures may be so far from the milking parlor as to make it impractical to drive lactating cows back and forth. Likewise, it may not be practical to place animals on small and scattered plots.

3. **It lessens refusal.** Less forage is refused as green chop than when grazed. This is attributed to the fact that it is easier for the caretaker to see leftover forage in a manger

than in the field, with the result that feed allowance of green chop is more nearly synchronized with animal needs than carrying capacity of pastures; hence, wastage is lessened with green chop.

4. **It lessens stress due to change.** There is always a a certain amount of change where grazing is involved — change from drylot feeds to green grass, and change from pasture to pasture. Even though green chop varies during the season, it usually makes for less change than conventional pasture grazing.

5. **It alleviates the need for fencing, water, and shade.** In areas where green chop is produced, there is no need for fence, water, or shade, because there are no animals therein.

FACTORS UNFAVORABLE TO GREEN CHOP

Like many other good things, there are **disadvantages** to green chop; among them, the following:

1. **Higher harvesting cost.** Green chop requires special equipment, and it must be harvested every day. Hence, it involves more cost for machinery and labor than where animals do their own harvesting.

2. **Green chop makes for late spring start.** Cows can be placed on pasture to graze at least 2 weeks earlier than green chopping can commence.

3. **Quantity of green chop varies.** Once green chopping begins, it is difficult to keep up with it. Some of it will likely have to be stored as silage or hay for later use.

4. **Manure and urine not animal-spread.** Where green chop is used, manure and urine are not animal-spread back on the land. In order to keep green chop production at a high level year after year, either manure or chemical fertilizer, or both, must be spread on the land.

5. **Bloat, overeating, toxicity, and hardware disease.** Bloat can be a problem with green chop, especially if it is high in legume. The problem of overeating and toxic plants may be worse with green chop than with pasture. Toxic plants are chopped and mixed with the feed so that animals cannot pick and choose to avoid them. Likewise, the hardware problem may be exaggerated with green chop, because pieces of wire may be chopped into bite-sized fragments.

COMPARISON OF GRAZING SYSTEMS

When an orchardgrass-Ladino clover sward was subjected to various grazing systems, highest yields of forage and acre returns were obtained with the more intensive management systems (Table 7–1). Green chopping and strip grazing produced highest forage yields and acre returns. Rotation grazing was next best, and continuous grazing was poorest.

The efficiency of supplying forage for dairy cattle by rotation and strip grazing, green chopping, and stored forage was compared in Wisconsin. As shown in Table 7–2, more of

TABLE 7–1
GRAZING METHODS RELATED TO RETURNS FROM ORCHARDGRASS-LADINO MIXTURE IN RELATIVE TERMS[1]

Grazing Method	Yield of Mixture	Selective Grazing	Milk or Gain per Animal	Herbage Wasted	Acre Returns	Productive Years
Continuous	100	High	100	50	100	Short
Rotational	140	Medium	95	35	140	Long
Strip	160	Low	90	20	150	Long
Green chop	160	Very low	85	5	160	Long

[1]From a paper entitled, "Pasture Production and Management," by Dr. John B. Washko, Department of Agronomy, Pennsylvania State University.

the forage grown was available for feeding under green chopping than under any other method. Forage loss was as follows: green chop, 2%; stored forage, 11%; strip grazing, 31%; and rotation grazing, 43%. The acreage required to feed a cow for 110 to 120 days was lowest under the stored feeding program and highest under rotation grazing. Milk production per acre from the forage was highest under stored feeding followed by green chopping, strip grazing, and rotation grazing in that order. While the stored feeding program was most efficient from the standpoint of requiring less acreage per cow and produced the highest milk yield on an acre basis, this system of feeding required the feeding of more concentrates for maximum milk production.

In a U.S. Department of Agriculture experiment at Beltsville, cost studies were run on rotation and strip grazing in comparison with green chopping on an orchardgrass-Ladino clover mixture.[2] Rotational grazing provided forage at lower cost than either strip grazing or green chopping. Rotational grazing cost less because it required no portable electric fences or labor to move them daily as strip grazing and there was no harvesting or handling of forage each day as under green chopping. The rotational and strip grazing methods under proper management were equally efficient in forage utilization and significantly better than green chopping because the grazing systems provided the same number of cows with feed for more days per acre.

TABLE 7–2
EFFICIENCY OF HARVESTING FORAGE BY DAIRY CATTLE—3-YEAR AVERAGE[1]

Type of Grazing	Loss	110–120 Days Acreage/Cow	Milk Lb/Acre	Actual Milk Lb/Acre	Level of Conc. Feeding/Day Lb
	(%)				
Rotation	43	1.14	2,411	2,757	4.5
Strip	31	0.76	4,292	5,044	5.2
Green chop (zero)	2	0.69	5,190	5,958	4.9
Stored forage	11	0.49	6,055	8,135	8.3

[1]Larsen, H., and R. Johnnes, "Green Feeding is Practical and Profitable," *Hoard's Dairyman*, Vol. 103, No. 11, pp. 589–600.

When rotation grazing and green chopping of Sudangrass were compared for dairy cattle feeding, the rotation system proved to be superior for milk production over a 6-year period in Delaware.[1] The cows on rotation grazing consumed more forage in a day than those fed green chop. They also selected the more digestible parts of each plant. The average digestible dry matter intake of a grazing cow was more than 5 lb per day greater with rotational grazing than with green chop. This additional forage consumption provided enough nutrients for 16 more pounds of 4% milk daily. In the green chopping system, the cows were fed both the nutritious and fibrous portions of the plant indiscriminately. Under grazing, the cow selected the more nutritious plant parts and that which she left as wastage was apparently of little value from the standpoint of nutrition and production.

Changes in net income of dairy producers as influenced by feeding their cattle under different pasture management systems were studied at Michigan. As indicated in Table 7–3 (next page), rotation and strip grazing and green chopping increased net income on a 27-cow herd when compared with continuous grazing. Also, as shown in Table 7–4 (next page), when herd size increased to 60 cows, higher income was obtained with the more intensive forage management systems. When compared with rotation grazing, highest net income was obtained by changing to stored feeding, followed by green chopping and strip grazing.

Many dairy producers feel that the bloat hazard is reduced by mowing alternate strips through the pasture, thereby allowing the animals to consume the dry forage along with the green grass.

[1]Baker, T. A., *et al.*, "Factors Affecting the Consumption of Sudan Grass by Dairy Cows," *Journal of Dairy Science*, Vol. 43, pp. 958-965

[2]Gordon, C. H., *et al.*, "A Comparison of the Relative Efficiency of Three Pasture Utilization Systems," *Journal of Dairy Science*, Vol. 42, pp. 1686-1697.

TABLE 7–3
CHANGES IN NET INCOME FOR 27-COW HERD
(10,000-LB AVERAGE)
UNDER FOUR PASTURE MANAGEMENT SYSTEMS[1]

Item	Continuous Grazing	Rotation Grazing	Strip Grazing	Green Chopping
Net income	$2,897	$3,248	$3,458	$3,223
Income change over continuous grazing . .	—	+351	+588	+326
Income change over rotation grazing	—	—	+237	−25

[1]Mich. Ag. Exp. Sta. Bull. 429.

TABLE 7–4
CHANGE IN NET INCOME ON 60-COW HERD
DIFFERENT FORAGE MANAGEMENT SYSTEMS[1]

Item	Change from Rotation Grazing to:		
	Strip Grazing	Green Chopping	Storage Feeding
Change in net income	+$514	+$720	+$818

[1]Mich. Ag. Exp. Sta. Bull. 429.

Cattle should not be allowed to become empty when they congregate in a drylot for shade or insect protection, then be allowed to gorge themselves suddenly on green forage.

Water and salt should be conveniently accessible at all times.

AVOIDING SUDDEN CHANGES

Changes from confinement to pasture, or from less succulent to more succulent pastures, should be made with care because (1) grass is laxative, and lactating cows may fall off in production, and (2) there may be bloat.

REDUCING BLOAT ON PASTURE

Bloat can cause producers serious losses. It occurs frequently on pastures that contain a high percentage of legumes, such as Ladino clover or immature alfalfa. It is much less serious where the pasture contains about 50% grass.

■ **Bloat prevention**—The incidence of bloat is lessened by (1) avoiding straight legume pastures and immature legumes, (2) feeding a course grass hay prior to turning onto lush pasture, (3) feeding dry forage along with pasture, (4) avoiding a rapid fill from an empty start, (5) keeping animals continuously on pasture after they are once turned out, (6) keeping salt and water conveniently accessible at all times, (7) avoiding frosted pastures, or (8) using poloxalene (Bloat Guard), oxytetracycline (Terramycin or Neo-Terramycin), or Laureth-23 (Enproal Bloat Blox) according to manufacturers' directions, including placing blocks containing these antifoaming agents in various parts of the pasture.

■ **Bloat treatment**—Where legume bloat is encountered, use poloxalene (Bloat Guard), oxytetracycline (antibiotic), or polyoxyethylene (23) lauryl ether (Laureth-23/Enproal Bloat Blox), according to the respective manufacturers' directions.

SUPPLEMENTING EARLY SPRING GRASS

Turning lactating cows to pasture when the first sprigs of green grass appear will usually make for a temporary deficiency of energy, due to (1) washy (high water content) grasses, and (2) inadequate forage for animals to consume. As a result, producers often encounter a drop in milk production.

If there is good reason why grazing cannot be delayed until there is adequate spring growth, it is recommended that early pastures be supplemented with grass hay (do not use a legume hay because it will accentuate looseness, which usually exists under such circumstances), preferably placed before them in a rack.

SUPPLEMENTING DRY PASTURE

Dry, mature grass characterizes (1) drought periods, and (2) fall-winter pastures. Such grasses are low in energy, in protein, in carotene, and in phosphorus, and perhaps certain other minerals. These deficiencies become more acute following frost and increase in severity as winter advances. This explains the often severe loss in milk production following the first fall freeze.

In addition to the deficiencies which normally characterize whatever plants are available, dry pasture may be plagued by a short supply of feed.

These deficiencies need to be corrected by proper and liberal feeding during such periods.

SEPARATING CLASSES AND AGES OF DAIRY CATTLE

It is important that different classes and ages of dairy cattle be kept in separate pastures, according to needs. The following groups should be sorted out:

1. **Lactating cows.** It is difficult to provide sufficient nutrients to meet the needs of heavy-milking cows because they lack the capacity to eat enough. This problem is accentuated when cows are on poor pasture because of the added energy expended in the work of grazing. For these reasons, lactating cows should always be accorded the best pastures.

2. **Dry cows.** Dry cows should be in healthy, vigorous condition. Most of their rations may consist of pasture plus such supplements as required—with emphasis on proteins, minerals, and vitamin A; and the kind and level of supplementation should be varied according to the quality and quantity of grass available.

3. **Heifers.** Because they are growing, heifers require liberal feeding, especially from the standpoint of both energy and proteins. A combination of grass, exercise, and proper supplementation offers the ideal way in which to condition and grow heifers.

EXTENDING THE GRAZING SEASON

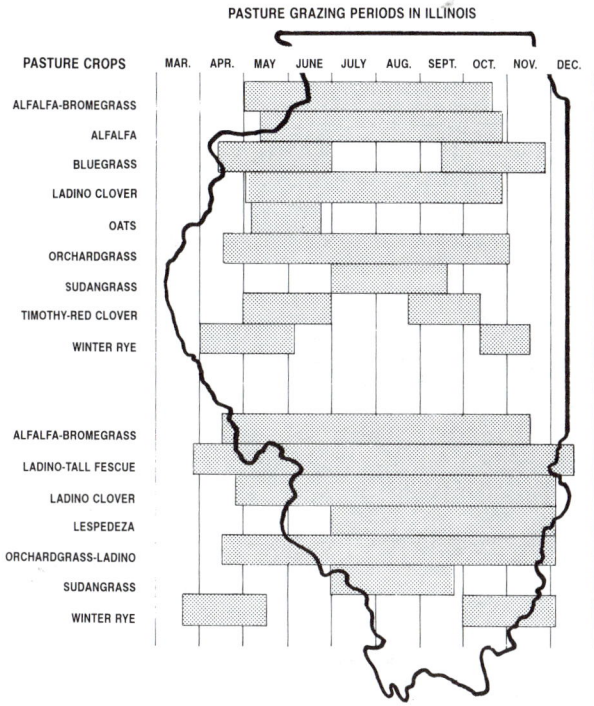

PASTURE GRAZING PERIODS IN ILLINOIS

PASTURE CROPS	MAR.	APR.	MAY	JUNE	JULY	AUG.	SEPT.	OCT.	NOV.	DEC.
ALFALFA-BROMEGRASS										
ALFALFA										
BLUEGRASS										
LADINO CLOVER										
OATS										
ORCHARDGRASS										
SUDANGRASS										
TIMOTHY-RED CLOVER										
WINTER RYE										
ALFALFA-BROMEGRASS										
LADINO-TALL FESCUE										
LADINO CLOVER										
LESPEDEZA										
ORCHARDGRASS-LADINO										
SUDANGRASS										
WINTER RYE										

Fig. 7–12. Suggested pasture crops for northern and southern Illinois, with grazing periods. The grazing season can be extended through the selection of species. (From *Illinois Forage Handbook*, Circ. 845, University of Illinois, p. 3.)

In practically all U.S. pasture areas, the grazing season can be extended by grazing earlier in the spring and later in the fall/winter, thereby lessening the amount of stored feed and supplemental protein needed for winter. From this standpoint, the southern states have a very real advantage over other areas. Through planting different crops at different seasons, progressive southern dairy producers are achieving year-round grazing. This fact, together with modern and more effective parasite and disease control measures, is the primary reason for the recent great expansion of all livestock production in the South.

Some Corn Belt dairy producers are approaching a 12-month dairy pasture by using two crops only, a legume (or legume-grass mix) and rye. In the North, where year-round grazing cannot be secured, it is merely recommended that dairy producers attain as long a grazing season as is possible, especially through arranging for early spring and late fall pastures (see Fig. 7–12).

In addition to lengthening the grazing season through the selection of species, earlier spring pastures can be secured by avoiding grazing too late in the fall and by the application of a nitrogen fertilizer in the fall or early spring. Nitrogen fertilizers will often stimulate the growth of grass so that it will be ready for grazing 10 days to 2 weeks earlier than unfertilized areas.

Fig. 7–13. Brown Swiss cows on pasture. (Courtesy, The Brown Swiss Cattle Breeders Assn. of the U.S.A., Beloit, WI)

QUESTIONS FOR STUDY AND DISCUSSION

1. List and discuss the advantages and disadvantages of producing milk from grass. How do you account for the current trend to confinement production by large, highly specialized operators?

2. Compute the monetary return that might reasonably be expected from an acre of pasture grazed by lactating cows vs the return from alternate cropping programs.

3. List six desirable characteristics of a good pasture.

4. List and discuss the classes of pasture. Will there be more irrigated pastures in the future?

5. List and discuss the factors affecting the value of pasture.

6. What are the primary factors which determine the specific grass or grass-legume mixture best adapted to and/or most common in each of the ten generally recognized U.S. pasture areas?

7. How does establishing a new pasture differ from improving or renovating an older pasture?

8. What practices are involved in efficient and profitable pasture management in the subhumid, humid, and irrigated areas of the United States?

9. Discuss the differences between, and the place of, the following seeded pasture grazing systems: continuous grazing, rotation grazing, intensive grazing, and strip grazing.

10. What factors are (a) favorable, and (b) unfavorable to green chop? Do you foresee more green chop used on dairy farms in the future?

11. Why are rotation grazing, strip grazing, and green chop grazing generally (a) more productive with high-producing dairy cows than with other classes of animals, and (b) more beneficial where parasite infections are heavy than where little or no parasite problems are involved?

12. Review the results of one experiment in which grazing systems were compared.

13. How would you reduce pasture bloat?

14. Why should cows on pasture be separated by classes and ages?

15. What are the primary differences between good management of milk cows on pasture vs good management of milk cows in confinement?

16. Select a specific dairy farm (either your home farm or one with which you are familiar). Then tell what pastures you would use for dairy cows, and justify your decision. Prepare a chart showing how you could advantageously extend the grazing season.

17. How would you improve upon the current pasture management practices followed on a dairy farm with which you are familiar?

SELECTED REFERENCES

Title of Publication	Author(s)	Publisher
Approved Practices in Pasture Management, Third Edition	M. H. McVickar	The Interstate Printers & Publishers, Inc., Danville, IL, 1974
Crop Production, Fifth Edition	R. J. Delorit L. J. Greub H. J. Ahlgren	Prentice-Hall, Inc., Englewood Cliffs, NJ, 1984
Feeds & Nutrition, Second Edition	M. E. Ensminger J. E. Oldfield W. W. Heinemann	The Ensminger Publishing Company, Clovis, CA, 1990
Feeds & Nutrition Digest	M. E. Ensminger J. E. Oldfield W. W. Heinemann	The Ensminger Publishing Company, Clovis, CA, 1990
Forage and Pasture Crops	W. A. Wheeler	D. Van Nostrand Company, New York, NY, 1950
Forage Crops	D. A. Miller	McGraw-Hill Book Company, New York, NY, 1984
Forages, Fourth Edition	M. E. Heath R. F. Barnes D. S. Metcalfe	The Iowa State University Press, Ames, IA, 1985
Grass Productivity	A. Voisin	Philosophical Library Inc., New York, NY, 1959
Grass to Milk	C. P. McMeekan	The New Zealand Dairy Exporter, Wellington, New Zealand, 1964
Grass, Yearbook of Agriculture, 1948	U.S. Department of Agriculture	Superintendent of Documents, Washington, DC, 1948
Intensive Grazing Management: Forage, Animals, Men, Profits	B. Smith P. Leung G. Love	The Graziers Hui, Kamuela, HI, 1986
Manual of the Grasses of the United States	A. S. Hitchcock	U.S. Department of Agriculture, Washington, DC, 1950
Pasture Book, The	W. R. Thompson	W. R. Thompson, State College, MS, 1950
Profitable Pasture Management	R. A. Chessmore	The Interstate Printers & Publishers, Inc., Danville, IL, 1979
Stockman's Handbook, The, Seventh Edition	M. E. Ensminger	Interstate Publishers, Inc., Danville, IL, 1992
Veld and Pasture Management in South Africa	N. M. Tainton	Shuter & Shooter, Pietermaritzburg, South Africa, 1984

Fig. 8–1. Alfalfa hay being converted into milk. Hay is the most important harvested forage for U.S. dairy cattle. (Courtesy, Ford New Holland, New Holland, PA)

Contents

CHAPTER

8

HAY AND OTHER DRY ROUGHAGES

Hay is forage harvested during the growing period and preserved by drying for subsequent use. Hays are made from legumes, grasses, and cereal crops. It ranks third among all livestock feeds, being exceeded only by pasture and corn. Average-quality hay runs 25 to 35% crude fiber and 45 to 55% TDN on an as-fed basis, whereas such concentrates as corn and wheat contain approximately 2 to 3% fiber and 80% TDN.

The object of haymaking is to (1) harvest the crop at the optimum stage of maturity which will provide the maximum yield of nutrients per acre without damage to the next crop, and (2) cure the crop properly by lowering the water content of the green herbage from 65–85% to 20% or less.

Hay is the most important harvested forage for U.S. dairy cattle. If fed hay alone, cows will consume 3 lb or more of it per 100 lb of liveweight daily.

Dairy cattle obtain approximately two-thirds of their feed from forages—pasture, hay, and silage. The proportion of concentrates to roughage is largely determined by the economics of the situation. When buying forage, this means the cost of energy and protein units per dollar. However, it is expected that forages will always remain the foundation of the dairy ration—and that means hay when it comes to winter feeding.

Of course, pasture is the leading forage utilized by dairy cows. But in most areas it is not available in sufficient quan-

tities on a year-round basis. Even in the pasture season, it is difficult to maintain grass in the quality and amounts needed for uniformly high milk production. Moreover, some dairy producers, particularly the very large ones, drylot the cows throughout the year. Thus, there is need for stored forage; and, for the most part, this means hay or silage. Some dairy producers use one or the other of these forages; others use both.

Drying, or making hay, is the most common method of preserving forage for storage, primarily because it is relatively easy to handle. It can be stored or transported long, chopped, pelleted, cubed, or baled into various types and sizes of bales. Modern equipment, including conditioners, hasten drying time; and automated systems facilitate handling.

The great capacity and specialized functions of the rumen allow the cow to use hay, and other forages, in large amounts. Bacteria and protozoa in the rumen break down and make available to the cow part of the nutrients in cellulose or fibrous material. But hay differs widely in nutrient content and palatability; and no amount of bacterial action can overcome deficiencies in either category.

In addition to the nutrients that it contains, and to its value in providing feed throughout the year, hay has other values. Dry feed is essential for the proper functioning of the digestive tract; it acts as a stimulant in moving the feed through the intestines, and it maintains the proper conditions in the rumen for the microbial action which plays such a vital role in the digestion of the fibrous portions of feeds. Hay is often used as a supplement to "washy" pastures and succulent silages. Also, it speeds along the development of the rumen function of the young calf, lessens the incidence of

Fig. 8–2. Many serious problems in high-production dairy herds have been traced to lack of hay in rations. Increased incidence of ketosis and displaced abomasums have been lessened by adding more hay to rations. (Courtesy, Ford New Holland, New Holland, PA)

ketosis and displaced abomasum in cows, and prevents a lowering of the fat content of the milk of lactating cows (unless it is finely ground). Also, and most important, good-quality hay is a hedge against high-concentrate prices; for when the price of such feeds increases disproportionately, increased amounts of hay may be fed and concentrates may be decreased, with a higher net return to the producer.

Despite its several advantages, hay has several shortcomings. It varies in nutrient content and palatability more then any other feed, because of differences in the (1) crops from which it is made, (2) stage of cutting, (3) handling, and (4) weather damage during curing. Moreover, not even ruminants can consume enough hay alone to meet the demands of high production; for example, on hay alone, dairy cows will produce only 50 to 70% as much milk as they would on a ration consisting of 50% concentrates.

During the curing process, the quality and feeding value of hay is decreased rapidly by rain, sun, bleaching, raking, handling when too dry, and storing with too much moisture. Studies conducted by the U.S. Department of Agriculture, at Beltsville, Maryland, revealed that the following losses were encountered in field-cured, second-cut alfalfa hay from the time of cutting to the time of feeding: leaves, 35%; dry matter, 20%; and proteins, 29%. Of course, it is recognized that losses also accrue when utilizing forage as pasture or silage; with pastures, there are trampling losses, and with silage, there are seepage losses.

In addition to the low nutrient content that characterizes poor-quality hay, a more serious loss may follow from feeding it. Recent experiments show that a part of the poor results obtained when feeding low-quality hay can be attributed to its failure to support maximum microflora in the rumen, with the result that the digestibility of crude fiber suffers. Hand in hand with the decline in rumen microflora activity, roughage consumption goes down. Of course, if they will not eat the hay, it will not do them any good. Since dairy cows normally obtain about 70% of their TDN from roughage, any decrease in hay consumption can be serious and is usually accompanied by a fall in milk production. This underscores the importance of high-quality hay for lactating cows.

An estimated 75% of all hay is fed on the farms on which it is produced, rather than being purchased. It is important, therefore, that dairy producers know how to produce good hay, as well as how to feed it, for most of them determine their own destiny from the standpoint of quality. For this reason, this section covers hay from production to feeding.

IMPORTANCE OF HAY

The importance of the nation's hay crop is attested to by the fact that the total area devoted to hay in the United States exceeds 60 million acres, the total production averages about 150 million tons, and the annual crop is worth approximately $10 billion—it is worth more than any other crop except corn and soybeans. On an air-dry tonnage basis, about three times as much hay is produced as silage.

HAY AS AN ENERGY SOURCE

Hay is an important source of energy for dairy cattle. Table 8–1 shows the percentage of total energy (TDN) intake provided to ruminants and horses by hay and other kinds of feed. As shown, hay is a more important source of energy for dairy cows than for any other class of farm animal. It is noteworthy, too, that better than one-half the total hay tonnage produced in North America is fed to dairy cattle, while beef cattle consume almost 40% of all hay produced. As increasing quantities of concentrates go to feed the world's hungry people, it is expected that dairy producers will depend even more on hay to meet a larger percentage of the total feed needs.

TABLE 8–1
PERCENTAGE OF ENERGY SUPPLIED BY HAY AND OTHER KINDS OF FEEDS[1]

Animal	Concentrates	Hay	Other Harvested Forages	Pasture	All Forage	Total
	(%)	(%)	(%)	(%)	(%)	(%)
Lactating cows	37.9	23.1	19.4	19.6	62.1	100
Other dairy cows	19.4	29.0	5.9	45.7	80.6	100
Finishing beef cattle	69.8	16.3	8.7	5.2	30.2	100
Other beef cattle	8.7	15.5	4.1	71.7	91.3	100
Sheep and goats	10.4	4.7	3.1	81.8	89.6	100
Horses and mules	20.6	18.3	10.2	50.9	79.4	100

[1]Based on USDA data. From paper entitled, "Hay Production, Preservation and Quality," by J. E. Baylor, The Pennsylvania State University, *Beef Cattle Science Handbook*, Vol. 13, p. 199, published by Agriservices Foundation, edited by M. E. Ensminger.

COMPARATIVE VALUE OF HAY

In dairy rations, hay is primarily a source of energy, but the legumes also serve as a source of protein.

Table 8–2 (next page) shows the dry matter (DM), crude protein (CP), and total digestible nutrients (TDN) in 100 lb of dry matter from corn grain, corn silage, and three different types of alfalfa hay (mature, midbloom, and early bloom). Note, too, that this table compares these crops on a per acre basis.

Of course, the economic comparisons in Table 8–2 are valid only at the stated prices of soybean meal and corn. However, in most practical feeding situations the comparisons are meaningful. It is noteworthy that, in terms of the economic value of the energy and protein provided by the different feeds listed in Table 8–2, early bloom alfalfa hay had a value nearly 95% (6.42 ÷ 6.76 × 100) that of corn grain. Note, too, that in terms of energy produced per acre, corn silage leads all other feeds.

TABLE 8–2
COMPARATIVE ECONOMICS OF THE NUTRIENTS
IN CORN GRAIN, CORN SILAGE, AND
ALFALFA HAY OF THREE QUALITIES[1]

Item	Corn Grain	Corn Silage	Alfalfa Hay		
			Mature	Mid-Bloom	Early Bloom
Analyses, DM basis	(%)	(%)	(%)	(%)	(%)
Dry matter (DM)	88.0	33.0	90.0	90.0	90.0
Crude protein (CP)	10.1	8.1	12.9	17.0	18.0
Total Digestible Nutrients (TDN)	90.0	70.0	50.0	58.0	60.0
Value of 100 lb DM	($)	($)	($)	($)	($)
CP value[2]	1.72	1.38	2.19	2.89	3.06
TDN value[3]	5.04	3.92	2.80	3.25	3.36
Total value	6.76	5.30	4.99	6.13	6.42
Total value/acre[2][3]	($)	($)	($)	($)	($)
16 tons silage or 100 bu grain	333.0	560.0	—	—	—
21 tons silage or 150 bu grain	500.0	735.0	—	—	—
5 tons hay	—	—	449.0	552.0	578.0
8 tons hay	—	—	719.0	883.0	924.0

[1]Adapted by the author from: *Haymaker's Handbook*, by J. E. Baylor, Professor of Agronomy, The Pennsylvania State University, and M. A. Balas, New Holland, Inc., published by Ford New Holland, Inc., New Holland, PA, 1987, p. 140, Table 17.1.

[2]44% soybean meal used as a standard protein source, priced at $150 per ton or $0.17 per pound of crude protein.

[3]Corn grain used as a standard TDN source, priced at $2.50 per bushel or $0.056 per pound of TDN.

KINDS OF HAY

Hays are made from legumes, grasses, and cereal crops. In terms of total tonnage produced annually, alfalfa accounts for approximately 57% of the U.S. hay crop. Many different kinds of hay make up the other 43% of the nation's hay supply; among them, the clovers, lespedeza, soybeans, and cowpeas, the cereal hays made from oats, barley, wheat, or rye, and grass hays made from Bermudagrass, prairie grass, redtop, Johnsongrass, orchardgrass, fescue, and timothy.

Legume hays are higher in protein, calcium, and carotene than grass hays; and they are usually more palatable. Of course, poor-quality legume hays—those cut at a late stage of maturity and exposed to weathering—are not as good as high-quality grass hays.

Alfalfa yields the highest tonnage per acre and has the highest protein content of the legume hays.

Clover is usually grown with timothy, as a grass-legume mixture. In comparison with alfalfa, clover-timothy mixed hays are lower in protein and not as high in quality. The lower quality is due to the fact that, generally speaking, at cutting time the timothy is at the right stage of maturity whereas the clover is overripe.

Lespedeza is an excellent hay for dairy cattle provided

(1) it is cured without weather damage, and (2) it is fine stemmed and free of foreign material.

Soybeans, cowpeas, and vetch are often made into hays and fed to dairy cattle. But they are not as valuable as alfalfa; generally, they are stemmy and difficult to cure. However, if they are cut at the proper stage and cured without loss of leaves, they make good feed.

Grass hays include prairie grass, redtop, Johnsongrass, orchardgrass, and timothy. Grass hays will grow under a wider range of conditions than alfalfa, but they yield less dry matter per acre. When cut at the usual stage of maturity, grass hays are less palatable than legume hays and lower in protein and mineral content. However, when grass hays are heavily fertilized and cut at an early stage of maturity, they are very palatable and about equal to alfalfa in protein content.

Cereal hays are made from oats, barley, wheat, and rye. When cut sufficiently early—in the flower stage, and before the milk stage—they retain much of their green color and have fair feeding value. They are low in protein; hence, they must be fed with a legume hay, grass silage, or a protein supplement. If cereal hays are cut too early, yield is reduced; if cut too late, they become fibrous and are of low feeding value.

Whenever feasible, it is recommended that a legume be grown for hay, for the reasons that, in comparison with grasses, legumes are (1) higher in protein, vitamins, and minerals; (2) higher yielding; and (3) nitrogen-fixing when inoculated, the bacteria (rhizobia) on their roots taking free atmospheric nitrogen from the air. However, a mixture of grasses and legumes is often preferred for reasons of palatability, ease in curing, erosion control, and lessening bloat.

More and more dairy producers are coming to appreciate the flexibility afforded by growing varieties of grasses and legumes that may be used three ways: for pasture, for hay, and for silage. With such flexibility, surplus pasture may be converted into hay, or if desired or the weather is not favorable for haymaking, the crop can be ensiled.

HAY QUALITY

Fig. 8–3. Hay quality can differ widely. At the left is a leafy, palatable, high-energy forage, rich in protein. At the right is a coarse, stemmy, unpalatable forage, low in nutrient content. (Courtesy, Ford New Holland, New Holland, PA)

Hay quality is the degree of excellence, or the productive worth, that hay possesses. It refers to the nutritive value of hay. For hay to be of superior quality, it must be high in four factors: (1) nutrients, (2) palatability (intake), (3) digestibility, and (4) efficiency of utilization.

The most accurate method of determining hay quality involves live animal experiments on the farm or ranch where the forage is to be fed. However, this is often too costly, slow, and impractical. Therefore, forage value is predicted by visual inspection, chemical analysis, and/or new methods such as near infrared analysis.

IMPORTANCE OF HAY QUALITY

Hay is feed. Thus, as with any feed, it is the end results from feeding hay—the value as determined by animals—that count. Generally speaking, dairy producers recognize that the feeding value of hay varies according to quality. However, it is doubtful that they realize just how much returns in milk production are affected by quality.

The quality of hay greatly affects its consumption. High-quality forage is more digestible and passes through the digestive tract more rapidly than low-quality forage; hence, animals will consume more of it.

Feeding trials at the University of Wisconsin, Madison, showed, conclusively, the effect of hay quality on milk production. Alfalfa hay at four stages of maturity—prebloom, early bloom, midbloom, and full bloom—was fed to lactating cows. Table 8–3 summarizes the results. It gives for the four different hays the crude protein (CP), neutral detergent fiber (NDF), and acid detergent fiber (ADF); the digestible dry matter (DDM); the dry matter intake (DMI); and the 4% fat corrected milk (FCM) produced. Note that with maturity of the hay the crude protein, DDM, DMI, and milk production decreased, while the NDF and ADF increased. The increase in NDF and ADF with maturity is as expected, because NDF is inversely correlated with intake, whereas ADF is highly correlated with digestibility.

At the Illinois Station, test cows ate 2.21 lb of early cut hay per 100 lb of body weight compared to 1.6 lb of late cut hay. This meant that cows on late cut hay ate 9 lb less per day than cows on hay cut in the bud stage. To make up this nutritional difference in intake, the cows on the late cut hay required 7 lb more grain daily for each cow. Hence, a 100-cow herd would require 700 lb more grain daily as a penalty for late cutting.

In feeding trials with lactating cows, Cornell workers compared alfalfa hay cut at two different stages of maturity—early bloom vs late bloom. They found that, in comparison with the late cut hay, the cows that were fed the early cut hay consumed 7 lb more of it per head per day and it was 16% more digestible and produced 12 lb more milk per day (see Fig. 8–4).

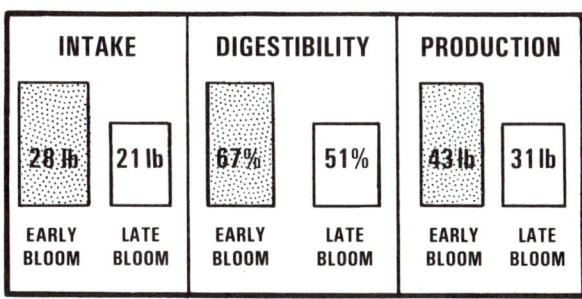

Fig. 8–4. Comparison of alfalfa hay cut at two different stages of maturity when fed to lactating cows (based on Cornell data).

Cornell University workers also compared the economic impact of feeding five different qualities of hay to a 120-cow dairy herd. In this study, half the dry matter was corn silage and the other half was hay. Everything else was held constant except for the hay quality. The results are reported in Table 8–4. Note that, as hay quality declined, the increase in purchased feed costs spiraled. The cost of purchased feed for the 120-cow dairy herd fed grass hay, the lowest hay quality, was $79.04 more per day, or $28,849.60 more per year, than for the cows fed the early cut legume. As dramatic as these results were, they do not tell the whole story. The higher quality hays also resulted in increased production and greater returns.

TABLE 8–3
EFFECT OF QUALITY OF ALFALFA HAY ON PERFORMANCE OF LACTATING COWS[1]

| Stage of Harvest | Composition | | | | | |
	CP	NDF	ADF	DDM	DMI	4% FCM
	(%)	(%)	(%)	(%)	(% BW[2])	(lb/day[3])
Prebloom	21.1	40.5	30.2	62.7	2.08	87.1
Early bloom . . .	18.9	42.0	33.0	61.6	1.97	77.2
Midbloom	14.7	52.5	38.0	54.8	1.48	66.2
Full bloom	16.3	59.5	45.9	52.9	1.42	64.7

[1]Kawas, J. R., N. A. Jorgensen, A. R. Hardie, and J. L. Danelon, *Journal of Dairy Science,* Abstract, Supplement 1, 1983, Vol. 66, p. 181; and Kawas, J. R., N. A. Jorgensen, and D. A. Rohwede, *Proceedings Wisconsin Forage Council Eighth Forage and Use Symposium,* 1984, p. 21, Tables 3 and 4.

[2]80% hay and 20% concentrate (DM basis).

[3]46% hay and 54% concentrate (DM basis).

TABLE 8–4
EFFECT OF FIVE DIFFERENT QUALITIES OF HAY ON THE DAILY FEED COST OF A 120-COW DAIRY HERD[1]

| Hay Quality | | | | |
Description	Percent Protein	Forage Fed	Concentrate Purchased	Purchased Feed Cost
	(%)	(tons DM)	(lb)	($)
Early cut legume	21	2.17	531	94.74
Legume	18	2.07	733	117.62
Mixed, mainly legume .	15.5	1.91	977	138.68
Mixed, mainly grass . .	12	1.84	1,158	160.84
Grass	10	1.77	1,278	173.78

[1]Adapted by the author from: *Haymaker's Handbook,* by J. E. Baylor, Professor of Agronomy, The Pennsylvania State University, and M. A. Balas, New Holland, Inc., published by Ford New Holland, Inc., New Holland, PA, 1987, p. 66, Table 9.2.

The Ohio Station compared hay containing 67% TDN to hay containing 61% TDN. The cows on the high-energy hay required approximately 7 lb less grain daily than the cows on the low-energy hay. When hay constituted 56% of the TDN in the ration, the cows on the low-energy hay needed 15 lb more grain per day than the cows on the high-energy hay.

The Wisconsin, Illinois, Cornell, and Ohio studies show that, in addition to the low nutrient content that characterizes poor-quality hay, a more serious loss may follow from feeding it. Studies show that part of the poor results obtained from feeding low-quality hay can be attributed to its failure to support maximum microflora in the rumen, with the result that the digestibility of the crude fiber suffers. Hand in hand with the decline in microflora activity, roughage consumption goes down. Of course, if animals won't eat feed, it won't do them any good.

CHARACTERISTICS OF HIGH-QUALITY HAY

High-quality hay is readily consumed and digested by animals, with the nutrients derived therefrom utilized efficiently in carrying out body functions.

The characteristics of high-quality hay, and the importance of hay quality, have long been recognized. The commonly used indicators of hay quality are visual inspection, chemical composition, and NDF, ADF and NIRS Analyses.

VISUAL INSPECTION

Fig. 8–5. James Tappan, Arizona Dairy Co., checking the quality of new-mown alfalfa hay. (Courtesy, James Tappan, Arizona Dairy Co., Higley, AZ)

Fortunately, hay quality and value can be estimated by certain characteristics. It is important therefore, that those who grow hay, and those why buy and sell hay, be acquainted with those recognizable characteristics of hay which indicate high palatability and nutrient content. If in doubt, the animals will tell them, for they like and thrive on high-quality hay.

The easily recognized characteristics of hay of high feeding value are:

1. It is made from plants cut at an early stage of maturity, thus assuring the maximum content of protein, minerals, and vitamins, and the highest digestibility.
2. It is leafy, thus giving assurance of high protein content.
3. It is bright green in color, thus indicating proper curing, a high carotene or provitamin A content, and palatability.
4. It is free from foreign material, such as weeds, stubble, etc.
5. It is free from must or mold or dust.
6. It is fine stemmed and pliable—not coarse, stiff, and woody.
7. It has a pleasing, fragrant aroma; it "smells" good enough to eat.

FEDERAL GRADES OF HAY OF HISTORIC INTEREST ONLY

Today, the federal grades of hay and straw are of historic interest only. They were created by the Agricultural Marketing Act of 1946, and they were terminated March 14, 1988. At the time the grades were discontinued, fewer than 50 requests per year for their use were being received by the U.S. Department of Agriculture.

The historic federal grades of hay were based entirely on subjective factors—leafiness, color, and presence or absence of foreign material. There were no indicators of nutrient content and projected animal productivity, such as crude protein, fiber, NDF, or ADF.

Thus, the demise of the 42-year-old federal grades of hay resulted because they did not adequately reflect feeding value. Progress had bypassed them.

CHEMICAL COMPOSITION

Visual estimates of hay quality are of value and should be used, but the most precise way to determine the nutrient value of hay is through chemical analysis. Analyses are not infallible, however; a Pennsylvania study revealed errors of as much as 5% in crude protein and 9% in TDN (energy) content of a forage, with evaluations made by trained individuals.

Visual inspection of hay is still needed. For example, hay cut at the right stage of maturity can become low-quality hay by poor hay-making practices and conditions, which only visual inspection can detect. Also, visual inspection is needed for (1) weed detection and color, (2) predicting palatability, and (3) detecting the effects of mold, rain damage, and brittleness. So, chemical analyses should supplement, but not replace, visual inspection. Also, it is recognized that any method of determining hay quality by means of chemical analyses is of value only if it is related to feeding value.

Fortunately, research has shown a high relationship between the chemical composition—especially the protein and fiber—of hay and its feeding value for animals. As a result, a growing number of states now have laboratories where, at a nominal charge, a quick check of moisture, protein, and/or crude fiber can be made. As hay matures, protein decreases (pounds of protein per acre decreased from 935 lb in early cut hay to 605 lb in late cut hay, according to a Cornell study) and fiber increases. Likewise, weathering lowers the protein and raises the fiber content since soluble nutrients are washed out by rain and leaves are lost during harvest. It is also noteworthy that palatability seems to be negatively correlated with crude fiber levels—the higher the fiber content, the lower the palatability; this is important, for if the animals won't eat, they can't produce. Cornell investigators found that cows ate 2 lb more of the early cut hay per day than of the late cut hay.

NDF, ADF, AND NIRS ANALYSES

Since 1865, fibrous materials traditionally have been analyzed by the *Weende proximate analysis* method. Although this method is still widely used, it is often supplemented with additional analyses. In a series of reports beginning in 1963,[1] Peter J. Van Soest proposed the *detergent analysis system*, better to evaluate the feeding value of fibrous materials; by using detergents, he separated the sample into two fibrous fractions: (1) a neutral detergent fibrous fraction (NDF), and (2) an acid detergent fibrous fraction (ADF). Further, he reported that, in comparison with traditional proximate analysis, NDF provided a better estimate of dry matter intake (consumption) by animals, and ADF provided a better estimate of the *in vivo* (inside the animal) dry matter digestibility. Van Soest partitioned forages into the following two fibrous fractions

1. **Neutral detergent fiber (NDF),** which is the cell wall material or plant structure in feed, is comprised of hemicellulose, cellulose, lignin, lignified N, and insoluble ash. This constituent is insoluble in neutral detergent and is only partially available to animals. The lower the NDF percentage, the more the animal will eat; it is inversely related to voluntary intake (consumption). Thus, a low percentage of NDF is desirable.

The NDF content is also positively correlated with eating time and rumination, and this may be related to the rate of particle size reduction. Additionally, NDF is related to the proper function and health of the rumen; roughage value (the amount, source, and physical form of dietary NDF) is associated with chewing time, saliva flow (buffering and pH in the rumen), rumen fermentation patterns, milk fat test, and total energy output.

2. **Acid detergent fiber (ADF),** which is the highly indigestible plant material in a forage, is comprised of cellulose, lignin, and insoluble ash. This constituent is insoluble in acid detergent. ADF differs from crude fiber in that it contains silica. Silica and lignin in plants are associated with low *in vivo* (in animal) digestibility. The lower the ADF, the more feed an animal can digest. Thus, a low ADF percentage is desirable.

But chemical determinations are slow! The laboratory determination of crude protein, neutral detergent fiber, acid detergent fiber, *in vitro* (in a test tube or other artificial environment) dry matter digestibility, mineral, and vitamin analyses may take up to 2 weeks. This prompted the search for a more rapid, yet reliable, method of assessing the nutritive value of hay. In 1976, Norris, *et al.*, indicated (*Journal of Animal Science,* 47:747–759) that a relatively new procedure, known as *near infrared reflectance spectroscopy,* which had been applied successfully to grain quality evaluation, could be used for predicting forage quality, also. Additional research confirmed the initial findings of Norris and developed data bases and procedures for quickly analyzing hays.

The near infrared reflectance spectroscopy (NIRS) is a non-consumptive instrumental method for fast, accurate, and precise evaluation of the chemical composition and associated feeding value attributes of forages and other feedstuffs. The instrument, known as a *near infrared analyzer,* produces infrared radiation over a given range of wavelengths and this radiation is focused onto the sample being tested. Because of the chemical structure of the sample material, certain combinations of infrared wavelengths are reflected and certain combinations are absorbed for each chemical characteristic tested, *e.g.*, energy values, crude protein, digestibility, minerals, NDF, and ADF. By using a system of filters and detectors, the instrument senses these reflected wavelengths and passes this information on to the computer. The computer sorts out the appropriate wavelength combinations and their relative magnitudes for each chemical characteristic and transforms these data into percentages.

The near infrared reflectance measures hay quality by comparing the energy reflected back from a hay sample with computerized standards established by conventional laboratory analysis of a large number of reference samples.

By using a portable near infrared analyzer in a mobile van, samples at hay markets can be evaluated in minutes, and hay can be bought and sold on the basis of known quality. Also, tests made at the farm by mobile units can provide accurate information needed to make decisions about balancing rations. Additionally, forage researchers are using NIRS to select experimental plant lines for improved nutrient quality.

The NIRS method of analysis has four main advantages: speed, simplicity of sample preparation, multiplicity of analysis with one operation, and nonconsumption of the sample (it can be analyzed again by the same or another procedure). With the NIRS method of analysis, it is possible to take a sample from a truckload of hay and provide, in less than 3 minutes, an analysis for crude protein, NDF, ADF, dry matter, lignin, and *in vitro* dry matter digestibility.

The chief disadvantages of the NIRS method are instrumentation requirements and costs, dependence on calibration procedures, complexity in the choice of data treatment, and lack of sensitivity for minor constituents.

On a voluntary basis, NIRS technology is being used currently and successfully in the NIRS Forage Research Project Network; and either NIRS or wet chemistry tests may be used in the laboratory certification program operated by the National Alfalfa Hay Test Association in conjunction with the American Forage and Grassland Council (AFGC) and the National Hay Association.

[1]Goldring, H. K. and P. J. Van Soest, *Forage Fiber Analyses,* Agriculture Handbook No. 379, ARS, USDA, 1970, p. 20 on which 12 papers are listed.

CORRECT SAMPLING NECESSARY

No forage test is any better than the sample taken. Thus, the most important single step in determining the chemical composition of hay is sampling. No matter how accurate the chemical analysis, a poor sampling technique can easily invalidate the results and lead to an erroneous conclusion. For instructions on correct sampling, and to determine if the state has a laboratory for analyzing forage samples, the dairy producer should see the local county agent or write to the state college of agriculture.

It is difficult to obtain a representative, meaningful sample of forage because of its bulky nature and variability within a given lot of hay as compared to most other crops. For the sample to be representative of a given lot of hay, it should have been taken from the same field (or at least it should have been produced under the same cultural conditions— same irrigation and fertility conditions, for example), same cutting, same stage of maturity, and all of it should have been baled within a 48-hour period using only one harvesting method. With conventional, rectangular bales, at least 20 bales should be sampled at random, by probing every third bale, for example. The probe, or core sampler, should be at least 3/8 in. in diameter. The center of either end of a rectangular bale may be probed by inserting the probe at a right angle to the face of the bale and to a depth of 12 to 18 in. The hay testing laboratory should be consulted relative to sampling large round bales and stacks.

Hay samples should be placed in a plastic bag or freezer carton; otherwise, the moisture content will not be meaningful.

EVALUATING TEST RESULTS

Test results can best be evaluated by comparing them with some standard. The testing laboratory may provide such information, possibly along with recommendations for applying the test results in balancing rations. For convenience, average crude protein and crude fiber values of some common hay crops are given in Table 8–5.

If a chemically analyzed sample runs higher in protein and lower in fiber than the average figures given in Table 8–5, it means that the sample is better than average-quality hay; conversely, if it is lower in protein and higher in fiber, the sample tested is below average quality. Of course, dairy producers should not settle for average protein or fiber content, for, on the whole, the vast majority of the U.S. hay crop is of low quality. For this reason, one should strive for the upper figures of the range given in Table 8–5. Certainly, poor-quality hay can be fed, and, under certain circumstances, it may even be economical and quite satisfactory— for example, for dry cows. However, when buying poor hay, the purchase price should be lowered accordingly; and the feed analysis should also be used as a basis of balancing the ration. By the same token, it is usually good business to pay a premium for high-quality hay. Some dairy producers are very wisely applying an escalator principle to hay purchases. They may pay $1.00 to $1.50 per ton for each 1% of protein above an agreed-upon figure; or they may dock the price by a corresponding amount if the content is lower. For example, if a vendor guarantees to deliver alfalfa with 15% crude protein and it is agreed that a $1.50 per ton premium will be paid for each 1% protein in excess of this figure, a $4.50 per ton premium would be added for alfalfa running 18% crude protein.

In some cases, hay is also purchased on the basis of moisture content; and the price drops as the moisture content increases—thereby discouraging selling water. For example, if 15% moisture is agreed upon, hay running 18% moisture would be docked 3% in price. Thus, if the base price of hay is $100.00 per ton, the price would be lowered to $97.00 per ton ($100.00 × 3% = $3.00; then, $100.00 – $3.00 = $97.00).

Others use crude fiber in the same manner—paying a premium for hay of low fiber content. Still others apply a more sophisticated evaluation, such as near infrared reflectance spectroscopy (NIRS), to arrive at the dollar value of hay.

MAKING QUALITY HAY

Fig. 8–6. Making quality hay in Colorado showing hay raking in the foreground and hay baling in the background. (Courtesy, Soil Conservation Service, Washington, DC)

TABLE 8–5
APPROXIMATE CHEMICAL COMPOSITION (MOISTURE-FREE) OF VARIOUS SUN-CURED HAYS

Kind of Hay	Crude Protein		Crude Fiber	
	Average	Range	Average	Range
	(%)	(%)	(%)	(%)
Alfalfa	15.5	12–19	29	22–36
Bromegrass	10.5	6–15	28	24–31
Ladino clover	18.5	16–21.5	22	18.5–23
Red clover	12.0	10.5–18.5	27	18–34
Lespedeza	13.0	11.5–14.5	27	22.5–32.5
Oat hay	5.0	4–6	28	26–32
Orchardgrass	8.1	6–14	30	26–31
Soybean	14.5	9–16.5	28	20.5–41
Timothy	6.5	5.5–9.5	30	28–31.5
Sudangrass	8.8	6.5–11	28	26–31.5

The object of haymaking is to (1) harvest the crop at the optimum stage of maturity, and (2) cure properly.

Hay quality begins with the soil and ends with the manager, with many intermediate factors affecting it along the line. Once forage is cut, opportunities to increase nutrient content are over; from that point on, quality can only be preserved.

There is no one best haymaking method or kind of equipment. These must necessarily vary with the size of the operation, the kind of hay, the climate of the area, the individual farm conditions and buildings, and the available labor and machinery and their cost. Yet, the principles of good haymaking and the objectives sought are the same everywhere.

GROWING FORAGE

Growing forage for hay has long been neglected. Average per acre yields are still well under one-half their potential. Little more than 1 acre of hay in 10 is fertilized on a regular basis; and the precious few acres that are fertilized get an average of only 12 to 15 lb per acre—a paltry amount compared to corn, which receives an average of about 200 lb of fertilizer per acre.

The steps in growing quality hay are:

1. **Match crop to soil.** Some forage crops will do better on certain soils than others. So, the crop should be fitted to the soil.

2. **Choose quality seed, and proven varieties and mixtures.** Most grass and legume seeds are extremely small and contain very little stored food material. Thus, it is important that good seed be used.

Also, recommended varieties should be selected. If there is a choice, consider simple grass-legume mixtures. Research has shown that in many areas and over a period of years such mixtures are frequently higher yielding and more persistent than legumes grown alone. Mixtures are also easier to harvest and cure as hay.

3. **Lime and fertilize.** To obtain top yields of hay, lime the soil (if needed) to a pH of 7, then use the right kinds and amounts of fertilizer for the species and soil. Legumes, such as alfalfa, need a nearly neutral soil to promote growth of nodule-forming bacteria. Fertilization should be based on soil test.

4. **Get good stand.** A good stand is important. In this connection, it is noteworthy that it takes about one-half million successfully established individual forage plants per acre for a productive stand.

5. **Irrigate where practical.** In arid and semiarid regions, irrigation is a must for hay production. Whether or not it will pay to irrigate hay crops in humid or semihumid areas, will depend on the water supply, irrigation costs, and returns.

6. **Control insects and diseases.** Forage insects destroy a lot of hay. But they can be controlled by using a proper combination of insecticides, cultural control practices (such as timely cutting), and biological control agents.

HARVESTING AT PROPER STAGE

Whether the crop is a grass or a legume, or a combination, the stage of maturity of the plants at the time of harvest affects digestibility, yield, and feeding value (see Fig. 8–7). Young, immature plants are high in protein and low in fiber or lignin. As hay crops mature, feeding value goes down and fiber content increases. Digestibility of the forage (TDN) declines about 0.5% each day cutting is delayed beyond the early bloom stage (Fig. 8–7); and the intake of forage decreases during this same period at more than 0.5% each day. Thus, in total, the feeding value of forage drops more than 1% for each day's delay after early bloom.

Forage dry weight yields increase until midbloom to late bloom stages (Fig. 8–7). Timothy and bromegrass fully headed, and red clover and alfalfa at full bloom, will give maximum yield of dry matter. However, maximum feeding value of first cutting forage is reached at least 10 days before the time of maximum dry weight yield (Fig. 8–7).

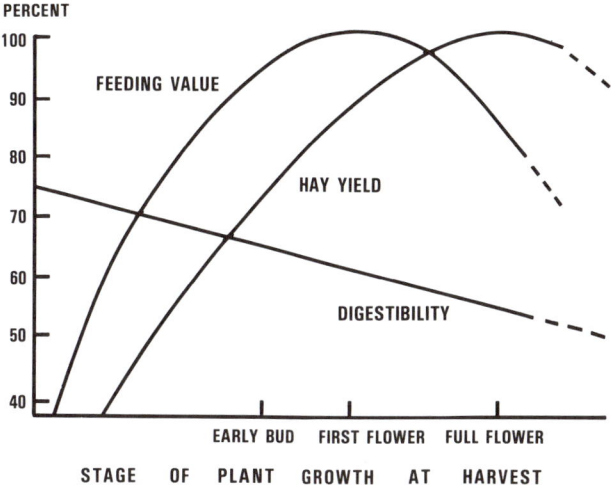

Fig. 8–7. Effect of advancing maturity on the feeding value, yield, and digestibility of alfalfa hay. Note that maximum feeding value per acre is reached 10 to 15 days before maximum yields.

Table 8–6 (next page) shows the primary changes in chemical constituents of alfalfa (a legume) and timothy (a grass) as affected by maturation. Note that, for both species, the DE, TDN, CP, lysine, Ca, and P decrease with maturity.

As shown in Fig. 8–7 and Table 8–6, advancing stage of maturity at harvest affects the chemical composition of hay adversely. More importantly, it has a direct and dramatic effect on animal production.

In Wisconsin, first-growth alfalfa-bromegrass hay was harvested each year for 3 years at four stages of maturity. Dry matter digestibility, digestible energy, and milk produc-

TABLE 8–6
EFFECT OF STAGE OF MATURITY AT HARVEST OF ALFALFA AND TIMOTHY ON COMPOSITION OF HAY[1]

Forage Crop	Stage of Maturity	DE[2]	TDN[3]	CP[4]	Ly-sine	Ca	P
		(Mcal/kg)	(%)	(%)	(%)	(%)	(%)
Alfalfa . . .	Prebloom	2.78	63	19.4	1.10	2.10	0.34
	First flower[5]	2.42	56	17.9	0.94	1.75	0.28
	Midbloom	2.29	52	16.0	0.90	1.50	0.25
	Full bloom	2.16	49	15.0	0.64	1.29	0.25
Timothy . .	Prehead	2.20	50	11.5	—	0.50	0.25
	Head	1.98	45	9.0	—	0.41	0.19

[1]Adapted by the author from: Rohweder, D. A. and R. Antoniewicz, *Alfalfa, the high quality hay for horses*, produced by Certified Alfalfa Seed Council, Inc., Woodland, CA, p. 6., Table 1.

[2]Digestible energy. Divide by 2.2 to obtain Mcal/lb.

[3]Total digestible nutrients. Improved harvest techniques can increase values 10%.

[4]Crude protein.

[5]First flower to 1/10 bloom.

tion all declined sharply with increasing maturity of the hay (Table 8–7).

TABLE 8–7
EFFECT OF STAGE OF MATURITY AT HARVEST OF ALFALFA-BROMEGRASS ON MILK PRODUCTION[1]

Stage of Maturity	Dry Matter Digestibility[2]	Digestible Energy[3]	4% Milk Production[4]
	(%)	(kcal/g)	(lb/day)
Vegetative	71.4	3.20	45
First flower	64.6	2.86	29
Full bloom	58.0	2.54	15
Green seed pod . .	55.2	2.43	4

[1]Adapted by the author from University of Wisconsin data. The alfalfa-brome was first cutting, made at Arlington, WI; and the data are an average of 3 years.

[2]Animal digestion trial data, values similar to total digestible nutrients.

[3]Animal energy digestibility × forage gross energy.

[4]Estimated for a 1,200-lb cow fed hay alone.

Stage of maturity also affects the vitamin content of hay. Carotene (precursor of vitamin A) and the B vitamins decrease as plants mature. Vitamin D content is the one exception—it increases as the forage is sun-cured.

Everything considered, there is a loss of about 1% in nutrient value for each day that hay harvest is delayed beyond the late vegetative stage of growth.

Table 8–8 gives guidelines relative to the proper forage-harvesting stage for maximum protein and minimum fiber.

TABLE 8–8
HAY CUTTING GUIDE

Kind of Hay	When to Cut
Alfalfa	Bud stage for first cutting; 1/10 bloom for second and later cuttings.
Alsike clover	Early bloom for 1/2 bloom stage.
Bermuda	When 16–18 in. tall, before lodging.
Birdsfoot trefoil	First flower to full bloom.
Bromegrass	Heads emerging.
Cowpeas	When pods are 1/2 to fully matured.
Crested wheatgrass	When the plants begin to head.
Crimson clover	From early bloom to 1/2 bloom.
Fescue	Boot to early head stage.
Grass-legume mixtures	When the legume is at the proper stage.
Johnsongrass, millet, Sudan-grass, sorghum hybrids . . .	40 in. height or early boot stage, whichever comes first.
Ladino clover	Few b,ooms to full bloom.
Lespedeza, annual	Early blossom.
Orchardgrass	Boot to early head stage.
Red clover	Late bud.
Sericea	When 12–15 in. high.
Small grains (oats, barley, wheat)	Boot stage to early dough stage.
Soybeans	Mid- to full-bloom and before bottom leaves begin to fall.
Sweet clover	Bud to very early flowering stage.
Timothy	Boot to early head stage.

CUTTING AND FIELD CURING HAY

Fig. 8–8. Hay mower-conditioner. This equipment reduces drying time and field losses. (Courtesy, Ford New Holland, New Holland, PA)

Fig. 8–9. Haymaking has gone modern! It is no longer a backbreaking, pitchfork job. This shows a pick-up baler with ejector—a bale loader. (Courtesy Ford New Holland, New Holland, PA)

Proper cutting and field curing of hay embraces all the steps from cutting to ready-for-packaging or storing. In modern hay making, it includes (1) cutting, curing in the swath and windrow, and raking or cocking; (2) reducing moisture content, while minimizing shattering, bleaching and fermenting; (3) reducing rain damage; and (4) considering chemical and preserving agents.

CUTTING/CURING IN THE SWATH AND WINDROW/RAKING OR COCKING

The common steps, methods, and equipment used in cutting and field curing hay are as follows:

1. **Cutting and curing in the swath or windrow.** Cutting, followed by curing in the swath or windrow, is the first step in haymaking, regardless of the subsequent method or type of equipment employed.

Any one of several types of mowers may be used, for all of them are designed to get the hay down. The most important thing is that the hay be cut at the proper stage of maturity (as noted in Table 8–8).

2. **Raking.** After the hay has wilted sufficiently in the swath, but while it is still tough and the leaves will not shatter, it should be windrowed. For this assignment, the side-delivery rake is preferred to the dump rake. The side-delivery rake rolls hay into fluffy, cylindrical windrows, which allows for good circulation of air; whereas dump rakes produce large windrows which are apt to remain damp underneath and bleach excessively on top. Where the hay crop is exceedingly heavy, the size windrow can be kept desirably small by limiting the width raked into each windrow.

If considerable shattering appears probable, it may be desirable to do the raking early in the morning when the dew makes the hay a bit tough.

Where windrowed hay is rained on, wait until the top half dries out, and then turn it upside down with the side-delivery rake (the use of the tedder for rewindrowing is not recommended because of excessive shattering).

3. **Cocking.** Formerly, well-made cocks, often adorned by hay caps, were considered a necessary part of good haymaking. However, this practice has greatly decreased, due primarily to higher labor costs and the advent of modern haymaking machinery. Today, the cocking of hay is confined almost entirely to use (a) in hot, arid regions where the leaves shatter if the hay is left in the swath or windrow for any appreciable length of time, and (b) as an emergency measure in order to protect hay when a storm is imminent.

REDUCING MOISTURE CONTENT/SHATTERING/ BLEACHING AND FERMENTING

Proper curing ensures that (1) the hay can be stored safely without heating excessively or becoming moldy; and (2) the maximum leafiness, green color, aroma, nutrient value, and palatability shall be retained. To the end that these desired objectives may be achieved, the following information is pertinent:

1. **Moisture content.** Freshly cut forage contains 75 to 80% moisture, whereas the maximum moisture content for safe hay storage is as follows:

> For loose hay—25% moisture.
> For baled hay—20 to 22% moisture (the lower figure for larger bales).
> For chopped hay—18 to 20% moisture.
> For cubes—16 to 17% moisture.

Hay of a higher moisture content than indicated should not be stored because (a) its value may be greatly lowered due to mold or to nutrient losses accompanying fermentation, and (b) of the ever-present danger of spontaneous combustion and a costly fire.

2. **Shattering losses.** Legume forages contain a larger proportion of leaves than do grasses, but unfortunately, the fine, thin legume leaves dry out more rapidly than the coarse stems to which they are attached. This results in considerable shattering losses, unless great care is taken. The importance of this condition is readily apparent when it is realized that in alfalfa, for example, 50% of the total weight of the plant is contained in the leaves, but the leaves contain 70% of the protein and 90% of the carotene content of the entire plant.

In field curing hay, losses from leaf shattering range from 2 to 5% for grass hay and 3 to 39% for legume hays, with as much as 15 to 20% for legume hays field cured under the most favorable conditions. Based on extensive experiments with field-cured alfalfa hay, the U.S. Department of Agriculture reported that leaf losses averaged 38.5% when none of the hay was wet; 47.3% when the hay was wet by 2 showers; and 74.5% when the hay was wet by 3 showers—and milk production per acre was 19.7% less per acre when cows were fed rain damaged, field-cured alfalfa hay in comparison with field-cured hay without damage by rain.

3. **Bleaching and fermenting losses.** In general, the carotene or provitamin A content of freshly cured hay is proportional to the greenness. With severe bleaching, more than 90% of the vitamin A potency may be destroyed.

Even under the best of conditions, there is an unavoidable loss through fermentation, especially losses in sugars, starch, and carotene. With good weather and proper curing methods, however, these losses will not be excessive.

REDUCING RAIN DAMAGE

The leaching losses from rain are less severe soon after mowing, but increase in severity as curing progresses. Also, repeated showers are more damaging than one heavy rain. Experimental studies have revealed that damaging rains may lower the feeding value of hay by one-fourth to one-third, or even more with severe exposure.

Losses from weather damage may be reduced (1) by using haymaking equipment that reduces the field drying time, (2) by understanding and using existing weather aides, and (3) by using proven chemical conditioning and preserving agents.

CONSIDER CHEMICAL CONDITIONING AND PRESERVING AGENTS

Chemical hay drying agents and preservatives are giving haymakers a big assist in lessening haymaking losses and improving hay quality. The big advantage of these products is that they speed up the haymaking process and reduce exposure to weather damage. In comparison with no treatment, the use of a desiccant, or drying agent, along with mechanical conditioning, can reduce the moisture content by an additional 2 to 10% during a 24-hour period. Adding a preservative to hay that is in the 25 to 35% moisture range will allow it to be baled and stored without undue heating.

■ **Chemical conditioners** — Chemical drying agents, which are sprayed on the crop at mowing time, break down the waxy cutin layer on the wall of the stem and allow moisture to escape, thereby promoting faster drying time, with the drying rate of the stems approaching that of the leaves. Several chemicals can be, and are, used for conditioning, including potassium carbonate, sodium carbonate, and sodium silicate. Also, methyl esters of fats, vegetable oils, or animal fats have been mixed with potassium carbonate in an attempt to increase the effectiveness of chemical conditioning.

Chemical conditioners are effective on legumes such as alfalfa, birdsfoot trefoil, and red clover, but, generally, they are not effective on grasses. Although they will reduce drying time on all cuttings of legumes, they are most effective on second and third cuttings and least effective on first and late autumn cuttings. This situation is attributed to the fact that conditioners work best when drying conditions are best (in the summer), and that first cutting has heavier yields and heavier swaths than later cuttings — conditions that hamper drying, because the moisture movement inside the swath is inhibited.

Studies show that drying agents are more effective as an addition to, and not as a substitute for, mechanical conditioners. The chemical of choice is applied at the time of cutting, by either of two techniques: (1) a spray boom mounted ahead of the reel; or (2) spray nozzles are mounted behind the reel, but in front of the conditioning rollers so that the rollers help distribute the spray.

■ **Preserving agents** — Under normal conditions, for safe baling, a moisture content of 20% or less is a must. Studies in many states have shown that, if properly treated with an adequate amount of the right preservative, alfalfa hay can be baled at 25 to 30% moisture, thereby speeding harvesting and lessening losses significantly. Preservatives act as fungicides and inhibit the growth and reproduction of the microorganisms that cause heating and molding in wet hay. A brief description of each of the common preservatives follows:

1. **Organic acids.** Propionic acid is the organic acid of choice. It is sometimes mixed with acetic acid, inorganic acids, formaldehyde, water, flavoring ingredients, and/or antioxidants. But, to be most effective, organic acid formulations should have at least 60% propionic acid; should be applied at the proper rate, depending on the moisture content of the hay; and must be uniformly distributed throughout the hay mass.

2. **Anhydrous ammonia.** When properly applied, anhydrous ammonia will stop bacteria and mold growth; and when applied to poor quality hay, it has the added advantage of increasing protein and digestibility. However, as a preservative it is not as effective as propionic acid. Also, unless large round bales are covered and/or contain less than 28% moisture, too much of the ammonia escapes. Excessive amounts of ammonia may cause animal disorders.

Where high-quality alfalfa hay is involved, which is already high in protein and digestibility, it is doubtful that the added expense of using ammonia can be justified.

3. **Bacterial inoculants.** Claims are made that most bacterial inoculants on the market will produce lactic acid, which acts as a fungicide and inhibits mold growth. More experimental work is needed, substantiating the effectiveness of bacterial inoculants as hay preservatives.

VALUE OF DIFFERENT CUTTINGS OF ALFALFA HAY

In many areas, there is a decided preference among livestock producers in favor of a certain cutting of alfalfa hay.

Generally, first cutting alfalfa hay is coarser stemmed and less leafy than later cuttings, and therefore of somewhat lower feeding value when the different cuttings are equally well cured. Also, the weather is often less favorable for curing the first cutting. Yet, dairy producers participating in the Wisconsin Forage Analysis Superbowl Contest, sponsored by the Wisconsin Forage and Grassland Council, report (*Holstein World*, March 10, 1986, p. 90) that they obtain higher milk production from feeding higher fiber first cutting alfalfa than from lower fiber subsequent cuttings. For unknown reasons, this appears to contradict the expected performance based on average chemical analysis, from feeding different cuttings of alfalfa hay. As shown in Fig. 8–10, on the average, each successive cutting is lower in crude fiber and higher in crude protein.

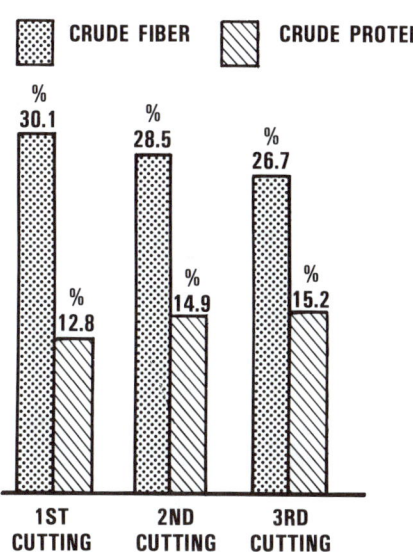

Fig. 8–10. Each successive cutting of alfalfa hay tends to be higher in nutritive value. Note that each successive cutting is lower in fiber and higher in protein. (Adapted by the author from *Atlas of Nutritional Data on United States and Canadian Feeds*, NRC-National Academy of Sciences, 1972)

HAYMAKING SYSTEMS

In haymaking, the term *system* refers to a team of processes and machines that does the work from field through feeding, saves crop nutrients, reduces manpower requirements, and eliminates drudgery. When each step is mechanized, it must be coordinated; otherwise, workers and machines end up waiting.

In recent years, automation has had great impact on haymaking. Some haymaking systems are completely mechanized from field to feeding.

There is no one best haymaking system for all conditions. Nevertheless, all good systems are fast, make handling easy, save labor and nutrients, and increase profits. Baling is the most popular hay-handling system in North America.

LONG, LOOSE HAY

The acreage harvested as long, loose hay has declined sharply in recent years, especially in the humid areas, because (1) of high labor cost, and (2) long hay is too bulky for mechanized feeding. Nevertheless, long, loose hay is still popular in many western areas where specialized handling equipment is used. Moreover, some of the newer systems of handling and self-feeding loose hay show promise.

The two common methods of handling long hay are:

1. **Loading with hay-loader directly from windrows.** In this method, cured hay is loaded on a truck or wagon directly from the windrow by means of a hay-loader. Usually the hay is then transported to a barn or stack where it is unloaded by fork or sling and moved away by hand. Sometimes it is chopped into the barn or other storage area.

2. **Hauling cured hay from windrows or cocks with buck rakes, sweep rakes, or sled.** In the West, much of the hay is cured in windrows or cocks and then transported by buck rakes, sweep rakes, or sleds to field stacks where it is stacked by hay stackers or other large mechanical devices. Then after going through a sweat in the stack, the hay is fed out as loose hay, or, if intended for market, it is baled.

Without doubt, this method results in the production of the highest percentage of good-quality hay of any known method, primarily because (a) more latitude is permissible in the moisture content when stacking than when baling or chopping, (b) the practice predominates in an area which normally has good haying weather, and (c) the method is prevalent on farms and ranches where haymaking is frequently a major enterprise and where the operators have the know-how to produce good hay.

CHOPPED HAY

Chopped dry hay fits into some feeding systems, particularly in the West.

For safe storage, the moisture of chopped hay should not exceed 18 to 20%.

Two common methods of chopping freshly cured hay follow:

1. **Field chopping cured hay directly from the windrow.** In this method, a field chopper gathers the cured hay from the windrow, chops it, and blows it into a truck or trailer. The chopped hay is then blown into the barn, stack, or other storage area.

Fig. 8–11. Field chopping cured hay from the windrow. (Courtesy, Gehl Company, West Bend, IN)

2. **Chopping into the barn or other storage area.** In this method, the cured hay is generally hauled from the windrow or cock to the barn or other storage area where it is chopped by a hay chopper or silage cutter and blown directly into the storage area. This method is slower and requires more labor than where field-cured hay is chopped directly from the windrow, but less expensive equipment is necessary.

PACKAGED HAY

Great strides have been made in hay packaging in recent years, characterized by the advent of round bales, large rectangular bales, loaflike stacks, and cubes.

Although large round bales and compressed stacks are better adapted to outside storage than small round bales, unrestricted access at feeding will result in excess wastage. Large rectangular bales are suitable for commercial marketing and long distance shipping.

Compressing hay into cubes and pellets makes for many advantages; among them, (1) completely mechanized handling; (2) high density, with more economical transportation and storage; (3) easier self-feeding and higher intake by animals; and (4) lower feeding losses.

BALES

The following choices of bales are available:

1. **Rectangular bales.** Conventional, small, rectangular packages (often called *square bales*), weighing 60 to 140 lb, were produced by the first baling machines; and they are still popular on farms and ranches that produce hay for their own use.

Today, hay is also being packaged in large rectangular bales, weighing from 1,000 to 2,000 lb, designed for custom operators or hay growers with large volumes of hay or straw.

Fig. 8–12. A baler making large rectangular bales which weigh up to 2,000 lb. Commercial hay producers like these big rectangular bales because they are easy to transport and store. (Courtesy, Ford New Holland, New Holland, PA)

2. **Large round bales.** Many makes and models of large round balers are on the market. All of them can be classified by the method of rolling the hay into a bale. One method is to pick up the hay from the windrow and roll it in a chamber, between a series of belts or chains. The other method is to roll the windrow on the ground, similar to rolling up a carpet. Both methods produce a rounded shape which gives weather protection. Large round bales range in weight from 850 to 2,000 lb, depending on the make of equipment.

Fig. 8–13. Round baler in operation. (Courtesy Ford New Holland, New Holland, PA)

STACKS

These are loaf-shaped (one system makes a circular stack), mechanically pressed haystacks. Long, loose hay is blown into a wagon and pressed down by a hydraulically operated canopy roof. Stacks range in size from 7 to 10 ft wide, 8 to 22 ft long, and 8 to 11 ft high, and weigh from 1 to 6 tons.

Fig. 8–14. This stacker produces tight, loaflike stacks of hay weighing up to one ton. (Courtesy, Hesston Corp., Hesston, KS)

CUBES (WAFERS)

Field cubers are machines that move across hayfields, pick up windrows of forage, and produce dense, high-quality forage cubes or wafers. Stationary cubers are used to produce similar cubes from loose haystacks or bales.

Hay cubes are of special interest to dairy producers because they have the advantages of pellets, without their disadvantages. Like finely ground forage that is pelleted, cube-feeding can be readily automated; and, in comparison with long hay, there is less transportation and storage cost. Besides, cubes will not lower the fat content of the milk as much as pellets when appreciable quantities of them are fed.

The Utah Station researchers compared cubed alfalfa hay with each (1) baled alfalfa, and (2) corn silage.[2] These studies showed that consumption of feed is higher when alfalfa hay is fed in cubed than in baled form, thereby increasing the potential for milk production. In one study, cubes resulted in a slight lowering in milk fat percentage; in the other study, cubes did not affect milk fat percentage.

PELLETS

Fig. 8–15. Hay cuber in operation, making cubes 1¼ in. square and 2 to 3 in. long. (Courtesy, Deere & Company, Moline, IL)

Pelleted forages are finely ground, then condensed.

The biggest deterrent to increased pelleting at the present time is the difficulty of processing chopped forage coarse enough so that it will not cause digestive disturbances or lower the fat content of milk. A minimum of a ¼-in. chop is recommended.

ARTIFICIAL DRYING

The use of forced air, either heated or unheated, for final drying is the most dependable way in which to preserve quality in hay. The application of heat provides faster drying, saves more leaves, and reduces losses of nutrients. Yet, artificial drying is on the decline in the United States because of the added cost in heating the air and the increased labor.

---

[2]Anderson, M. J., *et al.*, "Cubed Versus Baled Alfalfa for Dairy Cows," *Journal of Dairy Science*, Vol. 58, No. 1, 1975, p. 72.

MOW CURING

Mow curing, or drying, refers to the practice of curing partially dried hay—either long, chopped, or baled—in barn mows equipped with ventilation systems through which either unheated or heated air is forced.

Because of the relatively high cost involved in mow curing hay, in areas where poor haymaking weather generally prevails and field curing is hazardous, farmers may well consider the desirability of making silage.

ARTIFICIAL DEHYDRATORS

Artificial dehydrating refers to a process in which forage is taken from the field as soon as it is cut (or in some instances after wilting), put through a hay chopper or silage cutter, and dried in large driers of various types, following which it is finely ground. For the most part, this method of curing is limited to large commercial operations which process early cut alfalfa (or its leaves) and other legume and/or grass crops chiefly as a supplement for swine and poultry. Occasionally, artificially dehydrated hay is produced for other classes of animals, including dairy cattle, especially in those areas which rarely have good haymaking weather.

The most popular type of artificial dehydrator in use in this country is one that uses a high initial heat (1,200 to 1,400°F), and which is heated by gas or oil. In a good drier, the forage does not get sufficiently hot to be damaged, primarily because of the cooling effects produced when the water evaporates from the plant tissues, and because the forage is in contact with the hot air for only a few minutes.

Any burning of the leafy portions of the forage indicates that the temperature was too high or that the forage was overdried.

Except in special circumstances, it is seldom economical to feed artificially dehydrated hay to dairy cattle.

WAGON DRYERS

Wagon dryers were developed to reduce the high labor requirements of batch or platform drying. Essentially, they are batch dryers on wheels.

Some wagons are covered; they are known as *covered wagons* because of the outward appearance of the ballooned cover on the wagons during drying. The cover consists of a durable, lightweight material that will not leak water or air. Heat is conveyed under the cover. Other wagons are open and adapted to drying in a shed. In both types, the drying wagons are loaded in the field directly behind the baler, taken to the drying system where they are connected to a main air duct, and the hay is dried. After drying, they are taken to the storage place and unloaded. This procedure eliminates the unloading and reloading of wagons required in batch or platform drying.

Although good-quality hay can be produced by wagon drying, it requires more labor and handling than most other systems of haymaking. Additionally, drying costs money, both for equipment and heat.

STORING

Good hay should never be poorly stored. Naturally, the type of storage will vary from area to area. In the more arid sections where little rainfall comes during the fall and early winter, a good stack of loose or baled hay may provide entirely satisfactory storage. On the other hand, in high-rainfall areas, more expensive waterproof storage should be provided. At and between these two extremes, hay may be and is successfully stored in many different ways in different sections of the country.

Where different kinds and qualities of hay are produced or purchased, each kind and each quality should be stored in such manner that it will be accessible when needed. Otherwise, it may not be convenient to provide for variety in feeding and to feed some of the low-quality along with some of the high-quality hay.

BROWN HAY

Brown hay results when, due to inclement weather, hay wilted to about 50% moisture content is stocked or placed in storage. The damp mass soon ferments extensively, and heats (preferable not higher than 175°F). As a result of this action, the hay darkens in color, from dark brown to nearly black. Also, it becomes sweet, aromatic, and palatable. But as a result of the fermentation and heating to which it is subjected, the forage is lower in digestible protein, in total digestible nutrients, in vitamin content, and in feeding value.

Because of its lowered nutrient and feeding value and the danger of spontaneous combustion, therefore, the intentional making of brown hay is not recommended.

ADDITIVES AND PRESERVATIVES FOR HAY

Farmers in many countries of the world have traditionally added about 20 lb of salt per ton of new hay, at the time of stacking or putting it into the hay mow, in the belief that the salt would prevent the hay from molding and heating. Carefully controlled experiments have failed to substantiate claims that salt will prevent excess heating or sweating; nor has it prevented spontaneous combustion of hay. However, when salt is used in moderate amounts, it may improve the color, aroma, and palatability of poor-quality hay. It is recognized, too, that much higher levels of salt—quantities sufficiently high to harm animals—may prevent mold.

Over the years a number of products, both liquids and powders, said to preserve hay have become commercially available. While claims have been made that there will be no heating or molding when these materials are used, the results have been highly variable.

Recent studies of several state experiment stations have shown that propionic acid or a combination of propionic and other organic acids can be used successfully to preserve baled hay stored at 30% or less moisture. Anhydrous ammonia and ammonium isobutyrate have also been found to be effective in preventing heating and preserving the quality of high-moisture hay. In general, depending on the moisture content of the hay, application rates of 1 to 1.5% (20 to 30 lb/ton) of the additive are required for effective preservation.

Researchers are continuing to evaluate chemical hay preservatives and methods of application. But, based on presently available information, the following recommendations appear to be justified:

1. Generally, grass hay baled at less than 25% moisture, and legume hay at 20% moisture or less, will keep satisfactorily without a preservative.

2. For legume hay with between 20 to 25% moisture, apply at least ¼ lb of actual propionic acid per each 50-lb bale. This means that if the commercial hay preservative contains only 20% propionic acid, 1¼ lb of it should be added per each 50-lb bale.

3. For hay running 25 to 30% moisture, apply at least ½ lb of actual propionic acid per 50-lb bale.

4. Acid preservatives should be used only in situations where hay *must* be baled at high moisture (above 20% for legume hay, and above 25% for grass hay).

5. Complete and uniform coverage of the treated hay is essential for successful preservation. While other methods of application may be developed, the most practical method appears to be spraying it onto the hay as it enters the baling chamber.

SPONTANEOUS COMBUSTION

Wet hay ferments and generates heat. Sometimes this results in spontaneous combustion and fire, usually about a month to six weeks after storing. Here are the facts:

1. **Symptoms of heating.** The warning signals are hay that feels hot to the hands, strong burning odor, and visible vapor.

2. **Temperature of hot hay.** Hot spots may be located by probing the hay with a steel rod. Then the temperature of the hot spots may be tested with a thermometer (a dairy thermometer or other type) attached to a wire and dropped down a pipe. If the hay is over 140°F, it should be checked periodically during the day.

If the hay is 160°F, it should be checked hourly.

If the hay is 180°F, there are apt to be fire pockets; and it should be removed from a barn.

3. **Cooling hay.** Hay that is heating may be cooled by discharging through pipes into the hot areas either dry ice or liquid carbon dioxide.

4. **Removing hot hay.** When a fire is imminent and hot hay most be removed, it is important to have plenty of help on hand including the fire department. Then the hay should be removed cautiously and without wetting unless necessary.

5. **Precautions.** Never walk on hay that is heating—place planks over it, and do not breathe hot and noxious fumes.

BUYING AND SELLING HAY

Historically, most hay has been fed on the farms where it was produced. But this practice is changing. Today, about 29 million tons, or about 25% of the U.S. production, with a cash value of over $2 billion, is sold off the farm. The Northwestern states of Washington, Idaho, and Oregon, along with California and Texas, are the leading states in total value of hay sold. In California alone, the annual value of hay sold exceeded $425 million in the late 1980s. In four states—California, Arizona, New Mexico, and Washington—some 40% or more of the hay produced is sold. In California, nearly 70% of the hay produced is sold.

New hay markets are developing. As dairy producers become larger and more specialized, they prefer to care for animals and rely on other specialists to grow hay.

Additionally, considerable hay is exported, primarily to Japan, Korea, and Taiwan. In order to save space and cut down transportation costs, this generally requires a special hay package—pellets, wafers, or high-density bales.

HOW HAY IS SOLD

Fig. 8–16. Marketing hay is big business. (Courtesy, Ford New Holland, New Holland, PA)

New methods of marketing hay have made selling easier, but visual inspection is still the most common method of assessing quality and price when either selling or buying. However, the development of Near Infrared Reflectance Spectroscopy (NIRS) as a rapid, accurate, and precise method of measuring hay quality is having a major impact on the marketing of hay.

The traditional market channels for hay in the United States are as follows:

■ **Hay dealers or brokers**—Hay dealers or brokers are important suppliers of hay, as evidenced by the continued growth of the National Hay Association. In California, a high percentage of the hay is marketed through this channel. For the nation as a whole, dealers and brokers market 5 to 10% of the hay produced. They purchase hay from growers and sell it to consumers.

■ **Neighbor to neighbor**—This is the oldest market channel for hay, and it is still quite common. This is a disorganized market without any particular pricing structure. Hay is purchased on visual inspection.

■ **Associations and cooperatives**—These organizations vary in size, from very small to very large. The San Joaquin Hay Producers Association in California is one of the largest. Such associations normally purchase hay for cash and store it or move it to the consumer. This is considered a high-risk operation, because there are no futures markets that enable hedging protection.

■ **Auctions**—This is the most rapidly growing method of marketing hay. Originally, auctions simply brought together producers with loads of hay, an auctioneer, and an assembly of prospective buyers. Hay was sold by visual inspection, and the price determined by supply and demand. This approach is still widely used in many areas. Today, *Quality-Tested Hay Auctions* are becoming popular in some states, including Wisconsin and Minnesota.

■ **Contract**—This is an agreement between a hay grower and a livestock producer to supply hay of a specified quality at a prior agreed-upon price. It assures both the buyer and the seller an orderly market. Such eventualities as weather damaged hay should be covered in the contract.

Other approaches to marketing may have been developed. In several states, such as Oklahoma and Indiana, special computer assisted marketing systems, designed to help growers find buyers and to help buyers locate hay, have been developed. There is also a privately owned National Hay Exchange Corporation, which, by means of a computerized system, allows people to trade hay much easier, and in much less time, by matching buyer needs with seller offerings.

QUALITY-TESTED HAY AUCTIONS

In Wisconsin and Minnesota, two major hay producing and consuming states, the number of hay auctions has in-

Fig. 8–17. Automatic bale wagon stacking a loan. This type of equipment facilitates hay auctions. (Courtesy, Ford New Holland, New Holland, PA)

creased markedly in recent years. This is attributed to the introduction of *Near Infrared Reflectance Spectroscopy (NIRS)* equipment, usually in mobile units, as a rapid method of measuring hay quality prior to sale, plus the grading and selling of hay according to test results. These special auctions are known as *Quality-Tested Hay Auctions*.

The hay grading system used in Wisconsin and Minnesota is adapted from standards proposed by the American Forage and Grassland Council (AFGC) (see Table 8–9).

TABLE 8–9
FORAGE QUALITY STANDARDS FOR LEGUMES, GRASSES, AND LEGUME-GRASS MIXTURES[1]

Quality Standard	CP	ADF[2]	NDF[2]	DDM[3]	DMI[4]	RFV Index[5]
			— (% of DM) —			
Prime	>19	<31	<40	>65	>3.0	>151
1	17–19	31–35	40–46	62–65	3.0–2.6	151–125
2	14–16	36–40	47–53	58–61	2.5–2.3	124–103
3	11–13	41–42	54–60	56–57	2.2–2.0	102–87
4	8–10	43–45	61–65	53–55	1.9–1.8	86–75
5	<8	>45	>65	<53	<1.8	<75

[1]Standard assigned by Hay Marketing Task Force of AFGC.

[2]ADF = acid detergent fiber; NDF = neutral detergent fiber.

[3]Digestible dry matter (DDM, %) = 88.9 – (0.779 × ADF%).

[4]Dry matter intake (DMI, % of body weight) = 120 ÷ forage NDF (% of DMI).

[5]Relative feed value (RFV) calculated from DDM × DMI ÷ 1.29. Reference RFV of 100 = 41% ADF and 53% NDF.

Brief descriptions of the various factors included in the standards are as follows:

■ **Acid Detergent Fiber (ADF)** — Both animal and laboratory trials indicate that ADF is highly related to the digestibility of a forage. Factors which increase ADF content, such as increasing maturity, weathering, rain damage, and weeds, decrease digestibility.

■ **Neutral Detergent Fiber (NDF)** — Studies indicate that NDF is highly correlated with dry matter intake of the forage.

■ **Digestible Dry Matter (DDM)** — The accepted equation for predicting DDM of legumes, grasses, and legume-grass mixtures from ADF is:

$$DDM\ \% = 88.9 - (0.779 \times ADF\ \%).$$

■ **Dry Matter Intake (DMI)** — The amount of forage or feed DM an animal will consume is affected by how fast forages are digested and pass through the intestinal tract. The fiber fraction which appears to be most clearly related to the DMI of forages is Neutral Detergent Fiber (NDF). However, the exact NDF level in rations necessary to achieve optimum performance is uncertain. Wisconsin research indicates maximum feed intake in alfalfa-based dairy rations occurs when NDF is 1.2 lb per 100 lb of body weight.

$$DMI\ (\%\ of\ body\ weight) = \frac{120}{Forage\ NDF\ (\%\ of\ DM)}$$

■ **Relative Feed Value (RFV)** — Relative feed value is an index which combines important nutritional factors (potential intake and digestibility) into one number for a quick, easy and effective method of evaluating feeding value or quality. The formula for calculating RFV is the estimated digestibility and potential intake of a forage calculated from ADF and NDF fractions, respectively.

The calculation of RFV is made by multiplying DDM by DMI, then dividing by 1.29. The number derived from the RFV calculation has no units and is used only as an index for evaluating quality of hay or haylage made from legumes, grass, or legume-grass mixtures. The RFV concept should be used to evaluate quality only for those forages listed above.

$$RFV = \frac{DDM \times DMI}{1.29}$$

The RFV does not include crude protein (CP) because CP is influenced by factors unrelated to those affecting RFV. The CP should be considered, however, in pricing forages; and CP values are included in Table 8–9 to indicate the range of CP for forages of different qualities.

Various studies show that both visual appraisal and a forage test are necessary to assess properly the quality of a forage. For example, forage should be inspected for absence of mold and for the presence of good green color, as well as for foreign matter. However, descriptions of forage quality for marketing should include RFV.

In Wisconsin, it is authoritatively estimated that prices in Quality-Tested Hay Auctions have averaged about $1 per ton higher for each 1% increase in relative feed value (RFV). This means that a lot of hay with a RFV of 125 would be worth about $5 more per ton than hay with a RFV of only 120. Furthermore, Wisconsin surveys indicate that four out of five buyers had confidence in using RFV for pricing hay.

When summarizing the average price received per quality grade for the 2 years 1983–84 and 1984–85 (Table 8–10), Wisconsin workers noted a definite correlation between price and quality, especially for the better grades.

TABLE 8–10
PRICE VS QUALITY GRADE SUMMARY
QUALITY TESTED WISCONSIN HAY AUCTIONS

Grade	No. Lots	Average Prices
		($/ton)
Prime	95	$117.21
1	296	103.14
2	239	79.25
3	50	68.22
4	65	65.33
5	18	63.59

This new dimension is expected to have a major impact in the future on selling hay through auctions. For example, in a survey sent to buyers, 96% preferred having quality information available; and 99% preferred to purchase hay with feed quality information attached. It is noteworthy, too, that Wisconsin workers report that both hay sellers and buyers are enthusiastic about the program.

University of Minnesota workers report that the results of Quality Tested Hay Auctions in Minnesota have been similar to those reported in Wisconsin; namely, (1) improved hay quality, (2) higher prices for hay, and (3) increased demand.

More recently, workers at the universities of Wisconsin and Minnesota have incorporated a *relative feed value (RFV)* into the pricing scheme. Such a method allows for consideration of intake calculated from neutral detergent fiber (NDF) in addition to protein and digestibility or TDN.

Scientists generally agree that the uniform chemical analyses can improve the marketing and economics of alfalfa hay, provided the method is (1) accurate, (2) accepted by both buyers and sellers, and (3) used extensively, throughout the United States and worldwide.

Pricing hay based on quality as determined by analysis has been a major breakthrough in hay marketing. But factors other than feeding value as determined by analysis can also be important in pricing hay. Workers in the state of New Mexico, for example, report that weed-free alfalfa hay has sold for as much as $145 per ton as opposed to $100 per ton for weedy hay of otherwise similar quality. Clean hay, i.e., high-quality alfalfa without weeds, is higher in nutrients, so it is better for high-performance dairy cattle. Also, weeds may (1) cause digestive problems in animals, and (2) decrease palatability.

NATIONAL ALFALFA HAY QUALITY STANDARDS

While several states, including Oregon, Minnesota, and Wisconsin, did implement the new American Forage and Grassland Council (AFGC) standards in their hay marketing systems, these standards were never accepted by either the Federal Grain Inspection Service, or by much of the hay industry.

But the earlier efforts did lead to the formation of the National Alfalfa Hay Quality Committee and the development of national alfalfa hay quality standards. The intent of these standards is to have uniform testing throughout the United States of one hay crop, alfalfa, so both seller and buyer can receive accurate and interpretable results on any given lot of hay. While these latest standards are based on a minimum test of alfalfa hay for dry matter, crude protein (CP), ADF, and estimated digestible dry matter (EDDM) calculated from the ADF, test provisions are also included for a description sheet of visual characteristics.

Procedures for the national alfalfa hay quality standards are carefully spelled out and include: (1) specific sampling procedures using an improved hay core sampler, (2) suggested visual factors to be used with chemical analysis to describe the sample, (3) approved testing procedures using either NIRS or wet chemistry, (4) development of an acceptance of Acid Detergent Fiber (ADF) to predict estimated digestible dry matter, and (5) a voluntary laboratory certification program operated by the National Alfalfa Hay Test Association in conjunction with AFGC and the National Hay Association.

The bulky nature and the variability within a given lot of hay as compared to most other crops is well recognized.

Also, it is known that, in addition to the natural variation of hay, sample variation is influenced by sampling method, sampling equipment, and handling of the sample. Thus, it is important that the approved procedure be followed carefully.

SUMMARY OF HAY SELLING

At today's feed prices, hay (especially alfalfa) is a profitable crop to grow, both for direct utilization through livestock and as a cash crop *provided* it is of high quality and priced at its true value for feed. However, there is general agreement that if hay is to take its rightful place in the marketplace as a major source of protein and energy, many of the traditional selling arts must be replaced by developments of modern science.

The introduction of Near Infrared Reflectance Analysis and the development of standards and grades based on feeding value have alleviated some of the excuses for not marketing hay on the basis of analysis. Also, sampling procedures have been improved. So, systems of pricing hay consistent with the quality of the product sold are close at hand.

But other developments are necessary if hay is to become a leading cash crop, including:

1. **More markets.** There is need for more markets — such as auctions, dealers, and associations — to provide a ready market outlet for hay growers, and to attract both ample supplies of hay and buyers.

2. **Price protection for dealers while they have possession of the hay.** There is need for something similar to the futures market that will permit hedging and protection.

3. **More hay grower storage facilities.** There is need for additional grower storage so as to spread hay marketing over a longer period and avoid harvest market gluts that depress prices.

4. **Better hay market information.** There is need for hay growers and hay buyers to be better informed relative to supply, demand, and going prices.

FREEDOM FROM TOXIC RESIDUES

With the emphasis on residues in foods, it is important that hay be free from those residues which are prohibited. If milk (or products derived from it) are found to have residues, the blame cannot be shifted to the hay grower by the producer, unless there is a clear-cut case of fraudulent representation. The best assurance of freedom from such residue rests with the integrity of the hay growers, or those who represent them in selling their hay.

It is important that dairy producers and hay growers be well informed relative to (1) the chemicals which are banned, and (2) the conditions of application for those chemicals which are still permitted. In this way, disastrous financial effects from confiscation of market animals or products can be averted. So, *read the label*. Use proper pesticides and observe minimum days from last application to marketing. Wear protective rubber gloves, respirators, and coveralls when applying pesticides, especially those that are highly toxic. Wash before eating.

HAY SHRINKAGE

Hay buyers should figure hay shrinkage closely. Here is why: If a ton of hay containing 90% dry matter is bought for $75; 1,800 lb of dry matter have been purchased at this price. However, if the $75 per ton hay contains 80% dry matter, only 1,600 lb of dry matter have been purchased for this same price. Purchase of the high-moisture, 80% dry matter hay has resulted in a loss of 200 lb of hay, or $\frac{1}{9}$ of the dry matter, worth $8.33 ($\frac{1}{9}$ × $75). Thus, if the 90% dry matter hay is worth $75 per ton, then the 80% dry matter hay is worth only $66.67 ($75 − $8.33) per ton. If 1,000 tons of hay are involved, that is a loss of $8,330 (1,000 × $8.33).

In addition to moisture losses, newly harvested hay may be expected to lose about 5% weight from going through the sweat.

HAY FEEDING FUNDAMENTALS

Feeding is the end of the line for hay. No matter how carefully it has been grown, harvested, and stored, all that has gone before can be dissipated if it is improperly fed—unless hay feeding fundamentals are observed.

Ruminants, with their four stomach compartments and the help of microorganisms, can subsist largely, or entirely, on bulky, high-fiber forages which, because of their low energy per unit weight of dry matter, must be consumed in large quantities to supply their nutrient needs.

The economics of the situation—the relative price of forage and grain—calls for greater emphasis on forage accompanied by less grain feeding. With greater quantities of forage incorporated in rations, it is expected that performance—the production of milk—will decrease. However, maximum net returns, rather than just maximum production, will be the primary objective.

Increasingly, forage testing will be used in two ways: (1) to purchase hay on a quality basis, and (2) to balance rations more precisely.

MAXIMUM USE OF HOMEGROWN FORAGES

Although increased quantities of hay will be bought, especially by large dairies, most forages will continue to be homegrown for the following reasons: (1) Hay is bulky and costly to transport, unless it is cubed or pelleted; (2) hay crops have an important place in most crop rotations; and (3) hay crops can be grown on wasteland not suited to other crop productions.

LEGUME VS GRASS HAYS

With the higher cost of both protein supplements and grains, legume hays have a particularly valuable place in dairy cattle rations. Legumes are also a rich source of many minerals and vitamins.

HAY PREPARATION

Hay is fed as long hay or in processed form. The common methods of processing for dairy cattle are chopping and cubing.

Fig. 8–18. Alfalfa hay—long and corral-fed. Hay is a very important source of energy and fiber for U.S. dairy cows. (Courtesy, Holstein-Friesian Assn. of America, Brattleboro, VT)

Considerable hay is chopped in the West, for two reasons: (1) It facilitates handling, and (2) it lessens refusal and waste. Low-quality and coarse forages usually benefit more from chopping than high-quality forages.

Fine grinding is not desirable for ruminants; it results in reduced rumen acetate production and lowered milk fat percentage.

Cubing (1) makes automatic hay feeding feasible, (2) decreases nutrient losses, and (3) eliminates dust. Also, it narrows the spread between high- and low-quality forage; that is, the poorer the quality of the forage, the greater the advantage from cubing. This is so because such preparation assures complete consumption.

Cubes offer most of the advantages of pelleted forages, with few of the disadvantages. They alleviate fine grinding, and they facilitate automation in both haymaking and feeding, and they lower milk fat percentage only slightly, if at all.

HAY FEEDING SYSTEMS

Hay may be either (1) self-fed, or (2) limited fed.

Most hay is self-fed. With a manger full of hay in front of them, hay consumption is limited only by the capacity of animals—by the amount that they can hold. That is the reason that ruminants eat more cubes and pellets than long hay of equal quality.

Limited feeding of hay is accomplished either (1) by hand-feeding the hay and the concentrate allowances, or (2) by using a complete, mixed ration.

An increasing number of large commercial dairies are switching to complete rations. Most experiments and experiences have not shown any difference between mixed rations and the feeding of roughage and concentrates separately insofar as amount and efficiency of milk production are concerned. However, a mixed ration has the following advantages:

1. It makes for greater efficiency in feeding and lessens the sorting at the feed bunk.

2. Where the roughage is relatively unpalatable, a mixed ration forces consumption.

3. When concentrate consumption is to be limited, mixing with the roughage is desirable.

FEEDING HAY PACKAGES

Large round bales, large rectangular bales, and stacks are other alternatives to conventional, 60- to 140-lb, rectangular bales. There is little doubt that in many dairy operations, especially for dry cows and young stock, large hay packages can greatly decrease labor and result in similar animal performance. However, special attention needs to be paid to methods of feeding these big hay packages; otherwise, waste can easily wipe out any saving in labor.

Under an in-field storage system, the bales are dropped where they are made, and remain there until needed for fall or winter grazing. Little or no labor is involved in making hay except for mowing, windrowing, and baling; and there is no manure to haul. Cattle graze the grass growth which occurs subsequent to baling and consume the bales in the field. There is little or no labor in feeding. The cattle go to the feed, rather than necessitating that the feed be taken to them.

Fig. 8–19. Portable round bale, metal, feeding fence placed around a large bale. (Courtesy, Scranton Manufacturing Company, Inc.,

The following methods are used in grazing round bales or small stacks, along with regrowth in the field:

1. **Continuous access to all bales.** In this system, the cattle are given continuous access to all the bales in a field.

2. **Strip grazing bales and stacks.** In this system, the cattle are given access only to those bales or stacks which are to be consumed within a given period of time. This is accomplished by using an electric fence and cross-stripping the field containing the bales or stacks. Such a strip-grazing program will increase the number of cattle days by at least 35%.

Pennsylvania State University researchers showed that cows given access to grass regrowth and large round bales during October and November wasted 26% of the available standing grass and round bales. It should be noted, however, that considerable hay is always wasted, even under the best of circumstances.

Strip grazing will work for both bales and small haystacks. With stacks, it is especially important to limit-feed or strip-graze in order to avoid excess wastage.

3. **Other feeding methods.** Other systems require more investment than in-field grazing of bales or small stacks, but the additional numbers of cattle carried per acre may justify the increased cost. Among such systems are the following:

a. **Hay packages (large round bales or small stacks) placed in rows and grazed with an electric fence.** Usually, one side of an existing hayfield is used for such a storage and feeding area.

b. **Portable feeding gates (fences).** Portable feeding gates (or fences), usually made of metal, may be placed around large bales or stacks. Their main advantages are that the animals cannot gain access to any part of the stack except that exposed to them through the gate, and that the animals cannot trample the forage, since the gate protects the hay. Most feeding gates are designed so that the animals can push them toward the bale or stack as the hay is consumed.

c. **Feeding wagons.** There are several designs and sizes of self-feeding wagons, both commercial and homemade. They give much flexibility; wagons can be easily taken to the area where the round bales are stored, then pulled to various feeding locations.

d. **Three-sided feeder.** Bales may be placed on concrete and enclosed in a three-sided feeder.

The following points are pertinent to feeding large round bales and stacks:

1. **Frozen ground works best.** In-field grazing of big hay packages works best on frozen ground, because cattle eat more of the hay, and trample less of it than when on soft ground.

2. **Weed spots can result.** Small bales do not leave bare spots following grazing in the field. A sod readily pervades the space beneath the bales. However, when large bales or small stacks are left in the field throughout summer, fall, and most of the winter, bare, weedy spots are likely to appear.

3. **Dairy cows can use them, too.** Although most of the large bales and stacks are being used by beef cow-calf operators, they can also be used advantageously for dairy herds, especially for dry cows and young stock. Hay can be provided to a dairy herd in a three-sided feeder.

4. **Feeding space.** On a per cow basis, 18 to 24 in. of hay feeding space per cow should be allowed if all animals are to eat at the same time. Six to ten inches per cow will suffice when hay is available at all times.

BALING WIRE DANGER

Short pieces of baling wire are dangerous to cattle. If consumed, they are likely to pierce the stomach and damage the heart, probably killing the animal. Among cattle producers, this condition is known as *hardware disease*; among veterinarians, it is more technically known as *traumatic pericarditis*. Because of this hazard, it is urged (1) that bales of hay and straw be broken by pulling the whole wire off rather than by cutting, and (2) that all used baling wire be carefully folded and placed in a barrel (and later disposed of) rather than left on the ground and allowed to get mixed with either the feed or the bedding.

HAY FEEDING SCHEDULE

In general, animals may be given as much non-legume hay as they will consume, regardless of their previous ration. With legume hay, however, it is necessary that they be gradually accustomed to it; otherwise, looseness and scouring will likely result.

PROPORTION OF HAY TO CONCENTRATE

Cattle will eat 2 to 3 lb of hay per 100 lb of body weight if fed hay alone. Also, it is noteworthy that the higher the quality of the hay, the more of it they will eat, with the result that the grain requirement will be lessened.

The economics of the situation—the comparative price and quality of hay and concentrate—along with the management practices, will determine the proportion of hay to concentrate. Thus, during the period of low grain prices in relation to forage prices—from about 1950 to 1970—it was desirable to feed lactating cows high-energy rations and to maximize milk production. But the grain feeding binge ended with the world grain shortages and high-priced grains of the early 1970s. In the years ahead, with grain becoming more scarce and higher in price than forages, comparatively speaking, more forage and less grain will be fed to dairy cattle, and net returns will be more important than high production. Cattle will increasingly be "roughage burners." Dairy producers will rely upon the ability of the ruminant to convert coarse forage, grass, and by-product feeds, along with a minimum of concentrate, into palatable and nutritious food for human consumption, thereby competing less for humanly edible grains. Increasingly, the dairy cow of tomorrow will produce a maximum of milk on forage. More and more U.S. cereal grains will be used for human food, just as has been true, historically, in much of the rest of the world.

Ruminants can make the transition to more roughage with ease. For them, it is merely a "return to nature," for they evolved as consumers of forage.

The best buy (hay vs grain) may be determined by calculating the cost per pound of TDN and of protein in the hay and grain being compared. Then, the proportion of hay to concentrate can be varied accordingly. If hay is the best nutrient buy, feed more hay and less grain. On the other hand, if grain is the best buy, feed more grain and less hay.

RUMINANTS NEED HAY

Ruminants need some roughage. Hay fed early in life will develop the calf's rumen and prevent anemia.

Many serious problems in high-producing dairy herds have been traced to lack of hay in the ration; among them, increased incidence of ketosis and displaced abomasums. In addition to these maladies attributable to no-hay rations, it is noteworthy that milk fat percent can be as much as 1% less on an all-silage and concentrate ration. The explanation: The amount and length of hay in the ration affects cud chewing time and percent milk fat. Cud chewing time is the biological response that indicates several desirable factors—rumen muscle tone and function, the supply of saliva going into the rumen to buffer rumen acids, and rumen fermentation with the production of acetic acid. It is estimated that about 15 minutes of cud chewing time is needed per pound of dry matter eaten.

DIFFERENT QUALITIES OF HAY MAY BE USED

The type of ration which will be least costly and result in satisfactory performance will differ according to level of performance, age, etc. For example, the nutritive needs of a dry pregnant cow are much lower than those of a high-producing lactating cow. Thus, low-quality hay may be quite satisfactory for dry cows, whereas high-producing lactating cows should always have high-quality hay.

HAY WASTE AND REFUSAL

Dairy producers have commonly accepted 10% refusal as normal.

High-priced feeds and smaller margins are causing producers to scrutinize hay losses, and to do something about them. Feeding high-quality hay or the use of chopped hay in complete rations are the best ways in which to lessen waste and refusal.

SUPPLEMENTING THE HAY RATION

Hay is generally lower in energy and higher in fiber than most grains and concentrates. Legume hays have a high calcium content, but they vary in available phosphorus. If sun-cured properly, they are high in carotene and vitamin D, along with many of the B vitamins. A supplement should supply the nutrients that are most likely to be lacking in the hay; thus, supplements for alfalfa hay should be high in energy and phosphorus and low in fiber. Also, carotene or vitamin A should be provided if the hay has been bleached or has turned brown. Salt is lacking in hays and other natural feedstuffs and should be provided as a supplement, along with trace minerals that are deficient in the local area.

STRETCHING THE HAY SUPPLY

When hay is scarce and high in price, the supply of it for dairy cattle may be stretched. As the amount of hay fed is reduced, it must be replaced with other feeds so that the total ration is still balanced and fulfills all the nutrient requirements.

Most grains contain 75 to 80% TDN, while most medium- to good-quality hays contain 45 to 50% TDN. Hence, as a general rule of thumb, about 5 lb of grain equal 8 lb of hay, provided they are of comparable quality; for example, No. 2 hay and No. 2 corn. Thus, it follows that if corn can be bought for $100.00 per ton, hay should be bought for $62.50 per ton, or less.

If the price of hay is less than five-eighths the price of grain, relatively more of it should be fed; whereas, if the price of hay is higher than this, relatively more grain will make for cheaper production.

When hay is scarce and high in price, the hay supply for dairy cattle can be stretched as follows:

1. Feed only half to two-thirds the normal ration of hay, but be on the alert for digestive disturbances. With lactating cows, complete elimination of all the hay or other forages will result in low milk fat test and cow health problems; hence, a good rule of thumb for the dairy producer to follow is to feed at least 1.5% of the body weight of the cow as hay, or hay equivalent from other forages. For a 1,400-lb cow, this calls for a minimum of 21 lb of hay or hay equivalent. Under conditions of extreme shortages, hay consumption can probably be lowered to the 5 to 10 lb per cow range, but there is not general agreement on the minimum amount of hay necessary to maintain maximum feed intake and milk production.

2. Replace 1 lb of hay with 3 lb of silage. The basis for this substitution: hay is usually 90% dry matter, whereas silage runs about 30% dry matter (70% moisture); hence, 90% ÷ 30% = 3.

3. Replace 1 lb of hay with 4 lb of green chop. The basis for this substitution: green chop generally runs 22.5% dry matter (77.5% moisture); hence, 90% ÷ 22.5% = 4.

4. Replace each 2 lb of hay deleted with 1 lb of grain.

5. Make the maximum use of such feeds as cottonseed hulls, corncobs, straw, and grass aftermath in the ration for

(a) all but 5% of the alfalfa (or other legume) hay of grower rations; and (b) all of the "hottest" rations, adding such supplementary proteins, minerals, and vitamins as necessary to balance the ration.

6. Provide such supplementary proteins, minerals, and vitamins as necessary. This is especially important with gestating-lactating females or young, growing animals. This may be accomplished by (a) feeding some legume, either hay or silage; and/or (b) adding suitable protein, mineral, and vitamin supplements.

OTHER DRY ROUGHAGES

When hay is scarce and high in price, numerous other dry roughages may be fed to dairy cattle. Most of them are by-product feeds and crop residues; among them, corn residues, cereal grain straw and chaff, cotton crop residues, sorghum residues, and soybean residues. All of them are low in energy and protein, but high in fiber. Limited quantities of these low quality roughages are sometimes fed to lactating cows to stretch the hay supply and provide fiber. However, they are better suited for use in dry cow and replacement heifer rations. When such by-product feeds and crop residues are fed to lactating cows, dry cows, or replacement heifers, it is important that they be properly supplemented with energy, protein, minerals, and vitamins.

(For a detailed discussion of a great array of by-product feeds and crop residues, the reader is referred to *Feeds & Nutrition*, Chapter 12, By-product Feeds/Crop Residues. Dr. M. E. Ensminger is the senior author of *Feeds & Nutrition*, which is published by The Ensminger Publishing Company, Clovis, California.)

QUESTIONS FOR STUDY AND DISCUSSION

1. Why is hay the most important harvested forage for U.S. dairy cattle?

2. Discuss the value of hay for dairy cattle in the following roles: (a) speeding up the development of the rumen function of the young calf, and (b) lessening the incidence of ketosis and displaced abomasum.

3. Why does hay vary in nutrient content and palatability more than any other feed?

4. Discuss hay from each of the following standpoints:
   a. As an energy source for dairy cattle.
   b. As a protein source for dairy cattle.

5. What characteristics of alfalfa hay make it so popular as a dairy feed? Why should dairy producers who produce their own hay grow a legume whenever it is feasible?

6. Explain how low nutrient content alone does not account for the unfavorable results obtained when poor-quality hay is fed to lactating cows.

7. What are the easily recognized characteristics of good-quality hay?

8. What is the difference between, and the importance

of, neutral detergent fiber (NDF) and acid detergent fiber (ADF)? In what ways are they superior to proximate analysis?

9. What is near infrared reflectance spectroscopy (NIRS)? How is it used in forage analyses? What are its advantages and disadvantages?

10. Discuss each of the following points with reference to determining hay quality: (a) proper sampling, (b) evaluating the test results, and (c) using the results in buying hay and balancing rations.

11. List and discuss each of the steps in growing quality hay.

12. Describe the proper forage harvesting stage of alfalfa. How does the harvesting stage of alfalfa affect chemical composition and milk production of lactating cows?

13. Discuss reducing moisture content, shattering, bleaching, and fermenting in modern haymaking.

14. What chemical hay drying agents and preservatives are being used? How and why are they being used?

15. Is there any basis for dairy producers favoring a certain cutting of alfalfa hay?

16. Describe and discuss the place of each of the following haymaking systems: long hay, chopped hay, packaged hay, and artificial drying.

17. What prompted the development of big round bales and small mechanical stacks?

18. Why do dairy producers prefer cubes (wafers) to pellets for lactating cows?

19. List and describe the traditional market channels for hay in the United States.

20. What are the characteristics of quality-tested hay auctions?

21. Describe the hay quality standards for legumes, grasses, and legume-grass mixtures currently in use in Wisconsin and Minnesota—two great dairy states.

22. What developments in hay marketing are essential if hay is to become a leading cash crop?

23. When buying and selling hay, of what importance are the following: (a) freedom from toxic residues, and (b) hay shrinkage?

24. Discuss (a) hay preparation, (b) hay feeding systems, and (c) feeding hay packages.

25. How may a dairy producer lessen the hazard of hardware disease?

26. What factors determine the proportion of hay to concentrate in a ration?

27. Cite proof of the following statement: Ruminants need hay.

28. For what dairy animals may low-quality hay be used? Why is high-quality hay essential for lactating cows?

29. Should dairy producers accept a 10% hay waste and refusal as normal? If not, what can they do to lessen it?

30. When hay is high in price and scarce, as may happen after a long, hard winter and much snow in the northern part of the United States, how can dairy producers "stretch" their hay supplies?

31. What dry forages other than hay may be fed to dairy cattle? For what classes of dairy cattle are they best suited?

SELECTED REFERENCES

Title of Publication	Author(s)	Publisher
Crop Production, Fifth Edition	R. J. Delorit L. J. Greub H. L. Ahlgren	Prentice-Hall, Inc., Englewood Cliffs, NJ, 1984
Crop Quality, Storage, and Utilization	Edited by C. S. Hoveland	American Society of Agronomy, Crop Science Society of America, Madison, WI, 1980
Crops in Peace and War: Yearbook of Agriculture 1950–1951		U.S. Department of Agriculture, Washington, DC
Effect of Processing on the Nutritional Value of Feeds	National Research Council	National Academy of Sciences, Washington, DC, 1973
Feeds & Nutrition, Second Edition	M. E. Ensminger J. E. Oldfield W. W. Heinemann	The Ensminger Publishing Company, Clovis, CA, 1990
Feeds & Nutrition Digest	M. E. Ensminger J. E. Oldfield W. W. Heinemann	The Ensminger Publishing Company, Clovis, CA, 1990
Forage and Pasture Crops	W. A. Wheeler	Van Nostrand Reinhold Company, New York, NY, 1950
Forage Conservation in the '80s	Edited by C. Thomas	British Grassland Society, The Grassland Research Institute, Hurley, Maidenhead, Berkshire, UK, 1980
Forage Crops	D. A. Miller	McGraw-Hill Book Company, Inc., New York, NY, 1984
Forages, Fourth Edition	M. E. Heath R. F. Barnes D. S. Metcalfe	The Iowa State College Press, Ames, IA, 1985
Forages: Resources for the Future	J. E. Oldfield, Chairman of Task Force	Council for Agricultural Science and Technology, Ames, IA, 1986
Haymaker's Handbook	M. A. Balas J. E. Baylor	New Holland, Inc., New Holland, PA, 1987
Manual of the Grasses of the United States, Second Edition	A. S. Hitchcock; rev. by A. Chase	U. S. Department of Agriculture, Washington, DC
Plants for Man, Second Edition	R. W. Schery	Prentice-Hall, Inc., Englewood Cliffs, NJ, 1972
Stockman's Handbook, The, Seventh Edition	M. E. Ensminger	Interstate Publishers, Inc., Danville, IL, 1992

Fig. 9–1. Silage is especially important in all dairy regions of the U.S. that have humid climates and cold winters. (Courtesy, Holstein-Friesian Association of America, Brattleboro, VT)

Contents

CHAPTER

9

SILAGE/HAYLAGE/
HIGH-MOISTURE GRAIN

Silage is fermented forage plants. It is a very old method of preserving feed. Columbus found that the Indians used pits or trenches in which to store their grain, and, centuries earlier in the Old World, silos were used as a means of preserving both grain and green forage. The first tower silo built in the United States by other than Indians is said to have been erected by F. Morris in Maryland in 1876. Until about 1910, silage was generally thought of as a feed for dairy cows. Even today, most silage is used on dairy farms.

Silage making is one of the three common methods of utilizing forage crops, the other two methods being pasturing and haying. Pasturing is the least expensive of the three methods, but it is seasonal in nature. In the spring and early summer, forage plants generally grow faster than they can be utilized by normal grazing, and then become dormant in cold weather.

The surplus forage produced during the growing season may be preserved for feeding during the winter months and other periods of pasture scarcity by haymaking which, next to grazing, is the most efficient method during dry weather. But weather conditions are not always favorable for haymaking. Ensiling, on the other hand, can be done in inclement weather. Also, it has the added virtues of succulence and of preserving a higher proportion of the nutrients of the plant than can be accomplished in haymaking, although slightly greater cost may be involved than in normal field curing of hay.

The importance of silage in this country is attested to by the fact that about 120,000 tons are made annually. It is especially important in all dairy regions of the United States that have humid climates and cold winters. A wide variety of silo types continues to appear; and the use of preservatives has grown enormously since 1980.

Most of the silage is fed on the farms on which it is produced, rather than being purchased. It is important, therefore, that dairy producers know how to produce good silage, as well as how to feed it, for most of them determine their own destiny from the standpoint of quality. For this reason, this chapter covers silage from production to feeding.

Fig. 9–2. Corn, king of the silages for dairy cattle, being harvested with a forage harvester and blown into a trailer. (Courtesy, Gehl Company, West Bend, WI)

THE ENSILING PROCESS

The ensiling process refers to the changes which take place when forage or feed with sufficient moisture to cause fermentation is stored in a silo in the absence of air. The basic strategy of silage preservation is to exclude oxygen and to acidify the forage through bacterial fermentation. An understanding of these changes is likely to lead to the production of more high-quality silage.

The entire ensiling process requires 2 to 3 weeks, during which time the following aerobic (with air) and anaerobic (without air) activities predominate:

1. **Aerobic activity.** The living plant cells of the forage continue to respire, or breathe, consuming the oxygen of the silage-entrapped air, producing carbon dioxide and water,

and releasing energy or heat. Simultaneously, aerobic yeasts and molds thrive and multiply.

In good ensiling conditions and with proper management, the aerobic phase is very short, and silage temperatures seldom rise above 100°F. However, slow filling, inadequate packing of the forage, or leaky sealing around the silage, lengthens the aerobic phase, increases losses, and causes excessive heating.

2. **Anaerobic activity.** When the available oxygen of the entrapped air has been consumed, anaerobic bacteria multiply at a prodigious rate. Simultaneously, the molds and the yeasts die. Certain plant enzymes continue to function.

The combined anaerobic activity produces the following changes:

a. The nonstructural carbohydrates, especially sugars, are converted to lactic acid (the acid in sour milk), some acetic acid (the acid in vinegar), a small amount of other acids, alcohol, and carbon dioxide.

b. The plant proteins are broken down into peptides, ammonia, amino acids, amines, and amides. These non-protein nitrogen compounds are utilized less effectively by the animal than true protein.

c. When the acidity become high enough, the bacteria die and the silage stabilizes. This occurs when the silage pH reaches a low of 3.5 to 4.5 for corn and cereals, or 4.0 to 5.0 for grasses and legumes.

It is generally known that a number of factors affect the ensiling process and the quality of the silage produced; among them, the following:

1. **Forage composition and moisture content.** Proper fermentation requires sugars, oxygen-free conditions, moisture content in the correct range, and bacteria. The optimum moisture content for corn or cereal silages is 60 to 70%. Grass or legume silages should be in the range of 50 to 70%. However, with the right storage conditions and careful management, good silage can be made outside of these ranges.

In grasses, corn, and cereals, there are normally enough sugars for fermentation. Legumes with over 70% moisture are frequently low in sugars; this may limit fermentation and prevent the attainment of a stable pH. If the forage is harvested without rain damage and the silo is filled and sealed quickly, there is less respiration and more sugar available for fermentation.

Silage making is a preservation process; it cannot improve the quality of the forage. Thus, growing high-quality forage is the first requisite in making good silage.

2. **Harvest and storage management.** Grass and legume forages lose valuable nutrients during drying in the field. Rapid wilting and avoiding rain damage are two ways to increase the quality of the silage. Minimizing the time from cutting to ensiling also conserves valuable crop nutrients. Good packing of the forage is essential for keeping out air and allowing fermentation. This may require chopping the forage into sufficiently small particles and applying pressure to the top surface of the silage. Maintaining an airtight seal around the surfaces of the silage keeps out oxygen, promotes good fermentation, conserves nutrients, and reduces heating. Finally, careful management is needed after the silo is opened so as to minimize the exposure to air before the silage is fed to the animals.

3. **Preservatives.** A number of silage preservatives are currently available. These include (a) inoculates, which supply bacteria for a faster fermentation; (b) sugars to supplement the available plant sugars for fermentation; (c) enzymes which break down complex carbohydrates such as starch, cellulose, and hemicellulose to fermentable sugars; (d) acids which reduce the pH without fermentation and inactivate the enzymes which break down protein; and (e) aerobic inhibitors such as propionates and antioxidants which restrict the yeasts, molds, bacteria, and enzymes that degrade the forage in air. The use of additives can increase the preservation of crop nutrients and maintain a higher quality of silage. But, additives increase the cost of silage making, may not be reliably effective, and, in some cases, may be dangerous to handle.

The decision to use a silage additive should be based on (a) clear need to increase nutrient retention, and (b) scientific evidence that the additive is effective. Application rate is of critical importance in terms of the active ingredient supplied to each unit of forage. Proper application at the most appropriate point in the harvesting/storing process is necessary to achieve good mixing of the additive with the forage without slowing down the harvesting operation. Finally, the effectiveness of silage additives is increased when their use is accompanied by good management.

In summary, good management of ensiling includes (1) rapid drying to the proper moisture content, (2) chopping the forage to the proper particle size, (3) rapid filling of the silo, (4) good packing of the forage, (5) airtight sealing of the silo, (6) leaving the silo closed for at least 2 weeks to allow fermentation, and (7) minimizing the time of exposure to air while the silage is being fed to animals.

ADVANTAGES AND DISADVANTAGES OF SILAGE

Some of the **advantages** of silage are:

1. It retains a higher proportion of the nutrients of plants than can be accomplished by haymaking, even if the weather is satisfactory for the latter, chiefly because shattering and bleaching losses are held to a minimum. Thus, grass silage preserves 85% or more of the feed value of the crop, whereas haymaking under the best of conditions will preserve only 80%, and under poor conditions only 50 to 60%.

2. It makes possible the production of the maximum quantity of feed per acre of land and increases the dairy carrying capacity of the farm or ranch. Thus, corn, the chief U.S. silage crop, (a) yields more total digestible nutrients per acre than most other forage crops, and (b) has 30 to 50% higher feeding value as silage than when fed as grain and stover.

3. It is feasible to produce a top-quality feed during times of inclement weather when it would be impossible to cure the forage crop properly.

4. It is the most economical form in which the whole stalk of corn or sorghum can be processed and stored.

5. It requires less storage space per pound of dry matter than dry hay, even when the latter is baled or chopped. A cubic foot of silage contains two to three times

as much dry weight of feed as a cubic foot of long hay stored in the mow.

6. It practically eliminates the danger of loss by fire if stored within the recommended moisture range.

7. It is the most satisfactory and economical way in which to preserve a number of by-product feeds.

8. It improves the timeliness of harvest so as to maximize the quality of forage.

9. It is one of the best methods of controlling the European corn borer since the removal of cornstalks is required in making silage.

10. It helps to control weeds, which are often spread through hay or fodder.

11. It is the cheapest form in which a good succulent winter feed can be provided on most farms.

12. It is a better source of protein and of certain vitamins, especially carotene, and perhaps some of the unknown factors, than dried forage.

13. It is a very palatable feed and slightly laxative in nature.

14. It makes for less waste, the entire plant being eaten with relish, an important consideration with coarse stemmy forages.

15. It is without a peer from the standpoint of longtime storage, holding the feeding value of protein, carbohydrates, and carotene better than any other method of preservation, and providing a desirable backlog against drought or any other crop failure.

16. It may be completely mechanized as a feeding system, thereby eliminating much labor and time.

17. It offers many advantages over pasture, including (a) no fencing required, (b) approximately one-third more forage from the same acreage, (c) harvesting at optimum maturity, (d) more uniform quality, (e) little or no bloat, and (f) closer observation of animals that are confined to a lot or corral.

Some of the **disadvantages** of silage are:

1. It requires a silo or storage structure and other special equipment, for best results. In comparison with the simpler methods of field curing and storing hay, this is likely to mean higher costs—an important consideration with a small operator.

2. It possesses considerably less vitamin D than sun-cured hay.

3. It necessitates that 2 to 3 times as much tonnage be handled as when the same forage is dried for hay, due to the high water content.

4. It incurs an added expenditure when preservatives are necessary.

5. It may result in acidosis, milkfat depression, and other metabolic disorders, due to the short chop lengths required for good packing in the silo, if proper feeding management practices are not followed.

6. It may increase the need for more expensive dietary supplements because the degraded protein in silage may not be utilized effectively by the ruminant.

7. It may spoil during feeding, resulting in molds which may affect animal performance and contain toxins.

8. It may result in poor preservation if the moisture content is not right. With high-moisture content (above 65%), seepage occurs and a butyric fermentation may take place that results in an unpalatable and toxic feed. With low-moisture content (below 45%), excessive heating of the silage may occur, making the plant proteins indigestible and possibly leading to a silo fire.

9. It requires better management the year-round for stored silage than for stored hay.

SILOS

Silage may be stored in almost any kind of container. The main requisites of a good silo, regardless of kind, are:

1. That its size be in keeping with the number and kind of animals to be fed daily, the length of the feeding period, and the amount of forage available for ensiling.

2. That it exclude air from the stored material, including entrance of air around the doors of tower silos.

3. That the sidewalls be straight and smooth in order to prevent the formation of air pockets and allow for unimpeded packing.

4. That it be of proper dimension, thus making for better packing and less surface area to total mass exposed.

5. That it be properly reinforced. This point is especially important where direct cut grass silage is made, because it exerts from ½ to 2½ times as much pressure on the walls as does corn silage. Thus, tower silos which were originally built for corn or sorghum silage but which are to be filled with wet grass silage should be either (a) reinforced with extra bands placed around the lower part to strengthen the walls if an inspection reveals that the existing strength is not adequate, or (b) not filled to more than half capacity.

6. That adequate provision be made for the escape of surplus juices, by either a drain or a gravel bottom, along with positive containment to avoid surface pollution.

7. That it be conveniently located and accessible in all kinds of weather, from the standpoint of both filling and feeding.

Silos may be classified according to the five basic methods used for processing forages. Each method is associated with the shape and material of the structure, which also influences the efficiency of preserving the silage. The different shaped structures are also adapted to different methods of filling and unloading. Within each classification there are many variations of each type depending upon the manufacturer.

The kind of silo decided upon and the choice of construction material should be determined primarily by the cost and by the suitability to the particular needs of the farm.

Silos may be classified as follows:

1. Conventional upright (tower) silos
   a. Concrete stave
   b. Galvanized steel
   c. Wood stave
   d. Monolithic concrete (poured in place)
   e. Tile block
   f. Brick

2. Gastight (oxygen-limiting) silos
   a. Glass-lined structures

b. Concrete stave
c. Galvanized steel
d. Monolithic concrete

3. Pit silos
4. Horizontal silos
   a. Trench silos (belowground level)
   b. Bunker (aboveground level)

5. Temporary silos
   a. Enclosed stacks
   b. Open stacks
   c. Modified trench-stack silos
   d. Plastic or polyethylene bag silos
   e. Round bale bagged or wrapped silage

Some pertinent information relative to each main kind of silo is given in the discussion which follows, but it is not within the scope of this book to give detailed silo plans and specifications. The latter may be obtained from local authorities, from silo manufacturers, or from the state agricultural college.

CONVENTIONAL UPRIGHT (TOWER) SILOS

The upright or tower silo, which is sometimes referred to as the "watch tower of prosperity," is a cylinder built aboveground. Its round shape withstands pressure well and is adapted to good packing.

Fig. 9–3.   Upright (tower) silos made of concrete stave. (Courtesy, National Silo Association, Waterloo, IA; National Livestock Producer photo)

The tower silo is a permanent farm structure, and, as such, should be constructed to withstand long usage. Although tower silos are usually handy, they are, in their initial cost, generally the most expensive of all types. However, they have the following advantages: (1) durability, (2) minimum top and side spoilage, (3) convenience for feeding during periods of inclement weather, and (4) well adapted to automation (loading and unloading machinery).

Unloading mechanisms in tower silos include: (a) bottom unloaders, with elimination of doors; (2) center-core unloaders, with elimination of most of the doors; (3) top-unloaders; and (4) large diameter features (24 to 320 ft in diameter). These features are also available in gastight (oxygen-limiting) silos.

GASTIGHT (OXYGEN-LIMITING) SILOS

These silos resemble conventional tower silos, but they are more expensive because of their construction.

Sealed silos are designed for storage of wilted or even overwilted forage with as little as 40 to 55% moisture content or for the storage of high-moisture grain containing 22 to 30% moisture. These structures may be partly filled on widely separated dates, provided they are sealed between fillings. Packing and tramping of forage is not necessary, although distribution is desirable.

Practically all outside air is kept out of the oxygen-limiting silo, and carbon dioxide formed during fermentation is kept in a plastic breather bag and a pressure relief valve located in the top of some of these structures compensate for differences in inside and outside pressures without allowing outside air to contact the forage. Before each filling, the plastic breather bag should be checked for holes and flaws, the pressure relief valves should be inspected, the structure should be checked for leaks, door seals should be inspected, and unloaders should be checked for wear in order to prevent malfunctioning during unloading.

CAUTION: There is not sufficient oxygen inside a gastight (oxygen-limiting) silo to sustain life. So, never enter a filled sealed silo without proper life supporting equipment.

PIT SILOS

The pit silo is shaped like the tower silo, but inverted into the ground. It resembles a well or cistern. The walls of a pit silo may or may not be lined. Where the water table is low enough that the silo will not fill with water, such as in semiarid areas, the pit silo is very satisfactory.

In comparison with tower silos, pit silos have the following **advantages**: (1) they are never damaged by storm or fire, (2) they require less reinforcing, (3) they minimize silage losses because of not having doors, and (4) they avoid frozen silage. But they have the following **disadvantages**: (1) they are dangerous, due to the frequent presence of suffocating carbon dioxide gas; and (2) considerable work is involved in removing the silage, despite the development of a number of hoist devices.

CAUTION: Before entering a pit silo, it is recommended that a lighted candle or lantern be lowered into the silo. If the flame goes out, assume that the pit is dangerous to enter and replenish it with fresh air before entering.

HORIZONTAL SILOS

Only two types of horizontal silos will be discussed herein; namely, trench silos and bunker silos (or horizontal surface silos), both of which may be adapted to self-feeding.

TRENCH SILOS

The trench silo is a horizontal, trenchlike structure that can be built quickly and at low cost. It is most popular in areas where the weather is not too severe and where there is good drainage. The walls of a trench silo may or may not be lined, but for making good silage they should always be smooth, and there may or may not be a floor. A trench silo should be wider at the top than at the bottom, and the bottom should slope away from one end in order that excess juices will drain off.

Fig. 9–4. Two trench (horizontal) silos, side by each. One silo filled (left), and the other being filled. (Courtesy, Portland Cement Association, Skokie, IL)

Trench silos have the **advantages** of: (1) low initial cost; (2) low cost of filling machinery, for a blower is not necessary; (3) relative freedom from freezing; and (4) ease of construction. The chief **disadvantages** of trench silos in comparison with tower silos are the (1) larger area to seal, (2) higher spoilage losses, and (3) inconvenience in feeding during inclement weather. Because of shallowness, the forage should be packed very thoroughly in a trench silo by driving a tractor back and forth over it. When filling is completed, the top should be carefully sealed (1) by 3 to 6 in. of limestone or dirt, with or without a seeding of rye or winter wheat; (2) by wet straw, poor-quality hay, marsh grass, or sawdust; (3) by a mixture of dirt and straw; (4) by waterproof paper lapped with 12 in. at the joints and covered with dirt or straw; or (5) by polyethylene, plastic, aluminum, or other materials.

BUNKER SILOS

Aboveground horizontal silos are usually constructed with concrete floors and side walls of wood, concrete, or other materials. Bunker silos were originally intended for animals to self-feed directly, but skid loaders or front-end loaders on tractors are now the common methods of unloading. In warm periods, the exposed face of the silo may deteriorate rapidly due to the growth of yeasts and molds. Silo size and dimensions should be matched to the animal feed needs so as to obtain an unloading rate of 6 to 8 in. per day in warm weather.

TEMPORARY SILOS

Several kinds of aboveground temporary silos are used. Generally, this kind of storage is used to meet emergencies, to supplement permanent silos, or to ensile such by-product feeds as cannery refuse, pea vines, and beet tops or pulp. Aboveground temporary silos are low in cost, can be erected on short notice, require no special foundation, and can be set up on almost any level site convenient for filling and feeding.

The amount of spoilage in aboveground temporary silos can be kept to a minimum by having straight sides, considerable height, proper packing, and protection with fiber-reinforced paper, plastic, or other suitable material. Also, the use of propionic acid, formic acid, or other effective organic acids, applied to harvested materials at the time of chopping, is very effective in reducing spoilage in temporary silos. It is important that stacked material treated with organic acid be covered with plastic to prevent dilution by rain and snow.

The spoilage on the sides of temporary silos will vary from 4 to 20 in., with greater spoilage in grass silage than in easier-keeping corn and sorghum silage.

Perhaps most aboveground temporary silos can be classed as belonging to one of the following four kinds:

1. **Enclosed stacks.** These are built entirely aboveground, without trenches or holes. They are upright, are generally circular, and are enclosed by snow or picket fences, poles, wooden staves, heavy woven wire, or other materials. Most of them are lined with tar paper, plastic, or tough fiber-reinforced paper made especially for the purpose. Because of the relatively weak walls of such silos, their height should not be greater than twice their diameter unless poles are set at four to six points around their circumference and tied together at the top.

2. **Open stacks.** These are similar to enclosed stack silos, except that no supports or walls are used. As would be expected, greater spoilage is encountered in the open stack than in the enclosed stack, because of the greater evaporation and spoilage which accompanies the exposed sides. Less spoilage, percentagewise, occurs in stacks of considerable size—stacks that contain 500 to 1,000 tons or more silage.

3. **Modified trench-stack silo.** This silo, which is intermediate between a trench and a stack silo, is adapted to areas where the ground-water level is high. It is constructed by excavating a shallow trench 12 to 18 in. deep, by piling the excavated earth on either side of the trench to support the silage and to keep out surface water, by packing silage

thoroughly in and over the trench to a height of 10 to 15 ft, and by covering the stack with any one of the materials recommended for covering the trench silo. The modified trench-stack silo is designed to give greater protection and less spoilage than can be accomplished by open or closed stacks. Also, this type of silo is easier to feed from than a trench silo.

4. **Plastic silos.** Plastic (polyethylene) is now available for use as temporary silos, and for use as covers for trench, bunker, and tower silos, and as silo liners. If not punctured, it is nearly airtight. Plastic thicknesses range from 4 to 9 mils. The thicker grades have better tear and puncture resistance, and low permeability by both air and moisture; however, they cost more and are difficult to tie tightly. Thinner grade plastics are less costly, more pliable, and easier to seal.

The two common types of plastic silos are: (1) enclosed plastic bag or tube silos, and (2) round bale plastic covered silage.

■ **Enclosed plastic bag or tube silos** — These temporary silos are made of heavy plastic in the form of a tube into which forage is forced by a special machine (much like stuffing sausage). The machine needed to pack the tube is generally rented or owned cooperatively. The filled structure is 8 ft in diameter and about 100 ft long. Preservation of silage is excellent provided the ends are kept sealed and the plastic is not torn or damaged by rodents or other animals. To remove or self-feed the silage, the plastic is cut and folded back at one end to expose as much silage as needed each day. The plastic cannot be re-used.

Fig. 9–6. Plastic bags each of which stores and ensiles two 500 lb round bales. (Courtesy, American Polled Hereford Assn., Kansas City, MO)

2. **Plastic tubes.** These consist of several round bales stuffed by a machine into a long plastic tube which is then sealed at both ends. The filled plastic tube resembles an "Enclosed Plastic Bag or Tube Silo" described earlier, except that it consists of a row of round bales covered with plastic rather than long, continuous sausage-type silage material. Plastic tubes can be effective and time-saving, but the multiple bales stored in one package tend to increase the loss if the bag is torn, punctured, or opened for feeding. However, the tube can be easily tied off into one-bale (or more) segments for feeding.

3. **Sheet plastic.** Several round bales can be stacked under two sheets of plastic, with the plastic ends on the ground covered with soil, sand, or other effective sealing procedure. The hazard with this type of storage is that there are more possibilities for air leaks to develop, which may result in a large number of bales being spoiled.

Fig. 9–5. Plastic tube silos. (Courtesy, Montana State University, Bozeman)

■ **Round bale plastic covered silage** — The most common methods for using plastic material to produce round bale silage are:

1. **Individual bags.** Bags come in various lengths, diameters, and thicknesses. A tractor-mounted spear device is needed to lift the bale while applying the bag. Then the bale is placed in storage position before it is tied off. If possible, the bales should be stacked in cord-wood fashion to reduce exposed surface area. Then, a plastic cover over the entire stack may reduce storage damage.

Fig. 9–7. Several round bales of silage stacked under sheets of plastic. (Courtesy, Ford New Holland, New Holland, PA)

■ **Advantages and disadvantages of round bale plastic covered silage** — Round bale silage may serve as a supplement to, rather than a replacement for, other stored forages on most livestock farms. Some **advantages** are:

1. It does not require silo structures.
2. Hay-making equipment may be used to harvest it.
3. When silo capacity is lacking because of a surplus of forage, round bale silage can offer an effective method of storing excess forages.

Fig. 9–8. Round bale silage properly fed in a steel rack, thereby alleviating the energy for chopping and lessening the labor for feeding. (Courtesy, Ford New Holland, New Holland, PA)

4. The round bale silage system can be used to save a mowed field of hay when an anticipated rainstorm or extremely high humidity interfere with proper hay curing.
5. Round bale silage saves about one-third of the harvesting energy plus the fuel required for chopping silage.
6. Round baled silage can be self-fed if properly presented, thereby saving both labor and fuel by not requiring daily silage feeding.

But there are **disadvantages**; among them, the following:

1. Conditions associated with round bale silage are not optimum for fermentation.
2. Extreme care must be taken to eliminate air leaks.
3. The system requires prompt handling and storage of bales.
4. Machines for lifting and moving heavy, high-moisture bales must be available.
5. Plastic bags, storage tubes, or plastic sheets to cover group-stacked bales must be purchased.
6. Plastic is easily damaged, which can result in forage losses greater than in conventional silos.

Generally speaking, users of baled silage have liked the concept, but (1) not the machinery for bagging it, and (2) not the sometimes poor fermentation. Recently, automatic equipment has been developed that individually shrink-wraps big bales with tough plastic and provides an airtight seal; and additives/preservatives have been successfully added to large round bales.

SILAGE STORAGE LOSSES

Tight structures, good distribution and packing, and the proper use of plastic covers minimizes silage storage losses. Losses within type of silo storage vary widely based primarily on length of time and season of feedout; this is especially critical for silages continuously exposed to air. Silage losses also vary widely between kinds of silos, as shown in Table 9–1. Losses in trench and open stack silos are also influenced by depth; less surface is exposed in deeper silos.

TABLE 9–1
ESTIMATED (1) AVERAGE, AND (2) RANGE
OF SILAGE STORAGE LOSSES

Type of Silo	Percentage of Loss	
	Average	Range
	(%)	(%)
Gastight upright	5	1–10
Conventional upright	10	5–15
Horizontal (trench)	15	10–20
Open stack	25	15–30

Losses in the silo are of four types: (1) surface or top spoilage, (2) seepage, (3) gaseous, and (4) heating (browning reaction and spontaneous combustion).

Surface or top spoilage losses of 20% or more may occur in stack silos and in any uncovered bunk, trench, or pit silo. These losses can be reduced by the use of suitable protection, such as a plastic cover.

Seepage losses can be high in high-moisture silage stored in upright silos. The higher the silo, the greater the pressure and the higher the losses through seepage. The seepage carries soluble feed nutrients with it. Horizontal silos have less seepage loss than upright (tower) silos because of lower vertical pressure. Seepage losses can be reduced by wilting forages to less than 65% moisture before ensiling.

Gaseous losses are unavoidable so long as the plant material respires and there is subsequent fermentation. However, these losses can be minimized by avoiding entry of air into the silo, by having the pH decline rapidly, and by encouraging favorable fermentations.

Lowering the moisture without excluding the air may lead to heat damage, known as the browning reaction or Maillard reaction.

Spontaneous ignitions sometimes occur in low-moisture silage (haylage). For such losses to occur, there must be a build-up of temperature to the combustion point in the silo mass, combined with a low transfer of heat. These fires are very difficult, and usually impossible, to extinguish. The addition of water may build up pressure and lead to an explosion. Most silo fires should be allowed to burn.

HOW TO DETERMINE THE SIZE SILO TO BUILD

The size of silo to build should be determined by needs. With tower type and pit silos, this means (1) that the diameter should be determined by the quantity of silage to be fed daily, and (2) that the height (depth in a pit silo) should be determined by the length of the silage feeding period. Similar consideration should be accorded trench silos.

SIZE OF TOWER SILO

If the diameter is too great, the silage will be exposed too long before it is fed; and, unless a quantity is thrown away each day, spoiled silage will be fed.

The minimum recommended rate of removal of silage varies with the temperature. In most sections of the United States, it is desirable that a minimum of 1½ in. of silage be removed from tower silos daily during the winter feeding period, with the quantity increased to a minimum of 3 in. when summer feeding is practiced. Of course, the total daily silage consumption on any given farm or ranch will be determined by (1) the class and size of animals, (2) the number of animals, and (3) the rate of silage feeding.

Silo height should be determined primarily by the length of the intended feeding period. In general, however, the height should not be less than twice, nor more than 3½ times the diameter. The greater the depth, the greater the unit capacity. Extreme height is to be avoided because (1) of the excessive power required to elevate the cut silage material, and (2) of the heavier construction material required. Also, it is noteworthy that, with silos of larger diameters, more labor is required in carrying the silage to the silo door if silage is manually removed.

Table 9–2 may be used as a guide in computing the proper diameter of tower silo for any given farm.

TABLE 9–2
MAXIMUM DIAMETER OF TOWER SILO TO BUILD
IF SILAGE IS TO BE KEPT FRESH[1]

Inches of Silage Removed Daily	Total Silage Removed Daily with an Inside Diameter of:					
	10 Ft	12 Ft	14 Ft	16 Ft	18 Ft	20 Ft
	(lb)	(lb)	(lb)	(lb)	(lb)	(lb)
Summer: 3 in. daily will remove[2]	786	1,312	1,539	2,010	2,545	3,142
Winter: 1½ in. daily will remove[2]	393	656	770	1,005	1,272	1,571

[1]To convert pounds to kilograms multiply by 0.454.

[2]The pounds listed in each of the columns to the right are approximations based on an average constant weight of 40 lb of silage per cubic foot. Low moisture silages of 40–55% moisture content will weigh somewhat less than 40 lb per cubic foot.

Fig. 9–9 shows capacities of tower silos of different heights and diameters. It is based on well-eared corn silage harvested in the early dent stage, cut in ¼-in. lengths, well-tramped when filled, and with the silo refilled after settling for a day. Fig. 9–9 can be adapted for corn silage of different

stages of maturity and grain content, and for other kinds of silage, by applying the rules of thumb given in Table 9–3.

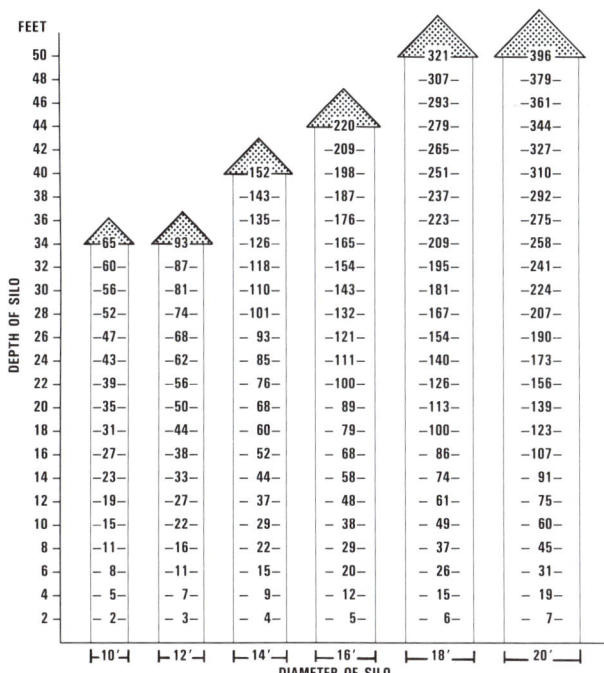

Fig. 9–9. Capacity in tons of settled corn silage in tower silos of varying sizes. See Table 9–5 (p. 157) for tabular material.

TABLE 9–3
EFFECT OF KIND OF SILAGE ON WEIGHT

Kind of Silage	Changes to be Made in the Number of Tons Shown in Table 9–5
Corn silage ensiled when less mature than usual	Add 5–10%.
Corn ensiled when dry or overripe	Deduct 5–10%.
Corn very rich in grain	Add 5–10%.
Corn with very little grain	Deduct 5–10%.
Sorghum silage	Use the same weights as used for corn silage of comparable grain and maturity.
Sunflower silage	Add 5–10%.
Grass silage	Add 10–15%.[1]

[1]For this reason, a stronger structure is necessary where grass silage is stored.

The following example will serve to illustrate how to determine the size tower silo to build:

Over a period of years, a dairy producer plans to winter 34 head of 425-lb replacement heifers on a ration of corn silage and protein supplement. There is a 240-day wintering period. No increase in heifer numbers is planned. What size tower silo should be built?

The answer is obtained as follows:

1. First, here are the silage requirements:

   a. Heifers weighing 425 lb on a ration of corn silage and protein supplement should receive about 30 lb of silage per head per day.

   b. 34 x 30 = 1,020 lb of silage required daily for the 34 heifers.

   c. 1,020 x 240 = 244,800 lb, or 122.4 tons, of silage required for the 240-day wintering period for the 34 heifers.

2. Next, here is the size silo to build:

   a. Table 9–2 shows that in order to remove 1,005 lb of silage daily (which is only slightly less than the 1,020 lb needed daily), with 1½ in. removed from the top of the silo each day, the diameter of the silo should not be greater than 16 ft.

   b. Fig. 9–9 can now be used as a guide in determining both the proper height (or depth) and diameter of the silo. Fig. 9–9 shows that a silo 16 ft in diameter and 27 ft high will hold 127 tons of silage, which would allow for 4.6 tons spoilage in excess of the required 122.4 tons. However, the height of a silo should not be less than twice the diameter. It appears best, therefore, to plan on a 14-ft diameter silo. As noted in Fig. 9–9, 34 ft of settled silage in a 14-ft diameter silo will provide 126 tons of silage, which would allow for 3.6 tons spoilage in excess of the required 122.4 tons. To allow for settling, an additional 4 to 6 ft should be added to the height, thus making a 38- to 40-ft height.

   c. The size silo to build to meet the needs outlined in this example, therefore, is one that is 14 ft in diameter and 38 to 40 ft high. Sufficient additional height, usually 4 to 6 ft, should be added to provide the necessary space required for the silage unloader.

SIZE OF TRENCH SILO

As in an upright silo, the cross-sectional area of a trench silo should be determined by the quantity of silage to be fed daily. The length is determined by the number of days of the silage feeding period. The only difference is that generally greater allowance for spoilage is made in the case of trench silos, though this factor varies rather widely.

Under most conditions, it is recommended that a minimum 4-in. slice be fed daily from the face (from the top to the bottom of the trench) of a trench silo during the winter months, with a somewhat thicker slice preferable during the summer months.

The dimensions, areas, and capacities given in Table 9–4 are based on the assumption that the silage weighs 35 lb per cubic foot,[1] which is an average figure for corn or sorghum silage. Thus, a trench silo 8 ft deep, 6 ft wide at the bottom, and 10 ft wide at the top has a cross-sectional area of 64 sq ft. This size silo will hold 747 lb of silage for each 4-in. slice, or 2,240 lb of silage for each 1-ft slice, or 112 tons in a trench 100 ft long.

TABLE 9–4
DIMENSIONS, CROSS-SECTIONAL AREA
OF TRENCH SILO, AND WEIGHT OF SILAGE
IN 4-INCH SLICE AND PER LINEAR FOOT[1]

Side Slope per Foot of Depth	Depth	Bottom Depth	Top Width		Cross-Sectional Area	Weight of Silage	
						4-Inch Slice	1-Foot Slice
(in.)	(ft)	(ft)	(ft)	(in.)	(sq ft)	(lb)	(lb)
3	4	5	7	0	24	280	840
4	4	6	8	8	29	338	1,015
5	4	7	10	4	33	385	1,155
3	6	6	9	0	45	525	1,575
4	6	7	11	0	54	630	1,890
5	6	8	13	0	63	735	2,205
3	8	6	10	0	64	747	2,240
4	8	7	12	4	77	898	2,695
5	8	8	14	8	91	1,062	3,185
3	10	6	11	0	85	992	2,975
4	10	8	14	8	113	1,318	3,955
5	10	10	10	4	142	1,657	4,970

[1]*Silos, Types and Construction*, Farmers' Bull. No. 1280, USDA, p. 55. To convert to metric, see the appendix section, "Weights and Measures."

For illustrative purposes, let us use the same example and silage requirements as were used in the section on "Size of Tower Silo," but this time determine the size trench silo to build rather than the size tower silo. Briefly, the requirements are for 1,020 lb of silage daily for a 240-day wintering period. As noted in Table 9–4, one day's feed or 1,020 lb of silage (1,062 lb to be exact) can be obtained in each 4-in. slice of a trench silo 8 ft wide at the bottom, 14 ft, 8 in. wide at the top, and 8 ft deep; or a 91 sq ft cross-sectional area. The cross-sectional area should not be larger than this if a 4 in. slice is to be removed daily in order to alleviate spoilage.

---

[1]Because the silage in trench silos is generally not so deep and well packed as the silage in tower silos, an average figure of 35 lb per cubic foot is used herein for trench silos and 40 lb for upright silos. With all types of silos — including aboveground and belowground types — the weight of a cubic foot of silage varies with the kind and maturity of material, moisture content, length of cut, rate of filling, and depth of the silo. Corn silage harvested when about 74% of the grain has passed the milk stage and containing approximately 70% moisture is considered average silage. Volume for volume, sorghum silage weighs about the same as corn silage. Grass, or grass-legume silage at 70% moisture is 10 to 15% heavier than corn silage, whereas grass haylage at 50% moisture is 10 to 15% lighter than corn silage of equal volume.

In order to obtain a 240-day feed supply, the filled trench must be 80 ft long (240 by 1/3 — the 1/3 representing 1/3 ft or 4 in.).

The size trench silo to build to meet the specified needs, therefore, is one that is 8 ft wide at the bottom, 14 ft, 8 in. wide at the top, 8 ft deep, and 80 ft long. In order to take care of spoilage and to provide a measure of safety, it is recommended that the actual length be from 85 to 90 ft.

About 8 ft for a trench silo is the most economical depth from the standpoint of cost and feeding. Of course, in filling it is desirable to pile silage 3 ft higher over the center of the trench and round it off. This provides for settling.

HOW TO ESTIMATE THE WEIGHT OF SILAGE IN A SILO

Sometimes, either (1) for inventory purposes, or (2) for purposes of buying or selling, dairy producers need to estimate the amount of silage remaining in a silo after part of it has been fed out. For a tower type silo, this may be done by referring to Table 9–5 and reading the accompanying directions.

Table 9–5 is for well-eared corn silage harvested in the early dent stage, cut in 1/4-in. lengths, well-tramped when filled, and with the silo refilled one time after settling for a day. The depth indicated in the left-hand column is the actual depth of the settled silage and not the height of the silo. As noted, silage is more compact and heavier as the depth increases.

Table 9–5 can be adapted for corn silage of different stages of maturity and grain content, and for other kinds of silages, simply by applying the rules of thumb given in Table 9–3.

To estimate the amount of average corn silage remaining in a tower silo after part of it has been fed out, proceed as follows:

1. Estimate the actual depth of silage left in the silo.

2. Estimate the original total depth of silage in the silo after settling 30 days.

3. Determine the feet of silage removed by subtracting the depth of silage left (1, above) from the original depth of silage in the silo (2, above).

4. Using Table 9–5, determine the original tonnage of silage contained in the silo.

5. Using Table 9–5, determine the amount of silage removed.

6. Determine the tonnage of silage remaining by subtracting the amount of silage removed (5, above) from the original tonnage (4, above).

As an example, let us assume that 10 ft of well-eared corn silage harvested in the early dent stage is left in a silo having a diameter of 14 ft and that, after settling, the entire depth of silage was 40 ft before feeding started. What tonnage of silage remains?

TABLE 9–5
GUIDE FOR ESTIMATING AMOUNT OF SILAGE IN A TOWER-TYPE SILO[1]

Depth of Settled Silage	Total Quantity of Settled Silage, from the Top to the Depth Indicated, in Silos Having a Diameter of:					
	10 Feet	12 Feet	14 Feet	16 Feet	18 Feet	20 Feet
(ft)	(tons)	(tons)	(tons)	(tons)	(tons)	(tons)
1	1	1	1	2	2	3
2	2	3	4	5	6	7
3	3	5	6	8	10	13
4	5	7	9	12	15	19
5	6	9	12	16	20	25
6	8	11	15	20	26	31
7	10	14	19	25	31	38
8	11	16	22	29	37	45
9	13	19	26	34	43	53
10	15	22	29	38	49	60
11	17	24	33	43	55	67
12	19	27	37	48	61	75
13	21	30	41	53	67	83
14	23	33	44	58	74	91
15	25	36	48	63	80	99
16	27	38	52	68	86	107
17	29	41	56	73	93	115
18	31	44	60	79	100	123
19	33	47	64	84	106	131
20	35	50	68	89	113	139
21	37	53	72	94	120	148
22	39	56	76	100	126	156
23	41	59	81	105	133	164
24	43	62	85	111	140	173
25	45	65	89	116	147	181
26	47	68	93	121	154	190
27	50	71	97	127	161	198
28	52	74	101	132	167	207
29	54	77	105	138	174	215
30	56	81	110	143	181	224
31	58	84	114	149	188	232
32	60	87	118	154	195	241
33	62	90	122	160	202	249
34	65	93	126	165	209	258
35	67	96	131	171	216	267
36	—	—	135	176	223	275
37	—	—	139	182	230	284
38	—	—	143	187	237	292
39	—	—	148	193	244	301
40	—	—	152	189	251	310
41	—	—	—	204	258	318
42	—	—	—	209	265	327
43	—	—	—	215	272	335
44	—	—	—	220	279	344
45	—	—	—	226	286	353
46	—	—	—	—	293	361
47	—	—	—	—	300	370
48	—	—	—	—	307	379
49	—	—	—	—	314	387
50	—	—	—	—	321	396

[1]This tabular material was used as a basis for Fig. 9–9. To convert to metric, see the appendix section, "Weights and Measures."

The answer is obtained as follows:

1. Estimated depth of silage remaining = 10 ft
2. Estimated original depth of settled silage = 40 ft
3. Feet of silage removed (40 – 10) = 30 ft
4. Original tonnage of silage before feeding as determined from Table 9–5; for a 14-ft diameter silo, 40 ft deep = 152 tons
5. Amount of silage removed as determined from Table 9–5; for a 140 ft diameter silo from which 30 ft has been removed = 110 tons
6. Estimated tonnage of silage remaining (152 – 110) = 42 tons

To estimate the amount of corn or sorghum silage in a trench silo (whether it is full or partly used), multiply the average width in feet by the depth in feet to get the cross-sectional area, then multiply by the length to get the volume and finally multiply the volume by 35 (the average weight of 1 cu ft of corn silage in a trench silo) in order to obtain the pounds of silage. For example, the amount of silage in a trench silo 8 ft wide at the bottom, 12 ft wide at the top, 8 ft deep, and 40 ft long is computed as follows:

1. $\dfrac{8 + 12 = 20}{2}$ = 10 ft; average width
2. 10 × 8 = 80 sq ft; cross-sectional area
3. 80 × 40 = 3,200 cu ft; volume of the silo
4. 3,200 × 35 = 112,000 lb, or 56 tons capacity

KINDS OF SILAGE

A great variety of crops can be and are made into silage. A rule of thumb is that crops that are palatable and nutritious to animals either as pasture, as green feed, or as dry forage also make palatable and nutritious silage. Likewise, crops that are unpalatable and nonnutritious as pasture, as green feed, or as dry forage also make unpalatable and nonnutritious silage. However, the palatability and nutrient value of some lowquality forages and crop residues can be improved by nutrient additives and fermentation.

Most silage in the United States is made from either corn or sorghum, with corn silage far in the lead—over 15 times as much corn silage as sorghum silage is made. In 1987, 80.6 million tons of corn silage and 5.4 million tons of sorghum silage were produced in the United States. At the present time, it is estimated that 75% of the nation's silage is made from corn and sorghum and 25% from grasses, legumes, and other feeds. In addition to the kinds of silage already mentioned, silage is made from sunflowers, the small grains, sugar beet tops, crop residues, wastes from food processing (sweet corn, green beans, green peas), root crops, and various vegetable residues.

Many of the leading silage crops are listed in Table 9–6, which also includes an estimate of the expected and potential production from these crops, along with the normal stage of maturity at harvest.

TABLE 9–6
LEADING SILAGE CROPS[1]

Crop	Expected Production[2]			Potential Production[3]			Stage for Ensiling
	Dry Matter Yield	Digestible Protein	TDN	Dry Matter Yield	Digestible Protein	TDN	
	(tons/A)	(lb/A)	(lb/A)	(tons/A)	(lb/A)	(lb/A)	
Corn	6.0	600	8,100	10.5	1,100	14,300	Hard dough or early glaze.
Grain sorghum	5.5	550	7,200	9.0	900	11,700	Soft to medium dough.
Forage sorghum	6.0	600	6,900	9.0	900	10,400	Soft to medium dough.
Sorghum-Sudangrass hybrids . .	4.5	540	5,000	8.0	960	8,800	Early bloom.
Sudangrass	3.5	420	3,900	6.0	720	6,600	Early bloom.
Alfalfa	5.0	1,500	6,500	9.0	2,700	11,700	Bud stage, first cut; early bloom, other cuts.
Legume-grass mixture	4.5	900	5,400	8.0	1,600	9,600	Legumes, early bloom.
Small grains	4.0	560	4,800	7.0	980	8,400	Boot to early head.

[1]From paper entitled, "Silage Production, Preservation, and Quality," by John E. Baylor, The Pennsylvania State University, *Dairy Science Handbook*, Vol. 9, 1976, p. 224, published by Agriservices Foundation, edited by M. E. Ensminger. To convert to metric: tons/A = 0.907 metric ton/0.4 hectare; lb/A = 0.454 kg/0.4 hectare.

[2]Expected production—Figures presented are better than average yields for the country. They are yields which might be expected for average or better farmers who are following recommended practices.

[3]Potential production—Figures presented are production yields attainable where good soils, excellent growing conditions, and top management are present. These silage yield equivalents have been produced on farms and experiment stations prior to this time.

The average composition of the silages produced from several of these crops is given in Table 9–7.

TABLE 9–7
COMPOSITION OF VARIOUS SILAGES

Type of Silage	Analyses On A Dry Matter Basis			
	Crude Protein	TDN	Ca	P
	(%)	(%)	(%)	(%)
Corn	8.3	68.0	0.31	0.27
Grain sorghum . . .	7.9	55.0	0.34	0.19
Forage sorghum . . .	9.2	57.9	0.30	0.24
Oats	10.0	57.0	0.47	0.33
Alfalfa	17.4	59.0	1.75	0.27

Fig. 9–10. Harvesting grain sorghum for silage on a southeast Kansas dairy farm. (Courtesy, A. O. Smith Harvestore Products, Arlington Heights, IL)

CORN AND SORGHUM SILAGE

For the United States as a whole, corn ranks first in importance as a silage crop. Generally more total digestible nutrients can be obtained from an acre of corn as silage—which will yield from 5 to 25 tons of forage per acre, with an average of about 14.5 tons—than can be obtained from an acre of any other crop. Also, corn ensiles easily without the aid of a preservative, and keeps almost indefinitely in a good silo, is highly palatable, is well adapted to mechanized feeding, and may be fed with little waste.

There are four kinds of corn silage; namely:

1. **The whole corn plant.** When at the peak of its nutritive value and right for ensiling, the whole corn plant contains 1½ times the nutrients of the ripened grain that the plant would have yielded. Also, in corn silage made from the whole crop more than 90% of the nutrients produced are saved.

2. **Ear corn silage.** The ensiled ears contain up to 68% of the nutrients of the entire corn plant.

3. **Corn stover silage.** The forage remaining after harvesting a grain crop. This accounts for about one-third of the total nutritive value of the crop.

4. **Shelled-corn silage.** This consists of the kernels only. At 70% dry matter (30% moisture), shelled-corn silage contains 61 to 66% of the nutrients in the whole crop (also see "High-Moisture Grain" later in this section).

The sorghums are more dependable and higher yielding than corn in certain areas, particularly in unirrigated, and relatively dry areas, of western and southwestern United States. Sorghum for silage is harvested with the same equipment as is used for corn silage. It should not be harvested for silage until the heads are soft to medium dough stage. Harvesting at this stage provides the highest yields of total feed material, enhances preservation, and makes silage that has good palatability.

On a dry-matter basis, corn silage contains an average of 8.3% crude protein, 68% total digestible nutrients, 0.31% Ca, and 0.27% P. Grain sorghum silage contains less protein and TDN than corn silage. Grass/legume silages contain more protein and less TDN than corn silage. The carotene content of corn silage is variable, but on the low side.

CORN AND SORGHUM RESIDUE SILAGE

Corn and sorghum residues—the forages that remain after harvesting a grain crop of corn or sorghum—may be used as cattle feed three ways: (1) grazed, (2) harvested (stacked or baled) and fed dry, or (3) ensiled and fed as silage. The discussion that follows will be limited to ensiling corn and sorghum residue.

When ensiled, cornstalks (stover) produce a product known as corn stover silage or cornstalk silage. When stalks are processed as silage, the use of a forage harvester equipped with a screen or a recutter-blower at the silo is necessary in order to chop the material finely. Fine chopping will ensure good packing and improve consumption by avoiding selectivity.

Where corn stover silage is made, the residue should be harvested as soon as possible after the grain is taken off, before the residue loses any moisture. At that time, the grain moisture will generally be under 30% and the refuse moisture will be above 48%. In an airtight silo, 40 to 45% moisture is very satisfactory for ensiling. In an unsealed or bunker silo, the moisture content should be 48 to 55% for proper lactic acid formation. Water may be added at the silo if necessary. As a precaution, some authorities recommend the addition of 56 lb of corn meal (or other finely ground grain) per ton of corn stover silage, as a means of providing readily-fermentable carbohydrates from which acids will form and act as a preservative. With husklage, the latter precaution is not necessary since there is sufficient grain remaining in the husk and cob.

Purdue University researchers fed corn stover silage supplemented with varying increments of energy for growing dairy heifers.[2] Twenty-eight yearling Holstein heifers were randomly assigned to each of four corn stover silage rations to determine the level of supplemental energy required to obtain growth response of 1.3 to 1.4 lb per day as recommended by the National Research Council (NRC). The four stover silage complete rations contained 0.37, 0.42, 0.47, and 0.52 Mcal estimated net energy (ENE) per pound of complete feed. Additions of increasing increments of ENE produced a linear increase in daily body weight gains of 0.75, 0.95, 1.28, and 1.5 lb for the four rations, respectively. Total dry matter and ENE intake increased linearly with increasing ENE concentration of the rations. The investigators concluded that, under the conditions of their study, a total ENE content of corn stover silage ration of approximately 0.5 Mcal per pound ration could have produced the growth response recommended by NRC for dairy heifers.

The biggest deterrent to harvesting stalklage, in either dry or ensiled form, is the cost—primarily for equipment. Rather than own such expensive equipment, which is used for a short period only, custom harvesting of stalklage is likely cheaper for most operators.

Husklage—the forage discharged from the rear of a corn combine, and consisting of the husks, cobs, and any grain carried through the combine—may also be ensiled. Ensiling husklage, along with recutting and adding water, results in increased cow consumption and less rejection of cobs.

Like corn, sorghum stover may either be grazed or harvested and stored either as dry feed or silage. Because the sorghum plant stays green late in the fall, good sorghum stover silage can be made without additional water.

GRASS/LEGUME (HAY CROP) SILAGE

Fig. 9–11. Grass silage being harvested from the windrow by a chopper. (Courtesy, Gehl Company, West Bend, WI)

Grass/legume (hay crop) silage refers to silage made from any of the green crops which might otherwise be grazed or dried and made into hay. This includes grasses (such as timothy or fescues), legumes (such as alfalfa or clovers), grass-legume mixtures, and cereal grains (such as oats). In practical operations, any adapted grasses and/or legumes may be used three ways: for grazing, for hay, or for silage.

Grass/legume silage can be produced in areas where the climate is too cool and the growing season too short for corn or sorghum silage.

Although grass and legume crops have been ensiled in Europe for hundreds of years, the practice did not become widely used in the United States until the 1930s. At that time, interest in hay crops for silage increased as a result of farmers (1) becoming aware of the field losses that occur in hay making, (2) being provided with the information necessary to make high-quality silages from grasses and legumes, and (3) having access to field choppers, which facilitated making silage from hay crops.

The New Hampshire Station compared hay crop silage and hay for lactating cows.[3] Holstein cows were divided into three groups for a 51-month continuous trial to study feed intake, production, profit, and dry cow responses when 43 to 44% of the forage dry matter was from either (1) wilted hay crop silage, (2) silage and hay (50:50), or (3) hay, all from the same field. The remainder of the forage for all cows was urea-treated corn silage. Common concentrate was fed to all cows *ad libitum* from calving to peak of lactation, then according to requirements until the end of lactation. Forages were fed *ad libitum* during lactation. Dry cows received a small amount of concentrate plus forage. The forage intake of hay was higher than the intake of hay crop silage, but this did not result in significantly more milk or income over feed cost. It was concluded that the decision to harvest the hay crop as wilted silage or as hay depends·on economics of earlier harvest and earlier regrowth with silage, and on its adaptability to blended complete rations rather than on differences in profitability or production response.

The following are the most important **advantages** of grass/legume silage:

1. It minimizes field, harvest, and storage losses of grass/legume forages (see Fig. 9–12).

2. It minimizes the dependence on favorable weather to harvest the crop.

3. It can be harvested with modern, efficient machinery and stored in large-volume structures.

4. It requires less supplemental feed than corn silage.

5. It can be handled and fed by mechanized methods, thereby reducing the labor requirements.

6. It kills weed seeds as a result of the fermentation.

[2]Colenbrander, V. F., *et al.*, "Corn Stover Silage Supplemented With Varying Increments of Energy for Growing Dairy Heifers," *Journal of Animal Science*, Vol. 33, No. 6, 1971, p. 1306.

[3]Holter, J. B., W. E. Urban, Jr., and H. A. Davis, "Haycrop Silage Versus Hay in a Mixed Ration for Lactating Cows," *Journal of Dairy Science*, Vol. 59, No. 6, p. 1087.

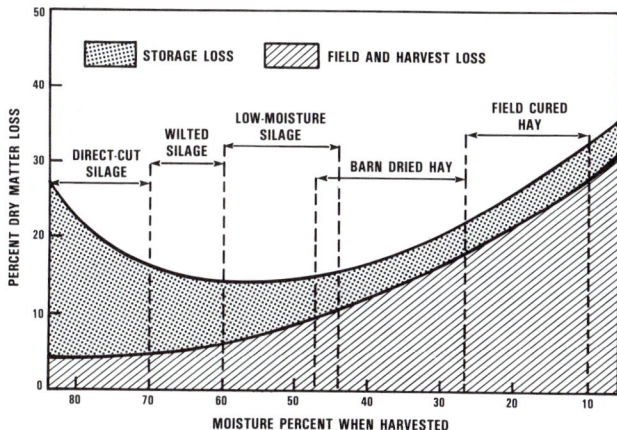

Fig. 9–12. Estimated total field, harvest, and storage loss when grass/legume forages are harvested at different moisture levels by alternative harvesting methods. (Courtesy, C. R. Hoglund, Professor Emeritus, Michigan State University, East Lansing)

Although grass/legume silage has important advantages, it also has the following **disadvantages**:

1. The initial investment costs for machinery, storage units, and feeding facilities are very high.

2. Silage-making machinery and storage and feeding facilities are highly specialized with the result that they have limited use for other purposes.

3. Inadequate fermentation may occur under certain conditions, resulting in poor-quality feed.

4. Storage and feeding losses may be high under poor management.

5. Like all silages, grass/legume silage is heavy, bulky, and costly to transport, thus its off-farm market value is limited.

6. Existing upright silos may not be in sufficiently good condition to store grass/legume silage.

7. Once grass/legume silage is removed from the silo, it must be fed within 12 to 24 hours to alleviate spoilage. Except for an oxygen-limiting silo, once feeding begins during warm weather, it is necessary to feed a minimum of 3 to 4 in. off the exposed surface daily to prevent spoilage in the silo.

Grass/legume silages are of three kinds based on moisture level:

1. Direct-cut silage, 70% moisture or above.
2. Wilted silage, 60 to 70% moisture.
3. Low-moisture silage (or haylage), 40 to 60% moisture. (Haylage is discussed in a separate section later in this chapter.)

DIRECT-CUT SILAGE

Direct-cut grass/legume silage is forage that is harvested and stored without field-drying, usually containing more than 70% moisture. Although direct-cut ensiling is the standard practice with mature corn and sorghum, it is not recommended for grass/legume crops because of (1) the difficulty in getting good preservation due to the high moisture content, and (2) the increased nutrient losses due to seepage.

The higher the moisture content of grass/legume forage the more critical the need for a low pH to obtain good preservation. Direct-cut silage is subject to excessive deterioration of protein, undesirable odor, excess loss of nutrients due to seepage, and deterioration of concrete stave silos. Also, when fed as the major forage, direct-cut silage usually results in lower consumption of dry matter and consequently lower animal production than wilted or low-moisture silage.

When making direct-cut silage, the following directions should be observed: (1) harvest at the proper stage of maturity; (2) avoid cutting when the forage is wet with dew or rain; (3) provide drainage for excess juice; (4) add an additive or preservative (coarsely ground cereal grain will absorb some of the excess moisture in addition to providing sugars for fermentation); (5) distribute evenly in the silo and pack thoroughly; and (6) cover with plastic or suitable material.

WILTED SILAGE

Today, a high percentage of grass/legume forage is dried to some degree prior to ensiling. Wilting gets rid of some water in the field, so less weight is handled. Also, in comparison with direct-cut silage, odor and seepage problems are reduced and additives or preservatives are usually not needed.

Authorities generally agree on the following rules for

Fig. 9–13. Alfalfa forage wilted to 65% moisture being field chopped for ensiling. (Courtesy, Ford New Holland, New Holland, PA)

making good wilted silage: (1) harvest at the proper stage of maturity; (2) allow the forage to wilt in the swath and/or windrow until the moisture reaches about 65%, which may take from 1 to 4 hours, or longer, depending on the weather, but which may be expedited by the use of a forage conditioner; (3) use a short cut (about 3/8 in.) on the forage harvester; (4) fill the silo rapidly and continuously; (5) distribute evenly and pack thoroughly; and (6) top off with 2 ft of unwilted material and cover with plastic or other suitable cover. No additive or preservative is needed with properly wilted silage, although it may be added if desired.

CORN OR SORGHUM SILAGE VS GRASS/LEGUME SILAGE

Frequently livestock producers are confronted with choosing between corn or sorghum silage and grass/legume silage. Under these circumstances, the following facts are pertinent:

1. Where adapted, corn or sorghum will generally produce a greater tonnage of feed per acre than grass/legume silage.

2. Good-quality corn or sorghum silage can be made more consistently and with greater ease than good-quality grass/legume silage.

3. Corn or sorghum silage may be more palatable than grass/legume silage, even when the latter is carefully preserved.

4. Grass/legume silage is generally higher in protein and carotene but lower in total digestible nutrients and vitamin D (wilted grass/legume silage is higher in vitamin D than unwilted) than corn or sorghum silage (generally, grass/legume silage contains about 90% as much TDN as corn silage, but it will equal corn silage in TDN when 150 lb of grain per ton has been added). Thus, grass/legume silage generally requires the addition to the ration of less protein supplement but more total concentrates than corn or sorghum silage.

5. Grass/legume silage can be produced in areas where the climate is too cool and the growing season too short for corn or sorghum silage.

6. The production of grass/legume silage will result in less soil washing than the production of corn or sorghum silage on lands subject to erosion.

7. Grass/legume silage will freeze next to the silo wall more than corn or sorghum silage, especially if it is ensiled when too wet (unwilted).

8. The silo can be kept working full time by using both grass/legume and corn or sorghum silage; ensiling the first cutting of grass/legume silage for summer feeding, and ensiling corn or sorghum silage for winter feeding.

9. In most of the major dairy states, alfalfa is the preferred forage and usually exceeds the grasses in protein yield and feeding value as a silage crop.

OTHER SILAGE CROPS

In the Northwest and North Central states, where the weather is cool and the growing season is short, sunflowers are sometimes grown for silage. Although they yield and ensile well, sunflower silage is neither as palatable nor as nutritious as corn, sorghum, or grass silage. Pound for pound, sunflower silage is about 80 to 85% as valuable as corn silage.

Throughout the United States, a great array of by-product feeds are ensiled, especially in the less expensive and temporary types of silos. Among such by-products are grain chaff, pea and bean vines, beet tops and pulp, sunflower hulls and chaff, potatoes, cannery refuse, cull and surplus fruits and vegetables, pulp and trimming wastes from market vegetables and fruits, wet brewers' and distillers' grains, almond hulls, and poultry litter. Sometimes Russian-thistles and other weeds are ensiled.

When potatoes, which contain about 80% moisture, are ensiled for cattle, it is recommended either (1) that 20 to 25 lb of dry hay, straw, or chaff be run through the ensilage cutter with each 100 lb of potatoes, or (2) that 1 ton of corn or sorghum silage be chopped with each 500 lb of potatoes. Frozen and sprouted potatoes should not be ensiled. Potato processing wastes (cull potatoes, off-flavor french fries and chips, etc.) can be ensiled in the same manner as unprocessed potatoes.

Either of the methods recommended for ensiling potatoes for cattle is equally adapted for the preservation of other high-moisture crops, such as apples, beets, pears, tomatoes, cauliflower, broccoli, kale, and trimming wastes from market vegetables — provided the added forage is in proportion to their respective moisture contents.

Cabbage, rape, and turnips should not be ensiled, as they make unsatisfactory, watery, foul-smelling silage.

COMBINING CROPS FOR SILAGE

Sometimes, in order to lower the moisture content, to alleviate the necessity of a preservative, and to assure better quality silage, forages of high sugar content are combined with forages of low sugar content. Thus, excellent silage can be made by mixing 1 ton of sorghum forage with each 3 tons of grass/legume silage material, or a ton of corn forage with each ton of grass/legume forage material (less sorghum forage is necessary than corn forage, because of the higher sugar content of the former).

At times such combination silage crops are even grown together; for example, corn and soybeans; millet or Sudangrass; and soybeans, oats, and peas.

A major difficulty in combining ensiling crops is that it is almost impossible to synchronize the stage of maturity of different crops so that they reach maximum yield and nutrient level at the same time.

RAIN-DAMAGED HAY SILAGE

Partly cured hay that has been rained upon, but is not moldy, may be salvaged as silage (although it will not be of high quality), provided it is finely chopped, distributed evenly, and packed in the silo thoroughly enough to squeeze out the air. It is recommended that it be placed in the bottom of the silo, and, preferably, that alternate loads of a green crop be mixed with it. Otherwise, satisfactory packing can be obtained by putting a few loads of greener-than-ordinary material on top of it.

FROSTED CROP SILAGE

Sometimes corn, sorghum, sunflowers, small grains, beans, and other crops, which may or may not have been intended for silage, are frosted before they reach the silage cutting stage. Corn that has been frosted before reaching maturity is commonly known as *soft corn*. Such frosted crops may be salvaged as silage. They should be cut at recommended moisture contents and ensiled according to directions. If they are too dry, water should be added.

Frosted crops, especially frosted sorghum, may be high in cyanide (HCN). (See *CAUTION* under the heading "Drought Stricken Crop Silage.")

DROUGHT STRICKEN CROP SILAGE

Sometimes corn or sorghum, or other crops, are drought stricken to the extent that little or no grain will be produced. Such crops may be harvested for silage and used as an energy source for ruminants. They should be cut and ensiled like any other silage crop. If they are too dry, water should be added.

Drought stricken crop silage may be used in the same manner as any other low-energy source. It is well-suited for dry dairy cows.

CAUTION: Danger of cyanide toxicity is much greater from sorghum than from corn. Drought stricken plants can accumulate cyanogenetic glycoside which hydrolizes to form free cyanide (HCN) The danger is increased when crops are grown on heavily nitrogen-fertilized soils or if any of the following have occurred: frosting, wilting, trampling, or hail. Any combination of these conditions can lead to a dangerous build-up or release of cyanide.

HARVESTING METHODS AND MACHINERY

There is no one best silage-making method or kind of equipment. These must necessarily vary with the kind of silage, the kind of silo, the size of operation, and the available labor and machinery and the cost.

Three principal kinds of machines are used for harvesting silage, Namely, field forage harvesters, row-crop binders, and stationary silo fillers.

Field forage harvesters, which were first developed around 1936, are more widely used than any other type of equipment for harvesting silage. Also, they tend to be concentrated in those states where the production of silage is most important. With different attachments, field forage harvesters can be used to harvest row crops for silage, grass silage as a standing crop or from the windrow, and hay from the windrow. Also, they can be used to harvest straw and other kinds of forage. Field choppers can even be adapted

for grinding and blowing high-moisture cob corn for ensiling. With appropriate attachments, a modern field harvester can be used to harvest all major ensiled crops. Thus, with a minimum of complementary machinery, a forage harvester can be the major piece of equipment in providing a completely ensiled ration for dairy cattle. But such equipment is expensive and may or may not be economical where a small operation is involved.

Field chopped forage is generally transported on wagons equipped with mechanical unloading devices or by means of dump trucks. Blowers and conveyors are used in filling both tower and horizontal silos. Frequently, trench silos are filled by dumping over the sides.

The use of row-crop binders reached a peak in 1942, following which they declined. Reduction in the numbers of these machines reflects the increased use of cornpickers, field forage harvesters, and grain combines. In addition to being used to harvest silage, row-crop binders are also used to harvest corn for grain, corn for fodder, and sorghum as bundle feed.

Beginning in 1951, the use of stationary silo fillers declined markedly. Today, few of them are used.

CHARACTERISTICS OF GOOD SILAGE

In order to make good-quality silage, dairy producers need to know what constitutes silage quality. They need to be acquainted with those recognizable characteristics of silage which indicate high palatability and nutrient content. The easily recognized characteristics of silage of high feeding value are:

1. **Odor.** It has a "clean," rather pleasing acid odor, in contrast to the foul or objectionable odor of poor silage.

2. **Taste.** The taste is pleasing, not bitter or sharp.

3. **Absence of mold and rot.** There is no visible mold, and it is not musty or slimy.

4. **Moisture and color.** It is uniform in moisture and color. Very high-moisture silage is likely to be dark colored, slimy textured, and have a disagreeable odor. Generally, green or brownish silage is good; tobacco brown, dark brown, caramelized, or charred silage indicate excessive heat; and black silage is rotten and should not be fed.

5. **Animal acceptance.** Animals like and thrive on good silage.

HOW TO MAKE GOOD SILAGE

In addition to using a sound silo of proper size, those who make good silage generally harvest at the proper stage of maturity, cut to proper length, control the moisture content of the forage, add a preservative only when needed, fill rapidly, distribute and tramp the forage in the silo, and seal or top-off the silo. Each of these factors will be discussed.

HARVEST AT PROPER STAGE OF MATURITY

Harvesting at the proper stage of maturity assures the maximum yield and nutrient content.

Fig. 9–14 shows the effect of stage of maturity of the corn plant on total dry matter accumulation.

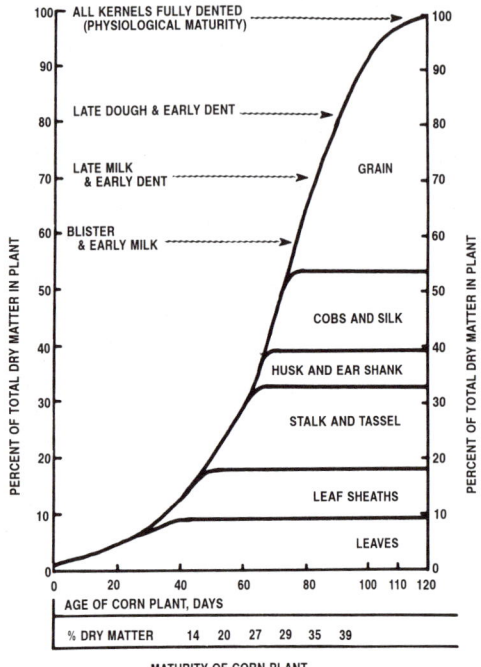

Fig. 9–14. Effect of maturity of corn plant on total dry matter accumulation. (Source: Iowa State University Special Report No. 48)

The *black layer test* can be applied quickly and easily to determine when to harvest corn for maximum yield and nutrient quality (see Fig. 9–15).

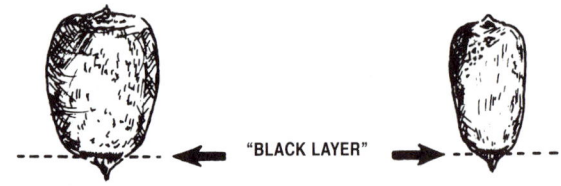

Fig. 9–15. Black layer near the tip of the kernel indicates that the grain is physiologically mature and ready for the silo.

When the grain reaches physiological maturity, several layers of cells near the tip of the kernel turn black, forming the *black layer*. This layer can be detected by removing several kernels from the middle of the ear, thence split them lengthwise or just cut off the tip, and look for the black layer near the tip. If the black layer is present, the grain is physiologically mature and ready for the silo.

At the black layer stage, usually the grains are dented and glazed, the lower four to six leaves of the corn plant are brown, and the plant contains 60 to 67% moisture. It can be cut 3 to 4 weeks past this stage with very little loss in dry matter or in feeding value.

Sorghum should be cut for silage when the seeds are hard.

Grass silage forages (grasses, legumes, and cereal crops) should be cut at the same stage at which they would make the best hay (see Table 8–8, p. 132, in Chapter 8, Hay).

The Tennessee Station researchers did not detect any significant difference in the milk production of Jersey cows fed corn silage at three different stages of maturity.[4] They concluded that selecting a high grain yielding corn silage and harvesting the crop when the dry matter content is between 30 and 35% to prevent seepage and spoilage losses may be as important as the physiological stage of maturity to determine the time of harvest.

CUT TO PROPER LENGTH

The length of the cut sections affects the packing and, hence, the quality of the silage. Also, the proper length of cut varies with the crop and the moisture content. Thus, for corn and sorghum crops, forage harvesters should be set to make a theoretical cut of ¼ to ⅜ in. If the knives are sharp and set up to the cutter bar, this will result in about 15% of the particles being 1½ in. and over, 25% of the particles being ¾ to 1½ in., and 60% being ⅛ to ¾ in. in length. Such a combination of particle size is necessary for high-quality feed. Grass silages should be more finely chopped than corn or sorghum silage. Also, wilted and dry forage and forage with hollow stems should be chopped more finely than forage of high-moisture content, thus permitting more thorough packing and eliminating most air pockets.

CONTROL THE MOISTURE CONTENT

Moisture content is one of the most important factors in determining quality of silage. Experimental work and practical experience have indicated that 60 to 67% is the best moisture content of a crop to be ensiled. However, low-moisture silage of 40 to 60% moisture is now being preserved successfully in (1) oxygen-limiting silos, and (2) tall, conventional silos that are properly topped-off with heavy, wet forage or sealed with a plastic cover.

Forage containing more than 60 to 67% moisture (1) is heavier and more costly to handle than is necessary; (2) is apt to produce slimy, putrid silage, due to the presence of butyric and other undesirable acids; (3) will have excessive leakage of the juices and some loss of nutrients, except carotene, from the silo; (4) will result in excessive deterioration in the silo walls due to the high acidity; and (5) will exert higher pressure on the silo walls—for the greater the moisture content, the greater the pressure.

If corn and sorghum are harvested at the stage recommended, their moisture content will be right. However, freshly cut grass and/or legume forage contains 75 to 80% moisture, which means that for proper ensiling its moisture content must be lowered by 10 to 15%

HOW TO LOWER THE MOISTURE CONTENT

The moisture content of silage material may be lowered by any one or a combination of the following methods: by

---

[4]Montgomery, M. J., *et al.*, "Effect of Maturity of Corn on Silage Quality and Milk Production," *Journal of Dairy Science*, Vol. 57, No. 6, 1974, p. 698.

conditioning and/or wilting, by adding dry hay or straw, by combining with corn or sorghum silage, or by adding a dry preservative of grain, dried molasses, or dried by-products of citrus or beets.

CONDITIONING-WILTING

This method is particularly applicable to the making of grass silage. Conditioning and/or wilting of grass silage increases the percentage of sugar in the forage, lessens the leakage of juice from the silo, lessens the pressure on the silo walls, and decreases the destructive action of the acids on the silo walls.

The needed 10 to 15% reduction in the moisture content of grass silage material can be accomplished by wilting for about 2 hours on a good drying day and up to 1 day or longer in slow drying weather.

The combination of conditioning and wilting is the method most commonly followed today. Excellent equipment is available for conditioning.

Excess drying should be avoided, as it will result in the forage becoming too dry for proper ensiling.

ADDING DRY HAY OR STRAW

The moisture content of any wet silage material can be lowered effectively by mixing dry hay or straw with it at the time of filling. Thus, during poor wilting weather, the moisture content of grass forage can be brought within the desired range by adding 5 to 20% hay or straw. Also, this is the standard method of lowering the moisture content when it is desired to ensile such high-moisture products as potatoes (see "Other Silage Crops").

Conditioning and wilting is the preferred method of lowering the moisture content of grass silage, rather than adding dry hay or straw, which reduce digestibility and energy content.

COMBINING WITH CORN OR SORGHUM SILAGE.

Sometimes the moisture content can be lowered sufficiently merely by mixing high water content crops with low water content crops (see sections entitled "Other Silage Crops" and "Combining Crops for Silage"). Simultaneously, usually more desirable bacterial action can be assured by this procedure.

ADDING A DRY PRESERVATIVE

Such dry preservatives as ground grain, corn-and-cob meal, dried molasses, and dried citrus meal, citrus pulp, or beet pulp will reduce the moisture content of freshly cut and unwilted forage, and, in turn, lessen the leakage (or seepage) from the silo.

HOW TO INCREASE THE MOISTURE CONTENT

If the crop is overripe and too dry when cut, or if it becomes overwilted, it will be necessary either to add water to the silo, or, perhaps preferably, to make it into hay.

Drier material may be used for silage by cutting shorter and packing more thoroughly. If necessary, water should be added or the dry material should be mixed with very green, freshly cut material by alternating loads.

HOW TO DETERMINE THE MOISTURE CONTENT

If corn or sorghum forage is harvested at the recommended stage (see section on "Harvest at Proper Stage of Maturity"), its moisture content will be satisfactory for silage making. On the other hand, wilting is always preferable with forages intended for grass silage.

With a little practice, dairy producers can usually determine when grass forage has wilted sufficiently and when the moisture content is between 60 to 67%. Here are some rule-of-thumb methods:

1. **The twist method.** Before chopping, the forage should be so well wilted that the stems may be twisted without breaking, but the limp leaves should show no signs of dryness. This test cannot be used for such coarse crops as sweet clover.

2. **The grab test (or squeeze method).** This test consists of taking a handful of the chopped forage and giving it a good hard squeeze for about 30 seconds, followed by opening the hand slowly, noting the condition of the ball of forage in the hand, and referring to the following guides:

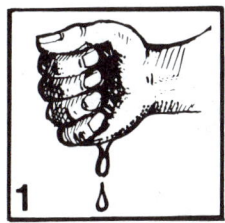

1. Juice runs freely or shows between the fingers. The crop contains 75 to 85% moisture and is too wet to make high-quality silage without treatment. Silages made from crops in this condition will lose large quantities of juice. When possible, wilt these crops. If they must be ensiled without wilting, use an effective chemical preservative or 200 lb of ground grain per ton of crop.

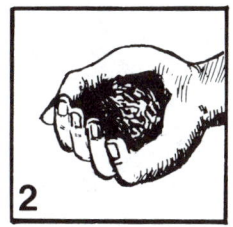

2. The ball holds its shape—the hand is moist. The crop contains 68 to 75% moisture. Some juices will escape from tower silos. Additional wilting in the field is desirable. Where this is not done, use a chemical preservative or 150 lb of ground grain per ton of crop, or layer with wilted crops. Odors will be strong without some treatment.

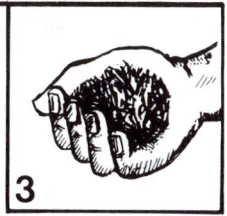

3. The ball expands slowly—no dampness appears on the hand. The crop contains 60 to 67% moisture. This is the best condition for ensiling legumes without treatment.

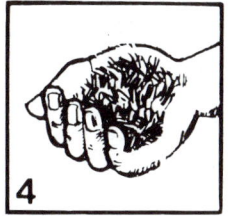

4. The ball springs out in the opening hand. The crop contains less than 60% moisture. Only very young crops wilted to this condition can be safely ensiled. Others are likely to mold in the silo unless layered with wet crops or placed in gastight silos.

Fig. 9–16. The grab test.

3. **The oven-drying method.** If in doubt or until more experience is obtained, the moisture content of a sample of any kind of forage may be obtained in about an hour's time by the following procedure:

    a. Weigh an empty tray on kitchen, bathroom, dairy barn, or other scales.

    b. Spread some of the forage in a thin layer on the tray.

    c. Weigh the tray and forage.

    d. Subtract "a" from "c," in order to obtain the weight of the green forage sample.

    e. Place the tray and sample in a preheated oven at 275°F. If the oven doesn't have a vent, leave the door slightly open.

    f. When the forage seems to be dry, weigh it again. Return it to the oven and reweigh at 5- to 10-minute intervals until the weight is nearly constant. Record final weight of the tray and dry forage.

    g. Subtract "a" from "f," in order to obtain the weight of the dry forage sample.

    h. Subtract "g" from "d," in order to obtain the weight of the water in the sample.

    i. Divide "h" by "d" (and multiply by 100), in order to obtain the percent of moisture in the forage.

4. **Other methods.** The heated-oil method and certain patented devices (such as forced air dryers) may be used for moisture determination. Recently, several electronic testers which give instantaneous moisture readings of forages have come on the market. Also, microwave ovens can be used for moisture determination, provided the forage is spread in a thin layer to prevent burning. The latter two methods are expensive, but fast.

SILAGE ADDITIVES AND PRESERVATIVES

Silage additives are products that provide supplemental nutrients which enhance the feeding value of silage.

Silage preservatives are products that enhance the keeping qualities of silage.

High-quality silage can be made without the use of additives or preservatives if good material is started with and all proven good practices are followed. But there are times when the ensiled material is either too wet or too dry; does not contain sufficient fermentable carbohydrates; is deficient in certain nutrients; is lacking in palatability; and/or the proven good practices cannot be followed. Under such circumstances, silage additives or preservatives may reduce silage losses and/or improve the feeding value of the silage.

Additives or preservatives should not be used as a substitute for a good silo or for proper chopping, packing, and sealing. Normally, additives or preservatives are neither needed nor added to corn or sorghum silages. But additives or preservatives may be very helpful if a grass/legume forage with over 70% moisture is ensiled.

A number of materials are available to incorporate in silage, with claims made that they will enhance the nutrient value, the preservation of the nutrients, and/or the palatability of the silage. *The bottom line when using any silage additive or preservative is how much it improves animal performance and net profit; and not whether it merely makes silage look better and smell better.*

Thorough testing of these materials would necessitate that each of them be used at several levels, with many kinds of silage, with each forage at various moisture contents, and under different storage conditions. Obviously, it would be highly impractical, if not impossible, to carry out such an extensive study. However, there is sufficient understanding of the process of silage formation, the requirements for the preservation of silage nutrients, and the mode of action of the ingredients used in various additives to make sound decisions as to whether they might be economically worthwhile. Also, some experimental testing has been done with certain additives.

In order to be effective, an additive or preservative should serve one or more of the following purposes.

1. Add nutrients.
2. Provide fermentable carbohydrates.
3. Furnish additional acids to increase acid conditions.
4. Inhibit undesirable types of bacteria and molds.
5. Reduce the amount of oxygen present, directly or indirectly.
6. Reduce the moisture content of the silage.
7. Absorb some nutrients which might otherwise be lost in seepage.
8. Reduce the fiber content of the forage through enzymatic action.

Four types of additives or preservatives are used in silage making: (1) feed additives, (2) acids, (3) fermentation aids, and (4) preservatives.

FEED ADDITIVES

Feed additives may be used to provide a readily available source of carbohydrates for fermentation into lactic acid, to reduce the moisture content, to provide needed nutrients, and/or to enhance palatability.

Feedstuffs used as silage additives include:

1. Corn-and-cob meal, ground corn, barley, or oats; applied in amounts varying from 100 to 300 lb per ton, depending on the moisture content of the crop.
2. Beet pulp, citrus pulp, chopped corncobs, or chopped hay to reduce seepage losses from the silo if moisture is high.
3. Molasses, either liquid or dehydrated, at rates of 40 to 80 lb per ton of green forage.
4. Dried whey, a product of the dairy industry, applied at the rate of 30 to 300 lb per ton, as a source of fermentable carbohydrate, protein, and minerals.
5. Nonprotein nitrogen (NPN) products such as urea and anhydrous ammonia.
6. Ground limestone.

■ **Grain and other feed ingredients**—Silage made from legumes or grasses may be improved under certain conditions by the addition of ground grain (corn, wheat, or barley), ground ear corn, beet pulp, citrus meal, citrus pulp, or other appropriate feed ingredient. The ground material will provide a readily available source of carbohydrates (sugar and starch) for bacterial fermentation and the production of acids; increase the feeding value of the silage, because 75 to 85% of the feeding value of the grain will be retained if the silo excludes air properly; and likely improve palatability. If the primary concern is to reduce the moisture content of the silage, cheaper materials, such as ground corn cobs, cottonseed hulls, oat hulls, or chopped straw or hay, may be more appropriate than ground grain. Dry, finely ground corn cobs will absorb nearly 200 lb of water per 100 lb of the cob material; dried beet pulp will absorb even more.

These materials may be added by feeding them into the blower from a properly adjusted hopper attachment.

When green forage is ensiled at a proper moisture content, there is usually no advantage to adding grain for the purpose of preservation or palatability.

■ Molasses (including cane or blackstrap, beet, corn, and citrus molasses) — Some green forages, such as legumes and certain grasses, are rather low in sugar content. Hence, adding molasses, which is high in sugar, may increase lactic and acetic acid production and improve silage quality and preservation. Also, molasses improves the palatability of silage and increases its nutritive value. For legumes, about 80 lb of molasses per ton are generally used, and for grasses about 40 lb per ton (molasses weighs 12 lb/gal). Addition of much less than these amounts, as an ingredient in mixed preservatives, is of little value. Much of the feeding value of the molasses is retained in the silage under good storage conditions and where there is no seepage loss.

Molasses may be added in either liquid or dehydrated form as the forage enters the blower.

When a grass and/or legume is wilted to 50 to 60% moisture content and adequately protected from air, an excellent feed with a good aroma and keeping quality can be obtained without the addition of molasses (sugar). Neither is the addition of sugar needed for corn silage.

■ Whey — Experiments and experiences indicate that adding dried whey to alfalfa silage or haylage slightly improves the quality and digestibility. There is also indication that adding dried whey to urea-treated corn silage may help reduce nitrogen losses from and improve the feeding value of the silage.

South Dakota Station workers studied corn silages containing either (1) 0.5% urea plus 1% dried whey, or (2) 0.5% urea; in a lactating trial, in a feeding trial with heifers 4 to 8 months of age, and in a digestion trial.[5] Adding 1% dried whey to urea-treated corn silage improved the feeding value of silage. This was evidenced by a 6.5% increase in milk production by lactating cows, a 7% increase in weight gains by growing heifers, a 2 to 3% increase in dry matter digestibility, and a 16% increase in apparent nitrogen digestibility by steers. While these improvements in feeding value are not large, they are sufficient to make adding dried whey to corn silage an economically sound practice under most farm conditions.

■ Urea, ammonia, and other NPN products — Urea, ammonia, and other NPN products can be added to corn or sorghum silage at the time of ensiling as a source of nonprotein nitrogen.

Urea increases the crude protein content of the silage and the amount of lactic and acetic acids produced. The addition of 10 lb of urea per ton of ensiled corn material will make for the following approximate increases on a dry matter basis: the crude protein from 8.3 to 12.3%, the lactic acid from 4.2 to 5.4%, and the acetic acid from 0.9 to 1.2%. Since the amount of nonprotein nitrogen that can be converted to microbial protein by the organisms in the rumen is limited, no more than 10 lb of urea should be added to a ton of ensiled corn material. The urea can be added by spreading it over the top of each load of chopped corn or it can be added to the chopped corn through the blower by commercially manufactured metering equipment.

More recently, ammonia-containing materials have been added to corn silage as a source of nonprotein nitrogen, including ammonia-water solutions, ammonia-mineral solutions, ammonia-mineral-molasses solutions, anhydrous ammonia gas, and cold flow ammonia. For dairy cattle, 5 lb of actual nitrogen (about 6 lb of ammonia) may be added per ton of wet silage. Ammonia treated corn silage has been found to contain increased concentrations of true protein, lactic acid, and acetic acid. Also, it may have a higher pH and be more stable than untreated silage when exposed to air. Special equipment is required to add ammonia or ammonia-containing materials.

■ What NPN-silage experiments show — Interest in adding nonprotein nitrogen (NPN) to corn silage has prompted a number of experiments.

Michigan Station researchers compared corn silages (34 to 37% dry matter) ensiled with (1) no additive; (2) 0.5% urea; (3) 0.75% urea; and (4) 0.75% urea + 0.17% $CaSO_4$ when fed as the only forage to lactating cows averaging 64.5 lb milk per day at the time treatments began.[6] Concentrates containing 8, 12, or 18% crude protein were fed at 1 lb per 3 lb of milk. The 0.5% urea silage was fed with each of three protein concentrates — 8, 12, and 18%. The other three silages were fed with 8% concentrate only. Cows fed low-protein rations (corn plus corn silage) consumed less feed and produced less milk than those fed rations to which 0.5 or 0.75% urea was added to the corn silage at ensiling time. Highest milk yields were obtained from the ration in which approximately 50% of the supplemental nitrogen came from urea added to corn silage and the remainder from soybean meal. Because of the slightly higher milk yields and the larger savings in the price of the concentrates fed, income over feed costs was about 10% greater for the urea-fed group than for the group fed all of its supplemental nitrogen as soybean meal. These studies clearly demonstrate a beneficial response from adding urea to corn silage at ensiling time

The Michigan Station workers also compared the feeding value of corn silage treated with an ammonia solution at

[5]Schingoethe, D. J., and G. L. Beardsley, "Feeding Value of Corn Silage Containing Added Urea and Dried Whey." *Journal of Dairy Science*, Vol. 58, No. 2, 1975, p. 196.

[6]Huber, J. T., and J. W. Thomas, "Urea-Treated Corn Silage in Low Protein Rations for Lactating Cows," *Journal of Dairy Science*, Vol. 54, No. 2, 1971, p. 224.

ensiling with that of urea-treated and control silages in dairy cattle rations.[7] On all-silage rations with no protein supplement, heifers ate more of the ammoniated than of the control silage. Lactating cows fed ammonia- and urea-treated silages showed higher milk yields than cows fed negative control rations with no added nitrogen. No significant differences in production were noted for cows fed control, urea-treated, or ammoniated silages at equal dietary nitrogen. This study demonstrated that the addition of an ammonia solution to corn silage results in a high-quality feed enriched with NPN which is equal to urea-treated silage in supporting milk yields of lactating cows. The lower cost of manufacturing ammonia than urea increases its potential as a silage additive. The Michigan data suggest that the ammonia solution exerts a buffering action which increases protein and lactic acid concentrations in ensiled material.

■ **Sulfur may be needed**—When corn silage reinforced with urea or ammonia is the major source of protein fed to ruminants, there may be inadequate sulfur for the rumen organisms to manufacture their own protein. In such cases, there is some experimental evidence that the addition of sulfur to achieve a nitrogen-sulfur ratio of less than 15:1 improves both the growth and milk production of cattle. The most practical way to provide additional sulfur is to add gypsum (calcium sulfate, $CaSO_4 2H_2O$) at the rate of 1.8 lb per ton of silage.

■ **Advantages and disadvantages of adding NPN**—Before adding NPN to corn silage, one should weigh carefully the advantages and disadvantages to accrue therefrom.

The **advantages** of adding NPN to corn silage are:

1. It is possible to raise the crude protein content by 4% on a dry basis—from 8.3 to 12.3% by the addition of 10 lb of urea per ton of ensiled corn material.
2. It (NPN) is effectively utilized by rumen bacteria.
3. It will not likely make for any intake or refusal problems (as can happen with urea in grain mixtures).
4. It alleviates or lessens the need for a protein supplement to balance rations.
5. It makes corn silage more stable. (There is less secondary fermentation or heating after removal from the silo.)

The **disadvantages** of adding NPN to corn silage are:

1. Seepage and ammonia gas losses occur. Even under ideal conditions, about 10% of the urea may be lost; and under average conditions, up to 30% of the urea may seep away or be lost as ammonia gas.
2. The expense of adding NPN occurs at ensiling time, thereby tying up capital.
3. Feed intake can be lowered if corn silage is too dry.
4. It necessitates care because ammonia gas, which is dangerous and causes irritation, can accumulate in the silo.
5. It lessens flexibility of feeding because the same amount of urea is contained in all the silage.
6. The monetary investment in NPN could be tied up in the silo for an extended period of time before any return on the investment.

---

[7]Huber, J. T., and O. P. Santana, "Ammonia-Treated Corn Silage for Dairy Cattle," *Journal of Dairy Science*, Vol. 55, No. 4, 1972, p. 489.

■ **Limestone**—Limestone (calcium carbonate) may be added at a level of 0.5 to 1% to corn silage to increase acid production. It neutralizes some of the initial acids as they are formed, allowing the lactic acid bacteria to perform longer and to produce more desirable acids.

Research has not shown any consistent increase in the nutritive value of silage treated with limestone.

The addition of limestone at ensiling time raises the naturally low calcium content of corn silage—a fact which should be considered when balancing rations. For lactating dairy cows fed high-alfalfa rations, this would be disadvantageous; but it can be rectified by increasing the phosphorus of the ration. Yet, experiments have shown that dairy cattle do not respond to the addition of limestone to silage.

ACIDS

Both inorganic (mineral) and organic acids may be used as additives. Mineral acids lower the pH immediately, while organic acids have a limited effect on lowering pH. Both mineral and organic acids limit microbial growth and help to stabilize silage.

The use of inorganic acids, such as hydrochloric, sulfuric and phosphoric, for forage preservation was pioneered by A. I. Virtanen, Finnish biochemist, in the 1920s. He discovered the AIV method, named from his initials, for preserving silage by acidification, for which he was awarded the Nobel Prize in Chemistry in 1945. His work was highly regarded in that area of the world because hay drying was difficult and dairying was, and still is, important.

■ **Inorganic acids**—Inorganic acids (hydrochloric acid, sulfuric acid, phosphoric acid) have been used as silage preservatives, almost entirely in Europe, in connection with the ensiling of high-moisture material. These acids substitute for the acids produced by bacterial action. However, they are very corrosive, causing problems in their application, including problems with the silo walls and silage handling equipment. Of the three acids, phosphoric is preferred because (1) it is less corrosive than sulfuric acid or hydrochloric acid, (2) it may enhance the phosphorus content of the silage, and (3) it increases the residual manure value from the silage. But phosphoric acid may introduce a problem of proper calcium-phosphorus ratio. This can result in some abnormal conditions and unsatisfactory performance in the animals to which it is fed.

When mineral acid preservatives are used, it is recommended that ground limestone or some other form of calcium or sodium carbonate be fed to animals at the rate of approximately 1 oz for each 10 lb of silage in order to neutralize the acid and prevent undesirable effects.

In general, the use of mineral acid preservatives is not considered as desirable as the use of molasses or grain, because (1) they produce a more sour and less palatable silage; (2) they may damage clothing, machinery, and/or masonry silo walls, due to their corrosiveness; and (3) they do not add to the nutrient value of the silage except by enhancing the preservation of carotene.

In general, the use of mineral acids has more disadvantages than advantages.

■ **Organic acids** — Propionic, acetic, lactic, citric, and formic acids are included in this group. They are used in a manner similar to inorganic acids, but they are much less corrosive and not so difficult to handle, although precautions must be taken.

Organic acids will enhance the preservation of forage without the loss of palatability. Also, they serve as mold inhibitors. Even so, like all additives, they cost money; hence, the economics of using them in making silage must be considered. When an organic acid is used, the following guidelines should be observed:

1. Add 1% of the acid to the forage in the field at the time of harvest or at the chopper.

2. Limit the presence of oxygen by using a sound, well-built silo.

3. Prevent dilution of organic acid treated silage by rain or snow by covering it with plastic when it is stored outside or in a temporary silo.

It appears that organic acids will find their greatest use in the preservation of high-moisture grain (see heading "High-Moisture Grain").

FERMENTATION AIDS

This group includes bacterial cultures, yeast cultures, and enzyme supplements. Controlled experiments support the claims made for some of these products, but not all of them. So, they should be purchased only from reputable sources that have valid research data to support the claims made for them.

■ **Bacterial cultures** — Silage additives containing cultures of acid-forming bacteria (*Lactobacillus*) are on the market. The basis for including these is to provide an inoculum or to increase the numbers of these bacteria and ensure rapid fermentation.

There are two schools of thought relative to the value of bacterial cultures as silage additives. The advocates claim that these products increase the dry matter, energy, and protein of the silage. Others report that the addition of such cultures is unnecessary and of questionable value because: (1) There are always sufficient numbers of these bacteria present on the ensiled material to bring about the proper fermentation; (2) the number of live bacteria present in these preparations cannot be guaranteed with accuracy; and (3) the number of bacteria provided through such additives is insignificant in comparison with numbers already present on the ensiled material.

■ **Yeast cultures** — Yeast cultures have also been included in certain silage additives. However, yeasts will sometimes grow in silage without an inoculum being added. When this happens, the silage has a yeasty odor and taste, which is considered undesirable. Yeast does have nutritional value, but because of the small quantity involved in additives, the contribution in this regard is minimal.

■ **Enzymes** — Cultures of molds, or of molds with other microorganisms, are sometimes added to silage to provide a source of enzymes. It is claimed that these enzymes improve the nutritive value of the silage by increasing its digestibility or digestible nutrient content. Although the enzyme activity of these preparations has not been measured experimentally, no doubt they vary considerably from batch to batch. Further, the quantity of enzymes added is insignificant compared to those already present in the silage.

PRESERVATIVES

This group includes antibiotics, salt, and sterilants. These products preserve silage by inhibiting microbial action or undesirable fermentations. All of them are of questionable value if air is properly excluded from the silage. If air is not excluded, they must be added at very high levels in order to be effective.

■ **Antibiotics** — Theoretically, antibiotics could preserve silage by selective action; by inhibiting undesirable microbial activity while allowing the desirable organisms to develop. So far, the results have been inconsistent.

■ **Salt (NaCl)** — At an appropriate level, salt inhibits certain microorganisms without preventing the action of bacteria which produce the desirable acids.

■ **Sterilants** — This group includes sulfur dioxide, sodium diacetate, sodium metabisulfite (sodium sulfite), sodium benzoate, and sodium nitrate. Sodium propionate and other organic acids have also been used as preservatives, because of their mold inhibiting properties. Each of these products appears to reduce carotene losses, improve the odor of silage, and/or lessen the production of toxic gases. But their effect on palatability is variable. The cost and inconvenience of application of these products may not justify their advantages.

SILAGE ADDITIVE AND PRESERVATIVE CONSIDERATIONS

A variety of silage preservatives is presently on the market; and, no doubt, new ones will follow. Many of these products have been inadequately tested. Yet, farmers are often in the position of having to decide whether an additive shall be used. In addition to understanding the silage-forming process and how different additives function, they should consider the following:

1. Additives and preservatives will not substitute for the proper exclusion of air.

2. Additives and preservatives do not produce new nutrients in silage, although they may aid in retention of those already present.

3. Additives and preservatives that add nutrients to the silage will be partially lost with any spoilage or seepage.

4. The cost of an additive or preservative is usually high in relation to the value of the silage.

5. Chemical analyses are of very limited use in evaluating silage additives and preservatives.

SILAGE ADDITIVE AND PRESERVATIVE RECOMMENDATIONS

When added to silage, the following materials will increase the amount of nutrients it contains:

1. Grain or grain by-products will increase total digestible nutrients and dry matter.
2. Molasses will increase the total digestible nutrients (TDN or energy) and may improve fermentation in legumes and certain grasses.
3. Urea or other NPN products will increase the nitrogen (crude protein).
4. Limestone will increase the calcium content.

Most of the arguments center around the use of the non-nutrient silage additives. A review of the literature indicates varying degrees of success from the use of such products. Some reports show positive effects from them while others show no effect. The most important consideration is whether the improved quantity and quality of forage from the use of an additive or preservative will offset its cost and application.

At the outset, it should be recognized that no silage additive or preservative can rectify mistakes that were made earlier—prior to incorporating the product. Neither can feeding more grain compensate for poor-quality silage.

Additives or preservatives are not essential to good silage formation when conditions of moisture and storage are right. Yet, under special circumstances they can be recommended for use. For example, molasses, grain, or grain by-products might be a wise addition to silage when conditions do not allow for proper wilting prior to ensiling, or when an *all-in-one* silage is being made. Urea may be an appropriate addition to an *all-in-one* silage or where increasing the protein content of the silage will simplify its feeding. It is doubtful that there is any justification for adding limestone unless this is a convenient method of calcium supplementation. The economy of most nutritive additives of this type depends largely on how well their nutrients are retained in the silage and the use made of them in balancing the rations.

When forages are stored at the proper moisture content, and when air is properly excluded, nutrient losses are low and a good-quality silage forms. Additives such as lactic acid bacteria, mold inhibitors, antibiotics, salt, enzymes, yeast cultures, and mineral acids can, therefore, do little if anything to improve the preservation of the silage or its feeding value. When high-moisture material is ensiled, grain is superior to any of these additives. When air is not properly excluded, none of these additives will correct the large fermentation and spoilage losses.

In short, there is no substitute for good management of forage crops for silage, with proper control of such factors as stage of maturity at harvest, harvesting methods, moisture content, fineness of chopping, distribution and packing, and exclusion of air.

In order to assess the value of a silage additive or preservative, it is recommended that the following criteria be applied:

1. Does the product lower the ensiling temperature?
2. Does the product increase aerobic stability?
3. Does the product increase dry matter and nutrient recovery from the silo?
4. Does the product improve feed value and animal performance, particularly when silage is a major ingredient of the ration?
5. Does the product make for sufficient benefits to offset costs and give a return on investment?

FILL RAPIDLY

Once silo filling is started, it should be rapid, so as to avoid spoilage before the silo is filled and sealed. Generally speaking, a silo should be filled in 2 days or less.

DISTRIBUTE FORAGE UNIFORMLY IN THE SILO

In order to avoid the presence of air pockets and spoilage, it is essential that any kind of chopped forage be distributed uniformly in the silo and that it be packed well. Proper silage distribution is obtained by keeping the material nearly level or slightly higher at the center. Silage distributing equipment is available for keeping the material in an upright silo level. These devices are very helpful, especially in silos of 14-ft or larger diameters.

Where corn, sorghum, and sunflower forage is harvested at a green, immature stage and cut into short lengths, tramping in an upright silo will not be necessary; but uniform distribution is very important. The only filling precaution under these conditions is to see that the top is carefully leveled and well packed and covered whenever filling is completed.

Grass silage (especially when wilted), hollow-stemmed forages, and forages that have matured or dried beyond the best silage stage should always be tramped well, especially near the wall.

Packing in a trench silo should be obtained by use of a tractor.

SEAL OR TOP-OFF THE SILO

Sealing or topping-off is necessary in order to avoid excess spoilage, especially with grass silage, which tends to dry out on the surface and to shrink away from the silo walls. This may be accomplished by carrying out one or more of the following procedures:

1. Leveling off the top and thoroughly tramping the last few feet, especially near the walls.
2. Topping-off the silo with 2 to 3 loads of wetter material.
3. Covering the top with plastic cut to fit the silo diameter and turned up against the silo wall a distance of 5 to 8 in.

FEEDING VALUE AND ECONOMY OF SILAGE

A common rule of thumb is that 3 lb of 70% moisture silage or 2 lb of 40% haylage are equivalent to 1 lb of hay of similar kind and quality; a difference due primarily to the higher water content of silage or haylage.

Many factors enter into any figures which propose to show the comparative economy of silage vs dry forages; among them, (1) the comparative yield of total digestible nutrients per acre, (2) the cost per ton for preserving and storing, (3) the relative nutrient and feeding value, (4) the distribution of labor, (5) the control of weeds, (6) the kind of haymaking weather, (7) the hazard of curing so much hay without it becoming overripe, (8) the price per ton, and (9) the machinery and efficiency of each method, etc., etc.

Numerous experiments have been conducted to determine the feeding value and economy of silage.

Cornell workers studied the effects of silage-based diets on feed intake, milk production, and body weight of dairy cows.[8] Lactating cows were assigned to three forage treatments: (1) corn silage *ad libitum*, (2) corn silage restricted plus hay crop silage, or (3) hay *ad libitum*. The cows on all three forage treatments had similar feed intakes, milk production, and body weights. There was no advantage from including hay with corn silage.

The University of Guelph, Guelph, Ontario, fed a group of 15 Holstein-Friesian heifers *ad libitum* from birth to 18 months of age on each of the following as the sole source of forage: (1) corn silage (35% dry matter) (2) blended corn silage (60% dry matter) and wilted hay crop silage (40% dry matter); and (3) blended corn silage (60% dry matter) and field-cured chopped hay (40% dry matter).[9] Blending of forage was done prior to feeding in a stationary auger feedlot mixer. Breeding commenced at the first estrus after 13 months of age and 770-lb weight. No differences occurred in intakes of forage or total dry matter, growth, age at first breeding, or conception. It was concluded that dairy heifers can be raised as successfully to 18 months on corn silage as the only forage as on combinations of corn and hay crop silage or corn silage and hay.

The New Hampshire Station studied corn silage with and without grass hay for lactating dairy cows.[10] Approximately 30 Holstein cows were assigned to three forage treatments of 0, 0.5, or 1.0 lb of hay per 100-lb body weight daily for the entire trial. Corn silage was fed *ad libitum*, except during the dry period, and concentrate was offered according to maintenance, growth, and production requirements. The investigators concluded that heavy feeding of corn silage will support high milk production adequately without creating any herd health problems. However, there is an advantage in forage dry matter intake and milk fat percentage when some hay crop is in the ration.

Maryland investigators studied the nutritive value of corn silage as influenced by grain content.[11] Two varieties of corn selected for a large difference in ear-to-stalk ratio were evaluated as silage with intake and digestibility trials with Holstein steers. In the second year, the same two varieties were evaluated with and without ears by the same test, using both steers and lactating cows. The effect of varietal difference in grain content and planting rate changed the digestible energy content by only two to three percentage units. Milk production was not significantly affected by variety but was lower when the ears were removed even though additional concentrates were supplied. It was concluded that variety and planting rate can influence yield per acre; however, the importance of these two factors on the energy content and intake of dry matter is questionable. Stage of maturity (or dry matter content) is more likely to influence cow performance as well as yield per acre and seepage losses.

At the Wisconsin Station,[12] high-energy and conventional corn silage harvested from alternate rows of corn for three consecutive years were compared in terms of field yields, utilization of nutrients for milk production, and digestibility. Three feeding trials involving 12 cows each were conducted to compare the feeding value of each silage in terms of milk production, milk composition, and body weight changes. The average daily milk production was slightly higher for the cows fed high-energy corn silage. Milk protein and solids-not-fat content increased while percent milk fat was depressed slightly in cows on high-energy corn silage. The depression in milk fat percentage was great enough to reduce the production of 4% FCM below that of cows fed conventional corn silage. There were no significant differences in body weight changes between rations. In each trial all animals gained weight at about equal magnitudes.

SILAGE POINTERS

Some additional pointers which may be of value to the dairy producer who is making or feeding silage follow.

COATING THE SILO

Since wet grass silage has a somewhat more corrosive action on concrete than does corn or sorghum silage, it may be desirable to apply a protective coating to the inside of concrete silos, whether of solid concrete or of stave construction. The problem is to find an effective, nontoxic, and economical coating. For information on the latest recommendations, the farmer should contact the local county agent, or write to the state college of agriculture.

[8]Belyea, R. L., *et al.*, "Effects of Silage Based Diets on Feed Intake, Milk Production, and Body Weight of Dairy Cows," *Journal of Dairy Science*, Vol. 58, No. 9, 1975, p. 1328.

[9]Grieve, D. G., *et al.*, "All Silage Forage Programs for Dairy Cattle. 1. Heifer Performance from Birth to Eighteen Months of Age," *Journal of Dairy Science*, Vol. 59, No. 5, 1976, p. 912.

[10]Holter, J. B., *et al.*, "Corn Silage With and Without Grass Hay for Lactating Dairy Cows," *Journal of Dairy Science*, Vol. 56, No. 7, 1973, p. 915.

[11]Hemken, R. W., *et al.*, "Nutritive Value of Corn Silage as Influenced by Grain Content," *Journal of Dairy Science*, Vol. 54, No. 3, 1971, p. 383.

[12]Jorgensen, N. A., *et al.*, "High Energy Corn Silage for Lactating Dairy Cattle," *Research Report 42*, The University of Wisconsin-Madison, June 1969.

NUTRIENT LOSSES IN LEAKAGE

Seepage losses vary with the moisture content, depth of silage, distribution of the silage, and the amount of nutrients in the seepage. Seepage losses may be as high as 14% of the dry matter stored.

The nutrient losses vary, but generally they are in proportion to the run-off. The nutrients lost in seepage from a 100-ton silo may equal the nutrients in ¾ ton or more of hay.

EXPOSURE TO AIR

Spoilage begins the moment silage is exposed to the air. Therefore, once the silo is opened for use, feed should be removed daily. In the wintertime, a minimum of 1½ in. of silage should be removed daily from a tower silo; in the summertime, 3 in.

Also, it should be realized that spoilage is likely to occur on the surface of the ensiled material if more than 2 days elapse between filling periods.

REMOVAL OF SILAGE FROM SILO

In the past, the common method of feeding silage was by hand. In the present era of bigness and automation, the removal and feeding of silage is being automated.

It is possible to achieve complete push button controlled feeding in an upright silo. With horizontal silos, the silage may be handled with a manure scoop, and sometimes a mechanical unloading wagon or truck — depending on distance from feedlot to silo.

Fig. 9–17.   Push button controlled silage feeding from upright silos. (Courtesy, Holstein-Friesian Assn. of America, Brattleboro, VT)

Fig. 9–18.   Lactating cows being bunk-fed grass silage from a self-unloading forage box. (Courtesy, Gehl Company, West Bend, WI)

Self-feeding from both tower and horizontal silos can be achieved, but this requires more management on the part of the operator.

MOLDY SILAGE

Moldy silage may be harmful. Any spoilage that causes animals to go off feed, or that upsets the metabolic processes, should not be fed.

Some conditions cause certain molds to produce toxins. The toxins are called *mycotoxins*, and the effects of the toxins on animals are called *mycotoxicoses*.

Mature ruminants appear to tolerate higher levels of mycotoxins than young ruminants.

One way in which to determine the potential toxicity of moldy silage is to feed it to some less valuable animals for at least 2 weeks. Observe the animals daily for signs of toxicity — such as reduced gain and going off feed. If no toxic effects are noticed, it is probably safe to feed the suspect silage to other animals. If ill effects are noticed, switch them to other feed immediately and dispose of the suspect silage by spreading it on the land and plowing it under.

SILAGE FOR SUMMER FEEDING

Some dairy producers use silage effectively as a summer feed. This practice is especially desirably in those areas where pastures dry up during the hot, dry months. It appears that more and more dairy producers will go to year-around silage feeding in a corral.

EFFECT OF SILAGE ON MILK ODOR AND FLAVOR

Silage sometimes affects the flavor and odor of milk, especially when ensiled too wet. This effect may be somewhat more pronounced with some silages than with others.

The dairy producer will do well, therefore, to feed all silages after, rather than before, milking.

DANGEROUS SILAGE GASES

Two types of toxic gases may be formed when making silage: (1) carbon dioxide (CO_2), and/or (2) nitrogen dioxide (NO_2).

Carbon dioxide forms soon after filling begins and continues until fermentation stops. It is a colorless, suffocating gas, which is heavier than air and tends to collect in low places.

Under drought conditions, corn, sorghum, and other grass species may accumulate higher than normal levels of nitrates. When ensiled, nitrates are converted to nitrites, then nitrites are converted to nitrogen oxide by bacteria and plant cells. As the nitrogen oxide comes in contact with air, it is oxidized to form nitrogen dioxide, a reddish brown-colored gas, which is heavier than air. This gas is highly toxic to both humans and farm animals.

Precautions against hazards caused by silage gases include (1) operating the blower for a 15-minute period if it is still connected, (2) swinging a piece of canvas, a tree branch, or a burlap bag vigorously so as to agitate the air and dilute gases that may be present, or (3) using proper life support equipment when entering an oxygen-limiting, or sealed, silo. Also, adequate provision for ventilation of the silo through the roof is essential.

A victim of silo gas should be moved into fresh air immediately, and artificial respiration should be applied. A physician should be called immediately.

HAYLAGE (LOW-MOISTURE SILAGE)

Fig. 9–19. Field chopping forage for haylage. (Courtesy, International Harvester Company, Chicago, IL)

This method of making silage is not new, contrary to common belief. It was developed many years ago at the Bacteriological Station at Crema, Italy.

Low-moisture grass/legume silage (or haylage), containing 40 to 60% moisture, is made with limited bacterial growth and fermentation. The term *oatlage* is sometimes used specifically to indicate low-moisture silage made from oats.

Fermentation is of minor concern in making low-moisture silage since little acid is produced and pH is not a useful criterion of quality. The most important factor is the establishment and maintenance of air-free conditions through fine chopping, rapid filling, and a good silo. Because of the difficulty of maintaining air-free conditions, stacks, bunkers, and trench silos are seldom used for storing low-moisture silage. Infiltration of air into the silage mass will result in growth of yeasts and molds and increase in temperature. Temperatures about 95°F for a few days will cause certain proteins to combine chemically with carbohydrates to form a product that is indigestible, termed *bound protein*. When more than 12% of the total protein is in the "bound" form, the silage has undergone excessive heating.

Properly made and stored low-moisture silage has a pleasant aroma and is a palatable, high-quality feed. Animals usually receive more dry matter and net feed value in low-moisture silage than in wilted silage made from the same cut. Low-moisture silage is increasing in popularity, especially as a dairy feed. It may be fed like wilted silage, with adjustment for difference in moisture content.

In a Wisconsin study,[13] low-moisture alfalfa-grass silage (haylage) was compared with wilted alfalfa-grass silage plus second crop hay in rations for dairy cows. Both low-moisture silage and wilted silage were stored in oxygen-limiting silos. Total dry matter consumption was higher for the low-moisture silage ration (P ≤ .01) and higher milk and fat corrected milk (FCM) production (P ≤.01) resulted. There was no significant difference in milk fat percentage. Forage dry matter consumption for cows fed low-moisture silage was 32.5 lb per day vs 30.5 lb for wilted silage plus hay. The FCM per day per cow fed the low-moisture silage treatment was 45.6 lb vs 42.2 lb for cows fed the wilted silage plus hay treatment. Total dry matter required per pound of FCM produced was similar for both rations. No detrimental effects were noted in the health of cows fed either forage ration.

In addition to excluding air by providing an airtight silo and by thorough packing, the following directions should be observed to make the best quality low-moisture silage: (1) harvest at the proper stage of maturity; (2) wilt in the swath and/or windrow until the moisture reaches 40 to 60% (35 to 40% for a bunker silo), with the required time determined by the weather, but which may be expedited by the use of a forage conditioner; (3) chop short, a ¼ in. cut is best; (4) fill silo rapidly and continuously; (5) add an additive or preservative if desired; (6) distribute silage evenly in the silo; (7) apply a top seal of forage containing 65 to 70% moisture, level, and tramp to remove air; and (8) crown the center slightly and cover with a plastic silo cap.

[13]Larsen, H. J., E. L. Jensen, and R. F. Johannes, "A Comparison of Haylage and Wilted Grass Silage Plus Hay in the Dairy Diet," *Research Report*, The University of Wisconsin-Madison, August 1971.

Fig. 9–20. Lactating cows eating haylage. (Courtesy, Ford New Holland, New Holland, PA)

HIGH-MOISTURE GRAIN

High-moisture grain refers to grain that is harvested at a moisture level of 22 to 40% and stored without drying.

Interest in storing and feeding high-moisture grain was prompted by the shift toward more field shelling of corn, instead of picking ear corn, because much shelled corn must be dried for safe storage whereas ear corn can be safely stored at moisture contents up to 24% without drying. Interest in high-moisture grain increased with high energy cost for drying. It takes approximately 1 gal of propane fuel to dry 4½ to 6 bu of corn, or 1 kilowatt hour of electricity to dry 10 to 12 bu of corn, with conventional high temperature drying to reduce the moisture content of wet grain 10 percentage units (i.e., to dry from 25% down to 15% moisture).

The **advantages** of high moisture grain in comparison with dried grain are:

1. It alleviates the high energy cost for drying grain.
2. It lessens field losses at harvest time.
3. It permits harvesting earlier, at higher moisture content, and usually during more desirable weather. As a result, it releases land for fall plowing in the North and for fall seeding of a second crop in the South.
4. It makes it practical to use later maturing, higher yielding varieties of corn and sorghum in the northern areas which often have early frost.
5. It usually requires less investment in processing equipment.
6. It improves the feeding value of grain for dairy cattle.

The **disadvantages** of high-moisture grain in comparison with dried grain are:

1. It requires a large inventory of high-moisture grain, which may increase capital requirements.
2. It limits market flexibility for the grain, since it must be fed to livestock.

3. It may result in higher storage losses than for dry grain if proper ensiling or acid-treatment is not followed.
4. It may freeze in the bunk in the winter, and flies may be a problem in the summer.

Today, considerable quantities of high-moisture ear corn, shelled corn, sorghum (milo), and small grains (wheat, barley, and oats), containing about 30% moisture, are stored and fed. Some high-moisture grains are planned for and intended. Others are the result of happenstances, such as crops planted late, early frost damage, or harvesting when wet.

HARVESTING

A number of field equipment combinations may be used to harvest high-moisture grains. Harvest cost plus product quality should be considered in selecting the most desirable combination.

High-moisture grain should be harvested when it reaches physiological maturity and the moisture is 22 to 40%; with ear corn, sorghum (milo), and small grains higher in moisture than corn at harvest time. At this stage, grains yield the maximum available nutrients per acre and preservation conditions are best. If grain is harvested when too wet and immature, dry matter yield per acre will be less. If the grain is too dry, mold will be high and fermentation will be low; thus, grain containing less than 22% moisture should be reconstituted by adding water.

STORAGE OF HIGH-MOISTURE GRAIN

There are three basic storage methods of high-moisture grains: (1) ensiling in sealed (airtight) storage, (2) ensiling in nonsealed storage, and (3) preservation with an organic acid or ammonia.

SEALED (AIRTIGHT) STORAGE

Fig. 9–21. High-moisture grain (corn) coming straight from the combine and going directly into oxygen-limited storage. (Courtesy, A. O. Smith Harvestore Products, Arlington Heights, IL)

High-moisture grain may be stored in an oxygen-limiting silo. It is not necessary to crack or grind grains before storing in this manner. Another advantage of sealed storage is that it can be unloaded from the bottom.

Sealed storage is the most popular method of storing high-moisture grains even though greater initial capital investment is required. This type of storage eliminates the 2 to 5% spoilage loss normally associated with unsealed storage.

UNSEALED STORAGE

Two types of unsealed high-moisture grain storage are used; (1) conventional upright silos made of concrete or steel, which are structurally adequate for storage of grass silage; or (2) horizontal silos. The grain should be ground into the storage unit, then firmly packed; otherwise, spoilage will result. With upright silos, about a 3 in. layer of the ground, high-moisture grain should be removed from the exposed surface each day during mild weather; with horizontal silos, 4 in. should be removed daily. Greater amounts should be removed from both types of storage units during warm weather to prevent spoilage.

PRESERVATION WITH AN ORGANIC ACID OR AMMONIA

Storage involving the treatment of high-moisture grain with an organic acid or ammonia to inhibit mold or spoilage is favored by many farmers because it alleviates artificial drying or the necessity to store in an airtight silo. Several different organic acids may be used—propionic, acetic, isobutyric, formic, benzoic, or a combination of these acids—but the most commonly used acids are propionic or propionic-acetic acid mixtures, marketed under various trade names. Anhydrous ammonia and other gaseous mixtures are also effective.

■ **Mode of action of organic acids**—High-moisture grain should be treated promptly after harvesting to avoid heating. When properly applied, propionic acid and other acid-type grain preservatives will kill most fungi and other microorganisms present on the outside of the treated grain. As the acid moves into the seed, it kills the embryo, thus nearly eliminating respiration and enzymatic activity. These actions prevent heating and retain nutrient content. While the exact mode of action is not known, the acid preservatives continue to inhibit growth of molds. The effect is probably due in part to the lower pH created by the acid; however, not all products which depress pH inhibit mold. To be effective, the treatment must provide continuing protection against mold and other microbial growth.

■ **Amount of organic acid to apply**—The amount of acid required to treat high-moisture grain depends on moisture content of the grain, length of storage desired, and the temperature. The recommended rate of application of propionic acid to high-moisture grain to provide protection for 1 year is shown in Table 9–8. These rates are for corn, but they are suitable for other high-moisture grains.

TABLE 9–8
AMOUNT OF 100% PROPIONIC ACID
REQUIRED FOR 1 YEAR OF STORAGE[1]

Percentage of Moisture Content of Grain	Percentage By Weight	Pounds Per Wet Bushel[2,3]	Pounds Per Ton[3]	Gallons Per Ton[3]
16–18	0.50	0.28	10	1.3
20	0.75	0.42	15	1.8
25	1.00	0.56	20	2.4
30	1.25	0.70	25	3.0
35	1.50	0.84	30	3.6

[1]The amounts of acid listed are for long term storage (1 year). For storage periods of 6 months or less, the amount of acid used could be reduced by ½.

[2]56 lb (25.5 kg) of high moisture corn.

[3]To convert to metric: lb/wet bushel = 0.45 kg/35.2 litres; lb/ton = 0.45 kg/0.907 metric ton; gal/ton = 3.78 litres/0.907 metric ton.

■ **Application guidelines**—

1. Check grain for moisture content so selected rate of acid application meets requirement for preservation.

2. Treat the grain immediately after harvesting so as to eliminate heating and mold development.

3. Make sure that all the grain is coated with the acid and that the application rate is correct.

4. Treat outdoors if possible; otherwise, provide adequate ventilation.

5. After treatment, flush equipment with water or untreated grain to prevent corrosion.

6. Observe safety guidelines at all times.

■ **Safe handling of grain preservatives**—When preservatives are applied, the following safety precautions should be observed by the user:

1. Follow the manufacturer's instructions.

2. Avoid storage of acids with fuels, lubricants, and pesticides. Store acids *only* in original container, tightly closed, with the bungs upright.

3. Use organic acids *only* on grain destined for animal feed.

4. If grain is treated in a building, adequate ventilation should be provided.

5. Acid should not be allowed to come in contact with skin or eyes. Protective gloves, goggles, respirators, or a face shield and protective clothing should be worn by the person applying it when there is a risk of contact.

6. A supply of water should be available to wash away any acid coming in contact with skin or eyes.

7. When contact occurs, drench and remove contaminated clothing immediately. Contaminated clothing should be washed before wearing.

8. Avoid breathing fumes during filling or when entering a freshly treated grain bin a day or two after treatment.

9. Organic acids are flammable so care should be exercised to avoid fire. Avoid smoking or having an open flame or heating device in treatment areas. Keep the acid away from any ignition source and maintain good ventilation in areas when it is being applied.

■ **Storage facility guidelines** — Nearly any weather-proofed facility can be used to store treated grains, with the precaution that the acid is very corrosive to steel, especially if galvanized. Concrete is also subject to corrosion, but to a lesser degree than steel. Among the facilities that may be, and are, used are the following:

1. Metal bins and buildings coated with acid-resistant paint, or linseed oil, or covered with polyethylene plastic.

2. Concrete silos or bins coated with acid-resistant paint or linseed oil, or covered with polyethylene plastic.

3. Wooden bins, cribs, and buildings. Except for nailhead corrosion, wooden bins or cribs are not affected by acids.

4. Quonset-type buildings, provided grain is not stored on a dirt floor and the metal is protected by acid-resistant paint, or linseed oil, or covered with polyethylene plastic.

Acid-treated grain stored in makeshift bins, piles, or bunkers should be provided with covers that will prevent moisture from entering the storage area.

■ **Handling and feeding** —

1. Since the acid is absorbed by the grain, protection is provided after removing from storage. Thus, treated grain can be mixed with dry grains and other feeds.

2. Treated grains can be transported. They have been successfully moved from one silo to another. However, movement from a large diameter to a smaller diameter silo is suggested for better packing.

3. Acid-treated high-moisture grain is readily accepted by cattle. In studies at the University of Wisconsin, dairy cows offered propionic acid-treated high-moisture ground ear corn consumed more dry matter than those offered similar ensiled corn or dried corn that had received no propionic acid treatment.

■ **Advantages and disadvantages of acid-treated grain** — In comparison with ensiled high-moisture grain, acid-treatment has the following advantages: (1) Removal from the silo is eliminated; (2) it can be stored in a barn or other temporary storage facility; (3) treated grain can be transported over long distances; and (4) large batches of a ration which include acid-treated high-moisture grain can be mixed without risk of spoilage.

The major disadvantages to acid-treated grain are: (1) It must be fed to livestock — it cannot be marketed for any other purpose; (2) it will not germinate, so it cannot be used for seed; (3) acids are corrosive to metal or concrete storage facilities; and (4) organic acids are costly, so for grain with over 30% moisture they may be uneconomical.

FEEDING VALUE OF HIGH-MOISTURE GRAIN

The feeding value of high-moisture grain is equal or slightly superior to that of dry grain, with some variation according to class and productivity of livestock.

Research studies have shown that the feeding value of properly ensiled or acid-treated high-moisture corn is equal to that of dry corn for lactating dairy cows; the milk yield and feed intake were similar. However, when high-moisture shelled corn supplies more than 50% of the total ration dry matter, depressed milk fat percentage may occur, with inadequate fiber given as the probable cause.

Some form of processing (*e.g.*, rolling or grinding) of high-moisture corn improves utilization. Whole kernels appearing in feces indicate incomplete digestion. Processed corn is higher in digestible, metabolizable, and net energy for dairy cows than rations containing whole shelled corn.

RECONSTITUTED GRAIN

Reconstituted grain is mature grain to which water has been added to raise the moisture content to 25 to 30% for storage.

BUYING AND SELLING HIGH-MOISTURE GRAIN

Most high-moisture grain is grown by livestock producers for livestock feed, or grown as a cash crop and sold at harvest to livestock producers. High-moisture grain is bought/sold at a lower figure than dry grain because of the water content. Also a greater quantity of it must be fed because of the water content. In most cases, the buying and selling transactions of high-moisture grains are based on an 87% dry matter and 13% moisture basis.

QUESTIONS FOR STUDY AND DISCUSSION

1. What is silage? How important is it as a U.S. feed?

2. Why do dairy producers incorporate more silage in their rations than beef cattle producers?

3. Since most silage is fed on the farms on which it is produced, how important is it that dairy producers know how to produce it?

4. Describe the ensiling process.

5. List and discuss three factors which may affect the ensiling process and the quality of silage produced.

6. List both the advantages and disadvantages of silage.

7. List and discuss the factors that dairy producers should consider in arriving at a decision as to whether they should use (a) pasture, (b) hay, (c) silage, or (d) any two or all three of these roughages.

8. Classify and describe the different kinds of silos. What are the advantages and disadvantages of each kind of silo?

9. Obtain the current price on a gastight (oxygen-limiting) upright silo. Then (a) compute the cost per ton for storage of corn silage therein, and (b) determine whether or not you can justify using such a silo for corn silage, filled on a once-per-year basis.

10. Describe and discuss the extent of each of the following losses in silage: (a) surface or top spoilage, (b) seepages, (c) gaseous, and (d) heating.

11. How would you go about determining the size of tower silo to build? How would you go about determining the size of trench silo to build?

12. How would you estimate the amount (weight) of silage (a) in a tower silo, (b) in a trench silo?

13. What accounts for corn ranking first in importance as a silage crop? List and describe the four kinds of corn silage.

14. Why do U.S. dairy producers use a high proportion of corn silage, whereas New Zealand producers rely largely on grass silage?

15. Under what circumstances should dairy producers make grass/legume (hay crop) silage rather than corn or sorghum silage?

16. Describe each of the following three kinds of grass/legume silage based on moisture content: (a) direct cut silage, (b) wilted silage, and (c) low-moisture silage (or haylage).

17. Describe each of the following silages: (a) combination silage crops, (b) rain-damaged hay silage, (c) frosted crop silage, and (d) drought stricken crop silage.

18. Discuss modern silage harvesting methods and machinery.

19. List and describe the easily recognized characteristics of good-quality silage.

20. Discuss the importance of each of the following practices from the standpoint of making good silage: (a) harvest at the proper stage of maturity, (b) cut to proper length, and (c) control of the moisture content. How would you (a) lower the moisture content or (b) increase the moisture content of silage?

21. Describe each of the following three methods for determining the moisture content of forage: (a) the twist method, (b) the grab test, and (c) the oven-drying method.

22. Define (a) silage additives, and (b) silage preservatives.

23. Describe the action of each of the following four types of additives or preservatives used in silage making: (a) feed additives, (b) acids, (c) fermentation aids, and (d) preservatives. List some products that are used in each of these categories.

24. In silage making, why is it important to (a) fill rapidly, (b) distribute forage uniformly in the silo, and (c) seal or top-off the silo?

25. Discuss the feeding value and economy of silage for dairy cattle, including what the experiments show.

26. Discuss the importance of each of the following silage pointers: (a) coating the silo, (b) nutrient losses in leakage, (c) exposure to air, (d) removal of silage from silo, (e) moldy silage, (f) silage for summer feeding, (g) effect of silage on milk odor and flavor, and (h) dangerous silage gases.

27. What is haylage? How does haylage compare with traditional silage as a feed for dairy cattle?

28. Define high-moisture grain. How is it harvested? How is it stored? How is it bought and sold?

29. What factors will determine whether a particular crop shall be utilized as pasture, hay, or silage?

SELECTED REFERENCES

Title of Publication	Author(s)	Publisher
Crop Production	R. J. Delorit L. J. Greub H. L. Ahlgren	Prentice-Hall, Inc., Englewood Cliffs, NJ, 1984
Crop Quality, Storage and Utilization	Edited by C. S. Hoveland	American Society of Agronomy, Crop Science Society of America, Madison, WI, 1980
Crops in Peace and War: Yearbook of Agriculture 1950-1951		U. S. Department of Agriculture, Washington, DC
Feeds & Nutrition, Second Edition	M. E. Ensminger J. E. Oldfield W. W. Heinemann	The Ensminger Publishing Co., Clovis, CA, 1990
Feeds & Nutrition Digest	M. E. Ensminger J. E. Oldfield W. W. Heinemann	The Ensminger Publishing Co., Clovis, CA, 1990
Forage Conservation in the 80s	Edited by C. Thomas	British Grassland Society, Hurley, Maidenhead, Berks, United Kingdom, 1979

Continued

SELECTED REFERENCES *(Continued)*

Title of Publication	Author(s)	Publisher
Forage Crops	D. A. Miller	McGraw-Hill Book Company, Inc., New York, NY, 1984
Forages, The Science of Grassland Agriculture, Third Edition	M. E. Heath R. F. Barnes D. S. Metcalfe	The Iowa State University Press, Ames, IA, 1985
Forages: Resources of the Future	J. E. Oldfield, Chairman of Task Force	Council for Agricultural Science and Technology, Ames, IA, 1986
Plants for Man, Second Edition	R. W. Schery	Prentice-Hall, Inc., Englewood Cliffs, NJ, 1972
Silage Additives USA	K. Bolsen	Chalcombe Publications, Marlow, Bottom, Marlow, Bucks, Great Britain, 1985
Stockman's Handbook, The, Seventh Edition	M. E. Ensminger	Interstate Publishers, Inc., Danville, IL, 1992

Fig. 9–22.   Tower silo, with Holstein cow in the foreground. (Courtesy, Milk Marketing, Inc., Strongsville, OH)

Fig. 9–23.   Swiss cows eating grass-legume silage. (Courtesy, Brown Swiss Cattle Breeders' Assn. of the U.S.A., Beloit, WI)

Fig. 10–1. A complete (all-in-one) ration being distributed by a self-unloading truck, at Arizona Dairy Co., Higley, Arizona. (Courtesy, Arizona Dairy Co., Higley, AZ)

Contents

CHAPTER

10

CONCENTRATES; SUPPLEMENTS; ADDITIVES

Concentrate feeds are those which are high in energy and low in fiber (less than 18%). Many different kinds of concentrate feeds are used in dairy cattle feeding. They may be classed as grains and other high-energy feeds; protein feeds; by-products, animal wastes, and crop residues; special feeds, supplements, and additives; or commercial feeds.

Three factors besides chemical composition are important in evaluating concentrates for milk cows—palatability, quality of milk produced, and cost. The most infallible way in which to appraise the first two factors is through actual feeding trials. Consideration of the third factor—cost—necessitates that dairy producers be keen students of values. They must change the formulations of their ration(s) in keeping with comparative feed prices, and do so without causing the animals to go off feed.

Corn and barley are the two chief grains used in dairy rations, although oats, sorghum grains, and wheat are also used when there is a price advantage. Corn is often used as ground ears (corn-and-cob meal), rather than as grain. In high-concentrate rations particularly, the fiber from the cobs may help to keep the digestive tract functioning normally. Grains make it feasible for dairy animals to produce at high levels. Even if forage is of high quality, it is difficult to get a production of more than 9,000 to 10,000 lb of milk during a 300-day lactation without feeding some grain. Today, many cows produce more than 20,000 lb of milk, and some herds now average more than this amount. Grains make it possible for superior cows to consume sufficient energy to sustain the high level of production of which they are genetically capable.

Oil meals are the major source of protein, with soybean meal and cottonseed meal most commonly used, along with some safflower meal, sunflower meal, copra meal, peanut meal, canola meal, and others. Urea is also used as a source of protein.

Mill by-products and sugar industry by-products are also extensively used in dairy rations.

GRAINS

Grains are seeds from cereal plants—members of the grass family, Gramineae.

Grains provide an excellent source of highly digestible energy for cattle that are either on high levels of production or unable to utilize forages (e.g., young calves). They furnish energy in concentrated form. For example, 5 lb of barley contains as much energy as 8 lb of hay or 25 lb of silage. However, several problems are inherent in the use of grains; among them, the following:

1. **High-concentrate rations may affect health and milk production.** Ruminants need some "roughage factor." On high-concentrate, low-roughage rations, numerous digestive disorders may occur, including displaced abomasum, acidosis, rumen parakeratosis, and low milk fat.

2. **Some grains must be processed before they can be fed.** The need for processing is primarily governed by the type of grain and the particular animal being fed. For example, grain fed to very young or very old animals and animals which do not thoroughly masticate their feed should be ground.

3. **Grains are more costly on a weight basis.** However, comparing the costs of grain to roughage, the energy content, digestibility, and other nutrients must be considered on a per unit feed basis. Thus, a relatively expensive grain containing large amounts of highly digestible energy may in reality be a better buy than a low-cost, low-quality roughage.

4. **Grains are extremely deficient in calcium and certain vitamins.** Most grains contain less than 0.1% calcium. Adequate amounts of phosphorus are generally present in grain, but the calcium to phosphorus ratio is highly imbalanced. Additionally, grains are deficient in certain vitamins; for example, vitamin A is low in all grains except fresh yellow corn.

The leading cereal grains fed to U.S. dairy cattle are barley, corn (maize), oats, rye, sorghum, and wheat. A brief discussion of each of these feeds follows. Also, the subject of "High-Moisture Grain" is covered. Additional concentrates and supplements are listed (1) in Chapter 12, Table 12–1, Feed Substitution Table for Dairy Cattle; and (2) in Chapter 21, Feed Composition Tables.

BARLEY

Barley is the leading dairy grain in western United States. Compared with corn, it contains somewhat more protein (crude protein: barley, 13%; corn, 9%) and fiber (due to the hulls) and somewhat less carbohydrate and fat. Like oats, the feeding value of barley is quite variable, due to the wide spread in test weight per bushel. For dairy cattle, barley is about 90% as valuable as corn.

When fed to dairy cattle, barley should always be steam rolled or ground.

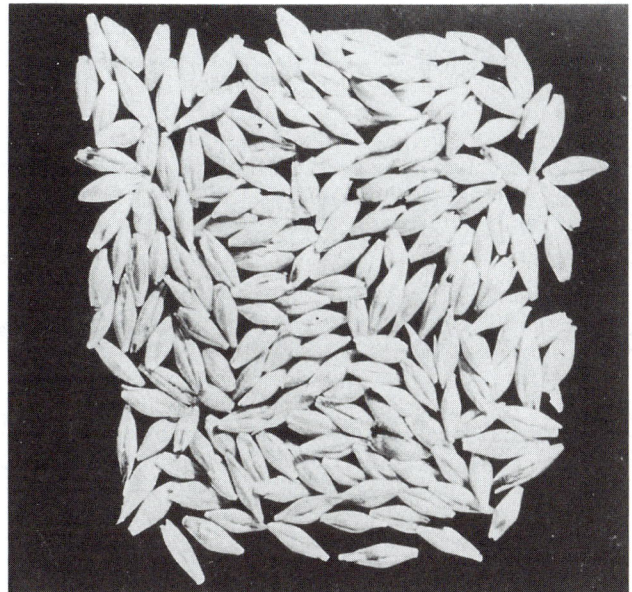

Fig. 10–2.   Barley, leading cereal grain for dairy cattle in western U.S. (Courtesy, J. I. Case Company, Racine, WI)

CORN (MAIZE)

Corn is the leading U.S. dairy feed. It is palatable, nutritious, and rich in energy-producing carbohydrate and fat, but it has certain very definite limitations. It lacks quality (being especially low in the amino acids lysine and tryptophan) and quantity of proteins (it runs about 9%), and it is deficient in minerals, particularly calcium.

Corn is higher in fat than barley or wheat (4% vs less than 2%). Fat not only contributes to the high-energy content of corn, but it also improves its palatability and feeding properties in general.

Corn may be fed to dairy cattle shelled (to young stock that masticate well), cracked, as corn-and-cob meal, or flaked.

OATS

Oats rank third in acreage among the cereal grains in the United States and second to corn in importance for stock feeding. About three-fourths the acreage of oats is in the North Central states.

Oats normally weigh 32 lb per bushel, but the best oats are heavier. The feeding value varies according to the hull content and test weight per bushel. On the average, oats contain about 30% hulls.

Oats are particularly valuable in calf starter rations and for getting animals on feed. They should be ground or rolled for cattle.

RYE

Rye thrives on poor sandy soils. In such areas, it is sometimes marketed through animals, although it generally sells at a premium for bread making or brewing.

Rye is similar to wheat in chemical composition, but it is less palatable. For the latter reason, it should be fed in combination with more relished feeds and limited to no more than one-third of the ration. When fed in limited amounts, the feeding value of rye is equal to corn. Because of the small, hard kernel, rye should always be processed.

SORGHUM

The grain sorghums—which include a number of varieties of earless plants, bearing heads of seeds—are assuming an increasingly important role in American agriculture. New and higher-yielding varieties have been developed and become popular. As a result, more and more grain sorghums are being fed to dairy cattle.

Fig. 10–3.   Corn, leading U.S. dairy feed. (Courtesy, USDA)

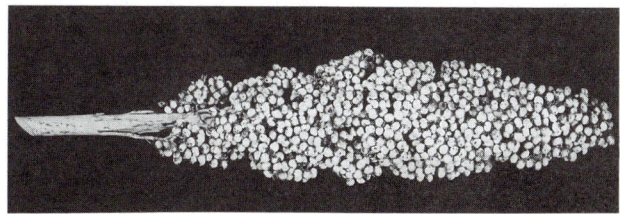

Fig. 10–4.   Head of grain sorghum. (Courtesy, USDA)

Kafirs and milos are among the more important sorghums grown for grain. Other less widely produced types include sorgo, feterita, durra, hegari, kaoliang, and shallu. These grains are generally grown in regions where climatic and soil conditions are unfavorable for corn.

Like white corn, the grain sorghums are low in carotene (vitamin A value). Also, they are deficient in other vitamins, and in proteins and minerals. Feeding trials show that the grain sorghums are worth about 90 to 95% as much as corn for dairy cattle, or about the same feeding value as barley. Sorghum should be ground or rolled for dairy cattle.

WHEAT

The total annual U.S. wheat tonnage is second only to corn. However, because it is produced mainly for the manufacture of flour and other human foods, it is generally too high in price to feed to dairy cattle. When the price is favorable or when the grain has been damaged by insects, frost, fire, or disease, it may be more profitable to market it through animals than for human consumption.

Compared with corn, wheat is higher in protein and carbohydrates, lower in fat, slightly higher in total digestible nutrients, and more palatable. Wheat, like white corn, is deficient in carotene. Pound for pound, it is equal to or up to 5% more valuable than corn for dairy cattle.

Fig. 10–5. Wheat, normally a bread grain, may be fed to dairy cattle when the price is favorable or the grain has been damaged. (Courtesy, J. I. Case Company, Racine, WI)

Since the kernels are small and hard, wheat should be ground coarsely or rolled for the most economical utilization by cattle.

HIGH-MOISTURE GRAIN

Feeding high-moisture corn or other wet grains to dairy cattle offers the following **advantages:**

1. Grain can be harvested 2 to 3 weeks earlier than normal, thereby reducing field losses and harvest problems associated with adverse weather.
2. Storage and handling losses are reduced.
3. It lends itself to automated feeding programs.
4. The expense of drying grain is eliminated.
5. It makes for a highly palatable feed.
6. It lessens the labor of processing/grinding grain.

High moisture shelled corn should be stored at a moisture content of 25 to 30%. Ground ear corn should contain 28 to 32% moisture for proper preservation. High moisture shelled corn and ear corn should be ground before storing in conventional silos. In airtight silos, shelled corn can be stored whole, then rolled upon removal from the silo. Propionic acid can be used effectively, according to label directions, to treat and preserve high-moisture corn or barley for dairy cattle.

■ **Moisture is important when buying feeds**—When buying grains, dairy producers should never lose sight of how much water they may be purchasing. Table 10–1 illustrates the relative value (dry matter purchased) when paying for corn on a 15.5% moisture basis while actually receiving corn of another moisture content. Thus, if producers were receiving 19% moisture corn and paying for 15.5% moisture, they would receive only 95.86% of the dry matter for which they paid. On the other hand, if corn is delivered with 7% moisture, while paying on a 15.5% moisture basis, producers would receive 110.06% of what they paid for.

TABLE 10–1
RELATIVE VALUE OF U.S. NO. 2 CORN (15.5% MOISTURE)
AS AFFECTED BY CHANGES IN MOISTURE

Moisture	DM Basis Multiplier	Moisture	DM Basis Multiplier
(%)		(%)	
0	1.1834	19	0.9586
1	1.1716	20	0.9467
2	1.1598	21	0.9349
3	1.1479	22	0.9231
4	1.1361	23	0.9112
5	1.1243	24	0.8994
6	1.1124	25	0.8876
7	1.1006	26	0.8757
8	1.0888	27	0.8639
9	1.0769	28	0.8521
10	1.0651	29	0.8402
11	1.0533	30	0.8284
12	1.0414	31	0.8166
13	1.0296	32	0.8047
14	1.0178	33	0.7929
15	1.0059	34	0.7811
16	0.9941	35	0.7691
17	0.9822	36	0.7574
18	0.9704		

■ **Moisture is important in formulating rations** — Careful feeders must constantly watch the moisture content of the feeds they buy, and the effect of moisture on their nutritional quality control. Most good feeders will readjust feeding formulas whenever moisture in a leading ingredient changes over 1%.

(Also see Chapter 9, section on "High-Moisture Grain.")

OTHER HIGH-ENERGY FEEDS

Although feed grains and their milling by-products comprise the vast majority of the energy feeds, numerous other feeds are routinely used to supply energy to dairy cattle. Seeds from plants other than *Gramineae* can be used effectively (for example, beans). Fats and oils provide an extremely concentrated source of energy. Molasses is a liquid energy feed that is highly palatable and digestible.

OTHER SEEDS

Seeds from plants other than cereal grains are used in dairy feeds when they are readily available and when the price is right. Legume seed, such as soybeans and peanuts, and whole cottonseed can be used for their energy content in addition to protein. Many types of seeds are by-products of cash crop enterprises, representing culls of processing or marketing. On occasion, a surplus of a certain seed generally used for human consumption may reduce the cost to a level where it becomes economically feasible to incorporate it in dairy feeds.

FATS AND OILS

Fat serves the following functions when added to dairy rations: it (1) increases the caloric density of the ration without lowering the forage (fiber) content; (2) controls dust; (3) lessens the wear and tear on feed mixing equipment; (4) facilitates pelleting of feeds; (5) increases palatability; (6) helps to homogenize and stabilize certain feed additives, especially those of a very fine particle size; and (7) increases the total amount of milk, butterfat, and SNF, but results in a slight decrease in the percentage of both butterfat and SNF.

Most forages and grains are low in lipids — they contain less than 2 to 3% fat. In general, dairy cows should be able to utilize 1 to 1½ lb of fat per day in addition to the fat present in natural feedstuffs. This means that about 3% more fat can be added to the total ration (forage plus concentrate), or that 5 to 6% fat can be added to the grain ration.

Added fat is especially effective in early lactation. Because of the increased caloric concentration provided by dietary fat and because high-producing cows are usually in negative energy balance during early lactation, fat is frequently added to the ration to increase the cow's energy intake and provide fatty acids to the udder. It may also be beneficial to provide supplemental fat when the capacity of the gastrointestinal tract limits energy intake.

An important consideration in the successful feeding of fats is that they may be used to provide increased energy without lowering the forage (fiber) intake. In early lactation,

substituting fat for a portion of the starch obtained in cereal grains in the ration of cows is a way in which to maintain high energy concentrations and high fiber intakes. The substitution of fat for grain alleviates the low milk fat syndrome caused by inadequate fiber and excessive grain. A ration containing a high proportion of forage helps maintain normal rumen function and provides an environment in which fat is less inhibitory to rumen fermentation and nutrient digestion.

The type of fat (saturated or unsaturated) added to the ration greatly influences the animal's nutrient utilization, milk production, feeding behavior, ration acceptability, the amount of fat that can be fed, and milk composition. Unsaturated fats are less desirable for dairy cows because of their inhibitory effects on rumen fermentation and digestion. Animal fats (which are more saturated) and blended animal-vegetable fats have generally given the most positive responses in animal performance.

When unsaturated fats are added to the ration, the calcium and magnesium contents of the cow's total ration DM may need to be increased by 20 to 30%, because added fat increases calcium and magnesium soap formation in the rumen and the excretion of these elements in the feces.

Because vegetable oils are high in unsaturated fat, they are less satisfactory than saturated fats as ration supplements. Whole seeds, such as cottonseed, soybeans, and sunflower seeds, have been used successfully, but the added fat derived from them should not exceed 1.0 lb/cow/day. Unsaturated fats that contain high levels of oleic acid apparently exceed the hydrogenation ability of rumen

Fig. 10–6. A complete (all-in-one) lactating ration containing whole cottonseed. Note the white cottonseeds. Whole cottonseed is a popular dairy feed in the West and Southwest. (Courtesy, Arizona Dairy Co., Higley, AZ)

microorganisms and result in the greatest milk fat depression. An increase in long-chain fatty acids in the ration (1) increases the secretion of milk and (2) inhibits the synthesis of short- and medium-chain fatty acids in mammary tissue.

Added ration fat, including feeding whole cottonseed and soybeans, decreases the protein content of milk by about 0.1%, primarily because of the lower casein content. Whole cottonseed also increases the proportion of long-chain fatty acids in milk.

In some cases, feeding rumen-protected fats that are resistant to microbial action in the rumen has increased milk fat percentage and the efficiency of milk production. However, if polyunsaturated fats are fed, this practice may result in milk with an increased proportion of polyunsaturated fatty acids and an increased susceptibility to oxidative rancidity. Limited data indicate that calcium salts of fatty acids (which are 82% fat) and prilled forms of saturated fats are effective sources of protected fat for dairy cows and that their use may allow the total fat in the ration to reach 7 to 8% of the dietary DM of the ration.

Until the rumen becomes functional, young dairy calves require some fat in the diet. A level of 10% fat in milk replacers appears to be sufficient to supply essential fatty acids and to carry fat-soluble vitamins, but insufficient to supply adequate energy for normal gains under optimum environmental temperatures. For veal production, a higher fat milk replacer (15 to 20% or more), will increase fat deposition in the carcass and is desirable. Also, 15 to 20% fat in milk replacers is needed for normal gains when calves are exposed to cold environmental temperatures.

MOLASSES

Molasses (including cane or blackstrap, beet, citrus, and wood molasses) is extensively used as a livestock feed. 1,605,000 metric tons are used for animal feeds in the United States, annually.

Cane and beet molasses are by-products of the manufacture of sugar from sugarcane and sugar beets, respectively. Citrus molasses is produced from citrus wastes. Wood molasses is a by-product of the manufacture of paper, fiberboard, and pure-cellulose from wood; it is an extract from the more soluble carbohydrates and minerals of the wood material. Cane or blackstrap molasses is by far the most extensively used type.

When used at levels of 10 to 15% of the ration, molasses has about three-fourths the energy value of corn. However, molasses has added value as an appetizer, to reduce dustiness of a ration, as a binder for pelleting, to stimulate rumen microbial activity, and as a source of unidentifiable factors. Also, cane molasses is a good source of certain minerals.

In hot, humid areas, molasses should be limited to 5% of the ration; otherwise, mold may develop. Where mustiness is a problem, it may be controlled by adding calcium propionate to the feed according to the manufacturer's directions.

Brix is a term used to express molasses quality, as reflected by the relative level of sugar present. It is arrived at by first determining specific gravity. Then by use of conversion tables, the degrees Brix, or level of sucrose present is obtained.

The different types of molasses may also be available in dehydrated form.

PROTEIN FEEDS

Protein supplements are feedstuffs that contain more than 20% protein or protein equivalent. At least 23 amino acids have been identified and may occur in combinations to form an almost limitless number of proteins.

Protein is essential for dairy cattle maintenance, growth, milk production, and the development of the fetus. Also, it is required for the formulation of enzymes and certain hormones that control or regulate chemical reactions in the body. The protein requirement is really a requirement for amino acids.

The protein composition of feeds, and the protein requirements of dairy cattle, may be expressed as crude protein, digestible protein, degraded intake protein, undegraded intake protein, and/or nonprotein nitrogen (NPN).

The amount of protein needed in the total ration of lactating cows is determined primarily by the amount of milk produced. Milk is a rich source of high-quality protein; so, as milk production increases, a substantial amount of dietary protein is necessary. Thus, a high-producing 1,320-lb cow yielding 88 lb of 3.5% protein milk daily secretes 3.08 lb of milk protein. A deficiency of protein results in lowered milk production and may depress the protein content of milk. Excess protein usually results in high-cost rations.

The amount of protein needed in the concentrate mix depends on the kind and quality of forage fed. As the amount of legume increases, the percentage of protein in the concentrate can be lowered. For most lactating cows, the total ration (forage plus grains and protein and energy supplements) should have 19% crude protein during the first one-third of lactation, lowered to 14% in midlactation and 12% during the dry period.

When more protein is fed than needed, the excess is used as a source of energy. Because protein feeds are generally more expensive than carbohydrate feeds, it usually is more economical to feed only the amount needed. Besides, a large excess of dietary protein may decrease the energy supply because excess protein must be deaminated to ammonia and, for the most part, transformed back into urea for excretion. Most cows fed good quality alfalfa hay (fed free-choice), along with grain fed according to production by one of the recommended systems, will not need any supplemental protein until they produce more than 50 to 60 lb of milk daily. As production increases above this amount, protein intake must be increased gradually, usually by decreasing the hay intake and replacing it with grain and protein concentrates. So long as the hay consumption does not fall below 12 to 15 lb daily, the supplemental concentrate need not contain more than 15 to 16% protein.

High-protein feeds are usually named and classified according to their origin and method of processing. On the basis of origin, they are usually grouped into two general categories as follows: (1) plant proteins, or (2) animal proteins.

PLANT PROTEINS

This group, which supplies the bulk of protein supplements for dairy cattle, includes the common oilseed by-products—soybean meal, cottonseed meal, linseed meal, peanut meal, safflower meal, sunflower meal, rapeseed meal (canola meal), and coconut (or copra) meal. They vary in protein content and feeding value, depending on the seed from which they are produced, the amount of hull and/or seed coat included, and the method of oil extraction used.

Protein quality is less important with ruminants because of microbial synthesis. In feeding mature cattle, a safe plan to follow is to provide a liberal supply of high-quality legume hay or lush young pasture along with the concentrates. Also, the quality of the proteins in a ration is likely to be higher if a variety of feeds is combined.

Even though they are not especially high in protein by comparison with other feedstuffs, the vegetative portions of many plants supply an extremely large portion of the protein in the total ration of dairy cattle, simply because these portions of feeds are consumed in large quantities. Needed protein not provided in these feeds is commonly obtained from one or more of the oilseed by-products.

Sometimes, the unprocessed seed is used to provide both a source of protein and a concentrated source of energy. The oil-bearing seeds are especially high in energy because of the oil that they contain.

Additional plant proteins are obtained as by-products from grain milling, brewing and distilling, and starch production. Most of these industries use the starch in grains and seeds, then dispose of the residue, which contains a large portion of the protein of the original plant seed.

Also, numerous good commercially manufactured protein supplements are available for dairy cattle.

OILSEED MEALS

Several rich oil-bearing seeds are produced for vegetable oils for human food (oleomargarine, shortenings, and salad oil), and for paints and other industrial purposes. In processing these seeds, protein-rich products of great value as dairy feeds are obtained; among them, soybean meal, coconut meal, cottonseed meal, linseed meal, peanut meal, rapeseed meal, safflower meal, sesame meal, and sunflower seed meal. Oil is extracted from these seeds by one of the following basic processes or modifications thereof: solvent extraction, hydraulic extraction, or expeller extraction.

SOYBEAN MEAL

Soybean meal has the highest nutritive value of any plant protein source. It is now the most widely used protein supplement in the United States.

Soybean meal is the ground residue (soybean oil cake or soybean oil chips) remaining after the removal of most of the oil from soybeans.

In the past, oil was extracted by the solvent, hydraulic, and expeller processes. But, today, almost all soybeans are solvent extracted. Soybean meal normally contains 41, 44, 48, or 50% protein, depending on the amount of hull removed. Because of its well-balanced amino acid profile, the protein of soybean meal is of better quality than other protein-rich supplements of plant origin. However, it is low in calcium, phosphorus, carotene, and vitamin D.

Fig. 10–7. Soybeans, the precious bean that originated in the Orient. Now the source of the major U.S. protein supplement—soybean meal. (Courtesy, J. C. Allen and Son, West Lafayette, IN)

COCONUT MEAL (COPRA MEAL)

This is the by-product from the production of oil from the dried meats of coconuts. The oil is generally extracted by either (1) the hydraulic process, or (2) the expeller process. Coconut meal averages about 21% protein content.

The lipid component of copra meal is very low in unsaturated fatty acids. Hence, dairy producers use copra meal to produce a pleasant-flavored, rather hard (highly saturated) butterfat. It appears that the maximum level of coconut meal in dairy rations is 3.3 to 6.5 lb per day. Higher amounts tend to produce tallowy butter.

COTTONSEED MEAL

The U.S. cotton crop ranks fourth in value, being exceeded only by soybeans, corn, and wheat. Among the oilseed meals, cottonseed meal ranks second in tonnage to soybean meal. The oil is extracted by (1) mechanical, (2) solvent, or (3) partially mechanically extracted and then solvent extracted process.

The protein content of cottonseed meal varies from about 22% in meal made from undecorticated seed to 95% in flour made from seed from which the hulls have been removed completely. Thus, by screening out the residual hulls, which are low in protein and high in fiber, the processor is able to make a cottonseed meal of the protein content desired—usually, 36, 41, 44, and 48%. The protein content of cottonseed meal varies with the geographical location in which it was grown. Meals manufactured from cottonseed produced on the West Coast generally contain higher protein levels than those produced throughout the rest of the United States.

LINSEED MEAL

Fig. 10–8. Flax seed, from which the oil is extracted, following which the residue is ground—producing linseed meal. (Courtesy, USDA)

Linseed meal is a by-product of flax, a fiber plant which antedates recorded history. In this country, most of the flax is produced as a cash crop for oil from the seed and the resulting by-product, linseed meal. Practically none of the U.S. flax crop is grown for fiber, for it is more economical to import if from those countries where cheaper labor is available.

Most of the nation's flax is produced in North Dakota, South Dakota, and Minnesota. Normally, an additional quantity of seed is imported and processed in our plants.

The oil is extracted from the seed by either of two processes: (1) the mechanical process (or so-called *old process*), or (2) the solvent process (or so-called *new process*).

Linseed meal is the finely ground residue (known as cake, chips, or flakes) remaining after the oil extraction. It averages about 35% protein content (33 to 37%). Linseed meal is lacking in carotene and vitamin D, and is only fair in calcium and the B vitamins. It is laxative; hence, it may be used to regulate the bowels.

Linseed meal has the reputation of importing "bloom" to the hair coats of cattle, and other animals, and is frequently used in fitting rations. About 7 lb per day is the maximum amount that can be fed to adult cattle.

PEANUT MEAL, AND PEANUT MEAL AND HULLS

Peanut meal, a by-product of the peanut industry, is ground peanut cake, the product which remains after the extraction of part of the oil of peanuts by pressure or solvents. It is a palatable, high-quality vegetable protein supplement used extensively in livestock and poultry feeds. Peanut meal ranges from 41 to 50% protein and from 4.5 to 8% fat. It is low in methionine, lysine, and tryptophan, and low in calcium, carotene, and vitamin D.

Peanut meal and hulls is ground peanut meal with added hulls, or the ground by-product remaining after extraction of part of the oil from whole or unshelled peanuts. Since about one-fourth of peanut meal and hulls consists of peanut hulls, it is high in fiber, averaging about 22.5%.

Since peanut meal tends to become rancid when held too long—especially in warm, moist climates—it should not be stored longer than 6 weeks in the summer or 2 to 3 months in the winter.

Fig. 10–9. Peanut crop. (Courtesy, USDA)

RAPESEED MEAL (CANOLA MEAL)

Rapeseed is grown extensively in Canada, where it has been renamed *canola*.

Canola was created from specially selected rapeseed by Canadian plant scientists in the 1970s. The old rapeseed was high in glucosinolate compounds, which, when fed to animals at high levels, made for palatability problems and lowered performance due to goitrogenic action. Canola changed this. The new canola is low in glucosinolates in the meal, and low in erucic acid (a long-chain fatty acid) in the oil. Canola is grown mainly in Canada, but it is increasing in the United States.

Canola meal averages about 36% crude protein and its amino acids compare favorably with soybean meal. When the price is favorable, canola meal may be used as a protein supplement for dairy cattle.

CAUTION: The lowering of the erucic acid and glucosinolate in rapeseed (canola) has proceeded at different rates in different countries. But the change is nearly complete in Canada, the leading nation, which produces about 20% of the world crop.

SAFFLOWER MEAL

A large proportion of the safflower seed is composed of hull—about 40%. Once the oil is removed from the seeds, the resulting product contains about 60% hulls and 18 to 22% protein. Various means have been tried to reduce this high hull content. Most meals contain seeds with part of the hull removed, thereby yielding a product of about 15% fiber and 40% protein. Safflower meal is not very palatable, but it can be used effectively when mixed with other feeds.

SESAME MEAL

Little sesame is grown in the United States despite the fact that it is one of the oldest cultivated oilseeds. The oil meal is produced from the entire seed. Solvent extraction yields higher protein (45%) but lower fat levels (1%) then either the screw press or hydraulic methods which produce meals containing about 38% protein and 5 to 11% oil. Excessive levels (greater than 7 lb per head per day) can produce soft butter.

SUNFLOWER MEAL

Sunflower meal—a newcomer to the oilseed industry in the United States—is rapidly gaining acceptance as a high-quality source of plant protein.

The oil meal varies considerably depending on the extraction process and whether the seeds are dehulled. Meal from prepressed solvent extraction of dehulled seeds contains about 44% protein, as opposed to 28% for whole seeds. Screw-pressed sunflower meal ranges from 28 to 45% protein. Sunflower hulls can be effectively used for roughage in dairy feeds.

ANIMAL PROTEINS

Animal proteins are derived from animal products. Milk and milk products, along with a limited amount of feather meal, constitute the bulk of animal proteins that are fed to dairy cattle.

NONPROTEIN NITROGEN FEEDSTUFFS (NPN)

Certain nonprotein nitrogen sources may be substituted for all or much of the supplemental protein required in most ruminant rations, provided such rations are adequate in minerals and readily available carbohydrates. Among such products are urea, ammoniated molasses, ammoniated beet pulp, ammoniated cottonseed meal, ammoniated citrus pulp, and ammoniated rice hulls.

The rumen microorganisms—which are a low form of plant life and are able to use inorganic compounds much like plants utilize chemical fertilizers—build proteins of high quality in their cells from sources of inorganic nitrogen that non-ruminants cannot use. Since the life-span of these microorganisms is short, further on in the digestive tract the ruminant digests the bacteria and obtains good protein therefrom. In ruminant nutrition, therefore, even such nonprotein sources of nitrogen as urea and ammonia have a protein replacement value. An exception is very young calves in which the rumen and its ability to synthesize are not yet well developed.

Only urea will be discussed in the section that follows, since it provides the great bulk of nonprotein nitrogen fed to dairy cattle.

UREA

Due to the increasing shortage and higher price of oilseed proteins, more and more urea is being fed to dairy cows.

Approximately 265,000 metric tons of urea are fed to ruminants annually in the United States, as a source of protein.

Urea is a white crystalline, odorless, nonprotein nitrogen compound of the formula N_2H_4CO. It is manufactured synthetically from the nitrogen of the air. When properly used, it is a safe, low-cost source of protein for dairy cattle; when improperly used, it becomes a hazard. It is recommended that urea in a concentrate mixture not exceed 2%, and that it not exceed 1% of the total hay and grain ration. It is further recommended that not more than one-third of the protein requirements be met through urea. Under experimental conditions, much higher quantities of urea have been fed to dairy cows. For example, beginning in 1958, Dr. A. I. Virtanen, the Finnish Nobel Prize winner, successfully fed dairy cows on a protein-free diet, utilizing small amounts of ammonium salts combined with urea as practically the sole source of nitrogen.[1]

Yet, it is emphasized that this was done under carefully controlled experimental conditions. Further, most state laws limit the amount of urea that can be put in commercial feeds.

One pound of 45% nitrogen (281% protein) urea provides as much protein value as 6.8 lb of 41% soybean meal or cottonseed meal (281 ÷ 41 = 6.8). However, urea does not provide energy, minerals, or vitamins, with the result that these must be provided through other sources when urea is substituted for protein meals.

When added to the dairy cattle ration, urea must be mixed thoroughly to ensure even distribution, and used according to directions. It may be toxic when improperly used, or when fed in too large amounts.

Normally, dairy producers limit urea to 1.5 to 2.0% of the concentrate ration. Higher levels (up to 2.75% of the concentrate) are unpalatable and depress appetite. However, the unpalatability may be alleviated by pelleting the urea with alfalfa.

METHODS OF FEEDING UREA

Urea is generally fed to dairy cattle by one of the following methods: (1) mixed in concentrate, (2) as a liquid supplement, or (3) mixed with silage.

UREA MIXED IN CONCENTRATE

Most of the urea fed to growing and lactating dairy cattle is incorporated into the concentrate portion of the ration. It

---

[1]*Science*, Vol. 153, No. 3744, Sept. 30, 1966, pp. 1603–1614.

can be supplied in a protein supplement or in the entire concentrate ration. Also, urea may be provided in a complete mixed feed.

Urea can be mixed in feed either as a powder or as an aqueous solution. Both methods are relatively simple and inexpensive. When urea is added as a powder, there is a chance that it will sift through the grain and be unevenly distributed, thereby increasing the chance of toxicity. However, if careful mixing procedures are followed, this hazard can be minimized. If urea is added to feed in aqueous form, uniform distribution throughout the feed can be assured. However, this method has two disadvantages: (1) special mixing equipment is required; and (2) once in solution, urea may be degraded during prolonged storage.

LIQUID SUPPLEMENTS

Liquid supplements combining molasses for energy and urea as a protein precursor are widely used. This type of supplement can also be used as a carrier for micronutrient and nonnutritive additives.

Several problems are inherent in feeding such liquid supplements. If they are to be effectively used, the urea must remain in solution or suspension. This means that the supplement must keep its chemical integrity throughout varying environmental temperatures over prolonged periods. Liquid supplements are extremely palatable; hence, there is a danger of over-consumption if intake is not monitored. Additionally, these supplements tend to be highly corrosive.

UREA MIXED WITH SILAGE

One way of feeding urea to dairy cattle is through the addition of urea to crops which are being ensiled. If chopped, whole plant corn is being ensiled at 35 to 40% dry matter, urea can be added at a level of 0.5% of wet material. This level should increase the crude protein level of the silage on a dry matter basis about five points. Urea levels higher than 0.5% can create palatability problems as well as storage problems. When the silage contains little or no grain, the amount of urea to be added should be reduced.

Silage tends to be rather variable in moisture, and this variability can affect the benefits of added urea. Hence, one should have a reasonable estimate of the moisture content of the material that is to be ensiled. Likewise, water in silage will create some leaching out of the urea. Also, ammonia is produced during the ensiling process, representing an additional loss of urea.

(Also see Chapter 9, section headed "Feed Additives — ■ Urea, ammonia, and other NPN products.")

BY-PRODUCTS, ANIMAL WASTES, AND CROP RESIDUES

Dairy cattle utilize a host of by-products, animal wastes, and crop residues. As the world grows more populous, increasingly the dairy industry will provide a practical and economical outlet for human-inedible feeds.

BY-PRODUCTS FROM PLANT PROCESSING

By-product feeds from plant processing are important in dairy rations. The milling, sugar, vegetable oil, and fermentation industries provide by-products of special significance to most commercial dairy rations, and for many home-mixed feeds. However, if dairy producers plan to incorporate a by-product feed in their rations, they should first determine the moisture content of the product, its relative feeding value, and the appropriate amount to feed. By-products are extremely variable in feeding value. For example, almond hulls can range from excellent to poor as a feedstuff. Additionally, some by-product feeds may contain pesticide residues that can be excreted in the milk. For the latter reason, producers should make sure that their feeds are free from environmental contaminants.

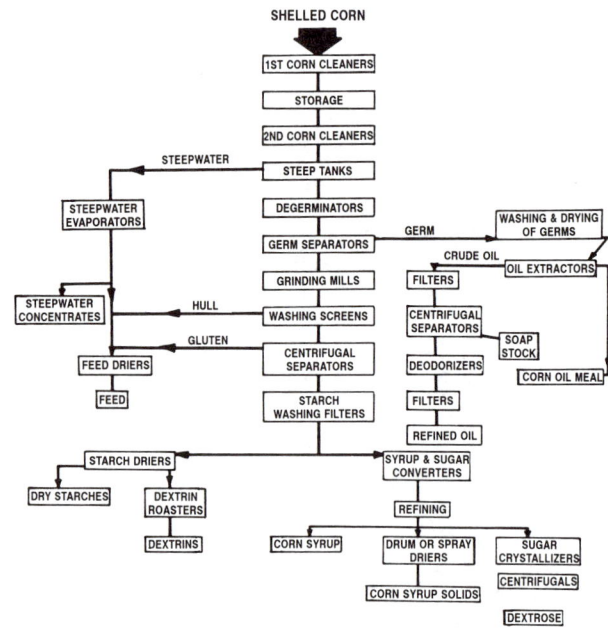

Fig. 10–10.   Wet corn milling. The by-products from this process are fed to dairy cattle and other animals.

Many of the residues and by-products from the vegetable industry can be used effectively in dairy rations. Such feeds as cannery wastes, pea vines and pods, and cottonseed hulls can be used as economic alternatives to the more traditional feeds.

Wheat bran and wheat middlings (or mill run) are the most widely used by-products of the milling industry. Wheat bran has always been considered to be an excellent dairy feed when the quality is high and the price is competitive. It is especially good as a conditioning feed and is widely used in fitting dairy animals for shows and sales. Wheat mill run (middlings) is usually more price competitive than wheat bran. When of high quality, it is an excellent feed, although it is not as high in digestibility or energy as the whole grains.

Beet pulp and beet molasses are used very extensively in dairy rations, where they add to the texture and palatability as well as serve as good sources of energy. Molasses reduces the dustiness of feeds and is an excellent carrier for minerals and vitamins, as well as other ingredients where uniform mixing is important. It is usually restricted to about 5% of the grain-concentrate mix. At higher levels, it may make the feed sticky and cause mechanical mixing and feeding difficulties. Also, excess molasses may be too laxative because of its high mineral and sugar content; hence, it may reduce the utilization of other less digestible ingredients.

When available at competitive prices, the dried or wet mash from the brewing and distilling industries, as well as some of the pharmaceutical fermentations, have been used extensively in dairy rations.

The vegetable oil industry extracts the oil from certain seeds, leaving a residue which is high in protein and beneficial in dairy rations. The oil meals are incorporated in dairy rations primarily as a source of protein. Soybeans are a very popular oil crop; and the extracted meal is one of the most common protein supplements in dairy rations. In cotton-growing areas, cottonseed meal is widely used in dairy rations. Other oil meals used in dairy rations include peanut meal, copra (coconut) meal, safflower meal, and several others of lesser importance. Some heating is desirable for most oil meals to destroy toxic or inhibitory materials which they contain. However, the heating must be controlled so that the digestibility of the proteins is not reduced to the point that it erases the benefits derived from eliminating the undesirable substances.

Additional information of interest to the dairy industry is presented herein relative to each of the following groups of by-products from plant processing: fruits and vegetables, fiber plants, sugar crops, brewing and distilling, and wood and paper products.

FRUIT AND VEGETABLE BY-PRODUCTS

Fruit and vegetable by-products may come from three sources: (1) cull, unmarketable or damaged, but wholesome, commodities; (2) crop residues left in the field; or (3) canning, juicing, or processing wastes. These products can be used successfully in many feeding programs. But problems involving their continued availability, storage, and handling must be considered, because many of them are highly perishable. Most of these by-product feeds are generally restricted to areas where processing and canning operations are located.

MISCELLANEOUS FIBROUS BY-PRODUCTS

The value of fiber in ruminant feeds has been well documented. Many of the oilseeds are hulled (decorticated) prior to processing. Cottonseed hulls and gin trash are widely used fibrous by-products. Soybeans, peanuts, and sunflowers are three additional oilseeds from which hulls are routinely removed in processing.

Hulls from the covered grains provide satisfactory roughage for dry cows. Rice, buckwheat, oats, and barley are commonly dehulled.

Hulls from many nuts, such as almonds and walnuts, can be used in dairy feeds when price and availability so warrant.

(Also see section on "Fibrous Feeds," which follows later in this chapter.)

SUGAR INDUSTRY BY-PRODUCTS

Two crops—sugarcane and sugar beets—account for the bulk of refined U.S. sugar, with sugar beets being the primary crop. Several by-product feeds are produced in various stages of processing these two crops for sugar.

SUGARCANE

In the processing of sugarcane, the white, crystalline sucrose is separated from the molasses and brown sugars. The molasses product, which is referred to as blackstrap molasses, is incorporated in many dairy rations.

After the juice has been extracted from the cane, the remaining by-product is known as *bagasse*. Bagasse is high in fiber, and the fiber is low in digestibility—only about 25%. Additionally, its TDN is extremely low, ranging from 20 to 25%. However, bagasse has been used effectively as a carrier of molasses, the combination of which yields a relatively high-fiber, high-energy feed.

SUGAR BEETS

The resulting beet pulp from the processing of sugar beets can be fed wet if used within a short time, or it can be ensiled or dried when long-term storage is desired. Molasses is often added to the dried beet pulp to increase the energy content; and, on occasion, beet pulp is ammoniated to provide a source of non-protein nitrogen.

Beet tops and crowns are relished by cattle. They can be fed fresh, dried, or ensiled.

BREWING AND DISTILLING INDUSTRY BY-PRODUCTS

Considerable quantities of grains are used in the brewing of beers and ales and in the distilling of liquors. After

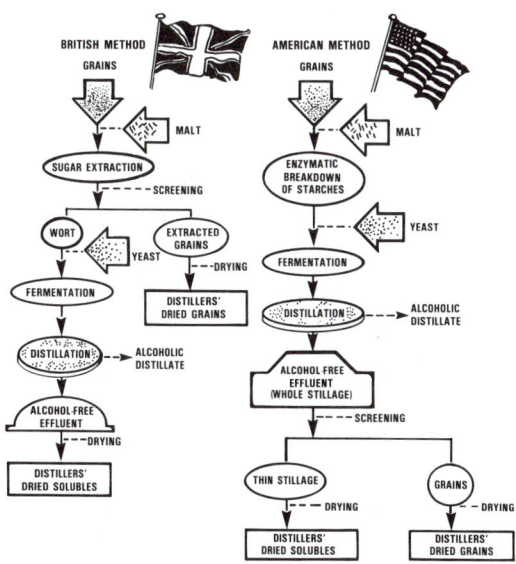

Fig. 10–11. British and American processes of distillation. The end product, distillers' dried grains, is fed primarily to dairy cattle as a source of protein.

processing, the remaining by-product can be readily adapted to dairy feeding programs.

Usually brewing and distilling by-products contain some of the B vitamins and other nutrients which are supplied by rumen fermentation.

WOOD AND PAPER INDUSTRY BY-PRODUCTS

Wood and by-products from the wood and paper industry offer an abundant source of potential feedstuff for ruminants. These by-products can be classified as follows: (1) milling and processing wastes, (2) treated wood scraps, (3) pulp scraps, (4) paper products, (5) wood sugar products, and (6) torula yeast. Since the digestibility and energy content of many of these products is low, their use must be restricted to those animals with low energy requirements, such as dry cows.

ANIMAL WASTES (MANURE)

Manure refers to a mixture of animal excrements (consisting of undigested feeds plus certain body wastes) and bedding. Where no bedding is included, it is commonly referred to as *pure manure*, *manure free of bedding*, or *manure exclusive of bedding*. The U.S. annual production of manure exclusive of bedding totals more than 1.5 billion tons.

Litter refers to the product formed when manure is mixed with bedding. The bedding is generally a highly absorbent material. A wide variety of materials is used for bedding, including straw, peanut hulls, wood shavings, sawdust, and bark.

Poultry wastes is a broad term which is sometimes applied to both cage layer wastes (where broken eggs, eggshells, and feathers are found in the manure) and broiler manure.

In mammals, urine and feces are voided separately; but in poultry the end products coming from the urinary and gastrointestinal tracts are mixed together and voided via the cloaca. Bedding with an absorbent material should be used for housed mammals in order to prevent the nitrogen in the urine from being lost through leaching. In poultry droppings, the fecal material prevents the nitrogen from leaching out, which is the primary reason that poultry manure has a higher nitrogen content than manure from four-footed animals.

Animal waste (manure) has nutritive value for ruminants because these animals are capable of utilizing nonprotein nitrogen and fiber. So, proper processing is important.

Broiler and layer litter has been successfully used as an ingredient of cattle feed for many years. However, wastes from all species may be, and are, used. Among the methods employed to process animal wastes prior to feeding are: deep-stacking, ensiling (fermentation), dehydration, and pelleting. The two most common and practical methods of processing are:

1. **Deep-stacking.** In this method, the litter is deep-stacked for several weeks, during which it generates temperatures of 160°F or higher, which render it free of any potentially pathogenic microorganisms that might be present. (Pathogenic bacteria do not grow at temperatures over 80°F, and they are killed at 145°F in a matter of minutes.) It

follows that there have been no documented animal health problems associated with feeding broiler or layer litter processed in this manner.

2. **Ensiling (fermentation).** Ensiling is a controlled fermentation process during which carbohydrates in the mixture are converted to lactic, acetic, and other acids. Once sufficient acids are produced, bacterial action ceases and the ensilage is stable. During the fermentation, heat is generated, thereby diminishing the hazard from certain pathogenic organisms that might be present.

Dehydration and pelleting of animal wastes are excellent processing methods as such. However, current energy costs make them uneconomical.

In December 1980, the U.S. Food and Drug Administration published a document leaving regulation of feeding animal waste to the individual states.

POULTRY WASTES

At the present time, most of the attention of animal wastes as feedstuffs has centered around poultry wastes because of its high nitrogen content.

Because poultry production is highly intensive, with many birds in a small area, waste disposal is a major problem. Most cage-layer operations produce manure free of bedding as the primary form of waste. Broiler operations, generally, produce litter because the birds are not caged; rather, they have access to the area of a pen or house, the floor of which is covered with bedding material.

On a moisture-free basis, cage-layer manure generally contains 25 to 35% crude protein and minimum fiber, while broiler litter contains somewhat less protein—about 18 to 30% and substantially more fiber due to the presence of absorbent materials.

TABLE 10–2
PERFORMANCE OF DAIRY HEIFERS FED
GRADED LEVELS OF TURKEY LITTER SILAGE[1]

Item	Treatment[2]			
	0%	15%	30%	45%
Initial weight . . . . (lb)	485	473	484	460
. . . . (kg)	220	215	220	209
Avg. daily gain . . . . (lb)	0.92	1.28	1.12	0.95
. . . . (kg)	0.42	0.58	0.51	0.43
Feed efficiency . . (feed/gain)	12.5	12.5	14.3	16.7

[1]Adapted by the author from "Turkey Litter Silage in Rations for Dairy Heifers," by D. L. Cross and B. F. Jenner, *Journal of Dairy Science*, Vol. 59, 1976, p. 919.

[2]Treatments represent amount of dry matter contributed by turkey litter silage to the total dry matter of the ration.

In a trial designed to test the effects of feeding different levels of turkey litter silage to dairy heifers, the South Carolina Experiment Station workers used a basal diet of corn silage and concentrate. Different levels of turkey litter silage were fed to heifers 8 to 12 months old. When turkey litter silage represented 15% of the total dry matter of the ration, heifers grew significantly faster than heifers fed the basal ration of corn silage-concentrate, but feed efficiency was the same. While the higher levels of turkey litter silage produced about the same rate of gain as the control, feed efficiency tended to decrease, though not statistically significant .

CROP RESIDUES

Crop residues are the portions of crops that are normally left in the field following harvest. Among such crop residues are: corn stalks and husklage, sorghum stalks, soybean refuse, small grain straws and chaff, and legume and grass seed straws. Crop residues must be fed to the right class of animals, and they must be properly supplemented.

Of all the crop residues, the residue of corn is produced in greatest abundance in the United States and offers the greatest potential for expansion in animal numbers. Corn usually produces an amount of residue equal to the quantity of the grain produced.

Generally speaking, crop residues may be grazed, processed as dry feed, or made into silage. The important thing to remember is that their relatively low value, in comparison with grains, necessitates low-cost harvesting, storing, and feeding. Also, they must be fed to the right class of animals — primarily dry cows, and they must be properly supplemented.

SPECIAL FEEDS, SUPPLEMENTS, AND ADDITIVES

Special feeds, as used herein, refers to feeds other than the usual concentrates and forages.

The term *supplement* refers to feedstuffs that are used to improve the value of basal feeds. Thus, supplements are products that provide an additional nutrient or nutrients. They can be used in large quantities, such as protein supplements, or in extremely small quantities, such as trace minerals.

An additive is a substance that is added to a basic feed, usually in small quantities, for the purpose of fortifying it with certain nutrients, stimulants, or medicines. In general, the term *feed additive* refers to a nonnutritive product that affects utilization of the feed or productive performance of the animal. However, in a legal sense, many nutritional supplements — for example, MHA (methionine hydroxy analog) — are considered to be additives.

SPECIAL FEEDS

A great array of special feeds are used by dairy producers. But only the following will be discussed herein: fiber, colostrum, milk replacers, calf starters, and hydroponics (sprouted grain).

FIBROUS FEEDS

Fiber is important in dairy rations. Excessive fiber levels limit intake and energy concentrations, while a shortage of fiber reduces rumen digestibility and milk fat test.

The amount of fiber to include in the ration of dairy cattle is influenced by the body condition and level of production of the animal, the type of fiber, the particle size, the amount of total DM consumed and its bulk density, the buffering capacity of the forage, the frequency of feeding, and the economics. Lactating cows that are fed to produce large amounts of milk, or young animals that are fed to achieve rapid growth, should receive more energy and less fiber than lower producing animals. Forages that are finely ground (processed to small particle size) are rapidly consumed and fermented in the rumen, which reduces (a) the animal's chewing time, (b) ruminal fluid, and (c) the acetate-to-propionate ratio in ruminal fluid. The result of these effects is a depression in milk fat percentage. Chopped alfalfa should average about $\frac{1}{4}$ in. in length to maintain a normal milk fat percentage. Feed factors, such as small particle size of forages, that reduce the pH of ruminal fluid, decrease the number and activity of fiber-degrading bacteria and cause a depression in fiber degradation. Feeding an insufficient amount of fiber or feeding forages that have a poor buffering capacity in the rumen may have undesirable effects on rumen fermentation, fiber degradation, and milk fat percentage that are similar to those caused by reducing the particle size of the forage.

So, the general recommendation is that lactating dairy cows should receive at least one-third of the total ration dry matter as long hay or as its DM equivalent in medium-to-coarse chopped silage or other forage. A minimum of 5 lb of forage dry matter measuring 1 to 2 in. in length will meet the fiber need of most lactating cows.

The values for neutral detergent fiber (NDF) and acid detergent fiber (ADF) are more accurate measures of the fiber component of feeds than are values for crude fiber. Yet, because both chemical and physical properties of feeds are involved in determining fiber quality and the energy value of feeds, there is currently no one fiber analysis that can accurately predict fiber quality and energy values for all feeds. NDF content is negatively correlated with dry matter intake and apparent digestibility of forages, but it is positively correlated with chewing time. ADF is more negatively correlated with apparent digestibility than is NDF. NDF and bulk density are positively related, which may explain the negative relationship between dry matter intake and the NDF content of the ration. According to University of Wisconsin researchers, NDF is a better predictor than ADF of dairy cow feed intake and milk production.

The optimum amount of NDF and ADF to include in the ration varies with the level of milk production and the type of forage that is fed to dairy cattle. A minimum of 21% of ADF and 28% of NDF is recommended for cows during the first 3 weeks of lactation. During times of high milk production, however, ADF and NDF contents of the ration are usually reduced to 19 and 25%, respectively, so that adequate dietary energy can be included to meet the cow's requirement. The ADF and NDF contents of the ration should be increased in later lactation to help prevent milk fat depression and because less energy is required for milk production. Seventy-five percent of the NDF in the ration should be supplied as forage.

(Also see section on "Miscellaneous Fibrous By-products," which is presented earlier in this chapter.)

COLOSTRUM

The subject of *colostrum* is fully covered in Chapter 12, in the section headed "Colostrum"; hence, the reader is referred thereto.

MILK REPLACERS

Milk replacers are usually classed as "special feeds," which, for the most part, are manufactured commercially, rather than home-mixed. The subject of *milk replacers* is fully covered in Chapter 12, in the section headed "Colostrum, Milk, and Milk Replacers Compared"; hence, the reader is referred thereto.

CALF STARTERS

The subject of *calf starters*, including the formulations of calf starters, is fully covered in Chapter 12, in the section headed "Calf Starters"; hence, the reader is referred thereto.

HYDROPONICS (SPROUTED GRAIN)

Hydroponics (or sprouted grain) is the growing of plants with their roots immersed in an aqueous solution containing the essential mineral nutrient salts, instead of soil. This means that sprouted grain for feed is produced with water and chemicals, without soil.

The Wisconsin Alumni Research Foundation chemically analyzed and compared the composition of oat grain and 5-day oat grass on a dry matter basis (see Table 10–3).

TABLE 10–3
COMPOSITION OF OAT GRAIN AND
5-DAY OAT GRASS, MOISTURE-FREE BASIS[1]

Constitutent	Oat Grain	Oat Grass
	(%)	(%)
Dry matter	100	100
Protein	15	21
Ether extract (fat)	4.21	5.2
Nitrogen-free extract	65.86	42.79
Fiber	11.71	26.11
Ash	3.22	3.9
Calcium	0.063	0.238
Phosphorus	0.360	0.509
	(mg/kg)	(mg/kg)
Carotene[2]	0.0	39.067
Vitamin E	17.95	48.87
Niacin	7.18	103.96
Riboflavin	1.96	22.29
Thiamin	3.14	12.86
Vitamin C	0.0	218.3

[1]Analyses by Wisconsin Alumni Research Foundation.

[2]Each mg of beta carotene was considered to be equivalent to 1,556 IU of vitamin A.

As shown in Table 10–3, the 5-day oat grass is a better source than oat grain of calcium, phosphorus, carotene, vitamin E, the B vitamins (riboflavin, thiamin, and niacin), and vitamin C. In addition to these comparative figures, however, the following facts are pertinent:

1. Supplemental quantities of calcium and phosphorus can be provided in many forms at a relatively low cost.

2. Sprouting greatly increases the carotene content; hence, if carotene, or vitamin A, is deficient, sprouted grains are a good supplemental source. However, supplemental vitamin A can be provided in a dry, stabilized form at a low cost.

3. Most rations are adequate in vitamin E. The B vitamins (riboflavin, thiamin, and niacin) are produced by the microorganisms in cattle, sheep, and horses; hence, supplemental quantities of them are not normally needed. Vitamin C is not required in the diet of farm animals.

The Michigan Station made a study of sprouted oats as a feed for dairy cows. They reported that the cost of sprouted oats was over four times that of the original oats or similar grains; and that there was a loss in nutrients during sprouting, a decrease in digestibility of sprouted oats, and no observed increase in milk production when sprouted oats were added to an adequate ration.[2]

Based on studies conducted by the different universities, sprouting results in an average loss of 83% of dry matter of the oat grain. One study showed a reduction in TDN from 75.7% in the oat grain to 70.2% for the sprouted oats. Also, the digestibility of dry matter, energy, protein, ether extract, and nitrogen-free extract was lower for the sprouted oats than for the oat grain.

In arriving at a decision whether or not to produce feeds hydroponically, consideration should be given to (1) the needs of different classes of livestock for each nutrient, and (2) the cost of supplying these nutrients in the form of sprouted grain.

Although sprouted grains are high-quality feeds from the standpoint of certain minerals and vitamins, the need for supplemental quantities of such nutrients in common rations for livestock is questionable.

Without doubt, sprouted grains will give an assist when added to poor rations—and the poorer the ration, the bigger the boost. However, with our present knowledge of nutrition, it should be recognized that balanced rations can be formulated without the added equipment and labor costs of sprouting grain.

SUPPLEMENTS

Mineral and vitamin supplementation is of paramount importance to *all* dairy rations. Imbalances, deficiencies, or excesses of minerals pose major problems. While toxicities of vitamins are rare, deficiencies are not; and in this era of highly refined scientific feeding, there can be no excuse for these occurrences.

[2]Report from *Quarterly Bulletin*, Vol. 44, No. 4, Michigan State University, May 1972, pp. 654-665.

MINERAL SUPPLEMENTS

The metabolic functions and interrelationships among the minerals are extremely varied and complex. An excessive amount of one mineral can create a deficiency of another. Additionally, several trace minerals have relatively narrow toxicity tolerances.

Almost all feeds contain at least limited amounts of the various minerals, but these levels are highly variable and often reflect the profile of the soil on which they are grown and the genetic variations among its plant species. In addition to the variability of mineral levels in individual feeds, the mineral requirements of animals are highly variable, depending on such factors as age, size, sex, type of production, and stage of production.

NEED FOR MINERAL SUPPLEMENTATION

Of the macrominerals demonstrated to be required by dairy cattle, only salt (sodium chloride), calcium, and phosphorus are routinely added to most rations. Two other macrominerals, magnesium and sulfur, are sometimes added to rations in specialized cases. Magnesium is sometimes provided in mineral mixes for cattle on pasture in areas where grass tetany is a problem. Sulfur is routinely added to rations containing urea, since this mineral element is a constituent of several amino acids and B vitamins.

Dairy producers should add trace minerals that are believed to be in short supply. This is usually accomplished through the use of a trace mineral mix.

Several products offering minerals in chelated form are on the market. Those selling chelated minerals generally recommend a smaller quantity of them (but at a higher price per pound) and extoll their "fenced-in" properties. When it comes to synthetic chelating agents, much needs to be learned about their selectivity toward minerals, the kind and quantity most effective, their mode of action, and their behavior with different species of animals and with varying rations.

GUIDELINES FOR MINERAL SUPPLEMENTATION

No single plan can be proposed as being the best for mineral supplementation. Rather, the dairy producer must tailor the supplement regimen to encompass the following considerations:

1. **Needs of the particular animal.** The age, sex, weight, and production parameters should be considered.
2. **Types of feed.** A high-concentrate ration will require a different mineral supplement than a high-roughage ration.
3. **Region from which the feeds were obtained.** The mineral content of the feed will reflect the mineral composition of the soil and the genetic makeup of the specific plant.
4. **Facilities.** Will the mineral supplement be mixed in the feed or offered free-choice? If the mineral mix is offered free-choice, containers protected from the elements—i.e., rain and wind—may have to be constructed.
5. **Cost of the minerals.** Minerals offered free-choice are utilized less efficiently because animals tend to either overeat or undereat. Additionally, some of the minerals will be lost due to silage and weather. By mixing the minerals in

the feed, the producer can ensure that there is a minimum of wastage. On the other hand, the requirements may not be known, and there are individual animal differences. Therefore, one must balance the cost of labor involved in mixing minerals in feed, along with assumed requirements, against the inefficiencies of free-choice supplementation of minerals.

Free-Choice Supplementation

Ad libitum (free-choice) supplementation of minerals is commonly practiced, but there are several serious flaws behind the rationale of this practice. It is based primarily upon the assumption that animals know their specific needs for minerals. That is to say, if an animal is deficient in calcium, it instinctively knows that it needs to consume the calcium supplement. This may be true in some animals, notably the rat, but recent research casts doubt on this practice in ruminants. Studies at both the Minnesota and Cornell Experiment Stations have shown great variation in the free-choice intake of minerals among cattle. It appears that the acceptability or palatability of the supplement plays a greater role in intake than the actual physiological need for the deficient mineral. Some animals consume minerals which they don't need, whereas others do not consume any minerals, even though they should. It follows that it is possible, even probable, that the producer has no idea of the mineral status of animals if they get their supplementation from mineral boxes. Therefore, an unthrifty animal, supposedly receiving adequate amounts of minerals, may, in fact, be suffering from one or more mineral deficiencies.

Supplementation of Mixed Feed

When cattle are fed a mixed feed, totally or in part, the needed minerals are usually incorporated in the ration in keeping with the known requirements. In general, the following recommendations are applicable to the mineral supplementation of mixed feeds:

1. Salt is usually incorporated in the ration at levels of 0.25 to 0.5%. If less salt is added, it can also be made available *ad libitum*.
2. Calcium and phosphorus are added as needed to balance the ration. Numerous calcium and phosphorus supplements are available at reasonable cost; and the wide variety enables producers to select one which fits their particular needs (see Table 10–4).
3. If cattle are housed in confinement where they receive little exposure to sunlight, careful attention must be given to providing adequate vitamin D, because vitamin D affects the assimilation and utilization of a number of minerals, especially calcium and phosphorus.
4. When the ration is suspected as deficient in one or more minerals, a trace mineralized salt or specific minerals should be added to the ration.

Supplementation on Pasture

Where dairy cattle are on pasture, minerals may be provided as follows:

1. **Where cattle are on liberal grain feeding.** Provide free access to a two-compartment mineral box, with (a) trace

mineralized salt in one side; and (b) a mixture of ⅓ trace mineralized salt (salt included for purposes of palatability), ⅓ dicalcium or defluorinated phosphate or steamed bone meal, and ⅓ ground limestone or oystershell flour in the other side.

2. **Where cattle are primarily on roughage (pasture, hay, and/or silage).** Provide free access to a two-compartment mineral box, with (a) trace mineralized salt in one side, and (b) a mixture of ⅓ trace mineralized salt (salt included for purposes of palatability); and ⅔ dicalcium or defluorinated phosphate or steamed bone meal in the other side.

CALCIUM AND PHOSPHORUS SUPPLEMENTS

Table 10–4 lists several sources of calcium and phosphorus and gives the typical analysis of each.

Where calcium alone is needed, ground limestone or oystershell flour are commonly used, either free choice or added to the ration in keeping with nutrient requirements. Where phosphorus alone is needed, monosodium phosphate or diammonium phosphate may be used.

VITAMIN SUPPLEMENTS

As with mineral supplements, careful consideration must be given to the vitamin supplementation of dairy feeds. While the requirements of vitamins are extremely small in comparison with energy and protein, the omission of a single required vitamin from the diet will produce specific deficiency symptoms, thereby reducing production. Moreover, the cost for vitamin supplementation constitutes a very small fraction of the total feed bill.

TABLE 10–4
TYPICAL ANALYSIS OF CALCIUM AND PHOSPHORUS SUPPLEMENTS (MOISTURE-FREE BASIS)[1]

Compound	Calcium Content (%)	Phosphorus Content (%)	Sodium Content (%)	Protein Equivalent N × 6.25 (%)	Fluorine Content (%)
Calcium compounds:					
Calcium carbonate	38.13	0.04	0.07	—	0.00
Limestone, ground	37.22	0.22	0.06	—	—
Oystershells, ground	36.27	0.10	0.21	0.7	—
Defluorinated phosphates manufactured from defluorinated phoshoric acid:					
Defluorinated phosphate	32.10	17.13	3.27	—	0.18
Diammonium phosphate	0.52	20.54	0.04	115.5	0.16
Dicalcium phosphate	22.67	19.00	1.61	—	0.10
Monoammonium phosphate	0.39	24.99	0.08	71.0	0.19
Monocalcium phosphate	18.80	21.27	0.06	—	0.14
Defluorinated phosphates manufactured from furnace phosphoric acid:					
Dicalcium phosphate	23.71	19.07	0.08	—	0.19
Disodium phosphate	—	22.32	32.00	—	—
Feed-grade phosphoric acid	0.18	27.84	0.23	—	0.25
Monocalcium phosphate	22.92	23.96	—	—	0.03
Monosodium phosphate	0.04	25.60	19.63	—	—
Sodium tripolyphosphate	—	25.38	31.23	—	0.03
Tricalcium phosphate	31.44	17.34	0.17	—	0.05
High-fluoride phosphate:					
Ground low-fluorine rock phosphate	36.00	14.00	—	—	0.45
Ground rock phosphate, raw	35.00	13.00	0.03	—	3.70
Soft rock phosphate	16.09	9.05	0.10	—	1.21
Packinghouse by-products:					
Bone charcoal (bone black)	34.08	15.85	—	—	—
Bone meal, steamed	27.31	12.40	0.42	19.5	0.07

[1]Adapted by the author from data provided especially for this book by the former International Feedstuffs Institute, now the Feed Composition Data Bank, National Agricultural Library, USDA, Beltsville, MD.

It is to be emphasized that subacute deficiencies can exist although the actual deficiency symptoms do not appear. Such borderline deficiencies are both the most costly and the most difficult with which to cope, going unnoticed and unrectified; yet they may result in poor and expensive milk production and impaired reproduction. Also, under farm conditions one will usually not find a vitamin deficiency which involves only a single vitamin. Instead, deficiencies usually represent a combination of factors, and usually the deficiency symptoms will not be clear-cut.

Formerly, a wide variety of feed ingredients was added to dairy rations for their vitamin content. But is was found that the vitamin concentration of feedstuffs varied tremendously, being affected by plant species and part (leaf, stalk, or seed), harvesting, storing, and processing. Generally speaking, vitamins are easily destroyed by heat, sunlight, oxidation, and mold growth. So, today, nutritionists rely on vitamin supplements, which in many cases are chemically pure sources that need to be used only in very minute amounts. In modern feed formulation, premixes often represent the commonsense approach to providing vitamins.

For adult dairy cattle, vitamins A, D, and E are of concern, with A being the one most likely to be deficient. Under ordinary circumstances, ruminants synthesize adequate B vitamins, and vitamins C and K. However, cobalt—a component of vitamin B-12—must be supplied in adequate amounts. Unless they are kept indoors, they usually receive sufficient exposure from direct sunlight to meet their needs for vitamin D. Young ruminants do not have fully functional rumens, with the result that the B vitamins and vitamin K should be provided in adequate amounts in feed until their rumens have developed. Also, recent evidence suggests that rumen microorganisms may not synthesize adequate amounts of niacin to meet the needs of high-producing cows in early lactation.

ADDITIVES

Many additives are used by dairy producers to increase milk production, affect milk composition, and/or improve feed efficiency; and new products are constantly evolving. Among such additives are the following; antibiotics, bovine somatotropin (BST), buffers, ionophores, and isoacids, (branched-chain fatty acids and valeric acid).

■ **Antibiotics**—Antibiotics, which are widely used in the diet of young dairy calves, are especially beneficial for calves exposed to adverse conditions of housing, sanitation, and disease. However, they should not be used as a substitute for good management and a clean, sanitary environment. The greatest benefits from feeding antibiotics accrue when calves are started on the antibiotic as soon as possible after birth, then continued on it during the milk or milk replacer feeding period. Antibiotics are mainly effective in increasing feed intake and growth, along with preventing diarrhea. Generally, antibiotics are fed at the following concentrations: In the milk replacer (dry basis), or in an equivalent amount of whole milk, 20 to 40 ppm; and in the starter ration, 10 to 20 ppm. For the prevention and control of disease, higher concentrations may be necessary: 50 to 100 ppm in the milk replacer, and 25 to 50 ppm in the starter.

Some studies have shown a slight increase in milk production when low levels of antibiotics are fed to lactating cows. However, the practice is not recommended because of (1) the presence of residual antibiotic in milk, and (2) the possibility of drug resistance in the animal.

Those using antibiotics should always read and follow the label directions on any antibiotic container before slaughtering animals or selling milk from cows treated with antibiotics.

■ **Bovine somatotropin (BST)**—Somatotropin is secreted naturally by the anterior pituitary gland, located in the skull at the base of the brain of all vertebrates. It is a peptide, and it is species-specific. Thus, bovine somatotropin (BST) and porcine somatotropin (PST) are distinctly different.

Many problems must be resolved before growth hormones become commercially available and widely used, including a practical method of administering somatotropin to animals. Since it is a protein, as is insulin, it cannot be taken orally in the feed because it would be digested in the gut.

Currently, scientists are working feverishly away perfecting and testing growth hormones. For dairy cattle, somatotropin is being tested as a stimulator of milk production, and as a growth promotant for replacement heifers.

The full impact of somatotropin on the dairy industry will not be felt until it has been used for a decade. It appears, however, that growth hormones will be one of the early significant commercial successes of DNA technology; and that they will usher in a new era in livestock production.

In 1985, Bauman and Eppard of Cornell University, and DeGeeter and Lanza of the Monsanto Company in St. Louis, reported responses of high-producing dairy cows to long-term treatment with pituitary somatotropin and recombinant somatotropin.[3] Cows injected with either type of somatotropin produced from 16 to 41% more 3.5% fat-corrected milk (FCM) than the control group. Cows injected with somatotropin consumed more feed on a body weight basis than the control group, but their relative feed energy efficiency was higher. The increased feed efficiency could be accounted for by the smaller proportion of total intake required to meet the cows' maintenance energy requirements.

Before the commercial use of bovine somatotropin can become a reality, a practical method of administering the product must be developed and FDA approval must be secured.

Some additional pertinent information relative to bovine somatotropin follows:

1. **Human safety.** BST is a natural protein produced by all cows, which is always present in milk. It is not a steroid or sex hormone, and it is not related to the steroid hormones which are used as growth promotants. There is no evidence that the BST normally found in milk is harmful when consumed by humans. Moreover, milk levels do not increase significantly following treatment. As a peptide, when the BST molecule is consumed by humans, it is broken down and inactivated in the digestive tract in the same manner as any other protein.

---

[3]Bauman, D. E., *et al.*, "Responses of High-Producing Dairy Cows to Long-Term Treatment with Pituitary Somatotropin and Recombinant Somatotropin," *Journal of Dairy Science*, Vol. 68, No. 6, 1985, p. 1352.

2. **How to administer.** Because BST is a peptide and would be broken down in the digestive tract, it cannot be fed. But daily injections do not seem practical. (In the experimental stage, it was injected daily, in the hip area.) So, a long lasting time-released injection or implant appears to be the answer.

3. **Increased feed efficiency and milk production.** In a commercial dairy and over a full lactation, bovine growth hormone may be expected to increase feed efficiency from 10 to 20%, and milk production by 10 to 25%. But scientists still do not know exactly how it works.

4. **Growth rate of heifers.** Bovine growth hormone will increase the growth rate of heifers by 8 to 10% and stimulate the development of secretory tissue in the mammary gland.

5. **No adverse effect on cows.** It appears that the growth hormone is a safe product to use on dairy animals; treated cows have about the same somatic cell count, disease incidence, post-calving behavior, and rebreeding performance as untreated cows. However, because of their higher production, treated cows will require greater feed intake and more intensive management and observation.

■ **Buffers (mineral salts)**—Buffers are used primarily to improve the feed intake, rumen function, milk production, milk composition, and health of lactating cows. When used for young calves, buffers have given inconsistent results; they will likely be most beneficial when the calf diet being fed results in higher than normal acidity.

The common buffers are: sodium bicarbonate ($NaHCO_3$), magnesium oxide (MgO), sodium bentonite, sodium sesquicarbonate, and calcium carbonate or limestone ($CaCO_3$). Buffers function to maintain the hydrogen ion concentration in the rumen, intestines, tissues, and body fluids, or to increase the rate of passage of liquids from the rumen, or both.

Buffers are of greatest benefit to cows in the following situations: (1) during early lactation; (2) when large amounts of rapidly fermentable carbohydrates are fed, especially when they are fed at infrequent intervals; (3) when fermented forage, primarily corn silage, is the major or only forage in the ration; (4) when the concentrate and forage are fed separately; (5) when the particle size of the total ration dry matter has been reduced by chopping, grinding, or pelleting to the extent that it increases the rate of ruminal fermentation and depresses salivary secretion and buffering capacity; (6) when cows are abruptly switched from high-forage to high-concentrate rations, especially during early lactation; (7) when the animal's milk fat content is low; or (8) when off-feed problems resulting from feeding rapidly fermentable feeds are encountered.

■ **Ionophores**—Ionophores are feed additives that change the metabolism within the rumen by altering the rumen microflora to favor propionic acid production. Currently, two ionophores—Bovatec (lasalocid) and Rumensin (monensin)—are FDA-approved for replacement heifers. Both are antibiotics. Feeding Bovatec or Rumensin to replacement heifers improves liveweight gains and the efficiency of feed utilization.

■ **Isoacids (branched-chain fatty acids and valeric acid)**—The isoacids provide three branch-chain fatty acids (isobutyric acid, isovaleric acid, and 2-methyl butyric acid)—the same fatty acids that are made by ruminant bacteria and are present naturally in the rumen of cattle. Isoacids are essential for the growth of some rumen organisms that digest fiber. The use of isoacids may boost milk production by 8 to 10%, or 4 to 6 lb per day, with little or no increase in feed consumption. The mode of action of isoacids is not entirely clear, but it appears to be due to enhancing fiber digestion and acetate production without stimulating insulin secretion. Not all dairy cattle benefit from the use of isoacids. Moreover, the benefits of a profitable response are delayed 30 to 60 days following the initiation of isoacids.

COMMERCIAL FEEDS

Fig. 10–12. A new and modern commercial feed mill of slip-form concrete construction. (Courtesy, Pennfield Corporation, Lancaster, PA)

Commercial dairy feeds are just what the name implies—feeds mixed by commercial manufacturers who specialize in the business. In 1988, 103.1 million tons of primary commercial feeds (complete feeds) were manufactured in the United States, and an additional 10 million tons of secondary feeds (supplements) were produced. Percentage-wise, 17.6% of the primary (complete) feeds were fed to dairy cattle.[4]

Several different types of commercial feeds are available for dairy cattle; among them, (1) complete lactating cow concentrates, (2) dry cow rations, (3) fitting rations, (4) growing or young stock rations, (5) calf starters, (6) milk replacer

---

[4]Source: American Feed Industry Association, Inc., Arlington, VA.

feeds, (7) protein supplements, and (8) mineral and vitamin premixes.

Enlightened dairy producers will know how to determine what constitutes the best in commercial feeds for their specific needs. They will not rely solely on how the feed looks and smells.

One of the important roles of commercial feed manufacturers is that of assuring a uniform mix of all ingredients, including those added in very minute amounts. Usually, commercial feed mills offer supplements which can be combined with grains, thereby coupling the advantages of both commercial and home mixing.

QUESTIONS FOR STUDY AND DISCUSSION

1. What are concentrates? How are they classified?

2. Define the term *grain*. List the leading grains that are fed to U.S. dairy cattle and give the pertinent characteristics of each.

3. What is high-moisture grain? Why is moisture important (a) when buying feeds, and (b) in formulating rations?

4. What functions do the addition of fats to feed serve?

5. Percentagewise, of the total ration, how much fat may be added to (a) lactating rations, and (b) milk replacers?

6. How and why is molasses used in dairy rations, in addition to serving as an energy feed?

7. Define (a) plant proteins, and (b) animal proteins. Give examples of each category that are commonly used in dairy cattle rations.

8. List the commonly used oilseed meals and give the pertinent characteristics of each.

9. Explain how ruminants are able to use nonprotein nitrogen. What are the three primary ways in which urea is fed?

10. Why will dairy producers feed more by-products, animal wastes, and crop residues as the world grows more populous?

11. Give pertinent information, along with examples, of each of the following by-products from plant processing: (a) fruits and vegetables, (b) fiber plants, (c) sugar crops, (d) brewing and distilling products, and (e) wood and paper products.

12. Differentiate between the terms *manure*, *litter*, and *poultry waste*.

13. How does the voiding of urine and feces differ between mammals and poultry? How does this difference affect the nitrogen content?

14. Describe each of the following ways of processing animal wastes: (a) deep-stacking, and (b) ensiling (fermentation).

15. What are crop residues? List the most common crop residues.

16. Define the terms *supplement* and *additive*.

17. What factors determine the amount of fiber to include in the ration of dairy cattle?

18. Of what importance are (a) neutral detergent fiber (NDF) and (b) acid detergent fiber (ADF) in dairy cattle rations.

19. Describe hydroponics. Would you recommend that a dairy producer buy special equipment and produce sprouted grain for lactating cows? Justify your answer.

20. Would you recommend that dairy producers provide mineral supplements to their herds? If so, would you (a) provide it free-choice, or (b) add it to the ration?

21. Where calcium alone is needed, what supplements may be used? Where phosphorus alone is needed, what supplements may be used?

22. Why did dairy producers move away from selecting and adding feed ingredients primarily because of their vitamin content?

23. What vitamins are most likely to be deficient in rations of adult dairy cattle? Under ordinary circumstances, what vitamins do ruminants synthesize?

24. Discuss the place of each of the following additives for dairy cattle: (a) antibiotics; (b) bovine somatotropin (BST); (c) buffers; (d) ionophores; and (e) isoacids.

25. What factors should be considered by a dairy producer in determining whether to use commercial feeds or home-mixed feeds?

SELECTED REFERENCES

Title of Publication	Author(s)	Publisher
Alternative Sources of Protein for Animal Production	National Research Council	National Academy of Sciences, Washington, DC, 1973
Animal Feeds	M. Gutcho	Noyes Data Corporation, Park Ridge, NJ, 1970
Animal Nutrition, Seventh Edition	L. A. Maynard, *et al.*	McGraw-Hill Book Company, New York, NY, 1979
Animal Nutrition, Third Edition	P. McDonald	Longman, London and New York, NY, 1981
Applied Animal Feeding and Nutrition, Third Edition	M. H. Jurgens	Kendall/Hunt Publishing Company, Dubuque, IA, 1974

(Continued)

SELECTED REFERENCES *(Continued)*

Title of Publication	Author(s)	Publisher
Basic Animal Nutrition and Feeding, Second Edition	D. C. Church W. G. Pond	John Wiley & Sons, New York, NY, 1982
Composition of Cereal Grains and Forages, Pub. 585	National Research Council	National Academy of Sciences, Washington, DC, 1958
Fats in Animal Nutrition	J. Wiseman	Butterworth, London, 1984
Feeds and Feeding, Third Edition	A. Cullison	Reston Publishing Company, Inc., Reston, VA, 1982
Feeds & Nutrition, Second Edition	M. E. Ensminger J. E. Oldfield W. W. Heinemann	Ensminger Publishing Co., Clovis, CA, 1990
Feeds & Nutrition Digest	M. E. Ensminger J. E. Oldfield W. W. Heinemann	Ensminger Publishing Co., Clovis, CA, 1990
Feeds for Livestock, Poultry and Pets	M. H. Gutcho	Noyes Data Corporation, Park Ridge, NJ, 1973
Handbook of Feedstuffs, The	R. Seiden W. H. Pfander	Springer Publishing Co., Inc., New York, NY, 1957
International Feed Nomenclature and Methods for Summarizing and Using Feed Data to Calculate Diets, An, Bull. 479	L. E. Harris J. M. Asplund E. W. Crampton	Agriculture Experiment Station, Utah State University, Logan, UT, 1968
Livestock Feeds and Feeding, Second Edition	D. C. Church	O & B Books, Inc., Corvallis, OR, 1984
Mineral Nutrition of Animals	V. I. Georvgievski, *et al.*	Butterworth, London, 1981
Mineral Nutrition of Livestock, The, Second Edition	E. J. Underwood	Food and Agriculture Organization of the United Nations, Rome, Italy, 1981
Nitrogen and Energy Nutrition of Ruminants	R. L. Shirley	Academic Press, Inc., Orlando, FL, 1986
Nonprotein Nitrogen in the Nutrition of Ruminants	J. K. Loosli I. W. McDonald	Food and Agriculture Organization of the United Nations, Rome, Italy, 1968
Processed Plant Protein Foodstuffs	Edited by A. M. Altschul	Academic Press, Inc., New York, NY, 1958
Processing and Utilization of Animal By-products	I. Mann	Food and Agriculture Organization of the United Nations, Rome, Italy, 1962
Protein Contribution of Feedstuffs for Ruminants	E. L. Miller I. H. Pike A. J. H. Zanes	Butterworth, London, 1982
Recent Advances in Animal Nutrition	W. Haresign D. J. A. Cole	Butterworth, London, 1985
Stockman's Handbook, The, Seventh Edition	M. E. Ensminger	Interstate Publishers, Inc., Danville, IL, 1992
Stockman's Handbook Digest	M. E. Ensminger	Interstate Publishers, Inc., Danville, IL, 1992
Urea and Non-protein Nitrogen in Ruminant Nutrition, Second Edition	Edited by H. J. Stangel	Nitrogen Division, Allied Chemical Corporation, Morristown, NJ, 1963
Urea as a Protein Supplement	Edited by M. H. Briggs	Pergamon Press, Inc., New York, NY, 1967
Use of Drugs in Animal Feeds, The, Pub. 1679	National Academy of Sciences	National Academy of Sciences, Washington, DC, 1969
Vitamins, The: Chemistry, Physiology, Pathology, Methods, Vols. I–III, Second Edition	Edited by W. H. Sebrell, Jr. R. S. Harris	Academic Press, Inc., New York, NY, 1967, 1968, 1971
Vitamins in Feeds for Livestock	F. C. Aitken R. G. Hankin	Commonwealth Agricultural Bureaux, Farnham Royal, Bucks, England, 1970
Vitamins and Hormones, Vol. XV	Edited by R. S. Harris G. F. Marrian K. V. Thimann	Academic Press, Inc., New York, NY, 1957

Fig. 11-1. Mixer wagon feeding cows in corrals. (Courtesy, Dr. Tom Schultz, Farm Advisor, Visalia, CA)

Contents

CHAPTER

11

FUNDAMENTALS
OF DAIRY
CATTLE NUTRITION

Dairy animals inherit certain genetic possibilities, but how well these potentialities develop depends upon the environment to which they are subjected; and the most important influence in the environment is nutrition. In turn, all feed comes directly or indirectly from plants which have their tops in the sun and their roots in the soil. Hence, we have the nutrition cycle as a whole—from the sun and soil, through the plant, thence to the animal, and back to the soil again.

From the above, it may be concluded that dairy nutrition is more than just feeding. Correctly speaking, *nutrition is the science of the interaction of a nutrient with some part of a living organism*. It begins with a knowledge of the fertility of the soil and the composition of plants; and it includes the ingestion of feed, the liberation of energy, the elimination of wastes, and all the syntheses essential for maintenance, growth, reproduction, and lactation.

A good understanding of nutrition is important because dairy animals are dependent upon nutrients for the processes of life. This chapter will be limited to a discussion of the fundamentals of dairy cattle nutrition.

PERSPECTIVE OF NUTRITION

The primary purpose of keeping dairy cattle is to transform feeds into milk. But this conversion must be done efficiently and economically. To do this, the principles of nutrition must be applied; and they must be augmented by superior breeding, good health, and competent management.

Like other sciences, nutrition does not stand alone. It draws heavily on the basic findings of chemistry, biochemistry, physics, microbiology, physiology, medicine, genetics, mathematics, and endocrinology, and, most recently, animal behavior, ecology, and biotechnology. In turn, it also contributes richly to each of these fields of scientific investigation.

BODY AND MILK COMPOSITION OF CATTLE

Table 11-1 shows that there is a wide range in the body composition of cattle according to age and nutritional state. Based on this table, together with other studies, the following conclusions relative to body composition may be drawn:

1. **Water.** On a percentage basis, the water content shows a marked decrease with advancing age, maturity, and

fatness. In cattle, the water content from conception to production changes as follows: embryo soon after conception (not shown in Table 11-1), 95%; newborn calf, 74%, 450-lb weaner, 69%; mature cow, 60%; and Choice grade steer, 53.5%.

2. **Fat.** The percentage of fat increases with growth and fattening.

3. **Fat and water.** As the percentage of fat increases, the percentage of water decreases.

4. **Protein.** The percentage of protein remains rather constant during growth, but decreases as the animal fattens.

On the average, there are 3 to 4 lb of water per 1 lb of protein in the body.

5. **Ash.** The percentage of ash shows the least change.

6. **Composition of gain.** The data presented in Table 11-1 clearly indicate that gain in weight does not provide an accurate measure of the actual gain in energy of the animal, because it tells nothing about the composition of gain.

TABLE 11-1
BODY AND MILK COMPOSITION OF CATTLE[1]

Cattle: Age or Status	Weight		Water	Fat	Protein	Ash
	(lb)	(kg)	(%)	(%)	(%)	(%)
Calf, newborn ..	70	31.8	74.4	2.5	19.0	4.1
Calf, weaner ...	450	204.5	69.0	9.0	18.0	4.0
Cow, breeding condition	1,100	500.0	60.0	18.0	17.5	4.5
Steer, Choice grade	1,050	477.3	53.5	26.0	17.0	3.5
Milk[2]	—	—	87.7	3.6	3.3	0.8

[1]Prepared by the author from numerous sources. Body composition ingesta-free (empty) basis.

[2]Cow's milk also contains about 4.7% lactose.

Also, the chemical composition of the body varies widely between organs and tissues and is more or less localized according to function. Thus, water is an essential of every part of the body, but the percentage composition varies greatly in different body parts; blood plasma contains 90 to 92% water; muscle, 72 to 78%; bone, 45%; and the enamel of the teeth, only 5%. Proteins are the principal constituents, other than water, of muscles, tendons, and connective tissues. Most of the fat is localized under the skin, near the kidneys, and around the intestines. But it is also present in the muscles (known as marbling in a carcass), bones, and elsewhere.

Table 11-1 does not reveal the very small amount of carbohydrates (mostly glucose and glycogen) present in the bodies of animals and found principally in the liver, muscles, and blood. Although these carbohydrates are very important in animal nutrition, they account for less than 1% of the body composition.

Table 11-1 also shows the composition of milk. It remains relatively constant, although certain feeding and management practices may alter its composition somewhat. For additional information on the production and composition of milk, the reader is referred to Chapter 18, Milk Secretion and Handling.

PHYSIOLOGY OF DIGESTION

In order fully to understand how the various nutrients are used in production, the dairy producer must first have a working knowledge of the anatomy of the digestive tract of the cow and how it functions. A nutritionally complete feed is not enough; the feed must be readily consumed, digested, and absorbed.

RUMINANT DIGESTIVE SYSTEM

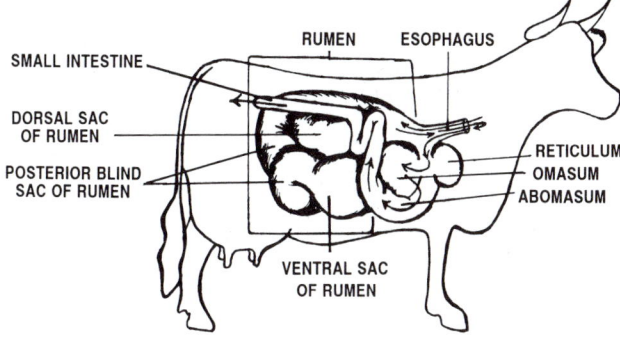

Fig. 11–2. Location and parts of the ruminant stomach (four compartments), with pathway of feeds indicated by arrows.

Cattle are ruminants. They differ from nonruminant animals (e.g., horse, pig, and human) in the following important ways:

1. **Mouth.** Ruminants have no upper incisor or canine teeth. Thus, they depend on the upper dental pad and lower incisors, along with the lips and tongue, for the prehension of feed.

2. **Four stomach compartments.** Ruminants possess four stomach compartments—rumen, reticulum, omasum, and abomasum (true stomach)—whereas monogastrics have one. Such a digestive system makes for two primary nutritional differences between ruminants and simple-stomached animals:

   a. **More space.** They have the necessary space for processing large quantities of bulky forages to provide their nutrients. The cow, for example, when compared to a human on a proportion-to-weight basis, has about nine times the digestive tract capacity.

   b. **More microorganisms.** The rumen provides a highly desirable environment for the enormous population of microorganisms. Typical counts of rumen bacteria range from 25 to 50 billion/milliliter, and typical counts of protozoa range from 200,000 to 500,000/milliliter. The number of rumen bacteria varies according to the nature of the diet, feeding regimen, time of sampling after feeding, species differences, individual animal differences, season, availability of green feed, and the presence or absence of ciliate protozoa.

   Rumen microorganisms serve two important functions:

   (1) They make it possible for ruminants to utilize roughage—to digest the fiber therein. They break down the cellulose and pentosans of feeds into usable organic acids, chiefly acetic, propionic, and butyric acid—commonly called the volatile fatty acids (VFA). These VFAs are largely absorbed through the rumen wall and provide the ruminant 60 to 80% of its energy needs. Microbial digestion is of great practical importance in the nutrition of ruminants; it is the fundamental reason why they can be maintained chiefly on roughages.

   (2) In exchange for their rumen-housing privileges, the microbes synthesize nutrients for their host, in a true type of symbiotic relationship. Rumen microbes synthesize, or manufacture, all the B complex vitamins and all the essential amino acids. The latter can even be made from nonprotein nitrogen compounds (NPN), such as urea or ammoniated products, or from proteins that are deficient in one or more of the amino acids. Finally, the microorganisms give their life to their host in payment for food and shelter, being digested further along in the gastrointestinal tract.

3. **Rumination.** A placid cow lying under a tree slowly chewing her cud conveys a special sense of contentment, symbolic of the tranquility of the countryside. But this activity, or phenomenon, which is peculiar to ruminants, is of great practical significance.

During rumination, the animal regurgitates and rechews a soft mass of coarse feed particles, called a *bolus*. Each bolus is chewed for about a minute, then swallowed again. Ruminants may spend 8 hours or more per day in rumination, the amount of time varying according to the nature of the diet. Coarse, fibrous diets result in more time ruminating. Rechewing does not improve digestibility. Rather, rumination has an important bearing on the amount of feed the animal can eat and utilize. Feed particle size must be reduced to allow passage of the material from the rumen. Because high-quality forages contain less fiber than low-quality forages, they require much less rechewing and pass out of the rumen at a faster rate; hence, they allow the cow to eat more.

4. **Eructation (belching of gas).** Substantially more gas is produced in digestion by ruminants than by simple-stomached animals. The microbial fermentation in the rumen results in the production of gases (primarily CO_2 and methane) which must be eliminated; otherwise, bloat results. Normally, these gases are expelled quite freely by eructation (belching) and, to a lesser extent, by absorption into the blood draining from the rumen, from which they are eliminated through exhaled air from the lungs.

5. **Stomach of the newborn.** When the calf is born, the rumen is small and the fourth stomach is by far the largest of the compartments. Thus, digestion in the young calf is more like that of a simple-stomached animal than that of a ruminant. The milk which the calf normally consumes bypasses the first two compartments by way of the esophageal groove and goes almost directly to the fourth stomach in which the rennin and other compounds for the digestion of milk are produced. If the calf gulps too rapidly, or gorges itself, the milk may go into the rumen where it is not digested properly and may cause upsets of the calf's digestive system. As the calf nibbles at hay, small amounts of material get into the rumen. When certain bacteria become established, the rumen develops and the calf gradually becomes a full-fledged ruminant.

CAPACITY OF THE DIGESTIVE TRACT

TABLE 11–2
PARTS AND AVERAGE CAPACITIES
OF DIGESTIVE TRACT OF CATTLE

Gastric Compartment	Capacity in Cattle	
	(gal)	(liter)
Rumen (paunch)	53.4	202
Reticulum (honeycomb)	2.0	8
Omasum (manyplies)	5.0	19
Abomasum (true stomach)	6.1	23
Subtotal	66.5	252
Small intestine	17.4	66
Cecum	2.6	10
Large intestine	7.4	28
Total	93.9	356

Due to the anatomic adaptations of the various livestock species, the relative importance of the various digestive organs is reflected in their respective capacities. In ruminants, the stomach (consisting of four compartments—rumen, reticulum, omasum, and abomasum) has the largest capacity (see Table 11–2).

PROCESS OF DIGESTION

Digestion, taken in a narrow sense of the word, can be defined as the process whereby proteins, fats, and complex carbohydrates are broken down into units that are of small enough size to be absorbed. This process is accomplished primarily through the action of digestive enzymes.

Enzymes are organic catalysts produced by certain cells within the body which speed biochemical reactions at ordinary body temperatures without being used up in the process. Enzymatic activity is responsible for most of the chemical changes occurring in feeds as they move through the digestive tract. A summary of the enzymes involved in the digestive process of animals is presented in Table 11–3.

TABLE 11–3
DIGESTIVE PROCESSES IN ANIMALS

Region	Secretion (Secreted By)	Enzyme	Enzyme Acts On, or Function	End Product of Digestion	Comments
Mouth	Saliva (salivary glands).	Amylase (ptyalin). Maltase.	Starch, dextrins. Maltose.	Maltose and dextrins. Glucose.	Saliva lubricates food. In ruminants, the buffer salts of saliva help control acidity in the stomach.
Rumen		Enzymes from microorganisms.	Cellulose, polysaccharides, starches, sugars, fats, proteins (urea).	Volatile fatty acids. Microbial protein. B vitamins. Vitamin K.	
Abomasum	Gastric juice and acids (chiefly HCl) (walls of stomach). Mucus.	Pepsin. Lipase (in carnivores). Amylase.	Protein. Fat.	Proteoses, polypeptides, peptides. Higher fatty acids and glycerol. Coating of stomach lining and lubrication of food.	
Nursing Animals	Gastric juice (walls of stomach).	Renin.	Milk protein (casein).	Coagulates milk protein.	
Duodenum (small intestine)	Pancreatic juice (pancreas).	Trypsin. Chymotrypsin. Amylopsin (amylase). Steapsin (lipase). Carboxypeptidase. Collagenase. Cholesterol esterase.	Proteins, proteoses, peptones, and peptides. Starch, dextrins. Fats. Peptides. Collagen. Cholesterol.	Peptones, peptides. Amino acids. Maltose, dextrins. Higher fatty acids and glycerol. Amino acids and peptides. Peptides. Cholesterol esterified with fatty acids.	Low in ruminants.
	Bile (liver).		Fats.	Emulsion of fats (soap, glycerol).	
Small intestine	Intestinal juice (secreted by intestinal wall).	Peptidase (erepsin). Sucrase (invertase). Maltase. Lactase. Polynucleotidase.	Peptides. Sucrose. Maltose. Lactose. Nucleic acid.	Amino acids and dipeptides. Glucose and fructose. Glucose. Glucose and galactose. Mononucleotides.	Very low in ruminants. Low in ruminants. High in young mammals.
Large intestine (cecum and colon)		Cellulase from microorganisms.	Cellulose, polysaccharides, starches, sugars.	Volatile fatty acids. Microbial protein. B vitamins. Vitamin K.	

PHYSIOLOGICAL PROCESSES

The discussion of the physiology of digestion will be divided into sections dealing with the regions of digestion instead of individual organs. These regions are the oral region, pharyngeal and esophageal region, gastric region, pancreatic region, hepatic region, and intestinal and cecum-colon region.

ORAL REGION

Three physical processes occur in the oral region of cattle—prehension, mastication, and the initiation of deglutition.

Prehension can be defined as the act of bringing food into the mouth. Numerous modes of prehension can be found in animals. The cow relies on structures of the mouth, such as the tongue, lips, and teeth.

Mastication is the act of chewing food. It involves the physical grinding and tearing of the food in addition to the admixture of saliva which lubricates the food as well as initiates a limited amount of enzymatic digestion. *Food that has been masticated and formed into a small compact ball for passage down the digestive tract is called a bolus.*

Deglutition is the act of swallowing. This process involves both voluntary and involuntary reflexes. Upon completion of mastication, the bolus is lifted by the tongue and moved to the back of the mouth. The bolus passes through the pharynx, causing a temporary inhibition of respiration by the reflex closure of the larynx, and finally down the esophagus to the gastric region.

TEETH

The teeth serve primarily as a mechanical aid for mastication. By tearing and grinding the food, they provide a means whereby a large surface area is created which can be exposed effectively to the digestive fluids of the tract. Cattle, being entirely herbivorous (plant eaters), do not require many teeth for the tearing of food; hence, they have no canine teeth, and incisors are found only on the lower jaw where they are used for shearing forages in prehension.

TONGUE

In cattle, the tongue is the primary structure for prehension. The tongue is elongated and covered with rough papillae, making it adapted to wrapping around grass and other forages. The cow then brings the forage into the oral cavity where it is sheared by the movement of the incisors against the dental pad.

Throughout the process of mastication, the tongue serves a threefold purpose. First, movement of the tongue transports the feed to the various areas of the mouth to be torn and ground. While doing this, the tongue is also mixing the feed with the various secretions of the mouth, ultimately forming a bolus. Secondly, the presence of taste buds on the tongue provides a neurological control for feed selection and intake. If the feed is bitter or unpalatable, impulses from the taste buds signal the animal to stop eating. Conversely, a desirable taste stimulates appetite. In the cow, the tip of the tongue contains a large number of taste buds while the middle portion has very few. The back portion of the tongue contains the highest density of taste buds. Finally, the tongue initiates the process of deglutition. When the bolus has been adequately prepared, the tongue moves it to the back of the mouth where neural receptors are stimulated and swallowing commences.

SALIVARY GLANDS

The salivary glands represent a network of accessory structures which are essential to digestion. Three pairs of salivary glands are of primary importance—parotid, submaxillary, and sublingual.

The uses of saliva in digestion are manifold, including the following:

1. **Lubricant.** These secretions act as aids in mastication, the formation of the bolus, and swallowing. Without this moisture, swallowing would be extremely difficult.

2. **Buffering capacity.** A large quantity of bicarbonate is secreted in saliva, thus serving as a buffer in the ingesta.

3. **Nutrients for rumen microorganisms.** Saliva contains considerable amounts of urea, mucin, phosphorus, magnesium, and chloride—all of which can be readily utilized by the bacteria and protozoa in the rumen.

4. **Prevention of frothing.** Gas can accumulate in the rumen and cause serious bloating if the eructation process is impaired. Saliva—acting as a surfactant—helps to prevent this problem.

5. **Taste.** Saliva solubilizes a number of the chemicals in the feed which, once in solution, can be detected by the taste buds.

6. **Protection.** The membranes within the mouth must be kept moist in order to remain viable. Saliva provides one means by which this is accomplished.

PHARYNGEAL AND ESOPHAGEAL REGION

The pharynx is the structure which controls the passage of air and feed. In this organ, the openings of the mouth, esophagus, posterior nares, Eustachian tubes, and larynx come together. During the act of swallowing, the opening into the larynx is reflexly closed by the arytenoid cartilages; and the epiglottis is passively folded over the opening of the larynx. This forces food into the esophagus, thus preventing any feed from passing into the respiratory tract.

The esophagus is a muscular tube extending from the pharynx to the cardia of the stomach. The musculature and innervation of the esophagus are such that peristaltic waves move the bolus. *Peristalsis is the coordinated contraction and relaxation of smooth muscles creating an unidirectional movement which pushes the bolus through the digestive tract.*

GASTRIC REGION

The primary differences in digestion among domestic livestock can be traced to the specialized development of the gastric region. The ruminant has been described as having four stomachs. In reality, the ruminant possesses a complex stomach consisting of four morphologically distinct compartments. These compartments are rumen, reticulum, omasum, and abomasum.

RUMEN AND RETICULUM

The rumen and reticulum are closely related as to physiological function and are often discussed together. The esophagus empties into the atrium ventriculi, a convex area formed by both the rumen and reticulum. The rumen is an extremely large compartment in the adult ruminant lined with a large number of papillae that increase the surface area for the churning of digested material and absorption. The reticulum, a structure which has an interior that very much resembles a honeycomb, acts as a collection compartment for foreign objects as well as an organ for digestion.

Ingested feed passes into these two compartments and is digested thoroughly through the action of various microorganisms (bacteria and protozoa) present in the rumen. The rumen, in effect, is a large physiological fermentation vat.

The microbes of the rumen digest carbohydrates to produce carbon dioxide and volatile fatty acids. Although a number of volatile fatty acids are produced, the vast majority of these end products are acetate, propionate, and butyrate. These products are then absorbed from the rumen and supply much of the energy required by the animal. Quite often, when high-concentrate rations are used, large quantities of lactic acid are produced and the pH of the rumen falls. Since most of the bacteria in the rumen are pH sensitive, any dramatic shift in pH will alter the proportions of the various types of microorganisms. When the ruminal pH drops too low, the animal goes off feed—a symptom of acute digestive problems.

Lipids are degraded by the ruminal microbes to fatty acids and glycerol. Glycerol is then primarily converted to propionate, with the long-chain fatty acids passing down to the small intestine for absorption.

Very few dietary proteins escape the degradation process of the rumen. The degree to which dietary protein is degraded is dependent on its solubility. A highly soluble protein will be rapidly degraded while a highly insoluble protein will probably leave the rumen relatively intact. Most dietary proteins are metabolized by the microorganisms and are incorporated as microbial protein. The microbes are then passed down the tract, which, through their own degradation, provide protein for the animal. With the degradation of the various dietary proteins, ammonia is produced in the rumen which can then either be absorbed through the rumen wall or provide nitrogenous precursors for the synthesis of bacterial protein. If the ration is high in sugars and starches, ammonia concentration is depressed.

Vitamin K and the B complex vitamins are all synthesized by the ruminal microbes. The young calf must obtain these vitamins from outside sources, but milk generally supplies the young calf's needs. Vitamin C is synthesized at the tissue level in ruminants. When young calves start to feed, the reticulo-rumen complex is not very well developed, and the population of microflora of the rumen is very small. An anatomic adaptation of the rumen and reticulum, called the esophageal groove, enables milk, through a series of reflex actions in the suckling calf, completely to bypass the rumen and reticulum and most of the omasum. Thus, milk rapidly passes into the last compartment of the gastric region—the abomasum. Here, the milk is coagulated through the action of the enzyme rennin, and digestion and absorption proceeds very much like that in the nonruminant. As the young calf grows older, it ingests increasingly larger quantities of solid feed which, in turn, stimulate the growth and development of the reticulum and rumen. Through its contact with the environment and other cattle, the young calf becomes naturally inoculated with microorganisms which benefit digestion.

Throughout the fermentation process in the rumen, various gases are produced and expelled through eructation (belching). Methane and carbon dioxide are the two gases produced most abundantly in the rumen. It has been estimated that methane constitutes 30 to 40% of the total rumen gas volume, and carbon dioxide 20 to 65%.

OMASUM

The omasum, or manyplies, is the next compartment for digestion. It contains numerous laminae (tissue leaves) that help grind ingesta. The exact physiological function of this compartment has not been fully elucidated, but many researchers feel that it serves to absorb water in addition to its function of grinding ingesta.

ABOMASUM

This compartment is analogous to the stomach of the nonruminant. It is the only compartment of the gastric region of the ruminant containing digestive glands. Digestive processes of this compartment are very similar to those of the stomach in the nonruminant.

PANCREATIC REGION

The pancreatic region involves the pancreas and the pancreatic duct—a duct leading from the pancreas to the small intestine.

The pancreas, an accessory organ of digestion, is a glandular structure that plays an essential role in the digestive physiology of animals. The pancreas—being both an endocrine and exocrine gland—serves two physiologically distinct functions. The endocrine function is that of the secretion of the hormones, insulin and glucagon. The exocrine function deals with the production and secretion of fluids that are necessary for digestion within the small intestine.

HEPATIC REGION

The hepatic region incorporates the liver, gallbladder, and bile duct.

In addition to the pancreas and salivary glands, the liver is an indispensable accessory organ of the gastrointestinal tract. From the stomach and small intestine, most of the absorbed nutrients travel through the portal vein to the liver—the largest gland in the body. The liver not only plays an important part in nutrient metabolism and storage, but also forms bile, a fluid essential for lipid absorption in the small intestine. The numerous psychological functions of the liver are:

1. Secretion of bile.
2. Detoxification of harmful compounds.
3. Metabolism of proteins, carbohydrates, and lipids.
4. Storage of vitamins.

5. Storage of carbohydrates.
6. Destruction of red blood cells.
7. Formation of plasma proteins.
8. Inactivation of polypeptide hormones.
9. Urea formation.

The primary role of the liver in digestion and absorption is the production of bile. Bile facilitates the solubilization and absorption of dietary fats and also aids in the excretion of certain waste products such as cholesterol and by-products of hemoglobin. The greenish color of bile is due to the end products of red blood cell destruction—biliverdin and bilirubin. Bile contains a number of salts resulting from the combination of sodium and potassium with bile acids. These salts combine with lipids in the small intestine to form micelles. *Micelles are colloidal complexes of monoglycerides and insoluble fatty acids that have been emulsified and solubilized for absorption.* When the micelle has been formed, the lipid can be digested and the resulting products (fatty acids and glycerol) can cross the mucosal barrier of the small intestine and enter the lymphatic system. Bile salts, however, do not travel with the lipid; rather, they are recycled into the enterohepatic circulation.

The volume of bile production is highly variable. An animal that has been starved produces little bile. Conversely, an animal that is fed a high-fat ration will produce substantial quantities of bile in order to keep up with absorptive requirements. Generally, the volume of bile is dependent on (1) blood flow, (2) nutritive state of the animal, (3) type of ration being fed, and (4) the enterohepatic bile salt circulation.

INTESTINAL AND CECUM-COLON REGION

The small intestine is divided anatomically into three sections—duodenum, jejunum, and ileum. The first segment, the duodenum, originates at the pyloric sphincter of the stomach and is closely attached to the body wall by a short mesentery. Both bile and pancreatic fluids are emptied into this segment. The next section is the jejunum. There is no clear demarcation between the jejunum and the ileum, but it is arbitrarily defined as the free border of the ileocecal fold.

Throughout the luminal surface of the small intestine lies an extensive network of fingerlike projections called villi. Each villus contains a lymph vessel called a lacteal and a series of capillary vessels. On the surface of the villi are a great number of microvilli which provide further surface area for absorption.

The small intestine terminates at the ileocecal valve—a sphincter that controls the flow of ingesta from the small intestine into the cecum and large intestine. This structure prevents the backflow of ingesta into the small intestine.

The cecum and colon in cattle are composed of several layers of muscle. There is a circular layer of muscle that forms the basic tube of the colon and facilitates movement. In addition to this layer of muscle, there are three strips of longitudinal muscle which form the *taenia coli*. These strips form a series of pouches or sacculations throughout the colon which are called *haustrae*. Ingesta is held in these saclike structures to facilitate the removal of water. Numerous mucous-secreting goblet cells can be found in the colon, but villi, such as the type that are found in the small intestine, are absent.

At the proximal end of the colon is a blind sac called the cecum. The cecum of the cow is not very well developed and plays a rather insignificant role in digestion. However, there is some absorption of volatile fatty acids in the cecum, and considerable amounts of water and electrolytes are absorbed in the colon.

CLASSIFICATION OF NUTRIENTS

Dairy cattle do not utilize feeds as such. Rather, they use those portions of feeds called *nutrients* that are released by digestion, then absorbed into the body fluids and tissues.

Nutrients are those substances, usually obtained from feeds, which can be used by the animal when made available in a suitable form to its cells, organs, and tissues. They include carbohydrates, fats, proteins, minerals, vitamins, and water. (More correctly speaking, the term *nutrient* refers to the more than 40 nutrient chemicals, including amino acids, minerals, and vitamins.) Energy is frequently listed with nutrients, since it results from the metabolism of carbohydrates, proteins, and fats in the body.

FUNCTIONS OF NUTRIENTS

Of the feed consumed, a portion is digested and absorbed for use by the animal. The remaining undigested portion is excreted and constitutes the major portion of the feces. Nutrients from the digested feed are used for a number of different body processes, the exact usage varying with the species, class, age, and productivity of the animal. All animals use a portion of their absorbed nutrients to carry on essential functions, such as body metabolism and maintaining body temperature and the replacement and repair of body cells and tissues. These uses of nutrients are referred to as *maintenance*. That portion of digested feed used for growth or the production of milk is known as *production requirements*. Another portion of the nutrients is used for the development of the fetus and is referred to as *reproduction requirements*.

Based on the quantity of nutrients needed daily for different purposes, nutrient demands may be classed as high, low, variable, or intermediate.

Requirements for milk are considered *high-demand uses*, whereas early gestation is a *low-demand use*. The last stages of pregnancy have variable requirements. Growth may be classed as intermediate in nutrient demands.

MAINTENANCE

Cattle, unlike machines, are never idle. They use nutrients to keep their bodies functioning every hour of every day, even when they are not being used for production.

Maintenance requirements may be defined as the combination of nutrients which are needed by the animal to keep its body functioning without any gain or loss in body weight or any productive activity. Although these requirements are relatively simple, they are essential for life itself. Cattle must have (1) heat to maintain body temperature, (2) sufficient energy to keep vital body processes functional, (3) energy for minimal movement, and (4) the necessary nutrients to repair damaged cells and tissues and to replace those which have become nonfunctional. Thus, energy is the primary nutritive need for maintenance. Even though the quantity of other nutrients required for maintenance is relatively small, it is necessary to have a balance of the essential proteins, minerals, and vitamins.

No matter how quietly cattle may be lying in a stall or in a pasture, they require a certain amount of fuel and other nutrients. The least amount on which an animal can exist is called its *basal maintenance requirement*.

There are only a few times in the normal life of cattle when only the maintenance requirement needs to be met. Such a status is closely approached by mature males not in service and by mature, dry non-pregnant females. Nevertheless, maintenance is the standard benchmark or reference point for evaluating nutritional needs.

Although nutrient needs are minimal during maintenance, it is noteworthy that one-third to one-half of the feed consumed by cattle as a whole is used to meet the maintenance requirement. Of course, on an individual basis, the higher the production of a lactating cow, the smaller the proportion of nutrients needed for maintenance (see Fig. 11–3).

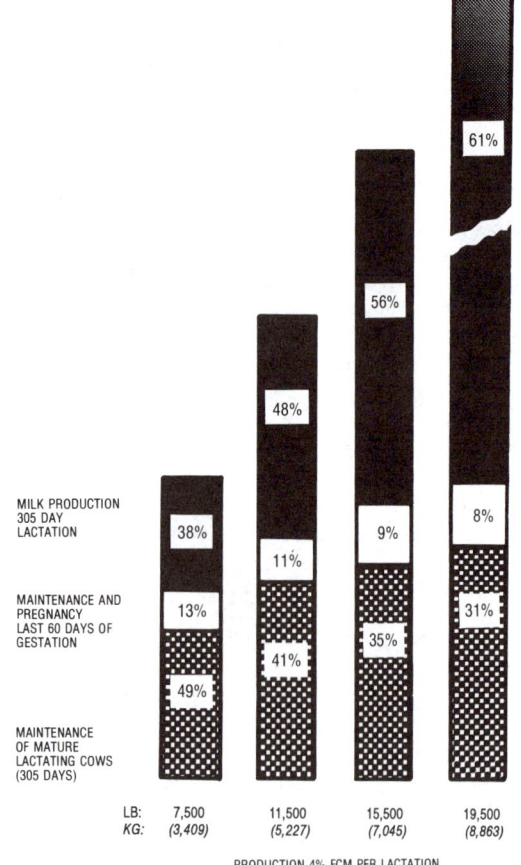

Fig. 11–3. Relative proportions of feed used by a 1,430-lb (*650 kg*) cow for (1) maintenance, (2) maintenance and pregnancy, and (3) milk production, at levels of 7,500, 11,500, 15,500, and 19,500 lb 4% FCM milk. Note that the percentage of feed used for maintenance decreases as production increases.

FACTORS AFFECTING MAINTENANCE REQUIREMENTS

Even though maintenance requirements might be considered as an expression of the nonproduction needs of cattle, there are many factors which affect the amount of nutrients necessary for this vital function; among them, (1) exercise, (2) weather, (3) stress, (4) health, (5) body size, (6) temperament, (7) individual variation, (8) level of production, and (9) lactation. The first four are *external factors*—they are subject to control to some degree through management and facilities. The others are *internal factors*—they are part of the animal itself. Both external and internal factors influence requirements according to their intensity. For example, the colder or hotter it gets from the most comfortable (optimum) temperature, the greater will be the maintenance requirements.

Other factors that affect the maintenance requirements are:

1. **Mature size of breed.** Larger breeds grow more rapidly than smaller breeds; hence, they have a higher maintenance requirement.

2. **Sex.** Young males gain more rapidly and have a higher maintenance requirement than young females.

GROWTH

Growth may be defined as the increase in size of bones, muscles, internal organs, and other parts of the body. It is the normal process before birth, and after birth until the animal reaches its full, mature size. Growth is influenced primarily by nutrient intake. The nutritive requirements become increasingly acute when young animals are under forced production, such as when heifers are bred to calve as two-year-olds.

Growth is the very foundation of animal production. Breeding females may have their reproductive ability seriously impaired if they have been improperly grown. Likewise, one cannot expect the most satisfactory yields of milk from dairy cows unless they were well developed during their growing period.

Generally speaking, organs vital for the maintenance of life—e.g., the brain, which coordinates body activities, and the gut, upon which the rest of the postnatal growth depends—are early developing; and the commercially more valuable parts, such as muscle, fat, and udder, develop later. However, not all gut is early developing; for example, the growth and functioning of the ruminant stomach is delayed.

The nutritive needs for growth vary with age, breed, sex, and health.

REPRODUCTION

Being born and born alive are the first and most important requisites of dairy production, for if cows fail to reproduce the dairy producer is soon out of business. A "mating of the gods," involving the greatest genes in the world, is of no value unless these genes result in (1) the successful joining of the sperm and egg, and (2) the birth of live offspring. It has been estimated that 20 to 50% of all cattle matings are infertile, that 25% of all cows culled from dairy herds are removed because of reproductive inefficiency, and that the overall average U.S. calf crop of all cattle (beef and dairy combined) is only 88%—the other 12% abort or are sterile. Certainly, there are many causes of reproductive failure, but scientists are agreed that nutritional inadequacies play a major role.

NUTRITIONAL REPRODUCTIVE FAILURE

Since dairy producers largely determine their own destiny when it comes to feeding, it is important that they know the causes of reproductive failure and how to rectify them.

Research gives ample evidence that the real cause of most reproductive failure is a deficiency of one or more nutrients just before or immediately following parturition—nutritive deficiencies during the critical period when life begins—a deficiency of energy, protein, minerals, and/or vitamins. Based on an extensive review of the voluminous literature, the author evolved with the following summary of the nutritional causes of reproductive failure in cattle:

1. **Overfeeding or underfeeding.** Overfeeding, accompanied by extremely high condition, or underfeeding, accompanied by emaciated and run-down condition, usually results in temporary sterility. Overfat females often experience birth difficulties. Excessive thinness results in low birth weights and weak young.

2. **Energy.** A low level of energy during the last third of pregnancy and immediately following parturition will have a marked adverse effect on rebreeding—fewer females will come in heat at the beginning of the breeding season, and fewer will conceive.

3. **Protein.** A low level of protein during gestation results in lowered reproduction, lighter birth weights, and delayed heat following parturition.

4. **Iodine.** A deficiency of iodine will cause impaired reproduction—weak or dead offspring at birth and big-necked (goitrous) calves.

5. **Vitamin A.** Low vitamin A will result in the birth of weak, malformed, partially blind, or dead young.

Most of the growth of the fetus occurs during the last third of pregnancy. Additionally, the female must store body reserves during pregnancy, because the demands for milk production are generally greater than can be supplied by the ration fed during early lactation. Hence, the nutrient requirements are very critical during this period, especially for young pregnant females.

It is also known that the ration exerts a powerful effect on sperm production and semen quality. There is abundant evidence that greater fertility of bulls exists under conditions where a well-balanced ration is provided.

LACTATION

Simply stated, milk production is a by-product of the reproductive process. The lactation requirements for moderate to heavy milk production are much more rigorous than the maintenance or pregnancy requirements. Fortunately, heavy milking cows can store up body reserves of certain nutrients before and during the pregnancy period, to be drawn upon following parturition. Here is how this phenomenon works: When properly fed before and during pregnancy, certain nutrient deposits are made in the body. Then, during lactation when the demands are greater than can be obtained from the feed, the cow draws from the stored body reserves. Thus, both calcium and phosphorus can be stored in the bones, then withdrawn during early lactation when milk production is at its peak. Of course, if there has not been proper body storage, something must "give"—and that something will be the mother, for nature ordained that growth of the fetus, and the lactation that follows, shall take priority over the maternal requirements. Hence, when there is a nutrient deficiency, the female's body will be deprived, or even stunted if she is young, before the developing fetus or milk production will be materially affected.

The nutrient needs for lactation depend on the amount and composition of milk secreted. Although high-producing cows require more total feed than low producers, they utilize proportionately more nutrients for milk production, and generally they return more net income over feed cost.

One of the most dramatic changes in the life cycle of cows is that which occurs at freshening, when a female suddenly makes the transition from nonlactating to lactating. In high-producing dairy cows, clinical milk fever at freshening is an indication of a very sudden lowering of the level of calcium in the blood serum, and putting it into the milk. If the reversal process is too drastic, the animal may be thrown into a blood calcium deficiency (hypocalcemia), followed by slowing of the heart and going into a coma.

CONDITIONING (FITTING)

Conditioning is what the name implies—the laying on of fat, especially in the tissues of the abdominal cavity and in the connective tissues just under the skin and between the muscles.

The objective of dairy beef producers is to finish animals to the degree of fleshing and carcass weight desired by the consumers, at a maximum of profit for their efforts. It is the normal feeding practice followed with dairy steers prior to slaughter, for the purpose of improving the flavor, tenderness, and quality of meat, better to meet consumer demands. Fattening is usually achieved through the use of high-energy feeds, carbohydrates and fats—a liberal allowance of grains.

Fitting is the conditioning of animals, usually for show or sale, through careful feeding, grooming, and exercising, to enhance their bloom and attractiveness. Fitting dairy animals for show or sale involves the application of similar principles and practices to those followed in fattening (conditioning) cattle for market. Animals intended for show or sale should be fed so as to achieve a certain amount of finish or bloom, but they should not be too fat. In general, most fitting rations are similar to the rations used in commercial fattening operations for animals of comparable ages, except that they are usually higher in protein content; experienced caretakers feel that they get more bloom by use of high-protein rations. Also, it is common practice to feed a palatable milk replacer to young animals that are being fitted for show or sale.

NUTRITIVE NEEDS OF DAIRY CATTLE

The first consideration in any dairy feeding program is to determine the nutritive needs for body maintenance, growth, pregnancy or reproduction, and milk production.

The body uses nutrients on a priority basis, with maintenance, growth, and pregnancy generally higher in priority than milk production. There are exceptions to this rule, as there are with most rules. Thus, some cows will produce milk at the expense of their maintenance and growth.

When cows are underfed, they cannot produce at their most profitable level. It is usually better, therefore, to feed them on the generous side than to feed so little that production drops to uneconomical levels. Of course, it is also unprofitable to waste nutrients. So, cows that put on flesh and fail to return it later in increased milk production should be fed somewhat different amounts and types of feeds than cows which have a higher priority for milk production.

ENERGY

Cows use energy for a variety of functions. A certain amount is used for body maintenance; heifers need energy for growth; pregnant cows need additional energy for development of the fetus; and lactating cows require energy to produce milk. For optimal milk production, minimum health disorders, and optimum reproductive efficiency, cows must not be too fat or too thin.

Lack of energy is the most common deficiency of dairy rations. In young animals, an insufficient supply of energy results in retarded growth and a delay in the onset of puberty. In lactating cows, it results in a decline in milk yields and a loss in liveweight; and severe and prolonged energy deficiency also depresses reproductive performance.

Most of the energy required is supplied by carbohydrates, although fats and protein are also used as energy. All cows, except low-producing ones—those producing less than 25 to 35 lb of milk per day, need some grain if they are to produce at top levels.

■ **Carbohydrates**—Carbohydrates are the major source of energy for dairy cattle. They constitute 50 to 80% of the dry matter of forages and grains. Three major categories of carbohydrates exist in feeds: (1) simple sugars (glucose or sucrose), (2) stored carbohydrates (starch), and (3) structural carbohydrate or fiber (cellulose and hemicellulose). Sugars are found in the cells of growing plants and in such feeds as molasses. Starch is the main component of grain. Cellulose and hemicellulose, which are classified as fiber, are made up of sugar molecules, as is starch, but they are bound together differently. Adult ruminants can digest fiber because the microbial population in the rumen breaks it down into usable products. However, lignin, which is not a true carbohydrate, is virtually indigestible.

■ **Fat**—Rations for baby calves, with large quantities of milk or milk replacer, may contain 15 to 35% fat in the dry matter consumed. Fat is mainly used in the rations of young calves, but it may be added to the ration of lactating cows to increase energy and reduce feed dustiness. In addition to the fat present in natural feedstuffs (generally less than 2 to 3%), dairy cows are able to utilize 1 to 1½ lb of fat per day; which

translates into about 5 to 6% added fat to the concentrate (grain) ration, or 3% added fat to the total mixed ration (grain and forage combined). Fat sources high in unprotected polyunsaturated fatty acids (soybean oil or corn oil) exert a greater negative effect on rumen microbes and depress fiber digestibility more than saturated fatty acid sources (tallow). Whole oilseeds (soybeans, sunflowers, or cottonseed) are good fat sources because they are slowly digested in the rumen, with the result that the oil is slowly released. Also, feeding rumen-protected fats that are resistant to microbial action in the rumen may increase milk fat percentage.

The NRC energy requirements for dairy cattle, which are presented in this book in Chapter 12, Tables 12–2 to 12–6, are expressed as total digestible nutrients (TDN), digestible energy (DE), metabolizable energy (ME), net energy for maintenance (NE_m), net energy for body gain (NE_g), and net energy for lactation (NE_{lc}). Separate energy values for each maintenance (NE_m) and gain (NE_g) are given because animals use energy for maintenance more efficiently than for growth. However, the efficiency of energy use by lactating cows for maintenance, pregnancy, and milk production is similar; so, only one energy value, net energy for lactation (NE_{lc}), is used for these functions (see Table 12–4).

The energy value of a feed may be separated into: (1) the losses that occur in digestion and metabolism, and (2) the net energy (NE) that is available to the animal for maintenance and production. The total energy in feed, which is determined by complete oxidation (burning) of the feedstuff and measurement of the heat produced, is known as *gross energy* and is expressed as calories. Common feedstuffs are similar in gross energy content, but differ in feeding value because of variations in digestibility. About 60% of the total energy in grain and 80% of the total energy in roughage is lost in feces, urine, gases, and heat. In the feed composition tables in Chapter 21 of this book, separate net energy values for each maintenance (NE_m) and gain (NE_g) are given.

In the NRC requirement tables which are reproduced in Chapter 12, Table 12–2 to 12–6, and in the Feed Composition Tables in Chapter 21 of this book, energy is also expressed as total digestible nutrients (TDN). TDN is comparable to digestible energy. It has been in use longer than the net energy system and more values are available for feedstuffs. TDN is computed as follows:

TDN = Digestible nitrogen-free extract (carbohydrate)
   + digestible crude fiber + digestible protein
   + (digestible ether extract × 2.25)

NE of lactation can be calculated from TDN as follows:

NE_{lc} (Mcal/lb DM) = (TDN, % of DM × .01114) − .054

The energy requirement for maintaining a lactating cow is affected by a number of factors, especially the following: (1) *body size*—the larger the animal, the higher the maintenance energy requirement; (2) *activity*—to support grazing activity, the maintenance allowance may be increased by 10% on good pasture and up to 20% on poor pasture; and (3) *cold temperature*—under severe winter conditions without access to dry shelter, the maintenance feed allowance may be increased up to 8%. Also, during the first lactation, when a heifer is still growing, her energy needs are about 20%

greater than a mature cow; and during the second lactation, her energy needs are 10% greater than a mature cow. The energy requirement for gestation is 30% of that required for maintenance alone, with most of the increase during the last 8 weeks of pregnancy.

Fig. 11–4. In comparison with confinement feeding, grazing activity may increase the maintenance allowance by 10% on good pasture and by up to 20% on poor pasture. (Courtesy, Milk Marketing, Inc., Strongsville, OH)

Table 12–4, in Chapter 12, includes allowances for liveweight changes during lactation. These values will aid the user in identifying the extent of dietary energy insufficiency during weight loss in early lactation and in estimating feed required to regain body weight in later lactation. The desired rate of liveweight gain will depend on the animal's body condition and stage of pregnancy.

When feed is restricted, a lactating cow will use the available energy for maintenance and reproduction at the expense of growth and milk production. Therefore, it is important to feed a ration that will fulfill the requirements for maintenance, growth, reproduction, and lactation.

For dry cows in good condition, medium quality forages will usually suffice as their primary feed. During the last 2 to 3 weeks prior to calving, concentrate feeding should be increased gradually to about 0.5% of their liveweight to allow the cow and the ruminal microorganisms to become adapted to the larger amounts of concentrate required in early lactation. Body weight losses after calving can be minimized by feeding as much of a properly balanced ration as the cow can safely use during the 10 to 12 weeks after calving. Proper feeding immediately after calving also helps to prevent ketosis.

The NRC maintenance requirements for growing replacement heifers are 12% higher than for beef cattle. Also, NRC (1) recommends that milk or milk replacer be fed to replacement calves for at least the first month of life, (2) states that longer periods (up to 2 months) of liquid feeding may be beneficial under some conditions because of decreased disease and death losses, and (3) recommends that veal calves be fed maximum *ad libitum* amounts of milk (or milk replacer).

The most common high energy dairy feeds are: barley, beet or sugar cane molasses, beet pulp, citrus pulp, corn, corn silage, fats, high-moisture corn, high quality legume forage (hay or silage), lush pasture, oats, sorghum grain, wheat, whole cottonseeds, and whole soybeans.

■ **NRC energy values of feed vs *Dairy Cattle Science* energy value of feeds** — In NRC's *Nutrient Requirements of Dairy Cattle*, Sixth Revised Edition, energy values of feeds for TDN, digestible energy (DE), and metabolizable energy (ME) (Chapter 12, Table 12–8) have been determined at the maintenance level of intake, whereas values for net energy for lactation are adjusted to three times (3X) the maintenance level (Chapter 12, Table 12–6). The NRC dairy cattle requirements for TDN, DE, and ME for animals fed above maintenance (lactating animals) have been increased to allow for the depression of digestibility of energy at these higher levels of intake. As long as formulators use both NRC energy values for feeds and NRC nutrient requirements, satisfactory rations will result.

In Chapter 21, Feed Composition Tables, in this book, the energy values of feeds are not identical to the energy values in NRC's *Nutrient Requirements of Dairy Cattle*, Sixth Revised Edition, due to varying conclusions from interpreting analytical information. However, the differences are minor and should not materially affect ration formulation. If, however, formulators are primarily interested in rations for dairy cattle, and if TDN is used as the energy measure, they may wish to refer to the NRC publication for the appropriate TDN values. As a convenience, Chapter 12, Table 12–8, of this book, contains TDN and NE_{lc} values from *Nutrient Requirements of Dairy Cattle*, Sixth Revised Edition, for the most common feeds used in dairy rations.

Note well: Net energy for lactation values (NE_{lc}) that conform to those used in *Nutrient Requirements of Dairy Cattle* may be calculated from TDN values presented in Chapter 21, Feed Composition Tables, by using the following equation:

$$NE_{lc} \text{ (Mcal/kg of DM)} = .0245 \times \text{TDN (\% of DM)} - .12$$

This equation is based on an average 4% reduction in digestibility for each multiple increase in intake over maintenance intake and assumes the intake will be 3X that of maintenance.

PROTEIN

Proteins are complex chemical structures which are made up of amino acids that are linked together in many different ways. Amino acids contain carbon, hydrogen, oxygen, nitrogen, and, in some cases, sulfur. There are 22 naturally occurring amino acids. The amino acids are supplied by the digestion of microbial protein and by feed protein that escapes microbial breakdown in the rumen. The proteins in practical energy sources supply some dietary protein that escapes rumen fermentation, and this protein, plus the microbial protein produced from supplemental NPN, may be enough to produce about 44 lb of milk per day. As milk production increases, a substantial amount of additional dietary protein from protein supplements must escape rumen fermentation to meet the cow's requirement for protein.

Protein is essential for dairy cattle for maintenance, growth, milk production, and the development of the fetus.

Also, it is required for the formulation of enzymes and certain hormones that control or regulate chemical reactions in the body. The protein requirement is really a requirement for amino acids.

The protein composition of feeds, and the protein requirements of dairy cattle, may be expressed as crude protein, digestible protein, degraded intake protein, undegraded intake protein, and/or nonprotein nitrogen (NPN).

1. **Crude protein.** Chemically, most proteins contain 16% nitrogen; so, crude protein is determined by finding the nitrogen content, then multiplying the result by 6.25 (100 ÷ 16 = 6.25). It is called *crude protein* because not all of the nitrogen in feeds is in the form of protein; rather, it is a combination of true protein and nonprotein nitrogen. Some feeds, particularly green roughages, contain one-third or more of their nitrogen as nonprotein nitrogenous substances such as amides, ammonium salts, amino acids, alkaloids, and other nitrogenous compounds. However, ruminal microorganisms make use of the various nitrogen sources for synthesis of microbial proteins which, in turn, are used by the cow. Consequently, for dairy cows and other ruminants, the amount of crude protein is about as good a measure for protein allowances as is the amount of true protein.

2. **Digestible protein.** This is the amount of crude protein consumed less the crude protein excreted in the feces. However, the term *apparent digestibility* is more accurate since it is recognized that a portion of the fecal nitrogen is derived from the animal and is not a feed residue.

3. **Degraded intake protein (DIP).** This refers to the intake crude protein that is broken down (degraded) by microorganisms in the rumen.

4. **Undegraded intake protein (UIP).** This is the crude protein that is not broken down in the rumen; instead, it is swept out of the rumen to the abomasum and small intestine for breakdown there and absorption as peptides and amino acids.

5. **Nonprotein nitrogen (NPN).** Feedstuffs which contain nitrogen in a form other than proteins or peptides are termed *nonprotein nitrogen (NPN)*. Nonprotein nitrogen compounds, such as urea and ammonium salts, have a crude protein value, but they do not supply any amino acids directly. The billions of microorganisms in the rumen convert nitrogen from NPN sources into amino acids for their growth and use. Then, the microbes pass into the small intestine where they are digested and release amino acids for absorption and utilization the same as amino acids released from the digestion of true proteins (composed of amino acids) in feeds.

■ **Amount of protein needed**— The amount of protein needed in the total ration of lactating cows is determined primarily by the amount of milk produced. Milk is a rich source of high-quality protein; so, as milk production increases, a substantial amount of dietary protein is necessary. Thus, a high-production 1,320-lb cow yielding 88 lb of 3.5% protein milk daily secretes 3.08 lb of milk protein. A deficiency of protein results in lowered milk production and may depress the protein content of milk. Excess protein usually results in high cost rations.

Fig. 11–5 depicts the increase in milk yield as the protein percentage of the total ration DM increases. At some point,

the value of increased milk will not exceed the cost of the additional protein, with this point determined by the relative milk price and the cost of the additional protein.

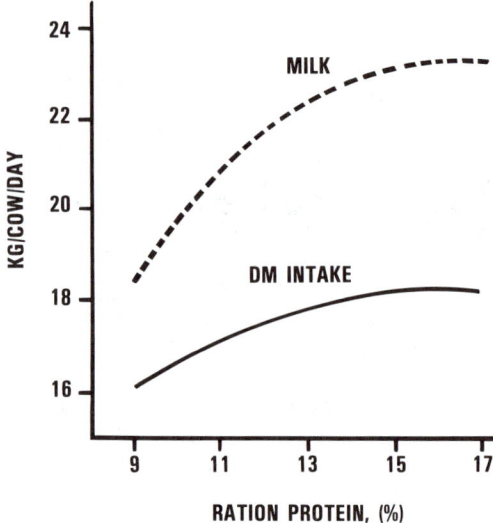

Fig. 11–5. This depicts the increase in milk yield as the protein present in the total ration DM increases. (Courtesy, J. W. Thomas, Ph.D., Professor Emeritus and Consultant, Department of Animal Science, Michigan State University,

The amount of protein needed in the concentrate mix depends on the kind and quality of forage fed. As the amount of legume increases, the percentage of protein in the concentrate can be lowered. For most lactating cows, the total ration (forage plus grains plus protein and energy supplements) should have 19% crude protein during the first one-third of lactation, lowered to 14% in midlactation and 12% during the dry period. The interaction of the protein and energy supply in the rumen of the high-producing cow is very important; so, an adequate supply of degradable intake protein (DIP) is essential to maximize both feed intake and ruminal digestibility. Moe, of the USDA, Agricultural Research Service, found that high-producing cows receiving 17% protein digested feed better than those fed 14% protein.[1]

■ **Excess protein**— When more protein is fed than needed, the excess is used as a source of energy. Because protein feeds are generally more expensive than carbohydrate feeds, it usually is more economical to feed only the amount needed. Besides, a large excess of dietary protein may decrease the energy supply because excess protein must be deaminated to ammonia and, for the most part, transformed back into urea for excretion. Most cows fed good-quality alfalfa hay (fed free-choice), along with grain fed according to production by one of the recommended systems, will not need any supplemental protein until they produce more than 50 to 60 lb of milk daily. As production increases above this amount, protein intake must be increased gradually, usually by decreasing the hay intake and

[1]*Research News*, ARS-USDA, p. 20, May 1977.

replacing it with grain and protein concentrates. So long as the hay consumption does not fall below 12 to 15 lb daily, the supplemental concentrate need not contain more than 15 to 16% protein.

DEGRADED INTAKE PROTEIN (DIP) AND UNDEGRADED INTAKE PROTEIN (UIP)

Approximately 60% of the crude protein in the typical dairy cow ration is broken down (degraded) by microbial digestion to ammonia. The rumen microbes must convert the ammonia to microbial protein in their own cells if the dairy animal is to receive any benefit. Fermentable energy must be available for the microorganisms to grow and synthesize the necessary amino acids. If rumen ammonia levels are excessively high, the ammonia is absorbed into the blood and either recycled or excreted in the urine as urea.

All feed protein sources are not degraded in the rumen to the same extent. The optimal ration will meet both the nitrogen requirement of rumen microorganisms for maximum synthesis of microorganism protein and allow for maximum escape or *bypass* of high quality feed protein for digestion in the small intestine. Protein synthesis by rumen microbes depends on feed intake, organic matter digestibility, feed type, protein level, and feeding system. Since 3.5 lb of microbial protein synthesis per day is near the maximum, the remainder of the protein must be derived from nondegraded (escape) protein sources. Young, fast-growing heifers and high-producing cows generally require additional nondegraded protein sources beyond their normal ration to meet total protein requirements; and the more rapid the growth and the higher the milk production, the greater the quantities of the undegradable protein needed. Brewers' grain, distillers' grain, corn gluten meal, fish meal, meat meal, and heat-treated soybeans are examples of feed with reduced rumen degradability that may be substituted in rations in which excess rumen ammonia exists and less than optimal amounts of quality protein (undegraded) pass into the small intestine.

UREA AND OTHER NPN PRODUCTS

Urea is a nonprotein nitrogen (NPN) compound, containing about 45% N, with a protein equivalent of 281% (45% N × 6.25), and with the chemical formula—

$$O = C \begin{smallmatrix} NH_2 \\ NH_2 \end{smallmatrix}$$

Using urea in the ration is similar to using degradable intake protein. It and other nonprotein nitrogen (NPN) compounds, such as ammonium salts, can be used to replace part of the protein required in dairy cattle rations after rumen function has become established. Studies have shown that when cattle are fed purified rations with only nonprotein nitrogen as a nitrogen (N) source, there is adequate microbial protein production for growing ruminants with a functional rumen to gain at about 65% of the level at which they gain when fed practical energy ingredients and protein supplements.[2] Also, studies show that there is adequate microbial

protein production for lactating cows to produce 8,800 lb of milk per lactation.[3]

The following guidelines should be observed for the successful use of urea in dairy rations:

1. All rations should be assessed for protein content before either supplemental NPN or natural protein is added to the ration. Protein may not be needed.

2. Feeds most successfully supplemented with NPN are high in energy, low in protein, and low in natural NPN (such as grains and corn silage).

3. Maximum amounts of urea to feed are:

    a. 1% urea in the grain mix; 0.5% urea in the total ration.

    b. 0.5% urea in corn silage (10 lb/ton). If 0.5% is added to corn silage, the amount in the grain should be no more than 0.5%. The addition of 10 lb of urea per ton of corn silage will increase the protein content from 8 to 12% on a dry matter basis (depending on losses incurred).

    c. 0.4 lb urea per head per day, with cows in early lactation limited to 0.2 lb of urea per head per day.

4. Urea is not palatable. So, it should be mixed thoroughly with the grain mix or silage. Molasses can improve acceptability.

5. If cattle have not been fed urea previously, a 7- to 10-day adjustment period in which the urea is gradually increased will help to maintain feed intake and production.

6. Total mixed rations and/or frequent feedings of feed containing urea maximizes utilization.

7. High levels of urea can be toxic; so, excessive intakes should be avoided. Urea should not be top-dressed; it should be mixed in the feed.

Urea can be used in a high-protein concentrate. A mixture of 87 lb of ground shelled corn and 13 lb of urea is equivalent in energy and crude protein to 100 lb of soybean meal.

A mixture of 56 lb shelled corn, 7 lb of urea, and 37 lb of soybean meal also equals 100 lb of soybean meal in total energy and protein equivalent, and can be used as a substitute for soybean meal.

■ **Other NPN products** — Several ammoniated products or ammonium salts are used successfully as sources of nitrogen. Monoammonium phosphate, which contains about 11% nitrogen (crude protein equivalent of 68.25%), is also used for phosphorus supplementation. Ammonia (cooled to form a liquid or in a water solution) may be added to corn silage at the rate of 7 lb (5 lb of nitrogen) per ton. Urea should not be fed in the concentrate when ammonia or other NPN has been added to the corn silage.

Since NPN products do not provide any energy, minerals, or vitamins, these nutrients must be provided through other sources.

(For additional discussion of NPN and urea, also see Chapter 10, sections headed "Nonprotein Nitrogen Feedstuffs (NPN)" and "Urea.")

---

[2]Oltjen, R. R., "Effects of Feeding Ruminants Nonprotein Nitrogen as the Only Nitrogen Source," *Journal of Animal Science*, 28:673, 1969.

[3]Virtanen, A. I., "On Nitrogen Metabolism in Milking Cows," *Federation Proceedings*, 28:232, 1969.

MINERALS

Minerals are inorganic elements, frequently found as salts with either inorganic elements or organic compounds.

Dairy cattle require at least 15 mineral elements. They are needed for both structural and regulatory functions. Minerals are needed for bone and teeth formation, and to maintain acid-base balance, water balance, and enzyme and hormone systems. They are also components of certain substances within the body, such as iron in hemoglobin. Additionally, a lactating cow needs minerals for the developing fetus and for milk production. Milk contains about 0.7% minerals; thus, a cow producing 20,000 lb of milk in a lactation secretes 140 lb of minerals per year.

The mineral elements that have been shown to be es-

sential for dairy animals and for which signs of deficiency have been described are generally classified as (1) major or macrominerals (those required in greatest quantities and present in animal tissues as higher levels), and (2) trace or microminerals (those required in smaller amounts and generally present in tissues at lower levels). Under practical feeding conditions, it is usually necessary to provide supplemental sources of several of these elements to meet the nutritional requirements of dairy cattle.

Mineral excesses should be avoided because of interaction with other minerals and possible toxicities and undesirable interactions. *The maximum tolerable level for a mineral element has been defined as that dietary level that, when fed for a limited period, will not impair animal performance and should not produce unsafe residues in human food derived*

**TABLE
DAIRY CATTLE**

Mineral	Conditions Usually Prevailing Where Deficiencies are Reported	Functions of Mineral	Deficiency Symptoms
Major or Macrominerals:			
SALT (NaCl, sodium and chloride)—The requirements for sodium and chlorine are commonly expressed as salt requirements because salt is an effective, economical way of supplementing rations with these elements.	Negligence; for salt is inexpensive. Deficiencies of sodium and chlorine may occur in dairy cattle because plants have low sodium contents, and because sodium needs increase during lactation and during periods of rapid growth.	Sodium (Na) functions in maintaining osmostic pressure, acid-base balance, and body-fluid balance; is involved in nerve transmission and active transport of amino acids; is required for cellular uptake of glucose through activation of the glucose carrier protein; and is a major cation of extracellular fluid and provides a majority of the alkaline reserve in plasma. Chlorine (Cl) is necessary for the activation of amylase; is essential for the formation of gastric hydrochloric acid; and is involved in respiration and regulation of blood pH, through the chloride shift.	Intensive craving of salt, manifested by the animals chewing and licking various objects, and by muscle cramps. Prolonged deficiency results in lack of appetite, unthrifty appearance, and decreased production. High-producing milk cows may collapse and die when salt defiency has been of long duration. It is noteworthy that when salt is omitted, sodium expresses its deficiency first.
CALCIUM (Ca)—Calcium is the most abundant mineral in the body. Most of the calcium in the body is found in the bones and teeth. It constitutes 2% of the body weight. In blood, calcium is found mostly in the plasma, with a controlled concentration of 10mg/100 ml.	A calcium deficiency may occur when the ration consists chiefly of dried mature grasses or cereal straws, and when cows are in heavy lactation. Osteomalacia may occur when there are high metabolic demands on calcium and phosphorus stores, such as occur during pregnancy and lactation.	Essential for bone formation, development of teeth, production of milk, transmission of nerve impulses, maintenance of normal muscle excitability (along with sodium and potassium), regulation of heart beat, movement of muscles, blood clotting (conversion of prothrombin to thrombin), and activiation and stabilization of enzymes (*i.e.*, pancreatic amylase).	A deficiency of calcium results in rickets in young animals and osteomalacia in older animals. Rickets may be caused by a deficiency of calcium, phosphorus, or vitamin D. It is characterized by improper calcification of the organic matrix of bones of young, growing animals. Thus, the bones are weak, soft, and lack density. Signs include swollen tender joints; enlargement of the ends of bones; and arched back; stiffness of the legs; and development of beads on the ribs. If the cause is not corrected, calves develop bowed and deformed legs. Also, rachitic bones are highly susceptible to fracture. Osteomalacia is the result of demineralization of the bones of adult animals. This condition is characterized by weak, brittle bones that may break when stressed. Milk fever (parturient paresis) in cows is caused by a disturbance in calcium metabolism, manifested by a marked drop in blood serum calcium, generally within 48 to 72 hours after calving.

from the animal. It is important not to exceed safe tolerances for dietary mineral elements in feeds

When dairy cattle are fed mixed feed, in part or totally, the needed minerals are usually incorporated in the ration in keeping with the known requirements. When animals are fed an unmixed ration or are on pasture, *ad libitum* supplementation of minerals is commonly practiced.

Tables 12–3, 12–4, and 12–5 of Chapter 12, Dairy Feeding Programs, show the daily calcium and phosphorus requirements for different classes of dairy animals; Table 12–6 shows the recommended content of major and trace minerals of rations for different classes of dairy animals; and Table 12–7 shows the maximum tolerable levels of certain elements. Table 11–4 presents in summary form the mineral requirements of dairy cattle. Chapter 21, Feed Composition Tables, presents the macro- and micromineral content of a great array of feeds.

In addition to the minerals listed and discussed in the narrative that follows, several other elements have been shown to be essential for one or more animal species; among them, arsenic, chromium, fluorine, lead, nickel, silicon, tin, and vanadium. Currently, however, these elements are not considered to be of practical importance in the feeding of dairy cattle.

DAIRY CATTLE MINERAL CHART

Table 11–4, Dairy Cattle Mineral Chart, presents in summary form the mineral elements that are known to be required by dairy cattle.

11–4
MINERAL CHART

Nutrient Requirements[1]		Recommended Allowances[1]	Practical Sources of the Mineral	Comments
Daily Nutrients/Animal	**Percentage of Ration**			
*A lactating cow producing 44 lb of milk daily, requires 21.3 to 23.3 g of sodium/cow daily.	*For lactating cows, a minimum of 0.43% sodium chloride (salt) in the ration. *For nonlactating dairy cattle, 0.25% sodium chloride (salt).	Free-choice feeding of salt. When included in a complete feed, 0.5% salt is recommended.	Salt should be available at all times. It should be both (1) self-fed, free-choice, and (2) mixed with other ration ingredients. Free access to salt in the form of loose, rock, coarse ground, or block salt. Cattle prefer loose salt to block salt, because it can be eaten more rapidly and with less effort. However, experiments with lactating cows have shown fully as good results with block salt as with loose salt even though smaller quantities were consumed. This means that the additional intake of loose salt over block salt does not appear to benefit cattle. Commercial mineral mixes (in block, or loose form) may contain ⅓ or more salt.	The salt requirements of cattle differ (1) between individuals, (2) according to whether milk is produced (being higher for lactating cows than for dry cows, because of the salt in the milk), (3) from season to season, (4) between block and loose salt (animals often consuming twice as much easy-to-get loose salt as block salt), and (5) according to the salt content of the soil, feed, and water (being higher when vegetable proteins are fed than when animal proteins are fed, higher on predominantly forage rations than on predominantly concentrate rations, and higher on lush early pasture than on more mature grasses). These are some of the reasons why free-choice feeding of salt is advocated.
*Variable, according to age, weight, and level of milk production (see Chapter 12, Tables 12–3, 12–4, and 12–5). Because true digestibilities of calcium in feedstuffs vary, the dietary calcium requirements shown in the tables may in some instances need to be adjusted.	*Variable according to age, weight, and level of milk production, ranging from 0.43% to 0.66% of a complete ration (see Chapter 12, Table 12–6). Because true digestibilities of calcium in feedstuffs vary, the dietary calcium requirements shown in the tables may in some instances need to be adjusted.	Free access to a calcium supplement, or a calcium supplement incorporated in the ration.	Legumes are high in calcium. Also, several of the oilseed meals are good sources of calcium. Sources of supplemental calcium include calcium carbonate, ground limestone, bone meal, dicalcium phosphate, defluorinated phosphate, monocalcium phosphate, and calcium sulfate. When both calcium and phosphorus need to be supplemented, they should be provided in a readily available and palatable form such as dicalcium phosphate, defluorinated phosphate, or bone meal.	Cows may be in negative calcium balance during early lactation, but the deficit is made up in late lactation and in the dry period. In addition to an adequate supply of calcium, proper utilization is dependent upon (1) a highly available source of the mineral, (2) a suitable ratio between calcium and phosphorus (somewhere between 1 and 2 parts of calcium to 1 part of phosphorus). Calcium-phosphorus ratios of 2:1 have been shown to be beneficial in reducing urinary calculi. When calculi problems are encountered, even higher levels of calcium may be advisable. Ratios between calcium and phosphorus of 7:1 have been reported to be satisfactory for cattle. Generally, when cattle receive at least ⅓ of a legume forage, ample calcium will be provided. But even nonlegume forages contain more calcium than cereal grains. Plants grown on calcium-rich soils are high in calcium. Calcium availability of 70% is generally assumed for all feedstuffs.

Continued

TABLE 11–4

Mineral	Conditions Usually Prevailing Where Deficiencies are Reported	Functions of Mineral	Deficiency Symptoms
Major or Macrominerals (continued):			
PHOSPHORUS (P)— Phosphorus has varied, but extremely important, biochemical physiological roles.	When cows are in heavy lactation. Semiarid regions are commonly associated with soils deficient in phosphorus. The phosphorus content of plants generally decreases markedly with maturity, with the result that deficiencies often occur in cattle subsisting for long periods on mature dried forage. High iron levels result in the formation of insoluble iron phosphate. Also, aluminum forms insoluble, unavailable phosphates.	Phosphorus is deposited in bones. It is also found in high concentrations in brain, muscle, liver, spleen, and kidneys. Phosphorus, as a component of phospholipids, influences cell permeability and is a component of myelin sheathing of nerves. Also, many energy transfers in cells involve the high-energy phosphate bonds in ATP. Phosphorus plays an important role in blood buffer systems. Activation of several B-vitamins (thiamin, niacin, pyridoxine, riboflavin, and pantothenic acid) to form coenzymes requires their initial phosphorylation. Phosphorus is also a part of the genetic materials DNA and RNA.	Phosphorus deficiencies in cattle are widespread. A deficiency of phosphorus results in decreased growth rates, in inefficient feed utilization, and in a depraved appetite (chewing of wood, soil, and bones—called *pica*); anestrus, low conception rate, and reduced milk production; low plasma phosphorus levels, and weak, fragile bones and stiffness of joints. A deficiency of phosphorus results in rickets in young cattle, and in osteomalacia, osteoporosis, and osteitis fibrosa in mature animals.
MAGNESIUM (Mg)— Magnesium is the fourth most abundant cation in the body.	When milk feeding of calves is prolonged without grain or hay. (Milk is rather low in magnesium.) When there is grass tetany, which is most likely to occur when cows in early lactation graze early spring pastures containing less than 0.2% magnesium.	Magnesium is required for skeletal development as a constituent of bone; plays an important role in neuromuscular transmission and activity; is required to activate many enzyme systems, including those involving ATP; and is required as a cofactor in decarboxylation and an activator of many peptidases. Approximately 65% of total body magnesium is contained in bone; the other 35% is distributed among various tissues and organs.	A magnesium deficiency causes grass tetany or grass staggers, characterized by anorexia, hyperemia, hyperirritability, convulsions, and death. Magnesium-deficient cattle exhibit loss of appetite and reduced dry matter digestibilties. Deficiencies in young cattle may result in defective bones and teeth.
POTASSIUM (K)— Potassium is the third most abundant mineral element in the body.	When cattle are fed high grain rations.	Essential for proper enzyme, muscle, and nerve function, rumen microorganism activity, and appetite.	Poor appetite and feed conversion, slow growth, stiffness, and emaciation.
SULFUR (S)—Sulfur is a component of protein, some vitamins, and several important hormones.	Cattle fed high-grain rations supplemented with nonprotein nitrogen. Also, sulfur deficiency may occur when dairy cattle are fed high corn silage rations, because corn silage is often low in sulfur.	Body functions that involve sulfur include protein synthesis and metabolism, fat and carbohydrate metabolism, blood clotting, endocrine function, and intra- and extracellular fluid acid-base balance. Sulfur has both structural and metabolic functions; it is found in virtually every tissue and organ of the body. Muscle has a fairly constant nitrogen to sulfur ratio of 15.3:1. The total body content of sulfur is approximatley 0.15%.	Depressed appetite, loss of weight, weakness, excessive salivation, watery eyes, dullness, emaciation, reduced milk production, and death. A lack of sulfur also results in a microbial population that does not utilize lactate.
Trace or microminerals:			
COBALT (Co)—The cobalt requirement of dairy cattle is actually a cobalt requirement of rumen microorganisms. The microbes incorporate cobalt into vitamin B–12, which is utilized by both microorganisms and animal tissues.	In cobalt-deficient soils where this element is not provided. Cobalt-deficient soils occur in many parts of the world, with large deficient areas in Australia, New Zealand, and along the southeast Atlantic Coast of the U.S.	The main function of cobalt is to serve as an integral part of vitamin B–12 (cobalamins). Vitamin B–12 is of importance in the metabolism of propionic acid, needed for the activity of the enzyme methylamalonyl-CoA isomerase. Vitamin B–12 is also a part of the enzyme that catalyzes the recycling of methionine from homocysteine after the loss of its labile methyl group. Vitamin B–12 is also needed for normal liver folate metabolism.	Loss of appetite and body weight, muscular wasting, severe anemia, decline of milk production, followed by death. In severe deficiency, the mucous membranes become blanched, the skin turns pale, a fatty liver develops, and the body becomes almost totally devoid of fat. With a deficiency of cobalt, there is usually a high mortality rate among calves.

(Continued)

Nutrient Requirements[1]		Recommended Allowances[1]	Practical Sources of the Mineral	Comments
Daily Nutrients/Animal	Percentage of Ration			
*Variable, according to age, weight, and level of milk production (see Chapter 12, Tables 12–3, 12–4, and 12–5).	*Variable according to age, weight, and type and level of milk production (see Chapter 12, Table 12–6).	Free access to a phosphorus supplement, or a phosphorus supplement added to the daily ration in keeping with the nutrient requirements.	Common sources of phosphorus are: dicalcium phosphate, defluorinated phosphate, bone meal, soft phosphate, sodium phosphate, ammonium polyphosphate, orthophosphates, metaphosphates, pyrophosphates, and tripolyphosphates. Oilseed meals and animal and fish products contain large amounts of phosphorus. Phytate phosphorus is not well utilized by nonruminants, but ruminants appear to use considerable quantities of this form of phosphorus.	Grains, grain by-products and high-protein supplements are fairly high in phosphorus; hence, rations high in such ingredients require little or no phosphorus supplementation. Calcium-phosphorus ratios of 2:1 are beneficial in reducing urinary calculi; and even higher levels of calcium may be necessary when urinary calculi is encountered. Ratios between calcium and phosphorus of 7:1 have been reported to be satisfactory for cattle.
*Young calves and growing cattle, 12–30 mg/kg body weight. Cows, 7–9 g/day during gestation and 21, 22, and 18 g/day during early, mid, and late lactation, respectively. Magnesium requirements are increased by feeding high levels of aluminum, potassium, phosphorus, or calcium; by younger cattle and magnesium-deficient cattle; and by high levels of milk production.	*0.25 to 0.3% dietary magnesium, with the supplemental magnesium provided in a readily available form such as magnesium oxide.		Commonly used feedstuffs vary widely in magnesium content and availability. Magnesium carbonate, oxide, and sulfate are good sources of supplemental magnesium.	Supplemental feeding of magnesium (20 g/day) reduces the incidence of grass tetany in many outbreaks. Availability values for magnesium as high as 70% have been observed in young milk-fed calves, but these decline to 30 to 50% in older calves. Magnesium absorption is lowest for young, highly succulent pastures, and increases with maturity.
	*For lactating cows, 0.9 to 1.0% of the ration. *For dry cows and young stock, 0.65% of ration dry matter. The needs for potassium vary with amounts of protein, phosphorus, calcium and sodium consumed.	1 to 1.25% of the ration for lactating cows. 0.7 to 1% of the total ration dry matter for dry cows and young stock.	Roughages usually contain ample potassium. Potassium chloride is the supplement of choice.	Grains often contain less than 0.5% potassium. Excessive levels of potassium have been found to interfere with magnesium absorption. Also, excessive levels of potassium, along with high levels of phosphorus, increase the incidence of phosphatic urinary calculi.
	*The NRC estimated minimum sulfur requirement for lactating cows is 0.2% of the ration; with the sulfur needs of dry cows and young stock calculated from the minimum protein requirement for these animals based on a nitrogen-to-sulfur ratio of 12:1.	The NRC suggested maximum level of sulfur in the ration is 0.4%.	Feeds high in protein are usually high in sulfur. The microbial population of the rumen has the ability to convert inorganic sulfur into organic sulfur compounds that can be used by the animal. So, either organic or inorganic sulfur can be utilized by cattle. Most feedstuffs provided to dairy cattle contain sufficient sulfur to meet their needs.	Sulfur increases copper requirements. Selenium can replace sulfur in some organic compounds. Sulfur requirements are primarily those involving amino acid nutrition.
	*0.1 ppm of ration dry matter, with a range of 0.07–0.11 ppm of dry matter.	Free access to a cobaltized mineral mixture in cobalt-defient areas; or administering a cobalt pellet. Supplements of 30 to 45 g of cobalt sulfate or 20 to 25 g of cobalt carbonate with 100 lb of salt have prevented cobalt deficiency problems.	A cobaltized mineral mixture may be prepared by adding cobalt at the rate of 0.2 oz/100 lb (*1.25 mg/kg*) of salt as cobalt chloride or cobalt sulfate, cobalt carbonate, cobalt oxide, or a good commercial mineral mixture or salt product may be used. Also, cobalt sulfate and cobalt oxide are effective as a drench; and a cobalt pellet (composed of cobalt oxide and finely divided iron) that lodges in the reticulum is an effective preventive.	Several good commercial cobalt-containing minerals are on the market. A vitamin B–12 injection will relieve a cobalt deficiency.

Continued

TABLE 11–4

Trace or microminerals: (continued)

Mineral	Conditions Usually Prevailing Where Deficiencies are Reported	Functions of Mineral	Deficiency Symptoms
COPPER (Cu)—Zinc and silver are antagonistic to copper absorption.	In copper-deficient areas (soils), as in Florida and the Coastal Plain region. On peat and muck soils, or where soil molybdenum levels are high. Deficiencies have occurred in calves kept on an exclusive milk diet for long periods.	Copper is necessary in hemoglobin formation, iron absorption from the small intestine, iron mobilization from tissue stores, and for the oxidation of iron, permitting it to bind with the iron transport—transferrin. Copper is essential in enzyme systems, hair development and pigmentation, bone development, reproduction, and lactation.	Emaciation, depigmentation (cattle turn yellowish) and loss of hair, stunted growth, anemia, and brittle and malformed bones. Also, heat periods are suppressed, and there may be depraved appetite and diarrhea. Young calves may have straight pasterns and stand forward on their toes. Low copper intake reduces the synthesis and activity of the copper-containing enzyme, tyrosinase, which is required for pigmentation of hair, wool, and feathers.
IODINE (I)—Two-thirds of ingested inorganic iodine is excreted by the kidneys.	In iodine-deficient areas (soils) where iodized salt is not fed (in northwestern U.S. and in the Great Lakes Region). Calves on an exclusive milk diet. Where feeds come from iodine-deficient areas. Substances that interfere with iodine metabolism. Rapeseed meal, soybean meal, and cottonseed meal have goitrogenic effects.	Inorganic iodine is taken up by the thyroid gland for the synthesis of thyroid hormones. Thyroid hormones have an active role in thermoregulation, intermediary metabolism, reproduction, growth and development, circulation, and muscle function.	Production of weak, goitrous, and dead calves. Goiter in the young, retarded growth and maturity, lowered metabolic rate, and increased water retention. Occasional borderline cases may survive; in these, the moderate thyroid enlargement disappears in a few weeks.
IRON (Fe)—Ferrous iron is absorbed to a much greater extent than ferric iron.	Calves on an exclusive milk ration (milk contains less than 10 ppm iron). Animals with excessive blood loss.	Iron has important biochemical functions in animals since it is a component of hemoglobin, myoglobin, cytochrome, and the enzymes catylase and peroxidase. Iron in these materials exists in porphyrin rings. Iron is involved in the transport of oxygen to cells and in cellular respiration.	Signs of lack of iron include anemia, reduced saturation of transferrin, listlessness, pale mucous membrane, reduced appetite and weight gain, and atrophy of the papillae of the tongue. Anemia and decreased growth in calves on exclusive milk diet.
MANGANESE (Mn)—Ruminants regulate manganese levels in blood and tissue via intestinal absorption.	In northwestern U.S. High-concentrate rations based on corn supplemented with nonprotein nitrogen.	Manganese is essential for normal reproduction in both males and females, for bone formation, and for the functioning of the central nervous system. Also, manganese is a preferred metal cofactor for many enzymes involved in carbohydrate metabolism and in mucopolysaccharide synthesis.	In males: impaired spermatogenesis, testicular and epididymal degeneration, sex hormone inadequacy, and sterility. In females: irregular and absent estrus, delayed conception, abortion, and deformed young at birth—crooked calves.
MOLYBDENUM (Mo)—Molybdenum is found in nearly all body cells and fluids.	Molybdenum toxicity occurs only occasionally in cattle and appears to be an area problem.	Molybdenum is a constituent of the enzymes xanthine oxidase, aldehyde oxidase, and sulfide oxidase; enzymes involved in the oxidation of purines and reduction of cytochrome C.	Molybdenum deficiencies have not been demonstrated in cattle. The NRC gives the maximum tolerable level of molybdenum as 6 ppm. Clinical signs of molybdenum toxicity in cattle are diarrhea, loss of appetite, anemia, ataxia, and bone malformation.

(Continued)

Nutrient Requirements[1]		Recommended Allowances[1]	Practical Sources of the Mineral	Comments
Daily Nutrients/Animal	Percentage of Ration			
	*NRC recommends a minimum of 10 ppm of copper in the ration. For presence of high levels of molybdenum and inorganic sulfate, increase the copper requirements.	Copper deficiency can be prevented by adding 0.25–0.5% copper sulfate to salt fed free-choice. Copper may also be injected as glycinate to meet the nutritional needs for the mineral.	Salt containing 0.25–0.5% copper sulfate.	Copper deficient cattle can be returned to normal by feeding 3 g of copper sulfate or blue vitriol every 10 days. An interesting interrelation exists between copper and molybdenum. An excess of molybdenum (in the presence of sulfate) causes a condition which can be cured only by administering copper. Excess copper is toxic; it accumulates in the liver, and death may result.
	*NRC recommends that lactating cows receive a dietary iodine concentration of 0.6 ppm, and that cows in last 2 months of gestation be fed 0.6 ppm of iodine. Young stock and dry cows should recieve 0.25 ppm iodine in the diet.	Free access to stabilized iodized salt containing 0.01% potassium iodide (0.0076% iodine).	Stabilized iodized salt containing 0.01% potassium iodide. Feed additives that supply iodine are: ethylenediamine dihydroiodide (EDDI), calcium iodate, cuprous iodide, potassium iodate, sodium iodate, potassium iodide, sodium iodide, and pentacalcium periodate.	The enlargement of the thyroid gland (goiter) is nature's way of trying to make enough thyroxin, when there is insufficient iodine in the feed. Eighty percent of hormonal iodine stored in the thyroid is thyroxin. The amount of iodine in milk is influenced by iodine intake, season, level of milk production, and use of iodine disinfectants.
	*100 ppm for calves, until 3 months of age, and 50 ppm for older cattle.		Levels of iron in common feed believed to be ample. Sources of supplemental iron in decreasing order of availability are: ferrous sulfate, ferrous carbonate, ferric chloride, and ferric oxide.	After calves are past 20 weeks of age, iron does not seem to be beneficial. About 30% of all calves are affected by prenatal iron defiency. In cattle, a majority of body iron is in the form of hemoglobin, with lesser amounts existing as protein-bound stored iron, myoglobin, and cytochrome.
	*40 ppm for lactating cows. *20 ppm for growing cattle and dry cows, with a range of 20–50 ppm. **Note well:** Requirements for manganese are increased by elevated dietary levels of calcium and phosphorus.	An intake of 40 ppm for mature breeding cattle and 20 ppm for growing-finishing cattle.	Most forages contain high levels of manganese. Manganous oxide, sulfate, and carbonate are good sources of supplemental manganese.	The managanese levels in pastures, grains, and forages are variable because of variations in plant species, soil types, soil pH, and fertilization practices. A deficiency of manganese exists in northwestern U.S. where it has been shown to cause *crooked calves*.
Requirements for molybdenum are not established. Because copper and sulfate alter molybdenum metabolism, arriving at the molybdenum requirement is impossible.	Requirements for molybdenum are not established.	As a feed additive, molybdenum is not cleared by the Food and Drug Administration.	Many feeds contain 6.8–13.6 mg/lb of ration dry matter.	Excess molybdenum may cause a copper deficiency. Sulfur, in the absence of molybdenum, also may cause a copper deficiency. Increasing copper level in ration to 1 g/head daily is effective in overcoming molybdenum toxicity in cattle.

(Continued)

Mineral	Conditions Usually Prevailing Where Deficiencies are Reported	Functions of Mineral	Deficiency Symptoms
Trace or microminerals: (continued)			
SELENIUM (Se)— Initially, interest in selenium was confined to the problem of toxicity in animals.	Low selenium forage and low vitamin E. It is an area problem, but it occurs in many parts of the U.S.	Selenium functions (1) as a component of glutathione peroxidase, an enzyme that destroys peroxides in tissues, and (2) intertwined with vitamin E in a mutual sparing effect.	A deficiency of selenium may cause white muscle disease; characterized by white muscle, heart failure, and paralysis evidenced by lameness or inability to stand. Depression of glutathione peroxidase in tissues of selenium-deficient animals may account for many of the manifestations of selenium deficiency. The NRC suggests that 2 mg/kg (*2 ppm*) dry weight ration is the maximum tolerable level for all species. Signs of toxicity include loss of appetite, loss of tail hair, sloughing of hoofs, and eventual death. Two types of selenium poisoning have been observed: (1) acute, blind staggers; and (2) chronic, alkali disease. Selenium toxicity can be counteracted by feeding some forms of sulfur. Toxic levels reported in South Dakota, North Dakota, Montana, Wyoming, Utah, Nebraska, Kansas, and Colorado.
ZINC (Zn)— Absorption of zinc occurs primarily from the abomasum and lower small intestine.	Zinc deficiencies have been reported in ruminants grazing forages low in zinc or high in compounds interfering with zinc utilization.	Zinc functions as both an activator and a constituent of several dehydrogenases, peptidases, and phosphates that are involved in nucleic acid metabolism, protein synthesis, and carbohydrate metabolism.	Deficiencies are characterized by decreased performance and listlessness, followed by development of swollen feet and a dermatitis that is most severe on the neck, head, and legs. Deficiencies may also result in vision impairment, excessive salivation, decreased rumen volatile fatty acid production, failure of wounds to heal normally, and impaired reproductive performance in both bulls and cows.

[1]As used herein, the distinction between *nutrient requirements* and *recommended allowances* is as follows: In nutrient requirements, no margins of safety are included intentionally; whereas in recommended allowances, margins of safety are provided to compensate for variations in feed composition, environment, and possible losses during storage or processing.

When preceded by an asterisk, the nutrient requirements were adapted by the author from *Nutrient Requirements of Dairy Cattle*, Sixth Revised Edition, Update 1989, published by the National Research Council-National Academy of Sciences, Washington, DC.

MAJOR OR MACROMINERALS

The major or macrominerals of importance in dairy cattle nutrition are: salt (sodium chloride), calcium, phosphorus, magnesium, potassium, and sulfur.

■ **Salt (sodium chloride [NaCl])**—Sodium and chlorine are usually provided in the form of common salt (NaCl). However, potassium chloride may be used as a source of chlorine, also. The current NRC recommendation is for a minimum of 0.43% sodium chloride in the total dairy ration, including that contributed by the feed (Table 12–6). Excessive levels of chlorine without sodium or potassium can contribute to an acidosis condition in dairy cattle.

■ **Calcium (Ca)**—Whole milk contains 0.12% calcium. The NRC subcommittee based the dietary calcium requirements for lactating and pregnant cows (Table 12–4) on 38% availability. Minimum calcium percentage for the complete ration (dry matter basis) recommended by the NRC for lactating cows varies from 0.43 to 0.66%, depending on level of milk production (Table 12–6). A deficiency of calcium may cause rickets, slow growth and poor bone development, easily frac-

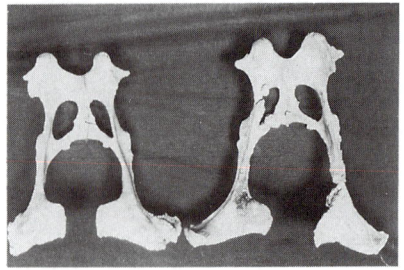

Fig. 11–6. Calcium deficiency. Lactating cows need calcium. Both hips of the cow shown above have been broken (knocked down) as a result of feeding a low-calcium ration. At lower left, the pelvis of a cow which has three breaks while the cow received a low-calcium ration. At lower right, the pelvis of the cow pictured above, showing the breaks involving both hip bones. (From *Fla. Ag. Exp. Sta. Tech. Bull. 262*, through the courtesy of R. B. Becker)

(Continued)

Nutrient Requirements[1]		Recommended Allowances[1]	Practical Sources of the Mineral	Comments
Daily Nutrients/Animal	**Percentage of Ration**			
	*NRC requirements for selenium for ruminants are 0.1 to 0.3 ppm, depending on the chemical form of selenium, the previous selenium status of the animal, and the amount of interfering or enhancing factors in the diet.	0.05–0.30 mg/kg *(0.1–0.3 ppm)* dry weight of ration.	In 1979, the FDA approved the addition of selenium as either sodium selenite or sodium selenate at the rate of 0.1 ppm complete feed for dairy cattle, beef cattle, and sheep. In 1987, FDA increased the allowance of selenium in complete feeds for cattle (dairy and beef), sheep, swine, chickens, turkeys, and ducks from 0.1 ppm to 0.3 ppm.	Selenium toxicity may occur when cattle consume feeds containing 10–30 ppm of selenium on a dry matter basis for an extended period. In Israel, in a series of experiments extending over 3 years, low doses of selenium injected intramuscularly reduced the incidence of retained placenta in cattle to half that of the controls. (Eger, S., *et al.*, "Effect of Selenium and Vitamin E on the Incidence of Retained Placenta," *Journal of Dairy Science*, Vol. 68, No. 8, Aug. 1985, p. 219.)
	*40 ppm of zinc are recommended for all classes and ages of dairy cattle.	*20–40 ppm zinc in the total feed (air-dry basis).	Feedstuffs vary widely in zinc concentrations, with legumes usually having higher concentrations than grasses, and with protein supplements of animal origin being higher than other protein supplements.	Mild zinc deficiency in cattle results in lowered weight gains without the development of a specific syndrome.

Mineral recommendations for all classes and ages of cattle: Provide free access to a two-compartment mineral box, with (1) salt (iodized salt in iodine-deficient areas) in one side and (2) dicalcium phosphate, defluorinated phosphate, or a mixture of ⅓ salt (salt added for purposes of palatability) and ⅔ steamed bone meal in the other side. Also, the mineral requirements may be met by using a good commercial mineral, in either block or loose form. If desired, the mineral supplement may be incorporated in the ration in keeping with the recommended allowances given in this table.

tured bones, reduced milk yield and increased incidence of milk fever. Feeding calcium at more than 0.95 to 1.0% (DM basis) in mixed rations may reduce dry matter intake and lower performance. The effects of variations in the calcium-to-phosphorus ratios have been overemphasized, as evidenced by studies showing that dietary calcium-to-phosphorus ratios of between 1:1 and 7:1 result in nearly equal performance, provided the animal's phosphorus intake meets its requirement.

■ **Phosphorus (P)** — Whole milk contains 0.09% phosphorus. The NRC subcommittee assumed a phosphorus availability from mixed rations fed to lactating cows of 45 to 50%. A deficiency of phosphorus may result in fragile bones, stiff joints, poor growth, low blood P (less than 4–6 mg/100 ml), depraved appetite (chewing wood, hair, and bones), and poor reproductive performance. Excessive phosphorus intakes may cause bone resorption, elevated plasma phosphorus levels, and urinary calculi.

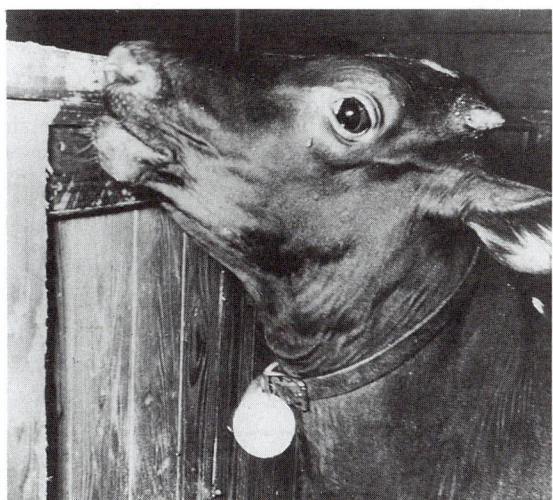

Fig. 11–7. Phosphorus-deficient calf chewing wood, a manifestation of depraved appetite. (Courtesy, Dr. S. E. Smith, Department of Animal Science, Cornell University, Ithaca, NY)

■ **Magnesium (Mg)**—Milk contains a substantial amount of magnesium (about 0.015%). Thus, when expressed as a percentage of the ration, the magnesium requirement increases with the cow's level of milk production. Under practical conditions, magnesium deficiencies may occur (1) when calves are fed an all-milk diet for extended periods, during which their body reserves of magnesium are depleted; or (2) when dairy cattle, especially older and lactating cows, are grazing lush, rapidly growing pastures that have been highly fertilized with nitrogen or potassium, or both, during cool seasons. Under conditions conducive to grass tetany and for high-producing cows in early lactation, the suggested requirement is 0.25 to 0.3% dietary magnesium, with the supplemental magnesium provided in a readily available form such as magnesium oxide. Magnesium toxicity is not known to be a practical problem in dairy cattle.

■ **Potassium (K)**—Milk contains about 0.15% potassium. The NRC minimum dietary potassium requirement for lactating cows is 0.9%, increased to 1% for high-yielding and early-lactation cows; for dry cows and young stock, it is 0.65%. Stress, especially heat stress, appears to increase the need for potassium, perhaps due to greater loss of potassium through sweat. The signs of relatively severe potassium deficiencies in lactating cows include a marked decrease in feed intake, loss in weight, decreased milk yield, pica, loss of hair glossiness, decreased pliability of hide, lower plasma and milk potassium, and higher hematocrit readings. Generally, forages contain considerably more potassium than is required by dairy cattle.

High levels of potassium (3% or above) in very lush forages grown on high potassium soils in cool weather appear to interfere with magnesium metabolism and utilization and are considered to be a factor in causing grass tetany of lactating cows.

■ **Sulfur (S)**—Milk contains 0.03% sulfur, much of which is in the form of the amino acids methionine and cystine. Sulfur is needed for microbial protein synthesis, especially when nonprotein nitrogen is fed. The NRC estimated minimum sulfur requirement for lactating cows is 0.2% of the ration; with the sulfur needs of other dairy cattle calculated from the minimum protein requirement for these animals based on a nitrogen-to-sulfur ratio of 12:1.

TRACE OR MICROMINERALS

The trace or microminerals of importance in dairy cattle nutrition are: cobalt, copper, iodine, iron, manganese, molybdenum, selenium, and zinc.

■ **Cobalt (Co)**—Normal cow's milk averages 0.38 to 1.04 mcg of cobalt/qt. Colostrum contains 4 to 10 times more cobalt than milk. Since cobalt is a component of vitamin B–12, ruminal microorganisms are able to synthesize this vitamin only when adequate cobalt is in the ration of the cow. NRC recommends a minimum of 0.1 ppm of cobalt in the total ration. Supplements of 30 to 45 g of cobalt sulfate or 20 to 25 g of cobalt carbonate with 100 lb of salt have prevented any cobalt deficiency problems. Also, a heavy pellet containing cobalt oxide and finely divided iron, which is administered orally and remains in the reticulo-rumen, will prevent cobalt deficiency for extended periods in cattle that graze cobalt-deficient pastures.

■ **Copper (Cu)**—Colostrum contains more copper than milk. The amount of copper in milk decreases with the length of lactation. Copper is needed for hemoglobin formation, although it is not actually contained in it. A deficiency of copper will result in anemia and bleaching of the hair. Black hair turns gray and red hair becomes yellow. NRC recommends a minimum of 10 ppm copper in the ration, with the caution that higher levels may be required for cattle grazing pastures or consuming feedstuffs that contain high levels of molybdenum or other interfering substances. Copper supplementation may be advisable under certain conditions, but it should be done with discretion. Excess copper is toxic and is a primary cause of oxidized flavor in milk.

■ **Iodine (I)**—About 10% of the iodine intake of lactating cows is normally excreted in milk. Iodine deficiency can be detected by analyzing milk or blood serum. Iodine concentration of less than 9.5 to 19 mcg/qt of milk or 37.8 mcg/qt of serum indicate iodine deficiency. Goiter (an enlargement of the thyroid gland) occurs in newborn calves if their mothers are fed iodine-deficient rations; necks of the calves are swollen and they are weak at birth or born dead. Much of the small amount of iodine in the body is contained in the thyroid gland as thyroxin and diiodotyrosine, both of which are contained in the protein thyroglobulin, a part of the thyroid hormone. The principal function of the thyroid gland is to regulate the metabolic rate. Many protein supplements (including soybean meal and cottonseed) are mildly goitrogenic because they reduce the availability of dietary iodine, and *Brassica* forages (cabbage, kale, rape) are highly goitrogenic. The NRC recommends that cows in lactation receive a dietary iodine concentration of 0.6 ppm, and that cows in the last 2 months of gestation be fed 0.6 ppm of iodine. When stabilized iodine is used, a level of 0.0076% in salt is adequate. The Northwest and Great Lakes regions are the most iodine-deficient areas of the United States. Lactating cows should not receive excessive dietary iodine because the resulting high iodine content of the milk is considered undesirable for humans. The use of iodine disinfectants as teat dips or udder washes can increase the iodine content of milk, but the main cause of high iodine levels in milk is dietary iodine.

■ **Iron (Fe)**—Iron is essential because it is a constituent in hemoglobin, the oxygen carrier in the blood. Cow's milk is low in iron—about 10 ppm. The iron requirements of a young calf are higher than those of a mature cow and are thought to be about 100 ppm until 3 months of age, and 50 ppm thereafter; as recommended by the NRC. The iron reserves of a newborn calf, which are primarily in the liver, are generally adequate to prevent serious anemia if calves are fed dry feeds at a few weeks of age. However, when calves are fed a milk diet exclusively for several weeks, they may develop iron deficiency anemia. Light colored veal is associated with low levels of muscle myoglobin and restricted iron intake. When veal calves are fed a dry ration, 40 ppm iron is sufficient to prevent severe anemia. The NRC gives an iron concentration of 1,000 ppm as the maximum tolerable level for cattle.

■ **Manganese (Mn)** — Manganese deficiency in dairy cattle is seldom a problem. In general, forages contain higher levels of manganese than grains. The manganese requirement for cattle is higher for reproduction than for growth. Little experimental work has been done on the manganese requirements of dairy cattle, but rations containing 40 ppm are recommended for lactating cows by the NRC. Trace mineralized salt and commercial mineral supplements usually contain manganese. Manganese toxicity in cattle is unlikely.

■ **Molybdenum (Mo)** — Molybdenum is an indispensable component of the enzyme xanthine oxidase, which is found in milk and distributed widely in animal tissue. Yet a deficiency of molybdenum has never been developed or observed in cattle. Molybdenum is known largely for its toxic characteristics; molybdenum toxicosis is a practical problem in grazing cattle in several areas of the world.

There is an antagonistic relationship between molybdenum and copper. Elevated dietary molybdenum increases both the animal's requirements for copper and the amount of copper that will cause toxicosis; increased dietary copper can reduce the toxic effect of molybdenum. Thus, the relative amounts of copper and molybdenum in the diet are important in determining the occurrence of molybdenum toxicosis. If the level of copper in the body is low, a lesser amount of molybdenum is toxic; as dietary copper increases, so does tolerance to molybdenum. High levels of both molybdenum and sulfur interfere with copper absorption. Molybdenum and sulfur also influence the metabolism of copper; added dietary molybdenum decreases metabolism of copper, whereas added dietary sulfur enhances it. The signs of molybdenosis have appeared in cattle that were fed about 6 ppm of molybdenum for several months. The NRC has set the maximum tolerable level of molybdenum for cattle for relatively short feeding periods at 10 ppm. *CAUTION:* As a feed additive, molybdenum is not approved by the Food and Drug Administration.

■ **Selenium (Se)** — Selenium, like molybdenum, was known for its toxic characteristics long before it was discovered to be an essential nutrient. However, research has firmly established the essentiality of selenium for ruminants; it is needed in trace amounts to prevent retarded growth, reproductive problems, retained placenta, white muscle disease — a condition that occurs in calves and lambs in selenium-deficient areas, and some mastitis problems. Also, it is closely associated with vitamin E; both selenium and vitamin E protect cells from the detrimental effects of peroxidation, but each takes a different approach. Vitamin E is present in the membrane components of the cell and prevents free-radical formation, whereas selenium functions throughout the cytoplasm to destroy peroxides. This explains why selenium will correct some deficiency symptoms of vitamin E, but not others. The current NRC recommended requirements for all cows and heifers are 0.3 ppm, which is the maximum level permitted by FDA. Deficient or toxic selenium areas are widely scattered throughout the United States and the world.

■ **Zinc (Zn)** — Milk generally contains about 4 ppm of zinc, but this level has been doubled by increasing the intake of zinc in the ration. Zinc is involved in several enzyme systems and is affected adversely when excess quantities of calcium are present. The NRC recommendation of 40 ppm of zinc for all classes and ages of dairy cattle is based upon limited data. Moderate excesses of zinc are not toxic to dairy cattle. Galvanized pipes and galvanized buckets, which are commonly used to provide water to cattle, contribute zinc along the way. Thus, it is unlikely that a zinc deficiency would occur under normal circumstances; so, zinc supplementation of dairy cattle may be considered as precautionary only.

VITAMINS

Vitamins are complex organic compounds that are required in minute amounts by one or more animal species for normal growth, production, reproduction, and/or health.

Dairy cattle, like other animals, require vitamins for optimum performance and health.

Vitamins are classified as fat-soluble or water-soluble. The fat-soluble vitamins include vitamins A, D, E, and K; the water soluble vitamins include the B vitamins and vitamin C.

Tables 12–3 and 12–4 of Chapter 12 present the daily vitamin A and vitamin D requirements for different classes of dairy animals; and Table 12–6 shows the recommended vitamin content of rations for different classes of dairy animals. Table 11–5 presents in summary form the vitamin requirements of all dairy cattle. Chapter 21, Feed Composition Tables, shows the fat-soluble vitamin and water-soluble vitamin content of feeds.

DAIRY CATTLE VITAMIN CHART

Table 11–5, Dairy Cattle Vitamin Chart (pp. 222–223), presents in summary form the vitamins that are known to be required by dairy cattle.

Vitamins Which May Be Deficient Under Normal Condtions	Conditions Usually Prevailing Where Deficiencies Are Reported	Functions of Vitamin	Deficiency Symptoms
Fat-Soluble Vitamins: **Vitamin A**—Vitamin A is found only in animals; plants contain the precursor—carotene. Vitamin A is the vitamin most likely to be of practical importance in dairy cattle.	Vitamin A deficiency is most likely to occur when dairy cattle are fed (1) high-concentrate rations; (2) bleached pasture or hay grown under drought conditions; (3) feeds that have had excess exposure to sunlight, air, and high temperature; (4) feeds that have been heavily processed or mixed with oxidizing materials such as minerals; and (5) feeds that have been stored for long periods of time. Dairy cattle particularly susceptible to vitamin A deficiency are: newborn calves deprived of colostrum; cattle that have been prevented from establishing or maintaining good liver stores through exposure to drought; cattle wintered without high-quality forage, and cattle exposed to stresses such as high temperatures or elevated nitrate intake.	Vitamin A functions as a component of the visual purple required for dim-light vision, and is essential for normal growth, reproduction, and maintenance of healthy epithelial tissue.	Signs of vitamin A deficiency include reduced feed intake, rough hair coat, edema of the joints and brisket, lacrimation, xerophthalmia, night blindness, slow growth, diarrhea, convulsive seizures, improper bone growth, blindness, low conception rates, abortion, stillbirths, blind calves, abnormal semen, reduced libido, and susceptibility to respiratory and other infections. Of these symptoms, only night blindness has proved unique to vitamin A deficiency. Clinical verification may include ophthalmoscopic examination, liver biopsy and assay, blood assay, testing spinal fluid pressure, conjunctival smears, and response to vitamin A therapy.
Vitamin D	Young calves kept indoors, especially in the wintertime. Cattle in northern U.S. on high silage and grain rations and a minimum of sun-cured hay.	Vitamin D is required for calcium and phosphorus absorption, normal mineralization of bone, and mobilization of calcium from bone.	Rickets in young calves, the symptoms of which are: decreased appetite, lowered growth rate, digestive disturbances, stiffness in gait, labored breathing, irritability, weakness, and occasionally, tetany and convulsions. Later, enlargement of the joints, slight arching of the back, bowing of the legs, and the erosion of the joint surfaces cause difficulty in locomotion. Posterior paralysis may follow fracture of vertebrae. In older animals with vitamin D deficiency, bones become weak and easily fractured, and posterior paralysis may accompany vertebral fractures. Vitamin D deficiency in the pregnant animal may result in dead, weak, or deformed calves at birth.
Vitamin E	Where soils are very low in selenium. When unsaturated fats are fed.	Vitamin E is an antioxidant. It has been widely used to protect and to facilitate the uptake and storage of vitamin A. In metabolism, it is linked closely with selenium. Some deficiency signs, particularly in white muscle disease, may respond to either selenium or vitamin E, or may require both.	Muscular dystrophy (commonly called white muscle disease) in calves 2 to 12 weeks of age; characterized by heart failure and paralysis varying in severity from slight lameness to inability to stand. Also, a dystrophic tongue is often seen in affected animals. A deficiency of vitamin E may be precipitated or accentuated by feeding unsaturated fats.
Vitamin K	When moldy sweet clover hay high in dicoumarol content is fed. Vitamin K deficiency results from the antagonistic action of dicoumarol that is formed in moldy sweet clover hay.	Vitamin K hastens blood clotting.	Sweet clover disease, characterized by prolonged blood clotting. Mild cases can be treated effectively with vitamin K.
Water-Soluble Vitamins: **B Vitamins**	When an antagonist is present. When ruminal synthesis is limited by lack of precursors or other problems.	Most of the established metabolic functions of B vitamins are important to cattle, as well as to other animals. Consequently, a physiological need for most B vitamins can be assumed for cattle of all ages. Vitamin B–12 is of special interest because of its role in propionate metabolism, and the practical incidence of vitamin B–12 deficiency as a secondary result of cobalt deficiency. Niacin has been reported to enhance protein synthesis by ruminal microorganisms.	Deficiency signs in young calves have been clearly demonstrated for thiamin, riboflavin, pyridoxine, pantothenic acid, biotin, nicotinic acid, vitamin B–12, and choline. Polioencephalomalacia in grain-fed cattle has been linked to thiaminase activity or production of a thiamin antimetabolite in the rumen. Affected animals have responded to intravenous adminstration of thiamin (2.2 mg/kg body weight).

[1]As used herein, the distinction between *nutrient requirements* and *recommended allowances* is as follows: In nutrient requirements, no margins of safety are included intentionally; whereas in recommended allowances, margins of safety are provided in order to compensate for variations in feed composition, environment, and possible losses during storage or processing.

11–5
VITAMIN CHART

Nutrient Requirements[1]		Recommended Allowances	Practical Sources of the Vitamin	Comments
Daily Nutrients/ Animal (or Injection)	Amount/Lb (or /kg) of Feed			
*Variable according to class, age, and level of milk production (see Chapter 12, Tables 12–3 and 12–4).	*Variable according to class, age, weight, and stage of production. It is higher for dry/pregnant cows and in early lactation (the first 3 weeks) than for lactation after 3 weeks or for growth (see Chapter 12, Table 12–4). On a dry ration basis, the vitamin A requirements are about as follows: *1. For first 3 weeks of lactation and for dry/pregnant cows, 1,816 IU/lb. *2. After first 3 weeks of lactation and for mature bulls, 1,453 IU/lb. *3. For calf milk replacer, 1,725 IU/lb. *4. For growing heifers and bulls, 999 IU/lb.	The vitamin A requirements listed in the 2 columns to the left are adequate under most practical conditions, but under certain stressful conditions such as low environmental temperature or exposure to infective bacteria, the requirements should be increased.	Stabilized vitamin A. Green pasture. Grass or legume silages. Yellow corn. Green hay not over 1 year old. The average carotene content of some common feeds is as follows: mg carotene per (lb) (kg) Legume hays (including alfalfa), avg. quality . . . . 9–14 20–31 Nonlegume hays, avg. quality . . . . 4–8 9–18 Dehydrated alfalfa meal, avg. quality . . . . 50–70 110–154 Yellow corn . . . . 0.8–1.0 1.8–2.2 Silages, corn, or sorghum 2–10 4–22	Carotene is rapidly destroyed by exposure to sunlight and air, especially at high temperatures. Hay over 1 year old, regardless of green color, is usually not an adequate source of carotene or vitamin A activity. Ensiling effectively preserves carotene, but the availability of carotene from corn silage may be low. The younger the animal, the quicker vitamin A deficiencies will occur. Mature animals may store sufficient vitamin A to last 6 months. When deficiency symptoms appear, they can be corrected (1) by increasing carotene intake through the introduction of high-quality forage, or (2) by supplying vitamin A in the feed or by injection.
*300 IU of vitamin D/100 lb of live weight.	*125 IU/lb (275 IU/kg) of dry ration. *Variable according to class, age, weight, and stage of production (see Chapter 12, Table 12–6).	Normally, dairy cattle receive sufficient vitamin D from exposure to direct sunlight or from sun-cured hay.	Exposure to direct sunlight. Sun-cured hay. Irradiated yeast.	Sun-cured alfalfa hay contains 300–1,000 IU/lb (661–2,204 IU/kg).
	*For young calves, 11 to 18 IU of vitamin E/lb of total dry feed.	Generally natural feeds supply adequate quantities of alpha-tocopherol for mature cattle, although muscular dystrophy in calves occurs in certain areas.	Alpha-tocopherol, added to the ration or injected intramuscularly. Commercial vitamin E supplements. Grains contain 6–15 mg vitamin E/lb (13–33 mg/kg).	The incidence of white muscle disease appears to be lower when the cows receive 2–3 lb (0.91–1.36 kg) of grain during last 60 days of pregnancy. Where supplemental vitamin E is needed, it may be added to the ration or injected intramuscularly.
			Vitamin K₁ is abundant in pasture and green roughage. Vitamin K₂ is synthesized in large amounts in the rumen. Either K₁ or K₂ effectively fulfill the vitamin K role in blood clotting mechanism.	Except when the dicoumarol content of hay is excessively high (as in moldy sweet clover hay), sufficient vitamin K is synthesized in the rumen of cattle.
	Chapter 12, Table 12–6, footnote 14, shows the minimum quantities of B vitamins that should be provided in milk replacers.	Usually, no dietary B vitamins need be supplied to cattle.	B vitamins are abundant in milk and many other feeds, and synthesis of B vitamins by ruminal microorganisms is extensive. Calves begin microbial synthesis of B vitamins very soon after the introduction of dry feed in the ration.	

*Where preceded by an asterisk, the nutrient requirements listed herein were taken from *Nutrient Requirements of Dairy Cattle*, Sixth Revised Edition, Update 1989, National Research Council-National Academy of Sciences, Washington, DC.

FAT SOLUBLE VITAMINS

Dairy cattle require fat-soluble vitamins A, D, E, and K. Generally, all classes of dairy cattle require a dietary source of vitamins A and E. Vitamin D must either be synthesized in the skin by action of ultraviolet radiation or be included in the ration. Rumen microbes synthesize adequate amounts of vitamin K to meet the needs of most dairy cattle with the exception of the young calf, whose rumen has not begun all of its functions.

Fortunately, under normal conditions, natural feeds furnish most fat-soluble vitamins or their precursors in adequate amounts. High-quality forages contain large amounts of vitamin A precursors, and vitamin E is abundant in most feeds. Vitamin D is found in large quantities in sun-cured forages. Additionally, cattle can store adequate reserves of the fat-soluble vitamins to meet their needs for several months. Yet, when dairy producers feed limited or low-quality forage, use high levels of ensiled forage, expose cattle to little sunlight, or use milk replacers for young calves, additional vitamins will probably be needed for optimum health and high performance.

■ **Vitamin A** — Vitamin A supplementation may be desirable (1) when poor-quality or limited amounts of forage are fed, (2) when feeding forage that has been stored for a long period and has lost its carotene through oxidation, or (3) when high levels of corn silage and low-carotene concentrates are fed.

A deficiency of vitamin A causes many problems. Some or all of the following symptoms may occur, depending on the length and severity of the deficiency: (1) night-blindness — a condition that is readily detected when animals are driven among obstacles in dim light, (2) watery eyes, (3) nasal discharge, (4) coughing, (5) diarrhea, (6) pneumonia, (7) lack of coordination, (8) staggering gait, (9) convulsive seizures, (10) complete blindness, (11) stratified keratinized epithelium, (12) increased susceptibility to infection, (13) loss of appetite, (14) emaciation, (15) rough hair coat, (16) scaly skin, (17) abortion, (18) shortened gestation period, (19) birth of dead, weak, or blind calves, and (20) retained placenta.

There are a number of indicators of vitamin A deficiency which may be used before clinical signs of deficiency become evident, one of the most sensitive of which in growing calves is the elevation of cerebrospinal fluid pressure.

The vitamin A requirements of cattle can be met by carotene in feeds, supplements of vitamin A in a stabilized form, or a combination of both. For cattle, 1 mg of carotene is considered to be equivalent to 400 IU (International Units) of vitamin A. The daily vitamin A requirements for different classes of dairy cattle are shown in Chapter 12, Tables 12–3 and 12–4; and Table 12–6 shows the recommended vitamin A content of rations for different classes of dairy animals. There are no allowances for vitamin A for milk production as intakes above those required for normal reproduction do not increase milk yield. **Note well:** The vitamin A requirements given in Tables 12–3, 12–4, and 12–6 are adequate under most practical conditions, but may be increased when animals are under certain stressful conditions such as low environmental temperature or exposure to infective bacteria.

Under most practical conditions, moderate excesses of vitamin A are not harmful. The NRC reports that the pre-sumed safe limit of vitamin A is 30,000 IU/lb of ration for both lactating and nonlactating cows.

■ **Vitamin D** — Cows fed sun-cured forage or exposed to sunlight do not need supplemental vitamin D. Even green forage, barn-cured hay, and silage have some vitamin D activity due to the irradiation of dead tissue of stems and leaves of growing plants. When animals are exposed to sunlight, vitamin D is synthesized by the skin in sufficient amounts for maintenance, growth, reproduction, and lactation. However, calves housed indoors need vitamin D supplementation due to lack of exposure to sunlight. Animal sources of vitamin D (called D_3) and plant sources (called D_2) are biologically equivalent in dairy cattle.

A vitamin D deficiency leads to a failure of bones to calcify normally, resulting in rickets in calves and osteomalacia in adults. Vitamin D deficiencies in calves kept indoors do occur, but deficiencies in mature cattle under normal conditions are extremely unlikely because exposure to sunlight provides adequate vitamin D. Some of the first signs of vitamin D deficiency rickets are decreases in the blood plasma concentrations of calcium or inorganic phosphorus, or both, and increases in serum phosphates.

Fig. 11–8. Calf with severe rickets. Note the bowed legs and swollen joints. Rickets may be caused by a lack of vitamin D, calcium, or phosphorus; or by an incorrect ratio of the two minerals. (Courtesy, Michigan State University, East Lansing)

■ **Vitamin E** — Vitamin E is an antioxidant associated with selenium. It stimulates the immune system and reduces the incidence of oxidized flavor when fed at high levels (400 to 1,000 mg/cow/day); and it may aid in protection against white muscle disease, caused by deficiency of selenium.

White muscle disease is characterized by (1) a weakening of the leg muscles, resulting in calves walking with a typical crossing of the hind legs; (2) relaxation of the pasterns and splaying of the toes; (3) impaired ability to suckle, due to the musculature of the tongue being affected; and (4) in advanced cases, the calf may be unable to hold up its head and to stand.

All green feeds are good sources of vitamin E. Cows on pasture or being fed green chop receive adequate vitamin E. However, the vitamin E content of dry feedstuffs decreases during storage.

One IU of vitamin E is defined as 1 mg of dl-alpha-tocopherol acetate. For young calves, the vitamin E requirements are estimated to range from 11 to 18 IU of vitamin E/pound of total feed (see Chapter 12, Table 12–6).

Vitamin E and selenium play a synergistic role in the nutrition of calves. Some deficiency signs, such as white muscle disease, may respond to either vitamin E or selenium; some deficiencies may require both. Also, the requirements for these nutrients may be influenced by the type of liquid in the diet.

■ **Vitamin K** — Vitamin K functions as a stimulant to blood coagulation. Either vitamin K_1 (phylloquinone) or vitamin K_2 (menaquinone) will meet the needs of cattle. Green, leafy materials of any kind, both fresh and dry, are good sources of vitamin K_1. Normally, vitamin K_2 is synthesized in large amounts in the rumen; so, dietary supplementation is not recommended.

When cows consume moldy sweetclover hay, which is high in dicoumarol, blood coagulation may be impaired, followed by generalized hemorrhaging. This syndrome, commonly called *sweetclover disease* or *sweetclover poisoning*, responds to treatment with vitamin K.

WATER-SOLUBLE VITAMINS

The water-soluble vitamins include biotin, choline, folacin (folic acid), inositol, niacin (nicotinic acid, nicotinamide), pantothenic acid (vitamin B–3), para-aminobenzoic acid (PABA), riboflavin (vitamin B–2), thiamin (vitamin B–1), vitamin B–6 (pyridoxine, pyridoxal, pyridoxamine), vitamin B–12 (cobalamins) and vitamin C (ascorbic acid, dehydroascorbic acid). However, a physiological need in cattle of all of these vitamins has not been demonstrated.

Until recently, it was assumed that dairy cattle with a functional rumen did not require supplemental B vitamins. The rumen microflora were believed to synthesize adequate amounts of these nutrients for the host's requirements. Besides, the B vitamins are relatively abundant in dairy feeds. But recent evidence suggests a need for supplemental niacin under certain conditions and possibly supplemental choline and thiamin in the case of mature cattle, for which microbial synthesis and quantities in feeds may be inadequate, especially during diseased conditions or periods of stress. It is assumed that dairy cattle of all ages have a physiological need for most of the B vitamins, especially biotin, choline, niacin, pantothenic acid, riboflavin, thiamin, vitamin B–6, and vitamin B–12. In young calves, deficiency signs have been demonstrated when there is inadequate intake of these vitamins, but, even without a functioning rumen, their needs for these B vitamins appear to be met when they are fed whole milk. When young calves are fed milk replacers, however, it is advisable to ascertain the adequacy of vitamin intakes until their rumens are functional. Chapter 12, Table 12–6, footnote

14, shows the minimum quantities of B vitamins that should be provided in milk replacers.

■ **Biotin** — A biotin deficiency in calves, characterized by paralysis of the hindquarters, has been produced. Signs of deficiency did not develop when synthetic milk was supplemented with 4.5 mcg of biotin/pound of feed and fed at 10% of liveweight.

■ **Choline** — Researchers have produced choline deficiency in calves by using a synthetic ration containing 15% casein. Within 6–8 days, the calves developed extreme weakness and labored breathing and were unable to stand. Supplementation of the ration with 236 mg of choline/quart of synthetic milk prevented the development of these signs. Adding choline to the ration may increase the percentage of milk fat in lactating cows.

■ **Niacin (nicotinic acid, nicotinamide)** — Niacin is required by the young preruminant calf. In order to prevent a niacin deficiency, it is recommended that niacin be added to milk replacers at a level of 2.6 ppm.

Some recent evidence suggests that rumen microorganisms may not synthesize adequate amounts of niacin to meet the needs of high-producing cows in early lactation. A summary of several studies of cows in early lactation shows a small increase in milk production and butterfat, along with a decrease in the incidence of ketosis, to the feeding of 6 g of niacin per cow per day. Although research results are not consistent, there appears to be significant benefits from feeding niacin to dairy cows with above average incidence of ketosis and that are overconditioned. Research studies indicate that supplementation of 6 to 12 g of niacin/day should begin 2 weeks before calving, then continue for 8 to 12 weeks after calving.

It is conjectured that the major reason for improvement in milk production that occurs with added niacin may be related to the role of niacin in carbohydrate and lipid metabolism and the resultant decrease in ketosis. Niacin may also influence rumen fermentation, as evidenced by greater microbial protein synthesis and increased levels of rumen propionate with niacin supplementation. When cows are fed heated soybean meal, rumen response to niacin is greater than it is when cows are fed unheated soybean meal.

■ **Pantothenic acid (vitamin B–3)** — Pantothenic acid deficiency in the calf is characterized by a scaly dermatitis around the eyes and muzzle, loss of appetite, diarrhea, weakness (unable to stand), and convulsions.

Pantothenic acid deficiency in animals with functioning rumens is unlikely due to microbial production of pantothenic acid.

■ **Riboflavin (vitamin B–2)** — Riboflavin deficiency in the calf is characterized by hyperemia (presence of blood) of the mucosa of the mouth, lesions in the corners of the mouth and along the edges of the lips, loss of hair — especially on the belly, and excess salivation.

A riboflavin deficiency in lactating cattle is unlikely because of the amounts of riboflavin that are present in feedstuffs and synthesized in the rumen.

■ **Thiamin (vitamin B–1)** — Thiamin deficiency in the calf may cause polioencephalomalacia, characterized by listlessness, muscular incoordination, progressive blindness, convulsions, and sudden death; which may be accompanied by diarrhea and dehydration. The condition is found primarily in cattle fed high-concentrate rations; and it has been linked to increased microbial thiaminase activity and the production of thiamin analogs in the rumen. Treatment consists of the IV or IM administration of thiamin at a rate of 1 mg/pound of liveweight.

■ **Vitamin B–6 (pyridoxine, pyridoxal, pyridoxamine)** — Vitamin B–6 deficiency has been produced in calves fed a synthetic diet. It is characterized by loss of appetite, cessation of growth; and after about 3 months, epileptic seizures in some, but not all, calves. Calves respond to vitamin B–6 therapy if it is initiated in the early stages of the disease.

■ **Vitamin B–12 (cobalamins)** — Vitamin B–12 deficiency has been produced in calves under 6 weeks of age by feeding them a diet containing no animal protein. Deficiency signs include poor appetite and growth, muscular weakness, and poor general condition. It has been suggested that the vitamin B–12 requirement for dairy cattle is between 0.15 and 0.3 g/pound liveweight.

Vitamin B–12 is of special interest in the mature ruminant because of its role in propionate metabolism and because of the incidence of B–12 deficiency as a secondary result of cobalt deficiency. Certain soils have insufficient cobalt to produce levels of the element in plants that are adequate to support optimum vitamin B–12 synthesis in the rumen.

WATER

Large amounts of water are essential if a cow is to produce to her maximum capacity. It is necessary for maintaining body fluids and proper ion balance; for digesting, absorbing, and metabolizing nutrients; for eliminating waste material and excess heat from the body; for providing a fluid environment for the fetus; and for transporting nutrients to and from body tissues.

The water that dairy cattle need is supplied by drinking, by water in the feed that they consume, and by metabolic water produced by the oxidation of organic nutrients.

Cows drink an average of 100 to 200 lb of water per day, with heavy producers drinking up to 300 lb per day (1 gal water = 8.33 lb). Cows need 4 to 5 lb of water for each pound of milk produced. The amount of water a cow will drink depends on her size and milk yield, the quantity of dry matter consumed, the temperature and relative humidity of the air, the temperature of the water, the quality of the water, and the amount of moisture in her feed.

Good-quality, clean water is of the utmost importance in a dairy cattle feeding program — a factor often neglected by many producers. Water troughs should be cleaned routinely to ensure that the water is free from dirt and pathogenic bacteria. A common drinking trough provides an excellent means of spreading parasites and disease.

It has long been known that cattle are sometimes poisoned when they drink lake water containing blue-green algae; thus, cattle should be prevented from drinking water with heavy algae growth.

In extremely cold weather, it is a good idea to have a tank heater to keep the water from freezing. Also, frequency of watering is important. Cows stabled in a stanchion-type barn produce 3.5 to 4% more milk if they have drinking cups available than if they are watered twice daily. Contrary to some opinions, cows do not produce more milk from softened water than from hard water.

Dairy cattle lose water from the body in saliva, urine, feces, and milk; through sweating; and by evaporation from body surfaces and the respiratory tract. The amount of water lost from the body of cattle is influenced by the activity of the animal, ambient temperature, humidity, respiratory rate, water intake, feed consumption, and other factors.

The amount of water required by the various types of dairy cattle varies considerably, as evidenced in Table 11–6. Several factors can affect the amount of water a particular animal will consume: (1) age, (2) body weight, (3) production, (4) weather (heat and humidity), and (5) type of ration. The intensity of production dramatically affects the water requirement. A cow that is not in lactation will have a daily water intake of about 90 lb. When she produces 20 to 50 lb of milk, this figure will increase to about 160 lb. When milk production reaches 80 lb of milk per day, water intake will reach close to 200 lb.

Fig. 11–9. Clear, clean water on a Missouri dairy farm. (Courtesy, *Holstein World*, Sandy Creek, NY)

TABLE 11–6
DAILY WATER CONSUMPTION OF DAIRY CATTLE

Age	Body Weight	Condition	Water Consumption
(weeks)	(lb)		(gal)
4	112	Growing	1.3–1.5
8	152	Growing	1.6–2.0
12	204	Growing	2.3–2.5
16	263	Growing	3.1–3.5
20	327	Growing	4.0–4.5
26	416	Growing	4.5–6.0
60	779	Growing	6–8
84	1,023	Pregnant	8–10
1–2 years	1,000–1,200	Fattening	8–9
2–8 years	1,200–1,600	Lactating	10–25
2–8 years	1,200–1,600	Grazing	4.5–9.0

QUESTIONS FOR STUDY AND DISCUSSION

1. Define *nutrition*. With what other sciences is nutrition intimately involved?

2. In what ways do age and nutritional state of cattle affect their body composition in water, fat, protein, and ash?

3. Describe the digestive system of ruminants.

4. Define (a) *rumination*, and (b) *eructation*. What roles do these two processes play in ruminant digestion?

5. How does the rumen of the newborn calf differ from that of an adult ruminant?

6. Compare the capacities of various parts of the digestive tract of the cow.

7. What are enzymes?

8. List and explain the functions of the major enzymes found in the digestive tract of the cow.

9. Define (a) *prehension*, (b) *mastication*, and (c) *deglutition*.

10. Why don't cows have need for canine teeth?

11. What functions does the tongue serve?

12. List the uses of saliva.

13. What is the pharynx? What is its role in digestion?

14. List and give the primary functions of each of the four compartments of the ruminant stomach.

15. The pancreas serves a two-fold purpose in the body. Discuss these roles.

16. List the primary functions of the liver.

17. What substances are absorbed in the colon?

18. What are nutrients?

19. What are the primary functions of nutrients? Discuss each of them.

20. Explain why the following situation applies: On an individual basis, the higher the production of a lactating cow, the smaller the proportion of nutrients needed for maintenance.

21. Discuss the importance of energy. What two classes of nutrients are used primarily for energy production? Which nutrient liberates the most energy?

22. What is TDN? How is it computed? What are its limitations?

23. What are the most common high energy dairy feeds?

24. What are proteins?

25. Of what significance is the finding that in high-producing cows bacterial synthesis of protein is insufficient to fulfill the amino acid needs?

26. Define *nonprotein nitrogen*. A hypothetical NPN product contains 92% nitrogen. What is its crude protein equivalent?

27. Explain why high-concentrate diets allow for better use of NPN than high-roughage diets.

28. Differentiate between macro- and microminerals. List the essential minerals under their appropriate division (macro or micro).

29. Explain how salt functions as a nutrient.

30. Discuss the importance of the calcium-phosphorus ratio and vitamin D in dairy rations.

31. What mineral is implicated with grass tetany?

32. Discuss the (a) functions, (b) deficiency symptoms, and (c) practical sources of each of the following minerals for dairy cattle: calcium, phosphorus, cobalt, iodine, and zinc.

33. Define *vitamin*. List the fat-soluble vitamins and give their functions and respective deficiency symptoms.

34. Why should newborn calves have access to water-soluble vitamins?

35. Describe the symptoms of vitamin E deficiency in calves.

36. Why is water so important for lactating cows?

SELECTED REFERENCES

Title of Publication	Author(s)	Publisher
Anatomy of the Domestic Animals, The	S. Sisson J. D. Grossman	W. B. Saunders Co., Philadelphia, PA, 1953
Anatomy and Physiology of Farm Animals	R. D. Frandson	Lea & Febiger, Philadelphia, PA, 1965
Animal Growth and Nutrition	Edited by E. S. E. Hafez A. I. Dyer	Lea & Febiger, Philadelphia, PA, 1969
Animal Nutrition, Second Edition	P. McDonald R. A. Edwards J. F. Greenhalgh	Oliver and Boyd, Edinburgh, Scotland, 1973

(Continued)

SELECTED REFERENCES (Continued)

Title of Publication	Author(s)	Publisher
Animal Nutrition, Second Edition	L. A. Maynard, *et al.*	McGraw-Hill Book Company, New York, NY, 1979
Animal Science, Ninth Edition	M. E. Ensminger	Interstate Publishers, Inc., Danville, IL, 1991
Bioenergetics and Growth	S. Brody	Reinhold Publishing Company, New York, NY, 1945
Dairy Cattle Feeding and Management, Sixth Edition	W. M Etgen P. M. Reaves	John Wiley & Sons, New York, NY, 1978
Dairy Cattle Feeding and Management	W. J. Miller	Academic Press, New York, NY, 1979
Dairy Cattle: Principles, Practices, Problems, Profits, Third Edition	D. L. Bath, *et al.*	Lea & Febiger, Philadelphia, PA, 1985
Digestive Physiology and Metabolism in Ruminants	Edited by Y. Ruckebusch P. Thivend	AVI Publishing Company, Inc., New York, NY, 1980
Digestive Physiology and Nutrition of Ruminants, Vols. 1 and 2	D. C. Church	O & B Books, Corvallis, OR, 1975
Dukes' Physiology of Domestic Animals, Eighth Edition	Edited by M. J. Swenson	Cornell University Press, Ithaca, NY, 1970
Energy Metabolism of Ruminants, The	K. L. Blaxter	Hutchinson Scientific and Technical, London, England, 1962
Fat-Soluble Vitamins, Vol. 9 of the *International Encyclopedia of Food and Nutrition*	Edited by R. A. Morton	Pergamon Press, New York, NY, 1970
Fat-Soluble Vitamins, The	Edited by H. F. DeLuca J. W. Suttic	University of Wisconsin, Madison, 1969
Feeds and Feeding, Second Edition	A. Cullison	Reston Publishing Company, Inc., Reston, VA, 1979
Feeds and Feeding, Twenty-second Edition	F. B. Morrison	The Morrison Publishing Company, Ithaca, NY, 1956
Feeds and Feeding, Abridged	F. B. Morrison	The Morrison Publishing Company, Ithaca, NY, 1956
Feeds & Nutrition, Second Edition	M. E. Ensminger J. E. Oldfield W. W. Heinemann	The Ensminger Publishing Company, Clovis, CA, 1990
Feeds & Nutrition Digest	M. E. Ensminger J. E. Oldfield W. W. Heinemann	The Ensminger Publishing Company, Clovis, CA, 1990
Fundamentals of Nutrition, Second Edition	L. E. Lloyd B. E. McDonald E. W. Crampton	W. H. Freeman and Company, San Francisco, CA, 1978
Livestock Feeds and Feeding, Second Edition	D. C. Church	O & B Books, Inc., Corvallis, OR, 1984
Nutrient Requirements of Dairy Cattle, Sixth Revised Edition, Update 1989	National Research Council	National Academy of Sciences, Washington, DC, 1989
Nutrients and Toxic Substances in Water for Livestock and Poultry	National Research Council	National Academy of Sciences, Washington, DC, 1974
Physiology of Digestion and Metabolism of the Ruminant	Edited by A. T. Phillipson	Oriel Press, Ltd., Newcastle upon Tyne, England, 1970
Principles and Practices of Feeding Dairy Cows	Edited by W. H. Broster R. H. Phipps C. L. Johnson	College of Estate Management, Reading University, England, 1986
Rumen and Its Microbes, The	R. E. Hungate	Academic Press, Inc., New York, NY, 1966
Science of Animals That Serve Humanity, The	J. R. Campbell J. R. Lasley	McGraw-Hill Book Company, New York, NY, 1985
Trace Elements in Human and Animal Nutrition, Fifth Edition	Edited by W. Mertz	Academic Press, Inc., New York, NY, 1987

Fig. 12–1.   Corral feeding. (Courtesy, The Brown Swiss Cattle Breeders' Assn. of the U.S.A., Beloit, WI)

Contents

CHAPTER

12

DAIRY

FEEDING

PROGRAMS

Feeding lactating dairy cattle differs from feeding other classes of farm animals. There is limited time between calvings in which to get whatever milk the cow will produce. Milk production reaches a peak 2 to 6 weeks after a cow freshens, then declines during the remaining portion of the lactation period. If feeding is limited, the rate of decline is more rapid than normal and total production is lowered. Thus, reducing the ration and concomitantly lowering production is almost certain to be economically unsound. By contrast, if the economics so warrant, a growing-finishing (meat) animal can be limited-fed for slower gains with market finish reached after a longer feeding period.

As labor became more scarce, and more expensive, dairy producers automated—they replaced part of the labor force with machines. Eventually, it became advantageous to divide the herd into corrals (groups) of cows producing at about the same level, with each corral of 50 to 100 cows managed as a unit. The corral concept led to changes in feeding practices, including group feeding based on the needs of a corral of cows, rather than the needs of an individual cow, and complete rations (grain and forage combined), with a different combination of feeds for each corral. Next came confinement, environmental control, and pollution control. Other innovations will follow.

This chapter deals with the mechanics of feeding dairy cattle: (1) evaluating and buying feeds, (2) feeding standards—ration formulation, and (3) feeding programs. Hence, each aspect is discussed separately within the chapter.

PART I
EVALUATING AND BUYING FEEDS

Profit is the ultimate criterion of success in any dairy operation, and the cost of nutrients is an important factor in determining success. The feed composition values presented in this and other books merely represent an averaging of an accumulation of data concerning the nutritive value of feeds. Considerable variation is inherent in the nutrient content of different samples of feeds. Thus, the successful dairy producer must recognize the value of a well-planned feed analysis program.

FEED ANALYSIS

Feeds are analyzed by physical, chemical, or biological procedures. Although physical evaluation may be the least accurate, it provides a quick and easy means of obtaining considerable information about the overall quality of a feed. The chemical procedure is more accurate than a physical evaluation, but it takes time. The biological method necessitates considerable time and expense, and the results are often variable.

PHYSICAL EVALUATION OF FEEDSTUFFS

In order to produce or buy superior feeds, dairy producers need to know what constitutes feed quality and how to recognize it. They need to be familiar with those recognizable characteristics of feeds which indicate high palatability and nutrient content. If in doubt, observation of the cattle consuming the feed will tell them; for cattle prefer and thrive on high-quality feed.

The physical evaluation of feedstuffs, especially forages, is based largely on sight and aroma appeal.

The easily recognizable characteristics of hay of high feeding value are given in Chapter 8, Hay and Other Dry Forages, in the section headed, "Characteristics of High-Quality Hay"; and the easily recognizable characteristics of silage of high feeding value are given in Chapter 9, Silage and Haylage, in the section headed, "Characteristics of Good Silage."

The easily recognizable characteristics of good grains and other concentrates are:

1. Seeds are not split or cracked.

2. Seeds are of low-moisture content—generally containing about 88% dry matter.

3. Seeds have a good color, characteristic of the species.

4. Concentrates and seeds are free from mold.

5. Concentrates and seeds are free from rodent and insect damage.

6. Concentrates and seeds are free from foreign material, such as iron filings.

7. Concentrates and seeds are free from rancid odor.

CHEMICAL ANALYSIS

A chemical analysis of a feed is only as good as the sample. If the sample is not representative of the entire batch, the evaluation is useless, no matter how extensively it is analyzed. This point bears emphasis, because feeds tend to be highly variable in composition.

Today, feeds are being analyzed routinely through highly sophisticated chemical procedures. Many agricultural experiment stations, as well as most large feed companies, have facilities to analyze feeds for both the prevention and diagnosis of nutritional problems.

A chemical analysis gives a solid foundation on which to start in the evaluation of feeds. Thus, feed composition tables serve as a basis for ration formulation and for feed purchasing and merchandising. Commercially prepared feeds are required by state law to be labeled with a list of ingredients and a guaranteed analysis. Although state laws vary slightly, most of them require that the feed label (tag) show in percent the minimum crude protein and fat, and the maximum crude fiber and ash. Some feed labels also include maximum salt, minimum TDN, and/or minimum calcium and phosphorus. These figures are the buyer's assurance that the feed contains the minimal amounts of the higher cost items—protein and fat; and not more than the stipulated amounts of the lower cost, and less valuable, items—the crude fiber and ash.

PROXIMATE ANALYSIS (WEENDE PROCEDURE)

For more than 100 years, feeds have been analyzed by a method developed by two scientists, Henneberg and Stohmann, at the Weende Experiment Station in Germany. This method is called the proximate analysis, or the Weende system, of feed analysis. Feeds are broken down into six components: (1) moisture, (2) ash, (3) crude protein, (4) ether extract, (5) crude fiber, and (6) nitrogen-free extract.

VAN SOEST ANALYSIS

The inadequacies of proximate analysis gave rise to the detergent system of feed analysis for estimating energy content of forages, developed by Peter J. Van Soest at the USDA Beltsville National Research facility, in the 1960s. By using detergents, the Van Soest system separates fibrous feeds into two fractions: a neutral detergent fibrous fraction and an acid detergent fibrous fraction.

■ **Neutral detergent fiber (NDF)**—Neutral detergents are used to separate the feed into two fractions: (1) neutral detergent solubles, representing the highly digestible portion of the feed and consisting of proteins, fats, and carbohydrates (along with nonprotein nitrogen, pectin, and soluble materials); and (2) neutral detergent fiber (NDF), representing the less digestible portion of the feed, consisting of plant cell walls, including lignin, cellulose, and hemicellulose.

NDF is closely related to feed intake because it contains all the fiber components that occupy space in the rumen and are slowly digested. Thus, the lower the NDF percentage, the more the animal will eat; it is inversely related to voluntary feed consumption. Hence, a low percentage of NDF is desirable. It is noteworthy that milk production of lactating cows is more highly correlated with the NDF portion of the ration than with the ADF in the ration.

■ **Acid detergent fiber (ADF)**—Acid detergent solutions are used to separate the feed into two fractions: (1) acid detergent solubles, containing the more readily digestible hemicellulose; and (2) acid detergent fiber (ADF), representing the less digestible portion of the feed and consisting of lignin (indigestible) and cellulose (digestible).

ADF is an indicator of forage digestibility because it contains a high proportion of lignin which is the indigestible fiber fraction. NDF will always be a higher number than ADF because ADF does not contain hemicellulose. The lower the ADF, the more feed an animal can digest. Hence, a low ADF percentage is desirable.

■ **Acid detergent lignin (ADL)**—Sulfuric acid may be used further to separate the ADF into (1) cellulose, which is digestible; and (2) lignin, which is indigestible.

BOMB CALORIMETRY

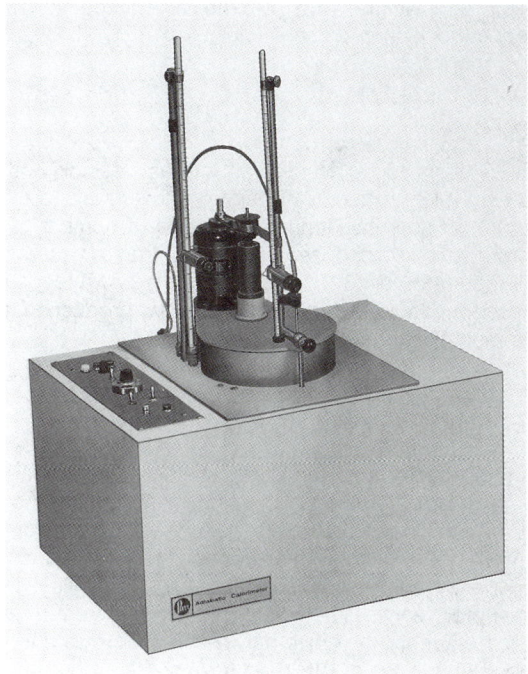

Fig. 12–2. Bomb calorimeter for the determination of gross energy. (Courtesy, Parr Instrument Company, Moline, IL)

When compounds are burned completely in the presence of oxygen, the resulting heat is referred to as gross energy or the heat of combustion. The bomb calorimeter is used to determine the gross energy of feed, waste products from feed (for example, feces and urine), and tissues.

The calorie is defined as the amount of heat required to raise the temperature of 1 g of water 1°C (precisely from 14.5° to 15.5°C).

Briefly stated, the bomb calorimeter works as follows: An electric wire is attached to the material being tested, so that it can be ignited by remote control; 2,000 g of water are poured around the bomb; 25 to 30 atmospheres of oxygen are added to the bomb; the material is ignited; the heat given off from the burned material warms the water; and a thermometer registers the change in temperature of the water. For example, if 1 g of feed is burned and the temperature of the water increases 1°C, 2,000 calories/gram are given off.

It is noteworthy that the determination of the heat of combustion with a bomb calorimeter is not as difficult or time-consuming as the chemical analyses used in arriving at TDN values.

BIOLOGICAL ANALYSIS

Quite often, biological assays are used in the analysis of micronutrients in feeds. There are two basic types of biological assays—(1) microbiological assays, and (2) the use of nutrient-deficient animals.

Biological assays tend to be laborious and time-consuming. Large numbers of samples are needed to produce statistically reliable results, and quite often data obtained from these assays are highly variable. The assay utilizing nutrient-deficient animals is particularly cumbersome because (1) the animals should be of approximately the same age, sex, and weight; and (2) time is required to induce deficient conditions in these animals.

FEEDS

Feed costs must be held as low as possible without depressing milk production if dairy producers are to obtain any return on their investment. They have a number of options from which to choose: (1) grow all their own feed, (2) buy all their feed, or (3) supplement homegrown feeds with purchased feeds. In order to get maximum production, producers must weigh all of their options.

HOMEGROWN VS PURCHASED FEEDS

A major factor in determining whether feeds should be homegrown or purchased is the system of farming. Over a period of time, dairy producers do those things which are most profitable to them, with some modification according to personal preference. Thus, profitability, with modification by personal preference, largely determines the system of farming. It follows that the choice between homegrown and purchased feeds for the strictly dairy farm or the crop and dairy farm combination is largely based on profitability.

Usually a combination of several factors determines the profits to accrue from homegrown or purchased feeds. Except for very large and intensive dairy operations, net returns are generally highest on those farms that produce most of their pasture, hay, and/or silage. However, on large specialized operations where land and labor are limited, highest net returns may accrue from the purchase of all feeds.

Generally speaking, part or all of dairy feeds are purchased under the following circumstances:

1. When land, water, and/or labor are limited.

2. Where the soil and climate of the farm are not suited to the production of a desired feed(s).

3. When feeds of comparable quality can be bought on a delivered basis more cheaply than they can be produced.

4. When storage facilities are limited, with the result that feeds produced on the farm cannot be held for yearlong use.

5. Where season-long financing for crop production cannot be secured, but where it is possible to "pay as you go," such as dairy producers using their monthly paycheck to buy feeds.

6. Where homegrown feeds need to be supplemented by certain nutrients, such as protein, minerals, and/or vitamins. This need exists on almost all dairy operations, for few of them produce all the nutrients needed for a balanced ration.

7. Where special feeds are desired, such as molasses, fats, etc.

8. Where a purchased feed can be substituted for a homegrown feed on an equal feeding value basis, but with the purchased feed acquired at a lower cost than the selling price of the homegrown product.

9. Where it is more profitable to use available capital, management, and labor for maximum animal production. This frequently applies to large dairies where profits may be maximized by increasing milk output rather than spending time and capital on raising crops.

10. Where the operator is not knowledgeable relative to crop production.

Generally speaking, forages are more apt to be homegrown than grains, primarily due to their greater bulkiness and higher cost of transportation. However, increased density of hay packages has made it economically feasible to transport such forages greater distances.

WHAT DAIRY PRODUCERS-FEED BUYERS SHOULD KNOW

Most purchased dairy feeds are bought by dairy producers themselves—by practical operators who subsequently feed them. So, no one has a greater incentive to purchase wisely and well than these producer-buyers. However, altogether too often they think only in terms of cost per bushel, per hundredweight, or per ton. Too often they merely consider the feed analysis or the guarantee based on averages, oblivious to such quality-affecting factors as variety, soil, and weather. Too often buying means haggling with the seller over the price of a feed, rather than considering the net returns. Too often when buying feeds these practical operators are competing with highly trained professionals with great expertise in buying—they are competing with buyers representing commercial feed companies or brokers. For these reasons, more and more livestock producers are purchasing formula feeds, rather than separate ingredients. By so doing, they are acquiring a service—ration formulation—as well as a feed.

Of course, price of feed is very important. For this reason, a sizable section in this chapter is devoted to "Best Buys in Feeds." Additionally, a number of complex and interrelated factors should be considered by the producer-buyer when purchasing feeds. Successful feed buying necessitates knowledge of all the factors that affect net returns, from the time a deal is made to buy the feed until the end product is marketed. Today, sophisticated dairy producers-feed buyers need to know the following:

1. They need to know the nutritive requirements of their cattle.

2. They need to know and speak the language of the feed industry.

3. They need to know production and economic trends.

4. They need to know business aspects, including sources of credit, interest rates, contracts, futures, and possible tax savings to accrue from purchasing feeds before the end of the year.

5. They need to know the different feed grades and quality classifications.

6. They need to know the restrictive use of certain feedstuffs, such as urea.

7. They need to know the associative, or additive, effects of certain feedstuffs.

8. They need to know the origin of the feed ingredients because of the direct and important relationship between the fertility of the soil and the climate of the area and the feed that it produces.

9. They need to know the local potential to grow certain feeds.

10. They need to know the longtime availability of feeds.

11. They need to know the moisture content of the feed ingredients.

12. They need to know transportation costs.

13. They need to know the storage capabilities of feeds.

14. They need to know feed shrinkage.

15. They need to know the risks; such as when wet hay ferments and generates heat in storage, resulting in a hazard of spontaneous combustion and fire, usually about a month or 6 weeks after storage.

16. They need to know what processing will be involved.

17. They need to know that certain feedstuffs affect the product produced; for example, when buying hay for dairy cows, buyers must avoid weeds that impart a strong flavor to milk.

18. They need to know about toxic residues.

19. They need to know the government regulations pertaining to the incorporation of feed additives, especially as to levels, drug combinations, and withdrawals.

20. They need to know the impact of foreign feed purchases.

OTHER FEED REQUISITES

In addition to the considerations already noted, it is important that all feeds—both bought and homegrown—meet the following requisites:

1. **Palatability.** Palatability is important. The producer should avoid mature, moldy, coarse, or weedy hay; finely ground hay or grain; and silages which are moldy, slimy, or too mature. It may be necessary to let the animals become accustomed to new feeds before judging their palatability. Some feeds which are unpalatable when first presented to cattle become quite palatable after they become accustomed to them.

2. **Variety.** Some variety in the ration is desirable, but palatability and nutritive content of individual feeds are more important than number of ingredients.

3. **Bulk.** Some bulk in the ration is desirable. Cows crave some dry forage and will eat bedding and other coarse material if sufficient fiber is not included in the ration. The proportion of grain to forage should be determined largely by comparative price. But the amount of bulk should not be decreased below that which will keep the animal on feed. This can be helped by processing the concentrate portion of the ration as coarsely as possible.

4. **Laxativeness.** Cows that receive average amounts of legume hay and/or silage seldom become constipated. Constipation can be corrected by feeding such feeds as alfalfa, wheat bran, linseed meal, or molasses. Lush green forage also has a laxative effect.

5. **Cost.** Cost is important, but net return is even more important; hence, it may well be said that it is net returns rather than cost per ton, or per bag, that count.

6. **Feeds affecting milk flavor.** Consumers want milk to taste like milk—not like silage, grass, or weeds.

Although feeds are not the only cause of milk flavors, they are major contributors. Feed flavors enter the milk through the digestive system, respiratory system, and by direct absorption. Research indicates that most feed flavors are detectable in the milk 20 minutes after the feed is consumed, and that they are usually most pronounced at the end of 2 hours.

Feed flavors that enter the milk through the respiratory system can usually be detected much sooner than those entering through the digestive system. For example, if a cow breathes air reeking with silage odors, these flavors can be detected in the milk almost immediately. Flavors that are directly absorbed by milk are less common, but they can be detected if the milk is left exposed for a long enough period.

The following control measures are recommended to alleviate feed flavors:

a. **Avoid sudden change to fresh, lush pasture.** Cows should be shifted from winter feeding, or old pasture, to new and lush pastures on a gradual basis. Also, cows should be taken out of pastures 2 to 3 hours before milking. For the same reason, freshly cut grass should not be fed immediately before milking.

b. **Control and avoid undesirable weeds.** When eaten by cows, many weeds will impart a strong flavor to milk; among them, wild onions, skunk cabbage, some members of the mustard family, bitterweed, carrot weed, ragweed, and others. It is easier to get rid of these weeds today than formerly, so they should be eliminated from pasture and hayfields utilized by milk cows.

c. **Feed silage after milking.** Silage flavor is both common and objectionable. It can be avoided by feeding all silages after milking, never before or during milking. Usually one will be safe if silage is not fed within 2 to 4 hours of milking time, but it is safer to feed it shortly after milking. This permits the flavor-causing material to pass through the cow's digestive system before the next milking.

If cows breathe the odor of silage, it will appear as flavor in the milk. Thus, silage should never be left in the mangers or feed alleys. In fact, it is preferable that it be fed in the corral, and not in the area where the cows are being milked.

7. **Poisonous plants and feeds.** Poisonous plants and feeds should be avoided. Dairy producers should know the poisonous plants common to the area, and avoid them. Also, the following poisons should be avoided: prussic acid, hydrocyanic acid, ergot, scabbed grain, smut on grain, spoiled or moldy feed, botulism, aflatoxins, excessive selenium, nitrates, lead, and mercury.

BEST BUY IN FEEDS

Feed prices vary widely. For profitable production, therefore, feeds with similar nutritive properties should be interchanged as price relationships warrant.

Purchase of feed nutrients at least cost, coupled with knowledge of the nutritive needs of animals and production results, provides a sound basis for arriving at the best buy in feeds.

Fig. 12–3. Stanchioned cows feeding in a stall barn. The owner's bottom line is: *cost of feed per hundred pounds of milk produced*. (Courtesy, Holstein-Friesian Assn. of America, Brattleboro, VT)

COST PER UNIT OF NUTRIENTS

One method of arriving at the best buy in feeds is to compute and compare the cost per unit of nutrients, based on feed composition. Where a chemical analysis of a specific feed is not available, feed composition tables, such as those in Chapter 21 of this book, may be used as good indicators. Thus, feed composition tables may serve as a basis of feed purchasing and merchandising, as well as for ration formulation.

The use of the cost per unit of nutrients method can best be illustrated by the examples that follow:

■ **Cost per pound of protein and TDN**—If 44% protein (crude) soybean meal is selling at $9.88 per 100 lb whereas 35% protein (crude) linseed meal sells for $6.25 per 100 lb, which is the better buy? Divide $9.88 by 44 to get 22.4¢ per pound of crude protein for the soybean meal. Then divide $6.25 by 35 and get 17.8¢ per pound of crude protein for the linseed meal. Thus, at these prices linseed meal is the better buy—by 4.6¢ (22.4 − 17.8 = 4.6) per pound of crude protein.

When buying energy feed, one can compare the cost per pound of total digestible nutrients (TDN). For example, if corn is priced at $3.63 per 100 lb and has a TDN of 91%, divide $3.63 by 91 and the result is 3.99¢ per pound of TDN. If milo with 86% TDN sells for $3.25 per 100 lb, divide $3.25 by 86, and the price is 3.78¢ per pound of TDN. Thus, milo would be the better buy by 0.21¢ (3.99 − 3.78 = 0.21) per pound of TDN.

■ **Cost per pound of phosphorus**—When buying a mineral supplement, the dairy producer may also check price against value received. For example, let's assume that the main need is for phosphorus, and that we wish to compare minerals, which we shall call brands "X" and "Y." Brand "X" contains 12% phosphorus and sells at $340 per ton or $17/cwt, whereas brand "Y" contains 10% phosphorus and sells at $320 per ton or $16/cwt. Which is the better buy?

COMPARATIVE VALUE OF BRANDS "X" AND "Y"
(based on phosphorus content alone)

Brand	Phosphorus (%)	Price/cwt ($)	Cost/lb Phosphorus ($)
"X"	12	17.00	1.42
"Y"	10	16.00	1.60

Hence, brand "X" is the better buy, even though it costs $1 more per hundred, or $20 more per ton.

One other thing is important when buying minerals. As a usual thing, the more scientifically formulated mineral mixes will have plus values in terms of trace mineral needs and balance.

■ **Factors other than price affect feeding value of feeds**—Of course, it is recognized that many other factors affect the actual feeding value of each feed, such as (1) palatability, (2) grade of feed, (3) preparation of feed, (4) ingredients with which each feed is combined, and (5) quantities of each feed fed. It follows that, from the standpoint of the producer, the most important measurement of a feed's usefulness is in terms of *net returns*, rather than cost per bag or cost per ton.

FEED SUBSTITUTION TABLE

The successful dairy producer recognizes that feeds of similar nutritive properties can and should be interchanged in the ration as price relationships warrant, thereby making it possible at all times to obtain a balanced ration at the lowest cost. Thus, (1) the cereal grains may consist of corn, barley, wheat, oats, and/or sorghum; (2) the protein supplements may consist of soybean, cottonseed, peanut, sunflower, and/or linseed meal; (3) the roughage may include many varieties of hays and silages; and (4) a vast array of by-product feeds may be utilized. The selection of alternative feeds has been made vastly easier and more accurate through the application of modern computer techniques.

Table 12–1, Feed Substitution Table for Dairy Cattle, is a summary of the comparative values of the most common U.S. feeds. In arriving at these values, two primary factors besides chemical composition and feeding value have been considered—namely, palatability and product quality.

In using this feed substitution table, the following facts should be recognized:

1. That, for best results, different ages and groups of animals within classes should be fed differently.

2. That individual feeds differ widely in feeding value. Barley and oats, for example, vary widely in feeding value according to the hull content and the test weight per bushel,

and forages vary widely according to the stage of maturity at which they are cut and how well they are cured and stored.

3. That nonlegume forages may have a higher relative value to legumes than herein indicated provided the chief need of the animal is for additional energy rather than for supplemented protein. Thus, the nonlegume forages of low value can be used to better advantage for wintering mature, dry cows than for young calves or lactating cows.

On the other hand, legumes may have a higher value relative to nonlegumes than herein indicated provided the chief need is for additional protein rather than for added energy.

4. That, based primarily on available supply and price, certain feeds—especially those of medium protein content, such as brewers' dried grains, corn gluten feed (gluten feed), distillers' dried grains, distillers' dried solubles, peanuts, and peas (dried)—may be used interchangeably as (a) grains and by-product feeds and/or (b) protein supplements.

5. That the feeding value of certain feeds is materially affected by preparation. Thus, wheat must be coarsely ground or rolled for cattle. The values herein reported are based on proper feed preparation in each case.

For these reasons, the comparative values of feeds shown in the feed substitution table are not absolute. Rather, they are reasonably accurate approximations based on average-quality feeds, together with experiences and experiments.

TABLE 12–1
FEED SUBSTITUTION TABLE FOR DAIRY CATTLE (AS-FED BASIS)

Feedstuff	Relative Feeding Value (lb for lb) In Comparison With The Designated (underlined) Base Feed Which = 100	Maximum Percentage of Base Feed (or comparable feed or feeds) Which It Can Replace For Best Results	Remarks
GRAINS, BY-PRODUCT FEEDS, ROOTS AND TUBERS:[1] (Low and Medium Protein Feeds)			
Corn, No. 2	_100_	_100_	The most important grain feed in the U.S. Grind coarsely or flake.
Almond hulls, dried, no shells	70–75	15–30	
Almond hulls and shell meal	35	15–20	
Apple pomace, air-dry	78	33⅓	Values given are for apple pomace with paper or rice hulls as press aids.
Bakery products, dried	110	15–30	
Bakery waste, not dried (30% water) . .	75	15–30	
Barley	90	25–100	The heavier the barley and the smaller the proportion of hulls, the higher the feeding value. Grind coarsely or roll for cattle. In Canada, where considerable barley is fed, it is often used as the only basal feed in the ration once animals are accustomed to it.
Beans (cull)	80	10	Best when cooked, but can also be fed raw. Beans should be ground. When cooked, 3–4 lb _(1.4–1.8 kg)_/head daily; when raw, 1–2 lb _(0.45–0.91 kg)_. Scouring may occur if they constitute more than 15% of total ration.
Beet pulp, dried	90	50	
Beet pulp, molasses, dried	90–95	50	
Beet pulp, wet	25	40	50% the value of corn silage. May compose 40% of ration on dry matter basis.

(Continued)

TABLE 12–1 *(Continued)*

Feedstuff	Relative Feeding Value (lb for lb) In Comparison With The Designated (underlined) Base Feed Which = 100	Maximum Percentage of Base Feed (or comparable feed or feeds) Which It Can Replace For Best Results	Remarks
GRAINS, BY-PRODUCT FEEDS, ROOTS AND TUBERS: (Continued)			
Brewers' dried grains	80	33⅓	Not very palatable; fed chiefly to dairy cattle.
Brewers' grains (wet)	13–15	33⅓	Grains usually come from barley. Best to haul and feed directly. Can be stored in silo if salt is added at rate of 25 lb *(11.4 kg)* per ton of grains.
Buckwheat	55–75	33⅓	Should be ground and mixed with other grains.
Carrots (cull)	10–15	20–25	Store 3–4 weeks before using; fresh carrots cause scouring. Feed whole or sliced.
Citrus pulp, dried	80–88	25–50	
Corn-and-cob meal	85–90	100	
Corn gluten feed (gluten feed)	85–90	50	
Distillers' dried grains	73–90	33⅓	Rye distillers' dried grains are of lower value than similar products made from corn or wheat. Distillers' dried grains are used chiefly for dairy cattle.
Distillers' dried solubles	73–90	33⅓	The chief difference between distillers' dried grains and distillers' dried solubles is the higher B vitamin content of the latter. Normally this is not important for dairy cattle.
Fat (animal or vegetable)	225	5	Fat has 203 megacalories energy/100 lb *(45.4 kg)* for maintenance and 127 megacalories for weight gain, as compared to 92 and 60, respectively, for corn.
Hominy feed	100	50	
Manure, cattle, without bedding	75	50	Approximately 80% of the total nutrients of feeds is excreted as animal manure. However, the feeding value of manure will vary according to (1) the nutritive value of the feeds initially fed, (2) the class, age, and individuality of the animal to which the feeds were initially fed, and (3) the handling and processing of the manure.
Manure, poultry (see poultry house litter)			
Molasses, beet	75	10–40	Value is highest when used as an appetizer. May be laxative if fed at levels above 6 lb *(2.7 kg)* daily.
Molasses, cane	75	10–40	Value is highest when used as an appetizer.
Molasses, citrus	65–75	10–40	
Molasses, wood	26–30	10–20	Unpalatable.
Oats	70–90	10–100	Valuable for young stock, and for breeding stock. Also, the feeding value of oats varies according to the test weight per bushel. Grind or roll for cattle.
Paunch, dried (also see "paunch-blood" under Protein Supplements of this table)	90	5–10	Dried paunch is not palatable, with the result that it depresses appetite. Rate of gain of young dairy stock is not affected, but feed efficiency is slightly lowered.
Peas (cull), dried	88	40	Because of lack of palatability, peas will lower feed intake if they constitute more than 20% of the total ration. Also, there is bloat hazard if they exceed 40% of the ration.
Pear waste, air-dry	75	40	
Potatoes (Irish), wet	20–25	85	When fed with alfalfa hay, they are worth about 80% as much per ton as corn silage. Do not feed frozen. Sunburned, decomposed, or sprouted potatoes should not make up more than 10% of potatoes fed.
Potatoes (Irish), dehydrated	88	50	Excellent source of energy, but deficient in protein, minerals, and vitamins.
Potatoes (sweet)	25	85	
Potatoes (sweet), dehydrated	95–100	50	Dehydrated sweet potatoes are more palatable than dehydrated Irish potaotes.
Poultry house litter	10–40	15–25	Poultry house litter may also be used as a protein source (see Protein Supplements, this table).

(Continued)

TABLE 12–1 *(Continued)*

Feedstuff	Relative Feeding Value (lb for lb) In Comparison With The Designated (underlined) Base Feed Which = 100	Maximum Percentage of Base Feed (or comparable feed or feeds) Which It Can Replace For Best Results	Remarks
GRAINS, BY-PRODUCT FEEDS, ROOTS AND TUBERS: (Continued)			
Prunes	62	15	Because of the laxative quality of prunes, they should be limited to 7% of the total ration.
Raisins (cull)	70	33⅓	
Raisin pulp	53	25	
Rice (rough rice)	80	100	
Rice bran	66⅔–75	33⅓	
Rice polishings	88	25	
Rye	96	33⅓	Not palatable when fed in large amounts.
Screenings, refuse	62–70	25–35	Should be finely ground in order to kill noxious weed seeds. Quality varies; good-quality screenings are equal to oats whereas poor-quality screenings resemble straw.
Sorghum (milo, kafir), grain	90–95	100	Varieties vary in protein content. Grind or roll for cattle.
Spelt and emmer	70–90	30–100	Similar to oats.
Wheat	100–105	50	Grind coarsely, or roll.
Wheat bran	70–90	25–33⅓	Bran is valuable for young dairy animals, for breeding animals, and for starting animals on feed.
Wheat-mixed feed (mill run)	95	33⅓	Sometimes fed to the breeding herd, or to young calves.
Wheat screenings	85	50	
Wood (cooked)	75–80	70	Wood products, which are largely cellulose and lignin, must be cooked before animals can digest them.
PROTEIN SUPPLEMENTS:			
Soybean meal (41%)	***100***	***100***	Slightly laxative effect.
Alfalfa or clover screenings	70–75	50	Grind finely to destory weed seeds.
Brewers' dried grains	55–65	50	Not very palatable. Fed chiefly to dairy cattle.
Copra meal (coconut oil meal), 21% . .	90–100	50	
Corn gluten feed (gluten feed)	65–75	50–100	
Corn gluten meal (gluten meal)	90–100	50	Somewhat unpalatable.
Cottonseed meal (41%)	100	100	
Distillers' dried grains	65–70	100	Rye distillers' grains are about 10% lower in protein than similar products made from corn or wheat. Low in palatability.
Distillers' dried solubles	70	100	
Feather meal (hydrolyzed; 84% protein)	175	50	Feather meal is unpalatable; hence, cattle must be accustomed to it gradually and it must be limited in quantity.
Legume screenings	75	75	Satisfactory, but less palatable than soybean or cottonseed meal.
Linseed meal (35%)	95	100	Linseed meal has laxative effect. Some cattle will not tolerate more than 5–8% linseed meal in the ration.
Paunch-blood feed (also see "paunch, dried" under Grains section of this table)	100	100	At slaughter, each bovine yields about 20 lb *(9.1 kg)* of paunch and 20 lb *(9.1 kg)* of blood. Dried paunch runs around 10% protein, dried blood around 80%, and a 50–50 mixture of the 2 products, around 45%.
Peanut meal (45%)	100	100	Peanut meal may become rancid if stored too long, especially in warm, moist climates.
Peas (cull), dried	65–75	50	

(Continued)

TABLE 12–1 *(Continued)*

Feedstuff	Relative Feeding Value (lb for lb) In Comparison With The Designated (underlined) Base Feed Which = 100	Maximum Percentage of Base Feed (or comparable feed or feeds) Which It Can Replace For Best Results	Remarks
PROTEIN SUPPLEMENTS: (Continued)			
Poultry house litter	50–55	25	Poultry house litter may also be used as an energy source (see Grains section of this table).
Rapeseed (canola) meal (37%)	88	75	Rapeseed meal should be limited to not more than 2 lb *(0.91 kg)* per cow. Canola meal may be used more liberally.
Safflower meal, well hulled (42%) . . .	92	100	
Safflower meal, with hulls (20%)	40–45	100	Safflower meal with hulls is unpalatable. Thus, it should be mixed with more palatable feeds.
Sesame meal	90–95	25	
Soybeans, whole	95–100	95	Soybean allowance should be limited to amount necessary to balance the ration. Larger amounts may be unduly laxative and cause cattle to go off feed.
Sunflower meal (39%)	95–100	100	If poorly hulled and lower protein content than 39%, feeding value will be lowered accordingly. It is well liked by cattle and keeps well in storage.
DRY FORAGES AND SILAGES:[2]			All the dry nonlegume forages listed herein are satisfactory when needed minerals and either a limited amount of legume hay or a protein supplement are supplied to balance the ration.
Alfalfa hay, all analyses	*100*	*100*	Does away with or lessens protein supplement requirements.
Alfalfa silage	33⅓–50	50–85	When alfalfa silage replaces corn silage, more energy feed must be provided but less protein.
Alfalfa straw	37	50	Feed with good hay.
Apple pomace silage	17–25	50–85	Usually fed as a substitute for corn or grass silage. 50% the value of corn silage. Sometimes fed out of a stack or trench silo.
Apples	17–25	50–85	Do not feed more than 25 lb *(11.4 kg)*/mature bovine. Danger of choking when fed whole. Relatively high handling cost.
Bagasse, dried; sugarcane or sorghum	10–20	5–10	
Barley hay	70	100	Avoid bearded varieties.
Barley silage	25–40	50–80	In silage, there is no problem with bearded varieties, which usually outyield beardless.
Barley straw	40	70	Of the cereal straws, barley ranks next to oat straw in feeding value. Feed to dry cows. Supplement daily with 5–6 lb *(2.3–2.7 kg)* alfalfa hay or 1–2 lb *(0.45–0.91 kg)* of 30–40% protein supplement.
Bean straw	34	50	Feed with good hay.
Beet tops, fresh	20	33⅓–50	Beet tops may be used for dry cows or older replacement heifers. Bloat may be a problem when tops are frozen. Tops are laxative. Add 2½ lb *(1.1 kg)* of ground limestone/ton of feed.
Beet top silage, sugar	17–25	33⅓–50	Feed 2 oz *(56.7 g)* of finely ground limestone or chalk with each 100 lb *(45.4 kg)* of tops, as calcium changes the oxalic acid to insoluble calcium oxalate.
Clover hay, crimson	90–100	100	Crimson clover hay has a considerably lower value if not cut at an early stage.
Clover hay, red	90–100	100	If the rest of the ration is adequate in protein, clover hay will be equal to alfalfa in feeding value; otherwise, it will be lower.
Clover straw	37	50	Feed with good hay.
Clover-timothy hay	80–90	100	Value of clover-timothy mixed hay depends on the proportion of clover present and the stage of maturity at which it is cut.
Corncobs, ground	70	90	Ground corncobs can be used as the only roughage for dry cows if properly supplemented with proteins, minerals, and vitamins.
Corn fodder	75	80–90	

(Continued)

TABLE 12-1 (Continued)

Feedstuff	Relative Feeding Value (lb for lb) In Comparison With The Designated (underlined) Base Feed Which = 100	Maximum Percentage of Base Feed (or comparable feed or feeds) Which It Can Replace For Best Results	Remarks
DRY FORAGES AND SILAGES: (Continued)			
Corn husklage (shucklage)	50	80–90	Highest and best use is for dry cows. It is slightly higher in energy and more palatable than corn stover.
Corn silage	33⅓–50	50–85	
Corn (sweet) silage, cannery waste . .	26–40	50–85	
Corn stover	45	70–90	Corn stover will meet the energy needs of dry cows, but is deficient in protein and low in phosphorus and vitamin A. Two acres of cornstalks will carry a cow 100–120 days.
Corn (sweet) stover	50	80–90	
Cottonseed hulls	66⅔	75	Use for dry pregnant cows. Supplement daily with 4–6 lb *(1.8–2.7 kg)* of good legume hay or 1–2 lb *(0.45–0.91 kg)* of a 30–40% protein supplement.
Cowpea hay	90–100	100	
Gin trash, cotton	75	75	
Grape pomace or meal	5–15	10–15	Pomace including stems is of little value as a feed.
Grass-legume mixed hay	80–90	100	Value depends on the proportion of legume present and the stage of maturity at which it is cut.
Grass-legume silage	32–47	50–85	Unless grain is added as a preservative, grass silage requires more energy feed, but less protein supplement than corn silage.
Grass silage	30–45	50–85	Grass silage must be supplemented with additional energy feeds, such as cereal grain or molasses, to be of the same value as corn silage.
Hop vine silage	20	50–75	It should be chopped when placed in the silo.
Hops, spent, dehydrated	80	50–65	Devoid of carotene; feed with legume hay.
Johnsongrass hay	70	100	
Lespedeza hay	80–100	100	Feeding value of lespedeza hay varies considerably with stage of maturity at which it is cut.
Mint hay	70–80	75	Cattle tire of mint hay when it is fed as the only roughage for extended periods.
Oat hay	75	100	
Oat silage	32–47	50–85	Must be chopped finely to exclude air from silo.
Oat straw	50	75	Oat straw is the best of the cereal straws. Use for dry cows. Supplement daily with 4–6 lb *(1.8–2.7 kg)* of good legume hay or 1–2 lb *(0.45–0.91 kg)* of 30–40% protein supplement.
Paper (newspaper; waste paper)	66⅔	50	Paper varies in feeding value in proportion to the cellulose (most paper is 60–90% cellulose) and lignin content. Magazine and bookstock papers are higher in cellulose and lower in lignin than newspapers; hence, of higher feeding value. Pelleting or cubing may increase the value of paper. *Caution:* Some newspapers contain heavy metals (boron, lead, barium, and antimony), sometimes used as a dye carrier in printer's ink, which may be toxic to animals. This is especially true of "funny" papers because of the quantity of heavy metals carried on the colored ink of the comics.
Pea straw	45–75	60–75	
Pea-vine hay	100–110	75–90	
Pea-vine silage	33⅓–50	50–85	Unless grain is added as a preservative, pea-vine silage requires more energy feed, but less protein supplement than corn silage.
Potato silage	25–30	50–75	About 75% the value of corn silage.
Prairie hay	65–70	100	
Reed canarygrass hay	70	100	
Rice straw	47	70	High levels of rice straw can be used if the straw is properly fortified.
Sawdust	75–80	70	Feeding value varies among species of trees. Digestibility is increased by cooking and other treatments.

(Continued)

TABLE 12–1 *(Continued)*

Feedstuff	Relative Feeding Value (lb for lb) In Comparison With The Designated (underlined) Base Feed Which = 100	Maximum Percentage of Base Feed (or comparable feed or feeds) Which It Can Replace For Best Results	Remarks
DRY FORAGES AND SILAGES: (Continued)			
Sorghum fodder	70	100	
Sorghum silage (grain varieties)	32–47	50–85	85–90% as valuable as corn silage. Must be supplemented in the same manner as corn silage.
Sorghum silage (sweet varieties)	25–30	50–85	Nearly equal to grain varieties in value per acre because of greater yield.
Sorghum (milo) stover	35	70–90	Can be grazed or harvested and stored either as dry feed or silage. About 2% higher in protein, but less palatable, than corn stover.
Soybean hay	85–90	50–75	Lower value than alfalfa hay, largely due to greater wastage in feeding. It may cause scouring when fed alone.
Sudangrass hay	70	100	
Sunflower silage	25–35	50–85	65–75% value of corn silage. Somewhat unpalatable and may cause constipation. Harvest for silage when ½–⅔ of heads are in bloom.
Sweet clover hay	100	100	Value of sweet clover hay varies widely. Moldy or spoiled sweet clover hay may cause sweet clover disease.
Timothy hay	70	100	
Vetch-oat hay	80–90	100	The higher the proportion of vetch, the higher the value.
Wheat hay	70	100	
Wheat straw	35	65	Of the cereal straws, wheat ranks third in nutritive value, behind oat straw and barley straw. Highest and best use is for dry cows. Supplement daily with 6 lb *(2.7 kg)* of alfalfa or 2 lb *(0.91 kg)* of a 30–40% protein supplement.

[1]Roots and tubers are of lower value than the grain and by-product feeds due to their higher moisture content.

[2]Silages are of lower value than dry forages due to their higher moisture content.

BUYING COMMERCIAL FEEDS

Commercial feeds are just what the term implies — instead of being home mixed, these feeds are mixed by commercial feed manufacturers who specialize in the business. In 1988, a total of 103.1 million tons of primary feeds (complete feeds) were manufactured in the United States, and an additional 10 million tons of secondary feeds (supplements) were produced; making for a total of 113.1 million tons of commercial feeds. *Primary feed is that which is mixed from individual ingredients, sometimes with the addition of a premix at a rate of less than 100 lb per ton of finished feed. Secondary feed is that which is mixed with one or more ingredients and a formula feed supplement (which is a primary feed)*; normally, the supplement is used at a rate of 300 lb or more per ton of finished feed, depending upon the protein content of the supplement and the percentage of protein content desired in the finished feed. The breakdown, percentagewise, by classes of livestock for which primary (complete) commer-

cial feeds were used in 1988, follow: poultry, 44.5%; beef and sheep, 17.6%; dairy, 16.9%; hogs, 14.2%; and all others, 6.8%.

Several different types of commercial feeds are available for dairy cattle; among them, (1) complete dairy concentrates, (2) dry cow rations, (3) fitting rations, (4) growing or young stock rations, (5) calf starters, (6) milk replacer feeds, (7) protein supplements, and (8) mineral and vitamin premixes.

The commercial feed manufacturer has the distinct advantages of (1) purchasing feed in quantity lots, making possible price advantages; (2) using computers for purchasing and least-cost formulating; (3) having the knowledge to manufacture medicated feeds; (4) having the knowledge and the facilities to manufacture specialty feeds, such as milk replacer; (5) processing and mixing economy control; (6) hiring scientifically trained personnel for use in determining the rations; and (7) controlling quality. Most dairy producers have neither the know-how nor the quantity of business to provide these services on their own. Because of these several advantages, commercial feeds are finding a place of increasing importance in dairy feeding.

HOME-MIXED VS COMMERCIAL FEEDS

The value of farm-grown grains—plus the cost of ingredients which need to be purchased in order to balance the ration, and the cost of grinding and mixing—as compared to the cost of commercial ready-mixed feeds laid down on the farm, should determine whether it is best to mix feeds at home or depend on ready-mixed feeds. Of course, the ultimate criterion for choosing between home-mixed and commercial feeds is which program will make for maximum returns to producers for their labor, management, and capital. Generally speaking, the use of commercial feeds makes it possible for the producer to have more animals and concentrate on production, whereas home mixing restricts animal numbers and necessitates that part of the time and capital be devoted to feed formulating and manufacturing.

The dairy producer has the following options from which to choose for home mixing feeds:

1. Purchase of a commercially prepared protein supplement (likely reinforced with vitamins and minerals), which may be blended with local or homegrown grain.

2. Purchase of a commercially prepared vitamin-mineral premix which may be mixed with an oil meal, and then blended with local or homegrown grain.

3. Purchase of individual ingredients (including vitamins and minerals) and mixing the feed from the ground up.

In summary, it may be said that there exist two good alternative sources of most feeds for the ration—home mixed or commercial—and the able manager will choose wisely between them.

HOW TO SELECT COMMERCIAL FEEDS

There is a difference in commercial feeds! That is, there is a difference from the standpoint of what dairy producers can purchase with their feed dollars. In particular, the smart operator will determine (1) the reputation of the manufacturer, and (2) the specific needs of the cattle.

The reputation of the manufacturer can be determined by (1) conferring with other producers who have used the particular products, and (2) checking on whether or not the commercial feed under consideration has a good record for meeting its guarantees. The latter can be determined by reading the bulletins and reports prepared by the respective state department in charge of monitoring feed quality and enforcing feed laws.

Feed requirements vary according to (1) the class, age, and productivity; and (2) whether the cattle are fed primarily for maintenance, growth, show-ring fitting, reproduction, or lactation. The wise producer will buy different formula feeds for different needs.

Fig. 12–4. This is the 170-ft commercial feed mill of Pennfield Feed-Grain, Lancaster, Pennsylvania. More than 1,000 tons of feed are handled daily in this mill. (Courtesy, Pennfield Feed-Grain, Lancaster, PA)

PART II
FEEDING STANDARDS—
RATION FORMULATION

Once dairy producers have become familiar with the various classes of feeds, they are in a position to make reasonable and responsible decisions about what feeds to include in their rations. In order to maximize production and profit, they must choose the feeds that are most economical for the particular animals to be fed. A high-protein, high-energy feed that is fed to low-producing or dry cows is unnecessarily expensive. Likewise, a low-cost, low-energy feed fed to high-producing cows will depress their potential for production and be an expensive feed.

FEEDING STANDARDS

Feeding standards are tables listing the amounts of one or more nutrients required by different species of animals for specific productive functions, such as growth and lactation. Most feeding standards are expressed in either (1) quantities of nutrients required per day, and/or (2) concentration in the ration; the first type is used where animals are provided a given amount of a feed during a 24-hour period, and the second is used where animals are provided a ration without limitation on the time in which it is consumed.

The most widely used feeding standards in the United States are those published by the National Research Academy (NRC) of the National Academy of Sciences. Today, the TDN system is gradually giving way to other energy evaluation systems, particularly net energy.

Although feeding standards are excellent and needed guides, there are still many situations where nutrient needs cannot be specified with great accuracy for animals. Also, in practical feeding operations, economy must be considered; for example, dairy producers are interested in obtaining that level of milk production which will make for the largest net returns in light of current feed costs and the market price of milk. Moreover, feeding standards tell nothing about the palatability, physical nature, or possible digestive disturbances of a ration. Neither do they give consideration to individual animal differences, management differences, and the effects of such stresses as weather, disease, parasitism, and surgery.

Thus, there are many variables that alter the nutrient needs and utilization of animals—variables that are difficult to include quantitatively in feeding standards, even when feed quality is well known.

NATIONAL RESEARCH COUNCIL (NRC) REQUIREMENTS[1]

The current recommended nutritive requirements for

---

[1]The National Research Council tables presented in this section were taken from *Feeds & Nutrition*, Second Edition, Chapter 20, for which they had been adapted. Dr. M. E. Ensminger is the senior author of *Feeds & Nutrition*.

dairy cattle are contained in *Nutrient Requirements of Dairy Cattle*, Sixth Revised Edition, 1989, prepared by the Subcommittee on Dairy Nutrition, National Research Council, and published by the National Academy Press, Washington, D.C., 1989. These are listed in Tables 12–2 to 12–6. It should be noted that the requirements listed in these tables do not allow for any margin of safety; that is, they do not provide for animal differences, feed differences, losses of certain nutrients in storage, and stresses. Accordingly, in the formulation of rations, margins of safety should be provided.

TABLE 12–2
DRY MATTER INTAKE REQUIREMENTS TO FULFILL NUTRIENT ALLOWANCES FOR MAINTENANCE, MILK PRODUCTION, AND NORMAL LIVE WEIGHT GAIN DURING MID- AND LATE LACTATION[1]

Live Wt.: (lb)		882	1,103	1,323	1,544	1,764
(kg)		*400*	*500*	*600*	*700*	*800*
FCM (4%)[2]		Percent Live Weight[3, 4]				
(lb/day)	*(kg/day)*	(%)	(%)	(%)	(%)	(%)
22	*10*	2.7	2.4	2.2	2.0	1.9
33	*15*	3.2	2.8	2.6	2.3	2.2
44	*20*	3.6	3.2	2.9	2.6	2.4
55	*25*	4.0	3.5	3.2	2.9	2.7
66	*30*	4.4	3.9	3.5	3.2	2.9
77	*35*	5.0	4.2	3.7	3.4	3.1
88	*40*	5.5	4.6	4.0	3.6	3.3
99	*45*	—	5.0	4.3	3.8	3.5
110	*50*	—	5.4	4.7	4.1	3.7
121	*55*	—	—	5.0	4.4	4.0
132	*60*	—	—	5.4	4.8	4.3

[1]Adapted by the author from *Nutrient Requirements of Dairy Cattle*, 6th rev. ed., update 1989, NRC, National Academy Press, p. 78, Table 6–1.

[2]4% fat-corrected milk (kg) = (0.4)(kg of milk) + (15)(kg of milk fat).

[3]The probable DMI may be up to 18% less in early lactation.

[4]DMI as a percentage of live weight may be 0.02% less per 1% increase in ration moisture content above 50% if fermented feeds constitute a major portion of the ration.

NOTE: The following assumptions were made in calculating the DMI requirements shown in this table:

1. The basic or reference cow used for the calculations weighed 1,323 lb *(600 kg)* and produced milk with 4% milk fat. Other live weights in the table and corresponding fat percentages were 882 lb *(400 kg)* and 5% fat; 1,103 lb *(500 kg)* and 4.5% fat; and 1,544 and 1,764 lb *(700 and 800 kg)* and 3.5% fat.

2. The concentration of energy in the ration for the reference cow was 1.42 Mcal of NE_{lc}/kg of DM for milk yields equal to or less than 22 lb *(10 kg)*/day. It increased linearly to 1.72 Mcal of NE_{lc}/kg for milk yields equal to or greater than 88 lb *(40 kg)*/day.

3. The energy concentrations of the rations for all other cows were assumed to change linearly as their energy requirements for milk production, relative to maintenance, changed in a manner identical to that of the 1,323-lb *(600-kg)* cow as she increased in milk yield from 22 to 88 lb *(10 to 40 kg)*/day.

4. Enough DM to provide sufficient energy for cows to gain 0.055% of their body weight daily was also included in the total. If cows do not consume as much DM as they require, as calculated from this table, their energy intake will be less than their requirements. The result will be a loss of body weight, reduced milk yields, or both. If cows consume more DM than what is projected as required from this table, the energy concentration of their ration should be reduced or they may become overly fat.

TABLE 12–3
DAILY NUTRIENT REQUIREMENTS OF GROWING DAIRY CATTLE AND MATURE BULLS[1]

Live Weight		Gain		Dry Matter Intake[2]		Energy						Protein			Minerals		Vitamins	
						TDN		DE	ME	NE$_m$	NE$_g$	CP	DIP[3]	UIP[4]	Ca	P	A	D
(lb)	(kg)	(lb)	(kg)	(lb)	(kg)	(lb)	(kg)	(Mcal)	(Mcal)	(Mcal)	(Mcal)	(g)	(g)	(g)	(g)	(g)	(1,000 IU)	(1,000 IU)
Growing Large-Breed Calves Fed Only Milk or Milk Replacer																		
88	40	0.4	0.2	1.06	0.48	1.37	0.62	2.73	2.54	1.37	0.41	105	—	—	7	4	1.70	0.26
99	45	0.7	0.3	1.19	0.54	1.54	0.70	3.07	2.86	1.49	0.56	120	—	—	8	5	1.94	0.30
Growing Large-Breed Calves Fed Milk Plus Starter Mix																		
110	50	1.1	0.5	2.87	1.30	3.22	1.46	6.42	5.90	1.62	0.72	290	—	—	9	6	2.10	0.33
165	75	1.8	0.8	4.37	1.98	4.90	2.22	9.78	8.98	2.19	1.30	435	—	—	16	8	3.20	0.50
Growing Small-Breed Calves Fed Only Milk or Milk Replacer																		
55	25	0.4	0.2	0.84	0.38	1.08	0.49	2.16	2.01	0.96	0.37	84	—	—	6	4	1.10	0.16
66	30	0.7	0.3	1.12	0.51	1.46	0.66	2.90	2.70	1.10	0.52	112	—	—	7	4	1.30	0.20
Growing Small-Breed Calves Fed Milk Plus Starter Mix																		
110	50	1.1	0.5	3.15	1.43	3.53	1.60	7.06	6.49	1.62	0.72	315	—	—	10	6	2.10	0.33
165	75	1.3	0.6	3.88	1.76	4.34	1.97	8.69	7.98	2.19	0.96	387	—	—	14	8	3.20	0.50
Growing Veal Calves Fed Only Milk or Milk Replacer																		
88	40	0.4	0.2	0.99	0.45	1.04	0.47	2.07	1.89	1.37	0.55	100	—	—	7	4	1.70	0.26
110	50	0.9	0.4	1.26	0.57	1.30	0.59	2.63	2.39	1.62	0.57	125	—	—	9	5	2.10	0.33
132	60	1.2	0.5	1.76	0.80	1.57	0.71	3.17	2.84	1.85	0.81	176	—	—	13	8	2.60	0.40
165	75	2.0	0.9	3.00	1.36	2.67	1.21	5.39	4.82	2.19	1.47	300	—	—	16	9	3.20	0.50
221	100	2.8	1.3	4.41	2.00	3.48	1.58	7.06	6.22	2.72	2.26	440	—	—	20	11	4.20	0.66
276	125	2.8	1.3	5.25	2.38	4.15	1.88	8.40	7.40	3.21	2.44	524	—	—	22	13	5.30	0.82
331	150	2.4	1.1	6.00	2.72	4.74	2.15	9.60	8.46	3.69	2.29	598	—	—	24	15	6.40	0.99
Large-Breed Growing Females																		
221	100	1.3	0.6	5.80	2.63	4.02	1.84	8.13	7.03	2.72	1.22	421	57	317	17	9	4.24	0.66
221	100	1.5	0.7	6.22	2.82	4.37	1.98	8.72	7.54	2.72	1.44	452	75	346	18	9	4.24	0.66
221	100	1.8	0.8	6.66	3.02	4.65	2.11	9.32	8.06	2.72	1.66	483	92	374	18	10	4.24	0.66
331	150	1.3	0.6	7.74	3.51	5.31	2.41	10.61	9.14	3.69	1.45	562	150	283	19	11	6.36	0.99
331	150	1.5	0.7	8.27	3.75	5.67	2.57	11.33	9.76	3.69	1.71	600	173	307	19	12	6.36	0.99
331	150	1.8	0.8	8.80	3.99	6.04	2.74	12.07	10.39	3.69	1.97	639	196	331	20	12	6.36	0.99
441	200	1.3	0.6	9.68	4.39	6.50	2.95	12.99	11.14	4.57	1.65	699	239	254	20	14	8.48	1.32
441	200	1.5	0.7	10.32	4.68	6.92	3.14	13.84	11.87	4.57	1.95	749	267	274	21	14	8.48	1.32
441	200	1.8	0.8	10.96	4.97	7.36	3.34	14.71	12.62	4.57	2.25	796	295	294	22	15	8.48	1.32
551	250	1.3	0.6	11.71	5.31	7.67	3.48	15.33	13.10	5.41	1.84	718	326	229	22	16	10.60	1.65
551	250	1.5	0.7	12.46	5.65	8.16	3.70	16.32	13.94	5.41	2.18	787	359	246	23	17	10.60	1.65
551	250	1.8	0.8	13.21	5.99	8.67	3.93	17.32	14.79	5.41	2.51	857	393	263	24	17	10.60	1.65
662	300	1.3	0.6	13.80	6.26	8.84	4.01	17.69	15.05	6.20	2.02	752	413	209	23	17	12.72	1.98
662	300	1.5	0.7	14.69	6.66	9.42	4.27	18.81	16.00	6.20	2.39	814	452	223	24	18	12.72	1.98
662	300	1.8	0.8	15.57	7.06	9.97	4.52	19.95	16.97	6.20	2.77	884	490	236	25	19	12.72	1.98
772	350	1.3	0.6	16.07	7.29	10.05	4.56	20.09	17.01	6.96	2.20	874	501	193	24	18	14.84	2.31
772	350	1.5	0.7	17.09	7.75	10.67	4.84	21.36	18.09	6.96	2.60	930	545	204	25	19	14.84	2.31
772	350	1.8	0.8	18.10	8.21	11.33	5.14	22.64	19.18	6.96	3.01	985	590	214	26	20	14.84	2.31
882	400	1.3	0.6	18.50	8.39	11.29	5.12	22.58	19.03	7.69	2.37	1,007	592	182	25	19	16.96	2.64
882	400	1.5	0.7	19.67	8.92	12.00	5.44	24.00	20.23	7.69	2.80	1,070	641	190	26	20	16.96	2.64
882	400	1.8	0.8	20.86	9.46	12.72	5.77	25.44	21.44	7.69	3.24	1,135	692	198	26	21	16.96	2.64
992	450	1.3	0.6	21.15	9.59	12.59	5.71	25.18	21.12	8.40	2.53	1,151	686	176	28	19	19.08	2.97
992	450	1.5	0.7	22.49	10.20	13.38	6.07	26.78	22.46	8.40	2.99	1,224	742	182	28	20	19.08	2.97
992	450	1.8	0.8	23.86	10.82	14.20	6.44	28.40	23.81	8.40	3.46	1,298	799	187	29	21	19.08	2.97

(Continued)

TABLE 12–3 *(Continued)*

Live Weight		Gain		Dry Matter Intake[2]		Energy							Protein			Minerals		Vitamins	
						TDN		DE	ME	NE_m	NE_g		CP	DIP[3]	UIP[4]	Ca	P	A	D
(lb)	(kg)	(lb)	(kg)	(lb)	(kg)	(lb)	(kg)	(Mcal)	(Mcal)	(Mcal)	(Mcal)		(g)	(g)	(g)	(g)	(g)	(1,000 IU)	(1,000 IU)
								Large-Breed Growing Females *(Continued)*											
1,103	500	1.3	0.6	24.10	10.93	13.98	6.34	27.96	23.32	9.09	2.69		1,311	785	175	28	20	21.20	3.30
1,103	500	1.5	0.7	25.64	11.63	14.88	6.75	29.74	24.81	9.09	3.18		1,395	848	179	28	20	21.20	3.30
1,103	500	1.8	0.8	27.18	12.33	15.79	7.16	31.55	26.32	9.09	3.68		1,480	913	182	29	21	21.20	3.30
1,213	550	1.3	0.6	27.39	12.42	15.48	7.02	30.95	25.67	9.77	2.84		1,490	891	180	28	20	23.32	3.63
1,213	550	1.5	0.7	29.15	13.22	16.47	7.47	32.95	27.33	9.77	3.37		1,587	963	183	28	20	23.32	3.63
1,213	550	1.8	0.8	30.96	14.04	17.51	7.94	34.99	29.02	9.77	3.90		1,685	1,035	185	29	21	23.32	3.63
1,323	600	1.3	0.6	31.11	14.11	17.13	7.77	34.24	28.23	10.43	3.00		1,694	1,007	193	28	20	25.44	3.96
1,323	600	1.5	0.7	33.19	15.05	18.26	8.28	36.50	30.09	10.43	3.55		1,805	1,088	194	28	21	25.44	3.96
1,323	600	1.8	0.8	35.26	15.99	19.40	8.80	38.79	31.98	10.43	4.11		1,919	1,170	195	29	21	25.44	3.96
								Small-Breed Growing Females											
221	100	0.9	0.4	5.31	2.41	3.68	1.67	7.35	6.34	2.72	0.91		386	38	249	15	8	4.24	0.66
221	100	1.1	0.5	5.82	2.64	4.01	1.82	8.03	6.92	2.72	1.16		422	59	275	16	8	4.24	0.66
221	100	1.3	0.6	6.31	2.86	4.37	1.98	8.71	7.51	2.72	1.40		458	80	300	17	9	4.24	0.66
331	150	0.9	0.4	7.30	3.31	4.90	2.22	9.78	8.39	3.69	1.09		529	129	222	17	10	6.36	0.99
331	150	1.1	0.5	7.94	3.60	5.31	2.41	10.63	9.12	3.69	1.39		575	156	243	18	11	6.36	0.99
331	150	1.3	0.6	8.58	3.89	5.76	2.61	11.50	9.86	3.69	1.69		622	185	263	19	11	6.36	0.99
441	200	0.9	0.4	9.35	4.24	6.09	2.76	12.16	10.38	4.57	1.26		578	217	201	19	13	8.48	1.32
441	200	1.1	0.5	10.14	4.60	6.59	2.99	13.19	11.25	4.57	1.60		648	251	217	20	13	8.48	1.32
441	200	1.3	0.6	10.94	4.96	7.12	3.23	14.23	12.14	4.57	1.95		718	286	232	20	14	8.48	1.32
551	250	0.9	0.4	11.55	5.24	7.28	3.30	14.57	12.36	5.41	1.41		629	305	185	21	15	10.60	1.65
551	250	1.1	0.5	12.52	5.68	7.89	3.58	15.78	13.38	5.41	1.80		682	346	197	21	16	10.60	1.65
551	250	1.3	0.6	13.49	6.12	8.51	3.86	17.01	14.43	5.41	2.20		753	389	209	22	16	10.60	1.65
662	300	0.9	0.4	13.98	6.34	8.53	3.87	17.06	14.38	6.20	1.56		761	395	176	22	16	12.72	1.98
662	300	1.1	0.5	15.15	6.87	9.24	4.19	18.48	15.57	6.20	1.99		824	445	184	23	17	12.72	1.98
662	300	1.3	0.6	16.32	7.40	9.97	4.52	19.92	16.79	6.20	2.43		888	495	192	23	17	12.72	1.98
772	350	0.9	0.4	16.69	7.57	9.86	4.47	19.71	16.50	6.96	1.71		909	490	173	23	17	14.84	2.31
772	350	1.1	0.5	18.08	8.20	10.67	4.84	21.35	17.87	6.96	2.18		985	548	178	23	18	14.84	2.31
772	350	1.3	0.6	19.51	8.85	11.51	5.22	23.03	19.28	6.96	2.66		1,062	608	183	24	18	14.84	2.31
882	400	0.9	0.4	19.80	8.98	11.29	5.12	22.58	18.77	7.69	1.84		1,078	592	177	24	18	16.96	2.64
882	400	1.1	0.5	21.48	9.74	12.26	5.56	24.50	20.36	7.69	2.35		1,169	661	181	24	19	16.96	2.64
882	400	1.3	0.6	23.20	10.52	13.23	6.00	26.45	21.98	7.69	2.87		1,263	730	183	25	19	16.96	2.64
992	450	0.9	0.4	23.46	10.64	12.90	5.85	25.80	21.27	8.40	1.98		1,276	706	191	27	18	19.08	2.97
992	450	1.1	0.5	25.49	11.56	14.02	6.36	28.04	23.12	8.40	2.52		1,387	786	193	28	19	19.08	2.97
992	450	1.3	0.6	27.56	12.50	15.17	6.88	30.33	25.01	8.40	3.08		1,500	867	194	28	19	19.08	2.97
								Large-Breed Growing Males											
221	100	1.8	0.8	6.17	2.80	4.32	1.96	8.66	7.48	2.72	1.42		448	65	401	18	10	4.24	0.66
221	100	2.0	0.9	6.55	2.97	4.59	2.08	9.16	7.92	2.72	1.60		475	79	433	19	10	4.24	0.66
221	100	2.2	1.0	6.90	3.13	4.83	2.19	9.67	8.36	2.72	1.79		501	93	465	20	11	4.24	0.66
331	150	1.8	0.8	7.94	3.60	5.51	2.50	11.03	9.52	3.69	1.64		576	155	364	20	12	6.36	0.99
331	150	2.0	0.9	8.38	3.80	5.82	2.64	11.63	10.03	3.69	1.85		607	172	393	21	13	6.36	0.99
331	150	2.2	1.0	8.80	3.99	6.11	2.77	12.22	10.55	3.69	2.07		639	190	422	22	13	6.36	0.99
441	200	1.8	0.8	9.77	4.43	6.68	3.03	13.34	11.48	4.57	1.84		709	241	333	22	15	8.48	1.32
441	200	2.0	0.9	10.28	4.66	7.01	3.18	14.02	12.06	4.57	2.08		745	262	359	23	15	8.48	1.32
441	200	2.2	1.0	10.78	4.89	7.36	3.34	14.71	12.66	4.57	2.33		782	284	385	24	16	8.48	1.32

(Continued)

TABLE 12–3 (Continued)

Live Weight		Gain		Dry Matter Intake[2]		Energy							Protein			Minerals		Vitamins	
						TDN		DE	ME	NE$_m$	NE$_g$		CP	DIP[3]	UIP[4]	Ca	P	A	D
(lb)	(kg)	(lb)	(kg)	(lb)	(kg)	(lb)	(kg)	(Mcal)	(Mcal)	(Mcal)	(Mcal)		(g)	(g)	(g)	(g)	(g)	(1,000 IU)	(1,000 IU)
	Large-Breed Growing Males (Continued)
551	250	1.8	0.8	11.62	5.27	7.78	3.53	15.58	13.37	5.41	2.03		843	325	305	24	17	10.60	1.65
551	250	2.0	0.9	12.19	5.53	8.18	3.71	16.35	14.03	5.41	2.30		885	350	329	25	18	10.60	1.65
551	250	2.2	1.0	12.79	5.80	8.58	3.89	17.13	14.70	5.41	2.57		927	375	352	26	18	10.60	1.65
662	300	1.8	0.8	13.52	6.13	8.91	4.04	17.80	15.22	6.20	2.21		863	408	281	25	19	12.72	1.98
662	300	2.0	0.9	14.18	6.43	9.33	4.23	18.66	15.96	6.20	2.51		934	436	302	25	19	12.72	1.98
662	300	2.2	1.0	14.84	6.73	9.77	4.43	19.53	16.70	6.20	2.80		1,004	464	323	26	20	12.72	1.98
772	350	1.8	0.8	15.48	7.02	10.01	4.54	20.02	17.06	6.96	2.38		885	490	261	26	20	14.84	2.31
772	350	2.0	0.9	16.23	7.36	10.50	4.76	20.98	17.88	6.96	2.70		956	522	280	26	20	14.84	2.31
772	350	2.2	1.0	16.98	7.70	10.98	4.98	21.94	18.70	6.96	3.02		1,027	554	298	27	21	14.84	2.31
882	400	1.8	0.8	17.55	7.96	11.14	5.05	22.27	18.91	7.69	2.55		955	572	244	26	21	16.96	2.64
882	400	2.0	0.9	18.39	8.34	11.66	5.29	23.32	19.80	7.69	2.89		1,001	608	260	27	21	16.96	2.64
882	400	2.2	1.0	19.23	8.72	12.19	5.53	24.39	20.71	7.69	3.24		1,056	644	277	28	22	16.96	2.64
992	450	1.8	0.8	19.73	8.95	12.28	5.57	24.56	20.78	8.40	2.71		1,074	656	230	29	21	19.08	2.97
992	450	2.0	0.9	20.66	9.37	12.86	5.83	25.72	21.76	8.40	3.08		1,125	696	245	29	22	19.08	2.97
992	450	2.2	1.0	21.61	9.80	13.45	6.10	26.89	22.75	8.40	3.44		1,176	736	259	29	23	19.08	2.97
1,103	500	1.8	0.8	22.05	10.00	13.47	6.11	26.92	22.69	9.09	2.87		1,201	742	220	29	21	21.20	3.30
1,103	500	2.0	0.9	23.11	10.48	14.09	6.39	28.19	23.76	9.09	3.25		1,257	786	233	29	22	21.20	3.30
1,103	500	2.2	1.0	24.14	10.95	14.73	6.68	29.47	24.84	9.09	3.64		1,314	830	346	29	23	21.20	3.30
1,213	550	1.8	0.8	24.56	11.14	14.69	6.66	29.38	24.66	9.77	3.02		1,336	831	213	29	21	23.32	3.63
1,213	550	2.0	0.9	25.71	11.66	15.39	6.98	30.76	25.82	9.77	3.43		1,399	879	225	29	22	23.32	3.63
1,213	550	2.2	1.0	26.88	12.19	16.07	7.29	32.16	27.00	9.77	3.84		1,463	927	236	30	23	23.32	3.63
1,323	600	1.8	0.8	27.25	12.36	15.99	7.25	31.95	26.71	10.43	3.17		1,483	923	211	29	21	25.44	3.96
1,323	600	2.0	0.9	28.55	12.95	16.74	7.59	33.47	27.97	10.43	3.60		1,554	976	221	29	22	25.44	3.96
1,323	600	2.2	1.0	29.86	13.54	17.51	7.94	34.99	29.25	10.43	4.03		1,624	1,029	231	30	23	25.44	3.96
1,433	650	1.8	0.8	30.19	13.69	17.33	7.86	34.67	28.86	11.07	3.32		1,643	1,020	212	29	21	27.56	4.29
1,433	650	2.0	0.9	31.64	14.35	18.17	8.24	36.33	30.24	11.07	3.77		1,722	1,078	222	29	22	27.56	4.29
1,433	650	2.2	1.0	33.10	15.01	19.01	8.62	38.00	31.63	11.07	4.22		1,801	1,137	230	30	23	27.56	4.29
1,544	700	1.8	0.8	33.43	15.16	18.79	8.52	37.59	31.14	11.70	3.46		1,820	1,124	219	29	22	29.68	4.62
1,544	700	2.0	0.9	35.06	15.90	19.71	8.94	39.40	32.64	11.70	3.93		1,907	1,187	227	29	22	29.68	4.62
1,544	700	2.2	1.0	36.67	16.63	20.62	9.35	41.23	34.16	11.70	4.40		1,996	1,252	235	30	23	29.68	4.62
1,654	750	1.8	0.8	37.02	16.79	20.37	9.24	40.73	33.59	12.33	3.60		2,015	1,235	232	29	22	31.80	4.95
1,654	750	2.0	0.9	38.85	17.62	21.37	9.69	42.73	35.23	12.33	4.09		2,114	1,305	239	29	23	31.80	4.95
1,654	750	2.2	1.0	40.68	18.45	22.38	10.15	44.74	36.89	12.33	4.58		2,213	1,376	246	30	23	31.80	4.95
1,764	800	1.8	0.8	38.72	17.56	21.30	9.66	42.59	35.12	12.94	3.74		2,107	1,303	216	29	22	33.92	5.28
1,764	800	2.0	0.9	40.59	18.41	22.34	10.13	44.67	36.83	12.94	4.25		2,210	1,377	221	29	23	33.92	5.28
1,764	800	2.2	1.0	42.51	19.28	23.40	10.61	46.76	38.55	12.94	4.76		2,313	1,451	227	30	23	33.92	5.28
	Small-Breed Growing Males
221	100	1.1	0.5	5.40	2.45	3.79	1.72	7.56	6.54	2.72	1.02		392	41	287	16	8	4.24	0.66
221	100	1.3	0.6	5.82	2.64	4.08	1.85	8.15	7.04	2.72	1.23		422	58	316	17	9	4.24	0.66
221	100	1.5	0.7	6.24	2.83	4.37	1.98	8.74	7.55	2.72	1.45		453	75	345	18	9	4.24	0.66
331	150	1.1	0.5	7.23	3.28	4.96	2.25	9.92	8.55	3.69	1.20		525	129	257	18	11	6.36	0.99
331	150	1.3	0.6	7.76	3.52	5.31	2.41	10.64	9.16	3.69	1.46		563	151	282	19	11	6.36	0.99
331	150	1.5	0.7	8.29	3.76	5.69	2.58	11.36	9.78	3.69	1.71		601	174	306	19	12	6.36	0.99

(Continued)

TABLE 12–3 *(Continued)*

Live Weight		Gain		Dry Matter Intake[2]		Energy							Protein			Minerals		Vitamins	
						TDN		DE	ME	NE$_m$	NE$_g$		CP	DIP[3]	UIP[4]	Ca	P	A	D
(lb)	(kg)	(lb)	(kg)	(lb)	(kg)	(lb)	(kg)	(Mcal)	(Mcal)	(Mcal)	(Mcal)		(g)	(g)	(g)	(g)	(g)	(1,000 IU)	(1,000 IU)
								**Small-Breed Growing Males** *(Continued)*											
441	200	1.1	0.5	9.11	4.12	6.09	2.76	12.18	10.45	4.57	1.37		630	213	232	20	13	8.48	1.32
441	200	1.3	0.6	9.70	4.40	6.50	2.95	13.02	11.17	4.57	1.66		699	241	252	20	14	8.48	1.32
441	200	1.5	0.7	10.34	4.69	6.95	3.15	13.87	11.90	4.57	1.96		751	268	273	21	14	8.48	1.32
551	250	1.1	0.5	11.00	4.99	7.21	3.27	14.41	12.31	5.41	1.53		648	296	210	21	16	10.60	1.65
551	250	1.3	0.6	11.73	5.32	7.70	3.49	15.38	13.14	5.41	1.86		718	328	228	22	16	10.60	1.65
551	250	1.5	0.7	12.48	5.66	8.18	3.71	16.35	13.97	5.41	2.19		787	361	245	23	17	10.60	1.65
662	300	1.1	0.5	12.99	5.89	8.31	3.77	16.64	14.15	6.20	1.68		707	378	193	23	17	12.72	1.98
662	300	1.3	0.6	13.85	6.28	8.86	4.02	17.74	15.09	6.20	2.04		754	415	207	23	17	12.72	1.98
662	300	1.5	0.7	14.73	6.68	9.44	4.28	18.85	16.04	6.20	2.41		814	453	221	24	18	12.72	1.98
772	350	1.1	0.5	15.13	6.86	9.46	4.29	18.91	16.01	6.96	1.82		823	461	180	23	18	14.84	2.31
772	350	1.3	0.6	16.12	7.31	10.08	4.57	20.15	17.06	6.96	2.22		877	503	191	24	18	14.84	2.31
772	350	1.5	0.7	17.11	7.76	10.72	4.86	21.41	18.13	6.96	2.62		932	547	203	25	19	14.84	2.31
882	400	1.1	0.5	17.42	7.90	10.63	4.82	21.25	17.91	7.69	1.96		947	545	171	24	19	16.96	2.64
882	400	1.3	0.6	18.54	8.41	11.33	5.14	22.64	19.08	7.69	2.39		1,010	594	180	25	19	16.96	2.64
882	400	1.5	0.7	19.71	8.94	12.04	5.46	24.06	20.27	7.69	2.82		1,073	644	189	26	20	16.96	2.64
992	450	1.1	0.5	19.91	9.03	11.84	5.37	23.70	19.87	8.40	2.10		1,083	634	166	28	19	19.08	2.97
992	450	1.3	0.6	21.21	9.62	12.63	5.73	25.26	21.18	8.40	2.55		1,155	689	174	28	19	19.08	2.97
992	450	1.5	0.7	22.56	10.23	13.43	6.09	26.84	22.51	8.40	3.01		1,227	744	180	28	20	19.08	2.97
1,103	500	1.1	0.5	22.67	10.28	13.14	5.96	26.29	21.93	9.09	2.23		1,233	726	167	28	19	21.20	3.30
1,103	500	1.3	0.6	24.17	10.96	14.02	6.36	28.04	23.39	9.09	2.71		1,315	788	173	28	20	21.20	3.30
1,103	500	1.5	0.7	25.69	11.65	14.91	6.76	29.81	24.87	9.09	3.20		1,398	851	177	28	20	21.20	3.30
1,213	550	1.1	0.5	25.73	11.67	14.55	6.60	29.08	24.12	9.77	2.36		1,400	825	174	28	19	23.32	3.63
1,213	550	1.3	0.6	27.47	12.46	15.52	7.04	31.05	25.75	9.77	2.87		1,495	895	178	28	20	23.32	3.63
1,213	550	1.5	0.7	29.24	13.26	16.52	7.49	33.03	27.40	9.77	3.39		1,591	966	181	28	20	23.32	3.63
1,323	600	1.1	0.5	29.22	13.25	16.07	7.29	32.14	26.50	10.43	2.48		1,590	933	187	28	19	25.44	3.96
1,323	600	1.3	0.6	31.22	14.16	17.18	7.79	34.35	28.32	10.43	3.02		1,699	1,012	190	28	20	25.44	3.96
1,323	600	1.5	0.7	33.25	15.08	18.30	8.30	36.59	30.17	10.43	3.57		1,810	1,091	192	28	21	25.44	3.96
								**Maintenance of Mature Breeding Bulls**											
1,103	500	—	—	17.40	7.89	9.57	4.34	19.15	15.79	9.09	—		789	472	161	20	12	21.20	3.30
1,323	600	—	—	19.96	9.05	10.98	4.98	21.95	18.10	10.43	—		905	573	155	24	15	25.44	3.96
1,544	700	—	—	22.40	10.16	12.33	5.59	24.64	20.32	11.70	—		1,016	670	148	28	18	29.68	4.62
1,764	800	—	—	24.76	11.23	13.63	6.18	27.24	22.46	12.94	—		1,123	764	142	32	20	33.92	5.28
1,985	900	—	—	27.06	12.27	14.88	6.75	29.76	24.53	14.13	—		1,227	854	135	36	22	38.16	5.94
2,205	1,000	—	—	29.28	13.28	16.10	7.30	32.20	26.55	15.29	—		1,328	943	129	41	25	42.40	6.60
2,426	1,100	—	—	31.44	14.26	17.31	7.85	34.59	28.52	16.43	—		1,426	1,029	122	45	28	46.64	7.26
2,646	1,200	—	—	33.56	15.22	18.46	8.37	36.92	30.44	17.53	—		1,522	1,113	115	49	30	50.88	7.92
2,867	1,300	—	—	35.63	16.16	19.60	8.89	39.21	32.32	18.62	—		1,616	1,196	108	53	32	55.12	8.58
3,087	1,400	—	—	37.68	17.09	20.73	9.40	41.45	34.17	19.68	—		1,709	1,277	102	57	35	59.36	9.24

[1]Adapted by the author from *Nutrient Requirements of Dairy Cattle*, 6th rev. ed., update 1989, NRC, National Academy Press, pp. 81–84, Table 6–2.

[2]The data for DMI are not requirements *per se*, unlike the requirements for net energy maintenance, net energy gain, and absorbed protein. They are not intended to be estimates of voluntary intake but are consistent with the specified dietary energy concentrations. The use of rations with decreased energy concentrations will increase dry matter intake needs; metabolizable energy, digestible energy, and total digestible nutrient needs; and crude protein needs. The use of rations with increased energy concentrations will have opposite effects on these needs.

[3]DIP – degraded intake protein.

[4]UIP – undegraded intake protein.

TABLE 12–4
DAILY NUTRIENT REQUIREMENTS OF LACTATING AND PREGNANT COWS[1]

Live Weight		Energy				Total Crude Protein	Minerals		Vitamins	
		TDN		DE	ME	NE$_{lc}$		Ca	P	A	D
(lb)	(kg)	(lb)	(kg)	(Mcal)	(Mcal)	(Mcal)	(g)	(g)	(g)	(1,000 IU)	(1,000 IU)

Maintenance of Mature Lactating Cows[2]

(lb)	(kg)	(lb)	(kg)	(Mcal)	(Mcal)	(Mcal)	(g)	(g)	(g)	(1,000 IU)	(1,000 IU)
882	400	6.90	3.13	13.80	12.01	7.16	318	16	11	30	12
992	450	7.54	3.42	15.08	13.12	7.82	341	18	13	34	14
1,103	500	8.16	3.70	16.32	14.20	8.46	364	20	14	38	15
1,213	550	8.75	3.97	17.53	15.25	9.09	386	22	16	42	17
1,323	600	9.35	4.24	18.71	16.28	9.70	406	24	17	46	18
1,433	650	9.94	4.51	19.86	17.29	10.30	428	26	19	49	20
1,544	700	10.50	4.76	21.00	18.28	10.89	449	28	20	53	21
1,654	750	11.07	5.02	22.12	19.25	11.47	468	30	21	57	23
1,764	800	11.60	5.26	23.21	20.20	12.03	486	32	23	61	24

Maintenance Plus Last 2 Months of Gestation of Mature Dry Cows[3]

(lb)	(kg)	(lb)	(kg)	(Mcal)	(Mcal)	(Mcal)	(g)	(g)	(g)	(1,000 IU)	(1,000 IU)
882	400	9.15	4.15	18.23	15.26	9.30	875	26	16	30	12
992	450	9.99	4.53	19.91	16.66	10.16	928	30	18	34	14
1,103	500	10.80	4.90	21.55	18.04	11.00	978	33	20	38	15
1,213	550	11.62	5.27	23.14	19.37	11.81	1,027	36	22	42	17
1,323	600	12.39	5.62	24.71	20.68	12.61	1,074	39	24	46	18
1,433	650	13.16	5.97	26.23	21.96	13.39	1,120	43	26	49	20
1,544	700	13.91	6.31	27.73	23.21	14.15	1,165	46	28	53	21
1,654	750	14.66	6.65	29.21	24.44	14.90	1,209	49	30	57	23
1,764	800	15.39	6.98	30.65	25.66	15.64	1,254	53	32	61	24

Fat	Energy								Total Crude Protein		Minerals				Vitamins	
	TDN		DE		ME		NE$_{lc}$				Ca		P		A	D
(%)	(lb)	(kg)	(Mcal/lb)	(Mcal/kg)	(Mcal/lb)	(Mcal/kg)	(Mcal/lb)	(Mcal/kg)	(g/lb)	(g/kg)	(g/lb)	(g/kg)	(g/lb)	(g/kg)		

Milk Production—Nutrients/2.2 lb or /kg of Milk of Different Fat Percentages

(%)	(lb)	(kg)	(Mcal/lb)	(Mcal/kg)	(Mcal/lb)	(Mcal/kg)	(Mcal/lb)	(Mcal/kg)	(g/lb)	(g/kg)	(g/lb)	(g/kg)	(g/lb)	(g/kg)		
3.0	0.616	0.280	0.56	1.23	0.49	1.07	0.29	0.64	35	78	1.24	2.73	0.76	1.68	—	—
3.5	0.662	0.301	0.60	1.33	0.52	1.15	0.31	0.69	38	84	1.35	2.97	0.83	1.83	—	—
4.0	0.708	0.322	0.64	1.42	0.56	1.24	0.34	0.74	41	90	1.46	3.21	0.90	1.98	—	—
4.5	0.755	0.343	0.69	1.51	0.60	1.32	0.35	0.78	44	96	1.57	3.45	0.98	2.13	—	—
5.0	0.801	0.364	0.73	1.61	0.64	1.40	0.38	0.83	46	101	1.68	3.69	1.04	2.28	—	—
5.5	0.847	0.385	0.77	1.70	0.69	1.48	0.40	0.88	49	107	1.78	3.93	1.10	2.43	—	—

Live Weight Change During Lactation—Nutrients/kg of Weight Change[4]

Weight																
loss	−0.990	−2.170	−4.34	−9.55	−3.75	−8.25	−2.25	−4.92	−145	−320	—	—	—	—	—	—
gain	1.030	2.260	4.52	9.96	3.88	8.55	2.32	5.12	145	320	—	—	—	—	—	—

[1]Adapted by the author from *Nutrient Requirements of Dairy Cattle*, 6th rev. ed., update 1989, NRC, National Academy Press, p. 84, Table 6–3.

[2]To allow for growth of young lactating cows, increase the maintenance allowances for all nutrients except vitamins A and D by 20% during the first lactation and 10% during the second lactation.

[3]Values for calcium assume that the cow is in calcium balance at the beginning of the last 2 months of gestation. If the cow is not in balance, then the calcium requirement can be increased from 25 to 33%.

[4]No allowance is made for mobilized calcium and phosphorus associated with live weight loss or with live weight gain. The maximum daily nitrogen available from weight loss is assumed to be 30 g or 234 g of crude protein.

TABLE 12–5
DAILY NUTRIENT REQUIREMENTS OF LACTATING COWS USING ABSORBABLE PROTEIN[1]

Live Weight		Fat	Milk		Live Weight Change		Dry Matter Intake		Energy					Protein		Minerals	
									TDN		NE$_{lc}$	NE$_{lcdm}$[2]		DIP[3]	UIP[4]	Ca	P
(lb)	(kg)	(%)	(lb)	(kg)	(lb)	(kg)	(lb)	(kg)	(lb)	(kg)	(Mcal)	(Mcal/lb)	(Mcal/kg)	(g)	(g)	(g)	(g)
Intake at 100% of the Requirement for Maintenance, Lactation, and Weight Gain
882	400	4.5	17.6	8.0	0.485	0.220	22.36	10.14	14.20	6.44	14.55	0.65	1.43	753	511	44	28
882	400	4.5	30.9	14.0	0.485	0.220	27.92	12.66	18.70	8.48	19.26	0.69	1.52	1,052	710	65	41
882	400	4.5	44.1	20.0	0.485	0.220	32.88	14.91	23.17	10.51	23.96	0.73	1.61	1,355	880	85	54
882	400	4.5	57.3	26.0	0.485	0.220	37.35	16.94	27.65	12.54	28.67	0.77	1.69	1,662	1,026	106	67
882	400	4.5	70.6	32.0	0.485	0.220	42.80	19.41	32.15	14.58	33.37	0.78	1.72	1,962	1,220	127	80
882	400	5.0	17.6	8.0	0.485	0.220	22.84	10.36	14.55	6.60	14.94	0.65	1.44	778	525	46	30
882	400	5.0	30.9	14.0	0.485	0.220	28.67	13.00	19.34	8.77	19.93	0.69	1.53	1,096	730	68	43
882	400	5.0	44.1	20.0	0.485	0.220	33.85	15.35	24.10	10.93	24.93	0.74	1.62	1,419	902	90	57
882	400	5.0	57.3	26.0	0.485	0.220	38.46	17.44	28.82	13.07	29.92	0.78	1.72	1,745	1,048	112	71
882	400	5.0	70.6	32.0	0.485	0.220	44.76	20.30	33.63	15.25	34.91	0.78	1.72	2,061	1,277	134	84
882	400	5.5	17.6	8.0	0.485	0.220	23.31	10.57	14.93	6.77	15.32	0.66	1.45	803	538	48	31
882	400	5.5	30.9	14.0	0.485	0.220	29.39	13.33	20.00	9.07	20.61	0.70	1.55	1,140	748	71	45
882	400	5.5	44.1	20.0	0.485	0.220	34.77	15.77	25.00	11.34	25.89	0.74	1.64	1,483	923	95	60
882	400	5.5	57.3	26.0	0.485	0.220	39.98	18.13	30.03	13.62	31.17	0.78	1.72	1,826	1,091	118	75
882	400	5.5	70.6	32.0	0.485	0.220	46.75	21.20	35.10	15.92	36.45	0.78	1.72	2,160	1,334	142	89
1,103	500	4.0	19.8	9.0	0.606	0.275	25.56	11.59	16.10	7.30	16.49	0.64	1.42	883	540	49	32
1,103	500	4.0	37.5	17.0	0.606	0.275	32.59	14.78	21.74	9.86	22.38	0.69	1.51	1,257	797	75	48
1,103	500	4.0	55.1	25.0	0.606	0.275	38.85	17.62	27.34	12.40	28.27	0.73	1.61	1,635	1,015	101	64
1,103	500	4.0	72.8	33.0	0.606	0.275	44.41	20.14	32.92	14.93	34.15	0.77	1.70	2,018	1,201	126	80
1,103	500	4.0	90.4	41.0	0.606	0.275	51.35	23.29	38.57	17.49	40.04	0.78	1.72	2,392	1,453	152	95
1,103	500	4.5	19.8	9.0	0.606	0.275	26.11	11.84	16.52	7.49	16.92	0.65	1.43	911	556	51	33
1,103	500	4.5	37.5	17.0	0.606	0.275	33.52	15.20	22.51	10.21	23.20	0.69	1.53	1,310	821	79	50
1,103	500	4.5	55.1	25.0	0.606	0.275	40.04	18.16	28.49	12.92	29.47	0.74	1.62	1,715	1,043	107	68
1,103	500	4.5	72.8	33.0	0.606	0.275	45.84	20.79	34.42	15.61	35.74	0.78	1.72	2,124	1,230	134	85
1,103	500	4.5	90.4	41.0	0.606	0.275	53.89	24.44	40.46	18.35	42.02	0.78	1.72	2,519	1,526	162	102
1,103	500	5.0	19.8	9.0	0.606	0.275	26.64	12.08	16.93	7.68	17.36	0.65	1.44	939	571	53	35
1,103	500	5.0	37.5	17.0	0.606	0.275	34.40	15.60	23.31	10.57	24.01	0.70	1.54	1,364	844	83	53
1,103	500	5.0	55.1	25.0	0.606	0.275	41.19	18.68	29.64	13.44	30.67	0.74	1.64	1,795	1,069	113	71
1,103	500	5.0	72.8	33.0	0.606	0.275	47.87	21.71	35.96	16.31	37.33	0.78	1.72	2,226	1,289	142	89
1,103	500	5.0	90.4	41.0	0.606	0.275	56.40	25.58	42.36	19.21	43.99	0.78	1.72	2,646	1,599	172	108
1,323	600	3.0	22.1	10.0	0.728	0.330	27.61	12.52	17.35	7.87	17.79	0.64	1.42	974	533	52	34
1,323	600	3.0	44.1	20.0	0.728	0.330	35.72	16.20	23.53	10.67	24.18	0.68	1.49	1,375	845	79	51
1,323	600	3.0	66.2	30.0	0.728	0.330	42.71	19.37	29.61	13.43	30.58	0.72	1.58	1,784	1,102	106	68
1,323	600	3.0	88.2	40.0	0.728	0.330	48.97	22.21	35.70	16.19	36.98	0.76	1.67	2,198	1,323	133	84
1,323	600	3.0	110.3	50.0	0.728	0.330	55.63	25.23	41.78	18.95	43.38	0.78	1.72	2,608	1,565	161	101
1,323	600	3.5	22.1	10.0	0.728	0.330	28.36	12.86	17.82	8.08	18.27	0.64	1.42	1,004	557	54	35
1,323	600	3.5	44.1	20.0	0.728	0.330	36.82	16.70	24.43	11.08	25.15	0.69	1.51	1,438	874	84	54
1,323	600	3.5	66.2	30.0	0.728	0.330	44.19	20.04	31.00	14.06	32.03	0.73	1.60	1,879	1,137	113	72
1,323	600	3.5	88.2	40.0	0.728	0.330	50.72	23.00	37.51	17.01	38.90	0.77	1.69	2,326	1,360	143	90
1,323	600	3.5	110.3	50.0	0.728	0.330	58.72	26.63	44.10	20.00	45.78	0.78	1.72	2,763	1,654	173	109

(Continued)

TABLE 12–5 (Continued)

Live Weight		Fat	Milk		Live Weight Change		Dry Matter Intake		Energy					Protein		Minerals	
									TDN		NE_{lc}	$NE_{lcdm}{}^2$		DIP^3	UIP^4	Ca	P
(lb)	(kg)	(%)	(lb)	(kg)	(lb)	(kg)	(lb)	(kg)	(lb)	(kg)	(Mcal)	(Mcal/lb)	(Mcal/kg)	(g)	(g)	(g)	(g)
Intake at 100%																	

Live Weight		Fat	Milk		Live Weight Change		Dry Matter Intake		TDN		NE_{lc}		$NE_{lcdm}{}^2$	DIP^3	UIP^4	Ca	P
(lb)	(kg)	(%)	(lb)	(kg)	(lb)	(kg)	(lb)	(kg)	(lb)	(kg)	(Mcal)	(Mcal/lb)	(Mcal/kg)	(g)	(g)	(g)	(g)
**Intake at 100% of the Requirement for Maintenance, Lactation, and Weight Gain** (Continued)																	
1,323	600	4.0	22.1	10.0	0.728	0.330	29.11	13.20	18.30	8.30	18.75	0.64	1.42	1,034	581	56	37
1,323	600	4.0	44.1	20.0	0.728	0.330	37.90	17.19	25.36	11.50	26.11	0.69	1.52	1,501	902	89	57
1,323	600	4.0	66.2	30.0	0.728	0.330	45.62	20.69	32.37	14.68	33.47	0.74	1.62	1,975	1,170	121	77
1,323	600	4.0	88.2	40.0	0.728	0.330	52.43	23.78	39.34	17.84	40.83	0.78	1.72	2,454	1,395	153	96
1,323	600	4.0	110.3	50.0	0.728	0.330	61.81	28.03	46.42	21.05	48.19	0.78	1.72	2,918	1,744	185	116
1,544	700	3.0	26.5	12.0	0.849	0.385	31.88	14.46	20.04	9.09	20.54	0.64	1.42	1,154	607	61	40
1,544	700	3.0	52.9	24.0	0.849	0.385	41.34	18.75	27.43	12.44	28.21	0.68	1.50	1,638	968	94	60
1,544	700	3.0	79.4	36.0	0.849	0.385	49.57	22.48	34.75	15.76	35.89	0.73	1.60	2,129	1,269	127	81
1,544	700	3.0	105.8	48.0	0.849	0.385	56.89	25.80	42.01	19.05	43.57	0.77	1.69	2,627	1,525	159	101
1,544	700	3.0	132.3	60.0	0.849	0.385	65.73	29.81	49.37	22.39	51.25	0.78	1.72	3,114	1,857	192	121
1,544	700	3.5	26.5	12.0	0.849	0.385	32.77	14.86	20.59	9.34	21.11	0.64	1.42	1,190	636	64	42
1,544	700	3.5	52.9	24.0	0.849	0.385	42.64	19.34	28.53	12.94	29.37	0.69	1.52	1,713	1,002	100	64
1,544	700	3.5	79.4	36.0	0.849	0.385	51.29	23.26	36.38	16.50	37.62	0.74	1.62	2,244	1,309	135	86
1,544	700	3.5	105.8	48.0	0.849	0.385	58.92	26.72	44.19	20.04	45.88	0.78	1.72	2,781	1,567	171	108
1,544	700	3.5	132.3	60.0	0.849	0.385	69.41	31.48	52.15	23.65	54.13	0.78	1.72	3,300	1,964	207	130
1,544	700	4.0	26.5	12.0	0.849	0.385	33.52	15.20	21.17	9.60	21.69	0.65	1.43	1,227	658	67	44
1,544	700	4.0	52.9	24.0	0.849	0.385	43.92	19.92	29.64	13.44	30.52	0.69	1.53	1,789	1,035	105	68
1,544	700	4.0	79.4	36.0	0.849	0.385	52.96	24.02	38.04	17.25	39.35	0.74	1.64	2,359	1,347	144	91
1,544	700	4.0	105.8	48.0	0.849	0.385	61.81	28.03	46.42	21.05	48.19	0.78	1.72	2,930	1,648	182	115
1,544	700	4.0	132.3	60.0	0.849	0.385	73.12	33.16	54.93	24.91	57.02	0.78	1.72	3,485	2,071	221	139
1,764	800	3.0	30.9	14.0	0.970	0.440	36.07	16.36	22.69	10.29	23.24	0.64	1.42	1,331	682	71	46
1,764	800	3.0	59.5	27.0	0.970	0.440	46.15	20.93	30.67	13.91	31.56	0.69	1.51	1,857	1,064	106	68
1,764	800	3.0	88.2	40.0	0.970	0.440	55.01	24.95	38.59	17.50	39.88	0.73	1.60	2,390	1,388	142	90
1,764	800	3.0	116.9	53.0	0.970	0.440	62.93	28.54	46.48	21.08	48.20	0.77	1.69	2,928	1,665	177	112
1,764	800	3.0	145.5	66.0	0.970	0.440	72.48	32.87	54.44	24.69	56.51	0.78	1.72	3,457	2,022	213	134
1,764	800	3.5	30.9	14.0	0.970	0.440	37.00	16.78	23.33	10.58	23.92	0.64	1.42	1,374	710	74	49
1,764	800	3.5	59.5	27.0	0.970	0.440	47.61	21.59	31.91	14.47	32.86	0.69	1.52	1,942	1,102	113	72
1,764	800	3.5	88.2	40.0	0.970	0.440	56.93	25.82	40.42	18.33	41.80	0.74	1.62	2,517	1,432	151	96
1,764	800	3.5	116.9	53.0	0.970	0.440	65.20	29.57	48.88	22.17	50.75	0.78	1.72	3,099	1,711	190	120
1,764	800	3.5	145.5	66.0	0.970	0.440	76.56	34.72	57.48	26.07	59.69	0.78	1.72	3,661	2,140	228	144
1,764	800	4.0	30.9	14.0	0.970	0.440	37.86	17.17	23.99	10.88	24.59	0.65	1.43	1,418	734	77	51
1,764	800	4.0	59.5	27.0	0.970	0.440	49.04	22.24	33.14	15.03	34.16	0.70	1.54	2,027	1,139	119	76
1,764	800	4.0	88.2	40.0	0.970	0.440	58.78	26.66	42.25	19.16	43.73	0.74	1.64	2,644	1,474	161	102
1,764	800	4.0	116.9	53.0	0.970	0.440	68.36	31.00	51.33	23.28	53.29	0.78	1.72	3,263	1,800	203	128
1,764	800	4.0	145.5	66.0	0.970	0.440	80.61	36.56	60.55	27.46	62.86	0.78	1.72	3,865	2,259	244	154
**Intake at 85% of the Requirement for Maintenance and Lactation**																	
882	400	4.5	44.1	20.0	−1.535	−0.696	25.62	11.62	18.72	8.49	19.41	0.76	1.67	1,066	687	85	54
882	400	4.5	57.3	26.0	−1.852	−0.840	30.91	14.02	22.58	10.24	23.41	0.76	1.67	1,310	931	106	67
882	400	4.5	70.6	32.0	−2.168	−0.983	36.18	16.41	26.44	11.99	27.41	0.76	1.67	1,554	1,187	127	80
882	400	5.0	44.1	20.0	−1.601	−0.726	26.70	12.11	19.51	8.85	20.23	0.76	1.67	1,118	720	90	57
882	400	5.0	57.3	26.0	−1.936	−0.878	32.30	14.65	23.62	10.71	24.47	0.76	1.67	1,377	987	112	71
882	400	5.0	70.6	32.0	−2.271	−1.030	37.93	17.20	27.69	12.56	28.72	0.76	1.67	1,635	1,255	134	84

(Continued)

Dairy Cattle Science

TABLE 12–5 *(Continued)*

Live Weight		Fat	Milk		Live Weight Change		Dry Matter Intake		Energy				Protein		Minerals	
									TDN		NE_{lc}	NE_{lcdm}[2]	DIP[3]	UIP[4]	Ca	P
(lb)	*(kg)*	(%)	(lb)	*(kg)*	(lb)	*(kg)*	(lb)	*(kg)*	(lb)	*(kg)*	(Mcal)	(Mcal/lb) *(Mcal/kg)*	(g)	(g)	(g)	(g)
									Intake at 85% of the Requirement for Maintenance and Lactation *(Continued)*							
882	*400*	5.5	44.1	*20.0*	−1.665	*−0.755*	27.78	*12.60*	20.31	*9.21*	21.05	0.76 *1.67*	1,169	761	95	60
882	*400*	5.5	57.3	*26.0*	−2.020	*−0.916*	33.71	*15.29*	24.63	*11.17*	25.54	0.76 *1.67*	1,443	1,042	118	75
882	*400*	5.5	70.6	*32.0*	−2.375	*−1.077*	39.65	*17.98*	28.97	*13.14*	30.03	0.76 *1.67*	1,717	1,323	142	89
1,103	*500*	4.0	55.1	*25.0*	−1.806	*−0.819*	30.14	*13.67*	22.03	*9.99*	22.83	0.76 *1.67*	1,286	810	101	64
1,103	*500*	4.0	72.8	*33.0*	−2.201	*−0.998*	36.76	*16.67*	26.86	*12.18*	27.83	0.76 *1.67*	1,590	1,134	126	80
1,103	*500*	4.0	90.4	*41.0*	−2.597	*−1.178*	43.35	*19.66*	31.69	*14.37*	32.84	0.76 *1.67*	1,894	1,458	152	95
1,103	*500*	4.5	55.1	*25.0*	−1.887	*−0.856*	31.49	*14.28*	23.02	*10.44*	23.85	0.76 *1.67*	1,350	864	107	68
1,103	*500*	4.5	72.8	*33.0*	−2.309	*−1.047*	38.54	*17.48*	28.16	*12.77*	29.18	0.76 *1.67*	1,674	1,205	134	85
1,103	*500*	4.5	90.4	*41.0*	−2.730	*−1.238*	45.58	*20.67*	33.30	*15.10*	34.52	0.76 *1.67*	1,998	1,546	162	102
1,103	*500*	5.0	55.1	*25.0*	−1.967	*−0.892*	32.83	*14.89*	23.99	*10.88*	24.87	0.76 *1.67*	1,414	917	113	71
1,103	*500*	5.0	72.8	*33.0*	−2.414	*−1.095*	40.31	*18.28*	29.46	*13.36*	30.53	0.76 *1.67*	1,758	1,275	142	89
1,103	*500*	5.0	90.4	*41.0*	−2.862	*−1.298*	47.78	*21.67*	34.91	*15.83*	36.19	0.76 *1.67*	2,103	1,633	172	108
1,323	*600*	3.0	66.2	*30.0*	−1.943	*−0.881*	32.44	*14.71*	23.68	*10.74*	24.56	0.76 *1.67*	1,399	860	106	68
1,323	*600*	3.0	88.2	*40.0*	−2.373	*−1.076*	39.60	*17.96*	28.93	*13.12*	30.00	0.76 *1.67*	1,728	1,223	133	84
1,323	*600*	3.0	110.3	*50.0*	−2.803	*−1.271*	46.79	*21.22*	34.18	*15.50*	35.44	0.76 *1.67*	2,057	1,585	161	101
1,323	*600*	3.5	66.2	*30.0*	−2.396	*−0.925*	34.05	*15.44*	24.87	*11.28*	25.79	0.76 *1.67*	1,476	924	113	72
1,323	*600*	3.5	88.2	*40.0*	−2.503	*−1.135*	41.76	*18.94*	30.52	*13.84*	31.63	0.76 *1.67*	1,830	1,308	143	90
1,323	*600*	3.5	110.3	*50.0*	−2.964	*−1.344*	49.48	*22.44*	36.16	*16.40*	37.48	0.76 *1.67*	2,184	1,692	173	109
1,323	*600*	4.0	66.2	*30.0*	−2.137	*−0.969*	35.65	*16.17*	26.06	*11.82*	27.01	0.76 *1.67*	1,552	988	121	77
1,323	*600*	4.0	88.2	*40.0*	−2.631	*−1.193*	43.92	*19.92*	32.08	*14.55*	33.27	0.76 *1.67*	1,932	1,393	153	96
1,323	*600*	4.0	110.3	*50.0*	−3.127	*−1.418*	52.19	*23.67*	38.12	*17.29*	39.52	0.76 *1.67*	2,311	1,798	185	116
1,544	*700*	3.0	79.4	*36.0*	−2.280	*−1.034*	38.06	*17.26*	27.81	*12.61*	28.83	0.76 *1.67*	1,669	1,054	127	81
1,544	*700*	3.0	105.8	*48.0*	−2.796	*−1.268*	46.68	*21.17*	34.11	*15.47*	35.36	0.76 *1.67*	2,064	1,489	159	101
1,544	*700*	3.0	132.3	*60.0*	−3.312	*−1.502*	55.30	*25.08*	40.40	*18.32*	41.88	0.76 *1.67*	2,458	1,924	192	121
1,544	*700*	3.5	79.4	*36.0*	−2.397	*−1.087*	40.02	*18.15*	29.24	*13.26*	30.30	0.76 *1.67*	1,761	1,131	135	86
1,544	*700*	3.5	105.8	*48.0*	−2.952	*−1.339*	49.28	*22.35*	36.01	*16.33*	37.32	0.76 *1.67*	2,186	1,591	171	108
1,544	*700*	3.5	132.3	*60.0*	−3.506	*−1.590*	58.54	*26.55*	42.78	*19.40*	44.34	0.76 *1.67*	2,611	2,052	207	130
1,544	*700*	4.0	79.4	*36.0*	−2.434	*−1.140*	41.96	*19.03*	30.65	*13.90*	31.78	0.76 *1.67*	1,853	1,208	144	91
1,544	*700*	4.0	105.8	*48.0*	−3.107	*−1.409*	51.86	*23.52*	37.90	*17.19*	39.28	0.76 *1.67*	2,308	1,694	182	115
1,544	*700*	4.0	132.3	*60.0*	−3.700	*−1.678*	61.78	*28.02*	45.14	*20.47*	46.79	0.76 *1.67*	2,764	2,180	221	139
1,764	*800*	3.0	88.2	*40.0*	−2.529	*−1.147*	42.23	*19.15*	30.85	*13.99*	31.98	0.76 *1.67*	1,871	1,176	142	90
1,764	*800*	3.0	110.3	*50.0*	−2.959	*−1.342*	49.41	*22.41*	36.10	*16.37*	37.42	0.76 *1.67*	2,200	1,538	169	107
1,764	*800*	3.0	132.3	*60.0*	−3.389	*−1.537*	56.58	*25.66*	41.34	*18.75*	42.86	0.76 *1.67*	2,529	1,900	196	124
1,764	*800*	3.5	88.2	*40.0*	−2.659	*−1.206*	44.39	*20.13*	32.44	*14.71*	33.62	0.76 *1.67*	1,973	1,261	151	96
1,764	*800*	3.5	110.3	*50.0*	−3.122	*−1.416*	52.10	*23.63*	38.08	*17.27*	39.46	0.76 *1.67*	2,327	1,645	181	114
1,764	*800*	3.5	132.3	*60.0*	−3.583	*−1.625*	59.82	*27.13*	43.70	*19.82*	45.31	0.76 *1.67*	2,682	2,028	211	133
1,764	*800*	4.0	88.2	*40.0*	−2.787	*−1.264*	46.55	*21.11*	34.00	*15.42*	35.25	0.76 *1.67*	2,075	1,346	161	102
1,764	*800*	4.0	110.3	*50.0*	−3.283	*−1.489*	54.82	*24.86*	40.04	*18.16*	41.51	0.76 *1.67*	2,455	1,751	193	122
1,764	*800*	4.0	132.3	*60.0*	−3.777	*−1.713*	63.06	*28.60*	46.08	*20.90*	47.76	0.76 *1.67*	2,835	2,156	225	142

[1]Adapted by the author from *Nutrient Requirements of Dairy Cattle*, 6th rev. ed., update 1989, NRC, National Academy Press, pp. 85–86, Table 6–4.

[2]NE_{lcdm} = net energy for lactation/kg of dry matter.

[3]DIP = degraded intake protein.

[4]UIP = undegraded intake protein.

TABLE 12-6
RECOMMENDED NUTRIENT CONTENT OF RATIONS FOR DAIRY CATTLE[1]

Cow Weight (lb)	(kg)	Fat (%)	Weight Gain (lb/day)	(kg/day)	Milk Yield (lb/day)	(kg/day)	Milk Yield (lb/day)	(kg/day)	Milk Yield (lb/day)	(kg/day)	Milk Yield (lb/day)	(kg/day)	Milk Yield (lb/day)	(kg/day)
882	400	5.0	0.485	0.220	15.4	7	28.7	13	44.1	20	57.3	26	72.8	33
1,103	500	4.5	0.606	0.275	17.6	8	37.5	17	55.1	25	72.8	33	90.4	41
1,323	600	4.0	0.728	0.330	22.1	10	44.1	20	66.2	30	88.2	40	110.3	50
1,544	700	3.5	0.849	0.385	26.5	12	52.9	24	79.4	36	105.8	48	132.3	60
1,764	800	3.5	0.970	0.440	28.7	13	59.5	27	88.2	40	116.9	53	147.7	67

	Lactating Cow Rations Milk Yield 1	Milk Yield 2	Milk Yield 3	Milk Yield 4	Milk Yield 5	Early Lactation (Weeks 0-3)	Dry, Pregnant Cows	Calf Milk Replacer	Calf Starter Mix	Growing Heifers and Bulls[2] 3-6 Mos.	6-12 Mos.	Over 12 Mos.	Mature Bulls	Maximum Tolerable Levels[3,4]
Energy:														
NE_{lc} (Mcal/lb)	0.64	0.69	0.74	0.78	0.78	0.76	0.57	—	—	—	—	—	—	—
NE_{lc} (Mcal/kg)	1.42	1.52	1.62	1.72	1.72	1.67	1.25	—	—	—	—	—	—	—
NE_m (mcal/lb)	—	—	—	—	—	—	—	1.09	0.86	0.77	0.72	0.64	0.52	—
NE_m (Mcal/kg)	—	—	—	—	—	—	—	2.40	1.90	1.70	1.58	1.40	1.15	—
NE_g (mcal/lb)	—	—	—	—	—	—	—	0.70	0.54	0.49	0.44	0.37	—	—
NE_g (Mcal/kg)	—	—	—	—	—	—	—	1.55	1.20	1.08	0.98	0.82	—	—
ME (mcal/lb)	1.07	1.15	1.23	1.31	1.31	1.27	0.93	1.72	1.41	1.18	1.12	1.03	0.91	—
ME (Mcal/kg)	2.35	2.53	2.71	2.89	2.89	2.80	2.04	3.78	3.11	2.60	2.47	2.27	2.00	—
DE (mcal/lb)	1.26	1.34	1.42	1.50	1.50	1.46	1.12	1.90	1.60	1.37	1.31	1.22	1.10	—
DE (Mcal/kg)	2.77	2.95	3.13	3.31	3.31	3.22	2.47	4.19	3.53	3.02	2.89	2.69	2.43	—
TDN (% of DM)	63.00	67.00	71.00	75.00	75.00	73.00	56.00	95.00	80.00	69.00	66.00	61.00	55.00	—
Protein equivalent:														
Crude protein (%)	12.00	15.00	16.00	17.00	18.00	19.00	12.00	22.00	18.00	16.00	12.00	12.00	10.00	—
DIP[5] (%)	7.80	8.70	9.60	10.30	10.40	9.70	—	—	—	4.60	6.40	7.20	—	—
UIP[6] (%)	4.40	5.20	5.70	5.90	6.20	7.00	—	—	—	8.20	4.40	2.10	—	—
Fiber content (minimum):[7]														
Crude fiber (%)	17.00	17.00	17.00	15.00	15.00	17.00	22.00	—	—	13.00	15.00	15.00	15.00	—
Neutral detergent fiber (NDF) (%)	28.00	28.00	28.00	25.00	25.00	28.00	35.00	—	—	23.00	25.00	25.00	25.00	—
Acid detergent fiber (ADF) (%)	21.00	21.00	21.00	19.00	19.00	21.00	27.00	—	—	16.00	19.00	19.00	19.00	—
Ether extract (minimum) (%)	3.00	3.00	3.00	3.00	3.00	3.00	3.00	10.00	3.00	3.00	3.00	3.00	3.00	—
Major or Macrominerals:														
Calcium (Ca) (%)	0.43	0.51	0.58	0.64	0.66	0.77	0.39[8]	0.70	0.60	0.52	0.41	0.29	0.30	2.0
Chlorine (Cl) (%)	0.25	0.25	0.25	0.25	0.25	0.25	0.20	0.20	0.20	0.20	0.20	0.20	0.20	
Magnesium (Mg)[9] (%)	0.20	0.20	0.20	0.25	0.25	0.25	0.16	0.07	0.10	0.16	0.16	0.16	0.16	0.5
Phosphorus (P) (%)	0.28	0.33	0.37	0.41	0.41	0.48	0.24	0.60	0.40	0.31	0.30	0.23	0.19	1.0
Potassium (K)[10] (%)	0.90	0.90	0.90	1.00	1.00	1.00	0.65	0.65	0.65	0.65	0.65	0.65	0.65	3.0
Sodium (Na) (%)	0.18	0.18	0.18	0.18	0.18	0.18	0.10	0.10	0.10	0.10	0.10	0.10	0.10	—
Sulfur (S) (%)	0.20	0.20	0.20	0.20	0.20	0.25	0.16	0.29	0.20	0.16	0.16	0.16	0.16	0.4

(Continued)

TABLE 12-6 (Continued)

	Lactating Cow Rations						Early Lactation (Weeks 0-3)	Dry, Pregnant Cows	Calf Milk Replacer	Calf Starter Mix	Growing Heifers and Bulls[2]				Maximum Tolerable Levels[3, 4]
	Milk Yield	Milk Yield	Milk Yield	Milk Yield	Milk Yield	Milk Yield					3-6 Mos.	6-12 Mos.	Over 12 Mos.	Mature Bulls	
Trace or Microminerals:															
Cobalt (Co) (ppm)	0.10	0.10	0.10	0.10	0.10	0.10	0.10	0.10	0.10	0.10	0.10	0.10	0.10	0.10	10
Copper (Cu)[11] (ppm)	10.00	10.00	10.00	10.00	10.00	10.00	10.00	10.00	10.00	10.00	10.00	10.00	10.00	10.00	100
Iodine (I)[12] (ppm)	0.60	0.60	0.60	0.60	0.60	0.60	0.60	0.25	0.25	0.25	0.25	0.25	0.25	0.25	50[13]
Iron (Fe) (ppm)	50.00	50.00	50.00	50.00	50.00	50.00	50.00	50.00	100.00	50.00	50.00	50.00	50.00	50.00	1,000
Manganese (Mn) (ppm)	40.00	40.00	40.00	40.00	40.00	40.00	40.00	40.00	40.00	40.00	40.00	40.00	40.00	40.00	1,000
Selenium (Se) (ppm)	0.30	0.30	0.30	0.30	0.30	0.30	0.30	0.30	0.30	0.30	0.30	0.30	0.30	0.30	2
Zinc (Zn) (ppm)	40.00	40.00	40.00	40.00	40.00	40.00	40.00	40.00	40.00	40.00	40.00	40.00	40.00	40.00	500
Vitamins:[14]															
Vitamin A (IU/lb)	1,453	1,453	1,453	1,453	1,453	1,453	1,816	1,816	1,725	999	999	999	999	1,453	29,964
Vitamin A (IU/kg)	3,200	3,200	3,200	3,200	3,200	3,200	4,000	4,000	3,800	2,200	2,200	2,200	2,200	3,200	66,000
Vitamin D (IU/lb)	454	454	454	454	454	454	454	545	272	136	136	136	136	136	4,540
Vitamin D (IU/kg)	1,000	1,000	1,000	1,000	1,000	1,000	1,000	1,200	600	300	300	300	300	300	10,000
Vitamin E (IU/lb)	7	7	7	7	7	7	7	7	18	11	11	11	11	7	908
Vitamin E (IU/kg)	15	15	15	15	15	15	15	15	40	25	25	25	25	15	2,000

[1]Adapted by the author from *Nutrient Requirements of Dairy Cattle*, 6th rev. ed., update 1989, NRC, National Academy Press, p. 87, Table 6-5.

[2]The approximate weight for growing heifers and bulls at 3-6 months is 331 lb *(150 kg)*; at 6-12 months, it is 551 lb *(250 kg)*; and at more than 12 months, it is 882 lb *(400 kg)*. The approximate average daily gain is 25 oz/day *(700 g/day)*.

[3]The maximum safe levels for many of the mineral elements are not well defined and may be substantially affected by specific feeding conditions. Additional information is available in Table 12-7 and in *Mineral Tolerance of Domestic Animals* (NRC, 1980).

[4]Vitamin tolerances are discussed in detail in *Vitamin Tolerance of Animals* (NRC, 1987b).

[5]DIP = degraded intake protein.

[6]UIP = undegraded intake protein.

[7]It is recommended that 75% of the NDF in lactating cow rations be provided as forage. If this recommendation is not followed, a depression in milk fat may occur.

[8]The value for calcium assumes that the cow is in calcium balance at the beginning of the dry period. If the cow is not in balance, then the dietary calcium requirement should be increased by 25 to 33%.

[9]Under conditions conducive to grass tetany, magnesium should be increased to 0.25 or 0.30%.

[10]Under conditions of heat stress, potassium should be increased by 1.2%.

[11]The cow's copper requirement is influenced by molybdenum and sulfur in the ration.

[12]If the ration contains as much as 25% strongly goitrogenic feed on a dry basis, the iodine provided should be increased by 2 times or more.

[13]Although cattle can tolerate this level of iodine, lower levels may be desirable to reduce the iodine content of milk.

[14]The following minimum quantities of B-complex vitamins are suggested per unit of milk replacer: Niacin (Nicotinic Acid, Nicotinamide, 2.6 ppm; Pantothenic Acid (Vitamin B-1), 13 ppm; Riboflavin (Vitamin B-2), 6.5 ppm; Vitamin B-6 (Pyridoxine, Pyridoxal, Pyridoxamine), 6.5 ppm; Folacin (Folic Acid), 0.5 ppm; Biotin, 0.1 ppm; Vitamin B-12 (Cobalamins, 0.07 ppm; and Choline, 0.26%. It appears that adequate amounts of these vitamins are furnished when calves have functional rumens (usually at 6 weeks of age) by a combination of rumen synthesis and natural feedstuffs.

Table 12–7 shows NRC's maximum tolerable dietary levels of certain elements.

Table 12–8, Composition of Feeds Commonly Used in Dairy Cattle Rations, consists of selected feeds from *Nutrient Requirements of Dairy Cattle*, Sixth Revised Edition, 1989, Table 7–1. In their feed compositions, the NRC committee assumed an average decrease of 4% per unit of dry matter intake above maintenance in calculating NE_{lc} values for feed ingredients, or an average discount of 8% based on their assumption that lactating cows are fed at 3X maintenance. For the convenience of those dairy producers and dairy nutritionists who wish to use these values, the author selected from *Nutrient Requirements of Dairy Cattle*, Sixth Revised Edition, 1989, Table 7–1, the feeds' most commonly used in dairy cattle rations, and herein reported them in Table 12–8. **Note well:** More complete feed compositions are presented in this book in Chapter 21, Feed Composition Tables.

TABLE 12–7
MAXIMUM TOLERABLE DIETARY LEVELS OF CERTAIN ELEMENTS[1]

Element	Maximum Tolerable Level	Element	Maximum Tolerable Level
	(ppm)		(ppm)
Aluminum	1,000[2]	Fluorine	40[4]
Arsenic:		Lead	30[3]
Inorganic	50	Mercury	2[3]
Organic	100	Molybdenum	10[5]
Bromine	200	Nickel	50
Cadmium	0.5[3]	Vanadium	50

[1]Adapted by the author from *Nutrient Requirements of Dairy Cattle*, 6th rev. ed., update 1989, NRC, National Academy Press, p. 88, Table 6–6.

[2]As soluble salts of high bioavailability. Higher levels of less soluble forms found in natural substances can be tolerated.

[3]Levels are based on human food residue considerations.

[4]As sodium fluoride or fluorides of similar toxicity. The maximum safe level of fluorine for growing heifers and bulls is lower than for other dairy cattle. Somewhat higher levels are tolerated when fluorine is from less available sources such as phosphates. Morphological lesions in cattle teeth may be seen when dietary fluoride for the young exceeds 20 ppm, but a relationship between the lesions caused by fluoride levels below the maximum tolerable levels and animal performance has not been established.

[5]Toxicity related to the dietary level of copper.

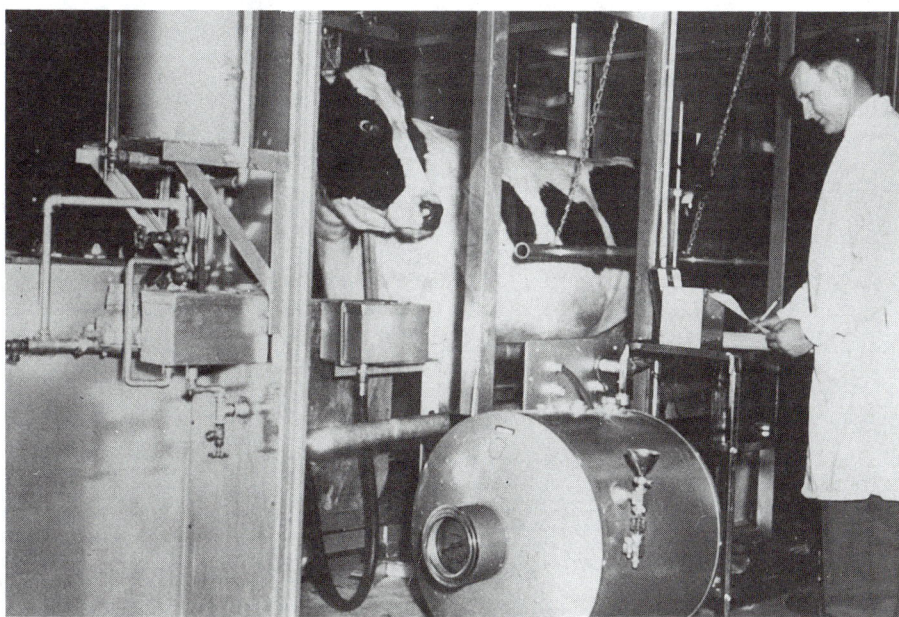

Fig. 12–5. A cow in an open circuit respiration chamber. The gas meter to the left of the chamber is used to measure the respiratory exchange of the cow. These data, plus the gas composition, provide the information needed to calculate the heat production (HP) of the cow. The HP of an animal consuming feed in a thermoneutral environment is composed of the heat increment (heat of fermentation plus heat of nutrient metabolism) plus heat used for maintenance (basal metabolism plus voluntary activity). (Courtesy, USDA)

Entry No.	Feed Name Description	International Feed Number[2]	Dry Matter	Values as Determined at Maintenance Intake			Production Growing Dairy Cattle		Production Lactating Cows	Production Growing Dairy Cattle		Production Lactating Cows	Crude Protein	Ether Extract	Total Ash	Crude Fiber
				TDN	DE	ME	NEM	NEG	NEL	NEM	NEG	NEL				
			(%)	(%)	(Mcal/kg)	(Mcal/kg)	(Mcal/kg)	(Mcal/kg)	(Mcal/kg)	(Mcal/lb)	(Mcal/lb)	(Mcal/lb)	(%)	(%)	(%)	(%)
	ALFALFA *Medicago sativa*															
1	HAY, SUN-CURED, EARLY VEGETATIVE	1–00–050	90	66	2.91	2.49	1.51	0.92	1.50	0.69	0.42	0.68	23.0	4.0	10.2	20.5
2	HAY, SUN-CURED, LATE VEGETATIVE	1–00–054	90	63	2.78	2.36	1.41	0.83	1.42	0.64	0.38	0.65	20.0	3.8	9.2	22.0
3	HAY, SUN-CURED, EARLY BLOOM	1–00–059	90	60	2.65	2.22	1.31	0.74	1.35	0.60	0.34	0.61	18.0	3.0	9.6	23.0
4	HAY, SUN-CURED, MIDBLOOM	1–00–063	90	58	2.56	2.13	1.24	0.68	1.30	0.57	0.31	0.59	17.0	2.6	9.1	26.0
5	HAY, SUN-CURED, FULL BLOOM	1–00–068	90	55	2.43	2.00	1.14	0.58	1.23	0.52	0.26	0.56	15.0	2.0	8.9	29.0
6	SILAGE, WILTED, 25–45% DRY MATTER	—	—	—	—	—	—	—	—	—	—	—	—	—	—	—
	(see similar maturity descriptions of hays)															
	ALMOND *Prunus amygdalus*															
7	HULLS	4–00–	90	59	2.60	2.18	1.27	0.70	1.33	0.58	0.32	0.60	2.7	3.6	7.6	11.0
	BARLEY *Hordeum vulgare*															
8	GRAIN	4–00–549	88	84	3.70	3.29	2.06	1.40	1.94	0.94	0.64	0.88	13.5	2.1	2.6	5.7
9	GRAIN, PACIFIC COAST	4–07–939	89	86	3.79	3.38	2.12	1.45	1.99	0.96	0.66	0.90	10.8	2.0	3.1	7.1
	BEET, SUGAR *Beta vulgaris altissima*															
10	PULP, DEHYDRATED	4–00–669	91	78	3.44	3.02	1.88	1.24	1.79	0.86	0.57	0.81	9.7	0.6	4.4	19.8
11	PULP W/MOLASSES, DEHYDRATED	4–00–672	92	78	3.44	3.02	1.88	1.24	1.79	0.86	0.57	0.81	10.1	0.6	6.1	16.5
	BLOOD															
12	MEAL	5–00–380	92	66	2.91	2.49	1.51	0.92	1.50	0.69	0.42	0.68	87.2	1.4	5.8	1.1
	BREWERS' GRAINS															
13	DEHYDRATED	5–02–141	92	66	2.91	2.49	1.51	0.91	1.50	0.69	0.41	0.68	25.4	6.5	4.8	14.9
	BROME *Bromus spp*															
14	FRESH, EARLY VEGETATIVE	2–00–892	34	74	3.26	2.85	1.75	1.13	1.69	0.80	0.51	0.77	18.0	3.7	10.7	24.0
	CASEIN															
15	DEHYDRATED (CATTLE)	5–01–162	91	89	3.92	3.51	2.20	1.52	2.06	1.00	0.69	0.94	92.7	0.7	2.4	0.2
	CITRUS *Citrus spp*															
16	PULP W/O FINES, DEHYDRATED	4–01–237	91	77	3.40	2.98	1.86	1.22	1.77	0.85	0.55	0.80	6.7	3.7	6.6	12.7
	(DRIED CITRUS PULP)															
	CORN, DENT YELLOW *Zea mays indentata*															
17	DISTILLERS' GRAINS, DEHYDRATED	5–28–235	94	86	3.79	3.38	2.12	1.45	1.99	0.96	0.66	0.90	23.0	9.8	2.4	12.1
18	EARS, GROUND (CORN AND COB MEAL)	4–28–238	87	83	3.66	3.25	2.03	1.37	1.91	0.92	0.62	0.87	9.0	3.7	1.9	9.4
19	GLUTEN, MEAL	5–28–241	91	86	3.79	3.38	2.12	1.45	1.99	0.96	0.66	0.90	46.8	2.4	3.4	4.8
20	GLUTEN, MEAL, 60% PROTEIN	5–28–242	90	89	3.92	3.51	2.20	1.52	2.06	1.00	0.69	0.94	67.2	2.4	1.8	2.2
21	GLUTEN W/BRAN (CORN GLUTEN FEED)	5–28–243	90	83	3.66	3.25	2.03	1.37	1.91	0.92	0.62	0.87	25.6	2.4	7.5	9.7
22	GRAIN, FLAKED	4–28–244	89	88	3.88	3.47	2.18	1.50	2.04	0.99	0.68	0.93	10.0	4.3	1.6	2.6
23	GRAIN, HIGH-MOISTURE	4–20–770	77	88	3.88	3.47	2.18	1.50	2.04	0.99	0.68	0.93	10.0	4.3	1.6	2.6
24	GRITS BY-PRODUCT (HOMINY FEED)	4–03–011	90	87	3.84	3.42	2.16	1.48	2.01	0.98	0.67	0.91	11.5	7.7	3.1	6.7
25	SILAGE, FEW EARS	3–28–245	29	62	2.73	2.31	1.38	0.80	1.40	0.63	0.36	0.64	8.4	3.0	7.2	32.3
26	SILAGE, WELL-EARED	3–28–250	33	70	3.09	2.67	1.63	1.03	1.60	0.74	0.47	0.73	8.1	3.1	4.5	23.7
	COTTON *Gossypium spp*															
27	HULLS	1–01–599	91	45	1.98	1.55	0.78	0.25	0.98	0.35	0.11	0.45	4.1	1.7	2.8	47.8
28	SEEDS, W/LINT	5–01–614	92	96	4.23	3.83	2.41	1.69	2.23	1.10	0.77	1.01	23.0	20.0	4.8	24.0
29	SEEDS, W/O LINT	5–01–	90	96	4.23	3.82	2.41	1.69	2.23	1.10	0.77	1.01	25.0	23.8	4.5	17.2
30	SEEDS, MEAL PREPRESSED, SOLV EXTD, 41% PROTEIN	5–07–872	91	76	3.35	2.93	1.82	1.19	1.74	0.83	0.55	0.79	45.6	1.3	7.0	14.1
31	SEEDS, MEAL PREPRESSED, SOLV EXTD, 44% PROTEIN	5–07–873	91	75	3.31	2.89	1.79	1.16	1.72	0.81	0.53	0.78	48.9	1.7	6.7	12.1
	FATS AND OILS (not exceeding 3% of diet)															
32	FAT, ANIMAL HYDROLYZED	4–00–376	99	177	7.30	7.30	5.84	5.84	5.84	2.65	2.65	2.65	—	99.5	—	—
	FESCUE, KENTUCKY 31 *Festuca arundinacea*															
33	FRESH, VEGETATIVE	2–01–902	29	67	2.91	2.49	1.51	0.92	1.50	0.69	0.42	0.68	14.5	5.5	9.9	24.6

12–8
CATTLE RATIONS (ON A 100% DRY MATTER BASIS)[1]

Entry No.	Neutral Detergent Fiber (%)	Acid Detergent Fiber (%)	Cellulose (%)	Lignin (%)	Macrominerals							Microminerals							Vitamins		
					Calcium (%)	Chlorine (%)	Magnesium (%)	Phosphorus (%)	Potassium (%)	Sodium (%)	Sulfur (%)	Cobalt (mg/kg)	Copper (mg/kg)	Iodine (mg/kg)	Iron (mg/kg)	Manganese (mg/kg)	Selenium (mg/kg)	Zinc (mg/kg)	A Activity (1,000 IU/kg)	D (1,000 IU/kg)	E (IU/kg)
1	38	28	22	5	1.80	0.34	0.26	0.35	2.21	0.22	0.33	0.10	11	0.19	253	45	0.37	24	80	1.9	—
2	40	29	23	7	1.54	0.34	0.24	0.29	2.56	0.15	0.31	0.09	9	0.18	227	34	0.35	27	81	—	—
3	42	31	24	8	1.41	0.38	0.33	0.22	2.52	0.14	0.28	0.16	11	0.17	192	31	0.34	25	56	2.0	26
4	46	35	26	9	1.41	0.38	0.31	0.24	1.71	0.12	0.28	0.36	14	0.16	134	28	0.32	23	46	2.0	11
5	50	37	28	10	1.25	0.35	0.31	0.22	1.53	0.11	0.27	0.33	14	0.13	150	37	0.29	25	26	2.0	11
6	—	—	—	—	—	—	—	—	—	—	—	—	—	—	—	—	—	—	—	—	—
7	25	20	14	6	0.23	—	0.13	0.11	0.53	0.02	0.11	0.30	11	—	301	21	—	24	—	—	—
8	19	7	5	2	0.05	0.18	0.15	0.38	0.47	0.03	0.17	0.10	9	0.05	85	18	0.22	19	1	—	25
9	21	9	—	—	0.06	0.17	0.14	0.39	0.58	0.02	0.16	0.10	9	—	97	18	0.11	17	—	—	30
10	54	33	31	2	0.69	0.04	0.27	0.10	0.20	0.21	0.22	0.08	14	—	329	38	—	10	—	0.6	—
11	44	25	22	3	0.61	—	0.16	0.10	1.78	0.53	0.42	0.23	16	—	207	27	—	10	—	—	—
12	—	—	—	—	0.32	0.30	0.24	0.26	0.10	0.35	0.37	0.10	11	—	4,064	6	0.80	5	—	—	—
13	46	24	18	6	0.33	0.17	0.16	0.55	0.09	0.23	0.32	0.08	23	0.07	266	40	0.76	30	0	—	29
14	56	31	27	3	0.50	—	0.18	0.30	2.30	0.02	0.20	0.08	11	—	200	142	—	27	184	—	—
15	0	0	0	0	0.67	—	0.01	0.90	0.01	0.01	—	—	4	—	15	5	—	30	—	—	—
16	23	22	18	3	1.84	—	0.17	0.12	0.79	0.09	0.08	0.16	6	—	378	7	—	15	—	—	—
17	43	17	12	5	0.11	0.08	0.07	0.43	0.18	0.10	0.46	0.09	48	0.05	223	23	0.48	35	1	—	—
18	28	11	9	2	0.07	0.05	0.14	0.27	0.53	0.02	0.16	0.31	8	0.03	91	14	0.09	14	2	—	20
19	37	9	8	1	0.16	0.07	0.06	0.50	0.03	0.10	0.39	0.08	30	—	423	8	1.11	29	7	—	34
20	14	5	4	1	0.08	0.10	0.09	0.54	0.21	0.06	0.72	0.05	29	0.02	313	7	0.92	35	14	—	26
21	45	12	—	—	0.36	0.25	0.36	0.82	0.64	1.05	0.23	0.10	52	0.07	471	26	0.30	72	3	—	14
22	9	3	2	1	0.03	0.05	0.14	0.29	0.37	0.03	0.12	0.05	4	—	30	5	0.08	14	1	—	25
23	9	3	2	1	0.02	0.05	0.14	0.32	0.35	0.01	0.14	0.05	4	—	30	6	0.08	18	1	—	25
24	55	13	10	2	0.05	0.06	0.26	0.57	0.65	0.09	0.03	0.06	15	—	75	16	0.11	3	—	—	—
25	53	30	23	5	0.34	—	0.23	0.19	1.41	—	0.08	—	—	—	—	—	—	—	5	—	—
26	51	28	24	4	0.23	—	0.19	0.22	0.96	0.01	0.15	0.06	10	—	260	30	—	21	18	0.1	—
27	90	73	59	24	0.15	0.02	0.14	0.09	0.87	0.02	0.09	0.02	13	—	131	119	—	22	—	—	—
28	44	34	24	10	0.21	—	0.46	0.64	1.00	0.01	0.26	—	9	—	151	19	—	33	—	—	—
29	37	26	12	14	0.12	—	0.41	0.54	1.18	0.01	—	—	11	—	108	14	—	36	—	—	—
30	26	19	12	6	0.22	0.04	0.55	1.21	1.39	0.04	0.34	0.82	20	—	223	23	—	69	—	—	—
31	28	21	13	7	0.17	0.04	0.55	1.00	1.39	0.04	0.34	0.82	20	—	223	23	10.00	69	—	—	—
32	—	—	—	—	—	—	—	—	—	—	—	—	—	—	—	—	—	—	—	—	—
33	—	—	—	—	0.51	—	—	0.37	—	—	—	—	—	—	—	—	—	—	—	—	—

(Continued)

TABLE 12–8

Entry No.	Feed Name Description	International Feed Number[2]	Dry Matter	Values as Determined at Maintenance Intake			Production Growing Dairy Cattle		Production Lactating Cows	Production Growing Dairy Cattle		Production Lactating Cows	Crude Protein	Ether Extract	Total Ash	Crude Fiber
				TDN	DE	ME	NEM	NEG	NEL	NEM	NEG	NEL				
			(%)	(%)	(Mcal/kg)	(Mcal/kg)	(Mcal/kg)	(Mcal/kg)	(Mcal/kg)	(Mcal/lb)	(Mcal/lb)	(Mcal/lb)	(%)	(%)	(%)	(%)
	FISH, MENHADEN *Brevoortia tyrannus*															
34	MEAL MECH EXTD	5–02–009	92	73	3.22	2.80	1.73	1.11	1.67	0.79	0.50	0.76	66.7	10.5	20.8	1.0
	FLAX *Linum usitatissimum*															
35	SEEDS, MEAL SOLV EXTD (LINSEED MEAL)	5–02–048	90	78	3.44	3.02	1.88	1.24	1.79	0.85	0.56	0.81	38.3	1.5	6.5	10.1
	MILK															
36	SKIMMED, DEHYDRATED (CATTLE)	5–01–175	94	85	3.75	3.34	2.10	1.43	1.96	0.95	0.65	0.89	35.8	0.9	8.4	0.2
	MOLASSES AND SYRUP															
37	BEET, SUGAR, MOLASSES, MORE THAN 48% INVERT SUGAR, MORE THAN 79.5° BRIX	4–00–668	78	75	3.31	2.89	1.79	1.16	1.72	0.81	0.53	0.78	8.5	0.2	11.3	—
38	CITRUS, SYRUP (CITRUS MOLASSES)	4–01–241	68	75	3.31	2.89	1.79	1.16	1.72	0.81	0.53	0.78	8.2	0.3	7.9	—
39	SUGARCANE, MOLASSES, DEHYDRATED	4–04–695	94	70	3.09	2.67	1.63	1.03	1.60	0.74	0.47	0.73	10.3	0.9	13.3	6.7
40	SUGARCANE, MOLASSES, MORE THAN 46% INVERT SUGAR, MORE THAN 79.5° BRIX (BLACK STRAP)	4–04–696	75	72	3.17	2.76	1.69	1.08	1.64	0.77	0.49	0.75	5.8	0.1	13.1	—
	OATS *Avena sativa*															
41	GRAIN	4–03–309	89	77	3.40	2.98	1.86	1.22	1.77	0.85	0.55	0.80	13.3	5.4	3.4	12.1
	PEANUT *Arachis hypogaea*															
42	KERNELS, MEAL SOLV EXTD (PEANUT MEAL)	5–03–650	92	77	3.40	2.98	1.86	1.22	1.77	0.85	0.55	0.80	52.3	1.4	6.3	10.8
	PINEAPPLE *Ananas comosus*															
43	PROCESS RESIDUE, DEHYDRATED (PINEAPPLE BRAN)	4–03–722	87	68	3.00	2.58	1.57	0.97	1.55	0.71	0.44	0.70	4.6	1.5	3.5	20.9
	RAPE *Brassica* spp															
44	SEEDS, MEAL SOLV EXTD	5–03–871	91	69	3.04	2.62	1.60	1.00	1.57	0.73	0.45	0.71	40.6	1.8	7.5	13.2
	SORGHUM *Sorghum bicolor*															
45	GRAIN, 8–10% PROTEIN	4–20–893	87	80	3.53	3.12	1.94	1.30	1.84	0.88	0.59	0.84	9.7	3.4	2.1	2.0
46	SILAGE	3–04–323	30	60	2.65	2.22	1.31	0.74	1.35	0.60	0.34	0.61	7.5	3.0	8.7	27.9
47	SILAGE, DOUGH STAGE	3–04–321	28	55	2.43	2.00	1.14	0.58	1.23	0.52	0.26	0.56	6.0	3.3	9.3	28.5
	SOYBEAN *Glycine max*															
48	HULLS	1–04–560	91	77	3.40	2.98	1.86	1.22	1.77	0.85	0.55	0.80	12.1	2.1	5.1	40.1
49	SEEDS, HEAT-PROCESSED	5–04–597	90	94	4.14	3.74	2.35	1.64	2.18	1.07	0.75	0.99	42.2	20.0	5.1	5.6
50	SEEDS, MEAL SOLV EXTD, 44% PROTEIN	5–20–637	89	84	3.70	3.29	2.06	1.40	1.94	0.94	0.64	0.88	49.9	1.5	7.3	7.0
	SUNFLOWER, COMMON *Helianthus annuus*															
51	SEEDS W/O HULLS, MEAL SOLV EXTD	5–04–739	93	65	2.87	2.45	1.47	0.88	1.47	0.67	0.40	0.67	49.8	3.1	8.1	12.2
	TIMOTHY *Phleum pratense*															
52	HAY, SUN-CURED, EARLY BLOOM	1–04–882	90	61	2.69	2.27	1.35	0.77	1.38	0.61	0.35	0.63	15.0	2.9	5.7	28.0
	TRITICALE *Triticale hexaploide*															
53	GRAIN	4–20–362	90	84	3.70	3.29	2.06	1.40	1.94	0.94	0.64	0.88	17.6	1.7	2.0	4.4
	UREA															
54	45% NITROGEN, 281% PROTEIN EQUIVALENT	5–05–070	99	0	0.00	0.00	0.00	0.00	0.00	0.00	0.00	0.00	281.0	0.0	—	0.0
	WHEAT *Triticum aestivum*															
55	BRAN	4–05–190	89	70	3.09	2.67	1.63	1.03	1.60	0.74	0.47	0.73	17.1	4.4	6.9	11.3
56	FLOUR BY-PRODUCT, LESS THAN 7% FIBER (WHEAT SHORTS)	4–05–201	88	73	3.22	2.80	1.73	1.11	1.67	0.79	0.50	0.76	18.6	5.2	4.9	7.7
57	FLOUR BY-PRODUCT, LESS THAN 9.5% FIBER (WHEAT MIDDLINGS)	4–05–205	89	69	3.04	2.62	1.60	1.00	1.57	0.73	0.45	0.71	18.4	4.9	5.2	8.2
58	GRAIN	4–05–211	89	88	3.88	3.47	2.18	1.50	2.04	0.99	0.68	0.93	16.0	2.0	1.9	2.9
	WHEY															
59	DEHYDRATED (CATTLE)	4–01–182	93	81	3.57	3.16	1.97	1.32	1.87	0.90	0.60	0.85	14.2	0.7	9.8	0.2

[1]Selected feeds from *Nutrient Requirements of Dairy Cattle*, 6th rev. ed., update 1989, NRC, National Academy Press, p. 90, Table 7–1.

[2]Some specific numbers have not been assigned by the USDA Feed Composition Data Bank.

(Continued)

Entry No.	Neutral Detergent Fiber (%)	Acid Detergent Fiber (%)	Cellulose (%)	Lignin (%)	Macrominerals							Microminerals							Vitamins		
					Calcium (%)	Chlorine (%)	Magnesium (%)	Phosphorus (%)	Potassium (%)	Sodium (%)	Sulfur (%)	Cobalt (mg/kg)	Copper (mg/kg)	Iodine (mg/kg)	Iron (mg/kg)	Manganese (mg/kg)	Selenium (mg/kg)	Zinc (mg/kg)	A Activity (1,000 IU/kg)	D (1,000 IU/kg)	E (IU/kg)
34	—	—	—	—	5.65	0.60	0.16	3.16	0.76	0.43	0.49	0.17	12	1.19	524	37	2.40	1.62	—	—	13
35	25	19	13	6	0.43	0.04	0.66	0.89	1.53	0.15	0.43	0.21	29	—	354	42	0.91	—	—	—	15
36	—	—	—	—	1.36	0.96	0.13	1.09	1.70	0.49	0.34	0.12	1	—	10	2	0.13	41	—	0.4	—
37	—	—	—	—	0.17	1.64	0.29	0.03	6.07	1.48	0.60	0.46	22	—	87	6	—	18	—	—	5
38	—	—	—	—	1.72	0.11	0.21	0.13	0.14	0.41	0.23	0.16	108	—	508	38	—	137	—	—	—
39	—	—	—	—	1.10	—	0.47	0.15	3.60	0.20	0.46	1.21	79	2.10	250	57	—	33	—	—	—
40	—	—	—	—	1.00	3.10	0.43	0.11	3.84	0.22	0.47	1.21	79	2.10	250	56	—	30	—	—	7
41	32	16	11	3	0.07	0.11	0.14	0.38	0.44	0.08	0.23	0.06	7	0.11	85	42	0.26	41	—	—	15
42	—	—	—	—	0.29	0.03	0.17	0.68	1.23	0.08	0.33	0.12	17	0.07	154	29	—	22	—	—	—
43	73	37	—	7	0.23	—	—	0.13	—	—	—	—	—	—	561	—	—	—	22	—	—
44	—	—	—	—	0.67	0.11	0.60	1.04	1.36	0.10	1.25	—	—	—	—	—	1.07	—	—	—	—
45	18	9	8	1	0.04	0.10	0.18	0.34	0.40	0.01	0.09	0.29	11	—	50	17	—	16	—	—	12
46	—	38	—	6	0.35	0.13	0.29	0.21	1.37	0.02	0.11	0.30	35	—	285	73	0.22	32	6	0.7	—
47	—	—	—	—	0.29	0.11	0.27	0.26	1.02	0.03	0.14	0.29	27	—	187	49	0.19	27	5	0.7	—
48	67	50	46	2	0.49	—	—	0.21	1.27	0.01	0.09	0.12	18	—	324	11	—	24	—	—	—
49	—	11	—	—	0.28	—	0.23	0.66	1.89	0.03	0.24	—	18	—	89	33	0.12	60	—	—	—
50	—	10	—	—	0.30	0.08	0.30	0.68	1.98	0.03	0.37	0.20	24	—	175	35	0.11	66	—	—	—
51	—	—	—	—	0.44	0.11	0.77	0.98	1.14	0.24	—	—	4	—	33	20	—	—	—	—	12
52	61	32	31	4	0.53	—	0.14	0.25	1.62	0.18	—	—	11	—	200	103	—	62	21	—	13
53	—	8	—	—	0.06	—	—	0.33	0.40	—	0.17	—	7	—	44	45	—	25	—	—	—
54	0	0	0	0	—	—	—	—	—	—	—	—	—	—	—	—	—	—	—	—	—
55	51	15	11	3	0.13	0.05	0.60	1.38	1.56	0.04	0.25	0.11	14	0.07	128	125	0.43	128	1	—	21
56	—	—	—	—	0.10	0.08	0.28	0.91	1.06	0.03	0.22	0.12	13	—	82	132	0.49	124	—	—	61
57	37	10	—	—	0.13	0.04	0.40	0.99	1.13	0.19	0.20	0.10	22	0.12	93	126	0.83	116	—	—	—
58	—	8	8	—	0.04	0.08	0.16	0.42	0.42	0.05	0.18	0.14	7	0.10	61	42	0.30	50	—	—	17
59	0	0	0	0	0.92	0.08	0.14	0.82	1.23	0.70	1.12	0.12	50	—	181	6	—	3	—	—	—

BALANCING RATION CONSIDERATIONS

When in confinement, cattle have access only to the feeds provided by dairy producers. Therefore, it is important to provide balanced rations.

Good producers should know how to balance rations. They should be able to select and buy feeds with informed appraisal; to check on how well their manufacturer, dealer, or consultant is meeting their needs; and to evaluate the results.

Ration formulation consists of combining feeds that will be eaten in the amount needed to supply the daily nutrient requirements of the animal. This may be accomplished by the method presented in the section that follows, but first the following pointers are necessary:

1. In computing rations, more than simple arithmetic should be considered, for no set of figures can substitute for experience and animal intuition. Formulating rations is both an art and a science—the art comes from animal know-how, experience, and keen observation; the science is largely founded on mathematics, chemistry, physiology, and bacteriology. Both are essential for success.

2. Before attempting to balance a ration, the following major points should be considered:

    a. Availability and cost of the different feed ingredients.

    b. Moisture content.

    c. Composition of the feeds under consideration. Feed composition tables (*book values*), or average analysis, should be considered only as guides, because of wide variations in the composition of feeds.

    d. Quality of feed. Numerous factors determine the quality of feed, including (1) stage of harvesting, (2) freedom from contamination, (3) uniformity, and (4) length of storage.

    e. Degree of processing of the feed.

    f. Soil analysis.

    g. The nutrient requirements and allowances.

3. In addition to providing a proper quantity of feed and to meeting the nutritive requirements, a well-balanced and satisfactory dairy ration should be:

    a. Palatable and digestible.

    b. Economical. Generally speaking, this calls for the maximum use of feeds available in the area, especially forages.

    c. So formulated as to nourish the billions of bacteria in the rumen in order that there will be satisfactory (1) digestion of forages; (2) utilization of lower quality and cheaper proteins and other nitrogenous products (thus, it is possible to use urea to constitute up to one-third of the total protein of the ration of ruminants, provided care is taken to supply enough carbohydrates and other nutrients to assure adequate nutrition for rumen bacteria); and (3) synthesis of B complex vitamins.

This means that rumen microorganisms must be supplied adequate (1) energy, including small amounts of readily available energy such as sugars or starches; (2) ammonia-bearing ingredients such as proteins, urea, and ammonium salts; (3) major minerals, especially sodium, potassium, and phosphorus; (4) cobalt and possibly other trace minerals; and (5) unidentified factors found in certain natural feeds rich in protein or nonprotein nitrogenous constituents.

    d. One that will enhance, rather than impair, the quality of the milk produced.

4. In addition to considering changes in availability of feeds and feed prices, ration formulation should be altered periodically to correspond to changes in weight and productivity of animals.

The ideal ration is one that will maximize production at the lowest cost. A costly ration may yield phenomenal production in cows, but the cost per pound of milk may make the ration economically infeasible. Likewise, the cheapest ration is not always the best since it may depress maximum production.

Therefore, the cost per unit of production is the ultimate determinant of what constitutes the best ration. Awareness of this fact separates a successful producer from the marginal or unsuccessful ones.

METHODS OF BALANCING RATIONS

In the sections that follow, three different methods of balancing rations are presented: (1) the square method, (2) the trial-and-error method, and (3) the computer method. In today's world, some of these are primarily of historic interest, but they present a progressive development of ration formulation technology. Despite the sometimes confusing mechanics of each system, if done properly, the end result of all three methods is the same—a ration that provides the desired allowance of nutrients in correct proportions economically (or at least cost), but, more important, so as to achieve the greatest net returns—for it is net profit, rather than cost, that counts. Since feed usually represent the greatest cost item in dairy production, the importance of balanced rations is evident.

An exercise in ration formulation follows for purposes of illustrating the application of each of these three methods:

1. **Square method,** applied to a calf starter.
2. **Trial-and-error method,** applied to a lactating cow ration.
3. **Computer method,** applied to a lactating cow ration.

SQUARE (OR PEARSON SQUARE) METHOD

The square method is a simple, direct, and easy way in which to figure proportions between two ingredients. It permits quick substitution of feed ingredients in keeping with market fluctuations, without disturbing the protein content.

In balancing rations by the square method, it is recognized that one specific nutrient alone receives major consideration. Correctly speaking, therefore, it is a method of balancing one nutrient requirement, with no consideration given to the other nutritive requirements.

To compute rations by the square method, or by any other method, it is first necessary to have available both feeding standards (see the nutrient requirement tables presented earlier in this chapter) and the feed composition tables in Chapter 21.

The following example shows how to use the square method in formulating a calf starter ration.

Example. *A dairy producer has calves to which it is desired to feed an 18% protein ration until they reach 12 weeks of age. Corn containing 8.9% protein is on hand. A 40% protein supplement, which is reinforced with minerals and vitamins can be bought. What percent of the ration should consist of corn and of the 40% protein supplement?*

Step by step, the procedure in balancing this ration is as follows:

1. Draw a square, and place the number 18 (desired protein level) in the center.

2. At the upper left-hand corner of the square, write *protein supplement* and its protein content (40); at the lower left-hand corner, write *corn* and its protein content (8.9).

3. Subtract diagonally across the square (the smaller number from the larger number), and record the difference at the corners on the right-hand side (40 – 18 = 22; 18 – 8.9 = 9.1). The number at the upper right-hand corner gives the parts of concentrate by weight, and the number at the lower right-hand corner gives the parts of corn by weight to make a ration with 18% protein.

```
Protein      40.0  ⟍  ⟋  9.1   Parts corn
Supplement         ⟍⟋
                   18
                   ⟋⟍
Corn          8.9  ⟋  ⟍  22.0  Parts corn
                             31.1  Total parts
```

4. To determine what percentage of the ration would be corn, divide the parts of corn by the total parts and multiply by 100: 22 ÷ 31.1 × 100 = 70.7% corn. The remainder, 29.3%, would be supplement.

TRIAL-AND-ERROR METHOD

In the example that follows, the trial-and-error method is used, with consideration given to energy and protein. Also, crude protein rather than digestible protein is used because (1) this is what feed manufacturers want to know as they plan feed formulas, and (2) this is what dairy producers see on the tag when they purchase feed. In most mixed feeds, approximately 80% of the total protein is digestible.

Example. *Let's assume that a dairy producer has a 1,433-lb cow producing 65 lb of milk testing 4% fat. The producer is feeding 14 lb of alfalfa hay and 40 lb of corn silage per day. Corn, oats, and soybean meal are available. What concentrate mix should the producer use to meet the needs of this lactating cow, from the standpoint of energy and protein?*

The available feeds have approximately the following composition (as-fed basis):

	TDN (%)	Crude Protein (%)
Alfalfa hay, all analyses	51	16.0
Corn silage, all analyses	18	2.2
Corn, all analyses	80	9.9
Oats, all analyses	69	11.9
Soybean meal, solv extd, 44%	76	44.4

Here are the steps in balancing this ration:

Step 1. The daily TDN and crude protein requirements of this cow (1,433 lb body weight, 65 lb of milk testing 4% fat) are:[2]

Requirements of cow for—

	TDN (lb)	Crude Protein (lb)	(g)
Maintenance	9.94	0.94	*428*
Milk production	20.9	5.87	*2,665*
Total	30.84	6.81	*3,093*

Step 2. The forage (14 lb alfalfa hay, 40 lb corn silage) is supplying:

TDN (lb)	Crude Protein (lb)
16.5	3.69

Step 4. Let's try out (that is why it is called the *trial-and-error method*) a grain mix of 700 lb corn, 280 lb oats, 10 lb monosodium phosphate, and 10 lb salt, and determine the amounts of TDN and crude protein in 1,000 lb of the grain mix:

	TDN (lb)	Crude Protein (lb)
Corn, 700 lb	560.0	69.3
Oats, 280 lb	193.2	33.3
Monosodium phosphate, 10 lb	—	—
Salt, 10 lb	—	—
Total	753.2	102.6
or in percent	75.3%	10.3%

Step 5. Divide the TDN needed from concentrate (16.5 lb) by the percent TDN in the mixture (75.3%). Thus, feeding 21.9 lb of the concentrate will meet the energy needs.

Step 6. Will this level of grain mix (21.9 lb) also meet the crude protein needs? By multiplying the pounds of concentrate mixture by the percent crude protein (21.9 × 10.3%), we find that the proposed concentrate would supply 2.26 lb of crude protein, whereas 3.69 lb are needed. Therefore, a high-protein supplement must be substituted for some of the homegrown grain.

Step 7. Let's substitute 175 lb of soybean meal for 175 lb of corn. Hence, the concentrate mix as now proposed will consist of:

	TDN (lb)	Crude Protein (lb)
Corn, 525 lb	420.0	52.0
Oats, 280 lb	193.2	33.3
Soybean meal, 175 lb	133.0	77.7
Monosodium phosphate, 10 lb	—	—
Salt, 10 lb	—	—
Total	746.2	163.0
or in percent	74.6%	16.3%

Step 8. By referring back to Step 3, we can divide the pounds of TDN and crude protein needed from the concen-

---

[2]From Table 12–4, "Daily Nutrient Requirements of Lactating and Pregnant Dairy Cattle."

trate by the percentage of TDN and crude protein found in the grain mix in Step 7. We find that 16.5 ÷ .746 = 22.1 lb needed to supply 16.5 lb TDN; and 3.69 ÷ .163 = 22.63 lb needed to supply 3.69 lb crude protein. Thus, we find that the following ration will supply the needed TDN (with a slight overage) and crude protein for a 1,433-lb lactating cow producing 65 lb of milk testing 4% fat:

	TDN (lb)	Crude Protein (lb)
Alfalfa hay, 14 lb	7.1	2.2
Corn silage, 40 lb	7.2	0.9
Concentrate mix (Steps 7 & 8), 22.63 lb . . .	16.9	3.7
Total	31.2	6.8

In many sections of the country, especially in grain deficient areas and on highly specialized dairies where little or no grain is grown, the dairy producer may find it most economical to purchase a commercial dairy feed to augment the roughage that is being fed.

COMPUTER-FORMULATED DAIRY RATIONS/DISKETTE[3]

Until the late 1970s, only those dairy producers with access to a large main-frame computer could formulate a ration using the computer. Usually this was limited to those associated with a university (Extension Service) or subscribing to a time-sharing system. Many rations were formulated by using a pencil, an eraser, paper, and a calculator. Balancing rations was time-consuming, and options were limited. Then came the microcomputer! By the mid-1980s, most dairy producers owned or had access to a microcomputer. Rations could be, and were, adjusted with changes in availability, price, composition, and moisture content of ingredients.

In the past, computer programs were written by software companies and universities to convert the tabular NRC requirements into equations that computed the animal requirements. With the publication of *Nutrient Requirements of Dairy Cattle*, Sixth Revised Edition, a new step was taken by the NRC committee. In addition to providing nutrient requirements for dairy cattle, as reproduced in this chapter in Tables 12–2 to 12–6, the committee published prediction equations. These equations may be easily translated into computer programs by programmers. Nutrient requirements for growing, lactating, or dry cows are generated by these equations.

Also, the NRC committee commissioned a software company, Microsoft® Corporation, to develop a FORTRAN program that is supplied with the dairy cattle publication in the form of a 5¼ in. diskette. This program will calculate nutrient requirements from the prediction equations and from specific information supplied by the producer or nutritionist. The results of the calculations may either (1) be displayed on the computer screen, or (2) be printed. The printout is in two parts: The first part lists the requirements for the major nutrients and for vitamins A and D and is useful to the producer or nutritionist in formulating rations. The second part is a very involved breakdown predicting utilization of dietary protein, information of value primarily to those who conduct research. Fig. 12–6 is a sample printout of the first page generated by this program.

```
NRC DAIRY (1988) REQUIREMENTS CALCULATED ON  3- 8-1989 AT 12:58
PREGNANT OR LACTATING CATTLE
ENERGY CONCENTRATION FED/NRC ASSUMED IS       1.000
LIVE WEIGHT IN LB IS                       1400.
MILK PRODUCTION IN KG IS                     75.
MILK FAT TEST % IS                            3.65
NUMBER OF DAYS PREGNANT IS                    0.
LACTATION NUMBER IS                           3
PROPORTIONAL FEED NEL/REQUIRED NEL IS         1.00
WEIGHT CHANGE IN LACTATION IS                  .750 LB
DRY MATTER INTAKE IS                         48.39  LB  OR     3.46  % LW
NEL NEEDED IS                                35.76  MCAL OR     .74  MCAL/LB
ME NEEDED IS                                 59.89  MCAL OR    1.24  MCAL/LB
DE NEEDED IS                                 69.11  MCAL OR    1.43  MCAL/LB
BASELINE TDN NEEDED IS                       34.56  LB  OR    71.41  % DM
CRUDE PROTEIN INTAKE NEEDED IS                7.902 LB  OR    16.33  % DM
UNDEGRADED INTAKE PROTEIN NEEDED IS           2.761 LB  OR     5.70  % DM
DEGRADED INTAKE PROTEIN NEEDED IS             4.678 LB  OR     9.67  % DM
INTAKE PROTEIN (IP) NEEDED IS                 7.439 LB  OR    15.37  % DM
CALCIUM NEEDED IS                              .285 LB  OR      .589 % DM
PHOSPHORUS NEEDED IS                           .181 LB  OR      .373 % DM
VITAMIN A NEEDED IS                         48263.  IU  OR   997.    IU/LB
VITAMIN D NEEDED IS                         19051.  IU  OR   394.    IU/LB
UNDEGRADED INTAKE PROTEIN IN IP IS                              37.11 % IP
```

Fig. 12–6. NRC dairy requirements calculated by using the data diskette supplied with *Nutrient Requirements of Dairy Cattle*, Sixth Revised Edition, 1989.

[3]This section on "Computer-Formulated Rations/Diskette" was prepared by L. M. Larsen, Ph.D., Consultant, Nutri-Systems, 426 E. Shields, Fresno, CA 93704.

The program runs on IBM-PC™ or PC-compatible computers. Various computer software companies have written the equations into their dairy ration formulation software, making the procedure of ration formulation more rapid and convenient. Fig. 12–7 is a computer printout of a ration for lactating cows formulated with the use of the NRC prediction equations.

An explanation of the numbered sections (columns) of Fig. 12–7 follows:

1. Feed cost per cow per day.

2. Input data used for computing the requirements.

3. Feedstuffs' amounts given in pounds per cow per day.

4. Nutrient composition of the ration on a 100% dry matter basis.

5. Nutrient names and units.

6. Daily intake of nutrients per cow per day.

7. The NRC requirements computed from the prediction equations.

8. Deficiencies (negative values) or excesses (positive

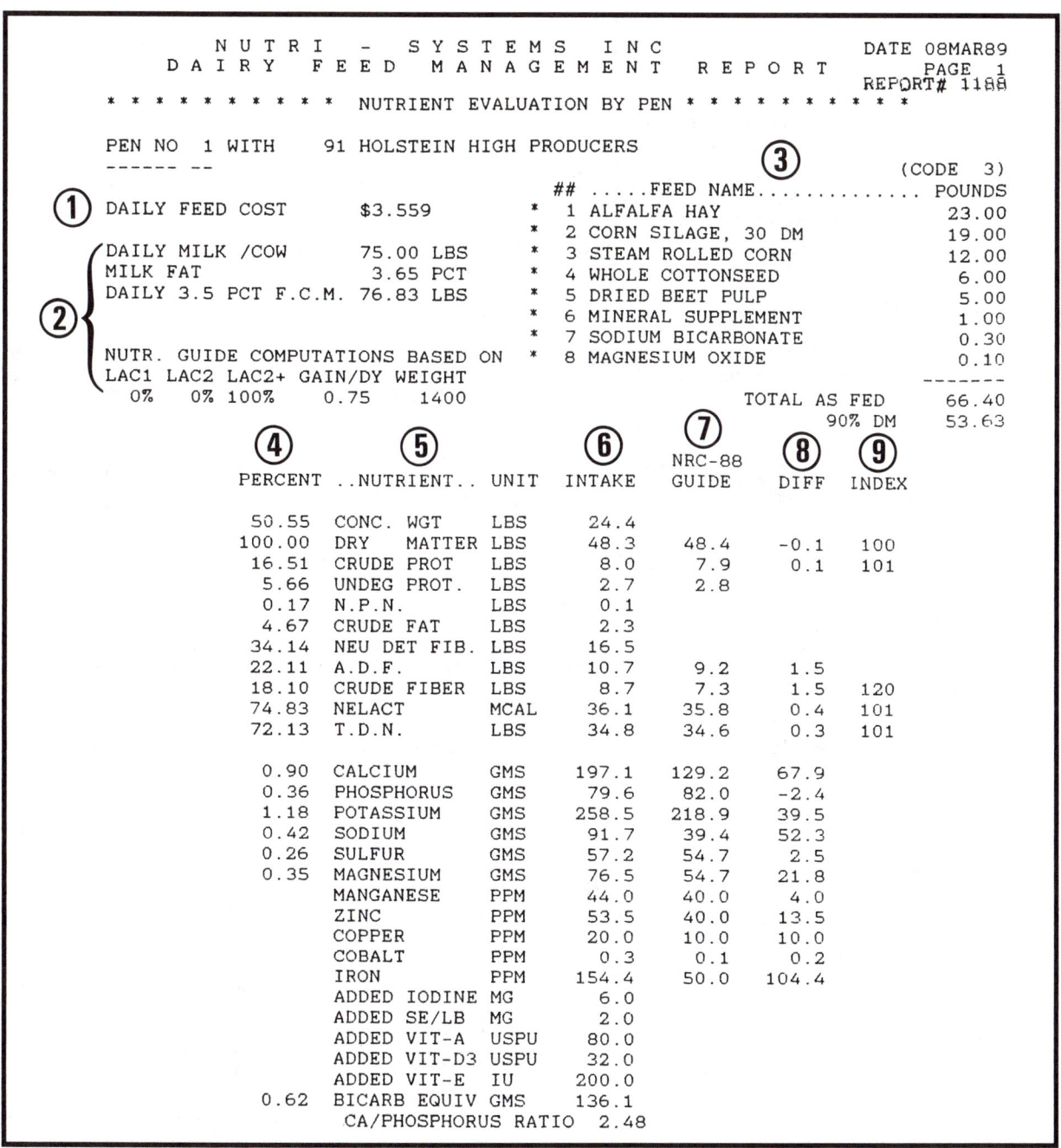

Fig. 12–7. Lactating dairy ration comparing nutrient intake (column 6) with the NRC requirement (column 7).

values). These values are computed by subtracting the requirements from the daily intake figures.

9. A computed index for dry matter, crude protein, NE_{lc}, and TDN. This value is computed as the percentage that nutrient intake is of the nutrient requirement.

(For a more detailed discussion of computer formulation of rations, see *Feeds & Nutrition*, of which Dr. M. E. Ensminger is the senior author, published by the Ensminger Publishing Company, Clovis, California, Chapter 19, section on "Computer Methods.")

PRECAUTIONS AND LIMITATIONS OF COMPUTER FORMULATION

When utilizing the computer in the formulation of least-cost rations, the dairy producer needs to consult an experienced nutritionist. The nutritionist can then interpret the printout from the computer and make any adjustments necessary to make the ration more realistic. It must be reiterated that the computer formulates rations objectively from the information that is fed into it. What comes out of the computer may be the best solution to the mathematical problem, but it may not be practical or realistic.

One of the major costs involved in computer formulations is the continual review and revision of the information that must go into the program. The user must be constantly updating costs in order to maximize the use of the computer.

The producer should be aware that radical changes in ration composition cannot be made without causing digestive disorders—especially in ruminant animals. Dairy animals need time to adapt to changes in rations, a fact which the computer does not consider.

When the computer formulates rations, it is using average values for the nutrient composition of the various feeds. We know that feeds can often vary in their nutrient composition so there is a good possibility that the chemical analysis of the formulated feed will not be the same as the formulated analysis. However, in most cases, this difference is not of sufficient magnitude to create problems.

Finally, it must be remembered that the results obtained from the computer are only as good as the person who feeds the information into the machine. If the data given to the computer are outdated or wrong, the ration that is formulated will be of little value.

FORMULATION WORKSHEET

When formulating rations, it is advisable to record the ration on a worksheet similar to that in Fig. 12–8. This worksheet is merely an example of the format that should be used. A similar sheet can be worked up for micronutrient composition of premixes for vitamins and minerals as well as for amino acids. The worksheet serves three purposes:

1. It provides a means of reviewing and double checking the calculations used to formulate the ration. If there is a gross error, it should become obvious when listed on the worksheet.

2. It can be used to organize mixing procedures. It is vital that the person mixing feed be able to refer to a worksheet on which can be recorded what has been mixed and what mixing order should be followed.

3. The worksheet can be filed for future reference. If any questions should arise when the feed is being used, the worksheet provides an orderly record of the content of the feed and its mixing.

MACRONUTRIENT WORKSHEET

Ration Number: _____   Date: _____

Ingredient	✔ If Mixed	Amount	Proximate Analysis				Energy	Minerals and Vitamins		
			Crude Fiber	Ether Extract (Fat)	N-free Extract	Crude Protein		Calcium (Ca)	Phosphorus (P)	Vitamin A
		(lb)	(lb)	(lb)	(lb)	(lb)	TDN = lb NE = Mcal ME = Mcal	(lb)	(lb)	(IU)
TOTAL										
NUTRIENT REQUIREMENTS										
NUTRIENT BALANCE (Total—Nutrient Requirements)										

Fig. 12–8. Formulation worksheet for macronutrients.

Each type of ration should be assigned a number for future reference, and the date of formulation and/or mixing should be recorded. In addition to listing the feed ingredients and their respective amounts, the nutrient requirement to be fulfilled by the ration should be listed on the worksheet immediately below the totals of various components of the feed. By subtracting the totals contained in the feed from the nutrient requirements, the person formulating the ration can then determine if there are any severe excesses or deficiencies in the ration.

PART III
FEEDING PROGRAMS

The dairy producer must put together the available feeds so as to achieve the most profitable production. At its best, developing a dairy ration involves combining the art and the science of feeding. For small herds, individual animal response may be satisfactory. With large commercial herds, the formulating of rations must be more precise, because small costs per cow become large costs when multiplied by many cows. Yet, the most sophisticated computer must be augmented by the good judgment of the manager if the rations are to be successful in meeting the nutrient needs of individual cows and of the herd as a whole. The dairy producer must always keep in mind that the best formula on paper is not always the best feed. A feed is of little value if it is not actually consumed.

Also, there should be a ration for every need—for lactating cows, dry cows, calves, replacement heifers, bulls, dairy beef, and show and sale animals.

FEED PREPARATION

Fig. 12–9. Stanchioned cows consuming long hay. (Courtesy, Holstein-Friesian Assn. of America, Brattleboro, VT)

Most grains for dairy cattle should be processed before feeding, although calves under six months of age can be fed whole corn. Fine grinding seems to be the least desirable method of processing, because of the poor palatability of the powdery texture. Coarse grinding is much better, but many producers prefer rolling to grinding. Even rolling can be undesirable if it leaves fine material. Flaking or wet rolling with the use of steam alleviates dust problems, especially when molasses is incorporated. Dairy producers should make certain that steam rolled grains have been dried to a normal moisture level of 10 to 15% before being weighed, regardless of whether the grain is purchased or of their own processing.

Pelleted concentrates are more compact and less dusty than ground grains. Also, cows will consume pelleted feeds faster than ground or flaked grain. Cows fed pelleted grain may produce slightly more milk, with a slightly lower butterfat content, than those fed unpelleted grain.

Cows produce as well on long hay as on chopped or ground hay. However, finely ground, pelleted hay affects the amount and proportion of volatile fatty acids in the rumen, with the result that the percentage fat content of the milk is lowered. Wafering or cubing, on the other hand, has little depressing effect on the fat content of the milk, and will increase intake and maintain or increase the milk production slightly. Both pelleting and wafering lessen the transportation, storage, and handling charges for hay compared to long or baled hay.

Additional information relative to feed preparation for dairy cattle is presented in Table 12–9.

TABLE 12–9
FEED PREPARATION FOR DAIRY CATTLE

Feed	Processing Methods	Comments
Grain	*Grinding* is the simplest and the most widely used grain processing method for dairy cattle. *Cracking, steam rolling,* and *pelleting* are the other popular procedures. *Exploding, extruding, flaking, micronizing, popping,* and *roasting* of high-moisture grain are preferable for high-producing lactating cows, but they are not widely used. *Dry or steam roll or grind coarsely* for dry cows, young stock, and low-producing cows. *Whole grain* (except for very hard seeds) may be fed to young calves under six months of age.	Butterfat is depressed unless the ration contains some threshold level of coarse material.
Forage	*Long hay or cubes.* Cubes lend themselves to automation, and lower milk fat percentage only slightly, if at all.	Finely ground or pelleted roughage will result in reduced rumen acetate production and lower milk fat percentage.

FEEDING LACTATING COWS

Few animal stresses are as great as those involved in the production of a large volume of milk. For each gallon of milk produced, 400 to 500 gal of blood must pass through the udder. Thus, if a cow is producing 10 gal (86 lb) of milk daily, 15 to 20 tons of blood course through the udder each 24 hours. This 10 gal of milk contains more than 3 lb of fat, more than 3 lb of protein, more than 4 lb of lactose (milk sugar), and more than ½ lb of minerals. All these must be supplied in the ration over and above the nutrients needed for the body processes, wastes, and energy to sustain all of the operation.

Also, producers realize greatest profits from feeding when cows convert the maximum proportion of their feed into milk. The nutrient requirements for production depend primarily on the amount and composition of the milk. These needs for cows of all sizes and levels of production are shown in Tables 12–2 to 12–6. Rations that fulfill these requirements, plus a margin of safety, can be formulated based on composition of feeds listed in Chapter 21, Feed Composition Tables. The primary concern in feeding lactating cows is to provide a ration adequate in energy, protein, fiber and roughage factor, salt, calcium, phosphorus, and vitamin A (carotene). When allowances for these nutrients are met, other minerals and vitamins usually are present in sufficient amounts, also.

Additional considerations in feeding lactating cows include palatability of the ration; physical form; protein and mineral content of concentrates; proportion of concentrate to roughage; relative prices of ingredients; voluntary feed intake; and frequency and regularity of feeding. Thus, the proper feeding of lactating cows necessitates that producers have sufficient knowledge relative to basic nutrient requirements and principles to plan an efficient feeding program, and the experience and management ability to apply it.

Dry matter consumption is very important in feeding dairy cows. The best ration formulation on paper will not make for profitable production if the cows either fail to eat it or are given insufficient amounts of it. Also, high-producing cows must consume very large amounts of a balanced ration if they are to produce to their maximum. For this reason, producers should adjust the feeding recommendations given in Table 12–2 to fit the needs of their cows, with consideration given to body weight and condition of cows, milk production, stage of lactation, weather and other environmental factors, and type and quality of feed.

THUMB RULES FOR FEEDING LACTATING COWS

The feed requirements of lactating cows are significantly influenced by the volume and composition of the milk that they produce. Although knowledge of the nutrient requirements of the animals and of the composition of feeds is essential in order to feed properly, the ability of the cows to consume sufficient volume of feed complicates adequate feeding. Table 12–10 may be used as a guide for dry matter intake; and the two sections that follow give some thumb rules relative to the amount and kind of forage and the amount and kind of concentrate to feed.

TABLE 12–10
DAILY DRY MATTER INTAKE GUIDELINES[1]

Live Wt.:	(lb)	900	1,100	1,200	1,300	1,500
	(kg)	409	499	545	590	681
Milk[2]			Percent of Body Weight[3]			
(lb/day)	(kg/day)	(%)	(%)	(%)	(%)	(%)
20	9.1	2.6	2.3	2.2	2.1	2.0
30	13.6	3.0	2.7	2.6	2.5	2.3
40	18.2	3.4	3.1	2.9	2.8	2.5
50	22.7	3.8	3.4	3.2	3.1	2.8
60	27.2	4.1	3.7	3.5	3.4	3.1
70	31.8	4.6	4.0	3.8	3.6	3.3
80	36.3	5.1	4.3	4.1	3.8	3.5
90	40.9		4.7	4.4	4.1	3.7
100	45.4		5.0	4.7	4.4	3.9

[1]Adapted by the author from: Linn, J. G., M. F. Hutjens, W. T. Howard, L. H. Kilmer, and D. E. Otterby, *Feeding the Dairy Herd*, Cooperative Extension Services, Universities of Illinois, Iowa State, Minnesota, and Wisconsin, 1988, p. 27, Table 19.

[2]Fat-corrected milk = (milk lb × 0.4) ÷ (fat lb × 15).

[3]Intakes may be up to 18% less for cows in early lactation.

AMOUNT AND KIND OF FORAGE TO FEED

The common thumb rules for forage feeding of lactating cows follow.

1. **Forage dry matter and intake.** The forage should constitute a minimum of 40% of the total dry matter of the ration and account for an intake of approximately 1.5% of the body weight daily.

2. **Acid detergent fiber (ADF).** The ADF should constitute 19% of the ration dry matter, increased to 21% during the first 3 weeks of lactation.

3. **Neutral detergent fiber (NDF).** The NDF should constitute 25% of the ration dry matter, increased to 28% during the first 3 weeks of lactation.

4. **Hay consumption.** If good quality hay only is fed, a cow will eat about 3 lb per 100 lb of body weight.

5. **Silage.** Depending on the moisture content, 2.5 to 4.5 lb of silage are equal to (and may replace) 1 lb of hay; the lower feeding value of silage is due to its high moisture content—hay runs 10 to 15% moisture, whereas silage runs 65 to 75% moisture.

6. **Hay/grain equivalent.** It takes about 3 lb of good hay to supply the same amount of usable energy as 2 lb of grain.

7. **Pasture (grass) consumption.** Cows will consume 100 to 200 lb of pasture per day; since pasture normally contains 70 to 85% moisture, that is 15 to 60 lb of dry matter per day.

8. **Yearly hay consumption.** Except for cows fed high grain rations, it takes 5 to 6 tons of hay (or an equivalent amount in dry matter from pasture or silage) to feed one cow for 1 year.

9. **Forage:concentrate ratio.** If forage is very high quality, cows will eat more of it, with the result that the grain requirement will be lessened. However, over and above

meeting the minimum forage requirement, the proportion of forage to concentrate should be determined primarily by the economics of the situation—that is, it should be decided on the basis of the relative price of available forage and concentrate, the milk production, and the net returns.

AMOUNT AND KIND OF CONCENTRATE (GRAIN) TO FEED

The common thumb rules for concentrate feeding of dairy cows follow.

1. **Amount of concentrate (grain).** The concentrate (grain) should constitute a maximum of 60% of the total dry matter of the ration and account for an intake of not to exceed 2.3% of the body weight daily. Table 12–11 can be used as a guide for feeding concentrate (grain) according to milk production.

TABLE 12–11
AMOUNT OF CONCENTRATE (GRAIN) TO FEED
BY PERIODS (1,400 LB *[636 KG]* COWS, 4% MILK)[1]

		Milk Production Ability of the Cow[2]			
Average Daily 1st Period . .	(lb)	50	60	80	90–100
Average Daily 1st Period . .	*(kg)*	*23*	*27*	*36*	*41–45*
Lactation Total	(lb)	10,000	12,000	15,000	18,000
Lactation Total	*(kg)*	*4,540*	*5,448*	*6,810*	*8,172*
Phase of Lactation		Grain to Milk Ratio			
1 (1st 10 weeks)		1:4	1:3	1:3	1:2.5
2 (2nd 10 weeks)		1:4	1:3	1:3	1:3
3 (last 24 weeks)		1:4	1:4	1:2.5	1:2.5
		Daily	Daily	Daily	Daily
4 (dry, 6–8 weeks)	(lb)	0–4	0–4	0–4	0–6
(dry, 6–8 weeks)	*(kg)*	*0–1.8*	*0–1.8*	*0–1.8*	*0–2.7*
Total grain (approximate)	(lb)	3,000	4,000	5,000	6,000
Total grain (approximate)	*(kg)*	*1,362*	*1,816*	*2,270*	*2,724*

[1]Adapted by the author from: Linn, J. G., M. F. Hutjens, W. T. Howard, L. H. Kilmer, and D. E. Otterby, *Feeding the Dairy Herd*, Cooperative Extension Services, Illinois, Iowa State, Minnesota, and Wisconsin, 1988, p. 26, Table 18.

[2]Ratios based on 100% dry matter basis, grain containing 80 Mcal, and forage 60 Mcal of NE_{lc} per 100 lb *(45 kg)*.

2. **Amount and kind of protein.** Feed protein according to requirements (19% in early lactation, decreased thereafter according to milk production, see Table 12–6). A low rumen degradable protein source is recommended for high-producing cows in early lactation. Limit urea to 0.4 lb per day, and preferably to 0.2 lb per day, in phases 1 and 2.

3. **Added fat.** In addition to the fat present in natural feedstuffs, lactating cows may be fed 1 to 1½ lb of *added fat* per day; which translates into about 6% added fat to the concentrate (grain) ration, or 3% added fat to the total mixed ration (grain and forage combined). Fats in oilseeds (soybeans or whole cottonseed) should be considered as added fat. When feeding added fat, increase the calcium to 0.9 to

1%, the magnesium to 0.3%, and the acid detergent fiber to 20%.

4. **Salt.** Include 1% salt in the concentrate (grain) mix, or 0.5% salt in the total ration (concentrate and forage combined); which will provide for a salt intake of 2 to 3 oz per cow per day.

5. **Calcium/phosphorus and trace minerals.** A calcium/phosphorus mineral source should constitute 1 to 2% of the grain mix, or be fed at a rate of 1 oz per 10 lb of milk. Trace minerals should be incorporated in the ration or self-fed in trace mineralized salt to meet the requirements.

6. **Vitamins.** Vitamins A, D, and E should be added to the ration to meet the requirements.

Any thumb rules, and even calculated values (calculated by hand or computer), are estimates to be used with some judgment by the feeder. The rule of the successful feeder is to increase concentrates so long as cows respond with extra milk at a profit. The commonly used profit indicator is the *milk-feed price ratio*, which is the pounds of 16% protein dairy concentrate equal in value to 1 lb of milk.

Generally, the cost of 16% protein dairy concentrate per pound is about 70% of the price received for milk; of course there are yearly, seasonal, and area variations. For the United States as a whole, the milk-feed price ratio was $1.31 in 1975 and $1.79 in 1986.[4] Thus, the cost per pound of 16% protein dairy concentrate was 76% the price received per pound of milk in 1975, and 56% in 1986.

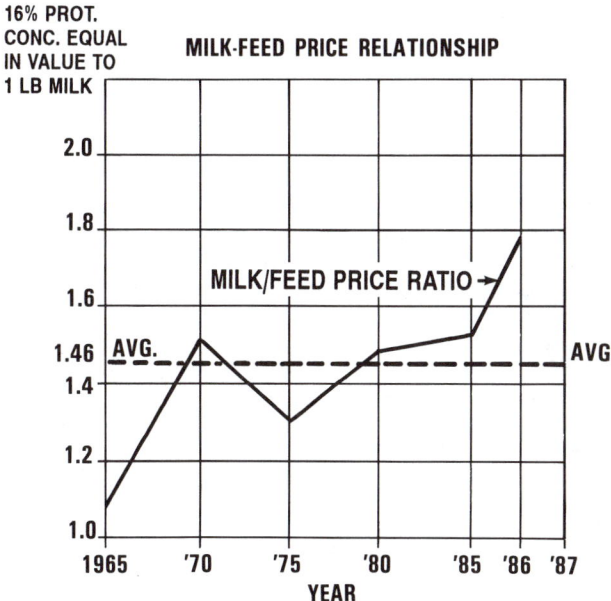

Fig. 12–10. Pounds of 16% protein dairy concentrate equal in value to 1 lb of milk. (Adapted by the author from USDA sources)

The amount of grain that it will pay to feed milking cows depends upon several factors: (1) quality of forage, (2) price of forage, (3) price and quality of concentrates, (4) price of milk, and (5) inherent producing ability of cows. Of these, the most important is the inherent milk-producing ability of the cow. To determine the inherent ability to produce milk, dairy

[4]USDA sources.

producers must rely on (1) the use of milk and fat production records of each cow, and (2) weighing and feeding the proper proportions of concentrates, with consideration given to the stage of lactation of the cows.

RATIONS FOR LACTATING COWS

Fig. 12–11.   Barn feeding a complete ration to lactating cows. (Courtesy, USDA)

Forages constitute the primary basis of most dairy rations, with alfalfa hay as the preferred roughage. Frequently, corn silage is used to replace part of the hay, but, regardless of the kind of forage, successful dairy producers balance the complete ration (forages and concentrates combined) so that all nutrients — energy, protein, minerals, and vitamins — are fed in sufficient amounts to meet the needs of the cow. For maximum intake, the complete ration must contain over 55% dry matter, except when pasture or green chop forage is fed.

Grain provides energy in concentrated form; for example, 5 lb of barley contain as much energy as 8 lb of hay or 25 lb of silage.

Cows fed an all-forage ration produce about half as much milk as cows fed concentrates in average amounts. This is because a cow's stomach simply is not big enough

to hold all the forage necessary to get the amount of energy needed. To provide the needed energy, grain must be added. Grain provides energy in concentrated form.

When fed an all-hay ration, cows can sustain a produc-

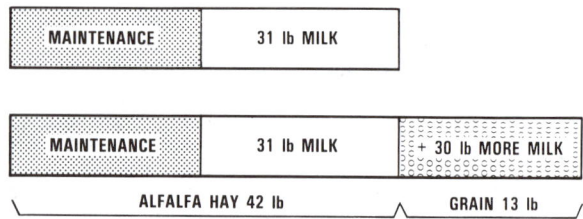

Fig. 12–12.   A mature Holstein cow can consume sufficient good-quality alfalfa hay to meet the energy required for maintenance plus about 31 lb of milk daily. An additional 13 lb of grain daily will allow the cow to produce 30 lb more milk. Thus, the milk:grain ratio for the additional grain is 2:3.

tion of 30 to 40 lb of milk daily, in addition to maintaining the other vital functions. The higher the quality of forage, the greater the amount that will be consumed, and the higher its percent digestibility. Cows can consume about 3% of their body weight in dry hay, which means that a 1,400-lb cow can consume more than 40 lb daily if hay is her only feed. As other feeds are introduced into the ration, hay consumption on a free-choice basis decreases.

Corn silage or hay-crop silage may replace hay at the rate of 2½ to 3 lb for each 1 lb of hay, based on dry matter and nutrient content. In tests with lactating cows, this was also the rate to which they voluntarily adapted when using various combinations of hay and silage. A daily silage allowance of 3 to 4.5% of body weight can be fed to a lactating cow without interfering with the total nutrient intake. This means that a 1,400-lb cow can be fed 40 to 60 lb of silage daily. Experience has shown that above these levels, total nutrient intake is decreased regardless of the types of feeds used. This means that corn silage can replace about one-third of the hay when it is economical to make a substitution.

Well-matured corn silage has about 30 to 45% of its weight in grain (dry basis); hence, on a moisture-free basis it is higher in energy than the hay that it replaces, but lower in protein. Table 12–12 shows what happens to the DM,

TABLE 12–12
NUTRIENTS SUPPLIED BY ALFALFA HAY AND CORN SILAGE (AS-FED BASIS)[1]

	DM	Energy				Total Protein		Calcium	Phos-phorus
		TDN		NE_lc					
		(lb)	(kg)	(Mcal/lb)	(Mcal/kg)	(lb)	(kg)	(g)	(g)
Forage A:									
Alfalfa hay (40 lb)[2]	36.4	20.8	9.44	22.0	48.5	7.2	3.27	269	36
Forage B:									
Alfalfa hay (25 lb)[2]	22.7	13.0	5.90	13.7	30.2	4.5	2.04	168	23
Corn silage (40 lb)[3]	12.0	8.8	4.00	8.0	17.6	1.0	0.45	4	9.6
Total forage	34.7	21.8	9.90	21.7	47.8	5.5	2.49	172	32.6

[1]Values from *Feeds & Nutrition*, by Ensminger *et al.*   [3]Mature corn silage.

[2]Early bloom hay.

energy, protein, and minerals contributed by the forage component of the ration when corn silage replaces about one-third of the alfalfa hay. Note that the DM and energy remain about the same, but that there is a decrease in the total protein, calcium, and phosphorus. Note, too, that there is little difference in energy values whether TDN or NE_{lc} are used as the measure, and that the Ca:P ratio decreases from 7.5:1 with the hay alone to 5.3:1 for the combination of hay and corn silage.

Alfalfa hay of good quality will supply much of the protein required by lactating dairy cows. Table 12–13 shows that the percentage total protein needed in the grain ration depends upon the type of roughage fed. As shown, with high quality legume hay, a 10% protein grain mix will suffice. With a low quality grass forage, however, the protein level of the grain mix should be much higher, up to 18 to 22%.

TABLE 12–13
PERCENT PROTEIN NEEDED IN GRAIN MIX

Forage Fed	Protein Needed in Grain Mix
	(%)
All legume	10–16
Mixed (part legume and part grass)	14–20
All grass	18–22

Since proteins are always the most expensive part of a ration, for practical reasons the dairy producer should not feed more of them than necessary. Nevertheless, the protein level should be in keeping with the production—the higher the yield, the higher the protein requirement. Generally speaking, this is best accomplished by varying the protein content of the grain mix according to the protein content of the roughage being fed.

The concentrate ration needed to supplement the available roughage on the dairy farm may be either home mixed or commercially manufactured. Home-mixing involves the mixing of homegrown grains and such purchased feeds as necessary to balance the ration. Commercially manufactured feeds are generally nutritionally complete, concentrate feeds. On small dairies where grain is home-grown or abundantly available locally, home mixing of concentrates is a widely used practice. However, there has been an increase in the use of commercial feeds on small dairies. Large specialized dairies purchase individual feed ingredients in bulk, then mix and feed total mixed rations.

Not all dairy producers, in the United States or in other countries, who home-mix feeds balance rations (1) on the basis of the chemical analyses of their feed ingredients, (2) with the use of computers, or (3) by combining the concentrates and forage into a complete ration. For these produc-

ers, Table 12–14 (next page), Feeding Guide for Lactating Cows, may serve as a useful guide. It shows how ingredients partitioned into four approximate protein levels (columns 2, 3, 4, and 5) may be combined to make concentrates suitable for feeding with three different qualities of roughages—excellent, medium, and poor.

Variations can and should be made in the rations listed in Table 12–14. Producers should give consideration to the supply of homegrown feeds, and to the availability and price of ingredients. Feeds of similar nutritive properties can and should be interchanged as price relationships warrant. Thus, the cereal grains may consist of corn, barley, wheat, oats, and/or sorghum; the protein supplements may consist of soybean, cottonseed, peanut and/or linseed meal; and a vast array of by-product feeds may be utilized.

Here is how to use Table 12–14: Let's assume that a producer has (1) medium-quality forage, and (2) both low- and medium-high protein (columns 2 and 4) ingredients from which to choose. How many pounds of each of the low- and medium-high protein ingredients will be required in 1,000-lb concentrate mix? Step by step, here is the answer:

1. Look under "Medium roughage—medium protein forage" 15–17% (column to the left).

2. Mix No. 6, containing 650 lb of low-protein ingredients (under column 2, under 12% ingredients) and 350 lb of medium-high protein ingredients (column 4, 18 to 28% protein), will meet the needs. The concentrates may be chosen from among those listed at the top of the respective columns of Table 12–14—the low-protein concentrates from column 2 (under 12%) and the medium-high protein concentrates from column 4 (18 to 28%).

Fig. 12–13. Corral feeding a complete ration to lactating cows. (Courtesy, James Tappan, Higley, AZ)

TABLE 12–14
FEEDING GUIDE FOR LACTATING COWS (AS-FED BASIS)[1]
Note: This shows how ingredients of four protein levels may be combined to make different concentrate mixes of approximate protein content to match three different qualities of roughages.

(1) Suggested Grain Mix, Based on Kind of Roughage Available	(2) Low-Protein (under 12%) Ingredients		(3) Low-Medium Protein (12–18%) Ingredients		(4) Medium-High Protein (18–28%) Ingredients		(5) High-Protein (Over 32%) Ingredients	
Feeds	(% protein)		(% protein)		(% protein)		(% protein)	
	Barley, all analyses . . . 11.7 Beet pulp w/molasses, dried 9.3 Corn-and-cob meal . . . 7.8 Corn #2 . . . . 8.9 Dairy feed, 12% . . . . . 12.0 Hominy feed . . . . 10.3 Molasses, cane* . . . . 4.3 Oats, all analyses . . . . 11.9 Rye, all analyses* . . . . 12.0 Sorghum (milo) . . . . . 10.1 Wheat, all analyses . . . 13.1		Dairy feed, 16% . . . . . 16.0 Wheat bran . . . . 15.5 Wheat middlings 16.4		Brewers' dried grains* . . 27.3 Copra (coconut) meal . . 21.3 Corn gluten feed . . . . . 23.0 Dairy feed, 18–24% . . 18–24 Distillers' dried grains . . 27.3 Malt sprouts . . . . 22.9 Peas, field* 23.2		Dairy feed, 32–34% . . 32–34 Corn gluten meal . . . . 60.8 Cottonseed meal* . . . . 41.2 Linseed meal . . . . . . 35.7 Peanut meal . . . . 49.0 Soybean meal 44.4	
	(lb)	(kg)	(lb)	(kg)	(lb)	(kg)	(lb)	(kg)
Excellent roughage—High-protein forage, 18%: (1) legume, or (2) legume and nonlegume mixed forages of *high quality*; consisting of dry forages and/or silage.								
Mix No. 1	1,000	454						
Mix No. 2	900	409					100	45
Mix No. 3	800	363			200	91		
Mix No. 4	850	386	100	45			50	23
Medium roughage—Medium-protein forage, 15–17%: (1) legume, or (2) legume and nonlegume mixed forages of *medium quality*; consisting of dry forages and/or silage.								
Mix No. 5	800	363					200	91
Mix No. 6	650	295			350	159		
Mix No. 7	700	318	100	45	100	45	100	45
Mix No. 8	Straight 16% dairy feed, or ½ Mix No. 9 and ½ 16% dairy feed							
Poor roughage—Low-protein forage, under 14%: nonlegume forage; consisting of dry forages and/or silage.								
Mix No. 9	700	318	300	136				
Mix No. 10	600	272			200	91	200	91
Mix No. 11	600	272	100	45	100	45	200	91
Mix No. 12	500	227	and 500 lb *(227 kg)* 32% dairy feed					

[1]The protein compositions in columns 2 to 5 may be obtained from Chapter 21, Feed Composition Tables.

Comments:

Add—To all rations (1) 1% iodized or trace-mineralized salt; (2) 1% steamed bone meal, dicalcium phosphate, or the equivalent (use monsodium phosphate or a high-phosphorus commercial mineral where alfalfa is fed liberally); (3) 1,000 IU of vitamin A/lb *(2,205 IU of vitamin A/kg)* of concentrate and, unless cows are in sunlight, add 150 IU of vitamin D/lb *(331 IU of vitamin D/kg)* of concentrate.

*Limitations—Wheat, not more than 50% of the ration; dried molasses beet pulp, 20%; molasses, 15%; peas and brewers' dried grains, 30%; rye, 10%; and cottonseed meal, 20% of the mix for calves, but as needed for mature cows.

FEEDING SYSTEMS

Traditional individual feeding of lactating cows in stanchioned barns or milking parlors is giving way to new feeding systems. Although the newer methods are not as effective as feeding cows individually, they are much more economical than feeding all cows in the herd the same amount of grain, regardless of production. Additionally, they make for considerable saving in labor and facilities.

PHASE FEEDING

Phase feeding is a feeding program that is divided into periods based on milk production, milk fat percentage, feed intake, and body weight. Fig. 12–14 illustrates the shape and relationship of curves for milk production, fat percentage, dry matter intake, and body weight. Based on these curves, four distinct feeding phases of lactating cows can be identified.

Producers should formulate rations to match each of these phases in order to optimize yield, minimize metabolic

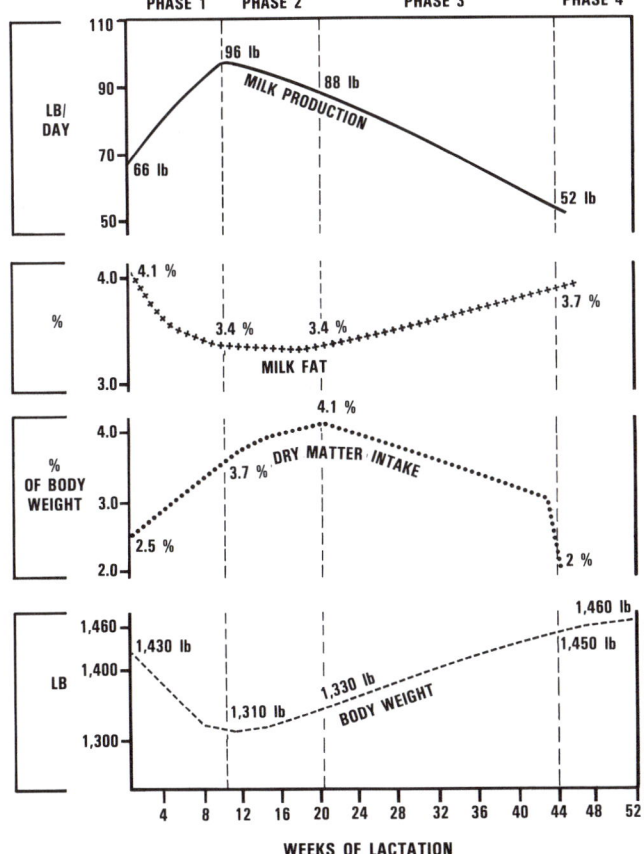

Fig. 12–14. Lactation cycle phases with corresponding changes in milk production, milk fat percentage, dry matter intake, and body weight. (*Source:* Linn, J. G., M. F. Hutjens, W. T. Howard, L. H. Kilmer, and D. E. Otterby, *Feeding the Dairy Herd*, Cooperative Ext. Services, Illinois, Iowa State, Minnesota, and

disorders, increase longevity, and increase profits. The four phases are:

1. **Phase 1, early lactation, 0 to 70 days postpartum.** During this period, milk production increases rapidly, peaking at 4 to 6 weeks after calving. But feed intake does not keep pace with nutrient needs (especially energy needs) for milk production, so body tissues are mobilized to meet these needs. During this phase, adjusting the cow to the milking ration is an important management practice. After calving, the grain should be increased by 1 to 1.5 lb daily to meet increased nutrient demands and minimize off-feed problems and acidosis. Excessive proportions of grain (over 60% of the total dry matter) can cause acidosis and low milk fat percentage. The fiber level in the total ration should not be less than 18% ADF, 28% NDF; and the forage should provide at least 21 percentage units of the NDF in the total ration. The physical form of the fiber is also important; normal rumination and digestion will be maintained if more than 50% of the forage is 1 in. in length or longer.

Protein content is critical during early lactation. Meeting or exceeding the crude protein requirements during this period helps to stimulate feed intake and permits efficient use of mobilized body tissue for milk production. Rations may need to contain 19% or more crude protein to meet the requirements during this phase. The type of protein (degradable and undegradable) and the amount of protein to feed will depend on ration ingredients, method of feeding, and milk production potential of the cow. A guideline followed by many producers is to feed 1 lb of soybean meal or equivalent protein supplement per 10 lb of milk above 50 lb. If urea is fed, it is best fed well mixed with corn silage or a part of the grain mix, and it should be limited to a maximum of 0.2 lb per cow daily.

When early lactation nutrient needs are not met, low peak production and ketosis may result. Low peak production presages low lactation production. If the grain intake is increased too rapidly or to too high a level, it may result in off-feed, acidosis, and displaced abomasum. To increase nutrient intake, (a) feed high quality forage, (b) include adequate protein in the ration, (c) increase grain intake at a constant rate after calving, (d) add 1 to 1.5 lb of fat/cow/day to the ration, (e) allow constant access to feed, and (f) minimize stresses.

2. **Phase 2, peak dry matter intake, second 10 weeks postpartum.** During this phase, cows should be fed to maintain peak milk production as long as possible. Feed intake is near maximum and can supply nutrient needs. Cows should be maintaining weight or making slight gains (see Fig. 12–14).

Grain intake may reach, but should not exceed, 2.3% of the cow's body weight (dry matter basis). High quality forage should be provided, with a minimum intake of 1.5% of the cow's body weight (dry matter basis) in order to maintain rumen function and normal fat test.

Potential problems during phase 2 include rapid drop in milk production, low fat test, silent heat periods (estrus not observed), and ketosis.

To maximize nutrient intake (a) feed forage and grain three or more times daily, (b) feed high quality feeds, (c) limit urea to 0.2 lb/cow/day, (d) minimize stress, and (e) use a total mixed ration.

3. **Phase 3, mid- to late-lactation, 140 to 305 days postpartum.** This is the easiest phase to manage. During this period, milk production is declining, the cow is pregnant, and nutrient intake will easily meet or exceed requirements. The level of grain feeding should be adequate to meet production requirements, and to begin to replace body weight lost during early lactation. Lactating cows require less feed to replace a pound of body tissue than dry cows; hence, it is more efficient to have cows gain body weight near the end of lactation than during the dry period. Young cows should receive additional nutrients for growth; 2-year-old heifers should receive 20% more than for maintenance, and 3-year-olds should receive 10% more.

■ **Phases 1, 2, and 3 guidelines/rations** — The following guidelines and example rations, which meet the nutrient requirements given in Table 12–6, may be used by dairy producers in developing a feeding program for lactating cows suitable for phases 1, 2, and 3.

a. **Net energy-lactation (NE_{lc}).** A minimum of 0.76 Mcal per pound of DM in early lactation, decreasing to 0.69 Mcal per pound in late lactation and 0.57 Mcal per pound during the dry period.

b. **Crude protein.** A minimum of 19% (DM basis) in early lactation, decreasing to 13% in late lactation and 12% during the dry period. Urea to be limited to a maximum of 0.4 lb per day or 1% of the grain mix.

c. **Forage dry matter.** A minimum of 1.5 lb per 100 lb body weight. High quality legume forage to be fed during early lactation.

d. **Fiber.** A minimum of 17% ADF in the DM during early lactation, increasing to 22% during the dry period. A minimum of 28% NDF during early lactation, increasing to 35% during the dry period.

e. **Minerals.** Include 0.5 to 1% trace mineralized salt in the grain mix; and approximately 1% calcium-phosphorus in the grain mix.

f. **Vitamins.** Incorporate in the ration sufficient vitamins A, D, and E to meet the requirements (see Table 12–6).

g. **Ration processing.** Forages and grains should not be chopped or ground too fine.

h. **Ration formulation.** Balance the ration to meet all requirements and feed it as a total mixed ration.

The example rations presented in Table 12–15 are suitable for the three phases.

TABLE 12–15
EXAMPLE RATIONS FOR VARIOUS MILK PRODUCTION PHASES, 1,350 LB *(613 KG)* COW, 3.8% FAT TEST[1]

Item		Phase 1	Phase 2	Phase 3
Milk . . . . (lb/day)		90	80	50
Milk . . . . *(kg/day)*		*40.9*	*36.3*	*22.7*
DM intake[2] . . . . (lb/day)		49	51	38
DM intake[2] . . . . *(kg/day)*		*22.2*	*23.2*	*17.3*

	As-Fed					
	(lb/day)	*(kg/day)*	(lb/day)	*(kg/day)*	(lb/day)	*(kg/day)*
Ration 1						
Alfalfa hay (88% DM), 140 RFV, 20% crude protein	28	*12.71*	34	*15.44*	27	*12.26*
Corn-oats[3]	21	*9.53*	24	*10.90*	16	*7.26*
Soybean meal, 44%	5	*2.27*				
Dical, 18% phosphorus	0.5	*0.23*	0.45	*0.20*	0.3	*0.14*
Salt, vitamins, trace mineralized	0.3	*0.14*	0.25	*0.11*	0.25	*0.11*
Weight change	−1.5	*−0.68*	—	—	+0.5	*+0.23*
Ration 2 (corn silage limit fed)						
Alfalfa hay, 140 RFV, 20% CP .	19	*8.63*	34	*15.44*	23	*10.44*
Corn silage (35% DM)	25	*11.35*	25	*11.35*	25	*11.35*
Corn-oats	18	*8.17*	12	*5.45*	10	*4.54*
Soybean meal, 44%	7.5	*3.41*	0.3	*0.14*	—	—
Dical, 18% phosphorus	0.45	*0.20*	0.5	*0.23*	0.3	*0.14*
Salt, vitamins, trace mineralized	0.3	*0.14*	0.25	*0.11*	0.25	*0.11*
Weight change	−1.2	*−0.54*	—	—	+0.5	*+0.23*
Ration 3 (hay limit fed)[4]						
All-grass hay, 113 RFV, 16% CP	10	*4.54*	10	*4.54*	10	*4.54*
Corn silage	41	*18.61*	70	*31.78*	57	*25.88*
Corn-oats	16	*7.26*	11	*4.99*	6	*2.72*
Soybean meal, 44%	11.5	*5.22*	8.2	*3.72*	4.5	*2.04*
Dical, 18% phosphorus	0.4	*0.18*	0.3	*0.14*	0.25	*0.11*
Limestone	0.4	*0.18*	0.3	*0.14*	0.15	*0.07*
Salt, vitamins, trace mineralized	0.3	*0.14*	0.25	*0.11*	0.25	*0.11*
Weight change	−1.4	*−0.64*	+0.7	*+0.32*	+0.5	*+0.23*
Ration 4						
All-grass hay, 113 RFV, 16% CP	23	*10.44*	32	*14.53*	24	*10.90*
Corn-oats	22	*9.99*	22	*9.99*	19	*8.63*
Soybean meal, 44%	8.5	*3.86*	3.5	*1.59*	1.1	*0.50*
Dical, 18% phosphorus	0.45	*0.20*	0.4	*0.18*	0.25	*0.11*
Limestone	0.2	*0.09*				
Salt, vitamins, trace mineralized	0.3	*0.14*	0.25	*0.11*	0.25	*0.11*
Weight change	−1.9	*−0.86*	—	—	+0.5	*+0.23*

[1]Source: Linn, J. G., M. F. Hutjens, W. T. Howard, L. H. Kilmer, and D. E. Otterby, *Feeding the Dairy Herd*, Cooperative Ext. Services, Illinois, Iowa State, Minnesota, and Wisconsin, 1988, p. 16, Table 6.

[2]Estimated average intake during the phase.

[3]85% corn–15% oats mix.

[4]Feed amounts may have to be limited during phase 2 and 3 to avoid over-conditioning.

4. **Phase 4, dry period, 45 to 60 days before parturition.** The dry phase is important. A good dry cow program can minimize metabolic problems at or immediately following calving and increase milk yield during the subsequent lactation.

Table 12–16 lists examples of dry cow rations.

TABLE 12–16
EXAMPLE DRY COW RATIONS, 1,400 LB *(636 KG)* DRY COW[1]

Forage	As-Fed	
	(lb/day)	*(kg/day)*
Grass forage		
Orchard grass hay, 12% crude protein	25.0	*11.35*
Corn	3.0	*1.36*
Soybean meal	0.5	*0.23*
Limestone	0.15	*0.07*
Trace mineralized salt and vitamins	0.1	*0.05*
Limited legume forage[2]		
Alfalfa hay, RFV 140, 20% crude protein	12.0	*5.45*
Corn silage	43.0	*19.52*
Monosodium phosphate	0.1	*0.05*
Trace mineralized salt and vitamins	0.1	*0.05*
Limited corn silage		
Alfalfa-grass hay, RFV 113, 16% crude protein	21.0	*9.53*
Corn silage	20.0	*9.08*
Dicalcium phosphate	0.1	*0.05*
Trace mineralized salt and vitamins	0.1	*0.05*

[1]Source: Linn, J. G., M. F. Hutjens, W. T. Howard, L. H. Kilmer, and D. E. Otterby, *Feeding the Dairy Herd*, Cooperative Ext. Services, Illinois, Iowa State, Minnesota, and Wisconsin, 1988, p. 17, Table 7.

[2]Ration contains excess energy as formulated and may over-condition cows in some situations.

Dry cows should be fed separately from lactating cows. Their ration should be formulated to meet their specific needs: body maintenance, fetal growth, and any additional body weight not replaced during phase 3. Daily dry matter intake should be near 2% of the cow's body weight; with forage intake a minimum of 1% of the body weight, and grain intake according to needs, but not to exceed 1% of body weight. One-half of 1% of the body weight in grain per day is usually sufficient in dry cow feeding programs. Dry cows should not be too fat. Feeding low quality forage, such as corn stalks or grass hay, is preferable to limit feeding. A protein level of 12% is sufficient during the dry period.

Some grain should be included in the dry cow ration starting 2 weeks before calving in order (1) to change the rumen bacteria from an all-forage digestion population to a mixed population of forage and grain digesters; and (2) to minimize the stress of ration changes following calving.

The calcium and phosphorus needs of dry cows should be met, but excesses should be avoided; sometimes, rations containing more than 0.6% calcium and 0.4% phosphorus have increased milk fever problems substantially. Trace minerals, including selenium, should be supplied in most dry cow rations. Also, adequate amounts of vitamins A, D, and E should be provided in the ration to lower milk fever, lessen retained placenta, and increase calf survival.

Potential problems during phase 4 include milk fever, displaced abomasum, retained placenta, fatty liver syndrome, and poor appetite, along with other metabolic disorders and diseases associated with the fat cow syndrome.

Key management precautions during the dry period include: (a) observing body condition of dry cows and adjusting energy feeding as necessary, (b) meeting nutrient requirements but avoiding excessive feeding, (c) changing the ration 2 weeks before calving to include grain and other small quantities of ingredients used in the lactating ration, (d) avoiding excess calcium and phosphorus intakes, and (e) limiting salt and other sodium-based minerals in the dry cow ration to reduce udder edema problems.

CHALLENGE FEEDING (LEAD FEEDING)

Challenge feeding, or lead feeding, refers to feeding the lactating cow so that she is challenged to reach her peak (summit) production level early in lactation.

Because of the strong relationship between peak (summit) milk yield and the total milk production for the entire lactation period, emphasis should be placed on attaining maximum yield between weeks 3 and 8. According to figures provided by Iowa State University extension dairy specialist Ron Orth, for every 5 lb increase in the summit peak, rolling herd average will increase 1,000 lb.

Preparation for challenge feeding begins during the dry cow period; (1) by having the dry cow in proper condition; and (2) by making the transition from dry cow ration to the lactating ration, thereby preparing the rumen bacteria.

After calving, challenge feeding calls for increasing the grain allowance several pounds per day above the cow's exact requirements at the time. The objective is to allow each cow to reach peak production at or near her genetic potential. The advantage of achieving maximum peak production will be maintained throughout the remaining months of lactation.

Calving time is a very traumatic experience for a high-producing cow. As a result, many cows have a depressed appetite for several days after calving. However, a skilled caretaker can get most of these cows on full feed very quickly without changing the ration markedly. But a heavy producer cannot consume sufficient energy to balance her energy output. Consequently, she will be forced to rely on body stores of fat and protein to supplement her ration. The objective in feeding the fresh cow is to keep her dependence on stored energy and protein as small and short in duration as possible. *Off-feed* is the greatest hazard and should be rigorously avoided.

During the period in which the concentrate is increased, a point will be reached when forage intake will be depressed. As this point is exceeded, each additional pound of concentrate will reduce forage intake by about $\frac{1}{2}$ lb of hay equivalent. This reduces the fiber content of the ration and may increase the incidence of milk fat depression, and rumen malfunction, but this is a temporary problem and should be viewed on the basis of its effect on the total lactation period. After peak milk flow is attained and challenging the cow ceases, forage intake will gradually increase and fat test will return to normal.

Challenge feeding helps a cow reach peak production earlier than she otherwise might, thus taking advantage of the fact that her system, at the time, is physiologically adapted to heavy production.

After peak production is reached, the amount of concentrate feed should be determined by a concentrate feeding guide based on body weight, milk production, and fat test (see Table 12–11).

CORRAL (GROUP) FEEDING

Individual feeding of lactating cows has largely given way to mechanized group feeding. The latter was developed for convenience and saving of labor, rather than for improved animal well-being or feed efficiency. Today, lactating herds with several hundred cows are common; and some herds number several thousand. In order to design a nutritional program for such large numbers that can be adapted to the specific needs of the cows, they are separated into groups according to production (and, therefore, nutritional needs).

When producers decide to go to group feeding, they must decide on the number of groups into which to divide the herd. To answer this question, consideration should be given to the following: (1) herd size; (2) types and costs of available feeds; (3) current type of housing, feeding, and milking system; and (4) overall economic integration of the operation—for example, labor, machinery, etc.

In large herds (more than 250 milking cows), a commonly used system is one in which a minimum of 5 groups are established: (1) high-production cows (about 90 lb of milk/head/day), (2) medium-production cows (about 65 lb of milk/head/day), (3) low-production cows (about 45 lb of milk/head/day), (4) dry cows, and (5) first calf heifers. More groups are desirable in very large herds if corrals and facilities are available. Because of feeding and social consideration, a maximum of 100 cows per group is advisable. With this program, there can be a maximum of two moves during the lactation cycle. In many cases, only one move is necessary; and in a few cases, no moves are required. This system allows each group to be fed according to need. The high-producing groups should be fed the highest quality ingredients at maximum levels. The middle-producing cows should be fed in such a way as to reduce feed cost, increase butterfat test, improve rumen function, and promote lactation persistency. The same holds true for the low-producing cows as for the medium producers except that considerable care must be exercised to avoid excessive fattening.

One of the problems inherent in group feeding concerns the behavioral adaptation of a newly introduced cow to a group. Group acceptance—*pecking order*—can pose occasional problems with a new cow, but the magnitude of the problem is usually not very great. One means of reducing this is to move several cows into a new group at the same time and just before feeding, rather than individually.

When group feeding programs are followed, grain is seldom fed in the milking parlor. This is commonly referred to as corral or bunk feeding since feeding generally takes place in bunks along the fenceline of the corrals or pens. Studies have demonstrated that cows fed their grain as a group in a common manger do as well as those fed individually in the milking parlor, but some cows may not always come into the parlor as easily when there is no grain to attract them. Some producers offer a minimum amount of feed in the parlor and the remaining amount in the corral with good success. The high producers seem to be more aggressive than the low producers; hence, they usually eat more when group fed.

Group feeding can be easily adapted to the use of complete feeds when the concentrates, roughages, and supplements are mixed into one feed rather than being fed separately. Some producers who use complete feeds prefer to feed dried roughages—especially long-stemmed hay—separately in order to enhance stimulation of the rumen and to facilitate mixing, because long hay does not lend itself to mixing in a mixer.

The **advantages** of group feeding and complete feed are:

1. It allows the producer to use special formulations which are particularly important to certain animals.

2. It eliminates the necessity of supplying minerals *ad libitum*.

3. It permits precise definition of the ration that is consumed.

4. It facilitates mechanized feeding; hence, less labor is required.

5. It eliminates the problems associated with the uncontrolled preferential consumption of a certain feedstuff.

6. It results in fewer digestive disturbances, such as displaced abomasums.

7. It eliminates the practice of feeding in the parlor.

8. It allows for maximum utilization of least-cost formulation.

9. It facilitates masking of certain unpalatable feeds, such as urea.

10. It is adaptable to conventional barn systems.

11. It allows the producer to fix a ratio of fiber to concentrate proportions in the ration.

12. It reduces the hazard of micronutrient deficiencies.

13. It provides the operator with daily feed consumption figures of the group, which can then be used to improve management decisions.

Among the **disadvantages** of group feeding and complete feeding are:

1. It necessitates specialized blending equipment to ensure thorough mixing.

2. It is uneconomical to divide small herds into groups to feed separate rations.

3. It is not applicable to grazing herds.

4. It is difficult to group animals in some barn designs.

5. Mismanagement can result in the so-called *fat cow syndrome* and related health problems, such as calving difficulties, poor reproduction, low production, low dry matter consumption, and metabolic disorders. In many cases, these problems do not become immediately obvious. Rather, they may take many months to develop.

■ **Automatic grain feeders**—The advent of electric and computerized grain feeders was heralded as a way in which producers could keep their cows in group housing but feed them individually. However, some producers report (1) that such feeders may result in overfeeding grain, accompanied by health problems and lower profits; and (2) that such feeders may not function properly when the grain ration contains adequate fiber content.

The following general types of mechanized grain-feeding systems are available:

1. **Free-choice, electronic grain feeders.** These units allow cows equipped with an identification unit (magnet, key, or chain) access to a feeding station. This system does not restrict access time or amount of grain consumed per feeder visit. Careful management is necessary to avoid digestive problems. The major advantages of this system are low initial investment and a simple design.

2. **Preset or computerized grain feeders.** This sys-

tem controls the maximum amount of grain that individual cows receive during a set period of time. Initial cost of this system varies widely depending on herd size and complexity of the system.

Regardless of which feeder is selected, successful adoption requires superior management. The entire daily grain allocation for individual cows and/or the herd can be fed through computer-operated feeders. Maximum daily grain capacity is generally 400 to 500 lb. It is recommended that a feeder accommodate 20 to 25 cows. Stall length, protection of the unit, and location of the unit relative to cow traffic patterns can affect the success of the unit.

■ **Example rations for group feeding** — Example rations for group feeding mature lactating cows at each of three milk production levels (90, 65, and 45 lb) are presented in Tables 12–17, 12–18, and 12–19; and example rations suitable for feeding dry cows are presented in Table 12–20 (p. 276).

TABLE 12–17
COMPLETE RATIONS FOR 1,300 TO 1,400 LB *(591 TO 636 KG)* COWS, IN EARLY LACTATION—HIGH-PRODUCTION GROUP (90-LB MILK, 3.6% FAT AVERAGE)[1]

Feeds	All Alfalfa		¾ Alfalfa ¼ Corn Silage		⅔ Alfalfa ⅓ Corn Silage		½ Alfalfa ½ Corn Silage		⅓ Alfalfa ⅔ Corn Silage		¼ Alfalfa ¾ Corn Silage	
	All Amounts on Dry Matter Basis											
	(lb)	(kg)	(lb)	(kg)	(lb)	(kg)	(lb)	(kg)	(lb)	(kg)	(lb)	(kg)
Alfalfa, 17% crude protein, 59% TDN	24.20	10.99	19.20	8.72	17.28	7.85	13.59	6.17	9.24	4.19	7.03	3.19
Corn silage, 8% crude protein, 68.7% TDN	—	—	6.40	2.91	8.63	3.92	13.59	6.17	18.51	8.40	21.10	9.58
Corn, 10% crude protein, 88% TDN	18.90	8.58	16.10	7.31	15.32	6.96	13.11	5.95	11.22	5.09	10.16	4.61
44% crude protein supplement, 49% crude protein, 86% TDN	5.50	2.50	6.60	3.00	7.00	3.18	7.83	3.55	8.82	4.00	9.34	4.24
Fat, 182% TDN	1.20	0.54	1.20	0.54	1.19	0.54	1.19	0.54	1.18	0.54	1.19	0.54
Dicalcium phosphate, 22% calcium, 19% phosphorus	0.42	0.19	0.52	0.24	0.51	0.23	0.49	0.22	0.46	0.21	0.44	0.20
Limestone, 34% calcium	—	—	0.07	0.03	0.14	0.06	0.25	0.11	0.41	0.19	0.48	0.22
Trace mineralized salt	0.25	0.11	0.25	0.11	0.25	0.11	0.25	0.11	0.25	0.11	0.25	0.11
Mineral-vitamin mix	0.08	0.04	0.08	0.04	0.08	0.04	0.09	0.04	0.10	0.05	0.11	0.05
	Ration Nutrient Information											
Dry matter	50.50	22.93	50.4	22.88	50.40	22.88	50.40	22.88	50.20	22.79	50.10	22.75
Crude protein	8.72	3.96	8.71	3.95	8.72	3.96	8.72	3.96	8.72	3.96	8.72	3.96
Crude protein (%)	17.26		17.29		17.29		17.30		17.36		17.40	
TDN	37.52	17.03	37.39	16.98	37.43	16.99	37.39	16.98	37.33	16.95	37.30	16.93
TDN (%)	74.29		74.19		74.26		74.18		74.36		74.446	
NE$_{lc}$ (Mcal)	38.72		38.66		38.72		38.72		38.72		38.72	
NE$_{lc}$ (Mcal/lb or kg)	0.767	1.69	0.767	1.69	0.768	1.69	0.768	1.69	0.771	1.70	0.773	1.70
Calcium (g)	208.00		206.00		206.00		206.00		205.00		205.00	
Calcium (%)	0.91		0.90		0.90		0.90		0.90		0.90	
Phosphorus (g)	103.00		114.00		114.00		114.00		114.00		114.00	
Phosphorus (%)	0.45		0.50		0.50		0.50		0.50		0.50	
Acid detergent fiber (%)	18.04		18.12		18.07		18.24		18.09		18.05	
Ether extract (%)	5.66		5.54		5.50		5.41		5.30		5.28	
Forage:grain ratio	48:52		51:49		51:49		54:46		55:45		56:44	

[1]Ration calculations based on 1989 NRC recommendations and on the use of either alfalfa or alfalfa and corn silage (CS) forage. Table 12–17 was prepared by D. E. Otterby, Professor, and J. G. Linn, Extension Animal Scientist, Department of Animal Science, University of Minnesota, St. Paul; with metric added by the author.

TABLE 12–18
COMPLETE RATIONS FOR 1,300 TO 1,400 LB *(591 TO 636 KG)* COWS, IN MID-LACTATION—MEDIUM-PRODUCTION GROUP (65-LB MILK, 3.6% FAT AVERAGE)[1]

Feeds	All Alfalfa		¾ Alfalfa ¼ Corn Silage		⅔ Alfalfa ⅓ Corn Silage		½ Alfalfa ½ Corn Silage		⅓ Alfalfa ⅔ Corn Silage		¼ Alfalfa ¾ Corn Silage		All Corn Silage	
	All Amounts on Dry Matter Basis													
	(lb)	*(kg)*	(lb)	*(kg)*	(lb)	*(kg)*	(lb)	*(kg)*	(lb)	*(kg)*	(lb)	*(kg)*	(lb)	*(kg)*
Alfalfa, 17% crude protein, 59% TDN . .	21.93	*9.96*	17.93	*8.14*	16.43	*7.46*	11.98	*5.44*	8.07	*3.66*	6.15	*2.79*	—	—
Corn silage, 8% crude protein, 68.7% TDN	—	—	5.98	*2.71*	8.20	*3.72*	—	—	—	—	—	—	—	—
Urea-corn silage, 12% crude protein, 70% TDN	—	—	—	—	—	—	11.98	*5.44*	16.16	*7.34*	18.46	*8.38*	26.39	*11.98*
Corn, 10% crude protein, 91% TDN . . .	17.12	*7.77*	14.29	*6.49*	13.23	*6.01*	13.25	*6.02*	12.48	*5.67*	11.80	*5.36*	6.28	*2.85*
44% CP supplement, 49% CP, 86% TDN	2.33	*1.06*	3.21	*1.46*	3.54	*1.61*	3.62	*1.64*	4.10	*1.86*	4.33	*1.97*	7.34	*3.33*
Dicalcium phosphate	0.34	*0.15*	0.33	*0.15*	0.33	*0.15*	0.35	*0.16*	0.36	*0.16*	0.35	*0.16*	0.36	*0.16*
Limestone	—	—	—	—	—	—	—	—	0.04	*0.02*	0.10	*0.05*	0.25	*0.11*
Trace mineralized salt	0.21	*0.10*	0.21	*0.10*	0.21	*0.10*	0.21	*0.10*	0.21	*0.10*	0.21	*0.10*	0.20	*0.09*
Mineral-vitamin mix	0.08	*0.04*	0.07	*0.03*	0.08	*0.04*	0.08	*0.04*	0.07	*0.03*	0.10	*0.05*	0.10	*0.05*
	Ration Nutrient Information													
Dry matter	42.00	*19.07*	42.00	*19.07*	42.00	*19.07*	41.50	*18.84*	41.50	*18.84*	41.50	*18.84*	40.90	*18.57*
Crude protein	6.60	*3.00*	6.60	*3.00*	6.60	*3.00*	6.60	*3.00*	6.60	*3.00*	6.60	*3.00*	6.60	*3.00*
Crude protein . . . . (%)	15.72		15.72		15.72		15.91		15.91		15.91		16.14	
TDN	29.90	*13.57*	29.85	*13.55*	29.84	*13.55*	29.56	*13.42*	29.73	*13.50*	29.70	*13.48*	29.60	*13.44*
TDN . . . . (%)	71.18		71.08		71.05		71.23		71.63		71.58		72.37	
NE$_{lc}$ . . . . (Mcal)	30.73		30.73		30.73		30.73		30.73		30.73		30.73	
NE$_{lc}$ . . . . (Mcal/lb or kg)	0.73	*1.61*	0.73	*1.61*	0.73	*1.61*	0.74	*1.63*	0.74	*1.63*	0.74	*1.63*	0.75	*1.65*
Calcium . . . . (g)	180.00		158.00		150.00		135.00		123.00		123.00		123.00	
Calcium . . . . (%)	0.95		0.83		0.79		0.72		0.65		0.65		0.66	
Phosphorus . . . . (g)	81.00		81.00		81.00		81.00		81.00		81.00		81.00	
Phosphorus . . . . (%)	0.42		0.42		0.42		0.43		0.43		0.43		0.44	
Acid detergent fiber . . . . (%)	19.00		19.57		19.78		19.73		19.59		19.69		19.03	
Ether extract . . . . (%)	3.51		3.36		3.34		3.00		2.84		2.74		2.95	
Forage:grain ratio	52:48		57:43		59:41		58:42		58:42		59:41		65:35	

[1]Ration calculations based on 1989 NRC recommendations and on the use of either alfalfa or alfalfa and corn silage (CS) forage. Table 12–18 was prepared by D. E. Otterby, Professor, and J. G. Linn, Extension Animal Scientist, Department of Animal Science, University of Minnesota, St. Paul; with metric added by the author.

Fig. 12–15. A complete ration fed along a fenceline manger. (Courtesy, James Tappan, Higley, AZ)

TABLE 12–19
COMPLETE RATIONS FOR 1,300 TO 1,400 LB *(591 TO 636 KG)* COWS,
IN LATE LACTATION—LOW-PRODUCTION GROUP (45-LB MILK, 3.8% FAT AVERAGE)[1]

Feeds	All Alfalfa		¾ Alfalfa ¼ Corn Silage		⅔ Alfalfa ⅓ Corn Silage		½ Alfalfa ½ Corn Silage		⅓ Alfalfa ⅔ Corn Silage		¼ Alfalfa ¾ Corn Silage		All Corn Silage	
	All Amounts on Dry Matter Basis													
	(lb)	*(kg)*	(lb)	*(kg)*	(lb)	*(kg)*	(lb)	*(kg)*	(lb)	*(kg)*	(lb)	*(kg)*	(lb)	*(kg)*
Alfalfa, 17% crude protein, 59% TDN . . .	26.14	*11.87*	21.53	*9.77*	19.74	*8.96*	14.99	*6.81*	10.46	*4.75*	7.98	*3.62*	—	—
Corn silage, 8% crude protein, 68.7% TDN	—	—	7.18	*3.26*	9.86	*4.48*	—	—	—	—	—	—	—	—
Urea-corn silage, 12% crude protein, 66% TDN	—	—	—	—	—	—	14.99	*6.81*	20.95	*9.51*	23.94	*10.87*	31.10	*14.12*
Corn, 10% crude protein, 91% TDN . . .	10.76	*4.89*	8.21	*3.73*	6.97	*3.16*	6.74	*3.06*	4.79	*2.17*	3.97	*1.80*	2.60	*1.18*
44% CP supplement, 49% CP, 81% TDN	—	—	—	—	0.34	*0.15*	0.13	*0.06*	0.63	*0.29*	0.94	*0.43*	2.20	*1.00*
Dicalcium phosphate, 22% Ca, 19% P .	0.27	*0.12*	0.26	*0.12*	0.25	*0.11*	0.30	*0.14*	0.31	*0.14*	0.31	*0.14*	0.30	*0.14*
Limestone, 34% calcium	—	—	—	—	—	—	—	—	—	—	—	—	0.12	*0.05*
Trace mineralized salt	0.19	*0.09*	0.19	*0.09*	0.19	*0.09*	0.19	*0.09*	0.19	*0.09*	0.19	*0.09*	0.18	*0.08*
Mineral-vitamin mix	0.06	*0.03*	0.06	*0.03*	0.06	*0.03*	0.07	*0.03*	0.08	*0.04*	0.09	*0.04*	0.20	*0.09*
	Ration Nutrient Information													
Dry matter	37.42	*16.99*	37.42	*16.99*	37.42	*16.99*	37.42	*16.99*	37.42	*16.99*	37.42	*16.99*	36.70	*16.66*
Crude protein	5.52	*2.51*	5.11	*2.32*	5.09	*2.31*	5.09	*2.31*	5.09	*2.31*	5.09	*2.31*	5.09	*2.31*
Crude protein . . . . (%)	14.75		13.65		13.60		13.60		13.60		13.60		13.87	
TDN	24.91	*11.31*	24.85	*11.28*	24.83	*11.27*	24.78	*11.25*	24.73	*11.23*	24.70	*11.21*	24.58	*11.16*
TDN . . . . (%)	66.52		66.40		66.35		66.22		66.10		66.02		66.98	
NE_{lc} . . . . (Mcal)	25.37		25.37		25.37		25.37		25.37		25.37		25.37	
NE_{lc} . . . . (Mcal/lb or kg)	0.677	*1.49*	0.678	*1.50*	0.678	*1.50*	0.678	*1.50*	0.678	*1.50*	0.678	*1.50*	0.691	*1.52*
Calcium . . . . (g)	195.00		173.00		165.00		148.00		129.00		117.00		96.00	
Calcium . . . . (%)	1.15		1.02		0.97		0.87		0.76		0.69		0.58	
Phosphorus . . . . (g)	64.00		64.00		64.00		64.00		64.00		64.00		64.00	
Phosphorus . . . . (%)	0.38		0.38		0.38		0.38		0.38		0.38		0.38	
Acid detergent fiber . . . . (%)	23.92		24.63		24.91		25.42		26.02		26.16		25.40	
Ether extract . . . . (%)	3.47		3.38		3.32		2.86		2.59		2.45		2.09	
Forage:grain ratio	70:30		77:23		79:21		80:20		84:16		85:15		85:15	

[1]Ration calculations based on 1989 NRC recommendations and on the use of either alfalfa or alfalfa and corn silage (CS) forage. Table 12–19 was prepared by D. E. Otterby, Professor, and J. G. Linn, Extension Animal Scientist, Department of Animal Science, University of Minnesota, St. Paul; with metric added by the author.

Fig. 12–16. A complete ration being fed from a self-unloading truck. (Courtesy, University of California, Davis)

TABLE 12–20
COMPLETE RATIONS FOR 1,300 TO 1,400 LB *(591 TO 636 KG)* DRY COWS[1]

Feeds	Alfalfa Corn Silage		Alfalfa-Grass Hay, Corn Stover		Alfalfa-Grass Hay		Oatlage		Urea-Corn Silage, Grass Hay		Grass Hay	
	All Amounts on Dry Matter Basis											
	(lb)	*(kg)*	(lb)	*(kg)*	(lb)	*(kg)*	(lb)	*(kg)*	(lb)	*(kg)*	(lb)	*(kg)*
Alfalfa, 17% CP, 59% TDN	11.09	5.03	—	—	—	—	—	—	—	—	—	—
Alfalfa-grass hay, 16.5% CP, 58% TDN	—	—	14.59	6.62	22.52	10.25	—	—	—	—	—	—
Grass hay, 12% CP, 60% TDN	—	—	—	—	—	—	—	—	13.50	6.13	17.93	8.14
Oatlage, 12.8% CP, 59% TDN	—	—	—	—	—	—	25.74	11.69	—	—	—	—
Corn silage, 8.7% CP, 68.7% TDN	12.66	5.75	—	—	—	—	—	—	—	—	—	—
Urea-corn silage, 12% CP, 66% TDN	—	—	—	—	—	—	—	—	12.28	5.58	—	—
Corn stover, 6.7% CP, 66% TDN	—	—	11.59	5.26	—	—	—	—	—	—	—	—
Corn, 10% CP, 88% TDN	—	—	—	—	3.47	1.58	1.09	0.49	—	—	5.46	2.48
44% CP supplement, 49.9% CP, 81% TDN . . .	—	—	—	—	—	—	—	—	—	—	0.53	0.24
Dicalcium phosphate, 22% Ca, 19% P	—	—	—	—	—	—	0.01	0.01	—	—	—	—
Limestone, 34% Ca	—	—	—	—	—	—	0.05	0.02	0.05	0.02	0.09	0.04
Monosodium phosphate, 22.5% P	0.05	0.02	0.10	0.05	0.03	0.01	—	—	—	—	—	—
Trace mineralized salt	0.07	0.03	0.07	0.03	0.07	0.03	0.07	0.03	0.07	0.03	0.07	0.03
Mineral-vitamin mix	0.03	0.01	0.02	0.01	0.02	0.01	0.06	0.03	0.06	0.03	0.06	0.03
	Ration Nutrient Information											
Dry matter	25.07	11.38	26.37	11.97	26.11	11.85	27.00	12.26	25.96	11.79	24.14	10.96
Crude protein	3.05	1.38	3.24	1.47	4.07	1.85	3.40	1.54	3.09	1.40	2.97	1.35
Crude protein . . . . (%)	12.18		12.29		15.58		12.60		11.92		12.31	
TDN	16.17	7.34	16.10	7.31	16.09	7.30	16.14	7.33	16.08	7.30	15.99	7.26
TDN . . . . (%)	64.49		61.08		61.64		59.78		61.98		66.27	
Calcium . . . . (g)	88.00		91.00		87.00		48.00		48.00		48.00	
Calcium . . . . (%)	0.78		0.77		0.73		0.39		0.41		0.44	
Phosphorus . . . . (g)	30.00		30.00		30.00		30.00		30.00		37.00	
Phosphorus . . . . (%)	0.27		0.25		0.25		0.25		0.27		0.34	
Acid detergent fiber . . . . (%)	30.16		37.31		28.96		33.48		35.05		31.32	
Forage:grain ratio	99:1		99:1		87:13		95:5		99:1		74:26	

[1]Ration calculations based on 1989 NRC recommendations. Table 12–20 was prepared by D. E. Otterby, Professor, and J. G. Linn, Extension Animal Scientist, Department of Animal Science, University of Minnesota, St. Paul; with metric added by the author.

When group feeding, first-calf heifers should be handled in a separate group and fed for both milk production and growth. Their nutrient requirements for milk production are similar to the requirements of their older counterparts producing milk at the same level, but, because of their growth, they should receive about 20% more nutrients than are required for maintenance.

Although variations can and should be made in Tables 12–17, 12–18, 12–19, and 12–20 rations, they are excellent guides. When milk yields and/or body weights differ from those used in these tables, a suitable computer program or hand calculations should be employed to obtain amounts to be fed, and to give consideration to costs of alternative feeds.

FEEDING ON PASTURE

On the average, a lactating cow on pasture spends 8 hours daily in harvesting, or grazing. In addition, she spends another 8 hours each day in rumination—the act of chewing the cud. This means that a cow spends two-thirds of her time eating. If she is a high producer, meeting her nutritive needs within this period of time calls for good pastures plus supplemental feeding.

Fig. 12–17. Lactating cows on pasture, with natural shade. (Courtesy, *Holstein World*, Sandy Creek, NY)

Problems of milk production are at a minimum during the early pasture season, when plant growth is lush. However, when the weather gets hot—the period known as the *summer slump*—it is a different story. High temperatures actually affect pasture growth more than the well-being of the cows. Many dairy producers have discontinued pasture grazing for two reasons: (1) It is difficult to keep milk production uniform when cows are on pasture, because of changing temperatures and pasture growth; and (2) with larger herds, it is not possible to have sufficient pasture in close proximity to headquarters.

Pasture of high quality has a value that is intermediate between concentrate and hay. Thus, on good pasture, it is possible to sustain a high level of production with less grain than is needed for conventional winter forage supplementation.

The following summer feeding program is recommended for most lactating cows on pasture:

1. Have good pastures and follow good pasture management. Practice a rotational system of grazing with cows kept within one area until forage is used, and then place them on another area. In this way pasture yields will be higher and cows will have a better quality feed during the entire pasture season. The length of rotation may vary from 1 to 3 days. Beyond 3 days, there will likely be some damage to the plants.

2. Consider ensiling the early pasture growth which is in excess of the amount the cows will consume, thereby avoiding overmature and/or wasted forage in the early season.

3. Feed hay and grain as a supplement to pastures when needed to obtain desired milk yield. If the pasture is of comparable quality to the forage that was fed in the winter season, the grain mix may be the same as was fed in the winter. On the other hand, if summer pastures are much better or much poorer than the quality of the winter roughage, the grain mix should be changed accordingly. Thus, the grain mix used when cows are on pasture should be formulated so as to balance the deficiencies of the grass, just the same as the winter concentrate should balance the deficiencies of the winter forage. Also, cows should be fed hay before turning them to pasture during periods when bloat is particularly hazardous, such as early and late in the season.

4. When pastures are poor and/or the weather is very hot, feed more grain and limit the hay, thereby providing needed added energy and avoiding the excess heat generated during the digestion of high-fiber rations. But do not reduce hay feeding so much as to cause a reduction in the fat content of the milk.

5. Consider supplementing summer pastures with silage. Succulence is of benefit in dairy rations; hence, if cows are accustomed to having silage during the winter, it may be continued to advantage in the summer when pastures are not doing well.

6. Provide adequate shade and water for cows on summer pasture.

FEEDING DRY COWS

Dry cows have three important needs: (1) regression of old milk producing cells in the udder and regeneration of new milk producing cells, (2) developing the unborn calf, and (3) storing body reserves for the next milking period. This necessitates that they be properly fed in late lactation and during the dry period.

Proper dry cow feeding is one of the most critical practices for successful dairy operations. It involves management practices (1) to have the cow dry 50 to 65 days, and (2) to have the cow in proper condition when beginning and finishing the dry period. Proper dry cow feeding begins during the last half to one-third of the lactation period when body reserves should be replaced. At drying off time, the cow's body condition should be classed as 2 or 3 on an arbitrary scale of 1 as too thin and 5 as too fat. At calving, the body condition should be in the 3.5 to 4 class. During the 60-day dry period, cows should be fed to gain 120–200 lb.

The following routine is recommended for dry cows:

1. Turn high-producing cows and cows milked 3 times daily dry 60 to 65 days before expected calving date. Turn cows more than 4 years of age dry 50 to 60 days before freshening. In many cases, dairy producers erroneously prolong the lactation of their high producers, subsequently lessening the period in which the cow can recuperate from the heavy demands of lactation. Records show that the highest total milk during a lactation is produced when cows are dry 50 to 60 days.

2. For cows producing low levels of milk at the end of lactation, the caretaker should stop grain feeding, limit water intake, and stop milking abruptly to hasten drying off. However, cows producing high levels of milk at the end of lactation may require continued milking for a short period during which they are fed a low-energy ration. This should reduce stress and facilitate drying off.

3. The dry period should be a time when the digestive tract has a chance to recondition itself following the rigors of high-concentrate feeds used during lactation. Concentrate

feeds put considerable stress on the rumen. Thus, roughage—especially hay—is extremely important during the dry period to promote healthy rumens. Ideally, the condition of the dry cow should be such that she will not have to gain much weight, since weight can be put on much more efficiently during the latter stages of lactation. Under proper dry cow management, small amounts of grain should only be required during the last 2 weeks of the dry period for rumen acclimation.

4. The 2 weeks prior to calving should be devoted to preparing the cow for parturition and the onset of lactation and for the adaptation of the rumen microflora to a radical change in the ration. If milk fever is a problem, as it is in most herds, it is desirable that the dry cow ingest less than 100 g of calcium and 40 g of phosphorus per day but still ingest enough to satisfy the requirements for these minerals. Additionally, supplemental vitamins A, D, and E, along with added selenium, should be provided. One commonly used procedure to adapt the ruminal microflora to the lactation rations is to feed the same ration, or percentage of the same ingredients, that the cows will receive during lactation. Grain levels should not exceed 8 lb per day. It is imperative that dry cows not be in a fat condition at calving, however.

On the first day after calving, cows should be fed the same amount of grain that they were getting before calving, followed by an increase of 2 to 3 lb per day, according to the cow's appetite. The experienced caretaker is in the best position to determine how much, and what, to feed each individual cow at, and after, calving time.

FEEDING CALVES

One of the most important phases of dairy production is that of feeding and managing dairy calves. Statistics reveal that more than 20% of the dairy calves die of sickness or disease before reaching maturity. With good management, these losses may be reduced to 3 to 5%. Many calf deaths are caused by faulty nutrition or poor housing and management.

Fig. 12–18. A bucket or bottle equipped with a plastic nipple is easy to clean and provides a convenient method of measuring the correct amount to feed the calf. (Courtesy, Milk Marketing, Inc., Strongsville, OH)

A carefully planned and executed feeding program is necessary to produce growthy, vigorous, and healthy calves. The following feeding program is recommended:

Day 1	Dam's colostrum
Day 2	Dam's colostrum
Day 3	Dam's colostrum
Day 4	Liquid feed of choice, introduce starter and water
Day 5 to weaning	Continue feeding program
Weaning to 12 weeks . .	Starter (up to 5 lb daily), introduce forage

DAIRY CALF PRODUCTION

Physiologically, the newborn calf is not a functioning ruminant. The abomasum, which represents the largest portion of the stomach of the newborn calf, is the primary functional unit of the gastric region. The sucking reflex allows colostrum or milk to bypasss the rudimentary rumen and reticulum via the esophageal groove directly into the abomasum where digestion is initiated. As the calf grows older, it consumes solid feedstuffs which serve as a mechanical stimulator on the other sections of the gastric region, thereby hastening their development. Additionally, the rumen becomes inoculated with microorganisms from the immediate environment.

Since milk is the primary product of dairy production, it is necessary to switch the young calf to cheaper feeds as expeditiously as possible. At the same time, it is important that the diet promote good health, growth, and development. Four feeds that are routinely fed to calves are (1) colostrum, (2) milk, (3) milk replacers, and (4) calf starters.

COLOSTRUM

Colostrum is the milk which is high in antibodies, and which is secreted by cows, and other mammalian females, for the first few days following parturition.

Colostrum (either dam's colostrum or mixed colostrum from first milking of older cows) should be fed to calves as soon after birth as possible (ideally within 15 minutes and certainly within 4 hours) to protect against disease. At the 4,800-cow Arizona Dairy, Higley, Arizona, every effort is made to remove calves from their mother *wet* and without nursing, so as to minimize infection from nursing. Then, as soon as possible, they are offered colostrum from a nippled bottle. If they fail to nurse naturally, the colostrum is hand-fed, using a specially designed tube to force it into the abomasum. This system, along with superior feeding, management, and sanitation at Arizona Dairy, has resulted in average calf losses of only 2.7% (1.4% at birth and during the first 24 hours, and 1.3% after 24 hours).

Early feeding of colostrum is necessary because:

1. Newborn calves have no antibodies to provide natural protection against disease until colostrum is received.

2. The calves' ability to absorb immunoglobulins (the

disease protecting component) is substantially reduced after 24 to 36 hours.

3. Calves may become infected with highly pathogenic bacteria immediately after birth.

Calves should receive a total of 10 to 12% of their birth weight as first milking colostrum, with half of this amount received 4 to 6 hours after birth.

Surplus colostrum can be frozen and stored for a period of 1 year or longer without losing its antibody value. It may be thawed, warmed to about 100°F, and fed as needed.

Excess colostrum is a very nutritious feed, but it has little or no immune benefit (antibody protection) to a calf beyond the first 24 hours of life. Colostrum contains about a third more solids than milk or reconstituted milk replacer, and is highly digestible. So, storage and subsequent feeding of excess colostrum is highly desirable. It may be fed fresh; frozen, then thawed prior to feeding; or stored as sour (fermented) colostrum. Since it is higher in solids than normal milk, it should be diluted 25 to 50% when fed to other than newborn calves, in order to avoid overfeeding and scours.

Naturally fermented, sour colostrum sometimes becomes putrid and unfit for consumption, especially during the summer months. The following organic acids have been used separately, and successfully, to acidify colostrum to a pH of 4.6:

Name of Acid	Concentration	Dilution
Formic acid	0.3% by weight	¼ cup acid/5 gal colostrum
Acetic acid	0.7% by weight	½ cup acid/4¼ gal colostrum
Propionic acid	1.0% by weight	1 cup per 6 gal colostrum

COLOSTRUM, MILK, AND MILK REPLACERS COMPARED

The composition of colostrum changes rapidly after calving. The first six milkings are higher in nutrients than normal milk or most reconstituted milk replacers. Table 12–21 shows the comparative composition and characteristics of colostrum, whole milk, and a reconstituted milk replacer.

Milk replacers vary in quality, so the buyer/user should study the feed tag. The best milk replacers contain 22% protein, all derived from milk products — skim milk powder, buttermilk powder, dried whole whey, de-lactosed whey, casein, and/or milk albumen. Chemically modified soy protein, soy isolates, and soy concentrates are good, but as plant proteins they are less digestible than milk protein. Meat solubles, fish protein concentrate, distillers' dried solubles, brewers' dried yeast, oat flour, and wheat flour are inferior as protein sources in milk replacers.

A good milk replacer powder should contain a minimum of 15% fat, and it may contain more than 20%. The higher fat level tends to reduce the severity of diarrhea and produce additional energy for growth. Good quality animal fats are preferable to most vegetable fats. However, soy lecithin, especially when homogenized, is an acceptable fat source and improves mixing qualities of the replacer.

The calf can use two carbohydrate sources in milk replacer: lactose (milk sugar) and dextrose. However, two other carbohydrate sources, starch and sucrose (table sugar), are not satisfactory and should be excluded from milk replacers.

Mastitic milk and discard milk (unmarketable milk from cows that were treated for mastitis, metritis, or other health problems) may be fed fresh in the same manner as whole milk, or they may be fermented or preserved with an organic acid. Milk collected for three to six milkings after antibiotic treatment will ferment normally. Extremely abnormal milk (bloody or watery) should not be fed.

Properly managed feeding of mastitic milk to calves will not increase mortality nor cause mastitis in these animals when they freshen. Neither will it cause diarrhea. Mastitic milk

TABLE 12–21
COMPOSITION AND CHARACTERISTICS OF COLOSTRUM, WHOLE MILK, AND RECONSTITUTED MILK REPLACER
(1 LB [0.45 KG] POWDER + 7 LB [3.18 KG] WATER)[1]

Item	First Milking	Second Milking	Second Day	Third Day	Whole Milk	Reconstituted Milk Replacer
Specific gravity .... (g/ml)	1.056	1.04	1.034	1.033	1.032	
Total solids .... (%)	23.9	17.9	14.0	13.6	12.9	12.5
Fat .... (%)	6.7	5.4	4.1	4.3	4.0	2.5
Nonfat solids .... (%)	16.7	12.2	9.6	9.5	8.8	11.25
Protein .... (%)	14.0	8.4	4.6	4.1	3.1	2.8
Lactose .... (%)	2.7	3.9	4.5	4.7	5.0	variable
Ash .... (%)	1.1	1.0	0.8	0.8	0.7	variable
Vitamin A .... (g/100 ml)	295.0	190.0	95.0	74.0	34.0	variable
Immunoglobulins .... (%)	6.0	4.2	1.0	—	—	0.0

[1]Source: Linn, J. G., M. F. Hutjens, W. T. Howard, L. H. Kilmer, and D. E. Otterby, *Feeding the Dairy Herd*, Cooperative Ext. Services, Illinois, Iowa State, Minnesota, and Wisconsin, 1988, p. 18, Table 8.

should not be fed to calves less than 2 days old, as the intestine is permeable to large protein molecules. Calves fed milk containing antibiotics should not be marketed for meat unless the required withdrawal period is observed prior to slaughter.

CALF STARTERS

A high quality, palatable calf starter should be offered when the calf is 4 days old, and not later than 10 to 12 days of age. The best starters are high in energy, contain 16 to 18% protein (20% if calves are weaned before 4 weeks of age), and are free of excessive fines. To encourage consumption, starters should consist of whole, coarsely ground, cracked, or rolled grains. Up to 5% molasses improves palatability and minimizes fines and dust. Whole grains, especially oats, can be fed with starter rations to calves up to 3 months of age. Calf starters should be fed until calves are about 12 weeks of age, with intake limited to 5 to 7 lb per calf daily.

Fig. 12–19. Calf starter and water (note two pails) provided to a calf in an individual pen. (Courtesy, Maddox Dairy, Riverdale, CA)

Many good commercial starters are on the market. Also, calf starters may be home-mixed. Table 12–22 presents examples of some good grain calf starters.

TABLE 12–22
GRAIN STARTER RATIONS FOR CALVES[1]

	Grain Starters[2]		
	1	2	3
Ingredients (air dry basis)			
Corn (cracked or coarse ground) . . . (%)	50.0	30.0	
Ear corn (coarse ground) . . . . (%)			50.0
Oats (rolled or crushed) . . . . (%)	22.0	18.0	
Barley (rolled or coarse ground) . . . (%)		20.0	21.0
Wheat bran . . . . (%)		8.0	
Soybean meal . . . . (%)	20.0	16.0	21.0
Molasses . . . . (%)	5.0	5.0	5.0
Dicalcium phosphate . . . . (%)	0.5	0.5	0.5
Limestone . . . . (%)	1.5	1.5	1.5
Trace mineralized salt and vitamins . (%)	1.0	1.0	1.0
Composition (dry matter basis)			
ADF . . . . (%)	7.0	6.9	9.1
Crude protein . . . . (%)	18.1	18.0	18.4
TDN . . . . (%)	80.0	78.8	78.0
Calcium (Ca) . . . . (%)	0.8	0.8	0.82
Phosphorus (P) . . . . (%)	0.48	0.56	0.47
Vitamin A . . . . (IU/lb)	1,000.0	1,000.0	1,000.0
Vitamin A . . . . *(IU/kg)*	*2,205.0*	*2,205.0*	*2,205.0*
Vitamin D . . . . (IU/lb)	150.0	150.0	150.0
Vitamin D . . . . *(IU/kg)*	*331.0*	*331.0*	*331.0*
Vitamin E . . . . (IU/lb)	11.0	11.0	11.0
Vitamin E . . . . *(IU/kg)*	*24.0*	*24.0*	*24.0*

[1]Source: Linn, J. G., M. F. Hutjens, W. T. Howard, L. H. Kilmer, and D. E. Otterby, *Feeding the Dairy Herd*, Cooperative Ext. Services, Illinois, Iowa State, Minnesota, and Wisconsin, 1988, p. 20, Table 11.

[2]Hay may be offered free choice with grain starters.

HAY OR SILAGE FOR CALVES

While calves may begin nibbling on good quality hay as early as 5 to 10 days of age, it is not necessary to feed forage before 8 to 10 weeks of age. If the housing and management system makes it inconvenient to provide forage, it may be desirable to incorporate a forage factor (more fiber) in the starter ration. Table 12–23 presents examples of suitable rations for calves not receiving hay or silage. Corn silage or pasture should not be fed before 3 months of age because of their high moisture content which can limit intake and growth. Low moisture haylage is acceptable if it is kept fresh.

TABLE 12–23
COMPLETE STARTER RATIONS FOR CALVES[1]

		Complete Starters		
		1	2	3
Ingredients (air dry basis)				
Corn (cracked or coarse ground) . . (%)		40.0	25.0	30.0
Oats (rolled or crushed) . . . . (%)		14.5	8.0	18.0
Beet pulp . . . . (%)			25.0	25.0
Alfalfa hay (ground) . . . . (%)			10.0	
Corn cobs (ground) . . . . (%)		15.0		
Soybean meal . . . . (%)		23.0	18.0	20.0
Molasses . . . . (%)		5.0	5.0	5.0
Dried whey . . . . (%)			7.0	
Dicalcium phosphate . . . . (%)		0.5	0.5	0.5
Limestone . . . . (%)		1.0	0.5	0.5
Trace mineralized salt and vitamins (%)		1.0	1.0	1.0
Composition (dry matter basis)				
ADF . . . . (%)		13.3	15.8	14.2
Crude protein . . . . (%)		18.3	18.0	18.2
TDN . . . . (%)		75.5	78.0	79.4
Calcium (Ca) . . . . (%)		0.63	0.72	0.58
Phosphorus (P) . . . . (%)		0.45	0.44	0.43
Vitamin A . . . . (IU/lb)		1,000.0	1,000.0	1,000.0
Vitamin A . . . . (IU/kg)		2,205.0	2,205.0	2,205.0
Vitamin D . . . . (IU/lb)		150.0	150.0	150.0
Vitamin D . . . . (IU/kg)		331.0	331.0	331.0
Vitamin E . . . . (IU/lb)		11.0	11.0	11.0
Vitamin E . . . . (IU/kg)		24.0	24.0	24.0

[1]Source: Linn, J. G., M. F. Hutjens, W. T. Howard, L. H. Kilmer, and D. E. Otterby, *Feeding the Dairy Herd*, Cooperative Ext. Services, Illinois, Iowa State, Minnesota, and Wisconsin, 1988, p. 20, Table 11.

WATER

Clean, fresh water in clean pails may be offered free-choice starting on day four. Calves fed limited liquid (such as when fed once-a-day) should receive supplemental water, especially during warm weather. Calves offered water during the liquid feeding period (birth to 4 weeks) tend to consume more starter and perform better than calves fed liquid only.

AMOUNT AND METHOD OF FEEDING, FREQUENCY OF FEEDING, AND AGE OF WEANING

Calves may be separated from their dams at birth, or within 12 to 24 hours after birth. In any case, they should receive their dam's colostrum for the first 3 days of life, following which they may be shifted to a liquid feed of the feeder's choice.

In order to obtain proper growth, calves must be provided adequate dry matter. For an 80- to 100-lb calf, this calls for 1 lb of dry matter (solids) daily from milk, surplus colostrum, or milk replacer, from birth to weaning at 4 weeks. Table 12–24 provides dairy calf feeding guidelines.

Milk or milk replacer may be fed by open pail, by nipple feeding from a pail or bottle, or by automated feeding equipment. Each method of feeding is satisfactory, provided it is accompanied by cleanliness and sanitation.

Most calf raisers feed twice daily, with half the amounts listed in Table 12–24 offered at each feeding. Weak or unthrifty calves may benefit from more frequent feedings. Calves need the same amount of dry matter daily, but liquid amounts may have to be reduced to avoid digestive upsets. Dry milk replacer can be added to whole or mastitic milk to increase solid content without increasing the volume of liquid fed. Once-a-day feeding of milk-fed calves has proven successful except when calves are housed in a very cold environment. If once-a-day feeding is practiced, calves should be checked for health and well being at least once in addition to feeding time.

TABLE 12–24
SUGGESTED RATES FOR FEEDING WHOLE MILK, FEEDING AND DILUTION RATES FOR COLOSTRUM,[1] AND RECONSTITUTED SCHEDULE FOR MILK REPLACERS; DAILY[2]

Body Weight		Whole Milk		Solids Content		Colostrum			Water			Replacer			Water	
(lb)	(kg)	(lb)	(kg)	(lb)	(kg)	(lb)	(kg)		(lb)	(kg)		(lb)	(kg)		(lb)	(kg)
50–60	22–27	5.0	2.3	0.6	0.27	3.6	1.6	+	1.2	0.54	0.6	0.27	+	4.0	1.8
61–70	28–31	5.5	2.5	0.7	0.32	4.2	1.9	+	1.4	0.65	0.7	0.32	+	5.0	2.3
71–80	32–36	6.5	2.9	0.8	0.36	4.8	2.2	+	1.6	0.73	0.8	0.36	+	5.5	2.5
81–90	37–40	7.0	3.2	0.9	0.41	5.4	2.4	+	1.8	0.82	0.9	0.41	+	6.0	2.7
91–100	41–45	8.0	3.6	1.0	0.45	6.0	2.7	+	2.0	0.91	1.0	0.45	+	7.0	3.2
101–110	46–50	9.0	4.1	1.1	0.50	6.6	3.0	+	2.2	1.00	1.1	0.50	+	8.0	3.6

[1]Solids content is assumed to be 16%.

[2]Source: Linn, J. G., M. F. Hutjens, W. T. Howard, L. H. Kilmer, and D. E. Otterby, *Feeding the Dairy Herd*, Cooperative Ext. Services, Illinois, Iowa State, Minnesota, and Wisconsin, 1988, p. 18, Table 10.

When calves are housed in hutches, and the weather is extremely cold, they should be fed a 20% fat replacer three times daily and the feed allowance should be increased by 1¼ to 1½ times in order to meet their increased energy requirements. Young calves that are doing poorly should be moved to warmer quarters.

Most producers wean calves between 4 and 8 weeks of age. When weaned at 3 weeks of age, calves may have slightly depressed growth rates, temporarily; however, by 12 weeks of age, their weights will be about the same as their later weaned mates. Weaning later than 8 weeks of age may lead to fat calves. A good practice is to wean according to starter intake—wean when the starter intake is 1 to 1½ lb/day. Starter intake can be encouraged by placing a little dry feed in the pail immediately after the liquid has been consumed. In general, early weaning (at 21 to 35 days of age) will reduce feed and labor costs. Calves eating less than 1 lb of starter per day or calves doing poorly should be fed liquid until performance improves and dry feed is consumed in satisfactory amounts.

Fig. 12–20. Group of four weaned Guernsey calves in a pen. Note the hay rack. (Courtesy, American Guernsey Assn., Reynoldsburg, OH)

VEAL CALF PRODUCTION

A veal calf is defined as a young bovine animal, usually not over 4 months of age, that has subsisted largely on milk or milk replacers, thus making the color of the lean meat light, grayish pink. The majority of veal calves are of dairy breeding, consisting of bull calves and heifer calves not retained as replacements. Producers receive a premium if lean meat of veal is light, grayish pink in color (due to reduced muscle myoglobin), characteristic of feeding milk, which is naturally low in iron. Some producers attempt to enhance the grayish pink color by restricting iron intake and exercise. Research has shown that a dietary iron concentration of 11.4 to 13.6 mg per pound of dry matter in milk replacers is sufficient for the well-being of veal calves and produces desirable grayish pink carcasses, with or without exercise.

A conversion rate of 10 lb of whole milk for 1 lb of body weight gain is normal. If milk replacer is used, the conversion rate is generally about 1.3 to 1.5 lb of dry replacer per pound of gain.

Profitable veal production depends on: (1) a low mortality rate, (2) economical housing, (3) plenty of inexpensive labor, and (4) an established market.

Two types of veal markets exist: (1) fancy veal, and (2) conventional veal. The choice of the type of veal production should be determined primarily by two factors: (1) the market price outlook, and (2) the availability of an abundance of grain and some high quality forage needed for raising conventional veal.

■ **Fancy veal**—Calves for fancy veal are marketed at 300 to 400 lb weight and 3 to 4 months of age. The meat must be very light colored, indicating that the calves have been fed milk and/or milk replacers, and that they have not been fed hay or grain.

■ **Conventional veal**—The production of conventional veal involves the same management and feeding practices used in raising heifer replacements. A good starter and milk or milk replacer normally are used from birth to 4 to 6 weeks of age. Then, a calf grower ration and a good quality forage may be fed until the calves are about 12 weeks old and weigh about 200 lb.

PREVENTING CALF SCOURS

To prevent calf scours, the producer must prevent primary infection of the newborn. This rests on strict sanitary measures and isolation, along with other preventive measures. Observance of the following practices will lessen the incidence of calf scours:

1. **Make certain that the calf gets first-milk colostrum.** Never assume that the calf has nursed. Many newborn calves never receive sufficient colostrum to protect them from calfhood diseases. So, colostrum should be fed, preferably by hand, soon after birth; large breed calves should receive a minimum of 2 to 3 qt, small breed calves should receive about 3 pt.

2. **Augment natural resistance with vitamins.** Supplementation with vitamins A, D, and E (oral or injectable) immediately after birth is helpful in increasing the calf's nat-

ural resistance to scours, especially if colostrum is low in vitamin A content.

3. **Avoid overfeeding and irregularity of feeding.** Overfeeding and irregularity are common causes of calf scours.

4. **Keep feeding utensils clean and sanitary.** Clean the feeding utensils thoroughly after each feeding, then store them upside down to drain and dry.

5. **Do not overcrowd.** In bedded areas, provide 24 to 28 sq ft per calf. In confined, elevated stalls, provide about 20 sq ft per calf.

6. **Provide adequate ventilation.** Provide a minimum of four air exchanges per hour in winter and 15 air exchanges per hour in the summer.

7. **Avoid damp, wet calves.** Provide plenty of dry bedding in maternity stalls; and provide adequate bedding and ventilation in the calf quarters.

■ **Use of electrolytes** — Feeding an oral electrolyte solution usually is beneficial when a calf has a mild case of scours (not off feed, not depressed, and no fever). Recommended treatment follows:

1. Delete or drastically reduce the amount of milk or milk replacer fed.

2. Feed only water containing an electrolyte for three to six feedings, depending on how soon the feces become firm. Frequent feeding of small volumes is advantageous. A 100-lb calf should consume about 5 qt (10% of body weight) daily.

Oral electrolyte solutions can be purchased commercially. If not readily available, a suitable electrolyte mixture can be made by combining the following ingredients:

> 4 tsp of table salt
> 3 tsp of baking soda
> ½ cup of "light" corn syrup
> 1 gal of water

FEEDING REPLACEMENT HEIFERS

Between weaning and calving (12 weeks to 2-year-olds), the nutrition of heifer replacements is often neglected. At its best, the feeding and management program during this period involves three distinct phases: (1) weaning (about 12 weeks of age) to 1 year; (2) 1 year to 2 months before calving at 2 years; and (3) 2 months before calving to calving.

■ **Replacement heifers, weaning (about 12 weeks of age) to 1 year** — During this period, replacement heifers may be fed forage free-choice and limited grain. The amount, and

Fig. 12–21.   Replacement heifers—group fed. (Courtesy, Milk Marketing, Inc., Strongsville, OH)

the protein content, of the grain mix needed will be determined by the quality of the forage being fed. Pasture can be used successfully in the feeding program of replacement heifers, provided it is supplemented with a grain mix and some dried forage, along with suitable minerals (incorporated in the grain mix or offered free-choice). Also, there should be access to clean, fresh water.

During the yearling stage, replacement heifers should not be overfed and become too fat. Overconditioning has an inhibitory effect on mammary secretory tissue development during the critical period of its maximum development between 3 and 9 months of age and results in lower milk production later in life. Overconditioning of heifers after 15 months of age does not affect mammary secretory tissue.

Table 12–25 (next page) lists some grower rations for 400-lb calves. If the protein content of the forage is good, little protein supplement will be required in the grain mix. Monensin or Lasalocid can be fed as directed on the label to heifers to improve growth and rate of gain.

TABLE 12–25
RATIONS FOR LARGE BREED DAIRY HEIFERS OF DIFFERENT WEIGHTS[1]

Weight	Rate of Gain (lb/day)	(kg/day)	Ration (As-fed)	(lb)	(kg)	Percent (%)
400 lb (182 kg)	1.7	0.8	Alfalfa hay, 140 RFV, 20% CP	8.5	3.9	
			Grain mix, 12.0% CP	3.0	1.4	
			coarse ground barley			71.0
			rolled or ground oats			23.0
			molasses			5.0
			trace mineral salt			0.5
			dicalcium phosphate			0.4
			vitamin premix			0.1
	1.7	0.8	Alfalfa-grass hay, 113 RFV, 16% CP	7.0	3.2	
			Grain mix, 14.5% CP	4.5	2.0	
			coarse ground corn			83.1
			soybean meal			15.5
			trace mineral salt			0.5
			dicalcium phosphate			0.8
			vitamin premix			0.1
	1.7	0.8	Orchardgrass hay	7.0	3.2	
			Grain mix, 16.5% CP	5.0	2.3	
			rolled or coarse ground barley			65.8
			molasses			4.0
			soybean meal			27.5
			trace mineral salt			0.5
			limestone			2.1
			vitamin premix			0.1
	1.7	0.8	Orchardgrass hay	4.5	2.0	
			Corn silage	12.0	5.4	
			Grain mix, 19.7% CP	2.5	1.1	
			coarse ground corn			13.5
			rolled or coarse ground oats			6.5
			soybean meal			74.8
			trace mineral salt			1.0
			limestone			4.0
			vitamin premix			0.2
700 lb (318 kg)	1.7	0.8	Alfalfa hay, 140 RFV, 20% CP	18.0	8.1	
			Grain mix	2.0	0.9	
			coarse ground corn			95.3
			trace mineral salt			1.5
			dicalcium phosphate			3.0
			vitamin premix			0.2
	1.7	0.8	Alfalfa hay, 140 RFV, 20% CP	10.0	4.5	
			Corn silage	15.0	6.8	
			Grain mix	1.0	0.5	
			ground ear corn			88.5
			trace mineral salt			4.0
			dicalcium phosphate			7.0
			vitamin premix			0.5

Weight	Rate of Gain (lb/day)	(kg/day)	Ration (As-fed)	(lb)	(kg)	Percent (%)
700 lb (318 kg)	1.7	0.8	Alfalfa-grass hay, 113 RFV, 16% CP	12.5	5.7	
			Grain mix	6.0	2.7	
			ground ear corn			95.9
			soybean meal			2.5
			trace mineral salt			0.6
			dicalcium phosphate			0.9
			vitamin premix			0.1
	1.7	0.8	Orchardgrass hay	4.0	1.9	
			Corn silage	30.0	13.6	
			Grain mix	2.5	1.1	
			soybean meal			91.8
			trace mineral salt			2.0
			dicalcium phosphate			1.0
			limestone			5.0
			vitamin premix			0.2
1,000 lb (454 kg)	1.7	0.8	Alfalfa hay, 113 RFV, 16% CP	13.0	5.9	
			Corn silage	35.0	15.9	
			Mineral-vitamin supplement	0.5	0.2	
			soybean meal			80.0
			trace mineral salt			12.0
			dicalcium phosphate			6.0
			vitamin premix			2.0
	1.7	0.8	Orchardgrass hay	10.0	4.5	
			Corn silage	38.0	17.3	
			Supplement	2.0	0.9	
			soybean meal			89.0
			trace mineral salt			3.0
			limestone			7.5
			vitamin premix			0.5
	1.7	0.8	Oatlage, 13% CP	50.0	22.7	
			Grain mix	4.0	1.8	
			coarse ground corn			93.9
			trace mineral salt			1.5
			limestone			3.0
			dicalcium phosphate			1.0
			vitamin premix			0.6
	1.7	0.8	Urea-corn silage, 0.5% urea	41.0	18.6	
			Grass hay	10.0	4.5	
			Supplement	0.5	0.2	
			soybean meal			64.0
			trace mineral salt			10.0
			limestone			24.0
			vitamin premix			2.0

[1]Source: Linn, J. G., M. F. Hutjens, W. T. Howard, L. H. Kilmer, and D. E. Otterby, *Feeding the Dairy Herd*, Cooperative Ext. Services, Illinois, Iowa State, Minnesota, and Wisconsin, 1988, pp. 20–21, Table 12.

■ **Replacement heifers, 1 year to 2 months before calving at 2 years**—If good quality forage is available, it may be the only feed required for heifers over 1 year of age. A suitable mineral mix should be provided on a free-choice basis. Heifers should gain 1.6 to 1.8 lb per day. If growth is not satisfactory, some grain should be provided. The rations shown in Table 12–25 indicate the amounts of feed when various forage and grain combinations are offered to 700- to 1,000-lb heifers. Heifers on good pasture require no grain or other forage. As pastures mature, dry out, or are heavily grazed, supplemental feed should be provided. Heifers that are seriously deficient in energy, phosphorus, or vitamin A may not exhibit estrus.

First estrus in heifers is dependent on size and weight, primarily weight. A general guide is that heifers will show their first estrus at 40% of their mature weight, which should be before 12 months of age. Heifers fed high planes of nutrition will show estrus at an earlier age than heifers grown at recommended rates, and underfeeding of heifers will delay estrus. Underfed or very slow growing heifers may ovulate, but estrous signs are often suppressed. Heifers in good condition and gaining weight at breeding time generally show more definite signs of estrus and have improved conception rates over heifers in poor condition and/or losing weight. Over-conditioned heifers require more services per conception than heifers of normal size and weight. Chapter 17, Table 17–3, shows normal heart girth measurement and weight of calves and heifers during the growing period.

■ **Two months before calving to calving**—The feeding of heifers during this period can affect milk production during the first lactation. During the last 2 months of gestation, heifers should make daily gains of about 2 lb per day, in comparison with 1.7 lb during early pregnancy. Heifers that are growing rapidly at calving time, and continuing to grow during the first lactation, are more persistent milkers than full-size heifers at calving.

The amount of grain to feed before calving will depend on forage quality and size and condition of the heifer. A good thumb rule is to feed grain at 1% of body weight starting about 6 weeks before calving. The ration should have adequate protein, minerals, and vitamins. Excess salt intake can contribute to udder edema and should be avoided the last 2 weeks before calving.

Well-grown heifers will have a minimum of problems at calving time. But plane of nutrition can affect ease of calving in two ways: (1) calf size, and (2) fatness of the dam. Fat heifers have a higher incidence of dystocia because of small pelvic openings and usually a larger than normal sized calf at birth. Underfed or poorly grown heifers will require more assistance at calving and have a higher death rate at calving than normal sized heifers.

FEEDING DAIRY BULLS

Bull calves raised for breeding purposes should be fed and handled much the same as heifers. But, since they grow slightly faster than heifers, they should receive somewhat more feed than heifers of the same age.

Older bulls should be kept in thrifty, vigorous condition, but they should not be permitted to become too fat. Mature bulls can be fed the same grain ration as the lactating cows. Depending on the quality of the roughage, usually about ½ lb of grain per 100 lb of body weight will suffice for the mature bull. Also, individual differences must be considered, for some bulls are easier keepers than others.

Fig. 12–22. A mature Guernsey bull in thrifty, vigorous condition. (Courtesy, American Guernsey Assn., Reynoldsburg, OH)

DAIRY BEEF

Dairy beef is beef derived from cattle of dairy breeding. Dairy heifers and steers finished at market weights comparable to the finished market weights of heifers and steers of beef breeding are commonly referred to as dairy beef. Generally speaking, there are two different finishing programs and market weights for dairy beef:

1. **High-energy ration; light market weights.** These animals are full fed a high-concentrate ration from about 300 lb to market weights of 750 to 950 lb.

2. **High-roughage; heavy market weights.** These animals are grown on a maximum roughage to 600- to 750-lb weight, following which the proportion of concentrate is increased; and they are marketed at weights of 1,150 to 1,400 lb.

FEEDING SHOW AND SALE ANIMALS

Dairy animals intended for show or sale should be fed so as to achieve a certain amount of finish or bloom, but they should not be too fat. Linseed meal, beet pulp, oats, barley, molasses, and wheat bran are popular feeds in a fitting and showing ration. Linseed meal, in particular, is used to impart a healthy bloom or shine to the hair. Likewise, good quality forages are always very important.

Fig. 12–23. Spunky Blevins, age 9, Mesa, Arizona, showing a registered Holstein senior heifer calf at the Arizona State Fair. (Courtesy, Arizona Dairy Co., Higley, AZ)

In fitting show and sale animals, it is important that the grain mixture be palatable and light, and that they not go off feed. Feeds are sometimes soaked before feeding to make them more palatable.

NUTRITIONAL DISORDERS OF DAIRY COWS

Milk production is a very intensive and demanding form of production, accompanied by much stress. Therefore, certain health problems are associated with it; among them, acidosis, bloat, displaced abomasum, fat cow syndrome, grass tetany, hardware disease, ketosis, mastitis, milk fever, retained placenta, and udder edema.

ACIDOSIS (LACTIC ACIDOSIS)

Acidosis is a metabolic disease of cattle.

When ruminants are switched from a high-roughage ration to a high-grain ration too rapidly, ruminal acidosis and atony may occur. The microflora of the rumen do not have time to adapt to the new ration, resulting in serious digestive disturbances. The pH of the rumen contents falls, and lactic acid production increases dramatically. Many of the microorganisms cannot live in this environment resulting in radical changes in the bacterial and protozoal populations. Rumen motility, then, becomes static. Afflicted animals show signs of weakness, diarrhea, and abdominal discomfort, and in many cases die.

Prevention consists in starting animals on a high-roughage ration and gradually reducing the roughage and increasing the grain; avoiding erratic feeding; and avoiding abrupt ration changes. Also, the addition of buffers to a high-grain ration aids in the prevention of acidosis.

Various treatments have been used with different degrees of success. Perhaps the most successful treatment consists in decreasing both the amount and kinds of feeds fed (lessening the total amount of the ration), then returning to a higher forage mix.

BLOAT

Bloat affects all ruminants, dairy cattle included. When dairy animals are first introduced to concentrates or lush legume pastures, bloat problems arise. Normally, the eructation process allows for the expulsion of gases that are produced in the rumen. In cases of bloat, frothing (the trapping of gas in the ingesta) prevents this process and intraruminal pressure builds, thereupon distending the left side of the abdomen until the animal is thrown off feed and goes down due to pain and the buildup of toxic metabolites.

Puncturing of the rumen of bloated cattle should be a last resort.

DISPLACED ABOMASUM

In recent years, dairy producers have encountered increasing numbers of cows with displaced abomasums. The practice of feeding dry cows liberal amounts of grain has been pointed to as one contributing factor. It has been suggested that the lack of bulk and rumen fill promotes flabby muscle tone of the rumen, thereby permitting the abomasum to become displaced. The feeding of some effective roughage is recommended.

FAT COW SYNDROME (FATTY LIVER SYNDROME)

Cows afflicted with fat cow syndrome have lowered levels of liver function due to enlarged livers infiltrated with fat. General herd signs include very fat cows in the dry cow group, decreased resistance to infection, increased incidence of metabolic diseases such as ketosis, reduced feed intake, and reduced milk production and body weight.

Normally, the amount of fat in the liver is quite low (*i.e.*, 1 to 2%), although it may increase to 4 to 10% precalving. However, cows with the fat cow syndrome may have more than 20% fat accumulation in the liver.

The incidence of the fat cow syndrome may be reduced by avoiding overconditioning of cows during late lactation and the dry period and by formulating rations that maximize feed intake after calving.

GRASS TETANY (HYPOMAGNESEMIA)

Grass tetany is a metabolic condition that affects cows (especially lactating cows) grazing lush pasture high in nitrogen, resulting in low absorption of magnesium, and is most common in the spring. Afflicted animals develop tetany, walk with a stiff gait, go into convulsions, and may die. During the danger period, cows grazing lush pasture should be supplemented with magnesium oxide.

HARDWARE DISEASE

The condition in which the collection of foreign objects irritates or punctures the reticulum is called *reticulitis* or *hardware disease*. Ruminants are grazers by nature and are sometimes indiscriminate in their selection of feed. Often, they will consume nails, pieces of wire, and other foreign objects. Due to the motility patterns of the gastric region, these objects tend to accumulate in the reticulum; and the presence of these sharp objects can pose serious problems, especially if the reticulum should be punctured.

KETOSIS

Ketosis is a metabolic disease characterized by a drop in milk production, hypoglycemia, ketonuria, and a rapid loss of weight. In general, the disorder develops within the first 30 days of lactation. While no preventative measures have proven to be 100% effective, the feeding of propylene glycol or sodium propionate has been successful in some cases; and the inclusion of niacin in the ration at a level of 6 g per head per day in the last 2 weeks of the dry period and in the fresh cow ration may aid in reducing the incidence of ketosis. Additionally, starting limited grain feeding during the latter part of the dry period is helpful in preventing ketosis.

If a cow comes down with ketosis, a glucose solution can be administered intravenously to promote rapid recovery.

MASTITIS

Mastitis is an infection of the mammary gland caused by any one of several bacterial organisms, most frequently *staphylococcus* or *streptococcus*. Symptoms vary with degree of inflammation. Acute cases show a swollen and painful udder, and frequently cause the cow to go off feed. Chronic cases result in slightly swollen udders and small flakes in the milk.

No feed is known to cause mastitis. However, the sudden addition of nutrients may result in a marked increase in milk production and cause more stress; in turn, this may cause subclinical cases. Also, feeding recommended levels of selenium and vitamin E may be helpful in preventing mastitis.

MILK FEVER

At or soon after calving (generally within 48 to 72 hours), a sharp decrease in blood calcium (hypocalcemia) occurs in some cows, resulting in loss of appetite, subnormal temperature, and an unsteady gait. This is followed by nervousness, and, finally, collapse or complete loss of consciousness. The head is usually turned back. The name *milk fever* is a misnomer, because the body temperature is below normal.

The triggering mechanism for this drop in blood calcium is the onset of lactation—an intensive mobilization of calcium.

Feeding practices involving dry cows can markedly reduce the incidence of milk fever. When certain cows are known to have a history of milk fever, excessive calcium intake during the dry period should be avoided. If the problem persists, some nutritionists recommend the limited feeding of a high-energy, low-calcium (less than 15 g of calcium per day) ration. After calving, the calcium levels should be raised rapidly to meet the high requirements of lactation.

Recent studies have indicated that the addition of certain anions (negatively charged ions) to the ration may reduce the incidence of milk fever by aiding calcium absorption and mobilization. But more experimental work is needed relative to this method.

RETAINED PLACENTA

Normally, the placenta is expelled within 3 to 6 hours after parturition. If it is retained as long as 12 hours after calving, competent assistance should be obtained.

Retained placenta occurs in about 10% of dairy cattle. It is more common following abnormally short or abnormally long pregnancies, among older cows, and following twinning. Experimentally, it has been found that a high incidence of retained afterbirth occurs when premature calving is induced by the administration of glucocorticoid drugs.

While infections such as brucellosis, vibriosis, and others have been associated with abortion and retained afterbirth, these are by no means the only causes. Its incidence increases with parturient hypocalcemia and appears to be related to the fat cow syndrome. Nutritionally, deficiencies of vitamin A, selenium, copper, and iodine have been incriminated. The prepartum injection of selenium at low doses has been shown to reduce the incidence of retained placenta. Also, it appears that fewer cases of retained placenta occur (1) when calves stay with their dams and nurse for 12 to 24 hours, and (2) when cows are kept on pasture year-round. Among cows which have previously retained the placenta, 20% are likely to do so again.

Calves born when the placenta is retained are likely to be weak. A retained placenta may cause pathological conditions resulting in uterine tissue destruction. This condition may or may not affect milk production, but it very likely will result in 5 to 10% lower fertility than for normal cows.

When a retained placenta is encountered, appropriate treatment should be administered by the veterinarian. The usual treatment consists of either antibiotics or sulfonamides, by direct infusion into the uterus or by other routes (or both).

It is seldom advisable to attempt removal of retained placenta. If the membranes are dragging on the ground, they should be cut off at the hocks. But never, never tie bricks or other objects to it. In most instances, the membranes will fall off by themselves in 1 to 2 weeks.

It is desirable to have all cows which have had retained afterbirth examined at about 30 days after calving. If pus is present, they may be treated with estrogenic hormones to induce heat and then the uterus can be infused with Lugol's (iodine) solution. Such examination and treatment may save considerable time with regard to the onset of normal cycles and may result in a higher conception rate.

UDDER EDEMA

Udder edema, characterized by excessive accumulation of fluid in the intercellular spaces of the udder and forward of it, is sometimes of serious magnitude before calving. The cause is not well understood, but a reduction of blood proteins at calving time and increased blood flow without compensatory lymph removal have been suggested. It appears that high intakes of sodium chloride or potassium chloride increase the severity of udder edema, and that restriction of the salt intake will reduce the severity.

Severe edema may reduce milk production and may be one of the causes of pendulous udder.

DRUGS AND PESTICIDES

Many drugs used in the treatment of cattle diseases, along with many pesticides, are excreted in milk. Such milk should be discarded to prevent the drugs from entering the human food supply. The presence of antibiotics, sulfas, and pesticides in milk is illegal. Dairy producers should follow a residue avoidance program.

TOXIC MATERIALS

Cattle are sometimes poisoned by toxic materials, especially lead, mercury, and mycotoxins. Hence, a summary relative to each of these potentially toxic materials follows:

■ **Lead**—Young calves are extremely susceptible to lead poisoning.

Important sources of toxic amounts of lead are lead-based paint (from discarded paint cans or peeling paint), used motor oil, discarded oil filters, storage batteries, and certain types of grease and linoleum.

Acute lead-toxicity symptoms include dullness, loss of appetite, abdominal pain, and constipation sometimes followed by diarrhea. In advanced stages, which may occur 2 to 3 days after a lethal dose, symptoms may include bellowing, staggering, snapping of eyelids, muscular twitching, frothing at the mouth, and convulsive seizures.

The diagnosis of lead poisoning in cattle is best made by simultaneously considering the clinical symptoms, the presence of a lead source, the lead content of the blood and feces, or the lead content of the soft tissues (especially the kidney cortex) if the animal dies. Blood levels may remain above normal (0.13 ppm Pb) for many weeks following a single oral dose. Kidney cortex values of 22.7 mg lead per lb (*50 mg/kg*) indicate death from lead poisoning.

If damage to tissue has been extensive, treatment is of little value.

■ **Mercury**—Mercury poisoning sometimes occurs in cattle. Seed grain, which has been treated with an organic mercury fungicide, is the most common source. Other sources are from overdosing of mercury-containing medicines, and from excessive absorption of liberally applied mercury skin ointments.

Toxicity and metabolism of mercury are greatly affected by the chemical form. In cattle, the organic mercury compounds (especially methyl mercury) are more toxic than the inorganic compounds. The kidneys and liver are the main deposition sites of both chemical forms; however, appreciable amounts of methyl mercury are found in the muscle and brain. Limited data with ruminants indicate that relatively small amounts of organic and inorganic mercury are secreted in the milk.

Different toxicity symptoms are produced from ingesting organic and inorganic mercury forms. Inorganic mercury compounds are very caustic. Their caustic action on the mucosal membranes of the alimentary tract results in a rapid development of gastroenteritis. Chronic mercurealism, in which small amounts of inorganic mercury are ingested over a longer period, is manifested in cattle by depression, loss of appetite, emaciation, and a stiff, stilted way that may progress to paresis. Alopecia, scabby lesions around the anus and vulva, pruritus, tenderness of gums and teeth, shedding of teeth, and chronic diarrhea are common symptoms in later development stages.

The alkyl mercuries (organic compounds) primarily affect the nervous system. Calves may appear normal for several days during daily ingestion of methyl mercury, then there is sudden and rapid onset of clinical symptoms, including ataxia, neuromuscular incoordination, headpressing, twitching of eyelids, tetanuslike spasms on stimulation, excessive salivation, recumbency, and inability to eat or drink. These are followed by tonic-clonic convulsions, with opisthotonos and death.

There is a dearth of data pertaining to the toxic dose of mercury to calves.

Treatment of mercury poisoning is not very satisfactory.

■ **Mycotoxins**—Dairy cattle, especially young calves, are susceptible to poisoning from the toxin produced by certain molds—especially aflatoxin, although it is not the only mycotoxin to be feared. Generally speaking, mature cattle appear to tolerate higher levels of mycotoxins and for longer periods of intake than simple-stomached animals. However, in addition to the effect of mycotoxins on the animal's health, milk may be contaminated by the residues of mycotoxins, or their metabolic products. Contaminated milk is of concern because aflatoxin has been clearly shown to be a carcinogen (tumor-producing).

Important cattle sources of aflatoxin (most studied of the group) are silage, corn and most other cereals, hay, and grasses. The mold can produce toxins on virtually any feed that will support life.

The prime cause of aflatoxin is moisture. Hence, proper harvesting, drying, and storage are important factors lessening contamination and toxin production. Propionic and acetic acids, and sodium propionate, will inhibit mold growth; hence, their use in preserving high-moisture grains is encouraged. Also, ultraviolet irradiation and anhydrous ammonia under pressure will reduce the toxicity of aflatoxins and, if continued long enough, will deactivate them entirely.

Molds affect cattle in a variety of ways, from decreased production to sudden death. Usually the first sign is loss of appetite and weight. A few animals will abort. With high intakes of mycotoxins, any one or a combination of the following symptoms may develop: liver damage, hyperkeratosis, a typical interstitial pneumonia, blood slimy scours, arched back, dry gangrene at the end of the tail or top of the hoof, hemorrhagic hepatitis, renal damage, lameness, and/or swollen legs.

When trouble strikes, the source of the mold should be removed immediately. Animals suffering from molds frequently respond to vitamin B injections. Iron therapy may be helpful, since hemorrhaging is a frequent problem.

Additional pertinent information relative to aflatoxin, most studied of the mycotoxins, follows:

1. Food and Drug Administration regulations do not per-

mit grain or feed containing more than 20 ppb of aflatoxin to be fed to animals.

2. The toxic symptoms from continued, long-term intake of aflatoxin by dairy cattle, and at selected levels, follow:

Age and Status	Aflatoxin Level (ppb)	Symptoms
Calves (milk fed)	200	Fatal
Lactating cows	20	Drop in milk yield; aflatoxin secreted in milk

QUESTIONS FOR STUDY AND DISCUSSION

1. How and why does the feeding of lactating cows differ from the feeding of other classes of farm animals?

2. Describe briefly each of the following methods of evaluating dairy feeds: (a) physical evaluation, (b) proximate analysis, (c) Van Soest analysis, (d) bomb calorimetry, and (e) biological analysis.

3. What major factors should be considered in determining whether dairy feeds shall be homegrown or purchased?

4. In addition to price of feed, list a minimum of eight other important things that a sophisticated dairy producer should know when buying feeds.

5. Discuss the importance of each of the following feed factors in dairy nutrition: palatability, variety, bulk, laxativeness, cost, and effect on milk flavor.

6. Explain how the cost per pound of protein and of TDN may be used to determine the best buy in feeds.

7. Of what value is a feed substitution table, like Table 12–1, to a dairy producer?

8. What are commercial feeds? What factors should a dairy producer consider when deciding whether to home-mix feeds or buy commercial feeds?

9. What are feeding standards? Of what value are the National Research Council (NRC) requirements? Why should one provide for added margins of safety when using them?

10. Why is dry matter intake of such practical importance for high-producing cows?

11. Three methods of balancing rations are presented: (a) square (Pearson square) method, (b) trial-and-error method, and (c) computer-formulated rations/diskette. Describe each method.

12. How should (a) grain and (b) forage be prepared for dairy cattle?

13. Under what sort of stress is the lactating cow placed? Discuss the importance of monitoring dry matter intake in dairy cows.

14. What general thumb rules should be followed when feeding (a) forages, and (b) concentrates?

15. What factors determine the amount of grain that it will pay to feed milking cows?

16. Why should a dairy producer be well informed relative to the milk-feed price relationship?

17. Discuss the economics of an all-forage ration versus a forage and concentrate ration for lactating cows.

18. Is there need for, or a better alternative to, Table 12–14 (or a similar table) for use in balancing lactating cow rations by producers who (a) home-mix their dairy rations, (b) are without chemical analyses of feed ingredients, (c) are without access to a computer, and (d) do not have processing and mixing equipment with which to combine their roughage(s) with their concentrates to make a complete ration?

19. Define phase feeding; and discuss the four lactation phases with corresponding changes in milk production, milk fat percentage, dry matter intake, and body weight.

20. Describe challenge feeding. Discuss its advantages and disadvantages.

21. Study the corral (group) feeding program presented in this chapter under the section headed "Corral (Group) Feeding," and analyze it from the following standpoints: (a) the number and types of groups; (b) minimizing the behavioral adaptation in moving animals to a new production group; (c) milking parlor vs corral feeding, or a combination of both; (d) the use of a complete feed vs feeding the concentrates and roughages separately; (e) the disadvantages of group feeding; (f) the use of automatic grain feeders; (g) the group feeding nutrient specification guidelines and example rations presented in Tables 12–17 to 12–20.

22. When on pasture, how does a lactating cow divide her time between harvesting and ruminating?

23. What two primary reasons have caused many dairy producers to discontinue pasture grazing of lactating cows?

24. Outline the routine for feeding and managing dry cows.

25. Outline and discuss a feeding program for young dairy calves, including colostrum, a starter ration, hay or silage, method of feeding, frequency of feeding, and age of weaning.

26. Why is it important that young calves get a good feeding of colostrum as soon after birth as possible?

27. What are veal calves? How are they fed? Why do consumers prefer grayish pink veal?

28. List recommended practices to lessen the incidence of calf scours.

29. Discuss the feeding of replacement heifers. What is the relationship of growth of dairy heifers to time of breeding?

30. Discuss the feeding of dairy bulls, dairy beef, and show and sale animals.

31. Discuss the cause, prevention, and treatment of each of the following health disorders of dairy cows: acidosis, bloat, displaced abomasum, fat cow syndrome, grass tetany, hardware disease, ketosis, mastitis, milk fever, retained placenta, and udder edema.

32. Why should dairy producers follow a drug and pesticide residue avoidance program?

SELECTED REFERENCES

Title of Publication	Author(s)	Publisher
Animal Nutrition	L. A. Maynard, *et al.*	McGraw-Hill Book Company, New York, NY, 1979
Applied Animal Feeding and Nutrition, Third Edition	M. H. Jurgens	Kendall/Hunt Publishing, Dubuque, IA, 1974
Basic Animal Nutrition and Feeding	D. C. Church W. G. Pond	O & B Books, Corvallis OR, 1974
Dairy Cattle: Principles, Practices, Problems, Profits, Third Edition	D. L. Bath, *et al.*	Lea & Febiger, Philadelphia, PA, 1985
Dairy Farm Management	T. Quinn	Delmar Publishers, Inc., Albany, NY, 1980
Digestive Physiology and Nutrition of Ruminants	D. C. Church	O & B Books, Corvallis, OR, 1974
Energy Metabolism of Ruminants, The	K. L. Baxter	Hutchinson & Co., Ltd., London, England, 1962
Feed Formulations, Third Edition	T. W. Perry	The Interstate Printers & Publishers, Inc., Danville, IL, 1982
Feeds and Feeding, Third Edition	A. Cullison	Reston Publishing Company, Inc., Reston, VA, 1982
Feeds and Feeding, Twenty-second Edition	F. B. Morrison	Morrison Publishing Co., Ithaca, NY, 1956
Feeds & Nutrition, Second Edition	M. E. Ensminger J. E. Oldfield W. W. Heinemann	The Ensminger Publishing Company, Clovis, CA, 1990
Feeds & Nutrition Digest	M. E. Ensminger J. E. Oldfield W. W. Heinemann	The Ensminger Publishing Company, Clovis, CA, 1990
Illinois-Iowa Dairy Handbook	Cooperative Extension Service	University of Illinois, Urbana, and Iowa State University, Ames, 1983
Large Dairy Herd Management	C. J. Wilcox, *et al.*	University Press of Florida, 1978
Livestock Feeds & Feeding, Second Edition	D. C. Church	O & B Books, Inc., Corvallis, OR, 1984
Nutrient Requirements of Dairy Cattle, Sixth Revised Edition	NRC, R. W. Henken, Chairman, Subcommittee, Dairy Cattle Nutrition	National Academy Press, Washington, DC, 1988, Update 1989
Principles and Practices of Feeding Dairy Cows	W. H. Broster, *et al.*	College of Estate Management, Reading University, England, 1986
Science of Providing Milk for Man, The	J. R. Campbell R. T. Marshall	McGraw-Hill Book Company, New York, NY, 1975
Stockman's Handbook, The, Seventh Edition	M. E. Ensminger	Interstate Publishers, Inc., 1992
United States-Canadian Tables of Feed Composition	J. H. Conrad, Chairman of Subcommittee	National Academy Press, Washington, DC, 1982

Fig. 13–1. Don't do it in the stream! Control pollution. (Courtesy, Soil Conservation Service, Washington, DC)

Contents

CHAPTER

13

DAIRY CATTLE BEHAVIOR AND ENVIRONMENT

A high-producing milk cow is a composite of unique behavioral traits blended with favorable environment and skillful management. A high yielder may be further described as having a ravenous appetite and aggressive eating habits; possessing an excellent temperament—she's neither bossy nor submissive, and she accepts zero flight (distance between cows); and maintaining a close relationship with her caretakers—she readily approaches good handlers. This behavioral characterization is exemplified by the world record milk champion, *Beecher Arlinda Ellen*, the great Holstein-Friesian cow which, in 1975, completed a 365-day production record of 55,661 lb of milk in twice-a-day milking in a milking parlor, with a peak of 194.5 lb, or 22.6 gal, per day. Upon monitoring *Ellen's* behavior, it was found that, during a typical 24-hour period, she spent 6 hours and 15 minutes eating and drinking, with the kinds and frequencies of consumption as follows: hay, 13 times; grain, 12 times; straw, 2 times; salt, 5 times; and water, 7 times. She spent 13 hours and 55 minutes lying down and resting, with about 2 hours more on her right side than on her left; and she spent an additional 30 minutes with her eyes closed (but cattle do not sleep in the ordinary sense of the word). The remaining 4 hours were devoted to ruminating while standing, eliminating, milking, self-grooming, interacting with calves in an adjacent pen and other cows outside her pen, and idling. To the ten-member Harold Beecher family of Rochester, Indiana, Beecher Arlinda Ellen was a personality and affectionately known as "Ellie."

A knowledge of cattle behavior—*cow sense*—is necessary in order to manage and handle dairy animals successfully.

This chapter is presented for the purpose of bridging the gap between the principles and application of dairy cattle behavior and environment.

Fig. 13–2. Contented cows! This requisite for high production represents a composite of unique behavioral characteristics, favorable environment, and skillful management. (Courtesy, The Brown Swiss Cattle Breeders' Assn., Beloit, WI)

ANIMAL BEHAVIOR

Animal behavior is the reaction of animals to certain stimuli, or the manner in which they react to their environment. Through the years, dairy cattle behavior has received less attention than the quantity and quality milk produced. But modern breeding, feeding, and management have brought renewed interest in behavior, especially as a factor in obtaining maximum production and efficiency. With the restriction, or confinement, of herds, many abnormal behaviors have evolved to plague those who raise them, including loss of appetite, pica, stereotyped movements, poor parental care, overaggressiveness, dullness, degenerate sexual behavior, and a host of other behavioral disorders. This has been due to a genetic time lag; dairy producers have altered the environment faster than the genetic makeup of animals.

This chapter is for the purpose of presenting some of the principles and applications of dairy cattle behavior.

CAUSES OF ANIMAL BEHAVIOR

Animal behavior is caused by, or is the result of, three forces: (1) genetic, (2) simple learning (training and experience), and (3) complex learning (intelligence).

GENETIC

We need to breed dairy cattle adapted to artificial environments. Confinement has not only limited space, but it has interfered with the habitat and social organization to which, through thousands of years of evolution, cattle became adapted and best suited. So, producers need to concern themselves more with the natural habitat of animals. Nature ordained that they do more than eat, sleep, and reproduce. We need to change dairy cattle through heredity and selection, as has been done in Israel, where they have selected intensively for docility for 35 years, with the result that the cows literally touch each other, with no antagonistic—or dominant—type response. (See subsequent section in this chapter headed "Breeding for Adaptation.")

SIMPLE LEARNING (TRAINING AND EXPERIENCE)

In general, the behavior of animals depends upon the particular reaction patterns with which they were born. These are called instincts and reflexes. They are unlearned forms of behavior.

Cattle learn by experience. However, the training is only as effective as the inherited neural pathways will permit. Several types of learning processes are known; among them are those that follow:

HABITUATION

Habituation is getting used to, or ignoring, certain stimuli.

Bunk breaking calves is an example. If calves 6 to 7 months of age are weaned without prior bunk breaking, then suddenly transferred to a corral where there is no milk or grass, and where their feed must be obtained from a bunk, it is a traumatic experience for them. This is so because (1) there is a mother-young separation reaction, (2) they get homesick (and animals do get homesick), and (3) there is a change in feed and water. On the other hand, if they have been bunk broken prior to weaning, they take to the new feed bunk in the feedlot because they are used to it.

CONDITIONING (OPERANT CONDITIONING)

Conditioning is the type of learning in which the animal responds to a certain stimulus. For example, upon hearing the rolling of a barn door a cow may lick her tongue and moo, even though she can see no feed; and upon hearing the rattle of a milk pail, she may let down her milk.

Artificial insemination techniques have been developed around the understanding of normal reproductive behaviors and the modifications of these behaviors. Semen collection routines are faced by behavioral responses that can change from impotence to optimum performance and high-quality semen. Proper stimulation of some bulls, for example, can increase sperm cell output by nearly 40%, compared to ejaculates after minimum stimulation.

Another example of conditioning is the use of an electric fence. When an electric fence is installed, the immediate instinct of cattle is to investigate—to touch it with their noses. Upon receiving a shock, they back off and let it alone. Thereafter, the electricity can be shut off for a considerable period of time before some animal again tests it.

Operant conditioning, or operant learning, refers to animal operation of some aspect of the environment to obtain access to feed or other animals. It is the learning of an act that has some consequence; i.e., one that operates the environment—like pressing a bar that supplies some feed or turns off a light.

Broadly speaking, training is operant conditioning—it is an attempt to modify an animal's behavior. There are two types of training: (1) reinforced training, usually with positive rewards, and (2) forced training in which the animal is compelled to do certain things.

INSIGHT LEARNING (REASONING)

Insight learning is the sudden adaptive reorganization of experience or sudden production of a new adaptive response not arrived at by overt trial-and-error behavior. It replaces trial-and-error. Of course, it is difficult to be certain in such cases that the animal did not have a similar type of problem before. Even so, the immediate application of past experience to a new situation is a noteworthy capacity.

The most important single factor to remember in training animals is that none of them (dogs included) can reason things out. An animal's mind functions by intuition, not logic. Moreover, it has no conscious sense of right and wrong. Thus, it is one of the trainer's tasks to teach an animal the difference between right and wrong—between good and bad. Although the animal cannot utilize pure reason, it can remember, and it has the ability to use the memory of one situation as it applies to another.

IMPRINTING (SOCIALIZATION)

This is a form of early social learning which has been observed in some species. The pioneering work in this field was done by the Austrian zoologist Lorenz, with goslings. He found that if a baby gosling was exposed immediately after hatching to some moving object, especially if the object emitted sound, it would adopt that object as its parent-companion. Further studies revealed that goslings would adopt any other moving object in the same manner—dogs, cats, humans, and so forth. Also, it was found that the same principle applies to other fowl and to mammals.

Apparently, inheritance controls the time and the length of the critical period when an individual can be imprinted, the type of object to which it can be imprinted, the tendency to respond to the first object to which it is exposed, and the permanence of the attachment to the object following imprinting.

MEMORY

Memory is the ability to remember or keep in mind; the capacity to retain or recall that which is learned or experienced.

The existence of dominance in cattle is evidence that cattle do remember (recognize) each other; otherwise, bunting and hooking would be promiscuous and continue without end.

COMPLEX LEARNING (INTELLIGENCE)

Complex learning (intelligence) is the capacity to acquire and apply knowledge—the ability to learn from experience and to solve problems. It is the ability to solve complex problems by something more than simple trial-and-error, habit, or stimulus-response modifications. In humans, we recognize this capacity as the ability to develop concepts, to behave according to general principles, and to put together elements from past experience into a new organization.

Animals learn to do some things, whereas they inherit the ability to do others. The latter is often called *instinct*.

Generally speaking, behavioral scientists are agreed that each species has its own special abilities and capacities, and that it should only be tested on these. For example, the dog, pig, and rat are more adept at solving a maze test than cattle. Hence, solving a maze in order to find food favors the scavengers (and the dog, the pig, and the rat are all scavengers) — they have connived for their food since the beginning of time.

Thus, each species is uniquely adapted to only one ecological niche. Moreover, a niche is filled by the particular species that can solve food finding therein, and that is best adapted under the conditions that prevail. It follows that intelligence comparisons between species are not meaningful, and that it is absurd to say that one species is smarter than another.

HOW CATTLE BEHAVE-BEHAVIORAL SYSTEMS

Animals behave differently, according to species. Also, some behavioral systems or patterns are better developed in certain species than in others. Ingestive and sexual behavior systems have been most extensively studied because of their importance commercially. Nevertheless, most cattle exhibit the following nine general functions or behavioral systems:

1. Agonistic behavior (combat).
2. Allelomimetic behavior.
3. Care-giving and care-seeking (mother-young) behavior.
4. Eliminative behavior.
5. Gregarious behavior.
6. Ingestive behavior (eating and drinking).
7. Investigative behavior.
8. Sexual behavior.
9. Shelter-seeking behavior.

AGONISTIC BEHAVIOR (COMBAT)

This type of behavior includes fighting, flight (distance between animals), and other related reactions associated with conflict. Among all species of farm mammals, males are more likely to fight than females. Nevertheless, females may exhibit fighting behavior under certain conditions. Castrated males are usually quite passive, which indicates that hormones (especially testosterone) are involved in this type of behavior. Thus, farmers have for centuries used castration as a means of producing docile males, particularly cattle, swine, and horses.

In combat, bulls paw the ground and bellow, followed by putting their heads together and butting.

Although young bulls raised together will seldom fight, a group of bulls may single out one individual and ride him to death, unless he is removed from the group.

Bringing together sexually mature strange bulls almost always results in a fight. Also, it is noteworthy that breeds of cattle differ in their agonistic behavior.

There is the hazard that bulls will be stifled as a result of fighting; hence, conditions that result in combat should be minimized.

High-yielding cows generally have excellent temperaments; and high-producing herds have tame cows, with zero flight distances.

ALLELOMIMETIC BEHAVIOR

Allelomimetic behavior is mutual mimicking behavior. Thus, when one member of a herd of cattle does something, another tends to do the same thing; and because others are doing it, the original individual continues.

In the wild state, this trait was advantageous in detecting the enemy, and in providing protection therefrom. Under domestication, animals are usually protected from predators. Nevertheless, the allelomimetic behavior still has important consequences.

Cows moving across a pasture toward a milking barn often display allelomimetic behavior. One cow starts toward the barn, and the others follow. Since the rest of the herd is following, the first cow proceeds on.

Because of stimulating and competing with each other, there is usually higher per animal feed consumption among a group of calves than by one calf alone.

Fig. 13–3. Allelomimetic behavior. They are competing with each other. (Courtesy, American Feed Industry Assn., Arlington, VA)

CARE-GIVING AND CARE-SEEKING (MOTHER-YOUNG) BEHAVIOR

The care-giving behavior is largely confined to females among domestic animals, where it is usually described as *maternal;* the care-seeking behavior is normal for young animals. This behavior begins shortly after birth and extends until the young are weaned. Care-giving and care-seeking vary widely among different species of farm animals.

Fig. 13–4. Care-giving and care-seeking behavior.

By the time the calf is 2 days old, the mother wanders more extensively, with the calf at her side. Soon, they rejoin the herd.

Recognition between mother and a calf is by smell (olfactory), sight (visual), and sound (auditory). Cows usually sniff their calves after being away for a time, and the calf recognizes its mother's call. The attachment of the mother to her calf is very strong. However, the calf accepts separation with less stress. Calves that are removed from their mothers during the critical period of an hour or so after birth, then resubmitted to their mothers, are frequently rejected.

Fig. 13–5. Baby-sitter! This shows a cow baby-sitting for three other cows (note four calves; her own and three others) that have gone for feed or water. (Courtesy, Dickinson Studio, Calhan, CO)

Nature ordained that cows seek isolation at calving time. So, where possible, they will hide out.

Following birth, the care-giving behavior of the new mother becomes evident almost immediately. She gets up and begins to dry her newborn calf by licking it. Simultaneously, some cows "talk" to their newborn. They may become quite concerned and nervous as their "baby" first attempts to stand, takes a few footsteps—and falters. Aided by its mother's licking and encouraged by her "talking," eventually the calf makes it to its unsteady feet and commences to search for a teat.

A newborn calf cannot see too well, but it can smell, touch, and taste. It associates everything that is good and that cares for it with its mother. This is the beginning of herd instinct.

If on pasture, the new mother usually hides her calf. During the first day or two, the calf sleeps a great deal, while the mother grazes nearby. But a mother takes great pains not to disclose the hiding place of her calf. At intervals, she returns to feed it. If it is necessary for her to leave her calf in order to get water or supplemental feed, she does not tarry much along the way. Frequently, where there are a number of newborn calves, the cows "baby-sit" for each other. Part of the cows will leave for feed or water, but one or two will remain behind and guard all the calves. Then, when the first cows to leave have returned, the "baby-sitters" will take their turn and depart. In this manner, there are older cows with the calves at all times.

When a calf in hiding is approached by a human, it will usually lie as close to the ground as possible, without any movement except for its eyes. If picked up, and if scared, it may bawl (cry) for its mother. If the mother hears the call, she will come running—often ready to fight. Frequently, other cows in the vicinity, especially if they have calves of their own, may join in the response. If the disturbed calf runs away, it will return to the area after the danger has passed.

Sometimes dairy producers desire to change the normal mother-young relationship of a cow to multiple suckling. This can be done provided the calves to be adopted are properly "mothered up." Thus, if four calves are to be raised by a cow (her own and three others), stable bonding can be achieved by applying her birth fluids to the calves being "grafted" and putting them with her within the critical period after the birth of her own calf.

If a calf is stillborn, or dies soon after birth, some cows will leave the place where the fetus lies, never to return. Others may return to their dead calves at frequent intervals over a period of several days, smelling it and mooing gently.

Dairy calves are normally removed from their mothers when they are from 1 to 4 days of age, with the result that the tie between the mothers and offspring is soon severed.

After weaning, the calf looks for care and shelter from the herd. Thus, if an animal is separated from the herd, it is stressed. It may even jump fences because of its strong instinct to rejoin the herd.

ELIMINATIVE BEHAVIOR

In recent years, elimination has become a most important phenomenon, and pollution has become a dirty word. Nevertheless, nature ordained that if animals eat, they must eliminate.

A full understanding of the eliminative behavior will make for improved animal building design and give a big assist in handling manure. Right off, it should be recognized that the eliminative behavior in farm animals tends to follow the general pattern of their wild ancestors; but it can be influenced by the method of management.

Cattle deposit their feces in a random fashion. Although cows can defecate while walking, with the result that their feces are scattered, generally they deposit their "chips" in neat piles. Most cows hump up to urinate, whereas bulls are inclined to stand squarely on all "fours."

GREGARIOUS BEHAVIOR

Gregarious behavior refers to the flocking, or herding, instinct of certain species. It is closely related to allelomimetic behavior. If animals imitate each other, they must stay together. If they stay together as a mobile group, they must use allelomimetic behavior to do so. All such behavior arises out of the process of social attachment.

Cattle tend to roam in groups of various sizes when a large herd is placed on a pasture. However, there is usually considerable space between the members of the herd. Moreover, on close observation it is evident that there are several small groups within a herd, each ranging from three to five head.

INGESTIVE BEHAVIOR (EATING AND DRINKING)

This type of behavior includes eating and drinking; hence, it is characteristic of animals of all species and all ages. It is very important because animals cannot live without feed and water. Moreover, for high production, animals must have aggressive eating habits; they must consume large quantities of feed.

The first ingestive behavior trait, common to all young mammals, is suckling.

Each species has its own particular method of ingesting feed. The natural feeding (grazing) position of cattle is heads down. In this position, they produce more saliva; and saliva aids digestion. When grazing, cattle wrap their tongues around grass, then jerk their heads forward so that the vegetation is cut off by the lower incisor teeth. (There are no upper incisor teeth, only the thick, hard dental pad.) When grazing, cattle also move their heads from side to side. This movement, aided by protuberant eyes and thin legs, gives them a continuous view of their entire surroundings, an essential for wild cattle in an environment containing dangerous predators. It is important that artificial feeding devices and arrangements not depart too far from this natural pattern,

Fig. 13–6. Ingestive behavior. (Courtesy, *The Dairyman*, Corona, CA)

because cows have a built-in antipathy to being more or less blindfolded while eating.

Cattle are ruminants. Thus, they regurgitate their feed and chew it again in a process known as *rumination.*

Rumination is the act of chewing the cud, characteristic of herbivorous animals with split hoofs. It involves regurgitation of ingesta from the reticulo-rumen, swallowing of regurgitated liquids, remastication of the solids accompanied by reinsalivation, and reswallowing of the bolus. Rumination occupies about 8 hours of the cow's time each day. (In addition, the harvesting or grazing time may take another 8 hours. This means that cows may work a 16-hour day.)

Fig. 13–7. Cows spend two-thirds of their time eating—in consuming feed and chewing their cud. (Photo by J. C. Allen and Son, Inc., West Lafayette, IN)

When the cow regurgitates, a soft mass of coarse feed particles, called a bolus, passes from the rumen through the esophagus in a fraction of a second. She chews each bolus for about 1 minute, then swallows the entire mass again. Originally, it was thought that the regrinding which occurred during rechewing helped the digestion by exposing a greater surface area to fiber-digesting microflora. But recent studies indicate that rechewing does not improve digestibility. Instead, rumination has an important effect on the amount of feed the animal can utilize. Feed particle size must be reduced to allow passage of the material from the rumen. It follows that high-quality forages require much less rechewing and pass out of the rumen at a faster rate; hence, they allow a cow to eat more. This concept is very important to the production of milk because a cow will eat only as much coarse material as she can grind up by ruminating.

INVESTIGATIVE BEHAVIOR

All animals are curious and have a tendency to explore their environment. Investigation takes place through seeing, hearing, smelling, tasting, and touching. Whenever an animal is introduced into a new area, its first reaction is to explore it. Experienced producers recognize that it is important to allow animals time for investigation before attempting to work them, either when they are placed in new quarters or when new animals are introduced into the herd.

Fig. 13–8. Investigative behavior. (Courtesy, Holstein-Friesian Assn. of America, Brattleboro, VT)

If they are not afraid, cattle investigate a strange object at close range. They proceed toward it with their ears pointed forward and their eyes focused directly upon it. As they approach the object, they sniff and their nostrils quiver. When they reach the object, sniffing is replaced by licking; and if the object is small and pliable, they may chew it or even swallow it.

Cattle exhibit investigative behavior when placed in a new pasture or in a new barn. As a result, if there is an open gate in a pasture or a hole in the fence, they usually find it, then proceed to explore the new area.

Calves are generally more curious than older cattle. Perhaps this is due to the fact that older animals have seen more objects, with the result that fewer things are new or strange to them.

SEXUAL BEHAVIOR

Reproduction is the first and most important requisite of dairy breeding. Without young being born and born alive, the other economic traits are of academic interest only. Thus, it is important that all those who breed dairy cattle should have a working knowledge of sexual behavior.

Sexual behavior involves courtship and mating. It is largely controlled by hormones, although males that are castrated after reaching sexual maturity (which are known as stags) usually retain considerable sex drive and exhibit sexual behavior. This suggests that psychological, or learned, as well as hormonal factors may be involved in sexual behavior.

Each animal species has a special pattern of sexual behavior. As a result, interspecies matings do not often occur.

Males in most species of farm animals detect females in heat by sight or smell. Also, it is noteworthy that courtship is more intense on pasture than under confinement, and that captivity has the effect of producing many distortions of sexual behavior compared to wild animals.

Today, livestock producers are attempting to control the sex life of animals, by bringing about ovulation at the time of choice of the owner, rather than of the female.

Experienced producers can usually detect in-heat cows through one or more of the following characteristic symptoms: (1) restlessness; (2) mounting other cows, and standing to be mounted by another cow (standing heat appears to be the best single indicator of the proper time to breed); (3) a noticeable swelling of the labia of the vulva; (4) an inflamed appearance about the lips of the vulva; (5) frequent urination; (6) switching and raising the tail; and (7) a mucous discharge. A day or two following estrus, a bloody discharge is sometimes seen. Dry cows and heifers usually show a noticeable swelling or enlargement of the udder during estrus, whereas in lactating cows a rather sharp decrease in milk production is often encountered. When kept alone, some cows become restless, walk the fence, and bawl when they are in heat. Some may even jump the fence, or go through it, as they attempt to find a bull.

A bull can often detect a cow that is coming in heat 24 to 48 hours before she will mate, at which time he will remain in her company. Courtship of the bull consists of following the in-heat cow, licking and smelling the external genitalia, with the head extended horizontally and the lip upcurled, and chin-resting, with the chin and throat resting on the cow's rump.

Fig. 13–9.   Courtship of the bull, showing chin-resting, in which the chin and throat are rested on the cow's rump.

SHELTER-SEEKING BEHAVIOR

Fig. 13–10.   Shelter-seeking behavior. Lactating cows under a shade and evaporative coolers. (Courtesy, J. Tappan, Arizona Dairy Co., Higley, AZ)

All species of animals seek shelter—protection from the sun, wind, rain and snow, insects, and predators.

Cattle are not as sensitive to extremes in temperature—heat and cold—as are swine. Nevertheless, they do seek shelter under natural conditions—this may consist of hills, valleys, timber, and other natural windbreaks; or they may even group closely together.

Cattle seem to be able to sense the coming of a storm, at which time they may race about and "act up." During a severe rain or snowstorm, they turn their rear ends to the storm and tend to drift away from the direction of the wind. By contrast, bison (buffalo) face a storm head on.

During the hot summer months, cattle seek either shade or a waterhole during the heat of the day. Then they graze in the cool of the evening or early morning. There are well-known breed differences in tolerance to heat.

SOCIAL RELATIONSHIPS OF CATTLE

Social behavior may be defined as any behavior caused by or affecting another animal, usually of the same species, but also, in some cases, of another species.

Social organization may be defined as an aggregation of individuals into a fairly well integrated and self-consistent group in which the unity is based upon the interdependence of the separate organisms and upon their responses to one another.

The social structure and infrastructure in the herd are of great practical importance. Some of the ideas on peck order (*bunt,* or *hooking* order) have had to undergo changes as a result of increased understanding of the social organization within the herd.

Older cows generally dominate the younger ones, and the heavier animals (usually the older animals) tend to dominate the lighter, for this reason, 2-year-old heifers should be segregated from older cows. However, among cows of similar age and breed, the more aggressive ones are most dominant. Also, cows with more seniority in the group and cows with horns tend to be of higher social rank. Aggression in cows appears to be ritualized, with most encounters taking place in the following sequence: approach, threat, and physical contact (or fighting).

Limited studies indicate a high relationship between social status of cattle and spacing, or social distance. The higher the social rank, the more likely cows are to be found near other members of the herd. Also, dominant cows tend to allow close approach by other cows more often than subordinates.

When moving from the paddock to the milking parlor, dairy cows travel in a consistent order. Mid-dominant cows tend to be in front of the group. However, the same individuals are seldom consistent leaders; instead, there is a pool of animals which tends to be in or near the lead. More consistency is found in the cows bringing up the rear of the moving herd. So, *rearship* is a more distinctive feature than *leadership.* The animals at the rear are usually the younger subordinate heifers. Also, in most herds there is a definite order in which cows enter the milking parlor; the mid-age, mid-dominant cows tend to be milked first, followed by the older cows. Social dominance orders (called *bunt order* in polled cows and *hook order* in horned cows) become more complex as herd sizes increase. Generally, not more than 100 dairy cows should be group-fed.

Fig. 13–11. When milk cows are moving to and from the milking parlor, bringing up the rear is more distinctive and consistent than leadership. Mid-dominant cows tend to be in front of the string, but their order changes. (Courtesy, USDA)

SOCIAL ORDER (DOMINANCE)

Within most groups of farm animals of the same species, there is a well-organized social rank. In chickens, in which it was first observed, the social rank order is called the *peck order*.

Thus, in most species of farm animals, the alpha animal in the herd or flock will be dominant over all other individuals and the omega animal will be subordinate to all. In between, some animals will be subordinate in some relationships and dominant in others. Moreover, once these relationships are established, they seldom change. The social order is usually important only in females, because mature male animals are seldom run together in groups.

When several cows are brought together to form a herd, there is a substantial period during which there is much butting and threat posturing in order to establish a dominance hierarchy. This is disturbing to a dairy herd and will result in reduced production. Usually the older and larger cows come out at the top of the hierarchy.

Fig. 13–12. Dominance. This shows a dominant cow attacking the neck of a subordinate. The latter submits and avoids a fight.

Once the social rank order is established, it results in a peaceful coexistence of the herd. Thereafter, when the dominant one merely threatens, the subordinate animal submits and avoids conflict. Of course, there are some pairs that fight every time they chance to meet. Also, if strange animals are introduced into such a group, social disorganization results in the outbreak of new fighting, as a new social rank order is established.

Among wild animals, social rank order is nature's way of giving mating priority to the top ranking males. Hence, they leave behind more of their progeny than do the less dominant males. Also, dominance establishes priority in feeding.

Social rank among dairy animals is of little consequence as long as they are on pasture, and if there is plenty of feed and water. But it becomes of very great importance when animals are placed in confinement. When cows are moved into limited quarters, social dominance decrees that replacement heifers be sorted out and fed separately, that young bulls be cared for in separate quarters, and that old cows with poor teeth be fed separately; otherwise, these animals will not get enough feed.

When self-feeders and central water tanks are used, care must be taken relative to providing both adequate space and proper placement; otherwise, submissive animals may find it difficult to get out of eye contact of dominant animals so as to eat and drink in peace.

Of course, social rank becomes doubly important if limited feeding is practiced. Under such circumstances, the dominant individuals crowd the subordinate ones away from the feed bunk, with the result that they may go hungry.

Several factors influence social rank; among them, (1) age—both young animals and those that are senile rank toward the bottom; (2) early experience—once a subordinate in a particular herd, usually always a subordinate; (3) weight and size; and (4) aggressiveness or timidity. Also, it is noteworthy that social rank is influenced by hormones.

In dairy confinement operations, social facilitation is of great practical importance. Dominants should be sorted out, and, if possible, grouped together. Of course, they will fight it out until a new social order is established. In the meantime, both feed efficiency and milk will suffer. But, as a result of removing the dominants, the feed intake of the rest of the animals will be improved, followed by greater feed efficiency, production, and profit. Among the more settled animals, social facilitation will become more evident. After the dominants have been removed, the rest of the animals will settle down into a new hierarchy, but within the limits of their dominance. Their interaction or social facilitation will be far more likely to have a calming effect on this group, to both the economic and practical advantages of the operator.

Dominance and subordination are not inherited as such, for these relations are developed by experience. Rather, the capacity to fight (agonistic behavior) is inherited, and, in turn, this determines dominance and subordination. Hence, when combat has been bred into the herd, such herds never have the same settled appearance and docility that is desired of high-production and intensive animals.

LEADER-FOLLOWER

Leader-follower relationships are important in cattle. The young follow their mothers; hence, they continue to follow their elders.

The leader is the cow that is usually at the head of a moving column and often seems to initiate a new activity. The leader may be small, but she is always intelligent.

Fig. 13–13. Leader-follower social relationship exhibited by a Jersey herd.

Right off, it is important to distinguish leader-follower relationships from dominance; in the latter, the herd is driven, rather than led. After the dominants have been removed from the herd, the leader-follower phenomenon usually becomes more evident. It is well known that the dominant animal is not necessarily the leader; in fact, it is very rarely the leader. It pays too much attention to other matters of dominance in its relationship within the herd, with the result that it does not develop the qualities of leadership.

When a string of cows moves from the pasture into the milking parlor, the dominant animals are generally in the middle of the procession; with the leader in front, and the subordinate ones bringing up the rear.

INTERSPECIES RELATIONSHIPS

Social relationships are normally formed between members of the same species. However, they can be developed between two different species. In domestication this tendency is important (1) because it permits several species to be kept together in the same pasture or corral, and (2) because of the close relationship between caretakers and animals. Such interspecies relationships can be produced artificially, generally by taking advantage of the maternal instinct of females and using them as foster mothers.

All sorts of bizarre interspecies relationships have been arranged—including cows raising pigs.

PEOPLE-ANIMAL RELATIONSHIPS

Social relationships can also be transferred to human beings. As a result, a young calf associates everything good or bad with humans. Unfortunately, this is the period in life during which calves are dehorned, castrated, branded, and

vaccinated; hence, it is no wonder that some cattle are hard to handle. In order to minimize the problem, calves should be worked as little as possible, with all such jobs done at one time.

Good dairy caretakers usually form a care-dependency relationship with the animals under their care, with the result that the cows readily come to them.

HOW CATTLE COMMUNICATE

Communication between cattle involves one individual giving some signal, which, on being received by another, influences its behavior.

SOUND

Sound communication is of special interest because it forms the fundamental basis of human language. The gift of language alone sets people apart from the rest of the animals and gives them enormous advantages in their adaptation to their environment and in their social organization.

Fig. 13–14. Communication by sound—a bellowing bull.

Sound is also an important means of communication among cattle. They use sounds in many ways; among them, (1) feeding, in sounds of hunger (bawling) by young; (2) distress calls like the bellowing of a bull; (3) sexual behavior and related fighting; (4) mother-young interrelations to establish contact and evoke care behavior; and (5) maintaining the group in its movements and assembly.

Cattle have a very acute sense of hearing, perceiving higher and fainter noises than the human ear.

SMELL

Cattle can smell at a greater distance than people. On a day with a 5-mile wind and a humidity of 75%, a cow can smell up to 6 miles away; as wind and humidity increase, she can smell even further.

In cattle, females in estrus secrete a substance that attracts males. Hence, bulls locate cows that are in heat by the sense of smell.

VISUAL DISPLAYS

When several strange cows are brought together, there is much threat posturing, as well as butting, in order to establish a dominance hierarchy. Also, bulls will strike a hostile stance prior to fighting.

Fig. 13–15. Agonistic behavior exhibited by a bull. Pawing the ground and bellowing are generally the first stage of combat in cattle.

Birds are noted for their sexual behavior in the act of courtship. Visual displays during courtship are less evident among cattle, but they do occur to some extent.

HOMING AND ORIENTATION (PATH-FINDING, NAVIGATION, MIGRATION)

Through sound, scent, or some sense of which we do not know, when cattle are moved to distant places, they often find their way back home.

NORMAL CATTLE BEHAVIOR

Dairy producers need to be familiar with behavioral norms of cattle in order to detect and treat abnormal situations — especially illness. Many sicknesses are first suspected because of some change in behavior — loss of appetite (anorexia); listlessness; labored breathing; posture; reluctance or unusual movement; persistent rubbing or licking; and altered social behavior, such as one animal leaving the herd and going off by itself — these are among the useful diagnostic tools.

Also, it is important to know how cattle see and sleep.

A summary of normal cattle behavior follows.

HEALTH

Some of the signs of good health are:

1. Contentment.
2. Alertness.
3. Eating with relish and cudding.

4. Sleek coat and pliable and elastic skin.
5. Bright eyes and pink membranes.
6. Normal feces and urine.
7. Normal temperature, pulse rate, and breathing rate.

Normal rectal temperature
    Average, 101.5°F
    Range, 100.4–102.8°F
Normal pulse rate
    60–70/min.
Normal breathing rate
    10–30/min.

SIGHT

The eyes of most animals are on the side of the head (the cat is an exception). This gives them an orbital, or panoramic, view — to the front, to the side, and to the back — virtually at the same time. Also, this is a rounded, or globular, type of vision. This leads to a different interpretation than that of the binocular type of vision of people.

The wide-set eyes of cattle enable them to have a large panoramic field of vision, even to the extent of seeing everything around them, with slight head movements. Only what is immediately behind their hindquarters is outside their field of view.

A cow does not see in color; she sees in shades of grays and blacks.

If a cow sees movement, her instinct is to escape; hence, movements around cattle should be made very quietly and slowly.

SLEEP

Normal behavior in sleep should be recognized, especially since it differs widely between species.

Cattle typically lie on the stomach or tilt to one side, with the forelimbs folded under the body; one hind limb extends forward, while the other protrudes toward the outside. Although cattle rest in this manner, they do not sleep in the sense that the term usually connotes. While lying down, they do shut their eyes for short periods of time.

Calves commonly spend up to ½ hour at a time with their heads turned back in the flank position.

ABNORMAL CATTLE BEHAVIOR

Abnormal behavior of domestic animals is not fully understood. As with human behavior disorders, more work is needed. However, we have learned from studies of captured wild animals that when the amount and quality, including variability, of the surroundings of an animal are reduced, there is increased probability that abnormal behavior will develop. Also, it is recognized that confinement of animals makes for lack of space; this often leads to unfavorable changes in habitat and social interactions.

Abnormal behavior in animals develops where there is a combination of confinement, excess stimulation, and forced production with a lack of opportunity to adapt to the situation.

Homosexual behavior is common among all species where adult mammals of one sex are confined together.

The *mean bull* complex is an example of abnormal behavior in cattle. Of course, there are inherited differences in the temperaments of cattle. Nevertheless, constant stress can change the temperament of an animal, just as it can in people. Thus, when a bull is kept for hand mating in a corral by which the cow herd passes each day, cows in heat, or coming in heat, stimulate his sexual behavior. Since he cannot respond naturally through coitus, he becomes a mean bull.

Pica (consumption of dirt, hair, bones, or feces) may develop, perhaps due to boredom, nutritional inadequacies, or physiological stress.

Milk cows may kick because they are in pain, they are frightened, or they have been mistreated.

APPLIED CATTLE BEHAVIOR

In the beginning of this chapter, it was stated that this presentation is for the purpose of bridging the gap between the principles and the application of dairy behavior. So, let us next turn to some practical applications of cattle behavior.

BREEDING FOR ADAPTATION

The wide variety of livestock in different parts of the world reflects a continuous process of natural and artificial selection which has resulted in the survival of animals well adapted to climate and other environmental factors. Among the examples are *Bos indicus* (Zebu) types of cattle in tropical areas, and *Bos taurus* cattle in temperate zones. Such adaptations relate to survival of the animals, but they do not necessarily entail maximum productivity of food for people. European cattle usually have much higher yields of milk than the breeds native to Africa or India. It is understandable, therefore, why there have been many attempts to introduce improved European dairy cattle into countries in which the productivity of native stock is low. But there are many problems in breed replacement, with the result that a large number of experimental introductions of new breeds have not been successful. Tropical Africa provides an example. Because of disease problems, poor resistance to high temperatures, and limited feed supplies, many of the attempts made by former colonial powers to improve the output of native stock by replacing them with the European breeds failed. Breed replacement or a crossbreeding system might seem to be a simple panacea for low productivity. However, unless associated with special provisions for subsequent importation of breeding stock and simultaneous improvement of the nutritional, parasitological, disease, and husbandry environments of the crossbreds, it is not likely to succeed.

Another example of breeding and selecting for a behavioral characteristic in dairy cattle pertains to milk stimulation and stripping. When dairy producers first modified the milking technique and went from hand milking to machine milking, they assumed the necessity of hand stripping. In recent years, many producers have given up hand stripping and selected cows capable of producing large quantities of milk by being milked by machine, without either stimulation before milking or stripping afterwards. This has been a selection for a behavioral characteristic—a low threshold to the milking stimulus. As a result, today a very large number of cows, especially within the Holstein-Friesian breed, will let down their milk effectively with no other stimulation than having the milking machine applied.

Cattle can be changed through heredity and selection. For example, in Israel, which has one of the highest average milk yields per cow of any nation in the world, the flight distance between cows approaches contiguousness in some herds, this is due to Israelis having selected intensively for docility for 35 years. In other words, the animals are literally touching each other, with no antagonistic or dominant-type response. This allows them to concentrate their animals even more than they had previously, thereby giving them a higher productivity per unit area. The only problem reported by Israelis is that estrus, or heat, in animals in close proximity is difficult to detect.

QUICK ADAPTATION—EARLY TRAINING

We need to breed and select animals that adapt quickly to an artificial environment—animals that not only survive, but that thrive, under the conditions that people impose upon them.

Also, early training and experience are extremely important. In general, young animals learn more quickly and easily than adults; hence, advance preparation for adult life will pay handsome dividends. The optimum time for such training varies according to species.

Both good and poor patterns of behavior and productive traits come from early experience. For example, feeding group-reared calves from nipple pails (rather than buckets) reduces the vice of suckling each other.

Stress can be reduced or avoided entirely if animals proceed through a graduated sequence of events leading to an otherwise noxious experience. If calves are properly started on feed, vaccinated, treated for parasites prior to weaning, the stress of subsequent weaning and movement to another location is minimized.

MILKING PARLOR DESIGN

Cows will enter milking parlors best when there is a continuous holding pen-milking parlor arrangement with no doors opening or closing.

A frequent fault in the design of many modern milking parlors is a failure to provide facilities that enable the cows about to leave the parlor to see those exiting. For the speedy evacuation of cows, those which have left should be visible to those ready to depart.

MANURE ELIMINATION

Body waste is a major concern; it is expensive, time-consuming, and a major pollution problem. But manure handling can be facilitated by an understanding and application of eliminative behavior.

Cattle are indiscriminate eliminators. Even so, this trait

can be used effectively. For example, if cattle are fed at the same time each day, feed is released from the rumen into the true stomach regularly; and the moment the latter happens, there is a gastro-colic reflex. When this happens and cattle are put under slight stress, they defecate. Knowing this, cattle can be moved to the defecating area at the right time.

COMPANIONSHIP

Companionship in animals is of great practical importance. Except for the cat, all domestic animals are highly social and have constant need for companionship.

If not too crowded, placing cattle together sometimes accomplishes two things: (1) greater feed consumption, due to the competition between them (social facilitation); and (2) a quieting effect.

ANIMAL ENVIRONMENT

Environment may be defined as all the conditions, circumstances, and influences surrounding and affecting the growth, development, and production of animals. The most important influences in the environment are the feed and quarters (space and shelter).

It is becoming increasingly difficult to define environment, because scientists continue to discover important new environmental factors. Primitive people recognized that the sun and fire provided both heat and light, that body heat could be conserved by draping the body with animal skins, and that trees and caves provided protection from the weather. Today, it is recognized that these, along with a host of other environmental factors affect animals and people.

The branch of science concerned with the relation of living things to their environment and to each other is known as ecology.

Dairy producers were not concerned with the effect of environment on animals so long as they grazed on extensive pastures. But rising feed and land costs, along with the concentration of animals into smaller spaces, changed all this.

People achieve environmental control through clothing, vacationing in resort areas, and air-conditioned homes and cars. In animals, environmental control involves space requirements, light, air temperature, relative humidity, air velocity, wet bedding, ammonia buildup, dust, odors, and manure disposal. Control or modification of these factors offers possibilities for improving animal performance. Although there is still much to be learned about environmental control, the gap between awareness and application is becoming smaller. Research on animal environment has lagged, primarily because it requires a melding of several disciplines—nutrition, physiology, genetics, engineering, and climatology. Those engaged in such studies are known as *ecologists*.

In the present era, pollution control is the first and most important requisite in locating a new dairy, or in continuing an old one. The location should be such as to avoid (1) the neighbors complaining about odors, insects, and dust; and (2) pollution of surface and underground water. Without knowledge of animal behavior, or without pollution control, no amount of capital, native intelligence, and sweat will make for a successful dairy enterprise.

HOW ENVIRONMENT AFFECTS DAIRY ANIMALS

Fig. 13–16. Production records are affected by management—by the art of caring for and handling lactating cows. In the dairy operation, management gives point and purpose to everything else.

This shows push-button feeding from silo to herd, and a paved corral with a shelter. Cows can get to the feed in all kinds of weather. (Courtesy, Portland Cement Association, Skokie, IL)

The effect of environment on dairy cattle was demonstrated in a classic experiment in New Zealand. (See Chapter 5, section headed "Records Affected By Environment.")

Among the environmental and other factors affecting milk production records are the following:

1. **Milking practices.** Good milking practices and a properly functioning machine are necessary for high milk production.

2. **Age of animal.** On the average, production increases each year from the time the first calf is born until the cow reaches five to eight years of age, after which it declines.

3. **Size of animals.** Within the same breed and age group, the larger animals, as measured by their capacity to consume more feed, usually produce more milk than the smaller animals.

4. **Season of freshening.** Cows that freshen in the spring and summer months usually produce less than cows freshening in the winter months. Of course, this variation differs among herds and areas.

5. **Calving interval.** Cows that calve within 12 to 14 months from last calving produce more for that lactation than cows calving at shorter intervals. Lifetime production will usually be reduced if calving intervals are longer than 12 to 14 months.

6. **Length of dry period.** Cows with a dry period of 6 to 8 weeks produce more during the following lactation than those cows with a dry period of less than 4 weeks' duration.

7. **Freedom from disease, parasites, and injury.** Any one of these may depress production, with the degree of depression determined by the severity of the ailment.

8. **Rate of maturity.** Some strains or families of cows mature at a slower rate than others.

9. **Yearly differences.** There are important yearly differences within a given area, primarily due to weather conditions and the general quality of feed available. Nevertheless, when an attempt is being made to evaluate the breeding worth of an individual animal, all of these factors may be playing an important part in the production records.

Over the long pull, selection provides a major answer to behavioral problems; we need to breed animals adapted to artificially made environments.

The following environmental factors are of special importance in any discussion of animal behavior-environment:

1. Feed and nutrition.
2. Water.
3. Weather.
4. Facilities.
5. Health.
6. Stress.

FEED

The most important influence in the environment is the feed. Animals may be affected by (1) too little feed, (2) rations that are too low in one or more nutrients, (3) an imbalance between certain nutrients, or (4) objection to the physical form of the ration—for example, it may be ground too finely.

Forced milk production and the feeding of forages and grains which are often produced on leached and depleted soils have created many problems in nutrition. These conditions have been further aggravated through the increased confinement of animals, many animals being confined to stalls or lots all or a large part of the year. Under these unnatural conditions, nutritional diseases and ailments have become increasingly common.

Also, nutritional reproductive failures plague dairy operations. Generally speaking, energy is more important than protein in reproduction. The level and kind of feed before and after parturition will determine how many females will show heat—and conceive. After giving birth, feed requirements increase tremendously because of milk production; hence, a lactating female needs a much greater feed allowance than during the pregnancy period. Otherwise, she will suffer a serious loss in weight, and she may fail to come in heat and conceive.

The following additional feed-environmental factors are pertinent:

1. **Regularity of feeding.** Animals are creatures of habit; hence, they should be fed at regular times each day, by the clock.

2. **Underfeeding.** Too little feed results in slow and stunted growth of young stock; in loss of weight, poor condition, and excessive fatigue of mature animals; and in poor reproduction, failure of some females to show heat, more services per conception, lowered young crop, light birth weights, and lowered milk production.

3. **Overfeeding.** Too much feed is wasteful. Besides, it creates a health hazard; there is usually lowered reproduction in breeding animals, and a higher incidence of digestive disturbances (bloat, founder, and scours)—and even death. Animals that suffer from mild digestive disturbances are commonly referred to as *off feed*.

4. **Deficiency of nutrient(s).** A deficiency of any essential nutrient required by a lactating female will lower milk production and feed efficiency, rather than make for significant changes in the composition of milk.

5. **Some feed ingredients and rations influence milk composition.** Some feeds reduce the fat percentage of milk. Among such feeds are cod-liver oil and other fish oils, certain pasturages (especially lush spring pastures), and pearl millet. Also, fine grinding of forage, too small an amount of roughage, or heated starch will lower the butterfat content of milk. On the other hand, such feeds as whole cottonseed, soybeans, and coconut oil result in a temporary increase in the fat content of milk.

The amounts of fat-soluble vitamins A, D, and E in milk are influenced by the levels of these particular vitamins in the ration, and in the case of vitamin D, exposure to sunlight is a factor also.

WATER

Fig. 13–17. Automatically supplied water tank for lactating cows.

Cattle can survive for a longer period without feed than without water. Water is one of the largest constituents in the animal body, ranging from 40% in very fat, mature cattle to 80% in newborn calves. Deficits or excesses of more than a few percent of the total body water are incompatible with health, and large deficits of about 20% of the body weight lead to death.

The total water requirement of dairy animals varies primarily with the weather (temperature and humidity); feed (kind and amount); age, and weight of animal; and the physiological state. The need for water increases with increased intakes of protein and salt, and with increased milk production of lactating animals. Water quality is also important, especially with respect to the content of salts and toxic compounds.

The water content of feeds ranges from about 10% in air-dry feeds to more than 80% in fresh, green forage. Feeds containing more than 20% water are known as *wet feeds*. The water content of feeds is especially important for animals which do not have ready access to drinking water.

Under practical conditions, the frequency of watering is best determined by the dairy animals, by allowing them access to clean, fresh water at all times.

WEATHER

Weather affects the maintenance requirements of animals. Extreme weather can cause wide fluctuations in dairy animal performance. The difference in weather impact from one year to the next, and between areas of the country, causes difficulty in making a realistic analysis of buildings and management techniques used to reduce weather stress. Weather may be modified by shelters. The research data clearly show that winter shelters and summer shades almost always improve production and feed efficiency. The issue is clouded only because the additional costs incurred by shelters have frequently exceeded the benefits gained by the improved performance, particularly in those areas with less severe weather and climate.

The maintenance requirements of animals increase as temperature, humidity, and air movements depart from the comfort zone. Likewise, the heat loss from animals is affected by these three items.

Cattle eat more just before a storm; possibly nature ordained this for the purpose of providing a reserve so that they may ride out the period of the storm. The ability to recognize the "sign language of animals" and to change the feeding program accordingly characterizes the experienced caretaker.

Butterfat percentage of milk varies with the season, being higher in the fall and winter and lower in the spring and summer. It may vary up and down seasonally by an average of 0.3 to 0.5%. Solids-not-fat also show a seasonable variation, with the low point in the spring and summer. The reasons for these changes are not known; it may be due to temperature and humidity, changes in body weight, or kinds and amounts of feeds.

Severe weather conditions usually decrease the total amount of milk produced and may influence the fat test either up or down. Temperatures above 85°F greatly affect cows, and the situation is accentuated when high temperatures are accompanied by high humidity.

In collaboration with the U.S. Department of Agriculture, Cornell University studied the effect of summer weather on performance of Holstein cows in three stages of lactation. They reported that, for all stages of lactation, 9% of the variation in milk yield, 13% in milk fat, 5% in feed intake, and 65% in rectal temperature were attributable to weather conditions.[1]

It is also noteworthy that cows calving in the fall months consistently produce more than those calving at other times of the year. Cows calving in the spring produce the least. This difference may be as much as 10 to 15%. This phenomenon may be due in part to temperatures; but more than likely, available feeds, including spring pastures to which fall-calving cows respond so well, may be a factor.

FACILITIES

Optimum facility environments can only provide the means for animals to express their full genetic potential of

Fig. 13–19.   Environmentally controlled dairy barn in Wisconsin, well insulated and ventilated (note fan). Also, note (1) automatic feed auger at top of rack for conveying chopped hay and silage, (2) automatic waterer at end of feed rack, and (3) curb along rack to encourage cows to stay near feed bunk while eating. (Courtesy, Babson Bros. Co., Oak Brook, IL)

Fig. 13–18.   Cow showing heat stress—tongue out and panting. (Courtesy, Dr. G. E. Higginbotham, Dairy Farm Advisor, Fresno/Madera Counties, CA)

[1]Maust, L. E., R. E. McDowell, and N. W. Hoover, "Effect of Summer Weather on Performance of Holstein Cows in Three Stages of Lactation," *Journal of Dairy Science*, Vol. 55, No. 8, p. 1133.

production, but they do not compensate for poor management, health problems, or improper rations.

With the shift to confinement structures and high-density production operations, building design and environmental control became more critical. Limited basic research has shown that animals are more efficient—that they produce and perform better, and require less feed—if raised under ideal conditions of temperature, humidity, and ventilation. In a study with lactating cows, researchers at the Florida Station found that controlled environment increased milk yield, fat percentage, fat yield, 4% fat-corrected milk, and conception rates. The primary reason for having livestock buildings, therefore, is to modify the environment. Properly designed barns and other shelters, shades, insulation, ventilation, and air-conditioning can be used to approach the environment that we wish.

However, the per head cost is much higher for environmentally controlled facilities. Thus, the decision on whether or not confinement and environmental control can be justified should be determined by economics. Will the cows in environmentally controlled quarters produce sufficiently more milk on less feed to justify the added cost? Of course, manure disposal and pollution control should also be considered.

There is still much to be learned about environmental control, but the gap between awareness and application is becoming smaller.

Environmentally controlled buildings are costly to construct, but they make for the ultimate in animal comfort, health, and efficiency of feed utilization. Also, like any confinement building, they lend themselves to automation, which results in a saving in labor; and, because of minimizing space requirements, they effect a saving in land cost. If they malfunction, however, they can cause animals to suffocate and result in large economic losses. Today, environmental control is rather common in poultry and swine housing; and it is on the increase with dairy cattle.

In hot climates, increased use is being made of shades, fans, sprinklers/sprayers/foggers, ventilation, and windbreaks.

Before an environmental system can be designed for animals, it is important to know their (1) heat production, (2) vapor production, and (3) space requirements. This information is as pertinent to designing livestock buildings as nutrient requirements are to balancing rations. This information is presented in Chapter 16, Dairy Cattle Buildings and Equipment.

Also, increasing attention needs to be given to other facility-related stress sources, such as space requirements, and the grouping of animals as affected by class, age, size, and sex.

HEALTH

Diseases and parasites (external and internal) are ever present animal environmental factors. Death takes a tremendous toll. Even greater economic losses in dairy cattle result from lowered milk production, retarded growth of young stock, poor feed efficiency, and in labor and drug costs. The signs of good health are summarized earlier in this chapter in the section headed, "Normal Cattle Behavior"; hence, the reader is referred thereto.

Any departure from the signs of good health constitutes a warning of trouble. Most sicknesses are ushered in by one or more signs of poor health—by indicators that tell expert caretakers that all is not well—that tell them that their animals will go off feed tomorrow, and that prompt them to do something about it today.

Among the signs of cattle ill health are lack of appetite (the animal does not eat or graze normally); listlessness; droopy ears; sunken eyes; humped-up appearance; abnormal dung—either very hard or watery dung suggests an upset in the water balance or some intestinal disturbance following infection; abnormal urine (repeated attempts to urinate without success or off-colored urine should be cause for suspicion); abnormal discharges from the nose, mouth, and eyes, or a swelling under the jaw; unusual posture—such as standing with the head down or extreme nervousness; persistent rubbing or licking; hairless spots, dull hair coat, and dry, scurfy, hidebound skin; pale, red, or purple mucous membranes lining the eyes and gums; reluctance to move or unusual movements; higher than normal temperature; labored breathing—increased rate and depth; altered social behavior such as leaving the herd and going off alone; and sudden drop in milk production. Disease affects milk secretion, in both total production and composition, with the degree of the effects determined by the kind and severity of disease. Mastitis will, for example, lower both the total production of milk and its composition.

STRESS

Fig. 13–20. Misting heat-stressed dairy cows from corral manger. (Courtesy, Dr. Tom Schultz, Farm Advisor, Visalia, CA)

Stress is defined as physical or psychological tension or strain. Stress of any kind affects animals. Among the external forces which may stress animals are previous nutrition, abrupt ration changes, change of water, space, level of production, number of animals together, changing quarters or mates, irregular care, transporting, excitement, presence of strangers, fatigue, previous training, illness, management, weaning, temperature, and abrupt weather changes.

Animals experience many periods of stress. For example, high-producing dairy cows are constantly under stress. As a result, they are usually very sensitive to any changes in their environment, even such things as a change in milkers, an unequal interval between milkings, or overmilking.

Studies have shown that the stressed cow will show a dramatic increase in the somatic cell count of milk from certain quarters, normally quarters which have shown infection in the past.

Separating a calf from its mother, followed in rapid succession by weaning, transporting many miles, going through an auction ring, and ending up in a corral without either milk or grass, where it is vaccinated, and sometimes even dehorned, is a traumatic experience and stress of the highest order. Such animals suffer from a lack of feed and water, digestive upsets, dehydration, and sometimes high fevers. The end result is lowered gains (usually shrinkage) and feed efficiency, illness, and sometimes even death. In order to get such environmentally stressed animals back on feed as soon as possible, it is important that the caretaker know what has happened to them and how to rectify the situation.

The weanling calf, being a ruminant, is dependent upon fermentation in the rumen for a major portion of its nutrients. Going without feed, whether voluntary or imposed, for 24 to 48 hours, will reduce rumen fermentation by as much as 70%. This is caused by a change in the ratio of rumen microorganisms which digest crude fiber, metabolize protein, and produce water-soluble vitamins. Thus, the ruminant loses its major source of energy, its ability to produce a balance of amino acids, and its primary source of vitamin C and the B vitamins. Since these vitamins are not stored in the body and are required on a daily basis, a vitamin deficiency develops very rapidly.

In addition to the rumen problem, the bacterial population of the lower digestive tract is altered, particularly if antibiotics have been administered to the stressed animal. This change paves the way for a rapid growth of E. coli and the development of scouring.

Thus, under stress, the animal suffers a severe malnutrition in terms of the major nutrients, along with an imbalance in blood electrolytes due to dehydration, acidosis, and scouring. This general breakdown in the normal state of the entire digestive system frequently results in the invasion of fungi, such as yeasts and molds.

Rations for animals under stress should be formulated to provide the best chance for the return of the digestive system to normalcy. To achieve this, the ration should provide a balance of nutrients ideally suited to the creation of normal rumen fermentation. This calls for the following:

1. **Restoration of fiber digestion.** The ideal feedstuffs for fiber-digesting organisms are those that contain highly digestible fiber, such as beet pulp, citrus pulp, soybean mill feed, and alfalfa meal. Also, *rumen stimulants*, such as those found in distillers' grains, molasses, and alfalfa meal, can be utilized to reduce the time required for normal rumen repopulation.

2. **High-quality proteins.** Since the ability to build amino acids in the usual way (by microorganisms) is reduced, nonprotein sources of nitrogen, such as urea, should be avoided. Thus, animals under stress should be fed oilseed or animal proteins.

3. **Restoration of flora in lower tract.** If left alone, the lower digestive tract will, in due time, restore its flora. But, since a speedy return to normal production is desired, some assist is usually desirable. The natural flora of the lower digestive tract include a large population of *Lactobacilli,* the presence of which creates a natural regulation of E. coli and a normal state of health in the lower digestive tract. The key to restoring the *Lactobacilli* is the use of organisms that will implant and grow in the tract.

Thus, the development of nutritional programs for animals under stress calls for the proper formulation of rations to correct the damage done by the stress. Since animals experience many stresses, the development of nutritional programs for periods of stress should receive as much attention as the medical treatment of ills. Also, and most important, stress should be kept to a minimum. For example, replacement heifers should be dehorned, vaccinated, treated for parasites, and on feed prior to movement to a new location.

In the life of an animal, some stresses are normal, and they may even be beneficial—they can stimulate favorable action on the part of an individual. Thus, we need to differentiate between stress and distress. Distress—not being able to adapt—is responsible for harmful effects. The trick is to manage stress so that it does not become distress and cause damage, and to recognize the warning signals of distress.

The principal criteria used to evaluate, or measure, the well-being or stress of people are: increased blood pressure, increased muscle tension, body temperature, rapid heart rate, rapid breathing, and altered endocrine gland function. In the whole scheme, the nervous system and the endocrine system are intimately involved in the response to stress and the effects of stress.

The principal criteria used to evaluate, or measure, the well-being or stress of cattle are: growth rate or production, efficiency of feed use, efficiency of reproduction, body temperature, pulse rate, breathing rate, mortality, and morbidity. Other signs of cattle well-being, any departure from which constitutes a warning signal are: contentment, alertness, eating with relish, cudding, sleek coat and pliable and elastic skin, bright eyes and pink eye membranes, and normal feces and urine.

Stress is unavoidable. Wild animals were often subjected to great stress; there were no caretakers to modify their weather, often their range was overgrazed, and sometimes malnutrition, predators, diseases, and parasites took a tremendous toll.

Domestic animals are subjected to different stresses than their wild ancestors, especially to more restricted areas and greater animal density. However, in order to be profitable, their stresses must be minimal.

CONTROL POLLUTION

Pollution is the major issue of the decade. Anything that defiles, desecrates, or makes impure or unclean the surroundings pollutes the environment and can have a detrimental effect on animal health and performance. Thus, gases, odorous vapors, and dust particles from animal wastes (feces and urine) in buildings directly affect the quality of the environment. Muddy lots and stray electrical voltage may also pollute the environment. For healthy and productive animals, each of these pollutants must be maintained at an acceptable level. Perhaps the most troublesome dairy cattle pollutants are manure and stray voltage.

Before constructing a dairy facility, the owner should become familiar with both state and federal regulations. The state regulations can be secured from the state water board. They differ from state to state, but most states require a catch basin (detention pond) sufficient to contain the runoff from a storm of the magnitude of the largest rainfall during a 48-hour period of the most recent ten years.

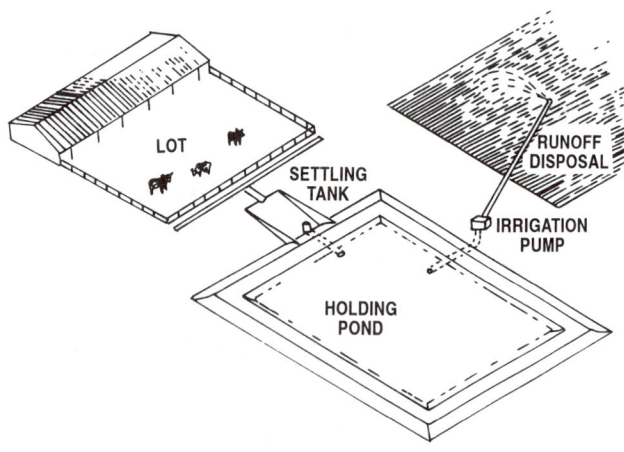

Fig. 13–21. A typical dairy corral runoff pollution control system. Solids are collected in the settling tank, then hauled. Liquid runoff is stored in a holding pond, then returned to the land through the irrigation system.

MANURE

Dairy operations located near centers of population are having an increasing number of complaints lodged against them because of manure and odor. Lawsuits, based on the nuisance law, have been filed against some of them.

No doubt, the manure pollution problem will persist. However, the energy crisis, accompanied by increased chemical fertilizer and feed prices, has caused manure to be looked upon as a resource and not a waste that presents a disposal problem. As a result, a growing number of American farmers are returning to organic farming—they are using more manure—the unwanted barnyard centerpiece of years gone by. They are discovering that they are just as good reapers of the land and far better stewards of the soil. Additionally, more and more poultry manure is being fed to ruminants.

Fig. 13–22. Alley manure scraper in a free stall barn. (Courtesy, Dr. J. Schingoethe, Dairy Science Department, South Dakota State University, Brookings)

In the future, as fertilizer and feed become increasingly scarce and expensive, the economic value of manure will increase.

From the standpoints of pollution control and of using manure as a fertilizer, it is important that the dairy producer (1) exercise certain precautions, and (2) know how much manure can be applied to the land.

STRAY VOLTAGE

Stray electrical voltage has caused serious problems on many dairy farms—affecting animal behavior and lowering milk production, although it may affect other animal species also. Contrary to popular belief, stray voltage is not new; it is as old as electricity itself. However, it has become a problem on many farms recently for two reasons: (1) There is more electrical load on today's farms; and (2) in the last 20 years we have used more equipment grounding for safety purposes.

Stray voltage is excessive voltage between two animal contact points. The conditions that cause stray voltage are, electrically, quite simple: If sufficient voltage is present, it may force a current through any available conductor, including a cow's body. Cows are good conductors because of their body design (the length from mouth to front and rear legs); cows bridge the gaps between electrically grounded objects and *true earth*. The cow does not feel the voltage as such; she feels the tingling current running through her body.

People seldom feel the current for several reasons. Usually, caretakers wear rubber-soled shoes when in the barn, whereas the bare-footed cow stands on concrete that is often wet. Also, humans have only two legs instead of four like the cow, and human's legs touch the floor near the same vicinity.

■ **Sources of stray voltage** — Any electrical condition which creates large enough voltage between two animal contact points may create a stray voltage problem. The source of stray voltage may be either *on-farm* or *off-farm*.

On-farm voltage problems stem from defective equipment, faulty wiring, bad connections, or having several 120-volt motors on the same line. On-farm stray voltage can be minimized by maintaining good electrical wiring systems that meet the requirements of the National Electric Code. Also, properly balanced 120-volt circuits and conversion of larger 120-volt meters to 240 volts will reduce the effect of secondary neutral voltage drops at the farm service entrance.

Off-farm voltage comes onto the farm through the electrical supplier's lines. Voltage will vary with the load and the natural grounding ability of the area. As usage increases, so may stray voltage. Heavier loads are seen at milking time and in the fall when grain dryers may be running on many farms.

■ **Signs of stray voltage** — One or more of the following signs may indicate that stray voltage exists in a dairy:

1. **Cows reluctant to enter the parlor.** When cows are subjected to stray voltage in the parlor stalls, they soon become reluctant to enter the parlor.
2. **Cows nervous in parlor.** Cows often dance or step around almost constantly while in the milking parlor.
3. **Uneven milk let-down, and milk-out.** When milk let-down and milk-out are uneven, more machine stripping is required and longer milking time becomes apparent.
4. **Increased mastitis.** When milk-out is incomplete, more mastitis is likely to occur; all that is required is the presence of an infectious bacteria. In turn, this will result in increased somatic count.
5. **Reduced feed intake in the parlor.** If cows encounter stray voltage while eating from the grain feeders, a reluctance to eat and reduced feed intake usually follow.
6. **Reluctance to drink water.** If stray voltage reaches the cows in stall barns through the water supply or metal drinking cups, the animals soon become reluctant to drink.
7. **Lowered milk production.** Each of the symptoms listed above is associated with stress and reduced feed intake, followed by a drop in daily milk production.

But detection of stray voltage is not easy! Other factors such as mistreatment of animals, milking machine problems, disease, sanitation, and nutritional disorders can create problems which manifest themselves in the seven symptoms mentioned above.

■ **Use voltmeter to monitor voltage** — The only sure method to determine if significant stray voltage is present is to have a qualified person perform a stray voltage survey, using approved equipment and monitoring the voltage through one, and preferably two, milkings. Point to point measurements between cow and contact points will determine if the voltage is actually getting to the cow. Generally, stray voltage is not constant throughout the day; so, readings should be taken over a long period.

Most milking machine company representatives, many power supplier employees, some milking equipment dealers, and some veterinarians and county extension agents have equipped themselves with suitable voltmeters and are pre-pared to lend assistance. *Someone familiar with electrical systems, wiring, and equipment should be present when measurements are made.*

POLLUTION LAWS AND REGULATIONS

Invoking an old law (the Refuse Act of 1899, which gave the Corps of Engineers control over runoff or seepage into any stream which flows into navigable waters), the U.S. Environmental Protection Agency (EPA) launched a program to control water pollution by requiring that all owners with cattle numbering 1,000 head or more the previous year must apply for a permit. The states followed suit; although differing their regulations, all of them increased legal pressures for clean water and air. Then followed the Federal Water Pollution Control Act Amendments, enacted by Congress in 1971, charging the EPA with developing a broad, national program to eliminate water pollution.

Owners/operators of dairy facilities with more than 1,000 animal units must apply. Animal units are computed as follows: multiply the number of mature dairy cattle by 1.4.

ANIMAL WELFARE/ANIMAL RIGHTS

In recent years, the behavior and environment of animals in confinement have come under increased scrutiny of animal welfare/animal rights groups all over the world. For example, in 1987 Sweden passed legislation designed (1) to phase out layer cages as soon as a viable alternative can be found; (2) to discontinue the use of sow stalls and farrowing crates; (3) to provide more space and straw bedding for slaughter hogs; and (4) to forbid the use of genetic engineering, growth hormones, and other drugs for farm animals except for veterinary therapy. Also, the law provides for fining and imprisoning violators.

Animal welfarists see many modern practices as unnatural, and not conducive to the welfare of animals. In general, they construe animal welfare as the well-being, health, and happiness of animals; and they believe that certain intensive production systems are cruel and should be outlawed. The animal rightists go further; they maintain that humans are animals, too, and that all animals should be accorded the same moral protection. They contend that animals have essential physical and behavioral requirements, which, if denied, lead to privation, stress, and suffering; and they conclude that all animals have the right to live.

Dairy producers know that the abuse of animals in intensive/confinement systems leads to lowered production and income — a case in which decency and profits are on the same side of the ledger. They recognize that husbandry that reduces labor and housing costs often results in physical and social conditions that increase animal problems. Nevertheless, means of reducing behavioral and environmental stress are needed so that decreased labor and housing costs are not offset by losses in productivity. The welfarist/rightists counter with the claim that the evaluation of animal welfare must be based on more than productivity; they believe that there should be behavioral, physiological, and environmental evidence of well being, too. And so the arguments go!

To all animal caretakers, the principles and application of animal behavior and environment depend on understanding; and on recognizing that they should provide as comfortable an environment as feasible for their animals, for both humanitarian and economic reasons. This requires that attention be paid to environmental factors that influence the behavioral welfare of their animals as well as their physical comfort, with emphasis on the two most important influences of all in animal behavior and environment—feed and confinement.

Animal welfare issues tend to increase with urbanization. Moreover, fewer and fewer urbanites have farm backgrounds. As a result, the animal welfare gap between town and country widens. Also, both the news media and the legislators are increasingly from urban centers. It follows that the urban views that are propounded will have greater and greater impact in the years ahead.

FOOD SAFETY AND DIET/HEALTH CONCERNS

Many food safety and diet/health concerns are unwarranted. American consumers are prone to over-react to rumors relative to their food. They care little about what they put on their backs, but they are greatly concerned about what goes into their stomachs.

America's food supply is the safest in the world! Nevertheless, there is need for constant vigilance and improvement, especially in animal products which are subject to all the hazards of other foods (spoilage, pesticides, toxicities), plus being capable of transmitting, or serving as passive carriers, of certain diseases to humans.

In colonial times, the dairy producer milked the cows; then, delivered the milk door-to-door to urban customers. If the milk was not acceptable, the matter was resolved quickly and on the spot, or the producer lost a customer. Today, the public expects the dairy team—farmers, processors, and retailers—to provide wholesome and safe dairy products free from disease agents, toxic substances, and pesticide and drug residues.

Uptake of pesticides by animals, leading to residues in animal products, can result either from direct application of pesticides to animals or from animals ingesting feeds carrying pesticide residues. Drug residues are caused by (1) producers failing to withdraw drugs from livestock far enough in advance of marketing products; (2) contaminated feed storage, mixing, and handling equipment; and/or (3) the wastes (feces and urine) of treated animals coming in contact with untreated animals. *Reading and following the directions on the label is the key to safe pesticide and drug use.*

Because the welfare of the nation is dependent upon the health of its people, dairy (and other) products are carefully monitored by various government agencies to assure consumers that they are wholesome and safe; and because of recognizing the importance of consumers in the safety of their food, the private sector may do additional testing. The agencies most responsible for this important work are:

1. The U.S. Department of Health and Human Services, including the following agencies: The Center for Disease Control, the Food and Drug Administration (FDA), and the National Institute of Health.

2. The U.S. Department of Agriculture, including the following agencies: the Agricultural Research Service, the Animal and Plant Health Inspection Service, the Cooperative State Research Service, the Federal Extension Service, the Labeling and Registration Section, and the Veterinary Service Division.

3. State and local government agencies.

4. International organizations engaged in health and/or nutrition activities, including the World Health Organization (WHO), and Food and Agriculture Organization (FAO).

5. Private industry groups such as the National Dairy Council.

6. Professional organizations, including dentists, dieticians, doctors, health educators, nurses, and public health workers.

7. Food processors and retailers.

QUESTIONS FOR STUDY AND DISCUSSION

1. Discuss the importance of the behavioral characteristics of the world record milk champion, *Beecher Arlinda Ellen*, from the standpoint of high production.

2. Define animal behavior.

3. Why has increased confinement of animals made for great interest in the subject of animal behavior?

4. Why has there been a genetic time lag; why have dairy cattle producers altered the environment faster than the genetic makeup of animals?

5. Define each of the following causes of animal behavior: (a) genetic, (b) simple learning, (c) complex learning. How do each of these causes of animal behavior affect dairy cattle?

6. List the nine general functions or behavioral systems that cattle exhibit.

7. Discuss the following behavioral systems as they pertain to cattle:
    a. Care-giving and care-seeking behavior.
    b. Eliminative behavior.
    c. Ingestive behavior.
    d. Sexual behavior.

8. How is gregarious behavior related to allelomimetic behavior?

9. Describe the social organization in a dairy cattle herd on pasture from the standpoints of (a) dominance and (b) spacing.

10. Describe the consistent order in which dairy cattle travel when moving from the paddock to the milking parlor.

11. How is the dominance hierarchy established when several dairy cows are brought together to form a herd? How is it maintained?

12. Explain the difference between dominance and leader-follower.

13. Discuss the importance of each of the following social relationships of dairy cattle: (a) interspecies relationships and (b) people-animal relationships.

14. Discuss how cattle communicate with each other, and the importance of each method.

15. Those who care for cattle need to be familiar with behavioral norms in order to detect and treat abnormal situations—especially illness. Describe a normal cow. Describe (a) sight and (b) sleep of cattle.

16. Discuss each of the following abnormal behaviors of cattle: (a) homosexual, (b) the *mean bull* complex, (c) pica, and (d) kicking when being milked.

17. List and discuss the significance of one example of the practical application of cattle behavior in each of the following areas:
   a. Breeding for adaptation.
   b. Quick adaptation-early training.

c. Milking parlor design.
d. Manure elimination.
e. Companionship.

18. Discuss how each of the following environmental factors affects dairy animals:
   a. Feed.
   b. Water.
   c. Weather.
   d. Facilities.
   e. Health.
   f. Stress.

19. Define pollution. Why has pollution become such a great issue?

20. Discuss each of the following troublesome pollutants which may plague dairy producers: (a) manure and (b) stray voltage.

21. Assume that you plan to establish a dairy herd of 1,000 lactating cows. With what pollution laws and regulations must you comply? How would you go about applying?

22. Discuss the animal welfare/animal rights issue.

23. Discuss food safety and diet/health concerns relative to dairy products.

SELECTED REFERENCES

Title of Publication	Author(s)	Publisher
Behavior of Domestic Animals, The, Third Edition	Edited by E. S. E. Hafez	The Williams and Wilkens Company, Baltimore, MD, 1975
Bibliography of Livestock Waste Management	J. R. Miner D. Bundy G. Christenbury	Office of Research and Monitoring, U.S. Environmental Protection Agency, Washington, DC, 1972
Biology of Stress In Farm Animals: an integrative approach	P. R. Wiepkema P. W. M. van Adrichem	Kluwer Academic Publishers, Hingham, MA, 1987
Concise Survey of Animal Behavior, A	E. K. Honore P. H. Klopfer	Academic Press, Inc., Harcourt Brace Jovanovich, Publishers, San Diego, CA, 1990
Development and Evolution of Behavior	Edited by L. R. Aronson, *et al.*	W. H. Freeman and Company, San Francisco, CA, 1970
Domestic Animal Behavior	J. V. Craig	Prentice-Hall, Inc., Englewood Cliffs, NJ, 1981
Effect of Environment on Nutrient Requirements of Animals	D. R. Ames, Chairman	NRC, National Academy Press, Washington, DC, 1981
Environmental and Functional Engineering of Agricultural Buildings	H. J. Barre L. L. Sammet G. L. Nelson	Van Nostrand Reinhold Co., New York, NY, 1988
Environmental Biology	P. L. Altman D. S. Dittmer	Federation of American Societies for Experimental Biology, Bethesda, MD, 1966
Environmental Control for Agricultural Buildings	M. L. Esmay J. E. Dixon	The AVI Publishing Company, Inc., Westport, CT, 1986
Environmental Management in Animal Agriculture	S. E. Curtis	Animal Environment Services, Mahomet, IL, 1981
Ethology, The Biology of Behavior, Second Edition	I. Eibl-Eibesfeldt	Holt, Rinehart and Winston, New York, NY, 1975

(Continued)

SELECTED REFERENCES (Continued)

Title of Publication	Author(s)	Publisher
Farm Animal Manures: an overview of their role in the agricultural environment	J. Azevedo P. R. Stout	Agricultural Publications, University of California, Berkeley, 1974
Guide to Environmental Research on Animals, A	R. G. Yeck, Chairman	NRC, National Academy of Science, Washington, DC, 1971
Health Issues Related to Chemicals in the Environment: A Scientific Perspective	A. L. Craigmill, Chairman	Council for Agricultural Sciences and Technology, Ames, IA, 1987
Impact of Stress, The, Proceedings	Edited by R. E. Moreng J. R. Herbertson	Colorado State University, Ft. Collins, 1986
Introduction to Animal Behavior, An,: ethology's first century, Second Edition	P. H. Klopfer	Prentice-Hall, Inc., Englewood Cliffs, NJ, 1974
Livestock Behavior, a practical guide	R. Kilgour C. Dalton	Westview Press, Boulder, CO, 1984
Livestock Environment, Proceedings, Second International Livestock Environment Symposium	D. S. Bundy, Planning Chairman	American Society of Agricultural Engineers, St. Joseph, MI, 1982
Mechanisms of Animal Behavior	P. Marler W. J. Hamilton	John Wiley & Sons, New York, NY, 1966
Organic Farming: current technology and its role in a sustainable agriculture	Edited by D. M. Kral	American Society of Agronomy, Madison, WI, 1984
Principles of Animal Behavior	W. N. Tavolga	Harper & Row, New York, NY, 1969
Principles of Animal Environment	M. L. Esmay	The AVI Publishing Company, Inc., Westport, CT, 1978
Readings in Animal Behavior	Edited by T. E. McGill	Holt, Rinehart and Winston, New York, NY, 1973
Safe and Effective Use of Pesticides, The	P. J. Marer	University of California Publications, Oakland, 1988
Scientific Aspects of the Welfare of Food Animals	F. H. Baker, Chairman	Council for Agricultural Science and Technology, Ames, IA, 1981
Social Hierarchy and Dominance	Edited by M. W. Schein	Dowden, Hutchinson & Ross, Inc., Stroudsburg, PA, 1975
Social Space for Domestic Animals	Edited by R. Zayan	Kluwer Academic Publishers, Hingham, MA, 1985
Social Structure in Farm Animals	G. J. Syme L. A. Syme	Elsevier Scientific Publishing Co., Amsterdam, The Netherlands, 1979
Stress Physiology in Livestock	M. K. Yousef	CRC Press, Inc., Boca Raton, FL, 1985
Structures and Environment Handbook		Midwest Plan Service, Iowa State University, Ames, 1972
Utilization, Treatment, and Disposal of Waste on Land, Proceedings	E. C. A. Runge, President of Society	Soil Science Society of America, Inc., Madison, WI, 1986

Fig. 14–1. Management gives point and purpose to everything else. (Courtesy, Holstein-Friesian Assn. of America, Brattleboro, VT)

Conents

CHAPTER

14

DAIRY CATTLE

MANAGEMENT

Management is the art of caring for, handling, or controlling. In the dairy operation, it gives point and purpose to everything else. It can make or break a dairy.

Higher levels of milk production per cow, bigger dairy herds, and increased facility and labor costs have focused attention on the need for more efficient management.

Innumerable management aspects are of great importance to dairy production. When disregarded, many of them will lessen production and make the enterprise unprofitable, no matter how good the breeding and feeding of the animals.

Pertinent facts relative to, along with methods of accomplishing, some important dairy management practices are covered in this chapter.

DAIRY MANAGER

The manager is the person who conducts, controls, or directs the business, hopefully at a profit.

The manager is the most important factor in the operation of a dairy enterprise. Unfortunately, this fact is often overlooked, primarily because of the accent on scientific advances, improved breeding and feeding, and automation.

The skill of the manager materially affects how well animals are bought and sold, the quality of the herd, the health of the animals, the efficiency of the ration, the stress of the animals, the performance of labor, the quality and price of the milk produced, the expression of the genetic potential of the herd, the public relations of the establishment, and the profit or loss. Indeed, dairy managers must wear many hats, and they must wear each of them well.

The bigger and the more complicated the dairy operation, the more competent the management required. This point bears emphasis because, currently, (1) bigness is a sign of the times, and (2) the most common method of attempting to "bail out" an unprofitable dairy is to increase its size. Although it is easier to achieve efficiency of equipment, labor, purchases, and marketing in big operations, bigness alone will not make for greater efficiency. Management is still the key to success.

REQUISITES OF SUCCESSFUL MANAGERS

The first and most important requisite of successful dairy managers is that they possess a great love for dairy cattle. This appears to be an inborn trait, for some people never acquire the natural ability to work with animals—no matter

how long or how hard they try. When such love exists, the animals are more docile and easier to handle, for the caretakers' feelings are relayed to them. Also, a great love for animals appears to be essential if caretakers are to feed and milk them regularly, cheerfully, and with enjoyment, without regard to long hours and Sundays or holidays; if they are to provide clean, dry bedding, despite the fact that a driving storm may make it necessary to repeat the operation the next day; if they are to serve as nursemaid to newborn or sick animals, though it may mean the loss of sleep and working with cold, numb fingers; and if they are to remain calm and collected, though striking an animal, or otherwise giving vent to their feelings, might appear to be warranted.

Also, successful dairy managers must be well versed relative to soils and crops, both of which must be considered whether feeds are homegrown or bought.

Finally, experience, industry, and good judgment are very necessary requisites for successful dairy managers—words that carry the same connotation in all industries, and that are self-explanatory.

DAIRY RECORDS

Production records are the most important management tool on the dairy farm. The primary purpose of dairy records is to give the producer detailed information on individual cows upon which to base daily management decisions—to allot concentrates to cows, to breed cows, to dry off cows, to cull cows, and to treat cows for diseases, parasites, and abnormalities. Records are also essential in order to evaluate the status of the dairy herd. A summation of the records on a monthly, semiannual, or annual basis allows the producer to determine the strengths, weaknesses, and profitability of the operation. Such summations make possible an informed evaluation of past management practices and long-range planning for the years ahead.

The desirable requisites of a dairy record-keeping system are that it be simple, complete, accurate, up to date, understandable, and require a minimum of time to keep.

MILK AND BUTTERFAT TESTING PROGRAMS

Individual cow records are a must in any progressive dairy production program. Producers use records as a guide for feeding, for locating and culling out the least profitable cows, and for maintaining a permanent, detailed record of each cow. Records necessitate that each cow be individually identified, and that there be milk and butterfat production records.

ALTERNATE DHI TESTING PLANS

The National Cooperative Dairy Herd Improvement (DHI) Program is a voluntary cooperative effort to improve the level and efficiency of milk production and increase dairy profits. It involves milk producers, local and state DHI organizations, extension services of land grant colleges and universities, and the U.S. Department of Agriculture.

The U.S. Department of Agriculture aids in conducting and distributing results of the sire evaluation phase of the DHI Program. It also coordinates, furnishes materials, provides statistical information, analyzes data, and researches various aspects of the program.

State and local Dairy Herd Improvement Associations (DHIA) conduct the program among producers, working through the Cooperative Extension Service in cooperation with the Federal Extension Service and Animal Science Research Service of the USDA.

Dairy cattle record-keeping plans may be either *official* or *unofficial,* with alternate choices under each grouping.

OFFICIAL DHI TESTING

This includes the Standard Dairy Herd Improvement Association (DHIA) and the Dairy Herd Improvement Registry (DHIR). Since both programs are official, a supervisor tests herds 4 to 12 times annually. Records from both programs are used in proving dairy sires.

DAIRY HERD IMPROVEMENT ASSOCIATION (DHIA)

This program, first adopted in 1926, is the most complete of all dairy production and record plans. More than half of the cows in the United States on production test are on this program. Both registered and grade cows can be enrolled.

In this program, a supervisor or tester employed by the local or state testing association visits the herd one day each month. The tester identifies all cows in the herd; weighs and takes representative samples of the milk from all animals in the herd for two consecutive milkings (three milkings on herds 3-times-daily-milking); then combines the milk samples and sends them to a central testing laboratory for analyses of components such as butterfat, protein, and somatic cell count (SCC). Records are obtained on an individual cow basis on monthly and accumulative records for milk, fat, and protein; amount and cost of feed, and income over feed cost; breeding dates, calving dates, dry dates, and other factors affecting productivity; and in some testing associations, somatic cells or the California Mastitis Test (CMT) is made as an aid in monitoring udder health.

The above information is fed into a computer, programmed to provide monthly summaries of (1) individual cows, and (2) the herd; and this information is sent to each producer (see Fig. 14–2 for example).

Fig. 14–2. Sample Day and Lactation Report used by the Mid-States Dairy Processing Center, Ames, IA.

ALTERNATE AM-PM TEST

Dairy producers have additional options which can reduce the cost and still maintain official records. The supervisor can weigh two milkings and sample only one of the milkings, or only one milking may be weighed and sampled. A milking time record is required if only one milking is weighed and sampled. The milking which is weighed and sampled is alternated from month to month.

The primary advantages of these programs is the lower cost and the reduced supervision time. They also reduce the amount of disruption in the parlor by reducing the sampling time. These programs are especially attractive to larger dairy herds. Supervisors may also find these programs beneficial since they allow additional herds to be tested.

DAIRY HERD IMPROVEMENT REGISTRY (DHIR)

This is the Standard DHIA record *plus* added requirements to satisfy the needs of breed associations. Among the latter, are *surprise tests,* made when the milk production of certain cows exceeds the breed average or another specified amount. Only registered dairy cows are eligible for DHIR records. The production records of herds enrolled in DHIR are transferred to the respective breed registries for official recording.

UNOFFICIAL DHI TESTING

These unofficial record plans are designed to aid in herd management at minimal cost. Among such plans are: Owner Sampler (OS), Weigh-A-Day-A-Month (WADAM), Milk Only Record (MOR) and Alternate AM-PM.

OWNER-SAMPLER RECORDS (OS)

Under the owner-sampler plan, the owner, rather than a supervisor, weighs and samples the milk. The samples are then tested at a central laboratory.

WEIGH-A-DAY-A-MONTH (WADAM)

In this program, the owner weighs the milk for each cow one day each month, and enters the weight and feeding information on the form provided. The information and forms are mailed to the supervisor, or a central office, where calculations are completed, following which summaries are returned to the owner.

MILK ONLY RECORDS (MOR)

This testing plan, which originated in North Carolina, is receiving considerable attention, especially in the southeastern states. These records involve milk weights recorded by the DHIA supervisor, without fat determination.

IDENTIFICATION

A requisite of any dairy record program is the identification of each animal in the herd. Daily management decisions concerning breeding, feeding, selection, calving, and culling depend on accurate identification of animals. Also, proper identification is necessary for keeping records of cows on official production testing programs and for registration of purebred cattle in the breed registry associations.

Identification of animals by appearance (size, color patterns, or other distinguishing marks) is usually satisfactory in a small herd of fewer than 50 animals. But most producers with herds of more than 50 cows have difficulty in recognizing each animal by its physical appearance if no other identification is present. So, some permanent system of identification is essential in larger herds. Also, positive identification of registered animals is necessary.

For registration, the broken-colored breeds (Ayrshires, Guernseys, and Holsteins) require sketches of color markings or photographs, while the Jersey and Brown Swiss breeds require permanent ear tattoos. These are listed on the registration papers. Other, and more usable, methods of identification include ear, leg, and tail tags; neck chains; hide brands; brisket tags; and marking paint. Each of the main methods of identification will be discussed briefly.

■ **Ear tags**—Ear tags are the most widely used means of identifying dairy animals, especially grades. They are made of steel, aluminum, nylon, or plastic. Ear tags are easily attached, but sometimes they are easily pulled out and lost, although the latter problem has been lessened by the new plastic and nylon types of tags which do not cut through the

Fig. 14–3. Ear tag, the most widely used means of identifying commercial dairy animals. (Courtesy, Brown Swiss Cattle Breeders' Assn. of the U.S.A., Beloit, WI)

ears as easily as the sharper metal varieties. Also, it is note-worthy that ear tags frequently rub and scratch the skin, thereby making openings for screwworm infestation—an important consideration in the South.

■ **Hide brands**—When properly applied, hide brands are permanent and easily read. Four methods of applying hide brands are:

1. **Freeze branding.** This method, which is increasing in popularity, was developed by the U.S. Department of Agriculture, at Washington State University. It makes use of a super-chilled (by dry ice or liquid nitrogen) copper branding "iron" which is applied to the closely clipped surface for about 20 seconds, thereby depigmenting the hair follicles, following which the hair grows out white. When properly done, this method is painless, permanent, and there is no hide damage.

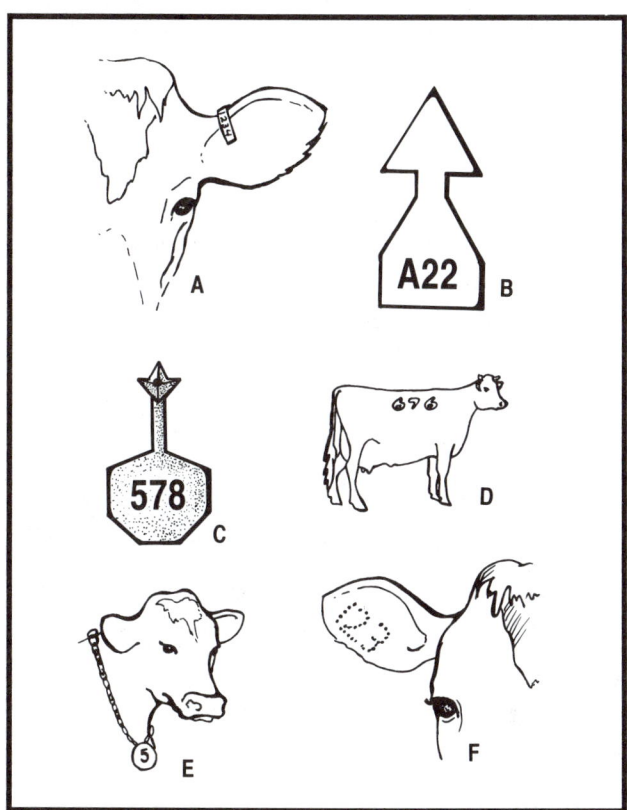

Fig. 14–4. Cattle identification methods: A—metal ear tag, showing proper placement; B—ear tag on which owner can write number; C—prenumbered ear tag; D—freeze brand; E—neck chain; F—tattoo.

On white cattle, deliberate overbranding (30 seconds or more) will produce a bald brand suitable for identification after clipping.

2. **Hot iron.** This is the traditional brand of the west. Irons are heated to a temperature that will burn sufficiently deep to make the scab peel, but which will not leave deep scar tissue. The proper temperature of the hot iron is indicated by a yellowish color. Branding is accomplished by placing the heated branding iron firmly against the body area

which it is desired to mark and by not allowing it to slip for the few seconds when the hide is burned. The branding iron should be kept free from dirt and adhering hair at all times.

Where electricity is available, the electric iron may be used; it keeps an even temperature, and if properly used, makes a clear, uniform brand.

3. **Branding fluids.** Branding fluids, which are less widely used in making hide brands, consist of caustic material applied by means of a cold iron. Best results are secured if the area is first clipped. The chemical method of producing hide brands is slower than the hot iron; the results are generally less satisfactory, particularly if the operator is inexperienced with the method; and the resulting brand is less permanent.

4. **Laser brand.** This type of brand is permanent, but it needs further development and experimental study.

A good hide brand, regardless of how it is applied, is one that is easily read, that is of simple design and yet cannot be easily changed or tampered with, that has no welds or thick points in the iron, and that interferes with the circulation as little as possible. Thick points mean deeper burning and slower healing, whereas small enclosed areas, such as a small "O," will slough out entirely.

■ **Neck chains or straps**—Neck chains or straps are the most common temporary identification of dairy cattle. Occasionally, they may be lost, but this is not particularly serious if the caretaker is on the alert and immediately replaces each one that is lost, without allowing several losses to accumulate before taking action. In rare instances, an animal will hang itself by the chain.

Neck chains or straps must be adjusted, for young animals grow, or animals change in condition.

■ **Tattoos**—Most purebred dairy cattle registry associations require that registered animals be individually tattooed. This method of marking consists of piercing the skin with instruments equipped with needle points which form letters or numbers. This operation is followed by rubbing indelible ink onto the freshly pierced area. It is well to disinfect the tattooing instrument carefully between each operation in order to alleviate the hazard of spreading warts to the pierced area, for warts make it impossible to read the tattoo.

Cattle must be confined in order to read tattoo numbers. Even then, tattoos are difficult to decipher on dark-skinned animals.

■ **Other identification**—Other identification marks used on the range include: (1) *buds* formed by making a strip incision through the nose; (2) *wattles* made by cutting down a strip of skin on the jaw bone; and (3) *dewlaps* formed by cutting a strip of skin on the brisket.

The U.S. Department of Agriculture requires that most cattle 2 years of age or older be backtagged or eartagged to identify the animals to their herd of origin before they are shipped across state lines.

South African officials are encouraging cattle raisers to keep records of the noseprints of cattle. Like human fingerprints, tiny ridges on cattle noses always form differing patterns. Of course, it is difficult to get a calf to hold still long enough to get a reliable print of its nose.

Various electronic devices are in different stages of research and development; among them—

1. **Radio transmitter in the second stomach.** The animal swallows a small radio transmitter enclosed in a ³⁄₄-in. x 2½-in. plastic capsule, which lodges in the second stomach. From there, it transmits a coded number when signaled by a receiving unit to do so. The transmitter can be retrieved at slaughter and reused.

2. **Transponder.** The transponder is a new technology. It can be used on livestock or machines for identification, tracking, and theft recovery. On dairy cows, it can be used to identify each individual for the grain feeder and the milking parlor. The transponder consists of an electromagnetic coil and microchip encapsuled in glass, varying from about the size of a grain of rice to much larger. The transponder has no power source of its own. A reader emits a magnetic field which activates the transponder so that it transmits its code number.

The transponder may be implanted just below the skin of the animal or on a cow's neckstrap.

In the future, transponders may replace other methods of animal identification.

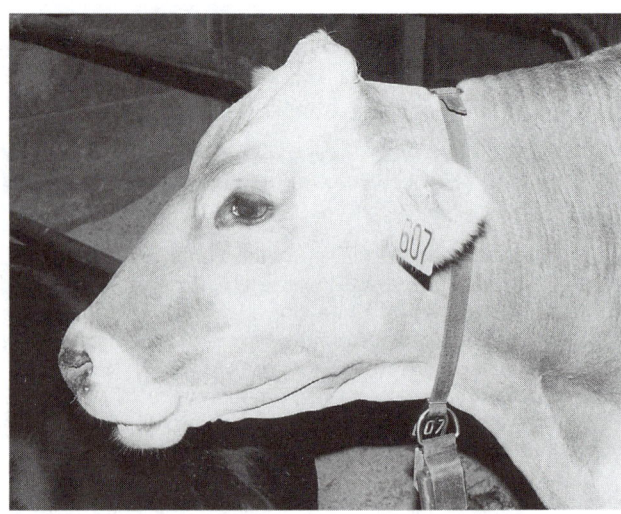

Fig. 14–5. Transponder on cow's neckstrap which identifies her for the grain feeder and milking parlor. (Courtesy, Dr. J. Schingoethe, South Dakota State University, Brookings)

MARKING PUREBRED DAIRY CATTLE

Table 14–1, Marking or Identifying Guide for Registered Dairy Cattle, summarizes the pertinent regulations of the registry associations relative to marking or identifying.

TABLE 14–1
MARKING OR IDENTIFYING GUIDE FOR REGISTERED DAIRY CATTLE

Breed	Association Rules Relative to Marking
Ayrshire	All calves must be tattooed before leaving individual pens or ties. Both ears may be used and the letters and numbers in the ears must be stated on the application for registration. Duplicate tattoos are not allowed in the same herd. The letters "I," "O," "Q," and "V" may be used only if accompanied by one or more other letters in the same ear. Tattoos may not exceed a total of 5 letters and numbers per line in each ear. The number shall be followed by the year letter designated by the Association.
Brown Swiss	Each animal must be tattooed in ear with indelible ink with such letters and numbers as the owner may select. No two animals in the same herd, of the same sex, shall have the same number. Both ears may be used. If only one ear is used, it is recommended that it be the left ear. All calves must be tattooed before leaving individual pens or ties.
Guernsey	The animal must be plainly tattooed in the ear with indelible ink or paste before application for registration is made. Both ears may be used, but it is not required. The tattoo may consist of either a series of numbers or a combination of letters and numbers selected by the owner but may not exceed a total of 6 numbers and letters. No two animals in the same herd and of the same sex can have the same tattoo. Vaccination tattoos are not acceptable as identification for registration.
Holstein-Friesian	All Holsteins should be identified by sketch (by use of the outline on the reverse side of the registry application form) or by photograph as soon after birth as possible. Ear tag idenfification may be used, but is not considered official for registration by the Association.
Jersey	All calves must be tattooed before leaving individual pens or ties. Both ears may be used and the letters and numbers in the ears must be stated on the application for registration. Tattoos must include at least one letter and one number. The letters "I," "O," "Q," and "V" may be used only if accompanied by an additional letter in the same ear. Tattoos may not exceed a total of 7 letters and numbers in each ear. The Association recommends that calves be tattooed in both ears with the same tattoo.
Milking Shorthorn	Milking Shorthorn cattle cannot be registered unless they have been tattooed in the ear with an individual identification number. An identification letter or initial may precede the number if desired. The application for registry of a calf must show whether the tattoo number appears in the right or left ear. Duplication of numbers for calves of same sex in the same herd is not permissible. Use of initial letters without a number is not sufficient—a number is required.

DEHORNING

Most dairy females are now dehorned, even in those herds that exhibit in the show-ring. The chief reasons for dehorning are:

1. Dehorned cattle are less likely to inflict injury upon other cattle or upon the attendants.

2. Less shed and feeding space is required for dehorned cattle, an important consideration when there is loose housing or group (corral) feeding.

3. Dehorned cows are quieter and easier to handle.

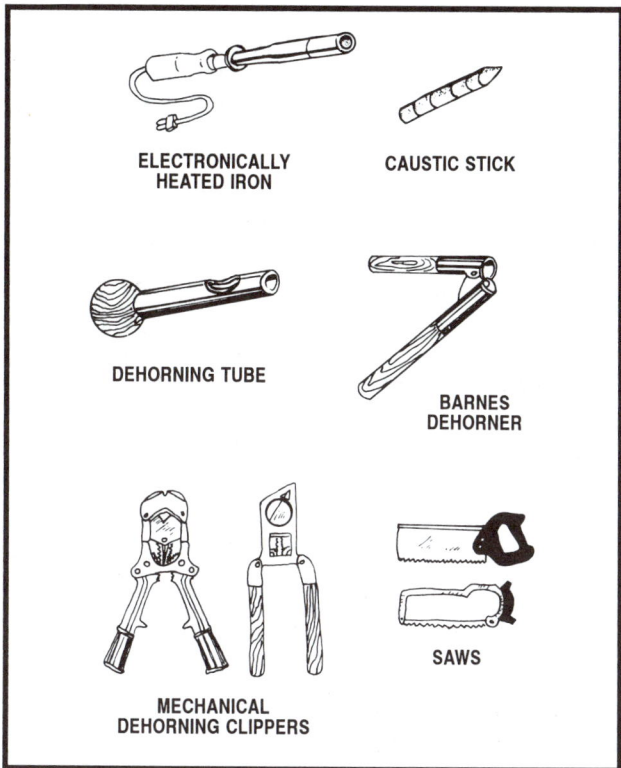

Fig. 14–6. Common instruments used for dehorning dairy cattle.

Early dehorning, preferably under two months of age, is recommended. Young calves are easier to handle, lose less blood, and suffer less setback. Also, the danger of infection and screwworm trouble is minimized when calves are dehorned at an early age.

The following general precautions should be observed in dehorning dairy cattle:

■ Dehorn calves before they leave the calf barn.

■ Thoroughly clean and disinfect all equipment before use.

■ Isolate animals for a few days following dehorning so that they can be easily observed and dealt with in case of an emergency.

■ Protect recently dehorned animals from flies and extremes of temperature.

■ Use a dehorning saw, clippers, or Barnes dehorner on older and larger animals, with the work done by an experienced person or a veterinarian.

■ Use a head gate or squeeze chute to secure older and larger animals.

Several methods of dehorning dairy animals are used, but the electrically heated iron and the use of caustic potash are most common. Other methods include the dehorning spoon and tube and the use of saws, clippers, or the Barnes dehorner.

ELECTRICALLY HEATED IRON

Where electricity is available, the electric hot iron may be used. It keeps an even temperature, without getting too hot or too cold.

The hot-iron method of dehorning consists of the application of a specially designed hot iron to the horn of young calves. The iron is fashioned with one end cupped out (bell-shaped) so that it fits over the small horn button. The hot iron should be applied for 15 to 20 seconds, or until a full ring of copper-colored burned tissue surrounds the small horn.

This system of dehorning is bloodless and may be used any time of the year, but it can be used on young calves only—preferably calves under 35 days of age.

CAUSTIC POTASH

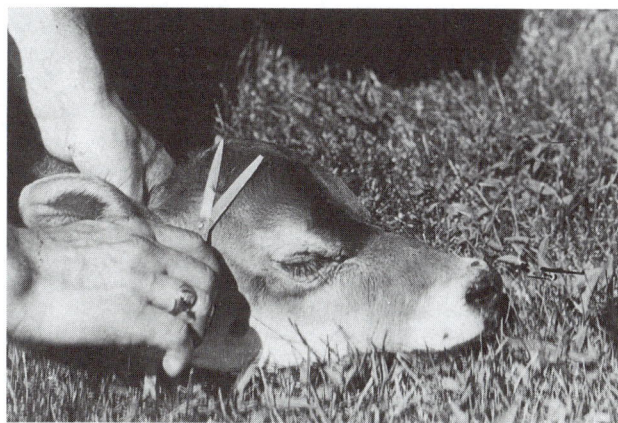

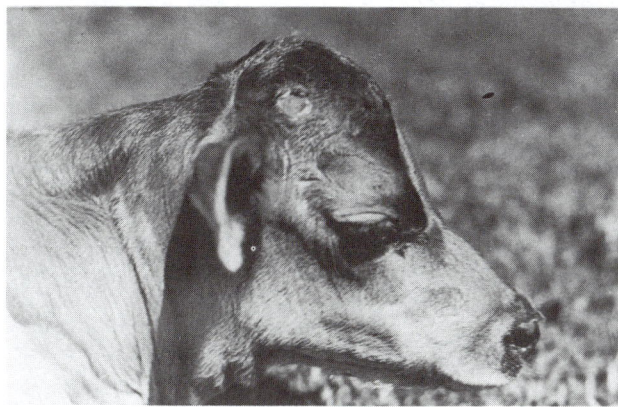

Fig. 14–7. Use of chemicals in dehorning. First the hair around the "button" is clipped (upper), following which caustic material should be rubbed over the little horn until blood appears (lower). (Courtesy, USDA)

Calves may be dehorned satisfactorily by the use of caustic potash (potassium hydroxide). This chemical can be purchased at almost any drug store, in stick, paste, or lacquer-base form. (The accompanying use of petroleum jelly is not necessary with the lacquer base.) This method really prevents horn growth and does not actually remove the horns. The treatment should be applied when the calf is from 3 to 20 days old and when only small buttons are present. After the hair around the buttons has been clipped or sheared closely, smear a ring of heavy grease or petroleum jelly around the clipped area to keep the caustic from running into the calf's eyes. Then rub the caustic material over the button or little horn until the blood appears. This should be done carefully; for, otherwise, some of the horn cells may not be destroyed and a scur may develop. If a caustic stick is used, it should be wrapped in a cloth or paper to protect the operator's hands from serious burns. Within a week or 10 days, the thick scab that appears over the horn buttons will drop off, and the calf will suffer little inconvenience. Calves treated with caustic should be isolated so that other animals do not lick the caustic, and they should be protected from rain for a day following the application, for the caustic may wash down and injure the side of the face or the eyes. Also, it is best not to turn calves back with their dams for a few hours following the application of the caustic.

DEHORNING SPOON AND TUBE

The dehorning spoon (or gouge) is a small instrument with which the horns of young calves can be gouged out. In the hands of an experienced operator it is both fast and effective. The use of the spoon leaves the head slightly rounded, and very seldom do scurs occur.

The dehorning tube is a newer instrument than the dehorning spoon. In comparison with the spoon, the tube is easier, faster, and less tiresome to use, and more certain to avoid regrowth.

Dehorning tubes come in four sizes, varying in diameter from ¾ in. for the smallest to 1⅛ in. for the largest. All four sizes should be available.

The steps and directions for using the dehorning tube are as follows:

1. Restrain the calf.
2. Select a sharp tube of proper size to fit over the base of the horn and include about ⅛ in. of skin all the way around.
3. Place the cutting edge straight down over the horn and then push and twist, first one way and then the other, until the skin has been cut through. A cut from ⅛- to ⅜-in. deep is required, the greater depth being necessary with older calves. Going deeper than necessary will cause excessive bleeding.
4. Turn the tube to about a 45° angle and rapidly shove and turn the cutting edge until the button comes off.

Either instrument can be used on calves up to 45 days of age.

SAWS; CLIPPERS; BARNES DEHORNERS

Saws or various forms of shears and clippers (including the Barnes-type dehorner) are used almost exclusively for dehorning animals over 3 months old. Whatever the instrument used, it is necessary to remove the horn with about ¼ in. of the skin around its base to make certain that the horn-forming cells are destroyed. The skin should then be allowed to grow over the wound. Dehorning of young animals can be done with greater ease, and there is less shock to the animals. However, some attention must be given to the season.

Ordinarily, clippers are satisfactory for removing the horns of younger cattle; but the hard, brittle horns of mature cattle can best be removed with a saw. With older animals, clippers are likely to sliver or crack the bone that forms the horn core. Moreover, the saw results in less loss of blood, for the action of the saw blade produces a lacerating of the blood vessels rather than a clean-cut cross section. On the other hand, the ragged wound made by a saw heals more slowly, and the operation is much slower.

While the dehorning operation with older cattle is being performed, it is necessary to have some device for confining or restraining animals. For this purpose, various types and arrangements of dehorning chutes, pinch gates, squeeze pens, and cattle stocks have been devised. Calves may be handled by throwing or snubbing them to a fence post and tying one side of the body against a strong fence or solid wall. Such methods are more difficult, however, for both handlers and animals.

REMOVING "EXTRA" TEATS

Heifer calves may be born with more than four teats. Usually the extra teats are located posterior to one or both rear teats, but they may be between the front and rear teats on one or both sides of the calf's udder. Since extra teats have no value and detract from the appearance of the udder and may interfere with milking, they should be removed when the calf is 1 to 2 months of age.

In most cases, extra teats can be removed by the caretaker, proceeding as follows:

1. Place the calf on its side, with the hind leg nearest the handler drawn forward.
2. Identify the extra teats to be removed. (If in doubt, have a veterinarian check before removing.)
3. Wash the udder with warm soap and water in order to remove foreign matter and lessen the chance of infection.
4. Wash the udder with a disinfectant; wipe dry.
5. Stretch extra teat and hold firmly with one hand.
6. Cut off the unwanted teat close to the udder, using curved, sharp, *disinfected* scissors. If necessary, trim the area around the base of the teat. Apply tincture of iodine to the cut area. If there is much bleeding, hold cotton to the site momentarily. In fly season, paint wounds with pine oil to repel flies and observe animals following the operation.
7. Have the veterinarian perform the operation if there is a double teat or when the extra teat is near the base of the normal teat.

When extra teats are properly removed, usually there is little bleeding, and the scar will be scarcely noticeable when the heifer freshens.

HOOF TRIMMING

Hoof troubles make cows reticent to walk for feed and water and result in a drop in milk production. Hence, regular inspection and care of the feet of all dairy animals is important, especially for the older and larger cows. It will reduce the incidence of foot rot, lameness, and other foot troubles.

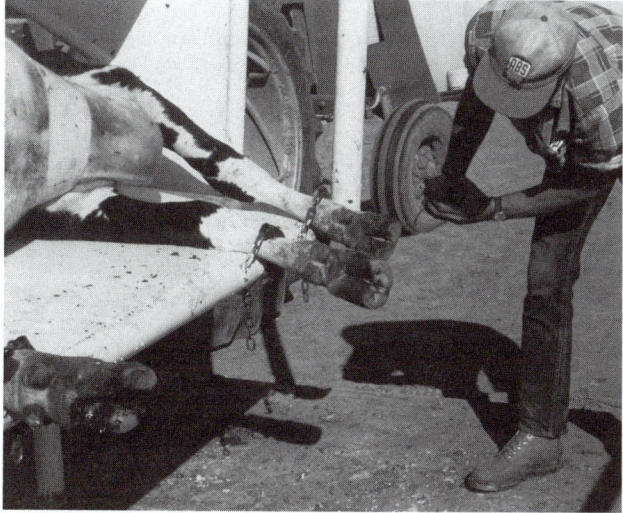

Fig. 14–8. Hoof trimming on a tilt table. (Courtesy, Dr. Tom Schultz, Farm Advisor, Visalia, CA)

Cows in confinement tend to grow long toes and build up excessive tissue on the soles of the feet. As a result, more weight is carried by the heels and the hocks, and the pasterns are subjected to extra stress. If these conditions are not corrected by proper hoof trimming, permanent damage may result in the form of crooked legs and weak pasterns, and the productive life of the cow will be shortened.

If the hoofs are trimmed properly, the animal stands squarely and walks properly, with each leg directly under the weight it supports.

The reason for and the technique of trimming hoofs are the same whether a cow is in a commercial production string or being fitted for show.

Also see Chapter 4, Fitting and Showing Dairy Cattle, Table 4–1, Training and Grooming Guide for Dairy Cattle.

MARKETING DAIRY HERD REPLACEMENTS

Most established dairy producers have surplus animals to sell—replacement heifers, cull cows and bulls, veal calves, and feeder animals for growing and finishing. Those establishing new herds, and to a more limited extent those maintaining herds, are on the buying end of the business. Dairy animals intended for slaughter are marketed through the same channels as beef animals sold for slaughter.

Dealers are the major outlet for dairy replacement stock, with most of their purchases made on the farm. Dealers resell replacement stock by either private treaty or auction. In addition to dealers, with some variation from area to area, replacement heifers are marketed through breed registry association sponsored sales (usually state or local sponsorship), cooperatives, artificial insemination associations, and dairy cattle sales associations.

In most areas, the following improvements would benefit both the seller and the buyer of replacement heifers:

1. Establishing and using uniform grades.
2. Reporting market information on prices.
3. Maintaining high health standards, and getting uniform health regulations from state to state.
4. Keeping production records on all cows from which replacement heifers will be sold.
5. Lowering the stress and disease losses which accompany the transportation of heifers.

TIME AND FREQUENCY OF FEEDING AND WATERING

Where complete rations are fed, cows are generally fed twice or three times daily. Where the forage and grain are fed separate, most dairy farms provide hay free-choice at all times, feed silage once or twice daily, and feed concentrates twice daily.

Hay may be put out in large amounts once daily, or in small amounts several times a day, depending on the type and preparation of the hay and labor situation. Cows will eat more if moderate amounts are fed at frequent intervals, but the labor requirement may be prohibitive. At least twice-a-day feeding of hay is desirable to reduce wastage. If wastage exceeds 10%, there is need for improvement in the manner of feeding or in the quality of the hay, or both.

Silage is usually fed once daily, because of the added cost of more frequent feeding. However, some of the automated systems do not have the added labor cost factor, with the result that they often feed twice daily, or more frequently. Silage should be fed soon after milking, so that any residual feed flavors will disappear before the next milking. Cows should not have access to silage, or other feeds that cause off-flavored milk, for at least 2 to 3 hours prior to milking.

Grain should be fed twice daily. It may be fed individually in the milking barn or milking parlor according to level of production. Where a milking parlor is used, feeding the concentrate therein serves as an inducement for the cows to enter the parlor. But high-producing cows are not in the milking parlor long enough to eat all the grain that they need; hence, part of the concentrate of high producers is generally fed in the manger along with the hay and/or silage. Group or corral feeding of grain has evolved, as a means of lowering labor costs. Under this system, lactating cows are grouped according to level of production. Grain is fed twice daily in the manger, right after milking, either (1) as a top dressing on the silage and/or hay, or (2) mixed with the silage and/or hay, with the grain allocation for each group or corral determined by the average level of production of each cow in the group.

Calves should be fed milk or milk replacer once or twice daily, at regular intervals. It is more harmful to overfeed than to underfeed a young calf.

Water should be clean, fresh, and available at all times. If there is too little water or if the cows must stand in line to get it, milk production will suffer. In cold weather, it may be necessary to protect water from freezing, as cows cannot get sufficient water by licking ice.

MANAGING DAIRY CALVES

The recommended practices in managing dairy calves follow, in summary form:

■ Dehorn anytime after 10 days of age, using an electric dehorner or caustic potash.

■ Remove extra teats before heifers are six months old; cut them off with clean scissors and disinfect with iodine.

■ Check for scours.

MANAGING REPLACEMENT HEIFERS

Fig. 14–9. Dairy herd replacements. (Courtesy, *Holstein World*, Sandy Creek, NY)

Good management of replacement heifers embraces the following principles and practices:

■ Separate bulls and heifers before 6 months of age; do not have over three month's difference in age of animals within a given group.

■ Treat for worms when the need is demonstrated.

■ Breed heifers at 15 to 18 months of age, but also consider weight and size.

■ Accustom bred heifers to milking barn procedure beginning about one month prior to calving.

MANAGING THE BULL

Many dairies rely on artificial insemination entirely, and do not keep mature bulls. Others keep one or more mature bulls.

All mature dairy bulls should be considered dangerous and handled as such. Safety pens and stalls are essential, and the bull should always be handled with a nose ring.

A satisfactory, but inexpensive, shelter should be provided; preferably in a separate barn, a little distance away from other animals. The stall should be about 12 ft square. It should open into a strongly fenced paddock to which the bull has access or is turned daily. The feed and watering arrangement should be such that the bull can be fed and watered without going into the pen.

The feeding of a herd bull should be such as to keep him in a thrifty, vigorous condition at all times.

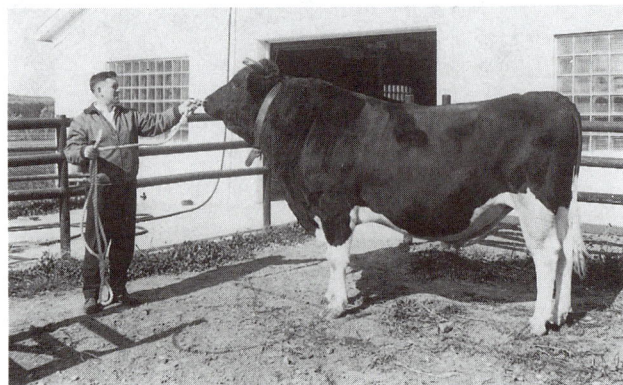

Fig. 14–10. Strong, safety bull pens are essential; and a bull should always be handled with a nose ring. (Courtesy, United Cooperative Farmers, Inc., Fitchburg, MA)

BEDDING CATTLE

Bedding is used primarily for the purposes of keeping animals clean and comfortable. But bedding has the following added values from the standpoint of the manure:

1. It soaks up the urine which contains about one-half the total plant food of manure.

2. It makes manure easier to handle.

3. It absorbs plant nutrients, fixing both ammonia and potash in relatively insoluble forms that protects them against losses by leaching. This characteristic of bedding is especially important in peat moss, but of little significance with sawdust and shavings.

The kind of bedding material selected should be determined primarily by (1) availability and price, (2) absorptive capacity, and (3) plant nutrient content. In addition, a desirable bedding should not be dusty, should not be excessively coarse, and should remain well in place and not be too readily kicked aside. Table 14–2 summarizes the characteristics of some common bedding materials.

In addition to the bedding materials listed in Table 14–2, many other products can be and are successfully used for this purpose, including leaves of many kinds, tobacco stalks, buckwheat hulls, processed manure (made by separating solid fibers from the liquid and water-soluble material in animal wastes), and shredded paper.

TABLE 14–2
WATER ABSORPTION CAPACITY OF BEDDING MATERIALS

Material	Lb of Water Per Lb of Bedding
Barley straw	2.10
Cocoa shells	2.70
Corn stover (shredded)	2.50
Corn cobs (crushed or ground)	2.10
Cottonseed hulls	2.50
Flax straw	2.60
Hay (mature, chopped)	3.00
Leaves (broadleaf)	2.00
(pine needles)	1.00
Oat hulls	2.00
Oat straw (long)	2.80
(chopped)	3.75
Peanut hulls	2.50
Peat moss	10.00
Rye straw	2.10
Sand	0.25
Sugar cane bagasse	2.20
Vermiculite[1]	3.50
Wheat straw (long)	2.20
(chopped)	2.95
Wood	
Dry fine bark	2.50
Tanning bark	4.00
Pine chips	3.00
sawdust	2.50
shavings	2.00
needles	1.00
Hardwood chips	1.50
shavings	1.50
sawdust	1.50

[1]This is a micalike mineral mined chiefly in South Carolina and Montana.

Fig. 14–11. Separated solids for bedding. (Courtesy, Dr. Tom Schultz, Farm Advisor, Visalia, CA)

Softwood (on a weight basis) is about twice as absorptive as hardwood, and green wood has only 50% the absorptive capacity of dried wood.

2. **Cut straw.** Cut straw will absorb more liquid than long straw, but chopped straws may be dusty.

3. **Fertility value.** From the standpoint of the value of plant food nutrients per ton of air dry material, peat moss is the most valuable bedding, and wood products the least valuable.

The minimum desirable amount of bedding to use is the amount necessary to absorb completely the liquids in manure. Per 24-hour confinement, the minimum daily bedding requirement, based on uncut wheat or oat straw, of a dairy cow is about 9 lb.

Under average conditions, about 500 lb of bedding are used for each ton of excrement.

REDUCING BEDDING NEEDS

In most areas, bedding materials are becoming scarcer and higher in price, primarily because (1) geneticists are breeding plants with shorter straws and stalks, (2) of more competitive and remunerative uses for some of the materials, and (3) the current trend toward more confinement rearing of livestock requires more bedding.

Dairy producers may reduce bedding needs and costs as follows:

1. **Collect liquid excrement separately.** Where the liquid excrement is collected separately in a cistern or tank, less bedding is required than where the liquid and solid excrement are kept together.

2. **Chop bedding.** Chopped straw, waste hay, fodder, or cobs will go further and do a better job of keeping animals dry than long materials. Chopped straw, for example, will soak up approximately 25% more moisture than long straw.

Naturally the availability and price per ton of various bedding materials vary from area to area, and from year to year. Thus, in the New England states shavings and sawdust are available, whereas other forms of bedding are scarce, and straws are more plentiful in the central and western states.

Table 14–2 shows that bedding materials differ considerably in their relative capacities to absorb liquid. Other facts of importance relative to certain bedding materials and bedding uses follow:

1. **Wood products (sawdust, shavings, tree bark, chips, etc.).** The suspicion that wood products will hurt the land is rather widespread, but unfounded. It is true that shavings and sawdust decompose slowly, but this process can be expedited by the addition of nitrogen fertilizers. Also, when plowed under, they increase soil acidity, but the change is both small and temporary.

3. **Use deep-bedding system.** A deep-bedding system—letting the bedding build up beneath; adding a light sprinkling of fresh bedding on top at intervals—will keep the animals warm and dry, and save in bedding.

4. **Ventilate quarters properly.** Proper ventilation lowers the humidity and keeps the bedding dry.

5. **Feed and water away from loafing quarters.** Cattle should be fed and watered in areas removed from their loafing quarters. With this type of arrangement, they defecate less in the bedded area.

6. **Consider rubber mats.** Rubber (either solid or foam rubber) bedding-replacers (or more correctly speaking, they are bedding-savers, for a limited amount of bedding is usually sprinkled over the top) are now available for use in stanchioned dairy barns. The life expectancy of solid rubber mats is 12 years; for foam rubber, about 4 years.

MANURE

The term manure refers to a mixture of animal excrements (consisting of undigested feeds plus certain body wastes) and bedding.

No doubt, the manure pollution problem, suspicioned or real, will persist. However, the energy crisis, accompanied by high chemical fertilizer prices, has caused manure to be looked upon as a resource. Therefore, planned manure management is an important part of modern dairy production. The collection, transport, storage, and use of manure must meet sanitary and pollution control regulations.

Modern dairy buildings and equipment should be designed to handle the manure produced by the animals that they serve; and this should be done efficiently, with a minimum of labor and pollution, so as to retrieve the maximum value of the manure, and make for maximum animal sanitation and comfort.

Table 14–3 shows the pounds of manure produced daily per 1,000 lb liveweight by different kinds of animals. These figures can be used to determine the tonnage production on a farm.

TABLE 14–3
DAILY PRODUCTION OF MANURE (FECES AND URINE)[1]

Kind of Animal	Production/Day/1000 Lb of Liveweight
	(lb)
Dairy cattle	85
Beef cattle	62
Sheep	36
Swine, breeders	50
Swine, feeders	69
Horses	50
Poultry	53
Humans	31

[1]Adapted by the author from *Agricultural Waste Management Field Manual*, USDA-SCS.

Table 14–4 shows the daily volume of manure produced by different kinds of animals. These figures can be used to determine storage needs in a lagoon or pit. A rule of thumb for under the floor pits is to figure that the pit will fill at the rate of 1 ft per month. Of course, the amount of the excreta is influenced by the composition of the ration; for example, high-silage rations produce more manure than high-grain rations.

TABLE 14–4
APPROXIMATE DAILY MANURE PRODUCTION, WITHOUT BEDDING[1]

Animal	Cu Ft/Day Solids and Liquids[2]	Gallons/Day[3]
1,000-lb cow	1½	11
1,000-lb steer	1	7½
10 head of sheep	½	4
10 head of hogs:		
50 lb	⅔	5
100 lb	1⅓	10
150 lb	2¼	17
200 lb	2¾	20½
250 lb	3½	26
1,000-lb horse	¾	5½

[1]Adapted by the author from *Michigan State University Circ. Bull. 231*.

[2]There are about 34 cu ft in a ton of manure.

[3]One cu ft = 7½ gal.

Fig. 14–12.   Lagoon for manure storage. (Courtesy, Dr. J. Schingoethe, South Dakota State University, Brookings)

DAILY MANURE PRODUCTION—STORAGE

Manure may be stored in a separate tank, in a nearby earthen dam lagoon, or it may be left to accumulate in a pit under slotted floors.

Storage capacity can be computed as follows:

Storage capacity = number of animals × daily manure
production × desired storage time
in days + extra water.

Example: *80 cows (1,000 lb each) × 1.5 x 120 days =
14,400 cu ft: 11 gal × 14,400 cu ft = 158,400-gal capacity.*

Water should be added to a pit (1) initially, and (2)
subsequently to replace evaporation losses from wastes.
Thus, if the manure is to be pumped, $\frac{1}{5}$ to $\frac{2}{5}$ of the storage
volume may be needed for the extra water. For irrigation,
there should be about 95% water and 5% manure. Water
should be kept to a minimum if the manure is to be field
spread with a tank wagon.

Generally 3 to 6 months' storage capacity is desirable.

AMOUNT, COMPOSITION, AND VALUE OF MANURE

The quantity, composition, and value of excrement produced vary according to species, weight, kind, and amount
of feed, and kind and amount of bedding. The author's computations are on a fresh manure (exclusive of bedding) basis.
Table 14–3 presents data by species per 1,000 lb liveweight,
whereas Table 14–5 gives yearly tonnage and value.

The data in Table 14–5 and Fig. 14–13 (next page) are
based on animals confined in stalls the year around. Actually,
the manure recovered and available to spread where desired
is considerably less than indicated because (1) animals are
kept on pasture and along roads and lanes much of the year,
where the manure is dropped, (2) of losses in weight which
often run as high as 60% when manure is exposed to the
weather for a considerable time, and (3) of losses in nitrogen
as NH_3 volatilizes with drying. Almost one-fourth of the total
nitrogen of cow manure may be lost in 12 hours of drying at
high temperature.

TABLE 14–5
QUANTITY, COMPOSITION, AND VALUE OF FRESH MANURE (FREE OF BEDDING)
EXCRETED BY 1,000 LB LIVEWEIGHT OF VARIOUS KINDS OF FARM ANIMALS

(1) Animal	(2) Tons Excreted/ Year/1,000 Lb Liveweight[1]	Composition and Value of Manure on a Tonnage Basis[2]						
		(3) Excrement	(4) Lb/Ton[3]	(5) Water	(6) N	(7) P_2O_5[4]	(8) K_2O[4]	(9) Value/ Ton[5]
			(lb)	(%)	(lb)	(lb)	(lb)	($)
Cow (beef or dairy)	12	Liquid	600	79	11.2	4.6	12.0	5.99
		Solid	1,400					
		Total	2,000					
Steer (finishing cattle)	8.5	Liquid	600	80	14.0	9.2	10.8	7.64
		Solid	1,400					
		Total	2,000					
Sheep	6	Liquid	660	65	28.0	9.6	24.0	13.48
		Solid	1,340					
		Total	2,000					
Swine	16	Liquid	800	75	10.0	6.4	9.1	5.65
		Solid	1,200					
		Total	2,000					
Horse	8	Liquid	400	60	13.8	4.6	14.4	7.05
		Solid	1,600					
		Total	2,000					

[1]*Manure Is Worth Money—It Deserves Good Care*, University of Illinois Circ. 595, 1953, p. 4.

[2]Columns 5, 6, 7, and 8 from *Farm Manures*, University of Kentucky Circ. 593, 1964, p. 5, Table 2.

[3]From *Reference Material for 1951 Saddle and Sirloin Essay Contest*, compiled by M. E. Ensminger, p. 43; data from *Fertilizers and Crop Production* by Van Slyke, published by Orange Judd Publishing Co.

[4]P_2O_5 can be converted to phosphorus (P) by dividing the figure given above by 2.29, and K_2O can be converted to potassium (K) by dividing by 1.2.

[5]Calculated on the assumption that nitrogen (N) retails at 25¢, P_2O_5 at 25¢, and K_2O at 17¢ per pound in commercial fertilizers.

TONS OF MANURE (FREE OF BEDDING) PRODUCED PER YEAR PER 1,000 LB WEIGHT

Fig. 14–13.   On the average, each class of stall-confined animals produces per year per 1,000 lb weight the tonnages shown above. (Drawing by R. F. Johnson)

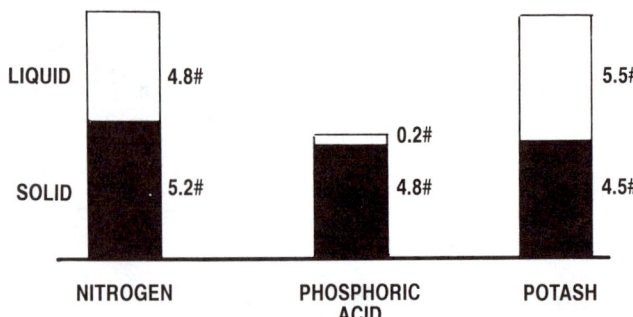

Fig. 14–14.   Distribution of plant nutrients between liquid and solid portions of a ton of average farm manure. As noted, the urine contains about half the fertility value of manure.

Fig. 14–15.   Liquid manure storage tank. (Courtesy, University of Minnesota, Crookston, MN)

About 75% of the nitrogen, 80% of the phosphorus, and 85% of the potassium contained in animal feeds are returned as manure. In addition, about 40% of the organic matter in feeds is excreted as manure. As a rule of thumb, it is commonly estimated that 80% of the total nutrients in feeds are excreted by animals as manure.

Naturally, it follows that the manure from well-fed animals is higher in nutrients and worth more than that from poorly fed ones. For example, steer manure produced by fattening cattle liberally fed on nutritious concentrates is more valuable than that produced from cattle wintered on hay.

The urine makes up 20% of the total weight of the excrement of horses, and 40% of that of hogs; these figures represent the two extremes in farm animals. Yet the urine, or liquid manure, contains nearly 50% of the nitrogen, 6% of the phosphorus, and 60% of the potassium of average manure; roughly one-half of the total plant food of manure (see Fig. 14–14). Also, it is noteworthy that the nutrients in liquid manure are more readily available to plants than the nutrients in the solid excrement. These are the reasons it is important to conserve urine.

The actual monetary value of manure can and should be based on (1) increased crop yields, and (2) equivalent cost of a like amount of commercial fertilizer. Numerous experiments and practical observations have shown the measurable monetary value of manure in increased crop yield. Table

14–5 (footnote 5), gives the equivalent cost of a like amount of commercial fertilizer. Of course, the cost of application of manure vs the cost of application of fertilizer must be considered.

Currently, we are producing manure (exclusive of bedding) at the rate of 1.25 billion tons annually (see Table 14–6). That is sufficient manure to add nearly 2/3 ton each year to every acre of the total land area (1.9 billion acres) of the United States.

Based on equivalent fertilizer prices (see Table 14–5) and livestock numbers (Table 14–6), the yearly manure crop is worth $6.7 billion.

Of course, the value of manure cannot be measured alone in terms of increased crop yields and equivalent costs of a like amount of commercial fertilizer. It has additional value for the organic matter which it contains, which almost all soils need, and which farmers and ranchers cannot buy in a sack or tank.

Also, it is noteworthy that, due to the slower availability of its nitrogen and to its contribution to the soil humus, manure produces rather lasting benefits, which may continue for many years. Approximately one-half of the plant nutrients in manure are available to and effective upon the crops in the immediate cycle of the rotation to which the application is

TABLE 14–6
TONNAGE AND VALUE OF MANURE (EXCLUSIVE OF BEDDING) EXCRETED BY U.S. LIVESTOCK[1]

Class of Livestock	Number of Animals on Farms[2]	Average Liveweight	Tons Manure Excreted/Year/ 1,000 Lb Liveweight[3]	Total Manure Production	Total Value of Manure[4]
		(lb)	(tons)	(tons)	($)
Cattle (beef and dairy; including steers) . . .	102,468,000	900	11	1,014,433,200	6,076,454,868
Sheep	10,328,000	100	6	6,196,800	83,532,864
Swine	50,960,000	200	16	163,072,000	92,356,800
Horses	8,519,000	1,000	8	68,152,000	480,471,600
				1,251,854,000	6,732,816,132

[1]In these computations, no provision was made for animals that died or were slaughtered during the year. Rather, it was assumed that their places were taken by younger animals, and that the population of each species was stable throughout the year.

[2]From USDA, *Agricultural Statistics 1987*, except horse numbers from American Horse Council, Inc., Washington, DC. The cattle and sheep figures are for Jan. 1, 1987; the swine figures are for Dec. 1, 1986; the horse numbers are for 1986. All figures are assumed averages throughout the year.

[3]*Manure Is Worth Money—It Deserves Good Care*, University of Illinois Circ. 595, 1953, p. 4.

[4]Computed on the basis of the value per ton given in the right-hand column of Table 14–5.

made. Of the unused remainder, about one-half, in turn, is taken up by the crop in the second cycle of the rotation; one-half of the remainder in the third cycle, etc. Likewise, the continuous use of manure through several rounds of a rotation builds up a backlog which brings additional benefits, and a measurable climb in yield levels.

Farmers and ranchers sometimes fail to recognize the value of this barnyard crop because (1) it is produced whether or not it is wanted, and (2) it is available without cost. Most of all, no one is selling it. Whoever heard of a traveling manure salesman?

MANURE GASES

When stored inside a building, gases from liquid wastes create a hazard and undesirable odors. Most (95% or more) of the gas produced by manure decomposition is methane, ammonia, hydrogen sulfide, and carbon dioxide. Several have undesirable odors or possible animal toxicity, and some promote corrosion of equipment. Table 14–7 gives some properties of the more abundant gases.

TABLE 14–7
PROPERTIES OF THE MORE ABUNDANT MANURE GASES[1]

Gas	Weight Air = 1	Physiologic Effect	Other Properties
CH_4 Methane	½	Anesthetic	Odorless, explosive
NH_3 Ammonia	⅔	Irritant	Strong odor, corrosive
H_2S Hydrogen sulfide . . .	1+	Poison	Rotten-egg odor, corrosive
CO_2 Carbon dioxide . . .	1⅓	Asphyxiant	Odorless, mildly corrosive

[1]*Beef Housing and Equipment Handbook*, Midwest Plan Service, Iowa State University, Ames, 1968, p. 10.

Animals and people can be killed (asphyxiated) because methane and carbon dioxide displace oxygen.

Most gas problems occur when manure is agitated or when ventilation fans fail.

No one should enter a storage tank, unless (1) the space over the wastes is first ventilated with a fan, (2) another person is standing by to give assistance if needed, and (3) wearing self-contained breathing equipment—the kind used for fire fighting or scuba diving.

It is important that maximum building ventilation be provided when agitating or pumping wastes from a pit. Also, an alarm system (loud bell) to warn of power failures in tightly enclosed buildings is important, because there can be a rapid buildup of gases when forced ventilation ceases.

MANURE USES

Historically, manure has been used as a fertilizer. But high feed prices and shortages of fossil fuels have made for new uses. Although it is expected that manure will continue to be used primarily as a fertilizer for many years to come, increasingly it will be recycled and used as a feed and converted into energy. Other manure-based products will continue to evolve, but it is expected that they will be of minor importance.

MANURE AS A FERTILIZER

With today's heavy animal concentration in one location, the question is being asked: How much manure can be applied to the land without depressing crop yields, making for salt problems in the soil, making for nitrate problems in feed, contributing excess nitrate to groundwater or surface streams, or violating state regulations?

Based on earlier studies in mid-western United States before the rise of commercial fertilizers, it would appear that one can apply from 5 to 20 tons of manure per acre, year after year, with benefit.

Fig. 14–16. Manure spreader. (Courtesy, Gehl Company, West Bend, WI)

Heavier applications can be made, but probably should not be repeated every year. With rates higher than 20 tons per annum, there may be excess salt and nitrate buildup. Excess nitrate from manure can pollute streams or groundwater and result in toxic levels of nitrate in crops. Without doubt the maximum rate at which manure can be applied to the land will vary widely according to soil type, rainfall, and temperature.

State regulations differ in limiting the rate of manure application. Missouri draws the line at 30 tons per acre on pasture, and 40 tons per acre on cropland. Indiana limits manure application according to the amount of nitrogen applied, with the maximum limit set at 225 lb per acre per year. Nebraska requires only one-half acre of land for liquid manure disposal per acre of feedlot, which appears to be the least acreage for manure disposal required by any state.

When a farmer has sufficient land, manure should be used at rates which supply only the nutrients needed by the crop rather than the maximum possible amounts suggested for pollution control.

PRECAUTIONS WHEN USING MANURE AS A FERTILIZER

The following precautions should be observed when using manure as a fertilizer:

1. Avoid applying waste closer than 100 ft to waterways, streams, lakes, wells, springs, or ponds.

2. Do not apply where percolation of water down through the soil is not good, or where irrigation water is very salty or inadequate to move salts down.

3. Do not spread on frozen ground.

4. Distribute the waste as uniformly as possible on the area to be covered.

5. Incorporate (preferably by plowing or discing) manure into the soil as quickly as possible after application. This will maximize nutrient conservation, reduce odors, and minimize runoff pollution.

6. Minimize odor problems by—

a. Spreading raw manure frequently, especially during the summer.

b. Spreading early in the day as the air is warming up, rather than late in the day when the air is cooling.

c. Spreading only on days when the wind is not blowing toward populated areas.

7. In irrigated areas, (a) irrigate thoroughly to leach excess salts below the root zone, and (b) allow about a month after irrigation before planting, to enable soil microorganisms to begin decomposition of manure.

MANURE AS A FEED

Recycling manure as a livestock feed is the most promising of the nonfertilizer uses. Various processing methods are being employed; and some manure is being fed without processing. More and more feedlot manure will be either (1) incorporated in a grower ration, or (2) fed to breeding herds during periods when pasture supplementation is beneficial, with the residues distributed over grazing areas where they would have fertilizing value.

Animal wastes contain several nutrients that are capable of being utilized when the material is recycled by feeding. Nitrogen, which is present in both protein and nonprotein forms, is a major constituent. Available energy is rather low. Fiber and ash are generally high. The high ash indicates that animal wastes are high in minerals; they are especially rich in phosphorus. Additionally, they contain certain vitamins synthesized in the digestive tract. The wastes processing the highest nutritive value are broiler litter, and layer waste.

One characteristic of all animal wastes is variability in composition due to diet regime, kind and amount of bedding, length of time before collecting, and processing method. The main difference in composition between raw and processed wastes is in the moisture content; many of the processed wastes are low in moisture.

The high fiber and considerable nonprotein nitrogen of animal wastes indicate that they are best suited for feeding to ruminants, since they possess a digestive tract capable of efficiently utilizing high fiber and nonprotein nitrogen. Also, because of their low energy content, they are best adapted for use in maintenance and gestating rations, rather than in lactating and growing rations.

Animal wastes processed by ensiling, dehydration, and other methods can be fed successfully to a wide range of animals. But, for best results the rations in which they become a part should be well balanced following their incorporation. Several workers have shown that the inclusion of too high levels of waste in a ration results in an excessive level of fiber and/or minerals, followed by lowered animal performance. Because of this limitation, not more than 10 to 20% waste should be included in high-energy rations, such as in cattle finishing rations. However, much higher levels (up to 80%) can be incorporated in rations of gestating cows.

■ **Poultry waste**—Nearly 100 million tons of poultry wastes (from layers, broilers, and turkeys) are produced annually. Because poultry production is highly intensive, with many birds in a small area, waste disposal is a major problem. Most cage-layer operations produce manure free of litter as the primary form of waste. Broiler operations, generally produce litter.

On a moisture-free basis, cage-layer manure generally contains 25 to 35% crude protein and minimal fiber, while broiler litter contains somewhat less protein—about 18 to 30% and substantially more fiber due to the presence of absorbent materials.

Poultry litter is the most collectable and the most nutritious of all animal wastes. It follows that many experiments have been conducted with it, involving feeding trials with different species. The results of numerous cattle experiments are summarized in Table 14–8. The mean values for waste-fed cattle reported therein were obtained by averaging all levels of feeding poultry wastes in the respective categories, though some of the levels were excessive. As shown in Table 14–8, the performance of cattle fed wastes was generally slightly lower than that of the controls that were fed traditional feed ingredients. But, on a dry-matter basis, animal wastes generally make for least cost rations and higher net returns.

Also, dried poultry litter has been fed successfully to dry and lactating cows.

Fig. 14–17. A manure burning electric generator plant. (Courtesy, Dr. Tom Schultz, Farm Advisor, Visalia, CA)

gas: their city cousins called it *sewer gas*.) However, it should be added that, due to capital and technical resources needed, for some time to come the production of methane

TABLE 14–8
PERFORMANCE OF CATTLE FED RATIONS CONTAINING POULTRY WASTES[1]

Species of Experimental Animal Used	Kind of Poultry Waste Studied	Performance of Experimental Animals		
		Criteria	Control Group	Waste-Fed Group
Cattle	Dehydrated layer waste	Daily gain . . . . , . . lb	2.35 *(1.07 kg)*	2.31 *(1.05 kg)*
		Daily feed dry-matter intake . . . lb	15.82 *(7.19 kg)*	15.44 *(7.02 kg)*
		Feed/gain ratio	7.81	7.72
Lactating Cows	Dehydrated layer waste	Milk yield . . . . lb/day	41.8 *(19.0 kg)*	38.94 *(17.7 kg)*
		Milk fat . . . . %	3.51	3.63
		Milk total solids . . . . %	12.04	12.01
Cattle	Poultry litter[2]	Daily gain . . . . lb	2.2 *(1.0 kg)*	1.91 *(0.87 kg)*
		Feed/gain ratio	10.18	11.58

[1]Adapted by the author from *Unidentified Resources as Animal Feedstuffs*, NRC, National Academy Press, Washington DC, 1983, pp. 132–144, Tables 35–41.

[2]Also, dried poultry litter has been fed successfully to dry and lactating dairy cows, to growing and breeding sheep, to growing swine, and to broilers.

MANURE AS A NONFEED ENERGY SOURCE

Manure can also serve as a source of nonfeed energy, which, of course, is not new. The pioneers burned dried bison dung, which they dubbed *buffalo chips*, to heat their sod shanties. In this century, methane from manure has been used for power in European farm hamlets when natural gas was hard to get. While the costs of constructing plants to produce energy from manure on a large-scale basis may be high, some energy specialists feel that a prolonged fuel shortage will make such plants economical. India now has many anaerobic digestion plants in operation.

Methane, of course, is usable like natural gas. There is nothing new or mysterious about this process. Sanitary engineers have long known that a family of bacteria produces methane when they ferment organic material under strictly anaerobic conditions. (Our grandparents called it *swamp*

by anaerobic digestion will likely be limited. If all animal manure were converted to energy, it has been estimated that it could produce energy equal to 10% of the petroleum requirements or 12.5% of our natural gas requirements.

SUMMARY OF MANURE

In the future, as fertilizer and feed become increasingly scarce and expensive, the economic value of animal manure will increase, and it will be looked upon as a resource and not as a waste that presents a disposal problem. More and more manure will be recycled and either (1) incorporated in a grower ration, or (2) fed to dry cows; and more and more manure will be used as a source of energy.

QUESTIONS FOR STUDY AND DISCUSSION

1. Define the terms *management* and *manager.*

2. Which is most important in a dairy operation—breeding, feeding, or management? Justify your answer.

3. Why are dairy records important?

4. Of the three milk and butterfat testing programs listed and discussed in this chapter, which would you recommend? Justify your choice.

5. For what purposes do progressive breeders use records as guides?

6. What method of individual animal identification would you recommend for (a) a commercial dairy and (b) a registered dairy herd? Justify your choices.

7. Discuss the present status and use of each of the following electronic devices for the identification of dairy cattle: (a) radio transmitter in the second stomach, and (b) transponder.

8. What method of dehorning would you recommend? Justify your choice.

9. Why should extra teats be removed from heifer calves? Give the procedure for removing extra teats.

10. Why, at what frequency, and how should hoofs be trimmed?

11. How may dairy herd replacements be marketed?

12. Discuss the time and frequency of feeding and watering dairy cows.

13. Outline pertinent management practices for (a) dairy calves, (b) replacement heifers, and (c) bulls.

14. What kind(s) of bedding do you recommend for dairy cattle?

15. List five ways in which dairy producers may reduce bedding needs.

16. Discuss each of the following pertaining to manure:
   a. The daily manure production from a dairy cow.
   b. The yearly production of manure from a dairy cow, along with its composition and value.
   c. Manure gases.
   d. Manure uses.

17. Would you recommend that a young person train to become a dairy manager?

SELECTED REFERENCES

Title of Publication	Author(s)	Publisher
Dairy Cattle: Principles, Practices, Problems, Profits, Third Edition	D. L. Bath, *et al.*	Lea & Febiger, Philadelphia, PA, 1985
Dairy Farm Management	T. Quinn	Delmar Publishers, Inc., Albany, NY, 1980
Illinois-Iowa Dairy Handbook	Cooperative Extension Service	University of Illinois, Urbana, and Iowa State University, Ames, 1983
Large Dairy Herd Management	C. J. Wilcox, *et al.*	University Press of Florida, 1978
Stockman's Handbook, The, Seventh Edition	M. E. Ensminger	Interstate Publishers, Inc., Danville, IL, 1992

Fig. 15–1. The cow doctor. (Courtesy, Bettmann Archive, New York, NY)

By
Dr. Robert F. Behlow, DVM, Professor and Extension Veterinarian,
North Carolina State University, Raleigh, North Carolina
and
Dr. M. E. Ensminger, President, Agriservices Foundation

Contents

---

[1]The material presented in this chapter is based on factual information believed to be accurate, but it is not guaranteed. When the instructions and precautions given herein are in disagreement with those of competent local authorities or manufacturers, always follow the latter.

CHAPTER

15

DAIRY CATTLE HEALTH, DISEASE PREVENTION, AND PARASITE CONTROL[1]

Contents   Page

with modern, rapid transportation facilities and the dense dairy population centers, the opportunities for animals to become infected are greatly increased compared with a generation ago.

It is conservatively estimated that annual U.S. losses from diseases, parasites, and pests of livestock and poultry are equivalent to about 15% of the cash receipts from marketings of livestock and livestock products.[2] Thus, with cash receipts of $76.218 billion in 1987, losses for that year on an equivalent basis of 15% would total $11.4 billion (76.218 × 15% = $11.4 billion).

Studies reveal that dairy producers suffer the following appalling losses:

1. Twelve percent of the cows that are bred never calve.
2. Calf losses from birth to weaning run 6%.
3. Ten percent of all calves (dairy and beef combined) are afflicted by calf scours, and 18% of all dairy calves so afflicted die.
4. Sterility and delayed breeding in dairy cattle make for an estimated yearly loss of $60 per cow, or a national total of $650 million.
5. Dairy herds average 10% breeding difficulty at any one time and 1.85 services per conception.
6. Retained placenta occurs in about 10% of the parturitions of dairy cattle.
7. Nearly 40% of all dairy cows have some form of mastitis, which causes a yearly loss of $90 to $250 per infected cow per year, depending on the severity of the infection, according to the National Mastitis Council.

In addition to the losses enumerated above, dairy producers may suffer even greater economic losses from decreased growth of young stock, lowered milk production, and increased production costs. Also, considerable cost is involved in keeping out diseases that do not exist in the United States; and quarantine of a diseased area may cause depreciation of land values or even restrict whole agricultural programs.

Dairy producers should also be well informed relative to the relationship of cattle diseases and parasites to other classes of animals and to humans, because many of them are transmissible between species. For example, over 200 different types of infectious and parasitic diseases can be transmitted from animals to people. Of most concern in the

---

[2]The author arrived at the 15% by two methods:

1. **Method #1:** A report prepared for the Council of Deans, Association of American Veterinary Medical Colleges, based on information available on disease losses as of Feb. 1, 1981, and including cattle (beef and dairy), sheep, swine, poultry, horses, and fish, showed animal disease losses of $10 billion per year. In 1980, cash receipts from marketing of total livestock and products amounted to $67.991 billion (Source: *Agricultural Statistics 1988*, USDA, p. 409, Table 581). So, a $10 billion disease/parasite loss was equivalent to 14.7% of the cash receipts from marketings of livestock and livestock products that year.

2. **Method #2.** *Losses in Agriculture*, Ag. Handbook No. 291, ARS, USDA, 1965, pp. 73–82, Tables 26–32, reported estimated annual U.S. losses from the more important diseases, parasites, and pests of livestock and poultry at $2.8 billion for the period 1951–60. During this same period (1951–60), cash receipts from marketings of total livestock and products averaged $17.76 billion (Source: *Agricultural Statistics 1962*, USDA, p. 567, Table 688). So, a $2.8 billion disease/parasite loss was equivalent to 15.8% of the average annual cash receipts for the 10-year period 1951–60.

Without doubt, one of the most serious menaces threatening the dairy industry is animal ill health, of which the largest loss is a result of the diseases that are due to a common factor transmitted from animal to animal. Today,

latter respect are such animal diseases as brucellosis (undulant fever), anthrax, Q fever, rabies, trichinosis, tuberculosis, and tularemia. Thus, rigid milk and meat inspection is necessary for the protection of human health. This is an added expense which the producer, processor, and consumer must share.

Fortunately, dairy cattle are milked two or three times daily, and milk production is a good barometer of the health of the cow. This daily observation and inspection enables the dairy producer to detect illness or disease in the early stage.

NORMAL TEMPERATURE, PULSE RATE, AND BREATHING RATE OF CATTLE

Table 15–1 gives the normal temperature, pulse rate, and breathing rate of cattle. In general, any marked and persistent deviations from these normals may be looked upon as a sign of animal ill health.

TABLE 15–1
NORMAL TEMPERATURE, PULSE RATE, AND BREATHING RATE OF CATTLE

Animal	Normal Rectal Temperature		Normal Pulse Rate	Normal Breathing Rate
	Average	Range		
	(°F)	(°F)	(rate/minute)	(rate/minute)
Cattle	101.5	100.4–102.8	60–70	10–30

Every dairy producer should have an animal thermometer, which is heavier and more rugged than the ordinary human thermometer. The temperature is measured by inserting the thermometer full length in the rectum, where it should be left a minimum of 3 minutes. Prior to inserting the thermometer, a long string should be tied to the end.

In general, infectious diseases are ushered in with a rise in body temperature, but it must be remembered that body temperature is affected by stable or outside temperature, exercise, excitement, age, feed, etc. It is lower in cold weather, in older animals, and at night.

The pulse rate indicates the rapidity of the heart action. The pulse of cattle is taken either on the outside of the jaw just above its lower border, on the soft area immediately above the inner dewclaw, or just above the hock joint. It should be pointed out that the younger, the smaller, and the more nervous the animal, the higher the pulse rate. Also, the pulse rate increases with exercise, excitement, digestion, and high outside temperature.

The breathing rate can be determined by placing the hand on the flank, by observing the rise and fall of the flanks, or, in the winter, by watching the breath condensate in coming from the nostrils. Rapid breathing due to recent exercise, excitement, hot weather, or poorly ventilated buildings should not be confused with disease. Respiration is accelerated in pain and in febrile conditions.

A PROGRAM OF DAIRY CATTLE HEALTH, DISEASE PREVENTION, AND PARASITE CONTROL

The operation of a dairy farm for maximum profit necessitates (1) good breeding, (2) good feeding, (3) good care and management, (4) good records, and (5) a good dairy health program. The primary objectives of the health program are (1) to prevent the introduction or occurrence of disease, and (2) to control the spread of infectious diseases and parasites.

The responsibility for the success of a dairy health program rests equally with the dairy caretaker and the veterinarian.

Dairy producers must provide good husbandry, nutrition, milking practices, sanitation, and records. Also, they should organize their work prior to the veterinarian's visitation, so that the veterinary work can be carried out in an efficient and organized manner, thereby effecting a saving in the veterinarian's time and the owner's expense. Dairy producers should inform their veterinarians of all treatments that were administered since the veterinarian's last visit, as well as any changes in feeding and management practices.

Veterinarians are responsible for providing an effective disease prevention program that meets the needs and requirements of each particular farm. Thus, they should make a visit routinely on a monthly basis, or more frequently in larger herds and at times of heavy breeding or trouble.

Although the exact herd health program will vary according to the specific conditions existing on each individual dairy farm, the basic principles will remain the same. Accordingly, the following program of dairy cattle health, disease prevention, and parasite control is presented with the hope that the dairy producer will use it (1) as a yardstick with which to compare their existing programs, or (2) as a guide so that they and their local veterinarians may develop a similar and specific program for each enterprise.

■ **General dairy cattle health**—The following general, overall health program should be observed on the dairy farm:

1. Provide plenty of exercise for pregnant cows, preferably by allowing them to graze in well-fenced pastures in which plenty of shade and water are available.

2. Keep lots and corrals well drained and as dry as practical, to prevent breeding places for foot-rot and parasites. Fence cattle out of pasture mud holes and ponds for the same reason.

3. If possible, divert drainage from adjacent farms and avoid common line fences which permit direct contact with the neighbor's cattle. Do not visit farms where infectious diseases exist, as the organisms may be brought home on shoes, clothing, or vehicles. For the same reason, feeds should not be bought from such farms, and one should not reuse feed bags.

4. Avoid areas on which cattle have over-wintered; and, preferably, use only those pastures that have not had cattle on them for one year or that have been plowed and renovated in the interim.

5. Follow a sound feeding program to prevent ketosis, milk fever, and nutritional diseases.

6. Prevent hardware disease by keeping wire, nails, and loose metals away from cows.

7. Test annually for tuberculosis, unless you are participating in a program which may permit less frequent testing.

8. The *milk ring test*, will serve as a screening test to monitor the herd against brucellosis.

9. Worm all calves and replacement heifers with a suitable vermifuge, as young animals are susceptible to parasite damage.

10. Use artificial insemination, unless (a) AI is not available or (b) there are compelling reasons to use a bull in natural service in a particular breeding program.

11. Lessen calving difficulties of first calf heifers by (a) feeding replacement heifers liberally and (b) breeding, when they reach adequate size, to a bull whose calves are small at birth.

12. When disease problems strike, isolate affected animals and obtain the services of a veterinarian. Remember that early diagnosis is the key to prompt treatment and response.

13. Isolate for 3 weeks all animals entering the herd from shows or sales.

■ **Infectious disease control** — Effective vaccines exist for the control of most common infectious diseases of dairy cattle.

Producers and their veterinarians should develop a vaccination program, then follow it routinely. Such a program will vary from herd to herd, depending upon (1) the location of the farm and (2) the disease situation on the farm and in the immediate area. Remember that the loss of just one animal, or the lowered milk production that usually accompanies a disease outbreak, will often exceed the cost of vaccinating the entire herd.

Also, some vaccinations can be used to control certain diseases once they have been accurately diagnosed.

■ **Reproductive disease control** — Reproduction is one of the most important parts of a dairy health program. Each day that a cow is "open" beyond 100 days following freshening costs the milk producer 50¢ to $1.00 per day. The following reproductive disease control program is recommended:

1. **Records.** Check all open cows for heat twice daily, and record heat periods. Cows remain in heat for only 14 to 16 hours. Hence, for highest conception, those found in heat in the morning should be bred that evening, and those found in heat in the evening should be bred the next morning. Also, keep accurate and complete breeding and calving records.

2. **Pregnancy examination.** A veterinarian or experienced technician should make a pregnancy examination anytime 35 days following service. Also, in problem herds, the veterinarian may perform an involution checkup 30 to 50 days after calving to determine if the ovaries and uterus are normal.

3. **Retained placenta; other genital abnormalities.** Cows with retained placenta should be examined and treated by the veterinarian 1 to 3 days following calving. Likewise, prompt treatment should be given for any abnormality of the genital organ, such as heat cycles of abnormal length or cystic ovaries.

■ **Modern milking** — A good milking program will lessen disease — especially mastitis, the bane of all milk produc-

ers — and increase profits from 5 to 25%. From a herd health standpoint, a modern milking program should embrace the following:

1. **Udder and teat protection.** Most mastitis is predisposed by injuries to the udder and teats. The best injury-preventive measure consists in using liberal amounts of clean, dry bedding.

2. **Sanitation.** Clean udders and teats are essential to the production of clean milk of high quality. This calls for (a) adequate space and a clean barn; (b) clean, dry corrals (free from low, wet, muddy, or swampy areas); and (c) clean, sterile milking equipment.

3. **Proper milking procedure.** First, the cow must be stimulated to release (let-down) her milk. This can be accomplished by the following:

a. Keeping the cow content in a quiet, peaceful atmosphere. Also, and most important, the milkers should love their work; otherwise, it is almost impossible to have this type of atmosphere.

b. Keeping regular milking hours.

c. Washing the teats and lower udder, which both stimulates and cleanses.

After stimulation, rapid and complete milking must follow immediately. Within about 45 seconds following the stimulation, pressure builds up in the cow's udder due to the action of the hormone oxytocin, which is released by the pituitary gland. This hormone action lasts for about 5 to 6 minutes. It causes the muscles to contract around the milk cells. So, within 1 minute, and not more than 1½ minutes, following the stimulation, attach the teat cups and begin milking.

Machine strip each teat when it is nearly collapsed and before milk flow has stopped. This is accomplished by pulling down gently on the teat cups and massaging each quarter with a gentle downward motion. Do not prolong machine stripping, as it will cause teat injury; about 20 seconds per cow is sufficient.

4. **Milking teat dip.** Use a postmilking teat dip (100 ppm iodophor or chlorine, or other sanitizing agent) to lessen the incidence of new infections during lactation.

5. **Milking machine repairs.** Keep milking machines — the most used machines on the dairy farm — in good operating condition at all times. Have them checked and serviced regularly; the vacuum and pulsations are of the greatest importance.

6. **Mastitis tests.** Quality milk cannot be produced if there is infectious mastitis in the herd. Clinical mastitis milk is unfit for human consumption because of (a) abnormal physical characteristics (flaky, lumpy, stringy, watery, or bloody); (b) high bacterial count; (c) poor flavor — usually salty; and (d) possible drug residues.

It has been said the producers themselves are responsible directly, or indirectly, for 90% of their mastitis troubles; however, most producers blame their milking machine. The three main routes through which mastitis comes are: (1) dirty, or poorly adjusted milking equipment; (2) poor milking practices; and (3) injuries to cows because of their surroundings.

Through the years, many different kinds of drug therapy have been used — including dyes, chemicals, sulfas, antibiotics, and nitrofurans. Many times such drugs have been effective, at least temporarily; in any event, acute cases of mastitis should be treated by a veterinarian.

In summary, it may be said that the producers themselves are unwittingly setting the stage for mastitis flare-ups in their herds, by providing the ideal conditions—poor milking practices, poor milking equipment, and improper surroundings. Through managed milking and sanitation, producers can reduce or eliminate mastitis.

■ **Calving time**—The following calving time care should be accorded:

1. When weather conditions permit, allow parturient cows to calve in a clean, uncontaminated, open pasture. During inclement weather or when pasture is not available, place the cows in isolated, roomy, light, well-ventilated maternity stalls—which should first be carefully and thoroughly disinfected (following the mixing directions printed on the disinfectant container) and provided with clean bedding for the occasion. After calving, all wet, stained, or soiled bedding should be removed and the floor sprinkled with lime; the afterbirth should be burned or buried deep in lime; and, if there has been trouble, the cows should be kept isolated until all discharges have ceased.

2. Unless the calves are born on a clean pasture away from possible infection, treat the navel cord of each newborn animal with tincture of iodine.

■ **Suckling calves**—

1. If the baby calves are confined to stalls, scrub stalls thoroughly twice each week with warm soap solution and disinfect the walls and feed bunks and/or mangers.

2. If you are in a brucellosis area, vaccinate all dairy heifer replacements against brucellosis at 2 through 6 months of age, observing the state regulations.

3. Vaccinate calves with blackleg and malignant edema bacterin at about 2 months of age and again at weaning time if these diseases are prevalent in the area.

■ **New stock**—

1. Vaccinate incoming calves against blackleg and malignant edema in areas that are endemic for these diseases.

2. Isolate newly acquired animals for a minimum of 3 weeks during which time they should be cared for by a separate caretaker.

3. While in isolation, test all newly acquired breeding animals for tuberculosis, brucellosis, leptospirosis, anaplasmosis, and Johne's disease; first, however, make every reasonable effort to ascertain that they come from herds which are known to be free from these and other diseases.

4. Spray newly acquired animals for lice control; and check them for internal parasites, and treat where indicated.

5. When possible, it is preferable to purchase virgin heifers from a disease control standpoint.

6. Thoroughly clean and disinfect the isolation stall after each animal(s) is removed and before a new animal(s) is placed therein. Disinfect with a hot 3% lye solution, followed by the use of another recommended disinfectant listed in Table 15–4, Disinfectant Guide.

■ **Sound practices**—The following practices are valuable adjuncts to the herd health program:

1. **Health records.** Each cow should be individually identified and have an individual health record. The latter should show all vaccinations, tests, past diseases, and treatments. This will assist the veterinarian in establishing a more accurate diagnosis when illness is encountered.

2. **Routine tests.** Routine tests should be conducted for brucellosis, tuberculosis, leptospirosis, and occasionally other diseases.

3. **Dehorning.** Dehorning prevents many injuries. Calves should be dehorned at an early age, when it is both easier and safer.

4. **Removal of extra teats.** Removal of extra teats imparts a better appearance to the udder and lessens some udder disease problems.

5. **Care of feet.** Sore feet markedly reduce production. So, the feet should be examined routinely, and trimmed and treated when necessary; thereby preventing many cases of lameness and foot disease.

6. **Parasite control.** Make routine examinations for both internal and external parasites, and treat as necessary. Parasites result in lowered feed efficiency and poor growth of young stock.

DISEASES OF DAIRY CATTLE

It is not intended that this book shall serve as a source of home remedies. Rather, enlightened dairy producers will institute a program designed to assure herd health, disease prevention, and parasite control. When animal disease troubles are encountered, they will not attempt to diagnose or treat. Instead, they will call upon their local veterinarian in exactly the same manner as they call upon their family doctors when human ill health is encountered.

This chapter is limited to nonnutritional diseases and ailments; mineral and vitamin deficiency diseases are covered in Tables 11–4 and 11–5, respectively, of Chapter 11, Fundamentals of Dairy Cattle Nutrition; and nutritional disorders of dairy cows are covered in Chapter 12, Dairy Feeding Programs.

ANTHRAX (SPLENIC FEVER, CHARBON)

Anthrax, also referred to as splenic fever or charbon, is an acute, infectious disease affecting all warm-blooded animals; but mature cattle are most susceptible. It usually occurs as scattered outbreaks or cases, but hundred of animals may be involved. Certain sections are known as anthrax districts because of the repeated appearance of the disease. Grazing animals are particularly subject to anthrax, especially when pasturing closely following a drought or on land that has been recently flooded. In the United States, most human infections of anthrax result from handling diseased or dead animals on the farm or from handling hides, hair, and wool in factories.

Historically, anthrax is of great significance. It was one of the first scourges to be described in ancient and Biblical literature; it marked the beginning of modern bacteriology, being described by Koch in 1876; and it was the first disease in which immunization was effected by means of an attenuated culture, Pasteur having immunized animals against anthrax in 1881. Vaccination of cattle in affected areas has reduced the incidence of anthrax to negligible proportions.

SYMPTOMS AND SIGNS[3]

The mortality is usually quite high. It runs a very short course and is characterized by a blood poisoning (septicemia). Victims present a picture of cerebral apoplexy (sudden staggering, difficult breathing, trembling, collapse, and a few convulsive movements) and die, frequently without showing any previous evidence of illness. During the course of the disease, the body temperature may reach 107°F; rumination ceases, milk production drops, and pregnant cows may abort.

CAUSE, PREVENTION, AND TREATMENT

The disease is identified by a microscopic examination of the blood in which will be found the typical large, rod-shaped organisms (*Bacillus anthracus*) causing anthrax. The bacillus can survive for years in a spore stage, resisting most destructive agents. As a result, it may remain in the soil for extremely long periods.

This disease is one that can largely be prevented by immunization. In the so-called anthrax regions, vaccination should be performed annually, usually in the spring, and well in advance of the time when the disease normally makes its appearance; and there should be adequate fly control by spraying animals during the insect season.

The nonencapsulated Stern-strain vaccine is used almost exclusively for livestock immunization. Vaccination should be done 2 to 4 weeks prior to the season when outbreaks may be expected. Animals should not be vaccinated within 60 days of slaughter.

Herds that are infected should be quarantined, and all milk and other products should be withheld from the market until the danger of disease transmission is past. The producer should never open the carcass of a dead animal suspected of having died from anthrax, because the organisms are infectious to humans; instead, the veterinarian should be summoned at the first sign of outbreak.

When the presence of anthrax is suspected or proved, all carcasses and contaminated material should be completely burned or deeply buried and covered with quicklime, preferably on the spot. This precaution is important because the disease can be spread by dogs, coyotes, buzzards, and other flesh eaters, and by flies and other insects.

When an outbreak of anthrax is discovered, all sick animals should be isolated promptly and treated. All exposed healthy animals should be vaccinated, pastures should be rotated, and a rigid program of sanitation should be initiated. Anthrax is a reportable disease, requiring quarantine. Hence, control measures will be carried out under the supervision of state or federal regulatory officials.

Treatment of affected animals, with massive doses of penicillin, may be effective if given soon enough.

---

[3]Currently, many veterinarians prefer the word *signs* rather than *symptoms*, but throughout this chapter the author accedes to the more commonly accepted terminology among dairy producers and includes the word symptoms.

BACILLARY HEMOGLOBINURIA (RED WATER DISEASE)

Bacillary hemoglobinuria, which is an acute, infectious disease, is often confused with other cattle diseases in which blood-colored urine is seen. The disease usually occurs in cattle that are pastured on meadows or irrigated lands where drainage is poor, and during the summer and early fall months. A mortality rate up to 100% occurs in untreated cases.

SYMPTOMS AND SIGNS

All ages are affected, but most losses occur in cows over one year of age.

The course of the disease is usually 2 days or less. Appetite, rumination, and milk flow suddenly cease. The animal hesitates to move and stands apart from the herd. The eyes are sunken, bloodshot, and may appear yellow. Breathing is rapid, and the temperature is high. Both the urine and the feces are usually blood tinged from destroyed red cells. It must be understood, however, that bloody urine (red water) may also be one of the symptoms in such conditions and diseases as lack of phosphorus, leptospirosis, Texas fever, plant poisoning, and anthrax. As red water disease can easily be confused with other conditions producing bloody urine, laboratory assistance is usually indicated in the event of unknown hemoglobinuria (bloody urine).

CAUSE, PREVENTION, AND TREATMENT

An anaerobic bacterium called *Clostridium hemolyticum* is the primary causative agent, and its toxin causes the blood breakdown. It is found in moist alkaline soils, especially in the low lying valley land of the Sierra Nevada, and the Coast and Cascade ranges in Nevada, California, and Oregon; and in Washington, Louisiana, Florida, Montana, Idaho, and Texas.

Vaccination about 2 weeks prior to the time of the previous annual outbreak is the best practical control measure. Also, the destruction of snails (by draining stagnant water and using bluestone) may aid in prevention.

Antiserum from hyperimmunized animals and blood transfusions, in conjunction with antibiotics, are the only known treatments for animals showing symptoms of the disease, but suitable medical treatment in the way of stimulants may be a valuable adjunct.

Since the disease may be confused with shipping fever, leptospirosis, and plant poisoning, it is essential that a veterinarian make the diagnosis.

BLACKLEG (BLACK QUARTER, EMPHYSEMATOUS GANGRENE, QUARTER ILL)

This is a very infectious, highly fatal disease of cattle, and less frequently of sheep and goats. The disease is widespread, especially in the western range states. It occurs at almost any season, predominating in the spring and fall months among pastured cattle; but it may occur in the winter in stabled cattle. Once prevalent in a community, the disease remains there as a permanent hazard, the infected territory being referred to as a *hot area*. It is seen most frequently in cattle ranging in age from 3 months to 2 years, but it may occur in older animals.

SYMPTOMS AND SIGNS

The incubation period is from 1 to 5 days, and its course is from 1 to 3 days. The first symptom noted is lameness, usually accompanied or followed by swellings of gas under the skin in the areas of the neck, shoulder, flanks, thighs, and breast, which crackle under pressure. High fever, loss of appetite, and severe depression accompany the symptoms. There are few recoveries; death is the usual termination, occurring within 3 days of the onset of symptoms.

Fig. 15–2. Heifer with blackleg, 6 hours before death. Note the lameness and the swelling over the neck and shoulder. (Courtesy, Veterinary Research Laboratory, Montana State University)

Animals that die from blackleg should not be cut open unless under the direction of a qualified veterinarian. The carcasses should be burned or deeply buried and the contaminated area disinfected. Eradication of blackleg from pastures is difficult if not impossible.

In the early stages of the disease, massive doses of antibiotics will sometimes save an animal. But a good immunization program is the key to preventing losses due to blackleg.

CAUSE, PREVENTION, AND TREATMENT

This disease is caused by an anaerobic bacterium, called *Clostridium chauvoei*, although it is often accompanied by other of the *Clostridia* genus. Infection is usually the result of wound contamination or ingestion of the organisms.

Calves should be vaccinated twice, 2 weeks apart, between 2 and 6 months of age. In high-risk areas, revaccination may be necessary at 1 year and every 5 years thereafter. A bacterin containing *Cl. chauvoei* and *Cl. septicum*, is a safe and reliable immunizing agent.

Parenteral and multiple local injections of penicillin will sometimes save an animal, provided they are given during the early stages of the disease. But a good immunization program is the key to preventing losses from blackleg.

BOVINE PULMONARY EMPHYSEMA (COW ASTHMA, PANTERS, LUNGERS, FOG FEVER, SKYLINE FEVER, SUMMER PNEUMONIA, GREEN GRASS POISONING, GRUNTS)

Bovine pulmonary emphysema usually occurs when cattle are abruptly changed from dry, mature feed to green, immature pasture. Typically, cases develop when cattle are moved from dry range to green mountain or irrigated pastures in the fall or late summer. It may also occur when cattle are first turned to lush pasture in the spring.

SYMPTOMS AND SIGNS

The disease is characterized by rapid and labored breathing (the animal forces air from the lungs, and may grunt with each breath). Affected animals may breathe through and froth at the mouth. The temperature remains normal or only slightly elevated, and the appetite is good.

CAUSE, PREVENTION, AND TREATMENT

The cause is unknown. It is noteworthy, however, that workers at Washington State University have produced the condition experimentally by giving cattle the essential amino acid tryptophan, leading to speculation that rapidly growing pastures may be high in tryptophan.

Prevention consists in avoiding sudden changes from dry or poor pasture to immature, green feed. Some dry hay should be fed while making the transition.

When bovine pulmonary emphysema strikes, call a veterinarian. Epinephrine, aminophyline, and corticosteroids are widely used. Remove affected animals from the offending pasture and handle quietly.

BOVINE RESPIRATORY DISEASE COMPLEX—BRDC (SHIPPING FEVER, HEMORRHAGIC SEPTICEMIA)

Bovine respiratory disease complex (BRDC) is an acute respiratory disease of cattle. The term *shipping fever* is losing favor because it is misleading and the term *hemorrhagic septicemia* is best reserved for the septicemic *Pasteurella* infections seen in cattle. The disease is most common in calves and following shipment. It occurs widely throughout the world, especially among thin and poorly nourished young animals that are subjected to shipment by truck or rail during periods of inclement weather, though it may occur in animals in good condition. The disease is a serious problem to both shippers and receivers of cattle.

SYMPTOMS AND SIGNS

The first sign of the disease (which may appear within 2 to 21 days after moving cattle) is a tired appearance and

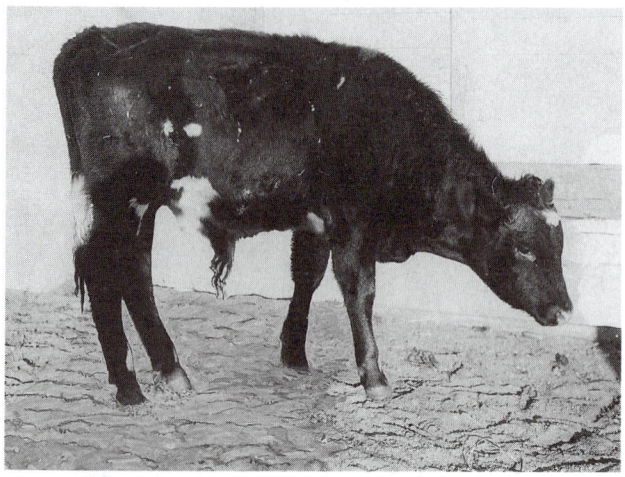

Fig. 15–3. An animal with shipping fever. The disease is most frequently associated with animals whose resistance has been lowered due to travel; hence, the name shipping fever. (Courtesy, USDA)

reduced appetite. The affected animal may show signs of depression, watery to slimelike nasal discharge, increased body temperature (rising to 105°–107°F), occasional soft or hacking cough, rapid breathing, loss of appetite, followed by loss of body weight and drop in milk production. In very acute forms, animals may die showing no symptoms. Death losses may be high in untreated cases.

Calves are more susceptible than older animals, but cattle of all ages are affected.

CAUSE, PREVENTION, AND TREATMENT

Bovine respiratory disease complex is caused from multiple infection due to the interaction of viruses and bacteria, accentuated by environmental conditions creating physical tension or stress. Change in weather and feed, overcrowding, hard driving, lack of rest, and improper shelter all help usher in the disease. The three viruses which cause most bovine respiratory infections are: IBR (infectious bovine rhinotracheitis), BVD (bovine virus diarrhea), and PI3 (parainfluenza 3). Other viruses which may cause respiratory problems include adenovirus, syncytial virus, rhinovirus, and rotavirus. Unfortunately, two or more of these organisms may infect a herd at the same time. But viruses are not the only agents of respiratory infection; bacteria can cause problems, too, especially in cattle already weakened by infections. For example, infection by *Pasteurella multocide* and *Pasteurella haemolytica* is thought to be a major cause of shipping fever (hemorrhagic septicemia). Other bacteria which may infect weakened cattle include *Haemophilus somnus* and species of *Salmonella*, *Pseudomonas*, and *Leptospira*.

As a preventive measure, one should eliminate as many as possible of the predisposing factors that lower the animal's vitality. Also, newly purchased animals should be isolated for 2 to 3 weeks before being placed in the herd.

Immunity against IBR, BVD, and PI3 can be achieved by administration of modified live or inactivated vaccines, in single or combination forms. The routes of administration of these vaccines are intramuscular (IBR, BVD, PI3), or in-

tranasal (IBR, PI3 only). Both intramuscular and intranasal vaccines provide adequate immunity.

Antibiotics and sulfa drugs are effective treatments if given early in the course of the disease. Treatment after BRDC develops is often disappointing.

BOVINE VIRUS DIARRHEA—BVD (MUCOSAL DISEASE)

Bovine virus diarrhea is not new, having first been described in 1946; but, in different periods of time and areas, it has been known as New York virus diarrhea, Indiana virus diarrhea, and mucosal disease. Improved methods of diagnosis and laboratory techniques now indicate that most, if not all, of the bovine virus diarrhea mucosal disease outbreaks are caused by the same organism.

The disease is widespread in the United States. The greatest dairy losses are in feed and milk production. Mortality is low, rarely exceeding 5%.

SYMPTOMS AND SIGNS

The incubation period is 7 to 9 days following exposure to the virus. The disease is characterized by high temperature (104° to 107°F) for 2 to 5 days, nasal discharge, rapid breathing, depression, and loss of appetite. Some animals make a prompt recovery. In other cases, signs persist, including nasal discharge and diarrhea. Sometimes blood flecks occur in the feces. Coughing, eye lesions, and lameness may affect 10% of the herd. In pregnant cows, abortions generally appear 3 to 6 weeks after infection; and, in lactating cows, a marked loss in milk production occurs.

CAUSE, PREVENTION, AND TREATMENT

As indicated by the name, the disease is caused by a virus.

The most effective preventive measures consist in avoiding contact with affected animals and in keeping away from contaminated feed and water. Also, all incoming animals should be isolated for at least 30 days. Once the disease makes its appearance, sick animals should be isolated and rigid sanitary measures should be initiated.

Where virus diarrhea is a constant problem, cows should be vaccinated. Immunity against BVD can be achieved by the intramuscular administration of modified live or inactivated vaccines. One vaccination should last a lifetime. Two precautions are important: (1) do not use the vaccine on pregnant cows because of possible abortions and birth defects and (2) do not vaccinate calves under 6 months of age because it may be ineffective due to the temporary immunity from colostrum of immune dams. Replacement dairy heifers should be vaccinated at 9 to 12 months of age.

Antibiotics or sulfonamides effectively combat the secondary bacterial invaders that accompany the disease. Administration of balanced electrolytes and fluid is indicated to rehydrate animals with diarrhea.

BRUCELLOSIS (BANG'S DISEASE)

Brucellosis, which occurs throughout the world, is an insidious (hidden) disease in which the lesions frequently are not evident. Although the medical term *brucellosis* is used in a collective way to designate the disease caused by the three different but closely related *Brucella* organism, the species names further differentiate the organisms as (1) *Br. abortus*, (2) *Br. suis*, and (3) *Br. melitensis.*

Brucellosis derives its name from a British Army surgeon, Sir David Bruce, who, in 1887, discovered the bacteria, later name *Brucella melitensis*. In cattle, it is called Bang's disease, after Professor Bang, noted Danish research worker, who, in 1896, first discovered the organism, *Brucella abortus*, responsible for bovine brucellosis, or contagious abortion, in cattle. In swine, it is Traum's disease, or infectious abortion, caused by *Brucella suis*. In goats, it is Malta fever, or abortion, caused by *Brucella melitensis*. In humans, it is Mediterranean fever, or undulant fever.

Although brucellosis remains a threat to the U.S. cattle industry, much progress has been made toward its eradication. The first national testing program, which was initiated in connection with the cattle reduction program necessitated by the drought of 1934, revealed a cattle infection level of 11.5%. Today, the infection level nationally is less than one-half of 1%.

Control and eradication of the disease are important in order to lessen economic loss. Additionally, we need to alleviate the danger of human infection, for brucellosis (undulant fever) is still an occupational risk for certain people.

Fig. 15–4. Cow with aborted fetus. Every case of abortion should be regarded with suspicion until proved noninfectious. (Courtesy, USDA)

SYMPTOMS AND SIGNS

Unfortunately, the symptoms of brucellosis are often rather indefinite. While abortion is the most readily observed symptom in cows, it should be borne in mind that not all animals that abort are effected with brucellosis and that not all animals affected with brucellosis will necessarily abort. On the other hand, every case of abortion should be regarded with suspicion until proved noninfectious.

The infected animal may prematurely give birth to a dead fetus, usually during the last third of pregnancy. On the other hand, the birth may be entirely normal; but the calf may be weak, or there may be retention of the afterbirth, inflammation of the uterus, and/or difficulty in future conception. The milk production is usually reduced. There may be abscess formation in the testicles of the male and swelling of the joints (arthritis). The observed symptoms in humans include weakness, joint pains, undulating (varying) fever, and occasionally orchitis (inflammation of the testes).

CAUSE, PREVENTION, AND TREATMENT

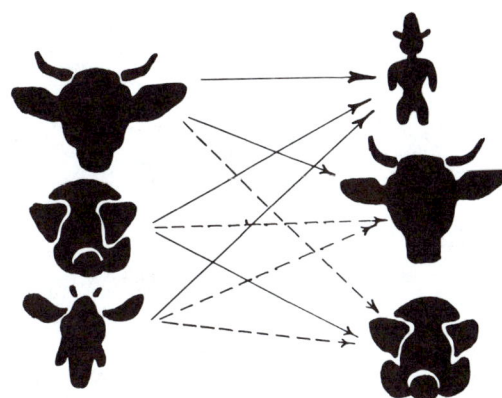

SOURCES OF INFECTION
Dotted lines indicate sometimes a source.

Fig. 15–5. Sources of brucellosis infection. (Drawing by R. F. Johnson)

The disease is caused by a bacteria called *Brucella abortus* in cattle, *Brucella suis* in swine, and *Brucella melitensis* in goats. The suis and melitensis types are seen in cattle, but the incidence is rare; swine are infected with both the suis and melitensis types; and horses may become infected with all three types.

Humans are susceptible to all three species of brucellosis. In most areas, the vast majority of undulant fever cases in humans are due to Brucella suis. The swine organism causes a more severe disease in human beings than the cattle organism, although not so severe as that induced by the goat type (*Brucella melitensis*). Fortunately, far fewer people are exposed to the latter, simply because of the limited number of goats and the rarity of the disease in goats in the United States. Dairy producers are aware of the possibility that human beings may contact undulant fever from handling affected animals, especially at the time of parturition; from slaughtering operations or handling raw meats from affected animals; or from consuming raw milk or other raw by-products from cows or goats, and eating uncooked meats infected with brucellosis organisms. The simple precautions of pasteurizing milk and cooking meat, however, make these foods safe for human consumption.

The *Brucella* organism is quite resistant to drying but is killed by the common disinfectants and by pasteurization. It is found in immense numbers in the various tissues of the aborted young and in the discharges and membranes from the aborted animals. It is harbored indefinitely in the udder and may also be found in the sex glands, spleen, liver, kidneys, bloodstream, joints, and lymph nodes.

Brucellosis appears to be commonly acquired through the mouth in feed and water contaminated with the bacteria, or by licking infected animals, contaminated feeders, or other objects to which the bacteria may adhere. Venereal transmission by infected bulls to susceptible cows through natural service may occur, but it is rare.

ROUTE OF BRUCELLOSIS GERMS IN THEIR ATTACK ON CATTLE

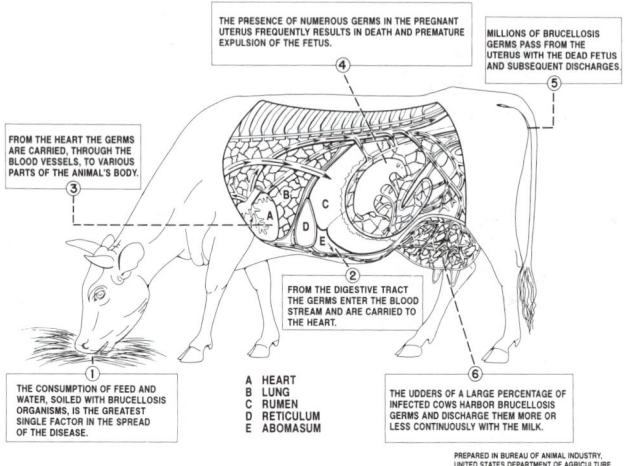

Fig. 15–6. Diagram showing how cattle become infected with the *Brucella* organism and its route of attack. (Courtesy, USDA)

Freedom from disease should be the goal of all control programs. Testing, the removal of infected animals, strict sanitation, proper and liberal use of disinfectants, isolation at the time of parturition, and the control of animals, feed, and water brought into the premises are the key to the successful control or eradication of brucellosis.

The nationwide cooperative federal-state brucellosis eradication program, which was initiated in 1934, has been very effective in reducing the incidence of bovine brucellosis in the United States. Four principles are involved: (1) finding infected animals and eliminating them from the herd; (2) vaccinating where there is a disease problem; (3) certifying brucellosis-free herds and areas—with the certification progressing from an individual herd, thence to an area or county, thence to a state; and (4) providing indemnity to farmers whose animals are condemned under the program.

In the United States, screening tests are used to locate infected herds. The *milk ring test*, using milk samples collected on each herd at the milk processing plant or creamery, is used in detecting infected dairy herds; and the *market cattle test*, based on blood samples at the time of slaughter, is used in detecting non-lactating infected herds. Supplemental tests, including acidified antigen tests such as the plate or card tests, may be employed to pinpoint infected animals in infected herds detected by a screening test.

The following tests are used for diagnosis of the disease in cattle:

1. **Agglutination tests,** of which there are two common methods:

a. **The tube or *slow* method,** in which a blood sample is taken from the jugular vein and used. This test is based on the phenomenon that the bloodstream of an infected animal contains an antibody, known as *agglutinin*. When blood serum containing this substance is brought in contact with a suspension of *Brucella* organisms (called an *antigen*), it causes the organisms to adhere to one another and form clumps. This action, known as *agglutination*, constitutes a simple test for diagnosing brucellosis in the living animal. Complete agglutination in dilutions of 1:100 and higher is positive.

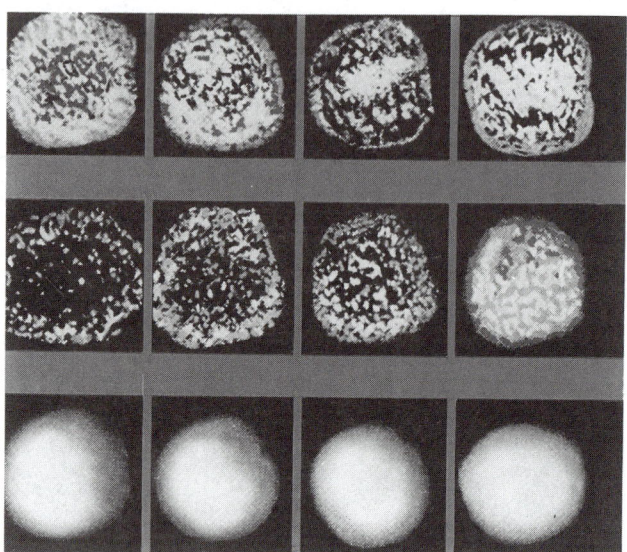

Fig. 15–7. Microscopic picture showing the blood (agglutination) test for diagnosis of brucellosis. Top row shows clumping (agglutination), indicating brucellosis. Center row shows complete clumping in the first three dilutions and partial clumping in the 1:200 dilution. Bottom row shows a negative test, indicating brucellosis-free.

b. **The plate or rapid test.** This is a rapid agglutination test which is done on a glass slide or plate. The antigen consists of specially selected strains of *B. abortus* stained with gentian violet and brilliant green.

2. **Milk ring test.** This is a modification of the agglutination test which is done with milk. The test involves mixing the antigen with fresh milk. The test depends on the fact that clumps of agglutinated organisms are carried to the surface by rising fat globules. A positive test is indicated by a purple cream layer with white milk below. The milk ring test is a highly efficient and accurate screening test for locating infected dairy herds.

3. **Card test.** This test involves the use of a disposable card in which blood serum or plasma is mixed with a buffered whole-cell suspension of *Br. abortus* (antigen), which reacts (agglutinates) with antibodies in the blood serum of animals infected with brucellosis. The card test detects infected animals that may be missed by the plate or tube agglutination tests. Also, it detects infected animals earlier, thereby making it possible to clear up an infected herd with fewer tests; it does not show nearly so many false reactions caused by nonspecific agglutinins as do the older tests; and it is more

likely to show Strain 19 vaccinated animals as negative (unless an actual infection is present), rather than reactors or suspects as the older tests frequently do.

Sound management practices, which include either buying replacement animals that are free of the disease or raising all females, are a necessary adjunct in prevention. Drainage from infected areas should be diverted or fenced off, and visitors (people and animals) should be kept away from animal barns and feedlots. Feeds should not be bought from farms that have infected animals, and one should beware of used feed bags. Animals taken to livestock shows and fairs should be isolated on their return and tested 30 days later.

Brucellosis cannot be eradicated by the use of Strain 19 alone, because (1) it is a live vaccine, and (2) it is only 65 to 75% effective under normal field conditions. Nevertheless, in problem herds and problem areas, it is recommended that dairy heifer calves be vaccinated with Strain 19 at 2 through 6 months of age. Overage vaccinations require longer to show a negative test, and, for this reason, often result in quarantines. A heifer vaccinated at 3 months of age will take about 40 days after vaccination to show a negative test, whereas a heifer vaccinated at 6 months of age will take about 70 days after vaccination to show a negative test, and an animal vaccinated at 8 months of age will require about 300 days to show a negative test.

To date, there is no known medicinal agent that is completely effective in the treatment of brucellosis in any class of farm animals. Thus, the dairy producer should not waste valuable time and money on so-called cures that are advocated by fraudulent operators.

(Also see later section in this chapter entitled, "Federal and State Regulations Relative to Disease Control.")

CALF DIPHTHERIA

Calf diphtheria is an acute, infectious disease of housed suckling calves and young cattle. The disease sometimes attacks these young animals as early as the third or fourth day after birth. If untreated, the mortality rate is very high. The disease is not to be confused with the diphtheria of humans, with which it has no relationship.

SYMPTOMS AND SIGNS

The affected animal shows difficulty in breathing, eating, and drinking. Drooling and swallowing movements may also be noted. Inspection of the mouth reveals yellowish crumbling masses and patches of dead tissue (diphtheritic membranes) on the borders of the tongue, adjacent to the molar teeth, and in the throat. Once established and unchecked, it will spread rapidly, eventually causing the death of the animal.

CAUSE, PREVENTION, AND TREATMENT

The alleged cause of this malady is the soil organism *Fusobacterium necrophorum*, the same organism that is often found in foot-rot. The organism gains entrance to the

tissues through wounds or eruptions in the mouth, and within 5 days after entrance will develop the symptoms noted.

Prevention consists of segregating the sick animals from the healthy ones and cleaning and disinfecting the quarters, not only after infection breaks out but before the calf is born. After outbreaks, all well animals should be checked daily.

Treatment consists of using sulfa drugs or broad spectrum antibiotics. The local application of a proteolytic enzyme for removal of the dead tissue is indicated.

CALF SCOURS (INFECTIOUS DIARRHEA, WHITE-SCOURS)

This is an acute, contagious, and often rapidly fatal disease of young or newborn calves. Most affected calves are less than 2 days of age. Outbreaks of calf scours are most common during fall, winter, and early spring.

Scours is the cause of more calf deaths than all other diseases combined. It is estimated that 10% of all calves in the United States are affected by the disease, and that 8% of beef calves and 18% of dairy calves so affected die. It is further estimated that calf scours cost the cattle industry $200 million annually.

SYMPTOMS AND SIGNS

Calf scours can vary from a mild to a severe disease. In the mild form, the main symptom is softer than normal feces. The severely affected calf initially appears depressed and has a lack of appetite. Then begins a severe diarrhea which consists of yellowish, foul-smelling, watery, or foamy feces. These calves can have a rough hair coat, sunken eyes, and appear emaciated. In very acute cases, death can occur before diarrhea is observed; however, death usually occurs 2 to 3 days after the onset of diarrhea. Some degree of associated pneumonia occurs more frequently in stabled dairy calves than in beef calves on the range.

It is sometimes difficult to distinguish the infectious dis-

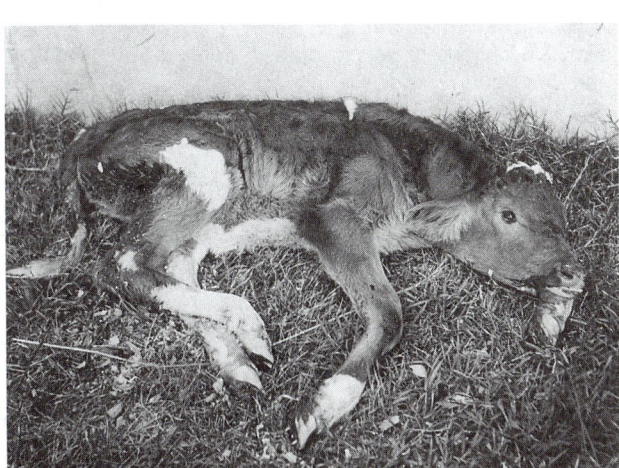

Fig. 15–8. Young calf with severe scours. Note sunken eyes, profuse salivation, and watery bowel discharge; which are characteristic symptoms. (Courtesy, University of North Carolina)

ease from diarrhea caused by other factors—such as overfeeding, irregular feeding, use of unclean utensils, too rapid changes in feed, or exposure to drafts and cold, damp floors. With the infectious type of scours, however, several calves are usually affected; and some animals may die quickly.

CAUSE, PREVENTION, AND TREATMENT

Enterotoxigenic K 99 plus *Escherichia coli*, rotavirus, and corona-like virus are the common causative agents. The *E. coli* cause diarrhea in calves from 1 to 3 days of age, the rotavirus and cornavirus cause diarrhea in calves from 5 to 15 days of age. Contributing causative factors include: insufficient colostrum, lack of antibodies in the colostrum of dams that have not been exposed to certain pathogens, stress due to weather, inferior milk replacers, inadequate housing and hygiene, and poor management.

To keep the disease away from the herd, one must prevent primary infection of the newborn. This rests on strict sanitary measures and isolation. The disease can be introduced by adding calves or adult animals from another herd. Calf diarrhea frequently occurs when a newly assembled herd begins to calve.

Weather conditions permitting, birth should preferably take place in the open on an uncontaminated, sun-exposed pasture. Otherwise, a clean, disinfected maternity stall should be provided, and the navel cord of the newborn calf should be treated with tincture of iodine. The newborn animal should be segregated from other animals and the contaminated quarters thoroughly cleaned and disinfected. Prevention should include proper feeding of pregnant cows and giving colostrum to calves that are subsequently to be raised on a milk replacer. When calf scours appear, infected animals should be segregated and the premises and feed containers should be thoroughly cleaned and disinfected.

The most effective preventive measure of calf scours involves the following three practices: (1) reduce the degree of exposure of newborn animals to the infectious agent, (2) provide resistance with adequate colostrum and optimal husbandry, and (3) increase the resistance of the newborn by vaccination of the dam 2 to 6 weeks before parturition to stimulate antibodies which are then passed on to the newborn through the colostrum.

Treatment of severely affected diarrheic calves should include discontinuing feeding milk for 24 to 48 hours; giving fluids orally and by injection to combat dehydration; administering gastrointestinal protectants; and giving antibiotics orally and by injection. The choice of the antibiotic should be made by the veterinarian, for in many areas the bacteria associated with calf diarrhea are resistant to many of the available drugs.

COW POX

Cow pox is an acute, infectious virus disease of the udder and teats of dairy cattle. It occurs throughout the United States, but it is not considered a serious disease.

SYMPTOMS AND SIGNS

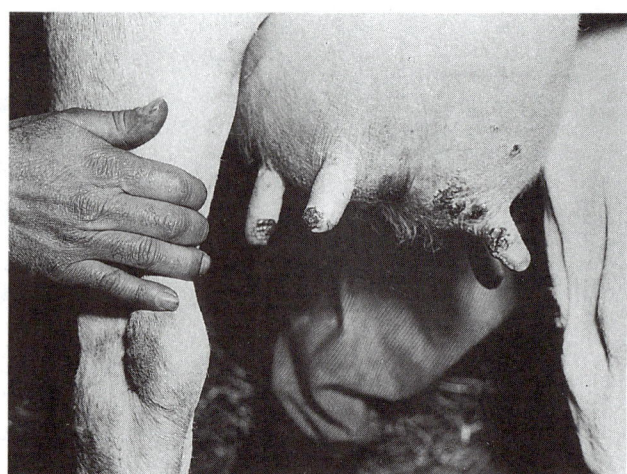

Fig. 15–9.   Cow pox, showing the scab stage. (Courtesy, College of Veterinary Medicine, University of Illinois)

Cow pox is characterized by reddish, painful spots that appear on the udder. In 24 to 48 hours, these change to yellowish white blisters. They then enter the pus stage, and in 10 to 12 days develop into scabs. In rare cases, the pox eruptions may appear on the body or limbs.

CAUSE, PREVENTION, AND TREATMENT

Cow pox is caused by a virus.

When an outbreak of cow pox occurs, the dairy producer should (1) isolate pox-infected cows and arrange for their care by a separate attendant, (2) have milkers wash their hands thoroughly as they go from cow to cow, (3) destroy flies and insects, and (4) vaccinate the herd.

To prevent cow pox (1) avoid contact with infected animals or milkers recently vaccinated against smallpox, (2) isolate new additions to the herd for 2 weeks, and (3) milk diseased cows last. Animals can be immunized by vaccination with a commercial smallpox vaccine, but the disease is not sufficiently prevalent or serious to justify widespread use of this practice.

There is no treatment that will destroy the virus or shorten the duration of the disease. A 3% orthophenylphenate solution applied as a wash to the udder and teats, or Whitfield's ointment, will tend to ease tissue sensitivity and soften the skin.

As a rule, one attack of cow pox confers a lasting immunity.

FESCUE FOOT (FESCUE TOXICITY)

Fescue is a valuable pasture grass that grows best in cool and cold weather. Under some condition, a fescue pasture may become toxic. Cattle that graze on such pastures do not perform well. In cold climates, they occasionally develop a crippling disease, known as fescue foot or fescue toxicity. Both dairy and beef cattle are susceptible; and this disease has been reported in sheep in Australia.

Most cases of fescue toxicity occur among cattle that graze pure stands of fescue during late fall and winter; and most toxic stands of fescue pasture are several years old. Fescue toxicity is more prevalent in animals suffering from malnutrition or parasitism.

SYMPTOMS AND SIGNS

There are variations in the severity of symptoms in cattle on toxic pastures. Some animals show no apparent lameness, whereas others show varying degrees of sloughing (necrosis) on the ends of their tails.

During the summer, cattle grazing toxic pastures show poor growth, increased temperatures, and increased pulse and respiratory rates. The main complaint that dairy producers make is the fact that the cattle are not doing as well as in previous years.

CAUSE, PREVENTION, AND TREATMENT

Fescue foot is associated with a fungus, which lives in the leaves, stems, and seed of the tall fescue plant and is not visible externally. The toxin is a vasoconstrictor that affects the blood vessels, which explains the extreme lameness found in the winter months. Occasionally, the circulation is closed to a degree that causes the entire foot to slough off. Such animals walk on stumps of bone.

The seeding of fungus-free fescue is the best way to prevent fescue foot. Toxic pastures should be renovated and some legume should be seeded with the fescue. It requires good pasture management, along with fertilization, to maintain a good fescue-legume pasture.

No mediation is effective for cattle with fescue foot. In severe cases where sloughing has occurred, the animals should be destroyed for humane reasons.

Cattle usually recover completely if they are removed from fescue pasture or fescue hay and are given other feed or pasture as soon as the first signs of the disease appears.

FOOT-AND-MOUTH DISEASE

This is a highly contagious disease of cloven-footed animals (mainly sheep, cattle, and swine) characterized by the appearance of watery blisters in the mouth (and in the snout in the case of hogs), on the skin between and around the claws of the hoof, and on the teats and udder. Fever, diminished rumination, and reduced appetite are other signs of the disease.

Humans are mildly susceptible, but very rarely infected, whereas the horse is immune.

Unfortunately, one attack does not render the animal permanently immune, but the disease has a tendency to recur, perhaps because there are several strains of the causative virus. The disease is not present in the United States, but there were at least nine outbreaks (some authorities claim 10) in this country between 1870 and 1929, each of which was stamped out by the prompt slaughter of every affected and exposed animal. No U.S. outbreak has occurred since 1929, but the disease is greatly feared. Drastic measures are exercised in preventing the introduction of the disease into

the United States, or, in the case of actual outbreak, in eradicating it.

Foot-and-mouth disease is constantly present in Europe, Asia, Japan, the Philippines, Africa, and South America. It has not been reported in New Zealand or Australia.

SYMPTOMS AND SIGNS

The disease is characterized by the formation of blisters (vesicles) and a moderate fever 3 to 6 days following exposure. These blisters are found on the mucous membranes of the tongue, lips, palate, cheeks, on the skin around the claws of the feet, and on the teats and udder.

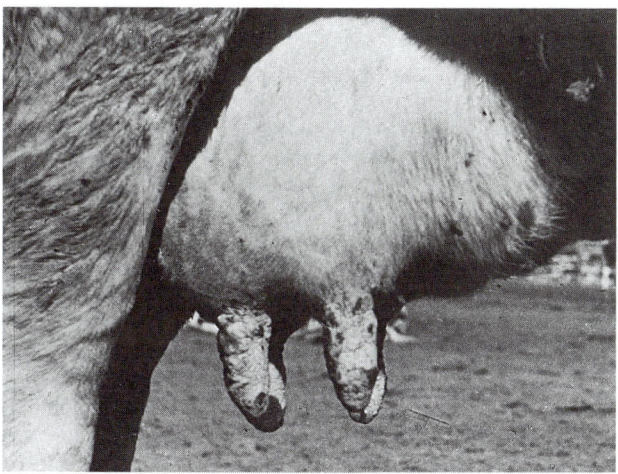

Fig. 15–10. Blisters (vesicles) on the teats of a cow with foot-and-mouth disease. (Courtesy, USDA)

Presence of these vesicles, especially in the mouth of cattle, stimulates a profuse flow of saliva that hangs from the lips in strings. Complicating or secondary factors are infected feet, caked udder, abortion, and great loss of weight. The mortality of adult animals is not ordinarily high, but the usefulness and productivity of affected animals is likely to be greatly damaged, thus causing great economic loss.

CAUSE, PREVENTION, AND TREATMENT

The infective agent of this disease is one of the smallest of the filterable viruses. In fact, it now appears that there are at least six strains of the virus. Infection with one strain does not protect against the other strains.

The virus is present in the fluid and coverings of the blisters, in the blood, meat, milk, saliva, urine, and other secretions of the infected animal. The virus may be excreted in the urine for over 200 days following experimental inoculation. The virus can also be spread through infected biological products and by the cattle fever tick.

Except for the nine outbreaks mentioned, the disease has been kept out of the United States by extreme precautions, such as quarantine at ports of entry and assistance with eradication in neighboring lands when introduction appears imminent. Neither live cloven-hoofed animals nor their fresh, frozen, or chilled meats can be imported from any country in which it has been determined that foot-and-mouth disease exists (meat imports from these countries must be canned or fully cured).[4]

In the United States, two methods have been applied in control: the slaughter method, and the quarantine procedure. Then, if the existence of the disease is confirmed by diagnosis the area is immediately placed under strict quarantine; infected and exposed animals are slaughtered and buried, with owners being paid indemnities based on their appraised value. Everything is cleaned and thoroughly disinfected.

In some countries, a third method of control is used—the use of a vaccine containing one or more immuno-types of the virus. However, the immunity produced by such vaccines lasts only 3 to 6 months. Vaccines have not been used in the outbreaks in the United States because they have not been regarded as favorable to rapid, complete eradication of the infection.

Fortunately, the foot-and-mouth disease virus is quickly destroyed by a solution of the cheap and common chemical sodium hydroxide (lye). Because quick control action is necessary, state or federal authorities must be notified the very moment the presence of the disease is suspected.

No effective treatment is known.

FOOTHILL ABORTION (EPIZOOTIC BOVINE ABORTION)

This is an infectious disease of cattle manifested primarily by abortion. It is epizootic in California, where it is known as *foothill abortion* because of its high prevalence in cows pastured in foothill terrain. The disease has been reported in western United States and in Europe.

SYMPTOMS AND SIGNS

Cows may abort when 3 to 6 months' pregnant, with the abortion rate frequently reaching 65%. Some calves are stillborn, others are weak.

CAUSE, PREVENTION, AND TREATMENT

The disease is caused by a virus (psittacoid virus).

The soft-bodied pajaroello tick, *Ornithodoros coriaceus*, is the vector.

Prevention consists in moving cattle out of tick-infested areas (dry-land brush areas) during the 3- to 6-month gestation period.

Immunizing agents and antibiotic medication have not been successful in preventing foothill abortion.

Animals which have aborted usually are immune and should be retained in the herd.

---

[4]Effective April 11, 1974, this rule was altered to permit dependent territories or possessions to be determined free of foot-and-mouth disease and rinderpest, regardless of the disease status of the mother country. Thus, dependent territories or possessions that are geographically separated from their mother country, such as colonies or former colonies, can be judged as to their livestock disease status by the same criteria previously applied to politically separate countries.

FOOT-ROT (FOUL-FOOT)

This disease is an inflammation of the hoofs of cattle, sheep, and goats; but cross infections of foot-rot between cattle and sheep do not occur. It is a potential hazard wherever animals of these species are kept, especially in wet, muddy areas.

SYMPTOMS AND SIGNS

A shrewd observer will first notice a reddening and swelling of the skin just above the hoof, between the toes, or in the bulb of the heel. As the infection progresses, lameness will be noted. If not arrested the infection will invade the soft tissue and cause a discharge of pus from the infected breaks in the skin. At this stage, a characteristic foul odor is present. Later, the joint cavities may be involved, and the animal may show fever and depression characteristic of a general infection. Affected animals lose weight, and, if lactating, produce less milk; and they may die if unattended.

CAUSE, PREVENTION, AND TREATMENT

Foot rot is a contagious, infectious disease caused by the organism *Bacterioides nodosus* in conjunction with *Fusobacterium necrophorum*. Possibly other organisms such as *Spirocheata penortha* and *Corynebacterium pyogenes* may be involved. Since the disease can be spread only by infected animals, early diagnosis followed by proper treatment will prevent spread of the disease and will result in considerable saving in time and expense. Walking animals over contaminated areas where infected animals have been is the principal means of spreading the disease, though the incidence is influenced by the weather and temperature. Cross infection between cattle and sheep does not occur.

Prevention of foot rot includes draining muddy pastures and segregating new animals. Foundation and replacement animals should be purchased from known clean sources. If animals from a questionable or unknown source, pass through a public market, or are transported in a public conveyance, they should be isolated on arrival, their hoofs should be trimmed, and then they should be walked through a 3% formalin foot bath, a 5% copper sulfate foot bath, or a mixed powdered copper sulfate and lime twice a day. Oral iodides have been beneficial as preventatives in some cases.

Control of foot rot in cattle is best achieved by alleviating muddy or abrasive walking areas, regular trimming of hooves, sanitation, isolating affected animals, and use of a suitable disinfectant.

Systemic and local treatment with antibiotics and sulfonamides is recommended. Other procedures that may speed recovery are cleaning the foot, applying a protective dressing, wiring the claws together, and removing the necrotic interdigital mass. Zinc methionine has been recommended for both treatment and prevention.

In advanced stages, best results are obtained by surgical amputation of the affected claw. Animals so treated soon walk on the one remaining healthy claw, as before.

INFECTIOUS BOVINE RHINOTRACHEITIS—IBR (RED NOSE)

Infectious bovine rhinotracheitis was first reported in the United States in 1950. Since then, it has occurred throughout the country. The main economic losses from the disease are in time, weight, milk production, and drugs. In dairies, IBR is most prevalent in large commercial operations.

SYMPTOMS AND SIGNS

Affected animals go off feed and lose weight; generally cough; may show pain in swallowing; usually slobber and show a nasal discharge; breathe rapidly, with difficulty, and in severe cases through the mouth; show severe inflammation of the nostrils and trachea; have a high fever, 104° to 107°F; and may remain sick for as long as a week. When the disease breaks out, 25 to 100% of the animals are affected. Death loss rarely exceeds 5%.

The IBR virus has a predilection for several different areas of the body. When in the eyes, it causes symptoms similar to pinkeye. When in the genital tract, it causes infectious postular vulvovaginitis (IPV).

Abortion due to the virus may occur in large dairy herds in which there is a recent history of IBR or of vaccination of pregnant animals against the disease.

CAUSE, PREVENTION, AND TREATMENT

The disease is caused by a virus.

Infectious bovine rhinotracheitis can be prevented by the use of a vaccine, of which there are two types:

1. The modified live virus vaccine provides a lasting immunity, but it should not be used on pregnant cows or on calves under 6 months of age.
2. The killed vaccine, which must be repeated.

There is no known treatment, but sulfonamides and antibiotics effectively combat the secondary bacterial invaders that accompany the disease.

JOHNE'S DISEASE (CHRONIC BACTERIAL DYSENTERY, PARATUBERCULOSIS)

This is a chronic, incurable, infectious disease seen chiefly in cattle; also found in sheep and goats, and more rarely in swine and horses. It resembles tuberculosis in many respects. The disease is very widespread, having been observed in practically every country where cattle are raised on a large scale. Apparently, it is increasing in the United States. It is one of the most difficult diseases to eradicate from a herd.

SYMPTOMS AND SIGNS

The disease seems to involve calfhood exposure with no evidence of infection for 6 to 18 months. At the end of this time, the animal loses flesh and displays intermittent diarrhea and constipation, the former becoming more prevalent. Af-

fected animals may retain a good appetite and normal temperature. The feces are watery but contain no blood and have a normal odor. The disease is almost always fatal, but with the animal living from a month to 2 years.

Upon autopsy, the thickening of the infected part of the intestines, covered with a slimy discharge, is all that is evident. This thickening prevents the proper digestion and absorption of food and explains the emaciation.

Fig. 15–11. Cow with Johne's disease. Note marked loss of flesh and tucked up flank. Diarrhea is a common symptom of the disease. (Courtesy, College of Veterinary Medicine, University of Illinois)

CAUSE, PREVENTION, AND TREATMENT

The disease is caused by the ingestion of a bacterium, *Mycobacterium paratuberculosis*. Inasmuch as this organism is acid-fast (that is, it retains certain dyes during a staining procedure), it resembles tuberculosis.

Effective prevention is accomplished by keeping the herd away from infected animals. If it is necessary to introduce new animals into a herd, they should be purchased from reputable breeders; and the owner should be questioned regarding the history of the herd. A new vaccine for Johne's has been approved by the USDA, but individual approval for each state is required for its use.

It must be borne in mind that apparently healthy animals can spread the disease. Testing at regular intervals of 3 to 6 months with *Johnin*, removing reactors, disinfecting quarters, and raising young stock away from mature animals should be practiced in infected herds. In using the Johnin test, however, it should be realized that it is not entirely accurate as a diagnostic agent. Some affected animals fail to react to the test.

No satisfactory treatment for Johne's disease has yet been found.

LEPTOSPIROSIS

Leptospirosis was first reported in cattle in the United States in 1944, although it had been found in dogs in the United States since 1939. Bovine leptospirosis has been reported in Europe, Australia, and the United States.

Human infections may be contracted through skin abrasions when handling or slaughtering infected animals, by swimming in contaminated water, through consuming raw beef or other uncooked foods that are contaminated, or through drinking unpasteurized milk.

SYMPTOMS AND SIGNS

In most herds, leptospirosis is a mild disease. However, the symptoms may vary from herd to herd, or even within a herd. In general, the symptoms noted in cattle are (1) high fever, (2) poor appetite, (3) abortion at any time, (4) bloody urine, (5) anemia, and (6) ropy milk.

All ages of cattle, and both sexes (including steers), are affected.

CAUSE, PREVENTION, AND TREATMENT

The disease is caused by several species of corkscrew-shaped organisms of the spirochete group; primarily *Leptospira pomona* in cattle, although *L. hardjo* and *L. grippotyphosa* are becoming more common.

The following preventive measures are recommended:

1. Blood test animals prior to purchase, isolate for 30 days, and then retest prior to adding them to the herd.
2. Do not allow animals to consume contaminated feed or water, or to breathe contaminated urinal mist.
3. Keep premises clean, and avoid used of stagnant water.
4. Control rodents.
5. Vaccinate susceptible animals annually if the disease is present in the area.

Where a herd is infected, the following control measures should be initiated:

1. Blood-test the herd and dispose of reactors either through (a) strict isolation, or (b) sale for slaughter only. Then retest every 30 days until two consecutive tests are obtained.

The same blood sample used in a brucellosis test may also be used for a leptospirosis test, simply by dividing the serum.

2. Spread the cattle over a large area; avoid congestion in a corral or barn.
3. Do not allow animals to drink from ponds, swamps, or slow-running streams, or to eat contaminated feed.
4. Clean and disinfect the premises; exterminate the rodents.

It should be recognized that carrier animals—animals that have had leptospirosis and survived—may spread the infection by shedding the organism in the urine. The infected urine may then either (1) be breathed as a mist in cow barns, or (2) contaminate feed and/or water and thus spread the infection. It is known that such recovered animals may remain carriers for 2 to 3 months or longer after getting over the marked symptoms. Fortunately, the organisms seldom survive for more than 30 days outside the animal. However, stagnant water and mild temperatures favor their survival.

Treatment, which should be prescribed by a veterinarian, may include blood transfusions, administration of selected antibiotics, and good care.

LISTERIOSIS (CIRCLING DISEASE)

Listeriosis is an infectious disease caused by the bacterium *Listeria monocytogenes*, which affects cattle, sheep, and goats; and it has been reported in swine, foals, and humans. Cattle of all ages are susceptible. One to 7% of the herd may be infected, and the mortality rate of infected animals is extremely high.

SYMPTOMS AND SIGNS

This disease affects the nervous system. Depression, staggering, circling, and strange, awkward movements are noted. One eye and one ear may be paralyzed. The animal may be seen holding a mouthful of hay for hours. There may be inflammation around the eye, and abortion may occur. The course of the disease is very short, with paralysis and death the usual termination. Positive diagnosis can be made only by laboratory examination of the brain.

CAUSE, PREVENTION, AND TREATMENT

Circling disease results from the invasion of the central nervous system by the causative bacteria, *Listeria monocytogenes*. The method of transmission is unknown. In an outbreak, affected animals should be segregated. If silage is being fed, discontinue that particular silage on a trial basis. Spoiled silage should be avoided, routinely.

A satisfactory therapeutic agent has not been found. Antibiotics and electrolytes give variable results.

LUMPY JAW AND WOODEN TONGUE

These two infections are chronic diseases affecting mainly the head of cattle—hence, the name *big head*. They occur most frequently in young cattle during the period of changing teeth. At one time, both of the conditions were referred to as actinomycosis, but now this term is used only for lumpy jaw. Actinobacillosis is the synonym for wooden tongue and soft tissue lesions.

SYMPTOMS AND SIGNS

Because of the area involved in these diseases, there is usually emaciation resulting from the difficulty encountered in chewing and swallowing.

Lumpy jaw only rarely attacks the soft tissues. It is usually confined to the bones of the lower jaw, although the upper jaw and nasal bones may be involved. The affected bone becomes enlarged and spongy and filled with creamy pus. As the disease progresses, inflamed cauliflower masses of tissue spread out and may appear on the surface, discharging pus or a foul odor. The surrounding flesh will also show inflammation, and the teeth may become loosened.

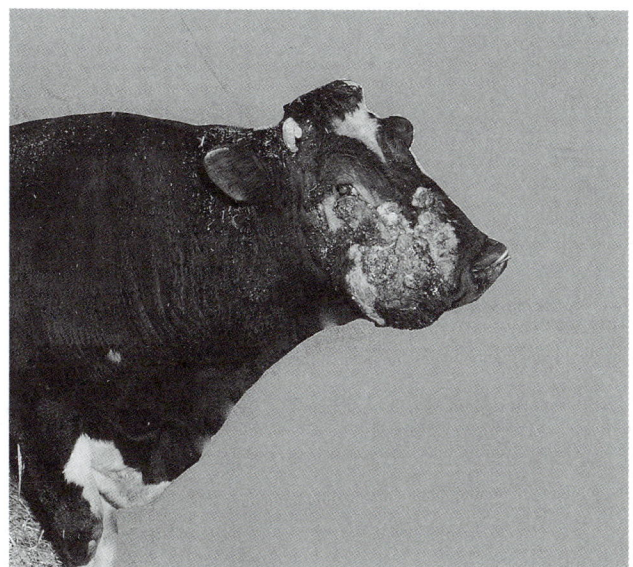

Fig. 15–12. Lumpy jaw (actinomycosis) in a Holstein bull, involving the upper jaw. Veterinarians were able to prolong the useful breeding life of this animal by massive antibiotic treatment. (Courtesy, College of Veterinary Medicine, University of Illinois)

Wooden tongue attacks chiefly the tissue in the throat area of cattle, but it may also involve the tongue, stomach, lungs, and lymph glands. The first lesion usually observed is a movable, tumorlike swelling about the size of a small egg under the skin in the infected area. The enlargements usually break open and discharge a light colored and very sticky pus. An involved tongue may or may not be ulcerated but will show an increase in size and hardness. The tongue may become quite immobile and may protrude from the mouth.

With this wooden tongue condition, there will be constant drooling, and the animals will lose weight and condition through inability to take feed. Although any chronic swelling in the region of the head should lead one to suspect the presence of infection, a positive diagnosis depends upon microscopic examination of the yellowish, granular pus material that will eventually discharge from the swelling.

CAUSE, PREVENTION, AND TREATMENT

Actinomyces bovis causes lumpy jaw; and *Actinobacillus lignieresei* causes wooden tongue. In each case, they may be assisted by secondary invaders. Both organisms lack invasive power, often being found in a normal oral cavity. They are thought to enter the tissue only by wound infections — for example, they may be carried in by the sharp awns of foxtail, barley, rye, bearded wheat, or oats.

Prevention consists of segregation and proper treatment or elimination of infected animals and the restricted feeding of material having sharp awns that might injure the animal's mouth. The latter precaution is important as the organism is a normal inhabitant of the mucous membranes of the mouth and nasal cavity of animals and humans.

Under some conditions, organic iodine appears to be an effective aid in the prevention and treatment of lumpy jaw in cattle. For prevention, add to the ration or salt 50 mg of ethylenediamine dihydriodide (EDDI) per head daily.

The veterinarian may, under certain conditions, (1) administer a water solution of an iodine salt of sodium or potassium, (2) prescribe an antibiotic, (3) resort to surgery, or (4) use x-ray therapy.

Treatment of lumpy jaw is not very satisfactory, but most cases of wooden tongue yield readily to treatment.

Sometimes treatment with organic iodide (EDDI) is effective. Add to the ration 250 to 500 mg/head/day for 2 to 3 weeks.

Superficial abscesses should be opened, drained, and swabbed with tincture of iodine.

MALIGNANT EDEMA (GAS GANGRENE)

This is an acute, generally fatal, toxemia of animals characterized by gangrene and emphysema around a wound. The incidence in a single herd may be high following castration, dehorning, or accidental wounds.

SYMPTOMS AND SIGNS

The affected animal goes off feed, breathes rapidly, and is profoundly depressed. A swelling forms around the wound. A gaseous and malodorous fluid exudes from the wound. In advanced stages of the disease, the animal is prostrated and often disoriented. There may or may not be a rise in temperature. Death occurs after a course of 12 to 48 hours. The mortality rate is high.

CAUSE, PREVENTION, AND TREATMENT

Malignant edema is caused by *Clostridium septicum* and related bacteria.

Since malignant edema is associated with contamination of wounds, the disease can be partially prevented by minimizing wounds and by castrating and dehorning under hygienic conditions.

Vaccination of young cattle with a vaccine containing *Cl. septicum* (for malignant edema) along with *Cl. chavoei* (for blackleg) at the time of the blackleg vaccination(s) will give some protection against malignant edema. Also, antibiotics may be administered for 4 to 5 days following surgery.

In the early stages of the disease, treatment with massive doses of antibiotics may be effective.

MASTITIS

Mastitis is an infectious inflammation or irritation in the udder which interferes with the normal flow of milk and/or its quality. It takes a heavier toll from the dairy industry than any other single disease. Losses are chiefly in decreased milk production and poor-quality milk.

Mastitis causes a yearly loss of $225 per afflicted cow, according to the National Mastitis Council.

Studies at Michigan State University have shown an overall average mastitis infection rate of 35% in dairies, with some individual operations running as high as 75%.

SYMPTOMS AND SIGNS

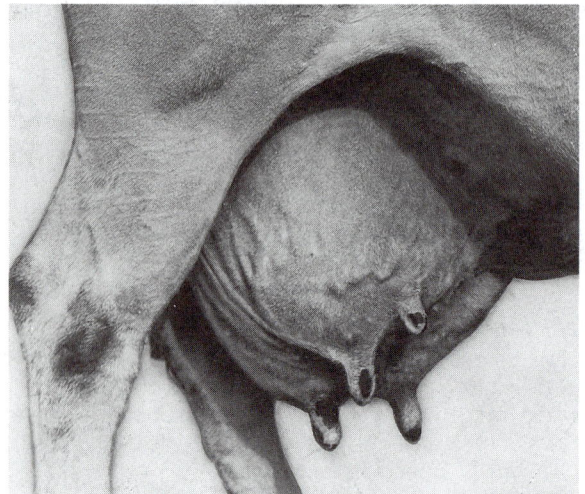

Fig. 15–13. Advanced mastitis. (Courtesy, USDA)

In acute mastitis, the udder is hot, very hard, and tender. The animal has an increase in temperature, refuses to eat, "loses its cud," and has dull eyes and a rough coat. Because of the soreness of the udder, the animal stands in an awkward position, and moves about and lies down with reluctance and difficulty. Milk is reduced greatly, and may become lumpy or watery. Abscesses may appear on the udder. Death often occurs in untreated, acute mastitis.

In chronic mastitis, the only symptom which may be noted is that the milk will be thick or lumpy.

CAUSE, PREVENTION, AND TREATMENT

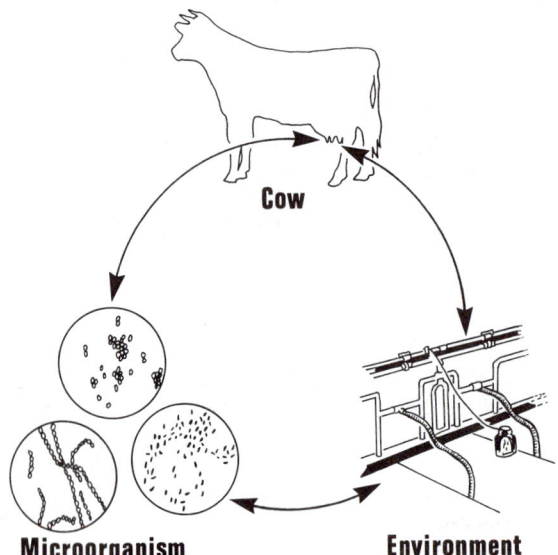

Fig. 15–14. Interrelationship of the three major factors involved in bovine mastitis.

Mastitis may be either infectious or noninfectious.

Infectious mastitis, resulting from the invasion of bacteria in the gland, may be from several different types of bacteria. Over 95% of all cases are caused by the following species of streptococci and staphylococci: *Streptococcus agalactiae, Streptococcus dysgalactiae, Streptococcus uberis,* and *Staphylococcus aurieus.* Infection with any of these organisms is usually chronic, with flareups occurring at regular intervals. No amount of drugs given to cows today can prevent another attack next month under the same conditions.

Noninfectious mastitis is the result of injury, chilling, bruising, or rough or improper milking.

The following preventive measures are recommended:

1. Milk all diseased cows last.
2. Use strip cup before milking. Promptly remove new cases from the milking line and segregate.

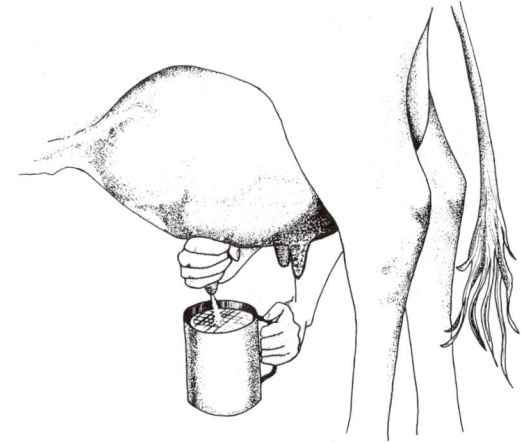

Fig. 15–15. Use the strip cup to obtain foremilk samples in order to spot new cases of mastitis.

3. Wash udders with clean individual towels placed in chlorine solution (with a strength of 200 ppm); then wring the towel dry and wipe.
4. Before milking each cow, wash hands with soap and water, disinfect them with chlorine solution, and wipe them dry, preferably with paper towels. Do not permit wet hand milking.
5. Milk properly, in a regular, rapid, and thorough manner.
6. If machines are used, remove the teat cups immediately when the milk flow stops.
7. Dip the ends of the teats in a chlorine solution after milking (chlorine with a strength of 200 ppm).
8. Do not milk on the floor.
9. Before sweeping the floor, sprinkle it with lime or superphosphate.
10. Use plenty of bedding for each cow.
11. Periodically check the milking machine, vacuum, and rubber parts. It is the most used machine on the dairy farm.
12. Keep the premises and equipment sanitary at all times. Use disinfectants.

13. Never make a cow hurry her movements; she may injure her udder or teats.

14. Use the utmost precaution in adding replacements to the herd. Preferably, raise them, or purchase heifers before they have freshened.

Although mastitis is usually apparent, it may be a "hidden" disease. Therefore, several different tests have been developed for detecting the presence of the causative microorganisms in lactating cows; among them: (1) *screening tests*, or *presumptive tests*, made either at the side of the cow or at the bulk tank, of which the California Mastitis Test (CMT) is the most widely used one; and (2) *specific laboratory tests designed to detect the causative organism*.

A reasonable goal, based on using the CMT, is to have at least 75% of the bucket milk samples score negative or trace. Less than 75% negative (–) and trace (T) bucket readings indicates a milking management problem. On an individual quarter-of-the-udder basis, 90% of the samples scoring negative or trace indicate a well-managed herd.

The veterinarian should be consulted relative to treatment.

Mastitis is usually treated by the intramammary injection of antibiotics, sulfa drugs, nitrofurans, or combinations of these drugs.

Acute cases should also be treated systemically by the veterinarian.

Local application of hot packs and udder massage will increase circulation in chronic cases; however, hot, acute, painful glands should be treated with cold applications.

Informative publications on mastitis are available from the National Mastitis Council, Inc., 1840 Wilson Boulevard, Suite 400, Arlington, VA 22201.

METRITIS

Metritis is an inflammation of the uterus, usually caused by various bacteria, which affects cattle, horses, sheep, and swine.

SYMPTOMS AND SIGNS

Metritis usually develops soon after the animal has given birth. It is characterized by a foul-smelling discharge from the vulva that becomes thick and yellow or white, and finally brownish or blood-stained. Also, there is chilling, high temperature, rapid breathing, marked thirst, loss of appetite, and lowered milk production. Pressure on the right flank may produce pain. The animal may lie down and refuse to get up. Affected animals may die in 1 to 2 days; or the acute infection may develop into a chronic form, producing sterility.

CAUSE, PREVENTION, AND TREATMENT

Metritis is caused by various types of bacteria, with *Escherichia coli* being the most common. Laceration at the time of calving, wounds caused by inexperienced operators, and/or retention of the afterbirth are the principal predisposing causes.

Preventive measures consist in alleviating as many of the predisposing factors as possible, including bruises and tears while giving birth, exposure to wet and cold, and the actual introduction of disease-causing bacteria during delivery or the manual removal of the afterbirth. Clean maternity stalls should be provided. If assistance at calving time becomes necessary, caretakers should first disinfect their hands and arms as well as the animal's external genitals.

Difficult parturition should be left to the veterinarian. Nothing is so distressing to the veterinarian as a history of long labor and well-meaning but ill guided attempts to remove a calf. Most cases are treated by introducing (in solution or tablets) an antibiotic or sulfa into the uterus.

NAVEL INFECTION (JOINT-ILL, NAVEL-ILL)

Navel infection is an infectious disease of newborn calves, foals, and lambs. It occurs less frequently in calves and lambs than in foals.

SYMPTOMS AND SIGNS

Navel infection is characterized by loss of appetite, by swelling, soreness and stiffness in the joints, by umbilical swelling and discharge, and by general listlessness.

Fig. 15–16. Calf with joint-ill. Note swelling in the joints. (Courtesy, College of Veterinary Medicine, University of Illinois)

CAUSE, PREVENTION, AND TREATMENT

Navel infection is caused by several kinds of bacteria.

The recommended preventive measures are sanitation and hygiene at mating and parturition and painting the navel cord of the newborn animal with iodine.

For treatment, the veterinarian may give a blood transfusion from the dam to the offspring; or administer a sulfa drug, an antibiotic, a serum, or a bacterin.

PINKEYE (KERATITIS)

This is the common name for an infectious disease that affects the eyes of cattle. It may be caused by several different infectious agents. Of the two most common forms of the disease, one is caused by a bacteria and the other by a virus. It attacks animals of any age, but is more common in young animals. It seems to become more virulent in certain years and in certain communities. The disease is widespread throughout the United States.

SYMPTOMS AND SIGNS

The first thing one may notice in bacterial pinkeye is the liberal flow of tears and the tendency to keep the eyes closed. There will be redness and swelling of the lining membrane of the eyelids and sometimes of the visible part of the eye. There

Fig. 15–17. Cow with pinkeye. Note eye discharge and the cloudiness or milkiness of the cornea or covering of the eyeball. (Courtesy, College of Veterinary Medicine, University of Illinois)

may also be a discharge of pus. Ulcers may form on the cornea. If unchecked, they may cause blindness and even loss of the eye. The attack may also be marked by slight fever, reduction in milk flow, and slight digestive upset.

In viral pinkeye, the eyeball itself is only slightly affected. Infectious bovine rhinotracheitis (IBR), a virus infection of the eyes of cattle, mainly affects the eyelids and the tissues surrounding the eyes. It causes a severe swelling of the lining of the lids.

CAUSE, PREVENTION, AND TREATMENT

■ **Bacterial pinkeye**—The most prevalent bacterial form of the disease is caused by *Moraxella bovis*. This organism produces a toxin which irritates and erodes the covering of the eye. Bacterial pinkeye occurs mainly during warm weather. Bright sunlight, wind, and dust may contribute to the cause of the disease. Cattle with white faces or lack of pigment around the eyes are rather susceptible. Transmission is mainly by flies and other insects that feed on eye dis-

charges of infected animals and then carry the infection to susceptible animals. Also, the disease is spread by direct contact, from animal to animal.

Prevention of bacterial pinkeye consists of the following: controlling face flies and other insects that feed around the eyes; good nutrition, including adequate vitamin A; and isolation of affected animals.

The most common treatment for bacterial pinkeye is the application of antibiotics or sulfa drugs to the affected eye as ointments, powders, or sprays; preferably, with treatment made twice daily. Foreign protein therapy, which is the subcutaneous or intramuscular injection of such things as sterile milk, has been used to treat pinkeye for years, with some success. Cortisone is sometimes combined with antibiotics and injected under the covering (at the outer edge) of the eyeball. The cortisone aids in reducing inflammation and pain and lessens the tears. Recovery is speeded up by keeping the infected animals in a dark barn. A commercially produced protective eye patch is now available. It completely covers the infected eye, holding the medication in place, protects the eye from insects and bright sunlight, and reduces the work and expense of handling and isolation. Held in place by a special adhesive, the eye patch drops off and decomposes after about 7 to 10 days.

■ **Virus pinkeye**—The best known virus infection of the eyes of cattle is caused by the "red nose" or infectious bovine rhinotracheitis (IBR) virus. It is much less common than bacterial pinkeye. When this organism infects the eyes of cattle, there may or may not be other signs of disease, such as respiratory infection, vaginitis, or abortion commonly associated with IBR. IBR conjunctivitis occurs most frequently in the winter, but it may be seen at other times of the year. The disease is highly contagious by direct and indirect contact of infected animals with susceptible animals.

IBR conjunctivitis may be prevented by proper vaccination of animals prior to onset of the disease. The herd should not be vaccinated once the disease appears; nor should pregnant cows be vaccinated. Affected animals should be isolated.

Treatment of IBR conjunctivitis is seldom of value, although antibiotics sometimes help reduce the secondary bacterial infection.

PNEUMONIA

Pneumonia is an inflammation of the lungs in which the alveoli (air sacs) fill up with an inflammatory exudate or discharge. The disease is often secondary to many other conditions. It is difficult to describe and classify, for the lung is subject to more forms of inflammation than any other organ in the body. It affects all animals. In cattle, it is seen most commonly as calf pneumonia, and frequently it accompanies shipping fever. If untreated, 50 to 75% of affected animals die.

SYMPTOMS AND SIGNS

The disease is ushered in by a chill, followed by elevated temperature. There is quick, shallow respiration, with discharge from the nostrils and perhaps from the eyes. A cough may be present. The animal appears distressed, stands with legs wide apart, drops in milk production, shows no appetite, and is constipated. There may be crackling noises with breathing, and gasping for breath may be noted. If the disease terminates favorably, the cough loosens and the appetite picks up.

Fig. 15–18. Calf with pneumonia. Note characteristic spread of front legs in an effort to ease breathing. (Courtesy, College of Veterinary Medicine, University of Illinois)

CAUSE, PREVENTION, AND TREATMENT

The causes are numerous. Many microorganisms found in other acute and chronic diseases, such as mastitis and metritis, have been incriminated; and pneumonia can be caused by a number of different viruses. One common cause that should be stressed is the inhalation of water or medicines that well-meaning but untrained persons give to animals in drenches. Also, it is generally recognized that changeable weather during the spring and fall, and damp barns, are conducive to pneumonia.

Prevention includes providing good hygienic surroundings and practicing good, sound husbandry.

Sick animals should be segregated and placed in quiet, clean quarters away from drafts.

Calves can be treated with a broad-spectrum antibiotic for 4 to 5 days. Secondary bacterial pneumonia may also be treated with sulfonamides or an antibiotic.

Q-FEVER (NINE-MILE FEVER)

Q fever was first recognized in Australia in 1935.

The disease has been identified in cattle in at least 35 states within the United States and, therefore, is recognized as endemic. An Ohio study indicates that the disease is more prevalent among large than among small dairy herds.

SYMPTOMS AND SIGNS

Coxiella burnetii produces only mild or inapparent illness in domestic animals, but affected animals serve as reservoirs of the organism. In cattle, herd to herd transmission has been demonstrated; and dairy farmers in areas where there is evidence of Q fever infection in cattle show serum antibodies against *C. burnetii*. Infection in a dairy herd may be followed by excretion of rickettsiae in the milk for as long as 32 months.

In humans, the disease manifests itself by acute onset, chills, prostration, and fever. Headache is pronounced in most cases. The fever is continuous and lasts from a few days to 2 or 3 weeks. It resembles influenza.

CAUSE, PREVENTION, AND TREATMENT

The causative organism of Q fever is *C. burnetii*. Ticks are the most important vector. People may acquire Q fever through the inhalation of contaminated dust (including tick feces). However, most persons become infected through exposure to livestock, or through the *ingestion* of their products (raw milk or meat of infected animals).

Prevention of Q fever in humans consists in avoiding ticks, exercising care when aiding animals at parturition, and pasteurizing milk properly.

Since milk-borne transmission of Q fever has occurred, pasteurization temperatures have been elevated slightly to ensure killing of the causative organism.

Vaccination appears to have some value in the control of the disease in both livestock and people.

Treatment with an antibiotic may reduce the duration of fever and illness, but the response to antibiotic therapy usually is not dramatic.

RABIES (HYDROPHOBIA, MADNESS)

Rabies is an acute infectious disease of all warm-blooded animals. It is characterized by deranged consciousness and paralysis, and terminates fatally. This disease is one that is far too prevalent, and, if present knowledge were applied, it could be controlled and even eradicated.

When a human being is bitten by a dog that is suspected of being rabid, the first impulse is to kill the dog immediately. This is a mistake. Instead, it is important to confine the animal under the observation of a veterinarian until the disease, if present, has a chance to develop and run its course. If no recognizable symptoms appear in the animal within a period of 2 weeks after it inflicted the bite, it is safe to assume that there was no rabies at the time. Death occurs within a few days after symptoms appear, and the dog's brain can be examined for specific evidence of rabies.

With this procedure, unless the bite is in the region of the neck or head, there will usually be ample time in which to administer the Pasteur treatment to exposed human beings. As the virus has been found in the saliva of a dog at least 5 days before the clinically recognizable symptoms, the bite of a dog should always be considered potentially dangerous until proven otherwise. In any event, when a human being is bitten by a dog, it is recommended that a physician be consulted immediately. Each year about 30,000 persons in the United States undergo the Pasteur treatment.

But not all animals that have bitten humans should be held for observation. Wild animals (skunks, raccoons, foxes, etc.) should be killed immediately and examined for evidence of rabies infection, because the signs of rabies in wild animals are variable and the duration of the virus excretion before clinical rabies develops may be longer than in dogs.

SYMPTOMS AND SIGNS

Less than 10% of the rabies cases appear in cattle, horses, swine, or sheep. The disease usually manifests itself in two forms: the furious, irritable, or violent form, or the dumb or paralytic form. It is often difficult to distinguish between the two forms, however. The furious type usually merges into the dumb form because paralysis always occurs just before death.

Fig. 15–19. Cow with rabies. Note the violent butting with the head; a characteristic of the furious form which is seen most often in cattle. At this stage, the animal is insane and is very dangerous, for it may attack and bite itself, other animals, or people. (Courtesy, Pitman-Moore, Indianapolis, IN)

The furious form is seen most often in cattle. In its early stages, the disease is marked by loss of appetite, cessation in milk secretion, anxiety, restlessness, and a change in disposition. This initial phase is followed by a stage of madness and extreme excitation indicated by a loud bellowing marked by a change in the voice, pawing of the ground, inability to swallow, and violent butting with the head. In all respects, the animal is insane and is very dangerous, for it may attack and bite itself or other animals and humans. On the fourth or fifth day, the animal becomes quieter and unsteady. This indicates approach of posterior paralysis. Loss of flesh is already very evident. On the sixth day, the animal may go into a coma and die.

CAUSE, PREVENTION, AND TREATMENT

Rabies is caused by a filtrable virus which is usually carried into a bite wound by the infected saliva of a rabid animal. The malady is generally transmitted to farm animals by dogs and certain wild animals, such as the fox and skunk.

Rabies can best be prevented by attacking it at its chief source, the dog. With the advent of an improved antirabies

vaccine for the dog, it should be a requirement that all dogs be immunized. This should be supplemented by regulations governing the licensing, quarantine, and transportation of dogs. For understandable reasons, the control of rabies in wild animals and bats is extremely difficult. In areas where rabies is present, all cattle should be vaccinated.

After the disease is fully developed in cattle, there is no known treatment. However, immediate use of a vaccine is recommended following known exposure of cattle, with the kind of vaccine determined by the veterinarian.

Persons bitten by a rabid animal should immediately report to the family doctor who will usually administer a vaccine. With severe bites, especially those around the head, antiserum is indicated.

TETANUS (LOCKJAW)

Tetanus is chiefly a wound infection disease that attacks the nervous system of cattle, swine, sheep, goats, and humans.

In the United States, the disease occurs most frequently in the South, where precautions against tetanus are an essential part of the routine treatment of wounds. The disease is worldwide in distribution.

SYMPTOMS AND SIGNS

The incubation period of tetanus varies from 1 to 2 weeks, but may be from 1 day to many months. It is usually associated with a wound but may not directly follow an injury. The first noticeable sign of the disease is a stiffness first observed about the head. The animal often chews slowly and weakly and swallows awkwardly. The third eyelid is seen protruding over the forward structure of the eyeball (called *haws*). Violent spasm or contractions of groups of muscles may be brought on by the slightest movement or noise. The animal usually attempts to remain standing throughout the course of the disease until close to death. If recovery occurs, it will take a month or more. In over 80% of the cases, however, death ensues, usually because of sheer exhaustion or paralysis of vital organs.

CAUSE, PREVENTION, AND TREATMENT

The disease is caused by an exceedingly powerful toxin (more than 100 times as toxic as strychnine) liberated by the bacterium *Clostridium tetani*. This organism is an anaerobe (lives in absence of oxygen) which forms the most hardy spores known. It may be found in certain soils, horse dung, and sometimes in human excreta. The organism usually causes trouble when it gets into a wound that rapidly heals or closes over. In the absence of oxygen, it then grows and liberates the toxin which follows up nerve trunks. Upon reaching the spinal cord, the toxin excites the symptoms noted above.

The disease can largely be prevented by reducing the probability of wounds, by general cleanliness, by proper wound treatment, and by vaccination with either a toxoid or an antitoxin in the so-called hot areas. Active immunization is achieved through two injections of toxoid at 2- to 4-week intervals, followed by annual booster injections.

When an animal has received a wound from which tetanus may result, short-term immunity can be conferred immediately by use of tetanus antitoxin, but the antitoxin is of little or no value after the symptoms have developed.

All perceptible wounds should be properly treated, and the animal should be kept quiet and, preferably, should be placed in a dark quiet corner free from flies. Supportive treatment is of great importance and will contribute towards a favorable course. This may entail artificial feeding. The animal should be placed under the care of a veterinarian.

TUBERCULOSIS

Tuberculosis is a chronic infectious disease of people and animals, which occurs worldwide. It is characterized by the development of nodules (tubercules) that may calcify or turn into abscesses. The disease spreads very slowly, and affects mainly the lymph nodes. There are three kinds of tuberculosis bacilli—the human, the bovine, and the avian (bird) types. Practically every species of animal is subjected to one or more of the three kinds, as shown in Table 15–2.

Fig. 15–20. Cow in the last stages of tuberculosis. Cattle are susceptible to all three kinds of tuberculosis. (Courtesy, USDA)

Many times an infected animal will show no outward physical signs of the disease. There may be a gradual loss of weight and condition and swelling of joints, especially in

TABLE 15–2
RELATIVE SUSCEPTIBILITY OF HUMANS AND ANIMALS TO THREE DIFFERENT KINDS OF TUBERCULOSIS BACILLI

Species	Susceptibility to Three Kinds of Tuberculosis Bacilli			Comments
	Human Type	Bovine Type	Avian (Bird) Type	
Humans	Susceptible	Moderately susceptible	Questionable	Pathogenicity of avian type for humans is practically nil.
Cattle	Slightly susceptible	Susceptible	Slightly susceptible	
Swine	Moderately susceptible	Susceptible	Susceptible	Ninety percent of all swine cases are due to the avian type.
Chickens	Resistant	Resistant	Very susceptible	Chickens only have the avian type.
Horses and mules	Relatively resistant	Moderately susceptible	Relatively resistant	Rarely seen in these animals in the U.S.
Sheep	Fairly resistant	Susceptible	Susceptible	Rarely seen in these animals.
Goats	Marked resistance	Highly susceptible	Susceptible	Rarely seen in these animals in the U.S.
Dogs	Susceptible	Susceptible	Resistant	Highly resistant.
Cats	Quite resistant	Susceptible	Quite resistant	Usually obtained from milk of tubercular cows.

In general, the incidence of tuberculosis is steadily declining in the United States, both in animals and humans. In 1917, when a thorough nationwide eradication campaign was first initiated, 1 cow in 20 had the disease; today, the number is about 1 in 1,000. Meanwhile, human mortality from tuberculosis dropped from 150 per 100,000 population in 1918 to 2 per 100,000 in 1988.

SYMPTOMS AND SIGNS

Tuberculosis may take one or more of several forms. Human beings get tuberculosis of the skin (lupus), of the lymph nodes (scrofula), of the bones and joints, of the lining of the brain (tuberculosis meningitis), and of the lungs. For the most part, tuberculosis in animals involves the lungs and lymph nodes. In cows, the udder becomes infected in chronic cases.

older animals. If the respiratory system is affected, there may be a chronic cough and labored breathing. Next to the lungs and lymph nodes, the udder is most frequently affected, showing increased size and swelling of the supra mammary lymph gland. Other seats of infection are the genitals, central nervous system, and the digestive system.

CAUSE, PREVENTION, AND TREATMENT

The causative agent is a rod-shaped organism belonging to the acid-fast group known as *Mycobacterium tuberculosis*. The disease is usually contracted by eating feed or drinking fluids contaminated by the discharges of infected animals.

With cattle, periodic testing and removal of the reactors is the only effective method of control. Also, cattle should not be pastured or housed with chickens. It is well to abide by the old adage, "once a reactor, always a reactor."

The test consists of the introduction of tuberculin—a standardized solution of the products of the tubercular bacillus—into an approved location on the animal.

There are three principal methods of tuberculin testing—the intradermic, subcutaneous, and ophthalmic. The first of these is the method now principally used. It consists of the injection of tuberculin into the dermis (the true skin).

Upon injection into an infected animal, tuberculin will cause a body reaction characterized by swelling at the site of injection. In human beings, the X-ray is usually used for purposes of detecting the presence of the disease, although the skin test may be used.

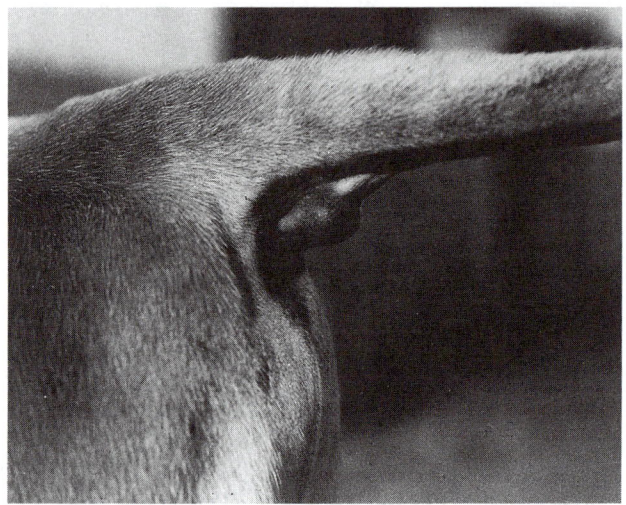

Fig. 15–21. A positive reaction (indicating the presence of tuberculosis) to the intradermic (into the true skin) tuberculin test in a cow. Reactors show a noticeable swelling, varying from the size of a pea to the size of a walnut, at the point of injection. The reading is made approximately 72 hours after injection. (Courtesy, College of Veterinary Medicine, University of Illinois)

As part of the federal-state tuberculosis eradication campaign of 1917, provision was made for indemnity payments on animals slaughtered.

Preventive treatment for both humans and animals consists of pasteurization of milk and creamery by-products and the removal and supervised slaughter of reactor animals. Vaccination against human tuberculosis, using B.C.G., is now being generally practiced in some foreign countries and being experimentally tried on some groups in this country. B.C.G. is the abbreviated designation of Bacille' Calmette-Guerin, after two French scientists who first prepared the vaccine. Although B.C.G. reduces the severity of the initial disease in cattle, it does not completely prevent infection. Also, vaccinated cattle react to the tuberculin test. It has, therefore, played little part in the control of bovine tuberculosis.

In human beings, tuberculosis can be arrested by hospitalization and complete rest along with drug therapy. But in animals this method of treatment is neither effective nor practical. Infected animals should be sent to slaughter.

All 50 states will accept for entry cattle meeting either of the following T.B. tests: (1) accredited herds tested within 12 months, or (2) individual negative tests within 30 days.

Cattle to be exported must be tested for tuberculosis and found free from the disease within 90 days of shipment.

(Also see later section in this chapter entitled, "Federal and State Regulations Relative to Disease Control.")

VAGINITIS (GRANULAR VAGINITIS, GRANULAR VENEREAL DISEASE)

This is an infectious disease which localizes in the cow's vulva and on the penis and prepuce of the bull, causing an inflammation of varying intensity. It occurs throughout the United States.

SYMPTOMS AND SIGNS

The tissue of the vagina is reddish, roughened, and granular in appearance. Economic losses are in terms of lower percentage calf crop and decreased milk production.

CAUSE, PREVENTION, AND TREATMENT

This condition is considered to be a response of the lymphatic tissue to an unknown irritant or antigen. It is non-specific, and not a disease in the classic sense.

Prevention consists in purchasing clean animals from clean herds and avoiding the use of bulls that have been exposed to the infection. Artificial insemination can be effectively employed in a control program. Also, in problem herds, vaccination 30 to 50 days prior to the breeding season should be considered.

Treatment of females is not indicated. The condition will clear up by itself in several weeks. Vaginitis in females is not directly related to fertility. However, the predisposing agents (viruses, bacteria, or fungi) may affect fertility.

The condition in bulls tends to be more persistent; and affected bulls may refuse to breed. They should be treated by massage of the anesthetized prolapsed penis and sheath with a suitable antibiotic ointment.

VIBRIONIC ABORTION (VIBRIO FETUS, VIBRIOSIS)

This is an infectious venereal disease of cattle which causes infertility and abortion. For diagnosis, laboratory methods must be used.

SYMPTOMS AND SIGNS

Infected herds are characterized by (1) abortions in the middle third of pregnancy, (2) several services per conception, and (3) irregular heat periods.

CAUSE, PREVENTION, AND TREATMENT

The disease is caused by the microorganism *Vibrio fetus*, which is transmitted at the time of coitus.

Prevention consists in avoiding contact with diseased

animals and contaminated feed, water, and materials. Also, vaccination, repeated annually, is effective in preventing the disease. Artificial insemination is a rapid and practical method of stopping infection from cow to cow.

Aborting cows should be isolated, and aborted fetuses and membranes should be burned or buried. Contaminated quarters should be thoroughly cleaned and disinfected.

Infected cows are treated by injecting drugs into the uterus and/or by allowing sexual rest.

WARTS (PAPILLOMATOSIS)

Warts, which are small tumors, are an infectious disease of cattle and other animals and humans. Young animals, under two years of age, are most often affected.

SYMPTOMS AND SIGNS

Warts are protruding growths on the skin, varying from very small to quite large, pendulous growths weighing several pounds. They may appear anywhere on the body, but they are especially common on the teats and/or around the head.

Although warts are a nuisance, their presence does not normally interfere with the animal's health. However, they damage the hide, making the leather derived therefrom weak in the affected area.

CAUSE, PREVENTION, AND TREATMENT

Warts are caused by a virus. It appears that each species of animals is attacked by a specific virus.

The following preventive measures are recommended:

1. Segregate "warty" cattle.
2. Clean and disinfect all exposed pens, stables, chutes, and rubbing posts.
3. Milk "warty" cows last.

The most common treatment among dairy producers consists of softening the wart with oil for several days, and then tying off the growth with thread or a rubber band or snipping it off with sterile scissors. The stump should then be treated with tincture of iodine. Wart vaccines help in some cases, but, generally speaking, they are more effective in prevention than in treatment. The veterinarian may resort to surgical removal of extremely large warts.

WINTER DYSENTERY (WINTER SCOURS)

This is an acute infectious disease of stabled cattle, both dairy and beef, most frequently occurring between the months of November and March.

SYMPTOMS AND SIGNS

It causes few death losses, but afflicted animals lose in condition; and, in lactating animals, there is a sharp reduction in milk flow.

The period of incubation is extremely short, varying from 3 to 5 days. A profuse watery diarrhea is the main symptom.

Often the feces are dark brown in color, and tend to become darker when intestinal hemorrhages occur. Usually the temperature remains normal, and the appetite is unchanged. Calves and young animals are least susceptible, but animals of all ages are affected. The seasonal incidence of the disease, the age and number of animals affected, together with the suddenness of the onset, are helpful in arriving at a correct diagnosis.

CAUSE, PREVENTION, AND TREATMENT

The cause is unknown, but the sudden onset and the rapidity of spread suggest a virus.

Prevention consists in isolating new or replacement animals. Also, any animal suffering from an acute attack of dysentery should be separated from the herd. Where the disease is encountered, rigid sanitation should be practiced.

Most cattle with winter dysentery do not require treatment since they recover within a few days. Severely afflicted animals may be given oral astringents along with fluid and electrolyte therapy.

PARASITES OF DAIRY CATTLE[5]

Dairy cattle are attacked by a wide variety of internal and external parasites.

Fig. 15–22. Calf with "bottle jaw." Generally this condition is indicative of a heavy infestation of several species of internal parasites. (Courtesy, School of Veterinary Medicine, Alabama Polytechnic Institute)

The prevention and control of parasites is one of the quickest, cheapest, and most dependable methods of increasing milk production with no extra cows, no additional feed, and little more labor. This is important, for, after all, the producer bears the brunt of reduced milk production, wasted

[5]From time to time, new insecticides and vermifuges are approved and old ones are banned or dropped. When parasitism is encountered, therefore, it is suggested that the dairy producer obtain from local authorities the current recommendation relative to the choice and concentration of the insecticide or vermifuge to use.

feed, and damaged hides. It is hoped that the discussion that follows may be helpful in (1) preventing the propagation of parasites, and (2) causing the destruction of parasites through the proper use of an effective anthelmintic or insecticide.

No specific suggestions concerning the choice of anthelmintics and insecticides for the control of parasites are given in the discussion that follows because of (1) the diversity of environments and management practices under which parasites occur, (2) the varying restrictions on the use of vermifuges and insecticides from area to area, and (3) the fact that registered uses of vermifuges and insecticides change from time to time. Information about what products are available and registered for use in a specific area can be obtained from the county extension agent, extension entomologist, or agricultural consultant.

ANAPLASMOSIS

Anaplasmosis is an infectious disease whose etiology and symptomatology are similar to cattle tick fever, except that more carriers are involved. It is caused by a minute parasite, *Anaplasma marginale*, which invades the red blood cells. The parasite is transmitted from infected to healthy animals by ticks, horseflies, stable flies, mosquitoes, deer flies, and probably by other biting insects.

DISTRIBUTION AND LOSSES CAUSED BY ANAPLASMOSIS

The disease is widely distributed in warm climates throughout the world. In the United States, it has been prevalent throughout the southern states, but it is slowly spreading to the northern states.

The mortality rate may vary from 2.5% to as high as 50 to 60%. The most severe losses are found in older animals and in hot weather.

LIFE HISTORY AND HABITS

In infected animals, the causative parasite, *Anaplasma marginale*, lives in the red blood cells. The parasite, and consequently the disease, may be transmitted from animal to animal by means of biting insects and by such mechanical agencies as needles, dehorning instruments, etc. Any animal that has once contracted the disease permanently retains the parasite in the blood, though no signs of ill health may be evident. Such animals are *carriers*, and are potential sources of danger to others.

In addition to carrier animals, there is another reservoir of anaplasmosis infection in the western range states—the wood tick. This insect is a biological vector, since the disease will overwinter in its body.

DAMAGE INFLICTED BY; SYMPTOMS AND SIGNS OF AFFECTED ANIMALS

The symptoms may be those of a mild, acute, or chronic condition. Calves usually have the mild type of infection, simply becoming "dumpy" for a few days and then appar-

ently recovering, though their blood remains the permanent abode of the parasite.

The more characteristic symptoms in mature animals include rapid, pounding heart action, labored and difficult breathing, rise in temperature (up to 107°F), dry muzzle, marked depression, tremors of the muscles, loss of appetite,

Fig. 15–23. Cow exhibiting typical symptoms of acute anaplasmosis. (Courtesy, USDA)

and a great reduction in the milk flow. Animals usually show yellowing of the eye and other mucous membranes and of the skin, as in jaundice. Depraved appetite, evidenced by the eating of bones or dirt, is not uncommon. Sick animals may also show brain symptoms and an inclination to fight. Unlike cattle tick fever, bloody urine is not common in anaplasmosis. In severe acute cases, death may follow in one to a few days. Recovery is usually very slow, and although no clinical symptoms remain, such animals continue as permanent carriers of the parasite.

PREVENTION, CONTROL, AND TREATMENT

Once an animal is infected with anaplasmosis, it becomes a permanent carrier, harboring the disease agent in its bloodstream for life. This animal then becomes a continuous source of infection in the herd.

Importing infected carrier cattle into clean areas and clean herds is the most common method of spreading the disease. Once an infected animal has been introduced, the disease is spread within the herd by insect vectors, primarily the wood tick and biting flies, or by mechanical means such as castration, ear tagging, dehorning, and vaccination. Infection can be transferred any time fresh blood is transferred from an infected to a noninfected animal.

Once an animal becomes a carrier, it is immune to subsequent exposure to the disease agent. However, carrier animals that have been treated and freed of infection become susceptible again.

Prevention in lightly infected areas consists in testing the herd and finding infected animals. Then, either removing the infected animals by culling them for slaughter or feeding an antibiotic. (Either Terramycin or Aureomycin, fed in a supplement, at a level of 5 mg of the antibiotic/pound body weight daily for 45 days.) Also, new additions to the herd should be tested to avoid reinfection. A new rapid card test, conducted at cowside, can be quickly made in the field to detect the presence of anaplasmosis. This test is approved by the U.S. Department of Agriculture.

In heavily infected areas, where there has been a high rate of infection for several years, the procedure, at present, it to live with the disease. Calves raised in such areas and herds will have a high rate of infection, develop a degree of immunity, and become carriers for life.

Vaccination with a killed vaccine with an adjuvant is available commercially in the United States; it reduces the severity of the infection in most cases. Vaccination of bulls is particularly recommended. The vaccine should be administered in two doses separated by an interval of 4 weeks, followed by a single annual booster injection.

Cattle that are acutely ill with anaplasmosis are often treated with an antibiotic (Terramycin or Aureomycin), but blood transfusion remains the most effective treatment. The chronically affected carrier animal can be treated effectively by the prolonged, high level feeding of an antibiotic with feed or water.

BLOWFLY

The flies of the blowfly group include a number of species that find their principal breeding ground in dead and putrefying flesh, although they sometimes infest wounds or unhealthy tissues of live animals and fresh or cooked meat. Black blowfly larvae frequently infest dehorning wounds during winter months and occasionally the navel of newborn animals. All the important species of blowflies except the flesh flies, which are grayish and have three dark stripes on their backs, have a more or less metallic luster.

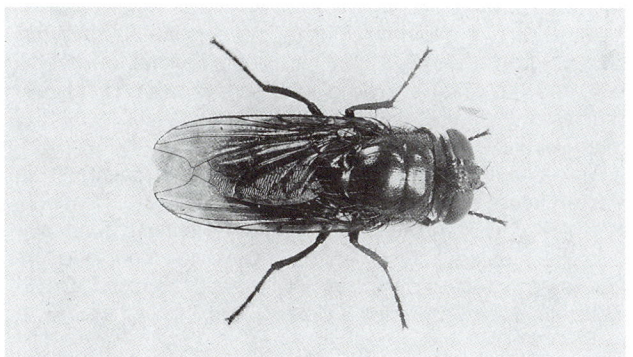

Fig. 15–24.   The black blowfly (*Phormia regina*). These flies are characterized by a metallic luster. (Courtesy, USDA)

DISTRIBUTION AND LOSSES CAUSED BY BLOWFLIES

Although blowflies are widespread, they present the greatest problem in the Pacific Northwest and in the South and southwestern states. Death losses from blowflies are not excessive, but they cause much discomfort to affected animals and lower production.

LIFE HISTORY AND HABITS

With the exception of the group known as gray flesh flies, which deposit tiny living maggots instead of eggs, the blowflies have a similar life cycle to the screwworm, except that the cycle is completed in about one-half the time.

DAMAGE INFLICTED BY; SYMPTOMS AND SIGNS OF AFFECTED ANIMALS

The blowfly causes its greatest damage by infesting wounds and the soiled hair of cattle. Such damage, which is largely limited to the black blowfly, is similar to that caused by screwworms. Infested animals rapidly become weak and fevered; and, although they recover, they may remain in an unthrifty condition for a long period.

PREVENTION, CONTROL, AND TREATMENT

Prevention of blowfly damage consists of eliminating the pest and decreasing the susceptibility of animals to infestation.

As blowflies breed principally in dead carcasses, the most effective control is effected by promptly destroying all dead animals by burning, deep burial, or sending to a rendering plant. The use of traps, poisoned baits, and electrified screens is also helpful in reducing trouble from blowflies. Suitable repellents, such as pine tar oil, help prevent the fly from depositing its eggs.

When animals become infested with blowfly maggots, their wounds should be treated twice weekly with a smear, dust, or pressurized spray of the proper insecticide, used according to manufacturer's directions.

BOVINE TRICHOMONIASIS

This is a protozoan venereal disease of cattle characterized by early abortions (usually between the second and fourth months of pregnancy) and temporary sterility. The protozoa that causes the disease, known as *Trichomonas foetus*, are one-celled, microscopic in size, and capable of movement. They are found in aborted fetuses, fetal membranes and fluids, vaginal secretions of infected animals, and the sheaths of infected bulls. Diagnosis can be confirmed microscopically. The infected bull is the source of the infection. On the other hand, the disease appears to be self-limiting in the cow.

DISTRIBUTION AND LOSSES CAUSED BY BOVINE TRICHOMONIASIS

This disease is being reported with increasing frequency throughout the United States and has become a serious problem in many herds. The economic loss is primarily due to the low percentage calf crops in infected herds.

LIFE HISTORY AND HABITS

The protozoa that cause the disease are one-celled microscopic organisms with three threadlike whips (flagella) at the front and one at the rear. The evidence indicates that the disease is spread from the infected to the clean cow by an infected bull at the time of service and that other types of contact infection do not occur. Following one or perhaps two abortions, cows appear to be immune to reinfection. Further than these facts, little is known of the life history and habits of *Trichomonas foetus*.

DAMAGE INFLICTED BY; SYMPTOMS AND SIGNS OF AFFECTED ANIMALS

No systemic disturbance is manifested by the infected bull. There may be some mucous discharge from the sheath, and the latter may be slightly inflamed. The only clinical evidence of infection is the transmission of the disease to the females serviced.

Infected cows frequently show a whitish vaginal discharge, and the following characteristic conditions usually exist when a herd is infected: (1) abortion in the first third of pregnancy, (2) uterine infections, (3) irregular heat periods, and (4) several services per conception.

Early abortions or erratic heat periods in individuals or herds that are known to be free of Bang's disease should lead one to suspect the presence of the trichomonad infection. Definite diagnosis of infection in the bull is made by means of microscopic examination of smears taken from the prepuce of the bull or the vagina of the cow.

Fig. 15–25.   Bull with trichomoniasis. This animal appeared normal but spread the disease during the breeding act. (Courtesy, College of Veterinary Medicine, University of Illinois)

PREVENTION, CONTROL, AND TREATMENT

Prevention lies in the use of clean bulls or artificial insemination. Slaughter, rather than treatment of bulls, is recommended. If natural service is essential, only young bulls should be used, as they lack susceptibility to trichomonad infection; all bulls over four years of age should be eliminated.

In cows, the disease appears to be self-limiting; that is, cows appear to acquire an immunity after about three months' sexual rest. Infected cows should be tested for a minimum of three months, and then bred by artificial insemination to a known clean bull.

CATTLE TICK FEVER (TEXAS FEVER)

This is an infectious protozoan disease of adult cattle caused by *Babesia bigemina*, which depends upon the tick, chiefly *Boophilus annulatus*, for its survival and transmission.

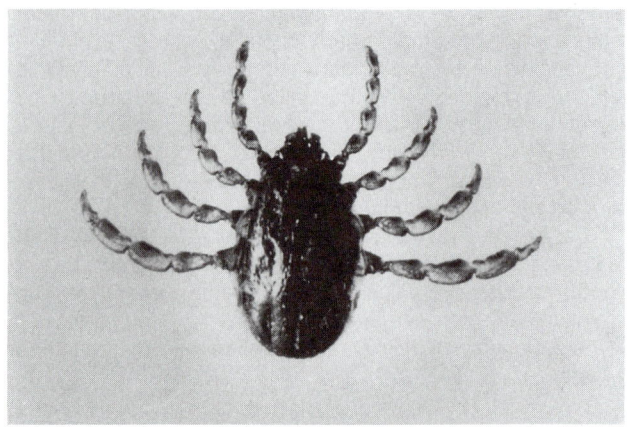

Fig. 15–26.   An adult cattle tick (*Boophilus annulatus*), chief transmitter of cattle tick fever. (Courtesy, USDA)

DISTRIBUTION AND LOSSES CAUSED BY CATTLE TICK FEVER

Prior to 1906, at which time a concerted effort was initiated to eradicate the cattle fever tick, this infectious disease of cattle and the parasite which transmits it were the most serious obstacles faced by the cattle industry in the 15 southern and southwestern states, representing a combined area of nearly one-fourth of the United States. At that time, conservative estimates placed the yearly losses at $40 million. Today, 99% of the formerly infested area has been freed (for the most part, the tick is confined to the Texas-Mexico border), and the once appalling losses have been practically eliminated. Cattle tick fever is not uncommon in Central and South America, however.

In addition to the serious death losses encountered in infected herds, the loss of blood—the only food of the cattle fever tick—results in serious damage. Infected young animals are stunted, mature animals are emaciated, and the milk flow of infected dairy animals is greatly reduced. Death occurs in about 10% of the chronic and 90% of the acute cases.

LIFE HISTORY AND HABITS

In 1889 and 1890, investigators of the USDA, Bureau of Animal Industry, established that (1) intracellular one-celled parasites, or protozoa, known as *Babesia* are the direct causative agents of the disease, and (2) cattle tick infestation is necessary in the transmission of the disease. Thus, for the first time in either human or veterinary medicine, the discovery was made that an intermediate biological carrier may transmit a disease. It is noteworthy that this pioneer work opened up an entirely new field in medical science, pointing the way for studies that later solved the problems of the spread of such dreaded diseases as malaria, yellow fever, Rocky Mountain spotted fever, typhus, and others.

The life history and habits of the protozoa which causes cattle tick fever are as follows: infected ticks, which have sucked blood from an infected cow, pass along the protozoa *Babesia (Piroplasma) bigemina* to their eggs. The female tick falls to the ground and deposits from 2,000 to 4,000 eggs. In 2 to 3 weeks, these eggs hatch into young ticks or larvae. The larvae climb on nearby vegetation to await the passing of cattle to which they attach themselves, biting and sucking blood from the host. In the latter process, the protozoa (*Piroplasma*) is passed into the blood of cattle — the protozoa of infected ticks having been passed into the eggs of the tick and through all stages of its growth.

DAMAGE INFLICTED BY; SYMPTOMS AND SIGNS OF AFFECTED ANIMALS

The incubation period of cattle tick fever is about 10 days. The disease is characterized by high temperature, rapid breathing, enlarged spleen, engorged liver, pale and yellow membranes, and red to black urine. Sometimes the symptoms subside only to reoccur at another time. Although immune, the recovered animals are permanent carriers of the disease. In infected areas, native cattle are either immune or only slightly affected.

Fig. 15–27. Cattle with cattle tick fever. The disease is characterized by high temperature, rapid breathing, enlarged spleen, engorged liver, pale and yellow membranes, and red to black urine. (Courtesy, USDA)

PREVENTION, CONTROL, AND TREATMENT

Prevention of the disease consists of avoiding contact with the cattle fever ticks, the only natural agent by which cattle tick fever is transmitted from animal to animal.

The most effective control measures are directed at the eradication of the fever ticks, either by killing them on the pastures or on the cattle. Pastures may be rendered tick-free by excluding all the host animals — cattle, horses, and mules — until all the ticks have died of starvation in 8 to 10 months.

Dipping or spraying with an approved insecticide is the most effective method of control.

Successful treatment of sick animals depends upon early recognition of the disease and prompt treatment. Agents commonly used include trypan blue, trypaflavine, and quinuronium sulfate.

Although immune, recovered animals are permanent carriers of the disease.

(Also see later section in this chapter entitled, "Federal and State Regulations Relative to Disease Control.")

COCCIDIOSIS

Coccidiosis — a parasitic disease affecting cattle, sheep, goats, swine, pet stock, and poultry — is caused by microscopic protozoan organisms known as coccidia, which live in the cells of the intestinal lining. Each class of domestic livestock harbors its own species of coccidia; hence, there is no cross infection between animals.

Cattle are known to be affected by 21 species of coccidia. But only *Eimeria bovis* and *E. zuerni* are important in the United States, with *E. zuerni* tending to cause the most serious infection.

DISTRIBUTION AND LOSSES CAUSED BY COCCIDIOSIS

The distribution of the disease is worldwide. Except in very severe infections, or where a secondary bacterial invasion develops, infested animals usually recover. The chief economic loss is in lowered growth of young stock. It is very severe in young dairy calves.

LIFE HISTORY AND HABITS

Infected animals may eliminate in their droppings thousands of coccidia organisms (in the resistant öocyst stage), daily. Under favorable conditions of temperature and moisture, coccidia sporulate to maturity in 3 to 5 days, and each öocyst contains eight infective sporozoites. The öocyst then gains entrance into an animal by being swallowed with contaminated feed or water. In the host's intestine, the outer membrane of the öocyst, acted on by the digestive juices, ruptures and liberates the eight sporozoites within. Each sporozoite then attacks and penetrates an epithelial cell, ultimately destroying it. While destroying the cell, however, the parasite undergoes sexual multiplication and fertilization with the formation of new öocysts. The parasite (öocyst) is then expelled with the feces and is again in a position to reinfect a new host.

The coccidia parasite abounds in wet, filthy surroundings; resists freezing and ordinary disinfectants; and can be carried long distances in streams.

DAMAGE INFLICTED BY; SYMPTOMS AND SIGNS OF AFFECTED ANIMALS

A severe infection with coccidia produces diarrhea, and the feces may be bloody. The bloody discharge is due to the destruction of the epithelial cells lining the intestines. Ensuing exposure and rupture of the blood vessels then produces hemorrhage into the intestinal lumen.

In addition to bloody diarrhea, affected animals usually show pronounced unthriftiness and weakness.

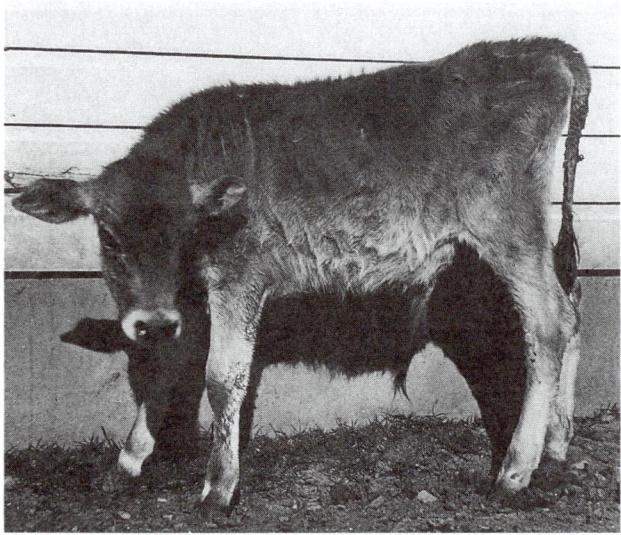

Fig. 15–28. Calf suffering from coccidiosis. The soiled tail is a typical symptom. (Courtesy, USDA, Regional Laboratory, Animal Disease Research, Auburn, AL)

PREVENTION, CONTROL, AND TREATMENT

Coccidiosis can be prevented by protecting animals from feed or water that is contaminated with the protozoa that causes the disease. Prompt segregation of affected animals is important and should be done if practical. Manure and contaminated bedding should be removed daily. Low, wet areas should be drained. If possible, segregation and isolation of animals by age should be used in controlling the disease. All precautions should be undertaken to keep droppings from contaminating the feed. Although the öocysts resist freezing and certain disinfectants and may remain viable outside the body for 1 or 2 years, they are readily destroyed by direct sunlight and complete drying.

Amprolium, a drug long used against coccidiosis of chickens, is approved for use in dairy calves.

If coccidiosis is suspected, the veterinarian should be consulted for treatment of affected animals and for advice on the management steps necessary to prevent further losses in the herd.

FLIES

Several kinds of flies live around dairy cattle; in and around buildings, in pastures, and on the animals. Whether feeding upon the animals or just annoying them, flies can reduce milk production and slow the growth of young stock. Additionally, flies are unsanitary and may contaminate milk.

Fig. 15–29 and Table 15–3 present pertinent information relative to flies that attack or annoy dairy cattle.

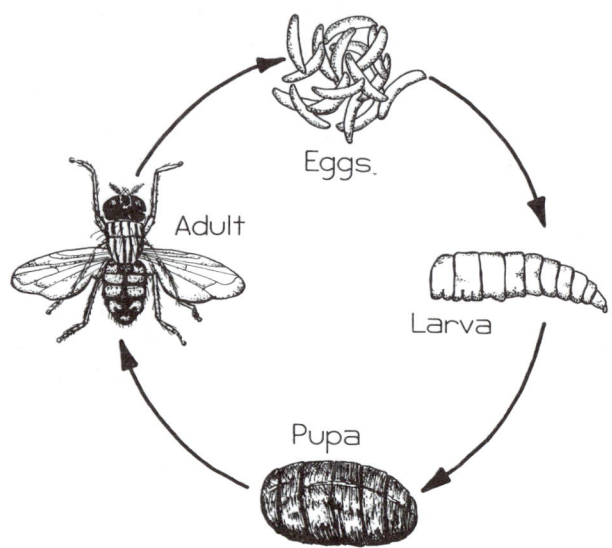

Fig. 15–29. The four stages in the life cycle of flies: egg, larva (maggot), pupa, and adult.

GASTROINTESTINAL WORMS

Many species of gastrointestinal worms may be found in the abomasum, small intestine, large intestine, lungs, and liver of cattle; among them, the following: stomach worms, bankrupt worms, medium brown stomach worms, nematodirus, threadworms, hookworms, cooperia, nodular worms, whipworms, lungworms, tapeworms, deer liver fluke, and coccidia.

When their presence is suspected, their diagnosis should be confirmed by the veterinarian, by microscopic examination of fresh manure samples. Such examination of the samples will show if parasite eggs or öocysts are present, and in what numbers.

DISTRIBUTION AND LOSSES CAUSED BY GASTROINTESTINAL WORMS

One or more species of internal parasites of cattle are found throughout the United States, but they may be especially severe in the South and on irrigated and permanent pastures.

The losses are in terms of lowered feed efficiency caused by disturbed digestion, lowered milk production, and some death losses. Young animals are more severely affected than mature cattle.

TABLE 15–3
FLIES OF DAIRY CATTLE AND THEIR CONTROL

Species of Flies	Distribution and Losses Caused by	Damage Inflicted; Symptoms and Signs of Affected Animals	Prevention, Control, and Treatment	Remarks
Several species of flies attack or annoy cattle. The most common ones follow:				
Biting midge *(Culicoides variipennis).* This fly is common in the U.S. It is a small fly that is usually brownish and hump-backed. They are blood-feeders.	Worldwide, in both warm and temperate climates. Milk production losses result from blood-feeding of the flies. Biting midges also transmit the virus that causes blue tongue in cattle.	Biting midges cause milk production losses, restless animals, and large, weeping, crusting lesions on the skin where large numbers of flies have fed.	Prevention and control consists of (1) stabilizing banks of bodies of water to reduce the habitat of the larvae, and (2) spraying animals with an approved insecticide according to manufacturer's directions.	Blue tongue in cattle can be prevented by the administration of a vaccine.
Black fly (buffalo gnat). These are small, dark colored flies, with a hump-backed appearance. They are blood-suckers.	Worldwide, in all climates. Losses result from irritation of animals, blood loss, and transmission of some animal diseases in the tropics.	Restlessness, milk production losses, and swollen, weeping lesions on the skin.	Black flies develop in well-aerated running water. Prevention consists of treatment of streams with insecticides.	Sometimes black flies occur in such large numbers that cattle are smothered. Irrigation structures downstream from dams are often sites of larval development.
Face fly *(Musca autumnalis)* was first found in this country in New York in 1953. It is a close relative of, and similar in appearance to, the housefly. Fig. 15–30	The face fly is more troublesome and causes greater losses than the horn fly. It attacks cattle that are in open fields and away from shade. The face fly causes direct irritation to eye tissues and aids in the transmission of pinkeye. The face fly is especially bad across southern U.S.	The face fly does not bite; but its habit of clustering around the eyes, mouth, and nostrils is extremely annoying to animals, interfering with their vision and breathing, and preventing normal grazing. Large populations force animals to leave pastures and seek relief in wooded areas and shelters.	Prevention consists in (1) scattering or removing fresh cow manure, and (2) spraying or dusting regularly, or self-treating with dust bags or back rubbers. Several insecticides are effective in the control of the face fly. One of these should be selected and applied in keeping with the manufacturer's directions.	When cattle enter a barn or darkened area, the fly leaves the animal's face and rests on fence posts, gates, sides of barns, etc. The adult fly hibernates in attics and other protected places during winter. The face fly lays its eggs in fresh manure, where the larvae develop.
Horn fly *(Haematobia irritans)*; also known as cattle fly, cow fly, or stock fly. This fly, which is about one-half the size of an ordinary housefly, is one of the most numerous and worst annoyances of cattle. Fig. 15–31	Throughout the U.S. An average of 4,000–5,000 flies per animal is not uncommon, and individual animals may support as many as 10,000 – 20,000 flies. Horn flies are more of a problem on grazing cattle than of cattle held in corrals. Horn flies produce irritation and worry, loss of blood, reduced vitality, and, in the South, sores that may become infested with screwworms. Lactating cows suffer lowered milk production.	Unlike the housefly and stablefly, the horn fly remains on cattle throughout the day and night. They usually feed twice a day, sometimes more frequently. Tormented cattle often refuse to graze during the day and seek protection by hiding in dark buildings, brush, or tall grass. Heavily infested cattle may also have rough, sore skin, and suffer an inevitable loss of condition.	In small pastures and where it is practical, spread fresh droppings with a spring-tooth harrow in order to hasten their drying. At frequent intervals, haul out cattle manure around barns, and spread thinly on the land, preferably with a manure spreader. The horn fly can be very effectively controlled in the adult stage by treating cattle. When sprays are used, only a small deposit of insecticide on the hair is sufficient. Also, complete coverage is not necessary because horn flies move about on the animal sufficiently to come in contact with insecticide deposited on almost any part of the animal's body. Thus, protection of cattle from attacks of horn flies can be accomplished by use of sprays, dusts, dust bags, dips, back rubbers, or pour-ons, used according to manufacturer's directions.	The horn fly is often found resting at the base of the horn; hence, the name. But it does not confine itself to this location on the animal. Horn flies may congregate by the hundreds or even thousands on the backs, shoulders, and bellies of cattle. The entire life cycle of the horn fly averages 9–12 days during the summer months, and the adult fly lives about 7 weeks.

(Continued)

TABLE 15-3 *(Continued)*

Species of Flies	Distribution and Losses Caused by	Damage Inflicted; Symptoms and Signs of Affected Animals	Prevention, Control, and Treatment	Remarks
Horseflies and deerflies are biting flies that attack cattle. The two most troublesome genera are *Tabanus* (horseflies) and *Chrysops* (deerflies). Fig. 15-32	Tabanids (horseflies) are found in all parts of the U.S., and large numbers may be expected wherever there are extended areas of permanently wet, undeveloped land and a mild climate. Generally, they are more of a problem to livestock than deerflies, but deerflies are often extremely annoying in the coastal areas of the South and the mountain areas of the West.	The bite from the slashing mouthparts of female horseflies and deerflies is very painful, and animals try to dislodge the fly with their tails or tongues or by stamping their feet. Heavily attacked animals stop grazing and tend to bunch together or seek shelter. Severe outbreaks can seriously affect milk production.	No really satisfactory method exists for controlling horseflies or deerflies. If possible, avoid pasturing cattle near swampy, wooded areas when these flies are numerous. Also, sheltering animals is often beneficial since tabanids do not ordinarily enter enclosures.	Horseflies and deerflies are also implicated in disease transmission because their habit of feeding on one animal and immediately attacking another can result in the direct transfer of pathogenic organisms that live in blood.
Houseflies *(Musca domestica)* are nonbiting flies that are common around barns and lots. Fig. 15-33	Houseflies become numerous both inside and outside barns and farm buildings. Houseflies are annoying to cattle and people, and they can spread human and animal diseases.	Although houseflies are nonbiting, they cause serious economic losses through annoyance of cattle and by disease transmission. Also, they create public health problems.	Insecticide alone will not control houseflies. Adequate sanitary measures, including proper disposition or handling of manure, are necessary to eliminate fly breeding areas. Spread manure thinly in fields so fly eggs and larvae will be killed by drying and heat. Several insecticides in fogs, mists, surface sprays, or baits may be used according to manufacturer's directions.	Houseflies breed in manure, garbage, and decaying vegetable matter. The eggs hatch after an incubation period of 12–36 hours. The larvae feed on the organic medium and grow to full size in 6–11 days.
Stablefly *(Stomoxys calcitrans)*, which is about the size of a housefly, is usually found in the vicinity of animals. Fig. 15-34	Stableflies are found in all parts of the U.S., but they are especially numerous in the central and southeastern states and in some coastal states. Large numbers can occur around barns where cows congregate and where there are accumulations of decaying plant material. Losses include: 1. Lowered milk production in dairy cattle, by as much as 50% in seasons when the number of flies becomes large. 2. Possible transmission of certain diseases and parasites.	Fly fighting and restlessness. In seeking natural protection, animals frequently resort to mudholes, brush, etc. Stableflies are known to transmit such diseases as anthrax and anaplasmosis.	Control of stableflies by direct application of insecticides to cattle is usually not satisfactory. They are best controlled by sanitation and by application of insecticides to the resting surfaces. Sanitation is the most effective method of controlling stableflies in such areas as corrals because it breaks the life cycle by removing the breeding sites. Corrals should be well drained, manure and decaying organic matter should be removed from inside and outside buildings and disposed of weekly or more often, if possible, by spreading it out to dry (this kills developing larvae). If manure cannot be spread, it should be placed in compact piles where the surface will dry quickly and become unattractive to females. With stableflies, insecticides should be used as a supplement to good sanitation, rather than as the principal method of control, because alone they may not do a satisfactory job. Residual sprays are the most effective method of treatment. They should be applied inside and outside barns and other farm structures where stableflies rest. The spray may be applied by spray gun or by fogging or misting devices. Care should be taken to prevent contamination of feed and drinking water, and animals should be removed from the area during the spraying. Application of insecticides to the cattle may afford only temporary relief. The preferred method of application is spraying since the insecticide should be applied to the legs and lower body of the animals.	As a blood-feeding fly, the stablefly is implicated in carrying disease, and high populations cause reduced milk production of lactating cows. The entire life cycle takes about 3 weeks, so several generations can develop during the summer. Although cattle feces are a suitable habitat for the development of stablefly larvae, they do not seem to be attractive to adult flies for egg laying unless they are mixed with straw, feed, or similar materials.

LIFE HISTORY AND HABITS

Most of the parasites listed above have similar life cycles, of which the common stomach worm is typical. Infected cattle carry the mature worms in the fourth stomach. Eggs from these worms are expelled with manure and develop on the pasture into infectious larvae. Then cattle become infected by eating grass infected with larvae. The latter develop into mature worms in the stomach and these again produce eggs that recontaminate the pasture.

DAMAGE INFLICTED BY; SYMPTOMS AND SIGNS OF AFFECTED ANIMALS

An animal with a light infection rarely shows any outward symptoms, and the symptoms are not specific. But infected animals generally show loss of milk production and weight, retarded growth, anemia, diarrhea, and/or lowered resistance to other diseases. With a heavy infection, there may be a swelling under the jaw (bottle jaw); and the parasites may even cause the death of the animals.

PREVENTION, CONTROL, AND TREATMENT

Preventive and control measures include (1) rotating pastures, (2) segregating calves from mature animals, (3) cross grazing with cattle and horses, (4) avoiding overstocking or overgrazing pastures since most of the infective larvae are on the bottom inch of grass, and (5) keeping feeders and waterers sanitary. In areas of constant exposure, routine treatment is recommended.

Treatment consists of administering the drug of choice according to manufacturer's directions, before calving, after calving, and when grazing is at its height. From the long list of drugs, dairy producers and their veterinarians can (1) select drugs for lactating cows which do not require discarding of milk, (2) rotate drugs to prevent parasitic resistance, and (3) choose the method of administration easiest to follow (although varying according to drug, they may be given as a drench, bolus, feed or mineral mix, paste, or injection).

GRUBS (HEEL FLIES, WARBLES)

Fig. 15–35. Adult female heel fly (*Hypoderma lineatum*) whose maggot stage is responsible for the common cattle grub. (Courtesy, USDA)

Cattle grubs are the maggot stage of insects known as heel flies, warble flies, or gadflies. Two species of cattle grubs are present in the United States. The northern cattle grub (*Hypoderma bovis*) occurs mainly in the north though it is found as far south as southern California, northern Arizona, Oklahoma, Tennessee, South Carolina, and Hawaii. The common cattle grub (*Hypoderma lineatum*) occurs throughout the 48 contiguous states and in Hawaii and Alaska. The cattle grub or heel fly is probably the most destructive insect attacking dairy animals.

Fig. 15–36. Dairy heifers running away from heel flies. Though the fly does not bite or sting, when it lays its eggs on the lower legs, it usually terrifies the animal, causing it to run with tail hoisted, seeking relief. (Courtesy, Livestock Conservation, Inc., Hinsdale, IL)

DISTRIBUTION AND LOSSES CAUSED BY CATTLE GRUBS

The species *Hypoderma lineatum* is widely distributed throughout the United States, whereas *Hypoderma bovis* is chiefly confined to the northern states.

The damage inflicted by cattle grubs affects dairy producers, packers, tanners, and, finally, consumers. The kinds of losses include the following:

1. **Decreased milk production or gains, mechanical injury, or even death.** Though the fly does not bite or sting, when it lays its eggs on the lower legs, it usually terrifies the animal, causing it to run with tail hoisted, seeking relief. It may run through fences, over cliffs, or become hopelessly bogged down in a mudhole or swamp. Milk production from dairy cows may be reduced from 10 to 25% during the period heel flies are laying their eggs.

2. **Carcass damage.** According to meat packers, about 35% of all cattle carcasses are damaged by grubs. The yellowish, water patches caused by the migration of the larvae under the skin are referred to by butchers as *pilled* or *licked beef*. Two to three pounds of "jellied" beef must be trimmed from the loins and ribs of each "grubby" animal, and the damaged cut of meat is devalued 2¢ per pound because of the ragged and unattractive appearance.

3. **Injury of hides.** Approximately one-third of all cattle hides produced in the United States are damaged by grubs. This loss is caused by the migration of the grub through the back, which leaves a scar in the most valuable part of the hide. According to trade custom, if a hide has as many as five grub holes, it is classed as grade No. 2 and is subject to a discount of 1¢ per pound. Commonly as many as 40,

and occasionally 100 or more, grub holes are found in a single hide. Hides of the latter quality are not considered worth tanning and are sold for by-products.

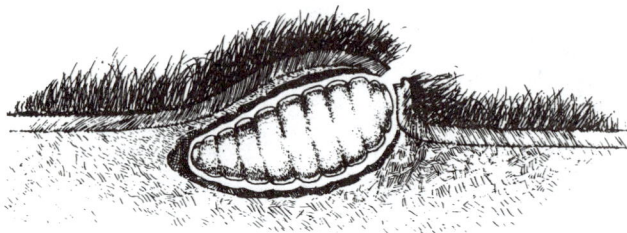

Fig. 15–37.   Hole in the hide on the back of an animal, with a grub inside. Note hide opening through which the grub obtains air and finally escapes. (Courtesy, USDA)

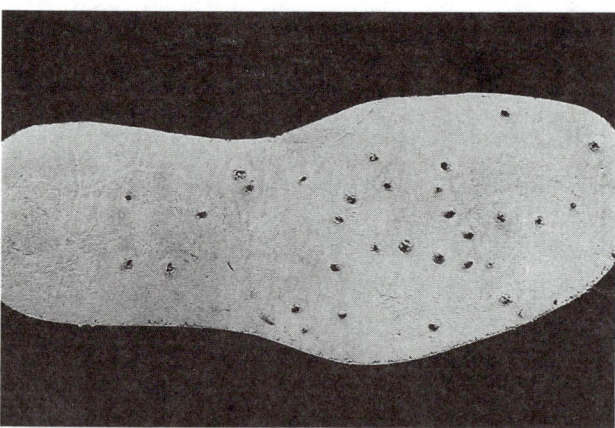

Fig. 15–38.   Shoe sole damaged by cattle grubs. About one-third of all cattle hides produced in the United States are damaged by grubs. (Courtesy, Livestock Conservation, Inc., Hinsdale, IL)

4.   **Shock to animals.** In certain older animals that have been previously sensitized, the breaking of a grub under its skin may cause a terrific reaction (anaphylaxis or allergic reaction). The area may be greatly swollen and form an abscess, and there may be such a general reaction that the animal may die from shock. At the first sign of shock, the local veterinarian should be summoned quickly, to administer appropriate stimulants.

LIFE HISTORY AND HABITS

Basically, the two species of cattle grubs have a similar life cycle. The female flies, called heel flies, attach their eggs to the hairs of the legs and bodies of cattle. The eggs hatch into larvae after about 3 days and enter the animals at the bases of the hairs. Once inside, the common cattle grub migrates from the point of entry to the gullet; the northern cattle grub migrates to the spinal column (both migrations take 2 to 4 months). After some additional months in the gullet or spinal column, grubs of both species migrate to the animal's backs, cut breathing holes in the hide, and remain there for about 6 weeks while they increase greatly in size.

The resultant swellings are often called *wolves* or *warbles*. Fully grown grubs leave the hide through the breathing holes and drop to the ground where they pupate. Then, in a few weeks, they transform to nonfeeding adult flies that emerge and mate. On bright sunny days the female then seeks cattle for egg laying, which causes the cattle to roam. The entire life cycle takes about 1 year, and the same stages are usually found at about the same time each year in any given area.

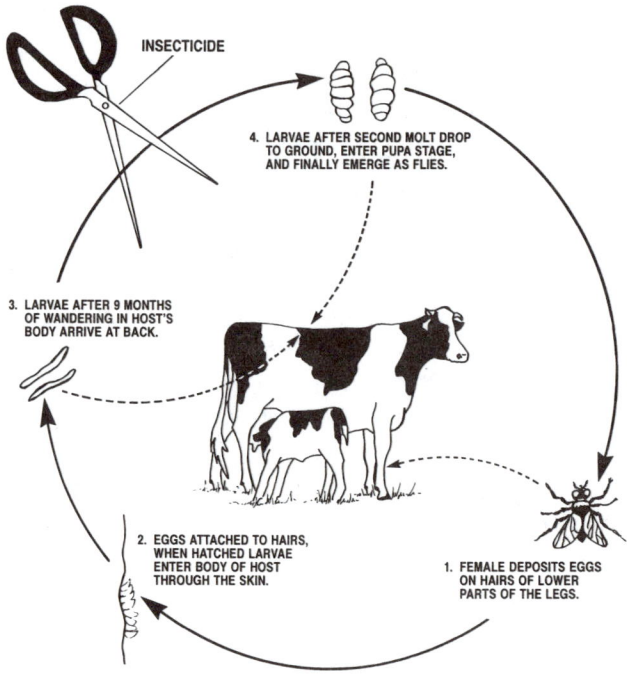

Fig. 15–39.   Diagram showing the life history and habits of cattle grubs. As noted (see scissors) effective control and treatment (cutting the cycle of the parasite) may be obtained by the application of a suitable insecticide (spray, dip, or dust). The first application should be made 25 to 30 days after grubs first appear in the back, with subsequent treatments at 30-day intervals thereafter as long as grubs are present in the back. Control may also be secured by the use of a systemic insecticide.

DAMAGE INFLICTED BY; SYMPTOMS AND SIGNS OF AFFECTED ANIMALS

The attack of the heel fly is unmistakable when, in the spring or early summer, cattle are seen madly running with their tails hoisted high over their backs in an attempt to escape. The presence of the grub (larva) in the back, usually from December to May, causes a characteristic swelling (and an opening in the skin, from which pus is discharged), which usually becomes conspicuous, so that a grubby back has a lumpy appearance.

PREVENTION, CONTROL, AND TREATMENT

Complete prevention of cattle grub damage within any given herd cannot be obtained unless all cattle grubs throughout the country are exterminated. This means a nationwide campaign in which all cooperate to eradicate the menace, farm by farm, county by county, and state by state.

Cattle grubs are most effectively controlled by applying a systemic insecticide to cattle as soon as possible after the activity of the heel flies ceases since those insecticides kill the young larvae in the animal's body. When the grubs are near the back or located in the back, treatments are less effective, and possible side effects are more likely. Side effects may also occur when there is a concentration of grubs in the gullet or spinal cord of treated cattle. A single treatment with a systemic insecticide should give excellent control of cattle grubs. For the correct timing, owners are advised to check with their local county agent or consultant. Systemics may be administered as sprays, dips, or as feed additives. Never use more than one systemic insecticide at a time, and always use a systemic in keeping with manufacturer's directions.

LICE

The louse is a small, flattened, wingless insect parasite of which there are several species.

Cattle are attacked by four species of bloodsucking lice—the short-nosed cattle louse (*Haematopinus eurysternus*), the cattle tail louse (*Haematopinus quadripertusus*), the long-nosed cattle louse (*Linognathus vituli*), and the little blue louse (*Solenopotes capillatus*), and by one species of biting louse (*Bovicola bovis*). Cattle lice will not remain on other farm animals, nor will lice from other animals infest cattle. Lice are always more abundant on weak, unthrifty animals and more troublesome during the winter months than during the rest of the year.

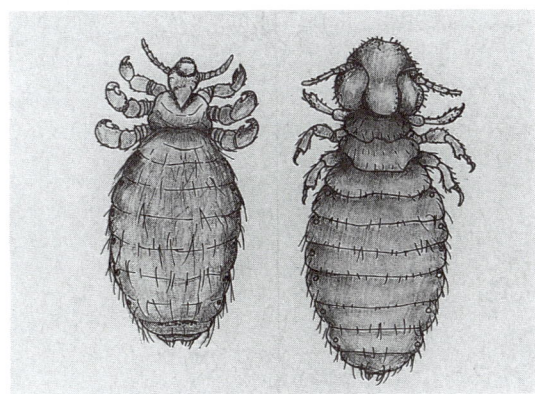

Fig. 15–40. Two species of cattle lice: (1) (left) hairy cattle louse—a bloodsucking louse; and (2) (right) chewing, or biting cattle louse.

DISTRIBUTION AND LOSSES CAUSED BY LICE

The presence of lice upon animals is almost universal, but the degree of infestation depends largely upon the state of animal nutrition and the extent to which the owner will tolerate parasites. The irritation caused by the presence of lice retards growth and milk production.

LIFE HISTORY AND HABITS

Lice spend their entire life cycle on the host's body. They attach their eggs or nits to the hair near the skin where they hatch in about 2 weeks. Two weeks later the young females begin laying eggs, and after reproduction they die on the host. Lice do not survive more than a week when separated from the host, but under favorable conditions eggs clinging to detached hairs may continue to hatch for 2 to 3 weeks.

DAMAGE INFLICTED BY; SYMPTOMS AND SIGNS OF AFFECTED ANIMALS

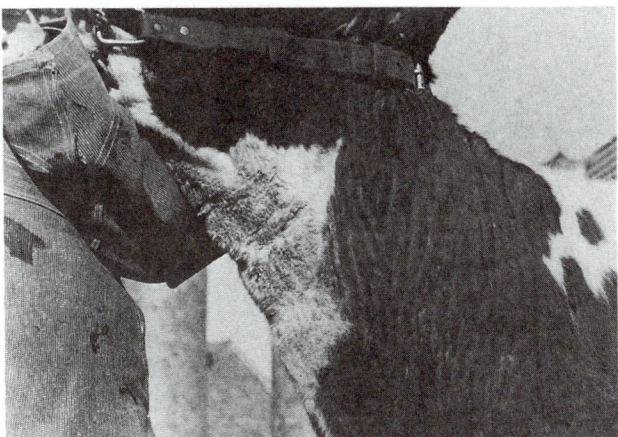

Fig. 15–41. Cow's neck heavily infested with short-nosed sucking lice. These pests seek the sheltered parts of the body on which to feed. (Courtesy, College of Veterinary Medicine, University of Illinois, Urbana)

Infestation shows up most commonly in winter in ill-nourished and neglected animals. There is intense irritation, restlessness, and loss of condition. As many lice are bloodsuckers, they devitalize their host. There may be severe itching and the animal may be seen scratching, rubbing, and gnawing at the skin. The hair may be rough, thin, and lack luster; and scabs may be evident. In cattle, favorite locations for lice are the root of the tail, on the inside of the thighs, over the fetlock region, and along the neck and shoulders. In some cases, the symptoms may resemble that of mange, and it must be kept in mind that the two may occur simultaneously.

With the coming of spring, when the hair sheds and the animals go to pasture, the problem of lice is greatly diminished.

PREVENTION, CONTROL, AND TREATMENT

Because of the close contact of domesticated animals, especially during the winter months, it is practically impossible to prevent entire herds from becoming slightly infested with the pests. Nevertheless, lice can be kept under control.

For effective control, all members of the herd must be treated simultaneously at intervals, and this is especially necessary during the fall months about the time they are placed in winter quarters. Cattle should be inspected for lice periodically throughout the winter and spring and re-treated when necessary. Approved insecticides applied by spraying are effective against lice. Dusting is less effective than spraying, but may be preferable when few animals are to be treated or during the winter months. The insecticide of choice should be applied according to the manufacturer's directions.

LIVER FLUKE

The liver fluke, *Fasciola hepatica*, is a flattened leaflike, brown worm, usually about an inch long. It affects cattle, sheep, goats, and other animals.

DISTRIBUTION AND LOSSES CAUSED BY LIVER FLUKES

The liver fluke is distributed throughout the world, wherever there are low-lying wet areas and suitable snails. In the United States, it is most common in some areas of the West and Southwest and in Florida.

Lowered milk production and feed inefficiency are the chief losses in dairy cattle. In addition, vast quantities of liver are condemned each year at the time of slaughter—an estimated 1,200 tons annually. In packinghouses, such livers are referred to as *fluky livers* or *rotten livers*.

LIFE HISTORY AND HABITS

Flukes reproduce by means of eggs which, after passing from the host, hatch into embryos equipped with cilia that enable them to move about. Upon encountering certain kinds of snails, they penetrate into the body of the intermediate host and develop into cercariae (flukes in the larval stage), which leave the snails and become encysted on the nearby vegetation. The encysted cercariae are then ingested by cattle during grazing. The fluke is liberated from the cyst, penetrates the intestinal wall, migrates about the abdominal cavity, and finally reaches the liver where maturity is attained 2 or 3 months after infestation.

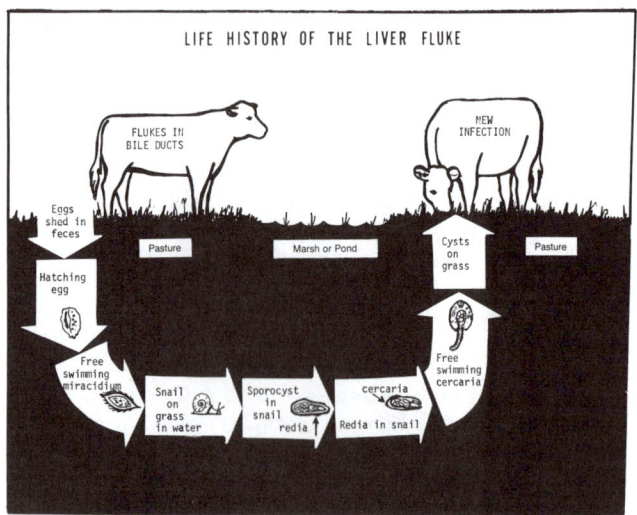

Fig. 15–42. Diagram showing the life history and habits of the liver fluke, *Fasciola hepatica*. When the eggs of the liver fluke hatch, they liberate larva called miracidium. The miracidium is ciliated and swims about in the water until it comes in contact with a suitable snail, into which it bores. Then it changes a number of times, finally becoming what is known as a redia. Large number of minute stages, shaped like tadpoles and known as cercariae, are produced in the redia. These eventually escape from the snail, swim about, and become encysted on grass or other vegetation.

DAMAGE INFLICTED BY; SYMPTOMS AND SIGNS OF AFFECTED ANIMALS

Infested cattle show anemia, as indicated by pale mucous membranes, digestive disturbances, loss of weight, and general weakness. As with most parasites, positive diagnosis consists of finding eggs in the feces by microscopic examination.

PREVENTION, CONTROL, AND TREATMENT

The following measures are recommended for the prevention and control of liver flukes:

1. Drainage or avoidance of wet pastures.
2. Where relatively small snail infested areas are involved, it may be practical to destroy the snail (carrier of liver flukes), preferably in the spring season, through—

   a. Applying 3 to 6 lb of copper sulfate (bluestone or blue vitrol) per acre of grassland, mixing and applying the small quantity of copper sulfate with a suitable carrier (such as a mixture of 1 part of the copper sulfate to 4 to 8 parts of either sand or lime), and

   b. Treating ponds or sloughs with 1 part of copper sulfate to 500,000 parts of water.

When copper sulfate is used in the dilutions indicated, it is not injurious to grasses and will not poison farm animals, but it may kill fish.

Snail infested pastures should not be used for making hay.

The recommended treatment, which should be under the direction of a veterinarian, consists in drenching according to manufacturer's directions with one of the approved wormers.

LUNGWORM

The lungworm, *Dictyocaulus viviparus*, is a white, threadlike worm 1½ to 3 in. long, found in the trachea and bronchi of cattle—especially calves.

DISTRIBUTION AND LOSSES CAUSED BY LUNGWORMS

Lungworms are distributed throughout the United States, especially on wet pastures.

The losses are chiefly in lowered feed efficiency, and in milk and meat production, there may be death losses.

LIFE HISTORY AND HABITS

In the bronchial tubes of the lungs, female lungworms produce large numbers of eggs. Usually, these hatch in the air passages and liberate larvae that are coughed up, swallowed, and eliminated in the feces. Sometimes the coughed-up eggs hatch in the stomach or intestines, but they may pass unhatched from the host, particularly when there is severe diarrhea.

Under favorable conditions, the larvae eliminated with the feces develop into the infective stage in about a week.

Then they crawl upon blades of grass, where they are ingested by grazing cattle. Thence they penetrate the intestinal wall and reach the lymph glands, from which they are eventually carried to the lungs.

Cattle lungworms mature in 3 to 4 weeks, at which time larvae appear in the feces. The worms live from 2 to 4 months in the host.

DAMAGE INFLICTED BY; SYMPTOMS AND SIGNS OF AFFECTED ANIMALS

Typical symptoms include coughing, labored breathing, loss of appetite, unthriftiness, and intermittent diarrhea. Death may follow, probably from suffocation or pneumonia.

PREVENTION, CONTROL, AND TREATMENT

Where lungworms are found, the following prevention and control measures are recommended:

1. Practice rigid sanitation.
2. Segregate calves from older animals where practical.
3. Keep calves on a good ration.
4. Avoid spreading infested manure on pastures.
5. Utilize dry pasture if possible.

Administer the drug of choice for lungworm treatment. Like all drugs, it should be given according to the manufacturer's directions.

MITES (MANGE, SCABIES)

Scabies in cattle, also known as scab, mange, or itch, is caused by mites living on or in the skin.

Each species of domesticated animals has its own peculiar species of mange mites, and, with the exception of the sarcoptic mites, the mites of one species of animals cannot live and propagate permanently on a different species. The sarcoptic mites are transmissible from one class of animals to another, and, in the case of the sarcoptic mite of the cow and horse, from animals to humans. Three of the more important species are the psoroptic, or common, scab mite (*Psoroptes equi* var. *bovis*); the sarcoptic scab mite (*Sarcoptes scabei* var. *bovis*); and the chorioptic, or symbiotic, scab mite (*Chorioptes bovis*). The sarcoptic form is most damaging, for, in addition to their tunneling, the mites secrete an irritating poison, which results in severe itching.

DISTRIBUTION AND LOSSES CAUSED BY MITES

Injury from mites is caused by bloodsucking and the formation of scabs and other skin affections. In a severe attack, the skin may be much less valuable for leather. Growth is retarded, and production of milk and meat is lowered.

The mites that attack cattle breed exclusively on the bodies of their hosts, and will live for only 2 or 3 weeks when removed therefrom. The female mite which produces sarcoptic mange—the most severe form of scabies—lays from 10 to 25 eggs during the egg-laying period, which lasts about 2 weeks. At the end of another 2 weeks, the eggs have hatched

and the mites have reached maturity. A new generation of mites may be produced every 15 days.

Mites are more prevalent during the winter months, when animals are confined and in close contact with each other.

DAMAGE INFLICTED BY; SYMPTOMS AND SIGNS OF AFFECTED ANIMALS

When the mite pierces the skin to feed on cells and lymph, there is marked irritation, itching, and scratching. Exudate forms on the surface, and this coagulates, crusting over the surface. The crusting is often accompanied or followed by the formation of a thick, tough, wrinkled skin. Frequently, there are secondary skin infections. The only certain method of diagnosis is to demonstrate the presence of mites.

Fig. 15–43.   A severe case of mange on a cow, caused by sarcoptic mites. Note the rough and wrinkled condition of the skin, with crusting over the surface. (J. W. McManigal, Agricultural Photographer, Horton, KS)

PREVENTION, CONTROL, AND TREATMENT

Prevention consists of avoiding contact with mangy animals or infested premises. Mange is a reportable disease; hence, infestations should be reported to proper livestock inspection authorities. The presence of certain species of cattle scab mites will result in the herd being quarantined. This prohibits their movement until they are inspected and found free of scabies.

Mites can be controlled by spraying or dipping infested animals with suitable insecticidal solutions, and by quarantine of affected herds. Only lime-sulphur is registered for use on lactating cows.

(Also, see later section in this chapter entitled, "Federal and State Regulations Relative to Disease Control.")

MOSQUITOES

Mosquitoes, particularly species of the genera *Aedes*, *Psorophora*, and *Culex*, are a severe nuisance to dairy cattle in many areas.

DISTRIBUTION AND LOSSES CAUSED BY MOSQUITOES

Mosquitoes are rather widely distributed, but they are most numerous in southeastern United States, especially in swampy regions that have permanent pools of water or that are exposed to frequent flooding. Sometimes they kill cattle, although this is rare.

LIFE HISTORY AND HABITS

Almost all female mosquitoes must take a blood meal before they can lay eggs. (The males do not suck blood, but feed on nectar and other plant juices.) Eggs are laid singly or in rafts on the surface of the water or on the ground in depressions that are flooded by tidal waters, seepage, overflow, or rainwater. The larvae and pupae are aquatic.

DAMAGE INFLICTED BY; SYMPTOMS AND SIGNS OF AFFECTED ANIMALS

Mosquitoes may occur in such abundance that cattle refuse to graze. Instead, they bunch together or stand neck deep in water to protect themselves from attack. Moreover, mosquitoes will annoy cattle day and night, so they can cause serious losses in milk production—or even death in extreme cases. Also, they may be disease carriers.

PREVENTION, CONTROL, AND TREATMENT

Mosquitoes can be controlled in several ways: (1) by elimination of breeding places, through providing fills, ditches, impoundments, improved irrigation methods, and other means of water manipulation; (2) by chemical destruction of larvae, through treating the relatively restricted breeding areas with proper larvicides; and (3) by chemical destruction of adults. Elimination of breeding sites is by far the most satisfactory and effective method of control. However, either this method or chemical destruction of larvae may not be economically practical if the breeding area is extensive. When the latter is the case, control can best be accomplished through group actions, such as mosquito abatement districts. The dairy producer can achieve some control by fogging or spraying the pasture, and some relief can be obtained by spraying the animals with insecticides.

RINGWORM

Ringworm, or barn itch, is a contagious disease of the outer layers of skin. It is caused by certain microscopic molds or fungi (*Trichophyton, Achorion,* or *Microsporon*). All animals are susceptible.

DISTRIBUTION AND LOSSES CAUSED BY RINGWORM

Ringworm is widespread throughout the United States. It is a contagious disease of the outer layer of skin caused by certain microscopic molds or fungi, which affect all animals. Though it may appear among animals on pasture, it is far more prevalent as a barn disease. It is unsightly, and affected animals may experience considerable discomfort; but the actual economic losses attributed to the disease are not too great.

LIFE HISTORY AND HABITS

The period of incubation for this disease is about 1 week. The fungi form seeds or spores that may live 18 months or longer in barns or elsewhere.

Ringworm is usually a winter disease, with recovery the following summer when the animals are on pasture.

DAMAGE INFLICTED BY; SYMPTOMS AND SIGNS OF AFFECTED ANIMALS

Fig. 15–44. Heifer with ringworm. The fungi causing these raised circular areas may be transmitted to humans. (Courtesy, College of Veterinary Medicine, University of Illinois, Urbana)

Round, scaly areas almost devoid of hair appear mainly in the vicinity of the eyes, ears, side of the neck, or the root of the tail. Crusts may form, and the skin may have a gray, powdery, asbestoslike appearance. The infected patches, if not checked, gradually increase in size. Mild itching usually accompanies the disease.

PREVENTION, CONTROL, AND TREATMENT

The organisms are spread from animal to animal or through the medium of contaminated fence posts, curry combs, and brushes. Thus, prevention and control consists of disinfecting everything that has been in contact with infected animals. The infected animals should also be isolated. Strict sanitation is essential in the control of ringworm.

The hair should be clipped, the scabs removed, and the area brushed and washed with soap. The diseased parts should be painted with tincture of iodine or salicylic acid and alcohol (1 part in 10) every 3 days, until cleared up. Certain proprietary remedies available only from veterinarians have proved very effective in treatment.

SCREWWORM

Wounds resulting from branding, castrating, and dehorning afford a breeding ground for screwworms, *Cochliomyia hominivorax*. Add to this the wounds from some types of vegetation, from fighting, and from bloodsucking insects, and ample places for propagation are provided.

DISTRIBUTION AND LOSSES CAUSED BY SCREWWORMS

Although the screwworm has been eradicated in the United States, it occasionally reappears.

LIFE HISTORY AND HABITS

The primary screwworm fly is bluish green in color, with three dark stripes on its back and reddish or orange color below the eyes. The fly generally deposits its eggs in shinglelike masses on the edges or the dry portion of wounds. From 50 to 300 eggs are laid at one time, with a single female being capable of laying about 3,000 eggs in a lifetime. Hatching of the eggs occurs in 11 hours, and the young whitish worms (larvae or maggots) immediately burrow into the living flesh. There they are fed and grow for a period of 4 to 7 days, shedding their skin twice during this period.

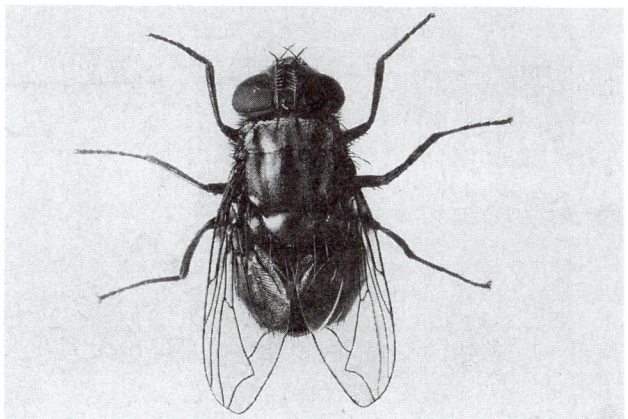

Fig. 15–45.   The screwworm fly (*Cochliomyia americana*). (Courtesy, USDA)

When these worms have reached their full growth, they assume a pinkish color, leave the wound, and drop to the ground, where they dig beneath the surface of the soil and undergo a transformation to the hard-skinned, dark-brown, motionless pupa. It is during the pupa stage that the maggot changes to the adult fly.

After the pupa has been in the soil from 7 to 60 days, the fly emerges from it, works its way to the surface of the ground, and crawls up on some nearby object (bush, weed, etc.) to allow its wings to unfold and otherwise mature. Under favorable conditions, the newly emerged female fly becomes sexually mature and will lay eggs 5 days later. During warm weather, the entire life cycle is usually completed in 21 days,

but under cold, unfavorable conditions the cycle may take as many as 80 days or longer.

DAMAGE INFLICTED BY; SYMPTOMS AND SIGNS OF AFFECTED ANIMALS

The injury caused by this parasite is inflicted chiefly by the maggots. The early symptoms in affected animals are loss of appetite and condition, and listlessness. Unless proper treatment is administered, the great destruction of many tissues kills the host in a few days.

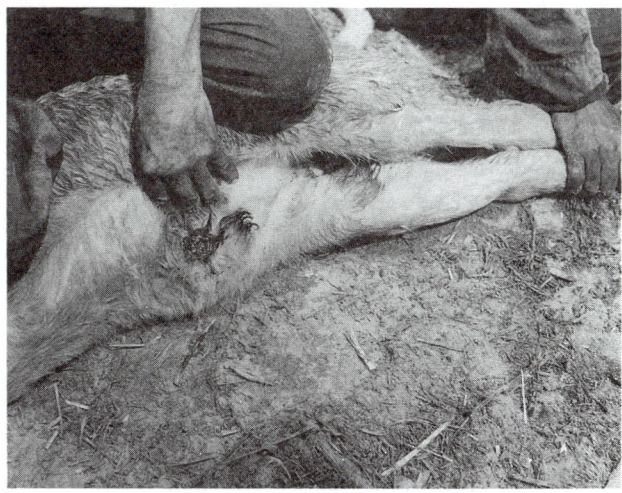

Fig. 15–46.   Screwworm in navel of calf. (Courtesy, USDA)

PREVENTION, CONTROL, AND TREATMENT

Prevention in infested areas consists mainly of keeping animal wounds to a minimum and of protecting those that do materialize.

In 1958, the U.S. Department of Agriculture initiated an eradication program. Screwworm larvae were reared on artificial media. Two days before fly emergence, the pupae were exposed to gamma irradiation at a dosage which caused sexual sterility but no other deleterious effects. Sterile flies were distributed over the entire screwworm-infested region in sufficient quantity to outnumber the native flies, at an average rate of 400 males per square mile per week. The female mates only once and, therefore, when mated with a sterile male does not reproduce. There was a decline in the native population each generation until the native males were so outnumbered by sterile males that no fertile matings occurred and the native flies were eliminated. This program has virtually eliminated all the losses caused by screwworms in the United States.

When maggots (larvae) are found in an animal, they should be removed and sent to the proper authorities for identification, and the animal should be treated with a proper insecticide.

(Also see later section in this chapter entitled, "Federal and State Regulations Relative to Disease Control.")

TICKS

The lone star tick (*Amblyomma americanum*), the Gulf Coast tick (*Amblyomma maculatum*), the Rocky Mountain wood tick (*Dermacentor andersoni*), the Pacific Coast tick (*Dermacentor occidentalis*), and the American dog tick (*Dermacentor variabilis*) are three-host species that attack cattle during the summer months. The black-legged tick (*Ixodes scapularis*) is also a three-host tick that is common in later winter and early spring. The winter tick (*Dermacentor albipictus*) is a one-host species found on cattle and horses in fall and winter. In addition, larvae and nymphs of the so-called *spinose* ear tick (*Octobius megnini*), a one-host species, attach deep in the ears of cattle and feed there for several months.

DISTRIBUTION AND LOSSES CAUSED BY TICKS

Ticks are widely distributed, especially throughout the southern part of the United States. But they are usually seasonal in their activities.

Ticks suck blood. They cause economic losses by transmitting diseases; by restlessness, anemia, and inefficient feed utilization; and by necessitating expensive treatments. Among the diseases transmitted to or produced in cattle by ticks are Texas fever, anaplasmosis, Q fever, tick paralysis, and piroplasmosis.

LIFE HISTORY AND HABITS

Generally, all species of ticks have similar stages of development. The females lay eggs that hatch into six-legged larvae (seed ticks). The larvae attach to a host, engorge on blood, and molt to eight-legged nymphs. The nymphs attach to a host, engorge on blood, and molt to eight-legged adults. Mating usually occurs on the host. The female then engorges fully, drops off the host, lays several thousand eggs, and dies.

DAMAGE INFLICTED BY; SYMPTOMS AND SIGNS OF AFFECTED ANIMALS

Generally speaking, injury to cattle from tick parasitism varies directly with numbers of parasites. Ticks feed exclusively on blood. Thus, when several hundred ticks feed, the host becomes anemic, unthrifty, and loses weight. In addition, some female ticks generate a paralyzing toxin. The spinose ear tick, commonly called the *ear tick*, takes up residence along the inner surfaces of the ears and in external ear canals, where it is extremely annoying. Cattle heavily parasitized by spinose ear ticks droop their heads, rub and shake their ears, and turn their heads to one side.

PREVENTION, CONTROL, AND TREATMENT

Because most species of ticks, except the ear tick, attach to the external surfaces of cattle, dipping and spraying are the most effective methods of control; however, dusts may be used. To treat animals for ear ticks, the chemical should be applied into the ears of the cattle.

CONTROL OF EXTERNAL PARASITES

Losses to the dairy industry due to insect pests can be considerably reduced by treating animals with insecticides that give economical control.

The cost of controlling external parasites on cattle with insecticides is usually very small compared with losses incurred when the infestations go uncontrolled. A good program for controlling external parasites of cattle should (1) be initiated during the early stages of infestation; (2) include use of good sanitation practices in addition to the application of insecticides (not all external parasites can be effectively controlled with insecticides); and (3) require use of insecticides in complete accordance with the labels and instructions.

No suggestions concerning the specific insecticides that should be used are presented in this book. Rather, a review follows relative to some of the types of formulations that are available, the type of treatments or applications with chemicals that may be used to control external parasites of cattle, and some of the restrictions and precautions to be observed when treating animals.

APPLICATION OF INSECTICIDES

The availability of an insecticide and the type of application(s) for which it was formulated are of prime importance, but the treatment of animals is dictated pretty much by the animal species, number of animals, available handling facilities, time or season, the target pest, management practices, and cost. The common methods of insecticide application are (1) spraying, (2) dipping, (3) back rubbers, (4) pour-on, (5) feed additives, (6) injectables, (7) insecticide impregnated ear tags, and (8) boluses.

USE INSECTICIDES SAFELY

Certain basic precautions must be observed when insecticides are to be used because, used improperly, they can be injurious to people, domestic animals, wildlife, and beneficial insects. Follow the directions and heed all the precautions on the labels.

■ **Selecting insecticides**—Always select the formulation and insecticide labeled for the purpose for which it is to be used.

■ **Storing insecticides**—Always store insecticides in original containers. Never transfer them to unlabeled containers or to food or beverage containers. Store insecticides in a dry place out of reach of children, animals, or unauthorized persons.

■ **Disposing of empty containers and unused insecticides**—The Environmental Protection Agency (EPA) has ruled that they consider triple rinsed containers as ordinary trash, not to be classified as a pesticide container. So, if a container is rinsed three times, the empty container can be disposed of as ordinary solid waste and sent to a landfill.

Always observe local regulations in the disposal of unused insecticides.

■ **Mixing and handling** — Mix and prepare insecticides in the open or in a well-ventilated place. Wear rubber gloves and clean dry clothing (respirator device may be necessary with some products). If any insecticide is spilled on you or your clothing, wash with soap and water immediately and change clothing. Avoid prolonged inhalation. Do not smoke, eat, or drink when mixing and handling insecticides.

■ **Applying** — Use only amounts recommended. Apply at the correct time to avoid unlawful residues in milk or meat. Avoid treating animals younger than specified on the label. Avoid retreating more often than label restrictions. Avoid drift on nearby crops, pastures, livestock, or other nontarget areas. Avoid prolonged contact with all insecticides. Do not eat, drink, or smoke until all operations have ceased and hands and face are thoroughly washed. Change and launder clothing after each day's work.

■ **In case of emergency** — If you accidentally swallow an insecticide, induce vomiting by taking one tablespoonful of salt in a glass of water. Repeat if necessary. Call a doctor.

■ **Withdrawal** — After treating animals with pesticides, observe the prescribed number of days between the last treatment and the use of milk or meat. Refer to the product labels for this information.

DISINFECTANTS

A disinfectant is a bactericidal or microbicidal agent that frees the object from infection (usually a chemical agent which destroys disease germs or other microorganisms, or inactivates viruses).

The high concentration of dairy animals and the continuous use of buildings often results in a condition referred to as *disease buildup.* As disease-producing organisms — viruses, bacteria, fungi, and parasite eggs — accumulate in the environment, disease problems can become more severe and be transmitted to each succeeding group of animals raised on the same premises. Under these circumstances, cleaning and disinfection become extremely important in breaking the life cycle. Also, in the case of a disease outbreak, the premises must be disinfected.

Under ordinary conditions, proper cleaning of barns removes most of the microorganisms, along with the filth, thus eliminating the necessity of disinfection.

Effective disinfection depends on five things:

1. Thorough cleaning before application.
2. The phenol coefficient of the disinfectant, which indicates the killing strength of a disinfectant as compared to phenol (carbolic acid). It is determined by a standard laboratory test in which the typhoid fever germ often is used as the test organism.

3. The dilution at which the disinfectant is used.
4. The temperature; most disinfectants are much more effective if applied hot.
5. Thoroughness of application, and time of exposure.

Disinfection must in all cases be preceded by a very thorough cleaning, for organic matter serves to protect disease germs and otherwise interferes with the activity of the disinfecting agent.

Sunlight possesses disinfecting properties, but it is variable and superficial in action. Heat and some of the chemical disinfectants are more effective.

The application of heat by steam, by hot water, by burning, or by boiling is an effective method of disinfection. In many cases, however, it may not be practical to use heat.

A good disinfectant should (1) have the power to kill disease-producing organisms, (2) remain stable in the presence of organic matter (manure, hair, soil), (3) dissolve readily in water and remain in solution, (4) be nontoxic to animals and humans, (5) penetrate organic matter rapidly, (6) remove dirt and grease, and (7) be economical to use.

The number of available disinfectants is large because the ideal universally applicable disinfectant does not exist. Table 15–4 (pp. 372–373) gives a summary of the limitations, usefulness, and strength of some common disinfectants.

When using a disinfectant, *always read and follow the manufacturer's directions.*

FEDERAL AND STATE REGULATIONS RELATIVE TO DISEASE CONTROL

Certain animal diseases are so devastating that no individual farmer or rancher could long protect privately owned herds against their invasion. Moreover, where human health is involved, the problem is much too important to be entrusted to individual action. In the United States, therefore, certain regulatory activities in animal disease control are under the supervision of various federal and state organizations. Federally, this responsibility is entrusted to the following agency:

Veterinary Service
Animal and Plant Health Inspection Service
U.S. Department of Agriculture
Federal Center Building
Hyattsville, MD 20782

FOOD AND DRUG ADMINISTRATION (FDA)

The FDA is charged with the responsibility of safeguarding American consumers against injury, unsanitary food, and fraud. It inspects and analyzes samples and conducts independent research on such things as toxicity (using laboratory animals), disappearance curves for pesticides, and long-range effects of drugs.

TABLE 15–4
DISINFECTANT GUIDE[1]

Kind of Disinfectant	Usefulness	Strength	Limitations and Comments
Alcohol (ethyl-ethanol, Isopropyl, Methanol)	Primarily as skin disinfectant and for emergency purposes on instruments.	70% alcohol—the content usually found in rubbing alcohol.	They are too costly for general disinfection. They are ineffective against bacterial spores.
Boric Acid[2]	As a wash for eyes, and other sensitive parts of the body.	1 oz in 1 pt water (about 6% solution).	It is a weak antiseptic. It may cause harm to the nervous system if absorbed into the body in large amounts. For this and other reasons, antibiotic solutions and saline solutions are fast replacing it.
Chlorines (sodium hypo-chlorate, chlormine-T)	Used for dairy equipment and as deodor-ants. They will kill all kinds of bacteria, fungi, and viruses, providing the concentration is sufficiently high.	Generally used at about 200 ppm for dairy equipment and as a deodorant.	They are corrosive to metals and neutral-ized by organic materials. Not effective against T.B. organisms and spores.
Cresols (many commercial products available)	A generally reliable class of disinfectant. Effective against brucellosis, shipping fever, and tuberculosis. Cresols give good results in foot baths.	Cresol is usually used as a 2–4% solution (1 cup/2 gal of water makes a 4% solution).	Effective on organic material. Cannot be used where odor may be absorbed, and, therefore, not suited for use around milk and meat.
Formaldehyde (gaseous disinfectant)	Formaldehyde will kill anthrax spores, T.B. organisms, and animal viruses in a 1–2% solution. It is often used to disinfect buildings following a disease outbreak. A 1–2% solution may be used as a foot bath to control foot-rot.	As a liquid disinfectant, it is usually used as a 1–2% solution. As a gaseous disinfectant (fumigant), use 1½ lb of potassium permanganate plus 3 pt of formaldehyde. Also, gas may be released by heating paraformaldehyde.	It has a disagreeable odor, destroys living tissue, and can be extremely poisonous. The bactericidal effectiveness of the gas is dependent upon having the proper relative humidity (above 75%) and temperature (above 86°F *[30°C]* and preferably near 140°F *[60°C]*).
Heat (by steam, hot water, burning, or boiling)	In the burning of rubbish or articles of little value, and in disposing of infected body discharges. The steam "Jenny" is effective for disinfection if properly employed, particularly if used in conjunction with a phenolic germicide.	10 minutes exposure to boiling water is usually sufficient.	Exposure to boiling water will destroy all ordinary disease germs but sometimes fails to kill the spores of such diseases as anthrax and tetanus. Moist heat is preferred to dry heat, and steam under pressure is the most effective. Heat may be impractical or too expensive.
Iodine[2] (tincture)	Extensively used as skin disinfectant, for minor cuts and bruises.	Generally used as tincture of iodine, either 2% or 7%.	Never cover with a bandage. Clean skin before applying iodine. It is corrosive to metals.
Iodophors (tamed iodine)	Primarily used for dairy utensils. Effective against all bacteria (both gram-negative and gram-positive), fungi, and most viruses.	Usually used as disinfectants at con-centrations of 50–75 ppm titratable iodine, and as sanitizers at levels of 12.5–25 ppm. At 12.5 ppm titratable iodine, they can be used as an antiseptic in drinking water.	Iodophors are combinations of iodine and detergents. They are inhibited in their activity by organic matter. They are quite expensive. They should not be used near heat.
Lime (quicklime, burnt lime, calcium oxide)	As a deodorant when sprinkled on manure and animal discharges; or as a disinfectant when sprinkled on the floor or used as a newly made "milk of lime" or as a whitewash.	Use as a dust; as "milk of lime"; or as a whitewash, but use fresh.	Not effective against anthrax or tetanus spores. Wear goggles, when adding water to quicklime.
Lye (sodium hydroxide, caustic soda)	On concrete floors; in milk houses because there is no odor; against microorganisms of brucellosis and the viruses of foot-and-mouth disease. In strong solution (5%), effective against anthrax and blackleg.	Lye is usually used as either a 2% or 5% solution. To prepare a 2% solution, add 1 can of lye to 5 gal of water. To prepare a 5% solution, add 1 can of lye to 2 gal of water. A 2% solution will destroy the organisms causing foot-and-mouth disease, but a 5% solution is necessary to destroy the spores of anthrax.	Damages fabrics, aluminum, and painted surfaces. Be careful, for it will burn the hands and face. Not effective against organism of T.B. or Johne's disease. It is relatively cheap. Lye solutions are most effective when used hot. Diluted vinegar can be used to neutralize lye.
Lysol (the brand name of a product of cresol plus soap)	For disinfecting surgical instruments and instruments used in dehorning, castrating, and tattooing. Useful as a skin disinfectant before surgery, and for use on the hands before castrating.	0.5–2.0%.	Has a disagreeable odor. Does not mix well with hard water. Less costly than phenol.

(Continued)

TABLE 15–4 *(Continued)*

Kind of Disinfectant	Usefulness	Strength	Limitations and Comments
Phenol (carbolic acid): 1. Phenolics—coal tar derivatives 2. Synthetic phenols	They are ideal general-purpose disinfectants. Effective and inexpensive. They are very resistant to the inhibiting effects of organic residue; hence, they are suitable for barn disinfection, and foot and wheel dip-baths.	Both phenolics (coal tar) and synthetic phenols vary widely in efficacy from one compound to another. So, note and follow manufacturer's directions. Generally used in a 5% solution.	They are corrosive, and they are toxic to animals and humans. Ineffective on fungi and viruses. Effective against bacteria, including T.B. organisms.
Quaternary ammonium compounds (QAC)	Very water soluble, ultrarapid kill rate, effective deodorizing properties, and moderately priced. Good detergent characteristics and harmless to skin.	Follow manufacturer's directions.	They can corrode metal. Not very potent in combating viruses. Adversely affected by organic matter; hence, they are of limited use for disinfecting livestock facilities. Not effective against T.B. organisms and spores. Not effective against anthrax and tetanus.
Sal soda	It may be used in place of lye against foot-and-mouth disease.	10½% solution (13½ oz/1 gal water)	
Sal soda and soda ash (or sodium carbonate)	They may be used in place of lye against foot-and-mouth disease.	4% solution (1 lb/3 gal water). Most effective in hot solution.	Commonly used as cleansing agents, but have disinfectant properties, especially when used as a hot solution.
Soap	Its power to kill germs is very limited. Greatest usefulness is in cleansing and dissolving coatings from various surfaces, including the skin, prior to application of a good disinfectant.	As commercially prepared.	Although indispensable to sanitizing surfaces, soaps should not be used as disinfectants. They are not regularly effective; staphylococci and the organisms which cause diarrheal disease are resistant.

[1]For metric conversions, see the Appendix of this book.

[2]Sometimes loosely classed as disinfectant but actually an antiseptic and practically useful only on living tissue.

U.S. DEPARTMENT OF AGRICULTURE (USDA)

The following four divisions of the U.S. Department of Agriculture have primary responsibilities in the area of animal and human health.

1. **The Animal and Plant Health Inspection Service.** This division is charged with maintaining the wholesomeness and safety of meats processed in packing plants that ship meat and meat products, including poultry and poultry products, interstate.

2. **The Labeling and Registration Section.** This section in the USDA has responsibility for the proper labeling and safe use of pesticides. Manufacturers of pesticides must present new products with their proposed labels for approval before they are authorized to sell them. The label must indicate, as a minimum, the following: the name of the product; the active and inactive ingredients, together with percentages of each, in the formulation; the pest(s) controlled; directions for use—including the method and rate of application; any restrictions to be observed in application and handling; and an antidote—if known.

3. **The Veterinary Service Division (VSD).** This division of the USDA is responsible for programs to control and eradicate (if possible) certain diseases of livestock, e.g., brucellosis, tuberculosis, and scabies.

4. **Stockyard Inspection.** With the advent of large public markets, public stockyards inspection was initiated. This is an addition to the regular inspection performed on animals by meat inspectors prior to slaughter. Among the principal disease for which inspections are made are: anthrax and scabies of cattle.

Not only are the incoming shipments of livestock inspected, but a reinspection is made of outgoing shipments. Tests for tuberculosis and brucellosis are accomplished, and dipping for scabies is performed before shipments are allowed to return to farms and ranches.

U.S. PUBLIC HEATH SERVICES (USPHS)

This section of the Department of Health and Human Services is concerned with the prevention and treatment of disease. It works in the areas of vector control, pollution control, and control of communicable diseases of people.

STATE VETERINARIANS, SANITARY COMMISSIONS, AND BOARDS

Most states have state veterinarians, or comparable officials, who direct the livestock sanitary and regulatory programs within their respective states. Livestock producers may secure the regulations applicable to the state in which they reside by writing their state department of agriculture.

QUARANTINE [6]

Many highly infectious diseases are prevented by quarantine from (1) gaining a foothold in this country or (2) spreading. By quarantine is meant (1) segregation and confinement of one or more animals in the smallest possible area to prevent any direct or indirect contact with animals not so restrained; or (2) regulating movement of animals at points of entry.

When an infectious disease outbreak occurs, drastic quarantine must be imposed to restrict movement out of an area or within areas. The type of quarantine varies from one involving a mere physical examination and movement under proper certification to the complete prohibition against the movement of animals, produce, vehicles, and even human beings.

FOREIGN DISEASE PROTECTION

Distance no longer provides a buffer against the invasion of foreign diseases. More than 90% of animals imported into the United States arrive by air. An airplane can outpace the development of clinical signs of diseases in an animal that has been exposed to infection just prior to shipment. This prompts great concern for epizootic diseases capable of crippling or destroying entire livestock populations. Such diseases still exist in Europe, Asia, Africa, and Latin America; among them are such dreaded diseases as rinderpest, contagious bovine pleuropneumonia, foot-and-mouth disease, hog cholera, Africa horse sickness, Africa swine fever, exotic Newcastle disease, African trypanosomiasis, East Coast fever, and piroplasmosis.

Fig. 15–47. Cow with foot-and-moth disease. This dreaded disease is capable of crippling the entire U.S. cattle industry. So, drastic measures are taken to prevent the introduction of the disease into the United States. (Courtesy, USDA)

---

[6]This portion from "Quarantine" to "State Indemnity Payments" was authoritatively reviewed by, and helpful suggestions were received from, Dr. M. A. Mixson, Chief Staff Veterinarian, Emergency Programs, Veterinary Services, Animal and Plant Health Inspection Service, U.S. Department of Agriculture, Washington, DC.

Today, there are stations at several entry points, where inspectors of the USDA's Animal and Plant Health Inspection Service (APHIS) inspect all animals to be imported into the United States. If no communicable diseases are found, the animals may be quarantined for a period of time, during which time they are treated for external parasites and subjected to various tests—e.g., cattle are tested for brucellosis. At the end of the quarantine period, if no communicable diseases are found, they are released to the purchaser.

FEDERAL QUARANTINE CENTER (HARRY S. TRUMAN ANIMAL IMPORT CENTER)

A Federal Quarantine Center was authorized in Public Law 91–239, signed by the President on May 6, 1970; and a 16.1-acre site for the Center was selected at Fleming Key, near Key West, Florida.

The quarantine center is designed to hold some 400 head of cattle, or other species in equivalent numbers, at one time for a 5-month quarantine period. This maximum security station enables American livestock producers to import breeding animals from all parts of the world, while at the same time safeguarding our domestic herds and flocks from such diseases as foot-and-mouth disease, rinderpest, piroplasmosis, and others.

INDEMNITY PAYMENTS

Where certain animal diseases are involved, the livestock producer can obtain financial assistance in eradication programs through federal and state sources.

Note well: Both federal and state indemnity payments are subject to change. So, for current regulations, the livestock producer should contact the local veterinarian or State Department of Agriculture.

FEDERAL INDEMNITY PAYMENTS

Information relative to indemnities paid to owners by the federal government for animals disposed of as a result of outbreaks of certain diseases is given in Chapter 1, Subchapter B, Title 9 of the Code of Federal Regulations, pertinent facts about which follow:

■ **Brucellosis and tuberculosis reactors**—Owners of cattle destroyed which are affected with brucellosis or tuberculosis may be paid indemnity by the USDA.

■ **Foot-and-moth disease, pleuropneumonia, rinderpest, and other contagious or infectious animal diseases which constitute an emergency and threaten the livestock industry of the country**—Under Title 9, Part 53, of the U.S. Code of Federal Regulations, the Secretary of Agriculture of the U.S. Department of Agriculture may declare a national emergency due to the existence of foot-and-mouth disease, rinderpest, contagious pleuropneumonia, or any other communicable disease of livestock which threatens the livestock industry of the country.

In order to reduce the cost of eradicating emergency disease to the livestock producer and to the state and federal government, it is essential that suspicious cases be promptly reported. If such a disease is suspected, the producer should promptly report the suspected case(s) to the practicing veterinarian and to state and federal animal health officials.

STATE INDEMNITY PAYMENTS

It is suggested that livestock producers secure the regulations applicable to the state in which they reside by writing to their state department of agriculture.

NATIONAL ANIMAL POISON CONTROL CENTER

Established in 1978 and maintained at the University of Illinois, Urbana-Champaign, the National Animal Poison Control Center hotline number is 217/333-3611. Recognizing that accidents do not wait for business hours, the Center is open 24 hours a day, every day of the week. The toxicology group is staffed to answer questions about known or suspected cases of poisoning or chemical contaminations involving any species of animal. It is not intended to replace local veterinarians or state toxicology laboratories, but to complement them.

The toxicologists at the center constantly update their files on chemicals, feed additives, human and veterinary drugs, pesticides, environmental contaminants, and plant and mold toxins. Their comprehensive file of information contains comparative species toxicity data, product ingredients, and recommended therapeutic and decontamination measures. The goal is a computer database containing 200,000 entries to facilitate quick and accurate responses to all types of poisoning/contamination incidents and inquiries.

Many times a proper treatment regime can be recommended over the telephone. When telephone consultation is inadequate or the problem is of major proportion, a team of veterinary specialists can arrive at the scene of a toxic contamination problem within a short time.

The cost of an investigation varies according to distance traveled, personnel time, and laboratory services required. Where consultation over the telephone is adequate, there is no charge to the veterinarian or producer.

QUESTIONS FOR STUDY AND DISCUSSION

1. Why are good dairy cattle health books of value to livestock producers?

2. Why should dairy producers by well informed relative to the relationship of cattle diseases and parasites to other classes of animals and to humans?

3. What is normal temperature, pulse rate, and respiration rate of cattle, and how would you determine each?

4. Select a specific dairy farm (either your own or one with which you are familiar) and outline (in 1, 2, 3, order) a program of dairy cattle health, disease prevention, and parasite control.

5. Assume that a specific contagious disease (you name it) has broken out in your herd. What steps would you take to meet the situation (list in 1, 2, 3 order; be specific)?

6. Give the cause, prevention, and treatment of each of the following diseases of dairy cattle: (a) bovine respiratory disease complex, (b) brucellosis, (c) calf diphtheria, (d) calf scours, (e) fescue foot, (f) foot-rot, (g) pneumonia, and (h) tuberculosis.

7. Since 1934, the United States has conducted a brucellosis control program. Yet, in 1990, after 56 years, about 0.5% of the nation's cattle were still infected with the disease. Has the program been a success or a failure?

8. How can milk and meat from brucellosis-infected cows be made safe for human consumption?

9. Sometimes dairy producers are accused of over-reacting in their efforts to keep foot-and-mouth disease out of this country as a means of preventing competition from foreign imports. Do you agree that cattle producers have overreacted?

10. Detail the measures that you would recommend to prevent mastitis.

11. Assume that a specific parasite (you name it) has become troublesome in your herd. What steps would you take to meet the situation (list in 1, 2, 3 order; be specific)?

12. Outline a control program for the most common species of flies that attack or annoy cattle.

13. Give the prevention, control, and treatment of each of the following parasites of dairy cattle: (a) bovine trichomoniasis, (b) coccidiosis, (c) gastrointestinal worms, (d) grubs, and (e) lice.

14. When mites are found in a herd, why is that herd quarantined? Who does the quarantining?

15. Explain (a) how screwworm flies are sterilized, and (b) how the screwworm control program worked.

16. If a child swallows an insecticide, what emergency steps should you take?

17. Assume that you have, during a period of a year, encountered cattle death losses from three different diseases (you name them). What kind of disinfectant would you use in each case?

18. Why are certain regulatory activities in animal disease control under the supervision of various federal and state agencies?

19. Why was the federal quarantine center, now known as the Harry S. Truman Animal Import Center, established?

20. When people are ill, they call the family doctor. Isn't it just as logical that a veterinarian be called when cattle are sick?

21. Discuss indemnity payments.

SELECTED REFERENCES

Title of Publication	Author(s)	Publisher
Animal Health, A Layperson's Guide to Disease Control, Second Edition	W. J. Greer J. K. Baker	Interstate Publishers, Inc., Danville, IL, 1992
Animal Health Livestock and Pets, 1984 Yearbook of Agriculture	Staff	U.S. Department of Agriculture, Washington, DC, 1984
Current Veterinary Therapy	Edited by J. L. Howard	W. B. Saunders Company, Philadelphia, PA, 1981
Diseases Transmitted from Animals to Man, Sixth Edition	Edited by W. T. Hubbert, *et al.*	Charles C. Thomas, Publisher, Springfield, IL, 1975
Disinfections, Sterilization, and Preservation	C. A. Lawrence S. S. Block	Lea & Febiger, Philadelphia, PA, 1968
Emerging Diseases of Animals	Veterinary Research Laboratory, Onderstepoort, South Africa	Food and Agriculture Organization of the United Nations, Rome, Italy, 1968
Hagan's Infectious Diseases of Domestic Animals, Sixth Edition	D. W. Bruner J. H. Gillespie	Cornell University Press, Ithaca, NY, 1973
Handbook of Pest Management in Agriculture, Vol. 1	D. Pimentel	CRC Press, Inc., Boca Raton, FL, 1981
Manual of Tropical Veterinary Parasitology	M. Shah-Fischer R. R. Say	CAB International, Wallingford, U.K., 1989
Merck Veterinary Manual, The, Sixth Edition	Edited by O. H. Siegmund	Merck & Co., Inc., Rahway, NJ, 1986
Nationwide System for Animal Health Surveillance, A	National Research Council	National Academy of Sciences, Washington, DC, 1974
State of Food and Agriculture, The	Staff	Food and Agriculture Organization of the United Nations, Rome, Italy, 1987–88
Stockman's Handbook, The, Seventh Edition	M. E. Ensminger	Interstate Publishers, Inc., Danville, IL, 1992

In addition to the above selected references, valuable publications on different subjects pertaining to animal disease, parasites, disinfectants, and poisonous plants can be obtained from the following sources:

1. Division of Publications, Office of Information, U.S. Department of Agriculture, Washington, DC, 20250.
2. Your state agricultural college.
3. Several biological, pharmaceutical, and chemical companies.

Fig. 16–1. Dairy. (Courtesy, D. R. Rush, Berkeley, CA)

Contents

CHAPTER

16

DAIRY CATTLE

BUILDINGS

AND EQUIPMENT

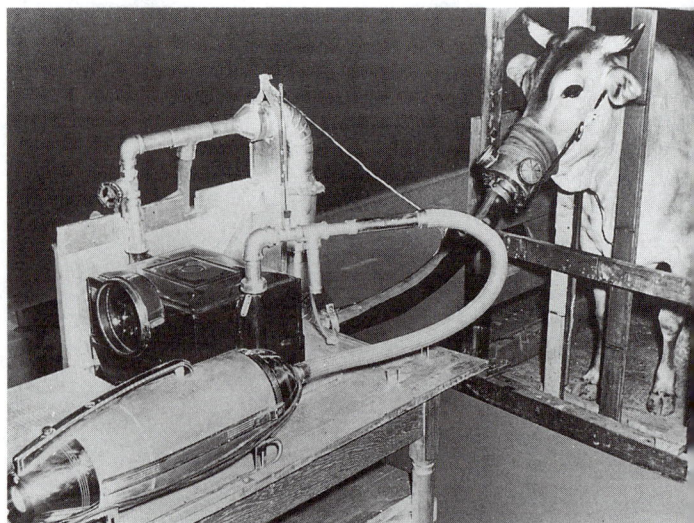

Fig. 16–2. Evaluating heat tolerance by respiratory volume. In the future, environmental control will be an important consideration in dairy buildings. (Courtesy, USDA)

Modern dairy cattle buildings and equipment should be designed to facilitate (1) freedom and individual comfort of cows, (2) automation and laborsaving, (3) herd health and sanitation, (4) bedding conservation, and (5) manure disposal.

Economical and successful dairy production depends largely upon the investment in practical and convenient buildings and equipment, as well as upon the care, feeding, and management of the herd. As would be expected in a country so big and diverse as the United States, there are wide differences in the systems and facilities of dairy production. A major difference exists between (1) the family-owned and operated farm herd, which usually relies on pastures in season and produces its own winter roughage; and (2) the large, commercial-type, drylot operation. In addition, facilities differ according to whether the operation is devoted to producing milk or raising heifer replacements, or a combination of both. Further facility differences exist between a strictly commercial enterprise and a purebred herd. Climate also makes for differences; from the animals being out in the open much of the time in the South and in California, to the need for warmth and a long winter feeding period in the North. Finally, there are differences due to meeting the health regulations of the particular area and market, the availability of materials and labor, and individual preferences.

ENVIRONMENTAL CONTROL

Nature gives dairy cattle an assist in environmental control through added hair growth for winter and shedding in the summer. Also, given an opportunity, animals seek out natural windbreaks in the winter and shades in the summer. But, further than this, they must rely on people to modify their environment.

The primary reason for having dairy buildings is to modify the environment. Properly designed barns and other shelters, shades, insulation, ventilation, and air conditioning can be used to approach the environment that we wish. Naturally, the investment in environmental control facilities must be balanced against the expected increased returns; and there is a point beyond which further expenditures for environmen-

tal control will not increase returns sufficiently to justify the added cost. This point of diminishing returns will differ between sections of the country, between animals of different ages (higher expenditures for environmental control can be justified for high-producing cows than for dry cows, for example), and between operators. Labor and feed costs will enter into the picture, also.

HEAT PRODUCTION OF CATTLE

The heat produced by cattle varies according to body weight, rate of feeding, environmental conditions, and degree of activity. However, Table 16–1 may be used as a guide.

TABLE 16–1
HEAT PRODUCTION OF CATTLE[1]

Heat Source	Heat Production, BTU/hr[2]			Heat Production, Kcal/hr		
	Temperature	Total	Sensible	Temperature	Total	Sensible
	(°F)			(°C)		
Cows, 1,000 lb (453.6 kg)	40	3,600	2,640	4	907.2	665.3
	70	3,000	1,550	21	756.0	390.6
Calves, 6–10 months	60	780	660	16	196.6	166.3
	80	720	420	27	181.4	105.8

[1]Adapted by the author from *Agricultural Engineers Yearbook,* St. Joseph, Michigan, ASAE Data Sheet D-249.2, p. 424.

[2]A British thermal unit (BTU) is the quantity of heat required to raise the temperature of 1 lb of water 1°F.

VAPOR PRODUCTION OF CATTLE

Most building designers are inclined to govern the amount of air change by the need for moisture removal. Table 16–2 gives the information necessary for determining the amount of moisture to be removed.

TABLE 16–2
VAPOR PRODUCTION OF CATTLE[1]

Vapor Source	Temperature		Vapor Production			
	(°F)	(°C)	(lb/hr)	(kg/hr)	(BTU/hr)	(Kcal/hr)
Cows, 1,000 lb (453.6 kg)	40	4	0.92	0.42	960	241.9
	70	21	1.38	0.63	1,450	365.4
Calves, 6–10 months	60	16	0.11	0.05	120	30.2
	80	27	0.29	0.13	300	75.6

[1]Adapted by the author from *Agricultural Engineers Yearbook*, St. Joseph, Michigan, ASAE Data Sheet D-249.2, p. 424.

The removal of such large quantities of water, especially in the winter when barns are closed, is a difficult problem for the designer to solve.

RECOMMENDED ENVIRONMENTAL CONTROL FOR DAIRY CATTLE

There is a paucity of experimental work on environmental control. Nevertheless, Table 16–3 represents the author's recommendations, based on experiences and experiments.

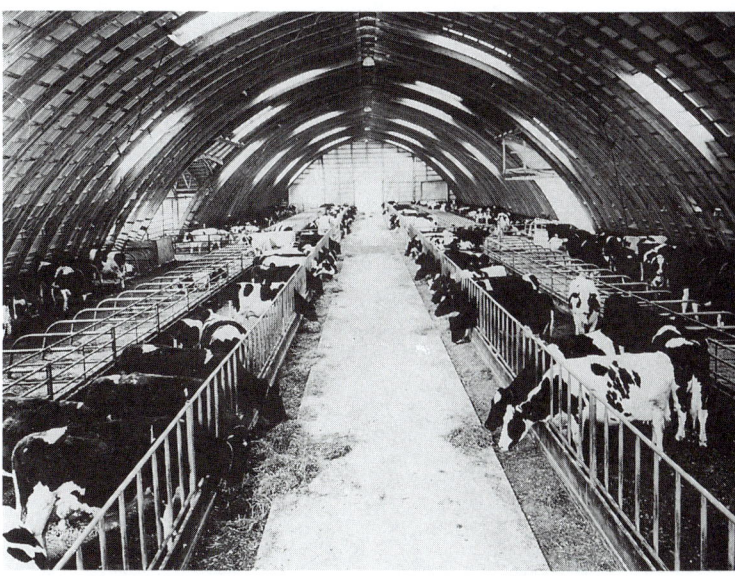

Fig. 16–3. Environmentally controlled dairy barn—a completely enclosed building with air circulation and with temperature, humidity, and light control. Environmentally controlled dairy barns are on the increase because they lend themselves to automation and make for the ultimate in animal comfort, health, and efficiency of feed utilization. (Courtesy, Ford New Holland, New Holland, PA)

PLANS AND SPECIFICATIONS

No attempt will be made in this chapter to present detailed building and equipment plans and specifications. Rather, it is desired merely to convey suggestions regarding some of the desirable features of buildings and equipment

TABLE 16–3
RECOMMENDED ENVIRONMENTAL CONDITIONS FOR CATTLE

Class of Animal	Temperature				Acceptable Humidity	Commonly Used Ventilation Rates[1]					Drinking Water			
	Comfort Zone		Optimum			Basis	Winter[2]		Summer		Winter		Summer	
	(°F)	(°C)	(°F)	(°C)	(%)		(cfm)	(m³/min)	(cfm)	(m³/min)	(°F)	(°C)	(°F)	(°C)
Dairy cows	40–70	5–21	50–60	10–15	50–75	1,000 lb (or 454 kg) per 100 lb (45 kg)	100	2.8	200	5.7	50	10	60–75	15–24
Dairy calves	50–75	10–24	65	17			10		25					

[1]Generally two different ventilating systems are provided: one for winter, and an additional one for summer. Hence, as shown in Table 16–3, the winter ventilating system in a dairy cow barn should be designed to provide 100 cfm (cubic feet/minute) for each 1,000-lb cow. Then, the summer system should be designed to provide an added 100 cfm, thereby providing a total of 200 cfm for summer ventilation.

In practice, in many buildings, added summer ventilation is provided by opening (1) barn doors, and (2) high-up hinged walls.

[2]Provide approximately 25% the winter rate continuously for moisture control.

CONFINEMENT ENVIRONMENTAL CONTROL

Confinement refers to limited quarters. Maximum environmental control necessitates a completely enclosed building with air circulation and temperature, humidity, and light control. Today, environmentally controlled confinement buildings are on the increase in the United States.

in use in various parts of the country. For detailed plans and specifications for a particular locality, the dairy producer should (1) study successful buildings and equipment on neighboring dairy farms; (2) consult the local county agricultural agent, vocational agriculture instructor, or lumber dealer; and/or (3) write to the state college of agriculture.

LOCATION OF THE DAIRY FARM HEADQUARTERS

Fig. 16–4. Well-located and attractive dairy farm headquarters. Note scenic views, access road, good drainage, and commanding view of the herd. (Courtesy, USDA)

When planning an entirely new dairy farm headquarters, the choice of location for the buildings is the first consideration. Likewise, when appraising the desirability of an existing headquarters, the same factors should be considered. These are:

1. **Central location.** From the standpoint of management, the most convenient location for the headquarters is near the center of the unit or in the middle of the long side. Normally, a natural dairy farm control center of this type makes for (a) the best accessibility to fields, for both equipment and stock, and (b) the most desirable overall visual supervision of operations. However, if the farm is deep, the location of the headquarters near the center of the unit may require a private all-weather lane or road leading from the highway to the headquarters, at added cost for construction and maintenance.

2. **Water supply.** Water must be available and plentiful. The availability of electricity and automatically operated pumps makes it possible to locate the headquarters farther from the source of water supply when it is desirable to do so to obtain other advantages. In irrigated areas, it is also desirable to be near an irrigation turnout.

3. **Roads.** It is preferable that the headquarters be located near an all-weather road or highway that is well maintained, but one should avoid having the farmstead on both sides of a road. Normally, a location along an all-weather road has better access to electric and telephone lines, the school bus, mail, religious and recreational facilities, and other services. Also, in irrigated areas, the irrigation turnout is usually near an all-weather road. When along dirt and gravel roads, the headquarters should be placed far enough away, taking into consideration the direction of the prevailing winds, to keep the dust from becoming a nuisance.

4. **Telephone and electricity.** If possible, the headquarters should be near well-maintained telephone and electric lines. Dairy farming is a business, and it is difficult to conduct any kind of business without access to a telephone. Likewise, electricity is essential for the operation of most modern utilities and automated equipment.

5. **Service facilities.** The farm should have convenient access to an established mail route, a school bus from a good school, delivery services, the church of preference, and recreational facilities of interest.

6. **Size and shape of area.** The area for the headquarters should be of adequate size, usually from 2 to 5 acres, and nearly square in shape. With further mechanization, smaller farmsteads are now needed than in the era when horses were extensively used. In general, the tendency is to have too large a farmstead; this adds to maintenance and weed-control costs and keeps valuable land out of production.

7. **Topography.** The topography should be high and level with no abrupt slopes. A relatively level area requires less site preparation, thus lowering building costs.

8. **Drainage.** The soil should be porous and the slope gentle, for this makes for dry corrals and lots. Dairy cattle health is much more easily maintained when the yards and buildings are well drained; and the work of the caretaker is more pleasant and easier under such conditions.

9. **Erosion control.** On those dairy farms in which soil erosion is a problem, it is desirable that the headquarters and fields be located to permit contour farming.

10. **Vegetation; windbreaks.** Natural shade trees for windbreaks, and a well-sodded area are valuable attributes. If a natural windbreak (hills or trees) is not available, wind protection may be provided by planting trees, or by utilizing the buildings themselves as windbreaks to protect yards, lots, and other open areas.

11. **Soil fertility.** A good fertile soil is desirable for the garden and yard.

12. **View.** A scenic view, especially from the living side of the house, makes for a "heap of living." Also, a location near the top of a slope may permit a commanding view of the rest of the farm.

The consideration of the above points in locating, planning, and improving the dairy farm headquarters can add materially to the convenience, comfort and pleasure of the operator, and to the profitable operation of the enterprise.

FARMSTEAD ARRANGEMENT

In planning a new dairy farm headquarters or in altering an old one, buildings, fences, lots, trees, etc., should be added according to an established master plan; for, once constructed, buildings are difficult and expensive to move. In general, for conservation of space and time, the barn and other service buildings should be located around a central court and should be so arranged that most of them can be seen from the house. In arriving at the best arrangement — which means the location and arrangement of individual buildings within the site — the farmstead cannot and should

Fig. 16–5. Well-arranged dairy farm headquarters. (Courtesy, United Co-operative Farms, Inc., Fitchburg, MA)

not be modeled after one popular pattern. Consideration should be given to the following pertinent factors:

1. **The house location comes first.** As the farm house is usually the headquarters or office of the dairy business as well as a home, its location is of greatest importance in farmstead arrangement. The house should be located (a) on a high area which is well drained away from the house and will command a view of other buildings as well as one or more scenic views; (b) where it is easily accessible; (c) in the direction of the prevailing winds, but sheltered from strong winds; (d) to obtain the maximum of sunlight in the North and a minimum of sunlight in the South; (e) with access of either the front or back door, but with best access to the front door; (f) where there is adequate yard which can be well landscaped; and (g) approximately 150 ft from the road.

2. **Orientation.** Fortunately, the farm or ranch headquarters need not be oriented with the compass. Although in general the farmstead plan will be developed to present the front to the road, most buildings can be turned, quarter-turned, or reversed, as may be necessary to take advantage of the prevailing winds, sunlight, view, etc. In general, dairy barns are placed with the long axis north and south, whereas, when possible, sheds in northern areas are faced to secure direct sunlight and yet to face away from the direction of the prevailing winds. In the South, sheds are usually oriented for maximum shade and storm protection.

3. **Direction of wind.** The house should be located on the windward side of the headquarters, with special consideration given to summer winds. Unless hills form a natural windbreak, it is desirable to arrange suitable tree plantings for this purpose. Usually, a tree windbreak is located 75 to 150 ft from the buildings to be protected, with three to seven rows of trees 20 to 75 ft wide.

4. **Efficiency.** The buildings should be located so as to require a minimum of walking when doing the chores. This means that those buildings in which the most time is spent — such as the dairy barn — should be closest to the house and the buildings should be near enough together to permit effi-

ciency of labor without making a fire hazard. Likewise, dairy barns should be convenient to lots and pastures.

5. **Corrals and lots.** The buildings and their adjacent corrals and lots should be arranged so that the buildings are accessible without walking through lots and corrals.

6. **Fire protection.** Buildings should be far enough apart so that fire will not spread easily from one building to another. In general, this means at least 100 ft apart in the case of large buildings. In acquiring added fire protection through spacing buildings further apart, one should avoid extreme distances that will mean inefficiency in operation; fire insurance is probably cheaper than labor.

7. **Appearance.** Careful attention to the dairy headquarters arrangement can add to the attractiveness of the entire unit. Manure piles and unsightly objects should not be visible from the main highway or house; shrubbery and trees should be planted to screen unsightly objects; fences and buildings should be repaired and painted regularly; and yards, driveways, and corrals should be kept free from rubbish, scattered farm machinery, etc.

8. **Gates and lanes.** The use of larger machinery has necessitated wider gates and lanes than have been commonplace in the past. Often the wider lanes can serve as pasture as well as roadway.

AUTOMATION

Fig. 16–6. Forage being fed by a self-unloading box. (Courtesy, Gehl Company, West Bend, WI)

Automation is a coined word meaning the mechanical handling of materials. Dairy producers automate to lessen labor and cut costs.

Modern dairy equipment has eliminated the pitchfork, bucket, and basket. Such chores as feeding, watering, bedding, barn cleaning, and milking have been mechanized. Dairy producers are using more self-unloading trucks and trailers, self-feeders, feed bunk augers and belts, laborsaving grain and forage processing equipment (cubes or wafers, etc.), automatic waterers, and manure disposal units. Automation of the dairy farm will increase.

SPACE REQUIREMENTS OF BUILDINGS AND EQUIPMENT FOR DAIRY CATTLE

One of the first and frequently one of the most difficult problems confronting the dairy producer who wishes to construct a building or item of equipment is that of arriving at the proper size or dimensions. Table 16–4 contains some conservative average figures, which it is hoped will prove helpful. In general, less space than indicated may jeopardize the health and well-being of animals, whereas more space may make the buildings and equipment more expensive than necessary.

TABLE 16–5
RECOMMENDED MINIMUM WIDTHS FOR SERVICE PASSAGES[1]

Kind of Passage	Uses	Minimum Width
		(ft)
Feed alley	For feed cart	4
Driveway	For wagon, spreader, or truck	9
Doors and gate	Drive-through	9
Doors and gate	Stall door or paddock	4

[1]For conversion to metric, see Appendix.

TABLE 16–4
SPACE REQUIREMENTS OF BUILDINGS AND EQUIPMENT FOR DAIRY CATTLE[1]

		Conventional Barn						Loose Housing		Maternity Stall	Holding Area in Preparation for Milking
		Standard Stanchion		Comfort Stalls		Tie Stalls		Group Housing	Freestall		
	Cow Size	Width	Length	Width	Length	Width	Length				
Lactating cows	Small cow	3'6"	4'8"	3'9"–4'0"	4'11"–5'0"	4'0"	5'2"	(sq ft/animal) 70	Small breeds: 3½' × 7'–7½' Large breeds: 4' × 7½'–8'	100' × 200'	(sq ft/cow) 15
	Medium cow	4'0"	5'4"	4'3"–4'6"	5'7"–5'8"	4'6"	5'10"				
	Large cow	4'6"	6'0"	4'9"–5'0"	6'2"–6'3"	5'0"	6'6"				

	Open Shed	Freestall	Corral (Lot) Space			Shade	Manger for Hay and/or Silage			Feed Bunk for Grain or Complete Ration			Water Space	
			Paved	Paved & Dirt	Dirt		Length/ Animal	Width (1 Sided)	Width (2 Sided)	Length/ Animal	Width (1 Sided)	Width (2 Sided)	Open Tank	Automatic Bowl
	(sq ft/ animal)		(sq ft/animal)			(sq ft/ animal)	(in.)	(in.)	(in.)	(in.)	(in.)	(in.)	(1 linear ft per)	
Lactating and dry cows	70	Same as for lactating cows	70	150	200	30–40	18	24 × 30	36	24–30	30	36	8–10 head	15 head
Replacement heifers	Small: 20–30 Large: 30–40	3½' × 4'7"	20–40	75–100	100–150	20–30	16	18 × 24	36	12–18	18–24	36	10 head	25 head

Calves:	Calf Stall	Calf Pens	Feed Boxes	Water, Automatic Cups
Individual stalls	3' wide, 6' long, 4' partitions	4' wide, 6' long, 4' partitions	8" wide, 10" long, 6" deep	Water bowl in each pen, or water pail
Group (maximum of 10 head)	. . . .	30 sq ft/calf, with stanchions	10" wide, 6" deep, with 1½ linear ft/calf	1 cup/5–8 calves

[1]For conversion to metric, see Appendix.

RECOMMENDED MINIMUM WIDTH OF SERVICE PASSAGES

The requirements for service passages given in Table 16–5 will be applicable to most dairy facilities.

STORAGE SPACE REQUIREMENTS FOR FEED AND BEDDING

The space requirements for feed storage for the dairy enterprise vary so widely that it is difficult to provide a suggested method of calculating space requirements applicable to such diverse conditions. The amount of feed to be stored depends primarily upon (1) length of pasture season, (2) method of feeding and management, (3) kind of feed, (4) climate, and (5) the proportion of feeds produced on the farm in comparison with feeds purchased. Normally, the storage capacity should be sufficient to handle all feed grain and silage grown on the farm and to hold purchased supplies. Forage and bedding may or may not be stored under cover. In those areas where weather conditions permit, hay and straw are frequently stacked in the fields or near the barns in baled or chopped form. Sometimes poled framed sheds or cheap covers of waterproof paper or wild grass are used for protection. Other forms of low-cost storage include temporary upright silos, trench silos, temporary grain bins, and open-wall buildings for hay.

TABLE 16–6
STORAGE REQUIREMENTS FOR FEED AND BEDDING[1]

Kind of Feed or Bedding	Pounds per Cubic Foot	Cubic Feet per Ton	Pounds per Bushel of Grain
Hay-straw:			
Loose			
Alfalfa	4.4–4	450–500	
Nonlegume	4.4–3.3	450–600	
Straw	3–2	670–1,000	
Baled			
Alfalfa	10–6	200–300	
Nonlegume	8–6	250–330	
Straw	5–4	400–500	
Chopped			
Alfalfa	7–5.5	285–360	
Nonlegume	6.7–5	300–400	
Straw	8–5.7	250–350	
Corn:			
15.5% moisture			
Shelled	44.8		56.0
Ear	28.0		70.0
Shelled, ground	38.0		48.0
Ear, ground	36.0		45.0
30% moisture			
Shelled	54.0		67.5
Ear, ground	35.8		89.6
Barley, 15% moisture	38.4		48.0
Ground	28.0		37.0
Flax, 11% moisture	44.8		56.0
Oats, 16% moisture	25.6		32.0
Ground	18.0		23.0
Rye, 16% moisture	44.8		56.0
Ground	38.0		48.0
Sorghum grain, 15% moisture	44.8		56.0
Soybeans, 14% moisture	48.0		60.0
Wheat, 14% moisture	48.0		60.0
Ground	43.0		50.0

[1]For conversion to metric, see Appendix.

Fig. 16–7. A pole-type barn for hay storage on the Harold Tollercup Dairy in Corona, California.

Table 16–6 gives the storage space requirements for feed and bedding. This information may be helpful to the individual operator who desires to compute the barn space required for a specific dairy enterprise. This table also provides a convenient means of estimating the amount of feed or bedding in storage.

DAIRY CATTLE BUILDINGS AND EQUIPMENT

Modern dairy facilities should provide for top performance, maximum sanitation and disease control, optimum environmental control, functional management and equipment, maximum labor efficiency, satisfactory manure disposal, high-quality milk, minimum housing and care cost per hundredweight of milk produced, and plans for possible expansion.

Prior to starting any new construction or remodeling of existing facilities, a thorough study should be made of the health regulations governing the market where milk is expected to be sold.

There is hardly any limit to the types of facilities and equipment used in caring for lactating cows, dry cows, calves, and replacement heifers. Likewise, there are numerous styles of facilities and equipment for storing, handling, and feeding hay, silage, and concentrates; there are different kinds of waterers; and three types of stanchions are in use — traditional stanchions, tie stalls, and comfort stalls.

HOUSING AND MILKING SYSTEMS

Two basic facility systems are used for lactating cows: (1) stall barns, and (2) loose housing. Thus, the starting point in designing lactating cow facilities is to select the system.

Many arguments are heard relative to the merits of each system. Experiments show that, with proper care and management, similar milk production can be achieved with either system. The stall barn allows the cows to be displayed to greater advantage, which is particularly important in the pure-bred herd; and, generally speaking, the cows are observed more frequently than in loose housing. On the other hand, loose housing requires less labor, saves bedding, results in less udder and leg injury, and usually costs less to construct because such items as expensive concrete work, stanchions, and water cups are omitted.

STALL BARNS

The stall barn consists of one or two rows of cows that are usually confined to stanchions, although tie or comfort-type stalls may be used. In tie stalls, each cow is individually tied with a strap or chain. This offers cows considerably more freedom than stanchion stalls, but requires more labor. Comfort stalls are a special type of tie stall, designed to give cows more freedom than is afforded by ordinary tie stalls. Each cow is secured with a strap or chain fastened to a curb at the front. Horizontal bars at the front force the cow to stand near the rear of the platform. Stalls are separated by fences made of pipe.

For the most part, concrete floors are used; and sometimes they are covered with rubber mats. The floor slopes into a gutter, which is usually 16 in. wide and 8 in. deep on the stall side and 6 in. deep in the alley side.

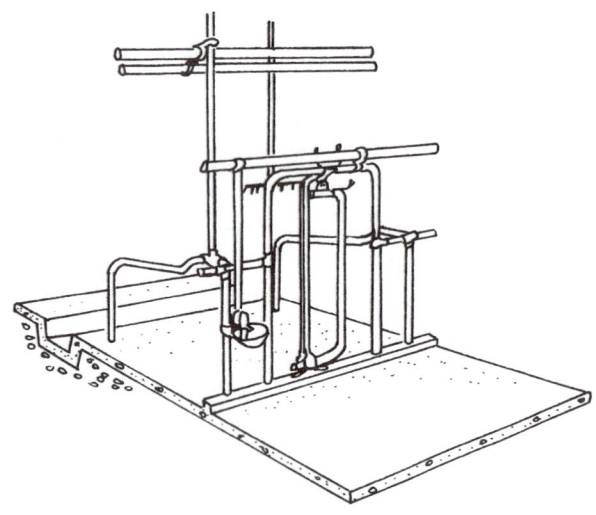

Fig. 16–8. Stanchion stall. (From *Dairy Housing and Equipment Handbook*, p. 4. Courtesy, Midwest Plan Service, Iowa State University, Ames)

Table 16–7 gives suggested stanchion-stall dimensions. Tie or comfort stalls should be 2 in. wider and 4 in. longer than the measurements given.

In stall barns, cows may be faced in (toward the center of the barn) or faced out (toward the walls). A time and motion study revealed that, in caring for a dairy herd, 60% of the time is spent behind the cows, 15% in front of the cows,

TABLE 16–7
SUGGESTED STANCHION-STALL DIMENSIONS[1]

Size of Cow	Stall Width	Stall Length
Small cow (Jersey)	3'6" to 3'10"	4'6" to 4'10"
Medium cow (Guernsey, Milking Shorthorn, Red Poll) . .	4'0" to 4'3"	5'0" to 5'4"
Large cow (Holstein, Brown Swiss)	4'4" to 4'6"	5'4" to 5'8"

[1]For conversion to metric, see Appendix.

Fig. 16–9. Stanchioned cows in stall barn. Note automatic watering cups. (Courtesy, Holstein-Friesian Assn. of America, Brattleboro, VT)

and 25% in other parts of the barn and in the milk house.[1] Since four times as much time is spent behind the cows as in front of them, it is obvious that this area should receive the most attention.

The advantages of facing the cows out are (1) it reduces labor, the wide middle alley in the center of the barn facilitates milking and cleaning, and (2) it alleviates splattered side walls.

Fig. 16–10. Stall barn, with cows facing out. (Courtesy, Holstein-Friesian Assn. of America, Brattleboro, VT)

[1]*Michigan Quarterly Bulletin*, Vol. 30, p. 15.

The advantages of facing the cows in are (1) it is easier to get them into their stalls, feeding is facilitated, because both rows of cows can be fed without backtracking, and (2) the sun's rays shine in the gutter where they are most needed.

During the winter months, the temperature in stall barns should not be permitted to drop below 40°F.

The standard milk house for a stall-type barn is located on the side of the barn opposite the feed room and silos, and it has provision for (1) cooling and storage of milk, (2) washing the utensils, (3) boiler or electric water heater, and (4) machinery for operating milkers, refrigeration units, etc.

LOOSE HOUSING

Loose housing is that system in which the herd is handled on a group basis except at milking time. It may be open lot or enclosed.

Currently, there is considerable interest in this system, primarily because it requires less labor than the stall system.

The following functional areas are involved in most loose housing systems:

1. **Resting or loafing area.** There are two rather distinct types of resting or loafing arrangements:

a. **Group housing.** In this system, the cows rest in a common area, with or without bedding, on a manure pack. About 60 to 70 sq ft of bedded area should be provided per cow. The bedding material should be deep to begin with; then each day the droppings should be removed and 10 to 12 lb of fresh bedding per cow added.

The reported advantage of group housing over individual stall housing is that the cost per cow is less.

b. **Free-stall housing.** This is a modification of the group housing system. It consists of individual open stalls, which are bedded. For small breeds, use 4- × 7-ft stalls; for large breeds, use 4- × 7½-ft stalls.

The reported advantages of free-stall housing over

Fig. 16–11. Cows bedded down in a resting or loafing barn. (Courtesy, Milk Marketing, Inc., Strongsville, OH)

Fig. 16–12. Suspended free-stalls. (Courtesy, Brown Swiss Cattle Breeders' Assn. of the U.S.A., Beloit, WI)

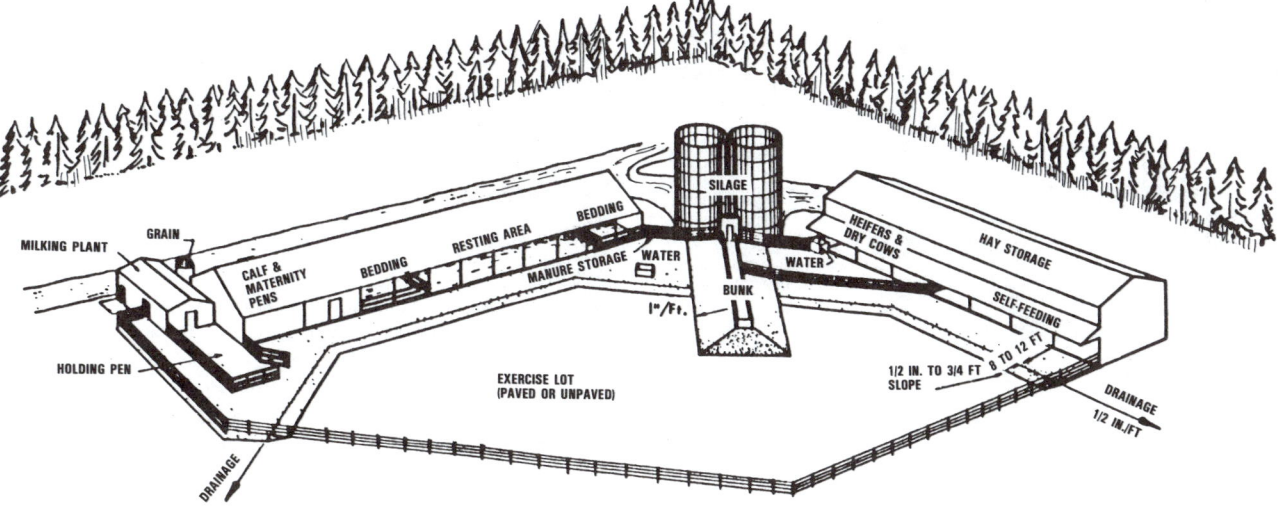

Fig. 16–13.   Layout of group housing. (From *Dairy Equipment Plans and Housing Needs.* Courtesy, Midwest Plan Service, Iowa State University, Ames)

Fig. 16–14. Open air free-stall housing, with milking parlor in the background. (Courtesy, Babson Bros. Co., Oak Brook, IL)

group housing are bedding costs may be reduced by as much as 75%; less labor is required to bed the cows; the cows are cleaner, with the result that less cow-washing time is required; and less space per cow is necessary.

2. **Calf and maternity quarters.** Housing for calves should be provided. Generally one individual calf facility for each ten cows will suffice. (See section headed, "Facilities for Calves.")

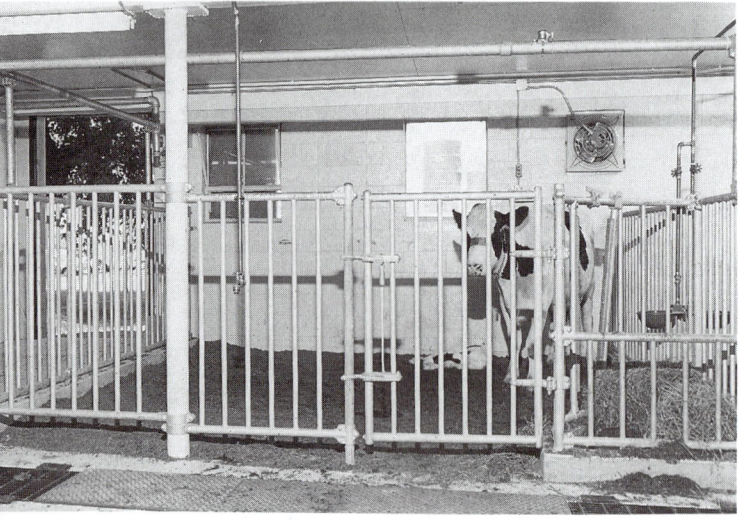

Fig. 16–15. Maternity stall at Dairy Research Unit, Ohio Agricultural and Development Center, Wooster, Ohio.

Also, box-type maternity and isolation stalls should be provided. It is recommended that 100 to 120 sq ft be allowed per stall, with one such stall per 20 milk cows in the herd. A deep stall (8 × 14 ft, for example) is preferred to a more nearly square stall (10 × 12 ft), because it is easier to keep clean. The temperature of the maternity stall should be about the same as is maintained for the milking herd.

3. **Feeding area.** Separate feeding and bedded areas are preferred. Also, more and more large dairies are feeding complete rations in fenceline feeders, with the herd separated into production groups.

Fig. 16–16. Cows eating in a separate feeding area, away from their bedding area. (Courtesy, Holstein-Friesian Assn. of America, Brattleboro, VT)

4. **Exercise yard.** This is usually a paved area which serves as a place for exercise. Approximately 100 sq ft per cow should be allowed. It should be designed so that it can be cleaned daily by scraping.

5. **Holding area.** This area is for the purpose of confining cows in preparation for milking. It should be paved, easy to clean daily, and funnelled to the parlor. About 15 to 20 sq ft per cow should be allowed.

6. **Milking parlor.** A loose housing system lends itself particularly well to the use of a milking parlor. (See section headed, "Milking Parlors.")

SHEDS

Sheds are versatile and widely used dairy cattle shelters throughout the United States. They are particularly suitable for housing dry cows and young stock.

Sheds usually open to the south or east, preferably opposite to the direction of the prevailing winds and toward the sun. They are enclosed on the ends and sides. Sometimes the front is partially closed, and in severe weather drop-doors may be used. The latter arrangement is especially desirable when the ceiling height is sufficient to accommodate a power manure loader.

So that the bedding will be kept reasonably dry, it is important that sheds be located on high, well-drained ground; that eave troughs and downspouts drain into suitable tile lines, or surface drains; and that the structures have sufficient width to prevent rain and snow from blowing to the back end. Sheds should be a minimum of 24 ft in depth, front to back, with depths up to 36 ft preferable. As a height of 8½ ft is necessary to accommodate some power-operated manure loaders, when this type of equipment is to be used in the shed, a minimum ceiling height of 10 ft is recommended. The extra 1½ ft allow for the accumulation of manure. Lower ceiling heights are satisfactory when it is intended to use a blade in cleaning the building.

The length of the shed can be varied according to needed capacity. Likewise, the shape may be either a single long shed or in the form of an L or T. The long arrangement permits more corral space. When an open shed is contemplated, thought should be given to feed storage and feeding problems.

Sometimes hayracks are built along the back wall of sheds, or next to an alley, if the shed is very wide or if there is some hay storage overhead. Most generally, however, hayracks, feed bunks, and watering troughs are placed outside the structure.

Fig. 16–17. A deep, open shed for dairy cows. (Courtesy, Kaiser Aluminum & Chemical Corp, Portsmouth, RI)

MILKING PARLORS

The milking parlor is for the purpose of improving labor efficiency, working conditions, and sanitation surrounding the milking operation. Parlors are of various designs. But all of them have many features in common, including —

1. An all-weather drive leading to the facility for bulk milk access.

2. A facility that (a) is well insulated, (b) can be heated to 50° to 60°F in the wintertime, (c) is well ventilated, (d) has sufficient size to permit herd expansion, and (e) has walls constructed of an easy-to-clean material.

Although there is a wide range of choices in milking parlors, the four most common ones are:

1. **Tandem-type (side-opening).** Where the cows stand single file in line and broadside to the operator's pit.

2. **Herringbone.** Where the cows stand in groups at an angle to the operator's pit. This parlor, which was developed in New Zealand, takes its name from the arrangement of the cows in the parlor.

3. **Polygon.** These are 4-sided herringbone parlors. The polygon was initially developed to maximize laborsaving capabilities of equipment used in mechanizing parlors. It appears best adapted for use in herds of more than 400 cows.

4. **Rotary (or carousel).** Where the parlor, which may incorporate either a tandem or herringbone principle, rotates around the operator.

Fig. 16–18 shows diagrams of seven different milking parlor arrangements, whereas Table 16–8 (next page) shows the capacities of milking parlors.

VARIOUS MILKING PARLORS

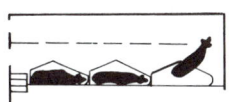

SINGLE SIDE OPENING

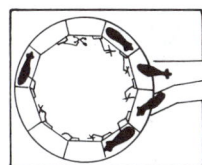

ROTARY TANDEM

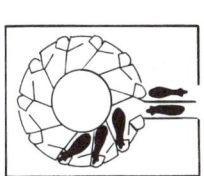

ROTARY HERRINGBONE

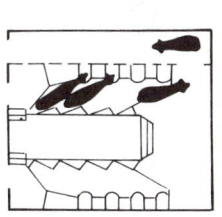

DOUBLE HERRINGBONE

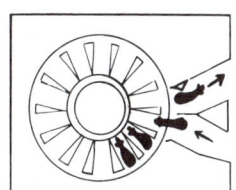

TURNSTYLE

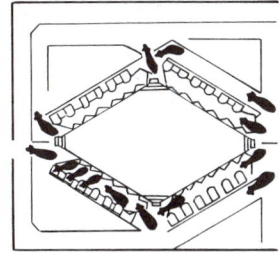

POLYGON HERRINGBONE

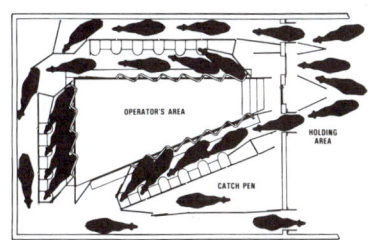

TRIGON

Fig. 16–18. Diagrams of various milking parlors.

TABLE 16–8
MILKING PARLOR CAPACITIES[1]
(HB = herringbone parlor; SO = side opening parlor)

Parlor Size	Cows per Hour	Cows per Work Hour	Pounds of Milk per Work Hour[2]
Double 3 HB	34	29	440
Double 4 HB	41	39	870
Double 6 HB	58	30	640
Double 8 HB	71	35	720
Double 10 HB	92	44	1,040
Double 2 SO	41	41	882
Double 3 SO	51	47	1,031
Rotary (turnstyle) ...	96	48	—
Polygon	134	67	1,566

[1]From *Dairy Housing and Equipment Handbook*, p. 22. Courtesy, Midwest Plan Service, Iowa State University, Ames. Data from a field survey.

MILKING EQUIPMENT

All commercial herds and many small herds are now milked by machine. In comparison with hand milking, machine milking (1) saves much labor, and (2) makes milking more acceptable to milkers.

(Also see Chapter 18, Milk Secretion and Handling, sections headed "Milking Equipment" and "Can and Bulk Systems of Handling.")

Fig. 16–19. Bucket system milking equipment in use on stanchioned cows in a stall barn. (Courtesy, Holstein-Friesian Assn. of America, Brattleboro, VT)

FACILITIES FOR LACTATING AND DRY COWS

Additional pertinent information relative to lots, housing, and feed and water facilities of lactating and dry cows follows:

■ Housing and lot facilities—

1. Open sheds may be used under a loose housing system. Free-stall housing saves bedding, keeps the cows cleaner, and saves labor. For small breeds, use 7- × 4-ft stalls; for large breeds, 7½- × 4-ft stalls.

2. Keep the temperature in stall barns 40°F or more; and in milking parlors 50° to 60°F during the winter.

3. Provide for proper ventilation—changing of air—in cold climates and during the winter months, with care taken to avoid direct drafts and coldness. A fan system installed according to manufacturer's directions is best. For each 1,000 lb of animal weight in a well-insulated barn, the fan should be capable of moving a minimum of 100 cu ft per minute (cfm).

4. Bed all cows except in dry climates.

5. Provide for efficient manure disposal, selecting the system best adapted for the particular dairy and area.

6. Provide the following minimum lot space per head:

Kind of Lot	Sq Ft/Animal
All paved	100
Paved and dirt	150
Dirt	200

7. Slope paved lots ¼ to ½ in./ft, and slope dirt lots ½ in. or more per foot.

8. Pave (rough finish) a 15- to 20-ft area around waterers, feed bunks and racks, and entrances to sheds.

■ Feed and water facilities—

1. Provide 30 linear in. per head of manger space for roughage feeding for the larger breeds of dairy cows, and 24 to 28 in. for the smaller breeds.

2. Make bunks and roughage racks 30 in. wide when cattle are fed from one side; 36 in. wide when feeding from both sides.

3. Provide adequate waterer space: (a) 1 linear ft of open tank per 8 to 10 head, or (b) one automatic bowl per 15 head.

4. Keep water temperature within the range of 35° to 80°F; warm it to 50°F in the winter.

FACILITIES FOR CALVES

Pertinent information relative to facilities and equipment for dairy calves follows:

■ Housing—

Some producers prefer to house calves in a warm barn, whereas other like cold barn housing. The important thing is to recognize that the preferred pen temperature is within the range of 50° to 75°F.

If a warm barn is desired, it should be heated. If a cold barn is used, the partitions should be solid in order to reduce drafts.

1. House calves separately (in individual pens or tie stalls)—from birth until at least 1 week after milk or milk substitute is discontinued. Thereafter, they may be raised in groups, with (a) a maximum of ten herd per group (preferably six to eight per group), and (b) a maximum age difference of 2 months between calves.

Fig. 16–20. Calf nursery at Dairy Research Unit, Ohio Agricultural and Development Center, Wooster, Ohio.

Fig. 16–21. Individual calf house or hutch; a 4- × 4-ft shed (with three tight walls), open to the south, with a 4- × 4-ft outside pen. (Courtesy, Ford New Holland, New Holland, PA)

Calf houses should be dry and well ventilated.

2. Provide a minimum of 24 sq ft pen space for individual calves; 2½- × 4-ft tie stalls; 30 sq ft for calves in groups without outside runs.

3. Solid partitions between individual pens reduce drafts and chilling. Front of pens should be wire or slatted.

4. Preferred pen temperature is within the range of 50° to 75°F.

■ **Feed and water**—

1. Feed boxes for individual calves should be 8 × 10 × 6 in. deep, and they should be removable so as to facilitate cleaning. Troughs for group feeding should be 10 in. wide and 6 in. deep, with 2 linear ft per calf; provide stanchions. Top of feed containers should be 20 in. from the floor, and feed containers should be located in a corner of the pen away from water.

Fig. 16–22. Calf feeding facilities and individual pens at Dairy Research Unit, Ohio Agricultural and Development Center, Wooster, Ohio.

2. Automatic drinking cups are preferred for both individual quarters and group pens (1 cup/5 to 8 calves). Where pails or tanks are used, keep them clean. Top of drinking cups for calves should be 20 in. from the floor.

3. Always locate water facilities at a corner away from the feed.

FACILITIES FOR REPLACEMENT HEIFERS

Pertinent information relative to facilities and equipment for replacement heifers follows:

■ **Housing and lots**—

1. Provide an open shed; allow 20 to 30 sq ft per head for small cattle, and 30 to 40 sq ft per head for large cattle. Bed sheds as needed.

Replacement heifers do not need a closed barn; the main requisite of their housing is that they be protected from drafts, rain, snow, and winds.

2. Provide the following lot space per head:

Kind of Lot	Sq Ft/Animal
All paved	50–75
Paved and dirt	75–100
Dirt	100–150

Fig. 16–23. Inside view of open-to-the-south heifer barn showing clear-span trusses and tombstone-design fenceline feeder; used for the bigger and older heifers that outgrow the first-stage hutches. (Courtesy, *Hoard's Dairyman*, Fort Atkinson, WI)

3. Slope paved (rough finish) lots ¼ to ½ in./ft, and dry lots ½ in. or more per foot.

4. Pave (rough finish) a 15- to 20-ft area around waterers, feed bunks and racks, and entrances to sheds.

5. Provide artificial shade in hot climates if natural shade is not available. Allow 20 to 30 sq ft per animal, and build shade 8 to 10 ft high.

■ **Feed and water facilities** —

Fig. 16–24. Feed bunk for group penned (with outside runs) heifers at Dairy Research Unit, Ohio Agricultural and Development Center, Wooster, Ohio.

1. Provide feed bunks that are 24 to 30 in. above the ground (to top of bunk; with height determined by size of cattle); 8 to 12 in. deep (12 in. deep for silage); and 18 to 24 in. wide when feeding from one side, 36 in. wide when feeding from two sides.

2. Allow the following amount of feeder space per head:

	Grain (in.)	Roughage (in.)
Small cattle	12	18
Large cattle	18	24

3. Allow 1 linear ft of water tank space for each 10 animals and one automatic watering bowl for each 25 animals. Water temperature may range from 35° to 80°F; warming to 50°F in the winter is desirable.

BULL BARN

The bull barn should be in near proximity to the dairy barn. Bulls need to be protected from mud and rain, but they should not be housed in a tight barn.

For safety reasons, the bull barn should be arranged so that (1) the bull can be fed and watered without entering the pen or stall, and (2) a safety breeding chute is built into the pen or lot, to eliminate danger at breeding time.

ISOLATION AND HOSPITAL FACILITIES

Sick animals and newly acquired animals should be isolated from the rest of the herd. A special barn away from the other buildings should be provided for this purpose.

DAIRY CATTLE EQUIPMENT

It is not proposed that all of the numerous types of dairy cattle equipment shall be described herein. Only those articles that are most common will be discussed. Suitable equipment saves feed and labor, and makes for increased production.

FEED BUNKS

Feed bunks are used for feeding (1) grain alone, (2) roughage, and/or (3) both grain and roughage. They may be either portable or stationary, and they may be located indoors or outdoors.

The bunk space per animal is determined by the size of animal and the number of animals that may eat at the same time. If all animals must eat at the same time, allow 30 in. of bunk space per cow and 24 in. per yearling.

If feed is always available (as when self-feeding hay or silage), 4 to 6 in. per animal will suffice.

Fenceline bunks, which are filled from one side by means of a power unloading wagon or truck, are very popular. They save labor. Moreover, they can be used with any size of operation and they are easily lengthened. A step next to the cow feeding side of the bunk will help keep the bunk clean by preventing the cows from standing along the bunk or backing up to it.

FENCELINE BUNK
TILT-UP CONCRETE

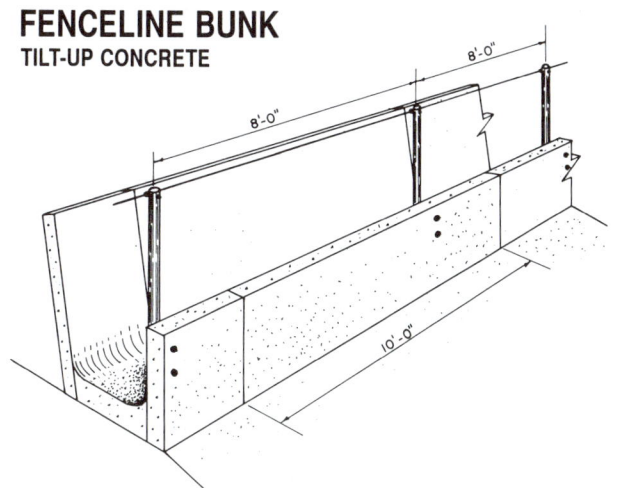

Fig. 16–25. Fenceline bunk. (From *Dairy Equipment Plans and Housing Needs*. Courtesy, Midwest Plan Service, Iowa State University, Ames)

HAYRACKS

Various sizes and designs of racks are used in feeding hay and other forages. In general, these structures are of two types: overhead racks, and low mangers. Overhead racks are easily moved, but they are difficult to clean; and often there is considerable wastage, especially when cows have horns and do considerable fighting. Low mangers result in less wastage of the leaves and other fine particles than overhead racks because the cattle must work down from the top.

HAY & SILAGE FEEDERS

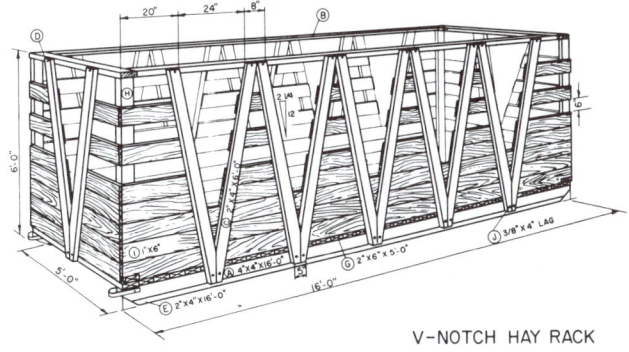

V-NOTCH HAY RACK

Fig. 16–26. A V-notch hayrack. (From *Dairy Equipment Plans and Housing Needs*. Courtesy, Midwest Plan Service, Iowa State University, Ames)

It is preferable that racks be of sufficient size so that one filling will last several days, thus lessening the labor requirements. A 15- to 20-ft area should be paved around hayracks. Portable racks should be mounted on runners or wheels, thus making it convenient to move them from place to place.

WATER FACILITIES

Water is cheap, abundant, and very important.

On the average, mature milk cows consume 12 to 15 gal of water per day. High-producing cows will consume two to three times this amount. Studies show that cows stabled in stanchion-type barns with water constantly available produce 3.5 to 4% more milk than those watered twice a day and over 10% more than those watered once per day. For this reason, as well as from a disease control standpoint, watering cups are used extensively in stall barns. A pressure system, in which the water is forced to the cups and let in by valve above the normal level of water in the cup, is preferred.

Tanks and troughs are used for outside watering. They should be of adequate size, and there should be provision for keeping animals out of them. Tanks should be surrounded by a 15- to 20-ft concrete apron, and provision should be made to pipe the overflow away from the vicinity of the tank. A suitable tank heater should be used where the installation is exposed to freezing temperatures.

Dairy cattle are frequently watered from ponds, springs, and streams. If properly managed and used, they may be quite acceptable.

SHADES

Fig. 16–27. Cows comfortably bedded down under a shade. (Courtesy, Holstein-Friesian Assn. of America, Brattleboro, VT)

Providing adequate shade to protect cattle from the sun is among the more important and widely used devices for improving the environment of cattle in hot climates.

The most satisfactory shades (1) provide 25 to 30 sq ft of shade per mature animal, (2) are 10 to 12 ft high, (3) are located with a north-south placement because they are drier—the sun can get underneath them to dry out the manure and urine—and (4) are open all around, so as to permit maximum air movement.

STOCKS

Cattle stocks may be used for trimming and treating hoofs, dehorning, drenching, ringing bulls, swinging injured animals, and restraining animals during surgical operations. The essential features of cattle stocks are (1) durability, (2) thorough restraint of the animals, (3) convenience for the operator, (4) a canvas sling to place under the animal to prevent it from lying down while in the stocks (the sling may be wound up on side rollers by means of turning rods), and (5) wooden sills that extend along either side at a height of 15 in. from the floor and on which the feet may be rested and tied while being trimmed or treated.

BREEDING RACK

Breeding racks are sometimes used by dairy producers who desire to breed young heifers to mature, heavy bulls.

LOADING CHUTE

The extensive use of trucks makes it desirable that the dairy farm be equipped with a chute for loading and unloading stock. Such equipment may be either portable or stationary. In the latter case, it is usually desirable to attach the chute to a corral.

The main essentials are (1) that the chute have proper height for the truck commonly served (or preferably have an adjustable height arrangement), (2) that it have adequate width to accommodate animals, and (3) that it have sufficient slope and cleating to the platform approach to prevent slipping. Most chutes are about 28 to 30 in. in width and 46 in. high.

OTHER DAIRY CATTLE EQUIPMENT

There is hardly any limit to the number of different articles of dairy cattle equipment, and the design of each. In addition to those already listed, the following equipment is used on most modern dairy farms:

1. **Feed carts.** Where feed cannot be fed from a self-unloading truck, feed carts should be used for handling grain, hay, and/or silage.

2. **Silage unloaders and loaders.** Two types of mechanical silo unloaders are available for upright or tower silos; one works from the top of the silo, the other works from the bottom. Also, silo loaders are available for trench silos.

3. **Elevators and conveyors.** Motor-driven elevators and conveyors are used to store baled hay and straw, bags of feed, and loose grain.

4. **Milk house equipment.** The equipment in the milk house should be (a) suitable for handling the grade of milk that is being produced, and (b) adequate for the amount of milk that is processed. This usually includes the following:
   a. **Washroom equipment.** The washroom generally contains a two-compartment wash vat, a storage place for milk cans and dairy utensils, and a cabinet for storing supplies and materials.

   b. **Boiler and heater.** Hot water may be supplied by a steam boiler, a steam generator, or an electric hot water heater. An electric steam sterilizing cabinet may be used where steam sterilization is required.

   c. **Milk cooler.** Proper cooling and storage of milk on the dairy farm requires facilities that will cool the milk quickly. This calls for the use of a mechanical refrigerator, of which many different types are on the market.

   d. **Milk cans and utensils.** If milk is handled in cans, rather than bulk, cans will be needed. Also, there is need for covered milk pails, milk strainers equipped with single use strainer pads, and a strip cup for each milker.

5. **Cleaning equipment.** The kind of cleaning equipment will vary according to the system used (stall vs loose housing), the design of the building, and the preference of the dairy producer. Among the different types of cleaning equipment are the following:
   a. **The manure spreader.** Manure may be stored on the spreader temporarily. But the manure spreader is primarily for spreading the manure on the land.

   b. **Litter carrier.** In some barns, the litter carrier is used for taking the manure from the barn to the storage area.

   c. **Mechanical gutter cleaner.** Several different makes of mechanical gutter cleaners are on the market. They are designed to remove the manure from the gutter and load it on a spreader or truck.

   d. **Mechanical manure loader.** A mechanical manure loader, mounted on a tractor and operated by tractor power, may be used for loading manure from piles or from loose housing areas.

   e. **Electric cow trainers.** A cow trainer consists of an electric fence control connected with a crossarm supported over the back of the cow. When the cow humps her back to defecate or urinate, the crossarm will shock her and cause her to step back and use the gutter. Most cows soon learn to step back before defecating or urinating, with the result that the stall platform remains cleaner.

PAVED LOTS

In those areas in which barnyard mud is a usual winter problem, a concrete lot floor for dairy cattle may constitute one of the most profitable improvements. A properly constructed paved lot results in: (1) a saving in time and labor; (2) less waste in feed and bedding; (3) greater conservation of manure; (4) greater sanitation and fewer diseases; and (5) more animal comfort, which means greater production of lactating cows and more growth of young stock.

The paved lot should be located where it will make for convenience in feeding and watering operations and will be sheltered from the prevailing winds and exposed to the sun. In most areas, these conditions are met by locating the paved lot adjacent to the building housing the animals and on the south or east side thereof.

With a paved lot, about 100 sq ft per mature animal will suffice. For cattle, the concrete should be about 5 in. thick. Subsurface drainage can be assured and proper footing obtained by placing the concrete slab on a well-tamped fill consisting of 3 to 6 in. of fine stone, gravel, or cinders. Adequate surface drainage may be provided by sloping the floor ½ in. per foot. In order to provide for expansion and contraction, the floor should be formed in sections about 10 ft square. An apron or cutoff wall should be added to prevent undermining of the floor. A low curb around the edge of the paving will also hold the manure; but some openings should be left to let rainwater escape. The floor should be finished with a wood float or broom, leaving an even, yet gritty, nonslip surface. The size of the paved lot will depend upon the number and size of animals kept and on management practices.

Hard-surfaced lots may be constructed with crushed limestone, gravel, paving brick, or concrete. Although concrete is the most costly, it is also the most permanent and satisfactory—all factors considered.

Properly constructed paved lots are easy to clean, and the manure is not trampled into the mud. It may be scooped up with a shovel or manure loader. On large lots, cleaning is facilitated by first bunching the refuse with a tractor equipped with a blade.

MANURE FACILITIES AND MANAGEMENT

Planned manure management is an important part of a modern dairy production program. The collection, transport, storage, and land application of manure must be compatible with sanitary milk production, housing systems, and pollution control. Likewise, manure should be handled so as to retain its highest value as a fertilizer.

Prior to construction, any proposed waste management system should be approved by the appropriate regulatory, public health, and milk market officials.

Manure can be handled in either of two ways:

■ **As a solid**, which runs 20 to 30% solids, and which refers to feces plus bedding, or feces after liquid separation.

■ **As a liquid**, which may be up to 15% solids, and which refers to feces, urine, and sometimes dilution water.

Manure can be hauled (1) daily and spread on available land, usually as a solid; or (2) from storage, either as a solid or liquid, and spread on cropland at a convenient time.

The average dairy cattle production of manure is shown in Table 16–9.

SOLID MANURE

Manure can be handled as a solid if it is mixed with bedding or if the liquids are allowed to evaporate or drain away.

Manure from a stall barn is usually loaded directly into a spreader with a barn cleaner and spread on the land daily.

Where long-term storage is planned, provision should be made for a period of 180 days or more. To calculate the size of storage areas, multiply the number of 1,000-lb cow units by 2.5. cu ft per day (this figure includes bedding), then multiply by the number of days of storage desired.

TABLE 16–9
DAILY DAIRY CATTLE MANURE PRODUCTION, SOLIDS AND LIQUIDS[1,2]
(Manure at 87.3% water and 62 lb/cu ft density)

Animal Size	Total Manure Production			Nutrient Content		
				N	P	K
(lb)	(lb/day)	(cu ft/day)	(gal/day)	(lb/day)		
150	12	0.19	1.5	0.06	0.010	0.04
250	20	0.33	2.4	0.10	0.020	0.07
500	41	0.66	5.0	0.20	0.036	0.14
1,000	82	1.32	9.9	0.41	0.073	0.27
1,400	115	1.85	13.9	0.57	0.102	0.38

[1]From *Dairy Housing and Equipment Handbook*, p. 37. Courtesy, Midwest Plan Service, Iowa State University, Ames

[2]To convert to metric, see Appendix.

Storage by stacking works best with manure containing bedding. It is well suited for use with stall barns and up to 80 cows in the herd. The investment in facilities is usually lower than with liquid storage systems. Moving stacked manure to the land requires a manure loader, a spreader, and a tractor.

LIQUID MANURE

Yearlong liquid storage of manure is a practical goal for dairy producers. It permits incorporating manure into the soil at the best time to preserve fertilizer value. Also, milkhouse and parlor wastes can go into a liquid manure storage.

In comparison with storing solid manure, storage of liquid manure is generally more costly. Also, odors can be a problem, especially when agitating and spreading; and labor requirements may interfere with field work.

Several types of storage are used for liquid manure: storage tank under the barn; outside, below ground storage tank; earthen storage basin; and above ground storage (silo).

Liquid manure is moved to storage by dropping through slots into a storage area below, automatic floor scrapers, tractor scrapers, barn cleaners, or by pumping from a hopper or tank into the storage structure.

SLOTTED ALLEYS

The scraping of alleys takes time and effort, and it can easily be neglected during busy seasons, resulting in undesirable conditions.

Slotted alleys eliminate the labor and cost of scraping and the cost of scraping equipment, because wastes pass directly through the slots into the storage area below. Manure does not build up on the floor. As a result, cows' feet remain comparatively clean, and cows track little manure into the milking parlor from the slotted holding area.

BARNYARD RUNOFF CONTROL

Precipitation that falls on or flows across manure-covered areas or manure stacks can cause severe pollution in streams, lakes, or ponds. This runoff must be kept from reaching usable private or public waters. Local regulations usually govern runoff control systems.

The best method of applying liquid manure to the land is through an irrigation system. Most irrigation systems can handle fluid waste with up to 4% solids, which are typical of lot runoff and effluent from a lagoon or milkhouse.

Fig. 16–28. Dairy waste water lagoon/holding pond. From here it goes to irrigation. Note ducks on the right and left. (Courtesy, Dr. Tom Schultz, Farm Advisor, Visalia, CA)

FENCES FOR DAIRY CATTLE

Good fences (1) maintain farm boundaries, (2) make livestock operations possible, (3) reduce losses to both animals and crops, (4) increase land values, (5) promote better relationships between neighbors, (6) lessen accidents from animals getting on roads, and (7) add to the attractiveness and distinctiveness of the premises.

The discussion which follows will be limited primarily to wire fencing, although it is recognized that such materials as rails, poles, boards, stone, and hedge have a place and are used under certain circumstances. Also, where there is a heavy concentration of animals, such as in corrals and feed yards, there is need for a more rigid type of fencing material than wire. Moreover, certain fencing materials have more artistic appeal than others; and this is an especially important consideration on the purebred establishment.

The kind of wire to purchase should be determined primarily by the class of animals to be confined. Table 16–10 is a suggested guide.

ELECTRIC FENCES

Where a temporary enclosure is desired as in rotation grazing, or where existing fences need bolstering from roguish or breachy animals, it may be desirable to install an electric fence, which can be done at minimum cost.

The following points are pertinent in the construction of an electric fence:

1. **Safety.** If an electric fence is to be installed and used, (a) necessary safety precautions against accidents to both persons and animals should be taken, and (b) the dairy producers should first check into the regulations of their own state relative to the installation and use of electric fences. *Remember that an electric fence can be dangerous.* Fence controllers should be purchased from a reliable manufacturer; homemade controllers may be dangerous.

2. **Charger.** The charger should be safe and effective (purchase one made by a reputable manufacturer). There are four types of chargers: (1) *the battery charger,* which uses a 6-volt hot shot battery; (2) *the inductive discharge system,* in which the current is fed to an interrupter device called a circuit breaker or chopper which energizes a current limiting transformer; (3) *the capacitor discharge system,* in which the power line is rectified to direct current and the current is stowed in the capacitor; and (4) *the continuous current type,* in which a transformer regulates the flow of current from the powerline to the fence.

3. **Wire height.** As a rule of thumb, the correct wire height for an electric fence is about three-fourths the height of the animal. Following are average fence heights above the ground for cattle and calves:

Cattle: 30 to 40 in.
Calves: 12 to 18 in.

TABLE 16–10
DAIRY FENCING CHART[1]

Kind of Stock	Recommended Woven Wire Height	Recommended Weight of Stay Wire	Recommended Mesh or Spacing Between Stays	Recommended Number of Barbed Wire Strands to Add to Woven Wire	Comments
	(in.)	(gauge)	(in.)		
Dairy cattle	47, 48, or 55	9 or 11	12	1 strand 2 in. to 3 in. above top of woven wire, with points 4 in. to 5 in. apart, to prevent animals from breaking down woven wire.	Fence for cattle lots should be constructed of wood, cable, pipe, or other strong material, and should be 60 in. high.
All farm animals	26 or 32	9 or 11 9 or 11	6 6	3 strands on top; 1 strand on bottom. 2 strands on top; 1 strand on bottom.	Cyclone fence, wood, pole, or other durable and attractive materials are usually used around the headquarters.

[1]To convert to metric, see Appendix.

Mixed livestock, three wires: 8, 12, and 32 in.

4. **Posts.** Either wood or steel posts may be used for electric fencing. Corner posts should be firmly set and well braced as required for any nonelectric fence so as to stand the pull necessary to stretch the wire tight. Line posts (a) need only be heavy enough to support the wire and withstand the elements, and (b) may be spaced 40 to 50 ft apart for cattle.

5. **Wire.** In those states where barbed wire is legal, new four point 12½ gauge hog wire is preferred. Barbed wire is recommended, because the barbs will penetrate the hair of animals and touch the skin, but smooth wire can be used satisfactorily. Rusty wire should never be used, because rust is an insulator.

6. **Insulators.** Wire should be fastened to the posts by insulators and should not come into direct contact with posts, weeds, or the ground. Inexpensive solid glass, porcelain, or plastic insulators should be used, rather than old rubber or necks of bottles.

7. **Grounding.** One lead from the controller should be grounded to a pipe driven into the moist earth. *An electric fence should never be grounded to a water pipe, because it could carry lightning directly to connecting buildings.* A lightning arrestor should be installed on the ground wire.

Fig. 16–29. Recommended height for electric fence for (A) cattle, and (B) calves. To convert to metric, see Appendix.

QUESTIONS FOR STUDY AND DISCUSSION

1. Make a critical study of your own dairy cattle facilities, or of a facility with which you are very familiar, and determine its (a) desirable, and (b) undesirable features.

2. Assume that cows in a completely environmentally controlled barn produce 1,000 lb more milk per head per year than cows in a conventional nonenvironmentally controlled unit. Also, assume that milk sells at $11/cwt. With a 100-cow herd, how much more money could you justify spending on an environmentally controlled unit than on a conventional unit?

3. What environmental conditions would you recommend for (a) mature dairy cows, and (b) calves?

4. Where would you get detailed plans and specifications for dairy buildings and equipment?

5. List, by rank, what you consider to be the three most important factors in (a) locating the dairy farm headquarters, and (b) farmstead arrangement. Justify your choices.

6. What determines the maximum that a dairy producer can afford to spend on automation?

7. One of the first and frequently one of the most difficult problems confronting the dairy producer who wishes to construct a building or item of equipment is that of arriving at the proper size or dimensions. In planning to construct new buildings and equipment for dairy cattle, what factors and measurements for buildings and equipment should be considered?

8. What factors should be considered in determining the storage space requirements for feed and bedding?

9. Assume that you have a purebred herd of 500 lactating cows for which you must construct new facilities. Assume further that you sell market milk plus 50 registered yearlings each year. Would you construct a loose housing system or a stall-barn system? Justify your decision.

10. What type milking parlor do you prefer? Justify your preference.

11. What brand of milking equipment do you prefer? Why do you favor it over other brands?

12. Would you recommend warm barn or cold barn housing for raising dairy calves in Minnesota or Wisconsin?

13. What kind of facilities would you recommend for raising replacement heifers?

14. For corral feeding, what kind of feed bunk would you recommend?

15. Give the general specifications for a lactating cow shade.

16. In planning a dairy farm, what consideration should be given to manure facilities and management, including (a) solid manure, (b) liquid manure, (c) slotted alleys, and (d) barnyard runoff control?

17. In the selection of woven wire fence, what is meant by the following number: 1155?

18. Under what circumstances should the use of electric fences be considered on the dairy farm?

SELECTED REFERENCES

Title of Publication	Author(s)	Publisher
Agricultural Engineers Yearbook	Edited by R. H. Hahn, Jr.	American Society of Agricultural Engineers, St. Joseph, MI, annually
Bibliography of Livestock Waste Management	J. R. Miner D. Bundy G. Christenbury	Office of Research and Monitoring, U.S. Environmental Protection Agency, Washington, DC, 1972
Dairy Housing II	R. L. Fehr, Planning Committee Chairman	American Society of Agricultural Engineers, St. Joseph, MI, 1983
Farm Builder's Handbook, Second Edition	R. J. Lytle	Structures Publishing Company, Farmington, MI, 1973
Farm Buildings	R. E. Phillips	Doane-Western, Inc., St. Louis, MO, 1981
Housing of Animals	A. Matson J. Daelemans J. Lambrecht	Elsevier, Amsterdam, The Netherlands, 1985
Livestock Waste Management and Pollution Abatement		American Society of Agricultural Engineers, St. Joseph, MI, 1971
Livestock Waste Management System Design Conference for Consulting and SCS Engineers		University of Nebraska Cooperative Extension Service, Lincoln, NE, 1973
Midwest Plan Service Handbooks, by species	Staff/Committee	Agricultural Engineering Department, Iowa State University, Ames, IA (pub. at intervals)
Practical Farm Buildings: A Text and Handbook, Second Edition	J. S. Boyd	The Interstate Printers & Publishers, Inc., Danville, IL, 1979
Stockman's Handbook, The, Seventh Edition	M. E. Ensminger	Interstate Publishers, Inc., Danville, IL, 1992
Structures and Environment Handbook		Midwest Plan Service, Iowa State University, Ames, IA, 1972
Ventilation of Agricultural Structures	M. A. Hellickson J. N. Walker	American Society of Agricultural Engineers, St. Joseph, MI, 1983

Plans and specifications for beef cattle buildings and equipment can also be obtained from your local county agricultural agent, your state college of agriculture, and materials and equipment manufacturers and dealers.

Fig. 16–30.   Dairy farm, with Jersey cows in the foreground. (Courtesy, Milk Marketing, Inc., Strongsville, OH)

Fig. 17–1. A group of 10 replacement heifers in a partially covered pen, with a feed bunk along the front. (Courtesy, Babson Bros. Co., Oak Brook, IL)

CHAPTER

17

DRY COWS; REPLACEMENT HEIFERS; DAIRY BEEF

Throughout the other chapters of this book, the accent is on lactating cows. Perhaps this is as it should be, for milk is the ultimate objective of dairying. This chapter is different! It is devoted to three important, but often neglected, phases of the dairy enterprise—dry cows, replacement heifers, and dairy beef. In order to underscore their importance, the business aspects are pointed up.

DRY COWS

Fig. 17–2. Dry cows in a corral. (Courtesy, Holstein-Friesian Assn. of America, Brattleboro, VT)

Dry cows are like banks. People deposit surplus money in banks, then draw out money from, or write checks on, their reserves as needed. Likewise, when properly fed, dry cows deposit nutrients in their bodies. Then, during lactation, when there is a negative balance and the demands are greater than can be obtained from the feed, they draw from these stored reserves in their bodies. For this reason, during her "vacation period" the dry cow must be fed a ration which will supply the nutrients needed to replenish her body reserves, provide for normal development of the fetus, and repair or rebuild the milk secreting tissue before another lactation starts.

(Also see Chapter 12, Part III, section on "Feeding Dry Cows.")

CONSEQUENCES OF INADEQUATE NUTRIENT RESERVES

Of course, if there has not been proper storage in the cow's body during the dry period, something must give; and that something may be:

1. **The health of the cow.** The warning signals of which are:

    a. Difficult calving and retained afterbirth.
    b. Milk fever or ketosis.

    c. Rebreedings.
    d. Cows "worn out" before 6 to 8 years of age.
    e. Young cows small or stunted.

2. **The calf.** More than half the fetal growth occurs during the last 2 months of lactation. Thus, deficiencies in the cow are often reflected in the calf, the warning signals of which are:

    a. Birth of calves that are small, weak, or stillborn.
    b. Calves that are difficult to raise—there is high mortality.

3. **The milk production.** The cow freshening in good condition starts off better and maintains a higher level of production than one in a run-down condition. Also, her milk is usually higher in solids. Cows with no dry period have their production level lowered by as much as 26%. It is reasonable to assume that cows with too short a dry period, or not fed properly during the dry period, will be affected in proportion to the degree of the short-changing.

WHAT ARE THE NEEDS?

Scientists and dairy producers agree that dry cow rations are more apt to be deficient in energy than in any other nutrient. It follows, therefore, that if the ration provides adequate energy, it will usually provide ample protein, also. Additionally, a good dry cow concentrate should provide the proper amount of minerals and vitamins.

Table 17–1 gives the daily energy requirements for dry pregnant cows (last 2 to 3 months of gestation).

TABLE 17–1
DAILY ENERGY (TDN AND NE$_{lc}$) REQUIREMENTS OF DRY PREGNANT COWS[1, 2]

Body Weight	Energy Required		Body Weight	Energy Required	
	TDN	NE$_{lc}$		TDN	NE$_{lc}$
(lb)	(lb)	(Mcal)	(lb)	(lb)	(Mcal)
882	9.15	9.30	1,433	13.16	13.39
992	9.99	10.16	1,544	13.91	14.15
1,103	10.80	11.00	1,654	14.66	14.90
1,213	11.62	11.81	1,764	15.39	15.64
1,323	12.39	12.61			

[1]Adapted by the author from *Nutrient Requirements of Dairy Cattle*, 6th rev. ed., Update, 1989, NRC, National Academy Press, p. 84, Table 6–3.

[2]To convert pounds to kilograms, multiply by 0.454.

HOW WELL ARE THE NEEDS BEING MET?

Most dairy producers bunk- or rack-feed the roughage—"all they will eat," without weighing it. Nevertheless, based on numerous studies, it is possible to estimate (1) how much roughage cows eat, and (2) how much energy they are getting from the roughage. Table 17–2 gives the TDN and hay equivalent values of some common roughages.

TABLE 17–2
TDN AND HAY EQUIVALENTS OF SOME ROUGHAGES

Feed	TDN	Hay Equivalent[1,2]
	(%)	(lb)
Alfalfa	44	1.6
Alfalfa	46	1.8
Alfalfa	48	2.0
Alfalfa	50	2.2
Legume silage, med. moisture, 35% dry matter	16.7	6.6
Grass silage, med. moisture, 35% dry matter . . .	15.6	7.04
Corn silage, dent stage, 30% moisture	18.5	5.94
Pasture or green chop:		
Legume, 20% dry matter	14.3	7.7
Grass, 20% dry matter	14.3	7.7

[1]Hay equivalent means the number of pounds of roughage cows will consume per 100 lb liveweight per day. Silages and pasture, which are on a wet basis, were equated to 1 lb of 50% TDN alfalfa hay.

[2]To convert pounds to kilograms, multiply by 0.454.

HOW MUCH CONCENTRATE SHOULD BE FED?

By referring to Tables 17–1 and 17–2, the concentrate needs for any given herd can be computed. Step by step, it consists in (1) determining the energy needs (estimate the average weight of the cows, then turn to Table 17–1), (2) calculating the proportion of the needs being met by the roughage (Table 17–2), and (3) arriving at the amount of concentrate to feed to meet the deficit (by subtracting the results of step 2 from step 1), then dividing by the percent TDN in the concentrate mix.

For purposes of illustrating, let us take two examples.

Example 1. *The dry cows in a given herd average 992 lb weight. They are receiving alfalfa hay testing 44% TDN. How much 70% TDN dry cow concentrate is needed?*

The answer:

Step	Lb
1. Table 17–1 shows that a 992-lb cow needs a daily TDN allowance of	9.99
2. Table 17–2 shows that a 992-lb cow self-fed alfalfa hay testing 44% TDN will consume 1.6 lb of it per 100 lb liveweight per day; hence, a 992-lb cow will consume	15.9
3. 15.9 lb of 44% TDN hay will provide total TDN in the amount of (15.9 × 44%)	7.0
4. The TDN deficit is 9.99 – 7.0	2.99
5. The daily allowance of 70% TDN dry cow ration needed to meet the deficit is 2.99 ÷ 70	4.3

Example 2. *The dry cows in a given herd average 1,544 lb weight. They are receiving a daily ration of 20 lb of 46% TDN alfalfa hay and 10 lb of 35% dry matter grass silage. How much 70% TDN dry cow concentrate is needed?*

The answer:

Step	Lb
1. Table 17–1 shows that a 1,544-lb cow needs TDN in the amount of	13.91

2. The hay and silage supply TDN as follows (see Table 17–2):	
Hay: 20 × 46%	9.2
Silage: 10 × 15.6%	1.56
Total	10.76
3. The TDN deficit is 13.91 – 10.76	3.15
4. The daily pounds of 70% TDN dry cow ration needed to meet the deficit are 3.15 ÷ 70	4.5

It is emphasized that the dry cow concentrate needs given above are minimum requirements, with no overage to provide for individual animal differences, feed differences, etc.

DRY COW PROGRAM

In England, the liberal feeding of dry cows is an old and well-established practice known as "steaming 'em up." Whether it is "steaming 'em up" in England or "more liberal feeding" in the United States, the important thing is that dry cows be well fed – that dairy producers recognize that feed plays an important part from the very day the cow is turned dry until the day she begins to milk.

The essential features of a dry cow program follow, although it is recognized that there are cow differences, depending on condition of the animal, level of production, size and age of cow, and quality of forage:

1. **Dry them off 60 days before calving.** Discontinue grain feeding and milking abruptly. Let *back pressure* in the udder build up. Examine the udder at intervals, but do not milk it out. At the end of 7 days, when the bag is soft and flabby, milk out what little secretion remains.

2. **Feed a special dry cow grain mix during this period.** The dry cow grain mix may contain 2 to 4% less protein than the lactating grain mix, because development of the fetus requires less protein than milk production. In order to lessen milk fever, the dry cow feed should also be (a) balanced for calcium and phosphorus (not to exceed a 2:1 calcium to phosphorus ratio in the total ration), and (b) fortified with vitamin D. Also, self-feed minerals in addition.

The level of feeding during the dry period should be determined by (a) the condition and size of the cow, and (b) the quality of roughage. The following guidelines, based on weight gains, may be used:

Weight of Cows	Desirable Dry Cow Gain
(lb)	(lb)
800–1,000	75–150
1,000–1,200	100–200
1,200–1,400	125–250

3. **Increase concentrate last 2 to 3 weeks of dry period.** Beginning 2 to 3 weeks before freshening, feed the cow about 4 lb of concentrate per day. Increase this by 1 lb each day until the cow is consuming 1 to 1.5 lb of concentrate per hundredweight of body weight. For a 1,200-lb cow, this will be 18 to 20 lb per day at calving. This will get the rumen, and the cow's appetite and eating habits, adjusted to feeding before freshening, thereby paving the way for liberal feeding immediately following freshening.

On the first day after freshening, feed the same concentrate level as before freshening. Then, increase the concentrate allowance by 2 to 3 lb per day according to the cow's appetite or maximum milk production, whichever comes first.

WHAT RESULTS MAY BE EXPECTED FROM PROPER FEEDING?

Proper dry cow feeding is not an expense; rather, it is an investment that pays dividends. Naturally, the net returns derived from feeding dry cows will vary according to the price of milk, the price of feed, the condition of the cows, the length of the dry period, etc. Nevertheless, here are some reasonable expectations:

1. Each pound of additional body weight gained during the dry period will make for 25 lb more milk (presently worth 11¢/lb) in the lactation ahead. Thin, dry cows use feed efficiently. On the average, they will make a pound of gain on 6 lb of added grain, at a cost of 40¢. In terms of dollars, this means that for every dollar spent for dry cow feed, a net return of $5.87 may be expected—definitely a good return on investment.

2. Michigan State University reports that, "for each 1,000 lb increase of milk per cow you gain $10 to $30 per cow above feed cost."[1] That is an average net return of $20 per cow. It is reasonable to expect that through proper dry cow feeding the productivity of the nation's cows could be increased by 1,000 lb per year, or $20 per cow. On a 100-cow herd, that is a net return of $2,000—decidedly profitable.

3. Other plus values, or goals that can be achieved from a good dry cow program include:

    a. A 10-month lactation period, a 60-day dry period, and a 12-month calving interval.

    b. At least 70% of the cows conceive at first breeding; not more than 1.3 breedings per conception.

    c. Not over 10% of the cows with reproductive difficulties at any given time.

    d. Fewer calving problems and retained afterbirth.

    e. A 90% calf crop.

    f. Fifty percent of the cows in the herd lasting until 8 years or older.

    g. A lower incidence of milk fever and ketosis.

REPLACEMENT HEIFERS

The average cow in the United States milks only 3.9 years before she breaks down or is sold because of poor production. To maintain a 100-cow dairy, therefore, 25 first-calf heifers must replace their elders on the milk line each year. But not all of today's heifers become tomorrow's cows. Statistics show that more than 30% of dairy calves die before reaching maturity,[2] and still others may be culled for one reason or another. Many of these deaths and culling losses are due to faulty nutrition.

Fig. 17–3. Replacement heifers in individual pens with feeders. (Courtesy, Dr. Tom Schultz, Farm Advisor, Visalia, CA)

To maintain *status quo* in a milking herd—with no provision for expansion whatsoever—each year the average dairy producer starts with at least three heifer calves for every 10 cows in the milking string. That is 33⅓% replacement per year.

Why the high turnover in lactating cows, and why the high calf mortality? Can anything be done to rectify the situation? Of course, many factors are involved. Potential is determined by heredity. But feed can either bring forth or inhibit this potential. More than any other environmental factor, feed determines age of puberty, age of breeding and freshening, and the productive life ahead. Also, feed can materially shorten the "freeloading" life of a heifer—the interval from birth to the milk line.

FEEDING DAIRY CALVES

Fig. 17–4. Guernsey calves tethered to individual hutches. (Courtesy, American Guernsey Assn., Reynoldsburg, OH)

---

[1] Hillman, D., Michigan State University Ext. Bull. E-423.

[2] *Hoard's Dairyman Feed Guide*, published by W. D. Hoard & Sons Co., Fort Atkinson, WI, 1965, p. 40.

Nature's way was for the cow to have sufficient, but only sufficient, milk for the calf. This is the beef system today, but the dairy cow has been developed to produce many times the milk needed for her calf. Also, the dairy producer wishes to use a calf raising system which will allow for the marketing of the greatest amount of salable milk, for milk is the most expensive feed that is fed to the calf. For these reasons, experiment stations and feed manufacturers have worked to perfect milk replacers and calf starters for growing calves.

The feeding and managing of dairy calves is fully covered in Chapter 12, Part III, of this book, in the several subsections under the heading "Feeding Calves;" so, the reader is referred thereto.

FEEDING REPLACEMENT HEIFERS

Fig. 17–5. Replacement heifers consuming liquid whey. (Courtesy, Dr. Tom Schultz, Farm Advisor, Visalia, CA)

The feeding and management of replacement heifers (from 12 weeks to 2 years old) is fully covered in Chapter 12, Part III, of this book, under the heading "Feeding Replacement Heifers; so, the reader is referred thereto.

NORMAL GROWTH OF HEIFERS

Well-grown heifers can be bred at 14 to 16 months of age. Table 17–3 can be used as a guide to determine if heifers have obtained sufficient growth to permit breeding at the age indicated. There is considerable data indicating that heifers calving at a respectively young age (around 24 months) are more productive and return more income during their lifetimes than heifers that calve at an older age (30 to 36 months). However, one must keep in mind the necessary size and maturity for calving and production. The author suggests that the figures given in Table 17–3 for Holsteins also be used for Brown Swiss, and that the figures given for Ayrshires also be used for Milking Shorthorns.

Fig. 17–6. Well grown group of replacement heifers in a corral. (Courtesy, Dr. Tom Schultz, Farm Advisor, Visalia, CA)

TABLE 17–3
NORMAL HEART GIRTH MEASUREMENT AND WEIGHT
OF CALVES AND HEIFERS DURING THE GROWING PERIOD[1,2]

Age	Holstein		Ayrshire		Guernsey		Jersey	
(mos)	(in.)	(lb)	(in.)	(lb)	(in.)	(lb)	(in.)	(lb)
Birth	31	96	29½	72	29	66	24½	56
1	33½	118	32	98	31½	90	29½	72
2	37	161	35½	132	34½	122	32½	102
3	40¼	213	38¾	179	38	164	35½	138
4	43½	272	42¾	236	41¼	217	38¼	181
5	47	335	45½	291	44¼	265	41½	228
6	50	396	48¼	340	47	304	44½	277
7	52½	455	51¼	408	49¾	362	47¼	235
8	54¾	508	53	447	51¾	410	49¼	369
9	57	559	55	485	53¾	448	51¾	409
10	58¾	609	57	526	55	486	53¼	446
11	60½	658	58	563	56¾	521	55	481
12	62½	714	59	583	58¼	549	56½	520
13	63¼	740	60¾	630	59¼	587	57½	540
14	64¼	774	62	666	60½	615	58½	565
15	65¼	805	63	703	61¾	640	59	585
16	66¼	841	64	731	62½	674	59¾	611
17	67¼	874	65¼	758	63½	696	60½	635
18	68½	912	66	781	65	727	61½	660
19	69¼	946	66½	813	65½	752	62½	687
20	70½	985	67½	841	66¼	780	63	712
21	71½	1,025	68½	885	67½	816	64	740

[1]Body weight for Holsteins and Jerseys from *USDA Tech. Bull. 1098* and *1099.* Heart girth measurements for these weights taken from *Res. Bull. 194* (1960), Nebr. Ag. Exp. Sta. Weights and heart girth measurements of Ayrshires and Guernseys calculated from data furnished by Professor H. P. Davis, University of Nebraska.

[2]To convert to metric, see Appendix.

As noted, heart girth measurements are included in Table 17–3. Thus, if no scales are available, an ordinary tape can be used to measure the distance around the body immediately behind the front legs (heart girth). Although this measurement is not precise, it is reasonably accurate and may be used for estimating body weights.

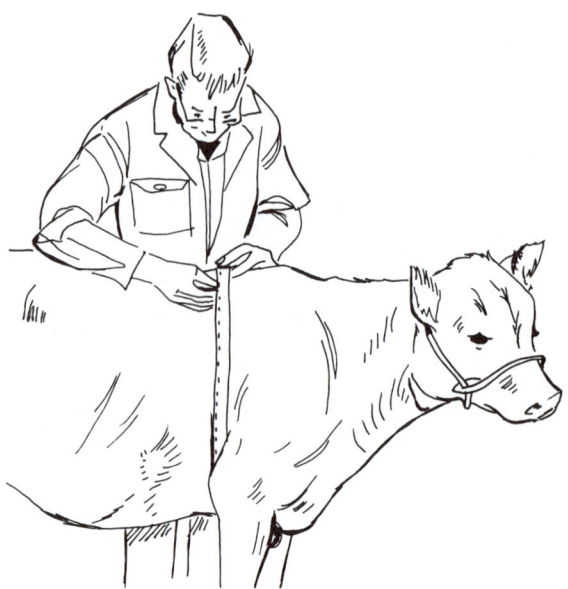

Fig. 17–7. How to tape measure a heifer. (From *Raising Dairy Calves and Heifers*, Farmers' Bull. No. 2176, USDA, p. 14)

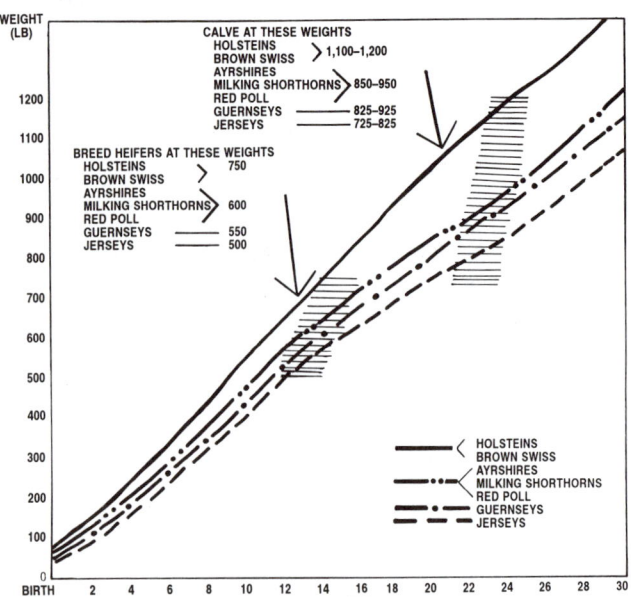

Fig. 17–8. Heifers that are well grown can be bred at the size and age shown, so as to calve at the size and age shown. To convert to metric, see Appendix.

LIBERAL FEEDING MAKES FOR EARLY SEXUAL MATURITY

The high cost of raising replacement heifers makes it imperative, for economic reasons, that breeders get females in reproduction as early in life as possible.

Research at several agricultural experiment stations has shown that well-fed replacement heifers develop faster sexually than those raised on a lower plane of nutrition. Heifers raised on a high plane of nutrition come into heat for the first time between 9 and 11 months of age, whereas those on a very low nutritional level require 18 to 20 months to reach puberty. In a long-term experiment designed to determine the effect of plane of nutrition on reproduction, Cornell University raised three groups of Holstein heifers from birth to first calving on three different nutritive levels—62% (low), 100% (medium), and 146% (high)—of the standard amount of total digestible nutrients (TDN). The 62% (low) TDN group showed first heat at 20.2 months of age, the 100% (medium) TDN group at 11.2 months, and the 146% (high) group at 9.2 months.[3] An extremely important finding in the Cornell study was that, regardless of plane of nutrition, the heifers (Holsteins) came into heat at about 600 lb bodyweight; showing that size and weight, not age, determine the time of sexual maturity. This points up the importance of liberal feeding for early breeding. Thus, most researchers and other knowledgeable dairy producers recommend *breeding according to size and not age* (see Fig. 17–8).

SIZE AND AGE AT FRESHENING AFFECT PRODUCTION

DHIA studies in Wisconsin and Pennsylvania showed that size and age at freshening affected lifetime production.[4] The Wisconsin study revealed (see Table 17–4) that, except for those calving at the very young age of less than 21 months, heifers freshening at an early age produced more butterfat (and, of course, more income) than those freshen-

TABLE 17–4
RELATION OF AGE OF FRESHENING TO BUTTERFAT PRODUCTION[1]

Age	Total Butterfat Production[2]
(months)	(lb)
18–21	1,870
22–23	1,930
24–25	1,910
26–27	1,810
26–29	1,760
28–31	1,720
32–33	1,580
34–35	1,540
36–42	1,490

[1]Chapman, A. B., and G. E. Dickerson, Proceedings, American Society of Animal Production, University of Wisconsin report, pp. 52–55.

[2]To convert pounds to kilograms, multiply by 0.454.

[3]Reid, J. T., *Influence of Early-Life Nutrition on the Performance of Dairy Cattle During Later Life*, Mim. Report.

[4]Chance, C. M., *Good Management of Dairy Heifers Pays*, University of Maryland Ext. Service Fact Sheet 180, p. 3.

ing at the ages most dairy producers are currently using 27 to 30 months. Also, from Table 17–4, it can be deduced that liberally fed heifers can have a full lactation by the time low-level heifers calve, thereby providing a possibility for increasing the milking life of a cow an additional year.

Early breeding of heifers makes for more profit than delayed breeding in three ways: (1) It lessens the time from birth to lactation and lowers the cost of carrying a nonproducing heifer, (2) lifetime production is greater (Table 17–5, next page), and (3) extensive studies reveal that within each breed the larger cows are usually the heavier milk producers. Based on a study of 2,833 Dairy Herd Improvement Association records, Cornell University reported that each 100 lb increase in body size of Holstein cows gave an average increase of 780 lb of milk each year; hence, the more growth and body development obtained on dairy heifers, the more milk they will produce.[5] Other reports indicate that within a breed an increase of 100 lb in size was correlated with increases in production of 200, 236, 390, and 570 lb of milk for the lactation, and that age of maturity affected production even more than size.[6] Early calving also permits faster genetic progress in the herd and helps maintain an established or desired seasonal calving schedule.

YOUNG GAINS ARE CHEAP GAINS

Young animals make more economical weight gains than older ones because (1) older animals require more feed for maintenance, and (2) the added weight of young animals is high in water content and lower in energy (70% water in a calf at birth vs 45% in a fat 2-year-old). Thus, it pays to give calves a good start in life and to keep them growing rapidly at an early age.

LIQUID FEEDING IS THE EXPENSIVE PERIOD

The most expensive period of calf raising is during the liquid feeding period. Hence, it is important to keep the liquid feeding period short and to change to dry feed as soon as possible to reduce costs. Milk or milk replacer can be discontinued at about one month of age if calves are eating enough dry feed to meet their dietary needs.

BUSINESS ASPECTS OF HEIFER PRODUCTION

Most people are in business to make money — and those who raise dairy heifer replacements are people. Thus, whether dairy producers are raising heifers for their own use or for sale to others, it should be done at a profit. This means that the business aspects must be considered; and the larger the business, the more important this becomes. The sections that follow are designed to point up the business aspects of raising dairy heifer replacements.

---

[5]Turk, K. L., and J. D. Burke, *Raising Dairy Calves and Heifers*, Cornell Ext. Bull. 761, p.25.

[6]*Journal of Dairy Science*, 41:747, 1958, and 44:515, 1961.

REPLACEMENT HEIFER COSTS

Fig. 17–9. Replacement heifers on improved pasture. Dairy producers can lower cost of raising replacement heifers to 24 months of age by about $120 per head by using good pasture. (Courtesy, Holstein-Friesian Assn. of America, Brattleboro, VT)

What does it cost to raise a replacement heifer? Are my costs in line? Most dairy producers would be hard pressed to answer these two questions. Yet, this information is needed in order for producers (1) to evaluate how well they are doing, and (2) to determine if it would be cheaper to buy or contract replacement heifers. Table 17–5 (next page) provides such guidelines for raising replacement heifers to 24 months of age.

■ **Feed costs** — The cost of feed accounts for 44% of the cost of raising replacements.

Feed costs (see Table 17–5, footnote 2) were based on going prices for the area. Naturally, feed costs vary from area to area, and according to prevailing prices.

■ **Livestock costs** — These costs account for 16% of the cost of raising replacement heifers to 24 months of age. Like feed costs, they also vary from area to area and from time to time.

■ **Fixed costs** — These costs account for 40% of the cost of raising replacement heifers.

■ **Total cost** — The bottom line of Table 17–5 shows that the total cost of raising replacement heifers to 24 months of age is $1,175.52. Based on this data, it would appear that the following conclusions relative to raising replacement heifers are justified:

1. Dairy producers cannot afford to raise heifers sired by poor-quality bulls.

2. It is costly to raise replacements, but it assures producers of the health status, age, breeding age, and pedigree of replacement heifers. Also, it eliminates transportation stress and the cost, time, and effort which would be required to find good replacements.

TABLE 17–5
REPLACEMENT HEIFER BUDGET[1]

	0–3 Months		3–12 Months		12–24 Months		0–24 Months	
	($ value)	(lb)	($ value)	(lb)	($ value)	(lb)	($ value)	(lb)
Feed Requirements:								
Forage (hay and equivalent)	5.50	200	101.75	3,700	305.25	11,100	412.50	15,000
Corn equivalent	11.25	252	25.00	560	10.00	229	46.25	1,041
Soybean meal	7.20	80	9.90	110	4.50	50	21.60	240
DiCal	1.10	5	5.50	25	6.60	30	13.20	60
Trace mineralized salt	0.21	3	1.05	15	1.75	25	3.01	43
Milk replacer	20.00	40					20.00	40
Total feed costs	45.26		143.20		328.10		516.56	
Livestock Costs:								
Bedding	500.00	250	17.00	850	22.00	1,100	44.00	2,200
Veterinary and medicine	8.00		6.00		8.00		22.00	
Breeding					25.00		25.00	
Power and fuel	4.00		8.00		7.00		19.00	
Supplies	2.35		1.55		15.50		19.40	
Overhead	2.00		6.00		8.00		16.00	
Interest	6.66		15.90		20.68		43.24	
Total livestock costs	28.01		54.45		106.18		188.64	
Fixed Costs:								
Buildings	17.50		52.50		70.00		140.00	
Equipment	12.75		38.25		51.00		102.00	
Livestock	3.44		1.13		3.75		8.31	
Labor charge	25.00		45.00		50.00		120.00	
Calf cost ($100.00)	100.00						100.00	
Total fixed costs	158.69		136.88		174.75		470.31	
Total of Above Costs	231.96		334.53		609.03		1,175.52	

[1]Source: *University of Wisconsin Bulletin A 2731.*

[2]Feed prices: Forage $55/T; corn; $2.50/bu; soybean meal, $180/T; dical, $22/cwt; salt, $7/cwt; milk replacer, $50/cwt; and bedding, $40T.

3. Producers cannot afford to delay first calving beyond 24 months of age. In comparison with a heifer calving at 30 months of age, a heifer that calves at 24 months of age and produced 12,000 lb of milk during her first lactation will produce an estimated 3,800 lb more milk by the time she reaches 56 months and 6,700 lb more by the time she is 84 months of age. Assuming feed costs of $2.19 per 100 lb of milk produced and selling at $9.50/cwt, the additional returns over feed costs at 56 and 84 months of age are $278 and $490 respectively, in favor of freshening at 24 months of age.

THE ECONOMICS OF USING A MILK REPLACER

Most recommended calf feeding programs call for the use of about 25 lb of milk replacer for each calf. Table 17–6 can be used in calculating in any area, and under any given pricing situation, the economics of using milk replacer instead of whole milk. When mixed with water as directed, each 25 lb of milk replacer will replace 225 lb of whole milk.

This means that when milk is selling at $11/100 lb, 25 lb of milk replacer has a milk equivalent value of $24.75. Since 25 lb of milk replacer costs $8.75, the dairy producer cannot afford not to use it in place of whole milk.

A calf that is fed whole milk for 2 to 3 months will consume 750 lb of it, or more. At $11/cwt, this amount of milk would have a sale value of $82.50. But this milk can be replaced with 83.25 lb of milk replacer, costing only $29.14, thereby effecting a saving of $53.36 per calf. Also, the pro-

TABLE 17–6
VALUE OF 25 POUNDS OF MILK REPLACER WITH RESPECT TO AN EQUAL AMOUNT OF FEED VALUE IN WHOLE MILK[1]

Milk Replacer	Whole Milk Equivalent	When Milk Price Is ($/100lb)						
		7.00	8.00	9.00	10.00	11.00	12.00	13.00
(lb)	(lb)	($)	($)	($)	($)	($)	($)	($)
25	225	15.75	18.00	20.25	22.50	24.75	27.00	29.25

[1]To convert pounds to kilograms, multiply by 0.454.

gram herein recommended makes it easy to wean calves at one month of age, thereby effecting a saving in labor. Of course, not every dairy producer feeds alike, even when using the whole milk system.

DAIRY BEEF

Dairy beef is just what the term implies — it is beef derived from cattle of dairy breeding or from dairy × beef crossbreds.

Dairy beef accounts for (1) about one-fourth of the beef consumed in this country, with these animals marketed as veal calves, cull dairy cows and bulls, and finished dairy heifers and steers; and (2) approximately 2% of farm income. Current economic conditions favor growing and finishing dairy steers and heifers, rather than marketing veal calves.

Today, dairy beef is extolled with pride.

But dairy beef has not always enjoyed status. Prior to about 1960, few self-respecting beef cattle producers would admit to finishing out dairy cattle. Given a choice between (1) topping the market with a uniform load of well-bred beef steers, even if they were fed at a loss, or (2) making money by feeding cattle of dairy breeding, most cattle feeders would have elected the first alternative—that is, they would have done so until recent years. They derived much satisfaction from topping the market, and they took pride in telling their neighbors about it. Likewise, meat packers were reluctant to have visitors to their coolers see yellow-finished carcasses, because of the yellow fat being indicative of dairy breeding or grass finish. Only the presence, suspicioned if not real, of goat carcasses was more humiliating to a packer. The near-contempt formerly evoked by cattle of dairy breeding was further evidenced on the nation's terminal markets by the names that were applied to them. Holsteins were known as *magpies*, and Jerseys were known as *yellow hammers*—terms which were neither endearing nor appetizing.

But time was! Today's cattle feeders are primarily concerned with rate and efficiency of gains, and net returns. As a result, most of them would just as soon feed steers of dairy breeding; some actually prefer them. Consumers demand beef that has a maximum amount of lean, with a minimum amount of waste fat, and which is tender and flavorful; and they could not care less whether it comes from a "critter" that was black, white-faced, roan, pink, yellow, or polka dot. As a result, more and more steers of dairy breeding are going the feedlot route, rather than as veal.

Fig. 17–10. Dairy beef in the making. Holstein steers on feed in a commercial feedlot near Corona, California. (Courtesy, Albers Milling Co., Los Angeles, CA)

HIGH-GROWTH THRUST ESSENTIAL

For a dairy beef program to be most successful, scientists and cattle producers in both Britain and the United States are agreed that the animals should have a high-growth potential, as evidenced by heavy birth weight and heavy weight at maturity. Since Holsteins are heavy at birth and mature out at around 1,400 lb, in comparison with mature weights of 1,000 to 1,200 lb of the European beef breeds, it can be readily understood that Holsteins are ideal when it comes to producing dairy beef.

(Also see Chapter 12, Part III, section on "Dairy Beef.")

DECIDE ON FEEDING PROGRAM

There is, of course, no one best system of producing dairy beef for any and all conditions. As is true in any type of cattle feeding program, the operator should make the best use of those feeds that are readily available at the lowest possible cost. Then, these feeds must be combined into satisfactory rations, with consideration given to both economy and probable market price of finished cattle of various weights, grades, and degrees of finish.

HIGH-ENERGY RATIONS; LIGHT MARKET WEIGHTS

If dairy steers are to be slaughtered at young ages and light weights, high-energy (low-roughage) rations are imperative. Under this system, usually young calves of either dairy or dairy × beef breeding are fed in confinement—generally in barns; are fed a milk replacer from one to four days of age to 200 to 300 lb; and are full-fed a high-concentrate ration from about 300 lb to market weight of 750 to 950 lb.

Crowding for market at an early age takes advantage of the fact that growth is generally most economical when most rapid, and that young gains are cheap gains. Also, experience shows that when Holstein calves are started on super-energy rations at around 300 to 350 lb weight and marketed under 950 lb, (1) there is excellent marbling with very little bark (outside fat), and (2) many of these animals will grade Choice, because of the superior marbling—despite some deficiencies in conformation.

HIGH-ROUGHAGE RATIONS; HEAVY MARKET WEIGHTS

If roughages are relatively more abundant and cheaper than concentrates, then it may be more remunerative to feed dairy beef more roughage and market at heavier weights—and with it to expect slower and less efficient gains. In any event, it is net returns that count, rather than rate of gain and pounds of feed required per pound of gain.

Under the high-roughage system, steers of dairy breeding are grown on maximum roughage to 600 to 750 lb weight, following which the ration of concentrate to roughage is increased. Most dairy steers fed according to this system are marketed at weights of 1,050 to 1,200 lb, grading Good or Commercial (most of them are too old to grade Standard, and lacking the necessary conformation and marbling to grade Choice).

DAIRY BEEF HAS GOOD POTENTIAL

There is ample evidence that male calves of the larger dairy breeds (Holsteins and Brown Swiss), along with dairy beef crosses, have the potential for producing acceptable beef with good feed efficiency, under a system of either (1) full-feeding from an early age on a high-energy ration, or (2) growing and finishing on a maximum of roughage and marketing at older ages and heavier weights. In the final analysis, therefore, the system selected should be determined by net returns. Both methods necessitate the rearing of young, 1- to 4-day-old calves to weights of around 300 lb, with such early rearing done by either a calf-raising specialist or by the cattle feeder who will do the ultimate finishing. Such calves must usually be obtained from over a wide area, and of variable ages and sizes; hence, they are difficult to come by. Also, death losses are frequently high and discouraging.

Both commercial cattle feeders and dairy producers are showing increased interest in producing dairy beef, with the result that there is competition between them. More dairy beef will be produced in the future.

QUESTIONS FOR STUDY AND DISCUSSION

1. Discuss possible consequences of inadequate nutrient storage in the cow's body during the dry period.

2. Outline a dry cow feeding, care, and management program.

3. What is the basis of the practice of increasing concentrates the last 2 to 3 weeks of the dry period?

4. Discuss the economics of putting weight on dry cows by grain feeding.

5. Establish reasonable goals to achieve in each of the following: calving interval, breedings per conception, percent reproductive difficulties, calving problems, percent calf crops, years in milk production, and incidence of milk fever and ketosis.

6. How do you account for an appalling 30% dairy calf loss from birth to maturity?

7. Under what circumstances may heart girth measurements of heifers be used to estimate weights?

8. What experimental evidence can you cite in support of the recommendation that heifers be bred according size rather than age?

9. Discuss each of the following:
   a. The relation of age at freshening to lifetime butterfat production.
   b. The effect of body size on milk production.

10. Analyze the following costs in raising replacement heifers: (a) feed costs, (b) livestock costs, (c) fixed costs, and (d) total costs.

11. Based on the cost of raising replacement heifers, what conclusions relative to raising replacement heifers are justified?

12. Discuss the economics of using a milk replacer rather than whole milk for calves.

13. Under what circumstances would you recommend that dairy producers feed out their surplus male calves?

14. Would you advocate a crossbreeding program in a dairy herd in which dairy beef is produced? If so, outline such a program.

15. Which dairy beef feeding program would you recommend — (a) high-energy rations, or (b) high-roughage rations? Discuss the economics of each.

16. Why not set up a cooperative to feed out in a central feedlot the surplus dairy calves from a number of dairies in a given area with each of the owners sharing in the profits?

SELECTED REFERENCES

Title of Publication	Author(s)	Publisher
Dairy Cattle, Principles, Practices, Problems, Profits, Third Edition	D. L. Bath, *et al.*	Lea & Febiger, Philadelphia, PA, 1985
Dairy Farm Management	T. Quinn	Delmar Publishers, Inc., Albany, NY, 1980
Nutrient Requirements of Dairy Cattle, Sixth Revised Edition, Update 1989	National Research Council	National Academy Press, Washington, DC, 1988
Principles and Practices of Feeding Dairy Cows	W. H. Broster, *et al.*	College of Estate Management, Reading University, England, 1986
Stockman's Handbook, The, Seventh Edition	M. E. Ensminger	Interstate Publishers, Inc., Danville, IL, 1992

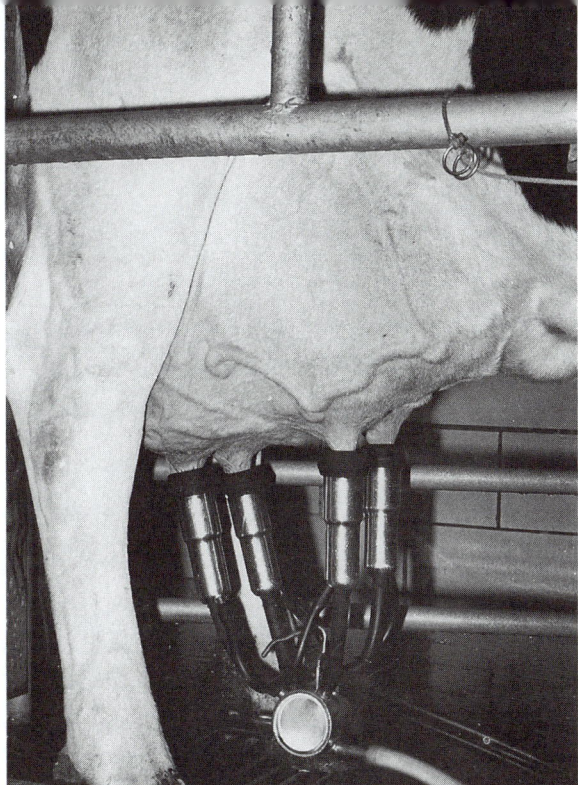

Fig. 18–1. Milking. (Courtesy, Holstein-Friesian Assn. of America, Brattleboro, VT)

Contents

CHAPTER

18

MILK SECRETION
AND HANDLING

Contents

Zoologically, cattle belong to the class Mammalia—warm-blooded, hairy animals that give birth to living young and suckle them for a variable period on a secretion from the mammary glands called milk.

The number of mammary glands and their position on the body is peculiar to each species. For example, the cow has four glands (quarters), each with a passageway (teat) to the outside, whereas the sow and the bitch generally have 10 or more mammary glands.

TABLE 18–1
NUMBER OF MAMMARY GLANDS IN SOME DOMESTICATED ANIMALS

Species	Number of Mammary Glands
Cattle	2 pair
Cat	4 pair
Dog	4–6 pair
Goat	1 pair
Guinea pig	1 pair
Horse	1 pair
Rabbit	3–5 pair
Rat	6 pair
Sheep	1 pair
Swine	4–9 pair (6 pair are most common)

The ability of dairy cattle to produce large amounts of milk is the principal reason why they are accorded a prominent place in American agriculture. It is important, therefore, that the physiology of milk production and the methods of *milk harvesting* be fully understood.

CHEMICAL COMPOSITION OF MILK

Contrary to popular belief, all milk is not alike. Chemically, it varies in composition by species (Table 18–2). Also, the composition of milk differs according to breeds. Table

TABLE 18–2
AVERAGE PERCENTAGE COMPOSITION OF MILK OF THE COW AND OTHER MAMMALS[1]

	Cow	Goat	Human	Sheep	Mare	Pig	Ass
Water	87.70	86.0	88.2	81.3	89.9	81.9	90.1
Fat	3.61	4.6	3.3	6.9	1.2	6.8	1.3
Lactose	4.65	4.2	6.8	5.2	6.9	5.5	6.5
Protein (N × 6.38)	3.29	4.4	1.5	5.6	1.8	5.1	1.6
Ash	0.75	0.8	0.2	1.0	0.3	0.7	0.5

[1]Source: Pearson, D., *The Chemical Analysis of Foods*, The Chemical Rubber Co., Cleveland, Ohio, 1971, p. 411, Table 12.1, with pig added by the author.

18–3 shows the average composition of milk of each of six major breeds of cattle. It is not claimed that these figures are true breed averages; rather, they give some indication of the levels of each component for each breed and the differences between breeds.

TABLE 18–3
COMPOSITION OF MILK FROM DIFFERENT BREEDS[1]

Breed	Fat	Protein	Lactose	Ash	Total Solids	SNF
	(%)	(%)	(%)	(%)	(%)	(%)
Ayrshire	4.1	3.6	4.7	0.7	13.1	8.52
Brown Swiss	4.0	3.6	5.0	0.7	13.3	8.99
Guernsey	5.0	3.8	4.9	0.7	14.4	9.01
Holstein	3.7	3.1	4.9	0.7	12.4	8.45
Jersey	5.1	3.9	4.9	0.7	14.6	9.21

[1]Source: *Dairy Guide*, The Ohio State University, Columbus.

Also, the composition of milk is greatly affected by both physiological and environmental factors, which will be discussed later in this chapter.

The first secretion of the mammary gland following parturition is known as *colostrum*. Colostrum is nature's product, designed to give young a good start in life. It is higher than milk in dry matter, protein, vitamins, and minerals (see Table 18–4). Additionally, it contains antibodies that give newborn animals protection against certain diseases.

TABLE 18–4
COMPOSITION OF COLOSTRUM FROM THE HOLSTEIN COW

Nutrient	Colostrum
	(%)
Total solids	23.9
Ash	1.11
Fat	6.7
Lactose	2.7
Protein	14.0

EXTERNAL STRUCTURE OF THE UDDER

The mammary glands, or udder, in the cow consist of four separate gland units, called quarters.

An extension of the udder, called the teat, hangs down from each quarter. It is hollow and more or less closed at the top and at the bottom. The bottom of the teat is closed by a circular (sphincter) muscle known as the teat canal, or streak canal. In about 40% of the cattle population, there are extra nonfunctional teats called supernumerary teats. These teats are normally cut off when the calf is very young (see Chapter 14, Dairy Cattle Management, section headed, "Removing Extra Teats").

Teats vary in shape from cylindrical to conical. Rear teats are usually shorter than the fore teats. Each teat in the cow has one streak canal that allows the milk to flow to it from the quarter to which it is connected. In some mammals, such as the mare, each teat contains more than one streak canal. Because the primary function of dairy cows is the production of milk, the size and shape of the teats are quite important. They should be of moderate size and located on the udder in such a way as to facilitate the use of milking machines. The sphincter in each teat should be tight enough to prevent the leakage of milk, and yet not so tight as to cause difficult milking. If the muscle is tight, the cow may be a "hard" milker. If it is loose, the cow is an "easy" milker.

The udder of the cow is located in the inguinal region and is covered with fine hair. But teats are generally devoid of hair. The right and left halves of the udder are divided by a longitudinal groove called the intermammary groove. Occasionally, a cow may have a distinct groove that separates the front and hindquarters, but this is generally considered to be undesirable.

The udder of a cow may weigh anywhere from 7 to 165 lb. Additionally, in many cases, it must also be able to support in excess of 80 lb of milk. The rear quarters secrete about 60% of the milk. The udder continues to grow in size until the cow is about six years of age.

SUSPENSION OF THE UDDER

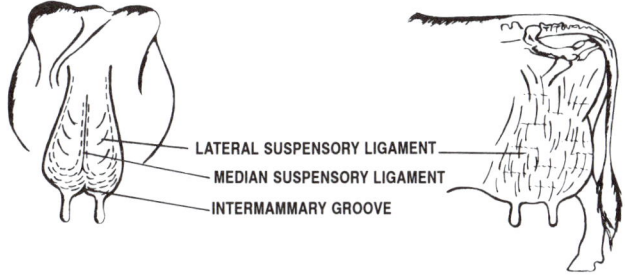

Fig. 18–2. How the udder is suspended.

A well-attached udder fits snugly against the abdominal wall in front and on the sides, and extends high between the thighs in the rear. *Breaking away* of the udder from the body occurs when the supporting ligaments weaken or stretch.

The primary supporting structures of the udder are (1) the skin, (2) the median suspensory ligament, and (3) the lateral suspensory ligament.

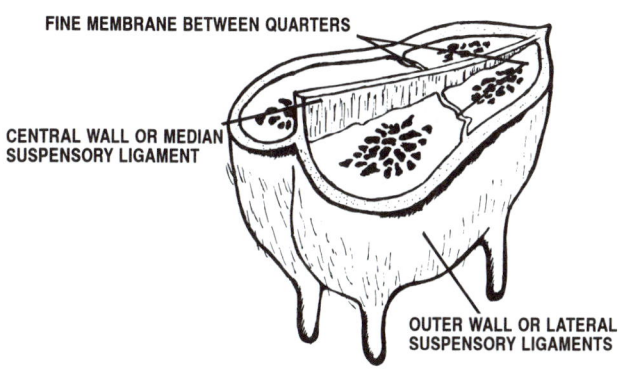

Fig. 18–3. Cross section of the udder.

The skin plays only a minor role in the support and stabilization of the udder. It is connected to loose connective aveolar tissue—cordlike tissue that keeps the surface of the forequarters close to the abdominal wall.

The median suspensory ligament is yellowish elastic tissue that separates the right and left halves of the udder. It connects the udder to the abdominal wall by way of a series of lamellae (plates) which are attached to the medial surface of the two halves of the udder. Since this tissue is elastic, it responds to the weight of milk in the udder. After milk ejection, the tissue can tighten up to accord more support to the udder.

In contrast to the median suspensory ligament, the lateral suspensory ligament is rather inflexible, consisting of white fibrous tissue. These ligaments surround the outer wall of the udder and are attached to the prepubic and subpubic tendons which are, in turn, attached to the ischium and the pubic bone. The intermammary groove is formed where the lateral suspensory ligament and the median suspensory ligament juncture.

INTERNAL ANATOMY OF THE UDDER

Milk is kept in the udder, and bacteria is kept out of the udder, primarily by the constriction of the streak canal. Within the teat is a duct with a capacity of 30 to 45 ml called the teat cistern. It is separated from the streak canal by a number of folds of tissue, generally four to eight in number, that radiate in several directions, called *Fürstenberg's rosettes*, serving as an additional means of preventing milk leakage.

The teat cistern is separated from the gland cistern by the cricoid fold. The gland cistern, capable of holding up to 400 ml of milk, acts as a collecting area for the mammary ducts. Branching from the gland cistern is an extensive, highly branched system of mammary ducts which may number anywhere from 8 to 50. Alveoli—the functional producers of milk—discharge their secretions into these ducts.

The dorsal part of the udder, containing a network of connective tissue in which little milk is produced, is rather hard and meaty.

ALVEOLI—THE PRODUCTION CENTER FOR MILK

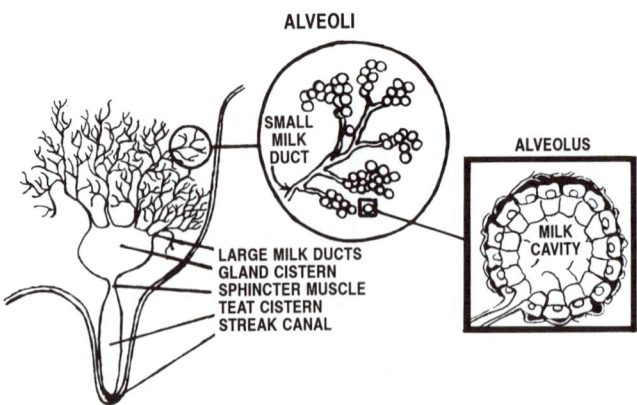

Fig. 18–4. Milk is produced and stored in the alveoli. By action on the musclelike cells surrounding the alveoli, the hormone, oxytocin, forces milk into the large ducts and cisterns of the udder. At milking time, it passes through the streak canal.

The basic milk-producing unit of the udder is a very small bulb-shaped structure with a hollow center called the alveolus. It is estimated that each cubic inch of udder tissue contains one million alveoli; hence, in total there are billions of alveoli in the udder. When an alveolus is filled with milk, it is about 0.1 to 0.3 mm in diameter.

Alveoli are lined with a single layer of epithelial cells which are responsible for secreting milk. Their functions are threefold: (1) remove nutrients from the blood, (2) transform these nutrients into milk, and (3) discharge the milk into the lumen. Each alveolus is surrounded by a network of capillaries from which nutrients are extracted and by a specialized type of muscle cell, called the myoepithelial cell, which is sensitive to the effects of oxytocin. When oxytocin is secreted into the blood, it stimulates contraction of these muscle cells, thereupon initiating milk ejection.

Groups of alveoli empty into a duct thereupon forming a functional unit called a lobule. Several lobules empty into another duct system forming a larger unit called a lobe. The ducts of lobes empty into what is referred to as a *galactophore*, which, in turn, empties into the gland cistern.

DUCTS

The ducts of the udder provide a storage area for milk and a means of transporting it to the outside. No milk secretion, *per se,* occurs within the ducts. The cells lining the cisterns and duct systems consist of two layers of epithelium. Myoepithelial cells are arranged in a longitudinal organization allowing the ducts to shorten and increase the diameter to facilitate the flow of milk.

INNERVATION (STIMULATION) OF THE UDDER

Milk secretion is regulated primarily through hormonal mechanisms. However, milk let-down is initiated largely through neural mechanisms.

The nervous system can be divided into two anatomical systems—the *somatic nervous system*, and the *autonomic nervous system*. The somatic nervous system enables the body to adapt to stimuli from the external environment. Various stimuli, such as touch, are perceived by specialized receptors within this system, and the body responds accordingly. The autonomic system involves the maintenance of homeostasis—the internal environment of the body. The autonomic system can be further divided into the *sympathetic autonomic nervous system* and the *parasympathetic autonomic nervous system*. The sympathetic system is generally associated with the traditional *fight or flight* response, and the parasympathetic system is usually associated with routine integration of normal activity.

In the udder, there is a network of afferent (sensory) and efferent (motor) nerves. Receptors in the udder are sensitive to touch, temperature, and pain. During preparation for milking, the washing and cleaning of the udder stimulates these receptors, and the process of milk ejection is initiated. Motor nerves transmit impulses from the brain and regulate blood flow and smooth muscle activity around the ducts and in the teat sphincter.

When the cow is startled or subjected to pain, the hormone adrenalin is released, and the sympathetic nervous system is stimulated. Blood vessels constrict so that blood can be shunted to other parts of the body (e.g., skeletal muscle) and milk ejection is slowed and production depressed.

There is no indication of parasympathetic innervation of the udder.

CIRCULATION IN THE UDDER

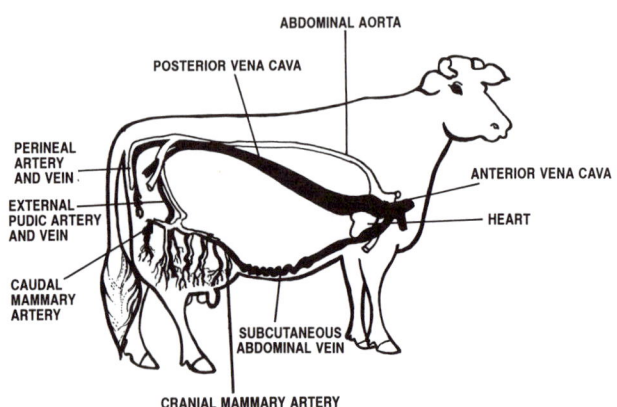

Fig. 18–5. Circulation in the udder.

Milk production places extremely high demands on the circulatory system. To produce 1 gal of milk, more than 500 gal of blood must pass through the udder. In the case of a low-producing cow, this ratio may increase to 1,000 to 1. The blood plasma of a lactating cow constitutes about 4.9% of the entire body weight, as opposed to 3.8% of the body weight for a nonlactating cow. Thus, it is imperative that the circulatory system within the udder be extensive and efficient.

Blood enters the udder through a pair of arteries called the external pudic arteries. These arteries enter the udder dorsal to the rear teats and branch to form the cranial (forward) and caudal (rear) mammary arteries. The cranial and caudal arteries branch into increasingly smaller vessels eventually forming a capillary system which surrounds each alveolus.

Blood is then picked up in the venules and anastomoses to form veins. These veins form a circle at the base of the udder. From this venous network, blood can travel through one of two routes. The first route is via the external pudic veins — veins which travel parallel to the external pudic arteries and empty into the vena cava where the blood then travels to the heart. The second route is through the subcutaneous abdominal veins, commonly called the milk veins. Blood in these veins travels anteriorly along the abdomen and eventually empties into the anterior vena cava.

The perineal veins travel through the udder carrying blood from the vulva and the anal region. However, blood from the udder does not enter this system.

LYMPH

Together with the circulatory system, the lymphatic system helps to regulate the proper fluid balance within the udder and combat infection. However, unlike the circulatory system, lymph — a colorless fluid drained from tissue — travels only in one direction, from the udder. This unidirectional movement is facilitated by the following mechanisms:

1. Blood capillary pressure.
2. Contraction of muscles surrounding the lymph vessels.
3. Valves that prevent the backflow of lymph.
4. Mechanical action of breathing.

Lymph travels from the udder to the thoracic duct and finally empties into the blood system in the anterior vena cava. Flow rates of lymph depend on the physiological status of the cow. A nonlactating cow will have a flow rate of from 15 to 250 ml per hour, while a lactating cow may have a flow rate of up to 2,600 ml per hour.

The onset of lactation initiates an increased blood flow to the udder, and sometimes the lymph system is unable to accommodate the increased volume of fluids. This fluid accumulation leads to a swelling of the udder called *udder edema*. This condition is particularly prevalent in first-calf heifers and high-producing cows because of the extreme changes taking place in the body in preparation for the high demands of lactation.

UDDER DEVELOPMENT

The development of the udder starts early in the growth of the fetus and continues to change under hormonal influences through prepuberty, puberty, pregnancy, and lactation.

EMBRYONIC AND FETAL DEVELOPMENT

The mammary glands develop from ectodermal tissue. At about 35 days of age, the fetus develops two parallel ridges on each side of its ventral midline. These ridges develop into a series of nodules which, in turn, sink into the dermis of the fetus at about 2 months to form the mammary buds, four of which are formed in the cow. A cordlike sprout, called the primary mammary cord, forms at the apex of each bud and becomes canalized at about 100 days. The lumen (chamber) formed by this canalization eventually develops into the gland cistern. The teat cistern develops at the end of the cord. Secondary sprouts branch off from the primary sprout and eventually divide into tertiary sprouts to form what is later to become the duct system.

At birth, the duct system is restricted to a relatively small area surrounding the gland cistern. The gland cistern and teat cistern are rather well developed, increasing only in size, not in morphology, as postnatal growth occurs.

BIRTH TO PUBERTY

During this period, there is an increase in the amount of connective tissue and fat in the udder. A limited amount of glandular development also occurs; but the mammary glands, for the most part, do not radically change in size or cellular function until puberty, pregnancy, and lactation.

From birth to puberty, thyroxin, growth hormone or somatotropic hormone (STH), and the corticosteriod hormones are involved in the general growth and development of the animal.

PUBERTY

Extensive changes begin to take place in the mammary glands at the onset of puberty. Follicle stimulating hormone (FSH) is secreted into the blood by the anterior pituitary, stimulating the growth and development of a follicle on the ovary. Thence, the newly functional ovary initiates estrogen secretion. Coinciding with the maturation of the follicle, luteinizing hormone (LH) is released by the anterior pituitary, thereupon triggering ovulation. After ovulation, a corpus luteum forms, which subsequently produces another female sex hormone, progesterone.

Estrogen is a potent stimulator of the development of the mammary duct system. Thus, with each recurring estrous cycle, more estrogen is released, and the duct system becomes more extensive.

PREGNANCY

Although the duct system of the mammary glands develops prior to pregnancy, the growth and development of alveoli do not. During the first 3 months of pregnancy, there is an extensive proliferation of the duct system. From about the third month of pregnancy, the lobule-alveolar system begins a rapid development under the influence of progesterone. Progesterone and estrogen are secreted at this stage of development by two sources — the ovary and the placenta.

Prolactin and growth hormone (STH) have also been implicated, along with estrogen and progesterone, in the development of the mammary system. Alveolar development starts around the gland cistern and branches out therefrom to the peripheral areas of the gland. By the fifth month of pregnancy, the lobular system is well formed.

PARTURITION AND LACTATION

Immediately prior to the birth of the calf, colostrum synthesis is initiated, and the udder begins to enlarge greatly. At this time, there is a large increase in the content of prolactin in the pituitary, as well as an increased activity of the adrenal cortex.

The functional development of the mammary glands is complete at the onset of the first lactation. The glands will continue to grow, however, until the cow is about 6 years old. Hence, production in each subsequent lactation should steadily increase to a peak at 6 years of age.

DRY COW

About 60 days before calving, cows are dried off to allow them to recuperate from the heavy demands of lactation and to help prepare for the next lactation. During this period, there is a degeneration and subsequent loss of alveolar epithelial cells called involution. However, the myoepithelial and connective tissues remain, and new alveolar epithelium develops prior to parturition.

MILK SYNTHESIS (GALACTOPOIESIS)

The efficiency of milk secretion becomes apparent when it is realized that a cow that produces 14,500 lb of milk during one year manufactures 523 lb of milk fat, 674 lb of milk sugar, 477 lb of milk protein, and 109 lb of minerals and vitamins, or a total of over 1,783 lb of food. (Since milk contains approximately 87.7% water and 12.3% solids, 14,500 lb of milk contains, 1,783 lb of solids—12.3% of 14,500.) This is equivalent to the carcass weight produced by 2½ steers in 18 months' time. Moreover, the cow is still alive and can repeat the productivity again and again, whereas the 2½ steers must be slaughtered or "spent."

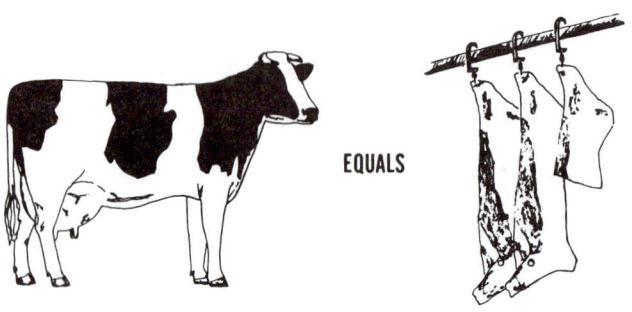

EQUALS

Fig. 18–6.   It takes 2½ steers 18 months' time (for each steer) to produce as much carcass weight as one cow produces in milk solids in 1 year. And the cow remains alive to do it over again!

The alveolus is, in effect, a milk factory. It has the ability to take nutrients from the blood and transform them into one of nature's most perfect foods. *Galactopoiesis is the term used to describe the biosynthesis of milk.*

For the first 1 to 2 hours after milking, there is no intramammary pressure, and milk secretion is at a minimum. However, once the body has had a chance to recuperate from the last milking and after the hormonal effects of milk ejection have subsided, galactopoiesis gets into full gear.

Milk is composed of the following components: (1) water; (2) fat; (3) solids-not-fat, such as protein, lactose, vitamins, and minerals; (4) sloughed-off secretory cells; and (5) bacteria.

CYTOLOGY (CELL STRUCTURE)

In order to understand fully the process of milk secretion, one must become familiar with the anatomy of the secretory cells. As is illustrated in Fig. 18–7, the main parts of the cell are the nucleus, endoplasmic reticulum, Golgi apparatus, mitochondria, and lysosomes.

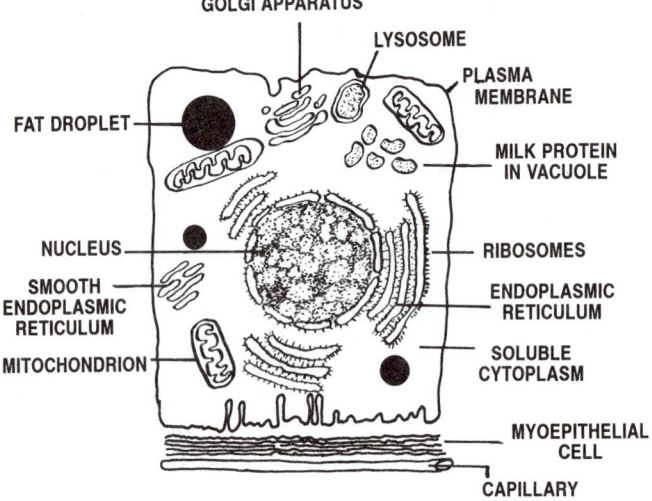

Fig. 18–7.   Cross section of a secretory cell.

NUCLEUS

The nucleus is the informational center for the cell. It contains all the genetic messages required for milk synthesis and cellular metabolism.

Two membranes surround the nucleus. The outer membrane is continuous with the endoplasmic reticulum and the Golgi apparatus. The inner membrane contains pores which allow material to pass from the nucleus to the cytoplasm.

Within the nucleus a number of nucleoli may be present, each of which carries genetic information.

CYTOPLASM

The cytoplasm is the area of the cell outside the nucleus, contained by the outer cell membrane. The cytoplasm contains numerous nutrients and structures which are responsible for carrying out the messages sent by the nucleus and for maintaining energy production needed to "drive" the cell.

MITOCHONDRIA

Often, the mitochondria are referred to as the *powerhouses of the cell*, and rightfully so, because most of the energy-liberating reactions occur in these organelles. As energy demands increase, the number of mitochondria likewise increases. Thus, mammary cells in nonlactating cows contain far fewer mitochondria than those in lactating cows.

LYSOSOMES

Lysosomes produce enzymes, called *lysozomes*, which break large molecules into smaller ones. When these enzymes leak out into the cell, the cell is destroyed and digested; hence, a membrane normally surrounds the lysosomes to prevent leakage.

When a cell becomes old and nonfunctional, the lysosomes provide one means of destroying the cell so another functional cell can take its place. Involution of the mammary gland during the drying off period, and in some cases of mastitis, has been linked to lysosomal activity.

ENDOPLASMIC RETICULUM

In the lactating cow, the endoplasmic reticulum is well developed and abundant. These organelles are responsible for the synthesis of milk protein. Attached to the outer surface of the membrane of the endoplasmic reticulum are a number of dense particles called *ribosomes*—structures in which most protein synthesis takes place.

The endoplasmic reticulum becomes smooth, lacking ribosomes, as it leads to the Golgi apparatus. These smooth channels act merely as a means for transferring material throughout the cytoplasm of the cell.

GOLGI APPARATUS

The Golgi apparatus in the secretory cells of lactating cows is an enlarged structure. Milk proteins are collected here for subsequent transport into the lumen of the alveolus.

CARBOHYDRATES IN MILK

Lactose, or milk sugar, is the most abundant form of carbohydrate in milk. *Lactose is a disaccharide—a compound sugar formed from one molecule of glucose and one molecule of galactose.* Glucose is found throughout the body and is one of the chief sources of energy. Most of the galactose in the body is produced in the milk, but some galactose is found in galactolipids, cerebrosides, and galactoproteins.

Lactose is synthesized solely from glucose. The secretory cell absorbs glucose and converts part of the supply to galactose, thereby providing the necessary units for lactose production. In the ruminant, most carbohydrates in the feed are broken down into the volatile fatty acids—acetic acid, propionic acid, and butyric acid. Acetic acid is used by the mammary gland primarily for the synthesis of milk fat. Propi-

onic acid is converted to glucose which is subsequently used for the production of lactose. Butyric acid is equally divided among lactose, fat, and casein production. Thus, a ration producing high quantities of propionic acid should increase the production of lactose.

Lactose synthesis is facilitated by an enzyme called lactose synthetase. It contains two subunits, which, when complexed, cause glucose and galactose to condense. The first subunit is called galactosyltransferase and is found in the Golgi bodies of the epithelium. The second subunit, lactalbumin, is produced by the ribosomes along the endoplasmic reticulum and moves to the Golgi apparatus to complex with galactosyltransferase.

The amount of lactose in milk is constant, ranging from 4.5 to 5% of the milk depending on the specific breed of cattle. Lactose in milk is largely responsible for the osmotic pressure exerted by milk. When the cell produces a large amount of lactose, fluids are drawn out into the milk to maintain a constant osmotic pressure.

MILK FAT

Only about 25% of the fatty acids found in milk fat are from dietary fat. The remaining portion is synthesized through numerous metabolic schemes. Triglycerides compose almost all of the milk fat. The fatty acids in these triglycerides are, for the most part, short-chain fatty acids (C_4 to C_{14}). These short-chain fatty acids are generally odoriferous, thereby creating the strong smells characteristic of certain types of cheeses. A triglyceride is formed by the combination of three fatty acids and one molecule of glycerol (see Table 18–5 and Fig. 18–8).

TABLE 18–5
PERCENTAGE OF FATTY ACIDS
IN THE TRIGLYCERIDES OF MILK FAT

Common Name	Scientific Name	Structure	Fatty Acid Content
			(%)
Saturated acids:			
Butyric	Butanoic	$CH_3(CH_2)_2COOH$	10.6
Caproic	Hexanoic	$CH_3(CH_2)_4COOH$	3.9
Caprylic	Octanoic	$CH_3(CH_2)_6COOH$	1.6
Capric	Decanoic	$CH_3(CH_2)_8COOH$	3.1
Lauric	Dodecanoic	$CH_3(CH_2)_{10}COOH$	3.4
Myristic	Tetradecanoic	$CH_3(CH_2)_{12}COOH$	10.2
Palmitic	Hexadecanoic	$CH_3(CH_2)_{14}COOH$	24.1
Stearic	Octadecanoic	$CH_3(CH_2)_{16}COOH$	10.7
Unsaturated acids:			
Palmitoleic	Hexadecenoic	$C_{16}H_{30}O_2$	1.0
Oleic	Octadecenoic	$C_{18}H_{34}O_2$	26.1
Linoleic	Octadecadienoic	$C_{18}H_{32}O_2$	3.2
Other			2.0

Almost all of the short-chain fatty acids are synthesized within the cell. Two mechanisms have been implicated in fatty acid synthesis and elongation. The first mechanism involves acetic acid—one of the volatile fatty acids produced in the rumen. In this mechanism, successive 2-carbon units are condensed until the proper length of the fatty acid has been attained. The second mechanism involves the use of 4-carbon molecules, β-hydroxybutyric acid, as the building blocks for fatty acid synthesis.

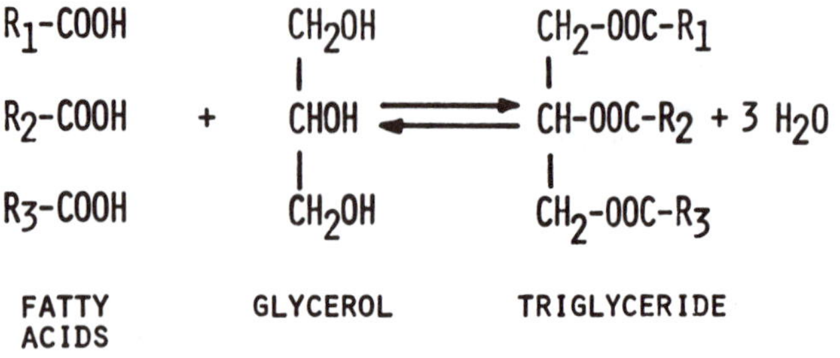

$$R_1\text{-COOH} \qquad CH_2OH \qquad CH_2\text{-OOC-}R_1$$
$$R_2\text{-COOH} \quad + \quad CHOH \rightleftharpoons CH\text{-OOC-}R_2 + 3\ H_2O$$
$$R_3\text{-COOH} \qquad CH_2OH \qquad CH_2\text{-OOC-}R_3$$

FATTY   GLYCEROL   TRIGLYCERIDE
ACIDS

Fig. 18–8. Synthesis of a triglyceride.

The acetic acid mechanism is used much more frequently than the β-hydroxybutyric acid mechanism. The acetic acid mechanism explains why forages are so important in maintaining fat test. When forages are digested, acetic acid is synthesized in large amounts in the rumen. When high-energy concentrates are fed, propionic acid is synthesized in large amounts at the expense of acetic acid. Since increased propionic acid levels increase the production of lactose and, therefore, increase milk yield, acetic acid levels are too low to maintain high fat test.

Long-chain fatty acids (C_{16} and C_{18}) are absorbed from the blood and incorporated into milk triglycerides. Occasionally, some β-oxidation of these long-chain fatty acids will occur, giving rise to C_{14} and C_{16} fatty acids.

Most of the glycerol used in the formation of triglycerides is synthesized in the cell from glucose. However, some glycerol is absorbed from the bloodstream as part of the triglycerides which are subsequently degraded in the cell.

Since fat droplets formed in the secretory cells are too large to pass through pores in cell membranes, the cells have developed a specialized mechanism for secretion called emeiocytosis (*cell vomiting*). In this mechanism, fat droplets migrate to the cell membrane leading to the lumen. The membrane then surrounds the droplet and the outer portions of the membrane break down, but the newly formed portion of the membrane that has surrounded the droplet prevents other cell contents from leaking out.

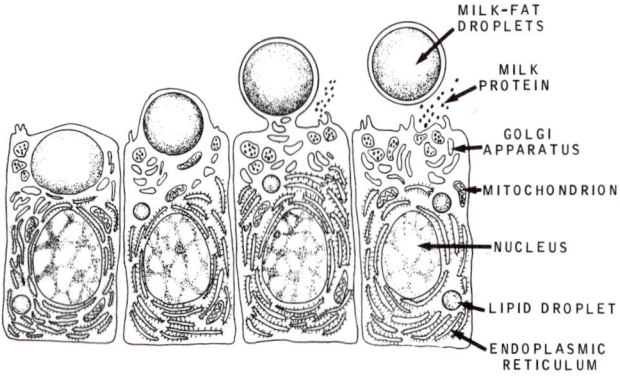

Fig. 18–9. Emeiocytosis, the process of excreting the fat droplet without exposure of the cytoplasm of the lumen of the cell.

MILK PROTEIN

The proteins that predominate in the milk protein fraction are α-casein, β-casein, α-lactalbumin, and β-lactoglobulin. These four proteins are found only in milk and constitute more than 90% of the milk protein. All of them are formed from free amino acid precursors in the blood. Casein is insoluble in water and forms micelles—small aggregates that are suspended in the milk. Kappa-casein prevents these micelles from forming curds. The primary function of casein is to provide a well-balanced protein for the calf. Alpha-lactalbumin and β-lactoglobulin are soluble. The role of α-lactalbumin in lactose production has been discussed previously. Beta-lactoglobulin is the predominant protein in whey.

A second group of proteins constitutes the remaining fraction of milk protein. This group contains immune globulins, serum albumin, and γ-casein. These proteins are not synthesized by mammary cells. Rather, they are absorbed from the blood intact and passed to the milk.

PROTEIN SYNTHESIS

The synthesis of milk protein is not a random process whereby a number of amino acids are joined together; rather, it is a detailed predetermined procedure. Within the cell, deoxyribonucleic acid (DNA) serves as the information center concerning the sequences of the various proteins to be synthesized in the cell. When DNA is decoded, amino acids are linked to form a specific protein which has its own particular physiological function.

In order for milk protein to be synthesized, all of its constituent amino acids must be available. If one amino acid is missing, the synthesis procedure is halted. When a particular amino acid is deficient, it is referred to as a limiting amino acid because it limits the synthesis of protein.

Deoxyribonucleic acid (DNA) is present in the nucleus and the mitochondria of the cell and acts as the genetic information source. It is composed of nucleotides containing (1) the purines, adenine and guanine; and (2) the pyrimidines, cytosine and thymine. Messages are relayed from DNA to the cytoplasm by RNA, whereupon the sequences of amino acids in protein synthesis are dictated from the sequences of the various nucleotides transcribed from the DNA.

Ribonucleic acid (RNA) acts as the messenger for DNA in the determination of the amino acid sequences of proteins. These nucleic acids consist of nucleotides containing the purines, adenine and guanine, and the pyrimidines, cytosine and uracil. There are three types of RNA, each one with a specialized function. *Messenger RNA (mRNA)* transcribes the sequences of DNA into a single strand of RNA. *Transfer RNA (tRNA)* are small molecules which act as carriers for the specific amino acids. The third type of RNA, *ribosomal RNA (rRNA)*, is a major component of the ribosomes within the cell, but the functions of this type of RNA are not clearly understood at the present time.

Protein synthesis involves a series of reactions which are specific for each protein. An outline of this procedure is as follows:

1. Messenger RNA transcribes the sequence *message* from DNA to form a template for protein synthesis. The sequences of the nucleotides in the DNA are the keys to the sequence pattern forming this template. Triplets of the purine and pyrimidine bases in the DNA form *codons* which correspond to specific amino acids and are signals which control protein synthesis. For example, the codon of adenosine, guanine, and guanine signals the incorporation of arginine.

2. A specific transfer RNA combines with each of the respective amino acids to form an aminoacyl complex. This reaction requires the expenditure of energy as shown in the following equation:

Amino acid + tRNA + ATP → Aminoacyl – tRNA + AMP + PP

There is at least one specific tRNA for each of the 20 amino acids.

3. The initiation of protein synthesis occurs when the ribosome (site of protein synthesis) recognizes a codon specific for initiation.

4. Once synthesis has been initiated, the protein is elongated through a series of successive additions of amino acids as determined by the mRNA template.

5. Eventually the procedure will be terminated when a codon specific for terminating protein synthesis is reached. Thereupon, the protein splits off of the ribosome.

VITAMINS AND MINERALS IN MILK

Vitamins and minerals are passed from the blood through the epithelial cells and into the milk through filtration. The cell membrane of the epithelial cell acts as a barrier to regulate osmotic pressure within the milk and also as a carrier system for particles that require active transport.

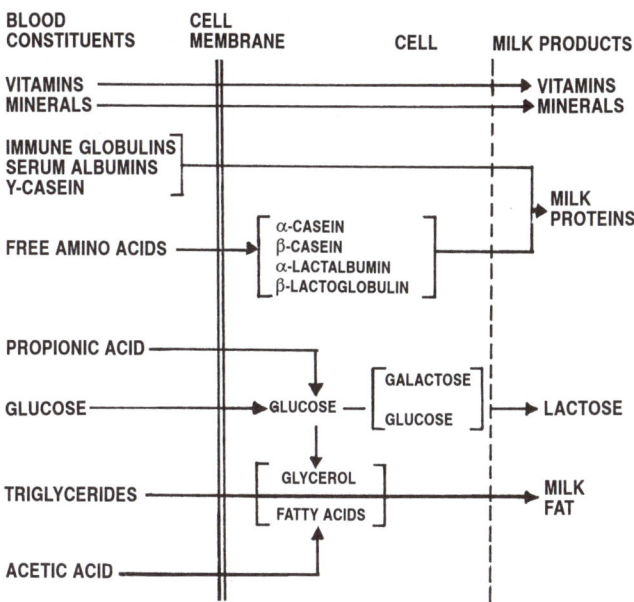

Fig. 18–10. Cellular metabolism.

Vitamins pass essentially unchanged from the blood into the milk. The concentration of the various vitamins, especially the fat-soluble vitamins, is dependent upon their respective concentrations in the blood.

Many of the minerals pass from the blood into the cells and are complexed with other compounds. About 75% of the calcium in milk is complexed with phosphate, citrate, and caseinate. More than 50% of the phosphorus in milk is attached to caseinate.

The primary minerals of milk are calcium, phosphorus, sodium, chlorine, and magnesium. In general, sodium and potassium concentrations are relatively constant as these minerals are involved in the regulation of osmotic pressure in the milk. However, when a cow has mastitis or is at the end of lactation, she may produce "salty" milk. This is due to the decrease in concentration of lactose and potassium in the milk and the relatively high concentrations of sodium and chlorine.

PHYSIOLOGICAL FACTORS AFFECTING AMOUNT AND COMPOSITION OF MILK

The variation in the butterfat composition of milk at the plant has puzzled dairy producers. And since the fat content of the milk has a bearing on the paycheck, it is an economic factor, too.

A number of physiological factors affect the amount and composition of milk; among them, the following:

1. **Breed and individual inheritance.** Variation in the ability of cows to produce total milk, fat, and solids-not-fat is an inherited characteristic. There is both a breed difference (Table 18–3) and an individual difference. In general, total milk production decreases and butterfat content increases by breeds in the following order: Holstein, Brown Swiss, Ayrshire, Guernsey, and Jersey.

Within the Holstein breed, a range in butterfat from 2.6 to 6% has been reported; and within the Jersey breed, from 3.3 to 8.4%. Similar variation between breeds and individuals exists in total milk production.

2. **Stage of lactation.** The greatest variation in the composition of milk takes place immediately following parturition, within the first 5 days after freshening. The secretory product known as colostrum, found in the udder at the time of calving and produced for a short time thereafter, is not milk as such. It contains more globulins, vitamins A and D, iron, calcium, magnesium, chlorine, and phosphorus than does milk; but it contains less lactose and potassium than milk.

Total milk production generally increases for the first month following freshening, then decreases gradually thereafter. Conversely, the butterfat test is usually higher toward the end of the lactation period than soon after freshening.

3. **Persistency.** This refers to the level at which milk production is maintained as lactation progresses. Generally speaking, following the peak lactation period, about a month after freshening, the total milk production each month is approximately 90% of that of the previous month (Fig. 18–11, next page).

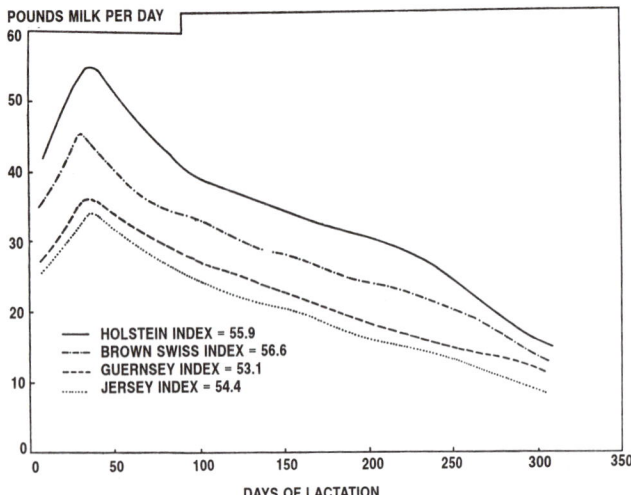

Fig. 18–11. Average lactation curves of four breeds of dairy cattle under Arkansas conditions. (From Ark. Exp. Sta. Bull. 678)

4. **Estrus; pregnancy.** Milk and butterfat production may fluctuate, usually downward, on the day of or the day following a heat period. Pregnancy seems to have little effect on milk composition. However, beginning about the fifth month of pregnancy, total production of gestating cows declines more rapidly than that of nonpregnant cows. It has been estimated that the energy requirement of the fetus is equivalent to about 400 to 600 lb of milk.

5. **Calving interval.** Research indicates that it is most profitable for cows to calve at 12-month, rather than longer, intervals. With an 8-week dry period, this means a lactation period of 10 months.

6. **First- and last-drawn milk.** The percentage of fat in last-drawn milk is higher than that in first-drawn milk. The reasons for this are not known.

7. **Age.** The age of a cow has a definite effect on production. Most cows reach maturity and maximum milk production at about 6 years of age, following which there is a decline in production. Records indicate that cows produce approximately 25% more milk at maturity than they do as 2-year-olds. Also, after passing their prime—6 years of age—butterfat gradually decreases with advancing age.

Age adjustment factors have been developed to standardize 305-day lactation records to a mature equivalent basis and to minimize environmental variation due to month of the year in which the record began. These factors remove, with considerable accuracy, environmental effects from age and month of calving in individual breeds and regions. Table 5–1 in Chapter 5 of this book shows the milk and fat age adjustment factors for cows in the United States calving in the month of May, by breed, for selected ages. A complete list of adjustment factors for milk and fat by breeds, by regions (and for the United States), by month of calving, and by age, is given in the following report: USDA-DHIA *Factors for Standardizing 305-Day Lactation Records for Age and Month of Calving,* ARS-NE-40, U.S. Department of Agriculture.

8. **Size.** Within a breed, large cows usually produce more milk than small cows. However, according to Brody of the Missouri Station, for each 100-lb increase in body weight, production increases only 70% of the proportional increase in body size.

ENVIRONMENTAL FACTORS AFFECTING AMOUNT AND COMPOSITION OF MILK

All animals, including dairy animals, are the result of two forces—heredity and environment. Because of this, the maximum development of dairy cattle characteristics of economic importance—particularly total milk production—cannot be achieved unless there are optimum conditions of environment. Among the environmental factors affecting amount and composition of milk are the following:

1. **Feed.** If milk cows are not fed, or if they do not eat, they will not produce. There are a number of ways in which feed may affect the quantity and/or composition of milk. Among them—

a. **Underfeeding.** Underfeeding usually refers to not providing sufficient energy. The degree of milk reduction resulting therefrom is dependent upon the extent of underfeeding and the length of time it exists.

b. **Challenge or lead feeding in early lactation.** One of the most critical periods for proper feeding is immediately following freshening. It is very difficult for high-producing cows to consume enough feed to supply the energy needed for production at this time. As a result, most cows lose weight during this period. The current system of increasing the concentrates, beginning 2 to 3 weeks before freshening until the cow is consuming 1 to 1.5 lb of concentrate per hundred pounds body weight at calving, is known as challenge or lead feeding. In this system, after freshening cows are fed to their inherited capacity for milk production as determined by profitability; in other words, at the point where the added milk produced does not pay for the added feed, it is time to discontinue further feed increases.

c. **Deficiency of nutrients.** A deficiency of any essential nutrient required by the cow will lower milk production and feed efficiency, rather than make for significant changes in the composition of milk.

d. **Some feed ingredients and rations influence feed composition.** Some feeds reduce the fat percentage of milk. Among such feeds are cod-liver oil and other fish oils, certain pasturages (especially lush spring pastures), and pearl millet. Also, fine grinding of forage, too small an amount of roughage, or heated starch will lower the butterfat content of milk. On the other hand, such feeds as whole cottonseed, whole soybeans, and coconut oil result in a temporary increase in the fat content of milk.

The amounts of fat-soluble vitamins A, D, and E in milk are influenced by the amounts of these particular vitamins in the ration; and in the case of vitamin D, exposure to sunlight is a factor, also.

2. **Length of dry period.** A dry period of approximately 60 days is recommended following each lactation period. This is important because it permits the cow's body to store up reserves so as to meet the rigorous demand of

the next lactation, and it permits proper involution and conditioning of the udder. A short dry period usually results in lower milk production.

3. **Condition at calving time.** Cows that are in a thin, run-down condition at calving time produce less milk than cows in good condition. Excessive condition will also lower milk production after freshening, but it should be added that this seldom happens in good producing dairy cows.

Cows in good flesh at calving time have been observed to start their lactation with 25% more milk production than those calving in poor condition. Generous feeding of thin cows following freshening may eliminate some of this difference, but it is questionable that thin, high-producing cows can ever consume enough to catch up.

4. **Frequency of milking.** Frequency of milking does result in more total milk produced; cows milked three times a day consistently produce more milk than those milked twice a day, and cows milked four times a day produce more milk than those milked three times daily. Also, it has been observed that cows milked more frequently are more persistent in their production throughout the lactation; that is, milk production declines less rapidly as lactation progresses. Of course, a decision as to whether or not it pays to milk more than twice daily will depend on whether the additional milk more than covers the added labor and other costs of obtaining it. In a limited number of herds, managed for intensive production, three daily milkings have been profitable.

Frequency of milking has no effect on butterfat percentage.

5. **Irregular feeding and milking.** Unequal intervals between milkings affects both quantity and composition of milk; more milk of slightly lower fat content is obtained following the longer intervals.

6. **Change of milkers.** High-producing dairy cows may be under stress, with the result that they are usually very sensitive to any changes, including that of the caretaker. Creating a pleasant, quiet, and comfortable environment causes a cow to perform more efficiently.

7. **Environmental temperature; season.** Butterfat percentage of milk varies with the season, being higher in the fall and winter and lower in the spring and summer. It may vary up and down seasonally by an average of 0.3 to 0.5%. Solids-not-fat also show a seasonable variation, with the low point in the spring and summer. The reasons for these changes are not known; it may be due to temperature and humidity, changes in body weight, or kinds and amounts of feeds may be reflected.

Severe weather conditions usually decrease the amount of total milk produced and may influence the fat test either up or down. Temperatures above 85°F greatly affect cows, and the situation is accentuated when high temperatures are accompanied by high humidity.

It is also noteworthy that cows calving in the fall months consistently produce more than those calving at other times of the year. Cows calving in the spring produce the least. This difference may be as much as 10 to 15%. This phenomenon may be due in part to temperature; but more than likely available feeds, including spring pastures to which fall-calving cows respond so well, may be a factor.

8. **Day-to-day variation.** Research has shown that day-to-day butterfat tests vary from 0.1 to 2%

9. **Disease.** Disease does affect milk secretion, in both total production and composition, with the degree of the effect determined by the kind and severity of disease. Mastitis will, for example, lower both the total production of milk and the composition thereof.

10. **Drugs.** Many types of drugs have been used in an effort to increase milk production and affect its composition. Most of them have no effect, so it is questionable that they can be used on a practical basis.

When added to the feed at certain levels, thyroprotein (thyroxin) stimulates the cow to produce more milk of a higher percentage fat. However, to be effective, it must be added at a specific time during the lactation period and cows must be fed more when they are receiving the drug.

Oxytocin will, on a temporary basis, increase yields of both milk and fat. This is because it permits greater release of milk from the udder. But it must be administered just after each milking in order to get the residual milk which makes its administration both expensive and time-consuming. Hence, it is not considered a practical procedure.

11. **Prepartum milking.** Prepartum milking is the practice of milking cows 10 days to 2 weeks before they are due to freshen. Those who follow this practice usually do so because they believe it will lessen congestion and swelling of the udder and belly of the cow. Among some, the feeling also persists that it will lessen the incidence of both mastitis and udder edema. It is known that prepartum milking will result in cows producing normal milk at the time of freshening, rather than colostrum. Thus, where prepartum milking is done, it is necessary to save (freeze) the early milk in order to have colostrum available for the newborn calf.

MILK EJECTION OR "LET-DOWN"

As has already been noted, the milk is stored in the alveoli. Before it is available to the calf or milker, it has to be forced from the alveoli into the larger ducts and cisterns. This process is known as the *let-down* of the milk. Here is how it works (see Fig. 18–12): when the udder (especially the teats) is stimulated by a calf or a milker, (1) impulses are conducted along the nerves to the posterior pituitary at the base of the cow's brain; (2) the posterior pituitary stores and releases the hormone oxytocin into the blood stream; (3) the blood transports oxytocin back to the udder; and (4) the oxytocin causes the smooth, musclelike cells surrounding each alveolus to contract, thereby forcing the milk out of them into the large ducts and cisterns of the udder.

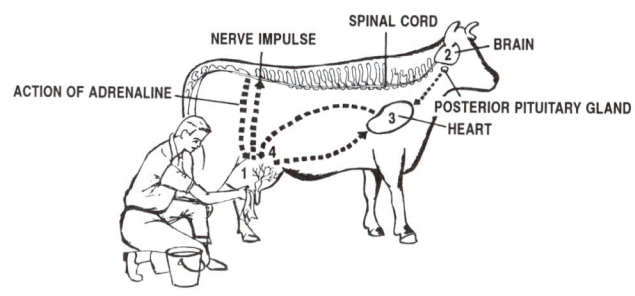

Fig. 18–12. Steps in milk let-down.

The stimulation of the udder lasts for a limited time only—less than 1 minute in a *fresh* cow, since oxytocin is destroyed in the bloodstream. Hence, once the *let-down* has occurred, it is important that the milk be removed within approximately 5 minutes to obtain the greatest amount. This is so because a second stimulation cannot be obtained soon after the first.

When a cow is frightened or angry (from being hit, chased, shouted at, barked at, or for other reasons), she may not let down her milk. This is because of an overriding hormone action. Upon such occasions, another hormone, adrenalin, is released into the bloodstream. This hormone interferes with the action of oxytocin by reducing blood circulation to the alveoli.

Hold up of milk may also be the result of a poorly operated milking machine or from a poor hand-milker.

MILKING THE COW; MANAGED MILKING

Milking is the act of removing milk from the udder. It is routinely carried out through three methods: (1) suckling by the calf, (2) hand-milking, or (3) machine milking.

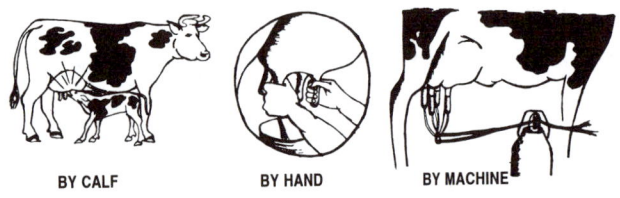

BY CALF   BY HAND   BY MACHINE

Fig. 18–13. Three ways to milk a cow.

SUCKLING

The fastest means of removing milk from the udder is through the use of the calf. The calf grasps the teat with its tongue and presses it against the soft palate on the roof of the mouth. Milk ejection is accomplished through the creation of a negative pressure in the mouth by the widening of the jaws and the retraction of the tongue. This action causes the streak canal to open, thereupon releasing milk from the udder. When the calf swallows, a positive pressure is created, which acts as a resting and a massage phase for the teat.

HAND-MILKING

Hand-milking is still widely used in the lesser developed countries where labor is cheaper than automation and in modern operations to milk out quarters which are infected or have been injured.

In hand-milking, the teat is grasped between the thumb and forefinger; then, by applying pressure with the other fingers, milk is forced from the teat cistern through the streak canal. Through this method, more milk can be obtained than by the use of a milking machine.

MACHINE MILKING

The history of machine milking in the United States goes back over 100 years. The first vacuum-type milking machine was patented by L. O. Colvin in 1865. As can be seen in Fig. 18–14, it was a rather crude apparatus. In 1878, Mrs. Anne Baldwin came out with her Hygienic Glove Milker, an apparatus that used a hand pump to provide milking action. In 1884, J. P. Martin devised a milking machine that had individual teat cups, connecting tubes, and vacuum pump. The pulsator was patented by Modestus Cushman in 1885. In 1892, the Mehring milker, powered by a hand pump and later by foot power, began to gain acceptance as a practical aid in milking. In 1903, an Australian, Alexander Gillies, developed the first prototype for what was eventually to develop into the modern milking machine. This machine has a source of vacuum, a collection receptacle, pulsator, hoses, and individual teat cups and liners.

Fig. 18–14. Historical development of the modern milking machine.

MILKING EQUIPMENT

Basically, there are two types of milking equipment: (1) the bucket system, and (2) the pipeline system.

In the bucket system, the milk is received directly into a nearby vacuumized portable bucket, which may be either of two types: (1) floor type, or (2) suspended type.

Conventional pipeline systems use a rigid heat-resistant glass or stainless sanitary pipe for carrying vacuum from the milk receiver to the individual milking units, and for carrying the milk from the units to the receiver. Pipeline milkers may be used in any of the following types of facilities: (1) stanchion barn, (2) herringbone milking parlor, (3) side-opening milking parlor, (4) walk-through milking parlor, or (5) rotary parlor.

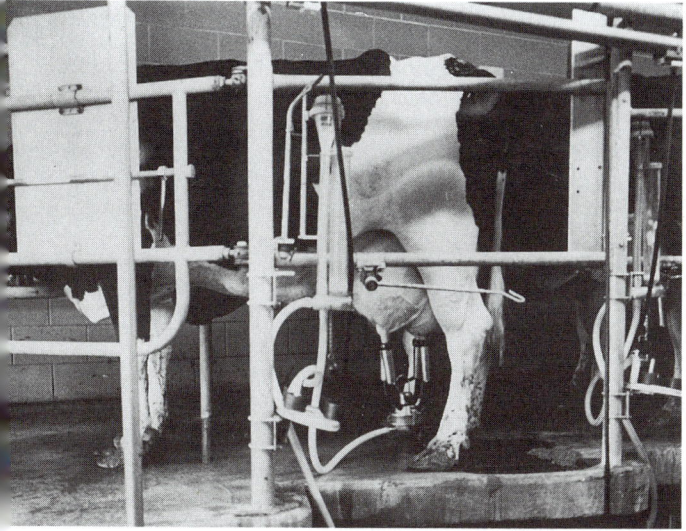

Fig. 18–15.   Modern herringbone milking parlor with large capacity pipeline milking system. (Courtesy, Babson Bros. Co., Oak Brook, IL)

Regardless of make, the mechanical milking systems can be separated broadly into three major parts: (1) vacuum supply, (2) pulsation, and (3) milking unit.

VACUUM SUPPLY

Two types of vacuum pumps are commonly used in milking operations—rotary and piston. Either type is readily adaptable to most milking operations; but it is generally recommended that the pump be of sufficient size to displace at least 25% more air than is required to operate the milking units and to lift and transport the milk to the cooling and storage area if a pipeline system is used. This additional displacement allows for decreased efficiency as the pump wears with age. The demands for a vacuum pump in a pipeline system are about 2.5 times as great as those for the bucket system.

PULSATION

Pulsators are installed in milking systems to direct the air flow leading to and from the milking unit. Two types of pulsators are currently available: (1) pneumatically controlled pulsators, and (2) electrically controlled pulsators. Pneumatic pulsators have a locking-screw valve which allows the milker to adjust the pulsation rate. Electric pulsators utilize a timing device to regulate pulsation rate.

Pulsation can be monitored by unit pulsators, pulsators placed on individual milking units, or by a master pulsator, a pulsator that controls several milking units.

Most milking machines synchronize all four teat cups, although some have two cups in the milking phase while the other two cups are in the resting phase. Before 1960, most machines used a milking time to collapse time ratio of 1:1. However, today most milking units use a ratio of 2:1 to 2.5:1 to allow for a faster flow rate. When ratios greater than 1:1 are used, the four teat cups must be in the same phase of milking.

MILKING UNIT

Each milking unit has four individual teat cups attached by hoses to either a suspended or a floor-type machine that houses a unit pulsator; or in cases where a master pulsator is used, a slave pulsator.

Each teat cup contains an inflation tube called a teat cup liner which is surrounded by a metal outer shell, generally of stainless steel. Teat cup liners come in a variety of bore sizes with the narrow bores generally causing less irritation to the teat than those with wide bores. This is because the top of narrow bore liners are lower on the teat, thereby avoiding contact with the sensitive, injury-prone area at the top of the teat. When teat liners lose their elasticity, teats become very susceptible to injury because the liner produces a sharp slapping action.

Fig. 18–16 illustrates the milking action within the teat cup. In the first phase of milking, called the expansion phase, pressure decreases in the space between the outer shell and the liner. This action dilates the teat canal and promotes milk flow. The pressure inside of the teat cistern is greater than the pressure outside of the teat and milk is forced through the streak canal. The massage phase, or nonmilking phase, is initiated when air is pumped into the space between the liner and the outer shell. The teat cup liner then collapses around the teat. This massaging action promotes circulation in the teat and allows the teat a brief moment of rest. If the massage phase is too short and the expansion phase too long, circulation is impaired, and the teat becomes injured due to congestion.

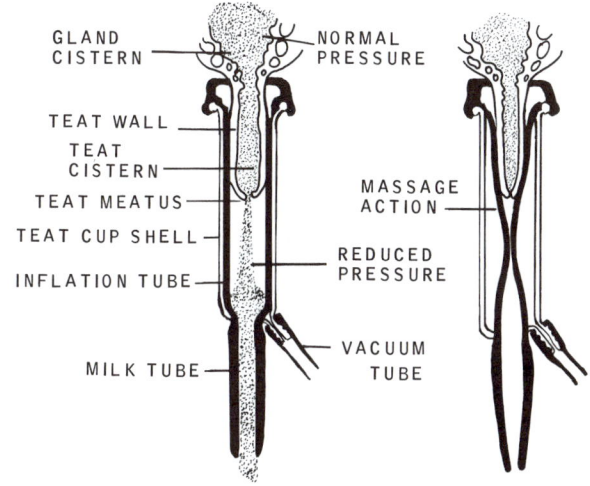

Fig. 18–16.   The mechanics of machine milking.

Milk flow rate is controlled by three factors:

1. **Pressure differential around the streak canal.** This pressure differential is determined by the pressure exerted by the milk in the udder and also by the pressure exerted by the vacuum pump and pulsator.

2. **Size of the streak canal.** More milk per unit time will flow through a larger opening.

3. **Tautness of the streak canal.** Several factors affect the tautness of the streak canal; among them, (a) intensity of stimulation for milk ejection, (b) age of the cow, and (c) effects of teat injury.

MILKING PROCEDURES

Milking is the most important single job to be done on the dairy farm. Some individuals are excellent milkers; others are very poor milkers—and this statement applies to both machine milking and hand-milking. Also, it is important that the cow be milked at regular times, preferably by the same milker; and that each milking be a pleasant experience. Cows like to be milked—if it is done properly.

The physiology of the discharge of milk is a delicate process, and it requires the close cooperation of the milker and the cow if it is to be successful. A managed milking program is made up of the following coordinated steps:

1. **Preparing the equipment.** Prior to milking, the equipment to be used in the milking process should be assembled and sanitized. Also, it should be checked and adjusted if necessary.

2. **Preparing the cow.** Under natural conditions, the cow is primed or stimulated by the suckling of the calf. This process can be simulated by washing the cow's teats and udder with warm water (120° to 130°F), then massaging and drying them with a paper towel. Following this process, remove two or three streams of milk from each quarter into a strip cup (never strip milk onto the floor) and examine for visible evidence of mastitis. Also, this (a) washes out any debris adhering to the end of the teat, and (b) enhances the let-down effect.

About 45 seconds after the priming stimulus, the udder becomes full and firm (especially in early lactation), and milk occasionally will leak from the teats. This is evidence that the cow has let down her milk and is ready for the next step.

3. **Attaching the teat cup and beginning.** About 1 minute after washing the udder, and not more than 1½ minutes, the teat cups should be attached and milking should begin. Most cows will milk out in 3 to 6 minutes, depending upon the amount of milk and the characteristics of the cow. Also, and most important, each quarter should be milked individually, because some quarters milk out faster than others.

4. **Stripping by machine.** When it is apparent that the cow is about milked out, she should be machine stripped. This consists of pulling down on the teat cups with one hand, and massaging the udder downward with the other. This process should not take over about 20 seconds.

5. **Removing the teat cup.** Both incomplete and overmilking should be avoided. The greatest cause of machine injury is leaving the teat cups on too long. Incomplete milking usually results because one or more quarters are more difficult to milk than the others.

As soon as the udder is empty, and before the teat cups crawl up, they should be removed, properly and gently. Then, dip the teats with a fresh disinfectant solution (100 ppm iodophor or chlorine, or other sanitizing agent). This will remove the milk from the ends of the teats and prevent the invasion of bacteria into the udder. Also, it will avoid attracting flies.

As soon as the teat cups have been removed from the udder of the cow, they should be cleaned. First, dip them in clean, cold water to remove milk inside the liners; then put them in a clean, warm, approved sanitizing solution. Change the solution after every five to seven cows.

6. **Cleaning up equipment.** After milking the last cow, all milking equipment should be thoroughly cleaned and put away.

7. **Milking time.** The actual milking time per cow will range from 3 to 6 minutes, with an average time of 3½ minutes for cows in mid-lactation. But additional time must be allowed for let-down, adjustments, and interval between cows.

The number of machines one milker can manage successfully depends upon the type of barn, the type of milking equipment, the ability of the milker, and the jobs the milker has other than milking

One milker should handle no more (preferably less) then the following number of units:

Type Milker	Units per Worker
Stanchion barns:	
Bucket	2
Pipeline	3
Milking parlor:	
Walk-through	3
Side-opening	3
Herringbone	4

With a three-in-line elevated parlor, one milker will average 18 to 25 cows per hour. With a four-in-line parlor, one milker will average 25 to 30 cows per hour. However, additional time must be allowed for bringing the cows in from the outside, setting up, cleaning up, as well as milking problem cows.

8. **Milking order.** Cows that have mastitis or a history of chronic mastitis are a source of infection to noninfected cows. Hence, it is well to milk *clean* cows first. A desirable milking order in stanchion barns is:

a. First-calf heifers that have been free of mastitis.

b. Older cows that have been free of mastitis.

c. Cows that have a previous history of mastitis, but which no longer show symptoms.

d. Cows with quarters producing abnormal milk.

CAN AND BULK SYSTEMS OF HANDLING

There are two characteristics of milk which make it ideal for the development of bacteria: (1) it is a well-balanced food in which bacteria thrive; and (2) as it comes from the cow, the temperature is ideal for bacterial growth. For these reasons, milk must be cooled to at least 50°F (preferably to 40°F) as soon as possible in order to inhibit bacterial growth.

Milk may be handled by either of two systems: the can system, or the bulk system. Until 1939, when the bulk system was first introduced in California, all milk was handled in cans. Today, most modern dairies use bulk tanks.

Generally speaking, the following advantages accrue to the use of bulk tanks, in comparison with cans: (1) a saving in labor, (2) less loss in milk, (3) alleviating 10-gallon cans, (4) higher butterfat tests (due to butterfat being left on lids of cans), (5) a saving in hauling costs, and (6) a premium paid by the plant.

DRYING OFF COWS

There are several methods of drying off cows, ranging from complete cessation of milking (see Chapter 12, Dairy Feeding Programs, under section on "Feeding Dry Cows") to intermittent milking. Perhaps in the final analysis each producer will do best by following his/her own procedure. Regardless of the method used, however, the grain should be reduced when it is decided to turn cows dry, with the tapering off of the grain beginning about a week before starting the actual drying off. Also, it is well to take cows off lush pastures or silage at that time, but they may be fed hay.

PRODUCE QUALITY MILK

Consumers and health departments all have a distinct interest in the quality of milk.

Quality milk can be produced only when the producer pays special attention to a number of factors:

1. **Health of the herd.** The herd should be free from diseases that might be spread to human beings through the milk. Bacteria in milk coming from cows must be eliminated. Mastitis is the most important herd health problem at the present time.

2. **Clean animals.** The milker should clean the flanks and udders of cows just prior to milking to prevent dirt from getting into the milk. Clean floors and bedding and a well-drained yard make the cleaning job easier.

3. **Clean equipment.** All milking equipment should be kept as clean and free from bacteria as possible. Bacteria grow in cracks and rough spots on equipment if it is not washed properly.

4. **Cool and store milk properly.** Proper cooling and storage of milk on the dairy farm require facilities which will cool the milk promptly from the in-the-pail temperature of about 90° down to 40°F, and then hold it at that temperature until it is collected. Bacteria will reproduce (divide) once every 30 minutes in 70° to 90°F temperature; thus, in 12 hours, one bacterium can reproduce 16 million. Cooling will control this growth.

5. **Keep barn and milk house clean.** The milking barn should be clean and have a concrete floor. Barn odors may be eliminated by ventilating the building well.

A milk room is important to the convenience of the operator, and an aid to the production of high-quality milk.

6. **Control flies.** Fly control measures are important to dairy producers. Flies add to the bacterial count of milk; cases are on record of flies carrying as many as 1¼ million bacteria. They can carry typhoid, dysentery, and other contagious diseases.

Breeding places for flies, such as manure piles and mudholes, should be eliminated.

7. **Control bacteria.** In summary, here is how the bacterial count in milk can be kept down:

   a. Rinse the utensils and equipment with hot water after cleaning so they dry off quickly.

   b. Remove all milkstone[1] from the equipment, as bacteria must have food.

   c. Cool the milk as quickly as possible to 40°F, as bacteria like high temperatures.

   d. Wash and sanitize with proper cleaning and sterilizing material.

   e. Have a well-lighted barn and milk house, as bacteria like darkness rather than light.

MILK FLAVOR

Most consumers base the quality of any product on its flavor, and milk is no exception. They want milk that "tastes good." The flavors most often found in milk and their causes and prevention are:

1. **Feed and weed flavors.** These have been covered under the section on "Other Feed Requisites" in Chapter 12, Dairy Feeding Programs, so repetition is unnecessary.

2. **Oxidized flavors.** These are sometimes described as cardboard flavors. Some causes of oxidized flavors are (a) metallic contamination from copper and iron, which may be alleviated by using stainless steel; (b) exposure to sunlight or just daylight; (c) foaming; and (d) drylot feeding. Feeding vitamin E to the milking herd will reduce or eliminate oxidized flavors.

3. **Rancid flavor.** This flavor is caused by a breakdown of the butterfat which releases strong-flavored acids. This action is caused by the enzyme lipase, which is present in all milk. The primary causes of rancid milk are (a) stripper cows (those well advanced in lactation); (b) excessive agitation of milk, due to high lifts and sharp turns in pipeline milking; and (c) slow cooling with foaming.

4. **Barny.** This flavor(s) is caused by dirty stables, poor ventilation, unclean milking, and unclean cows—all of which can be alleviated.

5. **Salty.** This flavor, which masks the slightly sweet flavor of milk, is caused by mastitis, stripper cows, or certain individual cows. Milk from cows that have mastitis, or from strippers, should not be marketed.

6. **Malty.** Malty flavor is primarily due to high bacteria count. The remedy is to keep bacteria out of milk as much as possible, and to prevent growth of those that do get into it. Clean and cold milk will practically eliminate malty flavor. Also, milk handlers should pick up all the milk and not leave any of it in the farm bulk tank.

7. **High-acid, sour milk.** This is due to very high bacterial count. In these days of mechanical refrigeration, there is no excuse for sour milk; simply cool it as rapidly as possible from the 90°F temperature of the milk pail to 40°F.

8. **Unnatural or foreign.** This refers to flavors that come from medicinal agents and disinfectants. The control of such off-flavors consists in (a) handling medicines and

---

[1]Milkstone is a complex mixture of milk and water minerals with entrapped fat, protein, soil particles, and microorganisms, plus cleaner and sanitizer resides. This film adheres tightly to the surface of milk-handling equipment and requires special acid treatment for removal.

disinfectants so that the flavor or odor from them will not get into the milk, and (b) using chemical sanitizers only in the concentrations indicated by the directions. Do not market milk from drug treated cows for at least 72 hours after last treatment, or longer if so prescribed on the drug label or by the veterinarian.

For good-tasting milk, the producer should keep it clean, keep it cold, feed silage after milking (not before), use good-quality feed, and not ship milk from problem cows.

DISEASES ASSOCIATED WITH LACTATION

Several diseases are directly associated with lactating cows; among them, mastitis, ketosis, milk fever, and udder edema. Mastitis, which is usually infectious, is covered in Chapter 15, Dairy Cattle Health, Disease Prevention, and Parasite Control. Ketosis, milk fever, and udder edema, which are nutritional in nature, are discussed in Chapter 12, Dairy Feeding Programs.

QUESTIONS FOR STUDY AND DISCUSSION

1. How many mammary glands are found in the following animals: (a) cow, (b) mare, (c) sow, and (d) goat?

2. Discuss the nutritional differences between milk from a human and milk from a cow.

3. Compare the nutrient content of colostrum and bovine milk.

4. What is the streak canal? Draw and label the various parts of the udder of the cow.

5. Which produces more milk, the rear or front quarters?

6. What is a "hard milker"?

7. What structures support the udder?

8. Diagram a cross section of the udder showing the duct system and the cisterns.

9. What is the alveolus? How does it function?

10. What is the function of the myoepithelial cells in the alveoli and ducts?

11. Differentiate between the somatic and autonomic nervous systems. Outline the neural controls of the udder.

12. Briefly outline the flow of blood into and out of the udder.

13. How is lymph movement facilitated?

14. Trace the development of the mammary system in the cow from the embryonic stage to the drying off of a lactating cow. What hormones influence this development?

15. Draw a cross section of a secretory cell and label its components. What are the functions of these components?

16. How is lactose formed and secreted into the milk?

17. Why is acetate important in milk fat synthesis?

18. What is emeiocytosis?

19. List the primary proteins found in the protein fraction of milk. What are their respective solubility characteristics?

20. Outline the synthesis of a protein.

21. List the primary minerals in milk.

22. List and discuss the physiological factors that affect (a) milk yield, and (b) milk composition.

23. List and discuss the environmental factors that affect (a) milk yield, and (b) milk composition.

24. Outline the process of milk ejection. What is milk *hold up*?

25. What three ways is milk removed from the udder? Which method yields the most milk?

26. Trace the history of milking machine development.

27. Discuss the various components of a machine milking system.

28. Outline the basic milking procedures.

29. What advantages generally accrue from the use of bulk tanks, in comparison with cans?

30. What method for drying off cows would you recommend?

31. What factors affect the quality of milk?

32. How are off-flavors produced in milk?

SELECTED REFERENCES

Title of Publication	Author(s)	Publisher
Biology of Lactation	G. H. Schmidt	W. H. Freeman and Co. Publishers, San Francisco, CA, 1971
Dairy Cattle: Principles, Practices, Problems, Profits, Third Edition	D. L. Bath, *et al.*	Lea & Febiger, Philadelphia, PA, 1985
Harvesting Your Milk Crop	C. W. Turner	Babson Bros. Co., Oak Brook, IL, 1973
Lactation—A Comprehensive Treatise, Vols. I, II, and III	B. L. Larson V. R. Smith	Academic Press, Inc., New York, NY, 1974
Science of Providing Milk for Man, The	J. R. Campbell R. T. Marshall	McGraw-Hill Book Company, New York, NY, 1975

Fig. 19–1. Milk truck in front of milking parlor. (Courtesy, Dr. Tom Schultz, Farm Advisor, Visalia, CA)

Contents

CHAPTER

19

MARKETING
MILK AND
DAIRY PRODUCTS

Contents

Marketing is that all-important end of the line; it is that which gives point and purpose to all that has gone before.

In our present system, the vast majority of the marketing of milk and dairy products is handled by specialists, usually under a myriad of complex regulations and controls. However, successful milk producers must understand milk markets and pricing systems, along with the factors affecting them, if they are to take full advantage of their financial opportunities. In the future, there will be increasing competition with the result that marginal dairy operations will, in all probability, be forced out of business. Thus, it is imperative that milk producers be aware of all the factors that influence the price that they will receive for their raw, unprocessed product.

MARKET IMPORTANCE OF MILK AND DAIRY PRODUCTS

The farm value of dairy products in 1990 was $20.2 billion. The total marketing bill—the cost of transporting, processing, and distributing dairy products, or the difference between consumer expenditures and farm value—that same year came to $39.7 billion. Upon being retailed, consumers spent $59.9 billion for these products (Table 19–1).

TABLE 19–1
FARM VALUE, MARKETING BILL, AND CONSUMER
EXPENDITURES FOR DAIRY PRODUCTS, 1990[1]

Item	Value
	(billion $)
Farm value	20.2
Marketing bill (cost for transporting, processing, and distributing)	39.7
Consumer expenditures (retail cost)	59.9

[1]Source: USDA

Other noteworthy statistics are (1) dairy products accounted for 12% and dairy cattle 2% of the cash income of the nation's farmers in 1989, and (2) customers spent about 11% of their food dollar for dairy products in 1989.

FARM PRODUCTION AND HANDLING OF MILK

Satisfactory milk marketing requires one basic ingredient—quality milk; and, ultimately, this means more income for the producer who takes pride and care in his/her operation.

The difference in price between Grade A milk and the lower grades is substantial. But it goes beyond this; quality can influence consumers' demand.

Buyers, consumers, and health departments share a common interest in the quality of milk marketed as fresh milk or used in manufacturing. In recent years, there has arisen a great interest by the public pertaining to the quality of the food we eat. Food must now be wholesome and free from contamination lest it be deemed unsafe or unfit for consumption.

Quality milk can be produced only when producers pay special attention to a number of factors; among them, herd health, the layout and structure of the barn and milk house, clean cows, care of the milking equipment, cooling and storage of milk, and transportation of milk to market.

COSTS INVOLVED IN THE PRODUCTION OF MILK

If milk producers are to be successful, they must weigh income of production against expenses incurred in production. A high-producing herd is desirable and much sought; but unless costs are held to a minimum, the operation may be an exercise in futility.

FEED

Feed is the largest single expenditure in dairying; it accounts for 45 to 65% of the cost of milk production. The producers must be well attuned to feed costs and new developments in dairy nutrition.

One commonly used indicator of the relative prosperity of dairy producers is the milk-feed price ratio. Table 19–2 gives the milk-feed (concentrate) price ratio for selected

TABLE 19–2
MILK-FEED (CONCENTRATE) PRICE RATIO, SELECTED YEARS[1]

Year	Milk-Feed (Concentrate) Price Ratio
1979	1:80
1980	1:76
1981	1:72
1982	1:83
1983	1:72
1984	1:65
1985	1:73
1986	1:79
1987	1:86
1988	1:58
1989	1:65
1990	1:71

[1]Milk-feed (concentrate) price ratio refers to the pounds of concentrate equal in value to 1 lb of milk. (Source: *Dairy Situation and Outlook Report*, USDA, ERS, DS-430, July 1991, p. 7)

years from 1979 to 1990, whereas Fig. 19–2 shows the milk-feed price ratio for 1965 to 1990. This ratio gives an indication of how much return producers are receiving for their feed inputs. In order to turn a profit, it is necessary that producers maintain a favorable milk-feed price ratio. Of course, the ratio fluctuates from season to season and from year to year, primarily in response to (1) availability and cost of feed, (2) demand for milk, (3) productivity of the dairy farm, and (4) imports. Also, it is recognized that the milk-feed price ratio does not tell the whole story; it does not indicate how energy, labor, and other managerial costs have increased in recent years.

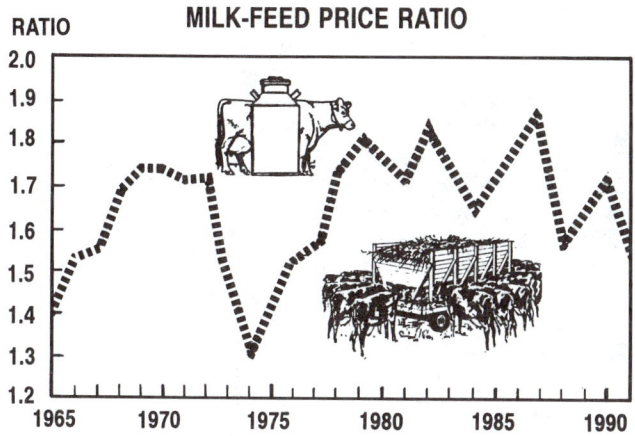

MILK-FEED PRICE RATIO

Fig. 19–2. Milk-feed price ratio, 1965 to 1990. The milk-feed ratio refers to the pounds of concentrate equal in value to 1 lb of milk. (Source: *Dairy Situation and Outlook Report*, USDA, ERS, DS-430, July 1991, p. 9)

LABOR

Dairy production is labor intensive. Fortunately, technology has developed new ways of reducing labor demands for milking. This decrease in labor requirements is reflected in Table 19–3.

TABLE 19–3
LABOR PER COW AND PER 100 POUNDS OF MILK, 1930–87[1]

| Year | Work-Hours | |
	Per Cow	Per 100 Lb Milk
	(hr)	(hr)
1930	147.1	3.26
1940	147.5	3.19
1950	125.5	2.36
1960	99.7	1.42
1970	68.1	0.70
1980	35.4	0.30
1985	21.2	0.16
1987	19.7	0.14

[1]*1989 Dairy Producer Highlights,* National Milk Producers Federation, Arlington, VA, p. 7. For milk cow enterprise only; does not include related dairy farm activities.

CAPITAL INVESTMENT

Numerous factors require the investment of large amounts of capital in dairy farms. Land, buildings, silos, machinery, milking equipment, and the animals themselves, all add up to a sizable sum. In addition to the cost of facilities and animals, depreciation and repairs of the physical plant, interest on loans, insurance, and taxes must be considered in the operating expenses.

The U.S. Department of Agriculture reported the following costs of production in 1989:[1]

1. Cost per dairy cow $1,207.00
2. Cost to produce 100 lb of milk . . . . $ 11.84

OTHER EXPENSES

Numerous other expenses are inherent in all dairy operations; among them, bedding, veterinary bills, utilities, general operational supplies, breeding fees, DHIA expenses, and transportation costs.

PRICES RECEIVED BY FARMERS FOR MILK

Milk production is seasonal in nature, and the economic principles of supply and demand dictate that prices decline when milk availability increases. Figs. 19–3 and 19–4 show the milk production and milk price fluctuations on a monthly basis, respectively. Milk production peaks in May and June

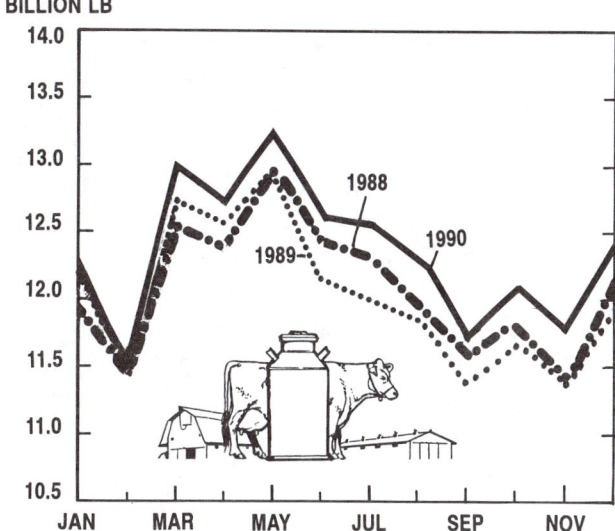

U.S. MILK PRODUCTION BY MONTHS
BILLION LB

Fig. 19–3. U.S. milk production by months. (From: *Dairy Situation and Outlook Report*, USDA, ERS, DS-429, April 1991, p. 61)

[1]*Dairy Background for 1990 Farm Legislation,* USDA, ERS, Commodity Economics Division, pp. 9 and 62.

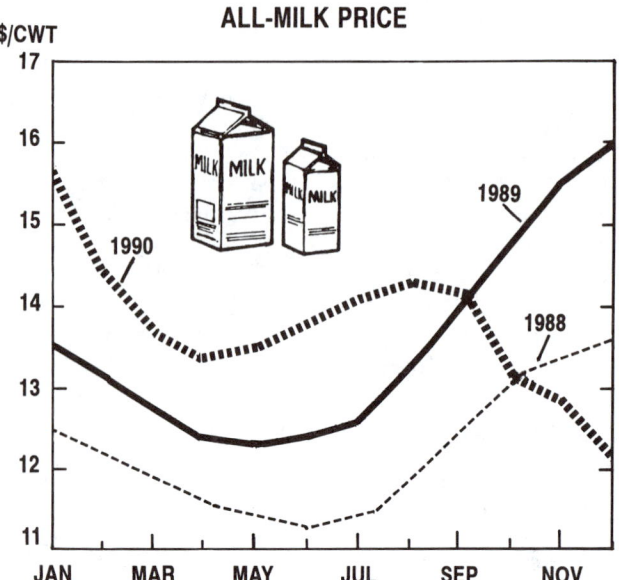

ALL-MILK PRICE

Fig. 19–4. U.S. milk prices by months on milk delivered to plants and dealers. (From: *Dairy Situation and Outlook Report,* USDA, ERS, DS-429, April 1991, p. 11)

and reaches its yearly low in November. Recently, milk price trends have been 1 month behind milk production, with the lower yearly prices occurring in April to June and the highest in November or December.

Table 19–4 gives a comparison of the prices received by

TABLE 19–4
PRICES RECEIVED BY FARMERS FOR HOGS, BEEF CATTLE, AND MILK, 1925–88[1]

Year	Prices Received			Index Numbers (1977 = 100)		
	Hogs	Beef Cattle	Milk Sold Wholesale[2]	Hogs	Beef Cattle	Milk Sold Wholesale
	($ per hundredweight)					
1925	10.91	6.53	2.38	28	19	25
1930	8.84	7.71	2.21	22	22	23
1935	8.65	6.04	1.72	22	18	18
1940	5.39	7.56	1.82	14	22	19
1945	14.00	12.10	3.19	35	35	33
1950	18.00	23.30	3.89	46	68	40
1955	15.00	15.60	4.01	38	45	41
1960	15.30	20.40	4.21	39	59	43
1965	20.60	19.90	4.23	52	58	43
1970	22.70	27.10	5.71	58	79	59
1975	46.10	32.20	8.75	117	94	90
1980	38.00	62.40	13.05	97	182	134
1985	44.00	53.70	12.75	112	156	131
1988	42.30	66.10	12.20	107	194	126

[1]Source: *Ag Prices Annual,* NASS, USDA.

[2]Average butterfat test.

farmers for milk, beef cattle, and hogs from 1925 to 1988. In the last 60 years, milk prices have risen steadily, while prices for hogs and cattle have tended to fluctuate with production cycles.

PARITY AND SUPPORT PRICES

Parity may be defined as a yardstick for measuring the relationship between the prices farmers receive for the products they sell and the prices they pay for the things they buy, including interest, taxes, and farm wage rates.

The price of a farm commodity may be said to be at parity when a given unit has the same purchasing power that it had in the base period. Thus, if dairy producers receive parity price per hundredweight for milk, they should be able to take the money and buy as much with it as they could back in 1910 to 1914, the base period. If the price of things farmers buy for their production program doubles, the parity price of farm commodities also doubles, to keep in line or equal parity.

The new parity formula, which became effective January 1, 1950, is still based on the years 1910 to 1914, but it reflects recent price relationships by using the latest 10-year averages.

In the Agricultural Act of 1949 and subsequent amendments to that Act, Congress established guidelines for the operation of the price support program. For many years, the minimum support price was 75% and the maximum was 90% of parity. Thus, the parity concept was important to milk producers until 1981, because the support price of milk was based thereon. However, legislation in 1981 departed from the parity concept for the first time and parity has not been used as a basis for establishing dairy price supports since then. The Food Security Act of 1985 set the support price of milk from 1986–1990. On January 1, 1990, the support price was $10.10 per cwt. The federal government makes substantial purchases of dairy products—through the CCC and other purchase programs—to support prices at announced levels.

HOW MILK IS SOLD BY DAIRY PRODUCERS

Milk producers sell most of their product in the form of whole milk. At one time, they marketed a considerable amount of their product as farm-separated cream, but the proportion of this product has been declining. In 1940, farm-separated cream accounted for 38% of their total marketing; today, marketing of cream is negligible.

In 1987, only 1.6% of the total milk production was used on farms, compared with 15% in 1950. Obviously, the point has been reached where little additional marketing of milk by farmers can be expected through decreased use on the farm. Future increases in milk marketings will have to come from increased production.

HOW MILK IS USED

Fig. 19–5 shows the milk supply, use, and stocks from 1975 to 1987. About 37.4% of the milk marketed by dairy farmers today is consumed in fluid form (Table 19–5, 36.6% sold by dealers plus 0.8% sold by producers directly). Fluid milk is retailed as pasteurized milk, homogenized milk, fortified milk (vitamin D), skimmed milk, flavored milk (whole milk with flavor added), or flavored milk drink (skimmed milk with flavor added).

In 1960, 23.9% of the milk supply was used for butter, whereas only 14.8% was so used in 1987. However, it is noteworthy that the decline in the use of milk for butter has slowed in recent years.

In 1987, farmers sold 98.4 lb out of every 100 lb of milk they produced. Of the milk which they sold, 89% of it was eligible for the fluid market, as Grade A; 11% of it was manufacturing milk, or Grade B.

MILK SUPPLY, USE, AND STOCKS

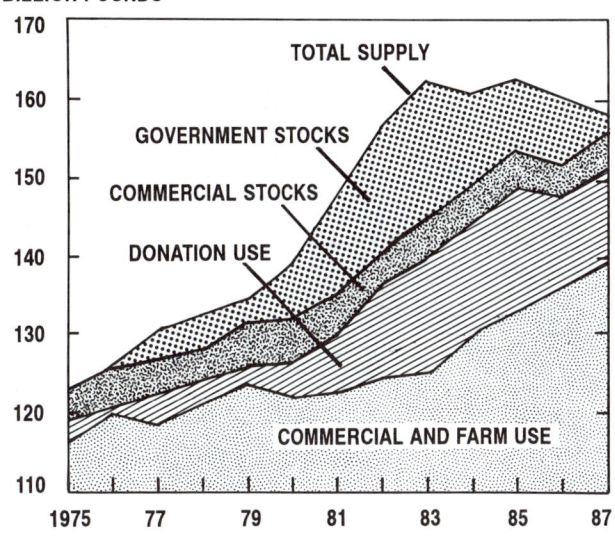

Fig. 19–5. Milk supply, use, and stocks, 1975 to 1987. (Source: *1988 Agricultural Chartbook*, Ag. Hdbk. No. 673, USDA, p. 91, Chart 198)

TABLE 19–5
UTILIZATION OF U.S. MILK SUPPLY, SELECTED YEARS 1965–87[1]

| Year | Fluid Milk and Cream | | Butter, Creamery | Cheese | Evaporated and Condensed Milk | Frozen Products | Other[2] | Used on Farms Where Produced[3] |
	Sold by Dealers	Sold by Producers Directly to Consumers						
	(percent of total milkfat supply)							
1960	41.4	1.7	23.9	10.9	4.4	7.7	2.6	7.4
1965	43.1	1.5	23.0	12.6	3.7	8.5	2.8	4.8
1970	42.8	1.5	20.4	16.6	2.8	9.4	3.1	3.4
1975	42.9	1.4	17.2	20.7	2.3	10.3	2.5	2.7
1980	38.4	1.1	17.7	26.4	1.6	9.3	3.7	1.8
1985	35.4	0.9	17.0	29.1	1.6	9.0	5.3	1.7
1987	36.6	0.8	14.8	30.0	1.5	9.4	5.3	1.6

[1]*1988 Milk Facts,* Milk Industry Foundation, p. 30. Supply of milkfat includes U.S. production, ingredient imports of milkfat and solids from sources outside the U.S., and net change in storage cream. Computations made by the Milk Industry Foundation, based on data from the USDA.

[2]Dry whole milk, creamed cottage cheese, and other miscellaneous uses.

[3]Milk fed to calves, consumed on farms as milk and cream, and used for farm-churned butter.

PER CAPITA CONSUMPTION

Table 19–6 shows the per capita consumption of each of the leading dairy products and the changes that have occurred since 1910.

TABLE 19–6
PER CAPITA CIVILIAN CONSUMPTION, MILK AND DAIRY PRODUCTS 1910–88[1]

Year	Fluid Milk Products				Whole Milk Equivalent of Butterfat	Manufactured Dairy Products						Total Human Use (Milk Equivalent of Butterfat[5])
	Fluid Whole Milk	Cream and Mixtures	Lowfat And Skim Milk[3]	Total Product		Butter	Total Cheese[4]	Cottage Cheese	Ice Cream	Evap. and Cond.	Nonfat Dry Milk	
						(pounds of product[6])						
1910						18.3	4.3	0.6	1.9	5.8		
1920						14.9	4.0	0.6	7.6	8.6	0.2	
1925						18.1	4.7	0.9	9.7	11.7	0.4	802
1930						17.6	4.7	1.2	9.8	13.6	1.3	819
1935						17.6	5.3	1.3	8.1	16.2	1.6	801
1940						17.0	6.0	1.9	11.4	19.3	2.2	819
1945						10.9	6.7	2.6	15.7	18.3	1.9	788
1950					321	10.7	7.7	3.1	17.2	20.5	3.7	740
1955	270	9	18	296	326	9.0	7.9	3.9	18.0	16.2	5.5	706
1960[2]	256	8	22	286	309	7.5	8.3	4.8	18.3	13.7	6.2	653
1965	252	7	32	292	285	6.4	9.6	4.7	18.5	15.6	5.6	625
1970	213	5	58	277	247	5.3	11.5	5.2	17.6	12.1	5.3	564
1975	184	5	78	267	223	4.7	14.5	4.7	18.5	8.9	3.3	539
1980	146	5	95	246	194	4.5	17.8	4.5	17.3	7.1	3.0	543
1985	120	6	107	239	186	4.9	22.6	4.1	18.0	7.9	2.2	593
1988	100	6	132	238	184	4.4	23.6	3.8	17.8	7.9	2.7	585

[1]Source: *1989 Dairy Producer Highlights,* National Milk Producers Federation, Arlington, VA, p. 17.

[2]Includes Alaska and Hawaii beginning with 1960.

[3]Includes buttermilk, flavored milk drinks, and yogurt.

[4]Natural equivalent.

[5]Includes the milk equivalent of butterfat for all fluid and dairy products and USDA donations.

[6]To convert pounds to kilograms, multiply by 0.454.

Fig. 19–5a. A mammoth cheese. (Courtesy, The Bettmann Archive, Inc., New York, NY)

Fig. 19–6 shows change in U.S. per capita dairy product sales from 1977 to 1987. It is noteworthy that declines in per capita consumption were registered in the following products: ice milk, butter, buttermilk, flavored milk and drinks, evaporated and condensed whole milk, creamed cottage cheese, whole milk, and nonfat dry milk. It is noteworthy, too, that yogurt and low fat cottage cheese led the increases. Increase in yogurt reflects the rising interest in health foods, whereas increased consumption of low fat cottage cheese reflect weight watching.

TERMINOLOGY OF MARKET REPORTS

Knowledgeable dairy producers and dairy manufacturers must follow market reports in order to determine the best channels through which to market their products, as well as to project future trends in supply and demand so that they may plan their programs accordingly. Table 19–7, Glossary of Terms Used in Federal-State Market News Reports, briefly describes the terms commonly associated with the marketing of commodities.

PERCENT CHANGE IN PER CAPITA SALES
1977–87

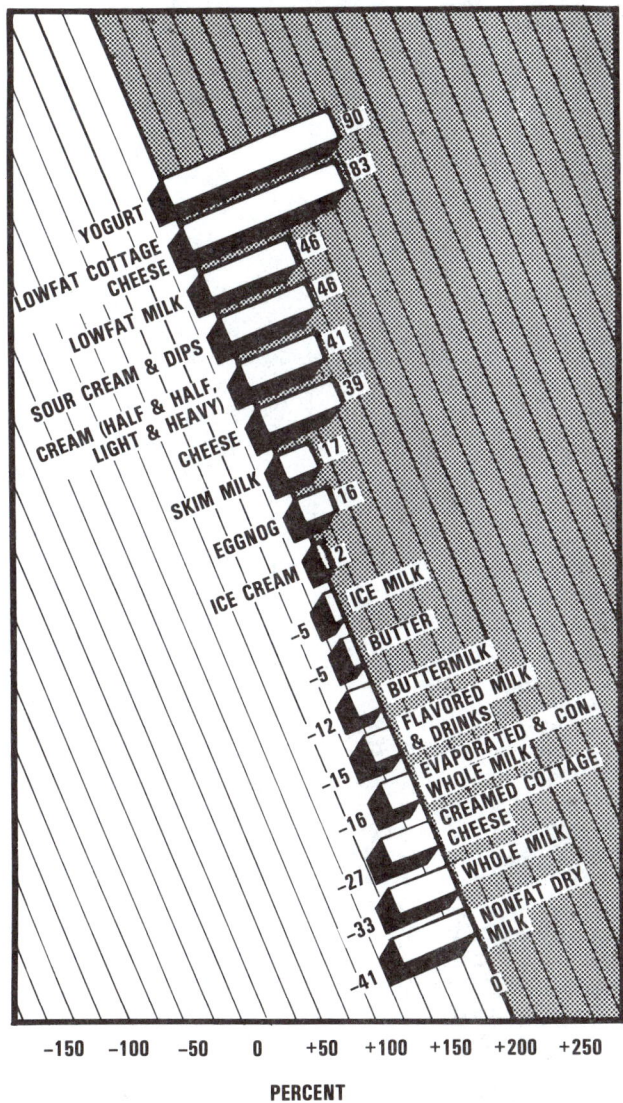

Fig. 19–6. Change in U.S. per capita consumption of milk and dairy products 1977 to 1987, based on sales. Bars to the left of the vertical line represent decreases; bars to the right represent increases. (Source: *1988 Milk Facts,* Milk Industry Foundation, Washington, DC, p. 16)

TABLE 19–7
GLOSSARY OF TERMS USED IN
FEDERAL-STATE MARKET NEWS REPORTS[1]

Terms	Definitions
Market	1. A geographic location where a commodity is traded. 2. The price, or price level, at which a commodity is traded. 3. To sell.
Market activity:	The pace at which sales are being made.
Active	Available supplies (offerings) are readily clearing the market.
Moderate	Available supplies (offerings) are clearing the market at a reasonable rate.
Slow	Available supplies (offerings) are not readily clearing the market.
Inactive	Sales are intermittent with few buyers or sellers.
Price Trend:	The direction in which prices are moving in relation to trading in the previous trading session.
Higher	The majority of sales are at prices measurably higher than the previous trading session.
Firm	Prices are tending higher, but not measurably so.
Steady	Prices are unchanged from previous trading session.
Weak	Prices are tending lower, but not measurably so.
Lower	Prices for most sales are measurably lower than the previous trading session.
Supply/offering:	The quantity of a particular item available for current trading.
Heavy	When the volume of supplies is above average for the market being reported.
Moderate	When the volume of supplies is average for the market being reported.
Light	When the volume of supplies is below average for the market being reported.
Demand:	The desire to possess a commodity coupled with the willingness and ability to pay.
Very good	Offerings or supplies are readily absorbed.
Good	Firm confidence on the part of buyers that general market conditions are good. Trading is more active than normal.
Moderate	Average buyer interest and trading.
Light	Demand is below average.
Very Light	Few buyers are interested in trading.
Mostly	The majority of sales or volume.
Undertone	Situation or sense of direction in an unsettled market situation.

[1]*Glossary of Terms Used in Federal-State Market News Reports,* Agricultural Marketing Service, USDA, June 1975.

MARKET CHANNELS FOR MILK AND DAIRY PRODUCTS

Milk moves from the farm to the consumer in the following three stages:

1. Assembly and transportation from farms to processing plants.
2. Processing and packaging or manufacturing into various dairy products.
3. Distribution of packaged milk and manufactured milk products to consumers.

Also, producers market their milk as (1) Grade A milk, or (2) manufacturing milk (Grade B).

DISTRIBUTION CHANNELS FOR FLUID MILK

Once milk is processed, it must be marketed rather rapidly because of its highly perishable nature. When consumers have a bad experience with milk, it is difficult to get them back into the habit of drinking milk. Hence, it is imperative that fluid milk be of high quality and possess desirable taste and packaging.

Milk is distributed in a number of ways: (1) home delivery, (2) retail stores, and (3) vending machines and self-service.

HOME DELIVERY

Throughout the first half of the 20th century, home delivery was the primary means of distributing milk. In fact, home delivery constituted one-fourth of all milk distributed from federal milk orders as late as 1966. However, in recent years this mode of distribution has dwindled — amounting to a mere 1% of the federal order markets in 1989.

RETAIL STORES

Today, the vast majority of milk is sold in retail stores. This is largely due to (1) increased mobility of the public

Fig. 19–8. Plastic milk bottles move through a high-speed gallon filler. (Courtesy, Carnation Co., Los Angeles, CA)

through the use of cars; (2) new technology in processing that increases the storage life of milk; (3) the widespread use of efficient refrigerators; (4) the increase in the methods of packaging milk (plastic and paper containers); (5) the increase in the number of supermarkets that offer low-priced milk; (6) the sales of milk by the retailer through the use of dairy industry personnel rather than wholesale dealers; and (7) increased costs involved with home deliveries.

Many stores today offer only milk brands that are processed by their own plants. Through this sales practice, the stores are able to reduce marketing costs, gain a better control of inventory, and predict sales volume more closely.

Other retail stores sell their own brand of milk at a lower price than the regional or national brands, but continue to offer the competitive brands so as to provide customers a wider choice. Through the use of private brands, the store has considerably more control over pricing and merchandising than with outside brands.

A third type of retail outlet is the milk store operated by a large dairy, integration from production through processing and marketing. As dairy operations become larger and more integrated in the future, the number of these specialty markets will likely increase.

Table 19–8 portrays the changes in fluid milk sales since 1971 in distribution methods, types of containers, and size of containers. As shown, the distribution of fluid milk through wholesale channels increased, while home delivery decreased; and the use of both plastic and gallon size containers increased dramatically.

Fig. 19–7. Buying milk in a retail store. (Photo by A. H. Ensminger)

TABLE 19–8
PERCENT OF FLUID MILK SOLD BY CONTAINER SIZES, TYPES, AND OUTLETS
FEDERAL ORDER MARKETS, NOVEMBER 1971–1990[1,2]

	1971	1973	1975	1977	1979	1981	1983	1984	1985	1986	1987	1988	1989	1990[3]
	(%)	(%)	(%)	(%)	(%)	(%)	(%)	(%)	(%)	(%)	(%)	(%)	(%)	
Distribution method:														
Home delivered	15	10	7	5	4	2	2	1	1	1	1	1	1	
Wholesale	85	90	93	95	96	98	98	99	99	99	99	99	99	
Supermarkets	N/A	N/A	N/A	43	46	50	50	50	53	54	54	55	54	
Dairy/convenience	N/A	N/A	N/A	10	10	10	10	11	9	9	9	9	10	
Military	N/A	N/A	N/A	2	1	1	1	1	1	1	1	1	1	
Schools	N/A	N/A	N/A	8	8	7	7	7	7	7	7	7	7	
Other	N/A	N/A	N/A	32	30	30	30	30	28	27	27	26	27	
Type of container:														
Glass	7	4	2	1	1	1	[4]	4	[4]	[4]	[4]	[4]	[4]	
Paper	78	71	67	58	49	42	38	36	34	33	33	32	31	
Plastic	15	25	31	41	50	57	62	64	65	66	67	67	69	
Metal cans	[4]	[4]	4	[4]	[4]	[4]	[4]	[4]	[4]	[4]	[4]	[4]	[4]	
Total	100	100	100	100	100	100	100	100	100	100	100	100	100	
Size of container:														
Gallon	29	37	43	49	53	57	58	59	60	61	60	61	61	
Half gallon	44	38	34	29	26	24	23	22	22	21	21	21	21	
Quart	10	8	7	6	5	5	5	5	5	5	5	5	4	4
Pint	1	1	1	1	1	1	1	1	2	2	2	2	2	2
Half-pint	11	10	11	11	11	10	10	10	9	8	10	9	10	9
Bulk—over 5 qts[5]	4	5	3	3	3	3	3	3	2	2	2	2	2	2
Other	1	1	1	1	1	[4]	4	[4]	[4]	[4]	[4]	[4]	[4]	[4]
Total	100	100	100	100	100	100	100	100	100	100	100	100		100

[1]Source: *1991 Milk Facts*, Milk Industry Foundation, Washington, DC, p. 19.

[2]November is considered representative of the annual average. Totals may not add due to rounding.

[3]Estimated by the Milk Industry Foundation.

[4]Less than 0.5%.

[5]Metal cans and plastic bag-in-box containers.

N/A = Not available.

VENDING AND SELF-SERVICE MACHINES

In areas where people congregate, vending machines containing milk can be effectively utilized. Little labor is involved, and the customer has ready access to the product.

HOW MILK IS PRICED AND REGULATED

Chaotic conditions in milk marketing, resulting from the breakdown of private controls and the serious economic plight of farmers during the depression years of the early 1930s, brought requests from organized producers and distributors for government control. Out of this evolved two forms of government controls — those established by the federal government, and those established by state governments; both were designed to bring more stability into the marketing of milk. Today, federal and state agencies, directly or indirectly, affect the pricing of milk marketed by dairy farmers in the United States.

FEDERAL MILK MARKETING ORDERS

Federal milk marketing orders were established, and are administered, by the Secretary of Agriculture under acts of Congress passed in 1933 and 1937. They are legal instruments, and they are very complex. However, stated in simple terms, they are designed to stabilize the marketing of fluid milk and to assist farmers in negotiating with distributors for the sale of their milk. Prices paid to farmers are controlled, but there is no direct control of retail prices.

Federal orders are not concerned with sanitary regulations. They regulate the handling and pricing of about 80% of the Grade A milk; and only Grade A milk is regulated by federal milk marketing orders. (Most of the remaining 20% of Grade A milk is regulated under state regulations.) Each of the 41 market orders has a market administrator and provisions for setting minimum farm prices and regulating transactions between farmers and milk dealers in their area.

Prices in other Grade A markets are influenced by prices established under federal orders or state control programs. Additionally, dairy support programs directly affect the prices of both manufacturing grade milk marketed by farmers and the milk farmers sell as farm-separated cream.

Table 19–9 shows the scope of the federal milk marketing order program from 1947 to 1988.

TABLE 19–9
SCOPE OF FEDERAL MILK MARKET ORDER PROGRAM, SELECTED YEARS[1,2]

	All U.S.		Federal Order Statistics							
	Milk Sold to All Plants and Dealers		Receipts in Federal Order Markets		Class I Sales in Federal Order Markets		Price per 100 Lb, 3.5% Milk			
Year	Total	Fluid Use	Quantity	As % of Milk Sold to Plants and Dealers	Quantity	Percent of U.S. Total Fluid Use	Class I	Blend	Number of Federal Order Markets[3]	Number of Handlers[3]
	(bil. lb)	(bil. lb)	(bil. lb)	(%)	(bil. lb)	(%)	($)	($)	(no.)	(no.)
1947	70.6	37.0	15.0	21.2	9.8	26.5	4.65	4.34	29	991
1950	74.2	39.2	18.7	25.1	11.0	28.1	4.51	3.93	39	1,101
1955	91.0	46.7	28.9	31.8	18.0	38.6	4.67	4.08	63	1,483
1960	103.9	50.9	44.8	43.1	28.8	56.6	4.88	4.47	80	2,259
1965	112.7	53.6	54.4	48.3	34.6	64.5	4.93	4.31	73	1,891
1970	110.0	50.3	65.1	59.6	40.1	79.7	6.74	5.95	62	1,588
1975	110.2	50.8	69.2	62.8	40.1	78.9	9.36	8.64	56	1,315
1980	124.7	50.8	84.0	67.4	41.0	80.7	13.77	12.86	47	1,091
1985	139.9	52.0	97.8	69.9	42.2	81.2	13.88	12.61	44	884
1988	142.1	54.5	100.1	70.4	43.1	79.1	13.42	12.14	42	796

[1]Sources: *Federal Milk Marketing Order Statistics*, AMS, and *Milk Production, Disposition and Income*, NASS, USDA.

[2]To convert pounds to kilograms, multiply by 0.454.

[3]At end of year.

An understanding of the terms used in milk marketing orders is helpful in comprehending what is behind the dairy producer's milk check. To this end, Table 19–10 is presented.

TABLE 19–10
GLOSSARY OF TERMS USED IN MILK MARKETING ORDERS[1]

Terms	Definitions
Adjusted uniform price	The price paid to a Grade A milk producer who has no daily base in any given month. (There are two exceptions which space limitation herein do not permit detailing.) This price is less than the uniform price by an amount equal to 25% of the difference between the uniform price and the excess price.
Base price	The price paid to Grade A producers for all milk shipped in a given month up to a volume equal to their daily base times the number of days in the month. Clearly, only producers with a daily base allotment receive the base price for any part of their monthly volume.
Basic formula price (BFP)	The BFP is used in computing Class I and II prices in federal order markets. The Minnesota-Wisconsin Series price for the second preceding month becomes the basic formula price in the current month.
Blend price	Net price per cwt received by the producer for a given month determined by dividing his/her total monthly milk payments (usually less shipping charges, assessments, dues, and retained funds) by the total volume (cwt) shipped in a given month. The blend price for the entire market is the total amount of money paid into the pool divided by the amount of milk shipped to the market.
Butterfat differential	Value accorded to the butterfat content of milk for the purpose of adjusting class prices based on a 3.5% butterfat content. In the Southern Michigan Order, all per cwt class prices are increased (decreased) by 0.113 times the 92 score Chicago butter price for that month for each 0.1% that the butterfat content is above (below) 3.5%.
Class I products	All skim milk and butterfat sold in fluid form (e.g., regular homogenized milk, 2% milk, half and half, buttermilk, low-fat milk, etc.).
Class II products (in orders w/3 classes)	All skim milk and butterfat sold as fluid cream, eggnog, yogurt (or any product containing 6% or more nonmilk fat which resembles these products), and cottage cheese.
Class III products	All skim milk and butterfat sold as cheese (except cottage cheese), butter, dry milk, and various other processed dairy products.
Class prices	Refers to the system under federal orders of valuing and pricing Grade A milk based on its ultimate end use. Some orders have two classes (I, II) while others, such as the Southern Michigan Order, have three classes (I, II, and III). Class I price is the price paid by handlers for milk used in Class I products; Class II price is the price paid by handlers for milk used in Class II products; Class III price is the price paid by handlers for milk used in Class III products.
Cooperative	An association of dairy farmers organized for the purposes of marketing and/or processing their milk under the provisions of the Capper-Volstead Act. The cooperative is owned and operated by the members through a Board of Directors and hired manager. The scope of activities carried on by dairy cooperatives is varied; their sphere of bargaining influence is often extensive. In many markets, cooperative associations are handlers when they market milk for their member producers.

(Continued)

TABLE 19-10 (*Continued*)

Terms	Definitions
Daily base	The volume of daily Grade A milk shipments for which a producer receives the base price. A base is usually earned each year by producing and selling grade A milk for the period of August 1–December 31. The daily base is calculated by dividing total shipments in this period by 122 days.
Equalization fund	A fund administered by market order authorities for the purpose of receiving handler milk payments and redistributing them to handlers in amounts calculated to allow each handler in a market-wide pool to pay a uniform set of prices to all producers in the marketing order area, regardless of their individual percentages of Class I, II, and III usage.
Excess price	The price producers receive for all Grade A milk shipped in a month in excess of their base allotments. The price is established at the Class III price which is the basic formula price.
Federal order market	A regulation issued by the Secretary of Agriculture which places certain requirements on the handling of milk in the area it covers. These requirements include the payment by handlers of established minimum prices for milk under a classified pricing plan and the pooling of milk payments for the purpose of paying all producers on the basis of a set of uniform or average prices.
Grade A milk	Milk produced on farms certified by the appropriate inspection authority to meet minimum structural and biological standards to ensure that all milk produced on these farms is of necessary purity for fluid consumption (i.e., a minimum of processing). Only milk of Grade A quality is included under a federal order.
Grade B price	The price which operators who produce Grade B (or manufacturing grade) receive. This is an unregulated price determined essentially in an open market environment receptive to supply and demand signals for milk of processing quality. The federal price support program does, however, act to keep this Grade B price above a certain parity-related level.
Handler	Milk processors-milk distributors who are subject to a federal order because they distribute milk in a regulated marketing area or because they process milk which may be sold in a regulated area.
Handler pool (individual handler pool)	A device for paying producers a uniform or average price for the milk they deliver to a milk handler. This price is calculated on the basis of *each handler's* use and receipts of milk and is paid to all producers served by that handler.
Location differential	An adjustment (no change or a reduction) of the price that handlers must pay into the pool for Class I milk received at their plant. The adjustment is based on the location of the plant and is designed to encourage handlers to buy and move milk efficiently in the process of meeting market needs. The differentials are designed so that all handlers can obtain their Class I milk needs at the same net price (price minus transportation costs) regardless of where the supplies originated.
Manufacturing grade milk (Grade B milk)	Milk produced under less controlled physical and biological conditions such that it is deemed suitable only for processing into dairy products (e.g., butter, powder, or cheese). Processing ensures that any physical or biological contamination is removed or rendered safe.
Market administrator	The official designated by the Secretary of Agriculture to administer a particular federal milk marketing order at the local level. The administrator and staff carry out the provisions of the order.
Market-wide pool	A device for computing and paying an average price to all producers selling to all handlers in a market order area. Total market-wide utilization and receipts are combined to achieve this purpose. (See also Handler pool.)
Minnesota-Wisconsin Series price	An estimate of the average price per cwt (3.5% butterfat) paid for manufacturing grade milk f.o.b. plants in Minnesota and Wisconsin in any given month, as reported by USDA through a monthly sampling of plants in those states. This is often called the M-W price or simply M-W.
Open base system	A plan under some federal orders whereby Grade A producers acquire a daily base of milk determined by the producers' shipments during certain months of the year. If the base is reestablished each year and if new producers are free to establish such a base allotment, the base system is called *open*; if acquisition is restricted (e.g., through the requirement of payment or waiting until increased usage or the removal of a producer with an existing base), the system is called *closed*. Two federal orders in the U.S. now use a closed base plant—called a Class I Base Plan.
Pool plant	A distributing plant, other than a producer-handler plant, which distributes on routes at least 50% of its producer milk receipts plus purchased fluid milk products as milk products. Specific regulations defining pool plants are complex, but, in general, the above definition is useful. A nonpool plant is a distributing plant which does not qualify as a pool plant.
Processor	As used in federal milk marketing orders, a handler.
Producer	For purposes of a discussion of pricing under federal order, a producer is a dairy farmer (other than a producer-handler) who is entitled to the protection and benefits of a milk order because he/she sell Grade A milk to handlers in a regulated market. (See also Producer-handler.)
Producer-handler	A person who operates a dairy farm and a milk plant from which fluid milk products are distributed in the marketing area and who receives fluid milk products only from his/her own production or by transfer from a pool plant.
Shipper	As used in federal milk marketing orders, a producer.
Skim milk	Whole milk after all the butterfat has been removed but still containing all solids-not-fat.
Super-pool premium (for Class I)	The difference between the federal order Class I price and the cooperative-negotiated higher Class I price. At times this has been over $1 per cwt. The total super-pool premium distributed among all cooperative members in the order is the per cwt premium times the Class I milk volume. This is not a part of a federal order.
Super-pool prices	The base, uniform, and adjusted uniform prices paid to cooperative members which are greater than their counterpart, federal order prices, because the total super-pool premium is allocated to these. The super-pool prices are computed by an escrow agent retained by the cooperatives participating in the super-pool and are, thus, not a part of a federal order.
Uniform price	The net value of all milk produced and sold in the market at class prices, divided by the total cwt of milk delivered. This is often referred to as the average price, market blend price, or weighted average price.

[1]*What's Behind the Grade A Dairyman's Milk Check?*, Ext. Bull. E-963, Cooperative Extension Service, Michigan State University.

STATE MILK CONTROL

Through state orders, 17 states have authority to set minimum farm prices and/or retail prices at the wholesale and retail levels. In some states, milk control commissions determine not only what farmers are to be paid but what price the stores can charge customers.

In setting minimum farm prices, state control agencies often operate in a manner similar to federal milk orders. Classified pricing principles are used, and prices are set for a particular market and not necessarily for the whole state. Retail prices are based on the cost of processing and distribution. It is noteworthy that fewer and fewer state milk commissions set retail milk prices. The foes of retail pricing point out that, on the average, retail price setting results in a lower price to the farmer than where retail prices are let alone.

Because of their inability to cope with out-of-state milk, state milk controls will likely decline in importance in the future; they will be replaced by federal milk orders.

COOPERATIVES

The practice of dealing separately with a large number of producers led to dissatisfaction in a number of cases. To rectify this situation, cooperatives were organized. These cooperative associations are of two general types:

1. Bargaining associations which do not handle any milk, but make all business arrangements.
2. Associations which process and distribute milk or assemble it for fluid use.

About 75% of the total deliveries of milk to plants and dealers in the United States is handled by cooperatives. In 1989, 260 U.S. cooperatives handled dairy products with a gross value of $19.3 billion.

In the last decade, a large number of small cooperatives have merged together to form regional organizations that offer a more powerful position for bargaining and more efficient use of marketing channels.

Cooperatives allow the dairy producers of a given area to integrate the various aspects of marketing. Procurement, assembling, marketing, and routing of milk are all handled by the cooperative. Thus, members of a cooperative are assured of a market for their milk.

OTHER REGULATORY PROGRAMS

Because of the essential nature of milk, plus the fact that it is easily contaminated and a favorable medium for bacterial growth, it is inevitable that numerous regulatory programs have evolved around it—federal, state, and local—some having been designed to control prices and assure a reasonably uniform flow of milk, and others for sanitary reasons.

SANITARY REGULATIONS

The sanitation of milk and dairy products is assured by the enforcement of sanitary regulations by federal, state, and local authorities.

There are more than 15,000 state, county, local, and municipal health and sanitation jurisdictions in the United States. Inspectors from these agencies regularly visit farms, plants, and stores, making sure dairy products keep their high quality. Unfortunately, from area to area, there are a bewildering number of different regulations, with the result that milk going to more than one city market is often subjected to duplication and confusion in inspection. Also, sanitary and health regulations have sometimes been used as barriers to keep milk out of a certain area for competitive reasons.

In 1923, the U.S. Public Health Service (USPHS) established an Office of Milk Investigations, and in 1924, the USPHS published its first Grade A pasteurized milk ordinance. Subsequently, this regulation has been revised several times.

Producers are issued permits allowing them to ship Grade A milk. The permit is revoked if either the bacteria count of raw milk exceeds 100,000 per milliliter or the cooling temperature exceeds 40°F in three of the last five samples.

The standard plate count of Grade A pasteurized milk may not exceed 20,000 per milliliter nor the coliform count 10 per milliliter in three of the last five samples or the processor's permit will be revoked.

In addition to cleanliness and freedom from mastitis, temperature is important in processing quality milk. Bacteria cannot reproduce effectively below 40°F; so, dairy farmers should cool milk below 40°F as quickly as possible. By law, all fluid milk sold for human consumption must be pasteurized; so, at dairy processing plants, milk is pasteurized to kill disease-causing organisms. It may be pasteurized at either (1) 145°F for 30 minutes, or (2) 161°F for 15 seconds. Additionally, milk should be refrigerated while in the store or in the home.

The Food and Drug Administration (FDA) is charged with inspecting dairy products and processing plants for contamination and adulteration.

Presently, many cooperatives and some milk dealers pay a premium to dairy farmers for producing high quality milk. A variety of premiums and penalty bases are in use. The following standards for milk quality for the highest premiums are proposed as reasonable goals:

1. Standard plate count (SPC) or plate loop count (PLC), less than 10,000 per ml.
2. Preliminary incubation count (PIC), less than 20,000 per ml.
3. Somatic cell count (SCC), less than 200,000 per ml.
4. Antibiotic and chemicals, no detectable levels.
5. Temperature, 40°F or lower.
6. Odors and flavors, none objectionable.
7. Acid degree value, 1.0 or below.
8. Milkfat, 3.25% or above.
9. Protein, 3.2% or above.
10. Farm inspection score, 90 or above.

STANDARDS AND GRADES

The U.S. Department of Agriculture has responsibility for the development of standards and grades for milk and dairy products.

The grades, classes, and uses of milk are shown in Fig. 19–9.

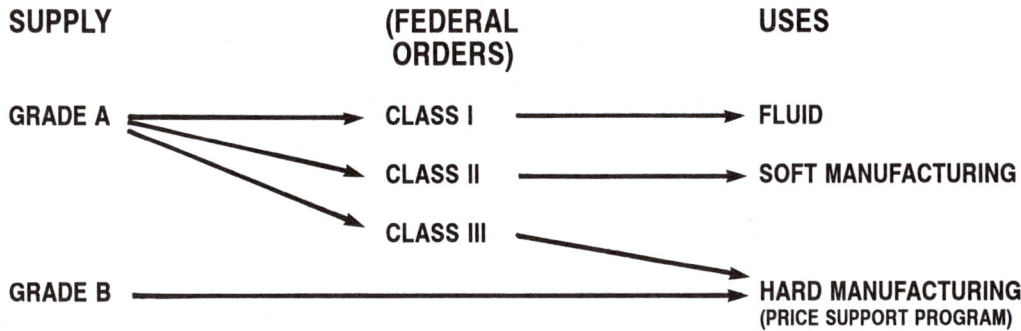

Fig. 19–9. Milk quality flows, showing grades, classes, and uses of milk.

About 90% of the nation's milk supply is Grade A, and about 45% of all Grade A milk that is sold is used for fluid milk products (beverage milk). Milk used for fluid products is designated (by federal orders) Class I. Most orders have two other classes: Class II includes milk used for soft products, including fluid cream, ice cream, cottage cheese, and yogurt; while Class III includes milk used for hard products, including cheese, butter, and nonfat dry milk.

The major dairy products for which the USDA has established grades, and the proportion graded are shown in Table 19–11.

It is expected that more and more dairy products will be federally graded.

TABLE 19–11
SELECTED DAIRY PRODUCTS GRADED BY USDA[1]

Product	Volume	Share of U.S. Production
	(million lb)	(%)
Butter	107.7	9.75
Cheese	5.3	0.1
Nonfat dried milk	16.2	1.53

[1]Source: USDA.

STATE TRADE PRACTICE LAWS

For almost 30 years, there has been considerable concern about competitive practices in the sale of fluid milk products and ice cream. Among the unfair trade practices sometimes observed or suspected in the marketing of milk are: discriminatory price cutting, secret rebates, loans, advertising rebates, furnishing and servicing equipment, and the giving of gifts and free signs.

Without doubt, more states will enact dairy fair trade practice laws, and this approach will be used by the state as a substitute for complete milk control.

METHODS OF PRICING OR PAYING FOR MILK

Economists refer to the different systems of paying for milk as *price plans*. These plans, which in actual practice generally involve two or more plants—for example, pricing based on (1) class, (2) grade, and (3) base surplus—are:

1. **Flat price plan.** This was the common method up to World War I. The milk producer was paid a uniform price for all milk sold, regardless of quality or the use made of it.

2. **Use classification plan.** Most marketing orders established two use classes—Class I and Class II.

Class I milk generally includes milk used in fluid form such as whole fluid milk, or milk for creamed drinks which must be made from milk approved by local health authorities. Generally speaking, Class I prices are 10 to 15% higher than Class II prices.

Class II milk usually includes milk in excess of fluid needs, which is used to make manufactured dairy products—primarily butter, nonfat dry milk, and cheese.

On some markets, a further division is made, primarily for milk going into cottage cheese, with the result that there are three classes of milk—Class I, Class II, and Class III.

3. **Blend prices.** When dealers buy according to classification prices, they may pay producers a blend price. The blend is an average of class prices weighted by the volume of milk in each class, usually quoted at a specific point and for a specific test of milk.

4. **Quality grade plan.** Frequently, the terms Grade A and Grade B (Grade B is usually called *Manufacturing Grade Milk*) are encountered in milk marketing. Although there may be some local variations in their use, Grade A usually refers to milk produced under conditions which make it acceptable for fluid use in a given market. Grade B often refers to milk produced under conditions which do not make it acceptable for fluid milk use—it is manufacturing milk.

The production of Grade A milk relative to that of Grade B milk has been increasing in recent years (see Fig. 19–10). In 1987, U.S. farmers marketed 139,058 million pounds of milk, of which 123,768 million pounds, or 89%, was Grade A, and 15,290 million pounds, or 11%, was Grade B.

GRADE A AND GRADE B MILK MARKETINGS

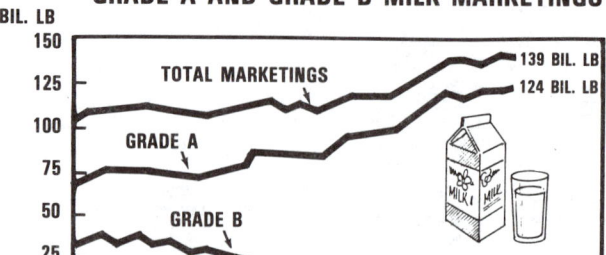

Fig. 19–10.   Grade A and Grade B milk marketings, 1960 to 1987. (Source: USDA)

5. **Base surplus plan.** The base surplus plan (or base rating plan) is designed to encourage that a uniform supply of milk be available. It compensates the producers who maintain a high fall production, when more milk is needed. The base period is established during the lowest production months, usually over a period of 3 to 6 months. Then, a producer's base is established by the average amount of milk delivered during the base period. The producer's base may be modified from time to time.

6. **Butterfat test price plan.** The butterfat test of milk affects the price. The common practice is to establish a price for 100 lb of milk of a specified butterfat test. Usually 3.5% butterfat is the basis for pricing, although several markets have established their base as high as 4% butterfat. Then, a price differential (per point or 0.1%) is set up for milk testing above or below this amount.

7. **Solids-not-fat price plan.** Today, the emphasis on the food value of milk is shifting from fat content to the other solids, especially protein. This is feasible because tests for solids-not-fat have been devised, and these are proving practical for field use. It is anticipated that this system of pricing milk will expand in the future.

On the average, whole milk contains about 2¼ lb of solids-not-fat for each pound of milk fat. Thus, milk testing 4% butterfat contains approximately 9 lb of solids-not-fat, to a total of 13 lb of solids per hundredweight.

8. **Component pricing plans.** Although various component pricing plans being formulated or adopted are not uniform, they continue to give price credit for butterfat in farm milk; in addition, they give credit for solids-not-fat, including protein. Some also involve end product pricing in which farm milk prices are based on the yield and market value of cheese and other dairy products that can be manufactured from the milk. Most also either (1) establish a maximum somatic cell count at which component premiums will be paid, or (2) pay a premium of 6 to 12¢ per hundredweight of milk for minimum somatic cell counts.

On the average, farm milk contains about 3.7% butterfat and 8.55% solids-not-fat, including 3.2% protein. A one point (0.1%) change in milk fat test is normally associated with a 0.4 point (0.04%) change in solids-not-fat. However, considerable variation in this average relationship does occur. Component pricing takes into consideration the value of variations

in solids-not-fat as well as fat; thus, its advocates feel that it would correct an inequity to dairy farmers under current butterfat differential pricing.

In 1988, the Federal Milk Marketing Administrator reported that 48% of all federal order milk was eligible for multiple component pricing (MCP), meaning that if the milk meets the minimum component, quality, and/or other requirements established by producers, cooperatives, or plants, payment may be based upon two or more components. If we include the non-federal order milk eligible for MCP, then approximately 60% of U.S. milk production is eligible for MCP. More and more component pricing will be implemented.

9. **Gallon or quart plan.** Occasionally, a producer supplies milk to a distributor on a per gallon or per quart basis. Since average milk weighs 2.15 lb to the quart and 8.6 lb to the gallon, 100 lb of milk would be equivalent to 46.5 qt or 11.6 gal. Thus, one can easily compute the possible returns from selling milk by different methods.

10. **Special milks.** Certain milks are sold under special labels. Among them are:

a. **Certified milk.** This is milk that is produced under special sanitary conditions prescribed by the American Association of Medical Commissioners. It is sold at a higher price than ordinary milk.

b. **Golden Guernsey milk.** Golden Guernsey milk is produced by owners of purebred Guernsey herds who comply with the regulations of the American Guernsey Cattle Association. Such milk is sold under the trade name *Golden Guernsey*, at a premium price.

c. **All-Jersey milk.** This is produced by registered Jersey herds whose owners comply with the regulations of the American Jersey Cattle Club. It is sold at a premium price under the trademark *All Jersey*.

PROFITABILITY OF DAIRY PROCESSING FIRMS

Like most businesses, dairy processing firms are owned by people; and most people want to make money, which is quite normal.

There are three ways of gauging the profitability of firms: (1) net return on sales, (2) return on assets, and (3) return on net worth. Table 19–12 gives these ratios for the fluid milk industry for 1984, 1988, 1989, and 1990.

TABLE 19–12
KEY PROFITABILITY AND EFFICIENCY RATIOS
OF THE FLUID MILK INDUSTRY[1]

	1984	1988	1989	1990
	(%)	(%)	(%)	(%)
Profitability				
Return on sales	1.0	1.0	1.0	0.8
Return on assets	4.1	4.4	4.4	2.9
Return on net worth	9.5	9.8	9.8	7.4
Efficiency				
Assets to sales	22.6	24.3	24.0	27.3
Sales to net working capital .	22.6	21.9	25.7	21.7

[1]Source: *1991 Milk Facts,* Milk Industry Foundation, Washington, DC, p. 10.

As shown in Table 19–12, the return on sales was about 1%. Also, the earnings on assets and earnings on net worth were low.

Of course, the volume of sales (the quantity of food and kindred products processed in a year) and the efficiency of operations make it possible for the fluid milk industry to operate on these comparatively small margins.

MANUFACTURED MILK PRODUCTS

Manufactured dairy products utilized about 62% of U.S. milk production.

The production of manufacturing grade milk is primarily centered in the Midwest and Great Lakes area. Thus, with the exception of ice cream making, the processing of manufactured dairy products—butter, nonfat dry milk powder, cheese, evaporated and condensed milk, and other products of minor importance—is concentrated near those areas of production.

NUMBER AND SIZE OF PLANTS

Since the 1940s, the number of milk manufacturing plants in the United States has been decreasing while the output per plant has been increasing (see Table 19–13). As shown, between 1948 and 1987, milk manufacturing plants became fewer and bigger. The statistics: During the 39-year period (1948 to 1987), 27.2 million more pounds of milk was manufactured, the number of plants decreased from 9,737 to 1,933—7,804, or 80%, and the plants became 7.5 times larger.

TABLE 19–13
TOTAL MILK MANUFACTURED AND NUMBER OF PLANTS MANUFACTURING MILK, 1948–87

Item	1948	1987	Change
			(%)
Total milk manufactured[1] . . . (million lb)	54.7	81.9	+ 50
Number of plants[2]	9,737	1,933	– 80
Average annual volume per plant (whole milk equivalent) . . . . . . (thousand lb)	5,622	42,392	+654

[1]USDA sources.

[2]*1988 Dairy Producer Highlights,* National Milk Producers Federation, p. 14.

USES OF MILK

The uses of milk have already been listed (see Table 19–5). A few pertinent points relative to each of the manufactured products will be presented in the sections which follow.

BUTTER

Butter is made from cream. As marketed, it consists of about 80% milk fat. The remainder is water, salt, and traces of other substances.

Fig. 19–11 shows the per capita consumption of butter and margarine, from 1910 to 1987. As noted, the per capita consumption of margarine surpassed butter in 1957. In 1987, the per capita consumption of butter was 4.7 lb, whereas the per capita consumption of margarine was 10.5 lb.

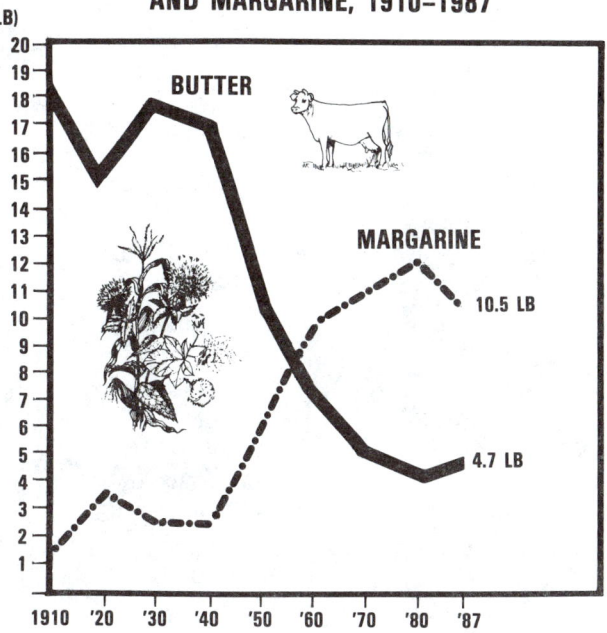

PER CAPITA CONSUMPTION OF BUTTER AND MARGARINE, 1910–1987

Fig. 19–11. The per capita consumption of butter and margarine from 1910 to 1987. (Source: USDA)

State and national laws and regulations bar additives to or changes in butter. Nevertheless, experiments are being conducted on low butterfat spreads. Besides their appeal to homemakers, these products would be better able to compete with oleomargarine.

Wisconsin, the leading state in butter production, accounted for 25% of U.S. butter production in 1990, followed by California, with 21%.

CASEIN

Casein, which is the major protein of milk, is found only in milk. It is obtained by acid or rennet coagulation of defatted milk. Casein contains a minimum of 80% crude protein. It gives milk its white color.

Casein is used as an ingredient of coffee whitener and whipped toppings, in baked goods, and as the main source of protein in the manufacture of meat analogs and in the protein supplementation of some meat products.

In 1987, the U.S. imported 108,136 metric tons of casein, which far exceeded imports of dried milk (1,301 metric tons), and of butter (905 metric tons).

CHEESE

Cheese is made by (1) exposing milk to specific bacterial fermentation, or (2) treating with enzymes, or both methods, to coagulate some of the proteins.

Milk can be, and is, processed into many different varieties of cheese. Some are made from whole milk, others from milk that has had part of the fat removed, and still others from skimmed milk. American types of cheese (Cheddar, Colby, washed curd, stirred curd, Monterey, and Jack) make up 60% of the nation's cheese output. The most important variety produced from skimmed milk is cottage cheese. Other important types of cheese are Italian (mostly soft varieties), Swiss, Muenster, brick, blue, and processed cheese.

Fig. 19–12. An assortment of cheeses. There are more than 2,000 different named cheeses. (Courtesy, United Dairy Assn., Rosemont, IL)

In 1987, 30% of all the milk used in manufactured dairy products was processed into cheese (exclusive of cottage cheese). The rising popularity of pizza in the United States accounts for much of the increase in cheese production and consumption in recent years.

The leading states in the production of cheese, excluding cottage cheese, by rank are: Wisconsin (with 32% of the total production in 1990), Minnesota, New York, California, Iowa, Pennsylvania, Missouri, Idaho, and South Dakota.

CONDENSED AND EVAPORATED MILK

The primary products within this category are evaporated milk and condensed milk packed in cans for consumer use, and condensed whole and skimmed milk shipped in bulk. Condensed and evaporated milk are manufactured by removing a major portion of the water from the whole milk in a machine called a vacuum pan. Condensed milk is further treated by the addition of large amounts of sugar.

CREAM

Cream is made by concentrating the fat portion of milk. Prior to the advent of the cream separator, this was accomplished by gravity separation. Today, it is done by passing milk through a cream separator. In commerce, whipping cream contains about 40% fat; coffee or table cream, 18 to 20%; and half-and-half, 12%.

DRIED MILK (WHOLE MILK, SKIMMED MILK, AND WHEY)

Among the dried milk products produced from milk are nonfat dried milk (skimmed milk), for both human food and animal feed; dried whey, for both human food and animal feed; and dried whole milk.

In 1987, the following quantities of these dried products were produced in the United States: 1,067 million pounds of dry whole milk, 1,059 million pounds of nonfat dried milk, and 1,034 million pounds of dried whey.

ICE CREAM AND SIMILAR FROZEN DESSERTS

Fig. 19–13. Meringue nut ice cream torte—a frozen dessert. (Courtesy, American Dairy Assn., Chicago, IL)

Currently, 99% of all frozen desserts in the United States consist of ice cream, ice milk, sherbet, and mellorine (made with a vegetable fat base). Other frozen desserts include frozen custard, frozen malted milk, artificially sweetened ice cream and ice milk, and water ices.

CULTURED MILK PRODUCTS

Numerous cultured and acidified milk products are sold as specialty dairy products. Among such products are yogurt, cultured buttermilk, cultured sour cream, and acidophilus milk.

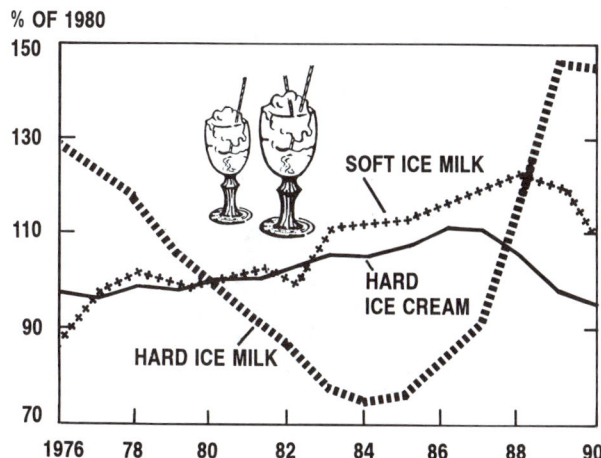

% OF 1980

Fig. 19–14.   U.S. frozen dessert sales 1976 to 1990. (Source: *Dairy Situation and Outlook Yearbook*, USDA, ERS, DS-431, August 1991, p. 7)

COMMODITY CREDIT CORPORATION (CCC)

In order to provide price supports for manufactured dairy products, the Commodity Credit Corporation (CCC) purchases butter, cheese, and nonfat dry milk for the various programs of the USDA.

Table 19–14 shows the CCC purchase prices for dairy products for selected dates from 1949 to 1989.

TABLE 19–14
COMMONDITY CREDIT CORPORATION
DAIRY PRODUCT PURCHASE PRICES AND SUPPORT PRICES, SELECTED DATES[1]

Effective Date	Support Price at National Standard Fat Test	Butter — Bulk Grade A New York	Cheddar Cheese — Grade A or Higher	Cheddar Cheese — Barrels Extra Grade	Nonfat Dry Milk Extra Grade Spray
	(¢/lb)	(¢/lb)	(¢/lb)	(¢/lb)	(¢/lb)
7/27/49	3.14	62.00	31.75		12.25
1/1/50	3.07	60.00	31.00		12.50
4/1/60	3.06	58.75	32.75		13.40
4/1/70	4.66	70.75	52.00		27.20
4/1/80	12.36	143.25	132.50	129.50	89.50
4/1/85	12.10	143.25	128.75	124.50	84.75
1/1/87	11.35	137.75	122.50	118.25	78.75
1/1/88	10.60	132.00	115.25	111.25	72.75
4/1/89	11.10	132.00	120.25	116.25	79.00

[1]Source: *Dairy Producer Highlights*, National Milk Producers Federation, Arlington, VA, p. 32.

Table 19–15 shows how the dairy products purchased by the CCC were utilized in 1987 and 1988.

TABLE 19–15
UTILIZATION OF CCC DAIRY PRODUCTS IN SELECTED YEARS[1, 2]

	Butter[3] 1987	Butter[3] 1988	Cheese[4] 1987	Cheese[4] 1988	Nonfat Dry Milk 1987	Nonfat Dry Milk 1988
	(million lb)	(million lb)	(million lb)	(million lb)	(million lb)	(million lb)
Domestic sales:						
For restricted use	0.0	0.0	0.0	0.0	0.0	0.0
For unrestricted use	0.1	0.1	21.9	0.2	70.8	1.1
Exchange for UHT milk	0.0	0.0	0.0	0.1	1.1	0.0
Manufacture of casein					0.0	0.0
Export sales:						
Noncommercial	31.0	41.1	18.0	24.2	387.9	117.4
Defense Department	4.7	3.7	1.3	2.8	3.5	2.7
Barter	0.0	0.0	0.0	0.0	0.0	0.0
Export Incentive Program	0.0	0.0	0.0	0.4	0.7	15.4
Domestic donations:						
Needy	58.8	66.6	416.9	129.9	99.8	61.8
Schools and institutions	102.2	110.0	212.6	205.1	49.2	41.1
Department of Defense	1.8	2.8	5.0	4.2	0.0	0.0
Veterans Administration	0.0	0.0	3.5	0.0	0.0	0.0
Bureau of Prisons	2.9	2.1	1.6	1.6	0.0	0.3
Foreign donations:						
Title II, PL–480	3.4	0.0	4.3	0.0	359.3	102.2
Section 416	46.2	0.0	12.4	0.0	63.3	0.0
Total Utilizations[2]	250.9	227.4	697.5	368.5	1,036.2	342.0

[1]Source: *Dairy Producer Highlights*, National Milk Producers Federation, Arlington, VA, p. 34.

[2]To convert pounds to kilograms, multiply by 0.454.

[3]Butter includes anhydrous milk fat and/or butter oil.

[4]Cheese amounts include Mozzarella cheese.

IMPORTS AND EXPORTS

Imports of a number of dairy products are restricted by specific import quotas; among such products are several types of cheese, butter, butteroil, butterfat mixtures, ice cream, frozen cream, nonfat dry milk, dried buttermilk and whey, evaporated milk, condensed milk, chocolate crumb, and animal feed with milk solids. Although not formally restricted, certain other dairy products may be limited by agreement between the United States and the exporting country.

As long as domestic prices are above world prices, and world supplies are ample, exporting countries will look to the United States as a possible market. As a result, import pressure will persist; yet, it is expected that imports of many commodities will continue to be limited by quotas.

Exports of dairy products are rather small. In 1987, U.S. exports of dairy products on a milk equivalent basis amounted to only 0.96% of the total U.S. milk supply. The three main U.S. dairy product exports in 1987, ranked in descending order of tonnage, were: nonfat dry milk, 311,852 metric tons; cheese, 19,560 metric tons; and butter, 7,472 metric tons. The exports of nonfat dry milk were greater than the exports of all other dairy products combined.

In recent years, imports of dairy products have exceeded exports in monetary value, resulting in a trade deficit of $221 million in 1985 to a high of $227 million in 1989. The trade deficit is pointed up in Fig. 19–15 and Table 19–16.

TABLE 19–16
DAIRY PRODUCT IMPORTS AND EXPORTS[1]

Year	Imports	Exports	Balance
	(million $)	(million $)	(million $)
1985	643	422	–221
1986	666	434	–232
1987	659	496	–163
1988	712	541	–171
1989	767	490	–277

[1]Source: *Dairy, Livestock, and Poultry: U.S. Trade and Prospects,* USDA, Foreign Agricultural Service, FDLP, 12–89, December, 1989, p. 2.

Exports of dairy products will continue to be influenced by the availability of surplus products and foreign policy. A more active role in meeting food deficiencies in the less developed countries of the world could increase total demand for dairy products and demand for export products.

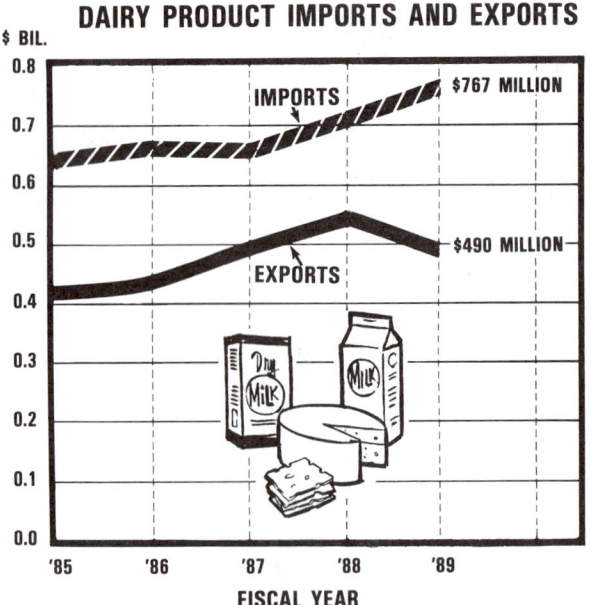

Fig. 19–15.   Dairy product imports and exports. (Source: *Dairy, Livestock, and Poultry: U.S. Trade and Prospects*, USDA, Foreign Agricultural Service, FDLP 12–89, December 1989, pp. cover and 2)

OUTLOOK

Per capita consumption of milk and most dairy products appears to have reached a rather stable stage. The most important factors affecting the future demand for dairy products will continue to be changes in population, income, consumer preferences, and new products.

It is expected that per capita consumption of various dairy products will follow the trend of recent years; products high in fat will decline in per capita consumption, while those low in fat will increase. The proportion of milk consumed in fluid form will decrease; per capita consumption of butter and evaporated milk will likely decline on a gradual basis; and low fat fluid milk, nonfat dried milk, low fat yogurt and frozen desserts, cheese, and sour cream dressings are likely to increase in per capita consumption.

Without doubt, new low fat dairy products and nondairy substitutes will replace some of the consumption of similar products higher in fat. The use of synthetic milk will increase.

Increasingly, the U.S. pricing system will (1) differentiate and price milk on a component basis, and (2) pay a premium for higher quality milk.

The operations involved in producing and marketing milk and dairy products will continue the current trend to bigness, fewer numbers, more mechanization, environmental control, and higher quality products.

QUESTIONS FOR STUDY AND DISCUSSION

1. Discuss the importance of the dairy industry in the United States, including its importance in providing jobs for Americans.

2. Table 19–1 shows that the marketing bill for dairy products exceeds the farm value of milk. Is this justified?

3. Discuss the main costs involved in the production of milk.

4. Of what value to a dairy producer is the milk-feed (concentrate) ratio?

5. Labor costs per cow and per 100 lb of milk have been lowered dramatically since 1925. How has this been accomplished?

6. Describe the relationship between milk production and prices with respect to seasonal fluctuations.

7. How stable are milk prices in comparison to hog and beef cattle prices?

8. Why has *home-use and fed* milk declined in recent years?

9. How is the milk marketed by farmers used today? What has caused per capita consumption of milk to decline in recent years?

10. Define: (a) inactive market activity, (b) steady price trend, (c) heavy offering, (d) very good demand, and (e) undertone.

11. What are the three stages of moving milk from the farm to the consumer?

12. Outline the distribution channels of milk. Why has home delivery of milk declined so drastically in recent years? Why has the use of both plastic and gallon size containers increased so dramatically in retail stores?

13. What caused government controls of milk to evolve?

14. What are federal milk marketing orders?

15. What roles do individual states play in regulating milk?

16. Should the states regulate milk prices, or should this be handled only through federal milk marketing orders?

17. Define each of the following: base price, blend price, Class I products, Class II products, Class III products, Grade A milk, and Grade B milk.

18. What is a cooperative dairy association? What roles does it fill in the dairy industry?

19. Discuss the sanitary regulations governing milk and dairy products.

20. Name the major dairy products for which the USDA has established grades.

21. Discuss the advantages and disadvantages of each method of paying for milk.

22. Analyze the profitability of dairy processing firms based on (a) return per dollar of sales, (b) return on assets, and (c) return on net worth. Are they doing all right?

23. What is meant by the term *manufactured milk products*?

24. Discuss the per capita consumption of butter and margarine since 1910.

25. How is cheese made? What is the leading state in the production of cheese? Does this leading state have a commanding lead in the production of cheese?

26. What are concentrated milk, condensed milk, and evaporated milk?

27. What is the chief difference between whipping cream, coffee cream, and half-and-half?

28. Name and describe three popular frozen desserts.

29. What is the function of the Commodity Credit Corporation in relation to the dairy industry?

30. Briefly discuss the role of imports and exports in the dairy industry in the United States.

31. Are you optimistic or pessimistic about the future of the U.S. dairy industry? With what dairy products will the increase and decrease in per capita consumption most likely occur?

SELECTED REFERENCES

Title of Publication	Author(s)	Publisher
Federal Milk Marketing Order Program, The, Mkt. Bull. No. 27		U.S. Department of Agriculture, Washington, DC, 1963
Fluid Milk Marketing	G. M Beal H. H. Bakken	Mimir Publishers, Madison, WI, 1956
Grade "A" Pasteurized Milk Ordinance, Pub. 229		U.S. Public Health Service, Washington, DC, 1965
Milk Production and Processing	H. J. Judkins H. A. Keener	John Wiley & Sons, Inc., New York, NY, 1963
Organization and Competition in the Dairy Industry, Tech. Study No. 3		National Commission on Food Marketing, Washington, DC, 1966
Organization and Competition in the Midwest Dairy Industries	S. W. Williams, *et al.*	Iowa State University Press, Ames, 1970

Fig. 19-16. Milk room, showing bulk holding tank for cooling milk, preferably to 40°F, prior to loading into a milk-collecting tanker. (Courtesy, National Milk Producers Federation, Arlington, VA)

Fig. 20–1. Dairy producer with computer. This computer is connected to the automated feeders. Twice a day the printout shows the feed consumed by each cow. Rations can be changed in keeping with the cow's production. (Courtesy, Brown Swiss Cattle Breeders' Assn. of the U.S.A., Beloit, WI)

Contents

CHAPTER

20

BUSINESS ASPECTS OF DAIRYING

With the increase in specialization and size of enterprises, the business aspects of dairy production have become more important. More capital is required, competent management is in demand, records are essential, computers have come in, and such things as tax management, estate planning, and liability require more attention. With this transition, dairy producers have become more sophisticated.

TYPES OF BUSINESS ORGANIZATIONS

The success of today's dairying is very dependent on the type of business organization. No one type of organization is superior under all circumstances; rather, each situation must be considered individually. The size of the operation, the family situation, and the objectives—all these, and more, are important in determining the best way in which to organize the dairy enterprise.

Three major types of business organizations are commonly found on fairy farms: (1) the sole proprietorship; (2) the partnership; and (3) the corporation. Among the factors which should be considered when deciding which business form best fits a given set of circumstances are the following:

1. Which type of organization is most likely to be looked upon favorably from the standpoint of more credit and capital?

2. How much capital will be required of each individual involved?

3. Are there tax advantages to be gained from the business organization?

4. Is expansion of the business feasible and facilitated?

5. Which type of organization reduces risks and liability most?

6. Which type of organization can be terminated most easily and readily?

7. Which type of ownership provides for the most continuity and ease of transfer?

8. What costs for legal and accounting fees are involved, in setting up the organization and in the preparation of the annual reports required by law?

9. Who will manage the business?

Most dairy enterprises are operated as sole proprietorships, not necessarily because this is the best type of organization, but with no effort to form some other type of organization it naturally results. Both the partnership and the corporation, which require special planning and effort to bring about, are well suited to the operation of large dairies.

PROPRIETORSHIP (INDIVIDUAL)

A sole proprietorship is a business which is owned and operated by one individual.

This is the most common type of business organization in U.S. farming—86.9% of the nation's farms are individually owned. Under the sole proprietorship, or individual ownership, one person controls the business. Proprietors may not provide all the capital used in the business; in fact, they usually do not. However, they have sole management and control of the operation, although this may be modified and delegated somewhat through lease agreements, contracts, etc. Basically, sole proprietors get all the profits of the business; likewise, they must absorb all the losses.

In comparison with other forms of organization, the sole proprietorship has two major limitations: (1) It may be more difficult to acquire new capital for expansion; and (2) not much can be done to provide for continuity and to keep the present business going as a unit, with the result that it usually goes out of existence with the passing of the owner.

PARTNERSHIP (GENERAL PARTNERSHIP)

A partnership is an association of two or more persons who, as co-owners, operate the business. About 10% of U.S. farms are partnerships.

Most dairy farm partnerships involve family members who have pooled land, machinery, working capital, and often their labor and management to operate a larger business than would be possible if each member limited the operation to one person's resources. It is a good way in which to bring a son or daughter, who is usually short on capital, into the business, yet keep the parent in active participation. Although there are financial risks to each member of such a partnership, and potential conflicts in management decisions, the existence of family ties tends to minimize such problems.

In order for a partnership to be successful, the enterprise must be sufficiently large to utilize the abilities and skills of the partners and to compensate them adequately in keeping with their respective contribution to the business.

A partnership has the following **advantages:**

1. **Combining resources.** A partnership often increases returns from the operation due to combining resources. For example, one partner may contribute labor and management skills, whereas another may provide the capital. Under such an arrangement, it is very important that the partners agree on the value of each person's contribution to the business, and that this be clearly spelled out in the partnership agreement.

2. **Equitable management.** Unless otherwise agreed upon, all partners have equal rights, regardless of financial interest. Any limitations, such as voting rights proportionate to investments, should be a written part of the agreement.

3. **Tax savings.** A partnership does not pay any tax on its income, but it must file an informational return. The tax is paid as part of the individual tax returns of the respective partners, usually at lower tax rates.

4. **Flexibility.** Usually, the partnership does not need outside approval to change its structure or operation – the vote of the partners suffices.

Partnerships may have the following **disadvantages:**

1. **Liability for debts and obligations of the partnership.** In a partnership, each partner is liable for all the debts and obligations of the partnership.

2. **Uncertainty of length of agreement.** A partnership ceases with the death or withdrawal of any partner, unless the agreement provides for continuation by the remaining partners.

3. **Difficulty of determining value of partner's interest.** Since a partner owns a share of every individual item involved in the partnership, it is often very difficult to judge value. This tends to make transfer of a partnership difficult. This disadvantage may be lessened by determining market values regularly.

4. **Limitations on management effectiveness.** This is due to personal differences among partners, and the responsibility of each partner for the acts of the other partners.

The above is what is known as a partnership or general partnership. It is characterized by (1) management of the business being shared by the partners, and (2) each partner being responsible for the activities and liabilities of all of the partners, in addition to his/her own activities within the partnership.

LIMITED PARTNERSHIP

A limited partnership is an arrangement in which two or more parties supply the capital, but only one partner is involved in the management. This is a special type of partnership with one or more *general partners* and one or more *limited partners.*

The limited partnership avoids many of the problems inherent in a general partnership and has become the chief legal device for attracting outside investor capital into farm ventures. As the term implies, the financial liability of each partner is limited to his/her original investment, and the partnership does not require, and in fact prohibits, direct involvement of the limited partners in management. In many ways, a limited partner is in a similar position to a stockholder in a corporation.

A limited partnership must have at least one general partner who is responsible for managing the business and who is fully liable for all obligations.

The **advantages** of a limited partnership are:

1. It facilitates bringing in outside capital.
2. It need not dissolve with the loss of a partner.
3. Interests may be sold or transferred.
4. The business is taxed as a partnership.
5. Liability is limited.
6. It may be used as a tax shelter.

The **disadvantages** of a limited partnership are:

1. The general partner has unlimited liability.
2. The limited partners have no voice in management.

CORPORATIONS

A corporation is a device for carrying out a farming enterprise as an entity entirely distinct from the persons who are interested in and control it. Each state authorizes the existence of corporations. As long as the corporation complies with the provisions of the law, it continues to exist – irrespective of changes in its membership.

Until about 1960, few U.S. farms were operated as corporations. In recent years, however, there has been increased interest in the use of corporations for the conducting of farm business. Even so, only about 1% of U.S. farms use the corporate structure.

From an operational standpoint, a corporation possesses many of the privileges and responsibilities of a real person. It can own property; it can hire labor; it can sue and be sued; and it pays taxes.

Separation of ownership and management is a unique feature of corporations. The owners' interest in a corporation is represented by shares of stock. The shareholders elect the board of directors who, in turn, elect the officers. The officers are responsible for the day-to-day operation of the business. Of course, in a close family corporation, shareholders, directors, and officers can be the same persons.

The major **advantages** of a corporate structure are:

1. It provides continuity despite the death of a stockholder.
2. It facilitates transfer of ownership.
3. It limits the liability of shareholders to the value of their stock.
4. It may make for some savings in income taxes.

The major **disadvantages** of a corporation are:

1. It is restricted to doing only what is specified in its charter.

2. It must register in each state.

3. It must comply with stipulated regulations which involve considerable paperwork and expense.

4. It is subject to the hazard of higher taxes.

5. It is possible to lose control.

FAMILY OWNED (PRIVATELY OWNED) CORPORATION

Still another type of corporation is family owned (privately owned). It enjoys most of the advantages of its generally larger outside investor counterpart, with few of the disadvantages. The chief **advantages** of the family-owned corporation over a partnership arrangement are:

1. **It alleviates unlimited liability.** For this reason, a lawsuit cannot destroy the entire business and all the individual partners with it.

2. **It facilitates estate planning and ownership transfer.** It makes it possible to handle the estate and keep the business in the family and going if one of the partners should die. Each of the heirs can be given shares of stock — which are easy to sell or transfer and can be used as collateral to borrow money — while leaving the management of the enterprise to those heirs interested in operating it, or even outsiders.

TAX-OPTION CORPORATION (SUBCHAPTER S CORPORATION)

Instead of paying a corporate tax, a corporation with no more than 35 stockholders may elect to be taxed as a partnership, with the income or losses passed directly to the shareholders, each of whom pays taxes on his/her share of the profits. This special type of corporation is variously referred to as a *tax-option* corporation, *subchapter S* corporation, pseudocorporation, or elective corporation.

For income tax purposes, the owners of a tax-option corporation are taxed as if they were a partnership. That is, income earned by the corporation passes through the corporation to the personal income tax returns of the individual shareholders. Thus, the corporation does not pay any income tax. Instead, the shareholders pay tax on their share of corporate income at their individual tax rate; and the shareholders report their share of long-term capital gains and receive their deductions therefor. Although each shareholder's portion of any corporate losses from current operations is deducted from their personal return, capital losses incurred by the corporation cannot be passed through to the shareholders.

Thus, there are some very real advantages to be gained from a subchapter S or tax-option corporation. However, in order to qualify as a subchapter S corporation, the following requisites must be met:

1. There cannot be more than 35 stockholders.

2. All stockholders must agree to be taxed as a partnership.

3. Nonresident aliens cannot own stock.

4. There can be only one class of stock.

5. Not more than 20% of the gross receipts of the corporation can be from royalties, rents, dividends, interest, or annuities plus gains from sale or exchange of stock and securities; and not more than 80% of the gross receipts can be from sources outside the United States.

CAPITAL REQUIREMENTS

Those thinking of becoming livestock operators inevitably ask: "How much money will it take, and what can I make?"

It takes a lot of capital to own and/or operate a modern livestock establishment. In 1987, U.S. farm assets — investments in land, improvements, machinery, equipment, animals, feed, and supplies — totalled $813.1 billion, while farm debt totaled $153.3 billion. Thus, in the aggregate, farmers had 81.2% equity in their businesses and 18.8% borrowed money (debts).

Perhaps agriculturists have been too conservative, for it is estimated that one-fourth to one-third of American farmers could profit from the use of more credit in their operations.

In 1987, the average U.S. farmer or rancher had more than $370,000 invested in land, machinery, livestock, working capital, and farm buildings. Some were much larger; investments of more than $1,000,000 per farm or ranch were not uncommon.

Another statistic which points up the enormity of capital needs is that it takes about $17.58 in farm assets to produce $1 of net farm income.[1]

Capital needs are determined primarily by (1) size of operation, (2) kind of dairy enterprise (milk production or replacement heifers), (3) ownership vs contract operation, and (4) location.

The U.S. Department of Agriculture reported[2] the following costs of production in 1989:

1. Cost per dairy cow $1,027.00
2. Cost to produce 100 lb of milk $11.84

A commercial dairy with 5,000 lactating cows requires far more capital than a family owned and operated dairy consisting of 50 lactating cows, although the per cow capital needs are generally less with increased size.

CREDIT IN AGRICULTURE

A big dairy operation necessitates both big money and knowledge of financing.

Credit may be defined as belief in the truth of a statement or in the sincerity of a person. In dairying, or in any other business transaction, credit means confidence that people will take care of their future obligations. Credit is the lifeblood of the dairy business. Without it, few large operations would be possible; for not many people are able to provide all the capital that they need.

---

[1]Based on 1987 figures, when farm assets were $813.1 billion and net farm income was $46.264 billion ($813.1 ÷ $46.26 = $17.58). Source, *Statistical Abstracts of the United States 1989,* Department of Commerce, p. 636, No. 1099.

[2]*Dairy Background for 1990 Farm Legislation,* USDA, ERS, Commodity Economics Division, pp. 9 and 62.

Most commercial lenders have guides and standards that set upper limits on the amount they will lend. Usually, to get credit on a mortgage for buying a dairy farm, the borrower is expected to make a down payment of 40 to 50% of the purchase price. Lenders usually will make loans on dairy animals and on new machinery for up to 80% of the purchase price.

Total farm investment in land, buildings, animals, and equipment has increased 4.8 times in 31 years, rising from $170 billion in 1956 to $813 billion in 1987. Farm debts have increased even more—they are 8.2 times larger than 31 years ago. The amount of debt owed by farmers and ranchers has risen from $18.8 billion in 1956 to $153.3 billion in 1987, and the trend shows no signs of letting up.

Credit is an integral part of today's dairy business. Wise use of it can be profitable, but unwise use of it can be disastrous. Accordingly, dairy producers should know more about it. They need to know something about the lending agencies available to them, the types of credit, and how to go about obtaining a loan.

TYPES OF CREDIT OR LOANS

Getting the needed credit through the right kind of loan is an important part of sound financial dairy farm management. The following three general types of agricultural credit are available, based on length of life and type of collateral needed:

■ **Short-term loans**—This type of loan is made for operating expenses and is usually for 1 year or less. It is used for the purchase of feed and for operating expenses; and it is repaid when milk is sold. Security, such as a chattel mortgage on the crop, may be required by the lender.

■ **Intermediate-term loans**—These loans are used to buy equipment and animals, for making land improvements, and for remodeling existing buildings. They are paid back in 1 to 7 years. Generally, they are secured by a chattel mortgage on animals and machinery.

■ **Long-term loans**—These loans are secured by mortgage on real estate and are used to buy land or make major improvements to farmland and buildings or to finance construction of new buildings. They may be for as long as 40 years. Usually they are paid off in regular annual or semiannual payments. The best sources for long-term loans are an insurance company, the Federal Land Bank, the Farm Home Administration, or an individual.

CREDIT SOURCES

Table 20–1 and Fig. 20–2 show where farmers borrow, the amount of loans from each source, and the percent of the total held by each type of lender.

TABLE 20–1
WHERE FARMERS BORROW (1989)[1]

Type and Source of Loan	Amount of Loan	Percent of Total
	(million $)	(%)
Real estate mortgage loans:		
Farm Credit System	26,059	35.0
Individuals and others	15,751	21.2
Commercial banks	15,263	20.5
Insurance companies	8,852	11.9
Farmers Home Administration	8,421	11.3
Total	74,346	100.0
Nonreal estate loans:		
Commercial banks	28,595	41.9
Farmers Home Administration	11,792	17.3
Individuals and others	11,760	17.2
Farm Credit System	9,120	13.4
Commodity Credit Corp.	7,000	10.2
Total	68,267	100.0
Total loans	142,613	
Percent real estate		52.1
Percent nonreal estate		47.9

[1]Data provided in a personal communication to the author from George D. Irwin, Deputy Director, Office of Financial Analysis, Farm Credit Administration, McLean, VA.

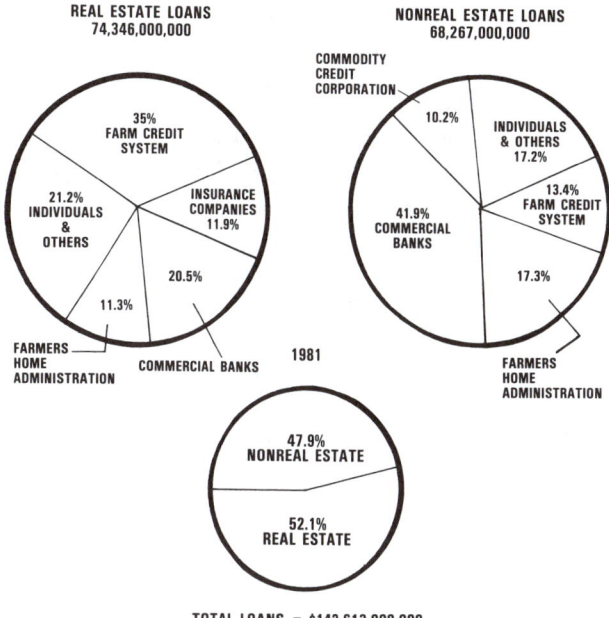

Fig. 20–2. Where farmers borrow (1989). (Source: Farm Credit Administration, Washington, DC)

CREDIT FACTORS CONSIDERED AND EVALUATED BY LENDERS

Potential money borrowers sometimes make their first big mistake by going in "cold" to see a lender, without adequate facts and figures, with the result that, to begin with, they have two strikes against getting the loan.

When considering and reviewing loan requests, the lender tries to arrive at the repayment ability of the potential borrower. Likewise, the borrower has no reason to obtain money unless it will make money.

Lenders need certain basic information in order to evaluate the soundness of a loan request. To this end, the following information should be submitted:

1. **Feasibility study.** Lenders are impressed with a borrower who has a feasibility study showing the current status of the enterprise, and achievable goals ahead—and how to get there.

In addition to spelling out the goals, assurance should be given of the necessary management skills to achieve them. Such an analysis of the present and projection into the future is imperative in big operations.

2. **The applicant, dairy establishment, and financial statement.** It is the borrower's obligation, and in his/her best interest, to present the following information to the lender:

a. **The applicant:**
   (1) Name of applicant and spouse; age of applicant
   (2) Number of children (minors; legal age)
   (3) Partners in business, if any
   (4) Years in area
   (5) References

b. **The dairy establishment:**
   (1) Owner or tenant
   (2) Location; legal description and county, and direction and distance from nearest town
   (3) Type of enterprise—market milk, heifer replacement, registered stock, or whatnot.

c. **Financial statement.** This document indicates the borrower's financial record and current financial position; potential ahead; and liability to others. Borrowers should always have sufficient slack to absorb reasonable losses due to such unforeseen happenstances as storms, droughts, diseases and poor markets, thereby permitting lenders to stay with them in adversity and to give them a chance to recoup their losses in the future. The financial statement should include the following:
   (1) Current assets:
      (a) Number and ages of cattle
      (b) Feed
      (c) Machinery
      (d) Cash. There should be reasonable cash reserves, to cut interest costs, and to provide a cushion against emergencies.
      (e) Bonds or other investments
      (f) Cash value of life insurance
   (2) Fixed assets:
      (a) Real property, with estimated value:
         i. Farm or ranch property
         ii. City property
         iii. Long term contracts

   (3) Current liabilities:
      (a) Mortgages
      (b) Contracts
      (c) Open account—to whom owed
      (d) Cosigner or guarantor on notes
      (e) Any taxes due
      (f) Current portion of real estate indebtedness due
   (4) Fixed liabilities—amount and nature of real estate debt:
      (a) Date due
      (b) Interest rate
      (c) To whom payable
      (d) Contract or mortgage

3. **Other factors.** Shrewd lenders usually ferret out many things; among them:

a. **The potential borrower.** Most lenders recognize that the potential borrower is the most important part of the loan. Lenders consider the borrower's—
   (1) Character
   (2) Honesty and integrity
   (3) Experience and ability
   (4) Moral and credit rating
   (5) Age and health
   (6) Family cooperation
   (7) Continuity, or line of succession

Lenders are quick to sense the *high-liver*—the person who lives beyond his/her means; the poor manager—the kind who would have made it except for hard luck, and to whom the hard luck happened many times; and the dishonest, lazy, and incompetent. In recognition of the importance of the person back of the loan, *key person* insurance on the owner or manager should be considered by both the lender and the borrower.

b. **Production records.** This refers to a good set of records showing efficiency of production. Such records should show milk production per cow, pounds of feed consumed per 100 lb of milk, the number of cows open longer than 100 days following calving, and the number of cases of mastitis/100 cows/month. Lenders will increasingly insist on good records.

c. **Progress with previous loans.** Has the borrower paid back previous loans plus interest; has the borrower reduced the amount of the loan, thereby giving evidence of progress?

d. **Profit and loss (P & L) statement.** This serves as a valuable guide to the potential ahead. Preferably, this should cover the previous 3 years. Also, most lenders prefer that this be on an accrual basis (even if the dairy producer is on a cash basis in reporting to the Internal Revenue Service).

e. **Physical plant:**
   (1) Is it an economic unit?
   (2) Does it have adequate water?
   (3) Is there adequate diversification?
   (4) Is the right kind of livestock being produced?
   (5) Is the farmstead neat and well kept?

f. **Collateral (or security):**
   (1) Adequate to cover loan, with margin

(2) Quality of security:
    (a) Grade of cattle
    (b) Type and condition of equipment
    (c) If feed storage is involved, adequate protection from moisture and rodents
    (d) Government participation
(3) Identification of security:
    (a) Ear tags, tattoo marks, or other identification of cattle
    (b) Serial numbers on equipment

4. **The loan request.** Dairy producers are in competition for money from urban businesses. Hence, it is important that their request for a loan be well presented and supported. The potential borrower should tell the purpose of the loan; how much money is needed, when it is needed, and what it is needed for; the soundness of the venture; and the repayment schedule.

CREDIT FACTORS CONSIDERED BY BORROWERS

Credit is a two-way street; it must be good for both the borrower and the lender. If borrowers are the right kind of persons and on a sound basis, more than one lender will want their business. Thus, it is usually well that borrowers shop around a bit — that they be familiar with several sources of credit and see what they have to offer. There are basic differences in length and type of loan, repayment schedules, services provided with the loan, interest rate, and the ability and willingness of lenders to stick by the borrower in emergencies and times of adversity. Thus, interest rates and willingness to loan are only two of the several factors to consider. Also, if at all possible, all borrowing should be done from one source; a one-source lender will know more about the borrower's operations and be in a better position to give assistance.

HELPFUL HINTS FOR BUILDING AND MAINTAINING A GOOD CREDIT RATING

Dairy producers who wish to build up and maintain good credit are admonished to do the following:

1. **Keep credit in one place, or in few places.** Generally lenders frown upon *split financing*. Borrowers should shop around for a creditor (a) who is able, willing, and interested in extending the kind and amount of credit needed; and (b) who will lend at a reasonable rate of interest, then stay with the borrower.
2. **Get the right kind of credit.** Do not use short-term credit to finance long-term improvements or other capital investments.
3. **Be frank with the lender.** Be completely open and aboveboard. Mutual confidence and esteem should prevail between borrower and lender.
4. **Keep complete and accurate records.** Complete and accurate records should be kept by enterprises. By knowing the cost of doing business, decision making can be on a sound basis.

5. **Keep annual inventory.** Take an annual inventory for the purpose of showing progress made during the year.
6. **Repay loans when due.** Borrowers should work out a repayment schedule on each loan, then meet payments when due. Sale proceeds should be promptly applied on loans.
7. **Plan ahead.** Analyze the next year's operation and project ahead.

BORROW MONEY TO MAKE MONEY

Dairy producers should never borrow money unless they are reasonably certain that it will make or save money. With this in mind, borrowers should ask, "How much should I borrow?" rather than, "How much will you lend me?"

CALCULATING INTEREST

The charge for the use of money is called interest. The basic charge is strongly influenced by the following:

1. The *basic cost* of money in the money market.
2. The *servicing costs* of making, handling, collecting, and keeping necessary records on loans.
3. The *risk* of loss.

Interest rates vary among lenders and can be quoted and applied in several different ways. The quoted rate is not always the basis for proper comparison and analysis of credit costs. Even though several lenders may quote the same interest, the effective or simple annual rate of interest may vary widely. The more common procedures for determining the actual annual interest rate, or the equivalent of simple interest on the unpaid balance, follow.

1. **Simple or true annual interest on the unpaid balance.** A $1,200 note payable at maturity (12 months) with 12% interest:

Interest paid . . . . $.12 \times \$1,200 = \$144$
Average use of the money . . . . $1,200 for the entire year
Actual rate of interest . . . . $\dfrac{\$144 \text{ (interest)}}{\$1,200 \text{ (used for 1 year)}} = 12\%$

2. **Installment loan (with interest on unpaid balance).**[3] A $1,200 note payable in 12 monthly installments with 12% interest on the unpaid balance:

Interest paid ranges from:

First month . . . . $\dfrac{.12 \times \$1,200}{12} = \$12$

    to

Twelfth month . . . . $\dfrac{.12 \times 100}{12} = \$1$

Total for 12 months is $78

---

[3]This method is used for amortized loans.

Average use of the money ranges from $1,200 for the first month down to $100 for the twelfth month, an average of $650 for 12 months.

Effective rate of interest . . . . $\dfrac{\$\,78}{\$650}$ = 12%

3. **Add-on installment loan (with interest on face amount).** A $1,200 note payable in 12 monthly installments with 12% interest on face amount of loan:

Interest paid . . . . .12 × $1,200 = $144

Average use of the money ranges from $1,200 for the first month down to $100 for the twelfth month, an average of $650 for 12 months.

Effective rate of interest . . . . $\dfrac{\$144}{\$650}$ = 22.15%

4. **Points and interest.** Some lenders now charge *points*. A point is 1% of the face value of the loan. Thus, if 4 points are being charged on a $1,200 loan, $48 dollars will be deducted and the borrower will receive only $1,152. But the borrower will have to repay the full $1,200. Obviously, this means that the actual interest rate will be more than the stated rate. But how much more?

Assume that a $1,200 loan is for 1 year and the annual rate of interest is 12%. Then the payment by the borrower of 4 points would make the actual interest rate as follows:

Interest . . . . .12 × $1,200 = $144
Average use of money . . . . . . $1,152 for one year
Effective rate of interest . . . $\dfrac{\$144\ (interest)}{\$1,152\ (used\ for\ 1\ year)}$ = 12.5%

5. **If interest is not stated, use this formula to determine the effective annual interest rate:**

Effective rate of interest =

$$\dfrac{\text{Number of payment periods} \times 2 \text{ in 1 year}[4] \times \text{Finance charges}[5]}{\text{Balance owed}[6] \times \text{Number of payments in contract plus 1}}$$

For example, a store advertises a refrigerator for $500. It can be purchased on the installment plan for $80 down and monthly payments of $35 for 12 months. What is the actual rate of interest if you buy on the time payment plan?

Effective rate of interest

$$\dfrac{2 \times 12 \times \$35}{\$420 \times (12 + 1)} = \dfrac{\$\ 840}{\$5,460} = 15.4\%$$

[4]Regardless of the total number of payments to be made, use 12 if the payments are monthly, use 6 if payments are every other month, or use 2 if payments are semiannual.

[5]Use either the time payment price less the cash price, or the amount you pay the lender less the amount you received if negotiating for a loan.

[6]Use cash price less down payment or, if negotiating for a loan, the amount you receive.

MANAGER

Fig. 20–3. Managers of large and successful dairies must wear many hats—and they must wear each of them well. (Courtesy, Holstein-Friesian Assn. of America, Brattleboro, VT)

According to Webster, *a manager is one who conducts business affairs with economy; and management is the act, or art, of managing, handling, controlling, or directing.*

Four major ingredients are essential to success in the dairy business: (1) good cattle, (2) good feeding, (3) good management, and (4) good records. A manager can make or break any dairy enterprise. Unfortunately, this fact is often overlooked in the present era, primarily because the accent is on scientific findings and automation.

In manufacturing and commerce, the importance and scarcity of top managers are generally recognized and reflected in the salaries paid to persons in such positions. Unfortunately, agriculture as a whole has lagged; and altogether too many owners still subscribe to the philosophy that the way to make money out of the dairy business is to hire a manager cheap, with the result that they usually get what they pay for—a "cheap" manager.

(Also, see Chapter 14, Dairy Cattle Management, section headed "Dairy Manager.")

TRAITS OF A GOOD MANAGER

There are established bases for evaluating many articles of trade, including hay and grain. They are graded according to well-defined standards. Additionally, we chemically analyze feeds and conduct feeding trials with them. But no such standard or system of evaluation has evolved for dairy managers, despite their acknowledged importance.

The author has prepared the "Dairy Manager Checklist," given in Table 20–2, which (1) students may use for guidance

TABLE 20–2
DAIRY MANAGER CHECKLIST

☐ CHARACTER—
   Absolute sincerity, honesty, integrity, and loyalty; ethical.

☐ INDUSTRY—
   Work, work, work; enthusiasm, initiative, and aggressiveness.

☐ ABILITY—
   Dairy know-how and experience; business acumen—including ability systematically to arrive at the financial aspects and convert this information into sound and timely management decisions; knowledge of how to automate and cut costs; common sense; organization; imagination; growth potential.

☐ PLANS—
   Sets goals, prepares organization chart and job description, plans work, and works plans.

☐ ANALYZES—
   Identifies the problem, determines the pros and cons, then comes to a decision.

☐ COURAGE—
   To accept responsibility, to innovate, and to keep on keeping on.

☐ PROMPTNESS AND DEPENDABILITY—
   A self-starter; has "T.N.T.," which means that assignments are done "today, not tomorrow."

☐ LEADERSHIP—
   Stimulates subordinates and delegates responsibility.

☐ PERSONALITY—
   Cheerful, not a complainer.

as they prepare themselves for managerial positions, (2) employers may find useful when selecting or evaluating a manager, and (3) managers may apply to themselves for self-improvement purposes. No attempt has been made to assign a percentage score to each trait, because this will vary among dairy establishments. Rather, it is hoped that this checklist will serve as a useful guide (1) to the traits of a good manager, and (2) to what the boss wants.

(Also see Chapter 14, Dairy Cattle Management, section headed "Requisites of a Successful Manager.")

ORGANIZATION CHART AND JOB DESCRIPTION

It is important that workers know to whom they are responsible and for what they are responsible; and the bigger and the more complex the operation, the more important this becomes. This should be written down in an organization chart and a job description. Samples of a job description and of an organization chart follow.

ORGANIZATION CHART OF BOSSIE DAIRY

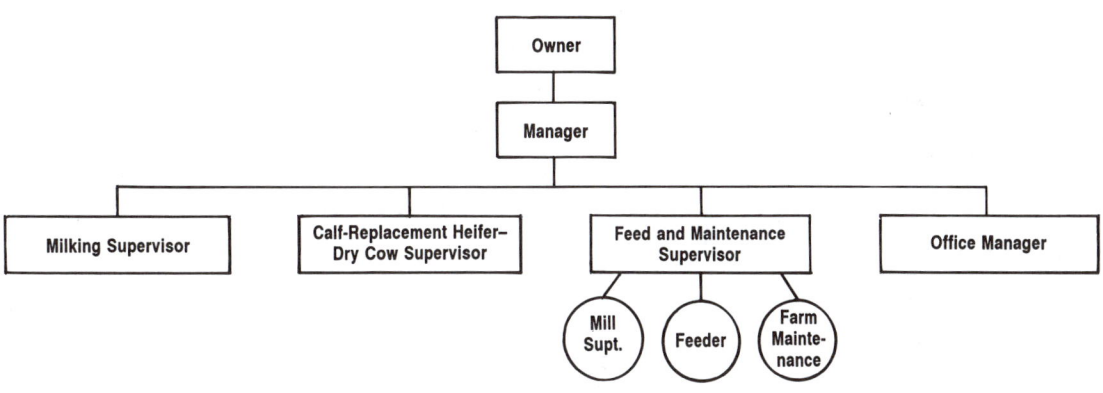

JOB DESCRIPTIONS OF BOSSIE DAIRY

Owner	Manager	Supervisor	Calf-Replacement Heifer-Dry Cow Supervisor	Feed and Maintenance Supervisor
Responsible for: 1. Making policy decisions. 2. Borrowing capital 3. Preparing long-term plan. 4. (List others.)	*Responsible for:* 1. Supervising all staff. 2. Budgets. 3. Buying replacement heifers, feed, equipment, and supplies. 4. Breeding cows, including A.I. work. 5. Marketing milk and cull cows. 6. Herd health. 7. (List others.)	*Responsible for:* 1. Milking and processing milk. 2. Housekeeping and sanitation of milkhouse. 3. Meeting milk production and conversion goals. 4. Milk production record keeping. 5. (List others.)	*Responsible for:* 1. Maternity barn. 2. Feeding, care, and management of replacement heifers. 3. Feeding and health of dry cows. 4. (List others.)	*Responsible for:* 1. Storing and processing feed; mixing rations. 2. Feeding. 3. Farm maintenance. 4. Manure disposal. 5. Crop production. 6. (List others.)

AN INCENTIVE BASIS FOR THE HELP

Big dairy establishments must rely on hired labor, all or in part. Good help—the kind that everyone wants—is hard to come by; it is scarce, in strong demand, and difficult to keep. And the agricultural work force situation is going to become more difficult in the years ahead. There is need, therefore, for some system that will (1) give a big assist in getting and holding top-flight help, and (2) cut costs and boost profits.

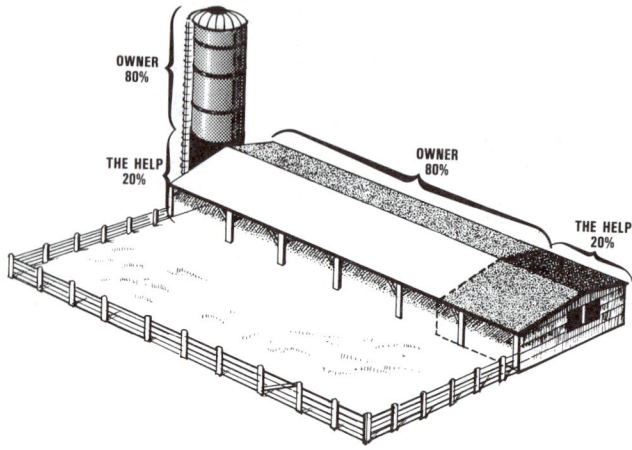

Fig. 20–4. A good incentive basis makes hired help partners in profit.

An incentive basis that makes hired help partners in profit is the answer (see Fig. 20–4).

Many manufacturers have long had an incentive basis. Executives are frequently accorded stock option privileges, through which they prosper as the business prospers. Common laborers may receive bonuses based on piecework or quotas (number of units, pounds produced). Also, most factory workers get overtime pay and have group insurance and a retirement plan. A few industries have a true profit-sharing arrangement based on net profits as such, a specified percentage of which is divided among employees. No two systems are alike. Yet, each is designed to pay more for labor, provided labor improves production and efficiency. In this way, both owners and laborers benefit from better performance.

Family-owned and family-operated farms have a built-in incentive basis; there is pride of ownership, and all members of the family are fully cognizant that they prosper as the business prospers.

Many different incentive plans can be, and are, used. There is no best one for all operations. The various plans in Table 20–3 are intended as guides only.

The incentive basis chosen should be tailored to fit the specific operation; with consideration given to kind and size of operation; extent of owner's supervision, present and projected productivity levels, mechanization, and other factors.

For most dairy operations, the author favors either a *production sharing* or a *production sharing and efficiency of production* type of incentive.

TABLE 20–3
INCENTIVE PLANS FOR DAIRY FARMS

Types of Incentives	Pertinent Provisions of Some Known Incentive Systems in Use	Advantages	Disadvantages	Comments
1. Bonuses	A flat, arbitrary bonus; at Christmas time, year-end, quarterly, or other intervals. A tenure bonus such as (1) 5–10% of the base wage or 2–4 weeks additional salary paid at Christmas time or year-end, (2) 2–4 weeks vacation with pay, depending on length and quality of service, or (3) $10 to $30/week set aside and to be paid if employee stays on the job a specified time.	It is simple and direct.	Not very effective in increasing production and profits.	
2. Equity-building plan	Employee is allowed to own a certain number of cows, or calves, which are usually fed without charge.	It imparts pride of ownership to the employee.	The hazard that the owner may feel that the employee accords his/her cows preferential treatment; suspicioned if not proved.	
3. Production sharing	Three bases are suggested; and one or all three might apply to any given dairy farm: (1) Calf raising: <table><tr><td>Death losses</td><td>Bonus/Month</td><td>or</td><td>Heifers</td></tr><tr><td>Under 5% . . . . . $15 to $30</td><td></td><td></td><td>raised to 1 month</td></tr><tr><td>Under 3% . . . . . $30 to $50</td><td></td><td></td><td>$5 to $10</td></tr></table>(2) Lactation averages: <table><tr><td>305 day M.E. lactation (lb milk)</td><td>Bonus/Worker ($)</td></tr><tr><td>Under 15,000 . . . .</td><td>0</td></tr><tr><td>15,000 to 15,999 . . . .</td><td>25</td></tr><tr><td>16,000 to 16,999 . . . .</td><td>30</td></tr><tr><td>17,000 to 17,999 . . . .</td><td>35</td></tr><tr><td>18,000 to 18,999 . . . .</td><td>40</td></tr><tr><td>19,000 to 19,999 . . . .</td><td>45</td></tr><tr><td>20,000 to 20,999 . . . .</td><td>50</td></tr></table>	It is an effective way to achieve higher production.	Net returns may suffer. For example, a higher milk production than is economical may be achieved by feeding more concentrated and expensive feeds than are practical. This can be alleviated by specifying the ration. If a high performance level already exists, further gains or improvements may be hard to come by.	Incentive payments for production above certain levels are more effective than paying for all units produced.

(Continued)

TABLE 20–3 (Continued)

Types of Incentives	Pertinent Provisions of Some Known Incentive Systems in Use	Advantages	Disadvantages	Comments
3. Production sharing (continued)	(3) Breeding efficiency:			

When Calving Interval Is (month)	Bonus for Cows Bred 60–100 Days After Calving ($)
Over 14	0
13.5 to 13.9	10
13.0 to 13.4	20
Under 13	30

Types of Incentives	Pertinent Provisions of Some Known Incentive Systems in Use	Advantages	Disadvantages	Comments
4. Profit sharing a. Percent of gross income b. Percent of net income	Here is a percent of net income basis, with an escalator arrangement. The *percent bonus* (column 3) can be varied as desired. Using line 2 as an example, column 5 is computed as follows: 20 + 10 = 30. Note that the 10 which is added on is taken immediately above the 30.	Net income sharing works better for managers, supervisors, and overseers than for common laborers because fewer hazards are involve in opening up the books to them. It is an effective way to get hired help to cut costs. It is a good plan for a hustler.	Percent of gross does not impart cost of production consciousness. Both (1) percent of gross income, and (2) percent of net income expose the books and accounts to workers who may not understand accounting principles. This can lead to suspicion and distrust. Controversy may arise (1) over accounting procedures—for example, from the standpoint of the owner a fast tax write-off may be desirable on new equipment but this reduces the net shared with the worker, and (2) because some owners are prone to overbuild and over equip, thereby decreasing net.	There must be prior agreement on what constitutes gross or net receipts, as the case may be, and how it is figured.

(1) Total Net Income ($)	(2) For Each Increase of ($)	(3) Bonus × Paid in % (%)	(4) Bonus = Paid in $ ($)	(5) Total Bonus ($)
1,000	1,000	1	10	10
2,000	1,000	2	20	30
3,000	1,000	3	30	60
4,000	1,000	4	40	100
5,000	1,000	5	50	150
6,000	1,000	6	60	210
7,000	1,000	7	70	280
8,000	1,000	8	80	360
9,000	1,000	9	90	450
10,000	1,000	10	100	550
11,000	1,000	11	110	660
12,000	1,000	12	120	780
13,000	1,000	13	130	910
14,000	1,000	14	140	1,050
15,000	1,000	15	150	1,200
16,000	1,000	16	160	1,360
17,000	1,000	17	170	1,530
18,000	1,000	18	180	1,710
19,000	1,000	19	190	1,900
20,000	1,000	20	200	2,100

Types of Incentives	Pertinent Provisions of Some Known Incentive Systems in Use	Advantages	Disadvantages	Comments

5. Production sharing and efficiency of production	Establish break-even point(s), then split profit(s) beyond this point(s) basis (1) 80% (owner) and 20% (help), or (2) use escalator arrangement, giving help greater percentage as profits rise.
	For example, the break-even points, above (below in feed efficiency) which profits would be split, might be—
	(1) In market milk production:
	    a. 15,000 lb milk/cow/year, and
	    b. 60 lb TDN consumed/cwt milk produced.
	(2) In a heifer replacement operation:
	    a. Death losses from birth to first calving, 10% and
	    b. Average age when joining the milk string, 24 months.
	Basis (1) udder health (for example, 20¢/cow/month for a somatic cell count of 500,000 to 600,000, graduated upward to 50¢/cow/month for a somatic cell count of 400,000 or less); or, (2) calving interval (for example, $20/month for a herd calving interval of 13.0 to 13.4 months, graduated upward to $50 bonus/month for a calving interval of 12.5 months or under).
		It embraces the best features of both production sharing and profit sharing without the major disadvantages of each. It (1) encourages high productivity and likely profits, (2) is tied in with prevailing prices, (3) does not necessitate opening the books, and (4) is flexible—it can be split between owner and employee on any basis desired, and the production part can be adapted to a sliding scale or escalator arrangement—for example, the incentive basis can be higher for each 1,000 lb milk increase above 19,000 lb than for each 1,000 lb milk increase above 16,000 lb.	It is a bit more complicated than some other plans and it requires more complete records.	When properly done, and all factors considered, this is a satisfactory incentive basis for a dairy enterprise.

INDIRECT INCENTIVES

Normally, we think of incentives as monetary in nature — as direct payments or bonuses for extra production or efficiency. However, there are other ways of encouraging employees to do a better job. The latter are known as indirect incentives. Among them are (1) good wages; (2) good labor relations; (3) adequate house plus such privileges as the use of the farm truck or car, payment of electric bill, use of a swimming pool, hunting and fishing, use of a horse, and furnishing milk; (4) good buildings and equipment; (5) vacation time with pay, time off, and sick leave; (6) group health insurance; (7) security; (8) the opportunity for self-improvement that can accrue from working for a top employer; (9) the right to invest in the business (10) an all-expense-paid trip to a short course, show, or convention; and (11) year-end bonus for staying all year. These indirect incentives will be accorded to the help of more and more establishments, especially the big ones.

HOW MUCH INCENTIVE PAY?

After (1) reaching a decision to go on an incentive basis, and (2) deciding on the kind of incentive, it is necessary to arrive at how much to pay. Here are some guidelines that may be helpful in determining this:

1. Pay the going base, or guaranteed, salary; then add the incentive pay above this.
2. Determine the total stipend (the base salary plus incentive) to which you are willing to go.
3. Before making any offers, always check the plan on paper to see (a) how it would have worked out in past years based on your records, and (b) how it will work out as you achieve the future projected production.

REQUISITES OF AN INCENTIVE BASIS

Owners who have not previously had experience with an incentive basis are admonished not to start with any plan until they are sure of both their plan and their help. Also, it is well to start with a simple plan; then a change can be made to a more inclusive and sophisticated plan after experience is acquired.

Regardless of the incentive plan adopted for a specific operation, it should encompass the following essential features:

1. A good owner (or manager) and good workers. No incentive basis can overcome poor managers. They must be good supervisors and fair to their help. Also, on big establishments, they must prepare a written organization chart and job description so the employees know (a) to whom they are responsible, and (b) for what they are responsible. Likewise, no incentive basis can spur employees who are not able, interested, and/or willing. This necessitates that they be selected with special care where they will be on an incentive basis. Hence, the three — good owner (manager), good employees, and good incentive — go hand in hand.
2. It must be fair to both employer and employees.
3. It must be based on and make for mutual trust and esteem.

Fig. 20–5. A good incentive program must compensate all members of the team. For example, without the cooperation of the milkers, no incentive program will succeed. The above picture shows the milker adjusting the electric pulsator to suit the individual cow. (Courtesy, Babson Bros. Co., Oak Brook, IL)

4. It must compensate for extra performance, rather than substitute for a reasonable base salary and other considerations (house, utilities, and certain provisions).
5. It must be as simple, direct, and easily understood as possible.
6. It should compensate all members of the team. For example, without the cooperation of the milkers, no dairy farm incentive program will succeed.
7. It must be put in writing, so that there will be no misunderstanding. For example, if some production-sharing plan is used in a market milk operation, it should stipulate the ration (or who is responsible for ration formulation).
8. It is preferable, although not essential, that workers receive incentive payments (a) at rather frequent intervals, rather than annually, and (b) immediately after accomplishing the extra performance.
9. It should give the hired help a certain amount of responsibility, from the wise exercise of which they will benefit through the incentive arrangement.
10. It must be backed up by good records; otherwise, there is nothing on which to base incentive payments.
11. It should be a two-way street. If employees are compensated for superior performance, they should be penalized (or, under most circumstances, fired) for poor performance. It serves no useful purpose to reward the unwilling, the incompetent, and the stupid. For example, no overtime pay should be given to employees who must work longer because of slowness or in order to correct mistakes of their own making. Likewise, if the reasonable break-even point on a market milk operation is 15,000 lb milk/cow/year and this production level is not reached because of obvious neglect (for example, poor milking), the employee(s) should be penalized (or fired).

RECORDS AND ACCOUNTS

The key to good business and management is records. The historian, Santayana, put it this way, "Those who are ignorant of the past are condemned to repeat it."

WHY KEEP RECORDS?

The chief functions of records and accounts are:

1. To provide profit and progress indicators. Production records on dairies are profit indicators and a way to measure progress.

2. To provide information from which the dairy business may be analyzed, with its strong and its weak points ascertained. From the facts thus determined, the operator may adjust current operations and develop a more effective plan of organization.

3. To provide a net worth statement, showing financial progress during the year.

4. To furnish an accurate, but simple, net income statement for use in filing tax returns.

5. To keep production records on the dairy and the crops.

6. To aid in making a credit statement when a loan is needed.

7. To keep a complete historical record of financial transactions for future reference.

Good records, properly analyzed and used, will increase net earnings and serve as a basis for sound management husbandry.

KIND OF RECORD AND ACCOUNT BOOK

The record forms will differ somewhat according to the type of enterprise. For example, with milk cows, cost per 100 lb of milk is the important thing, whereas in a heifer replacement program, it is percent of heifer calves dropped that are raised to six months of age plus growth rate. Net returns are important, but it is also necessary that records show all the items of cost and income—milk production and feed consumption of lactating cows; and mortality, feed cost, and growth rate of replacement heifers.

Dairy producers can make their own record book by ruling off the pages of a bound notebook to fit their specific needs, but the saving is negligible. Instead, it is recommended that they obtain a copy of a record book prepared for and adapted to their business. Such a book may usually be obtained at a nominal cost from the agricultural economics department of each state college of agriculture. Also, certain commercial companies distribute very acceptable record and account books at no cost.

KIND OF RECORDS TO KEEP

Most record and account books contain simple and specific instructions relative to their use. Accordingly, it is neither necessary nor within the realm of this book to provide such instructions. Instead, the comments made herein are restricted to the kind of records to keep.

At the outset, it should be recognized that the records should be easy to keep and should give the information desired to make a valuable analysis of the business. In general, the functions enumerated under the earlier section entitled, "Why Keep Records?" can be met by the following kinds of records:

1. **Annual inventory.** The annual inventory is the most valuable record that a producer can keep. It should include a list and value of real estate, cattle, equipment, feed, supplies, and all other property, including cash on hand, notes, bills receivable, and growing crops. Also, it should include a list of mortgages, notes, and bills payable. It shows the dairy producer what is owned and what is owed; whether he/she is getting ahead or going behind. The following pointers may be helpful relative to the annual inventory.

   a. **Time to take inventory.** The inventory should be taken at the beginning of the account year; usually this means December 31 or January 1.

   b. **Proper and complete listing.** It is important that each item be properly and separately listed.

   c. **Method of arriving at inventory values.** It is difficult to set up any hard and fast rule to follow in estimating values when taking inventories. Perhaps the following guides are as good as any:

      (1) **Real estate.** Estimating the value of farm real estate is, without doubt, the most difficult of all. It is suggested that the owner use either (a) the cost of the farm, (b) the present sale value of the farm, or (c) the capitalized rent value according to its productive ability with an average operator.

      (2) **Buildings.** Buildings are generally inventoried on the basis of cost less observed depreciation and obsolescence. Once the original value of a building is arrived at, it is usually best to take depreciation on a straight line basis by dividing the original value by the estimated life in terms of years. Most non-residential farm buildings are depreciated over a period of 31.5 years.

      (3) **Cows.** Cows are usually not too difficult to inventory because there are generally sufficient current sales to serve as a reliable estimate of value.

      (4) **Equipment.** The inventory value of equipment is usually arrived at by either of two methods: (a) the original cost less a reasonable allowance for depreciation each year, or (b) the probable price that it would bring at a well-attended auction.

      Under conditions of ordinary wear and reasonable care, it can be assumed that the general run of equipment (including automobiles and light general purpose trucks) will last about 5 years. Thus, with new equipment, the annual depreciation will be the original cost divided by five.

      (5) **Feed and supplies.** The value of feed and supplies can be based on market price.

   Two further points are important. Whatever method is used in arriving at inventory value (a) should be followed at both the beginning and the end of the year, and (b) should reflect the operator's opinion of the value of the property involved.

2. **Record of receipts and expenses.** Such a record

is essential to any type of well-managed business. To be most useful, these entries should not only record the amount of the transaction, but should give the source of the income or the purpose of the expense, as the case may be. In other words, they should show the producer from what sources the income is derived and for what it is spent.

The following kinds and arrangements of farm record books are commonly used for recording receipts and expenditures:

a. Those that devote a separate page to each enterprise; that is, a separate page is used for the market milk enterprise, another for replacement heifers, still another for crops (if there are such), and so on.

b. Those that provide for a record of receipts and expenses on the same page, using one column for receipts and another for expenses. This type is easy to keep, but very difficult to analyze from the standpoint of any particular enterprise.

c. Those that combine the features of both a and b above. The latter are more difficult to keep than the others, and may be confusing to the person keeping the record.

Household and personal accounts should be kept, but should be handled entirely separate from the dairy enterprise accounts because they are not farming expenses as such.

3. **Record of dairy and crop production.** A record of the production and sale of market milk, and of the yield of crops (if crops are grown) is most important, for the success of the dairy farm depends upon production. Additionally, the dairy producer should keep milk and butterfat records (DHIA); breeding records; herd health records, including mastitis records; and calf mortality records. Such records help in analyzing the farm business. They may be few or many, depending upon the wishes of the operator.

SUMMARIZING AND ANALYZING THE RECORDS

At the end of the year, the second or closing inventory should be taken, using the same method as was followed in taking the initial inventory. The final summary should then be made, following which the records should be analyzed. In the latter connection, producers should remember that the purpose of the analysis is not to prove that they have or have not been prosperous. They probably know the answer to this question already. Rather, the analysis should show actual conditions on the farm and point out ways in which these conditions may be improved.

Although producers can summarize and analyze their own records, there are many advantages in having the services of a specialist for this purpose. Such a specialist is in a better position to make a "cold" appraisal without prejudice, and to compare enterprises with those of other similar operators. Thus, the specialist may discover that, in comparison with other operators, the cows on a given farm are requiring too much feed to produce 100 lb of milk, or that the dairy enterprise is much less profitable than others have experienced. The local county agent can either render or recommend such specialized assistance. In some areas, it may consist in joining a cooperative farm record group or engaging the services of a consultant; in some states, such service is provided by the state agricultural college.

MILK PRODUCTION COSTS AND RETURNS

Table 20–4 shows milk production costs and returns, per cwt, for 1987–89. This record may be used by dairy producers —

1. To compare their own operation with these national averages — to see how well they are doing.
2. To serve as a guide for the preparation of detailed record forms for any dairy operation.

TABLE 20–4
MILK PRODUCTION COSTS AND RETURNS, PER CWT, 1987–89[1]

Item	1987	1988	1989
	($)	($)	($)
Cash receipts:			
Milk	12.51	12.23	13.55
Cull cows	1.18	1.24	1.27
Total	13.69	13.47	14.82
Variable cash expenses:			
Feed			
Concentrates (51 lb)	3.06	3.42	3.88
By-products (3.7 lb)	0.18	0.22	0.20
Hay (44 lb)	0.99	1.37	1.33
Silage (78 lb)	0.53	0.82	0.64
Pasture and other forage	0.06	0.07	0.07
Other			
Milk hauling and marketing	0.61	0.61	0.64
Artificial insemination	0.12	0.12	0.13
Veterinary and medicine	0.20	0.20	0.21
Livestock hauling	0.03	0.03	0.03
Fuel, lube, and electricity	0.23	0.22	0.24
Machinery and building repairs	0.38	0.38	0.39
Hired labor	0.96	0.99	1.04
DHIA fees	0.06	0.06	0.06
Dairy supplies	0.19	0.20	0.20
Dairy assessment	0.19	0.03	0.00
Total variable cash expenses	7.79	8.74	9.06
Fixed cash expenses:			
General farm overhead	0.70	0.81	0.95
Taxes and insurance	0.35	0.39	0.36
Interest	1.03	1.02	1.08
Total fixed cash expenses	2.08	2.22	2.39
Total cash expenses	9.87	10.96	11.45
Cash receipts less cash expenses	3.82	2.51	3.37
Capital replacement	1.59	1.70	1.65
Total, cash expenses and replacement . .	11.46	12.66	13.10
Net cash returns	2.23	0.81	1.72
Economic (full ownership) costs:			
Variable cash expenses	7.79	8.74	9.06
General farm overhead	0.70	0.81	0.95
Taxes and insurance	0.35	0.39	0.36
Capital replacement	1.59	1.70	1.65
Operating capital	0.06	0.08	0.09
Other nonland capital	0.58	0.70	0.77
Land	0.14	0.19	0.17
Unpaid labor (0.11 hr)	0.54	0.56	0.57
Total, economic costs	11.75	13.17	13.62
Residual returns to management and risk	1.94	0.30	1.20

[1]Source: *Economic Indicators of the Farm Sector, Cost of Production*, USDA, ERC, ECIFS9–1, Aug. 1990, p. 19.

ANALYZING A DAIRY BUSINESS— IS IT PROFITABLE?

Most people are in business to make money—and dairy producers are people. In some areas, particularly near cities and where the population is dense, land values may appreciate so as to be a very considerable profit factor. Also, a tax angle may be important. But neither of these should be counted upon. The dairy operation should make a reasonable return on the investment; otherwise, the owner should not be in the business.

For land and dairy cattle to be profitable, they should yield a return sufficient to the owner to (1) meet the interest payment on the investment, (2) retire a reasonable portion of the loan, and (3) provide satisfactory management return. But there is no more reason why large dairy holdings should be debt-free than there is for General Motors, or any other big corporation, to be debt-free.

A dairy owner or manager needs to analyze his/her business—to determine how well the enterprise is doing. With big operations, it is no longer possible to base such an analysis on the bank balance statement at the end of the year. In the first place, once per year is not frequent enough, for it is possible to go broke, without really knowing it, in that period of time. Secondly, a balance statement gives no basis for analyzing an operation—for ferreting out its strengths and weaknesses. In large dairy operations, it is strongly recommended that progress be charted by means of monthly or quarterly closings of financial records.

Also, dairy producers must not only compete with other dairy producers down the road, but they must compete with themselves—with their record last year and the year before. They must work ceaselessly at making progress, of improving the end product and lowering costs of production.

To analyze a dairy business, two things are essential: (1) good records, and (2) yardsticks, or profit indicators, with which to measure an operation. In addition to Table 20–4, profit indicators may be used as yardsticks.

Profit indicators are gauges for measuring the primary factors contributing to profit. In order for dairy producers to determine how well they are doing, they must be able to compare their own operations with something else; for example, (1) their own historical 5-year average, (2) the average for the United States or for their particular area, or (3) the top 5%. The author favors the latter, for high goals have a tendency to spur superior achievement.

DAIRY PROFIT INDICATORS

Many factors determine the profitableness of a dairy enterprise. Certainly, a favorable per-cow capital investment in land, buildings, equipment, and cows is a first requisite.

Tables 20–5 and 20–6 will serve as useful yardsticks for determining (1) how the dairy producer stacks up with the goals therein, and (2) where the producer is falling down.

TABLE 20–5
LACTATING COW PROFIT INDICATORS

Indicators	Production Goals	Achieved on Your Dairy
Investment/cow in land, buildings, equipment, and cows. Not more than	$4,500	
Milk production/cow/year	16,000 lb	
TDN consumed; pounds TDN/cwt milk. Not more than	52 lb	
Cows open longer than 100 days following calving. Not more than	10%	
Number cases of mastitis/month in lactating cows. Not more than	8%	

TABLE 20–6
HEIFER REPLACEMENT PROFIT INDICATORS

Indicators	Production Goals	Achieved on Your Dairy
Death losses from birth to first calving. Not more than	5%	
Average age when bred	12–14 months	
Average weight when bred:		
Jerseys	500 lb	
Guernseys	550 lb	
Ayrshires, Milking Shorthorn, Red Poll . . .	600 lb	
Holstein, Brown Swiss	750 lb	
Average age when joining the milking string	21–23 months	

BUDGETS IN THE DAIRY BUSINESS

A budget is a projection of records and accounts and a plan for organizing and operating ahead for a specified period of time. A short-term budget is usually for 1 year, whereas a long-term budget is for a period of years. The principal value of a budget is that it provides a working plan through which the operation can be coordinated. Changes in prices, droughts, and other factors make adjustments necessary. But these adjustments are more simply and wisely made if there is a written budget to use as a reference.

HOW TO SET UP A BUDGET

It is unimportant whether a printed form (of which there are many good ones) is used or one made up on an ordinary ruled 8½ x 11 in. sheet placed sidewise. The important things are (1) that a budget is kept, (2) that it be on a monthly basis, and (3) that the operator be comfortable with whatever forms or system used.

No budget is perfect. But it should be as good an estimate as can be made—despite the fact that it will be affected by such things as droughts, diseases, markets, and many other unpredictables.

A simple, easily kept, and adequate budget can be evolved by using forms such as Tables 20–7, 20–8, and 20–9.

TABLE 20–7
ANNUAL CASH EXPENSE BUDGET[1]

_____ for 19 ____
(name of dairy establishment)

Item	Total	Jan.	Feb.	Mar.	Apr.	May	June	July	Aug.	Sept.	Oct.	Nov.	Dec.
Heifer replacements													
Labor hired													
Feed purchased													
Gas, fuel, grease													
Taxes													
Insurance													
Interest													
Utilities													
etc.													
etc.													
etc.													
Total	60,000	5,000	6,000	4,000	5,000	5,000	6,000	4,000	5,000	5,000	6,000	4,000	5,000

[1]The Annual Cash Expense Budget should show the monthly breakdown of various recurring items—everything except the initial loan and capital improvements.

TABLE 20–8
ANNUAL CASH INCOME BUDGET[1]

_____ for 19 ____
(name of dairy establishment)

Item	Total	Jan.	Feb.	Mar.	Apr.	May	June	July	Aug.	Sept.	Oct.	Nov.	Dec.
Milk													
Cull cows													
Veal calves													
etc.													
etc.													
Total	100,000	12,000	7,000	7,000	7,000	7,000	7,000	12,000	7,000	7,000	7,000	7,000	13,000

[1]The Annual Cash Income Budget is just what the name implies—an estimated cash income by months.

TABLE 20–9
ANNUAL CASH EXPENSE AND INCOME BUDGET (CASH FLOW)[1]

_____ for 19 ____
(name of dairy establishment)

Item	Total	Jan.	Feb.	Mar.	Apr.	May	June	July	Aug.	Sept.	Oct.	Nov.	Dec.
Gross income	100,000	12,000	7,000	7,000	7,000	7,000	7,000	12,000	7,000	7,000	7,000	7,000	13,000
Gross expense	60,000	5,000	6,000	4,000	5,000	5,000	6,000	4,000	5,000	5,000	6,000	4,000	5,000
Difference	40,000	7,000	1,000	3,000	2,000	2,000	1,000	8,000	2,000	2,000	1,000	3,000	8,000
Surplus (+) or Deficit (–)	+	+	+	+	+	+	+	+	+	+	+	+	+

[1]The Annual Cash Expense and Income Budget is a cash flow budget, obtained from the first two forms. It is a money *flow* summary by months. From this can be ascertained when, and how much, money will need to be borrowed, the length of the loan, and a suitable repayment schedule. It makes it possible to avoid tying up capital unnecessarily, and to avoid unnecessary interest.

HOW TO FIGURE NET INCOME

Table 20–9 shows a gross income statement. But there are other expenses that must be taken care of before net profit is determined; namely:

1. **Depreciation on buildings and equipment.** It is suggested that the *useful life* of buildings and equipment be as follows, with depreciation accordingly: nonresidential buildings, 31.5 years; and machinery and equipment, 5 years. Sometimes, a higher depreciation, or amortization, is desirable because it produces tax savings and is protection against obsolescence due to scientific and technological developments.

2. **Interest on owner's money invested in farm and equipment.** This should be computed at the going rate in the area, say 10%.

Here is an example of how the above works:

Let us assume that on a given dairy establishment there was a gross income of $100,000 and a gross expense of $60,000, or a surplus of $40,000. Let's further assume that there are $40,000 worth of equipment, $30,000 worth of buildings, and $175,000 of the owner's money invested in farm and equipment. Here is the result:

Gross profit $40,000
Depreciation:
  Machinery, $ 40,000 @ 20% = $ 8,000
  Buildings, $ 30,000 @ 3.17% =   951
                                $ 8,951
  Interest $175,000 @ 10% = 17,500

                                     $26,451

Return to labor and management $13,549

Some people prefer to measure management by return on invested capital, and not wages. This approach may be accomplished by paying management wages first, then figuring return on investment.

ENTERPRISE ACCOUNTS

Where a dairy enterprise is diversified (for example, a farm selling market milk, replacement heifers, dairy beef, and crops), enterprise accounts should be kept—in this case four different accounts for four different enterprises. The reasons of keeping enterprise accounts are:

1. It makes it possible to determine which enterprises have been most profitable, and which least profitable.

2. It makes it possible to compare a given enterprise with competing enterprises of like kind, from the standpoint of ascertaining comparative performance.

3. It makes it possible to determine the profitableness of an enterprise at the margin (the last unit of production). This will give an indication as to whether to increase the size of a certain enterprise at the expense of an alternative existing enterprise when both enterprises are profitable in total.

COMPUTERS IN THE DAIRY BUSINESS

Accurate and up-to-the-minute records and controls have taken on increasing importance in all agriculture, including the dairy business, as the investment required to engage therein has risen and profit margins have narrowed. Today's successful dairy producers must have, and use, as complete records as any other business. Also, records must be kept current; it no longer suffices merely to know the bank balance at the end of the year.

Big and complex dairy enterprises have outgrown hand record keeping. It is too time-consuming, with the result that it does not allow management enough time for planning and decision making. Additionally, it does not permit an all-at-once consideration of the complex interrelationships which affect the economic success of the business. This has prompted a new computer technique known as linear programming.

Linear programming is similar to budgeting, in that it compares several plans simultaneously and chooses from among them the one likely to yield the highest returns. It is a way in which to analyze a great mass of data and consider many alternatives. It is not a managerial genie; nor will it replace decision-making managers. However, it is a modern and effective tool in the present age, when just a few cents per 100 lb of milk can spell the difference between profit and loss.

There is hardly any limit to what computers can do if fed the proper information. Among the difficult questions that they can answer for a specific dairy enterprise are:

1. **How is the entire operation doing so far?** It is preferable to obtain monthly progress reports; often making it possible to spot trouble before it is too late.

2. **What farm enterprises are making money; which ones are freeloading or losing?** By keeping records by enterprises—market milk production, replacement heifers, dairy beef, wheat, corn—it is possible to determine strengths and weaknesses; then either to rectify the situation or shift labor and capital to a more profitable operation. Through *enterprise analysis*, some operators have discovered that one part of the business may earn $10, or more, per hour for labor and management, whereas another may earn only $1 per hour, and still another may lose money.

3. **Is each enterprise yielding maximum returns?** By having profit, or performance, indicators in each enterprise, it is possible to compare these (a) with the historical average of the same dairy farm, or (b) with the same indicators of other similar establishments.

4. **How does this dairy enterprise stack up with its competition?** Without revealing names, the computing center (local, state, area, or national) can determine how a given dairy enterprise comparés with others—either the average, or the top (say, 5%).

5. **How to plan ahead.** By using projected prices and costs, computers can show what moves to make for the future—they can be a powerful planning tool. They can be used in determining when to buy feed, purchase replacement heifers, market cull cows, market dairy beef, etc.

6. **How can income taxes be cut to the legal minimum?** By keeping accurate record of expenses and figuring depreciations accurately, computers make for a saving in income taxes on most dairy establishments.

There are three requisites for linear programming a dairy establishment. These are:

1. Access to a computer.
2. Computer know-how, so as to set the program up properly and be able to analyze and interpret the results.
3. Good records.

Information on "how to Balance a Ration by Computer" is contained in Chapter 12, Dairy Feeding Programs.

DAIRY CATTLE CONTRACTS

With the increase in specialization and size of enterprises, the business aspects of dairy production have become more important. With this transition, the following types of contracts have evolved: (1) heifer replacement contracts, (2) cow rental contracts, and (3) cow pools.

HEIFER REPLACEMENT CONTRACTS

Contracting for heifer replacements evolved because some producers found that it was best to have specialists do this job for them. More specifically, here are the forces that are generally back of the movement:

1. Dairy producers who either do not wish or cannot afford to make the very considerable capital investment required for market milk production can, at a lower cost, utilize their land, facilities, and time for growing replacement heifers for other dairy producers.
2. Often those who grow replacement heifers under contract can arrange for the financing of the operation from the person for whom the heifers are being grown.
3. The desire of the producer to expand the milking herd, thereby crowding out the heifer replacement program.
4. High feed and land costs where the milking herd is maintained forcing a shift of heifer replacement production to more isolated and less expensive areas.
5. The desire of milk producers to obtain replacement heifers from genetically superior animals that they have developed in their herds. This can be achieved by placing their own calves out under a contractual arrangement.
6. The contract arrangement may be used as a means of lessening risks to each party, by limiting or apportioning losses.

The following types of heifer raising contracts are in use:

1. **The gain in weight or flat fee contract.** According to this plan, the contractor (the milk producer who is having heifers grown) pays the contractee (the heifer raiser) on the basis of (a) so many cents per pound of gain, or (b) a flat fee of so many dollars per month. Many variations of this plan exist.
2. **Option to purchase contract.** Under this type of contract, the grower purchases the calves, at anywhere from 4 days to 6 months of age, depending on the terms of the contract. But the original owner of the calves has the option to buy them back after they are bred. Because of changing economic conditions, the pricing arrangement is usually somewhat flexible, with prices of feed and milk being primary determining factors. If the original owner does not wish to exercise the option, the grower may sell them to the highest bidder.

Under this plan, the grower takes ownership of the calves, thereby providing a built-in incentive to do a good job.

3. **Co-op contracting heifers.** This type of co-op grows out surplus heifers belonging to members of the cooperative. Co-op members producing market milk transfer their calves, under contract, to farmers who are interested in growing them out cooperatively. The grower is paid on either a cost per pound of gain basis or so many dollars per month.

Naturally, the cost of raising dairy heifer replacements varies widely from area to area, regardless of whether they are raised on the farm where dropped or under contract. The primary factors accounting for these wide area variations are (1) cost of feed and labor, and (2) climate.

For some market milk producers, contracting for replacement heifers is advantageous. For others, there are advantages to on-the-farm raising, including the following:

1. It lessens the chance of introducing disease into the lactating herd.
2. Unless the milk producers place their own heifers out under contract, the source of replacements may be questionable and not up to the genetic standards of the herd in which they are to be used as replacements.
3. Contracting for heifers usually costs more than when they are home produced.
4. Contracting instead of raising heifer replacements alleviates a means of diversifying on the farm—the production of both market milk, and the raising of replacement heifers.
5. Home-raised replacements generally have less udder and breeding trouble than contract-raised heifers, with the result that they have a longer productive life in the herd.

COW RENTAL CONTRACTS

With the rapid growth in capital requirements in agriculture, some milk producers are now exploring alternate methods of gaining control of adequate resources. Lease agreements, long used for land and buildings, are now being used by some farmers for machinery and livestock. Several companies, as well as individuals, are now leasing cows to dairy producers.

A fair agreement makes good business sense. It distributes the income between the owner (lessor) and the milk producer (lessee) in keeping with the level of their respective contributions of resources and payments of costs. The owner and operator can arrive at a fair agreement by using a contributions analysis.

Lease or rental agreements can be advantageous for both the producer (lessee) and the owner (lessor).

■ **Advantages for dairy producers (lessees)**—Some potential advantages of dairy cow leasing for the producers are:

1. They may use their available capital for other and perhaps more effective purposes.
2. They may have use for additional capital over and above their borrowing capacity at a reasonable, or at least a profitable, rate of interest.

3. They may be able to gain control over a larger amount of capital than would be possible by simply borrowing capital.

4. They share risk with the owner (lessor), rather than assuming all the risk.

5. They may be able to generate an increased volume of business by employing more resources.

6. They may have a better chance to get started with an adequate amount of capital, should they be beginners.

7. They may obtain some cash flow advantages.

8. They may gain some income tax advantages.

■ **Advantages for owners (lessors)** — Some potential advantages of dairy cow leasing for owners (lessors) are:

1. They receive *rental* income, which may have income tax and/or social security advantages.

2. They can retain their breeding herd without having to provide labor and many other inputs.

3. They may have an opportunity for returns on a productive capital investment.

4. They may use the agreement as a vehicle to transfer ownership to the operator over a period of time.

5. They may elect to use such an agreement as a way to phase out of the business smoothly.

■ **Disadvantages for both dairy producers (lessees) and owners (lessors)** — Dairy cow leasing may have the following disadvantages for both the dairy producers and owners:

1. Both parties give up part of their control in the business.

2. Both parties give up part of their income and potential profit, by sharing returns.

3. Both parties run the risk of strained personal relationships.

The previous discussion has summarized briefly the potential advantages of dairy cow leasing for both dairy producers (lessees) and owners (lessors), along with some of the possible disadvantages for both parties. No generalization can be made as to which control method is better — owning or leasing; nor can generalizations be made as to the best type of cow lease. The choice depends on such factors as the availability and cost of capital, cow prices, lease terms, and dairy farmer goals.

The following example will illustrate how dairy producers can evaluate (1) the profitability of owning vs leasing, and (2) two different lease plans. Of course, the figures used are for illustrative purposes only. At the time an analysis and final decision is to be made, current cost and return figures for the area should be inserted.

■ **Three options** — Let us assume that a dairy producer has the following three alternatives from which to choose a particular method of acquiring additional cows:

1. **Cash purchase.** Cows can be purchased for $1,150 per head. Enough cash is on hand so that they can be purchased outright, with no credit needed. It is expected that the cows will be in the herd 4 years, following which they will be sold for $450 per cow.

2. **Lease Plan A.** with this plan, the producer can obtain possession of the cows in exchange for a monthly payment of $5 per $100 cow value. Assuming a $1,150 cow, this would mean monthly payments of $57.50 per cow. The first month's rent must be paid in advance and all subsequent payments are due at the beginning of each month.

All offspring belong to the dairy producer; and the lessor assumes responsibility for death losses, property taxes, and insurance. The contract expires after 4 years, following which the lessor regains possession of the cow.

3. **Lease Plan B.** Cows can be leased at a monthly rate of $6 per $100 cow value, or $69 for the $1,150 cow. The higher monthly payments under this plan than under Plan A give the dairy producer the right to buy the cow for $50 at the expiration of the 4-year lease. It is expected that the cow will be culled and sold for $450 at the end of the 4-year period. All other aspects of this plan are similar to Plan A.

The simplest way in which to analyze the three options is to budget the expected costs and returns for each alternative over the 4-year period and select the one having the greatest net benefit. Before itemizing and budgeting various cost and return variables, however, one should eliminate all variables common to all three alternatives. These variables will *not* affect the choice, but they do make the calculations more burdensome.

In the situation at hand, let us assume that the same basic cow is involved, regardless of the control alternative. Consequently, milk production and associated income can be ignored. Also, income derived from calves can be eliminated, since the dairy producer has right of ownership in these animals for all three alternatives. For the same reason, feed, bedding, breeding, veterinary service and medicine, power, labor, and equipment and building overhead costs can be omitted.

COW POOLS

A cow pool is defined as a business organization or cooperative which cares for and milks cows in a centralized location.

Cow pools are not new. The Walker-Gordon pool, located at Plansboro, New Jersey, was established in 1891; and it has operated continuously and successfully ever since.

Among the **advantages** ascribed to pools are: (1) young farmers with limited capital can get into the business by buying cows and putting them into pools, without having to own land and buildings; (2) reduced cost for overhead and equipment; (3) lower purchase price for milled dairy feeds; (4) greater labor efficiency is possible; (5) producers who pool their cows may use part of their labor more profitably in other enterprises; (6) farmers have more reasonable hours of work, and are not "tied down" to their milking chores every day of the year; (7) better production records are kept, with the result that the herd is upgraded; (8) a large supply of quality milk is provided, which can be readily marketed; (9) producers can qualify for a Grade A market without large capital outlay; and (10) lower milk hauling costs.

But pools have some **disadvantages**; among them are (1) the incidence of disease may be high, because of the concentration of animals; (2) farmers may lose control over the management of their herds, (3) nonagricultural interests may control the pool (and they may be impractical); (4) the commingling of cows belonging to different producers may create management problems; (5) inability of producers who go into the pool to use their present on-farm buildings and equipment for anything else; (6) established marketing procedures may be disrupted; (7) if a large number of producers of manufacturing milk are suddenly converted to the production of Grade A milk, a surplus of Grade A milk and depressed prices may follow for a period of time.

A careful weighing of both the advantages and the disadvantages would indicate that cow pools best serve either young or elderly dairy producers. Producers who are already selling Grade A milk may gain little from going into a cow pool.

The following types of cow pools are in operation:

1. **The Walker-Gordon-type of pool.** The cow owners in this pool care for and manage their herds in keeping with the rigid health and sanitation requirements specified by Walker-Gordon, in facilities owned by the latter. Owners remove cows to their home farms during the dry period.

Walker-Gordon provides the buildings and equipment; furnishes feed to the cow owners, milks the cows, and purchases, processes, and markets the certified milk. Manure is dehydrated and sold as a garden fertilizer under the trade name of *Bovung*.

2. **Custom cow pools.** This refers to pools that care for cows on a custom or contract basis. Two types of cow owners are involved: (a) dairy farmers who wish to be relieved of caring for their cows, and (b) nonfarmers, who buy cows as an investment. In comparison with small, individually operated dairies, cow pools generally provide more mechanization and know-how. Additionally, they appeal to nonfarm investors because they provide a *tax shelter*. There has been a trend toward investor-ownership of cows in pools, rather than farmer-ownership. It is expected that this trend will continue.

Some of the larger pools have on their staffs veterinarians who are responsible for the health of the herd, and nutritionists who are charged with the responsibility of formulating rations and obtaining maximum efficiency of production. Through custom or contract arrangement, pools sell the use of their facilities and services to cow owners, usually with a profit to each party.

Several different kinds of custom contracts exist; hence, there is no standard fee. One rather common arrangement consists in the cow owner and the pool entering into a 3-year contract with (a) the pool operator charging $30 to $60 per cow per year, payable in advance, for housing and equipment plus a management fee of 5% of that which remains after deducting all operating costs; and (b) the cow owner paying for all feed, labor, veterinary care, DHIA testing, breeding service, and marketing.

Where nonfarmers or investors are involved, a professional management firm may act as the representative of the owner. Their usual charge for a 10-year period is $1/9$ the value of each cow, payable in advance. Thus, if a cow is valued at $1,150, $1/9$ would be $127.78 per cow, or $12.78 per year per cow. Some management firms guarantee investors-dairy pro-

ducers a return of 1% per month on their investment. Where a guaranteed return exists, the management firm usually (a) gets all income in excess of 1% per month, plus the calves, and (b) agrees to replace death losses up to 25% of the herds owned by the investors.

3. **"Cowtel."** *Cowtels* differ from cow pools in that their herds are lotted and milked separately by ownership, whereas in a cow pool they are commingled. The main disadvantage of the cowtel is that it is more expensive to operate where each owner's herd is segregated. This problem can be lessened by specifying that the minimum-sized herd for separate penning shall be 30 cows (or whatever the number decided upon).

4. **Co-op cow pools.** Most cow pools are owned by individuals, partnerships, or corporations. However, there is increasing interest in cooperative cow pools, as a means of accommodating co-op members. The following two types of co-op cow pools exist:

a. **The Utah-type milking pool.** This type of pool operates in central and southern Utah. The cows are milked at a central location, according to one of the following three methods:

(1) All cows are kept on the producer's place and driven to the milking barn twice daily.

(2) All cows are corralled and housed on land belonging to the cooperative. The cooperative generally hires the milkers.

(3) A combination of the first two methods is followed; some cows are maintained on the producer's premises and driven back and forth, whereas others are maintained on the premises of the cooperative. In the Utah-type pools, an initial membership fee per cow is charged, plus a service charge per head per day.

b. **Co-op custom cow pools.** These are similar to custom cow pools (see point 2 under "Cow Pools") However, in the case of the co-op custom cow pool, the net profits are savings from the pool operation; hence, they are distributed to the producers in proportion to the number of cows placed in the pool by each member.

The three main factors affecting cow pool returns are (1) production per cow and milk price, (2) cow pool charges or fees, and (3) length of the cow's productive life, or depreciation rate. Studies indicate that a cow which does not produce 10,000 lb or more of milk per year is not likely to return much money in a pool!

FUTURES TRADING OF BUTTER IS HISTORIC

Butter futures were long traded at the Chicago Mercantile Exchange. But, with the decline in butter production and consumption, futures trading in butter was discontinued. It is noteworthy, however, that out of sentiment the "Merc" still has a spot market in butter every Friday morning from 10:00 to 10:03 a.m.—three minutes each week.

TAX MANAGEMENT AND REPORTING[7]

Good tax management and reporting consists in complying with the law, but in paying no more tax than is required. It is the duty of revenue agents to see that taxpayers pay the correct amount, and it is the business of taxpayers to make sure that they do not pay more than is required. From both standpoints, it is important that farmers should familiarize themselves with as many of the tax laws and regulations as possible.

Fig. 20–6. The cardinal principles of good tax management are: (1) keeping adequate records, and (2) conducting business affairs so that the taxes are no greater than necessary. (Courtesy, Holstein-Friesian Assn. of America, Brattleboro, VT)

The cardinal principles of good tax management are: (1) maintenance of adequate records, and (2) conduct of business affairs to the end that the tax required is no greater than necessary. Good tax management and good farm management do not necessarily go hand in hand, and may sometimes be in conflict. When the latter condition prevails, the advantages of one must be balanced against the disadvantages of the other to the end that there shall be the greatest net return.

---

[7]This section and the Estate Planning were prepared by the author's son, John J. Ensminger, LL.M., an attorney specializing in taxes and estate planning.

It is recognized that tax matters constitute a highly specialized and complex field, and that each farm will need separate considerations in appropriate planning. The recent rounds of federal tax legislation have made significant changes in the procedures dairy producers must use in accounting, as well as in their approaches to financial and estate planning. More than ever, it is important that they consult competent professionals before embarking upon any business operation involving animals. It is noteworthy that, if a dairy producer's return is to be audited, under the recently enacted Taxpayer Bill of Rights, the taxpayer is entitled to be represented at the audit by a representative. Though the IRS can require the taxpayer's attendance with a special summons, this is not likely to be used at the initial meeting.

Increasingly, as local governments must make up for decreased federal support, local tax matters become more important in planning; this also makes consultation with a specialist knowledgeable in state and local tax law, crucial for effective management.

Some tax pointers of particular interest to dairy producers follow.

FILE AN ESTIMATE OR FILE YOUR CURRENT RETURN

If your business is not a corporation and at least two-thirds of your gross income is from the business of farming, and if your tax year begins on January 1, you may elect to:

1. File an estimate of your tax and pay this amount by January 15 of the year following the close of your current tax year, then file your current return and pay any balance by April 15; or

2. File your return and pay any tax due for the year by the following March 1.

In addition, unlike most other taxpayers who must make payments equal to 90% of the current year's tax, farmers need only make payments equal to 66⅔% of the current year's tax.

KEEP ADEQUATE AND ACCURATE RECORDS AND ACCOUNTS

Farmers are not required to keep an elaborate set of records and accounts, such as would be necessary to reflect the income of a large automobile manufacturing company. On the other hand, where the accrual method is used, careful inventories must be kept, and farmers should have records which will support the figures that appear in these inventories. Even cash basis farmers must have evidence to support their income and expenditure figures.

Taxpayers should not forget that the burden of proof is on them, and not on the government. If taxpayers cannot prove income and expenditures, they are at the mercy of the government. In extreme cases, the government may actually recalculate income based on net worth, cash flow, or on bank deposits and withdrawals. If a large discrepancy is found, the taxpayer may be in serious trouble with the government; and the taxpayer's accountant may not be able to help out. The IRS will not accept the accountant's assertion that the taxpayer must have had the proper records in the past, or the accountant would not have signed the return. It is up to the taxpayer to keep and, as long as a tax year is open, retain the appropriate records. Records and accounts that should be kept include:

1. A summary of all farm business receipts and expenditures supported by deposit slips and checks, which should be annotated to identify the nature of the deposit or withdrawal.

2. A record of profits or losses on the sale of purchased animals.

3. A depreciation schedule for all farm buildings, machinery, purchased dairy or breeding animals, and other depreciable property.

4. All income and expenses of a personal nature that should be taken into account.

5. An animal record which should at least include the following:

    a. The date born or purchased.

    b. Ancestry.

    c. Purchase price.

    d. Designation as either *held for breeding* or *held for sale*.

    e. A record of the use of the animal — e.g., dates bred, etc.

    f. The amount of depreciation, if any, which was deducted for each animal on previous returns.

    g. The date sold, and reasons for sale.

    h. The sale price.

    i. Salvage value of breeding animals.

    j. Ratio of ordinary sales to capital gains.

Since a tax return is no more accurate than the information that goes on it, it is recommended that the above information be recorded in a suitable account or statement book as the year progresses, rather than trusting to memory or loose slips of paper.

SEPARATE THE FARM HOME FROM THE FARM BUSINESS

The farmhouse, and such expenses relating to it as fuel, insurance on the dwelling and contents, and expenses for groceries, clothing, etc., are not deductible. Likewise, only that part of an automobile that is chargeable to the farm business is deductible as a farm expense. It is generally wise to keep separate checking accounts for business and personal matters. This makes the process of an IRS audit, should one happen, considerably easier to deal with.

KEEP YEAR-TO-YEAR INCOME AS STEADY AS POSSIBLE

Though market prices affect year-to-year income on a dairy farm, good tax management consists in minimizing such fluctuations. Thus, a farmer with an income of $70,000 one year and $10,000 the next will usually pay more total tax than one with $40,000 for each of the two years.

The accrual basis of reporting tends to even up differences in year-to-year farm income. For example, farmers may sometimes find it desirable to withhold the sale of animals one year longer than normal. Under such circumstances, they face the possibility of having to pay tax on income from the sale of two years' production in one year unless they use the accrual basis of reporting. On the other hand, the farmer on the cash basis is also able to avoid undue fluctuations in income by offsetting unusually high income by purchasing in the year of that high income supplies for use during the following year, and by paying off expenses which might not normally be paid so quickly. (For example, in some states it is possible to pay two years' state taxes in one year.) Caution should be exercised in prepaying expenses because this type of tax planning may be challenged by the IRS if it causes a material distortion of income in a particular year.

SELECT THE BEST METHOD OF ACCOUNTING

Dairy producers may report on either the *cash receipts and disbursements (cash basis)* or the *accrual basis*. Also, they can use a combination of the two bases, a *hybrid* method of reporting. Thus, the accrual basis can be used for animals and the cash basis otherwise. While most taxpayers who produce, buy or sell merchandise must use inventories, and thus must use the accrual method, farmers must only do so to the extent that they are subject to inventory accounting requirements. If a return has never been filed, the choice is made on the first return.

If the accrual method is used, annual inventories must be kept, with taxes determined from increases in inventory and deductions given for decreases. Having elected a method, the farmer will not generally be allowed to change methods without the approval of the IRS This approval will not be granted unless the IRS and the taxpayer agree to the terms and conditions under which the change will be effected. With the general trend towards requiring accrual accounting, there is likely to be more resistance towards shifting from accrual to cash accounting than the reverse.

A description of each system follows.

CASH BASIS

Under this system, farm income includes all cash or value of merchandise or other property received during the tax year. It includes all receipts from the sale of items produced on the farm and profits from the sale of items that have been sold. It does not include proceeds from sales if the proceeds were not actually available during the tax year.

Allowable deductions include those business expenses incurred that were actually paid during the year, and depreciation on depreciable items.

1. **Drought relief.** After 1987, a farmer using the cash method who is forced to sell livestock as a result of drought conditions can defer income on the excess sales to the following year. The former limitations by which this treatment was restricted to certain animals were lifted by the Technical and Miscellaneous Revenue Act of 1988. The deferral of income is available only if the animals would not have been

sold but for the drought and the drought conditions resulted in the area being eligible for federal assistance.

2. **Feed purchases.** Purchasing feed for use in subsequent years by cash basis producers is an effective means of reducing current income, but the IRS will contest the deduction if (a) the payment is, in fact, a deposit, (b) there was no business purpose for it, or (c) it distorts income. Acceptable business purposes include guaranteeing prices and making certain of supply.

3. **Prepayments.** In 1986, Congress added a provision concerning *farming syndicates* which generally limited deductions for feed, seed, fertilizer or similar supplies to the taxable year in which the supplies are actually used or consumed. A farming syndicate is a publicly traded or limited partnership or other enterprise (but not a C corporation) engaged in the business of farming.

4. **Deferred payment contracts.** It is sometimes desirable to delay recognition of income from a sale of animals until a subsequent year. Though the IRS is prone to finding that income which has been earned has, in fact, been constructively received by a taxpayer, and thus includable in the taxpayer's income, there are certain procedures which can be used to delay recognition. The general procedure for such a deferral involves the receipt of the funds by a middle party, but not an agent. The dairy producer should not receive cash equivalents, such as negotiable notes or securities or letters of credit. If the farmer is to receive funds directly from the buyer, the contract should require a deferral of income. The contract should specify the terms under which the producer will receive payment. In the typical escrow agreement, there must be conditions, enforceable by both buyer and seller, which preclude the producer from receiving payment until a subsequent date. The producer cannot receive any present beneficial interest from the receipt of funds by the escrowee.

ACCRUAL BASIS

This system requires the keeping of complete annual inventories. Tax is paid on all income earned during the taxable year, regardless of whether payment was actually received, and on increases of inventory values of animals, crops, feed, dairy products, etc., at the end of the year as compared with the beginning of the year. All expenses incurred during the year's business are deducted from gross income regardless of whether payment is actually made, and deductions are made for any decrease in inventory values of animals, etc., during the year.

Four methods of inventorying are available to the accrual basis farmer or rancher.

1. **Cost.** Inventory items are valued at the actual cost of producing or purchasing them.

2. **The lower of cost or market value.** The comparison is made separately for each item in the inventory, not for the entire inventory. The entire stock should not be valued at cost and then at market, with the lower selected.

3. **Farm price.** Each item, raised or purchased, is valued at its market price less estimated direct cost of disposition. This method must be used for the entire inventory, except that animals may be inventoried by the next method.

4. **Unit livestock price.** Animals are classified according to kind and age, and a standard unit price is used for each animal within a class. All raised animals must be included in inventory under this method. Unit prices must reflect any costs required to be capitalized under the uniform capitalization rules. This method is usually chosen by large operations. Producers using the unit-livestock method are permitted to elect a simplified production method for determining costs required to be capitalized.

The third and fourth methods are unique to farmers and ranchers.

■ **Corporations and partnerships** – The Tax Reform Act of 1986 required most C corporations, partnerships in which one of the partners is a C corporation, and tax shelters to use the accrual method of accounting. An exception was made for farming businesses (unless they are tax shelters). Another exception applies to taxable entities with average annual gross receipts of less than $5 million. S corporations (in which there is generally no corporate-level tax) and partnerships can continue to use the cash method.

■ **Capitalization of inventory costs** – The Revenue Act of 1987 reduced the ability of many businesses to deduct expenses currently by requiring that a number of expenses associated with the production of inventory be capitalized. This applies to the direct costs of producing the inventory, as well as to certain indirect costs which are determined to be allocable to the inventory. An exception allowing current deductions for expenses was provided for taxpayers with gross receipts of $10 million or less. Also, an exception was provided for any plant or animal which is produced in a farming business and which has a preproductive period of two years or less. In 1988, Congress extended the exception so that only farms under the accrual method must use uniform capitalization. Because of the flux in this area, dairy producers who decided on an accounting method prior to 1988 should have their tax advisors reconsider these choices.

DISTINGUISH CAPITAL GAINS FROM ORDINARY INCOME

Except for the 28% tax cap on capital gains, there is no longer a difference in the tax rates applied to ordinary income and capital gains. However, political considerations are such that a differential appears possible in the future. If this should be the case, income reported as capital gains will be taxed at a lower rate. Thus, livestock held for sale in inventory, and livestock held for breeding purposes, may produce different tax effects when sold, even if the sale prices are the same. Nevertheless, developments in this area will have to come about before planning can be discussed.

SET UP DEPRECIATION SCHEDULES PROPERLY

Depreciation is estimated operating expense covering wear, tear, exhaustion, and obsolescence of property used in a farm business.

Depreciation may be taken on all farm buildings (except the dairy producer's personal residence), and on water systems, fences, machinery and equipment.

Those who file returns on a cash basis may also take depreciation on dairy cattle which were purchased, but they cannot take depreciation on animals they raised because all costs of raising are deducted as operating expenses. On the accrual basis, depreciation may be taken on purchased animals that are not included in inventory.

Taxpayers should list each building, and each piece of machinery on which depreciation is to be computed on the depreciation schedule. Such items as cows and small implements may be grouped together, but such groupings should be derived from totaling of a detailed individual list kept current in a permanent farm record book.

Depreciation is not available for inventory, which would include animals held for sale to customers. After 1986, depreciable property is placed in specific classes. Because the period over which property is amortized affects the overall tax revenues. Depreciation schedules which apply to dairy producers follow:

1. **Five-year property.** This includes automobiles and light, general-purpose trucks, certain technological equipment, and research and experimentation property.

2. **Ten-year property.** Single-purpose agricultural structures were originally recovered over 7 years, but after 1988 have a 10-year recovery period. A companion requirement limits recovery on such items to the 150% declining balance method (discussed below).

3. **27.5-year property.** This covers residential rental property.

4. **31.5-year property.** This covers nonresidential real property. This will include most farm buildings.

For property in the 5- and 10-year classes, depreciation was, prior to 1989, calculated on the double declining balance method, switching to the straight-line method at the time where depreciation is maximized. For the 27.5 and 31.5-year classes, the straight-line method is used. However, for personal property (i.e., nonreal property) placed in service in a farming business after 1988, the 150% declining balance must be used regardless of the recovery period.

For purchased animals, the price paid will generally determine the amount which can be depreciated. Inherited or gift animals can be depreciated. However, their value may have to be established by a qualified appraiser if the IRS contests the taxpayer's valuation.

If certain legislation being considered as this book goes to press is enacted, intangible property, such as goodwill and going concern value, may be depreciable over a 14-year period.

DO NOT OVERLOOK ADDITIONAL DEDUCTIONS

■ **Annual expensing** — The annual expensing limitation is $10,000 for property placed in service after 1986. However, this election is not available for taxpayers whose aggregate cost of qualifying property exceeds $210,000 (reduced dollar-for-dollar over $200,000). The amount which can be expensed is limited to taxable income derived from the trade or business. The repeal of the Investment Tax Credit and the longer recovery periods for most classes of property increases the value of this provision for the dairy farmer.

■ **Soil and water conservation** — Farmers can deduct soil and water conservation expenditures only if the expenditures are consistent with a conversation plan approved by the USDA or a comparable state agency. Such expenditures include treatment or movement of earth, such as leveling, terracing, or restoration of fertility; construction and protection of diversion channels, drainage ditches, and earthen dams; planting of windbreaks; etc. Though land clearing expenses are no longer deductible, ordinary maintenance, including brush clearing, remains deductible. Costs of fertilizing and other conditioning of land remain deductible. The amount deducted under this election cannot exceed 25% of the taxpayer's gross income from farming for the year. Part of the amount deducted may be recovered if the land is sold within ten years of the deduction.

■ **Education expenses** — Educational expenses, such as the cost of short courses, are deductible if they are taken to maintain or improve the skills of the person in conducting the operation, or, if the person is employed by a farming operation and they are taken as a requirement of continuing that employment. However, if taken to allow the person to enter another trade or business, such expenses will not be deductible.

■ **Pay children for farm work** — The farmer must be able to show that a true employer-employee relationship exists. To do so, children should, as much as possible, be treated as are other employees. They should be assigned definite jobs at reasonable, agreed-upon wages, and paid regularly.

TREAT LOSSES APPROPRIATELY

On the cash basis, no death deduction can be made for an animal that was born and raised on the farm, because the cost of raising the animal has been deducted already with operating expenses. On the accrual basis, when the value of an animal appears in the beginning-of-year inventory but not in the end-of-year inventory, the loss is automatically accounted for in the change in inventory value. Any money received from insurance or indemnity is entered as other farm income. Other death losses are listed on line 35, Part II of Form 1040 Schedule F, as "other deductions," with an explanation.

■ **Losses from destruction, theft, and condemnation** — Special treatment is available for certain gains or losses that are netted. The gains and losses can arise from the sale or exchange of property used in the trade or business, involuntary conversion or condemnation. If gains exceed losses, the net gain is treated as long-term capital gain. If losses exceed gains, the loss is ordinary. While this has limited significance as long as there is no tax differential between capital gain and ordinary income, the likelihood of a reintroduced capital gain preference makes the matter important to keep in mind.

■ **Passive activity losses** — Perhaps the most complicated addition to tax law in recent years was the passive activity loss concept, a development with many questions yet to be answered. Under this concept, all income and losses are divided between passive and nonpassive activities. A passive activity is one which involves the trade or business in which the livestock producer does not materially participate. Losses and credits from passive trade or business activities are disallowed to the extent they exceed aggregate passive income. Passive income does not include portfolio income (interest, dividends or royalties). However, rental activities are (if within the definition provided in the Internal Revenue Code) always passive.

■ **Material participation**—The IRS has provided seven exclusive tests for meeting the material participation requirement as to a particular activity:

1. **The 500 hours test.** The dairy owner participates more than 500 hours in the operation during the year. Obviously, full-time dairy producers will not have significant difficulties in meeting this requirement.

2. **Substantially all test.** The farmer's participation constitutes substantially all participation in the activity. Given the requirements for the care of animals, it is unlikely that this test is even necessary for producers, as they would then satisfy the first test in any case.

3. **The 100 hours test.** The individual participates for more than 100 hours and no other person participates for a greater number of hours. Again, this will not generally be relevant to dairy farmers. Nevertheless, a physician who owns animals and has a full-time employee to take care of them will often fail to be an active participant under this test.

4. **The related activities test.** The dairy farmer participates in a group of activities for more than 500 hours, more than 100 hours in each. This may apply where a producer has a number of operations, but only limited involvement in each.

5. **The 5 of 10 years test.** This allows a dairy farmer who has materially participated in the particular activity in the past to qualify as materially participating presently, even if his/her direct involvement has fallen off somewhat.

6. **Personal service activities.** This would apply to consultants involved in the dairy industry, but not to dairy producers running their own operations.

7. **Facts and circumstances test.** This test is, according to many experts, essentially similar to the 100 hours test.

Also, certain retired dairy farmers will qualify in the event of death during the year. Though these requirements will have no effect on the full-time farmer, they are important factors in terms of investment planning for anyone who is considering investments in rental real estate activities and other ventures.

■ **At-risk rules**—Another provision in the Internal Revenue Code limits losses to the extent that a taxpayer is at risk with respect to a particular activity. This means generally that a taxpayer is limited to the amount of his/her personal investment and the amount as to which he/she is personally liable. This provision specifically applies to farming, which includes livestock activities. The provision was designed principally to preclude losses from tax shelters and other leveraged investments where there may be no real chance that the taxpayer will have to cover the losses. Thus, it will seldom affect livestock producers whose credit is generally limited to the amount of collateral they can provide.

AVOID OPERATING THE BUSINESS AS A HOBBY

If an activity is not engaged in for profit, deductions are generally not available for the conduct of the activity except to the extent of income from it. (Actually, because the expenses of a hobby are limited to the extent of 2% of adjusted gross income, such expenses will not be fully deductible.) This requirement has often been applied when the IRS determines that a dairy operation is actually a hobby. Though the problem will generally not apply to full-time dairy farmers, others who devote a smaller amount of their time to an operation may find their activity is classified by the IRS as a hobby.

The general presumption for activities is that if an activity is profitable for 3 of the 5 consecutive years before the year being audited, it will be presumed to be engaged in for profit.

In determining whether a livestock operation is a business or a hobby, the IRS will examine the following factors:

1. **The manner in which the operator carries on the activity.** The more businesslike the conduct of the activity, the more likely it is to be recognized as a business. This includes the keeping of accurate records of income and expenses. If the operation is conducted in a manner similar to other profit-making livestock operations, it is more likely to be recognized as a business. If operating methods and procedures are changed because of losses, the impression is enhanced that the operation is a business. If the operation is typical of the other operations in the vicinity, it may indicate an attempt to fit into the livestock industry.

2. **The expertise of the operator and the employees.** A study of the industry and of other successful operations indicates a profit-making approach. If operators tend to ignore advice, they may have to establish that their expertise is even greater than that of their advisors.

3. **Time and effort spent in carrying out the operation.** The more time the owner devotes to the activity as a business and not as a recreational pursuit, the more likely the Service will find that the operation is a business. If the owner hires a full-time assistant to run day-to-day operations, he/she will be in a stronger position to argue that an attempt is being made to turn a profit. If the assistant is an inexperienced family member, the owner's position may, on the other hand, be weakened. Proper and rigid culling of herds will enhance the evidence for business conduct.

4. **The expectation that assets used in the activity will appreciate in value.** Even if current operations do not produce much income, the investment in land and buildings may support an argument that the owner has taken other businesslike factors into consideration. If the primary focus of the operation is breeding, it may take considerable time to get the necessary stock.

5. **Prior successes of the livestock producer.** The more experienced the producer and the more successful his/her prior livestock operations, the more he/she is likely to be seen as a serious business person. It may be important that the producer comes from a family of successful livestock producers.

6. **The operation's history of income and losses.** If losses are due to unforeseen circumstances (drought, disease, fire, theft, weather damages or other involuntary con-

versions, or from depressed markets), it may be possible to argue that there was nevertheless a profit motive in the operation.

7. **Occasional profits.** An occasional profit may indicate a profit motive if the investment or the losses of other years are comparatively small. The more speculative the venture, the more the livestock producer may be able to show that the losses were not due to a lack of profit intent.

8. **Financial status of the livestock producer.** The more the producer relies on the livestock operation, the more likely the producer is able to justify it as a business. If there are substantial profits from other sources, it may appear that the operation is nothing more than a private tax shelter. If this is the case, the producer may also have to worry about the effect of the limits on passive activity losses.

9. **Elements of recreation or pleasure.** Though having fun does not mean an operation is a hobby, the more the recreational element dominates the livestock producer's involvement, the more likely the livestock producer will have difficulty convincing the IRS that he/she is trying to make a profit. The presence of fishing holes, tennis courts, and guest houses may indicate that the producer has a country club (a different sort of business, but not a livestock operation).

PAY WITHHOLDING AND EMPLOYMENT TAXES

The Service has been increasingly aggressive in reclassifying independent contractors as employees because of its perception that many employers have been trying to avoid paying over withholding and employment taxes. Unfortunately, there is no one rule by which the status of someone working for a farmer can be determined. Generally, a worker is an employee (and not an independent contractor) if the farmer has the right to direct and control the worker's performance. The Service looks at no less than 20 factors to determine whether a worker is an employee. Thus, if the worker also serves others, basically controls how he/she accomplishes an assignment, may incur a loss in doing it, and furnishes his/her own supplies and training, he/she may be an independent contractor. Penalties for failure to make withholding and employment taxes are high, and farmers should consult professional tax advisors if they plan to treat regular workers as independent contractors.

ESTATE PLANNING

Human nature being what it is, most livestock producers shy away from suggestions that someone help plan the disposition of their property and other assets after they are gone. Also, they have a long-standing distrust of lawyers, legal terms, and trusts; and to them the subject of taxes on death seldom makes for pleasant conversation.

If a farmer has prepared a valid will, or placed the property in joint tenancy, the estate will be distributed as intended. If not, it goes to the heirs, according to the laws governing intestate (without a will) succession. The heirs are those persons whom the law appoints to succeed to the property in the event of intestacy, and are not necessarily the persons to whom the farmer would want to leave the property. These laws vary somewhat from state to state.

If no plans are made, estate taxes and settlement costs often run considerably higher than if proper estate planning is done. Today, dairying is big business; many have well over $1 million invested in land, animals and equipment. Thus, it is not a satisfying thought to those who have worked hard to build and maintain a good dairy during their lifetime to feel that the heirs will have to sell the facilities and animals to raise enough cash to pay estate and inheritance taxes. Therefore, dairy farmers should go to an estate planning specialist—a lawyer or company specializing in this work, or the trust department of a commercial bank. A discussion of some of the major considerations follows:

■ **Valuation can be based on farming use**—Owners of farms and small businesses have been granted an estate planning advantage by means of what is called *special use valuation.* Under this concept, a farm can escape valuation for estate tax purposes at the highest and best use. Thus, a farm located in an area undergoing development may be considerably more valuable to developers than it is as a farm. Nevertheless, if the family is willing to continue the farming or ranching use for 10 years, the farm can be included in the estate at its value as a farm. The aggregate reduction in fair market value cannot exceed $750,000.

In order to qualify for special use valuation, the decedent must have been a U.S. citizen or resident and the farm must be located in the United States. The farm must have been used by the decedent or a family member at the date of the decedent's death. A lease to a nonfamily member, if not dependent on production, will not satisfy this requirement. At least 50% of the value of the decedent's estate must consist of the farm and more than 25% of the estate must consist of the farm and real property. It may be possible to split up a farm and take the special valuation for only part of it, but this part must involve real property worth at least 25% of the estate.

The property must be passed to a qualified heir, including ancestors of the decedent, the spouse and lineal descendants, lineal descendants of the spouse or parents, and the spouse of any lineal descendant. Aunts, uncles and first cousins are excluded. Legally adopted children are included.

The property must have been owned by the decedent or a family member for 5 of the 8 years preceding the decedent's death and used as a farm in that period. The decedent or a family member must have participated in the farming operation for such a period prior to the decedent's death or disability.

■ **Electing special use valuation**—Though the procedures are clear as to how special use valuation is elected, the frequency with which mistakes are made indicates the importance of having a competent tax attorney or CPA firm prepare the estate tax return. A procedural failure denying the estate the considerable savings that can be gained by the election may give sufficient grounds for a malpractice suit against the return preparer.

■ **Recapture tax**—If the farm ceases to be operated by the heir or a family member within 10 years, an additional estate tax will be imposed and the advantage of the election will be substantially lost. Partition among qualified heirs will not bring about recapture. When heirs granted oil leases on a family farm, the portion of the land devoted to the oil rigs was subject to recapture. A recent change allows the surviving spouse of the decedent to lease a farm on a net cash basis to a family member without being subject to the recapture tax.

■ **Longer time to pay estate taxes**—Estates eligible for special use valuation may often be able to defer payment of estate taxes. Where more than 35% of an estate of a U.S. citizen or resident consists of a farm, the estate tax liability may be paid in up to 10 annual installments beginning as late as 5 years from when the tax might otherwise be due. Thus, a portion of the estate taxes is deferred as much as 15 years. For purposes of the 35% requirement, the residential buildings and improvements on them which are on the farm are considered to be part of the farming operation.

If more than 50% of the decedent's interest in the farm is disposed of in the deferral period, then the entire unpaid portion of the estate tax liability is accelerated. The transfer of the decedent's interest in a closely held business on the death of the original heir will not cause an acceleration if the transferee is a family member of the transferor.

■ **Use the gift tax exclusion for lifetime transfers**—The nontaxable gift tax exclusion remains at $10,000 per donee per year. A husband and wife who elect gift-splitting may jointly give $20,000 per recipient per year. These gifts may be in the form of interest in the farming operation.

■ **Plan with the unlimited marital deduction**—An unlimited deduction is permitted for the value of all property included in the gross estate that passes to the decedent's surviving spouse in the specified manner. Certain *terminable* interests do not qualify for such a deduction—that is, interests as to which of the surviving spouse's interests will terminate on the happening of some event. Surviving spouses may be given *qualified terminable interests*. The most common arrangement involves the surviving spouse receiving a lifetime interest in the farm, with the remainder passing on her/his death to others, perhaps the children of the decedent. No marital deduction is allowed if the surviving spouse is not a U.S. citizen, unless a specific trust arrangement is used.

■ **Consult a professional**—The preparation of wills, trusts, redemption agreements (if the farm is incorporated), partnership agreements, etc., requires consideration of the effects of federal and state tax law, as well as state law governing the various potential arrangements. Consequently, it is strongly advised that a competent professional be consulted in order to achieve an effective and cost-saving estate plan.

WILLS

A will is a set of instructions drawn up by or for an individual which details how he/she wishes the estate to be handled after death.

Despite the importance of a will in distributing property in keeping with the individual's wishes, about 50% of farmers and ranchers pass away without having written a will. This means the state law determines property distribution in such cases.

Every farmer/rancher should have a will. By so doing, (1) the property will be distributed in keeping with their wishes, (2) they can name the executor of the estate, and (3) sizable tax savings can be made by the way in which the property is distributed. Because technical and legal rules govern the preparation, validity, and execution of a will, it should be drawn up by an attorney. Wills can and should be changed and updated from time to time. This can be done either by (1) a properly drawn-up codicil (formal amendment to a will), or (2) a completely new will which revokes the old one.

The same attorney should prepare both the husband's and wife's wills so that a common disaster clause can be incorporated and the estate planning of each can be coordinated.

TRUSTS

A trust is a written agreement by which an owner of property (the trustor) transfers title to a trustee for the benefit of persons called beneficiaries. Both real and personal property may be placed in trust.

The trustee may be an individual(s), bank, or corporation, or a combination of two or three of these. Management skill should be considered carefully in choosing a trustee.

A trust can continue for any period of time set by the owner—for a lifetime, until the youngest child reaches age 21, etc. If the trust extends beyond a lifetime, there are limitations which should be explained by an attorney.

KINDS OF TRUSTS

Basically, there are two kinds of trusts, the *living* and the *testamentary*. The living or *inter vivos* trust is in essence an agreement between the trustor and the trustee and may be revocable or irrevocable.

The *revocable trust* can be terminated or altered; under it the trustor is concerned about the here and now, rather than only the hereafter. The trustor continues to make decisions, and can call off the whole arrangement (it is revocable) if it does not work out as expected. The revocable trust offers no special estate tax advantage; the assets of a revocable trust are included in the estate of the deceased creating the trust. However, it can be written in such a manner as to reduce substantially the estate taxes of the beneficiaries. Also, the revocable trust will eliminate the cost of probate—costs which may include executor's fees, attorney's fees, court costs, and appraisal fees.

The *irrevocable trust* cannot be amended, altered, revoked, or terminated. Under an irrevocable trust, the trustor must be willing to part with the trust property forever (irrevocably) and have nothing further to do with it and its administration. However, the irrevocable trust has many favorable aspects in estate planning; it will reduce estate taxes in both the estate of the trustor and the estate(s) of the life beneficiaries, and it avoids probate.

The *testamentary trust* is so-called because it is established under the provisions of the trustor's last will and testament. The testamentary trust does not become effective until after death of the trustor, followed by probate, there is no tax saving in the trustor's estate. However, the trust may be drafted to save estate taxes in the estates of the beneficiaries. A testamentary trust is useful when the heirs are minors or inexperienced in money matters.

PARTNERSHIP CONTRACT

Another logical step in the transfer of property is a partnership contract between the parents and their heirs(s) recorded in accordance with law. Appropriate counsel should be consulted in the preparation of such an agreement. Because of recently added *estate freeze* provisions, this approach to control of a farm may not provide the estate tax savings it could previously.

LIABILITY INSURANCE, WORKER'S COMPENSATION INSURANCE

Most farmers are in such financial position that they are vulnerable to damage suits. Moreover, the number of damage suits arising each year is increasing at an almost alarming rate, and astronomical damages are being claimed. Studies reveal that about 95% of the court cases involving injury result in damages being awarded.

Several types of liability insurance offer a safeguard against liability suits brought as a result of injury suffered by another person or damage to another's property.

Comprehensive personal liability insurance protects farm operators who are sued for alleged damages suffered from an accident involving their property or family. The kinds of situations from which a claim might arise are quite broad, including suits for injuries caused by animals, equipment, or personal acts.

Both worker's compensation and employer's liability insurance protect farmers against claims or court awards resulting from injury to hired help. Worker's compensation usually costs slightly more than straight employer's liability insurance, but it carries more benefits to the worker. An injured employee must prove negligence by the employer before the company will pay a claim under employer's liability insurance, whereas worker's compensation benefits are established by state law and settlements are made by the insurance company without regard to who was negligent in causing the injury. Conditions governing participation in worker's compensation insurance vary among the states.

LIVESTOCK INSURANCE

The ownership of a fine animal constitutes a risk; which means that there is a chance of financial loss. Unless the owner is in sufficiently strong financial position to assume this risk, the animal should be insured.

When insuring animals, the 10 pertinent points that follow should be considered, according to the American Live Stock Insurance Company:

1. Livestock mortality insurance is written for the purpose of protecting the actual investment of the livestock owner, not potential gain or profit.

2. A mortality policy cannot be construed in any way as a maintenance coverage; it does not include veterinarian or similar expenses.

3. Indemnity is payable only as a result of death loss.

4. Mortality coverage does not indemnify an insured against loss of an animal's ability to perform the functions for which it is kept.

5. Death from natural or accidental causes is included but mandatory slaughter by governmental authority or decree, or for expediency is not included.

6. The basis for valuing an animal should be actual sales price or fair and conservative appraisal by competent judges when no actual sales transaction has taken place. These values shall be subjected to acceptance by the company.

7. Mortality insurance is renewable only on evidence of reinsurability, both as to physical condition and market value.

8. Cancellation may only be effected by the insured, or by the company on notice given in conformation with whatever existing laws govern for the address of the insured as shown on the policy. Short rate basis if ordered by insured and *pro rata* basis if by company.

9. Policies may not be transferred from one insured to another unless agreed to through endorsement by company, nor may cover be switched from one animal to another unless agreed to by company.

10. Application subject to acceptance by company.

The rates of American Live Stock Insurance Company, Geneva, Illinois, which is rated A+ (Superior) by A. M. Best Company, the independent rating service, for cattle follow:

CATTLE—BEEF OR DAIRY—INSURANCE RATES

Conditions	Rate
Age limits, 3 mo–7 yr	
15-day term	$1.50/$100
(A 15-day policy may be endorsed "15 day cover to have its effective starting date at time of actual shipment.")	
1-mo term	$2.50/$100
2-mo term	$3.00/$100
3-mo term	$3.50/$100
6-mo term	$4.00/$100
1-yr term	$6.00/$100
Age exceptions—note added premium:	
Calves, 2–7 wk old	$4.00/$100
Calves, 7 wk–3 mo old	$2.00/$100
(This is an additional premium to be added to period coverage and considered earned in entirety when written.)	
Bulls past 6, up to eighth birthday, eligible for insurance after amount of cover has been confirmed by company. One dollar per hundred additional premium charged for each year or part over seventh birthday. *At ninth birthday, insurance not available.*	
Cows past 7 may be covered in same manner. *At tenth birthday annual insurance not available.*	

Generally, special stipulations and rates apply to (1) group (herd) insurance for cattle, (2) 4-H and FFA calves. For information relative to these, or other special types of coverage, the owner should make inquiry of a livestock insurance agent.

QUESTIONS FOR STUDY AND DISCUSSION

1. Why have the business aspects of dairy production become so important in recent years?

2. List and discuss each of the three major types of business organizations commonly found on dairy farms, including family owned and tax-option corporations.

3. Discuss the magnitude of U.S. farm assets vs U.S. farm debts. How much in farm assets is required to produce $1 in net farm income?

4. Assume that you are going to enter the dairy business, and that you have decided on the particular kind (with you making the decisions between market milk production, raising replacement heifers, producing dairy beef, etc.). What types of credit may be needed, how would you go about obtaining it, and from what source(s) would you hope to obtain the money?

5. What is a feasibility study? How would you go about making such a study or having such a study made?

6. In calculating interest, explain how *three points* would affect a $10,000 loan.

7. How may a student acquire the traits of a good manager?

8. Develop an organization chart for a dairy with which you are familiar.

9. Take your own dairy establishment, or one with which you are familiar, and develop a workable incentive basis for the help.

10. How may dairy producers use Table 20–4, Milk Production Costs and Returns, Per Cwt, 1987–89? How would you analyze a dairy business to determine whether it is profitable?

11. Develop a yearly budget for your own dairy establishment, or for one with which you are familiar.

12. How may computers be used, on a practical basis, for a dairy enterprise?

13. List the three requisites for linear programming a dairy establishment.

14. List the basic considerations for (a) a heifer replacement contract, (b) a cow rental contract, and (c) a cow pool.

15. Why does the Chicago Mercantile Exchange have a sentimental spot market for butter every Friday from 10:00 to 10:03 a.m. — three minutes each week?

16. What constitutes good tax management and reporting?

17. Explain the primary difference between (a) the cash basis, and (b) the accrual basis of accounting/tax reporting.

18. What is meant by a depreciation schedule? What property on a dairy would logically come under each of the following depreciation schedules: (a) 5-year property, (b) 10-year property, (c) 27.5-year property, and (d) 31.5-year property?

19. List and discuss briefly the seven exclusive tests which the IRS has provided for meeting the material participation requirement.

20. Discuss the importance of each of the following: (a) estate planning, (b) wills, (c) trusts, (d) liability insurance/worker's compensation insurance, and (e) livestock insurance.

SELECTED REFERENCES

Title of Publication	Author(s)	Publisher
Agricultural Statistics, 1988	Staff	U.S. Department of Agriculture, Washington, DC, 1988
Fact Book of Agriculture 1989	Staff	U.S. Department of Agriculture, Washington, DC, 1989
Introduction to Agribusiness Management, An, Second Edition	W. J. Wills	The Interstate Printers & Publishers, Inc., Danville, IL, 1979
Microcomputing in Agriculture	J. Legacy T. Stitt F. Reneau	Reston Publishing Company, Inc., Reston, VA, 1984
Spreadsheet Applications For Animal Nutrition and Feeding	R. J. Lane T. L. Gross	Reston Publishing Company, Inc., Reston, VA, 1985
Statistical Abstracts of the United States 1989	Staff	U. S. Department of Commerce, Washington, DC, 1989
Stockman's Handbook, The, Seventh Edition	M. E. Ensminger	Interstate Publishers, Inc., Danville, IL, 1992

Fig. 20–7. Dairy farm in northeastern United States. (Courtesy, United Cooperative Farmers, Inc.)

Fig. 20–8. Long, open-sided dairy barn with free stalls on a dairy with more than 3,000 lactating cows. (Courtesy, Maddox Dairy, Riverdale, CA)

Fig. 20–9. A 24-stall polygon herringbone milking parlor. (Courtesy, Arizona Dairy, Higley, AZ)

Fig. 21-1. Wheat straw in a reflux condenser being analyzed for cellulose and lignin. (Courtesy, USDA)

Contents

CHAPTER

21

FEED COMPOSITION
TABLES

In addition to the discussion that follows pertaining to composition of feeds, the reader is referred to Chapters 11 and 12 of this book.

Both nutritionists and dairy producers should have access to accurate and up-to-date composition of feedstuffs in order to formulate rations for maximum production and net returns. The ultimate goal of feedstuff analysis, and the reason for feed composition tables, is to be able to predict the productive responses of dairy cattle when they are fed rations of a given composition. In recognition of this need and its importance, the author spared no time or expense in compiling the feed composition tables presented in Chapter 21. At the outset, a survey of the industry was made in order to determine what kind of feed composition tables would be most useful, in both format and content. Secondly, it was decided to utilize, to the extent available, the feed compositions which, for many years, were compiled by Lorin Harris, Utah State University, now carried forward by the USDA, National Agricultural Library, Feed Composition Bank. These data were augmented by the author with feed compositions from the National Academy of Sciences, NRC, and from experimental reports, industries, and other reliable sources.

In Table 21–1, the commonly used dairy feeds are listed alphabetically, on both an As-Fed and Moisture-Free basis. Also, for each feed, the chemical analysis; digestible protein (ruminant); TDN, digestible energy, metabolizable energy, and net energy; and mineral and vitamin compositions are given. Values for each feed are presented in tabular form in a four-page spread.

Note well: In addition to the feed compositions presented in Table 21–1, should the need arise for additional feed compositions, and for additional mineral supplement compositions and vitamin supplement compositions, many more feeds are presented in *Feeds & Nutrition*, and *Feeds & Nutrition Digest*, books of which Dr. Ensminger is the senior author.

FEED NAMES

Ideally, a feed name should conjure up the same meaning to all those who use it, and it should provide helpful information. This was the guiding philosophy of the author when choosing the names given in the Feed Composition Table. Genus and species—Latin names—are also included. To facilitate worldwide usage, the International Feed Number of each feed is given. To the extent possible, consideration was also given to source (or parent material), variety or kind, stage of maturity, processing, part eaten, and grade.

Where feeds are known by more than one name, cross-referencing was used.

MOISTURE CONTENT OF FEEDS

It is necessary to know the moisture content of feeds in ration formulation and buying. Usually, the composition of a feed is expressed according to one or more of the following bases:

1. **As-Fed; A-F (wet, fresh).** This refers to feed as

normally fed to dairy cattle. As-Fed may range from near 0% to 100% dry matter.

2. **Air-Dry (approximately 90% dry matter).** This refers to feed that is dried by means of natural air movement, usually in the open. It may either be an actual or an assumed dry matter content; the latter is approximately 90%. Most feeds are fed in an air-dry state.

3. **Moisture-Free; M-F (oven-dry, 100% dry matter).** This refers to a sample of feed that has been dried in an oven at 221°F until all the moisture has been removed.

Formulas for adjusting moisture content from moisture-free to as-fed, or from as-fed to moisture-free, are given in *Feeds & Nutrition* and *Feeds & Nutrition Digest*, books of which Dr. Ensminger is the senior author.

CAROTENE

Where carotene has been converted to vitamin A, the conversion rate of the rat has been used as the standard value, with 1 mg of β-carotene equal to 1,667 IU of vitamin A.

PERTINENT INFORMATION ABOUT DATA

The information which follows is pertinent to Table 21–1.

■ **Variations in composition**—Feeds vary in their composition. Thus, actual analysis of a feedstuff should be obtained and used whenever possible, especially where a large lot of feed from one source is involved. Many times, however, it is either impossible to determine actual compositions or there is insufficient time to obtain such analyses. Under such circumstances, tabulated data may be the only information available.

■ **Feed compositions change**—Feed compositions change over a period of time, primarily due to (1) the introduction of new varieties, and (2) modifications in the manufacturing process from which by-products evolve.

■ **Biological value**—The response of animals when fed a feed is termed the biological value, which is a function of its chemical composition and the ability of the animal to derive useful nutrient value from the feed. The latter relates to the digestibility, or availability, of the nutrients in the feed. Thus, soft coal and shelled corn may have the same gross energy value in a bomb calorimeter but markedly different useful energy values when consumed by dairy cattle. Biological tests of feeds are more laborious and costly than chemical analyses, but they are much more accurate in predicting the response of animals to a feed.

■ **Where information is not available**—Where information is not available or reasonable estimates could not be made, no values are shown. Hopefully, such information will become available in the future.

■ **Calculated on a dry matter (DM) basis**—All data were calculated on a 100% dry matter basis (moisture-free), then converted to an as-fed basis by multiplying the decimal equivalent of the DM content times the compositional value shown in the table.

- **Fiber** — Four values relating to dietary fiber are given in the feed composition tables — crude fiber, neutral detergent fiber (NDF), acid detergent fiber (ADF), and lignin.

Crude fiber, methods for the determination of which were developed more than 100 years ago, is declining as a measure of low digestible material in the more fibrous feeds. The newer method of forage analysis, developed by Van Soest and associates of the U.S. Department of Agriculture, separates feed dry matter into two fractions: a neutral detergent fibrous fraction; and an acid detergent fibrous fraction. Also the amount of lignin in the ADF may be determined.

1. **Crude fiber (CF).** This fraction is an indicator of the relative indigestibility and bulkiness of the sample. It is the residue that remains after boiling a feed in a weak acid, and then in a weak alkali, in an attempt to imitate the process that occurs in the digestive tract. This procedure is based on the supposition that carbohydrates which are readily dissolved also will be readily digested by animals, and that those not soluble under such conditions are not readily digestible. Unfortunately, the treatment dissolves much of the lignin, a nondigestible component. Hence, crude fiber is only an approximation of the indigestible material in feedstuffs. Nevertheless, it is a rough indicator of the energy value of feeds. Also, the crude fiber value is needed for the computation of TDN.

2. **Neutral detergent fiber (NDF).** This is the fraction of the feed which is not soluble in neutral detergent. It consists of plant cell walls, including lignin, cellulose, and hemicellulose. NDF is closely related to feed intake because it contains all the fiber components that occupy space in the rumen and are slowly digested. The lower the NDF, the more forage the animal will eat; hence, a low percentage of NDF is desirable.

3. **Acid detergent fiber (ADF).** This is the fraction of the feed which is not soluble in acid detergent. It consists of cellulose (digestible) and lignin (indigestible). ADF is an indicator of forage digestibility because it contains a high proportion of lignin which is the indigestible fiber fraction. The lower the ADF, the more feed an animal can digest; hence, a low percentage of ADF is desirable.

4. **Lignin.** This fraction is essentially indigestible by all animals and is the substance that limits the availability of cellulose carbohydrates in the plant cell wall to rumen bacteria.

The acid detergent fiber procedure is used as a preparatory step in determining the lignin content of a forage sample. Hemicellulose is solubilized during this procedure, while the lignocellulose fraction of the feed remains insoluble. Cellulose is then separated from lignin by the addition of sulfuric acid. Only lignin and acid-insoluble ash remain upon completion of this step. This residue is then ashed, and the difference of the weights before and after ashing yields the amount of lignin present in the feed.

- **Nitrogen-free extract** — The nitrogen-free extract was calculated with mean data as: mean nitrogen-free extract (%) = 100 – % ash – % crude fiber – % ether extract – % protein.

- **Protein values** — Crude protein is determined by finding the nitrogen content and multiplying the result by 6.25. The nitrogen content of proteins averages about 16% (100 ÷ 16 = 6.25).

- **Energy** — The following four measures of energy are shown:

1. **TDN.** This value is given because there are more of them, and because it has been the standard method of expressing the energy value of feeds for many years. However, the following disadvantages are inherent in the TDN system: (a) Only digestive losses are considered — it does not take into account other important losses, such as those in the urine, gases, and increased heat production; (b) there is a poor relationship between crude fiber and NFE digestibility in certain feeds; and (c) it overestimates roughages in relation to concentrates when animals are fed for high rates of production, due to the higher heat loss per pound of TDN in high-fiber feeds.

2. **Digestible energy (DE).** Digestible energy is that portion of the gross energy in a feed that is not excreted in the feces. It is roughly comparable to TDN.

3. **Metabolizable energy (ME).** Metabolizable energy represents that portion of the gross energy that is not lost in the feces, urine and gas (mainly methane). It does not take into account the energy lost as heat, commonly called heat increment. As a result, it overevaluates roughages compared with concentrates, as does TDN and DE.

4. **Net energy (NE).** Net energy represents the energy fraction in a feed that is left after the fecal, urinary, gas, and heat losses are deducted from the GE. Because of its greater accuracy, net energy is being used increasingly in ration formulations, especially in computerized formulations for large operations.

Net energy is presented as: (a) net energy for maintenance (NE_m), (b) net energy for gain (NE_g), and (c) net energy for lactation (NE_{lc}).

- **Minerals** — The level of minerals in forages is largely determined by the mineral content of the soil on which the feeds are grown. Calcium, phosphorus, iodine, and selenium are well-known examples of soil nutrient–plant nutrient relationships.

- **Vitamins** — Generally speaking, it is unwise to rely on harvested feeds as a source of carotene (vitamin A value), unless the forage being fed is fresh (pasture or green chop) or of a good green color and not over a year old.

The author is very grateful to Lorin E. Harris, Ph.D., and Clyde R. Richards, Ph.D., Utah State University, Logan, for their interest and invaluable assistance in preparing the Feed Composition Tables for this book.

TABLE
COMPOSITION OF FEEDS, DATA EXPRESSED

Entry Number	Feed Name Description	International Feed Number	Moisture Basis: A-F (as-fed) or M-F (moisture-free)	Dry Matter (%)	Ash (%)	Crude Fiber (%)	Neutral Det. Fib. (NDF) (%)	Acid Det. Fib. (ADF) (%)	Lignin (%)	Ether Extract (Fat) (%)	N-Free Extract (%)	Crude Protein (%)
	ALFALFA (LUCERNE) *Medicago sativa*											
1	HAY, SUN-CURED, ALL ANALYSES	1–00–078	A-F	90	8.6	28.2	35.4	30.9	8.9	1.7	35.9	16.0
			M-F	100	9.5	31.2	39.2	34.2	9.8	1.9	39.7	17.7
2	HAY, PREBLOOM, SUN-CURED	1–00–054	A-F	90	8.3	20.7	38.3	29.6	5.5	4.1	36.4	20.2
			M-F	100	9.2	23.1	42.7	33.0	6.1	4.5	40.6	22.5
3	HAY, EARLY BLOOM, SUN-CURED	1–00–059	A-F	91	8.4	25.8	38.8	29.0	5.8	2.6	35.8	17.9
			M-F	100	9.2	28.5	40.7	32.0	6.4	2.9	39.6	19.8
4	HAY MIDBLOOM, SUN-CURED	1–00–063	A-F	91	7.8	25.5	43.2	33.4	6.7	3.3	37.4	17.1
			M-F	100	8.6	28.0	47.4	36.7	7.4	3.6	41.1	18.8
5	HAY, FULL BLOOM, SUN-CURED	1–00–068	A-F	91	7.1	27.3	45.0	35.2	6.9	3.1	37.9	15.5
			M-F	100	7.8	30.1	49.5	38.7	7.6	3.4	41.7	17.0
6	MEAL, DEHY, 15% PROTEIN	1–00–022	A-F	90	9.1	26.6	45.9	36.9	—	2.2	37.0	15.6
			M-F	100	10.0	29.4	51.0	41.0	—	2.5	40.9	17.3
7	MEAL, DEHY, 17% PROTEIN	1–00–023	A-F	92	9.7	24.0	41.3	31.5	9.7	2.8	37.8	17.4
			M-F	100	10.6	26.2	45.0	34.3	10.6	3.0	41.2	18.9
8	MEAL, DEHY, 20% PROTEIN	1–00–024	A-F	92	10.2	20.8	38.6	28.5	—	3.3	37.1	20.2
			M-F	100	11.1	22.7	42.0	31.0	—	3.6	40.5	22.1
9	MEAL, DEHY, 22% PROTEIN	1–07–851	A-F	93	10.2	18.3	36.3	26.0	—	4.1	38.1	22.2
			M-F	100	11.0	19.8	39.0	28.0	—	4.4	41.0	23.9
10	SILAGE, ALL ANALYSES	3–00–212	A-F	27	2.6	8.6	—	10.7	—	0.9	10.4	4.7
			M-F	100	9.4	31.7	—	39.2	—	3.2	38.2	17.4
11	SILAGE, PREBLOOM, WILTED	3–00–215	A-F	35	3.0	9.8	—	—	—	1.4	13.3	7.2
			M-F	100	8.6	28.4	—	—	—	4.0	38.3	20.7
12	SILAGE, EARLY BLOOM, WILTED	3–00–216	A-F	35	2.8	11.0	—	13.7	—	1.1	13.9	5.8
			M-F	100	8.2	31.8	—	39.6	—	3.2	40.2	16.7
13	SILAGE, MIDBLOOM, WILTED	3–00–217	A-F	38	3.0	11.9	—	—	—	1.2	15.6	6.4
			M-F	100	7.9	31.2	—	—	—	3.1	41.1	16.8
14	SILAGE, FULL BLOOM, WILTED	3–00–218	A-F	36	2.8	12.1	—	—	—	1.0	15.1	5.4
			M-F	100	7.7	33.2	—	—	—	2.7	41.5	14.9
	ALMOND *Prunus amygdalus*											
15	HULLS	4–00–359	A-F	90	6.1	13.5	28.8	25.2	—	2.9	63.8	4.1
			M-F	100	6.8	14.9	32.0	28.0	—	3.2	70.6	4.5
16	SHELLS, GROUND	1–07–754	A-F	92	4.0	39.6	—	—	—	3.7	40.0	4.4
			M-F	100	4.4	43.1	—	—	—	4.0	43.6	4.8
	APPLE *Malus* spp											
17	POMACE, WET	4–00–424	A-F	23	1.1	4.7	—	—	—	1.1	14.6	1.3
			M-F	100	5.0	20.7	—	—	—	4.7	64.0	5.6
18	POMACE, DEHY	4–00–423	A-F	89	3.0	16.2	—	—	—	4.3	61.2	4.4
			M-F	100	3.4	18.2	—	—	—	4.8	68.6	5.0
	BAHIAGRASS *Paspalum notatum*											
19	HAY, SUN-CURED	1–00–462	A-F	90	5.7	28.1	66.6	33.9	5.1	1.8	46.0	8.5
			M-F	100	6.3	31.2	73.9	37.7	5.7	2.0	51.1	9.5
	BAKERY WASTE											
20	DEHY (DRIED BAKERY PRODUCT)	4–00–466	A-F	91	3.7	1.2	—	1.6	—	10.9	65.3	10.1
			M-F	100	4.0	1.3	—	1.8	—	12.0	71.6	11.1
	BARLEY *Hordeum vulgare*											
21	GRAIN, ALL ANALYSES	4–00–549	A-F	88	2.4	5.0	16.8	10.7	1.5	1.7	67.7	11.7
			M-F	100	2.7	5.7	19.0	12.1	1.7	1.9	76.5	13.2
22	GRAIN, PACIFIC COAST	4–07–939	A-F	89	2.5	6.5	—	—	—	2.0	68.2	9.5
			M-F	100	2.8	7.3	—	—	—	2.2	76.9	10.8
23	GRAIN SCREENINGS	4–00–542	A-F	89	3.1	7.5	—	9.8	—	2.3	64.5	11.5
			M-F	100	3.4	8.4	—	11.0	—	2.6	72.6	13.0

21-1
AS-FED AND MOISTURE-FREE

Entry Number	Digestible Protein Ruminant (%)	TDN Ruminant (%)	Digestible Energy Ruminant (Mcal/lb)	(Mcal/kg)	Metabolizable Energy Ruminant (Mcal/lb)	(Mcal/kg)	Ruminant NE$_m$ (Mcal/lb)	(Mcal/kg)	Ruminant NE$_g$ (Mcal/lb)	(Mcal/kg)	Lactating Cows NE$_{lc}$ (Mcal/lb)	(Mcal/kg)	Calcium (Ca) (%)	Phosphorus (P) (%)	Sodium (Na) (%)	Chlorine (Cl) (%)
1	11.2	51	1.03	2.28	0.90	1.99	0.48	1.06	0.25	0.55	0.49	1.07	1.28	0.24	0.07	0.33
	12.4	56	1.14	0.52	0.99	2.20	0.53	1.18	0.28	0.61	0.54	1.18	1.42	0.26	0.08	0.37
2	15.3	54	1.25	2.75	1.02	2.24	0.57	1.26	0.34	0.74	0.58	1.27	1.34	0.30	0.10	0.31
	17.0	60	1.39	3.07	1.13	2.50	0.64	1.41	0.37	0.82	0.65	1.42	1.50	0.33	0.12	0.34
3	13.3	52	1.15	2.54	0.96	2.12	0.60	1.33	0.36	0.80	0.55	1.21	1.48	0.20	0.14	0.34
	14.7	58	1.27	2.80	1.06	2.35	0.67	1.47	0.40	0.88	0.61	1.33	1.63	0.22	0.15	0.38
4	12.0	52	1.12	2.46	0.94	2.07	0.52	1.14	0.28	0.62	0.51	1.13	1.27	0.22	0.11	0.34
	13.2	57	1.23	2.70	1.03	2.28	0.57	1.25	0.31	0.68	0.56	1.24	1.39	0.24	0.12	0.38
5	11.3	51	1.09	2.39	0.88	1.95	0.49	1.09	0.26	0.57	0.52	1.15	1.08	0.22	0.06	—
	12.5	56	1.19	2.63	0.97	2.14	0.54	1.20	0.29	0.63	0.58	1.27	1.19	0.24	0.07	—
6	10.9	53	1.07	2.35	0.89	1.97	0.54	1.20	0.31	0.68	0.56	1.23	1.24	0.22	0.07	0.44
	12.1	59	1.18	2.60	0.99	2.18	0.60	1.33	0.34	0.75	0.62	1.36	1.37	0.24	0.08	0.48
7	12.6	55	1.12	2.47	0.96	2.12	0.60	1.32	0.36	0.79	0.57	1.25	1.40	0.23	0.10	0.47
	13.7	60	1.22	2.69	1.05	2.32	0.65	1.44	0.39	0.86	0.62	1.37	1.52	0.25	0.11	0.52
8	15.1	57	1.07	2.36	0.93	2.06	0.58	1.28	0.34	0.75	0.59	1.30	1.59	0.28	0.11	0.47
	16.4	62	1.17	2.58	1.02	2.25	0.64	1.40	0.37	0.82	0.64	1.42	1.74	0.31	0.13	0.51
9	16.4	60	1.20	2.65	1.02	2.26	0.63	1.39	0.38	0.84	0.63	1.38	1.69	0.30	0.12	0.52
	17.7	65	1.29	2.85	1.10	2.43	0.68	1.49	0.41	0.90	0.67	1.49	1.82	0.33	0.13	0.56
10	3.4	16	0.28	0.61	0.27	0.59	0.16	0.36	0.09	0.20	0.16	0.35	0.48	0.07	0.04	0.11
	12.3	59	1.02	2.25	0.98	2.17	0.59	1.31	0.33	0.74	0.58	1.29	1.75	0.27	0.16	0.41
11	5.0	22	0.44	0.97	0.37	0.82	0.22	0.49	0.13	0.28	0.22	0.49	—	—	—	—
	14.5	63	1.27	2.79	1.08	2.37	0.64	1.40	0.37	0.82	0.64	1.42	—	—	—	—
12	3.5	18	0.39	0.86	0.33	0.72	0.17	0.36	0.08	0.17	0.18	0.40	—	—	—	—
	10.0	52	1.13	2.49	0.94	2.07	0.48	1.05	0.22	0.50	0.53	1.16	—	—	—	—
13	4.5	23	0.47	1.03	0.39	0.87	0.23	0.51	0.13	0.29	0.24	0.52	—	—	—	—
	11.8	61	1.22	2.70	1.03	2.28	0.61	1.34	0.35	0.76	0.62	1.37	—	—	—	—
14	3.6	21	0.43	0.94	0.36	0.79	0.20	0.45	0.11	0.24	0.21	0.47	—	—	—	—
	9.8	59	1.17	2.59	0.98	2.16	0.56	1.23	0.30	0.66	0.59	1.29	—	—	—	—
15	0.0	66	1.34	2.95	1.15	2.54	0.77	1.70	0.51	1.12	0.66	1.46	0.19	0.09	—	—
	0.0	73	1.48	3.27	1.28	2.82	0.85	1.88	0.56	1.24	0.73	1.61	0.21	0.10	—	—
16	1.1	49	0.98	2.16	0.80	1.76	0.46	1.02	0.23	0.51	0.50	1.11	—	—	—	—
	1.2	53	1.06	2.35	0.87	1.92	0.51	1.12	0.25	0.56	0.55	1.21	—	—	—	—
17	−0.4	17	0.33	0.73	0.29	0.64	0.17	0.38	0.11	0.24	0.17	0.37	0.02	0.02	—	—
	−1.7	74	1.46	3.21	1.27	2.79	0.76	1.67	0.48	1.06	0.74	1.62	0.10	0.10	—	—
18	−0.4	60	1.21	2.66	1.04	2.28	0.66	1.46	0.42	0.91	0.64	1.42	0.11	0.10	0.12	—
	−0.4	68	1.35	2.98	1.16	2.56	0.74	1.63	0.47	1.03	0.72	1.59	0.13	0.12	0.14	—
19	4.7	46	0.96	2.12	0.79	1.74	0.41	0.90	0.18	0.40	0.46	1.01	0.45	0.20	—	—
	5.2	51	1.07	2.36	0.88	1.93	0.45	1.00	0.20	0.45	0.51	1.13	0.50	0.22	—	—
20	6.3	82	1.62	3.57	1.43	3.16	1.00	2.20	0.70	1.54	0.85	1.88	0.14	0.22	1.02	1.47
	6.9	89	1.78	3.91	1.57	3.46	1.09	2.41	0.77	1.69	0.94	2.06	0.15	0.24	1.12	1.61
21	8.8	75	1.55	3.42	1.17	2.57	0.78	1.73	0.52	1.16	0.82	1.81	0.05	0.34	0.03	0.12
	10.0	85	1.75	3.86	1.32	2.91	0.89	1.95	0.59	1.31	0.93	2.05	0.06	0.39	0.03	0.13
22	7.1	75	1.51	3.32	1.34	2.96	0.81	1.77	0.54	1.20	0.76	1.67	0.05	0.34	0.02	0.15
	8.1	85	1.70	3.75	1.51	3.34	0.91	2.00	0.61	1.35	0.86	1.89	0.06	0.39	0.02	0.17
23	8.9	71	1.43	3.14	1.26	2.77	0.78	1.73	0.52	1.15	0.74	1.64	0.32	0.29	0.02	—
	10.0	80	1.60	3.53	1.42	3.12	0.88	1.94	0.59	1.30	0.84	1.84	0.36	0.33	0.02	—

TABLE 21–1

Entry Number	Feed Name Description	Moisture Basis: A-F (as-fed) or M-F (moisture-free)	Macro Minerals (Continued)			Micro Minerals						
			Magnesium (Mg)	Potassium (K)	Sulfur (S)	Cobalt (Co)	Copper (Cu)	Iodine (I)	Iron (Fe)	Manganese (Mn)	Selenium (Se)	Zinc (Zn)
			(%)	(%)	(%)	(ppm or mg/kg)	(ppm or mg/kg)	(ppm or mg/kg)	(%)	(ppm or mg/kg)	(ppm or mg/kg)	(ppm or mg/kg)
	ALFALFA (LUCERNE) *Medicago sativa*											
1	HAY, SUN-CURED, ALL ANALYSES	A-F	0.30	1.85	0.25	0.25	10.0	—	0.02	41.4	—	21.9
		M-F	0.33	2.05	0.28	0.28	11.0	—	0.02	45.8	—	24.3
2	HAY, PREBLOOM, SUN-CURED	A-F	0.19	2.25	0.48	0.26	10.2	—	0.02	42.2	—	33.5
		M-F	0.21	2.51	0.54	0.29	11.4	—	0.02	47.1	—	37.4
3	HAY, EARLY BLOOM, SUN-CURED	A-F	0.31	2.32	0.27	0.26	11.4	—	0.02	32.8	0.50	27.3
		M-F	0.34	2.56	0.30	0.29	12.6	—	0.02	36.2	0.55	30.2
4	HAY MIDBLOOM, SUN-CURED	A-F	0.32	1.42	0.26	0.36	16.1	—	0.02	55.1	—	28.1
		M-F	0.35	1.56	0.28	0.39	17.7	—	0.02	60.5	—	30.9
5	HAY, FULL BLOOM, SUN-CURED	A-F	0.25	1.42	0.27	0.21	9.0	—	0.02	38.5	—	23.7
		M-F	0.27	1.56	0.30	0.23	9.9	—	0.02	42.3	—	26.1
6	MEAL, DEHY, 15% PROTEIN	A-F	0.28	2.24	0.18	0.17	9.5	0.12	0.03	27.8	0.28	19.4
		M-F	0.31	2.48	0.20	0.19	10.5	0.13	0.03	30.8	0.31	21.4
7	MEAL, DEHY, 17% PROTEIN	A-F	0.29	2.38	0.23	0.30	8.6	0.15	0.04	31.0	0.34	19.3
		M-F	0.32	2.60	0.25	0.33	9.3	0.16	0.05	33.8	0.37	21.1
8	MEAL, DEHY, 20% PROTEIN	A-F	0.33	2.41	0.50	0.26	12.2	0.14	0.04	45.2	0.29	21.8
		M-F	0.36	2.63	0.55	0.28	13.3	0.15	0.04	49.4	0.31	23.8
9	MEAL, DEHY, 22% PROTEIN	A-F	0.31	2.40	0.30	0.31	9.8	0.17	0.04	36.4	0.53	19.5
		M-F	0.33	2.58	0.32	0.34	10.5	0.18	0.04	39.2	0.58	21.0
10	SILAGE, ALL ANALYSES	A-F	0.09	0.64	0.09	—	3.0	—	0.01	13.5	—	11.1
		M-F	0.33	2.35	0.31	—	11.1	—	0.03	49.7	—	40.7
11	SILAGE, PREBLOOM, WILTED	A-F	—	—	—	—	—	—	—	—	—	—
		M-F	—	—	—	—	—	—	—	—	—	—
12	SILAGE, EARLY BLOOM, WILTED	A-F	—	—	—	—	—	—	—	—	—	—
		M-F	—	—	—	—	—	—	—	—	—	—
13	SILAGE, MIDBLOOM, WILTED	A-F	—	—	—	—	—	—	—	—	—	—
		M-F	—	—	—	—	—	—	—	—	—	—
14	SILAGE, FULL BLOOM, WILTED	A-F	—	—	—	—	—	—	—	—	—	—
		M-F	—	—	—	—	—	—	—	—	—	—
	ALMOND *Prunus amygdalus*											
15	HULLS	A-F	—	0.48	0.10	—	—	—	—	—	—	—
		M-F	—	0.53	0.11	—	—	—	—	—	—	—
16	SHELLS, GROUND	A-F	—	—	—	—	—	—	—	—	—	—
		M-F	—	—	—	—	—	—	—	—	—	—
	APPLE *Malus* spp											
17	POMACE, WET	A-F	—	0.11	—	—	—	—	—	—	—	—
		M-F	—	0.47	—	—	—	—	—	—	—	—
18	POMACE, DEHY	A-F	0.06	0.43	0.02	—	—	—	0.03	7.2	—	—
		M-F	0.07	0.49	0.02	—	—	—	0.03	8.1	—	—
	BAHIAGRASS *Paspalum notatum*											
19	HAY, SUN-CURED	A-F	0.17	—	—	—	—	—	0.01	—	—	—
		M-F	0.19	—	—	—	—	—	0.01	—	—	—
	BAKERY WASTE											
20	DEHY (DRIED BAKERY PRODUCT)	A-F	0.16	0.40	0.02	1.22	11.0	—	0.02	65.0	—	17.8
		M-F	0.18	0.43	0.02	1.34	12.1	—	0.02	71.2	—	19.5
	BARLEY *Hordeum vulgare*											
21	GRAIN, ALL ANALYSES	A-F	0.13	0.46	0.15	0.17	7.6	0.04	0.01	16.0	0.16	39.3
		M-F	0.15	0.52	0.17	0.19	8.6	0.05	0.01	18.1	0.18	44.4
22	GRAIN, PACIFIC COAST	A-F	0.12	0.51	0.14	0.09	8.1	—	0.01	16.0	0.10	15.2
		M-F	0.14	0.58	0.16	0.10	9.1	—	0.01	18.0	0.11	17.1
23	GRAIN SCREENINGS	A-F	0.12	0.80	0.13	—	—	—	0.01	—	—	—
		M-F	0.14	0.90	0.15	—	—	—	0.01	—	—	—

(Continued)

Entry Number	Fat-Soluble Vitamins					Water-Soluble Vitamins								
	A	Carotene (Provitamin A)	D	E	K	B–12	Biotin	Choline	Folacin (Folic Acid)	Niacin	Pantothenic Acid (B–3)	Pyridoxine (B–6)	Riboflavin (B–2)	Thiamin (B–1)
	(IU/g)	(ppm or mg/kg)	(IU/kg)	(ppm or mg/kg)	(ppm or mg/kg)	(ppb or mcg/kg)	(ppm or mg/kg)	(ppm or mg/kg)	(ppm or mg/kg)	(ppm or mg/kg)	(ppm or mg/kg)	(ppm or mg/kg)	(ppm or mg/kg)	(ppm or mg/kg)
1	45.0	27.0	2	55.9	—	—	0.18	892	3.07	43	18.1	—	9.5	3.1
	49.8	29.9	2	61.9	—	—	0.20	988	3.40	48	20.1	—	10.5	3.4
2	300.2	180.1	—	—	—	—	—	—	—	—	—	—	—	—
	335.1	201.0	—	—	—	—	—	—	—	—	—	—	—	—
3	210.9	126.5	2	23.5	—	—	—	—	—	—	—	—	—	—
	233.0	139.8	2	26.0	—	—	—	—	—	—	—	—	—	—
4	50.5	30.3	1	—	—	—	—	—	—	—	—	—	9.6	—
	55.5	33.3	2	—	—	—	—	—	—	—	—	—	10.6	—
5	98.5	59.1	—	—	—	—	—	—	—	—	—	—	—	—
	108.4	65.0	—	—	—	—	—	—	—	—	—	—	—	—
6	124.1	74.5	—	81.9	9.59	—	0.25	1573	1.56	42	20.7	6.27	10.6	3.0
	137.3	82.3	—	90.6	10.61	—	0.28	1739	1.73	46	22.9	6.94	11.7	3.3
7	200.3	120.2	—	105.7	8.24	—	0.33	1369	4.37	37	29.7	7.18	12.9	3.4
	218.5	131.1	—	115.3	8.98	—	0.36	1494	4.77	40	32.4	7.83	14.1	3.7
8	265.4	159.2	—	143.3	14.19	—	0.35	1417	2.96	48	35.5	8.72	15.2	5.4
	289.7	173.8	—	156.4	15.50	—	0.39	1547	3.24	52	38.8	9.52	16.6	5.9
9	391.4	234.8	—	221.3	11.65	—	0.33	1605	5.15	50	39.0	8.28	17.6	5.9
	421.8	253.0	—	238.4	12.55	—	0.36	1729	5.55	54	42.0	8.92	19.0	6.3
10	42.2	25.3	—	—	—	—	—	—	—	—	—	—	—	—
	155.0	93.0	—	—	—	—	—	—	—	—	—	—	—	—
11	—	—	—	—	—	—	—	—	—	—	—	—	—	—
	—	—	—	—	—	—	—	—	—	—	—	—	—	—
12	—	—	—	—	—	—	—	—	—	—	—	—	—	—
	—	—	—	—	—	—	—	—	—	—	—	—	—	—
13	—	—	—	—	—	—	—	—	—	—	—	—	—	—
	—	—	—	—	—	—	—	—	—	—	—	—	—	—
14	—	—	—	—	—	—	—	—	—	—	—	—	—	—
	—	—	—	—	—	—	—	—	—	—	—	—	—	—
15	—	—	—	—	—	—	—	—	—	—	—	—	—	—
	—	—	—	—	—	—	—	—	—	—	—	—	—	—
16	—	—	—	—	—	—	—	—	—	—	—	—	—	—
	—	—	—	—	—	—	—	—	—	—	—	—	—	—
17	—	—	—	—	—	—	—	—	—	—	—	—	—	—
	—	—	—	—	—	—	—	—	—	—	—	—	—	—
18	—	—	—	—	—	—	—	—	—	—	—	—	—	—
	—	—	—	—	—	—	—	—	—	—	—	—	—	—
19	—	—	—	—	—	—	—	—	—	—	—	—	—	—
	—	—	—	—	—	—	—	—	—	—	—	—	—	—
20	7.0	4.2	—	40.9	—	—	0.07	917	0.19	26	8.2	4.29	1.4	2.9
	7.7	4.6	—	44.9	—	—	0.07	1005	0.20	28	9.0	4.70	1.5	3.2
21	3.4	2.0	—	23.2	0.22	—	0.15	1036	0.57	76	7.9	5.80	1.6	4.5
	3.8	2.3	—	26.2	0.24	—	0.17	1171	0.64	86	9.0	6.55	1.8	5.0
22	—	—	—	26.2	—	—	0.15	976	0.50	47	7.1	2.89	1.5	4.2
	—	—	—	29.6	—	—	0.17	1102	0.56	53	8.0	3.26	1.7	4.7
23	—	—	—	—	—	—	—	—	—	—	7.9	—	1.1	—
	—	—	—	—	—	—	—	—	—	—	8.9	—	1.2	—

TABLE 21–1

Entry Number	Feed Name Description	International Feed Number	Moisture Basis: A-F (as-fed) or M-F (moisture-free)	Chemical Analysis								
				Dry Matter	Ash	Crude Fiber	Neutral Det. Fib. (NDF)	Acid Det. Fib. (ADF)	Lignin	Ether Extract (Fat)	N-Free Extract	Crude Protein
				(%)	(%)	(%)	(%)	(%)	(%)	(%)	(%)	(%)
	BARLEY (Continued)											
24	MALT SPROUTS, DEHY	4–00–545	A-F	93	5.6	14.2	—	—	—	1.4	48.8	22.9
			M-F	100	6.1	15.3	—	—	—	1.5	52.6	24.6
25	HAY, SUN-CURED	1–00–495	A-F	88	6.6	23.6	—	—	—	1.9	48.5	7.8
			M-F	100	7.5	26.7	—	—	—	2.1	54.9	8.8
26	STRAW	1–00–498	A-F	91	6.7	37.9	77.5	51.1	6.9	1.7	41.1	4.0
			M-F	100	7.3	41.5	84.8	55.9	7.6	1.9	44.9	4.4
	BEAN											
27	SEEDS, NAVY	5–00–623	A-F	89	4.0	4.4	—	—	—	1.4	56.7	22.9
			M-F	100	4.5	5.0	—	—	—	1.5	63.4	25.6
	BEET, MANGEL *Beta vulgaris, macrorhiza*											
28	ROOTS, FRESH	4–00–637	A-F	11	1.1	0.8	—	—	—	0.1	7.7	1.3
			M-F	100	9.6	7.4	—	—	—	0.7	70.5	11.8
	BEET, SUGAR *Beta vulgaris, altissima*											
29	SILAGE, TOPS W/CROWNS	3–00–660	A-F	25	8.6	3.2	—	—	—	0.7	9.3	3.4
			M-F	100	34.0	12.8	—	—	—	2.9	37.0	13.4
30	PULP, WET	4–00–671	A-F	11	0.5	2.3	—	—	—	0.2	6.7	1.2
			M-F	100	4.7	21.3	—	—	—	2.1	60.8	11.2
31	PULP, DEHY	4–00–669	A-F	91	4.8	18.2	53.6	26.3	4.5	0.5	58.4	8.8
			M-F	100	5.3	20.1	59.0	29.0	5.0	0.6	64.3	9.7
32	PULP W/MOLASSES, DEHY	4–00–672	A-F	92	5.7	15.2	—	24.5	2.4	0.6	61.1	9.3
			M-F	100	6.2	16.6	—	26.6	2.6	0.6	66.5	10.1
	BERMUDAGRASS, COASTAL *Cynodon dactylon*											
33	HAY, SUN-CURED	1–00–716	A-F	91	6.3	27.0	69.2	34.6	—	2.0	43.9	11.7
			M-F	100	6.9	29.7	76.0	38.0	—	2.2	48.3	12.8
	BLOOD											
34	MEAL	5–00–380	A-F	91	5.3	1.0	—	—	—	1.3	3.1	80.5
			M-F	100	5.8	1.1	—	—	—	1.5	3.4	88.2
	BLUEGRASS, CANADA *Poa compressa*											
35	FRESH	2–00–764	A-F	31	2.8	8.3	—	—	—	1.2	13.8	5.3
			M-F	100	8.9	26.4	—	—	—	3.7	44.0	17.0
36	HAY, SUN-CURED	1–00–762	A-F	92	6.5	27.6	—	—	—	2.4	45.8	9.5
			M-F	100	7.0	30.1	—	—	—	2.6	49.9	10.3
	BLUEGRASS, KENTUCKY *Poa pratensis*											
37	FRESH, IMMATURE	2–00–777	A-F	31	2.9	7.8	17.1	9.0	—	1.1	13.7	5.4
			M-F	100	9.4	25.2	55.0	29.0	—	3.6	44.4	17.4
38	FRESH, EARLY BLOOM	2–00–779	A-F	35	2.5	9.6	22.8	11.2	—	1.4	15.7	5.8
			M-F	100	7.1	27.4	65.0	32.0	—	3.9	44.9	16.6
39	FRESH, MILK STAGE	2–00–782	A-F	42	3.1	12.7	28.6	16.0	—	1.5	19.8	4.9
			M-F	100	7.3	30.3	68.0	38.0	—	3.6	47.2	11.6
	BREWERS GRAINS											
40	WET	5–02–142	A-F	22	0.9	3.1	9.2	5.0	1.1	1.5	10.7	5.8
			M-F	100	3.9	14.1	42.0	23.0	5.0	6.8	48.7	26.4
41	DEHY	5–02–141	A-F	92	3.6	13.0	38.7	23.9	4.6	6.6	41.6	27.3
			M-F	100	4.0	14.1	42.0	26.0	5.0	7.1	45.2	29.6
	BROMEGRASS *Bromus spp*											
42	FRESH, IMMATURE	2–00–892	A-F	34	3.8	7.5	19.0	10.5	—	1.2	15.5	5.8
			M-F	100	11.4	22.1	56.0	31.0	—	3.7	45.8	17.1
43	FRESH, MATURE	2–00–898	A-F	56	3.7	18.5	40.3	24.6	—	1.2	29.0	3.6
			M-F	100	6.6	33.0	72.0	44.0	—	2.2	51.8	6.4
44	HAY, SUN-CURED	1–00–890	A-F	91	7.1	29.8	—	31.7	4.3	1.9	43.3	8.7
			M-F	100	7.8	32.9	—	34.9	4.7	2.1	47.7	9.5
45	HAY, MATURE, SUN-CURED	1–00–889	A-F	92	6.0	33.0	62.6	39.6	—	1.7	45.2	6.1
			M-F	100	6.6	35.9	68.0	43.0	—	1.8	49.1	6.6

(Continued)

Entry Number	Digestible Protein Ruminant (%)	TDN Ruminant (%)	Digestible Energy Ruminant (Mcal/lb)	(Mcal/kg)	Metabolizable Energy Ruminant (Mcal/lb)	(Mcal/kg)	Net Energy Ruminant NE_m (Mcal/lb)	(Mcal/kg)	Ruminant NE_g (Mcal/lb)	(Mcal/kg)	Lactating Cows NE_{lc} (Mcal/lb)	(Mcal/kg)	Calcium (Ca) (%)	Phosphorus (P) (%)	Sodium (Na) (%)	Chlorine (Cl) (%)
24	19.2	66	1.31	2.89	1.13	2.50	0.69	1.52	0.43	0.95	0.67	1.48	0.18	0.63	0.88	0.36
	20.7	71	1.41	3.11	1.22	2.69	0.74	1.63	0.47	1.03	0.72	1.60	0.19	0.68	0.95	0.39
25	4.3	50	0.99	2.17	0.79	1.73	0.45	0.99	0.22	0.49	0.50	1.10	0.21	0.25	0.12	—
	4.9	57	1.11	2.46	0.89	1.96	0.51	1.11	0.25	0.56	0.57	1.25	0.24	0.28	0.14	—
26	0.6	43	0.84	1.86	0.64	1.40	0.29	0.64	0.07	0.15	0.42	0.91	0.27	0.07	0.13	0.61
	0.7	47	0.92	2.03	0.70	1.54	0.32	0.70	0.08	0.16	0.45	1.00	0.30	0.07	0.14	0.67
27	20.2	76	1.53'	3.37	1.36	3.00	0.83	1.83	0.56	1.24	0.78	1.72	0.17	0.54	0.04	0.06
	22.5	85	1.71	3.77	1.52	3.36	0.93	2.05	0.63	1.39	0.87	1.93	0.19	0.61	0.05	0.06
28	0.9	9	0.18	0.39	0.15	0.34	0.09	0.21	0.06	0.14	0.09	0.20	0.02	0.02	0.07	0.16
	8.1	80	1.59	3.51	1.40	3.09	0.85	1.88	0.56	1.24	0.81	1.79	0.18	0.22	0.63	1.41
29	2.2	13	0.26	0.58	0.21	0.47	0.13	0.28	0.06	0.14	0.14	0.30	0.39	0.07	0.14	—
	8.7	52	1.04	2.29	0.84	1.86	0.50	1.11	0.25	0.55	0.55	1.20	1.56	0.28	0.54	—
30	0.7	8	0.16	0.35	0.14	0.30	0.08	0.17	0.05	0.11	0.08	0.17	0.10	0.01	0.02	—
	6.1	72	1.44	3.18	1.25	2.76	0.71	1.57	0.44	0.97	0.70	1.55	0.87	0.10	0.19	—
31	4.3	67	1.31	2.89	1.10	2.43	0.73	1.60	0.47	1.03	0.70	1.55	0.63	0.09	0.19	0.04
	4.7	74	1.44	3.18	1.21	2.68	0.80	1.76	0.52	1.14	0.77	1.70	0.70	0.10	0.21	0.04
32	6.1	69	1.39	3.05	1.21	2.67	0.73	1.61	0.47	1.04	0.71	1.55	0.56	0.09	0.48	—
	6.7	75	1.51	3.33	1.32	2.91	0.80	1.76	0.52	1.14	0.77	1.69	0.61	0.10	0.53	—
33	7.5	49	1.04	2.30	0.66	1.44	0.31	0.68	0.09	0.20	0.53	1.16	0.38	0.17	—	—
	8.2	54	1.15	2.53	0.72	1.59	0.34	0.75	0.10	0.22	0.58	1.28	0.42	0.18	—	—
34	57.2	61	1.20	2.65	0.99	2.19	0.63	1.38	0.38	0.84	0.60	1.33	0.29	0.25	0.32	0.30
	62.6	66	1.32	2.91	1.09	2.40	0.69	1.51	0.42	0.92	0.66	1.45	0.32	0.28	0.35	0.33
35	3.9	20	0.41	0.90	0.35	0.77	0.23	0.51	0.15	0.32	0.23	0.50	0.12	0.12	0.04	—
	12.5	65	1.30	2.86	1.11	2.44	0.74	1.63	0.47	1.03	0.72	1.60	0.39	0.39	0.14	—
36	4.1	57	1.14	2.52	0.97	2.13	0.61	1.35	0.37	0.81	0.61	1.35	0.28	0.24	0.10	—
	4.5	62	1.25	2.75	1.06	2.33	0.67	1.47	0.40	0.88	0.67	1.47	0.30	0.26	0.11	—
37	4.0	22	0.42	0.93	0.37	0.81	0.24	0.52	0.15	0.33	0.23	0.51	0.15	0.14	0.04	—
	12.9	72	1.38	3.03	1.19	2.61	0.77	1.70	0.49	1.08	0.75	1.64	0.50	0.44	0.14	—
38	4.2	24	0.49	1.07	0.42	0.93	0.26	0.57	0.16	0.36	0.25	0.55	0.16	0.14	0.05	—
	12.0	69	1.39	3.06	1.20	2.64	0.73	1.62	0.46	1.01	0.72	1.58	0.46	0.39	0.14	—
39	3.2	26	0.52	1.15	0.44	0.97	0.27	0.59	0.15	0.34	0.27	0.59	—	—	—	—
	7.6	62	1.24	2.73	1.05	2.31	0.63	1.39	0.37	0.81	0.64	1.41	—	—	—	—
40	4.2	15	0.32	0.70	0.27	0.60	0.18	0.40	0.12	0.26	0.17	0.38	0.06	0.12	0.06	0.03
	19.3	68	1.44	3.17	1.25	2.76	0.83	1.82	0.54	1.19	0.79	1.74	0.29	0.54	0.28	0.13
41	20.1	65	1.25	2.76	1.01	2.22	0.64	1.41	0.39	0.86	0.67	1.48	0.30	0.51	0.21	0.15
	21.8	71	1.36	2.99	1.09	2.41	0.69	1.53	0.42	0.93	0.73	1.61	0.33	0.55	0.23	0.17
42	4.7	25	0.50	1.11	0.44	0.97	0.24	0.53	0.15	0.33	0.24	0.52	0.20	0.13	0.01	—
	14.0	74	1.48	3.27	1.29	2.85	0.71	1.57	0.44	0.97	0.70	1.55	0.59	0.37	0.02	—
43	1.6	36	0.70	1.54	0.59	1.31	0.38	0.83	0.23	0.50	0.37	0.83	0.17	0.15	0.01	—
	2.9	65	1.25	2.75	1.06	2.33	0.67	1.47	0.40	0.88	0.67	1.47	0.30	0.26	0.02	—
44	4.8	48	1.12	2.48	0.95	2.09	0.50	1.11	0.27	0.60	0.50	1.11	0.32	0.15	0.03	—
	5.3	53	1.24	2.73	1.05	2.30	0.55	1.22	0.30	0.66	0.55	1.22	0.36	0.16	0.03	—
45	2.1	47	0.93	2.06	0.76	1.67	0.43	0.94	0.20	0.44	0.48	1.05	0.40	0.09	0.02	—
	2.3	51	1.02	2.24	0.82	1.81	0.46	1.02	0.21	0.47	0.52	1.14	0.43	0.09	0.02	—

TABLE 21–1

Entry Number	Feed Name Description	Moisture Basis: A-F (as-fed) or M-F (moisture-free)	Macro Minerals (Continued)			Micro Minerals						
			Magne-sium (Mg)	Potassium (K)	Sulfur (S)	Cobalt (Co)	Copper (Cu)	Iodine (I)	Iron (Fe)	Manga-nese (Mn)	Selenium (Se)	Zinc (Zn)
			(%)	(%)	(%)	(ppm or mg/kg)	(ppm or mg/kg)	(ppm or mg/kg)	(%)	(ppm or mg/kg)	(ppm or mg/kg)	(ppm or mg/kg)
	BARLEY (Continued)											
24	MALT SPROUTS, DEHY	A-F	0.17	0.25	0.79	—	5.9	—	0.02	29.4	0.42	58.4
		M-F	0.18	0.27	0.85	—	6.3	—	0.02	31.7	0.45	60.7
25	HAY, SUN-CURED	A-F	0.14	1.30	0.15	0.06	3.9	—	0.03	34.8	—	—
		M-F	0.16	1.47	0.17	0.07	4.4	—	0.03	39.4	—	—
26	STRAW	A-F	0.21	2.16	0.16	0.06	4.9	—	0.02	15.1	—	6.8
		M-F	0.23	2.37	0.17	0.07	5.4	—	0.02	16.6	—	7.4
	BEAN											
27	SEEDS, NAVY	A-F	0.13	1.31	0.23	—	9.9	—	0.01	21.1	—	—
		M-F	0.15	1.47	0.26	—	11.0	—	0.01	23.6	—	—
	BEET, MANGEL Beta vulgaris, macrorhiza											
28	ROOTS, FRESH	A-F	0.02	0.25	0.02	—	0.6	—	0.00	—	—	—
		M-F	0.20	2.30	0.20	—	5.5	—	0.00	—	—	—
	BEET, SUGAR Beta vulgaris, altissima											
29	SILAGE, TOPS W/CROWNS	A-F	0.27	1.45	0.14	—	—	—	0.01	—	—	—
		M-F	1.07	5.74	0.57	—	—	—	0.02	—	—	—
30	PULP, WET	A-F	0.02	0.02	0.02	—	—	—	0.00	—	—	1.1
		M-F	0.22	0.19	0.22	—	—	—	0.03	—	—	10.0
31	PULP, DEHY	A-F	0.26	0.18	0.20	0.07	12.5	—	0.03	34.2	—	0.7
		M-F	0.28	0.20	0.22	0.08	13.7	—	0.03	37.7	—	0.8
32	PULP W/MOLASSES, DEHY	A-F	0.15	1.63	0.39	0.21	14.7	—	0.02	18.4	—	5.1
		M-F	0.16	1.78	0.42	0.23	16.0	—	0.02	20.1	—	5.5
	BERMUDAGRASS, COASTAL Cynodon dactylon											
33	HAY, SUN-CURED	A-F	0.16	1.46	0.19	—	—	—	0.03	—	—	18.2
		M-F	0.17	1.61	0.21	—	—	—	0.03	—	—	20.0
	BLOOD											
34	MEAL	A-F	0.22	0.09	0.34	0.09	12.6	—	0.37	5.3	0.73	4.4
		M-F	0.24	0.10	0.37	0.10	13.8	—	0.41	5.8	0.80	4.8
	BLUEGRASS, CANADA Poa compressa											
35	FRESH	A-F	0.05	0.64	0.05	—	—	—	0.01	24.8	—	—
		M-F	0.16	2.04	0.17	—	—	—	0.03	79.1	—	—
36	HAY, SUN-CURED	A-F	0.30	1.73	0.12	—	—	—	0.03	84.9	—	—
		M-F	0.33	1.88	0.13	—	—	—	0.03	92.6	—	—
	BLUEGRASS, KENTUCKY Poa pratensis											
37	FRESH, IMMATURE	A-F	0.05	0.70	0.05	—	—	—	0.01	—	—	—
		M-F	0.18	2.27	0.17	—	—	—	0.03	—	—	—
38	FRESH, EARLY BLOOM	A-F	0.04	0.70	0.06	—	—	—	0.01	—	—	—
		M-F	0.11	2.01	0.17	—	—	—	0.03	—	—	—
39	FRESH, MILK STAGE	A-F	—	—	—	—	—	—	—	—	—	—
		M-F	—	—	—	—	—	—	—	—	—	—
	BREWERS GRAINS											
40	WET	A-F	0.03	0.02	0.07	0.02	4.9	—	0.01	9.0	—	23.2
		M-F	0.15	0.09	0.34	0.10	22.2	—	0.03	40.9	—	106.0
41	DEHY	A-F	0.15	0.09	0.30	0.08	21.7	0.07	0.02	37.2	—	27.3
		M-F	0.17	0.09	0.32	0.08	23.6	0.07	0.03	40.4	—	29.6
	BROMEGRASS Bromus spp											
42	FRESH, IMMATURE	A-F	0.06	1.46	0.07	—	—	—	0.01	—	—	—
		M-F	0.18	4.30	0.20	—	—	—	0.02	—	—	—
43	FRESH, MATURE	A-F	0.10	0.70	0.11	—	—	—	0.01	—	—	—
		M-F	0.18	1.25	0.20	—	—	—	0.02	—	—	—
44	HAY, SUN-CURED	A-F	0.09	1.49	0.18	—	—	—	0.02	—	—	—
		M-F	0.10	1.64	0.20	—	—	—	0.02	—	—	—
45	HAY, MATURE, SUN-CURED	A-F	0.08	1.75	—	—	—	—	—	—	—	—
		M-F	0.09	1.90	—	—	—	—	—	—	—	—

(Continued)

Entry Number	Fat-Soluble Vitamins					Water-Soluble Vitamins								
	A	Carotene (Provitamin A)	D	E	K	B–12	Biotin	Choline	Folacin (Folic Acid)	Niacin	Pantothenic Acid (B–3)	Pyridoxine (B–6)	Riboflavin (B–2)	Thiamin (B–1)
	(IU/g)	(ppm or mg/kg)	(IU/kg)	(ppm or mg/kg)	(ppm or mg/kg)	(ppb or mcg/kg)	(ppm or mg/kg)	(ppm or mg/kg)	(ppm or mg/kg)	(ppm or mg/kg)	(ppm or mg/kg)	(ppm or mg/kg)	(ppm or mg/kg)	(ppm or mg/kg)
24	—	—	—	3.7	—	—	4.09	1591	0.20	55	9.0	8.62	2.8	8.3
	—	—	—	4.0	—	—	4.40	1713	0.22	59	9.6	9.28	3.0	8.9
25	77.4	46.4	1	—	—	—	—	—	—	—	—	—	—	—
	87.5	52.5	1	—	—	—	—	—	—	—	—	—	—	—
26	3.5	2.1	1	—	—	—	—	—	—	—	—	—	—	—
	3.9	2.3	1	—	—	—	—	—	—	—	—	—	—	—
27	—	—	—	1.0	—	—	0.11	1017	1.29	24	2.4	0.30	1.7	6.3
	—	—	—	1.1	—	—	0.12	1136	1.44	27	2.7	0.33	1.9	7.1
28	0.2	0.1	—	—	—	—	—	—	0.17	3	1.0	0.43	0.4	0.3
	1.5	0.9	—	—	—	—	—	—	1.58	31	9.4	3.94	3.9	2.4
29	—	—	—	—	—	—	—	—	—	—	—	—	—	—
	—	—	—	—	—	—	—	—	—	—	—	—	—	—
30	—	—	—	—	—	—	—	—	—	—	—	—	—	—
	—	—	—	—	—	—	—	—	—	—	—	—	—	—
31	0.4	0.2	1	—	—	—	—	820	—	17	1.4	—	0.7	0.4
	0.4	0.2	1	—	—	—	—	902	—	18	1.5	—	0.8	0.4
32	0.4	0.2	—	—	—	—	—	814	—	16	1.5	—	0.7	—
	0.4	0.2	—	—	—	—	—	887	—	18	1.7	—	0.7	—
33	123.7	74.2	—	—	—	—	—	—	—	—	—	—	—	—
	136.2	81.7	—	—	—	—	—	—	—	—	—	—	—	—
34	—	—	—	—	—	44.3	0.09	780	0.10	31	2.3	4.41	2.0	0.3
	—	—	—	—	—	48.5	0.09	854	0.11	34	2.6	4.83	2.2	0.4
35	199.9	119.9	—	—	—	—	—	—	—	—	—	—	—	—
	637.6	382.5	—	—	—	—	—	—	—	—	—	—	—	—
36	378.4	227.0	—	—	—	—	—	—	—	—	—	—	—	—
	412.6	247.5	—	—	—	—	—	—	—	—	—	—	—	—
37	247.6	148.5	—	47.8	—	—	—	—	—	—	—	—	—	—
	803.4	481.9	—	155.0	—	—	—	—	—	—	—	—	—	—
38	163.4	98.0	—	—	—	—	—	—	—	—	—	—	—	—
	466.8	280.0	—	—	—	—	—	—	—	—	—	—	—	—
39	—	—	—	—	—	—	—	—	—	—	—	—	—	—
	—	—	—	—	—	—	—	—	—	—	—	—	—	—
40	—	—	—	5.5	—	—	—	—	—	—	—	—	—	—
	—	—	—	25.0	—	—	—	—	—	—	—	—	—	—
41	0.8	0.5	—	26.7	—	3.6	0.44	1651	0.22	44	8.2	1.03	1.5	0.6
	0.8	0.5	—	29.0	—	3.9	0.48	1792	0.24	47	8.9	1.11	1.6	0.7
42	259.6	155.7	—	—	—	—	—	—	—	—	—	—	—	—
	765.9	459.4	—	—	—	—	—	—	—	—	—	—	—	—
43	77.1	46.2	—	—	—	—	—	—	—	—	—	—	—	—
	137.5	82.5	—	—	—	—	—	—	—	—	—	—	—	—
44	50.3	30.2	—	—	—	—	—	—	—	—	—	—	—	—
	55.4	33.2	—	—	—	—	—	—	—	—	—	—	—	—
45	40.0	24.0	—	—	—	—	—	—	—	—	—	—	—	—
	43.5	26.1	—	—	—	—	—	—	—	—	—	—	—	—

TABLE 21-1

Entry Number	Feed Name Description	International Feed Number	Moisture Basis: A-F (as-fed) or M-F (moisture-free)	Chemical Analysis								
				Dry Matter	Ash	Crude Fiber	Neutral Det. Fib. (NDF)	Acid Det. Fib. (ADF)	Lignin	Ether Extract (Fat)	N-Free Extract	Crude Protein
				(%)	(%)	(%)	(%)	(%)	(%)	(%)	(%)	(%)
	BUCKWHEAT, COMMON *Fagopyrum sagittatum*											
46	GRAIN	4-00-994	A-F	88	2.1	10.6	—	14.9	—	2.4	61.5	11.1
			M-F	100	2.4	12.1	—	17.0	—	2.8	70.2	12.6
	CARROT *Daucus* spp											
47	ROOTS, FRESH	4-01-145	A-F	11	1.0	1.1	1.0	0.9	—	0.2	8.1	1.2
			M-F	100	8.4	9.5	9.0	8.0	—	1.3	70.7	10.0
	CASEIN											
48	ACID PRECIPITATED, DEHY	5-01-162	A-F	91	2.2	0.2	0.0	0.0	—	0.6	3.6	84.0
			M-F	100	2.4	0.2	0.0	0.0	—	0.7	3.9	92.7
	CITRUS *Citrus* spp											
49	PULP, SILAGE	4-01-234	A-F	21	1.2	3.3	—	4.2	—	2.1	12.8	1.5
			M-F	100	5.6	15.7	—	20.0	—	9.9	61.5	7.3
50	PULP W/O FINES, DEHY (DRIED CITRUS PULP)	4-01-237	A-F	91	6.0	11.6	20.9	20.0	—	3.4	63.9	6.1
			M-F	100	6.6	12.8	23.0	22.0	—	3.7	70.2	6.7
	CLOVER, ALSIKE *Trifoluim hybridum*											
51	FRESH	2-01-316	A-F	22	2.5	5.2	—	—	—	0.8	10.3	4.1
			M-F	100	9.3	23.3	—	—	—	3.6	45.7	18.1
52	HAY, SUN-CURED	1-01-313	A-F	88	7.6	26.3	—	—	—	2.4	39.1	12.4
			M-F	100	8.7	29.9	—	—	—	2.8	44.5	14.2
	CLOVER, CRIMSON *Trifolium incarnatum*											
53	FRESH	2-01-336	A-F	18	1.7	4.9	—	—	—	0.6	7.5	3.0
			M-F	100	9.5	27.7	—	—	—	3.3	42.6	17.0
54	HAY, SUN-CURED	1-01-328	A-F	88	7.8	28.1	—	—	—	2.0	35.2	14.7
			M-F	100	8.9	31.9	—	—	—	2.3	40.1	16.8
	CLOVER, LADINO *Trifolium repens*											
55	FRESH	2-01-383	A-F	18	1.9	2.5	—	—	—	0.9	8.1	4.4
			M-F	100	10.5	14.2	—	—	—	4.8	45.7	24.7
56	HAY, SUN-CURED	1-01-378	A-F	89	8.4	18.5	32.1	28.5	5.9	2.4	39.9	20.0
			M-F	100	9.4	20.8	36.0	32.0	6.6	2.7	44.7	22.4
	CLOVER, RED *Trifolium pratense*											
57	FRESH, EARLY BLOOM	2-01-428	A-F	20	2.0	4.6	8.0	6.2	—	1.0	8.3	3.8
			M-F	100	10.2	23.3	40.0	31.0	—	5.0	42.3	19.4
58	FRESH, FULL BLOOM	2-01-429	A-F	26	2.0	6.8	11.2	9.1	—	0.8	12.7	3.8
			M-F	100	7.8	26.1	43.0	35.0	—	2.9	48.6	14.6
59	HAY, SUN-CURED	1-01-415	A-F	88	6.7	27.1	49.5	36.2	8.8	2.5	39.2	13.0
			M-F	100	7.5	30.7	56.0	41.0	10.0	2.8	44.3	14.7
	COCONUT *Cocos nucifera*											
60	KERNELS W/COATS, MEAL MECH EXTD (COPRA MEAL)	5-01-572	A-F	92	6.4	12.1	—	18.3	—	6.8	45.0	21.2
			M-F	100	7.0	13.2	—	20.0	—	7.4	49.2	23.1
61	KERNELS W/COATS, MEAL SOLV EXTD (COPRA MEAL)	5-01-573	A-F	91	6.0	14.4	—	21.9	—	2.1	47.3	21.3
			M-F	100	6.6	15.8	—	24.0	—	2.3	52.0	23.4
	CORN, DENT YELLOW *Zea mays indentata*											
62	GRAIN, ALL ANALYSES	4-02-935	A-F	88	1.3	2.3	—	3.8	—	3.6	71.0	9.9
			M-F	100	1.5	2.6	—	4.3	—	4.1	80.7	11.2
63	GRAIN, HIGH MOISTURE	4-20-770	A-F	77	1.2	2.1	17.5	3.8	1.3	3.3	61.9	8.1
			M-F	100	1.6	2.7	22.9	4.9	1.7	4.3	80.8	10.6
64	GRAIN, FLAKED	4-28-244	A-F	89	0.9	0.6	8.0	2.7	—	2.0	75.4	9.9
			M-F	100	1.0	0.7	9.0	3.0	—	2.2	84.9	11.2
65	DISTILLERS GRAINS, DEHY	5-02-842	A-F	93	2.2	11.5	40.0	15.8	—	8.9	43.1	27.8
			M-F	100	2.4	12.3	43.0	17.0	—	9.5	46.2	29.7
66	DISTILLERS SOLUBLES, DEHY	5-28-237	A-F	93	7.2	4.6	21.4	6.5	—	8.6	45.2	27.4
			M-F	100	7.7	4.9	23.0	7.0	—	9.3	48.6	29.5
67	EARS, GROUND (CORN & COB MEAL)	4-28-238	A-F	87	1.7	8.2	24.4	9.6	—	3.2	65.7	7.8
			M-F	100	1.9	9.4	28.0	11.0	—	3.7	75.9	9.0

(Continued)

Entry Number	Digestible Protein Ruminant (%)	TDN Ruminant (%)	Digestible Energy Ruminant (Mcal/lb)	(Mcal/kg)	Metabolizable Energy Ruminant (Mcal/lb)	(Mcal/kg)	Ruminant NE$_m$ (Mcal/lb)	(Mcal/kg)	Ruminant NE$_g$ (Mcal/lb)	(Mcal/kg)	Lactating Cows NE$_{lc}$ (Mcal/lb)	(Mcal/kg)	Calcium (Ca) (%)	Phosphorus (P) (%)	Sodium (Na) (%)	Chlorine (Cl) (%)
46	8.0	63	1.26	2.77	1.09	2.40	0.65	1.43	0.41	0.90	0.63	1.40	0.10	0.33	0.05	0.04
	9.1	72	1.43	3.15	1.24	2.74	0.74	1.63	0.47	1.03	0.72	1.60	0.11	0.37	0.06	0.05
47	0.8	10	0.19	0.43	0.17	0.38	0.10	0.23	0.07	0.16	0.10	0.22	0.05	0.04	0.06	0.06
	7.3	84	1.69	3.72	1.50	3.30	0.91	2.00	0.61	1.35	0.86	1.89	0.40	0.35	0.48	0.50
48	81.5	81	1.56	3.44	1.22	2.68	0.82	1.81	0.55	1.22	0.78	1.72	0.61	0.82	0.01	—
	89.9	89	1.72	3.80	1.34	2.96	0.91	2.00	0.61	1.35	0.86	1.89	0.67	0.90	0.01	—
49	0.7	18	0.37	0.80	0.33	0.72	0.21	0.45	0.14	0.31	0.19	0.42	0.42	0.03	0.02	—
	3.5	88	1.76	3.87	1.57	3.46	0.29	2.18	0.68	1.50	0.92	2.04	2.04	0.15	0.09	—
50	2.7	75	1.41	3.10	1.12	2.46	0.74	1.63	0.48	1.06	0.77	1.71	1.69	0.12	0.07	—
	3.0	83	1.55	3.41	1.23	2.71	0.81	1.79	0.53	1.16	0.85	1.87	1.86	0.13	0.08	—
51	3.0	16	0.32	0.71	0.28	0.61	0.17	0.37	0.11	0.24	0.17	0.36	0.31	0.06	0.10	0.17
	13.5	71	1.42	3.14	1.23	2.72	0.76	1.67	0.48	1.06	0.74	1.62	1.36	0.29	0.45	0.77
52	8.3	51	1.01	2.24	0.85	1.86	0.49	1.08	0.26	0.58	0.51	1.13	1.14	0.22	0.40	0.68
	9.5	58	1.16	2.55	0.96	2.12	0.56	1.23	0.30	0.66	0.59	1.29	1.30	0.25	0.46	0.78
53	2.3	11	0.23	0.50	0.19	0.42	0.11	0.25	0.07	0.15	0.12	0.25	0.24	0.05	0.07	0.11
	13.1	65	1.29	2.84	1.10	2.42	0.65	1.43	0.38	0.85	0.65	1.44	1.38	0.29	0.40	0.61
54	10.2	50	1.00	2.21	0.84	1.84	0.54	1.19	0.31	0.68	0.55	1.21	1.23	0.19	0.34	0.55
	11.6	57	1.14	2.52	0.95	2.10	0.61	1.35	0.35	0.77	0.63	1.38	1.40	0.22	0.39	0.63
55	3.6	13	0.27	0.60	0.24	0.52	0.15	0.33	0.10	0.22	0.14	0.32	0.22	0.07	0.02	—
	20.2	76	1.53	3.37	1.34	2.95	0.86	1.89	0.57	1.25	0.82	1.80	1.27	0.42	0.12	—
56	15.4	58	1.16	2.55	0.99	2.18	0.58	1.27	0.34	0.75	0.58	1.28	1.30	0.30	0.12	0.27
	17.3	65	1.30	2.87	1.11	2.44	0.65	1.43	0.38	0.84	0.65	1.44	1.45	0.34	0.13	0.30
57	2.8	14	0.27	0.58	0.23	0.50	0.14	0.32	0.09	0.20	0.14	0.31	0.45	0.08	0.04	—
	13.9	69	1.34	2.96	1.15	2.54	0.73	1.61	0.46	1.01	0.72	1.58	2.26	0.38	0.20	—
58	2.7	17	0.34	0.75	0.29	0.64	0.17	0.38	0.10	0.22	0.17	0.38	0.27	0.07	0.05	—
	10.4	64	1.30	2.86	1.11	2.44	0.65	1.44	0.39	0.86	0.66	1.45	1.01	0.27	0.20	—
59	7.7	52	1.21	2.67	0.95	2.10	0.50	1.11	0.28	0.61	0.53	1.16	1.22	0.22	0.16	0.28
	8.7	59	1.37	3.02	1.08	2.38	0.57	1.26	0.31	0.69	0.60	1.31	1.38	0.25	0.18	0.32
60	17.6	74	1.52	3.34	1.33	2.94	0.84	1.85	0.57	1.25	0.80	1.75	0.19	0.60	0.04	—
	19.2	81	1.66	3.65	1.46	3.21	0.92	2.03	0.62	1.37	0.87	1.91	0.21	0.65	0.04	—
61	17.2	69	1.37	3.01	1.20	2.64	0.74	1.63	0.48	1.06	0.70	1.55	0.17	0.60	0.04	0.03
	18.8	75	1.50	3.31	1.31	2.90	0.81	1.79	0.53	1.16	0.77	1.70	0.19	0.66	0.04	0.03
62	8.1	80	1.63	3.59	1.26	2.78	0.86	1.90	0.59	1.31	0.82	1.81	0.05	0.28	0.01	0.05
	9.2	91	1.85	4.09	1.43	3.16	0.98	2.16	0.67	1.48	0.93	2.05	0.05	0.31	0.01	0.06
63	4.9	71	1.41	3.11	1.27	2.80	0.79	1.75	0.55	1.22	0.74	1.63	0.01	0.25	0.01	0.04
	6.4	92	1.85	4.07	1.66	3.66	1.04	2.28	0.72	1.59	0.96	2.12	0.02	0.33	0.01	0.05
64	6.2	78	1.57	3.45	1.41	3.09	0.88	1.94	0.49	1.34	0.83	1.82	—	—	—	—
	7.0	88	1.76	3.88	1.58	3.47	0.99	2.18	0.55	1.50	0.93	2.04	—	—	—	—
65	20.1	81	1.54	3.41	1.28	2.82	0.87	1.92	0.59	1.30	0.84	1.86	0.09	0.39	0.09	0.07
	21.5	87	1.65	3.64	1.37	3.02	0.93	2.05	0.63	1.39	0.90	1.99	0.10	0.42	0.09	0.08
66	21.5	80	1.55	3.42	1.34	2.96	0.92	2.03	0.63	1.40	0.86	1.89	0.30	1.30	0.23	0.26
	23.2	87	1.67	3.68	1.45	3.19	0.99	2.19	0.68	1.51	0.92	2.03	0.32	1.40	0.24	0.28
67	4.7	72	1.45	3.20	1.26	2.78	0.86	1.90	0.60	1.31	0.76	1.68	0.06	0.24	0.02	0.04
	5.5	83	1.68	3.70	1.46	3.21	1.00	2.20	0.69	1.52	0.88	1.95	0.07	0.27	0.02	0.05

TABLE 21–1

Entry Number	Feed Name Description	Moisture Basis: A-F (as-fed) or M-F (moisture-free)	Macro Minerals (Continued)			Micro Minerals						
			Magnesium (Mg)	Potassium (K)	Sulfur (S)	Cobalt (Co)	Copper (Cu)	Iodine (I)	Iron (Fe)	Manganese (Mn)	Selenium (Se)	Zinc (Zn)
			(%)	(%)	(%)	(ppm or mg/kg)	(ppm or mg/kg)	(ppm or mg/kg)	(%)	(ppm or mg/kg)	(ppm or mg/kg)	(ppm or mg/kg)
	BUCKWHEAT, COMMON *Fagopyrum sagittatum*											
46	GRAIN	A-F	0.10	0.45	0.14	0.05	9.5	—	0.01	33.7	—	8.8
		M-F	0.12	0.51	0.16	0.06	10.8	—	0.01	38.4	—	10.0
	CARROT *Daucus* spp											
47	ROOTS, FRESH	A-F	0.02	0.32	0.02	—	1.2	—	0.00	3.6	—	—
		M-F	0.20	2.80	0.17	—	10.4	—	0.00	31.5	—	—
	CASEIN											
48	ACID PRECIPITATED, DEHY	A-F	0.01	0.01	—	—	4.1	—	0.00	3.5	—	31.8
		M-F	0.01	0.01	—	—	4.5	—	0.00	3.9	—	35.1
	CITRUS *Citrus* spp											
49	PULP, SILAGE	A-F	0.03	0.13	0.00	—	—	—	0.00	—	—	3.3
		M-F	0.16	0.62	0.02	—	—	—	0.02	—	—	16.0
50	PULP W/O FINES, DEHY (DRIED CITRUS PULP)	A-F	0.15	0.71	0.17	0.17	5.0	—	0.03	6.6	—	13.7
		M-F	0.17	0.78	0.19	0.19	5.4	—	0.04	7.3	—	15.1
	CLOVER, ALSIKE *Trifoluim hybridum*											
51	FRESH	A-F	0.07	0.61	0.05	—	1.3	—	0.01	26.3	—	—
		M-F	0.32	2.70	0.22	—	6.0	—	0.04	117.1	—	—
52	HAY, SUN-CURED	A-F	0.40	1.95	0.17	—	5.3	—	0.02	60.5	—	—
		M-F	0.45	2.22	0.19	—	6.0	—	0.03	69.0	—	—
	CLOVER, CRIMSON *Trifolium incarnatum*											
53	FRESH	A-F	0.05	0.55	0.05	—	—	—	—	43.1	—	—
		M-F	0.29	3.10	0.28	—	—	—	—	245.8	—	—
54	HAY, SUN-CURED	A-F	0.25	2.10	0.25	—	—	0.06	0.06	183.3	—	—
		M-F	0.28	2.40	0.28	—	—	0.07	0.07	208.7	—	—
	CLOVER, LADINO *Trifolium repens*											
55	FRESH	A-F	0.09	0.33	0.02	—	—	—	0.01	12.7	—	—
		M-F	0.48	1.87	0.12	—	—	—	0.04	71.7	—	—
56	HAY, SUN-CURED	A-F	0.42	2.17	0.19	0.14	8.4	0.27	0.04	109.7	—	15.2
		M-F	0.47	2.44	0.21	0.16	9.4	0.30	0.05	123.1	—	17.0
	CLOVER, RED *Trifolium pratense*											
57	FRESH, EARLY BLOOM	A-F	0.10	0.49	0.03	—	—	—	0.01	—	—	—
		M-F	0.51	2.49	0.17	—	—	—	0.03	—	—	—
58	FRESH, FULL BLOOM	A-F	0.13	0.51	0.05	—	—	—	0.01	—	—	—
		M-F	0.51	1.96	0.17	—	—	—	0.03	—	—	—
59	HAY, SUN-CURED	A-F	0.34	1.60	0.15	0.14	18.8	0.22	0.02	95.2	—	32.5
		M-F	0.38	1.81	0.16	0.16	21.2	0.25	0.02	107.7	—	36.7
	COCONUT *Cocos nucifera*											
60	KERNELS W/COATS, MEAL MECH EXTD (COPRA MEAL)	A-F	0.30	1.65	0.34	0.13	16.7	—	0.07	70.1	—	48.5
		M-F	0.33	1.80	0.37	0.14	18.2	—	0.08	76.6	—	53.0
61	KERNELS W/COATS, MEAL SOLV EXTD (COPRA MEAL)	A-F	0.31	1.41	—	—	—	—	—	54.5	—	—
		M-F	0.34	1.55	—	—	—	—	—	59.8	—	—
	CORN, DENT YELLOW *Zea mays indentata*											
62	GRAIN, ALL ANALYSES	A-F	0.11	0.33	0.11	0.38	3.5	—	0.00	5.7	0.13	19.4
		M-F	0.13	0.37	0.13	0.43	4.0	—	0.00	6.4	0.14	22.0
63	GRAIN, HIGH MOISTURE	A-F	0.11	0.28	0.11	—	2.2	—	0.00	5.3	—	25.4
		M-F	0.15	0.37	0.14	—	2.9	—	0.00	6.9	—	33.2
64	GRAIN, FLAKED	A-F	—	—	—	—	—	—	—	—	—	—
		M-F	—	—	—	—	—	—	—	—	—	—
65	DISTILLERS GRAINS, DEHY	A-F	0.07	0.16	0.43	0.08	38.9	0.05	0.02	19.3	0.35	41.7
		M-F	0.07	0.18	0.46	0.08	41.7	0.05	0.02	20.7	0.38	44.7
66	DISTILLERS SOLUBLES, DEHY	A-F	0.60	1.70	0.37	0.17	77.9	0.08	0.05	72.0	0.37	88.0
		M-F	0.65	1.83	0.40	0.18	83.9	0.09	0.06	77.6	0.40	94.8
67	EARS, GROUND (CORN & COB MEAL)	A-F	0.12	0.46	0.14	0.27	6.8	0.02	0.01	19.9	0.07	12.1
		M-F	0.14	0.53	0.16	0.32	7.9	0.03	0.01	23.0	0.09	14.0

(Continued)

Entry Number	Fat-Soluble Vitamins					Water-Soluble Vitamins								
	A	Carotene (Provitamin A)	D	E	K	B–12	Biotin	Choline	Folacin (Folic Acid)	Niacin	Pantothenic Acid (B–3)	Pyridoxine (B–6)	Riboflavin (B–2)	Thiamin (B–1)
	(IU/g)	(ppm or mg/kg)	(IU/kg)	(ppm or mg/kg)	(ppm or mg/kg)	(ppb or mcg/kg)	(ppm or mg/kg)	(ppm or mg/kg)	(ppm or mg/kg)	(ppm or mg/kg)	(ppm or mg/kg)	(ppm or mg/kg)	(ppm or mg/kg)	(ppm or mg/kg)
46	—	—	—	—	—	—	—	439	—	18	11.5	—	4.7	3.7
	—	—	—	—	—	—	—	501	—	21	13.1	—	5.4	4.2
47	129.9	77.9	—	6.9	—	—	0.01	—	0.14	7	3.5	1.39	0.6	0.7
	1129.4	677.5	—	60.2	—	—	0.07	—	1.21	58	30.1	12.05	4.9	5.8
48	—	—	—	—	—	—	0.04	208	0.47	1	2.7	0.42	1.5	0.4
	—	—	—	—	—	—	0.05	229	0.52	1	2.9	0.47	1.7	0.5
49	—	—	—	—	—	—	—	—	—	—	—	—	—	—
	—	—	—	—	—	—	—	—	—	—	—	—	—	—
50	0.4	0.2	—	—	—	—	—	789	—	22	14.0	—	2.1	1.5
	0.4	0.2	—	—	—	—	—	867	—	24	15.4	—	2.3	1.6
51	—	—	—	—	—	—	—	—	—	—	—	—	4.4	2.0
	—	—	—	—	—	—	—	—	—	—	—	—	19.6	8.8
52	272.1	163.2	—	—	—	—	—	—	—	—	—	—	15.1	4.2
	310.1	186.0	—	—	—	—	—	—	—	—	—	—	17.2	4.8
53	—	—	—	—	—	—	—	—	—	—	—	—	—	—
	—	—	—	—	—	—	—	—	—	—	—	—	—	—
54	32.9	19.8	—	—	—	—	—	—	—	—	—	—	—	—
	37.5	22.5	—	—	—	—	—	—	—	—	—	—	—	—
55	96.2	57.7	—	—	—	—	—	—	—	—	—	—	4.2	—
	545.2	327.0	—	—	—	—	—	—	—	—	—	—	24.1	—
56	239.5	143.7	—	—	—	—	—	—	—	10	1.0	—	15.2	3.7
	268.7	161.2	—	—	—	—	—	—	—	11	1.1	—	17.0	4.2
57	81.5	48.9	—	—	—	—	—	—	—	—	—	—	—	—
	412.6	247.5	—	—	—	—	—	—	—	—	—	—	—	—
58	90.6	54.4	—	—	—	—	—	—	—	—	—	—	—	—
	345.9	207.5	—	—	—	—	—	—	—	—	—	—	—	—
59	40.5	24.3	—	—	—	—	0.09	—	—	38	9.9	—	15.7	2.0
	45.9	27.5	—	—	—	—	0.11	—	—	43	11.2	—	17.8	2.2
60	—	—	—	—	—	—	—	1046	1.08	27	6.1	—	3.3	0.8
	—	—	—	—	—	—	—	1143	1.18	30	6.6	—	3.6	0.9
61	—	—	—	—	—	—	—	1089	0.30	24	6.5	4.36	3.5	—
	—	—	—	—	—	—	—	1196	0.33	26	7.2	4.78	3.8	—
62	9.5	5.7	—	20.9	0.22	—	0.07	504	0.31	23	5.1	6.16	1.1	3.7
	10.8	6.5	—	23.8	0.25	—	0.08	573	0.35	26	5.8	7.01	1.2	4.2
63	—	—	—	—	—	—	—	—	—	—	—	—	—	—
	—	—	—	—	—	—	—	—	—	—	—	—	—	—
64	—	—	—	0.8	—	—	—	—	—	—	—	—	—	—
	—	—	—	0.9	—	—	—	—	—	—	—	—	—	—
65	5.2	3.1	—	—	—	0.3	0.41	1113	1.00	38	11.3	4.22	5.0	1.8
	5.6	3.3	—	—	—	0.3	0.44	1191	1.07	41	12.1	4.51	5.3	1.9
66	1.1	0.7	—	45.9	—	4.2	1.49	4751	1.34	124	23.3	9.41	15.1	6.8
	1.2	0.7	—	49.4	—	4.5	1.61	5116	1.45	133	25.0	10.14	16.3	7.3
67	5.3	3.2	—	17.5	—	—	0.03	357	0.24	17	4.2	5.97	0.9	2.9
	6.1	3.7	—	20.2	—	—	0.04	412	0.28	20	4.8	6.89	1.0	3.3

TABLE 21–1

Entry Number	Feed Name Description	International Feed Number	Moisture Basis: A-F (as-fed) or M-F (moisture-free)	Chemical Analysis								
				Dry Matter	Ash	Crude Fiber	Neutral Det. Fib. (NDF)	Acid Det. Fib. (ADF)	Lignin	Ether Extract (Fat)	N-Free Extract	Crude Protein
				(%)	(%)	(%)	(%)	(%)	(%)	(%)	(%)	(%)
	CORN, DENT YELLOW (Continued)											
68	GERM MEAL, WET MILLED, SOLV EXTD	5–28–240	A-F	92	3.9	12.2	—	—	—	1.5	53.5	20.7
			M-F	100	4.2	13.3	—	—	—	1.6	58.3	22.6
69	GLUTEN MEAL, 60% PROTEIN	5–28–242	A-F	90	1.7	1.8	12.6	4.5	—	2.1	23.7	60.8
			M-F	100	1.9	2.0	14.0	5.0	—	2.3	26.3	67.5
70	GLUTEN FEED	5–28–243	A-F	90	6.6	8.7	40.5	10.8	—	2.1	49.4	23.0
			M-F	100	7.4	9.7	45.0	12.0	—	2.4	55.0	25.6
71	GRITS (HOMINY GRITS)	4–03–011	A-F	90	2.8	4.8	49.5	11.7	—	6.5	65.8	10.3
			M-F	100	3.1	5.3	55.0	13.0	—	7.2	72.9	11.4
72	FODDER, W/EARS, W/HUSKS, SUN-CURED	1–02–775	A-F	90	4.9	32.6	—	—	—	7.4	29.5	15.6
			M-F	100	5.4	36.2	—	—	—	8.2	32.8	17.3
73	STOVER, W/O EARS, W/O HUSKS, SUN-CURED	1–02–776	A-F	85	6.1	29.3	57.0	33.2	—	1.1	43.2	5.4
			M-F	100	7.2	34.4	67.0	39.0	—	1.3	50.8	6.4
74	HUSKS, SUN-CURED	1–02–785	A-F	89	3.2	30.0	—	—	—	0.8	51.2	3.3
			M-F	100	3.6	33.9	—	—	—	0.9	57.9	3.7
75	LEAVES, SUN-CURED	1–02–788	A-F	87	8.5	24.0	65.3	42.1	6.2	2.6	44.4	7.6
			M-F	100	9.7	27.6	74.9	48.3	7.1	3.0	51.0	8.7
76	COBS, GROUND	1–02–782	A-F	90	1.6	32.2	80.1	31.5	—	0.6	52.7	2.8
			M-F	100	1.8	35.8	89.0	35.0	—	0.7	58.7	3.1
77	SILAGE, ALL ANALYSES	3–02–822	A-F	26	1.5	6.6	—	8.9	1.2	0.8	15.2	2.2
			M-F	100	5.6	25.1	—	34.0	4.4	3.2	57.8	8.3
78	EARS	4–07–740	A-F	70	1.6	7.3	—	—	—	2.5	51.5	7.1
			M-F	100	2.3	10.4	—	—	—	3.6	73.7	10.1
79	EARS W/HUSKS	4–02–839	A-F	43	1.5	4.8	—	—	—	1.6	31.5	4.0
			M-F	100	3.4	11.1	—	—	—	3.7	72.7	9.2
80	HUSKS (HUSKLAGE)	3–26–074	A-F	78	2.7	26.8	—	—	—	0.7	45.0	2.9
			M-F	100	3.4	34.3	—	—	—	0.9	57.7	3.7
	CORN, SWEET *Zea mays sacharatta*											
81	CANNERY RESIDUE	3–07–955	A-F	31	1.7	10.1	—	10.5	—	1.3	15.4	2.5
			M-F	100	5.4	32.7	—	34.0	—	4.3	49.6	8.0
	COTTON *Gossypium spp*											
82	SEEDS W/O LINT	5–13–749	A-F	91	4.7	19.4	—	—	—	20.4	24.5	21.8
			M-F	100	5.2	21.4	—	—	—	22.5	27.0	24.0
83	SEEDS, MEAL MECH EXTD, 36% PROTEIN	5–01–625	A-F	92	6.4	15.0	—	—	—	4.5	28.8	37.2
			M-F	100	7.0	16.3	—	—	—	4.9	31.3	40.5
84	SEEDS, MEAL MECH EXTD, 41% PROTEIN	5–01–617	A-F	93	6.1	11.9	25.9	18.5	5.6	4.7	28.9	41.0
			M-F	100	6.6	12.9	28.0	20.0	6.0	5.0	31.2	44.3
85	SEEDS, MEAL, PREPRESSED, SOLV EXTD, 41% PROTEIN	5–07–872	A-F	90	6.4	12.9	23.4	17.1	—	1.0	28.8	41.3
			M-F	100	7.0	14.2	26.0	19.0	—	1.2	31.9	45.7
86	SEEDS, MEAL SOLV EXTD, 41% PROTEIN	5–01–621	A-F	91	6.5	12.1	23.6	18.4	5.5	1.5	29.6	41.2
			M-F	100	7.1	13.4	26.0	20.2	6.0	1.6	32.5	45.4
87	SEEDS, MEAL SOLV EXTD, 46% PROTEIN	5–26–100	A-F	92	7.2	8.9	25.8	19.3	—	1.6	26.8	47.6
			M-F	100	7.8	9.6	28.0	21.0	—	1.8	29.1	51.7
88	SEEDS, MEAL SOLV EXTD, 48% PROTEIN	5–26–101	A-F	90	7.4	7.0	—	—	—	1.9	24.7	49.3
			M-F	100	8.2	7.7	—	—	—	2.1	27.4	54.7
89	SEEDS, LOW GOSSYPOL, MEAL SOLV EXTD	5–01–633	A-F	93	5.8	12.7	—	—	—	1.2	31.5	41.5
			M-F	100	6.3	13.7	—	—	—	1.3	34.0	44.8
90	SEEDS W/O HULLS, MEAL, PREPRESSED, SOLV EXTD, 50% PROTEIN	5–07–874	A-F	93	6.6	8.2	—	—	—	1.3	26.8	50.3
			M-F	100	7.1	8.8	—	—	—	1.4	28.8	54.0
91	SEEDS, MEAL SOLV EXTD, AMMONIATED	5–09–352	A-F	91	6.6	14.1	—	—	—	3.8	—	44.5
			M-F	100	7.3	15.5	—	—	—	4.2	—	48.9
	COWPEA, COMMON *Vigina sinensis*											
92	HAY, SUN-CURED	1–01–645	A-F	90	10.5	24.4	—	—	—	2.6	35.1	17.7
			M-F	100	11.7	27.1	—	—	—	2.9	38.8	19.6

(Continued)

Entry Number	Digestible Protein Ruminant (%)	TDN Ruminant (%)	Digestible Energy Ruminant (Mcal/lb)	(Mcal/kg)	Metabolizable Energy Ruminant (Mcal/lb)	(Mcal/kg)	Ruminant NE$_m$ (Mcal/lb)	(Mcal/kg)	Ruminant NE$_g$ (Mcal/lb)	(Mcal/kg)	Lactating Cows NE$_{lc}$ (Mcal/lb)	(Mcal/kg)	Calcium (Ca) (%)	Phosphorus (P) (%)	Sodium (Na) (%)	Chlorine (Cl) (%)
68	16.6	68	1.41	3.11	1.24	2.73	0.73	1.61	0.47	1.04	0.71	1.55	0.04	0.51	0.04	0.04
	18.1	74	1.54	3.39	1.35	2.98	0.80	1.76	0.52	1.14	0.77	1.69	0.04	0.55	0.04	0.04
69	54.5	—	—	—	—	—	—	—	—	—	—	—	0.07	0.45	0.05	0.09
	60.6	—	—	—	—	—	—	—	—	—	—	—	0.08	0.50	0.05	0.10
70	19.8	75	1.44	3.17	1.21	2.67	0.82	1.80	0.55	1.21	0.78	1.72	0.32	0.74	0.12	0.22
	22.0	83	1.60	3.52	1.34	2.96	0.91	2.00	0.61	1.35	0.87	1.91	0.36	0.82	0.14	0.25
71	7.3	84	1.69	3.73	1.29	2.85	0.88	1.95	0.61	1.34	0.91	2.00	0.05	0.51	0.08	0.05
	8.1	93	1.88	4.14	1.43	3.15	0.98	2.16	0.67	1.48	1.01	2.22	0.05	0.57	0.09	0.06
72	4.0	45	0.91	2.00	0.74	1.64	0.44	0.97	0.12	0.26	0.46	1.01	0.45	0.23	0.03	0.17
	4.4	50	1.01	2.22	0.83	1.82	0.49	1.08	0.13	0.29	0.51	1.12	0.50	0.25	0.03	0.19
73	2.3	51	1.01	2.23	0.85	1.87	0.51	1.12	0.29	0.63	0.52	1.15	0.49	0.08	0.06	—
	2.7	59	1.19	2.62	1.00	2.19	0.60	1.32	0.34	0.74	0.61	1.35	0.57	0.10	0.07	—
74	0.7	55	1.11	2.44	0.94	2.06	0.69	1.51	0.44	0.97	0.66	1.46	0.16	0.12	—	—
	0.8	62	1.25	2.75	1.06	2.33	0.78	1.71	0.50	1.09	0.75	1.65	0.18	0.14	—	—
75	3.0	53	1.07	2.35	0.90	1.98	0.55	1.22	0.32	0.71	0.56	1.23	0.60	0.22	—	—
	3.4	61	1.22	2.70	1.03	2.28	0.63	1.40	0.37	0.82	0.64	1.42	0.69	0.25	—	—
76	-0.4	44	0.91	2.00	0.74	1.62	0.39	0.87	0.17	0.37	0.44	0.96	0.11	0.04	—	—
	-0.4	50	1.01	2.23	0.82	1.80	0.44	0.96	0.19	0.42	0.49	1.07	0.12	0.04	—	—
77	1.0	18	0.34	0.75	0.30	0.66	0.19	0.42	0.12	0.26	0.17	0.37	0.08	0.07	0.01	0.05
	4.0	68	1.30	2.86	1.14	2.50	0.73	1.61	0.46	1.01	0.65	1.42	0.31	0.27	0.03	0.18
78	4.2	55	1.10	2.42	0.97	2.13	0.58	1.28	0.38	0.84	0.55	1.22	—	—	—	—
	6.0	78	1.57	3.46	1.38	3.04	0.83	1.83	0.54	1.20	0.79	1.75	—	—	—	—
79	2.2	31	0.66	1.45	0.57	1.27	0.33	0.74	0.21	0.47	0.32	0.71	0.04	0.12	0.00	—
	5.2	72	1.51	3.33	1.32	2.92	0.77	1.70	0.49	1.08	0.75	1.64	0.10	0.29	0.01	—
80	—	47	0.92	2.03	0.78	1.72	—	—	—	—	—	—	—	—	—	—
	—	60	1.18	2.60	1.00	2.20	—	—	—	—	—	—	—	—	—	—
81	1.2	22	0.42	0.92	0.36	0.79	0.24	0.53	0.15	0.34	0.23	0.51	0.10	0.24	0.01	—
	3.7	72	1.35	2.97	1.16	2.55	0.77	1.70	0.49	1.08	0.75	1.64	0.32	0.77	0.03	—
82	17.7	87	1.57	3.46	1.39	3.07	0.97	2.13	0.67	1.49	0.91	2.01	0.14	0.69	0.03	—
	19.4	95	1.72	3.80	1.53	3.38	1.06	2.34	0.74	1.63	1.00	2.21	0.16	0.76	0.03	—
83	29.4	66	1.32	2.91	1.15	2.52	0.73	1.62	0.47	1.05	0.71	1.56	0.18	0.95	0.04	—
	32.0	72	1.44	3.16	1.25	2.75	0.80	1.76	0.52	1.14	0.77	1.69	0.20	1.04	0.05	—
84	35.1	72	1.49	3.29	1.12	2.47	0.72	1.59	0.64	1.41	0.76	1.67	0.19	1.07	0.04	0.04
	37.9	77	1.61	3.56	1.21	2.67	0.78	1.72	0.69	1.52	0.82	1.81	0.21	1.16	0.05	0.05
85	35.4	72	1.51	3.33	1.14	2.51	0.70	1.55	0.64	1.41	0.76	1.67	0.16	1.07	0.04	0.06
	39.1	80	1.67	3.69	1.26	2.78	0.78	1.71	0.71	1.56	0.84	1.85	0.17	1.18	0.04	0.07
86	31.3	68	1.48	3.27	1.17	2.57	0.74	1.64	0.63	1.40	0.73	1.61	0.17	1.11	0.04	0.04
	34.5	75	1.63	3.60	1.28	2.83	0.82	1.80	0.70	1.54	0.80	1.77	0.19	1.22	0.05	0.05
87	41.1	71	1.42	3.13	1.25	2.74	0.77	1.70	0.51	1.12	0.74	1.63	—	—	—	—
	44.6	77	1.54	3.40	1.35	2.98	0.84	1.85	0.55	1.22	0.80	1.77	—	—	—	—
88	42.7	73	1.44	3.17	1.27	2.79	0.80	1.77	0.54	1.18	0.76	1.67	0.20	1.20	—	—
	47.3	81	1.59	3.51	1.40	3.09	0.89	1.96	0.60	1.31	0.84	1.85	0.22	1.33	—	—
89	35.5	68	1.36	3.01	1.19	2.62	0.71	1.57	0.45	1.00	0.69	1.52	—	—	—	—
	38.3	74	1.47	3.24	1.28	2.83	0.77	1.69	0.49	1.08	0.75	1.64	—	—	—	—
90	40.7	70	1.43	3.15	1.25	2.76	0.76	1.67	0.49	1.08	0.73	1.60	0.18	1.16	0.05	0.05
	43.8	75	1.54	3.38	1.35	2.97	0.81	1.79	0.53	1.16	0.78	1.72	0.19	1.24	0.06	0.05
91	—	66	1.30	2.88	1.07	2.36	0.66	1.46	0.43	0.94	0.70	1.54	—	—	—	—
	—	72	1.43	3.16	1.18	2.59	0.73	1.60	0.47	1.03	0.77	1.69	—	—	—	—
92	12.2	54	1.14	2.52	0.92	2.02	0.58	1.28	0.34	0.75	0.56	1.24	1.26	0.31	0.24	0.15
	13.4	60	1.26	2.78	1.02	2.24	0.64	1.41	0.38	0.83	0.62	1.38	1.40	0.35	0.27	0.17

TABLE 21–1

Entry Number	Feed Name Description	Moisture Basis: A-F (as-fed) or M-F (moisture-free)	Macro Minerals (Continued)			Micro Minerals						
			Magnesium (Mg)	Potassium (K)	Sulfur (S)	Cobalt (Co)	Copper (Cu)	Iodine (I)	Iron (Fe)	Manganese (Mn)	Selenium (Se)	Zinc (Zn)
			(%)	(%)	(%)	(ppm or mg/kg)	(ppm or mg/kg)	(ppm or mg/kg)	(%)	(ppm or mg/kg)	(ppm or mg/kg)	(ppm or mg/kg)
	CORN, DENT YELLOW (Continued)											
68	GERM MEAL, WET MILLED, SOLV EXTD	A-F	0.16	0.35	0.31	—	4.5	—	0.03	3.8	0.34	104.8
		M-F	0.17	0.38	0.34	—	4.9	—	0.04	4.1	0.37	114.2
69	GLUTEN MEAL, 60% PROTEIN	A-F	0.08	0.18	0.65	0.05	26.1	0.02	0.02	6.3	0.83	30.6
		M-F	0.09	0.20	0.72	0.05	29.0	0.02	0.03	7.0	0.92	34.0
70	GLUTEN FEED	A-F	0.33	0.57	0.21	0.09	47.1	0.07	0.04	23.1	0.27	64.6
		M-F	0.36	0.64	0.23	0.10	52.3	0.07	0.05	25.7	0.30	71.8
71	GRITS (HOMINY GRITS)	A-F	0.24	0.59	0.03	0.06	13.6	—	0.01	14.5	—	—
		M-F	0.26	0.65	0.03	0.06	15.1	—	0.01	16.1	—	—
72	FODDER W/EARS, W/HUSKS, SUN-CURED	A-F	0.26	0.84	0.13	—	6.9	—	0.01	61.4	—	—
		M-F	0.29	0.93	0.14	—	7.7	—	0.01	68.2	—	—
73	STOVER, W/O EARS, W/O HUSKS, SUN-CURED	A-F	0.34	1.24	0.15	—	4.3	—	0.02	115.9	—	—
		M-F	0.40	1.45	0.17	—	5.1	—	0.02	136.0	—	—
74	HUSKS, SUN-CURED	A-F	—	0.57	—	—	—	—	—	—	—	—
		M-F	—	0.65	—	—	—	—	—	—	—	—
75	LEAVES, SUN-CURED	A-F	—	1.56	—	—	—	—	—	—	—	—
		M-F	—	1.79	—	—	—	—	—	—	—	—
76	COBS, GROUND	A-F	0.06	0.78	0.42	0.12	6.6	—	0.02	5.6	—	—
		M-F	0.07	0.87	0.47	0.13	7.3	—	0.02	6.2	—	—
77	SILAGE, ALL ANALYSES	A-F	0.06	0.32	0.03	0.03	2.4	—	0.01	10.8	—	5.5
		M-F	0.22	1.22	0.12	0.10	9.2	—	0.02	41.1	—	21.2
78	EARS	A-F	—	—	—	—	—	—	—	—	—	—
		M-F	—	—	—	—	—	—	—	—	—	—
79	EARS W/HUSKS	A-F	0.05	0.21	0.06	—	—	—	0.00	—	—	—
		M-F	0.12	0.49	0.13	—	—	—	0.01	—	—	—
80	HUSKS (HUSKLAGE)	A-F	—	—	—	—	—	—	—	—	—	—
		M-F	—	—	—	—	—	—	—	—	—	—
	CORN, SWEET *Zea mays sacharatta*											
81	CANNERY RESIDUE	A-F	0.07	0.36	0.03	—	—	—	0.01	—	—	—
		M-F	0.24	1.15	0.11	—	—	—	0.02	—	—	—
	COTTON *Gossypium* spp											
82	SEEDS W/O LINT	A-F	0.32	1.11	0.24	—	49.0	—	0.01	11.1	—	—
		M-F	0.35	1.22	0.26	—	53.9	—	0.02	12.2	—	—
83	SEEDS, MEAL MECH EXTD, 36% PROTEIN	A-F	0.68	1.34	0.26	0.15	16.5	—	0.02	22.6	—	56.7
		M-F	0.74	1.46	0.28	0.16	17.9	—	0.02	24.5	—	61.7
84	SEEDS, MEAL MECH EXTD, 41% PROTEIN	A-F	0.53	1.33	0.40	0.63	18.5	—	0.02	22.3	—	61.8
		M-F	0.57	1.44	0.43	0.68	20.0	—	0.02	24.1	—	66.8
85	SEEDS, MEAL, PREPRESSED, SOLV EXTD, 41% PROTEIN	A-F	0.48	1.25	0.31	0.74	18.2	—	0.02	20.4	—	62.7
		M-F	0.53	1.38	0.34	0.82	20.2	—	0.02	22.5	—	69.4
86	SEEDS, MEAL SOLV EXTD, 41% PROTEIN	A-F	0.54	1.37	0.25	0.48	19.5	—	0.02	20.6	—	60.7
		M-F	0.59	1.51	0.27	0.53	21.4	—	0.02	22.7	—	66.7
87	SEEDS, MEAL SOLV EXTD, 46% PROTEIN	A-F	—	—	—	—	—	—	—	—	—	—
		M-F	—	—	—	—	—	—	—	—	—	—
88	SEEDS, MEAL SOLV EXTD, 48% PROTEIN	A-F	—	—	—	—	—	—	—	—	—	—
		M-F	—	—	—	—	—	—	—	—	—	—
89	SEEDS, LOW GOSSYPOL, MEAL SOLV EXTD	A-F	—	—	—	—	—	—	—	—	—	—
		M-F	—	—	—	—	—	—	—	—	—	—
90	SEEDS W/O HULLS, MEAL, PREPRESSED, SOLV EXTD, 50% PROTEIN	A-F	0.46	1.45	0.52	0.04	14.5	—	0.01	23.0	—	73.8
		M-F	0.50	1.56	0.56	0.05	15.6	—	0.01	24.8	—	79.4
91	SEEDS, MEAL SOLV EXTD, AMMONIATED	A-F	—	—	—	—	—	—	—	—	—	—
		M-F	—	—	—	—	—	—	—	—	—	—
	COWPEA, COMMON *Vigina sinensis*											
92	HAY, SUN-CURED	A-F	0.41	2.04	0.32	0.07	—	—	0.06	438.4	—	—
		M-F	0.45	2.26	0.35	0.07	—	—	0.06	485.1	—	—

(Continued)

Entry Number	Fat-Soluble Vitamins					Water-Soluble Vitamins								
	A	Carotene (Provitamin A)	D	E	K	B–12	Biotin	Choline	Folacin (Folic Acid)	Niacin	Pantothenic Acid (B–3)	Pyridoxine (B–6)	Riboflavin (B–2)	Thiamin (B–1)
	(IU/g)	(ppm or mg/kg)	(IU/kg)	(ppm or mg/kg)	(ppm or mg/kg)	(ppb or mcg/kg)	(ppm or mg/kg)	(ppm or mg/kg)	(ppm or mg/kg)	(ppm or mg/kg)	(ppm or mg/kg)	(ppm or mg/kg)	(ppm or mg/kg)	(ppm or mg/kg)
68	3.4	2.0	—	85.8	—	—	0.22	1586	0.20	39	4.2	—	3.8	4.5
	3.7	2.2	—	93.5	—	—	0.24	1728	0.22	42	4.6	—	4.1	4.9
69	—	—	—	14.6	—	—	—	—	—	—	—	6.39	—	—
	—	—	—	16.2	—	—	—	—	—	—	—	7.10	—	—
70	9.8	5.9	—	12.1	—	—	0.33	1514	0.27	70	13.6	13.93	2.2	2.0
	10.9	6.5	—	13.5	—	—	0.36	1684	0.30	78	15.1	15.49	2.5	2.2
71	15.4	9.2	—	—	—	—	0.13	1154	0.31	47	8.2	10.95	2.1	8.1
	17.0	10.2	—	—	—	—	0.15	1280	0.34	52	9.1	12.14	2.4	8.9
72	0.45	0.23	0.03	0.17	0.26	0.84	0.13	—	6.90	—	0.01	61.4	—	—
	0.50	0.25	0.03	0.19	0.29	0.93	0.14	—	7.70	—	0.01	68.2	—	—
73	6.3	3.8	1	—	—	—	—	—	—	—	—	—	—	—
	7.4	4.5	1	—	—	—	—	—	—	—	—	—	—	—
74	—	—	—	—	—	—	—	—	—	—	—	—	—	—
	—	—	—	—	—	—	—	—	—	—	—	—	—	—
75	—	—	5	—	—	—	—	—	—	—	—	—	—	—
	—	—	5	—	—	—	—	—	—	—	—	—	—	—
76	1.0	0.6	—	—	—	—	—	—	—	7	3.8	—	1.0	0.9
	1.2	0.7	—	—	—	—	—	—	—	8	4.2	—	1.1	1.0
77	—	—	—	—	—	—	—	—	—	—	—	—	—	—
	—	—	—	—	—	—	—	—	—	—	—	—	—	—
78	—	—	—	—	—	—	—	—	—	—	—	—	—	—
	—	—	—	—	—	—	—	—	—	—	—	—	—	—
79	—	—	—	—	—	—	—	—	—	—	—	—	—	—
	—	—	—	—	—	—	—	—	—	—	—	—	—	—
80	—	—	—	—	—	—	—	—	—	—	—	—	—	—
	—	—	—	—	—	—	—	—	—	—	—	—	—	—
81	6.9	4.2	—	—	—	—	—	—	—	—	—	—	—	—
	22.3	13.4	—	—	—	—	—	—	—	—	—	—	—	—
82	—	—	—	—	—	—	—	—	—	—	—	—	—	—
	—	—	—	—	—	—	—	—	—	—	—	—	—	—
83	—	—	—	15.1	—	—	1.12	2739	3.75	29	9.7	—	5.3	4.5
	—	—	—	16.4	—	—	1.21	2980	4.08	32	10.6	—	5.7	4.9
84	0.4	0.2	—	32.3	—	—	0.91	2753	2.45	35	10.2	5.00	5.2	7.1
	0.4	0.2	—	34.9	—	—	0.99	2974	2.65	38	11.0	5.41	5.6	7.6
85	—	—	—	—	—	—	0.55	2881	2.57	40	7.3	4.11	5.3	3.3
	—	—	—	—	—	—	0.61	3166	2.85	44	8.1	4.55	5.8	3.7
86	—	—	—	14.6	—	—	0.55	2780	2.55	41	13.7	5.41	4.7	7.3
	—	—	—	16.1	—	—	0.61	3058	2.81	45	15.1	5.95	5.2	8.0
87	—	—	—	—	—	—	—	—	—	—	—	—	—	—
	—	—	—	—	—	—	—	—	—	—	—	—	—	—
88	—	—	—	—	—	—	—	3316	—	51	15.5	—	6.0	—
	—	—	—	—	—	—	—	3674	—	56	17.1	—	8.6	—
89	—	—	—	—	—	—	—	—	—	—	—	—	—	20.3
	—	—	—	—	—	—	—	—	—	—	—	—	—	21.9
90	—	—	—	11.3	—	—	0.44	2962	0.93	45	14.3	6.29	4.9	8.2
	—	—	—	12.1	—	—	0.48	3184	1.00	48	15.4	6.76	5.3	8.8
91	—	—	—	—	—	—	—	—	—	—	—	—	—	—
	—	—	—	—	—	—	—	—	—	—	—	—	—	—
92	52.7	31.6	—	—	—	—	—	—	—	—	—	—	—	—
	58.3	35.0	—	—	—	—	—	—	—	—	—	—	—	—

TABLE 21–1

Entry Number	Feed Name Description	International Feed Number	Moisture Basis: A-F (as-fed) or M-F (moisture-free)	Chemical Analysis								
				Dry Matter	Ash	Crude Fiber	Neutral Det. Fib. (NDF)	Acid Det. Fib. (ADF)	Lignin	Ether Extract (Fat)	N-Free Extract	Crude Protein
				(%)	(%)	(%)	(%)	(%)	(%)	(%)	(%)	(%)
	DISTILLERS PRODUCTS (also see CORN)											
93	GRAINS, DEHY	5–02–144	A-F	93	1.5	12.8	—	—	—	7.4	43.5	27.3
			M-F	100	1.6	13.8	—	—	—	8.0	47.0	29.5
94	SOLUBLES, DEHY	5–02–147	A-F	92	6.2	3.4	—	—	—	8.9	44.8	28.8
			M-F	100	6.7	3.7	—	—	—	9.7	48.6	31.3
	FATS & OILS (not over 3% of diet)											
95	FAT, ANIMAL, HYDROLIZED	4–00–376	A-F	99	—	—	—	—	—	98.4	—	—
			M-F	100	—	—	—	—	—	99.2	—	—
96	FAT (LARD), SWINE	4–04–790	A-F	99	—	—	—	—	—	99.3	—	—
			M-F	100	—	—	—	—	—	100.0	—	—
97	OIL, SOYBEAN	4–07–983	A-F	100	0.3	—	—	—	—	95.0	7.3	1.4
			M-F	100	0.3	—	—	—	—	95.5	7.3	1.4
	FESCUE, MEADOW *Festuca elatior*											
98	HAY, SUN-CURED	1–01–912	A-F	88	7.9	28.0	74.0	43.8	—	2.4	41.0	8.2
			M-F	100	9.0	32.0	84.6	50.0	—	2.7	46.9	9.4
	FESCUE, TALL (ALTA) *Festuca arundinacea*											
99	HAY, SUN-CURED	1–05–694	A-F	89	5.9	32.6	61.7	35.6	—	2.0	41.4	7.2
			M-F	100	6.6	36.6	69.3	40.0	—	2.2	46.5	8.1
	FISH, ANCHOVY *Engraulis ringen*											
100	MEAL MECH EXTD	5–01–985	A-F	92	14.7	1.0	—	—	—	4.1	6.7	65.4
			M-F	100	16.0	1.1	—	—	—	4.5	7.3	71.1
	FISH, MENHADEN *Brevoortia*											
101	MEAL MECH EXTD	5–02–009	A-F	92	19.1	0.9	—	—	—	9.6	0.8	61.2
			M-F	100	20.9	1.0	—	—	—	10.5	0.8	66.8
	FISH, WHITE *Gadidae, Lophildae, Rajidae* (families)											
102	MEAL MECH EXTD	5–02–025	A-F	91	23.1	0.5	—	—	—	4.7	0.1	62.6
			M-F	100	25.4	0.6	—	—	—	5.1	0.1	68.8
	FLAX, COMMON *Linum usitatissimum*											
103	SEEDS, GROUND	5–02–042	A-F	96	5.0	6.3	—	—	—	35.9	30.7	17.8
			M-F	100	5.2	6.6	—	—	—	37.5	32.1	18.6
104	SEEDS, MEAL SOLV EXTD, 35% PROTEIN (LINSEED MEAL)	5–26–090	A-F	90	5.8	8.9	—	—	—	1.7	38.2	35.7
			M-F	100	6.4	9.9	—	—	—	1.9	42.3	39.6
105	SEED SCREENINGS, MEAL MECH EXTD	5–02–054	A-F	91	8.6	11.2	22.9	15.5	6.4	7.3	39.6	24.8
			M-F	100	9.4	12.2	25.0	17.0	7.0	8.0	43.3	27.1
	GRAPES *Vitis spp*											
106	POMACE, DEHY (MARC)	1–01–208	A-F	90	7.5	27.9	48.1	49.2	31.8	7.6	35.3	12.1
			M-F	100	8.3	30.9	53.2	54.4	35.2	8.4	39.0	13.4
	LESPEDEZA, COMMON *Lespedeza striata*											
107	HAY, SUN-CURED, ALL ANALYSES	1–08–591	A-F	89	4.7	28.4	—	—	—	2.5	39.4	13.8
			M-F	100	5.3	32.0	—	—	—	2.8	44.4	15.6
108	HAY, PREBLOOM, SUN-CURED	1–20–881	A-F	89	6.4	22.8	—	—	—	2.7	43.1	14.3
			M-F	100	7.2	25.5	—	—	—	3.0	48.3	16.0
109	HAY, EARLY BLOOM, SUN-CURED	1–20–882	A-F	91	5.5	29.1	—	—	—	3.6	37.9	14.8
			M-F	100	6.0	32.0	—	—	—	4.0	41.7	16.3
110	HAY, MIDBLOOM, SUN-CURED	1–02–554	A-F	91	4.5	26.2	—	—	—	2.3	46.5	11.4
			M-F	100	4.9	28.8	—	—	—	2.5	51.2	12.6
111	HAY, FULL BLOOM, SUN-CURED	1–20–887	A-F	89	5.0	27.4	—	—	—	1.9	42.2	12.8
			M-F	100	5.6	30.7	—	—	—	2.1	47.3	14.3
	MEAT											
112	MEAL RENDERED	5–00–385	A-F	94	28.1	2.7	—	—	—	9.1	3.3	50.7
			M-F	100	29.9	2.9	—	—	—	9.7	3.5	54.0
113	W/BLOOD, MEAL RENDERED (TANKAGE)	5–00–386	A-F	92	21.1	1.8	—	—	—	8.7	0.1	60.5
			M-F	100	22.9	2.0	—	—	—	9.4	0.1	65.6

(Continued)

Entry Number	Digestible Protein Ruminant (%)	TDN Ruminant (%)	Digestible Energy Ruminant (Mcal/lb)	Digestible Energy Ruminant (Mcal/kg)	Metabolizable Energy Ruminant (Mcal/lb)	Metabolizable Energy Ruminant (Mcal/kg)	Ruminant NEm (Mcal/lb)	Ruminant NEm (Mcal/kg)	Ruminant NEg (Mcal/lb)	Ruminant NEg (Mcal/kg)	Lactating Cows NElc (Mcal/lb)	Lactating Cows NElc (Mcal/kg)	Calcium (Ca) (%)	Phosphorus (P) (%)	Sodium (Na) (%)	Chlorine (Cl) (%)
93	18.9	78	1.61	3.54	1.43	3.16	0.93	2.06	0.65	1.43	0.87	1.92	0.12	0.54	0.05	0.05
	20.4	84	1.74	3.85	1.56	3.43	1.02	2.24	0.70	1.55	0.95	2.09	0.13	0.59	0.05	0.06
94	24.0	68	1.39	3.07	1.22	2.69	0.77	1.70	0.51	1.12	0.74	1.63	0.24	1.35	0.45	—
	26.0	75	1.52	3.34	1.33	2.93	0.84	1.86	0.56	1.22	0.80	1.77	0.26	1.47	0.49	—
95	—	223	4.46	9.84	4.31	9.49	2.95	6.49	2.23	4.91	2.43	5.35	—	—	—	—
	—	225	4.50	9.92	4.34	9.57	2.97	6.55	2.25	4.95	2.45	5.39	—	—	—	—
96	—	—	—	—	—	—	—	—	—	—	—	—	—	—	—	—
	—	—	—	—	—	—	—	—	—	—	—	—	—	—	—	—
97	-1.8	193	3.86	8.51	3.70	8.15	2.40	5.30	1.79	3.94	2.09	4.61	—	—	—	—
	-1.8	194	3.88	8.55	3.71	8.19	2.42	5.32	1.80	3.96	2.10	4.63	—	—	—	—
98	4.6	53	1.06	2.35	0.90	1.98	0.55	1.20	0.32	0.70	0.56	1.22	0.33	0.25	—	—
	5.3	61	1.22	2.68	1.02	2.26	0.62	1.38	0.36	0.80	0.64	1.40	0.37	0.29	—	—
99	3.6	48	0.95	2.10	0.78	1.72	0.44	0.96	0.21	0.47	0.48	1.06	0.35	0.21	0.05	—
	4.0	54	1.07	2.36	0.88	1.93	0.49	1.08	0.24	0.53	0.54	1.19	0.39	0.24	0.06	—
100	57.3	72	1.45	3.19	1.27	2.81	0.79	1.73	0.52	1.14	0.75	1.65	3.74	2.48	0.88	1.00
	62.3	79	1.57	3.47	1.39	3.05	0.85	1.88	0.56	1.24	0.81	1.79	4.07	2.70	0.95	1.08
101	49.6	67	1.33	2.94	1.16	2.55	0.75	1.65	0.49	1.08	0.72	1.58	5.19	2.88	0.41	0.55
	54.1	73	1.45	3.21	1.26	2.79	0.82	1.81	0.53	1.18	0.79	1.73	5.67	3.14	0.45	0.60
102	58.3	70	1.40	3.09	1.23	2.71	0.78	1.71	0.51	1.13	0.74	1.63	6.60	3.98	0.46	—
	64.0	77	1.54	3.39	1.35	2.98	0.85	1.88	0.56	1.24	0.81	1.79	7.25	4.37	0.51	—
103	13.9	—	—	—	—	—	—	—	—	—	—	—	0.28	0.55	—	—
	14.5	—	—	—	—	—	—	—	—	—	—	—	0.29	0.58	—	—
104	30.3	70	1.41	3.10	1.24	2.73	0.75	1.65	0.49	1.08	0.72	1.58	0.40	0.82	0.14	0.60
	33.6	78	1.56	3.43	1.37	3.02	0.83	1.83	0.54	1.20	0.79	1.75	0.44	0.91	0.15	0.66
105	20.4	54	1.07	2.37	0.90	1.98	0.52	1.14	0.28	0.62	0.54	1.19	0.42	0.56	0.14	—
	22.3	59	1.17	2.59	0.98	2.16	0.56	1.24	0.31	0.68	0.59	1.30	0.46	0.62	0.15	—
106	1.6	24	0.48	1.06	0.30	0.66	—	—	—	—	—	—	0.52	0.15	0.08	0.01
	1.7	27	0.53	1.17	0.33	0.73	—	—	—	—	—	—	0.58	0.17	0.09	0.01
107	5.9	44	0.98	2.15	0.80	1.77	0.53	1.16	0.29	0.65	0.54	1.19	0.78	0.25	—	—
	6.7	50	1.10	2.42	0.91	2.00	0.59	1.30	0.33	0.73	0.61	1.35	0.88	0.29	—	—
108	7.2	49	1.04	2.30	0.87	1.92	0.55	1.21	0.32	0.70	0.56	1.24	1.03	0.20	—	—
	8.0	55	1.17	2.57	0.97	2.15	0.62	1.36	0.35	0.78	0.63	1.38	1.16	0.22	—	—
109	10.8	52	1.03	2.28	—	—	—	—	—	—	—	—	1.07	0.22	—	—
	11.9	57	1.13	2.50	—	—	—	—	—	—	—	—	1.17	0.24	—	—
110	6.1	52	1.06	2.34	0.89	1.95	0.54	1.18	0.30	0.66	0.55	1.22	1.07	0.17	—	—
	6.7	57	1.17	2.58	0.98	2.15	0.59	1.30	0.33	0.73	0.61	1.35	1.18	0.19	—	—
111	8.5	53	1.05	2.32	0.88	1.94	0.51	1.12	0.28	0.62	0.53	1.17	1.02	0.18	—	—
	9.5	59	1.18	2.60	0.99	2.18	0.57	1.26	0.31	0.69	0.60	1.31	1.14	0.21	—	—
112	43.8	67	1.20	2.64	0.89	1.95	0.52	1.15	0.28	0.62	0.69	1.53	8.61	4.58	1.05	1.11
	46.7	71	1.28	2.81	0.94	2.08	0.56	1.23	0.30	0.66	0.74	1.63	9.18	4.88	1.11	1.18
113	52.8	67	1.34	2.96	1.17	2.58	0.74	1.62	0.48	1.05	0.71	1.56	5.87	3.09	1.67	1.73
	57.3	73	1.46	3.21	1.27	2.80	0.80	1.76	0.52	1.14	0.77	1.69	6.37	3.36	1.81	1.88

(Continued)

TABLE 21–1

Entry Number	Feed Name Description	Moisture Basis: A-F (as-fed) or M-F (moisture-free)	Macro Minerals (Continued)			Micro Minerals						
			Magne-sium (Mg)	Potassium (K)	Sulfur (S)	Cobalt (Co)	Copper (Cu)	Iodine (I)	Iron (Fe)	Manga-nese (Mn)	Selenium (Se)	Zinc (Zn)
			(%)	(%)	(%)	(ppm or mg/kg)	(ppm or mg/kg)	(ppm or mg/kg)	(%)	(ppm or mg/kg)	(ppm or mg/kg)	(ppm or mg/kg)
	DISTILLERS PRODUCTS (also see CORN)											
93	GRAINS, DEHY	A-F	0.09	0.20	0.46	0.09	47.9	—	0.03	35.0	—	—
		M-F	0.10	0.21	0.49	0.10	51.7	—	0.03	37.8	—	—
94	SOLUBLES, DEHY	A-F	0.53	1.97	—	0.20	71.6	—	0.03	67.1	—	138.0
		M-F	0.58	2.14	—	0.21	77.9	—	0.03	69.7	—	150.0
	FATS & OILS (not over 3% of diet)											
95	FAT, ANIMAL, HYDROLIZED	A-F	—	—	—	—	—	—	—	—	—	—
		M-F	—	—	—	—	—	—	—	—	—	—
96	FAT (LARD), SWINE	A-F	—	—	—	—	—	—	—	—	—	—
		M-F	—	—	—	—	—	—	—	—	—	—
97	OIL, SOYBEAN	A-F	—	—	—	—	—	—	—	—	—	—
		M-F	—	—	—	—	—	—	—	—	—	—
	FESCUE, MEADOW *Festuca elatior*											
98	HAY, SUN-CURED	A-F	0.44	1.61	—	0.12	—	—	—	21.4	—	—
		M-F	0.50	1.84	—	0.14	—	—	—	24.5	—	—
	FESCUE, TALL (ALTA) *Festuca arundinacea*											
99	HAY, SUN-CURED	A-F	0.20	2.12	—	—	—	—	—	—	—	—
		M-F	0.23	2.38	—	—	—	—	—	—	—	—
	FISH, ANCHOVY *Engraulis ringen*											
100	MEAL MECH EXTD	A-F	0.25	0.72	0.78	0.17	9.1	3.14	0.02	11.0	1.36	105.0
		M-F	0.27	0.78	0.84	0.19	9.9	3.41	0.02	11.9	1.47	114.2
	FISH, MENHADEN *Brevoortia*											
101	MEAL MECH EXTD	A-F	0.15	0.70	0.56	0.15	10.3	1.09	0.06	37.0	2.15	144.2
		M-F	0.17	0.77	0.61	0.17	11.3	1.19	0.06	40.4	2.34	157.5
	FISH, WHITE *Gadidae, Lophildae, Rajidae* (families)											
102	MEAL MECH EXTD	A-F	0.18	0.45	—	3.36	4.1	—	0.03	9.4	1.61	69.1
		M-F	0.20	0.49	—	3.69	4.5	—	0.03	10.3	1.77	75.9
	FLAX, COMMON *Linum usitatissimum*											
103	SEEDS, GROUND	A-F	—	—	—	—	—	—	—	—	—	—
		M-F	—	—	—	—	—	—	—	—	—	—
104	SEEDS, MEAL SOLV EXTD, 35% PROTEIN (LINSEED MEAL)	A-F	1.37	0.39	—	—	—	—	—	—	—	—
		M-F	1.52	0.43	—	—	—	—	—	—	—	—
105	SEED SCREENINGS, MEAL MECH EXTD	A-F	0.39	0.77	0.23	—	—	—	0.01	—	—	—
		M-F	0.43	0.84	0.25	—	—	—	0.01	—	—	—
	GRAPES *Vitis spp*											
106	POMACE, DEHY (MARC)	A-F	0.09	0.82	—	—	—	—	—	36.8	—	21.9
		M-F	0.10	0.91	—	—	—	—	—	40.7	—	24.2
	LESPEDEZA, COMMON *Lespedeza striata*											
107	HAY, SUN-CURED, ALL ANALYSES	A-F	0.20	1.21	0.21	—	7.1	—	0.02	99.6	—	21.3
		M-F	0.23	1.37	0.24	—	8.0	—	0.02	112.3	—	24.0
108	HAY, PREBLOOM, SUN-CURED	A-F	0.21	1.07	—	—	—	—	0.03	159.5	—	—
		M-F	0.24	1.20	—	—	—	—	0.03	178.4	—	—
109	HAY, EARLY BLOOM, SUN-CURED	A-F	0.23	1.00	—	—	—	—	0.36	232.5	—	—
		M-F	0.25	1.10	—	—	—	—	0.40	255.5	—	—
110	HAY, MIDBLOOM, SUN-CURED	A-F	0.22	0.94	—	—	—	—	0.03	210.6	—	—
		M-F	0.24	1.04	—	—	—	—	0.03	231.9	—	—
111	HAY, FULL BLOOM, SUN-CURED	A-F	0.20	0.93	—	—	—	—	0.03	135.2	—	—
		M-F	0.23	1.04	—	—	—	—	0.03	151.4	—	—
	MEAT											
112	MEAL RENDERED	A-F	0.25	0.55	0.46	2.25	9.6	—	0.05	11.8	0.51	74.3
		M-F	0.27	0.58	0.49	2.40	10.2	—	0.05	12.6	0.54	79.2
113	W/BLOOD, MEAL RENDERED (TANKAGE)	A-F	0.36	0.55	0.70	0.15	38.8	—	0.21	19.2	—	—
		M-F	0.39	0.60	0.76	0.17	42.1	—	0.23	20.8	—	—

(Continued)

Entry Number	Fat-Soluble Vitamins					Water-Soluble Vitamins								
	A	Carotene (Provitamin A)	D	E	K	B–12	Biotin	Choline	Folacin (Folic Acid)	Niacin	Pantothenic Acid (B–3)	Pyridoxine (B–6)	Riboflavin (B–2)	Thiamin (B–1)
	(IU/g)	(ppm or mg/kg)	(IU/kg)	(ppm or mg/kg)	(ppm or mg/kg)	(ppb or mcg/kg)	(ppm or mg/kg)	(ppm or mg/kg)	(ppm or mg/kg)	(ppm or mg/kg)	(ppm or mg/kg)	(ppm or mg/kg)	(ppm or mg/kg)	(ppm or mg/kg)
93	13.0	7.8	—	30.5	—	—	—	2645	—	47	11.9	6.00	6.6	2.5
	14.0	8.4	—	32.9	—	—	—	2858	—	51	12.9	6.48	7.1	2.6
94	1.9	1.1	—	—	—	2.9	2.84	4992	—	143	25.3	8.66	11.3	6.9
	2.0	1.2	—	—	—	3.1	3.09	5425	—	155	27.5	9.42	12.3	7.5
95	—	—	—	—	—	—	—	—	—	—	—	—	—	—
	—	—	—	—	—	—	—	—	—	—	—	—	—	—
96	—	—	—	22.8	—	—	—	—	—	—	—	—	—	—
	—	—	—	23.0	—	—	—	—	—	—	—	—	—	—
97	—	—	—	—	—	—	—	—	—	—	—	—	—	—
	—	—	—	—	—	—	—	—	—	—	—	—	—	—
98	105.8	63.4	—	118.6	—	—	—	—	—	—	—	—	—	—
	120.9	72.5	—	135.6	—	—	—	—	—	—	—	—	—	—
99	30.8	18.5	—	—	—	—	—	—	—	—	—	—	—	—
	34.6	20.7	—	—	—	—	—	—	—	—	—	—	—	—
100	—	—	—	3.7	—	214.5	0.20	3700	0.16	81	10.0	4.71	7.3	0.5
	—	—	—	4.0	—	233.2	0.21	4023	0.17	88	10.9	5.12	8.0	0.6
101	—	—	—	6.8	—	122.0	0.18	3112	0.15	55	8.6	3.80	4.8	0.6
	—	—	—	7.4	—	133.2	0.20	3398	0.17	60	9.4	4.15	5.3	0.6
102	—	—	—	8.9	—	84.3	0.08	4295	0.35	59	9.9	5.30	9.1	1.7
	—	—	—	9.8	—	92.6	0.09	4719	0.38	65	10.9	5.83	10.0	1.8
103	—	—	—	—	—	—	—	—	—	—	—	—	—	—
	—	—	—	—	—	—	—	—	—	—	—	—	—	—
104	—	—	—	5.9	—	—	—	1216	2.85	30	—	9.93	2.9	9.4
	—	—	—	6.5	—	—	—	1346	3.15	33	—	10.99	3.2	10.4
105	—	—	—	—	—	—	—	—	—	—	—	—	—	—
	—	—	—	—	—	—	—	—	—	—	—	—	—	—
106	—	—	—	—	—	—	—	253	—	18	3.1	—	2.2	—
	—	—	—	—	—	—	—	279	—	20	3.4	—	2.5	—
107	73.9	44.3	—	—	—	—	—	—	—	—	—	—	8.7	—
	83.3	50.0	—	—	—	—	—	—	—	—	—	—	9.8	—
108	—	—	—	—	—	—	—	—	—	—	—	—	—	—
	—	—	—	—	—	—	—	—	—	—	—	—	—	—
109	—	—	—	—	—	—	—	—	—	—	—	—	—	—
	—	—	—	—	—	—	—	—	—	—	—	—	—	—
110	—	—	—	—	—	—	—	—	—	—	—	—	—	—
	—	—	—	—	—	—	—	—	—	—	—	—	—	—
111	—	—	—	—	—	—	—	—	—	—	—	—	—	—
	—	—	—	—	—	—	—	—	—	—	—	—	—	—
112	—	—	—	0.9	—	75.2	0.12	1980	0.39	56	6.0	4.23	5.2	0.2
	—	—	—	1.0	—	80.1	0.13	2110	0.42	60	6.4	4.51	5.5	0.2
113	—	—	—	—	—	89.4	—	2203	1.54	38	3.2	—	2.2	0.3
	—	—	—	—	—	97.1	—	2391	1.67	41	3.5	—	2.4	0.4

(Continued)

TABLE 21–1

Entry Number	Feed Name Description	International Feed Number	Moisture Basis: A-F (as-fed) or M-F (moisture-free)	Chemical Analysis								
				Dry Matter	Ash	Crude Fiber	Neutral Det. Fib. (NDF)	Acid Det. Fib. (ADF)	Lignin	Ether Extract (Fat)	N-Free Extract	Crude Protein
				(%)	(%)	(%)	(%)	(%)	(%)	(%)	(%)	(%)
	MEAT (Continued)											
114	W/BLOOD & BONE, MEAL RENDERED (TANKAGE)	5–00–387	A-F	93	28.2	2.2	—	—	—	12.7	3.1	46.6
			M-F	100	30.4	2.4	—	—	—	13.7	3.3	50.2
115	W/BONE, MEAL RENDERED	5–00–388	A-F	93	28.0	2.4	—	—	—	10.0	2.6	50.4
			M-F	100	30.0	2.6	—	—	—	10.7	2.8	54.0
	MILK											
116	FRESH (CATTLE *bos taurus*)	5–01–168	A-F	12	0.8	—	—	—	—	3.6	4.7	3.3
			M-F	100	6.2	—	—	—	—	29.5	37.6	26.7
117	DEHY (CATTLE *Bos taurus*)	5–01–167	A-F	95	5.4	0.2	—	—	—	26.3	38.1	25.3
			M-F	100	5.6	0.2	—	—	—	27.6	39.9	26.6
118	SKIMMED, FRESH (CATTLE *Bos taurus*)	5–01–170	A-F	10	0.7	—	—	—	—	0.1	5.8	3.0
			M-F	100	6.9	—	—	—	—	1.0	60.6	31.2
119	SKIMMED, DEHY (CATTLE *Bos taurus*)	5–01–175	A-F	94	8.0	0.2	0.0	—	—	1.1	51.6	33.3
			M-F	100	8.4	0.2	0.0	—	—	1.2	54.8	35.4
	MILLET, FOXTAIL *Setaria italica*											
120	FRESH	2–03–101	A-F	29	2.5	9.2	—	—	—	0.9	13.4	2.8
			M-F	100	8.6	32.0	—	—	—	3.1	46.7	9.6
121	GRAIN	4–03–098	A-F	89	3.4	7.4	—	—	—	4.1	63.0	11.4
			M-F	100	3.8	8.3	—	—	—	4.6	70.6	12.8
122	HAY, SUN-CURED	1–03–099	A-F	87	7.5	25.7	—	—	—	2.5	43.7	7.5
			M-F	100	8.6	29.6	—	—	—	2.9	50.3	8.6
	MILLET, PROSO (BROOMCORN, HOG MILLET) *Panicum miliaceum*											
123	GRAIN	4–03–120	A-F	90	2.9	5.3	—	14.9	3.2	3.6	66.3	11.6
			M-F	100	3.3	6.0	—	16.6	3.6	4.0	73.9	12.9
	MOLASSES AND SYRUP											
124	BEET, SUGAR, MOLASSES, MORE THAN 48% INVERT SUGAR, MORE THAN 79.5 DEGREES BRIX	4–00–668	A-F	78	8.9	—	—	—	—	0.2	62.2	6.6
			M-F	100	11.4	—	—	—	—	0.2	79.9	8.5
125	CITRUS, SYRUP (CITRUS MOLASSES)	4–01–241	A-F	67	5.1	—	—	—	—	0.2	55.7	5.8
			M-F	100	7.6	—	—	—	—	0.3	82.7	8.5
126	SUGAR CANE, MOLASSES, DEHY	4–04–695	A-F	94	12.5	6.3	—	—	—	0.9	65.0	9.7
			M-F	100	13.3	6.7	—	—	—	0.9	68.8	10.3
127	SUGAR CANE, MOLASSES, MORE THAN 46% INVERT SUGAR, MORE THAN 79.5 DEGREES BRIX (BLACKSTRAP)	4–04–696	A-F	74	9.8	0.4	—	0.3	0.2	0.2	59.7	4.3
			M-F	100	13.2	0.5	—	0.4	0.3	0.2	80.2	5.8
	NAPIERGRASS *Pennisetum purpureum*											
128	PREBLOOM, FRESH	2–03–158	A-F	20	1.7	6.7	14.0	9.0	—	0.6	9.5	1.8
			M-F	100	8.6	33.0	70.0	45.0	—	3.0	46.7	8.7
129	LATE BLOOM, FRESH	2–03–162	A-F	23	1.2	9.0	17.3	10.8	—	0.3	10.8	1.8
			M-F	100	5.3	39.0	75.0	47.0	—	1.1	46.8	7.8
	OATS *Avena sativa*											
130	GRAIN, ALL ANALYSES	4–03–309	A-F	89	3.1	10.7	26.4	14.2	2.7	4.7	58.9	11.9
			M-F	100	3.4	11.9	29.6	15.9	3.0	5.2	66.1	13.3
131	GRAIN, LIGHT, LESS THAN 27 LB/BU (34.7 KG/HL)	4–03–318	A-F	91	4.2	14.4	—	—	—	4.5	55.7	11.9
			M-F	100	4.6	15.9	—	—	—	4.9	61.5	13.1
132	GRAIN, PACIFIC COAST	4–07–999	A-F	91	3.8	11.2	—	—	—	5.0	61.8	9.1
			M-F	100	4.2	12.3	—	—	—	5.5	68.0	10.0
133	CEREAL BY-PRODUCT (FEEDING OAT MEAL, OAT MIDDLINGS)	4–03–303	A-F	91	2.3	3.6	—	—	—	6.4	63.7	14.8
			M-F	100	2.5	4.0	—	—	—	7.0	70.2	16.3
134	GROATS	4–03–331	A-F	90	2.1	2.5	—	—	—	6.2	63.0	15.8
			M-F	100	2.4	2.8	—	—	—	6.9	70.3	17.6
	ORANGE *Citrus sinensis*											
135	PULP W/O FINES, DEHY (DRIED ORANGE PULP)	4–01–254	A-F	88	3.7	8.4	18.5	14.1	—	1.7	66.9	7.5
			M-F	100	4.2	9.6	21.0	16.0	—	1.9	75.9	8.5

(Continued)

Entry Number	Digestible Protein Ruminant (%)	TDN Ruminant (%)	Digestible Energy Ruminant (Mcal/lb)	(Mcal/kg)	Metabolizable Energy Ruminant (Mcal/lb)	(Mcal/kg)	Ruminant NE$_m$ (Mcal/lb)	(Mcal/kg)	Ruminant NE$_g$ (Mcal/lb)	(Mcal/kg)	Lactating Cows NE$_{lc}$ (Mcal/lb)	(Mcal/kg)	Calcium (Ca) (%)	Phosphorus (P) (%)	Sodium (Na) (%)	Chlorine (Cl) (%)
114	40.2	63	1.44	3.17	1.26	2.79	0.90	1.99	0.62	1.36	0.85	1.86	11.16	5.41	—	—
	43.2	68	1.55	3.42	1.36	3.00	0.97	2.14	0.67	1.47	0.91	2.00	12.01	5.82	—	—
115	45.8	66	1.32	2.91	1.14	2.51	0.71	1.57	0.45	1.00	0.69	1.52	10.00	4.94	0.72	0.75
	49.1	71	1.41	3.11	1.22	2.70	0.76	1.68	0.49	1.07	0.74	1.63	10.72	5.30	0.77	0.80
116	3.2	16	0.32	0.71	0.30	0.65	0.19	0.42	0.14	0.30	0.18	0.40	0.12	0.09	0.05	0.11
	25.5	128	2.61	5.75	2.40	5.29	1.53	3.37	1.11	2.46	1.46	3.22	0.93	0.75	0.38	0.92
117	20.7	113	2.27	4.99	2.09	4.61	1.27	2.81	0.92	2.02	1.17	2.57	0.89	0.70	0.36	1.48
	21.8	119	2.38	5.24	2.20	4.84	1.34	2.95	0.96	2.12	1.22	2.70	0.93	0.74	0.38	1.55
118	2.5	9	0.18	0.39	0.16	0.35	0.10	0.22	0.07	0.16	0.10	0.21	0.13	0.10	0.04	0.05
	26.0	92	1.86	4.11	1.66	3.66	1.06	2.33	0.74	1.63	1.00	2.20	1.31	1.04	0.47	0.54
119	30.0	80	1.43	3.15	1.07	2.37	0.69	1.52	0.43	0.95	0.83	1.82	1.28	1.02	0.51	0.90
	31.8	85	1.52	3.35	1.14	2.51	0.73	1.62	0.46	1.01	0.88	1.94	1.36	1.09	0.54	0.96
120	1.7	18	0.35	0.77	0.30	0.65	0.17	0.37	0.09	0.21	0.17	0.38	0.09	0.05	—	—
	5.9	63	1.22	2.69	1.03	2.27	0.58	1.29	0.33	0.72	0.61	1.33	0.32	0.19	—	—
121	7.5	76	1.45	3.19	1.28	2.82	0.75	1.65	0.49	1.09	0.71	1.58	—	0.41	—	—
	8.4	85	1.62	3.57	1.43	3.16	0.84	1.85	0.55	1.22	0.80	1.77	—	0.46	—	—
122	3.7	51	1.00	2.19	0.83	1.83	0.46	1.01	0.24	0.52	0.49	1.08	0.29	0.16	0.09	0.11
	4.3	59	1.15	2.52	0.95	2.10	0.53	1.16	0.27	0.60	0.56	1.24	0.33	0.18	0.10	0.13
123	7.9	74	1.47	3.25	1.31	2.88	0.75	1.66	0.50	1.09	0.72	1.59	0.03	0.30	—	—
	8.8	82	1.64	3.62	1.45	3.21	0.84	1.85	0.55	1.22	0.80	1.77	0.03	0.34	—	—
124	3.6	61	1.20	2.64	1.04	2.29	0.70	1.54	0.47	1.04	0.65	1.43	0.12	0.03	1.16	1.28
	4.6	78	1.54	3.38	1.33	2.94	0.90	1.98	0.60	1.33	0.83	1.83	0.16	0.03	1.48	1.64
125	2.0	51	1.01	2.22	0.86	1.89	0.57	1.26	0.38	0.83	0.52	1.15	1.18	0.09	0.28	0.07
	3.0	75	1.49	3.29	1.27	2.81	0.85	1.87	0.56	1.23	0.78	1.71	1.76	0.13	0.41	0.11
126	5.3	66	1.36	2.99	1.18	2.60	0.70	1.55	0.44	0.97	0.68	1.51	1.04	0.42	0.19	—
	5.7	70	1.44	3.17	1.25	2.75	0.74	1.64	0.47	1.03	0.73	1.60	1.10	0.45	0.20	—
127	0.6	60	1.22	2.68	1.12	2.46	0.77	1.70	0.53	1.18	0.64	1.41	0.74	0.08	0.16	2.26
	0.8	81	1.63	3.60	1.50	3.31	1.04	2.28	0.72	1.58	0.86	1.89	1.00	0.11	0.22	3.04
128	1.0	11	0.22	0.49	0.19	0.41	0.11	0.25	0.06	0.13	0.12	0.26	0.12	0.08	0.00	—
	5.0	55	1.11	2.44	0.91	2.01	0.55	1.21	0.29	0.64	0.58	1.28	0.60	0.41	0.01	—
129	0.8	12	0.24	0.54	0.20	0.44	0.13	0.28	0.07	0.15	0.13	0.29	0.08	0.07	0.00	—
	3.6	53	1.06	2.34	0.87	1.91	0.55	1.21	0.29	0.64	0.58	1.28	0.35	0.30	0.01	—
130	9.2	69	1.36	3.00	1.19	2.62	0.80	1.77	0.54	1.19	0.70	1.55	0.08	0.34	0.05	0.09
	10.3	77	1.53	3.37	1.33	2.94	0.90	1.98	0.60	1.33	0.79	1.74	0.09	0.38	0.06	0.10
131	8.2	59	1.25	2.76	1.08	2.38	0.70	1.55	0.45	0.99	0.68	1.50	—	—	—	—
	9.1	66	1.38	3.04	1.19	2.62	0.78	1.71	0.50	1.10	0.75	1.66	—	—	—	—
132	6.8	70	1.41	3.11	1.24	2.73	0.76	1.68	0.50	1.11	0.73	1.61	0.10	0.31	0.06	0.12
	7.5	78	1.55	3.42	1.36	3.00	0.84	1.85	0.55	1.22	0.80	1.77	0.10	0.34	0.07	0.13
133	10.9	86	1.72	3.79	1.55	3.42	0.95	2.10	0.67	1.47	0.89	1.95	0.07	0.44	0.09	0.05
	12.1	95	1.90	4.18	1.71	3.78	1.05	2.32	0.73	1.62	0.98	2.15	0.08	0.48	0.10	0.06
134	11.1	87	1.72	3.80	1.46	3.21	1.02	2.24	0.72	1.58	0.88	1.94	0.08	0.43	0.05	0.08
	12.3	98	1.92	4.24	1.63	3.59	1.14	2.50	0.80	1.77	0.98	2.17	0.08	0.48	0.06	0.09
135	5.9	79	1.45	3.19	1.28	2.83	0.71	1.58	0.47	1.02	0.69	1.51	0.62	0.10	—	—
	6.7	89	1.64	3.62	1.45	3.21	0.81	1.79	0.53	1.16	0.78	1.72	0.71	0.11	—	—

(Continued)

TABLE 21–1

Entry Number	Feed Name Description	Moisture Basis: A-F (as-fed) or M-F (moisture-free)	Macro Minerals (Continued)			Micro Minerals						
			Magnesium (Mg)	Potassium (K)	Sulfur (S)	Cobalt (Co)	Copper (Cu)	Iodine (I)	Iron (Fe)	Manganese (Mn)	Selenium (Se)	Zinc (Zn)
			(%)	(%)	(%)	(ppm or mg/kg)	(ppm or mg/kg)	(ppm or mg/kg)	(%)	(ppm or mg/kg)	(ppm or mg/kg)	(ppm or mg/kg)
	MEAT (Continued)											
114	W/BLOOD & BONE, MEAL RENDERED (TANKAGE)	A-F	—	—	0.26	—	—	—	—	—	0.26	—
		M-F	—	—	0.28	—	—	—	—	—	0.28	—
115	W/BONE, MEAL RENDERED	A-F	1.02	1.33	0.25	0.18	1.5	1.32	0.07	13.3	0.26	94.3
		M-F	1.09	1.43	0.27	0.19	1.6	1.41	0.07	14.3	0.28	101.1
	MILK											
116	FRESH (CATTLE *bos taurus*)	A-F	0.01	0.14	0.04	0.00	0.1	—	0.00	—	—	2.3
		M-F	0.10	1.13	0.32	0.01	0.8	—	0.01	—	—	23.0
117	DEHY (CATTLE *Bos taurus*)	A-F	0.09	1.05	0.31	0.01	0.9	—	0.02	0.4	—	21.7
		M-F	0.09	1.10	0.32	0.01	0.9	—	0.02	0.5	—	22.8
118	SKIMMED, FRESH (CATTLE *Bos taurus*)	A-F	0.01	0.12	0.03	0.01	1.1	—	0.00	0.2	—	4.9
		M-F	0.12	1.29	0.32	0.11	11.6	—	0.01	2.3	—	51.0
119	SKIMMED, DEHY (CATTLE *Bos taurus*)	A-F	0.12	1.60	0.32	0.11	11.7	—	0.00	2.1	0.12	38.5
		M-F	0.13	1.70	0.34	0.12	12.4	—	0.00	2.3	0.13	40.9
	MILLET, FOXTAIL *Setaria italica*											
120	FRESH	A-F	—	0.56	—	—	—	—	—	—	—	—
		M-F	—	1.94	—	—	—	—	—	—	—	—
121	GRAIN	A-F	—	0.31	—	—	—	—	0.01	—	—	—
		M-F	—	0.35	—	—	—	—	0.01	—	—	—
122	HAY, SUN-CURED	A-F	0.20	1.69	0.14	—	—	—	—	120.1	—	—
		M-F	0.23	1.94	0.16	—	—	—	—	138.1	—	—
	MILLET, PROSO (BROOMCORN, HOG MILLET)											
123	GRAIN	A-F	0.16	0.43	—	—	—	—	0.01	—	—	—
		M-F	0.18	0.48	—	—	—	—	0.01	—	—	—
	MOLASSES AND SYRUP											
124	BEET, SUGAR, MOLASSES, MORE THAN 48% INVERT SUGAR, MORE THAN 79.5 DEGREES BRIX	A-F	0.23	4.73	0.46	0.36	16.8	—	0.01	4.5	—	14.0
		M-F	0.29	6.07	0.60	0.47	21.6	—	0.01	5.8	—	18.0
125	CITRUS, SYRUP (CITRUS MOLASSES)	A-F	0.14	0.09	0.16	0.11	72.8	—	0.04	40.9	—	92.4
		M-F	0.21	0.14	0.23	0.16	108.0	—	0.05	60.7	—	137.1
126	SUGAR CANE, MOLASSES, DEHY	A-F	0.44	3.40	0.43	1.15	74.9	—	0.02	54.1	—	31.2
		M-F	0.47	3.60	0.46	1.21	79.4	—	0.03	57.3	—	33.0
127	SUGAR CANE, MOLASSES, MORE THAN 46% INVERT SUGAR, MORE THAN 79.5 DEGREES BRIX (BLACKSTRAP)	A-F	0.31	2.98	0.35	1.18	48.9	1.56	0.02	43.7	—	15.6
		M-F	0.42	4.01	0.47	1.59	65.7	2.10	0.03	58.8	—	20.9
	NAPIERGRASS *Pennisetum purpureum*											
128	PREBLOOM, FRESH	A-F	0.05	0.27	0.02	—	—	—	—	—	—	—
		M-F	0.26	1.31	0.10	—	—	—	—	—	—	—
129	LATE BLOOM, FRESH	A-F	0.06	0.30	0.02	—	—	—	—	—	—	—
		M-F	0.26	1.31	0.10	—	—	—	—	—	—	—
	OATS *Avena sativa*											
130	GRAIN, ALL ANALYSES	A-F	0.14	0.40	0.21	0.06	6.0	0.11	0.01	35.8	0.22	34.9
		M-F	0.16	0.45	0.23	0.06	6.7	0.13	0.01	40.1	0.24	39.2
131	GRAIN, LIGHT, LESS THAN 27 LB/BU (34.7 KG/HL)	A-F	—	—	—	—	—	—	—	—	—	—
		M-F	—	—	—	—	—	—	—	—	—	—
132	GRAIN, PACIFIC COAST	A-F	0.17	0.38	0.20	—	—	—	0.01	38.0	0.08	—
		M-F	0.19	0.42	0.22	—	—	—	0.01	41.8	0.08	—
133	CEREAL BY-PRODUCT (FEEDING OAT MEAL, OAT MIDDLINGS)	A-F	0.14	0.50	0.22	0.05	5.2	—	0.04	43.8	—	139.5
		M-F	0.16	0.55	0.24	0.05	5.7	—	0.04	48.3	—	153.8
134	GROATS	A-F	0.11	0.35	0.20	—	6.0	0.11	0.01	27.8	—	0.0
		M-F	0.13	0.39	0.22	—	6.7	0.12	0.01	31.0	—	0.1
	ORANGE *Citrus sinensis*											
135	PULP W/O FINES, DEHY (DRIED ORANGE PULP)	A-F	—	—	—	—	—	—	—	—	—	—
		M-F	—	—	—	—	—	—	—	—	—	—

(Continued)

Entry Number	Fat-Soluble Vitamins					Water-Soluble Vitamins								
	A	Carotene (Provitamin A)	D	E	K	B–12	Biotin	Choline	Folacin (Folic Acid)	Niacin	Pantothenic Acid (B–3)	Pyridoxine (B–6)	Riboflavin (B–2)	Thiamin (B–1)
	(IU/g)	(ppm or mg/kg)	(IU/kg)	(ppm or mg/kg)	(ppm or mg/kg)	(ppb or mcg/kg)	(ppm or mg/kg)	(ppm or mg/kg)	(ppm or mg/kg)	(ppm or mg/kg)	(ppm or mg/kg)	(ppm or mg/kg)	(ppm or mg/kg)	(ppm or mg/kg)
114	—	—	—	0.8	—	104.4	0.07	2067	0.57	58	4.8	—	5.0	0.2
	—	—	—	0.9	—	112.4	0.08	2225	0.62	63	5.2	—	5.4	0.2
115	—	—	—	0.9	—	118.4	0.10	2049	0.37	51	5.5	5.86	4.7	0.2
	—	—	—	0.9	—	126.9	0.11	2195	0.40	55	5.9	6.28	5.0	0.2
116	—	—	—	—	—	—	—	904	—	1	8.4	—	1.7	0.3
	—	—	—	—	—	—	—	7311	—	10	68.0	—	13.8	2.4
117	—	—	0	—	—	—	0.38	—	—	8	22.7	4.71	19.6	3.8
	—	—	0	—	—	—	0.40	—	—	9	23.8	4.94	20.6	3.9
118	—	—	—	—	—	—	—	—	—	1	3.5	—	2.0	0.4
	—	—	—	—	—	—	—	—	—	12	36.9	—	20.8	4.6
119	—	—	0	9.1	—	50.9	0.33	1394	0.62	11	36.4	4.10	19.1	3.7
	—	—	0	9.6	—	54.1	0.35	1480	0.66	12	38.6	4.35	20.3	3.9
120	—	—	—	—	—	—	—	—	—	—	—	—	—	—
	—	—	—	—	—	—	—	—	—	—	—	—	—	—
121	—	—	—	—	—	—	—	—	—	33	—	—	1.1	3.8
	—	—	—	—	—	—	—	—	—	37	—	—	1.2	4.3
122	86.9	52.1	—	—	—	—	—	—	—	—	—	—	—	—
	100.0	60.0	—	—	—	—	—	—	—	—	—	—	—	—
123	—	—	—	—	—	—	—	438	—	24	10.9	—	3.3	7.5
	—	—	—	—	—	—	—	488	—	27	12.2	—	3.7	8.3
124	—	—	—	4.0	—	—	—	827	—	41	4.5	—	2.3	—
	—	—	—	5.1	—	—	—	1062	—	53	5.8	—	2.9	—
125	—	—	—	—	—	—	—	—	—	27	17.2	—	6.2	—
	—	—	—	—	—	—	—	—	—	40	25.5	—	9.2	—
126	—	—	—	5.2	—	—	—	—	—	—	—	—	—	—
	—	—	—	5.5	—	—	—	—	—	—	—	—	—	—
127	—	—	—	5.4	—	—	0.69	764	0.11	36	37.4	4.21	2.8	0.9
	—	—	—	7.3	—	—	0.92	1027	0.15	49	50.3	5.67	3.8	1.2
128	—	—	—	—	—	—	—	—	—	—	—	—	—	—
	—	—	—	—	—	—	—	—	—	—	—	—	—	—
129	—	—	—	—	—	—	—	—	—	—	—	—	—	—
	—	—	—	—	—	—	—	—	—	—	—	—	—	—
130	0.2	0.1	—	14.9	—	—	0.27	967	0.39	14	9.9	2.53	1.4	6.0
	0.2	0.1	—	16.8	—	—	0.30	1084	0.44	16	11.1	2.84	1.5	6.8
131	—	—	—	—	—	—	—	—	—	—	—	—	—	—
	—	—	—	—	—	—	—	—	—	—	—	—	—	—
132	—	—	—	20.2	—	—	—	917	—	14	11.7	—	1.2	—
	—	—	—	22.2	—	—	—	1009	—	16	12.8	—	1.3	—
133	—	—	—	23.7	—	—	0.22	1157	0.46	24	17.6	—	1.7	7.0
	—	—	—	26.1	—	—	0.24	1277	0.51	26	19.4	—	1.9	7.7
134	—	—	—	14.8	—	—	—	1132	0.51	10	13.8	1.00	1.2	6.5
	—	—	—	16.5	—	—	—	1264	0.57	11	15.4	1.12	1.3	7.2
135	—	—	—	—	—	—	—	—	—	—	—	—	—	—
	—	—	—	—	—	—	—	—	—	—	—	—	—	—

(Continued)

TABLE 21-1

Entry Number	Feed Name Description	International Feed Number	Moisture Basis: A-F (as-fed) or M-F (moisture-free)	Chemical Analysis								
				Dry Matter	Ash	Crude Fiber	Neutral Det. Fib. (NDF)	Acid Det. Fib. (ADF)	Lignin	Ether Extract (Fat)	N-Free Extract	Crude Protein
				(%)	(%)	(%)	(%)	(%)	(%)	(%)	(%)	(%)
	ORCHARDGRASS *Dactylis glomerata*											
136	FRESH, ALL ANALYSES	2-03-451	A-F	26	2.6	6.4	13.9	—	—	1.6	11.3	3.9
			M-F	100	10.0	24.8	54.0	—	—	6.4	43.8	15.0
137	HAY, SUN-CURED	1-03-438	A-F	89	6.5	31.0	64.1	36.0	—	2.8	39.7	9.4
			M-F	100	7.3	34.7	71.8	40.3	—	3.1	44.4	10.5
138	HAY, FULL BLOOM, SUN-CURED	1-03-427	A-F	93	8.3	33.4	58.5	34.3	4.3	2.9	39.5	8.7
			M-F	100	8.9	36.0	63.1	37.0	4.6	3.1	42.6	9.4
	PANGOLAGRASS *Digitaria decumbens*											
139	HAY, SUN-CURED, 29-42 DAYS' GROWTH	1-26-214	A-F	91	7.3	32.8	66.0	39.0	5.5	1.8	—	6.5
			M-F	100	8.0	36.0	73.0	43.0	6.0	2.0	—	7.1
	PAPER											
140	WASTE	1-26-072	A-F	92	0.5	63.4	86.5	73.6	22.1	3.6	23.9	0.6
			M-F	100	0.5	68.9	94.0	80.0	24.0	3.9	26.0	0.7
	PEA *Pisum* spp											
141	SEEDS	5-03-600	A-F	89	2.9	5.5	—	—	—	1.1	56.7	23.2
			M-F	100	3.2	6.1	—	—	—	1.2	63.4	26.0
142	SILAGE, VINES W/O SEEDS, W/PODS	3-03-596	A-F	25	2.2	7.3	—	—	—	0.8	11.0	3.2
			M-F	100	9.0	29.8	—	—	—	3.3	44.9	13.1
	PEANUT *Arachis hypogaea*											
143	HAY, SUN-CURED	1-03-619	A-F	91	8.2	30.3	—	37.2	—	3.3	39.1	9.9
			M-F	100	9.0	33.4	—	41.0	—	3.6	43.1	10.9
144	HULLS (PODS)	1-08-028	A-F	91	3.8	57.2	69.0	60.8	21.8	2.1	20.7	7.3
			M-F	100	4.2	62.9	75.8	66.8	23.9	2.3	22.8	8.0
145	SEEDS W/O HULLS, MEAL MECH EXTD (PEANUT MEAL)	5-03-649	A-F	93	5.0	6.2	13.2	5.6	1.0	5.6	26.7	49.2
			M-F	100	5.4	6.7	14.2	6.1	1.1	6.0	28.8	53.1
146	SEEDS W/O HULLS, MEAL SOLV EXTD (PEANUT MEAL)	5-03-650	A-F	93	5.8	7.7	—	—	—	2.2	27.9	49.0
			M-F	100	6.3	8.3	—	—	—	2.4	30.1	52.9
	PEARL MILLET *Pennisetum glaucum*											
147	FRESH	2-03-115	A-F	21	1.9	6.5	—	—	—	0.6	9.7	2.1
			M-F	100	9.2	31.1	—	—	—	2.9	46.8	10.1
148	SILAGE	3-20-903	A-F	30	—	11.4	—	—	—	—	—	2.8
			M-F	100	—	38.0	—	—	—	—	—	9.2
	PINEAPPLE *Ananus comosus*											
149	CANNERY RESIDUE, DEHY (PINEAPPLE BRAN)	4-03-722	A-F	87	3.0	18.2	63.5	32.2	—	1.3	60.5	4.0
			M-F	100	3.5	20.9	73.0	37.0	—	1.5	69.5	4.6
	POTATO *Solanum tuberosum*											
150	TUBERS, FRESH	4-03-787	A-F	24	1.1	0.6	—	—	—	0.1	19.5	2.2
			M-F	100	4.8	2.4	—	—	—	0.4	83.2	9.3
151	TUBERS, DEHY	4-07-850	A-F	91	7.2	2.1	—	2.7	—	0.5	73.3	8.1
			M-F	100	7.9	2.3	—	3.0	—	0.5	80.5	8.9
152	CANNERY RESIDUE, DEHY	4-03-775	A-F	89	3.0	6.5	—	—	—	0.3	71.5	7.4
			M-F	100	3.4	7.3	—	—	—	0.4	80.5	8.4
153	TUBERS, SILAGE, BOILED	4-03-767	A-F	23	1.5	0.7	—	—	—	0.1	19.1	1.9
			M-F	100	6.5	3.2	—	—	—	0.4	81.7	8.2
	RAPE *Brassica napus*											
154	FRESH	2-03-867	A-F	17	2.1	2.4	—	—	—	0.6	8.5	2.9
			M-F	100	12.6	14.7	—	—	—	3.8	51.2	17.6
155	SEEDS, MEAL MECH EXTD	5-03-870	A-F	92	6.9	12.0	—	—	—	7.3	30.1	35.6
			M-F	100	7.5	13.1	—	—	—	7.9	32.7	38.7
156	SEEDS, MEAL SOLV EXTD, 34% PROTEIN	5-26-082	A-F	90	7.0	13.0	—	—	—	2.5	33.5	34.0
			M-F	100	7.8	14.4	—	—	—	2.8	37.2	37.8
	RAPE, SUMMER *Brassica napus, annua*											
157	SEEDS, MEAL PREPRESSED, SOLV EXTD	5-08-135	A-F	92	7.2	9.3	33.1	16.6	—	1.1	33.9	40.5
			M-F	100	7.8	10.1	36.0	18.0	—	1.2	36.8	44.0

(Continued)

Entry Number	Digestible Protein Ruminant (%)	TDN Ruminant (%)	Digestible Energy Ruminant (Mcal/lb)	(Mcal/kg)	Metabolizable Energy Ruminant (Mcal/lb)	(Mcal/kg)	Ruminant NE$_m$ (Mcal/lb)	(Mcal/kg)	Ruminant NE$_g$ (Mcal/lb)	(Mcal/kg)	Lactating Cows NE$_{lc}$ (Mcal/lb)	(Mcal/kg)	Calcium (Ca) (%)	Phosphorus (P) (%)	Sodium (Na) (%)	Chlorine (Cl) (%)
136	2.7	17	0.35	0.77	0.30	0.66	0.19	0.42	0.12	0.26	0.18	0.41	0.09	0.05	0.03	—
	10.5	67	1.35	2.98	1.16	2.56	0.73	1.61	0.46	1.01	0.72	1.58	0.37	0.18	0.13	—
137	5.7	51	1.23	2.71	0.96	2.11	0.50	1.09	0.27	0.59	0.52	1.15	0.34	0.23	0.01	0.37
	6.4	57	1.38	3.03	1.07	2.36	0.56	1.22	0.30	0.66	0.58	1.29	0.38	0.26	0.02	0.41
138	5.0	50	1.02	2.25	0.83	1.82	0.46	1.01	0.22	0.49	0.50	1.10	—	—	—	—
	5.4	54	1.10	2.42	0.89	1.97	0.49	1.09	0.24	0.53	0.54	1.19	—	—	—	—
139	—	41	0.82	1.80	0.64	1.41	0.32	0.71	0.10	0.23	0.41	0.89	0.45	0.21	—	—
	—	45	0.90	1.98	0.70	1.55	0.35	0.78	0.11	0.25	0.45	0.98	0.46	0.23	—	—
140	—	25	0.92	2.02	0.75	1.66	—	—	—	—	—	—	0.09	0.06	—	—
	—	27	1.00	2.20	0.82	1.80	—	—	—	—	—	—	0.10	0.07	—	—
141	19.0	77	1.56	3.44	1.39	3.07	0.87	1.91	0.59	1.31	0.81	1.79	0.12	0.41	0.04	0.05
	21.2	87	1.75	3.85	1.56	3.44	0.97	2.14	0.67	1.47	0.91	2.00	0.14	0.46	0.05	0.06
142	1.9	14	0.28	0.61	0.23	0.51	0.13	0.29	0.07	0.15	0.14	0.31	0.32	0.06	0.00	—
	7.7	57	1.13	2.49	0.94	2.07	0.53	1.18	0.28	0.61	0.57	1.25	1.31	0.24	0.01	—
143	5.7	48	0.95	2.10	0.78	1.71	0.39	0.85	0.16	0.36	0.45	0.99	1.12	0.14	—	—
	6.3	53	1.05	2.32	0.86	1.89	0.43	0.94	0.18	0.40	0.49	1.09	1.23	0.16	—	—
144	1.7	16	0.33	0.72	0.14	0.32	-0.16	-0.36	-0.38	-0.83	0.11	0.24	0.24	0.06	0.12	—
	1.9	18	0.36	0.79	0.16	0.35	-0.18	-0.39	-0.41	-0.91	0.12	0.26	0.26	0.07	0.13	—
145	45.5	81	1.54	3.40	1.47	3.24	0.86	1.89	0.58	1.28	0.81	1.78	0.20	0.56	0.12	0.03
	49.1	37	1.66	3.67	1.59	3.50	0.92	2.04	0.63	1.38	0.87	1.92	0.22	0.61	0.13	0.03
146	42.3	73	1.43	3.15	1.29	2.84	0.78	1.72	0.51	1.13	0.74	1.63	0.36	0.61	0.03	0.03
	45.7	79	1.54	3.40	1.39	3.07	0.84	1.86	0.55	1.22	0.80	1.76	0.39	0.66	0.03	0.03
147	1.3	13	0.25	0.56	0.21	0.47	0.12	0.27	0.07	0.15	0.13	0.28	—	—	—	—
	6.3	62	1.22	2.68	1.02	2.26	0.59	1.29	0.33	0.72	0.61	1.34	—	—	—	—
148	—	18	0.35	1.70	0.30	0.65	0.17	0.38	0.10	0.21	0.18	0.40	—	—	—	—
	—	59	1.18	2.60	0.99	2.18	0.57	1.27	0.32	0.70	0.60	1.33	—	—	—	—
149	0.7	64	1.27	2.81	1.11	2.45	0.70	1.55	0.46	1.00	0.67	1.49	0.20	0.11	—	—
	0.8	73	1.46	3.23	1.27	2.81	0.81	1.78	0.52	1.15	0.77	1.71	0.23	0.13	—	—
150	1.4	19	0.38	0.84	0.34	0.74	0.20	0.45	0.14	0.30	0.19	0.43	0.01	0.06	0.02	0.07
	5.7	81	1.62	3.58	1.43	3.16	0.87	1.92	0.58	1.28	0.83	1.82	0.04	0.24	0.09	0.28
151	2.9	75	1.42	3.14	1.26	2.77	0.85	1.88	0.58	1.28	0.84	1.85	0.07	0.20	0.01	0.36
	3.2	83	1.56	3.45	1.38	3.04	0.94	2.06	0.64	1.40	0.92	2.03	0.08	0.22	0.01	0.40
152	5.7	79	1.47	3.23	1.30	2.86	0.73	1.60	0.47	1.04	0.70	1.53	0.14	0.23	—	—
	6.4	90	1.65	3.64	1.46	3.22	0.82	1.80	0.53	1.18	0.78	1.73	0.16	0.25	—	—
153	0.4	18	0.35	0.78	0.31	0.68	0.17	0.38	0.11	0.24	0.17	0.37	—	—	—	—
	1.8	75	1.51	3.33	1.32	2.91	0.73	1.61	0.46	1.01	0.72	1.58	—	—	—	—
154	2.4	13	0.25	0.54	0.21	0.47	0.12	0.26	0.07	0.16	0.12	0.26	0.25	0.07	—	—
	14.5	79	1.47	3.24	1.28	2.83	0.71	1.57	0.44	0.97	0.70	1.54	1.47	0.43	—	—
155	30.4	71	1.41	3.11	1.24	2.73	0.79	1.75	0.53	1.16	0.75	1.66	0.66	1.04	—	—
	33.0	77	1.53	3.38	1.35	2.97	0.86	1.90	0.57	1.26	0.82	1.81	0.72	1.14	—	—
156	—	—	—	—	—	—	—	—	—	—	—	—	—	—	—	—
	—	—	—	—	—	—	—	—	—	—	—	—	—	—	—	—
157	34.6	70	1.40	3.08	1.22	2.69	0.75	1.65	0.49	1.07	0.72	1.58	0.66	0.93	—	—
	37.6	76	1.52	3.34	1.33	2.93	0.81	1.79	0.53	1.17	0.78	1.72	0.72	1.01	—	—

TABLE 21–1

Entry Number	Feed Name Description	Moisture Basis: A-F (as-fed) or M-F (moisture-free)	Macro Minerals (Continued)			Micro Minerals						
			Magnesium (Mg)	Potassium (K)	Sulfur (S)	Cobalt (Co)	Copper (Cu)	Iodine (I)	Iron (Fe)	Manganese (Mn)	Selenium (Se)	Zinc (Zn)
			(%)	(%)	(%)	(ppm or mg/kg)	(ppm or mg/kg)	(ppm or mg/kg)	(%)	(ppm or mg/kg)	(ppm or mg/kg)	(ppm or mg/kg)
	ORCHARDGRASS *Dactylis glomerata*											
136	FRESH, ALL ANALYSES	A-F	0.06	0.74	—	0.06	2.5	—	0.00	28.5	—	5.3
		M-F	0.24	2.88	—	0.21	9.8	—	0.01	110.4	—	20.6
137	HAY, SUN-CURED	A-F	0.16	2.68	0.23	0.34	12.9	—	0.01	162.7	—	32.0
		M-F	0.18	3.00	0.26	0.38	14.5	—	0.02	182.3	—	35.8
138	HAY, FULL BLOOM, SUN-CURED	A-F	—	—	—	—	—	—	—	—	—	—
		M-F	—	—	—	—	—	—	—	—	—	—
	PANGOLAGRASS *Digitaria decumbens*											
139	HAY, SUN-CURED, 29–42 DAYS' GROWTH	A-F	0.14	—	—	—	—	—	—	—	—	—
		M-F	0.15	—	—	—	—	—	—	—	—	—
	PAPER											
140	WASTE	A-F	—	—	—	—	—	—	—	—	—	—
		M-F	—	—	—	—	—	—	—	—	—	—
	PEA *Pisum* spp											
141	SEEDS	A-F	0.12	0.95	—	—	—	—	0.01	2.9	—	23.0
		M-F	0.14	1.06	—	—	—	—	0.01	3.2	—	25.7
142	SILAGE, VINES W/O SEEDS, W/PODS	A-F	0.10	0.34	0.06	—	—	—	0.00	—	—	—
		M-F	0.39	1.40	0.25	—	—	—	0.01	—	—	—
	PEANUT *Arachis hypogaea*											
143	HAY, SUN-CURED	A-F	0.44	1.25	0.21	0.07	—	—	—	—	—	—
		M-F	0.49	1.38	0.23	0.08	—	—	—	—	—	—
144	HULLS (PODS)	A-F	0.15	0.87	0.09	0.11	16.2	—	0.03	62.5	—	21.9
		M-F	0.17	0.95	0.10	0.12	17.8	—	0.03	68.7	—	24.1
145	SEEDS W/O HULLS, MEAL MECH EXTD (PEANUT MEAL)	A-F	0.26	1.16	0.22	0.11	15.4	0.07	0.03	25.5	—	33.0
		M-F	0.28	1.25	0.24	0.12	16.6	0.07	0.03	27.6	—	35.6
146	SEEDS W/O HULLS, MEAL SOLV EXTD (PEANUT MEAL)	A-F	0.27	1.16	0.31	—	—	—	—	—	—	—
		M-F	0.30	1.25	0.33	—	—	—	—	—	—	—
	PEARL MILLET *Pennisetum glaucum*											
147	FRESH	A-F	—	—	—	—	—	—	—	—	—	—
		M-F	—	—	—	—	—	—	—	—	—	—
148	SILAGE	A-F	—	—	—	—	—	—	—	—	—	—
		M-F	—	—	—	—	—	—	—	—	—	—
	PINEAPPLE *Ananus comosus*											
149	CANNERY RESIDUE, DEHY (PINEAPPLE BRAN)	A-F	—	—	—	—	—	—	0.05	—	—	—
		M-F	—	—	—	—	—	—	0.06	—	—	—
	POTATO *Solanum tuberosum*											
150	TUBERS, FRESH	A-F	0.03	0.51	0.02	—	6.7	—	0.00	9.8	—	—
		M-F	0.14	2.17	0.09	—	28.4	—	0.01	41.7	—	—
151	TUBERS, DEHY	A-F	0.11	1.97	0.08	—	—	—	—	2.3	—	2.0
		M-F	0.12	2.16	0.09	—	—	—	—	2.5	—	2.2
152	CANNERY RESIDUE, DEHY	A-F	—	—	—	—	—	—	—	—	—	—
		M-F	—	—	—	—	—	—	—	—	—	—
153	TUBERS, SILAGE, BOILED	A-F	—	—	—	—	—	—	—	—	—	—
		M-F	—	—	—	—	—	—	—	—	—	—
	RAPE *Brassica napus*											
154	FRESH	A-F	0.01	0.56	0.11	—	1.4	—	0.00	7.7	—	—
		M-F	0.06	3.37	0.68	—	8.1	—	0.02	46.0	—	—
155	SEEDS, MEAL MECH XTD	A-F	0.50	0.83	—	—	6.8	—	0.02	55.3	0.96	43.2
		M-F	0.54	0.90	—	—	7.4	—	0.02	60.2	1.04	47.0
156	SEEDS, MEAL SOLV EXTD, 34% PROTEIN	A-F	—	—	—	—	—	—	—	—	—	—
		M-F	—	—	—	—	—	—	—	—	—	—
	RAPE, SUMMER *Brassica napus, annua*											
157	SEEDS, MEAL PREPRESSED, SOLV EXTD	A-F	—	—	—	—	—	—	—	—	—	—
		M-F	—	—	—	—	—	—	—	—	—	—

(Continued)

Entry Number	Fat-Soluble Vitamins					Water-Soluble Vitamins								
	A	Carotene (Provitamin A)	D	E	K	B–12	Biotin	Choline	Folacin (Folic Acid)	Niacin	Pantothenic Acid (B–3)	Pyridoxine (B–6)	Riboflavin (B–2)	Thiamin (B–1)
	(IU/g)	(ppm or mg/kg)	(IU/kg)	(ppm or mg/kg)	(ppm or mg/kg)	(ppb or mcg/kg)	(ppm or mg/kg)	(ppm or mg/kg)	(ppm or mg/kg)	(ppm or mg/kg)	(ppm or mg/kg)	(ppm or mg/kg)	(ppm or mg/kg)	(ppm or mg/kg)
136	137.1	82.2	—	112.3	—	—	—	—	—	—	—	—	—	1.9
	531.8	319.0	—	435.6	—	—	—	—	—	—	—	—	—	7.3
137	28.9	17.3	—	170.7	—	—	—	—	—	—	—	—	6.1	2.6
	32.4	19.4	—	191.1	—	—	—	—	—	—	—	—	6.8	2.9
138	—	—	—	—	—	—	—	—	—	—	—	—	—	—
	—	—	—	—	—	—	—	—	—	—	—	—	—	—
139	—	—	—	—	—	—	—	—	—	—	—	—	—	—
	—	—	—	—	—	—	—	—	—	—	—	—	—	—
140	—	—	—	—	—	—	—	—	—	—	—	—	—	—
	—	—	—	—	—	—	—	—	—	—	—	—	—	—
141	1.2	0.7	—	3.0	—	—	0.18	547	0.22	31	27.8	1.97	1.8	4.6
	1.3	0.8	—	3.3	—	—	0.20	612	0.25	34	31.1	2.21	2.0	5.2
142	77.2	46.3	—	—	—	—	—	—	—	—	—	—	—	—
	315.0	189.0	—	—	—	—	—	—	—	—	—	—	—	—
143	52.6	31.5	—	—	—	—	—	—	—	—	—	—	8.8	—
	58.0	34.8	—	—	—	—	—	—	—	—	—	—	9.7	—
144	1.3	0.8	—	—	—	—	—	—	—	—	—	—	—	—
	1.5	0.9	—	—	—	—	—	—	—	—	—	—	—	—
145	—	—	—	2.4	—	—	0.33	1975	0.66	173	47.6	6.12	9.1	5.7
	—	—	—	2.6	—	—	0.36	2132	0.71	186	51.4	6.61	9.8	6.2
146	—	—	—	2.9	—	—	—	1896	—	178	36.8	5.95	5.3	—
	—	—	—	3.2	—	—	—	2049	—	192	39.8	6.43	5.7	—
147	63.0	37.8	—	—	—	—	—	—	—	—	—	—	—	—
	304.2	182.5	—	—	—	—	—	—	—	—	—	—	—	—
148	3.0	—	—	—	—	—	—	—	—	—	—	—	—	—
	10.0	—	—	—	—	—	—	—	—	—	—	—	—	—
149	78.4	47.0	—	—	—	—	—	—	—	—	—	—	—	—
	90.0	54.0	—	—	—	—	—	—	—	—	—	—	—	—
150	—	—	—	—	—	—	—	—	—	17	—	—	0.5	1.2
	—	—	—	—	—	—	—	—	—	74	—	—	2.0	5.0
151	—	—	—	—	—	—	0.10	2622	0.61	33	20.0	14.12	1.0	—
	—	—	—	—	—	—	0.11	2879	0.66	37	22.0	15.50	1.1	—
152	—	—	—	—	—	—	—	—	—	—	—	—	—	—
	—	—	—	—	—	—	—	—	—	—	—	—	—	—
153	—	—	—	—	—	—	—	—	—	—	—	—	—	—
	—	—	—	—	—	—	—	—	—	—	—	—	—	—
154	—	—	—	—	—	—	—	—	—	—	—	—	—	—
	—	—	—	—	—	—	—	—	—	—	—	—	—	—
155	—	—	—	18.8	—	—	—	6532	—	155	9.0	—	3.0	1.8
	—	—	—	20.4	—	—	—	7103	—	168	9.8	—	3.3	1.9
156	—	—	—	—	—	—	—	—	—	—	—	—	—	—
	—	—	—	—	—	—	—	—	—	—	—	—	—	—
157	—	—	—	—	—	—	—	—	—	—	—	—	—	—
	—	—	—	—	—	—	—	—	—	—	—	—	—	—

TABLE 21–1

Entry Number	Feed Name Description	International Feed Number	Moisture Basis: A-F (as-fed) or M-F (moisture-free)	Chemical Analysis								
				Dry Matter	Ash	Crude Fiber	Neutral Det. Fib. (NDF)	Acid Det. Fib. (ADF)	Lignin	Ether Extract (Fat)	N-Free Extract	Crude Protein
				(%)	(%)	(%)	(%)	(%)	(%)	(%)	(%)	(%)
	REDTOP *Agrostis alba*											
158	FULL BLOOM, FRESH	2–03–891	A-F	26	1.8	6.6	16.6	—	—	0.9	14.8	2.1
			M-F	100	7.0	25.1	64.0	—	—	3.5	56.3	8.1
159	HAY, SUN-CURED	1–03–885	A-F	92	6.0	28.4	—	—	—	2.8	47.4	7.4
			M-F	100	6.6	30.9	—	—	—	3.1	51.5	8.1
	RICE *Oryza sativa*											
160	GRAIN, GROUND (GROUND ROUGH RICE, GROUND PADDY RICE)	4–03–938	A-F	89	5.3	8.6	—	—	—	1.6	65.9	7.5
			M-F	100	6.0	9.7	—	—	—	1.8	74.1	8.4
161	BRAN W/GERMS (RICE BRAN)	4–03–928	A-F	91	11.3	11.9	28.0	25.7	3.6	13.5	41.0	13.0
			M-F	100	12.5	13.1	30.9	28.4	4.0	14.9	45.2	14.3
162	GROATS, POLISHED (POLISHED RICE)	4–03–942	A-F	89	0.5	0.4	14.2	0.9	—	0.5	80.3	7.0
			M-F	100	0.6	0.4	16.0	1.0	—	0.5	90.6	7.9
163	GROATS (BROWN RICE)	4–03–936	A-F	88	1.0	0.8	—	—	—	1.8	77.2	7.4
			M-F	100	1.2	0.9	—	—	—	2.0	87.6	8.4
164	HULLS	1–08–075	A-F	92	19.0	38.9	71.9	62.3	9.6	1.0	30.3	3.0
			M-F	100	20.6	42.2	78.0	67.6	10.4	1.1	32.9	3.2
	RYE *Secale cereale*											
165	DISTILLERS GRAINS, DEHY	5–04–023	A-F	92	2.3	12.3	—	—	—	6.0	48.3	23.0
			M-F	100	2.5	13.4	—	—	—	6.5	52.6	25.1
166	GRAIN, ALL ANALYSES	4–04–047	A-F	87	1.6	2.2	—	—	—	1.5	70.0	12.0
			M-F	100	1.9	2.5	—	—	—	1.7	80.1	13.8
167	SILAGE, MOLASSES ADDED	3–04–021	A-F	24	1.8	8.4	—	—	—	0.7	10.4	2.4
			M-F	100	7.6	35.4	—	—	—	3.0	43.9	10.1
168	SILAGE, WILTED	3–08–601	A-F	30	2.5	10.8	—	—	—	1.0	12.5	3.5
			M-F	100	8.2	35.6	—	—	—	3.3	41.3	11.6
	RYEGRASS *Lolium* spp											
169	HAY	1–04–057	A-F	88	7.1	25.3	56.3	37.0	—	1.8	46.2	7.5
			M-F	100	8.1	28.8	64.0	42.0	—	2.1	52.5	8.5
	SAFFLOWER *Carthumus tinctorius*											
170	SEEDS	4–07–958	A-F	93	3.0	23.6	—	37.2	—	30.8	20.9	14.9
			M-F	100	3.2	26.3	—	40.0	—	33.1	22.4	16.0
171	SEEDS, MEAL SOLV EXTD, 20% PROTEIN	5–26–095	A-F	92	4.6	32.2	—	39.6	—	1.1	32.7	21.6
			M-F	100	5.0	34.9	—	43.0	—	1.2	35.5	23.4
172	SEEDS, W/O HULLS, MEAL MECH EXTD	5–08–499	A-F	91	6.5	12.8	—	—	—	6.0	23.9	42.0
			M-F	100	7.2	14.0	—	—	—	6.6	26.2	46.1
173	SEEDS, W/O HULLS, MEAL SOLV EXTD, 42% PROTEIN	5–26–094	A-F	92	6.5	14.6	—	19.2	—	1.3	26.3	42.7
			M-F	100	7.2	16.0	—	21.0	—	1.5	28.8	46.7
	SESAME *Sesamum indicum*											
174	SEEDS	5–08–509	A-F	95	5.8	10.6	—	—	—	44.1	10.5	23.7
			M-F	100	6.1	11.2	—	—	—	46.6	11.1	25.0
175	SEEDS, MEAL MECH EXTD	5–04–220	A-F	93	10.3	5.6	15.8	15.8	—	8.7	23.0	45.0
			M-F	100	11.2	6.1	17.0	17.0	—	9.4	24.8	48.6
176	SEEDS, MEAL SOLV EXTD, 44% PROTEIN	5–26–096	A-F	92	13.1	6.8	—	—	—	1.4	25.8	45.0
			M-F	100	14.2	7.4	—	—	—	1.5	28.0	48.9
	SORGHUM *Sorghum bicolor*											
177	FODDER W/HEADS, SUN-CURED	1–07–960	A-F	90	8.9	25.6	—	—	—	2.0	47.4	6.2
			M-F	100	9.9	28.4	—	—	—	2.2	52.6	6.9
178	STOVER W/O HEADS, SUN-CURED	1–04–302	A-F	92	8.9	29.9	—	39.9	—	1.6	46.8	4.4
			M-F	100	9.7	32.6	—	43.6	—	1.8	51.1	4.9
179	GRAIN, ALL ANALYSES	4–04–383	A-F	90	1.8	2.6	16.2	8.1	1.2	2.7	71.6	11.5
			M-F	100	1.9	2.8	18.0	9.0	1.3	2.9	79.5	12.8
180	GRAIN, LESS THAN 9% PROTEIN	4–08–138	A-F	89	2.1	2.2	16.0	8.0	—	2.9	72.4	8.9
			M-F	100	2.4	2.5	18.0	9.0	—	3.3	81.8	10.1

(Continued)

Entry Number	Digestible Protein Ruminant (%)	TDN Ruminant (%)	Digestible Energy Ruminant (Mcal/lb)	(Mcal/kg)	Metabolizable Energy Ruminant (Mcal/lb)	(Mcal/kg)	Ruminant NE_m (Mcal/lb)	(Mcal/kg)	Ruminant NE_g (Mcal/lb)	(Mcal/kg)	Lactating Cows NE_lc (Mcal/lb)	(Mcal/kg)	Calcium (Ca) (%)	Phosphorus (P) (%)	Sodium (Na) (%)	Chlorine (Cl) (%)
158	1.2	16	0.33	0.72	0.28	0.61	0.16	0.36	0.10	0.21	0.17	0.37	0.16	0.10	0.01	—
	4.4	62	1.24	2.73	1.04	2.30	0.62	1.38	0.36	0.80	0.64	1.40	0.62	0.37	0.05	—
159	3.7	50	1.02	2.25	0.84	1.86	0.50	1.11	0.27	0.59	0.53	1.17	0.39	0.20	0.06	0.06
	4.0	54	1.11	2.45	0.92	2.02	0.55	1.21	0.29	0.04	0.58	1.27	0.43	0.22	0.07	0.07
160	4.0	68	1.35	2.98	1.18	2.61	0.71	1.57	0.46	1.01	0.68	1.51	0.07	0.32	0.06	0.07
	4.5	76	1.52	3.35	1.33	2.93	0.80	1.76	0.52	1.14	0.77	1.70	0.07	0.36	0.07	0.08
161	8.6	64	1.10	2.42	0.99	2.18	0.62	1.38	0.38	0.84	0.53	1.18	0.07	1.44	0.03	0.07
	9.5	71	1.21	2.67	1.09	2.40	0.69	1.52	0.42	0.93	0.59	1.30	0.08	1.59	0.04	0.08
162	3.6	78	1.59	3.51	1.44	3.17	0.90	1.98	0.62	1.37	0.85	1.87	0.02	0.11	0.01	0.04
	4.0	88	1.80	3.96	1.63	3.58	1.01	2.24	0.70	1.55	0.96	2.11	0.03	0.13	0.02	0.04
163	3.2	78	1.56	3.43	1.39	3.07	0.83	1.82	0.56	1.23	0.78	1.71	0.03	0.20	0.02	0.07
	3.6	88	1.76	3.89	1.58	3.47	0.94	2.06	0.64	1.40	0.88	1.94	0.04	0.23	0.03	0.07
164	0.1	11	0.27	0.60	0.16	0.35	−0.26	−0.57	−0.47	−1.04	0.08	0.17	0.11	0.10	0.02	0.07
	0.2	12	0.30	0.65	0.17	0.38	−0.28	−0.62	−0.51	−1.13	0.08	0.19	0.12	0.10	0.02	0.08
165	13.8	54	1.08	2.38	0.90	1.99	0.45	1.00	0.22	0.49	0.50	1.09	0.15	0.48	0.17	0.05
	15.0	59	1.18	2.59	0.98	2.17	0.49	1.09	0.24	0.53	0.54	1.19	0.16	0.52	0.18	0.05
166	8.4	73	1.42	3.12	1.18	2.60	0.80	1.75	0.54	1.18	0.74	1.63	0.06	0.31	0.02	0.03
	9.6	84	1.62	3.57	1.35	2.97	0.91	2.01	0.61	1.35	0.85	1.86	0.07	0.36	0.03	0.03
167	1.3	11	0.25	0.55	0.20	0.45	0.13	0.28	0.07	0.15	0.14	0.30	—	—	—	—
	5.6	48	1.05	2.31	0.85	1.88	0.54	1.20	0.29	0.63	0.58	1.27	—	—	—	—
168	2.1	14	0.32	0.70	0.26	0.57	0.17	0.36	0.09	0.19	0.17	0.38	—	0.07	—	—
	6.8	48	1.05	2.30	0.85	1.88	0.54	1.20	0.29	0.63	0.58	1.27	—	0.23	—	—
169	3.8	53	1.06	2.34	0.89	1.97	0.54	1.19	0.31	0.68	0.55	1.22	—	—	—	—
	4.3	60	1.21	2.66	1.02	2.24	0.61	1.35	0.35	0.78	0.63	1.38	—	—	—	—
170	7.2	83	1.07	2.36	1.19	2.62	0.93	2.06	0.64	1.42	0.87	1.92	0.24	0.57	0.06	—
	7.7	89	1.15	2.53	1.27	2.81	1.00	2.21	0.69	1.52	0.94	2.06	0.26	0.61	0.06	—
171	17.4	46	0.87	1.92	0.86	1.90	0.51	1.12	0.27	0.59	0.40	0.89	0.31	0.61	—	—
	18.9	50	0.95	2.09	0.94	2.06	0.55	1.21	0.29	0.65	0.44	0.96	0.34	0.66	—	—
172	38.0	70	1.40	3.08	1.22	2.70	0.76	1.67	0.50	1.10	0.72	1.60	—	—	—	—
	39.5	77	1.53	3.37	1.34	2.96	0.83	1.83	0.55	1.20	0.79	1.75	—	—	—	—
173	36.7	66	1.15	2.53	0.97	2.15	0.58	1.27	0.33	0.74	0.56	1.24	0.38	1.08	—	—
	40.1	72	1.26	2.77	1.06	2.35	0.63	1.39	0.37	0.81	0.61	1.35	0.41	1.18	—	—
174	19.2	99	1.97	4.34	1.80	3.96	—	—	—	—	—	—	0.97	0.72	—	—
	20.3	104	2.08	4.59	1.90	4.19	—	—	—	—	—	—	1.02	0.76	—	—
175	38.5	71	1.41	3.12	1.24	2.73	0.75	1.66	0.49	1.08	0.72	1.59	2.01	1.36	0.05	0.07
	41.6	76	1.53	3.37	1.34	2.95	0.81	1.79	0.53	1.16	0.78	1.72	2.17	1.46	0.05	0.07
176	39.4	69	1.38	3.04	1.13	2.48	0.71	1.56	0.46	1.01	0.79	1.75	2.01	1.28	—	—
	42.8	75	1.50	3.30	1.23	2.70	0.77	1.70	0.50	1.10	0.86	1.90	2.18	1.39	—	—
177	2.4	51	1.02	2.24	0.84	1.86	0.51	1.12	0.28	0.61	0.53	1.17	0.56	0.17	0.02	—
	2.6	56	1.13	2.49	0.94	2.06	0.56	1.24	0.31	0.68	0.59	1.30	0.62	0.19	0.02	—
178	0.7	47	0.92	2.02	0.74	1.64	0.40	0.89	0.18	0.39	0.42	0.93	0.37	0.10	—	—
	0.7	51	1.00	2.21	0.81	1.79	0.44	0.97	0.19	0.43	0.46	1.02	0.40	0.11	—	—
179	7.1	67	1.34	2.96	1.17	2.58	0.53	1.18	0.30	0.66	0.55	1.21	0.05	0.32	0.03	0.08
	7.9	75	1.49	3.29	1.30	2.87	0.59	1.31	0.33	0.73	0.61	1.35	0.06	0.35	0.03	0.09
180	5.3	75	1.50	3.30	1.33	2.93	0.78	1.73	0.53	1.16	0.74	1.64	0.03	0.27	0.04	—
	6.0	84	1.69	3.73	1.50	3.31	0.89	1.95	0.59	1.31	0.84	1.85	0.03	0.31	0.05	—

TABLE 21–1

Entry Number	Feed Name Description	Moisture Basis: A-F (as-fed) or M-F (moisture-free)	Macro Minerals (Continued)			Micro Minerals						
			Magnesium (Mg)	Potassium (K)	Sulfur (S)	Cobalt (Co)	Copper (Cu)	Iodine (I)	Iron (Fe)	Manganese (Mn)	Selenium (Se)	Zinc (Zn)
			(%)	(%)	(%)	(ppm or mg/kg)	(ppm or mg/kg)	(ppm or mg/kg)	(%)	(ppm or mg/kg)	(ppm or mg/kg)	(ppm or mg/kg)
	REDTOP *Agrostis alba*											
158	FULL BLOOM, FRESH	A-F	0.07	0.62	0.04	—	—	—	0.01	—	—	—
		M-F	0.25	2.35	0.16	—	—	—	0.02	—	—	—
159	HAY, SUN-CURED	A-F	0.20	1.74	0.23	0.13	3.6	0.09	0.02	207.7	—	—
		M-F	0.22	1.89	0.25	0.15	3.9	0.10	0.02	225.5	—	—
	RICE *Oryza sativa*											
160	GRAIN, GROUND (GROUND ROUGH RICE, GROUND PADDY RICE)	A-F	0.13	0.47	0.05	—	—	—	—	18.0	—	15.0
		M-F	0.14	0.53	0.05	—	—	—	—	20.2	—	16.9
161	BRAN W/GERMS (RICE BRAN)	A-F	0.85	1.69	0.18	1.38	11.0	—	0.02	337.6	—	37.4
		M-F	0.94	1.87	0.20	1.53	12.1	—	0.02	372.4	—	41.3
162	GROATS, POLISHED (POLISHED RICE)	A-F	0.09	0.23	0.08	0.85	5.4	—	0.00	29.6	—	13.7
		M-F	0.10	0.26	0.09	0.96	6.1	—	0.00	33.4	—	15.4
163	GROATS (BROWN RICE)	A-F	0.08	0.30	0.04	0.73	3.7	—	0.00	20.3	—	14.3
		M-F	0.09	0.34	0.05	0.82	4.2	—	0.00	23.0	—	16.2
164	HULLS	A-F	0.41	0.64	0.08	2.05	3.1	—	0.01	295.0	—	22.0
		M-F	0.45	0.69	0.09	2.22	3.4	—	0.01	320.1	—	23.9
	RYE *Secale cereale*											
165	DISTILLERS GRAINS, DEHY	A-F	0.17	0.07	0.44	—	—	—	—	18.4	—	—
		M-F	0.18	0.08	0.48	—	—	—	—	20.0	—	—
166	GRAIN, ALL ANALYSES	A-F	0.12	0.46	0.15	—	7.5	—	0.01	72.0	—	28.1
		M-F	0.14	0.52	0.17	—	8.6	—	0.01	82.3	—	32.2
167	SILAGE, MOLASSES ADDED	A-F	—	—	—	—	—	—	—	—	—	—
		M-F	—	—	—	—	—	—	—	—	—	—
168	SILAGE, WILTED	A-F	—	0.56	—	—	—	—	—	—	—	—
		M-F	—	1.85	—	—	—	—	—	—	—	—
	RYEGRASS *Lolium* spp											
169	HAY	A-F	—	—	—	—	—	—	—	—	—	—
		M-F	—	—	—	—	—	—	—	—	—	—
	SAFFLOWER *Carthumus tinctorius*											
170	SEEDS	A-F	0.34	0.74	0.06	—	10.0	—	0.03	1.1	—	30.0
		M-F	0.36	0.79	0.06	—	10.7	—	0.04	1.2	—	32.2
171	SEEDS, MEAL SOLV EXTD, 20% PROTEIN	A-F	0.32	0.74	0.20	—	9.6	—	0.04	17.7	—	39.6
		M-F	0.35	0.80	0.22	—	10.4	—	0.05	19.2	—	43.0
172	SEEDS, W/O HULLS, MEAL MECH EXTD	A-F	—	—	—	—	—	—	—	—	—	—
		M-F	—	—	—	—	—	—	—	—	—	—
173	SEEDS, W/O HULLS, MEAL SOLV EXTD, 42% PROTEIN	A-F	1.18	1.18	0.34	1.83	80.6	—	0.09	36.6	—	168.5
		M-F	1.29	1.29	0.38	2.00	88.0	—	0.10	40.0	—	184.0
	SESAME *Sesamum indicum*											
174	SEEDS	A-F	—	—	—	—	—	—	—	—	—	—
		M-F	—	—	—	—	—	—	—	—	—	99.6
175	SEEDS, MEAL MECH EXTD	A-F	0.46	1.25	0.33	—	—	—	0.01	47.7	—	99.6
		M-F	0.50	1.35	0.35	—	—	—	0.01	51.5	—	107.5
176	SEEDS, MEAL SOLV EXTD, 44% PROTEIN	A-F	—	—	—	—	—	—	—	47.5	—	—
		M-F	—	—	—	—	—	—	—	51.6	—	—
	SORGHUM *Sorghum bicolor*											
177	FODDER W/HEADS, SUN-CURED	A-F	0.27	1.12	—	—	—	—	—	—	—	—
		M-F	0.30	1.24	—	—	—	—	—	—	—	—
178	STOVER W/O HEADS, SUN-CURED	A-F	—	1.10	—	—	—	—	—	—	—	—
		M-F	—	1.20	—	—	—	—	—	—	—	—
179	GRAIN, ALL ANALYSES	A-F	0.14	0.35	0.15	0.28	9.7	—	0.01	9.8	—	42.4
		M-F	0.16	0.38	0.17	0.31	10.8	—	0.01	10.9	—	47.1
180	GRAIN, LESS THAN 9% PROTEIN	A-F	—	0.35	—	0.07	9.7	0.02	0.00	15.4	—	13.7
		M-F	—	0.40	—	0.08	11.0	0.03	0.00	17.4	—	15.4

(Continued)

Entry Number	Fat-Soluble Vitamins					Water-Soluble Vitamins								
	A	Carotene (Provitamin A)	D	E	K	B–12	Biotin	Choline	Folacin (Folic Acid)	Niacin	Pantothenic Acid (B–3)	Pyridoxine (B–6)	Riboflavin (B–2)	Thiamin (B–1)
	(IU/g)	(ppm or mg/kg)	(IU/kg)	(ppm or mg/kg)	(ppm or mg/kg)	(ppb or mcg/kg)	(ppm or mg/kg)	(ppm or mg/kg)	(ppm or mg/kg)	(ppm or mg/kg)	(ppm or mg/kg)	(ppm or mg/kg)	(ppm or mg/kg)	(ppm or mg/kg)
158	66.9	40.1	—	—	—	—	—	—	—	—	—	—	—	—
	254.4	152.6	—	—	—	—	—	—	—	—	—	—	—	—
159	6.1	3.7	—	—	—	—	—	—	—	—	—	—	—	—
	6.6	4.0	—	—	—	—	—	—	—	—	—	—	—	—
160	—	—	—	14.0	—	—	—	926	0.25	40	7.1	—	0.7	—
	—	—	—	15.7	—	—	—	1041	0.28	45	8.0	—	0.8	—
161	—	—	—	60.4	—	—	0.43	1230	2.20	299	22.8	13.24	2.6	22.4
	—	—	—	60.7	—	—	0.47	1357	2.42	330	25.2	14.61	2.8	24.8
162	—	—	—	3.5	—	—	—	901	0.15	15	3.5	0.39	0.6	0.7
	—	—	—	4.0	—	—	—	1017	0.17	17	3.9	0.45	0.6	0.7
163	—	—	—	10.3	—	—	0.09	—	0.19	43	10.7	7.00	0.6	2.9
	—	—	—	11.7	—	—	0.10	—	0.21	49	12.1	7.94	0.7	3.3
164	—	—	—	7.5	—	—	—	—	—	28	7.9	0.07	0.5	2.2
	—	—	—	8.1	—	—	—	—	—	31	8.6	0.08	0.6	2.4
165	—	—	—	—	—	—	—	—	—	17	5.2	—	3.3	1.3
	—	—	—	—	—	—	—	—	—	18	5.7	—	3.6	1.4
166	0.1	0.1	—	14.5	—	—	0.06	419	0.62	14	7.5	—	1.7	4.1
	0.2	0.1	—	16.6	—	—	0.06	479	0.71	16	8.5	—	1.9	4.7
167	—	—	—	—	—	—	—	—	—	—	—	—	—	—
	—	—	—	—	—	—	—	—	—	—	—	—	—	—
168	—	—	—	—	—	—	—	—	—	—	—	—	—	—
	—	—	—	—	—	—	—	—	—	—	—	—	—	—
169	175.8	105.5	—	—	—	—	—	—	—	—	—	—	—	—
	199.9	119.9	—	—	—	—	—	—	—	—	—	—	—	—
170	—	—	—	—	—	—	—	—	—	—	—	—	—	—
	—	—	—	—	—	—	—	—	—	—	—	—	—	—
171	—	—	—	0.9	—	—	—	1541	—	12	36.2	474.4	2.2	—
	—	—	—	1.0	—	—	—	1673	—	13	39.3	515.0	2.4	—
172	—	—	—	—	—	—	1.42	2608	0.44	22	86.6	—	4.0	18.2
	—	—	—	—	—	—	1.56	2860	0.49	24	95.0	—	4.4	20.0
173	—	—	—	0.6	—	—	1.56	3156	1.47	21	38.2	10.71	2.3	4.2
	—	—	—	0.7	—	—	1.70	3447	1.60	23	41.7	11.70	2.5	4.6
174	—	—	—	—	—	—	—	—	—	—	—	—	—	—
	—	—	—	—	—	—	—	—	—	—	—	—	—	—
175	0.7	0.4	—	—	—	—	—	1533	—	19	5.9	12.45	3.4	2.8
	0.8	0.5	—	—	—	—	—	1655	—	20	6.4	13.44	3.6	3.0
176	—	—	—	—	—	—	—	1517	—	—	6.3	—	3.7	—
	—	—	—	—	—	—	—	1649	—	—	6.8	—	4.0	—
177	—	—	—	—	—	—	—	—	—	—	—	—	—	—
	—	—	—	—	—	—	—	—	—	—	—	—	—	—
178	—	—	—	—	—	—	—	—	—	—	—	—	—	—
	—	—	—	—	—	—	—	—	—	—	—	—	—	—
179	2.0	1.2	—	—	—	—	0.26	686	0.22	47	10.2	5.41	1.2	4.5
	2.2	1.3	—	—	—	—	0.29	762	0.24	52	11.3	6.00	1.4	5.0
180	—	—	—	2.2	—	—	0.29	763	0.22	48	12.8	4.63	1.3	4.4
	—	—	—	2.5	—	—	0.32	862	0.25	54	14.4	5.23	1.5	5.0

(Continued)

TABLE 21–1

Entry Number	Feed Name Description	International Feed Number	Moisture Basis: A-F (as-fed) or M-F (moisture-free)	Chemical Analysis								
				Dry Matter	Ash	Crude Fiber	Neutral Det. Fib. (NDF)	Acid Det. Fib. (ADF)	Lignin	Ether Extract (Fat)	N-Free Extract	Crude Protein
				(%)	(%)	(%)	(%)	(%)	(%)	(%)	(%)	(%)
	SORGHUM (Continued)											
181	GRAIN, 9–12% PROTEIN	4–08–139	A-F	89	1.9	2.4	—	—	—	2.7	72.2	9.8
			M-F	100	2.1	2.6	—	—	—	3.1	81.1	11.0
182	GRAIN, MORE THAN 12% PROTEIN	4–08–140	A-F	89	2.3	1.8	—	—	—	1.5	71.8	11.6
			M-F	100	2.6	2.0	—	—	—	1.7	80.7	13.0
183	DISTILLERS GRAINS, DEHY	5–04–374	A-F	94	4.3	12.1	—	—	—	8.3	38.3	30.8
			M-F	100	4.6	12.9	—	—	—	8.8	40.8	32.9
184	SILAGE, DOUGH STAGE	3–04–321	A-F	29	2.5	8.3	19.2	10.9	1.9	0.9	15.1	2.3
			M-F	100	8.5	28.6	66.2	37.4	6.5	3.1	52.0	7.9
185	SILAGE, MATURE	3–04–322	A-F	32	2.3	7.8	16.3	8.5	2.1	0.9	18.2	2.7
			M-F	100	7.1	24.3	51.0	26.6	6.5	2.8	57.1	8.6
	SORGHUM, JOHNSONGRASS *Sorghum halepense*											
186	HAY, SUN-CURED	1–04–047	A-F	91	7.7	30.4	—	—	—	2.0	43.7	6.7
			M-F	100	8.6	33.6	—	—	—	2.2	48.3	7.5
	SORGHUM, SORGO *Sorghum bicolor saccharatum*											
187	SILAGE	3–04–468	A-F	29	2.4	7.0	18.5	11.0	1.7	0.7	16.7	1.9
			M-F	100	8.3	24.4	64.0	38.0	6.0	2.5	58.1	6.8
	SORGHUM, SUDANGRASS *Sorghum bicolor sudanense*											
188	FRESH	2–04–489	A-F	19	1.5	5.8	—	7.9	1.0	0.6	8.6	2.3
			M-F	100	8.0	30.9	—	42.0	5.3	3.2	45.8	12.1
189	FRESH, IMMATURE	2–04–484	A-F	18	1.5	6.7	9.9	5.2	—	0.6	6.8	2.6
			M-F	100	8.2	36.7	55.0	29.0	—	3.2	37.5	14.4
190	FRESH, MATURE	2–04–487	A-F	30	2.4	10.6	—	—	—	0.5	14.5	1.6
			M-F	100	8.1	35.8	—	—	—	1.7	48.9	5.5
191	HAY, SUN-CURED	1–04–480	A-F	91	10.7	26.2	60.2	20.4	35.5	1.6	41.7	10.9
			M-F	100	11.8	28.7	66.0	22.4	39.0	1.7	45.8	12.0
192	SILAGE	3–04–499	A-F	23	2.1	7.9	16.4	1.2	9.7	0.7	9.9	2.6
			M-F	100	9.2	34.0	71.0	5.0	42.0	2.9	42.9	11.1
	SOYBEAN *Glycine max*											
193	HAY, SUN-CURED	1–04–558	A-F	89	7.2	30.6	—	35.7	—	2.3	35.0	14.1
			M-F	100	8.0	34.3	—	40.0	—	2.5	39.3	15.8
194	HULLS	1–04–560	A-F	91	4.6	36.2	59.4	42.4	1.8	2.0	37.0	10.8
			M-F	100	5.1	40.0	65.6	46.8	2.0	2.2	40.9	11.9
195	SEEDS	5–04–610	A-F	92	5.1	5.4	—	9.2	—	17.2	25.9	38.4
			M-F	100	5.6	5.8	—	10.0	—	18.7	28.1	41.7
196	SEEDS, MEAL MECH EXTD, 41% PROTEIN	5–04–600	A-F	90	6.0	6.0	—	—	—	4.7	30.4	42.9
			M-F	100	6.7	6.7	—	—	—	5.2	33.8	47.7
197	SEEDS, MEAL SOLV EXTD, 44% PROTEIN	5–20–637	A-F	89	6.4	6.2	12.5	8.9	—	1.5	30.6	44.4
			M-F	100	7.2	7.0	14.0	10.0	—	1.7	34.3	49.8
198	SEEDS W/O HULLS, MEAL SOLV EXTD, 49% PROTEIN	5–20–638	A-F	90	6.1	3.7	6.6	6.2	—	1.2	29.8	49.0
			M-F	100	6.8	4.1	7.4	6.9	—	1.4	33.2	54.6
199	SILAGE	3–04–581	A-F	30	3.0	9.0	—	12.0	—	0.8	12.2	5.2
			M-F	100	9.9	29.9	—	40.0	—	2.6	40.5	17.1
	SUGAR CANE (BAGASSE) *Saccharum officinarum*											
200	DEHY	1–04–686	A-F	91	2.9	42.3	78.8	54.5	12.8	0.7	43.8	1.4
			M-F	100	3.1	46.5	86.5	59.8	14.0	0.8	48.0	1.6
	SUNFLOWER, COMMON *Helianthus annuus*											
201	SEEDS	5–08–330	A-F	94	3.7	22.7	—	—	—	32.3	14.4	20.9
			M-F	100	4.0	24.1	—	—	—	34.4	15.3	22.2
202	SEEDS W/O HULLS, MEAL MECH EXTD, 41% PROTEIN	5–26–097	A-F	92	6.7	13.2	—	—	—	7.5	23.9	40.7
			M-F	100	7.3	14.3	—	—	—	8.2	26.0	44.2
203	SEEDS W/O HULLS, MEAL SOLV EXTD, 44% PROTEIN	5–26–098	A-F	93	7.7	11.0	—	—	—	2.9	24.6	46.8
			M-F	100	8.3	11.8	—	—	—	3.1	26.5	50.3

(Continued)

Entry Number	Digestible Protein Ruminant (%)	TDN Ruminant (%)	Digestible Energy Ruminant (Mcal/lb)	*(Mcal/kg)*	Metabolizable Energy Ruminant (Mcal/lb)	*(Mcal/kg)*	Ruminant NEm (Mcal/lb)	*(Mcal/kg)*	Ruminant NEg (Mcal/lb)	*(Mcal/kg)*	Lactating Cows NElc (Mcal/lb)	*(Mcal/kg)*	Calcium (Ca) (%)	Phosphorus (P) (%)	Sodium (Na) (%)	Chlorine (Cl) (%)
181	6.1	73	1.42	*3.14*	1.11	*2.44*	0.73	*1.62*	0.48	*1.06*	0.76	*1.67*	0.03	0.27	0.02	—
	6.9	82	1.60	*3.53*	1.24	*2.74*	0.82	*1.82*	0.54	*1.19*	0.85	*1.88*	0.04	0.30	0.02	—
182	7.6	69	1.39	*3.06*	1.22	*2.69*	0.76	*1.67*	0.50	*1.11*	0.72	*1.59*	0.03	0.29	0.04	—
	8.6	78	1.56	*3.44*	1.37	*3.02*	0.85	*1.88*	0.56	*1.24*	0.81	*1.79*	0.03	0.32	0.05	—
183	24.7	78	1.56	*3.45*	1.39	*3.06*	0.85	*1.88*	0.57	*1.27*	0.80	*1.77*	0.15	0.69	0.05	—
	26.3	83	1.67	*3.68*	1.48	*3.26*	0.91	*2.00*	0.61	*1.35*	0.86	*1.89*	0.16	0.74	0.05	—
184	0.9	16	0.31	*0.68*	0.24	*0.53*	0.16	*0.35*	0.08	*0.18*	0.16	*0.35*	—	—	—	—
	3.0	56	1.06	*2.34*	0.82	*1.81*	0.54	*1.19*	0.29	*0.63*	0.54	*1.19*	—	—	—	—
185	1.0	17	0.33	*0.72*	0.27	*0.58*	0.16	*0.35*	0.08	*0.17*	0.16	*0.36*	—	—	—	—
	3.0	53	1.03	*2.26*	0.83	*1.83*	0.50	*1.10*	0.25	*0.54*	0.51	*1.12*	—	—	—	—
186	3.0	51	1.01	*2.23*	0.84	*1.85*	0.48	*1.06*	0.25	*0.56*	0.51	*1.13*	0.80	0.27	0.01	—
	3.3	56	1.12	*2.46*	0.93	*2.04*	0.53	*1.18*	0.28	*0.61*	0.57	*1.25*	0.89	0.30	0.01	—
187	0.5	17	0.34	*0.74*	0.28	*0.62*	0.16	*0.35*	0.09	*0.19*	0.17	*0.37*	0.10	0.06	0.04	0.02
	1.7	58	1.17	*2.58*	0.98	*2.15*	0.55	*1.22*	0.30	*0.66*	0.58	*1.29*	0.35	0.21	0.15	0.06
188	1.5	13	0.25	*0.56*	0.22	*0.47*	0.13	*0.29*	0.08	*0.18*	0.13	*0.28*	0.09	0.08	—	—
	8.3	68	1.35	*2.98*	1.15	*2.53*	0.71	*1.56*	0.44	*0.96*	0.68	*1.50*	0.49	0.44	—	—
189	1.8	12	0.25	*0.54*	0.21	*0.47*	0.14	*0.30*	0.09	*0.19*	0.13	*0.29*	0.08	0.08	0.00	—
	9.9	68	1.39	*2.98*	1.16	*2.56*	0.74	*1.63*	0.47	*1.03*	0.72	*1.60*	0.43	0.41	0.01	—
190	0.8	19	0.36	*0.79*	0.30	*0.67*	0.16	*0.35*	0.08	*0.18*	0.17	*0.37*	0.09	0.06	—	—
	2.1	65	1.21	*2.67*	1.02	*2.25*	0.53	*1.17*	0.28	*0.61*	0.57	*1.25*	0.32	0.21	—	—
191	4.7	51	1.00	*2.20*	0.82	*1.82*	0.48	*1.07*	0.25	*0.55*	0.52	*1.14*	0.47	0.28	0.01	—
	5.2	56	1.10	*2.42*	0.90	*1.99*	0.53	*1.17*	0.28	*0.61*	0.57	*1.25*	0.51	0.31	0.02	—
192	1.6	13	0.26	*0.58*	0.22	*0.48*	0.13	*0.28*	0.07	*0.15*	0.13	*0.30*	0.12	0.05	0.01	—
	6.8	57	1.14	*2.51*	0.94	*2.08*	0.55	*1.21*	0.29	*0.64*	0.58	*1.28*	0.50	0.21	0.02	—
193	9.5	49	1.11	*2.45*	0.86	*1.90*	0.52	*1.14*	0.29	*0.63*	0.55	*1.21*	1.13	0.22	0.10	0.13
	10.7	55	1.25	*2.75*	0.97	*2.13*	0.58	*1.28*	0.32	*0.70*	0.62	*1.36*	1.26	0.24	0.11	0.15
194	6.5	69	1.20	*2.65*	1.03	*2.26*	0.75	*1.66*	0.50	*1.09*	0.72	*1.59*	0.45	0.19	0.03	—
	7.2	77	1.33	*2.92*	1.13	*2.50*	0.83	*1.84*	0.55	*1.21*	0.80	*1.75*	0.49	0.21	0.03	—
195	34.5	84	1.69	*3.72*	1.52	*3.34*	0.93	*2.04*	0.64	*1.41*	0.86	*1.90*	0.25	0.60	0.00	0.03
	37.5	92	1.83	*4.04*	1.65	*3.63*	1.01	*2.22*	0.70	*1.53*	0.94	*2.07*	0.27	0.65	0.00	0.03
196	36.6	77	1.53	*3.38*	1.37	*3.01*	0.86	*1.89*	0.58	*1.29*	0.81	*1.78*	0.26	0.61	0.18	0.07
	40.7	85	1.70	*3.76*	1.52	*3.35*	0.95	*2.10*	0.65	*1.43*	0.90	*1.98*	0.29	0.68	0.20	0.08
197	37.8	76	1.45	*3.19*	1.17	*2.59*	0.79	*1.74*	0.53	*1.16*	0.75	*1.66*	0.35	0.64	0.03	—
	42.4	85	1.62	*3.57*	1.32	*2.90*	0.89	*1.95*	0.59	*1.30*	0.85	*1.87*	0.40	0.71	0.04	—
198	42.4	78	1.51	*3.32*	1.10	*2.42*	0.72	*1.59*	0.47	*1.03*	0.79	*1.75*	0.25	0.63	0.00	0.07
	47.3	87	1.68	*3.70*	1.22	*2.70*	0.81	*1.78*	0.52	*1.15*	0.88	*1.95*	0.28	0.70	0.00	0.08
199	3.2	16	0.32	*0.71*	0.26	*0.58*	0.15	*0.33*	0.07	*0.16*	0.16	*0.36*	0.40	0.13	0.03	—
	10.6	53	1.07	*2.35*	0.87	*1.92*	0.50	*1.10*	0.25	*0.54*	0.54	*1.20*	1.32	0.44	0.09	—
200	−1.5	43	0.88	*1.93*	0.69	*1.53*	0.37	*0.82*	0.15	*0.32*	0.40	*0.88*	0.47	0.26	0.04	—
	−1.7	48	0.96	*2.12*	0.76	*1.68*	0.41	*0.90*	0.16	*0.36*	0.44	*0.96*	0.51	0.29	0.04	—
201	16.7	78	1.56	*3.44*	1.38	*3.05*	0.89	*1.95*	0.60	*1.33*	0.83	*1.83*	0.16	0.67	0.02	—
	17.8	83	1.66	*3.66*	1.47	*3.25*	0.94	*2.08*	0.64	*1.41*	0.88	*1.95*	0.17	0.71	0.02	—
202	36.5	68	1.38	*3.04*	1.13	*2.48*	0.63	*1.38*	0.42	*0.92*	0.75	*1.66*	—	—	—	—
	39.7	74	1.50	*3.30*	1.23	*2.70*	0.68	*1.50*	0.45	*1.00*	0.82	*1.80*	—	—	—	—
203	42.1	65	1.22	*2.70*	1.01	*2.23*	0.59	*1.30*	0.34	*0.74*	0.63	*1.40*	—	—	—	—
	45.3	70	1.32	*2.90*	1.09	*2.40*	0.64	*1.40*	0.36	*0.80*	0.68	*1.50*	—	—	—	—

(Continued)

TABLE 21–1

Entry Number	Feed Name Description	Moisture Basis: A-F (as-fed) or M-F (moisture-free)	Macro Minerals (Continued)			Micro Minerals						
			Magnesium (Mg)	Potassium (K)	Sulfur (S)	Cobalt (Co)	Copper (Cu)	Iodine (I)	Iron (Fe)	Manganese (Mn)	Selenium (Se)	Zinc (Zn)
			(%)	(%)	(%)	(ppm or mg/kg)	(ppm or mg/kg)	(ppm or mg/kg)	(%)	(ppm or mg/kg)	(ppm or mg/kg)	(ppm or mg/kg)
	SORGHUM (Continued)											
181	GRAIN, 9–12% PROTEIN	A-F	0.15	0.34	0.14	0.07	9.7	0.02	0.00	15.4	—	13.7
		M-F	0.17	0.38	0.16	0.08	10.9	0.03	0.00	17.3	—	15.4
182	GRAIN, MORE THAN 12% PROTEIN	A-F	0.17	0.34	0.16	—	—	—	0.01	—	—	—
		M-F	0.19	0.38	0.18	—	—	—	0.01	—	—	—
183	DISTILLERS GRAINS, DEHY	A-F	0.18	0.36	0.17	—	—	—	0.01	—	—	—
		M-F	0.19	0.38	0.18	—	—	—	0.01	—	—	—
184	SILAGE, DOUGH STAGE	A-F	—	—	—	—	—	—	—	—	—	—
		M-F	—	—	—	—	—	—	—	—	—	—
185	SILAGE, MATURE	A-F	—	—	—	—	—	—	—	—	—	—
		M-F	—	—	—	—	—	—	—	—	—	—
	SORGHUM, JOHNSONGRASS *Sorghum halepense*											
186	HAY, SUN-CURED	A-F	0.31	1.22	0.09	—	—	—	0.05	—	—	—
		M-F	0.35	1.35	0.10	—	—	—	0.06	—	—	—
	SORGHUM, SORGO *Sorghum bicolor saccharatum*											
187	SILAGE	A-F	0.08	0.32	0.03	—	9.0	—	0.01	17.6	—	—
		M-F	0.27	1.12	0.10	—	31.1	—	0.12	61.0	—	—
	SORGHUM, SUDANGRASS *Sorghum bicolor sudanense*											
188	FRESH	A-F	0.07	0.40	0.02	0.03	6.7	—	0.00	15.3	—	—
		M-F	0.35	2.14	0.11	0.13	35.9	—	0.02	81.4	—	—
189	FRESH, IMMATURE	A-F	0.06	0.39	0.02	—	—	—	0.00	—	—	—
		M-F	0.35	2.14	0.11	—	—	—	0.02	—	—	—
190	FRESH, MATURE	A-F	—	—	—	—	—	—	—	—	—	—
		M-F	—	—	—	—	—	—	—	—	—	—
191	HAY, SUN-CURED	A-F	0.34	1.90	0.06	0.12	28.6	—	0.02	69.5	—	34.6
		M-F	0.37	2.08	0.06	0.13	31.4	—	0.02	76.3	—	38.0
192	SILAGE	A-F	0.10	0.60	0.01	0.06	8.5	—	0.00	22.8	—	—
		M-F	0.42	2.61	0.06	0.27	36.6	—	0.01	98.8	—	—
	SOYBEAN *Glycine max*											
193	HAY, SUN-CURED	A-F	0.71	0.92	0.25	0.08	8.0	0.22	0.03	94.3	—	21.5
		M-F	0.81	1.04	0.28	0.08	9.0	0.24	0.03	105.8	—	24.1
194	HULLS	A-F	—	1.15	0.08	0.11	16.1	—	0.03	9.9	—	21.8
		M-F	—	1.27	0.09	0.12	17.8	—	0.03	11.0	—	24.1
195	SEEDS	A-F	0.27	1.66	0.22	—	18.2	—	0.01	36.4	0.11	56.9
		M-F	0.29	1.80	0.24	—	19.8	—	0.01	39.6	0.12	61.8
196	SEEDS, MEAL MECH EXTD, 41% PROTEIN	A-F	0.26	1.79	0.33	0.18	21.7	—	0.02	31.3	0.10	57.2
		M-F	0.29	1.98	0.37	0.20	24.1	—	0.02	34.8	0.11	63.6
197	SEEDS, MEAL SOLV EXTD, 44% PROTEIN	A-F	0.27	1.98	0.41	1.38	19.9	—	0.02	31.6	0.49	50.5
		M-F	0.31	2.22	0.42	1.55	22.3	—	0.02	35.5	0.55	56.6
198	SEEDS W/O HULLS, MEAL SOLV EXTD, 49% PROTEIN	A-F	0.37	1.79	0.41	2.69	13.5	0.15	0.01	49.5	—	51.1
		M-F	0.41	1.99	0.46	3.00	15.0	0.17	0.01	55.2	—	56.9
199	SILAGE	A-F	0.12	0.39	0.09	—	2.9	—	0.01	42.7	—	10.3
		M-F	0.40	1.28	0.31	—	9.6	—	0.03	141.3	—	34.0
	SUGAR CANE (BAGASSE) *Saccharum officinarum*											
200	DEHY	A-F	0.08	0.34	0.09	—	—	—	0.02	—	—	—
		M-F	0.08	0.37	0.10	—	—	—	0.02	—	—	—
	SUNFLOWER, COMMON *Helianthus annuus*											
201	SEEDS	A-F	0.37	0.68	0.28	—	23.5	—	0.01	21.9	—	68.6
		M-F	0.39	0.72	0.29	—	25.0	—	0.01	23.3	—	73.0
202	SEEDS W/O HULLS, MEAL MECH EXTD, 41% PROTEIN	A-F	—	—	—	—	—	—	—	—	—	—
		M-F	—	—	—	—	—	—	—	—	—	—
203	SEEDS W/O HULLS, MEAL SOLV EXTD, 44% PROTEIN	A-F	—	—	—	—	—	—	—	—	—	—
		M-F	—	—	—	—	—	—	—	—	—	—

(Continued)

Entry Number	Fat-Soluble Vitamins					Water-Soluble Vitamins								
	A	Carotene (Provitamin A)	D	E	K	B–12	Biotin	Choline	Folacin (Folic Acid)	Niacin	Pantothenic Acid (B–3)	Pyridoxine (B–6)	Riboflavin (B–2)	Thiamin (B–1)
	(IU/g)	(ppm or mg/kg)	(IU/kg)	(ppm or mg/kg)	(ppm or mg/kg)	(ppb or mcg/kg)	(ppm or mg/kg)	(ppm or mg/kg)	(ppm or mg/kg)	(ppm or mg/kg)	(ppm or mg/kg)	(ppm or mg/kg)	(ppm or mg/kg)	(ppm or mg/kg)
181	—	—	—	1.3	—	—	0.29	762	0.22	48	12.8	4.60	1.3	4.4
	—	—	—	1.5	—	—	0.32	857	0.25	54	14.4	5.17	1.5	5.0
182	—	—	—	—	—	—	—	—	—	—	—	—	—	—
	—	—	—	—	—	—	—	—	—	—	—	—	—	—
183	—	—	—	—	—	—	0.31	805	—	—	—	—	—	—
	—	—	—	—	—	—	0.33	858	—	—	—	—	—	—
184	—	—	—	—	—	—	—	—	—	—	—	—	—	—
	—	—	—	—	—	—	—	—	—	—	—	—	—	—
185	—	—	—	—	—	—	—	—	—	—	—	—	—	—
	—	—	—	—	—	—	—	—	—	—	—	—	—	—
186	58.8	35.3	—	—	—	—	—	—	—	—	—	—	—	—
	64.9	38.9	—	—	—	—	—	—	—	—	—	—	—	—
187	20.4	12.2	—	—	—	—	—	—	—	—	—	—	—	—
	70.7	42.4	—	—	—	—	—	—	—	—	—	—	—	—
188	57.1	34.3	—	—	—	—	—	—	—	—	—	—	—	—
	304.6	182.7	—	—	—	—	—	—	—	—	—	—	—	—
189	59.9	35.9	—	—	—	—	—	—	—	—	—	—	—	—
	329.2	187.5	—	—	—	—	—	—	—	—	—	—	—	—
190	—	—	—	—	—	—	—	—	—	—	—	—	—	—
	—	—	—	—	—	—	—	—	—	—	—	—	—	—
191	—	—	—	—	—	—	—	—	—	—	—	—	—	—
	—	—	—	—	—	—	—	—	—	—	—	—	—	—
192	40.6	24.3	—	—	—	—	—	—	—	—	—	—	—	—
	175.4	105.2	—	—	—	—	—	—	—	—	—	—	—	—
193	53.1	31.8	1	26.3	—	—	—	—	—	—	—	—	—	—
	59.5	35.7	1	29.5	—	—	—	—	—	—	—	—	—	—
194	—	—	—	6.6	—	—	—	588	—	25	13.4	1.70	3.6	1.6
	—	—	—	7.3	—	—	—	649	—	27	14.8	1.88	4.0	1.8
195	1.5	0.9	—	33.7	—	—	0.38	2931	—	23	16.0	11.04	2.9	11.3
	1.6	1.0	—	36.6	—	—	0.42	3184	—	24	17.4	12.00	3.2	12.2
196	0.4	0.2	—	6.5	—	—	0.33	2623	6.39	31	14.3	7.22	3.4	3.9
	0.4	0.2	—	7.3	—	—	0.36	2916	7.10	34	15.8	8.02	3.8	4.3
197	—	—	—	3.0	0.22	2.0	0.36	2706	0.69	26	13.8	5.90	3.0	6.6
	—	—	—	3.4	0.25	2.2	0.41	3036	0.77	29	15.5	6.62	3.4	7.4
198	—	—	0	3.3	—	2.0	0.38	2772	0.59	24	14.1	5.59	2.9	3.5
	—	—	0	3.7	—	2.2	0.42	3089	0.66	27	15.7	6.23	3.3	3.9
199	52.2	31.3	—	—	—	—	—	—	—	—	—	—	—	—
	172.9	103.7	—	—	—	—	—	—	—	—	—	—	—	—
200	—	—	—	—	—	—	—	—	—	—	—	—	—	—
	—	—	—	—	—	—	—	—	—	—	—	—	—	—
201	—	—	—	—	—	—	—	—	—	—	—	—	3.3	0.4
	—	—	—	—	—	—	—	—	—	—	—	—	3.5	0.5
202	—	—	—	—	—	—	—	—	—	—	—	—	—	—
	—	—	—	—	—	—	—	—	—	—	—	—	—	—
203	—	—	—	—	—	—	—	—	—	—	—	—	—	—
	—	—	—	—	—	—	—	—	—	—	—	—	—	—

(Continued)

TABLE 21–1

Entry Number	Feed Name Description	International Feed Number	Moisture Basis: A-F (as-fed) or M-F (moisture-free)	Chemical Analysis								
				Dry Matter (%)	Ash (%)	Crude Fiber (%)	Neutral Det. Fib. (NDF) (%)	Acid Det. Fib. (ADF) (%)	Lignin (%)	Ether Extract (Fat) (%)	N-Free Extract (%)	Crude Protein (%)
	SUNFLOWER, COMMON (Continued)											
204	SILAGE, LATE BLOOM	3–04–732	A-F	26	3.9	9.9	—	—	—	0.5	9.5	2.1
			M-F	100	15.0	38.2	—	—	—	2.1	36.5	8.2
205	SILAGE, MILK STAGE	3–04–733	A-F	21	2.1	6.2	—	—	—	1.3	9.5	2.1
			M-F	100	10.0	29.4	—	—	—	5.9	45.0	9.7
206	SILAGE, MATURE	3–04–735	A-F	26	2.6	10.2	9.2	8.6	—	1.1	9.9	2.2
			M-F	100	10.0	39.1	35.3	33.2	—	4.3	38.0	8.6
	SWEET CLOVER, YELLOW *Melilotus officinalis*											
207	HAY, SUN-CURED	1–04–754	A-F	89	7.6	28.9	—	—	—	1.9	36.4	13.7
			M-F	100	8.6	32.6	—	—	—	2.2	41.1	15.4
	TIMOTHY *Phleum pratense*											
208	HAY, SUN-CURED, ALL ANALYSES	1–04–893	A-F	91	4.6	30.3	63.7	36.4	—	2.4	47.3	6.8
			M-F	100	5.1	33.2	70.0	40.0	—	2.6	51.8	7.4
209	HAY, PREBLOOM, SUN-CURED	1–04–881	A-F	89	6.3	28.2	56.0	29.4	2.8	2.5	39.8	12.4
			M-F	100	7.1	31.6	62.7	32.9	3.1	2.8	44.6	13.9
210	HAY, EARLY BLOOM, SUN-CURED	1–04–882	A-F	89	5.1	30.3	54.4	30.4	3.8	2.6	41.7	9.5
			M-F	100	5.7	33.9	61.0	34.1	4.3	2.9	46.8	10.7
211	HAY, MIDBLOOM, SUN-CURED	1–04–883	A-F	89	5.6	30.3	58.0	33.6	4.4	2.3	42.1	8.6
			M-F	100	6.3	34.0	65.3	37.8	4.9	2.6	47.4	9.7
212	HAY, FULL BLOOM, SUN-CURED	1–04–884	A-F	89	4.6	31.1	60.5	33.8	—	2.7	43.4	6.8
			M-F	100	5.2	35.1	68.0	38.0	—	3.0	48.9	7.7
213	SILAGE, ALL ANALYSES	3–04–922	A-F	34	2.4	12.0	—	—	—	1.2	15.1	3.6
			M-F	100	7.1	35.0	—	—	—	3.4	44.0	10.5
214	SILAGE, WILTED	3–04–930	A-F	42	2.9	14.2	—	—	—	1.3	18.7	4.4
			M-F	100	7.0	34.2	—	—	—	3.2	45.0	10.7
	TOMATO *Lycopersicon esculentum*											
215	POMACE, DEHY	5–05–041	A-F	92	6.8	25.0	50.4	46.5	10.5	9.8	29.3	21.0
			M-F	100	7.4	27.2	54.8	50.5	11.4	10.7	31.9	22.9
	TREFOIL, BIRDSFOOT (DEERVETCH) *Lotus corniculatus*											
216	FRESH	2–20–786	A-F	19	2.2	4.1	9.5	—	—	0.8	8.5	3.7
			M-F	100	11.2	21.2	49.4	—	—	4.0	44.3	19.3
217	HAY, SUN-CURED	1–05–044	A-F	91	6.7	29.3	42.8	32.8	—	1.9	38.9	13.9
			M-F	100	7.4	32.3	47.0	36.0	—	2.1	42.9	15.3
	TRITICALE *Triticale hexaploide*											
218	GRAIN	4–20–362	A-F	89	1.8	3.0	11.9	—	—	1.5	67.3	15.4
			M-F	100	2.0	3.3	13.3	—	—	1.7	75.7	17.3
219	SILAGE, HEAD EMERGING	4–20–	A-F	30	—	—	—	11.4	—	—	—	3.0
			M-F	100	—	—	—	38.0	—	—	—	10.0
	TURNIP *Brassica rapa*											
220	ROOTS, FRESH	4–05–067	A-F	9	0.8	1.1	4.0	3.1	—	0.2	5.9	1.2
			M-F	100	8.7	11.5	44.0	34.0	—	1.9	64.8	13.1
	UREA											
221	45% NITROGEN, 281% PROTEIN EQUIVALENT	5–05–070	A-F	99	—	—	0.0	0.0	—	—	—	281.7
			M-F	100	—	—	0.0	0.0	—	—	—	285.0
	VETCH *Vicia spp*											
222	HAY, SUN-CURED	1–05–106	A-F	89	7.8	24.8	42.7	29.4	—	2.7	35.1	18.4
			M-F	100	8.8	27.9	48.0	33.0	—	3.0	39.6	20.7
	WHEAT *Triticum aestivum*											
223	BRAN	4–05–190	A-F	89	5.9	10.0	40.9	12.0	2.6	4.0	53.6	15.5
			M-F	100	6.7	11.2	45.9	13.5	3.0	60.2	17.5	13.5
224	DISTILLERS GRAINS, DEHY	5–05–193	A-F	93	3.0	11.8	—	—	—	6.7	40.2	31.6
			M-F	100	3.3	12.6	—	—	—	7.2	43.0	33.9
225	FLOUR, LESS THAN 1.5% FIBER	4–05–199	A-F	88	1.5	0.9	—	—	—	1.7	70.5	13.7
			M-F	100	1.7	1.0	—	—	—	1.9	79.9	15.5

(Continued)

Entry Number	Digestible Protein Ruminant (%)	TDN Ruminant (%)	Digestible Energy Ruminant (Mcal/lb)	(Mcal/kg)	Metabolizable Energy Ruminant (Mcal/lb)	(Mcal/kg)	Ruminant NE$_m$ (Mcal/lb)	(Mcal/kg)	Ruminant NE$_g$ (Mcal/lb)	(Mcal/kg)	Lactating Cows NE$_{lc}$ (Mcal/lb)	(Mcal/kg)	Calcium (Ca) (%)	Phosphorus (P) (%)	Sodium (Na) (%)	Chlorine (Cl) (%)
204	0.9	11	0.24	0.52	0.19	0.41	0.09	0.19	0.02	0.05	0.11	0.24	—	—	—	—
	3.6	43	0.91	2.01	0.72	1.58	0.33	0.72	0.09	0.19	0.43	0.94	—	—	—	—
205	1.0	11	0.24	0.53	0.20	0.44	0.10	0.21	0.04	0.09	0.11	0.24	—	—	—	—
	4.7	51	1.13	2.48	0.93	2.06	0.45	0.99	0.20	0.44	0.51	1.12	—	—	—	—
206	0.5	10	0.24	0.52	0.19	0.41	0.06	0.14	0.00	0.00	0.10	0.21	—	—	—	—
	1.9	38	0.91	2.01	0.72	1.58	0.24	0.53	0.00	0.01	0.37	0.81	—	—	—	—
207	9.5	49	0.98	2.16	0.81	1.78	0.49	1.07	0.26	0.57	0.51	1.13	1.44	0.24	0.08	0.33
	10.7	55	1.11	2.44	0.91	2.01	0.55	1.21	0.29	0.64	0.58	1.28	1.63	0.27	0.09	0.37
208	2.8	53	1.06	2.35	0.85	1.88	0.50	1.10	0.27	0.58	0.51	1.13	0.38	0.17	0.03	0.49
	3.0	58	1.17	2.57	0.93	2.06	0.55	1.20	0.29	0.64	0.56	1.24	0.41	0.19	0.03	0.53
209	7.5	57	1.15	2.53	0.97	2.15	0.57	1.25	0.33	0.73	0.58	1.27	0.41	0.36	0.06	—
	8.4	64	1.28	2.83	1.09	2.41	0.64	1.40	0.37	0.82	0.64	1.42	0.45	0.40	0.07	—
210	5.4	52	1.03	2.27	0.86	1.89	0.52	1.14	0.29	0.63	0.54	1.18	0.46	0.25	0.09	—
	6.0	58	1.16	2.55	0.96	2.12	0.58	1.28	0.32	0.71	0.60	1.33	0.51	0.29	0.10	—
211	5.0	53	1.06	2.33	0.88	1.95	0.51	1.13	0.28	0.63	0.53	1.18	0.32	0.20	0.08	—
	5.6	59	1.19	2.62	1.00	2.19	0.58	1.28	0.32	0.71	0.60	1.33	0.36	0.23	0.10	—
212	3.5	51	1.03	2.26	0.86	1.89	0.48	1.07	0.26	0.57	0.51	1.13	0.36	0.21	0.11	0.55
	4.0	58	1.16	2.55	0.97	2.13	0.55	1.20	0.29	0.64	0.58	1.27	0.41	0.24	0.12	0.62
213	2.0	20	0.41	0.90	0.34	0.75	0.20	0.44	0.11	0.25	0.21	0.46	0.19	0.10	0.04	—
	5.8	59	1.19	2.62	1.00	2.20	0.58	1.29	0.32	0.71	0.60	1.33	0.57	0.29	0.11	—
214	2.2	24	0.49	1.07	0.41	0.89	0.24	0.53	0.13	0.29	0.25	0.55	0.23	0.12	0.08	0.28
	5.4	58	1.17	2.57	0.97	2.15	0.58	1.28	0.32	0.71	0.60	1.33	0.56	0.29	0.20	0.66
215	11.9	60	1.21	2.66	1.03	2.27	0.61	1.35	0.37	0.81	0.61	1.35	0.39	0.55	—	—
	12.9	66	1.31	2.89	1.12	2.47	0.67	1.47	0.40	0.88	0.67	1.47	0.43	0.60	—	—
216	2.8	13	0.26	0.58	0.23	0.50	0.16	0.35	0.10	0.22	0.15	0.33	0.34	0.05	0.02	—
	14.6	68	1.36	2.99	1.17	2.57	0.81	1.79	0.53	1.16	0.78	1.72	1.74	0.26	0.11	—
217	9.6	54	0.91	2.01	0.83	1.84	0.55	1.22	0.32	0.69	0.57	1.25	1.54	0.21	0.06	—
	10.6	59	1.01	2.22	0.92	2.03	0.61	1.34	0.35	0.77	0.62	1.37	1.70	0.23	0.07	—
218	11.1	75	1.44	3.17	1.27	2.80	0.75	1.65	0.49	1.09	0.71	1.57	0.04	0.30	0.01	—
	12.5	84	1.62	3.56	1.43	3.15	0.84	1.86	0.56	1.22	0.80	1.77	0.04	0.34	0.01	—
219	—	17	0.33	0.73	0.27	0.60	0.16	0.34	0.08	0.17	0.17	0.37	—	—	—	—
	—	55	1.10	2.43	0.91	2.00	0.52	1.14	0.26	0.58	0.56	1.23	—	—	—	—
220	0.9	8	0.16	0.34	0.14	0.31	0.09	0.19	0.06	0.13	0.08	0.18	0.06	0.03	0.01	0.06
	9.8	85	1.70	3.74	1.51	3.33	0.94	2.06	0.64	1.40	0.88	1.94	0.64	0.32	0.10	0.65
221	—	—	—	—	—	—	—	—	—	—	—	—	—	—	—	—
	—	—	—	—	—	—	—	—	—	—	—	—	—	—	—	—
222	13.4	55	1.09	2.41	0.92	2.04	0.55	1.22	0.32	0.71	0.56	1.24	1.21	0.30	0.46	—
	15.1	62	1.23	2.72	1.04	2.30	0.62	1.38	0.36	0.80	0.64	1.40	1.36	0.34	0.52	—
223	12.0	63	1.26	2.78	1.09	2.40	0.67	1.48	0.42	0.93	0.64	1.41	0.13	1.16	0.06	0.05
	13.3	70	1.42	3.12	1.22	2.70	0.75	1.66	0.48	1.05	0.72	1.58	0.14	1.30	0.06	0.06
224	26.5	78	1.52	3.35	1.35	2.96	0.81	1.79	0.54	1.19	0.77	1.70	0.11	0.58	—	—
	28.4	84	1.63	3.59	1.44	3.18	0.87	1.92	0.58	1.28	0.83	1.82	0.12	0.63	—	—
225	12.6	87	1.57	3.46	1.40	3.09	0.77	1.69	0.51	1.13	0.73	1.61	0.04	0.29	—	—
	14.3	99	1.78	3.92	1.59	3.51	0.87	1.92	0.58	1.28	0.83	1.82	0.04	0.33	—	—

Entry Number	Feed Name Description	Moisture Basis: A-F (as-fed) or M-F (moisture-free)	Magne-sium (Mg) (%)	Potassium (K) (%)	Sulfur (S) (%)	Cobalt (Co) (ppm or mg/kg)	Copper (Cu) (ppm or mg/kg)	Iodine (I) (ppm or mg/kg)	Iron (Fe) (%)	Manga-nese (Mn) (ppm or mg/kg)	Selenium (Se) (ppm or mg/kg)	Zinc (Zn) (ppm or mg/kg)
	SUNFLOWER, COMMON (Continued)											
204	SILAGE, LATE BLOOM	A-F	—	—	—	—	—	—	—	—	—	—
		M-F	—	—	—	—	—	—	—	—	—	—
205	SILAGE, MILK STAGE	A-F	—	—	—	—	—	—	—	—	—	—
		M-F	—	—	—	—	—	—	—	—	—	—
206	SILAGE, MATURE	A-F	—	—	—	—	—	—	—	—	—	—
		M-F	—	—	—	—	—	—	—	—	—	—
	SWEET CLOVER, YELLOW *Melilotus officinalis*											
207	HAY, SUN-CURED	A-F	0.39	1.35	0.42	—	8.8	—	0.02	95.4	—	—
		M-F	0.44	1.53	0.47	—	10.0	—	0.02	107.7	—	—
	TIMOTHY *Phleum pratense*											
208	HAY, SUN-CURED, ALL ANALYSES	A-F	0.11	1.43	0.11	0.07	4.3	0.03	0.01	45.2	—	15.5
		M-F	0.12	1.57	0.12	0.08	4.7	0.04	0.01	49.5	—	17.0
209	HAY, PREBLOOM, SUN-CURED	A-F	0.10	2.72	0.12	—	23.0	—	0.02	79.5	—	59.8
		M-F	0.11	3.05	0.13	—	25.8	—	0.02	89.0	—	67.0
210	HAY, EARLY BLOOM, SUN-CURED	A-F	0.11	2.14	0.12	—	57.1	—	0.02	91.8	—	55.3
		M-F	0.13	2.41	0.13	—	64.0	—	0.02	103.0	—	62.0
211	HAY, MIDBLOOM, SUN-CURED	A-F	0.12	1.61	0.12	—	14.3	—	0.01	49.9	—	38.2
		M-F	0.13	1.82	0.13	—	16.0	—	0.02	56.1	—	43.0
212	HAY, FULL BLOOM, SUN-CURED	A-F	0.10	1.77	0.12	—	25.7	—	0.01	82.4	—	47.9
		M-F	0.11	2.00	0.13	—	29.0	—	0.01	93.0	—	54.0
213	SILAGE, ALL ANALYSES	A-F	0.05	0.58	0.05	—	1.9	—	0.00	30.9	—	—
		M-F	0.15	1.69	0.13	—	5.5	—	0.01	90.2	—	—
214	SILAGE, WILTED	A-F	0.06	0.70	0.06	—	2.3	—	0.01	37.5	—	—
		M-F	0.15	1.69	0.15	—	5.5	—	0.01	90.2	—	—
	TOMATO *Lycopersicon esculentum*											
215	POMACE, DEHY	A-F	0.18	3.34	—	—	30.0	—	0.42	47.1	—	—
		M-F	0.20	3.63	—	—	32.6	—	0.46	51.2	—	—
	TREFOIL, BIRDSFOOT (DEERVETCH) *Lotus corniculatus*											
216	FRESH	A-F	0.08	0.63	0.05	0.09	2.5	—	0.01	16.0	—	6.0
		M-F	0.40	3.26	0.25	0.49	12.8	—	0.03	82.9	—	31.1
217	HAY, SUN-CURED	A-F	0.46	1.74	0.23	0.10	8.4	—	0.02	26.0	—	69.9
		M-F	0.51	1.92	0.25	0.11	9.3	—	0.02	28.7	—	77.2
	TRITICALE *Triticale hexaploide*											
218	GRAIN	A-F	0.23	0.51	—	0.08	8.3	—	0.01	42.5	—	31.2
		M-F	0.26	0.57	—	0.09	9.3	—	0.01	47.8	—	35.1
219	SILAGE, HEAD EMERGING	A-F	—	—	—	—	—	—	—	—	—	—
		M-F	—	—	—	—	—	—	—	—	—	—
	TURNIP *Brassica rapa*											
220	ROOTS, FRESH	A-F	0.02	0.26	0.04	—	2.0	—	0.00	3.9	—	2.7
		M-F	0.20	2.82	0.43	—	21.3	—	0.01	42.7	—	29.4
	UREA											
221	45% NITROGEN, 281% PROTEIN EQUIVALENT	A-F	—	—	—	—	—	—	—	—	—	—
		M-F	—	—	—	—	—	—	—	—	—	—
	VETCH *Vicia spp*											
222	HAY, SUN-CURED	A-F	0.24	1.88	0.13	0.32	8.8	0.44	0.04	53.9	—	—
		M-F	0.27	2.12	0.15	0.36	9.9	0.49	0.05	60.8	—	—
	WHEAT *Triticum aestivum*											
223	BRAN	A-F	0.58	1.23	0.22	0.08	11.0	0.07	0.02	114.9	0.64	94.6
		M-F	0.65	1.38	0.25	0.08	12.4	0.07	0.02	129.0	0.72	106.2
224	DISTILLERS GRAINS, DEHY	A-F	—	—	—	—	—	—	—	15.0	—	—
		M-F	—	—	—	—	—	—	—	16.1	—	—
225	FLOUR, LESS THAN 1.5% FIBER	A-F	—	0.05	—	—	—	—	—	—	—	—
		M-F	—	0.06	—	—	—	—	—	—	—	—

(Continued)

Entry Number	Fat-Soluble Vitamins					Water-Soluble Vitamins								
	A	Carotene (Provitamin A)	D	E	K	B–12	Biotin	Choline	Folacin (Folic Acid)	Niacin	Pantothenic Acid (B–3)	Pyridoxine (B–6)	Riboflavin (B–2)	Thiamin (B–1)
	(IU/g)	(ppm or mg/kg)	(IU/kg)	(ppm or mg/kg)	(ppm or mg/kg)	(ppb or mcg/kg)	(ppm or mg/kg)	(ppm or mg/kg)	(ppm or mg/kg)	(ppm or mg/kg)	(ppm or mg/kg)	(ppm or mg/kg)	(ppm or mg/kg)	(ppm or mg/kg)
204	—	—	—	—	—	—	—	—	—	—	—	—	—	—
	—	—	—	—	—	—	—	—	—	—	—	—	—	—
205	—	—	—	—	—	—	—	—	—	—	—	—	—	—
	—	—	—	—	—	—	—	—	—	—	—	—	—	—
206	—	—	—	—	—	—	—	—	—	—	—	—	—	—
	—	—	—	—	—	—	—	—	—	—	—	—	—	—
207	145.8	87.4	2	—	—	—	—	—	—	—	—	—	—	—
	164.6	98.8	2	—	—	—	—	—	—	—	—	—	—	—
208	39.8	23.8	2	57.6	—	—	0.06	741	2.09	31	7.2	—	9.2	1.5
	43.5	26.1	2	63.1	—	—	0.07	811	2.29	34	7.9	—	10.1	1.7
209	186.1	111.6	—	—	—	—	—	—	—	—	—	—	—	—
	208.4	125.0	—	—	—	—	—	—	—	—	—	—	—	—
210	78.0	46.8	—	11.6	—	—	—	—	—	—	—	—	—	—
	87.5	52.5	—	13.0	—	—	—	—	—	—	—	—	—	—
211	79.0	47.4	2	—	—	—	—	—	—	—	—	—	—	—
	88.9	53.3	2	—	—	—	—	—	—	—	—	—	—	—
212	—	—	—	—	—	—	—	—	—	—	—	—	—	—
	—	—	—	—	—	—	—	—	—	—	—	—	—	—
213	51.3	30.8	—	—	—	—	—	—	—	—	—	—	—	—
	149.6	89.8	—	—	—	—	—	—	—	—	—	—	—	—
214	—	—	—	—	—	—	—	—	—	—	—	—	—	—
	—	—	—	—	—	—	—	—	—	—	—	—	—	—
215	—	—	—	—	—	—	—	—	—	—	—	—	6.1	11.3
	—	—	—	—	—	—	—	—	—	—	—	—	6.7	12.3
216	—	—	—	—	—	—	—	—	—	—	—	—	—	—
	—	—	—	—	—	—	—	—	—	—	—	—	—	—
217	217.8	130.6	—	—	—	—	—	—	—	—	—	—	14.6	6.2
	240.4	144.2	—	—	—	—	—	—	—	—	—	—	16.1	6.8
218	—	—	—	—	—	—	—	457	—	—	—	—	0.4	—
	—	—	—	—	—	—	—	514	—	—	—	—	0.5	—
219	—	—	—	—	—	—	—	—	—	—	—	—	—	—
	—	—	—	—	—	—	—	—	—	—	—	—	—	—
220	—	—	—	—	—	—	—	92	0.26	7	1.7	—	0.6	0.7
	—	—	—	—	—	—	—	1009	2.84	72	19.0	—	6.5	7.1
221	—	—	—	—	—	—	—	—	—	—	—	—	—	—
	—	—	—	—	—	—	—	—	—	—	—	—	—	—
222	—	—	—	—	—	—	—	—	—	—	—	—	—	—
	—	—	—	—	—	—	—	—	—	—	—	—	—	—
223	4.4	2.6	—	14.3	—	—	0.38	1232	1.77	197	28.0	10.34	3.6	8.4
	4.9	2.9	—	16.0	—	—	0.42	1383	1.98	221	31.4	11.61	4.0	9.4
224	1.8	1.1	—	—	—	—	—	—	—	56	8.2	—	3.7	2.0
	2.0	1.2	—	—	—	—	—	—	—	60	8.7	—	4.0	2.1
225	—	—	—	—	—	—	—	972	—	26	13.0	—	0.8	5.8
	—	—	—	—	—	—	—	1102	—	29	14.7	—	0.9	6.6

TABLE 21–1

Entry Number	Feed Name Description	Inter-national Feed Number	Moisture Basis: A-F (as-fed) or M-F (moisture-free)	Chemical Analysis								
				Dry Matter	Ash	Crude Fiber	Neutral Det. Fib. (NDF)	Acid Det. Fib. (ADF)	Lignin	Ether Extract (Fat)	N-Free Extract	Crude Protein
				(%)	(%)	(%)	(%)	(%)	(%)	(%)	(%)	(%)
	WHEAT (Continued)											
226	GERM MEAL	5–05–218	A-F	88	4.3	3.1	—	4.4	—	8.5	48.1	24.4
			M-F	100	4.9	3.5	—	5.0	—	9.6	54.4	27.6
227	GLUTEN	5–05–221	A-F	90	1.8	3.2	—	—	—	2.0	19.8	63.4
			M-F	100	2.0	3.6	—	—	—	2.2	21.9	70.3
228	GRAIN, ALL ANALYSES	4–05–211	A-F	89	1.8	2.6	—	7.1	—	1.8	69.7	13.1
			M-F	100	2.0	2.9	—	8.0	—	2.0	78.4	14.7
229	GRAIN, HARD RED SPRING	4–05–258	A-F	88	1.7	2.6	37.9	11.0	—	1.8	67.4	14.2
			M-F	100	1.9	2.9	43.3	12.6	—	2.1	76.9	18.2
230	GRAIN, HARD RED WINTER	4–05–268	A-F	89	1.7	2.6	24.8	3.9	0.9	1.6	69.8	12.8
			M-F	100	2.0	2.9	28.0	4.4	1.0	1.8	78.8	14.5
231	GRAIN, SOFT RED WINTER	4–05–294	A-F	88	1.9	2.3	—	—	—	1.6	71.2	11.4
			M-F	100	2.1	2.6	—	—	—	1.8	80.5	12.9
232	GRAIN, SOFT WHITE WINTER	4–05–337	A-F	90	1.5	2.3	12.6	3.6	—	1.5	75.0	10.2
			M-F	100	1.6	2.6	14.0	4.0	—	1.7	82.9	11.3
233	GRAIN, SOFT WHITE WINTER, PACIFIC COAST	4–08–555	A-F	89	1.9	2.5	—	—	—	1.9	72.9	10.0
			M-F	100	2.1	2.8	—	—	—	2.2	81.7	11.2
234	GRAIN SCREENINGS	4–05–216	A-F	89	3.2	5.3	—	—	—	2.8	64.1	13.3
			M-F	100	3.6	6.0	—	—	—	3.2	72.3	15.0
235	HAY, SUN-CURED	1–05–172	A-F	89	7.0	25.7	60.5	36.5	—	2.0	46.4	7.7
			M-F	100	7.9	29.0	68.0	41.0	—	2.2	52.3	8.7
236	MIDDLINGS, LESS THAN 9.5% FIBER	4–05–205	A-F	89	4.7	7.7	32.9	8.9	—	4.3	55.7	16.4
			M-F	100	5.3	8.7	37.0	10.0	—	4.9	62.7	18.5
237	MILL RUN, LESS THAN 9.5% FIBER	4–05–206	A-F	90	5.1	8.2	—	9.9	—	4.1	57.4	15.1
			M-F	100	5.7	9.1	—	11.0	—	4.6	63.9	16.7
238	RED DOG, LESS THAN 4% FIBER	4–05–203	A-F	88	2.4	2.9	—	—	—	3.4	64.0	15.6
			M-F	100	2.7	3.3	—	—	—	3.8	72.5	17.6
239	SHORTS, LESS THAN 7% FIBER	4–05–201	A-F	88	4.4	6.4	—	—	—	4.6	56.5	16.5
			M-F	100	5.0	7.2	—	—	—	5.2	63.9	18.7
	WHEY, CATTLE *Bos taurus*											
240	FRESH	4–08–134	A-F	7	0.7	—	0.0	0.0	—	0.3	5.1	0.9
			M-F	100	9.4	—	0.0	0.0	—	4.3	73.9	13.2
241	CONDENSED	4–01–180	A-F	54	5.0	0.3	—	—	—	0.4	41.2	6.9
			M-F	100	9.3	0.5	—	—	—	0.8	76.6	12.8
242	DEHY	4–01–182	A-F	93	8.8	0.2	0.3	0.2	—	0.8	70.2	13.3
			M-F	100	9.4	0.2	0.3	0.2	—	0.8	75.3	14.2
243	LOW LACTOSE, DEHY (DRIED WHEY PRODUCT)	4–01–186	A-F	93	15.4	0.2	0.0	0.0	—	1.0	60.0	16.7
			M-F	100	16.5	0.2	0.0	0.0	—	1.1	64.3	17.9
244	ALBUMIN (DRIED MILK ALBUMIN)	5–01–177	A-F	92	29.1	0.9	—	—	—	1.0	12.8	48.4
			M-F	100	31.6	0.9	—	—	—	1.1	13.9	52.5
	YEAST, BREWERS *Saccharomyces cerevisiae*											
245	DEHY	7–05–527	A-F	93	6.5	3.0	—	3.7	—	0.9	38.8	43.6
			M-F	100	7.0	3.2	—	4.0	—	1.0	41.7	47.1
246	IRRADIATED, DEHY	7–05–529	A-F	94	6.2	6.2	—	—	—	1.1	32.4	48.1
			M-F	100	6.6	6.5	—	—	—	1.2	34.5	51.2
	YEAST, TORULA *Torulopsis utilis*											
247	DEHY	7–05–534	A-F	93	8.0	2.5	—	3.7	—	1.6	31.5	49.6
			M-F	100	8.6	2.7	—	4.0	—	1.7	33.8	53.3

(Continued)

Entry Number	Digestible Protein Ruminant (%)	TDN Ruminant (%)	Digestible Energy Ruminant (Mcal/lb)	(Mcal/kg)	Metabolizable Energy Ruminant (Mcal/lb)	(Mcal/kg)	Ruminant NE$_m$ (Mcal/lb)	(Mcal/kg)	Ruminant NE$_g$ (Mcal/lb)	(Mcal/kg)	Lactating Cows NE$_{lc}$ (Mcal/lb)	(Mcal/kg)	Calcium (Ca) (%)	Phosphorus (P) (%)	Sodium (Na) (%)	Chlorine (Cl) (%)
226	22.9	83	1.66	3.66	1.50	3.30	0.96	2.11	0.67	1.48	0.89	1.95	0.06	0.95	0.02	0.06
	26.0	94	1.88	4.14	1.69	3.73	1.08	2.38	0.76	1.67	1.00	2.21	0.06	1.07	0.03	0.07
227	55.8	81	1.61	3.55	1.44	3.18	0.90	1.99	0.62	1.37	0.84	1.85	0.06	0.23	0.06	—
	61.8	90	1.79	3.94	1.60	3.52	1.00	2.20	0.69	1.52	0.93	2.05	0.07	0.25	0.07	—
228	10.5	77	1.54	3.40	1.28	2.82	0.88	1.93	0.60	1.33	0.81	1.79	0.05	0.35	0.06	0.08
	11.7	87	1.73	3.82	1.44	3.17	0.98	2.17	0.68	1.49	0.92	2.02	0.06	0.39	0.06	0.09
229	10.0	78	1.56	3.43	1.39	3.07	0.87	1.91	0.60	1.31	0.81	1.78	0.04	0.37	0.02	0.08
	11.4	89	1.78	3.91	1.59	3.50	0.99	2.18	0.68	1.50	0.92	2.04	0.05	0.42	0.02	0.09
230	8.8	78	1.57	3.45	1.40	3.09	0.88	1.94	0.61	1.34	0.82	1.81	0.04	0.37	0.02	0.05
	9.9	88	1.77	3.90	1.58	3.49	1.00	2.19	0.69	1.51	0.93	2.05	0.05	0.42	0.02	0.06
231	8.6	78	1.57	3.45	1.40	3.09	0.89	1.95	0.61	1.35	0.83	1.82	0.05	0.36	0.01	0.07
	9.7	89	1.77	3.91	1.59	3.50	1.00	2.21	0.69	1.52	0.94	2.06	0.06	0.40	0.01	0.08
232	6.4	80	1.60	3.54	1.44	3.16	0.91	2.00	0.62	1.38	0.84	1.86	—	0.40	0.02	—
	7.0	89	1.77	3.91	1.59	3.50	1.00	2.21	0.69	1.52	0.93	2.06	—	0.44	0.02	—
233	6.2	79	1.58	3.48	1.41	3.11	0.88	1.95	0.61	1.34	0.82	1.82	0.09	0.31	0.01	—
	7.0	88	1.77	3.89	1.58	3.48	0.99	2.18	0.68	1.50	0.92	2.04	0.10	0.35	0.01	—
234	9.6	63	1.26	2.79	1.10	2.42	0.63	1.39	0.39	0.86	0.62	1.37	0.13	0.34	0.05	—
	10.8	71	1.43	3.14	1.24	2.73	0.71	1.57	0.44	0.97	0.70	1.54	0.14	0.38	0.05	—
235	3.8	49	0.94	2.07	0.79	1.74	0.45	0.99	0.22	0.49	0.54	1.19	0.13	0.18	0.19	—
	4.3	56	1.06	2.33	0.89	1.96	0.51	1.11	0.25	0.56	0.61	1.34	0.15	0.20	0.21	—
236	12.4	74	1.39	3.07	1.16	2.57	0.78	1.72	0.52	1.15	0.79	1.73	0.13	0.89	0.01	0.04
	14.0	83	1.57	3.45	1.31	2.89	0.88	1.94	0.59	1.30	0.88	1.95	0.15	1.00	0.01	0.04
237	11.4	71	1.46	3.21	1.14	2.52	0.76	1.68	0.50	1.11	0.76	1.68	0.10	1.02	—	—
	12.6	79	1.62	3.58	1.27	2.87	0.85	1.87	0.56	1.23	0.85	1.87	0.11	1.13	—	—
238	12.9	77	1.47	3.23	1.39	3.06	0.82	1.80	0.55	1.22	0.77	1.70	0.06	0.51	0.01	0.14
	14.6	87	1.66	3.67	1.57	3.47	0.92	2.04	0.63	1.38	0.87	1.92	0.07	0.58	0.02	0.16
239	12.9	76	1.47	3.24	1.25	2.75	0.85	1.88	0.58	1.28	0.79	1.75	0.09	0.80	0.03	0.05
	14.6	86	1.66	3.66	1.41	3.11	0.96	2.12	0.66	1.45	0.90	1.98	0.10	0.91	0.03	0.06
240	0.6	7	0.13	0.29	0.12	0.26	—	—	—	—	—	—	0.06	0.05	—	—
	8.8	94	1.88	4.15	1.70	3.74	—	—	—	—	—	—	0.81	0.71	—	—
241	4.5	47	0.94	2.07	0.84	1.85	0.56	1.24	0.39	0.86	0.52	1.15	0.39	0.47	—	—
	8.4	87	1.74	3.84	1.56	3.43	1.04	2.30	0.72	1.60	0.97	2.13	0.72	0.88	—	—
242	8.9	76	1.51	3.33	1.28	2.83	0.87	1.92	0.59	1.30	0.78	1.71	0.86	0.76	0.62	0.07
	9.5	82	1.62	3.57	1.38	3.03	0.93	2.06	0.63	1.40	0.83	1.83	0.92	0.82	0.66	0.08
243	11.5	74	1.40	3.09	1.14	2.52	0.75	1.66	0.49	1.08	0.77	1.69	1.49	1.11	1.44	1.03
	12.4	79	1.50	3.31	1.22	2.70	0.81	1.78	0.52	1.15	0.82	1.81	1.60	1.18	1.54	1.10
244	41.8	59	1.28	2.83	1.11	2.44	0.76	1.67	0.50	1.09	0.73	1.60	10.86	4.03	—	—
	45.4	64	1.39	3.07	1.20	2.65	0.82	1.81	0.54	1.18	0.79	1.74	11.79	4.37	—	—
245	39.0	73	1.46	3.21	1.28	2.83	0.81	1.79	0.54	1.19	0.77	1.69	0.14	1.36	0.07	0.07
	41.9	78	1.57	3.45	1.38	3.04	0.87	1.92	0.58	1.28	0.83	1.82	0.15	1.47	0.08	0.08
246	—	72	1.43	3.16	1.26	2.77	—	—	—	—	—	—	0.78	1.42	—	—
	—	76	1.53	3.37	1.34	2.95	—	—	—	—	—	—	0.83	1.51	—	—
247	45.1	72	1.49	3.29	1.27	2.81	0.82	1.81	0.55	1.21	0.78	1.71	0.55	1.61	0.01	0.02
	48.5	78	1.60	3.53	1.37	3.02	0.88	1.95	0.59	1.30	0.84	1.84	0.59	1.73	0.01	0.02

Entry Number	Feed Name Description	Moisture Basis: A-F (as-fed) or M-F (moisture-free)	Macro Minerals (Continued)			Micro Minerals						
			Magne-sium (Mg)	Potassium (K)	Sulfur (S)	Cobalt (Co)	Copper (Cu)	Iodine (I)	Iron (Fe)	Manga-nese (Mn)	Selenium (Se)	Zinc (Zn)
			(%)	(%)	(%)	(ppm or mg/kg)	(ppm or mg/kg)	(ppm or mg/kg)	(%)	(ppm or mg/kg)	(ppm or mg/kg)	(ppm or mg/kg)
	WHEAT (Continued)											
226	GERM MEAL	A-F	0.25	0.94	0.27	0.12	9.2	—	0.01	132.5	0.46	119.4
		M-F	0.28	1.06	0.30	0.14	10.4	—	0.01	149.9	0.52	135.1
227	GLUTEN	A-F	0.04	0.02	0.95	0.05	11.6	0.06	0.01	18.1	3.75	38.5
		M-F	0.04	0.02	1.00	0.05	12.8	0.06	0.01	20.1	4.16	42.6
228	GRAIN, ALL ANALYSES	A-F	0.14	0.14	0.18	0.44	5.8	0.09	0.01	41.5	0.26	31.4
		M-F	0.15	0.15	0.20	0.50	6.5	0.10	0.01	46.7	0.29	35.2
229	GRAIN, HARD RED SPRING	A-F	0.14	0.36	0.15	0.12	6.0	—	0.01	37.0	0.26	37.9
		M-F	0.16	0.41	0.17	0.14	6.8	—	0.01	42.2	0.30	43.3
230	GRAIN, HARD RED WINTER	A-F	0.12	0.43	0.14	0.15	5.1	—	0.00	30.4	0.29	35.2
		M-F	0.13	0.49	0.15	0.16	5.7	—	0.00	34.3	0.33	39.8
231	GRAIN, SOFT RED WINTER	A-F	0.10	0.41	0.11	0.10	7.0	—	0.00	33.4	0.04	42.1
		M-F	0.11	0.46	0.12	0.12	8.0	—	0.00	37.8	0.05	47.7
232	GRAIN, SOFT WHITE WINTER	A-F	—	—	0.12	0.14	7.1	—	0.00	36.2	0.05	27.1
		M-F	—	—	0.13	0.15	7.8	—	0.00	40.0	0.05	30.0
233	GRAIN, SOFT WHITE WINTER, PACIFIC COAST	A-F	0.13	0.45	0.16	—	—	—	0.01	—	—	—
		M-F	0.15	0.51	0.18	—	—	—	0.01	—	—	—
234	GRAIN SCREENINGS	A-F	0.21	0.81	0.20	1.25	2.3	—	0.01	28.9	0.60	38.9
		M-F	0.24	0.91	0.22	1.41	2.6	—	0.01	32.5	0.68	43.9
235	HAY, SUN-CURED	A-F	0.11	0.88	0.19	—	—	—	0.02	—	—	—
		M-F	0.12	1.00	0.22	—	—	—	0.02	—	—	—
236	MIDDLINGS, LESS THAN 9.5% FIBER	A-F	0.34	0.98	0.17	0.50	15.9	0.11	0.01	114.0	0.74	96.9
		M-F	0.38	1.10	0.19	0.57	17.9	0.12	0.01	128.3	0.83	109.1
237	MILL RUN, LESS THAN 9.5% FIBER	A-F	0.48	1.20	0.30	0.21	18.5	—	0.01	104.1	—	—
		M-F	0.53	1.34	0.34	0.23	20.6	—	0.01	115.8	—	—
238	RED DOG, LESS THAN 4% FIBER	A-F	0.18	0.52	0.24	0.12	6.3	—	0.01	52.1	0.32	65.0
		M-F	0.21	0.59	0.27	0.13	7.1	—	0.01	59.1	0.37	73.7
239	SHORTS, LESS THAN 7% FIBER	A-F	0.27	0.93	0.21	0.11	11.5	—	0.01	114.1	0.48	102.4
		M-F	0.31	1.05	0.23	0.12	13.0	—	0.01	129.1	0.54	115.9
	WHEY, CATTLE *Bos taurus*											
240	FRESH	A-F	—	0.19	—	—	—	—	0.00	0.2	—	—
		M-F	—	2.75	—	—	—	—	0.03	3.2	—	—
241	CONDENSED	A-F	—	—	—	—	—	—	—	—	—	—
		M-F	—	—	—	—	—	—	—	—	—	—
242	DEHY	A-F	0.13	1.11	1.04	0.11	46.5	—	0.02	5.9	—	3.2
		M-F	0.14	1.19	1.11	0.12	49.9	—	0.02	6.3	—	3.4
243	LOW LACTOSE, DEHY (DRIED WHEY PRODUCT)	A-F	0.21	2.95	1.07	—	7.0	9.85	0.03	8.0	0.05	7.9
		M-F	0.23	3.16	1.15	—	7.5	10.55	0.03	8.6	0.06	8.4
244	ALBUMIN (DRIED MILK ALBUMIN)	A-F	—	—	—	—	—	—	—	—	—	—
		M-F	—	—	—	—	—	—	—	—	—	—
	YEAST, BREWERS *Saccharomyces cerevisiae*											
245	DEHY	A-F	0.24	1.69	0.43	0.51	38.4	0.36	0.01	6.7	0.91	39.0
		M-F	0.26	1.82	0.46	0.54	41.3	0.38	0.01	7.2	0.98	41.9
246	IRRADIATED, DEHY	A-F	—	2.14	—	—	—	—	—	—	—	—
		M-F	—	2.28	—	—	—	—	—	—	—	—
	YEAST, TORULA *Torulopsis utilis*											
247	DEHY	A-F	0.14	1.92	0.55	0.03	11.9	2.50	0.01	9.3	—	99.5
		M-F	0.15	2.06	0.59	0.03	12.8	2.69	0.01	10.0	—	107.0

(Continued)

| Entry Number | Fat-Soluble Vitamins | | | | | Water-Soluble Vitamins | | | | | | | | |
| | A | Carotene (Provitamin A) | D | E | K | B–12 | Biotin | Choline | Folacin (Folic Acid) | Niacin | Pantothenic Acid (B–3) | Pyridoxine (B–6) | Riboflavin (B–2) | Thiamin (B–1) |
	(IU/g)	(ppm or mg/kg)	(IU/kg)	(ppm or mg/kg)	(ppm or mg/kg)	(ppb or mcg/kg)	(ppm or mg/kg)	(ppm or mg/kg)	(ppm or mg/kg)	(ppm or mg/kg)	(ppm or mg/kg)	(ppm or mg/kg)	(ppm or mg/kg)	(ppm or mg/kg)
226	—	—	—	141.2	—	—	0.22	3062	2.12	68	18.6	9.97	6.0	23.1
	—	—	—	159.7	—	—	0.24	3465	2.40	77	21.0	11.28	6.8	26.2
227	—	—	0	34.1	—	73.1	0.00	577	0.74	74	5.8	2.26	0.7	0.9
	—	—	0	37.7	—	81.0	0.00	640	0.82	82	6.4	2.51	0.7	1.0
228	—	—	—	15.5	—	0.9	0.10	918	0.43	59	11.3	3.74	1.3	4.3
	—	—	—	17.4	—	1.0	0.11	1032	0.49	66	12.7	4.20	1.4	4.8
229	—	—	—	12.7	—	—	0.11	1010	0.41	56	9.6	5.11	1.4	4.2
	—	—	—	14.4	—	—	0.13	1153	0.46	64	11.0	5.83	1.6	4.8
230	—	—	—	11.1	—	—	0.11	1004	0.38	53	10.1	3.01	1.3	4.5
	—	—	—	12.5	—	—	0.12	1133	0.43	60	11.4	3.40	1.5	5.1
231	—	—	—	15.6	—	—	—	892	0.41	53	10.1	3.21	1.5	4.7
	—	—	—	17.7	—	—	—	1009	0.46	60	11.4	3.63	1.7	5.3
232	—	—	—	30.9	—	—	—	—	—	62	11.2	4.79	—	—
	—	—	—	34.2	—	—	—	—	—	69	12.3	5.29	—	—
233	—	—	—	—	—	—	—	973	—	46	11.1	—	1.1	—
	—	—	—	—	—	—	—	1090	—	52	12.4	—	1.2	—
234	—	—	—	—	—	—	—	869	0.43	58	11.3	—	0.9	6.3
	—	—	—	—	—	—	—	980	0.49	65	12.7	—	1.0	7.2
235	126.3	75.8	1	—	—	—	—	—	—	—	—	—	15.1	—
	142.3	85.4	2	—	—	—	—	—	—	—	—	—	17.0	—
236	5.1	3.1	—	23.8	—	—	0.24	1246	1.24	95	17.8	9.14	2.0	14.2
	5.8	3.5	—	26.9	—	—	0.27	1403	1.39	107	20.0	10.29	2.3	15.9
237	—	—	—	31.9	—	—	0.31	1005	1.08	116	13.7	11.09	2.1	15.2
	—	—	—	35.5	—	—	0.34	1118	1.20	129	15.2	12.33	2.4	17.0
238	—	—	—	37.4	—	—	0.11	1453	0.82	46	13.3	5.40	2.2	21.8
	—	—	—	42.4	—	—	0.12	1648	0.93	52	15.0	6.12	2.5	24.7
239	5.1	3.1	—	36.0	—	—	—	1697	1.51	105	21.9	—	4.1	19.5
	5.8	3.5	—	40.7	—	—	—	1920	1.71	119	24.8	—	4.6	22.1
240	—	—	—	—	—	—	—	—	—	1	5.3	—	1.4	0.3
	—	—	—	—	—	—	—	—	—	14	76.7	—	20.3	4.3
241	—	—	—	—	—	—	—	—	—	3	11.8	—	14.2	2.6
	—	—	—	—	—	—	—	—	—	5	22.0	—	26.5	4.8
242	—	—	—	0.2	—	18.9	0.35	1790	0.85	11	46.2	3.21	27.4	4.0
	—	—	—	0.2	—	20.3	0.38	1921	0.91	11	49.6	3.45	29.4	4.3
243	—	—	—	—	—	35.9	0.50	4096	0.89	18	74.5	4.48	47.6	5.0
	—	—	—	—	—	38.4	0.54	4387	0.96	19	79.8	4.79	50.9	5.4
244	—	—	—	—	—	—	—	—	—	2	7.3	—	8.8	0.7
	—	—	—	—	—	—	—	—	—	2	7.9	—	9.6	0.7
245	—	—	—	2.1	—	1.1	1.04	3847	9.69	443	81.5	36.67	34.1	85.2
	—	—	—	2.3	—	1.1	1.12	4134	10.41	476	87.6	39.40	36.6	91.6
246	—	—	—	—	—	—	—	—	—	—	—	—	18.5	—
	—	—	—	—	—	—	—	—	—	—	—	—	19.7	—
247	—	—	—	—	—	4.0	1.19	2981	25.66	512	107.5	34.48	47.7	6.8
	—	—	—	—	—	4.3	1.27	3203	27.58	550	115.6	37.06	51.3	7.3

Fig. 21–2. Two of the most important crops for dairy cattle—alfalfa and corn, being grown in contour strips to prevent erosion on this Wisconsin farm. (Courtesy, USDA, Soil Conservation Service)

Fig. A–1. Electronic grain feeding system. Modern dairying involves the use of weights and measures and other useful information contained in the Appendix. (Courtesy, David J. Schingoethe, South Dakota State University, Brookings)

Contents

SECTION I—ANIMAL UNITS

An animal unit is a common animal denominator, based on feed consumption. It is assumed that one mature cow represents an animal unit. Then, the comparative (to a mature cow) feed consumption of other age groups or classes of animals determines the proportion of an animal unit which they represent. For example, it is generally estimated that the ration of one mature cow will feed five mature ewes, or that five mature ewes equal one animal unit.

The original concept of an animal unit included a weight stipulation—an animal unit referred to a 1,000-lb cow, with or without calf at side. Unfortunately, in recent years, the 1,000-lb qualification has been dropped. Certainly, there is a wide difference in the daily feed requirements of a 900-lb cow and of a 1,500-lb cow; and feed requirements increase with higher levels of lactation.

APPENDIX

521

Also, the period of time to be grazed has an effect on the total carrying capacity. For example, if an animal is carried for 1 month only, it will take one-twelfth of the total feed required to carry the same animal 1 year. For this reason, the term *animal unit months* is becoming increasingly important. So, in addition to the weight factor, the time factor has a distinct bearing on the ultimate carrying capacity of a tract of land; and, in lactating cows, level of production is also a factor.

Table A–1 gives the animal units of different classes and ages of livestock.

TABLE A–1
ANIMAL UNITS

Type of Livestock	Animal Units
Chickens:	
75 layers or breeders	1.0
325 replacement pullets to 6 months of age	1.0
650 7-week-old broilers	1.0
Turkeys:	
35 breeders	1.0
40 turkeys raised to maturity	1.0
75 turkeys raised to 6 months of age	1.0
Cattle:	
Cow, with or without unweaned calf at side, or heifer 2 years old or older	1.0
Bull, 2 years old or older	1.3
Young cattle, 1 to 2 years	0.8
Weaned calves to yearlings	0.6
Horses:	
Horse, mature	1.3
Horse, yearling	1.0
Weanling colt or filly	0.75
Sheep:	
5 mature ewes, with or without unweaned lambs at side	1.0
5 rams, 2 years old or over	1.3
5 yearlings	0.8
5 weaned lambs to yearlings	0.6
Goats—7	1.0
Swine:	
Sow	0.4
Boar	0.5
Pigs to 200 lb	0.2
Rabbits—56	1.0
Fish—259	1.0

SECTION II—WEIGHTS AND MEASURES

Increasingly, dairy producers and those who counsel with them need to use the metric system. Hence, they need to have a working knowledge of it, along with conversion tables.

The basic metric units are the *meter* (length/distance), the *gram* (weight), and the *liter* (capacity). The units are then expanded in multiples of 10 or made smaller by 1/10. The prefixes, which are used in the same way with all basic metric units, follow:

"milli-"	=	1/1000	"deca-"	=	10
"centi-"	=	1/100	"hecto-"	=	100
"deci-"	=	1/10	"kilo-"	=	1,000

METRIC VS U.S. CUSTOMARY

LENGTH
1 CENTIMETER = 0.4 INCH

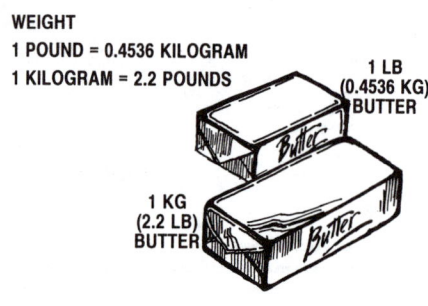

WEIGHT
1 POUND = 0.4536 KILOGRAM
1 KILOGRAM = 2.2 POUNDS

VOLUME
1 QUART = 0.946 LITER
1 LITER = 1.057 QUART

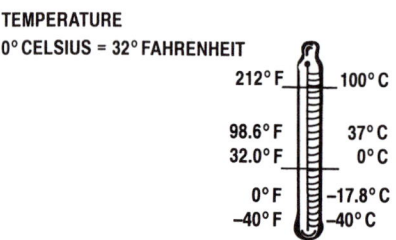

TEMPERATURE
0° CELSIUS = 32° FAHRENHEIT

212° F	100° C
98.6° F	37° C
32.0° F	0° C
0° F	–17.8° C
–40° F	–40° C

Fig. A–2. Some direct conversions and readings.

The following tables will facilitate conversion from metric units to U.S. customary, and vice versa:

Table A–2 Weights and Measures—
  Weight
  Length
  Surface/Area
  Volume

Table A–3 Temperature

Table A–2 is a conversion table, from metric system to U.S. customary system, and vice versa.

TABLE A–2
WEIGHTS AND MEASURES

Weight

Unit	Is Equal To	
Metric system:	**(metric)**	**(U.S. customary)**
1 microgram (mcg)	0.001 mg	
1 milligram (mg)	0.001 g	0.015432356 grain
1 centigram (cg)	0.01 g	0.15432356 grain
1 decigram (dg)	0.1 g	1.5432 grains
1 gram (g)	1,000 mg	0.03527396 oz
1 decagram (dkg)	10 g	5.643833 dr
1 hectogram (hg)	100 g	3.527396 oz
1 kilogram (kg)	1,000 g	35.274 oz; 2.2046223 lb
1 ton	1,000 kg	2,204.6 lb; 1.102 tons (short); 0.984 ton (long)
U.S. customary:	**(U.S. customary)**	**(metric)**
1 grain	0.037 dr	64.798918 mg; 0.64798918 g
1 dram (dr)	0.063 oz	1.771845 g
1 ounce (oz)	16 dr	28.349527 g
1 pound (lb)	16 oz	453.5924 g; 0.4536 kg
1 hundredweight (cwt)	100 lb	
1 ton (short)	2,000 lb	907.18486 kg; 0.907 (metric) ton
1 ton (long)	2,200 lb	1,016.05 kg; 1.016 (metric) ton
1 part per million (ppm)	0.4535924 mg/lb; 0.907 g/ton; 0.0001%; 0.00013 oz/gal	1 mcg/g; 1 mg/l; 1 mg/kg
1 percent (%) (1 part in 100 parts)	1.28 oz/gal; 8.34 lb/100 gal	10,000 ppm; 10 g/l

Weight Conversions

U.S. Customary to Metric		Multiply By	Metric to U.S. Customary		Multiply By
To Change			**To Change**		
grains	to milligrams	64.799			
ounces	to grams	28.35	grams	to ounces	0.035
pounds	to grams	453.6			
pounds	to kilograms	0.454	kilograms	to pounds	2.205
tons	to metric tons	0.9	metric tons	to tons	1.102

Weight—Unit Conversion Factors

To Change		Multiply By	To Change		Multiply By
milligrams/pound	to grams/ton	2	milligrams/gram	to milligrams/pound	453.6
grams/pound	to grams/ton	2,000	milligrams/kilogram	to milligrams/pound	0.4536
pounds/ton	to grams/ton	453.6	micrograms/kilogram	to grams/pound	0.4536
parts per million	to milligrams/pound	0.4536	grams/ton	to grams/pound	0.0005
parts per million	to percent . . . . . move decimal 4 places to left		grams/ton	to pounds/ton	0.0022
milligrams/pound	to parts per million	2.2046	grams/ton	to percent	0.00011
			percent	to grams/ton	9,072.0
parts per million	to grams/ton	0.907	grams/ton	to parts per million	1.1

(Continued)

TABLE A–2 *(Continued)*

Length

Unit	Is Equal To	
Metric system:	**(metric)**	**(U.S. customary)**
1 millimicron (mμ)	0.000000001 m	0.000000039 in.
1 micron (μ)	0.000001 m	0.000039 in.
1 millimeter (mm)	0.001 m	0.0394 in.
1 centimeter (cm)	0.01 m	0.3937 in.
1 decimeter (dm)	0.1 m	3.937 in.
1 meter (m)	1 m	39.37 in.; 3.281 ft; 1.094 yd
1 hectometer (hm)	100 m	328.08 ft; 19.8338 rd
1 kilometer (km)	1,000 m	3,280.8 ft; 0.621 mi
U.S. customary:	**(U.S. customary)**	**(metric)**
1 inch (in.)	1 in.	25 mm; 2.54 cm
1 hand*	4 in.	10.16 cm
1 foot (ft)	12 in.	30.48 cm; 0.305 m
1 yard (yd)	3 ft	0.914 m
1 fathom** (fath)	6.08 ft	1.829 m
1 rod (rd), pole, or perch	16.5 ft; 5.5 yd	5.029 m
1 chain	792 in.; 66 ft; 22 yd	20.116 m
1 furlong (fur.)	220 yd; 40 rd	201.168 m
1 mile (mi)	5,280 ft; 1,760 yd; 320 rd; 8 fur.	1,609.35 m; 1.609 km
1 knot or nautical mile	6,080 ft; 1.15 land miles	1.85 km
1 league (land)	3 mi (land)	4.827 km
1 league (nautical)	3 mi (nautical)	4.827 km

Length Conversion

U.S. Customary to Metric		Metric to U.S. Customary	
To Change	**Multiply By**	**To Change**	**Multiply By**
inches . . . . to millimeters	25.4	millimeters . . . . to inches	0.04
inches . . . . to centimeters	2.54	centimeters . . . . to inches	0.4
feet . . . . to centimeters	30.5	centimeters . . . . to feet	0.033
feet . . . . to meters	0.305	meters . . . . to feet	3.3
yards . . . . to meters	0.914	meters . . . . to yards	1.1
miles . . . . to kilometers	1.609	kilometers . . . . to miles	0.6

*Used in measuring height of horses.

**Used in measuring depth at sea.

(Continued)

TABLE A–2 *(Continued)*

Surface/Area

Unit	Is Equal To	
Metric system:	**(metric)**	**(U.S. customary)**
1 square millimeter (mm^2)	0.000001 m^2	0.00155 in.^{2}
1 square centimeter (cm^2)	0.0001 m^2	0.155 in.^{2}
1 square decimeter (dm^2)	0.01 m^2	15.5 in.^{2}
1 square meter (m^2)	1 centare (ca)	1,550 in.^{2}; 10.76 ft^2; 1.196 yd^2
1 are (a)	100 m^2	119.6 yd^2
1 hectare (ha)	10,000 m^2	2.47 acres
1 square kilometer (km^2)	1,000,000 m^2	247.1 acres; 0.386 mi^2
U.S. customary:	**(U.S. customary)**	**(metric)**
1 square inch (in.^{2})	1 in. × 1 in.	6.452 cm^2
1 square foot (ft^2)	144 in.^{2}; 0.111 yd^2	0.093 m^2
1 square yard (yd^2)	1,296 in.^{2}; 9 ft^2	0.836 m^2
1 square rod (rd^2)	272.25 ft^2; 30.25 yd^2	25.29 m^2
1 rood	40 rd^2	10.117 a
1 acre	43,560 ft^2; 4,840 yd^2; 160 rd^2; 4 roods	4,046.87 m^2; 0.405 ha
1 square mile (mi^2)	640 acres; 1 section	2.59 km^2; 259 ha
1 township	36 sections; 6 miles square	

Surface/Area Conversions

U.S. Customary to Metric		Metric to U.S. Customary	
To Change	**Multiply By**	**To Change**	**Multiply By**
square inches . . . . to square centimeters	6.452	square centimeters . . . . . . to square inches	0.155
square feet . . . . to square centimeters	929.1	square centimeters . . . . . . to square feet	0.001
square feet . . . . to square meters	0.09	square meters . . . . to square feet	10.764
square yards . . . . to square meters	0.836	square meters . . . . to square yards	1.196
square miles . . . . to square kilometers	2.6	square kilometers . . . . to square miles	0.4
acres . . . . to hectares	0.4	hectares . . . . to acres	2.5

Weights/Measures/Unit Area

Unit	Is Equal To
Volume per unit area:	
1 l/ha	0.107 gal/acre
1 gal/acre	9.354 l/ha
Weight per unit area:	
1 kg/cm^2	14.22 lb/in.^{2}
1 kg/h	0.892 lb/acre
1 lb/in.^{2}	0.0703 kg/cm^2
1 lb/acre	1.121 kg/ha
Area per unit weight:	
1 cm^2/kg	0.0703 in.^{2}/lb
1 in.^{2}/lb	14.22 cm^2/kg

(Continued)

TABLE A–2 *(Continued)*

Volume

Unit		Is Equal To	
Metric system—liquid and dry:		(U.S. customary—liquid)	(U.S. customary—dry)
1 milliliter (ml)	0.001 liter	0.271 dram (fl)	0.061 in.³
1 centiliter (cl)	0.01 liter	0.338 oz (fl)	0.61 in.³
1 deciliter (dl)	0.1 liter	3.38 oz (fl)	
1 liter (l)	1,000 cc	1.057 qt; 0.2642 gal	0.908 qt
1 hectoliter (hl)	100 liters	26.418 gal	2.838 bu
1 kiloliter (kl)	1,000 liters	264.18 gal	1,308 yd³

Unit		(ounces)	(cubic inches)	(metric)
U.S. customary—liquid:				
1 teaspoon (t)	60 drops	⅛	. . . .	5 ml
1 dessert spoon	2 t			
1 tablespoon (T)	3 t	0.5	. . . .	15 ml
1 fl oz		1	1.805	29.57 ml
1 gill (gi)	0.5 c	4	7.22	118.29 ml
1 cup (c)	16 T	8	14.44	236.58 ml; 0.24 litres
1 pint (pt)	2 c	16	28.88	0.47 litres
1 quart (qt)	2 pt	32	57.75	0.95 litres
1 gallon (gal)	4 qt	8.34 lb	231	3.79 litres
1 barrel (bbl)	31.5 gal			
1 hogshead (hhd)	2 bbl			
U.S. customary—dry:				
1 pint (pt)	0.5 qt	. . . .	33.6	0.55 litres
1 quart (qt)	2 pt	. . . .	67.2	1.1 litres
1 peck (pk)	8 qt	. . . .	537.61	8.81 litres
1 bushel (bu)	4 pk	. . . .	2,150.42	35.24 litres

Unit		(metric)	(U.S. customary)
Metric system—solid:			
1 cubic millimeter (mm³)		0.001 cc	
1 cubic centimeter (cc)		1,000 mm³	0.061 in.³
1 cubic decimeter (dm³)		1,000 cc	61.023 in.³
1 cubic meter (m³)		1,000 dm³	35.315 ft³; 1.308 yd³

Unit		(U.S. customary)	(metric)
U.S. customary—solid:			
1 cubic inch (in.³)		. . . .	16.387 cc
1 board foot (fbm)		144 in.³	2,359.8 cc
1 cubic foot (ft³)		1,728 in.³	0.028 m³
1 cubic yard (yd³)		27 ft³	0.765 m³
1 cord		128 ft³	3.625 m³

Volume Conversions

U.S. Customary to Metric			Multiply By	Metric to U.S. Customary			Multiply By
To Change				To Change			
ounces (fluid)	to cubic centimeters	. . . .	29.57	cubic centimeters	. . . . . .	to ounces (fluid)	. . . .	0.034
ounces	to milliliters	. . . .	29.57	milliliters		to ounces	0.034
quart	to liters	. . . .	0.946	liters		to quarts	1.057
cubic inches	to cubic centimeters	. . . .	16.387	cubic centimeters	. . . . . .	to cubic inches	0.061
cubic yards	to cubic meters	. . . .	0.765	cubic meters		to cubic yards	1.308

TEMPERATURE

Table A–3 and Fig. A–3 show Fahrenheit/Centigrade conversions.

CALORIES

TABLE A–3
TEMPERATURE

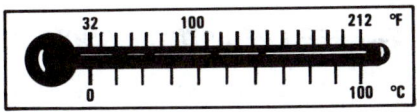

Fig. A–3. Fahrenheit-Centigrade scale for direct conversion reading.

One Fahrenheit (F) degree is 1/180 of the difference between the temperature of melting ice and that of water boiling at standard atmospheric pressure. One Fahrenheit degree equals 0.556°C.

Once Centigrade (C) degree is 1/100 of the difference between the temperature of melting ice and that of water boiling at standard atmospheric pressure. One Centigrade degree equals 1.8°F.

To Change	To	Do This
Degrees Fahrenheit	Degrees Centigrade	Subtract 32, then multiply by 0.556 (5/9)
Degrees Centigrade	Degrees Fahrenheit	Multiply by 1.8 (9/5) and add 32

One calorie is the amount of heat required to raise one gram of water, one degree centigrade at sea level.

Unit	Is Equal To
1 kcal	1,000 cal
1 mcal	1,000 kcal
1 therm	1,000 kcal or 1 mcal

WEIGHTS AND MEASURES OF COMMON FEEDS

In calculating rations and mixing concentrates, it is usually necessary to use weight rather than measures. However, in practical feeding operations it is often more convenient for the farmer or rancher to measure the concentrates. Table A–4 will serve as a guide in feeding by measure.

TABLE A–4
WEIGHTS AND MEASURES OF COMMON FEEDS

Feed	Approximate Weight	
	Pounds per Quart	Pounds Per Bushel
Alfalfa meal	0.6	19
Barley	1.5	48
Beet pulp, dried	0.6	19
Brewers' grains, dried	0.6	19
Buckwheat	1.6	50
Buckwheat bran	1.0	29
Corn, husked ear	—	70
Corn, cracked	1.6	50
Corn, shelled	1.8	56
Corn meal	1.6	50
Corn-and-cob meal	1.4	45
Cottonseed meal	1.5	48
Cowpeas	1.9	60
Distillers' grain, dried	0.6	19
Fish meal	1.0	35
Gluten feed	1.3	42
Linseed meal, old process	1.1	35
Linseed meal, new process	0.9	29
Meat scrap	1.3	42
Milo (grain sorghum)	1.7	56
Molasses feed	0.8	26
Oats	1.0	32
Oats, ground	0.7	22
Oat middlings	1.5	48
Peanut meal	1.0	32
Rice bran	0.8	26
Rye	1.7	56
Sorghum, grain	1.7	56
Soybeans	1.8	60
Tankage	1.6	51
Velvet beans, shelled	1.8	60
Wheat	1.9	60
Wheat bran	0.5	16
Wheat middlings, standard	0.8	26
Wheat screenings	1.0	32

SECTION III—GESTATION TABLE

The dairy producer who has information relative to breeding dates can easily estimate parturition dates from Table A–5. Consideration should also be given to the breed differences indicated in Table A–6.

TABLE A–5
GESTATION TABLE FOR COWS

Date Bred	Date Due, 282 Days	Date Bred	Date Due, 283 days
Jan. 1	Oct. 11	July 5	Apr. 14
Jan. 6	Oct. 16	July 10	Apr. 19
Jan. 11	Oct. 21	July 15	Apr. 24
Jan. 16	Oct. 26	July 20	Apr. 29
Jan. 21	Oct. 31	July 25	May 4
Jan. 26	Nov. 5	July 30	May 9
Jan. 31	Nov. 10	Aug. 4	May 14
Feb. 5	Nov. 15	Aug. 9	May 19
Feb. 10	Nov. 20	Aug. 14	May 24
Feb. 15	Nov. 25	Aug. 19	May 29
Feb. 20	Nov. 30	Aug. 24	June 3
Feb. 25	Dec. 5	Aug. 29	June 8
Mar. 2	Dec. 10	Sept. 3	June 13
Mar. 7	Dec. 15	Sept. 8	June 18
Mar. 12	Dec. 20	Sept. 13	June 23
Mar. 17	Dec. 25	Sept. 18	June 28
Mar. 22	Dec. 30	Sept. 23	July 3
Mar. 27	Jan. 4	Sept. 28	July 8
Apr. 1	Jan. 9	Oct. 3	July 13
Apr. 6	Jan. 14	Oct. 8	July 18
Apr. 11	Jan. 19	Oct. 13	July 23
Apr. 16	Jan. 24	Oct. 18	July 28
Apr. 21	Jan. 29	Oct. 23	Aug. 2
Apr. 26	Feb. 3	Oct. 28	Aug. 7
May 1	Feb. 8	Nov. 2	Aug. 12
May 6	Feb. 13	Nov. 7	Aug. 17
May 11	Feb. 18	Nov. 12	Aug. 22
May 16	Feb. 23	Nov. 17	Aug. 27
May 21	Feb. 28	Nov. 22	Sept. 1
May 26	Mar. 5	Nov. 27	Sept. 6
May 31	Mar. 10	Dec. 2	Sept. 11
June 5	Mar. 15	Dec. 7	Sept. 16
June 10	Mar. 20	Dec. 12	Sept. 21
June 15	Mar. 25	Dec. 17	Sept. 26
June 20	Mar. 30	Dec. 22	Oct. 1
June 25	Apr. 4	Dec. 27	Oct. 6
June 30	Apr. 9		

TABLE A–6
GESTATION VARIATION BETWEEN DAIRY BREEDS

Breed	Average Gestation Length
	(days)
Ayrshire	278.7
Brown Swiss	290.8
Guernsey	284.0
Holstein	278.9
Jersey	279.3

[1]Breeding Committee, American Dairy Science Association.

SECTION IV—BREED REGISTRY ASSOCIATIONS

A breed registry association consists of a group of breeders banded together for the purposes of (1) recording the lineage of their animals, (2) protecting the purity of the breed, (3) encouraging further improvement of the breed, and (4) promoting the interest of the breed. A list of the breed registry associations is given in Table A–7. No claim is made that all breed registries are listed therein.

TABLE A–7
DAIRY BREED REGISTRY ASSOCIATIONS

Breed	Association	Adddress
Ayrshire	Ayrshire Breeders' Assn.	2 Union Street Brandon, VT 05733
Brown Swiss	The Brown Swiss Cattle Breeders' Assn. of the U.S.A.	Box 1038 Beloit, WI 53511
Guernsey	The American Guernsey Cattle Assn.	P.O. Box 666 Reynoldsburg, OH 43068–0666
Holstein-Friesian	Holstein-Friesian Assn. of America	1 Holstein Place Brattleboro, VT 05302–0808
Jersey	The American Jersey Cattle Club	6486 E. Main Street Reynoldsburg, OH 43068–2362
Milking Shorthorn	The American Milking Shorthorn Society	P.O. Box 449 Beloit, WI 53511

SECTION V—DAIRY MAGAZINES

The dairy magazines publish news items and informative articles of special interest to dairy producers. Also, many of them employ field representatives whose chief duty it is to assist in the buying and selling of animals. A list of dairy magazines is given in Table A–8. No claim is made that all dairy magazines are listed therein.

TABLE A–8
DAIRY MAGAZINES

Breed	Title	Address
General	*California Agribusiness Dairyman*	1185 W. Hedges, Fresno, CA 93728
	Dairy Contact	11802 124th Street, Suite 214, Edmonton, Alta. T5L 0M3 Canada
	Dairy Illustrated	1880 Country Farm Drive, Naperville, IL 60540
	Dairy Journal	Tulare, CA 93275
	Dairyman, The	P.O. Box 299, Sandy Creek, NY 13145
	Dairymen's Digest	P.O. Box 5040, Arlington, TX 76005
	Farm & Dairy	Box 38, Salem, OH 44460
	Hoard's Dairyman	38 W. Milkwaukee Avenue, Fort Atkinson, WI 53538–0801
	Sunbelt Dairyman, The	P.O. Box 843, Franklin, IN 37064
Ayrshire	*Ayrshire Digest*	2 Union Street, Brandon, VT 05733
	Canadian Ayrshire Review	P.O. Box B.P. 188, Ste-Anne de Bellevue, Quebec, H9X 1C0 Canada
Brown Swiss	*Brown Swiss Bulletin, The*	P.O. Box 1038, Beloit, WI 53511
Guernsey	*Guernsey Breeders Journal*	P.O. Box 27410, Columbus, OH, 43227
Holstein	*Arizona Holstein Sun*	Box 365, Marana, AZ 85238
	California Holstein News	1177 West Hedges, Fresno, CA 93728
	Holstein Journal	335 Lesmill Road, Don Mills, Ont. M3B 2V1 Canada
	Holstein World	P.O. Box 299, Sandy Creek, NY 13145
	Texas Holstein News	Route 1, Buda, TX 78610
Jersey	*Canadian Jersey Breeder*	343 Waterloo Avenue, Guelph, Ont. N1H 3K1 Canada
	Jersey Journal	6486 E. Main Street, Reynoldsburg, OH 43068–2362
Milking Shorthorn	*Journal of the Milking Shorthorn and Illawarra Breeds*	P.O. Box 449, Beloit, WI 53511

SECTION VII—STATE COLLEGES OF AGRICULTURE

U.S. dairy producers can obtain a list of available bulletins and circulars, and other information regarding livestock by writing to (1) their state agricultural college (land-grant institution), or (2) the Superintendent of Documents, Washington, DC; or by going to the local county extension office (farm advisor) of the county in which they reside. Canadian dairy producers may write to the Department of Agriculture of their province or to their provincial university. A list of U.S. land-grant institutions and Canadian provincial universities follows:

TABLE A–9
U.S. LAND-GRANT INSTITUTIONS AND CANADIAN PROVINCIAL UNIVERSITIES

State	Address
Alabama	School of Agriculture, Auburn University, Auburn, AL 36830
Alaska	Department of Agriculture, University of Alaska, Fairbanks, AK 99701
Arizona	College of Agriculture, The University of Arizona, Tucson, AZ 85721
Arkansas	Division of Agriculture, University of Arkansas, Fayetteville, AR 72701
California	College of Agricultural and Environmental Sciences, University of California, Davis, CA 95616
Colorado	College of Agricultural Sciences, Colorado State University, Fort Collins, CO 80521
Connecticut	College of Agriculture and Natural Resources, University of Connecticut, Storrs, CT 06268
Delaware	College of Agricultural Sciences, University of Delaware, Newark, DE 19711
Florida	College of Agriculture, University of Florida, Gainesville, FL 32611
Georgia	College of Agriculture, University of Georgia, Athens, GA 30602
Hawaii	College of Tropical Agriculture, University of Hawaii, Honolulu, HI 96822
Idaho	College of Agriculture, University of Idaho, Moscow, ID 83843
Illinois	College of Agriculture, University of Illinois, Urbana-Champaign, IL 61801

(Continued)

TABLE A–9 *(Continued)*

State	Address
Indiana	School of Agriculture, Purdue University, West Lafayette, IN 47907
Iowa	College of Agriculture, Iowa State University, Ames, IA 50010
Kansas	College of Agriculture, Kansas State University, Manhattan, KS 66506
Kentucky	College of Agriculture, University of Kentucky, Lexington, KY 40506
Louisiana	College of Agriculture, Louisiana State University and A&M College, University Station, Baton Rouge, LA 70803
Maine	College of Life Sciences and Agriculture, University of Maine, Orono, ME 04473
Maryland	College of Agriculture, University of Maryland, College Park, MD 20742
Massachusetts	College of Food and Natural Resources, University of Massachusetts, Amherst, MA 01002
Michigan	College of Agriculture and Natural Resources, Michigan State University, East Lansing, MI 48823
Minnesota	College of Agriculture, University of Minnesota, St. Paul, MN 55101
Mississippi	College of Agriculture, Mississippi State University, Mississippi State, MS 39762
Missouri	College of Agriculture, University of Missouri, Columbia, MO 65201
Montana	College of Agriculture, Montana State University, Bozeman, MT 59715
Nebraska	College of Agriculture, University of Nebraska, Lincoln, NE 68503
Nevada	The Max C. Fleischmann College of Agriculture, University of Nevada, Reno, NV 89507
New Hampshire	College of Life Sciences and Agriculture, University of New Hampshire, Durham, NH, 03824
New Jersey	College of Agriculture and Environmental Science, Rutgers University, New Brunswick, NJ 08903
New Mexico	College of Agriculture and Home Economics, New Mexico State University, Las Cruces, NM 88003
New York	New York State College of Agriculture, Cornell University, Ithaca, NY 14850
North Carolina	School of Agriculture, North Carolina State University, Raleigh, NC 27607
North Dakota	College of Agriculture, North Dakota State University, State University Station, Fargo, ND 58102
Ohio	College of Agriculture and Home Economics, The Ohio State University, Columbus, OH 43210
Oklahoma	College of Agriculture and Applied Science, Oklahoma State University, Stillwater, OK 74074
Oregon	School of Agriculture, Oregon State University, Corvallis, OR 97331
Pennsylvania	College of Agriculture, The Pennsylvania State University, University Park, PA 16802
Puerto Rico	College of Agricultural Sciences, University of Puerto Rico, Mayaguez, PR 00708
Rhode Island	College of Resource Development, University of Rhode Island, Kingston, RI 02881
South Carolina	College of Agricultural Sciences, Clemson University, Clemson, SC 29631
South Dakota	College of Agriculture and Biological Sciences, South Dakota State University, Brookings, SD 57006
Tennessee	College of Agriculture, University of Tennessee, P.O. Box 1071, Knoxville, TN 37901
Texas	College of Agriculture, Texas A&M University, College Station, TX 77843
Utah	College of Agriculture, Utah State University, Logan, UT 84321
Vermont	College of Agriculture, University of Vermont, Burlington, VT 05401
Virginia	College of Agriculture, Virginia Polytechnic Institute and State University, Blacksburg, VA 24061
Washington	College of Agriculture, Washington State University, Pullman, WA 99163
West Virginia	College of Agriculture and Forestry, West Virginia University, Morgantown, WV 26506
Wisconsin	College of Agricultural and Life Sciences, University of Wisconsin, Madison, WI 53706
Wyoming	College of Agriculture, University of Wyoming, University Station, P.O. Box 3354, Laramie, WY 82070

Canada	Address
Alberta	University of Alberta, Edmonton, Alberta T6H 3K6
British Columbia	University of British Columbia, Vancouver, British Columbia V6T 1W5
Manitoba	University of Manitoba, Winnipeg, Manitoba R3T 2N2
New Brunswick	University of New Brunswick, Fredericton, New Brunswick E3B 4Z7
Ontario	University of Guelph, Guelph, Ontario N1G 2W1
Quebec	Faculty d'Agriculture, L'Universite Laval, Quebec City, Quebec G1K 7D4; and Macdonald College of McGill University, Ste. Anne de Bellevue, Quebec H9X 1C0
Saskatchewan	University of Saskatchewan, Saskatoon, Saskatchewan S7N 0W0

SECTION VII—POISON INFORMATION CENTERS

With the large number of chemical sprays, dusts, and gases now on the market for use in agriculture, accidents may arise because of operators being careless in their use. Also, there is always the hazard that a child may eat or drink something that may be harmful. Centers have been established in various parts of the country where doctors can obtain prompt and up-to-date information on treatment of such cases, if desired.

Local medical doctors have information relative to the Poison Information Centers of their area, along with some of the names of their directors, telephone numbers, and street numbers. When calling any of these centers, one should ask for the "Poison Information Center." If this information cannot be obtained locally, call the U.S. Public Health Service at Atlanta, Georgia; or Wenatchee, Washington.

Also, the *National Poison Control Center* is located at the University of Illinois, Urbana-Champaign. It is open 24 hours a day, every day of the week. The *hot line* number is: (217) 333-3611. The toxicology group is staffed to answer questions about known or suspected cases of poisoning or chemical contaminations involving any species of animal. It is not intended to replace local veterinarians or state toxicology laboratories, but to complement them. Where consultation over the telephone is adequate, there is no charge to the veterinarian or producer. Where telephone consultation is inadequate or the problem is of major proportions, a team of veterinary specialists can arrive at the scene of a toxic or contamination problem within a short time. The cost of a personal visitation varies according to the distance traveled, personnel time, and laboratory services required.

Fig. A–4. Records.

Fig. A–5. Brown Swiss cow.

Fig. A–6. Calf feeder stanchions. (Courtesy, Dr. Tom Schultz, Farm Advisor, Visalia, CA)

Fig. A–8. Lactating cows in free stall barn. (Courtesy, Maddox Dairy, Riverdale, CA)

Fig. A–7. Corral shade. (Courtesy, Dr. Tom Schultz, Farm Advisor, CA)

Fig. A–9. Guernsey cows at feed bunk. (Courtesy, American Guernsey Assn., Reynoldsburg, OH)

INDEX

G

H